Intermediate Algebra

SECOND EDITION

Barbara A. Poole
North Seattle Community College

Prentice Hall
Englewood Cliffs, New Jersey 07632

To my family,
Don, Julie, Jim, Lisa, Daniel, and Michael

Editor-in-chief: Timothy Bozik
Managing editor: Jeanne Hoeting
Editorial/production supervision: Nicholas Romanelli
Interior design: Judith A. Matz-Coniglio
Cover design: Jeannette Jacobs
Manufacturing buyers: Paula Massenaro/Trudy Pisciotti
Editorial assistants: Nancy Bauer/Audra Walsh

Printed in the United States of America
10 9 8 7 6 5 4 3 2 1

ISBN 0-13-075326-2

Prentice-Hall International (UK) Limited, *London*
Prentice-Hall of Australia-Pty. Limited, *Sydney*
Prentice-Hall Canada Inc., *Toronto*
Prentice-Hall Hispanoamericana, S.A., *Mexico*
Prentice-Hall of India Private Limited, *New Delhi*
Prentice-Hall of Japan, Inc. *Tokyo*
Simon & Schuster Asia Pte, Ltd., *Singapore*
Editora Prentice-Hall do Brasil, Ltda., *Rio de Janeiro*

Contents

10

Relations and Functions *705*

11

Exponential and Logarithmic Functions *763*

Final Examination *821*

Appendices *829*

A: More on Sets *829*

B: Tables *833*

Bibliography *836*

Answers *837*

Index *884*

Preface

The purpose of this book is to provide my students with an algebra text that they can read and understand. Explanations are written carefully in language that is familiar to the general population as well as to those students for whom English is a second language. Real-world applications of algebra are used throughout the book, to illustrate to students why algebra has a practical importance to them.

The following features of the text will make it easier for students to learn algebra:

Realistic word problems: Real-life situations are used to show the many applications of algebra. Geometry, one of the most common applications of mathematics, is used extensively. Word problems are interspersed throughout the book. Worked-out solutions to most odd-numbered word problems appear in the back of the book.

Definitions: The most important definitions are given twice—once where the concept is introduced in the text but also again in the chapter summary (a convenient reference for students preparing for an examination).

Historical notes: Labeled ''Did You Know?,'' these notes appear throughout the book, increasing students' understanding and appreciation of the *spirit* of mathematics.

Study hints: These appear periodically to help students improve their study skills and test preparation.

Objectives: Each section of the text begins with a list of objectives specifying the concepts to be learned in that section. The objectives are useful for the instructor preparing a lesson as well as for the student previewing or reviewing material.

Step-by-step examples: Each example is worked out in a clear and step-by-step manner. Often, brief explanations are given adjacent to the steps of the example.

Exercises: Examples are accompanied by exercises related directly to them, in the adjacent margin, allowing students to practice new skills *immediately* after learning them. Answers to these margin exercises appear at the end of the section, for immediate feedback. Each section is followed by a problem set, carefully graded for difficulty, and paired so that odd- and even-numbered problems are very similar. Finally, each chapter contains additional exercises that can be used as extra drill or to prepare for a test.

Pretests: These appear at the beginning of each chapter to allow students to assess their knowledge of the chapter.

Writing in mathematics: Each section contains ''Reading and Writing'' problems to help in the development of verbal skills.

Calculator problems: These appear in some sections to show how a calculator is useful for solving more complicated problems. There are also problems that illustrate the use of an advanced scientific (graphing) calculator.

Keystrokes for the calculator: Instructions on the use of the scientific calculator appear throughout the book.

Challenge problems: At the end of each section there are challenge problems to motivate students.

Critical thinking problems: Labeled "Think About It," critical thinking problems appear periodically to fortify students' thinking abilities. Some of these problems may also be used for cooperative learning.

Checkup problems: These problems provide a review of all prior sections and are intended to help prepare students for the next section. For example, since reciprocal is used to define perpendicular lines, reciprocals are reviewed in the previous section's Checkup Problems.

Summaries: A summary occurs at the end of each chapter to reinforce concepts and preparation for tests; the summaries are also a resource of definitions and problem-solving techniques.

Cooperative learning exercises: Each chapter contains a set of exercises for review and group discussion. Many of these exercises are critical thinking problems.

Chapter tests: Each set of cooperative learning exercises is followed by a chapter test that can be used as a practice tool.

Cumulative reviews: Cumulative review exercises at the ends of chapters 2, 4, 6, 8, and 10 provide additional review of previously covered topics to help students prepare for examinations.

Final examination: A comprehensive Final Examination is provided at the end of the book (see page 821).

Preceding some problems, the symbol 🖩 indicates a problem where use of a calculator is appropriate; the symbol * indicates a more challenging exercise.

Supplementary Materials

For the student the following supplements are provided by the publisher. Ask your instructor if they are available.

Student Solutions Manual	0-13-075359-9
Demo 3.5 IBM Interactive Algebra Tutor	0-13-503830-0
3.5 IBM Interactive Algebra Tutor	0-13-075357-0
Demo 5.25 IBM Interactive Algebra Tutor	0-13-503822-7
5.25 IBM Interactive Algebra Tutor	0-13-075383-1
Demo MAC Interactive Algebra Tutor	0-13-503814-6
MAC Interactive Algebra Tutor	0-13-075391-2
Demo Lecture Video	0-13-503848-0
Lecture Video Series	0-13-293267-9

For the instructor the following supplements are available from the publisher:

Instructor's Manual with Tests	0-13-075334-3
Instructor's Solutions Manual	0-13-075342-4
Test Item File	0-13-075367-X
Demo PH Test Manager 2.0	0-13-075763-2
3.5 PH Test Manager 2.0	0-13-075409-9
5.25 PH Test Manager 2.0	0-13-075417-X
Demo MakeTest for the Macintosh	0-13-075771-3
MakeTest for the Macintosh	0-13-075425-0

Acknowledgments

I wish to thank my husband, Donald, my daughter, Julia, and my son, James, and his wife, Lisa, for their encouragement and support. Although my grandson Daniel is too young to lend encouragement, he provided many diversions from writing!

My colleague at North Seattle Community College, Vicky Ringen, offered me many helpful suggestions. I also thank the college for its support of this project.

Finally, I extend my gratitude and appreciation to Sudhir Kumar Goel of Valdosta University for his accuracy check of the proofs during production and to the following reviewers of the manuscript for their helpful comments and suggestions:

James D. Blackburn, Tulsa Junior College (Tulsa, OK)
Marilyn Carlson, University of Kansas (Lawrence, KS)
Deanne Christianson, University of the Pacific (Stockton, CA)
James Coleman, New Community College of Baltimore (Baltimore, MD)
Joyce Huntington, Walla Walla Community College (Walla Walla, WA)
Glenn Jacobs, Greenville Technical College (Greenville, SC)
Robert C. Jolly, St. Philip's College (San Antonio, TX)
Phyllis H. Jore, Valencia Community College East (Orlando, FL)
Linda Kyle, Tarrant County Junior College (Hurst, TX)
Shelby Morgan, North Lake College (Irving, TX)
Steven Terry, Ricks College (Rexburg, ID)

BARBARA A. POOLE

STUDY HINTS

Success in mathematics often depends on your attitude toward it. Recognize that mathematics is widely used in the real world and in your other classes. Most people can succeed in it if they are willing to spend the time on it. You should plan to spend a minimum of two hours studying mathematics outside class for each hour that you spend in the class. The following are other suggestions for being successful in this course.

Read the objectives at the beginning of each section. These help you to focus on the material to be learned.

Study the section carefully. Work the margin exercises as you study. These help you to determine if you have understood the section. It may be helpful to highlight important concepts.

At the end of each section, study it as if you were going to have a quiz on it. This helps to prepare you for the chapter test. Also, what you learn in the next section may be based on the information from the preceding section or sections.

If you are taking a lecture class, read the section the day before the lecture on it. This will help you to understand the lecture and ask questions in class.

Some students, particularly those with learning disabilities, find that using blue overlays on the page helps them to focus on the material. They may also use colored pens for operation symbols as they do their homework. Some students also like to use note cards with important concepts on one side and sample problems on the other side.

With the right attitude and proper study habits, you *can* be successful in this course. Good luck!

CHAPTER 1

Real Numbers

Pretest

Find the following.

1. $|-17.4|$

2. $\left|\frac{11}{8}\right|$

3. Replace the ? with > or <: -15.7 ? -22.8.

Add.

4. $(-9) + (-8)$

5. $7.8 + (-13.6)$

6. $-\frac{3}{7} + \frac{8}{21}$

Subtract.

7. $-8 - 29$

8. $-33.4 - (-42.1)$

9. $-\frac{5}{6} - \frac{9}{8}$

Multiply.

10. $(7.8)(-9.6)$

11. $-\frac{8}{3}\left(\frac{9}{24}\right)$

12. $(-8)(-7)(-3)$

Divide.

13. $\frac{-48}{-6}$

14. $\frac{-200}{-25}$

15. $\frac{81}{-9}$

16. $\frac{-5}{7} \div \left(\frac{-21}{8}\right)$

17. $\frac{-3}{4} \div \frac{9}{16}$

18. $\frac{0}{15}$

Find the reciprocal.

19. $\frac{-7}{8}$

20. 12

21. Evaluate when $x = -5$, $y = -1$, and $z = -2$: $4xy - 7yz$.

Multiply.

22. $-4(3x - 8)$

23. $5(7y - 9)$

Combine like terms.

24. $-8x - 5y - 14x + 3y$

25. $6a - 12b - 9a - 7b$

Simplify.

26. $15a - (9a + 8)$

27. $3(x - 8) - 5(x - 7)$

28. $4x - 3[4 - 7(2x - 3)]$

29. $\frac{6^2 - 9(8)}{9}$

For Problems 30–33, assume that all variables represent nonzero real numbers.

Perform the operation indicated. Write with positive exponents.

30. $(-3x^4y^{-3})(-9x^5y^{-6})$

31. $\frac{-24x^{-2}y^5}{16x^{-8}y^9}$

Simplify and write with positive exponents.

32. $(-3x^{-4}y^5)^{-3}$

33. $\left(\frac{-x^{-4}y^5}{3x^7y^{-1}}\right)^{-4}$

34. Write in scientific notation: 75,400,000.

35. Write in decimal notation: 5.2×10^{-6}.

1.1 BASIC DEFINITIONS

The material in this chapter may look familiar to you. However, study it carefully. Understanding algebra begins with understanding Chapter 1.

OBJECTIVES

1. *Use roster and set-builder notation to name sets*
2. *Identify whole numbers, natural numbers, integers, and rational numbers*
3. *Find the absolute value of a number*
4. *Determine which of two numbers is smaller*
5. *Find the opposite or additive inverse of a number*

1 Set Notation

A ***set*** is a collection of objects. The objects in the set are called ***elements*** of the set.

The set containing the numbers -4, 0, and 5 can be named $\{-4, 0, 5\}$. This method of naming sets is called ***roster notation.*** Some important sets are shown in roster notation as follows. Since the elements of the sets cannot be counted, they are infinite sets as indicated by the three dots.

The ***natural numbers*** are the set of counting numbers.

$$\text{Natural numbers} = \{1, 2, 3, 4, \ldots\}$$

The ***whole numbers*** consist of zero and the natural numbers.

$$\text{Whole numbers} = \{0, 1, 2, 3, 4, \ldots\}$$

To solve some problems and symbolize debt, people began to use negative numbers. The ***integers*** are as follows:

$$\text{Integers} = \{\ldots, -3, -2, -1, 0, 1, 2, 3, \ldots\}$$

We may picture a set of numbers by using a drawing called a ***number line.*** Some integers are shown here on a number line.

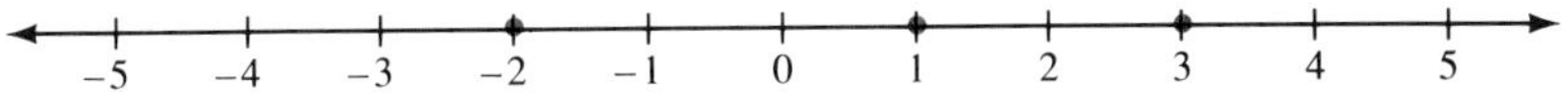

Positive numbers are placed to the right of zero. Negative numbers are shown to the left of zero. Zero is neither positive nor negative.

When we write a letter such as b or x, the letter may stand for any number. We call these letters ***variables.*** Variables can also be used to define sets of numbers. For example, the set of all natural numbers between 2 and 7 can be written

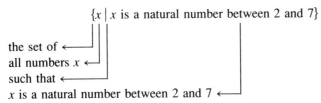

$$\{x \mid x \text{ is a natural number between 2 and 7}\}$$

the set of ←
all numbers x ←
such that ←
x is a natural number between 2 and 7 ←

This is called ***set-builder notation.*** The set could have been written in roster notation as follows.

$$\{3, 4, 5, 6\}$$

EXAMPLE 1 Write the set containing the even natural numbers less than 9 using both roster and set-builder notation.

Roster notation:

$$\{2, 4, 6, 8\}$$

Set-builder notation:

$$\{x \mid x \text{ is an even natural number less than 9}\}$$ ■

□ **DO EXERCISE 1.** Check your answers at the end of this section.

□ **Exercise 1** Write the set containing the five whole numbers less than 5 in both roster notation and set-builder notation.

□ **Exercise 2** Consider the numbers: 4, 0, $\frac{7}{8}$, $\frac{11}{5}$, -6. Find:

a. The natural numbers

b. The whole numbers

c. The integers

d. The rational numbers

Another important set of numbers is the rational numbers. We can use set-builder notation to describe them.

The ***rational numbers*** are of the form a/b, where a and b are integers and b is not zero.

$$\text{Rational numbers} = \left\{ \frac{a}{b} \;\middle|\; a \text{ and } b \text{ are integers and } b \neq 0 \right\}$$

The following are examples of rational numbers.

$$\frac{3}{4},\quad 0,\quad -\frac{8}{7},\quad 1.5,\quad \text{and}\quad 6$$

Notice that 1.5 can be written as $\frac{3}{2}$ and 6 can be written as $\frac{6}{1}$.

Irrational numbers are numbers such as $\sqrt{2}$ that are not rational because they cannot be written as the ratio of two integers. We study them in a subsequent chapter.

2 The Real Numbers

The ***real numbers*** are composed of the rational and irrational numbers. There is only one real number for every point on the number line.

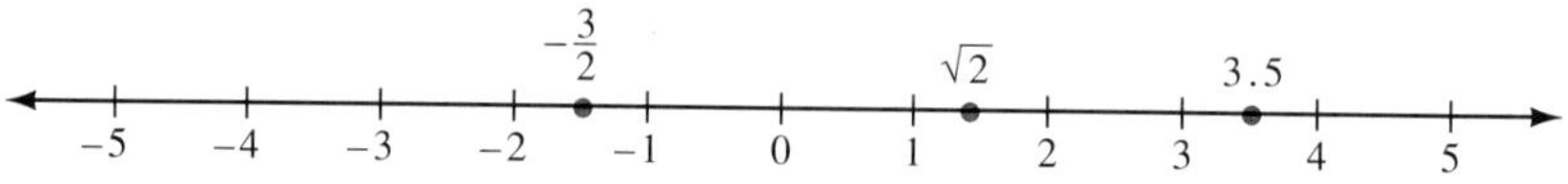

Notice that the natural numbers are a subset of the whole numbers (if a number is a natural number, it is also a whole number). Similarly, the whole numbers are a subset of the integers (every whole number is an integer), the integers are a subset of the rational numbers, and the rational numbers are a subset of the real numbers. The following chart shows the relationship of these sets of numbers.

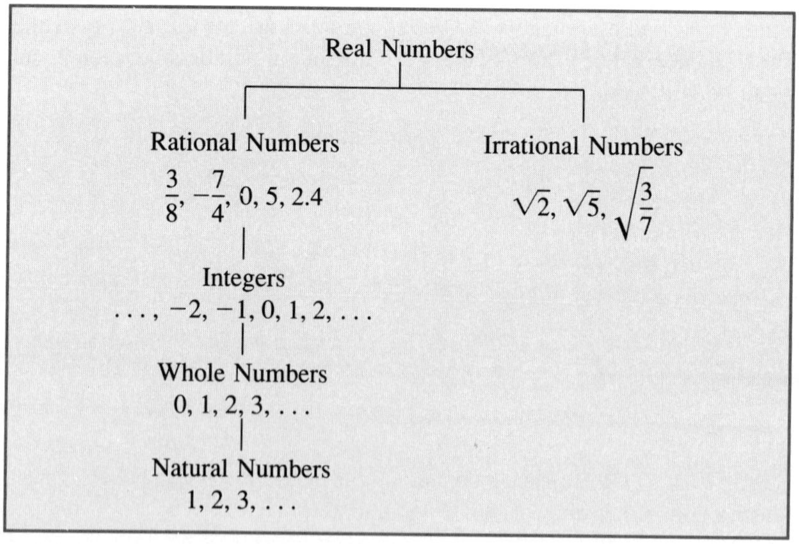

EXAMPLE 2 Consider the numbers 7, -2, $\frac{3}{8}$, 0. The natural number is 7. The whole numbers are 7 and 0. The integers are 7, -2, and 0. The rational numbers are 7, -2, $\frac{3}{8}$, and 0. ■

□ **DO EXERCISE 2.**

3 Absolute Value

The ***absolute value*** of a number is its distance from zero on the number line. Absolute value is denoted by | |.

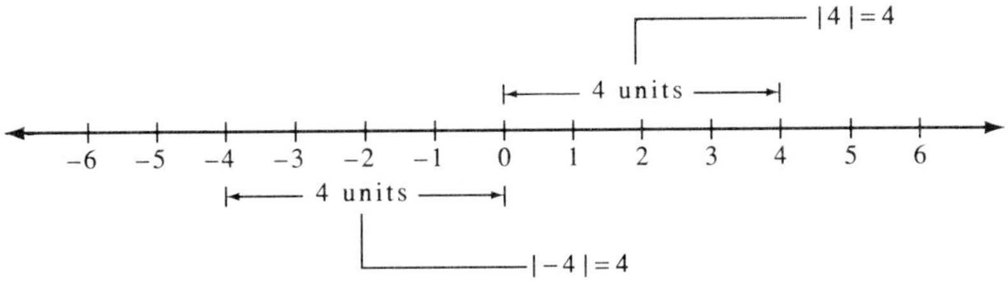

EXAMPLE 3 Find the following.

a. $|5| = 5$ Since the distance of 5 from 0 is 5.

b. $|-3| = 3$ Since the distance of -3 from 0 is 3.

c. $\left|-\frac{7}{8}\right| = \frac{7}{8}$

d. $-|6| = -6$ Since $|6| = 6$.

e. $-|-4| = 4$ Since $|-4| = 4$. ■

□ DO EXERCISE 3.

The absolute value of a number may be found with a graphing calculator. Since instructions for these calculators vary, consult your manual.

□ **Exercise 3** Find the following.

a. $|8|$

b. $|-7.1|$

c. $|0|$

d. $|-6|$

e. $\left|\frac{3}{4}\right|$

f. $\left|-\frac{7}{5}\right|$

g. $-|8|$

h. $-|4|$

i. $-|-2|$

j. $-|-9|$

4 Inequalities

If a number is to the right of another number on the number line, it is the greater of the two numbers. If a number is to the left of another number, it is the smaller of the numbers. The symbol $>$ stands for "greater than" and $<$ means "less than." Statements such as $4 < 6$ and $-1 > -2$ are called ***inequalities.***

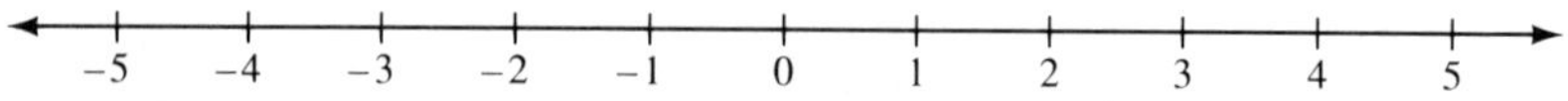

EXAMPLE 4

a. $5 > 3$ Since 5 is to the right of 3 on the number line.

b. $2 > -1$

c. $-1 > -3$ Since -1 is to the right of -3 on the number line.

d. $0 < 3$ Since 0 is to the left of 3 on the number line.

e. $-4.5 < -2.6$

f. $-\frac{1}{5} > -\frac{3}{4}$ Since $-\frac{3}{4}$ is to the left of $-\frac{1}{5}$ on the number line. ■

□ **DO EXERCISE 4.**

All positive real numbers are greater than zero and all negative real numbers are less than zero.

> If x is a positive real number, then $x > 0$.
>
> If x is a negative real number, then $x < 0$.

□ **Exercise 4** Replace the ? with $>$ or $<$.

a. $3 \ ? \ -1$

b. $8 \ ? \ 5$

c. $-4 \ ? \ -6$

d. $-3.8 \ ? \ 0$

e. $\frac{7}{8} \ ? \ \frac{3}{4}$

f. $-\frac{7}{9} \ ? \ -\frac{2}{3}$

5 Opposites or Additive Inverses

For every positive real number there is a corresponding number to the left of zero on the number line called its ***opposite.*** Similarly, for every negative real number there is a number to the right of zero on the number line which is its opposite.

The opposites of the positive real numbers are the negative real numbers. Similarly, the opposites of the negative real numbers are the positive real numbers. Zero is neither positive nor negative. The opposite of zero is zero. The sum of a number and its opposite is zero. Therefore, a number and its opposite are often called ***additive inverses.***

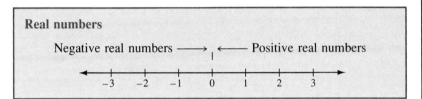

> **Real numbers**
>
> Negative real numbers ⟶ | ⟵ Positive real numbers
>
> 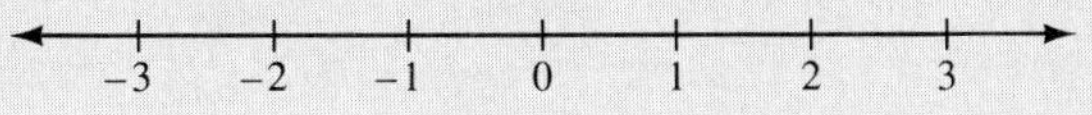
>

EXAMPLE 5

a. Find the opposite of 5.
The opposite of 5 is negative 5.

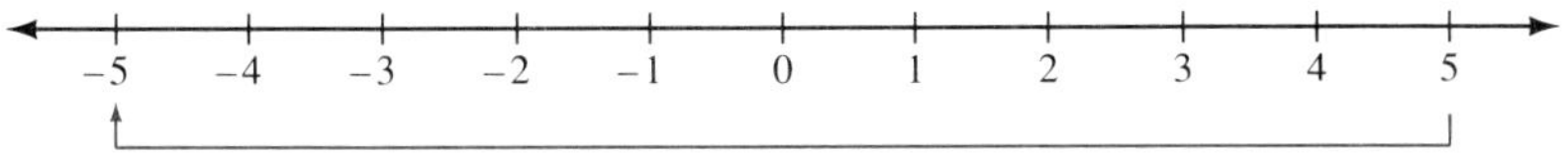

b. Find the opposite of −3.
The opposite of −3 is 3.

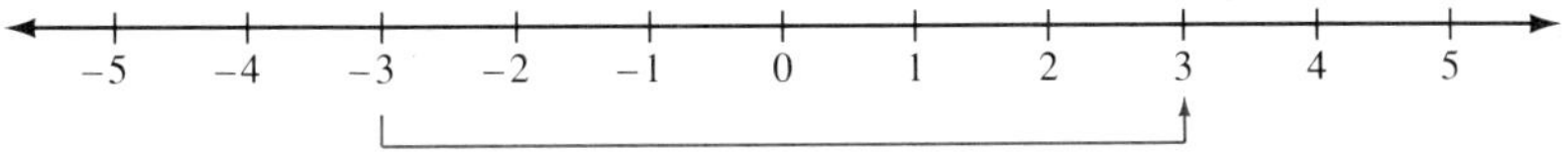

c. Find the opposite of $-\frac{9}{5}$.
The opposite of $-\frac{9}{5}$ is $\frac{9}{5}$. ■

□ **DO EXERCISE 5.**

$a \leq b$ and $b \geq a$ are also inequalities. $a \leq b$ is read "a is less than or equal to b." $b \geq a$ is read "b is greater than or equal to a."

We can use the definition of opposite to give a definition of absolute value that is often used in mathematics.

> For any real number a,
>
> $$|a| = \begin{cases} a & \text{if } a \geq 0 \\ -a & \text{if } a < 0 \end{cases}$$

For example,

$$|7| = 7 \quad \text{and} \quad |-3| = -(-3) = 3$$

□ **Exercise 5** Find the opposite.

a. −8

b. 6.7

c. $-\frac{5}{3}$

d. $\frac{3}{7}$

Answers to Exercises

1. {0, 1, 2, 3, 4}; $\{x \mid x$ is a whole number less than $5\}$

2. a. 4 **b.** 4, 0 **c.** 4, 0, −6 **d.** 4, 0, $\frac{7}{8}$, $\frac{11}{5}$, −6

3. a. 8 **b.** 7.1 **c.** 0 **d.** 6 **e.** $\frac{3}{4}$ **f.** $\frac{7}{5}$ **g.** −8

h. −4 **i.** −2 **j.** −9

4. a. > **b.** > **c.** > **d.** < **e.** > **f.** <

5. a. 8 **b.** −6.7 **c.** $\frac{5}{3}$ **d.** $-\frac{3}{7}$

Problem Set 1.1

NAME ____________________

DATE ____________________

CLASS ____________________

In all the problem sets, do the odd problems. If you need additional practice, do the even problems. Answers to the odd problems are given at the back of the book.

Consider the numbers: $-\frac{3}{7}, -2, 3, \frac{9}{5}, 0$. *Find the following.*

1. The natural numbers

2. The whole numbers

3. The integers

4. The rational numbers

Consider the numbers: $-8, 3, -\frac{5}{4}, 6.2, 0, 5\frac{3}{7}, 4, 9.1$. *Find the following.*

5. The whole numbers

6. The natural numbers

7. The rational numbers

8. The integers

Write each set in roster notation.

9. The set of all letters in the word "April"

10. The set of letters in the word "Mexico"

11. The set of even whole numbers less than 12

12. The set of odd natural numbers between 4 and 14

13. The set of even integers

14. The set of odd integers

Write each set in set-builder notation.

15. $\{0, 1, 2, 3, 4, 5, 6\}$

16. $\{8, 9, 10, 11, \ldots\}$

17. The set of natural numbers

18. The set of whole numbers

19. The set of all real numbers less than -4

20. The set of all real numbers greater than 10

Find the following.

21. $|9|$

22. $|-11|$

23. $|-2.1|$

24. $|25|$

25. $\left|-\frac{3}{4}\right|$

26. $\left|\frac{11}{5}\right|$

27. $-|7|$

28. $-|-3|$

29. $-\left|-\frac{9}{5}\right|$

30. $-|3.8|$

31. $|-18|$

32. $-|-20|$

Replace the ? with > or <.

33. 15 ? 10

34. 20 ? 24

35. 9 ? 0

36. 15 ? 0

37. $-6 \ ? \ 0$

38. $-11 \ ? \ 0$

39. $-7 \ ? \ -12$

40. $-4 \ ? \ -2$

41. $-16 \ ? \ -5$

42. $-8 \ ? \ -25$

43. $-15.38 \ ? \ 0.784$

44. $-12.57 \ ? \ -6.354$

45. $-2 \ ? \ -6$

46. $0 \ ? \ -1$

47. $3.8 \ ? \ 3.7$

48. $-2.1 \ ? \ -3.2$

49. $-\frac{3}{4} \ ? \ -\frac{1}{2}$

50. $-\frac{2}{3} \ ? \ -\frac{3}{8}$

Find the opposite.

51. 7

52. 15

53. -9.4

54. 11.2

55. $-\frac{1}{5}$

56. $\frac{11}{4}$

57. 15

58. -45

59. The set of counting numbers is also called the ___________ numbers.

60. Numbers of the form *a*/*b*, where *a* and *b* are integers and *b* is not zero, are called ___________ numbers.

61. The ___________ ___________ of a number is its distance from zero on the number line.

62. The ___________ of 8 is -8.

Think About It

True or false?

* 63. Every natural number is an integer.

* 64. Every rational number is a whole number.

* 65. Every integer is a natural number.

* 66. Every whole number is a rational number.

* 67. Some real numbers are irrational numbers.

* 68. Some integers are whole numbers.

* 69. Arrange the following numbers from smallest to largest:

$$-1.15, \ \frac{2}{3}, 0.09, \ \frac{87}{100}, \ 0.74, \ \frac{5}{8}, \ 0.235, \ -\frac{9}{5}$$

* 70. Arrange the following numbers from largest to smallest:

$$\frac{1}{5}, \ -2.1, \ 0, \ \frac{3}{8}, \ -\frac{7}{10}, \ 0.19, \ -\frac{4}{5}$$

1.2 ADDITION AND SUBTRACTION

In this section we review addition and subtraction of signed numbers. We also review the commutative law of addition and the associative law of addition.

OBJECTIVES

1. *Add signed numbers*
2. *Subtract signed numbers*
3. *Identify examples of the commutative and associative laws of addition*

1 Addition of Signed Numbers

> To add two real numbers with the same sign, add their absolute values and give the answer the common sign.

EXAMPLE 1 Add.

a. $9 + 3 = (+9) + (+3) = +12 = 12$

We may associate signed numbers with money.

$$\$9 + \$3 = \$12$$

b. $(-6) + (-2) = -8$ Add $|-6|$ and $|-2|$ and give the answer the common sign.

If we spend \$6 and we spend \$2, we have spent \$8.

c. $(-3.2) + (-5.8) = -9.0$

d. $-\frac{1}{5} + \left(-\frac{1}{10}\right) = -\frac{1}{5} \cdot \frac{\mathbf{2}}{\mathbf{2}} + \left(-\frac{1}{10}\right)$ Multiply by $\frac{2}{2}$ to get the least common denominator.

$$= -\frac{2}{10} + \left(-\frac{1}{10}\right) = -\frac{3}{10}$$

e. $-\frac{1}{3} + \left(-\frac{1}{4}\right) = -\frac{1}{3} \cdot \frac{\mathbf{4}}{\mathbf{4}} + \left(-\frac{1}{4}\right) \cdot \frac{\mathbf{3}}{\mathbf{3}}$ Multiply each fraction by 1 to get the least common denominator.

$$= -\frac{4}{12} + \left(-\frac{3}{12}\right) = -\frac{7}{12}$$ ■

□ **DO EXERCISE 1.**

□ **Exercise 1** Add.

a. $7 + 8$

b. $(-4) + (-5)$

c. $-9 + (-6)$

d. $-2.1 + (-0.4)$

e. $-\frac{1}{6} + \left(-\frac{5}{12}\right)$

f. $-\frac{5}{2} + \left(-\frac{2}{3}\right)$

When we add two signed numbers, it may be necessary to *subtract*.

> To add two real numbers with different signs, subtract the smaller absolute value from the larger. Give the answer the sign of the number with the larger absolute value.

□ **Exercise 2** Add.

a. $3 + (-8)$

b. $(-7) + 4$

c. $9 + (-6)$

d. $-9.3 + 9.3$

e. $-7.1 + 3.2$

f. $10.1 + (-4)$

g. $\frac{7}{6} + \left(-\frac{1}{3}\right)$

h. $-\frac{2}{3} + \frac{3}{5}$

□ **Exercise 3** Subtract.

a. $6 - 2$

b. $-5 - 7$

c. $-9 - (-10.1)$

d. $-4 - (-8.3)$

e. $-7 - (-7)$

f. $-\frac{5}{6} - \frac{5}{12}$

g. $-\frac{3}{8} - \left(-\frac{1}{4}\right)$

h. $\frac{1}{2} - \frac{3}{5}$

EXAMPLE 2 Add.

a. $(-3) + (+8) = +(8 - 3) = +5 = 5$ — $|+8| > |-3|$, so the answer is positive.

b. $(+2) + (-9) = -(9 - 2) = -7$ — $|-9| > |+2|$, so the answer is negative.

If we have \$2 and we spend \$9, we are in debt \$7.

c. $(+10.4) + (-4.6) = +(10.4 - 4.6) = +5.8 = 5.8$

d. $-3.2 + 4.9 = +(4.9 - 3.2) = +1.7 = 1.7$

e. $-\frac{5}{4} + \frac{3}{8} = -\left(\frac{5}{4} - \frac{3}{8}\right) = -\left(\frac{10}{8} - \frac{3}{8}\right) = -\frac{7}{8}$

f. $\frac{1}{3} + \left(-\frac{5}{7}\right) = -\left(\frac{5}{7} - \frac{1}{3}\right) = -\left(\frac{15}{21} - \frac{7}{21}\right) = -\frac{8}{21}$

g. $-6 + 6 = 0$ — If a number is added to its opposite, the result is zero.

h. $3.5 + (-3.5) = 0$ ■

□ DO EXERCISE 2.

2 Subtraction of Signed Numbers

To subtract one real number from another, add the opposite of the second real number to the first real number.

$$a - b = a + (-b) \quad \text{where } a \text{ and } b \text{ are real numbers}$$

EXAMPLE 3 Subtract.

a. $3 - (+8) = 3 + (-8) = -5$ — *Add* the opposite of 8.

If we have \$3 and we spend \$8, we are in debt \$5.

b. $-4 - (+7) = -4 + (-7) = -11$ — *Add* the opposite of 7.

c. $-6 - (-1) = -6 + (+1) = -5$ — *Add* the opposite of -1.

If we have a debt of \$6 and someone takes away a debt of \$1, we have gained \$1 and now owe \$5.

d. $-3.4 - (-7.2) = -3.4 + (+7.2) = 3.8$

e. $-\frac{1}{2} - \frac{3}{4} = -\frac{1}{2} + \left(-\frac{3}{4}\right) = -\frac{2}{4} - \frac{3}{4} = -\frac{5}{4}$ ■

□ DO EXERCISE 3.

3 The Commutative and Associative Laws of Addition

For any real numbers a and b,

$a + b = b + a$ ***Commutative law of addition***

The commutative law of addition says that we may add in any order. Notice that the *order of the letters* is different on each side of the equal sign.

Parentheses tell us to do the operation inside them first.

$$6 + (5 + 3) = 6 + 8 = 14$$

For any real numbers a, b, and c,

$a + (b + c) = (a + b) + c$ ***Associative law of addition***

The associative law of addition says that when we are only doing additions, we may move the grouping symbols. The grouping does not affect the sum. Notice that the *order of the letters does not change.*

EXAMPLE 4 Is the commutative or associative law of addition illustrated by each equation?

a. $7 + 5 = 5 + 7$ Commutative law of addition

The order of the numbers changes.

$$\left.\begin{array}{l} 7 + 5 = 12 \\ 5 + 7 = 12 \end{array}\right\}$$ Notice that the result is the same.

b. $3 + (8 + 9) = (3 + 8) + 9$ Associative law of addition.

The order of the numbers does not change. We do the calculations inside the parentheses first but we get the same result.

$$3 + (8 + 9) = 3 + 17 = 20$$

and

$$(3 + 8) + 9 = 11 + 9 = 20$$ ■

□ **DO EXERCISE 4.**

□ **Exercise 4** Is the commutative or associative law of addition illustrated by each equation?

a. $3 + 4 = 4 + 3$

b. $7 + (8 + 9) = (7 + 8) + 9$

c. $2 + (8 + 9) = (8 + 9) + 2$

CALCULATOR

You may want to use a calculator to add or subtract larger signed numbers. If your calculator has a $\boxed{+/-}$ key, signed numbers may be added or subtracted as follows. (If you have a graphing calculator, consult your manual.)

	ENTER	DISPLAY
$-73 + 84$	$\boxed{73}\ \boxed{+/-}\ \boxed{+}\ \boxed{84}\ \boxed{=}$	11
$-27 - 36$	$\boxed{27}\ \boxed{+/-}\ \boxed{-}\ \boxed{36}\ \boxed{=}$	-63
$24 - (-12) - 48$	$\boxed{24}\ \boxed{-}\ \boxed{12}\ \boxed{+/-}\ \boxed{-}\ \boxed{48}\ \boxed{=}$	-12

DID YOU KNOW?

Brahmagupta, a seventh-century Hindu mathematician and astronomer, used both negative and positive numbers.

Answers to Exercises

1. a. 15 **b.** -9 **c.** -15 **d.** -2.5 **e.** $-\frac{7}{12}$ **f.** $-\frac{19}{6}$

2. a. -5 **b.** -3 **c.** 3 **d.** 0 **e.** -3.9 **f.** 6.1 **g.** $\frac{5}{6}$ **h.** $-\frac{1}{15}$

3. a. 4 **b.** -12 **c.** 1.1 **d.** 4.3 **e.** 0 **f.** $-\frac{5}{4}$ **g.** $-\frac{1}{8}$ **h.** $-\frac{1}{10}$

4. a. Commutative law of addition **b.** Associative law of addition **c.** Commutative law of addition

Problem Set 1.2

NAME ______________________

DATE ______________________

CLASS ______________________

Add.

1. $7 + 8$

2. $9 + 8$

3. $-5 + (-7)$

4. $-3 + (-10)$

5. $4 + (-2)$

6. $8 + (-3)$

7. $-5 + 5$

8. $-3 + 3$

9. $-15 + 8$

10. $14 + (-7)$

11. $-6 + (-4)$

12. $-7 + (-5)$

13. $-3.2 + 8.5$

14. $-9.1 + 6.4$

15. $-3.8 + 0$

16. $0 + (-10.7)$

17. $-\frac{1}{3} + \frac{3}{4}$

18. $-\frac{3}{8} + \frac{2}{5}$

19. $\frac{1}{2} + \left(-\frac{1}{2}\right)$

20. $-\frac{7}{9} + \frac{7}{9}$

21. $-\frac{2}{3} + \left(-\frac{1}{6}\right)$

22. $-\frac{11}{5} + \left(-\frac{3}{10}\right)$

23. $-5.2 + (-3.8)$

24. $-7.6 + (-8.5)$

Subtract.

25. $-3 - 7$

26. $-2 - 5$

27. $8 - 9$

28. $10 - 12$

29. $3 - (-8)$

30. $2 - (-7)$

31. $15 - (-20)$

32. $18 - (-25)$

33. $9.1 - 10.2$

34. $6.5 - 7.8$

35. $-10.1 - 11.3$

36. $-19.1 - 3.6$

37. $-\frac{1}{2} - \frac{1}{7}$

38. $-\frac{3}{8} - \frac{5}{16}$

39. $-\frac{1}{2} - \left(-\frac{3}{4}\right)$

40. $-\frac{2}{5} - \left(-\frac{1}{5}\right)$

41. $(-8.4) - (-7.1)$

42. $(-3.5) - (-4.3)$

43. $-74 - 32$

44. $-28 - 16$

45. $32 - 95$

46. $54 - 88$

47. $\frac{7}{8} - \frac{11}{9}$

48. $\frac{3}{7} - \frac{5}{8}$

49. $\frac{6}{5} - \left(-\frac{8}{15}\right)$

50. $\frac{3}{8} - \left(-\frac{7}{24}\right)$

51. The temperature in Anchorage at 5 A.M. was -9°F. By 2 P.M. the temperature had increased 22°. What was the temperature at 2 P.M.?

52. Jolene paid \$75 on a debt of \$238. How much does she still owe on the debt?

53. The elevation of the Hawaiian mountain Mauna Loa is 13,680 feet. The elevation of Death Valley is -282 feet. Find the difference in elevation between the two points.

54. The elevation of the Dead Sea is -1299 feet. The elevation of Death Valley is -282 feet. What is the difference in elevation between the two points?

Is the commutative or associative law of addition illustrated by each equation?

55. $7 + 9 = 9 + 7$

56. $15 + 34 = 34 + 15$

57. $5 + (7 + 6) = (5 + 7) + 6$

58. $6 + (3 + 4) = (6 + 3) + 4$

59. $(3 + 8) + 9 = 9 + (3 + 8)$

60. $8 + 2 = 2 + 8$

61. $9 + 6 = 6 + 9$

62. $8 + (3 + 2) = (8 + 3) + 2$

63. $5 + (3 + 2) = (3 + 2) + 5$

64. $15 + 12 = 12 + 15$

65. John owed \$58 on his credit card account. He charged an additional amount of \$38. What negative number represents the amount that John owes?

66. Karen had \$35 in her checking account. She wrote a check for \$79. What negative number represents the amount that she is overdrawn?

67. To add two real numbers with different signs, ____________ the smaller absolute value from the larger. Give the answer the sign of the number with the ____________ absolute value.

68. To subtract one real number from another real number, add the ____________ of the second real number to the first real number.

69. The ____________ law of addition states that we may add real numbers in any order.

70. The grouping does not affect the sum according to the ____________ law of addition.

Add or subtract as indicated.

71. $-85.174 + (-64.938)$

72. $-36.874 + (-225.683)$

73. $-97.582 + 68.85$

74. $-123.89 + 54.328$

75. $-224.631 + 365.842$

76. $-55.901 + 78.324$

77. $-84.2 - 73.85$

78. $-96.784 - 43.29$

79. $17.2 - (-18.55)$

80. $36.8 - (-15.738)$

81. $-84.25 - (-36.814)$

82. $-174.1 - (-202.05)$

83. $7.342 - 6.985 - 8.41$

84. $-3.184 + 10.12 - (-16.5)$

85. $-16.28 - (-8.5431) - 14.3$

86. $15.935 - 18.72 - 54.3715$

87. Kevin owed $571.55 on his credit card. He used it to buy clothing costing $157.85. What negative number represents the amount that Kevin owes on his credit card?

88. Susan borrowed $314.75 from her friend. She paid back $285.34. What negative number represents the amount Susan owes her friend?

* **89.** $6 + (-4) + (-5)$

* **90.** $8 + (-16) - (-3)$

* **91.** $-5 - (-8) + (-2)$

* **92.** $-(-9) + 10 - 12$

* **93.** $-\frac{5}{3} + \frac{3}{4} - \left(-\frac{1}{6}\right)$

* **94.** $\frac{5}{8} - \frac{3}{4} - \left(-\frac{8}{5}\right)$

95. $-3 - |5| - |-9|$

96. $5 - (-11) + |-8|$

1.3 MULTIPLICATION AND DIVISION

In this Section we review multiplication and division of real numbers. When we multiply numbers, we find their product. Finding a product is usually called *multiplying*. When we divide one number by another number, we find the quotient. Finding the quotient is usually termed *dividing*.

1 Like Signs

> To multiply or divide two real numbers with the same sign, multiply or divide their absolute values. Give the answer a positive sign.

EXAMPLE 1 Multiply or divide.

a. $(-3) \cdot (-4) = +12 = 12$

b. $(7) \cdot (2.1) = +14.7 = 14.7$

c. $(-18) \div (-6) = +3 = 3$

d. $\dfrac{-20}{-5} = +4 = 4$

e. $-\dfrac{3}{5} \cdot -\dfrac{2}{7} = +\dfrac{6}{35} = \dfrac{6}{35}$ Multiply numerators and multiply denominators.

f. $0 \cdot 7 = 0$ Notice that any number times zero is zero. ■

Parentheses are often used to indicate multiplication. For example

$$(-3) \cdot (-2) = (-3)(-2) = -3(-2)$$

□ DO EXERCISE 1.

2 Different Signs

> To multiply or divide two real numbers with different signs, multiply or divide their absolute values. Give the answer a negative sign.

EXAMPLE 2 Multiply or divide.

a. $4(-2) = -8$

b. $-15 \div 3 = -5$

c. $3.2(-1.8) = -5.76$

d. $\dfrac{-50.2}{2} = -25.1$

e. $\dfrac{7}{9} \cdot (-4) = -\dfrac{28}{9}$ In algebra we often leave the answer as an improper fraction. ■

□ DO EXERCISE 2.

3 Commutative and Associative Laws of Multiplication

The commutative and associative laws also hold for multiplication of real numbers.

OBJECTIVES

1 *Multiply or divide numbers with like signs*

2 *Multiply or divide numbers with different signs*

3 *Identify examples of the commutative and associative laws of multiplication*

4 *Identify division problems that are not defined*

5 *Find the reciprocal of a number*

6 *Find other names for a rational number*

□ **Exercise 1** Multiply or divide.

a. $(-5)(-6)$

b. $(-40) \div (-10)$

c. $5.5(1.1)$

d. $72 \div 8$

e. $-4.2 \div (-2.1)$

f. $-\dfrac{11}{5} \cdot -\dfrac{2}{3}$

g. $6.2(0)$

□ **Exercise 2** Multiply or divide.

a. $(-2)7$

b. $18 \div (-9)$

c. $21.7 \div (-7)$

d. $8(-9)$

e. $\dfrac{-12}{4}$

f. $-\dfrac{8}{7} \cdot \dfrac{3}{5}$

□ **Exercise 3** Is the commutative or associative law of multiplication illustrated by each equation?

a. $3 \cdot 4 = 4 \cdot 3$

b. $(9 \cdot 8) \cdot 2 = 9 \cdot (8 \cdot 2)$

c. $(7 \cdot 6) \cdot 4 = 4 \cdot (7 \cdot 6)$

□ **Exercise 4** Divide.

a. $\frac{0}{6}$

b. $\frac{7}{0}$

c. $-\frac{8}{0}$

d. $\frac{0}{2}$

For any real numbers a and b,

$a \cdot b = b \cdot a$, ***Commutative law of multiplication***

For any real numbers a, b, and c

$a \cdot (b \cdot c) = (a \cdot b) \cdot c$, ***Associative law of multiplication***

The commutative law of multiplication says that we may multiply in any order. The associative law of multiplication tells us that the grouping does not affect the product. Notice that the order of the letters is different on each side of the equal sign for the commutative law, whereas for the associative law, the order of the letters remains the same.

EXAMPLE 3 Is the commutative or associative law of multiplication illustrated by each equation?

a. $7 \cdot (3 \cdot 4) = (7 \cdot 3) \cdot 4$ Associative law of multiplication.

b. $(4 \cdot 5) \cdot 6 = 6 \cdot (4 \cdot 5)$ Commutative law of multiplication. ■

□ **DO EXERCISE 3.**

4 Division by Zero

Division may be defined so that the quotient a/b is the number (if it exists) that when multiplied by b gives a.

Division by zero is undefined.

We may not divide by zero since the result of the division would have to be some number, say c. Let $a/0 = c$. Then $a = 0 \cdot c$ by the definition of division. But any number we substitute for c gives $a = 0$. So a would have to be zero. Is $0/0$ defined?

Suppose that $0/0 = c$. Then $0 = 0 \cdot c$ and c can be any number.

$$\frac{0}{0} = 4 \qquad \text{Since } 0 = 0 \cdot 4.$$

$$\frac{0}{0} = -3 \qquad \text{Since } 0 = 0 \cdot (-3).$$

Both 4 and -3 and other numbers could be substituted for c. The answer is not unique, so *we do not divide by zero*. We may divide zero by a nonzero number, however.

EXAMPLE 4 Divide.

a. $\frac{3}{0}$ Undefined.

b. $\frac{0}{5} = 0$ Zero cookies divided among 5 boys gives each boy 0 cookies! ■

□ **DO EXERCISE 4.**

5 Reciprocals

Two numbers whose product is 1 are called ***reciprocals*** or multiplicative inverses of each other. To find the reciprocal of a number, invert the number. Zero does not have a reciprocal, since $1/0$ is undefined.

Every nonzero real number a has a reciprocal $1/a$.

EXAMPLE 5

Number	*Reciprocal*
$\frac{3}{5}$	$\frac{5}{3}$
$\frac{-8}{7}$	$\frac{7}{-8}$
$6 = \frac{6}{1}$	$\frac{1}{6}$
0	No reciprocal ■

□ DO EXERCISE 5.

Division of Fractions

To divide two fractions, multiply by the reciprocal of the divisor. For any real numbers a and b, $b \neq 0$,

$$a \div b = \frac{a}{b} = a \cdot \frac{1}{b}$$

To ***factor*** a number means to write the number as a multiplication of numbers. We often factor the numerator or denominator (or both) of a fraction in order to simplify it. We also recall that any number divided by itself is 1.

EXAMPLE 6 Divide.

a. $\frac{3}{4} \div \frac{5}{7} = \frac{3}{4} \cdot \frac{\mathbf{7}}{\mathbf{5}} = \frac{21}{20}$

b. $\frac{-7}{2} \div \frac{-11}{9} = \frac{-7}{2} \cdot \frac{\mathbf{9}}{\mathbf{-11}} = \frac{63}{22}$

c. $\frac{2}{3} \div \frac{8}{5} = \frac{2}{3} \cdot \frac{\mathbf{5}}{\mathbf{8}} = \frac{2}{3} \cdot \frac{5}{4 \cdot 2}$ Factor a denominator.

$= \frac{2 \cdot 5}{3 \cdot 4 \cdot 2} = \frac{2 \cdot 5}{2 \cdot 3 \cdot 4}$ Use the commutative law of multiplication in the denominator.

$= \frac{2}{2} \cdot \frac{5}{3 \cdot 4} = \frac{5}{12}$ Eliminate $\frac{2}{2}$ or 1. ■

□ DO EXERCISE 6.

6 Finding Other Names for a Rational Number

Your answer to a problem involving the division of integers may be equivalent to the answer shown in the book because we may change the signs on a fraction, as shown.

$$+\frac{-1}{+5} = -\frac{+1}{+5} \quad \text{Since } -\frac{+1}{+5} = -1\left(\frac{+1}{+5}\right) = \frac{-1}{+5} \quad \text{or} \quad +\frac{-1}{+5}.$$

Also,

$$+\frac{+1}{-5} = -\frac{+1}{+5}$$

□ **Exercise 5** Find the reciprocal.

a. $\frac{2}{3}$

b. -8

c. $\frac{-9}{5}$

d. $\frac{10}{7}$

□ **Exercise 6** Divide.

a. $\frac{5}{3} \div \frac{6}{7}$

b. $-\frac{3}{8} \div \frac{1}{9}$

c. $\frac{9}{8} \div \frac{3}{1}$

d. $-7 \div -\frac{14}{5}$

□ **Exercise 7** Find two other names for the numbers by changing signs.

a. $\frac{-2}{5}$

b. $\frac{-8}{5}$

c. $-\frac{11}{3}$

d. $\frac{6}{-7}$

Every fraction has three signs. There is a sign on a fraction and a sign on the numerator and a sign on the denominator. We may change any *two* of these signs at the *same time*. The result is an equivalent fraction.

For any real numbers a and b, $b \neq 0$,

$$\frac{-a}{b} = -\frac{a}{b} = \frac{a}{-b}$$

Remember that a number without a sign is positive.

EXAMPLE 7 Find two other names for the number by changing signs.

$$\frac{-9}{+7}$$

$$\frac{-9}{+7} = +\frac{-9}{+7}$$

Hence, changing two signs, we obtain

$$+\frac{-9}{+7} = -\frac{+9}{+7} = +\frac{+9}{-7}$$ ■

□ **DO EXERCISE 7.**

You may want to use a calculator to help you with this section.

CALCULATOR

	ENTER	DISPLAY
15(−7)	[15] [×] [7] [+/−] [=]	−105
−128 ÷ (−16)	[128] [+/−] [÷] [16] [+/−] [=]	8
54 ÷ 0	[54] [÷] [0] [=]	Error message: the calculator may blink

Answers to Exercises

1. a. 30 **b.** 4 **c.** 6.05 **d.** 9 **e.** 2 **f.** $\frac{22}{15}$ **g.** 0

2. a. −14 **b.** −2 **c.** −3.1 **d.** −72 **e.** −3 **f.** $-\frac{24}{35}$

3. a. Commutative law of multiplication **b.** Associative law of multiplication **c.** Commutative law of multiplication

4. a. 0 **b.** Undefined **c.** Undefined **d.** 0

5. a. $\frac{3}{2}$ **b.** $\frac{1}{-8}$ **c.** $\frac{5}{-9}$ **d.** $\frac{7}{10}$

6. a. $\frac{35}{18}$ **b.** $-\frac{27}{8}$ **c.** $\frac{3}{8}$ **d.** $\frac{5}{2}$

7. a. $-\frac{+2}{+5}$ or $-\frac{2}{5}$, $+\frac{+2}{-5}$ or $\frac{2}{-5}$ **b.** $-\frac{+8}{+5}$ or $-\frac{8}{5}$, $+\frac{+8}{-5}$ or $\frac{8}{-5}$

c. $+\frac{-11}{+3}$ or $\frac{-11}{3}$, $+\frac{+11}{-3}$ or $\frac{11}{-3}$ **d.** $-\frac{+6}{+7}$ or $-\frac{6}{7}$, $+\frac{-6}{+7}$ or $\frac{-6}{7}$

Problem Set 1.3

NAME ____________________

DATE ____________________

CLASS ____________________

Multiply.

1. $(-3)(-2)$

2. $(-5)(-4)$

3. $(-7)(-5)$

4. $(-6)(-8)$

5. $(-7)(-9)$

6. $(-8)(-9)$

7. $(-5)(4)$

8. $(-3)(6)$

9. $8(-7)$

10. $6(-9)$

11. $4(-2)$

12. $7(-1)$

13. $5(-1.3)$

14. $7(-3.6)$

15. $-\frac{3}{4}\cdot\frac{8}{7}$

16. $-\frac{1}{9}\cdot\frac{2}{5}$

17. $-\frac{17}{3}\cdot\left(-\frac{9}{8}\right)$

18. $-\frac{7}{15}\cdot\left(-\frac{5}{6}\right)$

19. $0\cdot\frac{5}{8}$

20. $8(0)$

21. $(-3)(4)(7)$

22. $6(-1)(4)$

23. $(8)(-2)(-3)$

24. $(-1)(4)(-5)$

Divide, if possible.

25. $15 \div (-5)$

26. $20 \div (-2)$

27. $-16 \div (-4)$

28. $-18 \div (-6)$

29. $\frac{35}{-7}$

30. $\frac{48}{-6}$

31. $\frac{-72}{-8}$

32. $\frac{-64}{-8}$

33. $\frac{-7.2}{2.4}$

34. $\frac{-10.8}{-3.6}$

35. $\frac{0}{7}$

36. $\frac{0}{5}$

37. $\frac{8}{0}$

38. $\frac{4}{0}$

Find the reciprocal.

39. $\frac{2}{7}$

40. $\frac{5}{9}$

41. $\frac{-1}{3}$

42. $\frac{-5}{8}$

43. $\frac{11}{2}$

44. $\frac{15}{7}$

45. 12

46. 16

47. -9

48. -7

Divide.

49. $-\frac{3}{8} \div \frac{5}{11}$

50. $-\frac{1}{5} \div \frac{7}{12}$

51. $\frac{9}{8} \div \left(-\frac{7}{4}\right)$

52. $\frac{10}{3} \div \left(-\frac{1}{5}\right)$

53. $-\frac{3}{7} \div \left(-\frac{9}{14}\right)$

54. $-\frac{5}{8} \div \left(-\frac{15}{16}\right)$

55. $-81.48 \div 4.2$

56. $-52.91 \div 3.7$

57. $-60.5 \div (-12.1)$

58. $-183.4 \div (-26.2)$

59. $-36 \div 0.9$

60. $78 \div (-3.9)$

61. $-\frac{7}{8} \div \frac{5}{8}$

62. $-\frac{3}{5} \div \frac{5}{11}$

63. $-\frac{2}{9} \div \left(-\frac{4}{7}\right)$

64. $-\frac{8}{3} \div \left(-\frac{7}{9}\right)$

65. $\frac{9}{8} \div \left(-\frac{3}{16}\right)$

66. $\frac{6}{5} \div \left(-\frac{2}{15}\right)$

67. $\frac{-8}{12 - 12}$

68. $\frac{-7}{25 - 25}$

Find two other names for the numbers by changing signs.

69. $\frac{-3}{8}$

70. $\frac{-9}{5}$

71. $\frac{11}{-7}$

72. $\frac{1}{-8}$

73. $-\frac{5}{4}$

74. $-\frac{2}{3}$

75. According to the ___________ law of multiplication, grouping does not affect the product.

76. Division by zero is ___________ ___________.

77. To find the reciprocal of a number, ___________ the number.

78. To divide two fractions, we multiply by the ___________ of the divisor.

Multiply. Round answers to the nearest thousandth.

79. $(-6.24)(-5.73)$

80. $(-14.357)(-6.94)$

81. $14.8(-26.798)$

82. $4.95(-65.312)$

83. $-16.34(0.847)$

84. $(-0.037)(-25.4)$

85. $-732.83(-521.4)$

86. $(-36.95)(424.538)$

Divide. Round answers to the nearest thousandth.

87. $\frac{7.324}{-3.928}$

88. $\frac{-5.147}{6.59}$

89. $\frac{-16.345}{-29.48}$

90. $\frac{-25.74}{-9.856}$

91. $\frac{-8.91}{5.734}$

92. $\frac{0.738}{-9.23}$

93. $\frac{-35{,}000}{-15.165}$

94. $\frac{-27{,}400}{45.105}$

95. $\frac{-6600}{3.82}$

96. $\frac{-7450}{-6.94}$

Multiply.

* **97.** $\left(\frac{3}{5}\right)\left(\frac{-10}{7}\right)\left(\frac{-14}{9}\right)$

* **98.** $\left(\frac{-7}{8}\right)\left(\frac{-24}{9}\right)\left(\frac{27}{6}\right)$

* **99.** $\left(\frac{-9}{5}\right)\left(\frac{3}{77}\right)(0)$

* **100.** $\left(\frac{101}{8}\right)(0)\left(\frac{5}{99}\right)$

Divide.

* **101.** $\frac{\frac{-10}{17}}{\frac{-12}{5}}$

* **102.** $\frac{\frac{-33}{23}}{\frac{55}{4}}$

* **103.** $\frac{\left|\frac{-7}{10}\right|}{\frac{-8}{15}}$

* **104.** $\frac{\frac{9}{16}}{\left|\frac{-12}{7}\right|}$

1.4 COMBINING TERMS AND SIMPLIFYING EXPRESSIONS CONTAINING GROUPING SYMBOLS

OBJECTIVES

1 *Evaluate expressions*

2 *Use the distributive laws to multiply expressions*

3 *List the terms and the coefficients of each term of an expression*

4 *Factor expressions*

5 *Combine like terms*

6 *Simplify expressions containing grouping symbols*

1 Evaluating Expressions

Evaluations are very important in the applications of mathematics to business, science, and other fields. Recall from Section 1.1 that a *variable* is a letter that may stand for any number.

We sometimes indicate multiplication by writing several variables or a number and several variables together. If the expression contains an addition or subtraction, we agree that we will *do all multiplications first.*

EXAMPLE 1

a. Evaluate the expression $5x + xy$ when $x = -2$ and $y = -4$.

$$5x + xy = 5(\mathbf{-2}) + (\mathbf{-2})(\mathbf{-4}) = -10 + 8 = -2$$

b. What is the weight w (in pounds) of a man of height h of 70 inches if height and weight are related by the formula $w = \frac{11}{2}h - 220$?

$$w = \frac{11}{2}h - 220$$

$$w = \frac{11}{2}(\mathbf{70}) - 220 \qquad \text{Substitute 70 for } h.$$

$$w = 385 - 220 = 165$$

The man weighs 165 pounds. ■

□ DO EXERCISE 1.

□ **Exercise 1** Evaluate the following expressions when $x = 4$ and $y = -3$.

a. $xy - 3x$

b. $6y - 3xy$

2 The Distributive Laws

The distributive laws tell us one way that we can evaluate an expression.

Distributive Law of Multiplication over Addition

$$a(b + c) = ab + ac$$

for any real numbers a, b, and c.

Distributive Law of Multiplication over Subtraction

$$a(b - c) = ab - ac$$

for any real numbers a, b, and c.

EXAMPLE 2

a. $3(4 + 5) = 3 \cdot 4 + 3 \cdot 5 = 12 + 15 = 27$ Use a distributive law.

We may also combine numbers within the parentheses and then multiply.

$$3(4 + 5) = 3(9) = 27$$

b. $7(8 - 2) = 7 \cdot 8 - 7 \cdot 2 = 56 - 14 = 42$ Use a distributive law.

or $7(8 - 2) = 7(6) = 42$ ■

□ DO EXERCISE 2.

□ **Exercise 2** Evaluate in two ways.

a. $2(3 + 5)$

b. $9(8 - 6)$

□ **Exercise 3** Multiply.

a. $3(x - 7)$

b. $x(y + 8)$

□ **Exercise 4** List the terms and the coefficients of each term.

a. $x - 4$

b. x

c. $x - y + 1$

d. $\frac{y}{7} - 3$

When a variable is contained within the parentheses, we must use the distributive laws to multiply out the expression.

EXAMPLE 3

a. $-4(x + 3) = -4 \cdot x + (-4) \cdot (3) = -4x - 12$

b. $a(x - 5) = ax - 5a$ ■

□ DO EXERCISE 3.

3 Terms and Coefficients

A ***term*** is a number or the product of a number and a variable or variables. In a mathematical expression, the terms occur at each plus or minus sign.

The ***numerical coefficient*** of a variable in a term is a number by which the variable is multiplied. If no number appears by the variable, the numerical coefficient is understood to be either $+1$ or -1, depending on whether the variable is preceded by a plus sign or a minus sign. The numerical coefficient is often simply called the coefficient. If a term contains only a number, the numerical coefficient is that number.

EXAMPLE 4

Expression	*Terms*	*Coefficients of Each Term*
$x + 4y - 2$	$x, 4y, -2$	$1, 4, -2$
y	y	1
$\frac{x}{5} - y$	$\frac{x}{5}, -y$	$\frac{1}{5}, -1$ ■

□ DO EXERCISE 4.

Like terms differ only in their numerical coefficients. The numerical coefficients may also be the same.

For example, x and $4x$ are like terms.

4 Factoring

Since multiplication is commutative, the distributive laws may be reversed and written as follows:

For any real numbers a, b, and c

$$ba + ca = (b + c)a$$

$$ba - ca = (b - c)a$$

This allows us to factor some expressions. Factoring is the reverse of multiplication. This type of factoring will help us combine like terms.

EXAMPLE 5

a. $8\mathbf{x} + 9\mathbf{x} = (8 + 9)\mathbf{x}$

b. $5\mathbf{y} - 8\mathbf{y} = (5 - 8)\mathbf{y}$

c. $\mathbf{b} + 7\mathbf{b} = 1\mathbf{b} + 7\mathbf{b} = (1 + 7)\mathbf{b}$ ■

□ DO EXERCISE 5.

5 Combining Like Terms

Only like terms may be combined.

EXAMPLE 6 Combine like terms.

a. $5x + 3x + 4y$

$= \mathbf{5x} + \mathbf{3x} + 4y = (5 + 3)x + 4y = \mathbf{8x} + 4y$ Factor out x.

b. $7x + y - 2x$

$= \mathbf{7x} - \mathbf{2x} + y$ Use a commutative law.

$= (7 - 2)x + y$ Factor out x.

$= \mathbf{5x} + y$

c. $6 - 8 + 4y$

$= \mathbf{6} + (\mathbf{-8}) + 4y$ Use the definition of subtraction.

$= \mathbf{-2} + 4y$

d. $4x - 5y + x + 3y$

$= 4x + x - 5y + 3y$ Use a commutative law.

$= 4x + x + (-5y) + 3y$ Definition of subtraction.

$= (4 + 1)x + (-5 + 3)y$ Recall that $x = 1x$.

$= 5x - 2y$

e. $-4x + 6 - 5x - 8$

$= -4x - 5x + 6 - 8$

$= (-4 - 5)x + 6 - 8$ Factor out x.

$= -9x - 2$ ■

□ DO EXERCISE 6.

□ Exercise 5 Factor.

a. $7x + 8x$

b. $9y - 5y$

c. $5a - 6a$

d. $b - 4b$

□ Exercise 6 Combine like terms.

a. $2y - 3x + 4y$

b. $6x + 5y - 8x$

c. $9 - 5y + 7 - 3y$

d. $8y - 6x - 9y - x$

□ **Exercise 7** Simplify by writing with the least number of symbols.

a. $-(x + 8)$

b. $-(3y - 2z + 4)$

□ **Exercise 8** Simplify by writing with the least number of symbols.

a. $a - (4a + 7)$

b. $2x - (5 - 6x)$

6 Simplifying Expressions Containing Grouping Symbols

Three different sets of ***grouping symbols*** may be used. They are parentheses, ( ), brackets, [ ], and braces, { }. To *simplify* an expression often means to write the expression with the least number of symbols.

Recall that $-x = -1x$. This gives us the following property, since $-1x$ means $-1 \cdot x$.

The Property of −1

For any number *a*,

$$-a = -1 \cdot a$$

When a negative sign precedes an expression within grouping symbols, the negative sign can be interpreted as a -1. Then apply the distributive law to simplify the expression.

EXAMPLE 7 Simplify by writing with the least number of symbols.

a. $-(x + 3) = -1(x + 3) = -1x + (-1)(3) = -x - 3$

b. $-[4x + y - 2] = -1[4x + y - 2] = (-1)4x + (-1)y - (-1)(2) = -4x - y + 2.$ ■

□ **DO EXERCISE 7.**

Notice that when grouping symbols are preceded by a negative sign, the signs of all the terms within the grouping symbols are changed.

$$-[4x + y - 2] = -[+4x + y - 2] = -4x - y + 2$$

All signs within grouping symbols are also changed when the grouping symbol is preceded by a minus sign.

EXAMPLE 8

a. $2x - (3x + 4) = 2x - 3x - 4$

$= (2 - 3)x - 4$ Factor out x.

$= -x - 4$

b. $5a + 3 - (7a - 1) = 5a + 3 - 7a + 1$

$= 5a - 7a + 3 + 1$ Use a commutative law.

$= (5 - 7)a + 3 + 1$ Factor out a.

$= -2a + 4$ ■

□ **DO EXERCISE 8.**

We may have more than one set of grouping symbols in an expression. One way to remove them is to *work out the innermost grouping symbols first.* If the grouping symbols are preceded by both a subtraction sign and a number, treat this as a negative number. Use the distributive property to multiply the negative number times the expression inside the grouping symbols. Combine terms, if possible, after removing each grouping symbol.

EXAMPLE 9 Simplify by writing with the least number of symbols.

a. $5x - 4(x - 2)$

$= 5x - 4(x) - (-4)2$ — Use a distributive law.

$= 5x - 4x + 8$

$= x + 8$

b. $-[8 - (3x + 4)]$

$= -[8 - 3x - 4]$ — Remove innermost grouping symbols.

$= -[4 - 3x]$ — Combine terms.

$= -4 + 3x$

c. $-3[-7 + (5x - 2)]$

$= -3[-7 + 5x - 2]$ — A plus sign preceding parentheses does not change the signs within it.

$= -3[-7 - 2 + 5x]$ — Use a commutative law.

$= -3[-7 + (-2) + 5x]$ — Definition of subtraction.

$= -3[-9 + 5x]$ — Combine like terms.

$= 27 - 15x$ — Use a distributive law.

d. $-\{2[x - 3(3x + 4)]\}$

$= -\{2[x - 9x - 12]\}$ — Remove the innermost grouping symbols.

$= -\{2[(1 - 9)x - 12]\}$ — Factor out x.

$= -\{2[-8x - 12]\}$ — Combine like terms.

$= -\{-16x - 24\}$ — Use a distributive law.

$= 16x + 24$

e. $3x - 2[4 - (2x + 7)]$

$= 3x - 2[4 - 2x - 7]$

$= 3x - 2[4 - 7 - 2x]$ — Use a commutative law.

$= 3x - 2[4 + (-7) - 2x]$ — Definition of subtraction.

$= 3x - 2[-3 - 2x]$

$= 3x + 6 + 4x$ — Use a distributive law.

$- 7x + 6$ ■

□ **DO EXERCISE 9.**

□ **Exercise 9** Simplify by writing with the least number of symbols.

a. $3y - 5(2y - 1)$

b. $6x + (4x - 7)$

c. $2[3 - (8x - 5)]$

d. $-3[4a - 2(3a + 6)]$

e. $8y - \{3[y - (7y + 2)]\}$

f. $-\{3x - 2[8 - x + (2x + 10)]\}$

Answers to Exercises

1. a. -24 **b.** 18

2. a. $2(3 + 5) = 2(8) = 16$
$2(3 + 5) = 2 \cdot 3 + 2 \cdot 5$
$= 6 + 10 = 16$

b. $9(8 - 6) = 9(2) = 18$
$9(8 - 6) = 9 \cdot 8 - 9 \cdot 6$
$= 72 - 54 = 18$

3. a. $3x - 21$ **b.** $xy + 8x$

4.

	Terms	*Coefficients*		*Terms*	*Coefficients*
a.	$x, -4$	$1, -4$	**b.**	x	1
c.	$x, -y, 1$	$1, -1, 1$	**d.**	$\frac{y}{7}, -3$	$\frac{1}{7}, -3$

5. a. $(7 + 8)x$ **b.** $(9 - 5)y$ **c.** $(5 - 6)a$ **d.** $(1 - 4)b$

6. a. $6y - 3x$ **b.** $-2x + 5y$ **c.** $16 - 8y$ **d.** $-y - 7x$

7. a. $-x - 8$ **b.** $-3y + 2z - 4$

8. a. $-3a - 7$ **b.** $8x - 5$

9. a. $-7y + 5$ **b.** $10x - 7$ **c.** $16 - 16x$ **d.** $6a + 36$
e. $26y + 6$ **f.** $-x + 36$

Problem Set 1.4

NAME ______________________

DATE ______________________

CLASS ______________________

Evaluate when $a = -1$, $b = 2$, and $c = -3$.

1. $a - ab$

2. $ac + b$

3. $bc + 2ab$

4. $3ab - ac$

5. $2ac - 3bc$

6. $5ac + ab$

Multiply.

7. $2(x + 3)$

8. $3(x + 4)$

9. $5(x - 2)$

10. $6(2y - 3)$

11. $-4(3x + 5)$

12. $-7(5x + 1)$

13. $-8(2y - 7)$

14. $-9(3y - 1)$

15. $5a(x + 2y - z)$

16. $4b(x - 3y + 2z)$

List the terms and coefficients of each term.

17. $-2y + 4$

18. $5y - z$

19. $7 - \frac{y}{4}$

20. $\frac{x}{5} + 2$

21. $4x - 6y + 10$

22. $3x + 2y - 6$

Factor.

23. $12x + 7x$

24. $9x + 8x$

25. $15y - 4y$

26. $11y - 7y$

27. $-5a + 9a$

28. $-4b + 7b$

29. $6x - 11x$

30. $7y - 20y$

31. $-8a - 9a$

32. $-10b - 17b$

33. $5x - 3x - 8x$

34. $-7a + 8a - 4a$

35. $\frac{1}{2}x - \frac{3}{8}x$

36. $\frac{7}{5}y + \frac{4}{9}y$

Combine like terms.

37. $8x - 2x$

38. $9x + 3x$

39. $4a + 8a$

40. $3y - 5y$

41. $7x + 2x$

42. $8z - 3z$

43. $6y - 10y$

44. $4x - 5x$

45. $-2y - 3y$

46. $-z - 5z$

47. $8y - 2y + 3y$

48. $9x + x - 2x$

49. $-2x - 3x + x$

50. $-5y + 2y - 3y$

51. $8a - 2b + 2a + 4b$

52. $7b + 3c - 8b + c$

53. $9x + 4 - 5x - 7$

54. $4y + 1 - 6y + 2$

55. $3z - 4 + 5z - 1$

56. $8a - 7 - 9a - 2$

Simplify by writing with the least number of symbols.

57. $-(x + 2)$

58. $-(x + 3)$

59. $-(y - 2)$

60. $-(y - 7)$

61. $-(2a - 3b + c)$

62. $-(5x + y - z)$

63. $3 + (5a - 2)$

64. $7 + (2x - 3)$

65. $6x - (3x + 4)$

66. $7y - (2y + 3)$

67. $8a - (5 - 2a)$

68. $9b - (3 - 4b)$

69. $3(x + 2) - 4(x - 1)$

70. $5(x - 3) - 2(x - 1)$

71. $8a - 4(2a - 3)$

72. $7y - 2(5y + 1)$

73. $3[4x - (3x - 2)]$

74. $-2[3y + (5y - 1)]$

75. $-4[3 - 2(6y - 4)]$

76. $5[4a - 3(2a + 1)]$

77. $-\{4[x - 2(3x + 4)]\}$

78. $-\{5y - [7 - 2y + (y - 3)]\}$

79. $-\{2 + 4[x - 3(x - 2)]\}$

80. $2\{x - 3[2 - (x + 4)]\}$

81. $-[3x - 2(4x - 5)]$

82. $-3[x + 4(2x - 1)]$

83. The number by which a variable is multiplied is called the __________ of the variable.

84. Terms that differ only in their numerical coefficients are called __________.

85. The numerical coefficient of $-x$ is __________.

86. To __________ an expression often means to write the expression with the least number of symbols.

Evaluate when $a = -6.4$, $b = 3.72$, and $c = -5.098$. Round answers to the nearest thousandth.

87. $ab + 3ac$

88. $5ac - bc$

89. $4ac - 2ab$

90. $6ab + 2ac$

Combine like terms.

91. $7.29x - 4.78y + 5.7x + 3.092y$

92. $8.934y - 6.4z - 12.05y - 7.83z$

Evaluate when $a = -\frac{1}{2}$, $b = \frac{5}{8}$, $c = -\frac{3}{4}$, and $d = -\frac{1}{3}$.

* **93.** $\dfrac{ab}{c + d}$

* **94.** $\dfrac{bc}{a - d}$

* **95.** $\dfrac{4cd}{a - b}$

* **96.** $\dfrac{ad}{-3b + c}$

1.5 EXPONENTS AND ORDER OF OPERATIONS

1 Exponential Notation

We want to find a shorter notation for such numbers as

$$3 \cdot 3 \cdot 3 \cdot 3$$

Factors of a number are numbers that are multiplied together to give that number. The number

$$3 \cdot 3 \cdot 3 \cdot 3$$

has four factors that are all the same, so we write it as 3^4. The number 3 is called the ***base*** and 4 is called the ***exponent.***

> exponent
>
> 3^4
>
> base

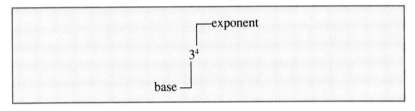

The exponent is also called the ***power.***

Any number of factors may be written in this way.

> $$\underbrace{a \cdot a \cdot a \cdots a \cdot a}_{n \text{ factors}} = a^n$$
>
> where n is an integer greater than 1.
>
> **1.** a^n is read "a to the nth power."
> **2.** a^3 is read "a-cubed" or "a to the third power."
> **3.** a^2 is read "a-squared" or "a to the second power."

Writing numbers such as $3 \cdot 3 \cdot 3 \cdot 3$ as 3^4 is called writing the number in ***exponential notation.***

EXAMPLE 1 Write in exponential notation.

a. $6 \cdot 6 \cdot 6 = 6^3$

b. $yy = y^2$

c. $4y \cdot 4y \cdot 4y = (4y)^3$

d. $5 \cdot x \cdot x \cdot x = 5x^3$

e. $(-2y) \cdot (-2y) \cdot (-2y) = (-2y)^3$ ■

□ DO EXERCISE 1.

Caution: a^n does not mean n times a. 2^3 is $2 \cdot 2 \cdot 2$ or 8, *not* $3 \cdot 2$ or 6.

We may reverse the procedure.

OBJECTIVES

1 *Rewrite expressions with or without whole-number exponents*

2 *Simplify expressions using the order of operations agreement*

□ **Exercise 1** Write in exponential notation.

a. $5 \cdot 5 \cdot 5$

b. $(-2x) \cdot (-2x) \cdot (-2x) \cdot (-2x)$

c. $-3 \cdot y \cdot y$

□ **Exercise 2** Evaluate the exponential expression.

a. 2^3

b. $(-3)^3$

c. $(-5y)^2$

d. $(3y)^4$

e. $\left(\frac{1}{2}\right)^4$

f. $(0.1)^3$

□ **Exercise 3** Write without exponents.

a. 5^1

b. $(-4)^0$

c. $(3x)^0$, $x \neq 0$

d. a^0, where $a \neq 0$

EXAMPLE 2 Evaluate the exponential expression.

a. $8^2 = 8 \cdot 8 = 64$

b. $(-2b)^3 = (-2b)(-2b)(-2b)$

$= (-2)(-2)(-2) \cdot b \cdot b \cdot b$

$= -8b^3$

c. $\left(\frac{1}{4}\right)^3 = \frac{1}{4} \cdot \frac{1}{4} \cdot \frac{1}{4} = \frac{1}{64}$

d. $(0.2)^4 = (0.2)(0.2)(0.2)(0.2)$

$= 0.0016$

e. $-(2x)^2 = -(2x)(2x) = -4x^2$

f. $(-2x)^2 = (-2x)(-2x) = 4x^2$ ■

□ **DO EXERCISE 2.**

One and zero are also used as exponents. Consider the following:

$$3 \cdot 3 \cdot 3 = 3^3$$

$$3 \cdot 3 = 3^2$$

We are dividing the left side of the first expression by 3 to get the left side of the second expression. Hence, continuing the pattern, the next expressions are

$$3 = 3^1$$

and

$$1 = 3^0$$

This means that $3 = 3^1$ and $1 = 3^0$. Exponents of 0 and 1 are defined as follows.

$a^1 = a$	for any number a
$a^0 = 1$	for any number a except 0

EXAMPLE 3 Write without exponents.

a. $4^1 = 4$

b. $9^0 = 1$

c. $(-3y)^0 = 1$ if $y \neq 0$ ■

□ **DO EXERCISE 3.**

2 Order of Operations

When we work examples that contain more than one operation, we must know which operation to do first. The following agreement has been established.

> If grouping symbols are present: do the operations within grouping symbols first, in the order given below. If there are no grouping symbols:
>
> 1. Evaluate exponential expressions working from left to right.
> 2. Perform multiplication and division in the order in which they occur, working from left to right.
> 3. Add or subtract in order from left to right.
> 4. If there is a fraction in the expression, do all work in the numerator and in the denominator, then simplify, if possible.

EXAMPLE 4 Simplify by writing with the least number of symbols.

a. $(-3)^2 + 6 - 3^2 = 9 + 6 - 9$ Evaluate exponential expressions.

$= 6$

Caution: Notice that $(-3)^2 = (-3)(-3) = +9$ but $-3^2 = -(3)(3) = -9$.

b. $5 - 2(3 + 4) = 5 - 2(7)$ Do the operations within grouping symbols.

$= 5 - 14$ Multiply before subtracting.

$= -9$

c. $3^2 + \frac{10}{5} = 9 + \frac{10}{5}$ Evaluate the exponential expression.

$= 9 + 2$ Divide before adding.

$= 11$

d. $12 \div 3(2) = 4(2) = 8$ Remember to divide and multiply in order from left to right.

e. $\frac{9 - 2 \cdot 7}{4(3) + 2} = \frac{9 - 14}{12 + 2} = \frac{-5}{14}$ Do all work in the numerator and denominator.

f. $16 \div 4(2) - 5 + 1 = 4(2) - 5 + 1$

$= 8 - 5 + 1$

$= 4$ ■

□ DO EXERCISE 4.

□ **Exercise 4** Simplify by writing with the least number of symbols.

a. $7 + 3(8 - 11)$

b. $2(8) - 5^2$

c. $18 \div 6(4)$

d. $\frac{6 - 2(5 + 3)}{3^2 - 4}$

e. $4 + 7 - (-4)^2$

CALCULATORS

Many scientific calculators do automatically some of the order of operations. To see if your calculator has this feature, do the following exercise.

$$8 + 3(7)$$

ENTER	DISPLAY
[8]	8
[+]	8
[3]	3
[×]	3
[7]	7
[=]	29

If the display reads 29, your calculator has the order of operations built into it. If the display reads 77, you can get the correct result, 29, by using the order of operations.

ENTER	DISPLAY
[3]	3
[×]	3
[7]	7
[+]	21
[8]	8
[=]	29

The following are other examples.

EXPRESSION	BASIC CALCULATOR ENTER	SCIENTIFIC CALCULATOR ENTER	DISPLAY
$-5 - 8(-4)$	[8] [×] [4] [=]		32
	Therefore, $(-8)(-4) = 32$.		
	[32] [−] [5] [=]		27
		[5] [+/−] [−] [8] [×] [4] [+/−] [=]	27
$\frac{18 - 2(3)}{4}$	[2] [×] [3] [=]		6
	[18] [−] [6] [=]		12
	[12] [÷] [4] [=]	[18] [−] [2] [×] [3] [=] [÷] [4] [=]	3
$5^2 + 9(-8)$	[5] [×] [5] [=]		25
	[9] [×] [8] [=]		72
	Therefore $9(-8) = -72$.		
	[72] [−] [25] [=]		47
	Therefore, $25 - 72 = -47$.	[5] [x^2] [+] [9] [×] [8] [+/−] [=]	−47

A scientific calculator was used for these examples.

EXPRESSION	SCIENTIFIC CALCULATOR ENTER	DISPLAY
$5 - (6 - 9)$	[5] [−] [(] [6] [−] [9] [)] [=]	8
$-6(3)^4 - 27$	[6] [+/−] [3] [y^x] [4] [=] [−] [27] [=]	−513

DID YOU KNOW?

The famous mathematician and philosopher, René Descartes, introduced our present exponential notation in his book, *Discours de la methode,* in 1637. Mathematicians wrote out a term, for example $x \cdot x \cdot x$, until Descartes began using the condensed notation, x^3. Descartes said that he did his best thinking by staying in bed until noon.

Answers to Exercises

1. a. 5^3 **b.** $(-2x)^4$ **c.** $-3y^2$

2. a. 8 **b.** -27 **c.** $25y^2$ **d.** $81y^4$ **e.** $\frac{1}{16}$ **f.** 0.001

3. a. 5 **b.** 1 **c.** 1 **d.** 1

4. a. -2 **b.** -9 **c.** 12 **d.** -2 **e.** -5

Problem Set 1.5

NAME ______________________

DATE ______________________

CLASS ______________________

Write in exponential notation.

1. $7 \cdot 7 \cdot 7 \cdot 7$

2. $3 \cdot 3$

3. $x \cdot x \cdot x \cdot x \cdot x$

4. $yyyyyy$

5. $2a \cdot 2a \cdot 2a$

6. $5b \cdot 5b \cdot 5b \cdot 5b$

Evaluate.

7. 2^3

8. 3^4

9. $(-1)^3$

10. $(-1)^4$

11. $(-3y)^2$

12. $(4x)^3$

13. $(2xy)^3$

14. $(-3xy)^2$

15. 2^1

16. $(-5)^1$

17. $(-3)^0$

18. 9^0

19. $\left(\frac{7x}{9}\right)^0$

20. $\left(\frac{3y}{5}\right)^0$

Simplify by writing with the least number of symbols.

21. $7 - 3(8)$

22. $5 - 4(-6)$

23. $8 - (3 - 9)$

24. $1 - [2 - (-3)]$

25. $8^2 - 3(2)$

26. $7^2 + 5(3)$

27. $3 + (4 - 2^2)$

28. $5 - (3^2 - 7)$

29. $5 \div 5(6)$

30. $12 \div 6(3)$

31. $8 - \frac{15}{5}$

32. $2 - \frac{16}{8}$

33. $\frac{5(-3) + 1}{-8}$

34. $\frac{15 - 7(3)}{3}$

35. $\frac{3(7 - 5) - 2^2}{4}$

36. $\frac{2(3 - 6) - 4^2}{5}$

37. $\frac{8(3) + 2(5)}{-6}$

38. $\frac{3(7) - 2(4)}{2}$

39. $\frac{10 - 6(8 + 2)}{3^2(2) + 2}$

40. $\frac{7(3 - 4) + 5}{2 + 2^2(3)}$

41. $\frac{6 \div 3(2)}{4^2}$

42. $\frac{8(2) \div 4}{2^2}$

43. $\frac{7 + 3(4)}{10 + 3(3)}$

44. $\frac{8 - 2(6)}{2(5) - 6}$

45. $12 - 4(7 - 9)$

46. $15 - 3(6 - 10)$

47. $32 \div (-4) \div (-2)$

48. $45 \div (-9) \div \left(-\frac{1}{5}\right)$

49. $4^3 + 18 \cdot 24 - (23 + 15 \cdot 4)$

50. $2^5 + 21 \cdot 16 - 34 + 8^2$

51. $5 - 4(8) - 11$

52. $5 - [4(8) - 11]$

53. $[3(6 - 8)]^2$

54. $-3(5 - 9)^2$

55. $4^3 - 8^2 - 78$

56. $94 - 2^4 - 9^2$

57. $30 - 4^4 \div (-2)$

58. $5 \times 10^2 - 7000$

59. In the exponential expression 2^7, 2 is called the ____________ and 7 is called the ____________.

60. The number 5^4 is read "five to the fourth ____________."

61. The number 4^3 may be read "four ____________" or "four to the third ____________."

62. The number 6^2 may be read "six ____________" or "six to the second ____________."

Write without exponents. Round answers to the nearest thousandth.

63. $(3.42)^5$

64. $(2.31)^4$

65. $(-4.09)^2$

66. $(-3.57)^3$

67. $(-2.7)^5$

68. $(-4.9)^4$

69. $(-0.73)^3$

70. $(0.788)^3$

71. $(0.57)^5$

72. $(-0.35)^5$

73. $(-3.28)^4$

74. $(5.29)^2$

Simplify by writing with the least number of symbols.

75. $6.74 - (3.4)(2.8)$

76. $(5.9)(-6.7) + 3.52$

77. $(2.8)^2 - (4.2)(3.6)$

78. $(7.5)(-2.8) + (0.7)^2$

79. $\dfrac{12.34}{(3.2)(4.4)}$

80. $\dfrac{(-7.4)(3.9)}{6.2}$

81. $\dfrac{(7.5)(-4.2) - 3.29}{-5.7}$

82. $\dfrac{14.76 + (5.9)(-6.3)}{-3.4}$

83. $\dfrac{(5.5)^2 - (6.8)(-3.2)}{(-4.2)^3}$

84. $\dfrac{(-2.2)^3 + (-3.7)(4.5)}{(6.9)^2}$

85. $\dfrac{16.35 - (4.2)(7.4)}{(3.3)(2.7)}$

86. $\dfrac{19.48 + (2.2)(3.1)}{4.4 - 7.8}$

Write in exponential notation.

* 87. $(-4)(-4)(-4)$

* 88. $(-6y)(-6y)(-6y)(-6y)$

Write without exponents.

* 89. $\left(-\frac{1}{4}\right)^3$

* 90. $-\left(\frac{1}{8}\right)^2$

* 91. -2^4

* 92. $(-2)^4$

Simplify by writing with the least number of symbols.

* 93. $-\frac{3}{4}(-12) + \frac{3}{7}(-14) - \frac{2}{5}(-15)$

* 94. $6\left[2 + \frac{3}{4}(-12)\right] - 8 \cdot \frac{5}{2}$

* 95. $\dfrac{\frac{5}{6} \cdot 27 + 4}{\frac{4}{3} \cdot 9 - 2}$

* 96. $\dfrac{\frac{3}{4} \cdot 12 - 3}{8 - \frac{8}{5} \cdot 20}$

1.6 MORE ABOUT EXPONENTS

1 Negative Integer Exponents

Recall that we explained the definitions for a^1 and a^0 by expanding a power of 3 and showing the division of the right side of the equation by 3.

$$3^2 = 3 \cdot 3$$
$$3^1 = 3$$
$$3^0 = 1$$

Continuing the pattern, we have

$$3^{-1} = \frac{1}{3}$$
$$3^{-2} = \frac{1}{3^2}$$

This series suggests the following definition.

> For any positive integer, *n*,
>
> $$a^{-n} = \frac{1}{a^n}, \qquad a \neq 0$$

EXAMPLE 1 Rewrite with positive exponents.

a. $5^{-4} = \dfrac{1}{5^4}$

Caution: Do not confuse the negative exponent with the sign of the result.

$$5^{-4} = \frac{1}{5^4} \quad \text{but} \quad 5(-4) = -20$$

b. $x^{-3} = \dfrac{1}{x^3}, \quad x \neq 0$

c. $(-3x)^{-2} = \dfrac{1}{(-3x)^2}$

$$= \frac{1}{9x^2}, \quad x \neq 0 \quad ■$$

□ **DO EXERCISE 1.**

We may reverse the rule.

EXAMPLE 2 Rewrite with negative exponents.

a. $\dfrac{1}{3^5} = 3^{-5}$

b. $\dfrac{1}{(-x)^3} = (-x)^{-3}, \quad x \neq 0 \quad ■$

□ **DO EXERCISE 2.**

OBJECTIVES

1 *Rewrite expressions with or without negative exponents*

2 *Multiply and divide using the rules for exponents*

3 *Raise to powers using the rules for exponents*

□ **Exercise 1** Rewrite with positive exponents.

a. 7^{-2}

b. $(-2y)^{-3}, \quad y \neq 0$

□ **Exercise 2** Rewrite with negative exponents.

a. $\dfrac{1}{2^4}$

b. $\dfrac{1}{(-2y)^7}, \quad y \neq 0$

□ **Exercise 3** Multiply and write with positive exponents. Assume that all variables represent nonzero real numbers.

a. $4^5 \cdot 4^{-7}$

b. $x^{-3}x^{-8}$

c. $(-4x^{-2})(8x^9)$

d. $(-5y^7)(-2y^5)$

e. $(3ab^3)(-4a^{-6}b^{-5})$

f. $(8xy^{-3})(9x^3y^5)$

2 Multiplication and Division

Recall that

$$a^5 \cdot a^{-3} = a \cdot a \cdot a \cdot a \cdot a \cdot \frac{1}{a \cdot a \cdot a}$$

$$= a \cdot a \cdot \frac{a \cdot a \cdot a}{a \cdot a \cdot a} = a^2 \qquad \text{Eliminate } \frac{a \cdot a \cdot a}{a \cdot a \cdot a} \text{ or } 1.$$

Hence

$$a^5 \cdot a^{-3} = a^2.$$

Notice that we may get the result above by adding the exponents. This is true for all exponential expressions with the same base.

> **Product Rule for Exponents**
>
> To multiply exponential expressions with the same base, we keep the base and add the exponents. For any integers m and n, and any nonzero real number a,
>
> $$a^m \cdot a^n = a^{m+n}$$

EXAMPLE 3 Multiply and write with positive exponents. Assume that all variables represent nonzero real numbers.

a. $7^3 \cdot 7^2 = 7^{3+2} = 7^5$ Do not multiply the bases.

b. $x^{-2} \cdot x^{-4} = x^{-2+(-4)} = x^{-6} = \dfrac{1}{x^6}$

c. $2x^{-4}y(-3x^2y^3) = 2(-3)x^{-4}x^2y^1y^3$ Recall that $y = y^1$.

$= -6x^{-4+2} \cdot y^{1+3}$ Multiply coefficients.

$= -6x^{-2}y^4$ Use the product rule.

$= -\dfrac{6y^4}{x^2}$ ■

□ **DO EXERCISE 3.**

When dividing exponential expressions with the same base,

$$\frac{a^4}{a^2} = \frac{a \cdot a \cdot a \cdot a}{a \cdot a} = \frac{a \cdot a}{a \cdot a} \cdot a \cdot a = a^2$$

We may simplify by subtracting the exponents.

> **Quotient Rule for Exponents**
>
> To divide exponential expressions with the same base, keep the base and subtract the exponent of the denominator from the exponent of the numerator. For any integers m and n, and any nonzero real number a,
>
> $$\frac{a^m}{a^n} = a^{m-n}$$

EXAMPLE 4 Divide and write with positive exponents. Assume that all variables represent nonzero real numbers.

a. $\dfrac{8^9}{8^4} = 8^{9-4} = 8^5$

b. $\dfrac{8^3}{8^{-4}} = 8^{3-(-4)} = 8^{3+4} = 8^7$ Do not divide the bases.

c. $\dfrac{x^7}{x^4} = x^{7-4} = x^3$

d. $\dfrac{y^{-2}}{y^{-5}} = y^{-2-(-5)} = y^{-2+5} = y^3$

e. $\dfrac{12x^3y^9}{3x^5y^6} = 4x^{3-5}\,y^{9-6}$ Divide the coefficients and use the quotient rule for exponents.

$= 4x^{-2}y^3 = \dfrac{4y^3}{x^2}$

f. $\dfrac{5x^7y^4}{10x^2y^4} = \dfrac{5}{10}x^{7-2}y^{4-4} = \dfrac{1}{2}x^5y^0$ Recall that $y^0 = 1$.

$= \dfrac{x^5}{2}$ ■

□ DO EXERCISE 4.

□ **Exercise 4** Divide and write with positive exponents. Assume that all variables represent nonzero real numbers.

a. $\dfrac{x^8}{x^5}$

b. $\dfrac{8^3}{8^7}$

c. $\dfrac{15x^4y^2}{3x^{-5}y^{-6}}$

d. $\dfrac{-3xy^{-5}}{9xy^{-2}}$

3 Raising Powers to Powers

$$(a^3)^2 = a^3 \cdot a^3 = (a \cdot a \cdot a)(a \cdot a \cdot a) = a^6$$

Power Rule for Exponents

To raise exponential expressions to a power, keep the base and multiply the exponents. For any nonzero real number *a*, and any integers *m* and *n*,

$$(a^m)^n = a^{mn}$$

EXAMPLE 5 Simplify by raising to powers. Write with positive exponents.

a. $(x^3)^5 = x^{3(5)} = x^{15}$

b. $(3^{-5})^4 = 3^{(-5)(4)} = 3^{-20} = \dfrac{1}{3^{20}}$ ■

Caution: Do not confuse multiplication with raising to powers.

$$a^3 \cdot a^2 = a^{3+2} = a^5$$

$$(a^3)^2 = a^{3(2)} = a^6$$

□ DO EXERCISE 5.

□ **Exercise 5** Simplify by raising to powers. Write with positive exponents. Assume that all variables represent nonzero real numbers.

a. $(2^5)^3$

b. $(x^{-4})^{-2}$

□ **Exercise 6** Simplify by raising to powers. Write with positive exponents. Assume that all variables represent nonzero real numbers.

a. $(ab)^5$

b. $(a^2b^3)^5$

c. $(2x^2y^5)^2$

d. $(-3xy^{-5})^4$

e. $(-x^2y^{-7})^{-2}$

f. $(7x^{-3}y^{-4})^2$

Notice that $(2^3b^2)^3 = (2^3b^2) \cdot (2^3b^2) \cdot (2^3b^2) = 2^9b^6$.

Raising a Product to a Power

A product of exponential expressions to a power may be simplified by raising each factor to the given power. For any nonzero real numbers *a* and *b* and any integer *m*,

$$(ab)^m = a^m b^m$$

EXAMPLE 6 Simplify by raising to powers. Write with positive exponents. Assume that all variables represent nonzero real numbers.

a. $(ab)^3 = a^{1(3)}b^{1(3)} = a^3b^3$

b. $(a^3b^4)^2 = a^{3(2)}b^{4(2)} = a^6b^8$

c. $(3x^4y^{-2})^3 = 3^3(x^4)^3y^{(-2)3}$ Raise each factor to the power.

$$= 3^3x^{12}y^{-6} = \frac{27x^{12}}{y^6}$$

d. $(-2xy^{-3})^{-5} = (-2)^{-5}(x^1)^{-5}(y^{-3})^{-5}$

$$= (-2)^{-5}x^{-5}y^{15} = \frac{y^{15}}{(-2)^5x^5} = \frac{y^{15}}{-32x^5}$$ ■

□ **DO EXERCISE 6.**

Raising a Quotient to a Power

Fractions may be raised to powers by raising the numerator to the power and the denominator to the power. For any integer *m* and any nonzero real numbers *a* and *b*,

$$\left(\frac{a}{b}\right)^m = \frac{a^m}{b^m}$$

EXAMPLE 7 Simplify by raising to powers. Write with positive exponents. Assume that all variables represent nonzero real numbers.

a. $\left(\frac{a}{b}\right)^6 = \frac{a^{1(6)}}{b^{1(6)}} = \frac{a^6}{b^6}$

b. $\left(\frac{x^{-3}}{y^2}\right)^5 = \frac{x^{(-3)(5)}}{y^{2(5)}} = \frac{x^{-15}}{y^{10}} = \frac{1}{x^{15}y^{10}}$

c. $\left(\frac{a^3b^{-2}}{2b^5}\right)^2 = \frac{(a^3b^{-2})^2}{(2b^5)^2}$

$= \frac{a^{3(2)}b^{(-2)(2)}}{2^2b^{5(2)}} = \frac{a^6b^{-4}}{4b^{10}} = \frac{a^6}{4b^{14}}$ ■

□ **DO EXERCISE 7.**

Definitions and Rules for Exponents

$a^{-n} = \frac{1}{a^n}, \quad a \neq 0 \qquad a^m \cdot a^n = a^{m+n}$

$a^1 = a \qquad \frac{a^m}{a^n} = a^{m-n}$

$a^0 = 1, \quad a \neq 0 \qquad (a^m)^n = a^{mn}$

$(ab)^m = a^m b^m$

$\left(\frac{a}{b}\right)^m = \frac{a^m}{b^m}$

DID YOU KNOW?

Archimedes, born about 287 B.C., was one of the greatest mathematicians in history. He stated the rule, $a^m \cdot a^n = a^{m+n}$ in his work *Sand Reckoner*. In this work he also estimated the number of grains of sand needed to fill the universe.

□ **Exercise 7** Simplify by raising to powers. Write with positive exponents. Assume that all variables represent nonzero real numbers.

a. $\left(\frac{a^4}{b^{-2}}\right)^{-5}$

b. $\left(\frac{3x^{-2}y^8}{y^3}\right)^2$

Answers to Exercises

1. a. $\frac{1}{7^2}$ **b.** $\frac{1}{(-2y)^3}$

2. a. 2^{-4} **b.** $(-2y)^{-7}$

3. a. $\frac{1}{4^2}$ or $\frac{1}{16}$ **b.** $\frac{1}{x^{11}}$ **c.** $-32x^7$ **d.** $10y^{12}$ **e.** $\frac{-12}{a^5b^2}$
f. $72x^4y^2$

4. a. x^3 **b.** $\frac{1}{8^4}$ **c.** $5x^9y^8$ **d.** $\frac{-1}{3y^3}$

5. a. 2^{15} **b.** x^8

6. a. a^5b^5 **b.** $a^{10}b^{15}$ **c.** $4x^4y^{10}$ **d.** $\frac{81x^4}{y^{20}}$ **e.** $\frac{y^{14}}{x^4}$ **f.** $\frac{49}{x^6y^8}$

7. a. $\frac{1}{a^{20}b^{10}}$ **b.** $\frac{9y^{10}}{x^4}$

Problem Set 1.6

NAME ____________

DATE ____________

CLASS ____________

Assume that all variables represent nonzero real numbers.

Rewrite with positive exponents.

1. 4^{-3}

2. 5^{-2}

3. y^{-4}

4. a^{-7}

5. $(-3x)^{-1}$

6. $(-2y)^{-5}$

Rewrite with negative exponents.

7. $\frac{1}{3^4}$

8. $\frac{1}{7^3}$

9. $\frac{1}{(-3)^5}$

10. $\frac{1}{(-2)^8}$

11. $\frac{1}{(6y)^7}$

12. $\frac{1}{(8y)^2}$

Multiply and write with positive exponents.

13. $3^8 \cdot 3^4$

14. $2^5 \cdot 2^3$

15. $6^{-8} \cdot 6^2$

16. $5^{-4} \cdot 5^2$

17. $x^{-3} \cdot x^{-2}$

18. $y^{-5} \cdot y^{-4}$

19. $a^3 \cdot a^{-6}$

20. $b^5 \cdot b^{-2}$

21. $(-2x)(-3x^2)$

22. $(3y)(-5y^3)$

23. $(7x^2y^3)(-3x^5y)$

24. $(-2x^4y)(8xy^7)$

Divide and write with positive exponents.

25. $\frac{4^5}{4^2}$

26. $\frac{7^3}{7^2}$

27. $\frac{x^9}{x^{-2}}$

28. $\frac{y^5}{y^{-3}}$

29. $\frac{a^{-2}}{a^{-4}}$

30. $\frac{b^{-5}}{b^{-3}}$

31. $\frac{10^{-2}}{10^3}$

32. $\frac{10^{-5}}{10^2}$

33. $\frac{-4x^2}{2x}$

34. $\frac{9x^5}{-3x}$

35. $\frac{-25x^{-3}y^5}{5x^2y^2}$

36. $\frac{18x^8y^{-2}}{-3x^3y^5}$

37. $\frac{-2x^{-4}y^5}{8x^{-6}y^8}$

38. $\frac{16x^7y^{-3}}{-20x^9y^{-5}}$

39. $\frac{x^{-5n}}{x^{7n}}$

40. $\frac{y^{-8p}}{y^{6p}}$

41. $\frac{a^{4p}}{a^{-9p}}$

42. $\frac{b^n}{b^{-2n}}$

43. $\frac{x^{-2t}}{x^{-3t}}$

44. $\frac{x^{-7q}}{x^{-5q}}$

45. $\frac{-8x^{-2}y^4}{-12x^7y^{-6}}$

46. $\frac{-14x^3y^8}{-21x^{-5}y^{-4}}$

47. $\frac{-32a^{-9}b^6}{24a^3b^7}$

48. $\frac{48p^5q^{-4}}{36p^{-2}q^{-8}}$

49. $\frac{5^{2n+2}}{5^{n+3}}$

50. $\frac{9^{n+4}}{9^{3n-7}}$

Simplify by raising to powers. Write with positive exponents.

51. $(2^3)^4$ **52.** $(3^5)^2$ **53.** $(x^{-3})^{-4}$ **54.** $(x^{-2})^{-5}$

55. $(7^{-4})^3$ **56.** $(5^{-2})^7$ **57.** $(3x^2y^3)^2$ **58.** $(2x^3y^5)^3$

59. $(8x^{-2}y)^2$ **60.** $(9xy^{-3})^2$ **61.** $(-x^3y^{-5})^{-5}$ **62.** $(-x^4y^{-8})^{-3}$

63. $\left(\frac{a^3}{b^{-5}}\right)^{-6}$ **64.** $\left(\frac{a^{-5}}{b^6}\right)^{-7}$ **65.** $\left(\frac{2x^{-3}y^4}{x^5}\right)^3$ **66.** $\left(-\frac{3xy^{-3}}{x^2y^4}\right)^{-2}$

67. $\left(\frac{-2x^4y^{-2}}{xy^3}\right)^2$ **68.** $\left(\frac{3a^5b^{-3}}{a^{-2}b}\right)^{-3}$ **69.** $(7^4)^x$ **70.** $(3^5)^y$

71. $(4^{2p})^{6q}$ **72.** $(10^{5p})^{3q}$ **73.** $\left(\frac{-5x^6y^{-3}}{7x^{-2}y^5}\right)^{-2}$ **74.** $\left(\frac{3x^4y^7}{8x^{-3}y^8}\right)^{-1}$

75. $\left(\frac{8^{-1}x^{-3}y^4}{4^{-2}x^5y^{-6}}\right)^{-3}$ **76.** $\left(\frac{6^{-2}a^{-5}b^8}{3^{-4}a^7b^{-2}}\right)^{-4}$ **77.** $\left(\frac{-300a^7b^{-2}}{5a^3b^{-4}}\right)^8$ **78.** $\left(\frac{128p^{-3}q^{-1}}{-4p^{-5}q^{-6}}\right)^6$

79. $\left(\frac{5x^{-4}y^3}{20x^2y^5}\right)^{-9}$

80. To multiply exponential expressions with the same base, we keep the base and ____________ the exponents.

81. To divide exponential expressions with the same base, keep the base and ____________ the exponents.

82. To raise exponential expressions to a power, keep the base and ____________ the exponents.

83. Fractions may be raised to powers by raising the ____________ to the power and the ____________ to the power.

Simplify by raising to powers and evaluating. Round answers to the nearest hundredth.

84. $[(4^5)]^2$

85. $[(5^2)]^4$

86. $[(2^3)]^{-2}$

87. $[(3^{-2})]^{-3}$

88. $[(3.5)^2]^3$

89. $[(4.9)4]^2$

Multiply and write with positive exponents.

* **90.** $\left(\frac{1}{2}\right)^3\left(\frac{1}{2}\right)^4$

* **91.** $\left(\frac{1}{5}\right)^7\left(\frac{1}{5}\right)^{-2}$

Simplify and write with positive exponents.

* **92.** $\dfrac{2^{-1}x^4(x^3)^{-1}}{2^3x^{-2}}$

* **93.** $\dfrac{3^2y^4(y^{-2})^{-1}}{3^4y^{-3}}$

* **94.** $\left[\dfrac{(-x^2)^6y^2}{2x^{10}}\right]^2$

* **95.** $\left[\dfrac{(a^2b)^2c^4}{a(bc)^3}\right]^3$

* **96.** $\left[\left(\dfrac{x^{-2}}{y^4}\right)^{-2}\cdot\left(\dfrac{x^3}{y^{-2}}\right)^3\right]^{-1}$

* **97.** $\left[\left(\dfrac{3a}{b^{-2}}\right)^{-1}\cdot\left(\dfrac{a^4}{b^3}\right)^{-2}\right]^3$

Think About It

* **98.** Write with a positive exponent $\dfrac{1}{a^{-n}}$.

* **99.** Show that $\left(\dfrac{a}{b}\right)^{-n}=\left(\dfrac{b}{a}\right)^n$.

1.7 SCIENTIFIC NOTATION

In the application of mathematics to the sciences we often find very large and very small numbers. For example, the distance from Earth to Mars is about 220,000,000 miles.

1 Scientific Notation

Scientific notation is a way of writing these numbers using exponents.

> A number is in *scientific notation* if it is a number between 1 and 10 times an integer power of 10. Powers of 10 are also in scientific notation.
>
> $$a \times 10^n$$
>
> where $1 \le a < 10$ and n is an integer is in scientific notation.

The following numbers are in scientific notation.

$$3.25 \times 10^4 \quad \text{and} \quad 2.8 \times 10^{-3}$$

EXAMPLE 1

a. Write 752 in scientific notation.
We want a number between 1 and 10, so we write 7.52. But now we have moved the decimal point two places to the left, which is the same as *dividing* by 100, so we must *multiply* 7.52 by 100 or 10^2 to give an equivalent number.

$$752 = 7.52 \times 10^2$$

b. Write 0.0025 in scientific notation.
We write 2.5 to get a number between 1 and 10, but we have moved the decimal point three places to the right. This is the same as multiplying by 1000, so we must divide 2.5 by 1000 or multiply by $\frac{1}{1000}$ or 10^{-3}. Hence

$$0.0025 = 2.5 \times 10^{-3} \quad ■$$

□ **DO EXERCISE 1.**

OBJECTIVES

1. *Write numbers in scientific notation*
2. *Write numbers in decimal notation*
3. *Calculate with numbers in scientific notation*

□ **Exercise 1** Write in scientific notation.

a. 3540

b. 7,450,000

c. 0.028

d. 0.000045

CALCULATOR

Numbers in scientific notation may be entered on some calculators with a key labeled

[EXP] or [EE] or [SCI]

NUMBER	ENTER
534,000	[5.34] [EXP] [5]
0.00000083	[8.3] [EXP] [7] [+/−]

□ Exercise 2 Write in decimal notation.

a. 6.98×10^5

b. 2.34×10^7

c. 3.28×10^{-3}

d. 4×10^{-5}

□ Exercise 3 Calculate. Write answers in scientific notation.

a. $(6.34 \times 10^{12})(2.1 \times 10^{-14})$

b. $(3.8 \times 10^{-4})(2 \times 10^6)$

c. $\dfrac{1.59 \times 10^9}{3 \times 10^7}$

d. $\dfrac{4.2 \times 10^{-3}}{1.68 \times 10^{-7}}$

2 Decimal Notation

We may convert from scientific notation to decimal notation.

EXAMPLE 2

a. Write 5.38×10^4 in decimal notation.

$$5.38 \times 10^4 = 5.38 \times 10{,}000 = 53{,}800$$

b. Write 4.57×10^{-1} in decimal notation.

$$4.57 \times 10^{-1} = 4.57 \times \frac{1}{10} = \frac{4.57}{10} = 0.457$$

Dividing by 10 is the same as moving the decimal point one place to the left. ■

□ DO EXERCISE 2.

3 Multiplying and Dividing

When we calculate with very large or very small numbers, it is helpful to use scientific notation.

EXAMPLE 3 Calculate. Write answers in scientific notation.

a. $(3.4 \times 10^{17})(7.1 \times 10^{-12})$

$$(\mathbf{3.4} \times 10^{17})(\mathbf{7.1} \times 10^{-12})$$
$$= (\mathbf{3.4})(\mathbf{7.1}) \times 10^{17} \times 10^{-12}$$
$$= 24.14 \times 10^5$$
$$= 2.414 \times 10 \times 10^5$$

Change 24.14 to scientific notation.

$$= 2.414 \times 10^6$$

Recall that $10 = 10^1$.

b. $\dfrac{7.8 \times 10^{-9}}{2 \times 10^{-7}}$

$$\frac{7.8 \times 10^{-9}}{2 \times 10^{-7}} = \frac{7.8}{2} \times \frac{10^{-9}}{10^{-7}}$$
$$= 3.9 \times 10^{-9-(-7)}$$
$$= 3.9 \times 10^{-2}$$ ■

□ DO EXERCISE 3.

EXAMPLE 4 The minimum distance of the planet Jupiter from the Earth is 368,000,000 miles. How long would it take a rocket, traveling at 2900 miles per hour to reach Jupiter?

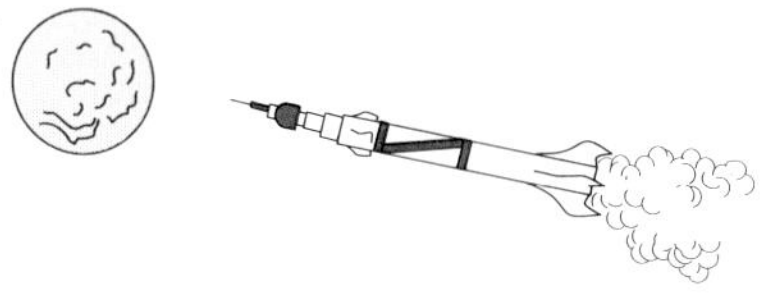

We must divide the number of miles to Jupiter by the speed of the rocket.

$$\frac{368,000,000}{2900}$$

Write each number in scientific notation. Then use the rules for exponents.

$$\frac{368,000,000}{2900} = \frac{3.68 \times 10^8}{2.9 \times 10^3}$$

$$= \frac{3.68}{2.9} \times \frac{10^8}{10^3}$$

$$\approx 1.3 \times 10^5$$

It takes the rocket approximately 1.3×10^5 hours to reach Jupiter. ■

□ **DO EXERCISE 4.**

□ **Exercise 4** A computer can do one addition in 1.4×10^{-7} second. How long would it take the computer to do a billion (10^9) additions? Give the answer in seconds and in minutes.

Answers to Exercises

1. a. 3.54×10^3 **b.** 7.45×10^6 **c.** 2.8×10^{-2} **d.** 4.5×10^{-5}

2. a. 698,000 **b.** 23,400,000 **c.** 0.00328 **d.** 0.00004

3. a. 1.3314×10^{-1} **b.** 7.6×10^2 **c.** 5.3×10 **d.** 2.5×10^4

4. 1.4×10^2 seconds or 140 seconds or approximately 2.3 minutes

Problem Set 1.7

NAME ______________

DATE ______________

CLASS ______________

Write in scientific notation.

1. 3720 **2.** 283 **3.** 50,400 **4.** 25

5. 0.0134 **6.** 0.00237 **7.** 0.138 **8.** 0.00029

9. 39,000,000 **10.** 5,700,000,000,000 **11.** 0.0000000023

12. 0.0000098 **13.** 0.00000000005 **14.** 0.000000000007

Write in decimal notation.

15. 4.52×10^2 **16.** 3.80×10^3 **17.** 6.69×10 **18.** 1.28×10^4

19. 3.356×10^{-2} **20.** 4.19×10^{-3} **21.** 2.03×10^{-1} **22.** 7.15×10^{-4}

23. 5.23×10^8 **24.** 7.19×10^9 **25.** 9.408×10^{-7} **26.** 3.685×10^{-10}

Calculate. Write answers in scientific notation.

27. $(3.2 \times 10^4)(7.8 \times 10^2)$

28. $(5.4 \times 10^6)(2.9 \times 10^3)$

29. $(3.1 \times 10^{-3})(4 \times 10^8)$

30. $(9.2 \times 10^{-7})(3 \times 10^4)$

31. $(2.18 \times 10^{-6})(3.7 \times 10^{-4})$

32. $(4.86 \times 10^{-2})(1.9 \times 10^{-5})$

33. $\dfrac{6.3 \times 10^7}{2.1 \times 10^5}$

34. $\dfrac{7.8 \times 10^9}{2.6 \times 10^3}$

35. $\dfrac{3.4 \times 10^3}{6.8 \times 10^5}$

36. $\dfrac{1.9 \times 10^6}{7.6 \times 10^8}$

37. $\dfrac{7.5 \times 10^{-3}}{2.5 \times 10^{-4}}$

38. $\dfrac{3.0 \times 10^{-6}}{6.0 \times 10^{-8}}$

39. The distance from the earth to the sun is 93,000,000 miles. Write this number in scientific notation.

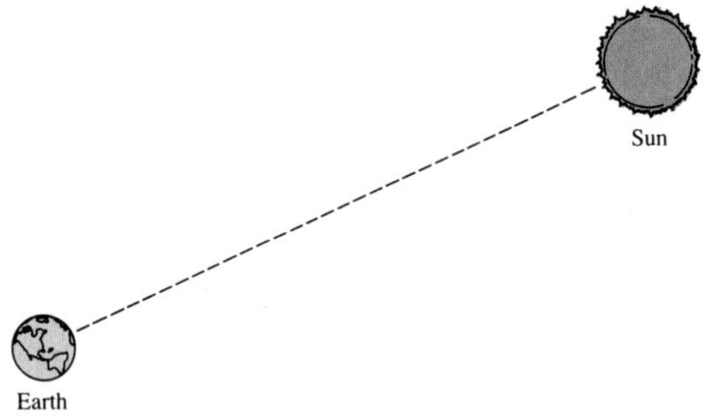

40. A certain computer can do one addition in 2×10^{-7} second. Write this number in decimal notation.

41. The wavelength of a light is 6.6×10^{-5} centimeter. Write this number in decimal notation.

42. There are 300,000 words in the English language. Write this number in scientific notation.

43. The distance light travels in 1 year is 5.87×10^{12} miles. How far does light travel in 32 years? Write your answer in scientific notation.

44. Saturn is 8.86×10^{8} miles from the sun. How long does it take a space probe, traveling at 3.1×10^{3} miles per hour, to reach the sun from Saturn? Write your answer in scientific notation.

45. A pulsating radio source, CP 1919, is 9.392×10^{16} miles from earth. If there are 5.87×10^{12} miles in a light year, how many light years is CP 1919 from Earth?

46. In 1990, it was estimated that the world's richest woman had assets of \$10,720,000,000. If she spent \$2,500,000 per year, how many years would it take her to spend her the amount of her assets in 1990?

47. The mass of an electron is 9.11×10^{-28} gram. What is the mass of 2 million electrons?

48. In 1989, 7,610,000 barrels of oil per day were produced in the United States. At this rate, how many barrels of oil were produced in 360 days?

49. A number is in ____________ if it is a number between 1 and 10 times an integer power of 10.

50. To write 397 in scientific notation, we must move the decimal point two places to the ____________ and multiply by 10^{2}.

51. The number 0.0035 is in scientific notation if we move the decimal point three places to the ____________ and multiply by 10^{-3}.

52. To write 2.75×10^{3} in decimal notation, we multiply 2.75 by 1000, which moves the decimal point three places to the ____________.

Calculate. Give answers in scientific notation. Round answers to the nearest tenth.

53. $\dfrac{7.38 \times 10^{-4}}{16.94 \times 10^{-3}}$

54. $\dfrac{8.934 \times 10^{5}}{2.74 \times 10^{2}}$

55. $\dfrac{3.7 \times 2.5 \times 10^4}{4.32 \times 10^{-2}}$

56. $\dfrac{7.2 \times 6.8 \times 10^{-3}}{9.46 \times 10^4}$

57. $\dfrac{5.4 \times 10^6 \times 16.73 \times 10^{-3}}{7.81 \times 10^3 \times 4.5 \times 10^{-2}}$

58. $\dfrac{6.28 \times 10^{-5} \times 4.9 \times 10^4}{5.5 \times 10^8 \times 3.64 \times 10^{-6}}$

Think About It

First write each number in scientific notation. Calculate. Write answers in scientific notation.

59. $\dfrac{0.00008}{2000}$

60. $\dfrac{90{,}000}{0.03}$

61. $\dfrac{0.002 \times 2600}{0.000013}$

62. $\dfrac{0.015 \times 700}{0.03}$

63. $\dfrac{320{,}000}{0.08 \times 20{,}000}$

64. $\dfrac{400}{1000 \times 800{,}000}$

65. $\dfrac{0.018 \times 30{,}000}{200 \times 0.0006}$

66. $\dfrac{0.0003 \times 46{,}000}{0.000111}$

67. $\dfrac{760{,}000 \times 0.02}{0.00018 \times 800}$

68. $\dfrac{32 \times 0.0035}{160 \times 2500}$

Chapter 1 Summary

Section 1.1

A ***set*** is a collection of objects. The objects in the set are called ***elements*** of the set.

The ***natural numbers*** are the set of counting numbers.

$$\text{Natural numbers} = \{1, 2, 3, 4, \ldots\}$$

The ***whole numbers*** consist of zero and the natural numbers.

$$\text{Whole numbers} = \{0, 1, 2, 3, 4, \ldots\}$$

The ***integers*** are as follows:

$$\text{Integers} = \{\ldots, -3, -2, -1, 0, 1, 2, 3, \ldots\}$$

The ***rational numbers*** are of the form a/b, where a and b are integers and b is not zero.

$$\text{Rational numbers} = \left\{\frac{a}{b} \,\middle|\, a \text{ and } b \text{ are integers and } b \neq 0\right\}$$

Irrational numbers are numbers such as $\sqrt{2}$ that are not rational because they cannot be written as the ratio of two integers.

The ***real numbers*** are composed of the rational and irrational numbers.

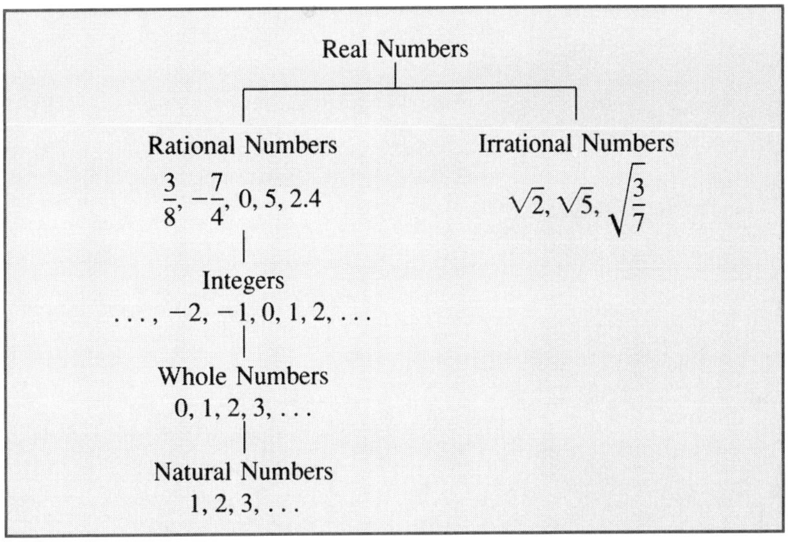

The ***absolute value*** of a number is its distance from zero on the number line. Absolute value is denoted by $|\ \ |$.

If a number is to the right of another number on the number line, it is the greater of the two numbers. If a number is to the left of another number, it is the smaller of the numbers. The symbol $>$ stands for "greater than" and $<$ means "less than." Statements such as $4 < 6$ and $-1 > -2$ are called ***inequalities.***

All positive real numbers are greater than zero and all negative real numbers are less than zero.

If x is a positive real number, then $x > 0$.

If x is a negative real number, then $x < 0$.

The ***opposites*** of the positive real numbers are the negative real numbers.

Similarly, the *opposites* of the negative real numbers are the positive real numbers. The *opposite* of zero is zero.

$a \leq b$ and $b \geq a$ are also inequalities. $a \leq b$ is read "a is less than or equal to b." $b \geq a$ is read "b is greater than or equal to a."

For any real number a,

$$|a| = \begin{cases} a & \text{if } a \geq 0 \\ -a & \text{if } a < 0 \end{cases}$$

Section 1.2

To add two real numbers with the same sign, add their absolute values and give the answer the common sign.

To add two real numbers with different signs, subtract the smaller absolute value from the larger. Give the answer the sign of the number with the larger absolute value.

To subtract one real number from another, add the opposite of the second real number to the first real number.

$$a - b = a + (-b) \qquad \text{where } a \text{ and } b \text{ are real numbers}$$

Commutative law of addition: $a + b = b + a$ for any real numbers a and b.

Associative law of addition: $a + (b + c) = (a + b) + c$ for any real numbers a, b, and c.

Section 1.3

To multiply or divide two real numbers with the same sign, multiply or divide their absolute values. Give the answer a positive sign.

To multiply or divide two real numbers with different signs, multiply or divide their absolute values. Give the answer a negative sign.

Commutative law of multiplication: $a \cdot b = b \cdot a$ for any real numbers a and b.

Associative law of multiplication: $a \cdot (b \cdot c) = (a \cdot b) \cdot c$ for any real numbers a, b, and c.

Division by zero is undefined.

Two numbers whose product is 1 are called ***reciprocals*** or multiplicative inverses of each other. To find the reciprocal of a number, invert the number. Zero does not have a reciprocal, since 1/0 is undefined.

Every nonzero real number a has a reciprocal $1/a$.

To divide two fractions, multiply by the reciprocal of the divisor. For any real numbers a and b, $b \neq 0$,

$$a \div b = \frac{a}{b} = a \cdot \frac{1}{b}$$

To ***factor*** a number means to write the number as a multiplication of numbers. For any real numbers a and b, $b \neq 0$,

$$\frac{-a}{b} = -\frac{a}{b} = \frac{a}{-b}$$

Section 1.4

If an expression contains an addition or subtraction and multiplications all multiplications must be done first.

Distributive law of multiplication over addition:

$$a(b + c) = ab + ac \qquad \text{for any real numbers } a, b, \text{ and } c$$

Distributive law of multiplication over subtraction:

$$a(b - c) = ab - ac \qquad \text{for any real numbers } a, b, \text{ and } c$$

A ***term*** is a number or the product of a number and a variable or variables.

The ***numerical coefficient*** of a variable in a term is a number by which the variable is multiplied. The numerical coefficient is often simply called the ***coefficient.***

Like terms differ only in their numerical coefficients. The numerical coefficients may also be the same.

Three sets of ***grouping symbols*** are parentheses, ( ), brackets, [ ], and braces, { }.

The property of −1: For any number a, $-a = -1 \cdot a$.

To remove grouping symbols from an expression, work out the innermost grouping symbols first.

Section 1.5

A number written in ***exponential notation*** is shown below. The exponent is also called the ***power.***

$$3^4$$

exponent (4); base (3)

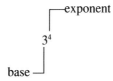

$a \cdot a \cdot a \cdot \cdots \cdot a \cdot a = a^n$, where n is an integer greater than 1.

a^n is read "a to the nth power."

a^3 is read "a-cubed" or "a to the third power."

a^2 is read "a-squared" or "a to the second power."

$a^1 = a$ for any number a.

$a^0 = 1$ for any number a except 0.

Order of operations agreement

If grouping symbols are present: do the operations within grouping symbols first, in the order given below. If there are no grouping symbols:

1. Evaluate exponential expressions working from left to right.
2. Perform multiplication and division in the order in which they occur, working from left to right.
3. Add or subtract in order from left to right.
4. If there is a fraction in the expression, do all work in the numerator and in the denominator, then simplify, if possible.

Section 1.6

For any positive integer n,

$$a^{-n} = \frac{1}{a^n} \qquad a \neq 0$$

Product rule for exponents: For any integers m and n, and any nonzero real number a,

$$a^m \cdot a^n = a^{m+n}$$

Quotient rule for exponents: For any integers m and n, and any nonzero real number a,

$$\frac{a^m}{a^n} = a^{m-n}$$

Power rule for exponents: For any nonzero real number a, and any integers m and n,

$$(a^m)^n = a^{mn}$$

Raising a product to a power: For any nonzero real numbers a and b, and any integer m,

$$(ab)^m = a^m b^m$$

Raising a quotient to a power: For any integer m, and any nonzero real numbers a and b,

$$\left(\frac{a}{b}\right)^m = \frac{a^m}{b^m}$$

Section 1.7

A number is in ***scientific notation*** if it is a number between 1 and 10 times an integer power of 10. Powers of 10 are also in scientific notation.

$a \times 10^n$ where $1 \leq a < 10$ and n is an integer is in scientific notation

Chapter 1 Additional Exercises (Optional)

NAME ______________________

DATE ______________________

CLASS ______________________

Section 1.1

Consider the numbers: $-4, \frac{11}{8}, -\frac{3}{5}, 0, 9$. *Find the following.*

1. The whole numbers **2.** The natural numbers **3.** The rational numbers **4.** The integers

Find the following.

5. $-|7|$ **6.** $-|-9|$ **7.** $|-4|$ **8.** $-|3|$

9. $-|-10|$ **10.** $-|-8|$ * **11.** $|-4| + 3$ * **12.** $|5| - |-2|$

Find the opposite.

13. -8 **14.** $-\frac{11}{3}$ **15.** $-\frac{3}{5}$ **16.** 12

Section 1.2

Add or subtract, as indicated.

17. $-\frac{3}{5} - \left(-\frac{2}{3}\right)$ **18.** $-\frac{3}{7} + \left(-\frac{6}{5}\right)$ **19.** $-\frac{3}{4} + \frac{1}{16}$

20. $-\frac{1}{5} - \frac{1}{10}$ **21.** $2.6 - (-5)$ **22.** $-7.1 + 3.2$

23. $-5 + 3 - 10 + 8$ **24.** $6 - 9 + 12 - 4$ * **25.** $|-3| - 4 + 7 - |6|$

* **26.** $|5| + 2 - |8| + 4$ * **27.** $|8 - 9| - |3 + 6|$ * **28.** $|9 - 2| + |3 - 5|$

Is the commutative or associative law of addition illustrated by the following equations?

29. $6 + (4 + 3) = (6 + 4) + 3$ **30.** $5 + (6 + 2) = (6 + 2) + 5$

31. $3 + (2 + 8) = (2 + 8) + 3$ **32.** $7 + (9 + 4) = (7 + 9) + 4$

Section 1.3

Multiply.

* **33.** $(-3.2)(-4.8)(2)$

* **34.** $(6.7)(-1.3)(5)$

35. $(-3)(-5)(-2)$

36. $(7)(9)(-3)$

Divide.

37. $-18 \div (-6)$

38. $24 \div (-4)$

39. $-\frac{2}{5} \div \frac{8}{5}$

40. $-\frac{10}{9} \div \left(-\frac{5}{6}\right)$

Find the reciprocal.

41. $\frac{1}{2}$

42. $\frac{-1}{8}$

43. 6

44. -4

Section 1.4

Evaluate when $a = -1$, $b = 3$, $c = -2$, and $d = 0$.

* **45.** $\frac{ad - bc}{ab}$

* **46.** $\frac{bd}{ab - ac}$

47. $\frac{bcd}{abc - c}$

48. $\frac{bd - ac}{abd - b}$

List the terms of the expression.

49. $x - 2y + 4$

50. $\frac{x}{4} - 8$

51. $x + 3y - 2$

52. $3x - 4y - z$

Simplify.

53. $5 - 4[x - (2x + 3)]$

54. $3 + 2[y - 3(y - 4)]$

55. $-3\{2[y - 3(3y + 1)]\}$

56. $-4\{x - [2x + (3x - 4)]\}$

Section 1.5

Evaluate.

57. $(-2xy)^3$

58. $(3xy)^2$

59. $(4xy)^2$

60. $(-2x)^2$

61. $(-5y)^2$

62. $(-xy)^3$

Simplify.

63. $\dfrac{17 - 3 \div 3}{8 + 3(2)}$

64. $\dfrac{4(3) \div 6}{3(8) - 4}$

65. $5 + 3(4 - 2)^2$

66. $6 + 8(3 + 4)^2$

Section 1.6

Simplify and write with positive exponents. Assume that variables represent nonzero real numbers.

67. $(3x^{-2})^2$

68. $(-2y^3)^{-2}$

69. $(-3^{-2})^{-3} \cdot (-3^2)^{-1}$

70. $(2^{-3})^{-2} \cdot (2^4)^{-1}$

* **71.** $\dfrac{(2x^2y^4)^{-3}}{(3x^{-2}y^3)^{-2}}$

* **72.** $\dfrac{(a^{-4}b^{-1})^{-3} \cdot (a^5b^{-2})^2}{a^7b^{-5}}$

Section 1.7

73. Light travels at about 186,000 miles per second. Write this number in scientific notation.

74. A pound of sugar contains about 2,260,000 grains. Write this number in scientific notation.

75. Saturn is about 8.86×10^8 miles from the sun. Write this number in decimal notation.

76. The velocity of light in a vacuum is about 3×10^{10} centimeters per second. Write this number in decimal notation.

COOPERATIVE LEARNING

1. Find two numbers whose absolute value is $\frac{11}{4}$.

2. Which number is greater, -25.8 or -37.4? Explain.

3. Dynagem Company and its subsidiaries had assets of \$6,540,000 and liabilities of \$8,900,000. ***Net worth*** is defined as assets minus liabilities. What was the company's net worth? If the net worth was distributed equally between the company and its four subsidiaries, what was the net worth of Dynagem Company?

4. Explain why zero does not have a reciprocal.

5. Use the formula $F = \frac{9}{5}C + 32$ where F is degrees Fahrenheit and C is degrees Celsius to convert -20 degrees Celsius to degrees Fahrenheit.

6. Write two factors whose product is $81y - 72$.

7. Simplify: $7p - \{p - 4 - [p - 3 - (p - 2) - 1] - p - 5\}$.

8. Simplify and write with positive exponents. Assume that all variables represent nonzero real numbers.

$$\left(\frac{-a^{-3}b^{-7}}{2a^{12}b^{-14}}\right)^3$$

9. The average daily increase in the world's population is about 270,000. What is the average yearly increase in the population? Write the answer in scientific notation.

STUDY HINT

Preparing for a Test

There is a practice test at the end of each chapter. Try to do the practice test without looking back in your book. Place a * by each problem that you cannot do correctly. When you have finished the test, look in the book to find how to do the starred problems. Study each one until you can do it without looking in the book or at your notes. Some students find that studying in groups is very helpful. Teaching a concept to someone is one of the best ways to learn it yourself.

Show your work, neatly, on the test. Often, your instructor will give partial credit for the correct method of working a problem. You will usually make fewer errors if you show your work, carefully.

Work the easier problems, first. Then complete the more difficult problems. Check your work.

Some students have math anxiety and get nervous about tests. Usually, if you know the material very well, you will have more confidence and be less nervous. Also, many schools have counselors who give classes on test anxiety. Check to see if this service is available at your school.

Chapter 1 Practice Test

NAME ______

DATE ______

CLASS ______

Find the following.

1. $|-5.3|$

2. $\left|\frac{5}{6}\right|$

3. Replace the ? with > or <: $-9.3 \ ? \ -7.1$

Add.

4. $(-5) + (-2)$

5. $6.3 + (-9.4)$

6. $-\frac{2}{3} + \frac{3}{4}$

Subtract.

7. $-9 - 7$

8. $-14.7 - (-15.5)$

9. $-\frac{1}{3} - \frac{1}{7}$

Multiply.

10. $(-2.3)(-7.4)$

11. $-\frac{9}{5}\left(\frac{7}{18}\right)$

12. $(-9)(-8)(-2)$

Divide.

13. $\frac{-18}{-6}$

14. $\frac{-50}{10}$

15. $\frac{64}{-8}$

16. $\frac{-2}{3} \div \left(\frac{-9}{7}\right)$

17. $\frac{-20}{3} \div \frac{10}{9}$

18. $\frac{0}{4}$

Find the reciprocal.

19. $\frac{3}{8}$

20. -4

1. ______
2. ______
3. ______
4. ______
5. ______
6. ______
7. ______
8. ______
9. ______
10. ______
11. ______
12. ______
13. ______
14. ______
15. ______
16. ______
17. ______
18. ______
19. ______
20. ______

21. ______________________

22. ______________________

23. ______________________

24. ______________________

25. ______________________

26. ______________________

27. ______________________

28. ______________________

29. ______________________

30. ______________________

31. ______________________

32. ______________________

33. ______________________

34. ______________________

35. ______________________

36. ______________________

21. Evaluate when $x = 3$, $y = -2$, and $z = -1$: $3xy - 4yz$.

Multiply.

22. $7(2x - 4)$

23. $-2(3y - 4)$

Combine like terms.

24. $7b - 4c - 9b + c$

25. $8x \quad 4y \quad 7x \quad 3y$

Simplify.

26. $9a - (3a + 4)$

27. $4(x - 2) - 3(x - 2)$

28. $4x - 2[3 - 5(2x + 1)]$

29. $\dfrac{3^2 - 2(7)}{4}$

For problems 30–33, assume that all variables represent nonzero real numbers. Perform the operation indicated. Write with positive exponents.

30. $(-2x^2y^{-2})(-4x^3y^{-5})$

31. $\dfrac{-16x^{-4}y^3}{12x^{-5}y^8}$

Simplify and write with positive exponents.

32. $(-4x^{-3}y^2)^{-2}$

33. $\left(\dfrac{-x^{-2}y^3}{2x^5y^{-1}}\right)^2$

34. Write in scientific notation: 8,325,000.

35. Write in decimal notation: 4.7×10^{-3}.

36. The mass of a hydrogen atom is 1.673×10^{-24} gram. What is the mass of 3000 hydrogen atoms?

CHAPTER 2

Linear Equations and Inequalities

Pretest

Solve.

1. $x - 21 = 35$

2. $-\frac{3}{4}x = -\frac{21}{12}$

3. $3.7y - 21.4 = -5.4y + 42.3$

4. $7x - 3(2x - 1) = -(x + 8)$

5. The perimeter of a rectangular parking area is 84 meters. If the length is 4 meters more than the width, find the dimensions of the parking area.

6. Find three consecutive odd integers such that the sum of the first and the third is 178.

7. Anton invested \$12,500 for 1 year and earned \$790 interest. Some of the money was invested at 6% and the remainder at 7%. How much was invested at 7%?

8. Solve $5y - 3x = 7$ for y.

9. Solve and graph $3x + 8 \geq 5x - 2$.

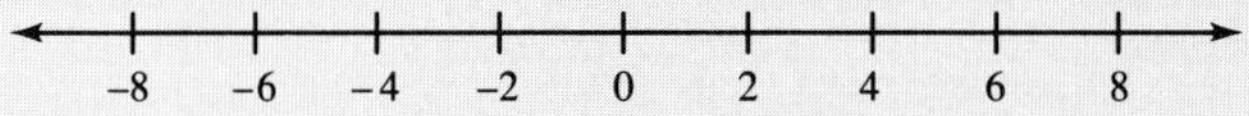

Solve.

10. $15x < -75$

11. $-2(x - 7) \leq -5(2x + 1)$

12. $|6x + 4| = 16$

13. $|7 - 12y| = -2$

14. $|x - 5| = |x - 2|$

15. A typist can be paid in two ways:

Method 1: $200 plus $6 per hour

Method 2: $8 per hour

If the job takes n hours, for what values of n does Method 1 give the typist better wages?

Solve and graph.

16. $|3x - 9| < 12$

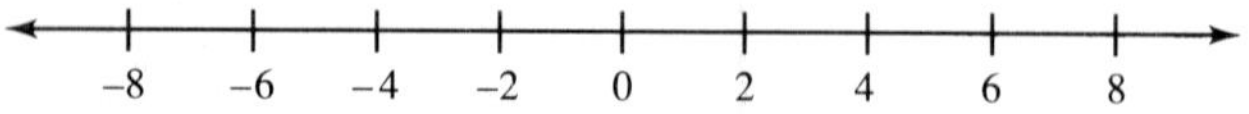

17. $|5x - 4| \geq 6$

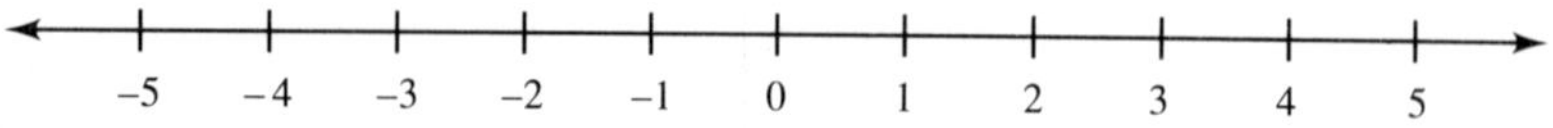

2.1 SOLVING LINEAR EQUATIONS

In Sections 2.2 and 2.3 we solve applied problems using linear equations. In this section we show how to solve the equations.

OBJECTIVES

1 *Determine if an equation is false, true, or conditional*

2 *Determine if a given value of a variable is a solution of an equation*

3 *Solve equations using the addition property*

4 *Solve equations using the multiplication property*

5 *Solve equations using both the addition and multiplication properties*

6 *Solve equations containing parentheses*

7 *Solve equations containing fractions or decimals*

1 True, False, and Conditional Equations

An ***equation*** is a statement which says that two mathematical expressions, such as $x + 4$ and 6, represent the same number. Some equations are always false, some are always true, and other equations are sometimes true and sometimes false, depending on the value assigned to the variable. We call the latter equations ***conditional.***

EXAMPLE 1 Determine if the equations are true, false, or conditional.

a. $4 + 3 = 8$ is false.

b. $7 + 8 = 15$ is true.

c. $x - 2 = 6$ is conditional. We have not assigned a value to the variable x.

d. $3(x - 4) = 3x - 12$ is true for any value of the variable x.

e. $2(x + 1) = 2x + 3$ is false for any value of x. ■

□ **DO EXERCISE 1.**

A ***linear equation in one variable*** is an equation with an exponent of 1 on the variable in any term. The equation $x + 4 = 10$ is linear. The equation $x^2 - 2x + 3 = 0$ is not linear because the exponent is 2 on a variable in one of the terms. Terms containing only numbers are sometimes considered to contain the variable $x^0 = 1$, for example, $3 = 3x^0 = 3(1) = 3$. For purposes of the definition of a linear equation, we will assume that terms containing only numbers do not include a variable. We will learn more about linear equations in Chapter 3.

2 Solutions of Equations

A ***solution of an equation*** is a number that gives us an equation which is true when we substitute that number for the variable. The set of all solutions is called the ***solution set.***

EXAMPLE 2 Is 3 a solution for the following equation?

$$5x - 4 = 11$$

Yes. If we substitute 3 for x, we get a true equation.

$5x - 4$	11
$5(3) - 4$	11
$15 - 4$	11
11	11

■

□ **DO EXERCISE 2.**

□ **Exercise 1** Determine if the equations are true, false, or conditional.

a. $9 - 2 = 7$

b. $x + 3 = 7$

c. $9 + 7 = 17$

d. $8 + x = x$

e. $3(x - 4) = 3x + 13$

f. $2x - 8 = 2(x - 4)$

□ **Exercise 2** Are the given values of the variables solutions for the equations?

a. $2x - 5 = 3, \quad x = 4$

b. $3x - 8 = 15, \quad x = 7$

c. $9 - x = 15, \quad x = -6$

d. $5 - 4x = 12, \quad x = -4$

□ **Exercise 3** Solve.

a. $x - 7 = 4$

b. $x + \frac{2}{3} = \frac{3}{4}$

c. $y - 19.1 = 89.3$

d. $15 = 28 + x$

3 The Addition Property of Equality

To ***solve*** an equation means to find all solutions of the equation. An equation may sometimes be solved by adding the same number to both sides.

Addition Property of Equality

If a, b, and c are real numbers and

$$a = b$$

then

$$a + c = b + c$$

EXAMPLE 3 Solve.

a. $x - 9 = -17$

$x - 9 + \mathbf{9} = -17 + \mathbf{9}$ Add 9 to each side of the equation.

$x = -8$

Check $x - 9 = -17$

$-8 - 9$	-17
-17	-17

Substitute -8 for x.

The solution is -8.

b. $\frac{3}{4} = \frac{1}{4} + x$

$\frac{3}{4} + \left(-\frac{\mathbf{1}}{\mathbf{4}}\right) = \frac{1}{4} + \left(-\frac{\mathbf{1}}{\mathbf{4}}\right) + x$ Add $-\frac{1}{4}$ to each side of the equation.

$\frac{2}{4} = x$

$\frac{1}{2} = x$ Simplify the fraction.

The number $\frac{1}{2}$ checks, so it is the solution. ■

□ **DO EXERCISE 3.**

4 The Multiplication Property of Equality

It is also possible to solve some equations by multiplying both sides of the equation by the same nonzero number.

Multiplication Property of Equality

If a, b, and c are real numbers and

$$a = b$$

then

$$ac = bc$$

EXAMPLE 4 Solve.

$$\frac{3x}{4} = 7$$

$$\mathbf{\frac{4}{3}} \cdot \frac{3x}{4} = 7 \cdot \mathbf{\frac{4}{3}} \qquad \text{Multiply both sides of the equation by } \tfrac{4}{3}.$$

$$x = \frac{28}{3}$$

Check

$\frac{3x}{4} = 7$	
$\frac{3}{4} \cdot \frac{28}{3}$	7
7	7

The solution is $\frac{28}{3}$. ■

In algebra, we usually give the solution as a simplified improper fraction.

□ **DO EXERCISE 4.**

We may divide both sides of the equation by the same nonzero number since this is the same as multiplying both sides by the reciprocal of the number.

EXAMPLE 5 Solve.

$$5x = 3$$

$$\frac{5x}{5} = \frac{3}{5} \qquad \text{Divide both sides of the equation by 5; this is the same as multiplying both sides by } \tfrac{1}{5}.$$

$$x = \frac{3}{5}$$

The answer $\frac{3}{5}$ checks, so it is the solution. ■

□ **DO EXERCISE 5.**

5 The Addition and Multiplication Properties of Equality

We may need to use both the addition and the multiplication properties of equality to solve equations. When both properties are used, it is usually easier to use the addition property first.

□ **Exercise 4** Solve.

a. $\frac{2x}{5} = -10$

b. $9x = 15$

□ **Exercise 5** Solve.

a. $-3.2y = -9.6$

b. $-\frac{x}{8} = \frac{5}{8}$

□ **Exercise 6** Solve.

a. $5x - 4 = 3$

b. $-8 + 2x = 6$

□ **Exercise 7** Solve.

a. $5y - 8 = -8 + 3y - y$

b. $6x - 15 = -7 - 4x - x$

EXAMPLE 6 Solve.

$$4x - 5 = 11$$

$$4x - 5 + \mathbf{5} = 11 + \mathbf{5} \quad \text{Use the addition property.}$$

$$4x = 16$$

$$\frac{4x}{4} = \frac{16}{4} \quad \text{Use the multiplication property, which allows us to divide by 4.}$$

$$x = 4$$

The number 4 checks, so it is the solution. ■

□ **DO EXERCISE 6.**

If there are like terms on one or both sides of an equation, we combine them first. Then if there are variables on both sides of the equation, we use the addition property to get the terms containing the variables on one side of the equation and all other terms on the other side. Terms containing variables may be on either side of the equation.

EXAMPLE 7 Solve.

$$3x - 2 + 4x = 7 - x + 15$$

$$7x - 2 = -x + 22 \quad \text{Combine like terms.}$$

$$7x = -x + 22 + 2 \quad \text{Add 2 to both sides.}$$

$$7x = -x + 24 \quad \text{Combine like terms.}$$

$$7x + x = 24 \quad \text{Add } x \text{ to both sides.}$$

$$8x = 24 \quad \text{Combine like terms.}$$

$$\frac{8x}{8} = \frac{24}{8} \quad \text{Divide both sides by 8.}$$

$$x = 3$$

Check $3x - 2 + 4x = 7 - x + 15$

$3x - 2 + 4x$	$7 - x + 15$
$3(3) - 2 + 4(3)$	$7 - 3 + 15$
$9 - 2 + 12$	$7 - 3 + 15$
19	19

The solution is 3. ■

□ **DO EXERCISE 7.**

6 Equations with Parentheses

Many equations with parentheses may be solved by first removing the parentheses. We remove the parentheses by using the distributive laws. Also, remember that to solve an equation, the sign on the variable must be positive in the last step.

EXAMPLE 8 Solve.

$20 - (y + 5) = 9(2y + 8)$	
$20 - y - 5 = 18y + 72$	Use a distributive law.
$15 - y = 18y + 72$	Combine like terms.
$-y = 18y + 72 + (-15)$	Add -15 to both sides.
$-y = 18y + 57$	Combine like terms.
$-19y = 57$	Add $-18y$ to both sides.
$\frac{-19y}{-19} = \frac{57}{-19}$	Divide both sides by -19 to give $+y$.
$y = -3$	

Check $20 - (y + 5) = 9(2y + 8)$

$20 - (-3 + 5)$	$9[2(-3) + 8]$
$20 - (2)$	$9[-6 + 8]$
18	18

The solution is -3. ■

□ **DO EXERCISE 8.**

7 Equations Containing Fractions

When an equation contains fractions or decimals we clear them first by using the multiplication property. We multiply both sides of the equation by the *least common denominator* (LCD).

EXAMPLE 9 Solve.

a.

$\frac{6y}{5} + \frac{3y}{10} = \frac{16}{5}$	
$\mathbf{10}\left(\frac{6y}{5} + \frac{3y}{10}\right) = \mathbf{10}\left(\frac{16}{5}\right)$	Multiply both sides by the LCD of 10.
$10\left(\frac{6y}{5}\right) + 10\left(\frac{3y}{10}\right) = 10\left(\frac{16}{5}\right)$	Use a distributive law.
$12y + 3y = 32$	Simplify.
$15y = 32$	
$y = \frac{32}{15}$	

The solution is $\frac{32}{15}$.

b.

$0.08x - 0.03x = 0.65$	
$\mathbf{100}(0.08x - 0.03x) = \mathbf{100}(0.65)$	Multiply both sides by 100 to clear the decimals.
$100(0.08x) - 100(0.03x) = 100(0.65)$	Use a distributive law.
$8x - 3x = 65$	
$5x = 65$	
$x = 13$	

The solution is 13.

□ **Exercise 8** Solve.

a. $5x - (2x - 10) = 25$

b. $3(y - 4) - (2y - 3) = -4$

□ **Exercise 9** Solve.

a. $\frac{2x}{3} + \frac{3x}{4} = 7$

b. $0.7y + 1.1y = 5.4$

c. $\frac{5x}{3} - 2 = \frac{6}{5}$

$\mathbf{15}\left(\frac{5x}{3} - 2\right) = \mathbf{15}\left(\frac{6}{5}\right)$ Multiply both sides by the LCD of 15.

$15\left(\frac{5x}{3}\right) - 15(2) = 15\left(\frac{6}{5}\right)$ Use a distributive law.

$25x - 30 = 18$ Simplify.

$25x = 48$ Add 30 to each side.

$x = \frac{48}{25}$

The solution is $\frac{48}{25}$. ■

□ **DO EXERCISE 9.**

Suggestions for Solving Linear Equations

1. Multiply on both sides of the equation to clear fractions or decimals.
2. Remove parentheses, using the distributive laws.
3. Combine like terms on each side of the equation, if possible.
4. Get the terms containing the variable on one side of the equation and all other terms on the other side by using the addition property.
5. Combine like terms, if possible.
6. Use the multiplication property to solve for the variable.

Answers to Exercises

1. a. True **b.** Conditional **c.** False **d.** False **e.** False **f.** True

2. a. Yes **b.** No **c.** Yes **d.** No

3. a. 11 **b.** $\frac{1}{12}$ **c.** 108.4 **d.** -13

4. a. -25 **b.** $\frac{5}{3}$

5. a. 3 **b.** -5

6. a. $\frac{7}{5}$ **b.** 7

7. a. 0 **b.** $\frac{8}{11}$

8. a. 5 **b.** 5

9. a. $\frac{84}{17}$ **b.** 3

Problem Set 2.1

NAME

DATE

CLASS

Solve and check, using the addition property of equality.

1. $x + 3 = 15$

2. $x + 8 = 20$

3. $y - 15 = 27$

4. $y - 32 = 55$

5. $-16 = 7 + x$

6. $-25 = 4 + x$

7. $x - 5.3 = -8.9$

8. $x - 6.8 = -7.3$

9. $z - \frac{5}{8} = \frac{3}{4}$

10. $z - \frac{5}{9} = \frac{1}{18}$

Solve and check, using the multiplication property of equality.

11. $\frac{2x}{3} = 4$

12. $\frac{3x}{5} = 15$

13. $\frac{8x}{7} = -12$

14. $\frac{7x}{4} = -14$

15. $16 = 8x$

16. $18 = 6x$

17. $3x = -11$

18. $4x = -21$

19. $-9x = -180$

20. $-5x = -250$

Solve and check, using both the addition and multiplication properties.

21. $3x - 15 = 60$

22. $5x - 7 = 43$

23. $6x + 5 = -13$

24. $8x + 4 = -60$

25. $-70 = -5 + 5x$

26. $-30 = -3 + 3x$

27. $3y - 15 = 15 - 3y$

28. $7z - 1 = 23 - 5z$

29. $5y - 9 = -9 + 2y$

30. $3a - 8 = -8 - 5a$

31. $8 - 5b = b - 16$

32. $5 - 4x = x - 13$

33. $12x + 7 = 7x - 2$

34. $2x + 3 = x - 4$

35. $3y + 7 - 2y = 4y - 2$

36. $3y - 6y + 3 = y - 4$

37. $19x - 16 + 14x = 21x + 4$

38. $4x + 5 = 12x - 15x - 2$

39. $-\frac{7}{6}x + \frac{1}{3} = -9$

40. $-\frac{9}{4}y + 2 = -\frac{91}{4}$

41. $\frac{m}{2} + \frac{m}{3} = 5$

42. $\frac{x}{5} - \frac{x}{4} = 1$

43. $7x - 2x + 4 - 5 = 3x - 5 + 6$

44. $12y - 15y - 8 + 6 = 4y + 6 - 1$

45. $10 - 8x - 4 - 3x + 4 = -9x + 4 - 4x$

46. $7 - 6x - 10 - 2x + 3 = -5x + 3 - 2x$

47. $10(3x + 2) = 80$

48. $9(5x - 2) = 27$

49. $2(a - 5) = 5a$

50. $3(b - 4) = -6$

51. $5x - (2x - 10) = 25$

52. $8y - (3y - 5) = 40$

53. $3(x + 5) = 2x - 1$

54. $4(x - 4) = 20x + 8$

55. $3(2b + 1) = 2(b - 2) + 5$

56. $4(z - 1) = 6 - 2(z + 3)$

57. $0.6s - 0.3(5s + 2) = 0.4 - 0.5s$

58. $0.3(2t + 1) - 0.2(t - 2) = 0.5$

59. $6x - 4(3 - 2x) = 5(x - 4) - 10$

60. $4p - 3(4 - 2p) = 2(p - 3) + 6p + 2$

61. $\frac{1}{4}(16x + 8) - 34 = -\frac{1}{2}(8x - 16)$

62. $\frac{1}{3}(12x + 48) - 40 = -\frac{1}{4}(24x - 144)$

63. $\frac{1}{2}(x - 4) - \frac{1}{4}(x + 1) = \frac{1}{4}(2x - 3)$

64. $\frac{1}{6}(y + 8) - \frac{1}{9}(y - 2) = \frac{4y}{9}$

65. A statement which says that two mathematical expressions such as $x - 3$ and 2 represent the same number is an ____________.

66. Equations that are sometimes true and sometimes false depending on the value assigned to the variable are called ____________.

67. A ____________ equation in one variable is an equation with an exponent of 1 on the variable in any term.

68. A ____________ of an equation is a number that gives us an equation which is true when we substitute that number for the variable.

69. We may ____________ the same number to both sides of an equation.

70. We may ____________ or ____________ both sides of an equation by the same nonzero number.

Solve. Round answers to the nearest hundredth.

71. $6.834x + 4.592 = 29.578$

72. $7.934y - 5.328 = 10.54$

73. $2.932y - 7.4 = 4.396 - y$

74. $y + 3.395 = 6.47y - 18.485$

75. $8.24x + 3.5(2.07 - 3.3x) = -10.2$

76. $7.65 - 3.4x = 7.28(5.02 - 6.78x)$

Solve.

* **77.** $-[x - (4x + 2)] = 2 + (2x + 7)$

* **78.** $-11y - (5 - 6y) = -(6 - 3y) + 1$

* **79.** $\dfrac{a-4}{4} - \dfrac{a+1}{8} = \dfrac{2a-3}{8}$

* **80.** $\dfrac{x+8}{2} - \dfrac{x-2}{3} = \dfrac{4x}{3}$

81. $6[5 - 3(4 - x)] - 4 = 10[3(5x - 4) + 8] - 52$

Think About It

* **82.** Consider the equation

$$\frac{5}{8}y + \frac{7}{6}y = \frac{43}{3}$$

Suppose that you multiplied both sides by 48, rather than the LCD, 24. Would you get the correct answer? Explain.

Checkup

The following problems provide a review of some of Section 1.1.

Consider the numbers $-\frac{11}{5}, -1, 0, 3, \frac{15}{4}$*. Find the following.*

83. The whole numbers

84. The natural numbers

85. The rational numbers

86. The integers

Replace the ? with < or >.

87. 0 ? −10

88. −7 ? −15

89. −8 ? −3

90. −27 ? −9

Word Problems

When you study biology, chemistry, physics, business, or other fields, you will be doing word problems specific to that field. To solve word problems, be sure to do all of the steps listed in Section 2.2. Do not omit the step of writing a variable to represent the unknown. This helps you to organize your thoughts and answer the question posed in the word problem.

Many word problems are done by type. The more that you practice each type of problem, the more confidence you will gain in your ability to do them!

STUDY HINT

2.2 APPLIED PROBLEMS

OBJECTIVES

1. *Write mathematical expressions for phrases*
2. *Translate sentences into equations and solve the equations*
3. *Solve word problems about perimeter*
4. *Solve word problems involving consecutive integers*
5. *Solve word problems involving percent*

1 Writing Mathematical Expressions for Phrases

Algebra is used to solve applied problems in the real world. Often the solution of a problem can be simplified if we can write an equation and solve it. To write the equation, we must be able to translate phrases into mathematical expressions. Some translations are shown below. We let x represent "a number."

EXAMPLE 1

Phrase	***Mathematical Expression***	
Addition		
A number plus 5	$x + 5$	
4 more than a number	$x + 4$	
A number increased by 3	$x + 3$	
The sum of a number and 8	$x + 8$	
16 added to a number	$x + 16$	
Subtraction		
A number decreased by 7	$x - 7$	
15 minus a number	$15 - x$	
7 less than a number	$x - 7$	
Subtract 3 from a number	$x - 3$	Notice that x
9 subtracted from a number	$x - 9$	is written first.
Multiplication		
Five times a number	$5x$	
The product of 9 and a number	$9x$	
Double x	$2x$	
One-fourth of a number	$\frac{1}{4}x$	"Of" indicates multiplication.
Twice a number	$2x$	
Division		
A number divided by 7	$\frac{x}{7}$ ■	

□ **DO EXERCISE 1.**

□ **Exercise 1** Write mathematical expressions for the phrases.

a. A number increased by 7

b. Subtract 5 from a number

c. A number decreased by 16

d. The product of −9 and a number

e. Ten divided by a number

□ **Exercise 2** Translate to an equation.

a. The sum of a number and 3 is 10.

b. Double a number and decrease it by 5. The result is 12.

c. A number divided by 9, minus the number, is 15.

d. If 2 is subtracted from a number, the result is the product of the number and 5.

e. Four times a number, minus 8, is 17.

f. Four times the result of subtracting 8 from a number is 17.

□ **Exercise 3** If 6 is subtracted from a number, the result is −4. What is the number?

2 Solving Word Problems

Once we have learned to translate phrases into symbols, we use this skill to write equations. We also use the fact that the word "is" means =.

EXAMPLE 2 *Sentence*	*Equation*
Twice a number, plus 8, is 26.	$2x + 8 = 26$
Three times a number, increased by 2, is the same as the product of 5 and the number.	$3x + 2 = 5x$
A number divided by 7, plus the number, is 25.	$\frac{x}{7} + x = 25$
Five times a number, minus 6, is 11.	$5x - 6 = 11$
Five times the result of subtracting 6 from a number is 11. ■	$5(x - 6) = 11$

□ **DO EXERCISE 2.**

We are now ready to solve word problems.

Steps in Solving a Word Problem

1. Read the problem carefully. Decide what questions are to be answered.
2. Choose a variable to represent the unknown quantity. If it is helpful, make a drawing.
3. Write an equation. Sometimes we may need to remember a formula that is not given. Some geometric formulas are given on the inside cover of this book.
4. Solve the equation.
5. Decide if the questions have been answered and check the answer in the original word problem.
6. Write the answer in a sentence.

EXAMPLE 3 Eight plus five times a number is seven times the number. What is the number?

Variable Let x represent the number.

Eight plus five times a number is seven times the number.

Equation $8 + 5x = 7 \cdot x$

Solve $8 + 5x = 7x$

$8 = 7x - 5x$ Add $-5x$ to both sides of the equation.

$8 = 2x$

$4 = x$ Divide both sides by 2.

Check Five times 4 is 20. If we add 8 to 20, we get 28. This equals 7 times 4, which is also 28.

The number is 4. ■

□ **DO EXERCISE 3.**

3 Perimeter Problems

EXAMPLE 4 The perimeter of a rectangle is 96 centimeters. What are its dimensions if the length is 14 centimeters more than the width?

Variable If w represents the width in centimeters, $w + 14$ represents the length in centimeters. We used w to represent the width in centimeters so that when we solve the problem for w, it is clear that it is the width.

The formula for the perimeter of a rectangle is as follows:

$$P = 2L + 2W$$

Drawing

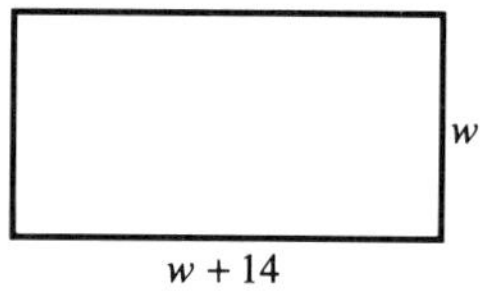

Equation $P = 2L + 2W$

Solve $96 = 2(\mathbf{w + 14}) + 2\mathbf{w}$ Substitute for L and W.

$96 = 2w + 28 + 2w$ Use a distributive law.

$96 = 2w + 2w + 28$

$96 = 4w + 28$

$96 - 28 = 4w$ Add -28 to both sides of the equation.

$68 = 4w$

$17 = w$

$w + 14 = 31$

The answers check.

The dimensions of the rectangle are 17 centimeters and 31 centimeters. ■

□ DO EXERCISE 4.

4 Consecutive Integer Problems

Consecutive integers are integers that follow one another in order, such as 2 and 3. Two numbers are ***consecutive even integers*** if they are even and one number is 2 larger than the other. For example 4 and 6 are consecutive even integers. ***Consecutive odd integers*** are odd numbers and one number is 2 larger than the other. Three and 5 are examples of consecutive odd integers.

□ **Exercise 4** The perimeter of a rectangle is 68 meters. If the length is 2 meters more than 3 times the width, find the dimensions of the rectangle.

□ **Exercise 5** Find three consecutive integers whose sum is 78.

□ **Exercise 6** At the end of the year, the price of a new car was reduced by 12%. If the sale price of the car was $9504, what was the original price of the car?

EXAMPLE 5 Find three consecutive integers such that the sum of the first and third is 146.

Variable Let x represent the first integer. Then $x + 1$ is the second integer and $x + 2$ is the third integer.

The sum of the first and third is 146.

Equation $x + (x + 2) = 146$

Solve

$$x + x + 2 = 146$$

$$2x + 2 = 146$$

$$2x = 146 - 2 \quad \text{Add } -2 \text{ to both sides.}$$

$$2x = 144$$

$$x = 72 \quad \text{First integer.}$$

$$x + 1 = 73 \quad \text{Second integer.}$$

$$x + 2 = 74 \quad \text{Third integer.}$$

Check The sum of 72 and 74 is 146.
The answers check in the word problem.

The integers are 72, 73, and 74. ■

□ **DO EXERCISE 5.**

5 Percent Problems

EXAMPLE 6 To buy more inventory for his business, Carlos took a 17% cut in salary for the year 1985. If his new salary was $20,750, what was his original salary?

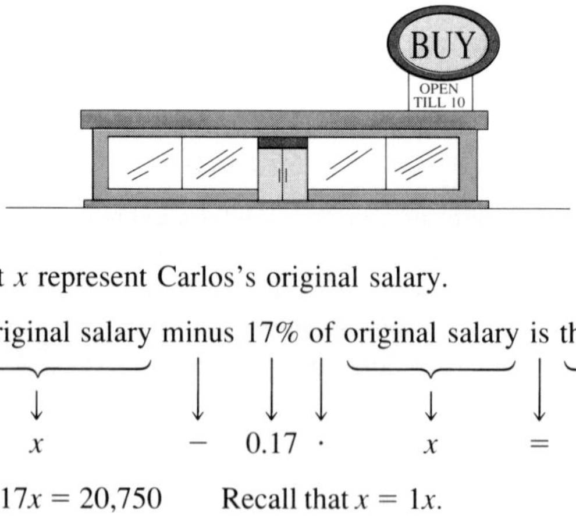

Variable Let x represent Carlos's original salary.

The original salary minus 17% of original salary is the new salary.

Equation $x - 0.17 \cdot x = 20{,}750$

Solve

$$x - 0.17x = 20{,}750 \quad \text{Recall that } x = 1x.$$

$$0.83x = 20{,}750$$

$$x = 25{,}000 \quad \text{Divide both sides by 0.83. You may want to use a calculator.}$$

Check 17% of 25,000 is 4250. If we subtract 4250 from 25,000, we get 20,750. The answer checks.

Carlos's original salary was $25,000. ■

□ **DO EXERCISE 6.**

EXAMPLE 7 26.88 is what percent of 84?

Variable Let x represent the percent number.

26.88 is what percent of 84?

↓ ↓ ↓ ↓ ↓ ↓

Equation $26.88 = x \quad \% \quad \cdot 84$

$26.88 = x(0.01)(84)$ Substitute 0.01 for %.

$26.88 = 0.84x$

$32 = x$ Divide both sides by 0.84

Check $32\%(84) = 0.32(84) = 26.88$. The answer checks.

26.88 is 32% of 84. ■

□ **DO EXERCISE 7.**

DID YOU KNOW?

The mathematical symbols used in this algebra book became popular and widely used about four centuries ago.

MATH SYMBOL	DATE SYMBOL CAME TO BE USED		
×	1631	<	1631
·	1631	≠	1739
÷	1659	x^3	1637
( )	1556	y^{-3}	1659
=	1557	√	1525
>	1631		

□ **Exercise 7** What percent of 62 is 14.26?

Answers to Exercises

1. **a.** $x + 7$ **b.** $x - 5$ **c.** $x - 16$ **d.** $-9x$ **e.** $\frac{10}{x}$

2. **a.** $x + 3 = 10$ **b.** $2x - 5 = 12$ **c.** $\frac{x}{9} - x = 15$
d. $x - 2 = 5x$ **e.** $4x - 8 = 17$ **f.** $4(x - 8) = 17$

3. $x - 6 = -4$
$x = 2$
The number is 2.

4. $2(3w + 2) + 2w = 68$
$6w + 4 + 2w = 68$
$8w = 64$
$w = 8$
$3w + 2 = 26$
The width is 8 meters and the length is 26 meters.

5. $x + (x + 1) + (x + 2) = 78$
$3x + 3 = 78$
$3x = 75$
$x = 25$
$x + 1 = 26$
$x + 2 = 27$
The integers are 25, 26, and 27.

6. $x - 0.12x = 9504$
$0.88x = 9504$
$x = 10{,}800$
The original price was $10,800.

7. $x\% \cdot 62 = 14.26$
$x(0.01)(62) = 14.26$
$0.62x = 14.26$
$x = 23$
23% of 62 is 14.26.

Problem Set 2.2

NAME

DATE

CLASS

Write mathematical expressions for the phrases.

1. The sum of a number and 5

2. Six subtracted from a number

3. A number increased by 10

4. A number decreased by 15

5. Subtract 8 from a number

6. A number subtracted from 2

7. The product of a number and -5

8. The product of 9 and a number

9. One-fifth of a number

10. Five-thirds of a number

11. A number divided by 8

12. Fifteen divided by a number

13. Four times a number, minus 7

14. Twice a number, plus 25

15. Four times the result of subtracting 7 from a number

16. Twice the sum of a number and 25

17. A number divided by 24, minus 10

18. Three divided by a number, plus 18

Solve.

19. If 9 is subtracted from a number the result is -4. Find the number.

20. If 15 is subtracted from a number the result is -6. Find the number.

21. If 16 is added to 5 times a number, the result is 13 times the number. What is the number?

22. If 9 is subtracted from a number divided by 4, the result is -3. What is the number?

23. If $\frac{3}{4}$ of a number is subtracted from 8, the result is 14. Find the number.

24. If $\frac{3}{2}$ of a number is added to 14, the result is 38. Find the number.

25. The perimeter of a garden is 128 meters. If the length is 3 times the width, find the dimensions of the garden.

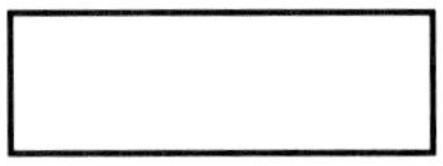

26. If the length of a house is twice the width and the perimeter is 250 feet, find the dimensions of the house.

27. The width of a rectangle is 20 centimeters less than the length. If the perimeter is 248 centimeters, find the dimensions of the rectangle.

28. The width of a rectangle is 26 inches less than the length. If the perimeter is 156 inches, find the dimensions of the rectangle.

29. The sum of three consecutive integers is 102. Find the integers.

30. Find three consecutive integers such that the sum of the second and third is 79.

31. Find two consecutive even integers such that 7 times the smaller is 6 times the larger.

32. Find two consecutive odd integers such that 2 times the first plus 3 times the second is 111.

33. The price of a house increased 8%. If the new price was $77,760, what was the original cost of the house?

34. The price of a lot increased 6% over last year's price. If the lot now costs $19,080, what was the price of the lot last year?

35. The cost of rent and electricity for a home was \$6740 for a year. If rent was \$4660 more than electricity, what was the cost of electricity?

36. The cost of room and board and tuition at State Tech was \$3420. If tuition was \$180 less than room and board, what was the cost of tuition?

37. The price of a new car was reduced by 16%. If the sale price of the car was \$11,760, what was the original price of the car?

38. For the company to stay in business, workers took a 12% cut in salary. If Knute's new salary was \$20,240, what was his original salary?

39. What percent of 73 is 24.82?

40. What percent of 45 is 24.3?

41. Eighteen less than 7 times a number is 8 more than 6 times the number. Find the number.

42. If 15 is subtracted from 8 times a number, the result is 5 times the number plus 3. Find the number.

43. An electrician cuts a piece of wire 26 feet long into three pieces. The second piece is 2 feet longer than the first piece. The third piece is twice as long as the first piece. Find the length of each piece of wire.

44. A farmer cuts a 55-foot piece of rope into three pieces. The second piece is 3 feet shorter than the first piece. The third piece is twice as long as the second piece. What is the length of each piece of rope?

45. Find two consecutive odd integers such that 3 times the first minus 12 is 1 less than twice the second.

46. Find two consecutive even integers such that if 20 is added to the first, the result is 16 less than twice the second.

47. The price of a dress was reduced 33%. If the sale price was \$43.55, what was the original price of the dress?

48. A washing machine was on sale for \$276. It was marked "25% off." What was the original price of the washing machine?

49. 151.04 is what percent of 236?

50. 109.62 is what percent of 378?

51. Integers that follow one another in order, such as 4 and 5, are called ____________ integers.

52. ____________ integers are even and one number is 2 larger than the other.

53. The perimeter of a rectangle is 72.14 meters. If the length is 3.5 meters more than the width, what are the dimensions of the rectangle?

54. The width of a rectangle is 15.84 inches less than the length. If the perimeter is 281.888 inches, find the dimensions of the rectangle.

55. The price of a house increased 8.5%. If the new price was $280,000, what was the original cost of the house?

56. Manuel received a 7.8% increase in his salary. If his new salary was $34,927.20, what was his original salary?

* 57. Three times a number minus 7 is twice the result of subtracting 3 from a number, plus 8. Find the number.

* 58. Four times the result of subtracting 5 from a number is 7 times the number, minus 4. Find the number.

* 59. The perimeter of a rectangle is twice the length plus 84 feet. What are its dimensions if the width is 6 feet less than the length?

* 60. The length of a rectangle is 3 meters less than twice the width. The perimeter is 4 times the width, plus 2 meters. Find the dimensions of the rectangle.

Think About It

* 61. Jennifer is 4 years older than Tom. In 12 years the sum of their ages will be 40. How old is Jennifer now?

62. An average score of 90 or above in a geology class gives a 4.0 grade. A student has grades of 97, 86, and 82 on three exams. Find the score on the fourth exam that will give the student a 4.0 for the course.

Checkup

The following problems provide a review of some of Section 1.2.

Which of the commutative or associative laws of addition is illustrated by each equation?

63. $3 + 8 = 8 + 3$

64. $7 + 5 = 5 + 7$

65. $2 + (9 + 7) = (2 + 9) + 7$

66. $8 + (4 + 3) = (8 + 4) + 3$

67. $(3 + 5) + 6 = 6 + (3 + 5)$

68. $9 + (1 + 5) = (1 + 5) + 9$

69. $4 + (8 + 2) = (4 + 8) + 2$

70. $7 + (3 + 4) = (3 + 4) + 7$

71. $6 + (5 + 7) = (6 + 5) + 7$

72. $(3 + 9) + 1 = 1 + (3 + 9)$

2.3 MORE APPLIED PROBLEMS

In everyday life we often calculate the value of a set of coins or the interest on a savings account. In this section we discuss coin and interest problems and additional number problems.

1 Number Problems

EXAMPLE 1 If the sum of two numbers is 20 and twice the larger is three times the smaller, find the numbers.

Notice that if we knew that the sum of two numbers was 20 and one of them was 4, the other number would be $20 - 4$. We use a similar idea to write mathematical expressions for the numbers.

20 represents the sum of the numbers.

Variable Let x represent smaller number.

Then $20 - x$ represents the larger number.

Equation Twice the larger is three times the smaller.

$$2 \quad (20 - x) = 3 \cdot x$$

$40 - 2x = 3x$ — Use a distributive law.

$40 = 5x$ — Add $2x$ to each side of the equation.

$8 = x$ — Divide each side of the equation by 5.

$20 - x = 20 - 8 = 12$

Check The sum of 12 and 8 is 20. Twice the larger is $2(12) = 24$. This is the same as $3(8) = 24$.

The numbers are 8 and 12.

Caution: Notice that the larger number was $20 - x$, *not* $x - 20$. ■

□ **DO EXERCISE 1.**

OBJECTIVES

1 *Solve number problems*

2 *Solve coin problems*

3 *Solve interest problems*

□ **Exercise 1** The sum of two numbers is 15. One number is four times the other. Find the numbers.

□ **Exercise 2** Jon has three times as many nickels as quarters. The value of the coins is \$2.40. How many of each does he have?

2 Coin Problems

EXAMPLE 2 A collection of 16 dimes and nickels is worth \$1.25. How many of each type of coin is in the collection?

16 represents the total number of coins.

Variable Let d represent the *number* of dimes.

Then $16 - d$ represents the *number* of nickels.

It is helpful to make a chart.

	(Value of each coin)	·	(number of coins)	=	(total value)
Dimes	0.10	·	d	=	$0.10d$
Nickels	0.05	·	$(16 - d)$	=	$0.05(16 - d)$

Equation $\begin{pmatrix}\text{Value of}\\\text{dimes}\end{pmatrix} + \begin{pmatrix}\text{value of}\\\text{nickels}\end{pmatrix} = \begin{pmatrix}\text{total value}\\\text{of all the coins}\end{pmatrix}$

$$0.10d + 0.05(16 - d) = 1.25$$

Solve Multiply each term by 100 to clear the decimals.

$$10d + 5(16 - d) = 125$$

$$10d + 80 - 5d = 125 \quad \text{Use a distributive law.}$$

$$5d + 80 = 125 \quad \text{Combine terms.}$$

$$5d = 125 - 80 \quad \text{Add } -80 \text{ to each side.}$$

$$5d = 45$$

$$d = 9$$

$$16 - d = 16 - 9 = 7$$

Check Nine coins plus 7 coins equals 16 coins. $0.10(9) + 0.05(7) = 1.25$. The answer checks.

There are 9 dimes and 7 nickels in the collection. ■

□ **DO EXERCISE 2.**

3 Interest Problems

To solve simple interest problems we use the following formula.

$$I = Prt$$

where I is the simple interest, P is the principal (amount invested), r is the interest rate, and t is the time in years.

EXAMPLE 3 Mrs. Rich invested \$28,000 for 1 year and earned \$2120 interest. If part of the money was invested at 8% interest per year and part at 7% interest per year, how much was invested at each rate?

28,000 represents the total principal invested.

Variable Let x represent the principal invested at 8%.

Then $28{,}000 - x$ represents the principal invested at 7%.

P	$\cdot$	r	$\cdot$	t	$=$	I
x	$\cdot$	0.08	$\cdot$	1	$=$	$0.08x$
$(28{,}000 - x)$	$\cdot$	0.07	$\cdot$	1	$=$	$0.07(28{,}000 - x)$

Equation

$$\begin{pmatrix}\text{Interest earned}\\ \text{at } 8\%\end{pmatrix} + \begin{pmatrix}\text{interest earned}\\ \text{at } 7\%\end{pmatrix} = \begin{pmatrix}\text{total}\\ \text{interest}\end{pmatrix}$$

$$0.08x + 0.07(28{,}000 - x) = 2120$$

$$8x + 7(28{,}000 - x) = 212{,}000 \quad \text{Multiply by 100.}$$

$$8x + 196{,}000 - 7x = 212{,}000$$

$$x = 16{,}000$$

$$28{,}000 - x = 28{,}000 - 16{,}000 = 12{,}000$$

Check \$16,000 + \$12,000 = \$28,000. The interest earned at 8% is 0.08(16,000) = 1280. The interest earned at 7% is 0.07(12,000) = 840. \$1280 + \$840 = \$2120. The answer checks.

The amount invested at 8% was \$16,000 and the amount invested at 7% was \$12,000. ■

□ DO EXERCISE 3.

□ Exercise 3 The total interest for 1 year earned on \$15,000 was \$1270. Part was invested at 7% per year and part at 9% per year. How much was invested at each rate?

Answers to Exercises

1. Let x represent the smaller number.
   Then $4x$ represents the larger number.
   $$x + 4x = 15$$
   $$5x = 15$$
   $$x = 3$$
   $$4x = 4(3) = 12$$
   The numbers are 3 and 12.

2. Let q represent the number of quarters.
   Then $3q$ represents the number of nickels.
   $$0.25q + 0.05(3q) = 2.40$$
   $$25q + 5(3q) = 240$$
   $$25q + 15q = 240$$
   $$40q = 240$$
   $$q = 6$$
   $$3q = 18$$
   Jon has 6 quarters and 18 nickels.

3. Let x represent the principal invested at 7%.
   Then $15{,}000 - x$ represents the principal invested at 9%.
   $$0.07x + 0.09(15{,}000 - x) = 1270$$
   $$7x + 9(15{,}000 - x) = 127{,}000$$
   $$7x + 135{,}000 - 9x = 127{,}000$$
   $$-2x = -8000$$
   $$x = 4000$$
   $$15{,}000 - x = 11{,}000$$
   The amount invested at 7% is \$4000 and the amount invested at 9% is \$11,000.

Problem Set 2.3

NAME ______________________

DATE ______________________

CLASS ______________________

1. The sum of two numbers is 16. Three times the larger number is 9 times the smaller number. Find the numbers.

2. The sum of two numbers is 25. Three times the smaller number is twice the larger number. Find the numbers.

3. The sum of two numbers is 1. Three times the larger number, minus the smaller number, is 55. Find the numbers.

4. The sum of two numbers is 3. Four times the smaller, plus the larger, is -48. Find the numbers.

5. Susan has a total of 29 quarters and nickels. They are worth \$4.45. How many of each coin does she have?

Bank

6. A collection of 35 coins has a value of \$6.65. If the coins are dimes and quarters, how many of each type is in the collection?

7. Ken has 4 times as many dimes as nickels. If the value of the coins is \$6.75, how many of each coin does he have?

8. There are twice as many quarters as nickels in a coin collection. If the value of the coins is \$9.90, how many of each type are in the collection?

9. Kevin invested \$14,000 for 1 year and earned \$900 interest. If part of the money was invested at 6% interest per year and part at 7% interest per year, how much was invested at each rate?

10. Susan earned \$222 interest for 1 year on an investment of \$3000. Some of the money was invested at 8% interest per year and the remainder was invested at 7% interest per year. How much did she invest at each rate?

11. Interest of \$630 was earned on \$9000 invested in two accounts for 1 year. If one account earned 9% per year and the other account earned 6% per year, how much was invested in each account?

12. If \$12,000 is invested in two accounts for 1 year, the interest earned is \$985. How much is invested in each account if one account earns 8% per year and the other account earns 9% per year?

13. The total interest earned for 1 year on $8000 was $600. If part of the money was invested at 10% interest per year and part was invested at 6% interest per year, how much was invested at each rate?

14. A total of $5000 was invested for 1 year and the money earned $430 interest. Some of the money was invested at 7% interest per year and the remainder at 11% interest per year. How much was invested at each rate?

15. The sum of Juan and Maria's ages is 71 and the difference in their ages is 3 years. If Juan is older, how old is Maria?

16. The sum of Mark and Christine's ages is 42. If Mark is 5 times as old as Christine, how old is Mark?

17. The sum of two numbers is 45. Twice the smaller, minus 6 gives the larger number. Find the larger number.

18. The sum of two numbers is 21. Twice the smaller, minus 3, gives the larger number. Find the smaller number.

19. Karen has 3 times as many quarters as dimes. If the value of the coins is $10.20, how many quarters does she have?

20. Robert has $4.95 in nickels and quarters. If he has 6 times as many nickels as quarters, how many nickels does he have?

21. A collection of 47 coins has a value of $3.90. If the coins are dimes and nickels, how many dimes are in the collection?

22. There are 36 nickels and quarters in a coin collection. If the value of the collection is $8.20, how many nickels are in the collection?

23. Dr. Minh earned $495 interest on two accounts in 1 year. If 3 times as much was invested at 8% interest per year as was invested at 9% interest per year, how much was invested at 9%?

24. Four times as much money was invested in an account earning 9% interest per year than was invested in an account earning 6% interest per year. If $147 interest was earned on the accounts in 1 year, how much was invested at 9%?

25. The sum of two numbers is 15. Three times the smaller, plus 8, is 3 less than 4 times the larger. Find the smaller number.

26. The sum of two numbers is 44. If 2 is subtracted from 5 times the larger, the result is 6 times the smaller, minus 13. Find the larger number.

27. The measure of one angle of a triangle is twice the measure of a second angle. The measure of the third angle is 5° more than the measure of the first angle. The sum of the measures of the angles of a triangle is 180°. Find the measure of each angle of the triangle.

28. The measure of one angle of a triangle is 10° more than the measure of a second angle. The measure of the third angle is 20° more than the measure of the second angle. The sum of the measures of the angles of a triangle is 180°. Find the measure of each angle of the triangle.

29. John has a collection of 79 pennies and nickels worth $1.99. How many pennies are in the collection?

30. Linda has 80 quarters and pennies worth $2.48. How many quarters does she have?

31. Roberto has twice as many dimes as nickels in his bank. There are also quarters in the bank and it contains 48 coins. If the value of the coins is $5.50, how many dimes are in the bank?

32. April keeps pennies, nickels, and dimes in a glass jar. The number of nickels in the jar is one more than the number of pennies, and the total number of coins is 71. If the value of the coins is $4.67, how many dimes are in the jar?

33. An investor earned $980 interest on three accounts in 1 year. The total amount invested was $13,000, and $3000 more was invested at 8% interest per year than at 6% interest per year. The principal in the third account earned 9% interest per year. How much was invested at 8% interest per year?

34. Ken invested $14,000. Twice as much was invested at 5% interest per year as was invested at 7% interest per year. He also invested some money at 4% interest per year. How much money did he invest at 5% interest per year if his total annual interest was $710?

35. To solve interest problems, we use the formula ____________.

36. Rhonda earned $5110 interest on two accounts in 1 year. If 3 times as much was invested at 8.5% as was invested at 9.5%, how much was invested at 9.5%?

37. Twice as much money was invested in an account earning 7.5% as was invested in an account earning 6.5%. If $13,760 interest was earned on the accounts in 1 year, how much was invested at 7.5%?

* 38. How many pounds of chocolates worth $3.20 per pound should be mixed with chocolates worth $2.00 per pound to make 8 pounds of chocolates worth $2.75 per pound?

* **39.** Turkey sells for \$3.85 per pound and roast beef sells for \$4.90 per pound. If John bought $\frac{1}{2}$ pound more roast beef than turkey and spent \$19.95, how much of each did he buy?

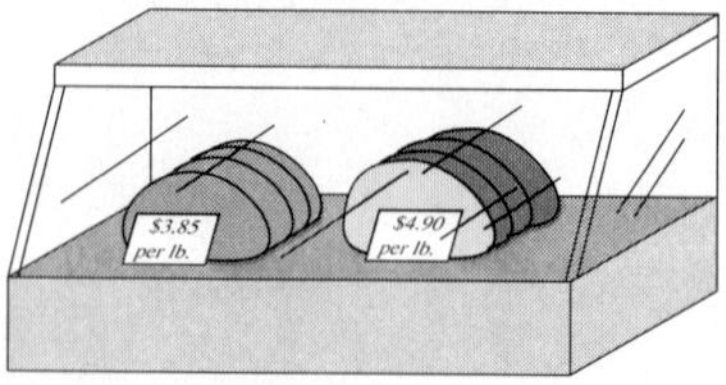

* **40.** If \$4000 is invested at 8% and \$5500 is invested at 9%, at what rate should \$3000 be invested to give a total annual income of \$1025?

* **41.** An investment of \$8000 is made at a certain interest rate, and \$5000 is invested at twice this interest rate. If the total interest earned is \$1080, at what rate is the \$5000 invested?

* **42.** A number divided by 7, minus 4, is the same as 3 times the result of subtracting 15 from the number, plus 1. Find the number.

* **43.** Twenty-five subtracted from twice the result of subtracting 18 from a number is the same as the number divided by 8, less 16. What is the number?

Checkup

The following problems provide a review of some of Section 2.1 and will help you with the next section.

Solve.

44. $3x + 6 = 18$

45. $5x - 2 = 23$

46. $38 = 7y + 3$

47. $22 = 3y - 2$

48. $4x - 5 = 6x - 5$

49. $8y - 16 = -4y + 20$

50. $3(y - 4) = -8y - 21$

51. $5(z - 2) + 3 = 4(2z - 5) - 5$

52. $\frac{x}{2} + \frac{x}{4} = 6$

53. $\frac{y}{3} - \frac{y}{9} = 1$

54. $15y - 3y - 9 = 5y + 7 - 2$

55. $33 - 5x - 18 = -10x + 39 - x$

2.4 FORMULAS

OBJECTIVE

1 *Solve formulas for a given letter*

1 Many word problems can be solved if we know a relationship between certain variables. This relationship is often called a ***formula.*** Recall that the formula for the perimeter P of a rectangle of length L and width W is

$$P = 2L + 2W$$

We may know the perimeters and the widths of many rectangles and want to find the lengths. We use our techniques for solving equations to solve the equation for L. Remember that to solve for L we must isolate it on one side of the equation.

EXAMPLE 1 Solve $P = 2L + 2W$ for L.

$$P = 2L + 2W$$

$$P + (\mathbf{-2W}) = 2L + 2W + (\mathbf{-2W})$$ Add $-2W$ to both sides of the equation.

$$P - 2W = 2L$$

$$\frac{P - 2W}{2} = L$$ Divide both sides of the equation by 2.

Notice that if the perimeter P is 40 feet and the width W is 10 feet, then

$$L = \frac{P - 2W}{2} = \frac{40 - 2(10)}{2} = \frac{40 - 20}{2} = 10$$

If we want to find many lengths, given each perimeter and width, this method is easier than solving the original formula $P = 2L + 2W$ for L. ■

□ **DO EXERCISE 1.**

□ **Exercise 1** Solve $P = 2L + 2W$ for W

EXAMPLE 2 Solve the interest formula $I = Prt$ for t.
We want to isolate t on one side of the equation.

$$I = Prt$$

$$\frac{I}{Pr} = t$$ Divide both sides of the equation by Pr. ■

□ **DO EXERCISE 2.**

□ **Exercise 2** Solve the formula for the volume of a rectangular solid $V = LWH$ for H.

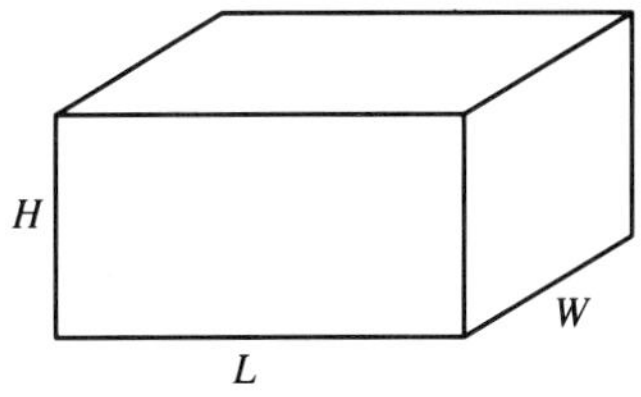

When we want to solve a formula containing fractions for a certain letter, it is usually easier to multiply first by the least common denominator on both sides of the formula to clear fractions.

□ **Exercise 3** Solve the formula for the volume of a cone,
$V = \frac{1}{3}\pi r^2 h$, for h.

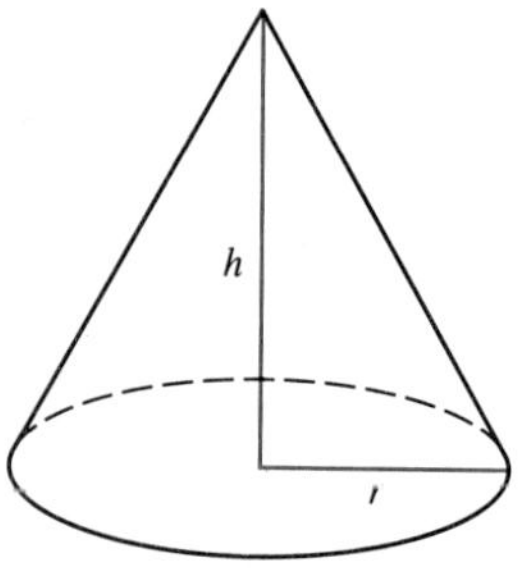

EXAMPLE 3 Solve the formula for the area of a triangle $A = \frac{1}{2}bh$ for h.

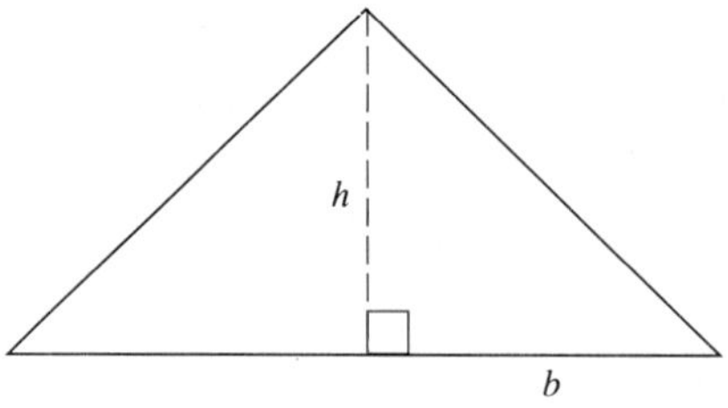

$$A = \frac{1}{2}bh$$

$$\mathbf{2}A = \mathbf{2}\left(\frac{1}{2}bh\right) \qquad \text{Multiply both sides by 2 to clear the fraction.}$$

$$2A = bh \qquad \text{Simplify.}$$

$$\frac{2A}{b} = h \qquad \text{Divide both sides by } b. \quad ■$$

□ **DO EXERCISE 3.**

A ***constant*** is a letter that stands for a specific number. Usually, letters from the beginning of the alphabet such as a, b, and c are used for constants. Letters such as x, y, and z from the end of the alphabet are used for variables.

□ **Exercise 4** Solve for the given letter.

a. $by + 3 = d$ for y

b. $6 - by = a$ for y

EXAMPLE 4 Solve $ax + b = c$ for x.

$$ax + b = c$$

$$ax + b + (\mathbf{-b}) = c + (\mathbf{-b}) \qquad \text{Add } -b \text{ to both sides.}$$

$$ax = c - b$$

$$x = \frac{c - b}{a} \qquad \text{Divide both sides by } a. \quad ■$$

□ **DO EXERCISE 4.**

Answers to Exercises

1. $W = \frac{P - 2L}{2}$

2. $H = \frac{V}{LW}$

3. $h = \frac{3V}{\pi r^2}$

4. a. $y = \frac{d - 3}{b}$ **b.** $y = \frac{a - 6}{-b}$

Problem Set 2.4

NAME
DATE
CLASS

Solve.

1. $D = L - N$ for N
(a discount formula)

2. $D = L - N$ for L

3. $A = P + Prt$ for r
(amount of a loan formula)

4. $A = P + Prt$ for t

5. $d = rt$ for t
(a distance formula)

6. $d = rt$ for r

7. $P = RB$ for R
(a percentage formula)

8. $P = RB$ for B

9. $A = LW$ for L
(area of a rectangle)

10. $A = LW$ for W

11. $C = 2\pi r$ for r
(circumference of a circle)

12. $C = 2\pi r$ for π

13. $I = Prt$ for P
(interest formula)

14. $I = Prt$ for r

15. $V = LWH$ for L
(volume of a rectangular solid)

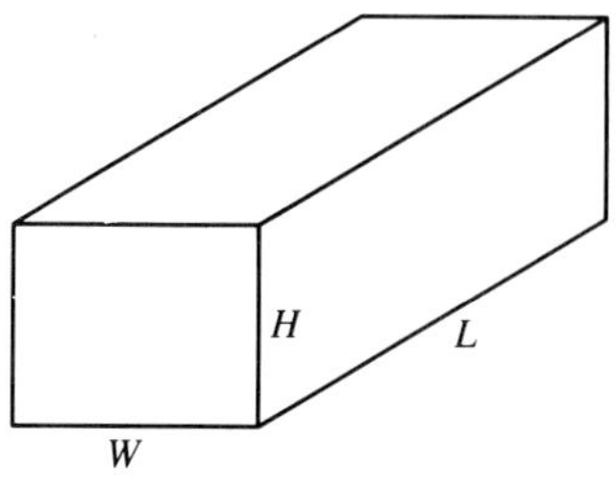

16. $V = LWH$ for W

17. $A = \frac{1}{2}bh$ for b

(area of a triangle)

18. $V = \frac{1}{3}\pi r^2 h$ for r^2

19. $A = \frac{h}{2}(b + c)$ for b
(area of a trapezoid)

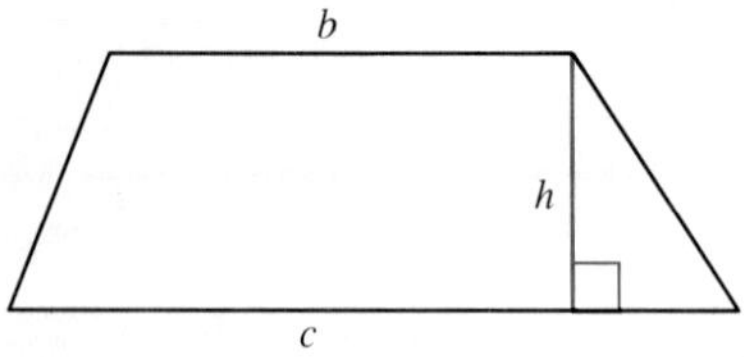

20. $A = \frac{h}{2}(b + c)$ for c

21. $R = \frac{C - S}{n}$ for S
(depreciation formula)

22. $R = \frac{C - S}{n}$ for C

23. $9 + by = c$ for y

24. $8 - ax = d$ for x

25. $a + c = -bx$ for x

26. $b + d = ay$ for y

27. $15 - gx = h$ for x

28. $-32 - ky = j$ for k

29. $3x + 4y = 7$ for y

30. $5x - 6y = 11$ for y

31. $7x - 2y = 5$ for y

32. $9x - 3y = 14$ for y

33. $6x - 3y = 24$ for y

34. $5x - 8y = 16$ for y

35. $Ax + By = C$ for y
(equation of a line)

36. $Ax + By = C$ for x

37. $c^2 = a^2 + b^2$ for a^2
(Pythagorean theorem for right triangles)

38. $c^2 = a^2 + b^2$ for b^2

39. The formula for the volume V of a sphere of radius r is $V = \frac{4}{3}\pi r^3$. Solve this formula for r^3.

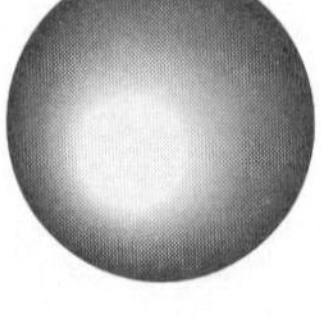

40. The height s in feet at time t in seconds of a projectile launched upward from the ground at a velocity of v feet per second is given by the equation $s = vt - 16t^2$. Solve this formula for v.

41. In Problem 5 we solved the formula $d = rt$ for t. Use this new formula to find the time it takes to travel a distance d of 90 miles at a rate r of 25 miles per hour.

42. In Problem 16 we solved the formula $V = LWH$ for W. Use it to find the width W of a rectangular solid with a volume V of 9000 cubic inches, a length L of 30 inches, and a height H of 20 inches.

43. In Problem 17 we solved the formula $A = \frac{1}{2}bh$ for b. Use this new formula to find the base of a triangle with an area A of 41 square inches and a height h of 7 inches.

44. In Problem 3 we solved the formula $A = P + Prt$ for r. Use this new formula to find the rate of interest r if the principal P was \$1400, the time t was 1 year, and the amount A paid back was \$1526.

45. A ____________ is a relationship between certain variables.

46. A letter that stands for a specific number is a ____________.

47. Solve $C = 2\pi r$ for r if $C = 46.472$ and $\pi = 3.14$.

48. Solve $V = LWH$ for L if $V = 255.808$, $W = 5.71$, and $H = 6.4$.

49. Solve $A = \frac{1}{2}bh$ for b if $A = 162.344$ and $h = 89.2$.

50. Solve $A = \frac{h}{2}(b + c)$ for h if $A = 70.516$, $b = 8.253$, and $c = 12.487$.

Solve.

* **51.** $p = \dfrac{S}{S + F}$ for F

* **52.** $R = \dfrac{C - S}{n}$ for S

* **53.** $V = \frac{1}{6}h(b + 4M + B)$ for M

* **54.** $m = \frac{C(100 - p)}{100 - d}$ for p

Use the formulas inside the front cover of your book to do the following problems.

* **55.** If the surface area and radius of a right circular cylinder are given, find a formula for the height.

* **56.** If the area and both bases of a trapezoid are given, find a formula for the height.

Checkup

The following problems provide a review of some of Sections 1.4 and 2.1 and will help you with the next section.

Simplify.

57. $3x - 4(2x + 1)$

58. $5(x - 2) + 3(2x - 5)$

59. $-5[2a + 3(4a - 1)]$

60. $-\{3 - 2[x - (4x - 5)]\}$

61. $4\{2x - [7 - (5x + 3)]\}$

62. $-\{8x - 3[4x - 2(5 - x)]\}$

Solve.

63. $x + 8 = 17$

64. $15 = y - 9$

65. $5y + 4 = 4y - 7$

66. $9y - 3 = 10y + 4$

2.5 THE ADDITION PROPERTY OF INEQUALITY

1 Recall from Section 1.1 that if a number a is to the right of another number b on the number line, we say that a is greater than b, written $a > b$. We also learned in Chapter 1 that if a number a is to the left of another number b on the number line, we say that a is less than b, written $a < b$.

"Greater than or equal to" is written $\geq$. "Less than or equal to" is written $\leq$.

$a > b$	means	a is greater than b.
$a < b$	means	a is less than b.
$a \geq b$	means	a is greater than or equal to b.
$a \leq b$	means	a is less than or equal to b.

An ***inequality*** is like an equation with the equal sign replaced by $<$, $>$, $\leq$, or $\geq$. We use many of the same techniques for solving inequalities that we used for solving equations. *An inequality usually has many solutions.* The set of all solutions is called the ***solution set.***

We solved some equations by using the addition property of equality. There is an addition property of inequality that helps us solve inequalities.

Addition Property of Inequality

The same number may be added to both sides of an inequality. If

$$a < b$$

then

$$a + c < b + c$$

The property is also true for $\leq$, $>$, and $\geq$.

EXAMPLE 1 Solve and graph $x + 2 > 4$.

$$x + 2 > 4$$

$$x + 2 + (-2) > 4 + (-2)$$ Use the addition property to add -2 to both sides.

$$x > 2$$

The solutions are all numbers greater than 2. The solutions may also be shown in the set-builder notation of Section 1.1 as follows.

$$\{x \mid x > 2\}$$

This is read "the set of all x such that x is greater than 2." We may graph these solutions on the number line. We have graphed $x > 2$ below. The left parenthesis indicates that 2 is not included; 2 is not in the solution set. In some books ○ is used instead of (or).

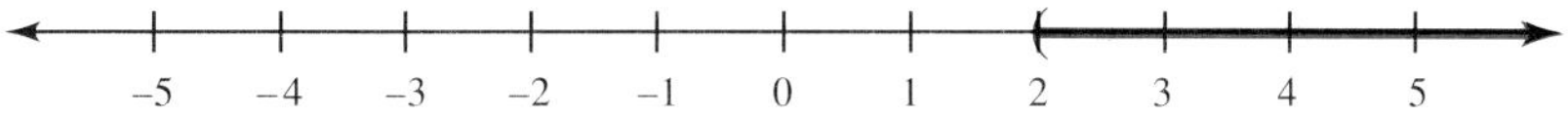

■

□ **DO EXERCISE 1.**

OBJECTIVE

1 *Solve inequalities using the addition property*

□ **Exercise 1** Solve and graph.

a. $x + 3 > 2$

b. $x + 4 < 6$

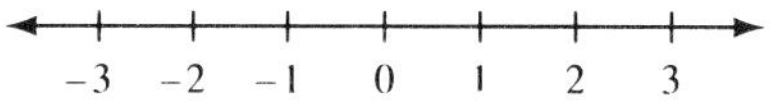

□ **Exercise 2** Solve and graph.

a. $5x + 2 \geq 4x - 1$

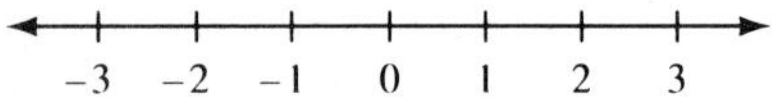

b. $-x + 3 \leq -2x + 5$

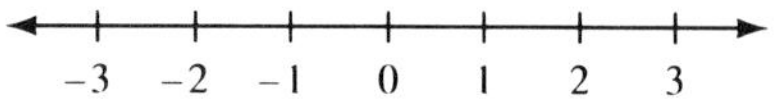

□ **Exercise 3** $A = \{1, 3, 5, 7, 9\}$ and $B = \{-3, -1, 0, 1, 3, 5\}$. Find $A \cap B$.

We may need to use the addition property of inequality more than once to solve the inequality. Our objective is to isolate the variable on one side of the inequality.

EXAMPLE 2 Solve $3y + 3 \leq 2y + 1$.

$$3y + 3 \leq 2y + 1$$

$$3y + 3 + (\mathbf{-3}) \leq 2y + 1 + (\mathbf{-3}) \qquad \text{Add } -3 \text{ to both sides.}$$

$$3y \leq 2y - 2$$

$$3y + (\mathbf{-2y}) \leq 2y - 2 + (\mathbf{-2y}) \qquad \text{Add } -2y \text{ to both sides.}$$

$$y \leq -2$$

The solution set is $\{y \mid y \leq -2\}$. To graph the inequality we use a square bracket at -2, to show that -2 is also a solution. In some books ● is used instead of [or].

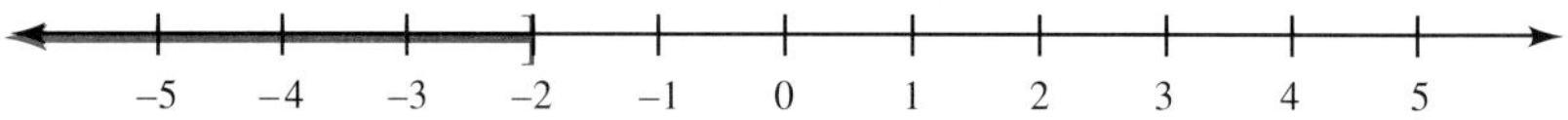

■

□ **DO EXERCISE 2.**

Compound Inequalities

When a given problem contains two or more inequality symbols we have a ***compound inequality.*** Compound inequalities are formed by two or more inequalities that are joined with the word *and* or the word *or*. To solve compound inequalities, we find the set of all solutions.

The ***intersection*** of two sets A and B is the set of all elements that are common to A and B. The intersection of sets A and B is written $A \cap B$ and may be shown as follows:

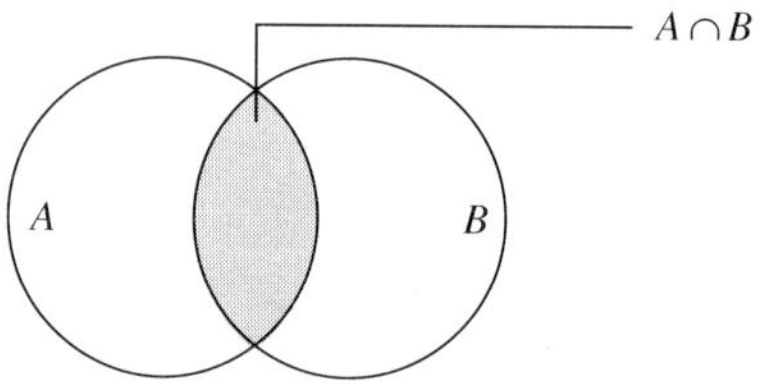

In set-builder notation,

$$A \cap B = \{x \mid x \text{ is an element of } A \text{ and } x \text{ is an element of } B\}$$

EXAMPLE 3 $A = \{1, 2, 3, 4\}$ and $B = \{2, 4, 6\}$. Find $A \cap B$.

The numbers 2 and 4 are common to both sets. Therefore,

$$A \cap B = \{2, 4\}$$ ■

□ **DO EXERCISE 3.**

When two or more inequalities are joined by the word *and* to make a compound inequality, the solution set is the *intersection* of the solution sets of the individual inequalities.

$a < x < b$ means x is greater than a and x is less than b

Notice that the inequality symbol points to the smaller number, so $a < x$ and $x > a$ are equivalent statements. The inequality $a < x < b$ may also be written $b > x > a$.

EXAMPLE 4 Graph $-3 < x \leq 2$.

This inequality means that $x > -3$ *and* $x \leq 2$. The solutions must satisfy both inequalities. The top graph shows $x > -3$ and the middle graph shows $x \leq 2$. The bottom graph shows the numbers common to the two graphs.

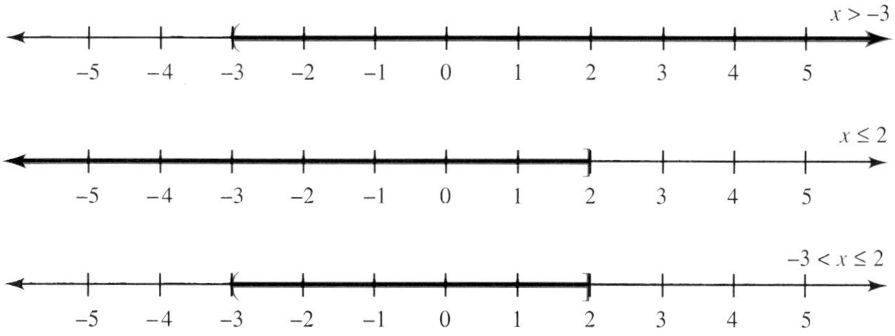

■

□ **DO EXERCISE 4.**

We may need to solve the inequality before we graph it.

EXAMPLE 5 Solve and graph $-7 < x - 5 \leq -2$.
The inequality may be solved in two ways.

a. $-7 < x - 5 \leq -2$
The inequality means that $x - 5 > -7$ and $x - 5 \leq -2$. Solve each of these inequalities.

$$x - 5 > -7 \quad \text{and} \quad x - 5 \leq -2$$
$$x - 5 + \mathbf{5} > -7 + \mathbf{5} \qquad x - 5 + \mathbf{5} \leq -2 + \mathbf{5}$$
$$x > -2 \quad \text{and} \quad x \leq 3$$

b. We may also solve the inequality by adding 5 to each part of it.

$$-7 < x - 5 \leq -2$$
$$-7 + \mathbf{5} < x - 5 + \mathbf{5} \leq -2 + \mathbf{5}$$
$$-2 < x \leq 3$$

The solution set is $\{x \mid -2 < x \leq 3\}$. The inequality also means that $x > -2$ and $x \leq 3$. The graph follows.

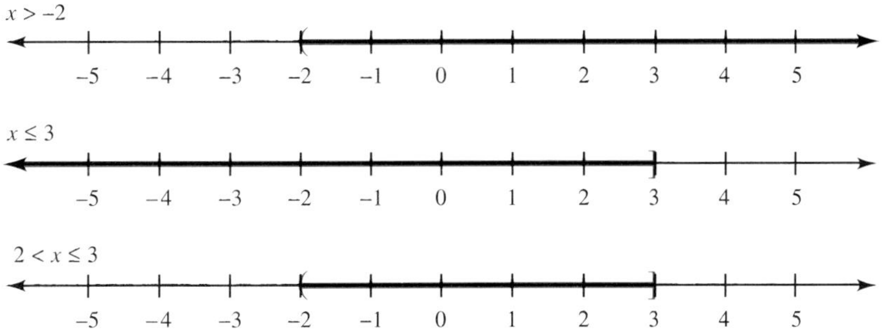

■

□ **DO EXERCISE 5.**

□ **Exercise 4**

a. $-2 \leq x < 1$

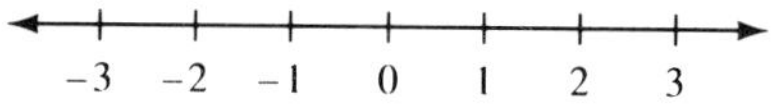

b. $-3 \leq x \leq 2$

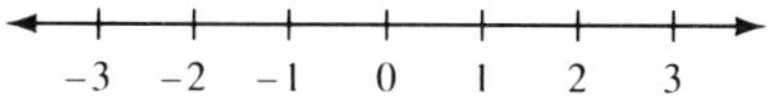

□ **Exercise 5** Solve and graph.

a. $-3 \leq x + 2 < 5$

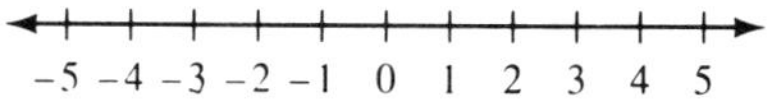

b. $-8 < x - 5 < -1$

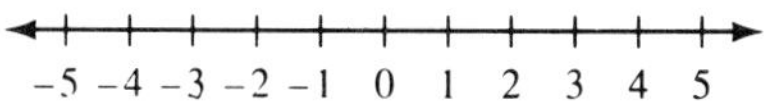

□ **Exercise 6** Solve and graph $x + 2 < 5$ and $x - 8 > 2$.

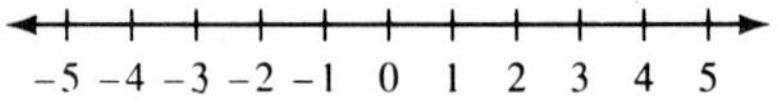

If two sets have no common elements, we say their intersection is the ***empty set,*** denoted by $\varnothing$. The intersection of the following two sets is empty.

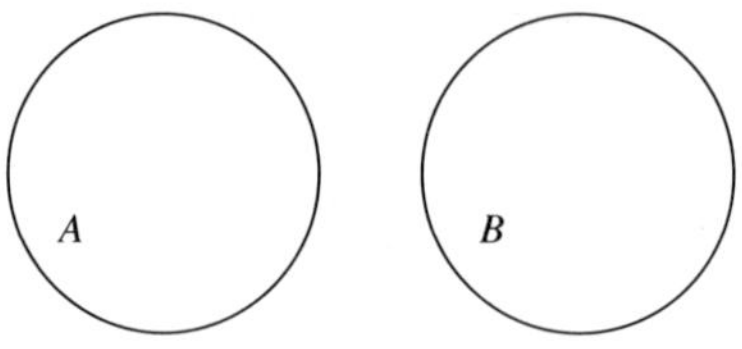

EXAMPLE 6 Solve and graph $x + 4 > 6$ and $x + 5 < 2$.

$x + 4 > 6$	and	$x + 5 < 2$
$x + 4 + (-4) > 6 + (-4)$		$x + 5 + (-5) < 2 + (-5)$
$x > 2$		$x < -3$

It is not possible for a number to be both greater than 2 and less than −3. The intersection is empty. The solution set is the empty set, $\varnothing$. The graph is as follows:

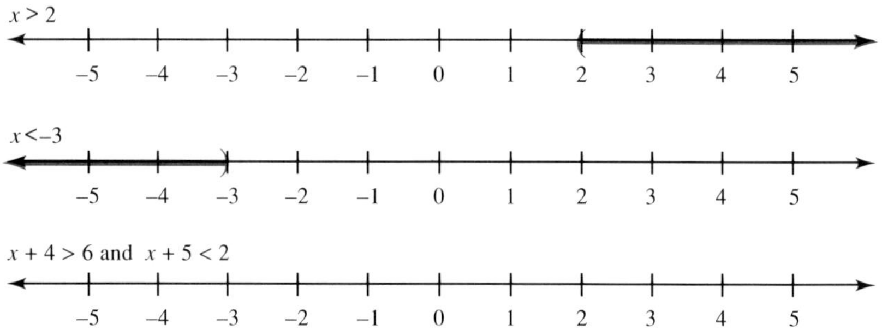

■

□ **Exercise 7** $A = \{2, 4, 6, 8\}$ and $B = \{6, 7, 8, 9\}$. Find $A \cup B$.

□ **DO EXERCISE 6.**

The ***union*** of two sets A and B is the set of all elements that are in A or in B or both. The union of sets A and B is written $A \cup B$ and may be shown as follows:

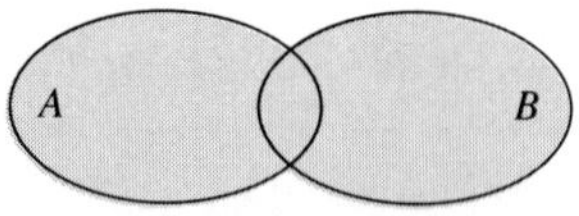

In set-builder notation,

$$A \cup B = \{x \mid x \text{ is an element of } A \text{ or } x \text{ is an element of } B\}$$

EXAMPLE 7 $A = \{1, 2, 3, 4\}$ and $B = \{3, 5, 7\}$. Find $A \cup B$.

The numbers that are in A or in B or in both are 1, 2, 3, 4, 5, and 7. Therefore,

$$A \cup B = \{1, 2, 3, 4, 5, 7\} \quad ■$$

Caution: Do not repeat elements in a set.

□ **DO EXERCISE 7.**

In Example 5 the solutions were true for both inequalities making up the compound inequality. In the following example, the solutions are true for *one or the other* parts of the compound inequality. The solution set is the union of the solutions of the given inequalities.

EXAMPLE 8 Solve and graph $x + 5 < 3$ or $x - 4 \geq -3$.

There is no alternative method for solving this inequality.

$$x + 5 < 3 \quad \text{or} \quad x - 4 \geq -3$$

$$x + 5 + (-5) < 3 + (-5) \qquad x - 4 + 4 \geq -3 + 4$$

$$x < -2 \quad \text{or} \quad x \geq 1$$

The solution set is $\{x \mid x < -2 \text{ or } x \geq 1\}$.

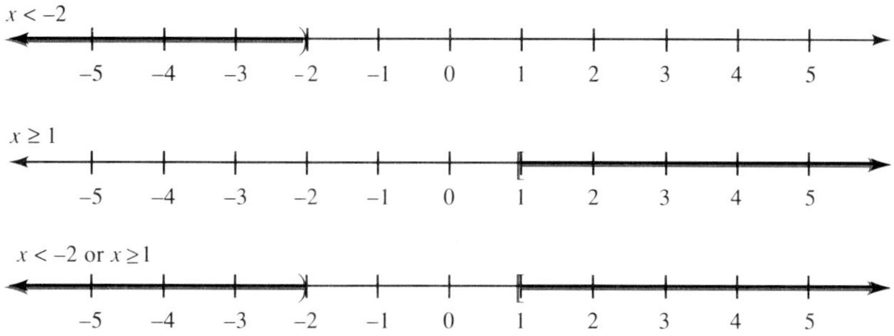

We have shown on our graph all numbers less than -2 or greater than or equal to 1. ■

□ **DO EXERCISE 8.**

DID YOU KNOW?

Thomas Harriot invented the symbols < and >. He may have gotten the idea for these symbols while looking at markings on the back of an Indian he saw while he was surveying land in Virginia. Harriot was sent to America by Sir Walter Raleigh.

□ **Exercise 8** Solve and graph.

a. $x - 4 < -7$ or $x + 1 > 2$

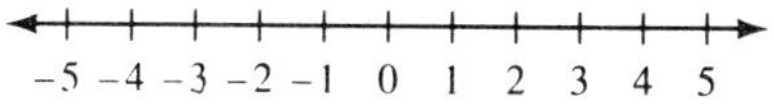

b. $x - 2 \geq 1$ or $x + 3 < -1$

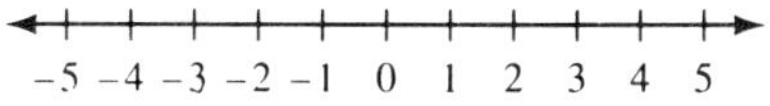

Answers to Exercises

1. a. $\{x \mid x > -1\}$; 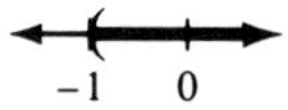

b. $\{x \mid x < 2\}$; 

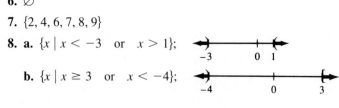

2. a. $\{x \mid x \geq -3\}$; 

b. $\{x \mid x \leq 2\}$; 

3. $\{1, 3, 5\}$

4. a. 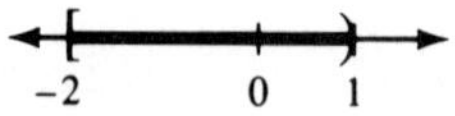

b. 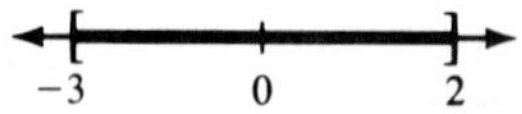

5. a. $\{x \mid -5 \leq x < 3\}$; 

b. $\{x \mid -3 < x < 4\}$; 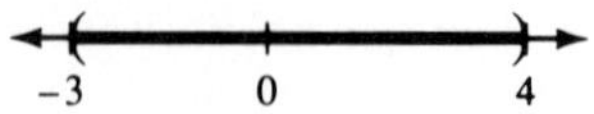

6. $\varnothing$

7. $\{2, 4, 6, 7, 8, 9\}$

8. a. $\{x \mid x < -3 \quad \text{or} \quad x > 1\}$; 

b. $\{x \mid x \geq 3 \quad \text{or} \quad x < -4\}$; 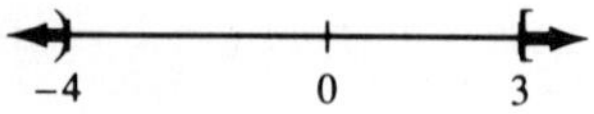

Problem Set 2.5

NAME

DATE

CLASS

Solve and graph.

1. $x + 2 > 5$

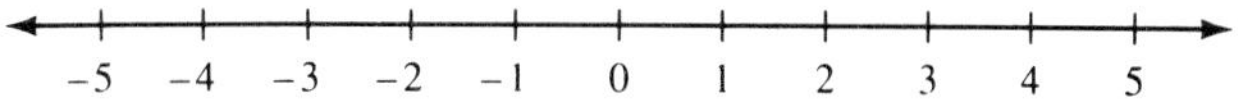

2. $x - 4 > -2$

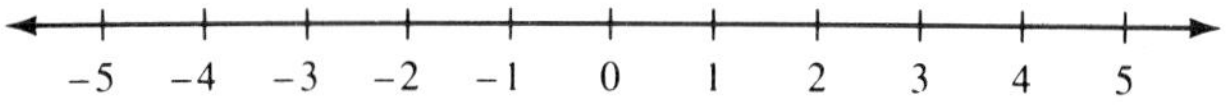

3. $x - 3 \leq 1$

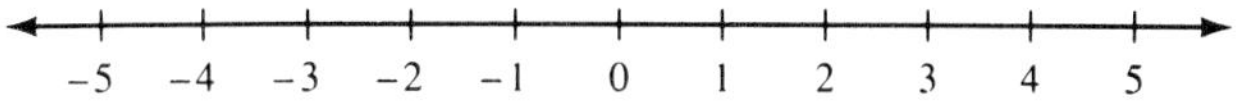

4. $x + 1 \leq 2$

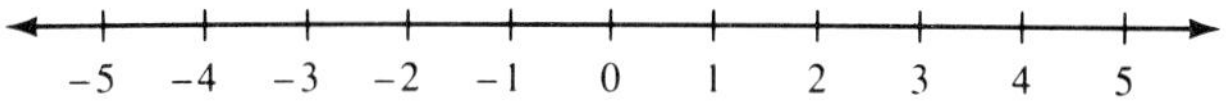

5. $4x - 1 < 3x + 3$

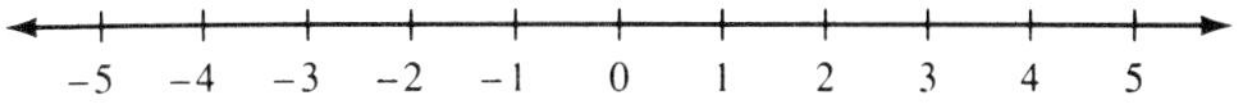

6. $8x - 9 < 7x - 7$

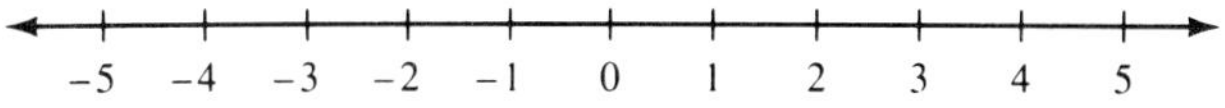

7. $2x - \frac{2}{3} \geq x - \frac{7}{6}$

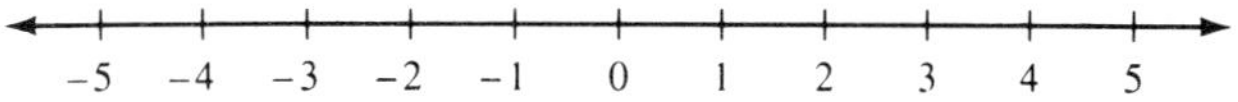

8. $6x + \frac{1}{5} \geq 5x - \frac{3}{5}$

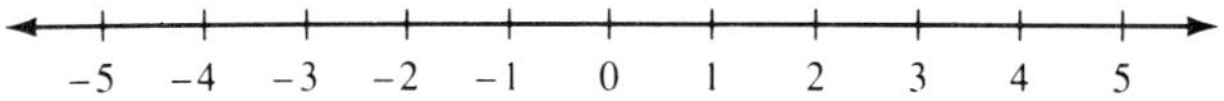

9. $-3 \leq x + 2 \leq 1$

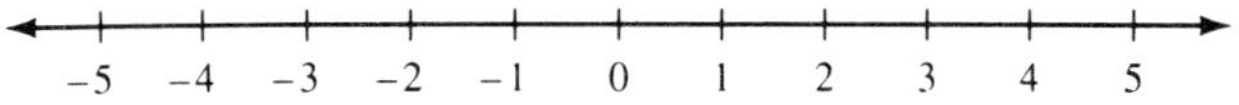

10. $2 < x + 4 < 6$

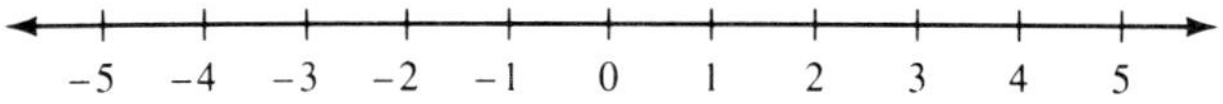

11. $7 < x + 4 \leq 9$

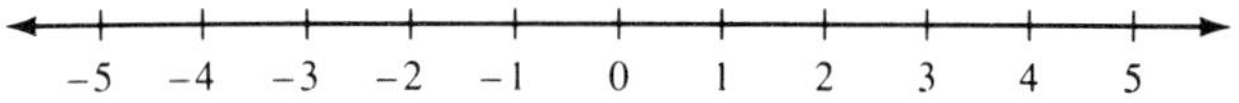

12. $-5 \leq x - 3 < 1$

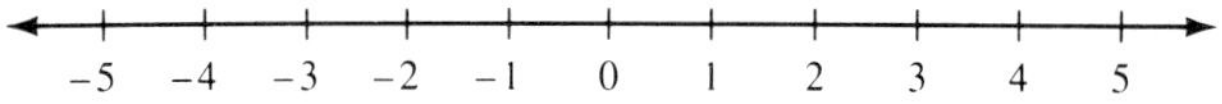

13. $x + 6 < 3$ or $x - 7 > -5$

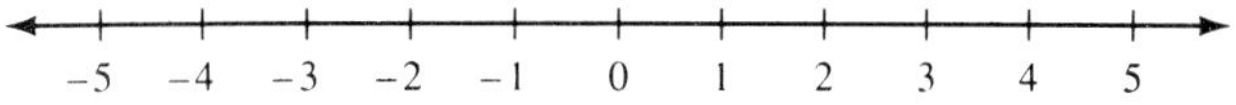

14. $x - 2 \leq -2$ or $x - 3 \geq 1$

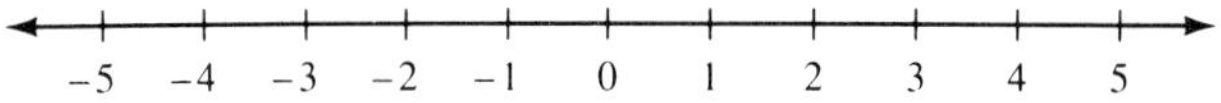

15. $x - 1 \leq 0$ or $x + 3 > 7$

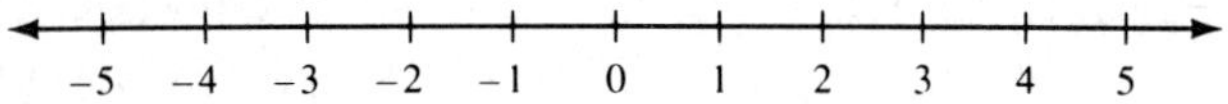

16. $x - 5 < -8$ or $x - 2 \geq 1$

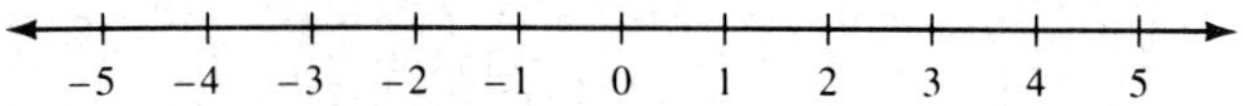

17. $x + 2 < 0$ or $x - 4 > 1$

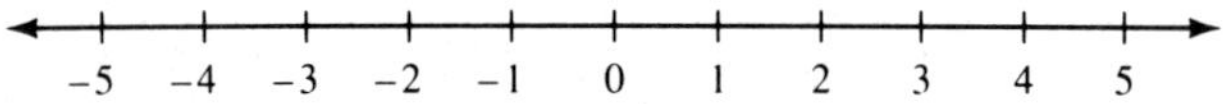

18. $x - 3 < -5$ or $x + 1 > 2$

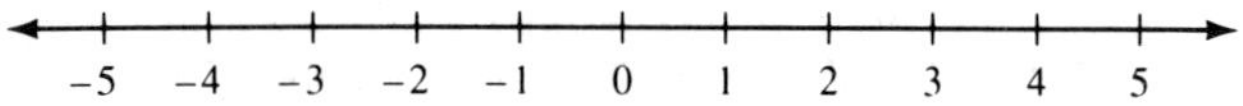

19. $-3 \leq x + 5 \leq -1$

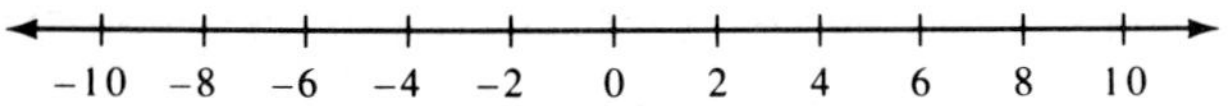

20. $0 < x + 3 \leq 7$

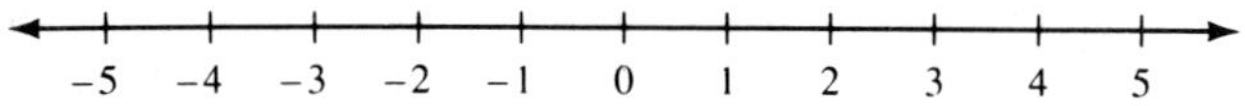

21. $x + 4 < 2$ or $x - 1 \geq 2$

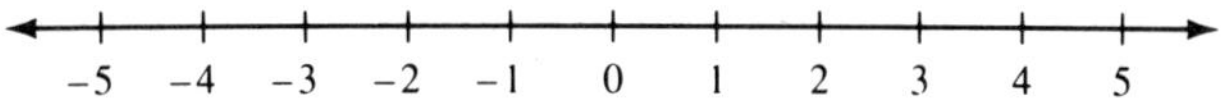

22. $x + 1 \leq 3$ or $x + 2 \geq 5$

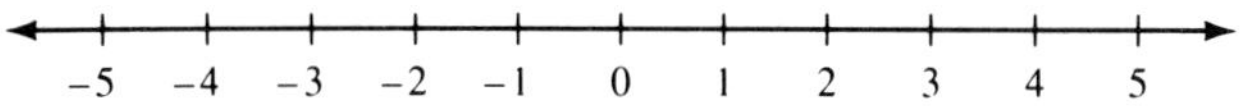

23. $-5 < x - 2 < 3$

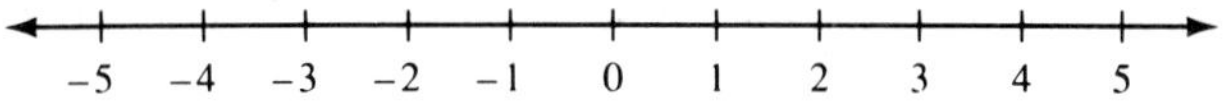

24. $-3 < x + 5 < 1$

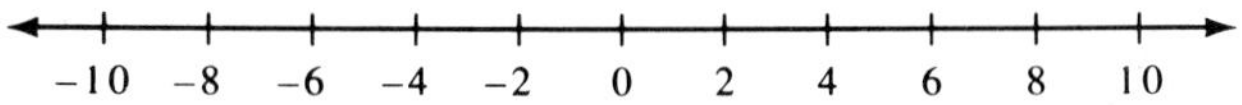

25. $x + 6 > 8$ and $x - 2 < -4$

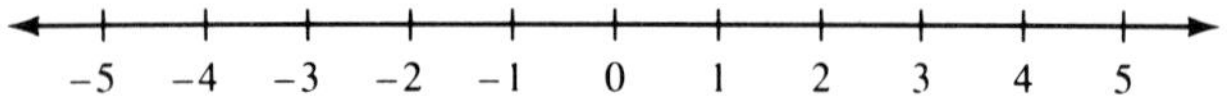

26. $x + 4 < 2$ and $x - 3 > 7$

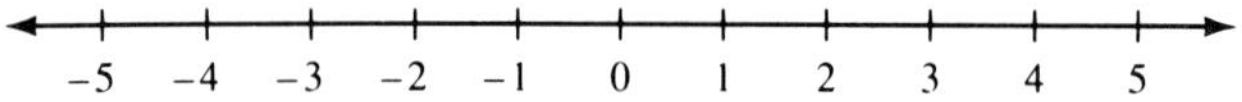

27. $-7.8 < x - 3.3 < 2.2$

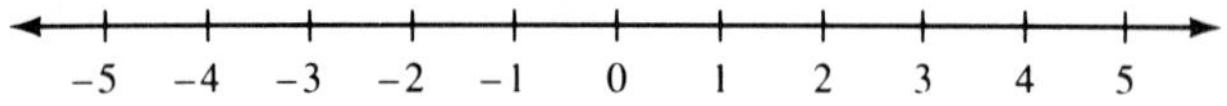

28. $-4 \leq x - 1.5 < 3.5$

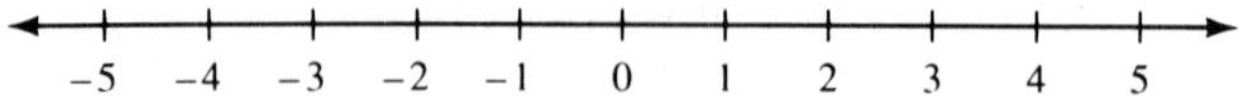

29. $-18 \leq x - 12 < -7$

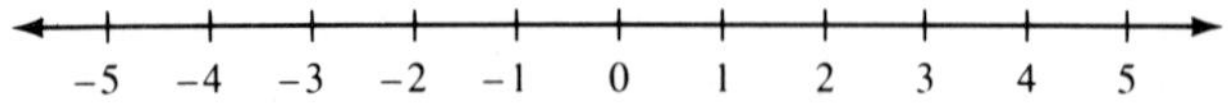

30. $8 < x + 12 \leq 17$

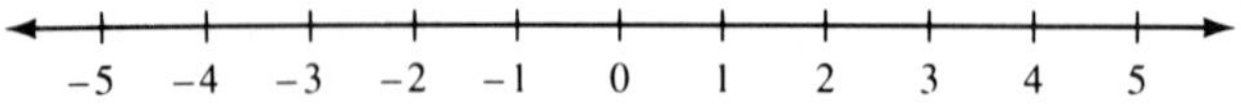

31. $x - 18 < -22$ or $x + 31 > -29$

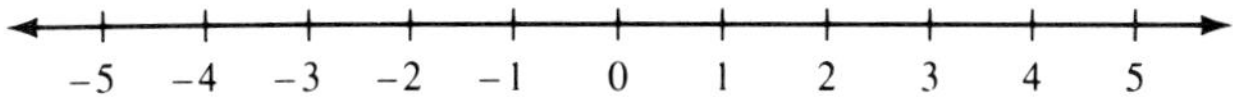

32. $x + 17 < 21$ or $x - 45 > -39$

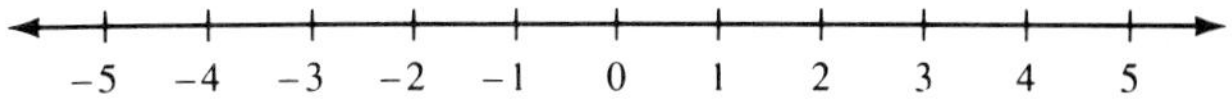

33. $x + 4.7 \leq -9.2$ or $x - 6.5 > 0$

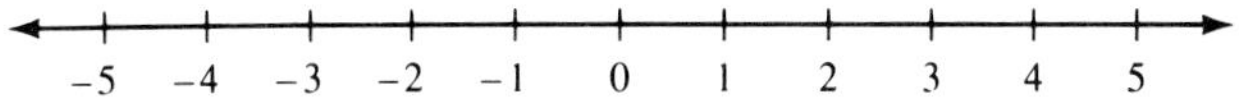

34. $x - 3.9 < 2.1$ or $x - 8.5 \geq 0$

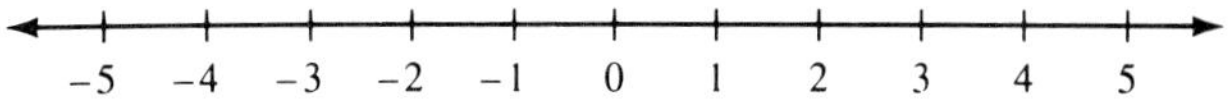

35. $x + 7 \geq 0$ and $x + 9 \leq 0$

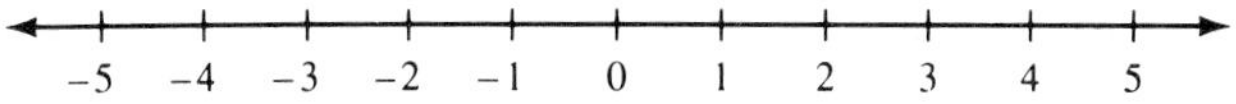

36. $x - 3 \geq -2$ and $x + 4 \leq -2$

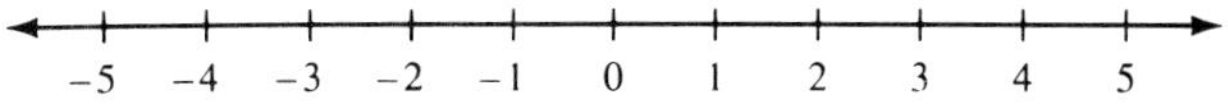

37. If a number a is to the right of another number b on the number line, a is ____________ than b.

38. An inequality usually has ____________ solutions.

39. The same number may be ____________ to both sides of an inequality.

40. An inequality that contains two or more inequality symbols is called a ____________ inequality.

Solve.

41. $7.8x - 8.2154 < 6.8x + 9.34$

42. $5.247x + 3.91 > 4.247x + 2.0435$

43. $-6.347 \leq x + 7.28 < 3.8254$

44. $9.28 > x - 4.9567 > -2.392$

45. $x - 9 < -2.01357$ or $x - 2.71845 > 6.3472$

46. $x - 2.0345 < 4.071$ or $x + 3.2187 \geq 7.192$

* 47. $-\frac{3}{8} < x + \frac{1}{12} \leq \frac{5}{16}$

* 48. $-\frac{15}{9} \leq x - \frac{4}{7} < \frac{8}{21}$

* **49.** $x - 114 < -118$ or $x + 116 > 113$

* **50.** $x + 204 < 814$ or $x - 317 > 580$

Checkup

The following problems provide a review of some of Sections 1.5 and 2.1 and will help you with the next section.

Simplify.

51. $9 \div 3(2)$

52. $21 \div 7(2)$

53. $\dfrac{3(4) + 5(4)}{2^2}$

54. $\dfrac{6(5) + 2^2}{5^2 + 2(4)}$

Solve.

55. $7x + 3 = 4x + 9$

56. $5x - 8 = 7x + 10$

57. $2x - 4 = 6x + 11$

58. $5x + 9 = 3x - 2$

59. $7x + 5 = 3x - 2$

60. $9x - 3 = 6x + 4$

2.6 THE MULTIPLICATION PROPERTY OF INEQUALITY

OBJECTIVES

1. *Solve inequalities using the multiplication property*
2. *Solve inequalities using both the addition and the multiplication properties*
3. *Solve compound inequalities using both the addition and multiplication properties*
4. *Solve word problems involving inequalities*

1 Using the Multiplication Property

Notice that	$6 > 4$	is a true inequality.
	$\mathbf{3}(6) > \mathbf{3}(4)$	Multiply both sides by positive 3.
	$18 > 12$	is a true inequality.
Consider	$2 < 5$	is a true inequality.
	$\mathbf{-4}(2) > \mathbf{-4}(5)$	Multiply both sides by -4 and reverse the inequality symbol.
	$-8 > -20$	is a true inequality.

Multiplication Property of Inequality

We may multiply both sides of an inequality by the same *positive* number without reversing the inequality symbol. If

$$a < b$$

then

$$ac < bc \qquad \text{if } c \text{ is positive}$$

If we multiply both sides of an inequality by the same *negative* number, we must *reverse* the inequality symbol. If

$$a < b$$

then

$$ac > bc \qquad \text{if } c \text{ is negative}$$

The property also holds for $>$, $\leq$, and $\geq$.

If we divide both sides of an inequality by a number, this is the same as multiplying both sides by the reciprocal of the number. Therefore, we may *divide both sides of an inequality by the same positive number.* However, if we *divide both sides of the inequality by the same negative number, we must reverse the inequality* symbol in order to get an equivalent inequality.

EXAMPLE 1 Solve.

a. $\dfrac{x}{3} > 8$

$\mathbf{3}\left(\dfrac{x}{3}\right) > \mathbf{3}(8)$ Multiply both sides by 3.

$x > 24$

The solution set is $\{x \mid x > 24\}$.

□ Exercise 1 Solve.

a. $\frac{x}{3} < 5$

b. $\frac{x}{-7} \geq 2$

c. $-4x < 40$

d. $-3y \geq 18$

b. $\frac{y}{-2} \leq 4$

$-2\left(\frac{y}{-2}\right) \geq -2(4)$ Multiply by -2 and reverse the inequality symbol.

$y \geq -8$

The solution is any number greater than or equal to -8. The solution set is $\{y \mid y \geq -8\}$.

c. $-8x < 24$

$\frac{-8x}{-8} > \frac{24}{-8}$ Divide by -8 and reverse the inequality symbol.

$x > -3$

The solution set is $\{x \mid x > -3\}$. ■

□ DO EXERCISE 1.

2 Using Both the Addition and the Multiplication Properties

Both the addition and the multiplication properties must be used to solve some inequalities. It is easier to use the addition property first to get all the terms with variables on one side of the inequality and all the other terms on the other side. Then use the multiplication property to solve the inequality.

EXAMPLE 2 Solve.

a. $8y + 16 \leq 12 + 5y$

$8y + 16 + (\mathbf{-16}) \leq 12 + 5y + (\mathbf{-16})$ Add -16 to both sides.

$8y \leq -4 + 5y$

$8y + (-5y) \leq -4 + 5y + (-5y)$ Add $-5y$ to both sides.

$3y \leq -4$

$\frac{3y}{3} \leq \frac{-4}{3}$ Divide by 3.

$y \leq \frac{-4}{3}$

The solution set is $\{y \mid y \leq -\frac{4}{3}\}$.

b. $3(x + 2) - 5x < x$

$3x + 6 - 5x < x$ Use a distributive law.

$-2x + 6 < x$ Combine like terms.

$-2x + 6 + 2x < x + 2x$ Add $2x$ to each side.

$6 < 3x$

$\frac{6}{3} < \frac{3x}{3}$ Divide by 3.

$2 < x$

The solution set is $\{x \mid 2 < x\}$.

This result is equivalent to $x > 2$. You may find it easier to graph in this form. Note that we may isolate the variable on either side. ■

Steps in Solving a Linear Inequality in One Variable

1. Combine like terms and remove parentheses on each side of the inequality, if necessary.
2. Use the addition property of inequality to get the terms containing the variable on one side of the inequality and all other terms on the other side.
3. Combine like terms, if possible.
4. Use the multiplication property of inequality to isolate the variable on one side. *Remember to reverse the inequality symbol if both sides of the inequality are multiplied or divided by a negative number.*

□ **DO EXERCISE 2.**

3 Compound Inequalities

To solve many *compound* inequalities, we must use both the addition and the multiplication properties of inequality.

EXAMPLE 3 Solve.

a.

$$-2 < 3x + 4 \leq 7$$

$$-2 + (\mathbf{-4}) < 3x + 4 + (\mathbf{-4}) \leq 7 + (\mathbf{-4}) \qquad \text{Add } -4.$$

$$-6 < 3x \leq 3$$

$$\frac{-6}{3} < \frac{3x}{3} \leq \frac{3}{3} \qquad \text{Divide by 3.}$$

$$-2 < x \leq 1$$

The solution set is $\{x \mid -2 < x \leq 1\}$.

b.

$$3x + 4 < -5 \quad \text{or} \quad -2x < 3x - 15$$

$$3x + 4 + (-4) < -5 + (-4) \quad \text{or} \quad -2x + (-3x) < 3x - 15 + (-3x)$$

$$3x < -9 \quad \text{or} \quad -5x < -15$$

$$\frac{3x}{3} < \frac{-9}{3} \quad \text{or} \quad \frac{-5x}{-5} > \frac{-15}{-5}$$

$$x < -3 \quad \text{or} \quad x > 3$$

The solution set is $\{x \mid x < -3 \text{ or } x > 3\}$. ■

□ **DO EXERCISE 3.**

□ **Exercise 2** Solve.

a. $3 + 4y \leq 15$

b. $2x - 7 > 4x + 9$

c. $2(x - 3) < 5x + 3$

d. $4(y - 3) + 2 \geq 6$

□ **Exercise 3** Solve.

a. $-5 < 4x + 3 < 15$

b. $6 \geq -2x - 8 > -4$

c. $2x + 5 < -7$ or $9 \geq -3x$

□ **Exercise 4** Andrew must have an average of at least 90 to get an A in his history class. There will be three tests and he got scores of 95 and 84 on his first two tests. What score must he get on his last test to get an A?

4 Applied Problems

Inequalities are used to solve many problems in the real world.

EXAMPLE 4 Maria must have an average of at least 90 to get an A in her chemistry class. There are four tests and her first three test scores were 90, 81, and 92. What score on the last test will give her an A in the class?

Let x represent Maria's score on the last test

Since there are four tests, Maria's average will be

$$\frac{90 + 81 + 92 + x}{4}$$

Her average must be at least 90 (greater than or equal to 90).

$$\frac{90 + 81 + 92 + x}{4} \geq 90$$

$$\frac{263 + x}{4} \geq 90$$

$$4\left(\frac{263 + x}{4}\right) \geq 4(90) \quad \text{Multiply by 4 to clear fractions.}$$

$$263 + x \geq 360$$

$$x \geq 360 - 263 \quad \text{Add } -263 \text{ to both sides.}$$

$$x \geq 97$$

A score of 97 will give Maria an A in the class. ■

□ **DO EXERCISE 4.**

Answers to Exercises

1. a. $\{x \mid x < 15\}$ **b.** $\{x \mid x \leq -14\}$ **c.** $\{x \mid x > -10\}$
d. $\{y \mid y \leq -6\}$

2. a. $\{y \mid y \leq 3\}$ **b.** $\{x \mid x < -8\}$ **c.** $\{x \mid x > -3\}$ **d.** $\{y \mid y \geq 4\}$

3. a. $\{x \mid -2 < x < 3\}$ **b.** $\{x \mid -7 \leq x < -2\}$
c. $\{x \mid x < -6 \text{ or } x \geq -3\}$

4. 91

Problem Set 2.6

NAME

DATE

CLASS

Solve.

1. $\frac{x}{5} < 7$

2. $\frac{y}{8} \geq 2$

3. $\frac{x}{-2} \leq 9$

4. $\frac{x}{-3} > 4$

5. $28 > 4x$

6. $27 \leq 3y$

7. $-9y > 36$

8. $-5x \leq 45$

9. $2 + 7y \leq 16$

10. $3 + 4x > 23$

11. $2 \geq 8 - 5x$

12. $3.4 < 3.2x - 9.4$

13. $\frac{1}{3} + 5x < \frac{3}{4}$

14. $\frac{1}{4} - 3y < \frac{5}{8}$

15. $x + 4(2x - 1) < 5x$

16. $-2(y - 4) > -3(y + 1)$

17. $-3(x - 6) \leq 2x - 5$

18. $-2(x + 4) > 6x + 8$

19. $12 \leq 2x + 4 < 16$

20. $-2 < 5y + 3 \leq 8$

21. $-12 < -4x + 8 < 24$

22. $-3 \leq -3y + 6 \leq 15$

23. $-6x + 1 \geq -11$ or $5x > 15$

24. $-3x > 9$ or $2x + 5 \geq 7$

25. Linda wants to earn $1540 per year in interest. What amount must she invest at 7% to earn at least $1540?

26. What amount must be invested at 9% interest to earn at least $1620 in 1 year?

27. A diet must include three foods. It must include twice as many grams of protein as fruit and 5 grams of fat. If this diet contains at least a total of 50 grams of food, how many grams of fruit should be included?

28. Amy must take at least 40 units of medicine per day. The medicine comes in red pills of 3 units each and blue pills of 4 units each. If she must take twice as many red pills as blue pills, how many of each pill does she need?

29. A plumber can be paid in two ways:

Method 1: $600 plus $5 per hour

Method 2: $8 per hour

If the job takes n hours, for what values of n does Method 1 give the plumber the better wages?

30. Tom has twice as many quarters as nickels, and he has at least 21 coins. At least how many nickels does he have?

31. Six times a number is between -18 and 12. The number is between what values?

32. One-fourth of a number added to 6 gives a result of at least 9. The number must have at least what value?

33. A craftsman sold rings worth $800 in June, $1500 in July, and $2300 in August. What is the minimum amount of sales he must make in September to earn $5200 during this 4-month period?

34. Stacie earned $3250, $1830, and $2780 working for 3 months as a sales representative. What is the minimum amount she must earn the next month in order to earn at least $9000 during this 4-month period?

35. The Sportsmans Club wants to collect at least 2125 pounds of aluminum cans during their collection drives. On the first two drives they collected 820 pounds and 537 pounds. What is the minimum number of pounds of cans that they must collect on the last drive?

36. Hassan wants to save at least $3000 in a 5-month period. The first 4 months he has saved $625, $340, $280, and $587. What is the minimum amount he must save during the fifth month?

37. Rent a Wreck rents cars for $18 per day or fraction of a day and unlimited mileage. What is the maximum number of days that Bangahn can rent a car if she wants to spend less than $340?

38. Bryan's Electrical Service charges $32 per hour or fraction of an hour. Cynthia wants to spend less than $850 to wire her house. What is the maximum number of hours that the electrical service can work on her house?

39. The owner of an ice cream store rents the store for $800 per month plus 6% of the total sales during the month. The owner wishes to earn a minimum of $2000 per month. Find the minimum sales that will enable the owner to achieve her goal.

40. Dungeness Public Utility District charges 5 cents per kilowatthour of electricity. How many kilowatthours of electricity can the Jergens family use if they want their electric bill to be less than $209?

41. Multiplying both sides of an inequality by the same ____________ number does not reverse the inequality symbol.

42. If we multiply or divide both sides of an inequality by the same ____________ number, we must reverse the inequality symbol.

Solve. Round answers to the nearest hundredth.

43. $6.8315y \leq 15.0923$

44. $-7.2957z > 4.37742$

45. $7.58 + 4.352y \geq 26.1175$

46. $-15.322 < 8.61 - 5.983x$

47. $-3(2.45x - 7.235) \leq 16.56$

48. $2(3.895 - 6.289y) > -13.5926$

Solve.

* 49. $-\frac{3}{4}(x + 6) + \frac{9}{2}(2x - 5) < 0$

* 50. $\frac{10}{3}(3p - 1) \geq \frac{5}{2}(2p - 3)$

* 51. $5(x - 1) < 4x - (x + 3)$ or $3(x + 1) - 1 > -(x - 2) + 3x$

* 52. $y - (2y - 1) > y - 5$ or $5 + y - (2y + 3) > 9 - 2y$

Think About It

Determine whether the statement is true or false. If it is false, give a counterexample.

* 53. For any real numbers a, b, c, and d, if $a < b$ and $c < d$, then $ac < bd$.

* 54. For any real numbers a, b, c, and d, if $a < b$ and $c < d$, then $a + c < b + d$.

Checkup

The following problems provide a review of some of Section 1.1 and will help you with the next section.

Find the following.

55. $|-3|$

56. $|6|$

57. $-\left|\frac{3}{4}\right|$

58. $\left|-\frac{7}{8}\right|$

59. $-|-2.4|$

60. $-|10.5|$

61. $-\left|-\frac{12}{5}\right|$

62. $-\left|\frac{23}{7}\right|$

63. $-|15.6|$

64. $-|-42.8|$

2.7 ABSOLUTE VALUE EQUATIONS AND INEQUALITIES

1 Distance

Notice that the distance between -2 and 3 on the number line is 5 units.

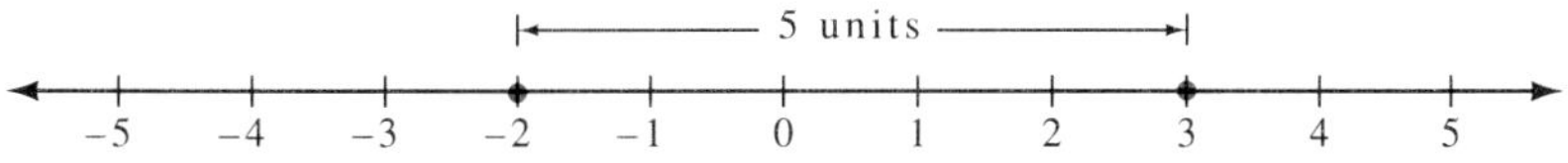

To find the ***distance*** between two numbers on the number line, we may subtract one number from the other and find the absolute value of the result. The distance, d, between -2 and 3 is

$$d = |-2 - 3| = |-5| = 5.$$

The order in which we subtract is not important since we are taking the absolute value of the result. Notice that if we had not used absolute value we would get a negative distance which does not make sense.

> The distance, d, between any real numbers a and b is
>
> $$d = |a - b|$$

EXAMPLE 1 Find the distance between the following numbers.

a. 7 and -2

$$d = |7 - (-2)| = |9| = 9$$

b. -15 and -8

$$d = |-15 - (-8)| = |-7| = 7 \quad ■$$

□ **DO EXERCISE 1.**

2 Equations

Recall from Section 1.1 that the *absolute value* of a number is its distance from zero on the number line. So the equation $|x| = 3$ means "find all the numbers whose distance from zero is 3." We can see from the number line below that there are two such numbers, 3 and -3.

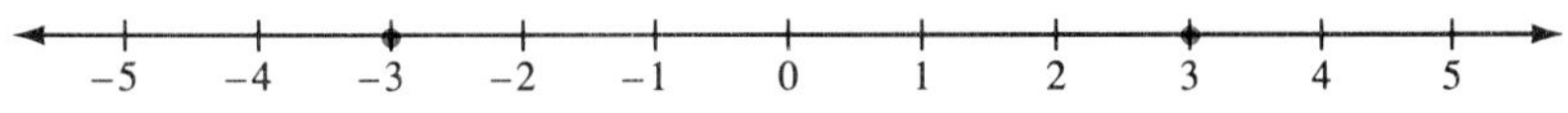

$$\text{If } |x| = 3$$

$$\text{then } x = 3 \quad \text{or} \quad x = -3$$

The solutions are 3 and -3. The solution set is $\{3, -3\}$. We may also want to solve an equation such as $|3x - 4| = 7$. The following is a general rule for solving equations with absolute value.

> To solve an equation of the form $|ax + b| = c$, where $a \neq 0$ and c is a *positive* number, solve the two equations
>
> $$ax + b = c \quad \text{or} \quad ax + b = -c$$

OBJECTIVES

1 *Find the distance between two points on the number line*

2 *Solve equations containing absolute value symbols*

3 *Solve and graph inequalities containing absolute value symbols*

□ **Exercise 1** Find the distance between the numbers.

a. 15 and -3

b. -25 and -9

c. -8 and 6

d. 55 and 17

□ **Exercise 2** Solve.

a. $|x| = 7$

b. $|x| = 15$

c. $|2x + 3| = 5$

d. $|y - 7| = 25$

□ **Exercise 3** Solve.

a. $|3x + 4| = -5$

b. $|2x - 7| = 0$

EXAMPLE 2 Solve.

a. $|x| = 9$

$$x = 9 \qquad \text{or} \qquad x = -9$$

The solution set is $\{9, -9\}$.

b. $|3x - 4| = 2$

The $|3x - 4|$ represents the distance between $3x$ and 4. The equation says that this distance must be 2.

$$|3x - 4| = 2$$

$$3x - 4 = 2 \qquad \text{or} \qquad 3x - 4 = -2$$

$$3x = 6 \qquad\qquad 3x = 2 \qquad \text{Add 4.}$$

$$x = 2 \qquad\qquad x = \frac{2}{3}$$

The solution set is $\{2, \frac{2}{3}\}$.

Notice that when $x = 2$, then $3x = 6$ and the distance between $3x$ and 4 is the distance between 6 and 4, which is 2. When $x = \frac{2}{3}$, $3x = 2$ and the distance between $3x$ and 4 is the distance between 2 and 4, which is 2.

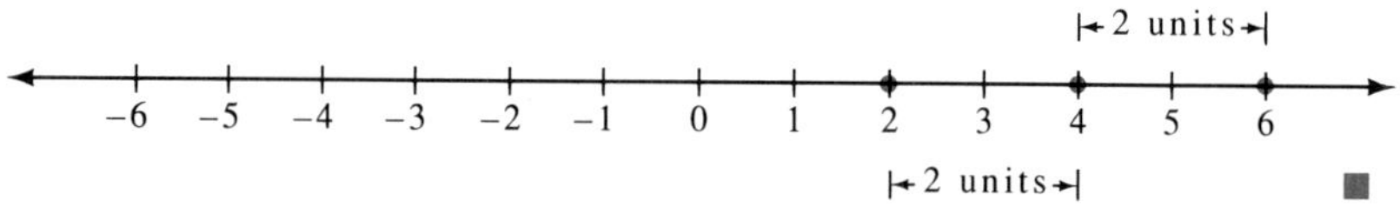

□ **DO EXERCISE 2.**

There are some special cases of absolute value equations.

EXAMPLE 3 Solve.

a. $|3p - 2| = -7$

The absolute value of an expression can never be a negative number. Therefore, the equation has *no solution*. The solution set is $\varnothing$.

b. $|5x - 4| = 0$

The expression $5x - 4$ will equal 0 only if

$$5x - 4 = 0$$

$$5x = 4$$

$$x = \frac{4}{5}$$

The solution is $\frac{4}{5}$. The solution set is $\{\frac{4}{5}\}$. ■

Caution: Equations containing one set of absolute value symbols have two solutions except in the special cases shown in Example 3.

□ **DO EXERCISE 3.**

There are also absolute value equations that contain two absolute value expressions. For two expressions to have the same absolute value they must be equal or be opposites of each other.

To solve an equation of the form

$$|ax + b| = |cx + d|$$

solve the two equations

$$ax + b = cx + d \quad \text{or} \quad ax + b = -(cx + d)$$

EXAMPLE 4 Solve: $|z + 4| = |2z - 3|$.

$$z + 4 = 2z - 3 \quad \text{or} \quad z + 4 = -(2z - 3)$$

$$4 + 3 = 2z - z \qquad z + 4 = -2z + 3$$

$$7 = z \qquad z + 2z = 3 - 4$$

$$3z = -1$$

$$z = -\frac{1}{3}$$

The solution set is $\{7, -\frac{1}{3}\}$. ■

□ DO EXERCISE 4.

3 Inequalities

Since absolute value represents the distance of a number from zero on the number line, the inequality $|x| < 3$ has as solutions all those numbers whose distance from zero is less than 3. The solutions are all numbers between -3 and 3.

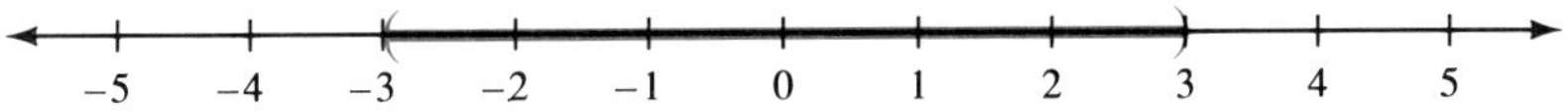

If $|x| < 3$, then $-3 < x < 3$. The solution set is $\{x \mid -3 < x < 3\}$.

To solve the inequality $|ax + b| < c$, where $a \neq 0$ and c is a positive number, solve the following inequality:

$$-c < ax + b < c$$

The rule also holds if $<$ is replaced by $\leq$.

EXAMPLE 5 Solve and graph.

a. $|x| < 2$

If $|x| < 2$, then

$$-2 < x < 2$$

The solution set is $\{x \mid -2 < x < 2\}$.

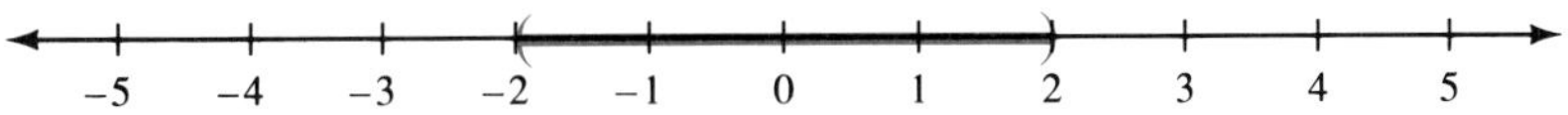

□ Exercise 4 Solve.

a. $|x + 6| = |2x + 5|$

b. $|5x - 2| = |x - 1|$

□ **Exercise 5** Solve and graph.

a. $|x| < 4$

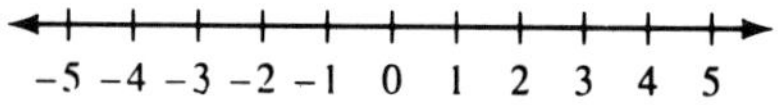

b. $|y| \leq 1$

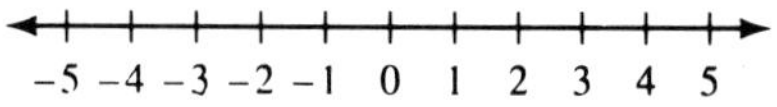

c. $|x - 3| \leq 12$

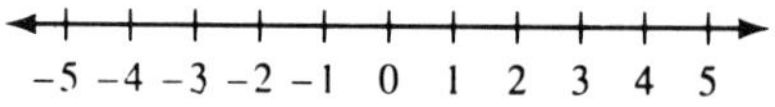

d. $|2y - 4| < 5$

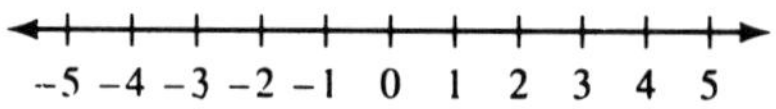

b. $|4x - 3| \leq 13$

$$-13 \leq 4x - 3 \leq 13$$

$$-10 \leq 4x \leq 16 \qquad \text{Add 3.}$$

$$-\frac{10}{4} \leq \frac{4x}{4} \leq \frac{16}{4} \qquad \text{Divide by 4.}$$

$$-\frac{5}{2} \leq x \leq 4 \qquad \text{Simplify.}$$

The solution set is $\{x \mid -\frac{5}{2} \leq x \leq 4\}$. The graph follows.

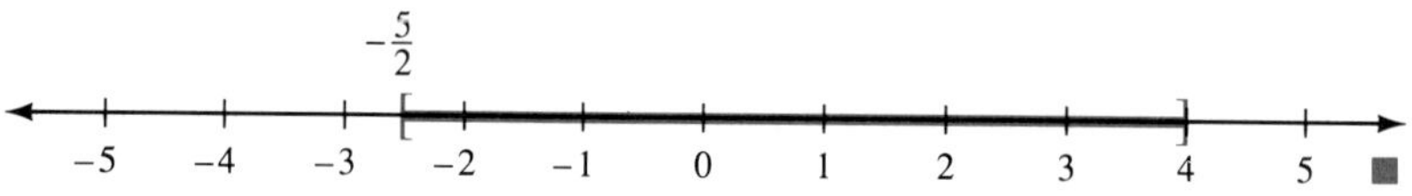

■

□ **DO EXERCISE 5.**

To solve absolute value inequalities containing the greater than symbol $>$, we again recall that absolute value represents the distance between a number and zero. Therefore, to solve the equation $|x| > 3$, we need to find all numbers whose distance from zero is greater than 3. The graph of the solutions is as follows:

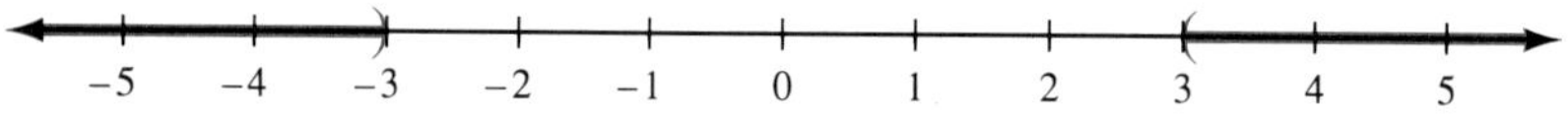

If $|x| > 3$, then $x < -3$ or $x > 3$. The solution set is $\{x \mid x < -3$ or $x > 3\}$.

> To solve the inequality $|ax + b| > c$, where $a \neq 0$ and c is a positive number, solve the following inequality.
>
> $$ax + b < -c \quad \text{or} \quad ax + b > c$$
>
> The rule also holds when $>$ is replaced by $\geq$.

EXAMPLE 6 Solve and graph.

a. $|x| \geq 2$

If $|x| \geq 2$, then

$$x \leq -2 \quad \text{or} \quad x \geq 2$$

The solution set is $\{x \mid x \leq -2$ or $x \geq 2\}$.

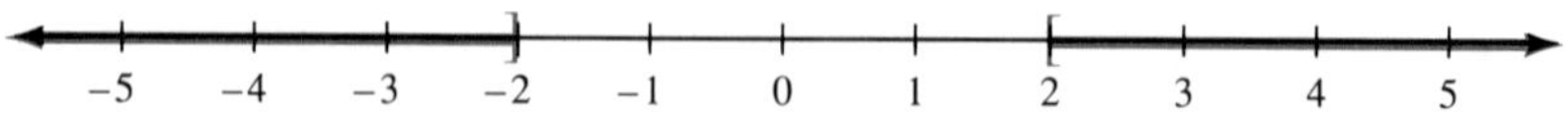

b. $|4x - 1| > 3$

$$4x - 1 < -3 \quad \text{or} \quad 4x - 1 > 3$$

$$4x < -2 \quad \text{or} \quad 4x > 4 \qquad \text{Add 1.}$$

$$x < -\frac{1}{2} \quad \text{or} \quad x > 1 \qquad \text{Divide by 4.}$$

The solution set is $\{x \mid x < -\frac{1}{2} \text{ or } x > 1\}$. The graph follows.

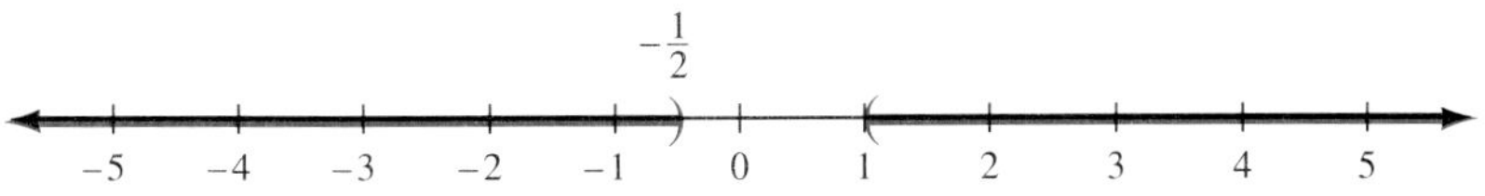

□ **DO EXERCISE 6.**

There are special cases of absolute value inequalities.

EXAMPLE 7 Solve.

a. $|x| \geq -7$
This is true for all real numbers.
The solution set is $\{x \mid x \text{ is a real number}\}$.

b. $|3y + 5| < -4$
There is no number whose absolute value is less than -4. There is *no solution* to this inequality.
The solution set is $\varnothing$. ■

□ **DO EXERCISE 7.**

Suggestions for Solving Absolute Value Equations and Inequalities, where $a \neq 0$ and c is a positive number

	Solution Is Found by Solving
$\lvert ax + b\rvert = c$	$ax + b = c$ or $ax + b = -c$
$\lvert ax + b\rvert < c$	$-c < ax + b < c$
$\lvert ax + b\rvert > c$	$ax + b < -c$ or $ax + b > c$

The rules also hold when $<$ is replaced by $\leq$ and $>$ is replaced by $\geq$.

□ **Exercise 6** Solve and graph.

a. $|x| > 4$

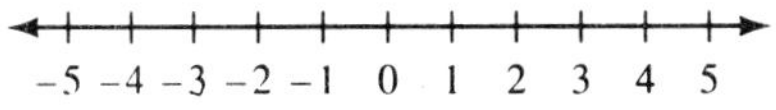

b. $|y| \geq 1$

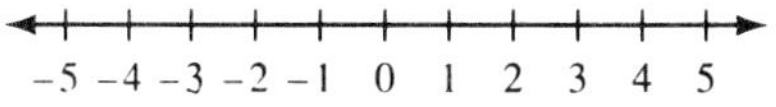

c. $|x + 2| > 5$

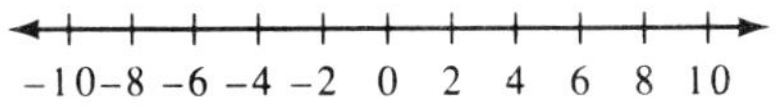

d. $|2y - 1| \geq 5$

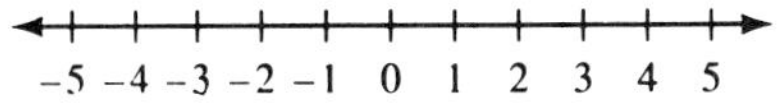

□ **Exercise 7** Solve.

a. $|5x| \geq -2$

b. $|4x - 2| < -9$

Answers to Exercises

1. a. 18 **b.** 16 **c.** 14 **d.** 38

2. a. $\{7, -7\}$ **b.** $\{15, -15\}$ **c.** $\{1, -4\}$ **d.** $\{-18, 32\}$

3. a. $\varnothing$ **b.** $\left\{\frac{7}{2}\right\}$

4. a. $\left\{1, -\frac{11}{3}\right\}$ **b.** $\left\{\frac{1}{4}, \frac{1}{2}\right\}$

5. a. $\{x \mid -4 < x < 4\}$;

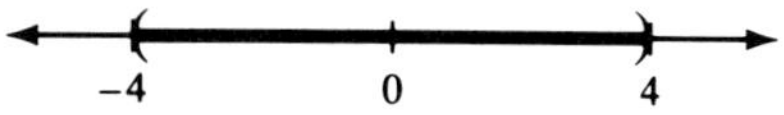

b. $\{y \mid -1 \le y \le 1\}$;

c. $\{x \mid -9 \le x \le 15\}$;

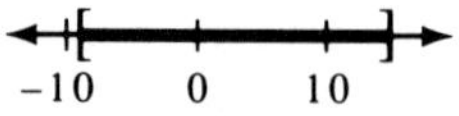

d. $\left\{y \mid -\frac{1}{2} < y < \frac{9}{2}\right\}$;

6. a. $\{x \mid x < -4 \text{ or } x > 4\}$;

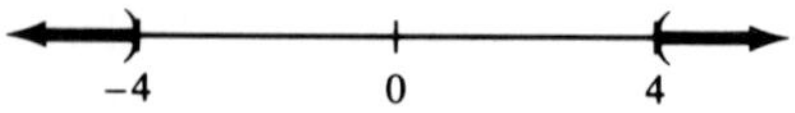

b. $\{y \mid y \le -1 \text{ or } y \ge 1\}$;

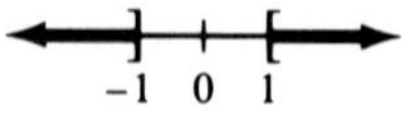

c. $\{x \mid x < -7 \text{ or } x > 3\}$;

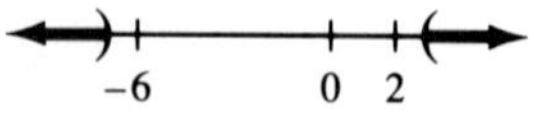

d. $\{y \mid y \le -2 \text{ or } y \ge 3\}$;

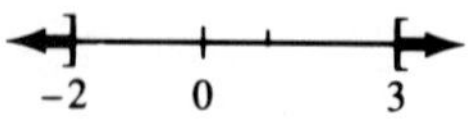

7. a. $\{x \mid x \text{ is a real number}\}$ **b.** $\varnothing$

Problem Set 2.7

NAME

DATE

CLASS

Find the distance between the numbers.

1. 12, 9

2. 25, 20

3. −9, −16

4. −15, −26

5. −51, 25

6. −75, 40

7. 0, −8

8. −10, 0

Solve.

9. $|x| = 8$

10. $|x| = 25$

11. $|y + 1| = 9$

12. $|x + 4| = 15$

13. $|2x + 5| = 3$

14. $|3x - 2| = 7$

15. $|5x - 1| = 9$

16. $|4x - 8| = 16$

17. $|4x - 5| = 13$

18. $|5y - 1| = 26$

19. $\left|\frac{1}{2}x + 3\right| = 4$

20. $\left|\frac{2}{3}p - 1\right| = 2$

21. $|x| = -5$

22. $|y| = -8$

23. $|5x - 2| = 0$

24. $|3x + 7| = 0$

25. $|7x - 8| = -4$

26. $|9x + 20| = -11$

27. $|2q + 4| = |3q - 1|$

28. $|5a + 7| = |4a + 3|$

29. $|2x + 5| = |x - 3|$

30. $|4x + 6| = |5x - 2|$

31. $|4x - 8| = |2x|$

32. $|4x - 3| = |3x - 4|$

Solve and graph.

33. $|x| < 5$

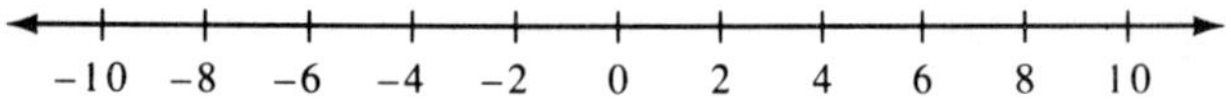

34. $|y| \leq 7$

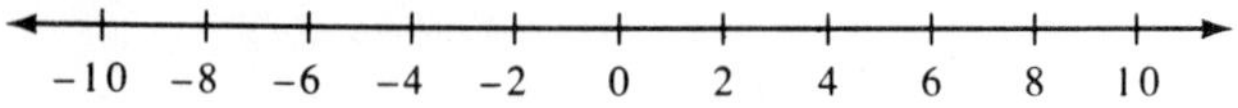

35. $|x + 5| \leq 8$

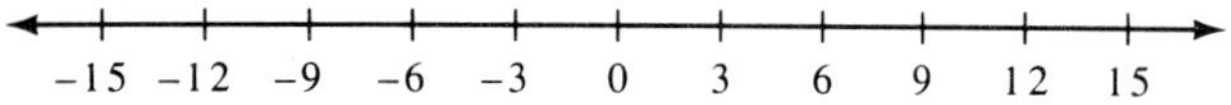

36. $|y - 3| < 2$

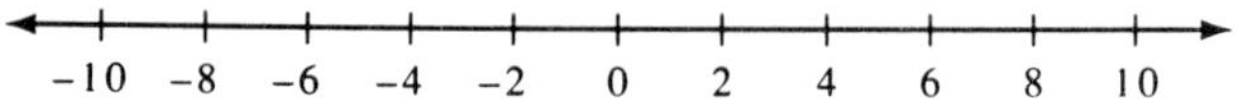

37. $|3x - 7| \leq 2$

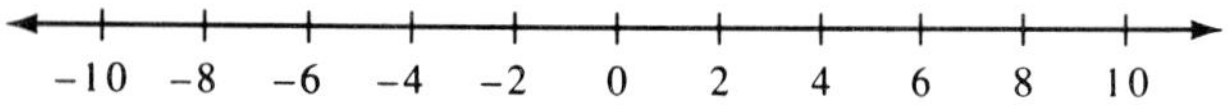

38. $|2x + 8| < 4$

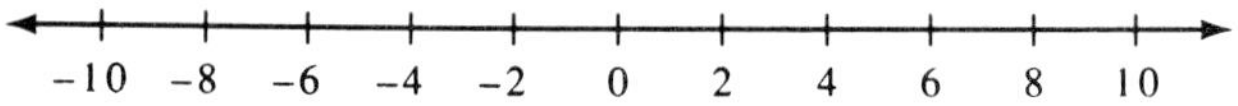

39. $|x| \geq 5$

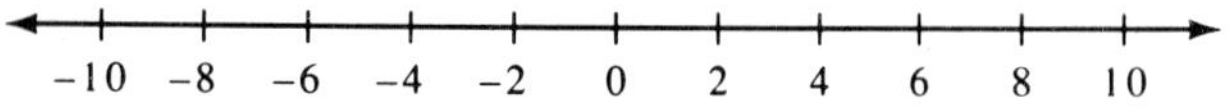

40. $|y| > 6$

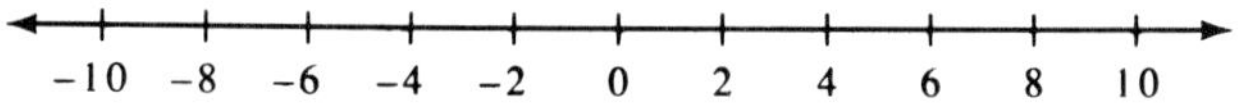

41. $|x - 2| > 4$

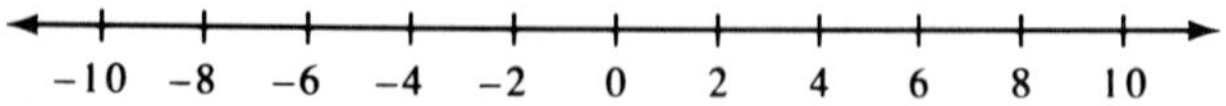

42. $|y + 3| \geq 1$

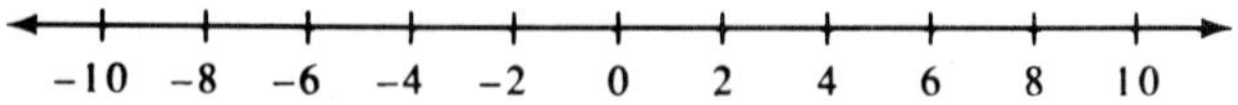

43. $|5x - 5| \geq 10$

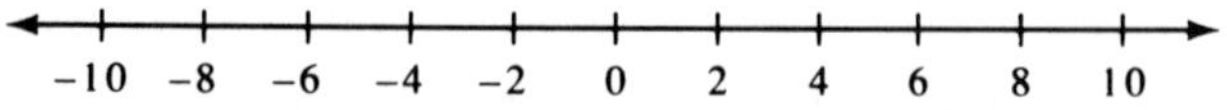

44. $|6x - 6| \geq 12$

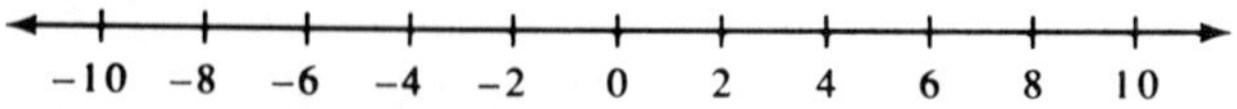

45. $|2x - 3| < 4$

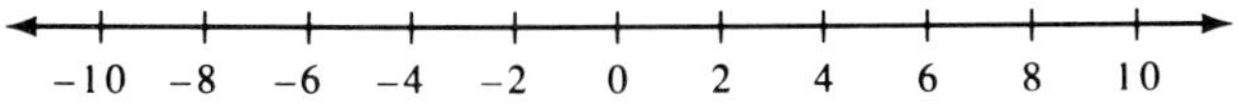

46. $|3x - 2| \leq 2$

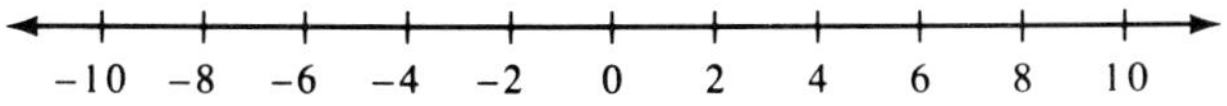

47. $|4x + 1| \geq 5$

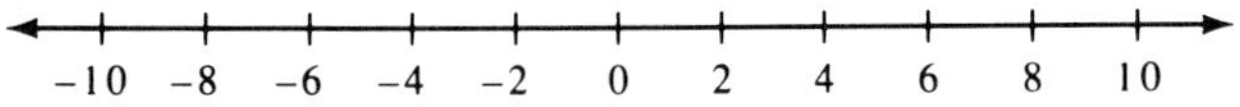

48. $|3x + 6| > 9$

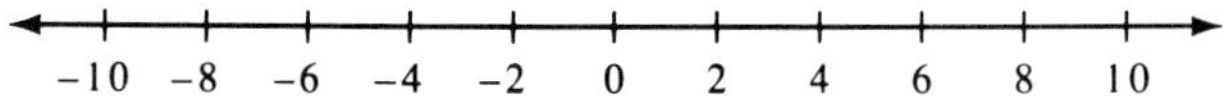

49. $|8y - 3| \leq 11$

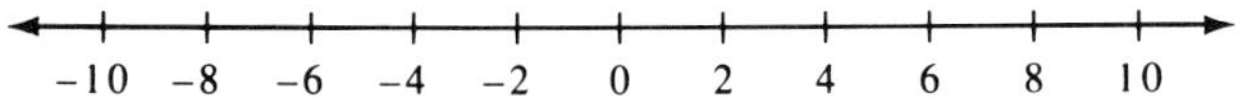

50. $|6y - 1| < 7$

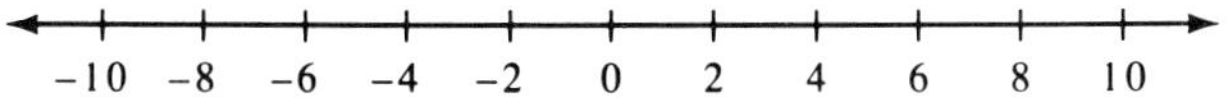

51. $|3x + 2| > -5$

52. $|7x - 1| > -12$

53. $|x| < -3$

54. $|x| < -8$

55. $\left|\frac{4 - 3y}{4}\right| > 2$

56. $\left|\frac{7 - 2x}{2}\right| > \frac{5}{2}$

57. $|7x - 6| < -2$

58. $|5x + 3| < -9$

59. $|x + 4| \geq -3$

60. $|3x - 2| \leq -4$

61. To find the distance between two numbers on the number line, we subtract one number from the other and find the ____________ of the result.

62. Absolute value inequalities usually have ____________ solutions.

Solve. Round answers to the nearest hundredth.

63. $|5.1342x + 7.894| = 1.34756$

64. $|2.189x - 0.8534| = 0.46$

65. $|4.9327x - 8.194| < 7.59064$

66. $|0.178x + 0.3954| \leq 1.3922$

67. $|0.0679x + 0.4692| \geq 1.05993$

68. $|3.247x - 4.28| > 2.8634$

* 69. $|x| - |3x - 1| = 0$

* 70. $|4x - 6| + |2x| = 0$

* 71. $|x + 6| > 0$

* 72. $|x - 5| < 0$

* 73. $|3x - 7| \leq 0$

* 74. $|6x + 5| \leq 0$

Checkup

The following problems provide review of some of Section 1.4.

Evaluate when $x = -2$, $y = 4$, and $z = -1$.

75. $xy - z$

76. $xz + y$

77. $3yz - xy$

78. $5xz + yz$

79. $3xy + 2yz$

80. $5xz - 4xy$

Chapter 2 Summary

Section 2.1

An ***equation*** is a statement which says that two mathematical expressions represent the same number.

Equations that are sometimes true and sometimes false, depending on the value assigned to the variable, are called ***conditional.***

A ***linear equation in one variable*** is an equation with an exponent of 1 on the variable in any term.

A ***solution*** of an equation is a number that gives an equation which is true when that number is substituted for the variable. The set of all solutions is called the ***solution set.***

Addition property of equality: If a, b, and c are real numbers and

$$a = b$$

then

$$a + c = b + c$$

Multiplication property of equality: If a, b, and c are real numbers and

$$a = b$$

then

$$ac = bc$$

Solving linear equations

1. Multiply on both sides of the equation to clear fractions or decimals.
2. Remove parentheses, using the distributive laws.
3. Combine like terms on each side of the equation, if possible.
4. Get the terms containing the variable on one side of the equation and all other terms on the other side by using the addition property.
5. Combine like terms, if possible.
6. Use the multiplication property to solve for the variable.

Section 2.2

Solving a word problem

1. Read the problem carefully. Decide what questions are to be answered.
2. Choose a variable to represent the unknown quantity. If it is helpful, make a drawing.
3. Write an equation. Sometimes we may need to remember a formula that is not given. Some geometric formulas are given on the inside cover.
4. Solve the equation.
5. Decide if the questions have been answered and check the answer in the original word problem.
6. Write the answer in a sentence.

Consecutive integers are integers that follow one another in order. Two numbers are ***consecutive even integers*** if they are even and one number is two larger than the other.

Consecutive odd integers are odd numbers and one number is two larger than the other.

Section 2.3 *Interest formula:*

$$I = Prt$$

where I is the simple interest, P is the principal (amount invested), r is the interest rate, and t is the time in years.

Section 2.4 A ***formula*** is a relationship between variables.

A ***constant*** is a letter that stand for a specific number.

Section 2.5 $a > b$ means that a is greater than b.

$a < b$ means that a is less than b.

$a \geq b$ means that a is greater than or equal to b.

$a \leq b$ means that a is less than or equal to b.

An inequality usually has many solutions. The set of all solutions is called the ***solution set.***

Addition property of inequality:

$$\text{If } a < b$$
$$\text{then } a + c < b + c.$$

The property is also true for $\leq$, $>$, and $\geq$.

Compound inequalities are formed by two or more inequalities that are joined with the word *and* or the word *or*.

The ***intersection*** of two sets is the set of all elements that are common to both sets.

$a < x < b$ means that x is greater than a and x is less than b.

If two sets have no common elements, their intersection is the empty set, denoted by $\varnothing$.

The ***union*** of two sets is the set of all elements that are in one or the other of the sets or in both of them.

Section 2.6 *Multiplication property of inequality:*

If $a < b$, then $ac < bc$, if c is *positive*.

If $a < b$, then $ac > bc$, if c is *negative*.

The property also holds for $>$, $\leq$, and $\geq$.

Solving a linear inequality in one variable

1. Combine like terms and remove parentheses on each side of the inequality, if necessary.
2. Use the addition property of inequality to get the terms containing the variable on one side of the inequality and all other terms on the other side.
3. Combine like terms, if possible.
4. Use the multiplication property of inequality to isolate the variable on one side. Remember to reverse the inequality symbol if both sides of the inequality are multiplied or divided by a negative number.

Section 2.7

The distance d, between any real numbers a and b is

$$d = |a - b|$$

Solving absolute value equations and inequalities, where $a \neq 0$ and c is a positive number

	Solution Is Found by Solving:
$\lvert ax + b \rvert = c$	$ax + b = c$ or $ax + b = -c$
$\lvert ax + b \rvert < c$	$-c < ax + b < c$
$\lvert ax + b \rvert > c$	$ax + b < -c$ or $ax + b > c$

The rules also hold when $<$ is replaced by $\leq$ and $>$ is replaced by $\geq$.

To solve an equation of the form

$$|ax + b| = |cx + d|$$

solve the two equations

$$ax + b = cx + d \quad \text{or} \quad ax + b = -(cx + d)$$

Chapter 2 Additional Exercises (Optional)

NAME ______

DATE ______

CLASS ______

Section 2.1

Solve.

1. $x + \frac{1}{8} = -\frac{5}{18}$

2. $y - \frac{3}{10} = \frac{4}{15}$

3. $-\frac{7x}{8} = \frac{21}{16}$

4. $\frac{11x}{9} = \frac{22}{27}$

5. $x - \frac{3}{4} = \frac{1}{12}$

6. $x + \frac{1}{14} = -\frac{4}{21}$

7. $\frac{2}{3}y = \frac{3}{5}$

8. $\frac{7}{8}y = \frac{16}{5}$

9. $3y - 7y + 2 = y - 8$

10. $2x + 3 = 5x - 7x - 9$

11. $0.5(x - 2) + 0.8x = 1.6$

12. $0.6t + 0.25 - 0.5 = 0.1t$

* **13.** $\frac{2p}{3} + \frac{1}{3} - \frac{p}{4} + \frac{1}{4} = -2$

* **14.** $\frac{19x}{6} - \frac{2}{3} + \frac{5x}{3} = \frac{5}{6}$

Section 2.2

15. Five times the result of subtracting 4 from a number is 70. Find the number.

16. Thirty minus twice the result of subtracting 2 from a number is 22. Find the number.

17. If the length of a rectangular room is 4 feet more than the width and the perimeter is 52 feet, find the dimensions of the room.

18. The sum of three consecutive integers is 129. Find the integers.

19. The price of a car decreased 5%. If the new price of the car is $11,400, what was the original price of the car?

20. The cost of rent and utilities for 1 month was $492. If rent was 5 times as much as utilities, what was the cost of rent?

21. What percent of 57 is 21.66?

22. What percent of 84 is 57.12?

Section 2.3

23. The sum of two numbers is 27. The larger number minus 3 times the smaller number is -9. Find the numbers.

24. The sum of two numbers is 36. Twice the smaller number plus the larger number is 48. Find the numbers.

25. A collection of 29 coins has a value of \$2.10. If the coins are nickels and dimes, how many of each coin are in the collection?

26. A collection of 24 nickels and quarters has a value of \$1.80. How many nickels are in the collection?

27. Andrea earned \$600 interest in 1 year on two accounts. If twice as much was invested at 7% as was invested at 6%, how much was invested at each rate?

28. The total interest earned for 1 year on \$7000 was \$490. If part of the money was invested at 8.5% and part was invested at 5%, how much was invested at each rate?

Section 2.4

Solve.

29. $S = \pi r^2 + 2\pi rh$ for h

30. $A = \frac{1}{2}h(b_1 + b_2)$ for b_2

31. $3y + 2x = 4$ for y

32. $7y - 3x = 5$ for y

33. $5y - 3x = 8$ for x

34. $6y - 2x = 9$ for x

35. $-2y + 4x = 7$ for y

36. $-8y - 3x = 2$ for y

Section 2.5

Solve.

37. $6x + 2 \leq 5x - 7$

38. $7x - 5 \geq 6x + 8$

39. $3x - \frac{3}{4} > 2x + \frac{1}{5}$

40. $1.3x + 0.6 < 2.3x + 1.4$

41. $8 < x + 3 \leq 9$

42. $2 \leq x + 5 \leq 4$

43. $x + 6 < 4$ or $x - 5 > 4$

44. $x + 3 \leq 2$ or $x - 4 \geq 1$

Section 2.6

Solve and graph.

45. $-\frac{7}{2}x \geq 14$

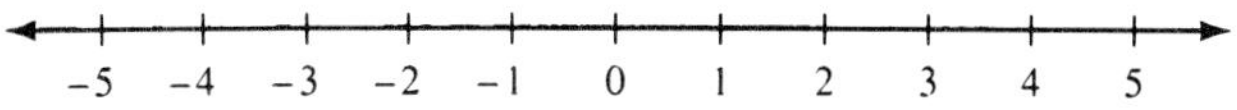

46. $-\frac{6}{5}x \leq -6$

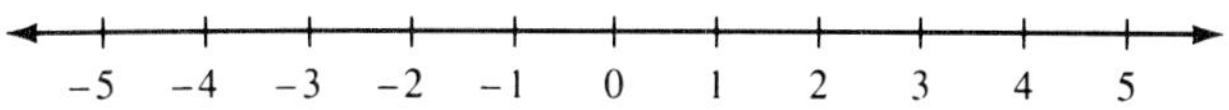

47. $-6x < -18$

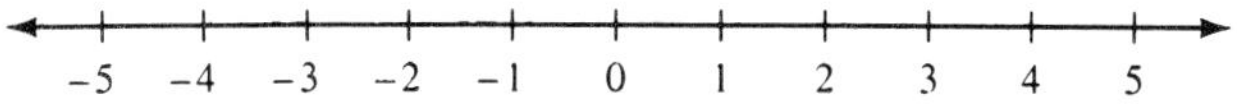

48. $-3x > 9$

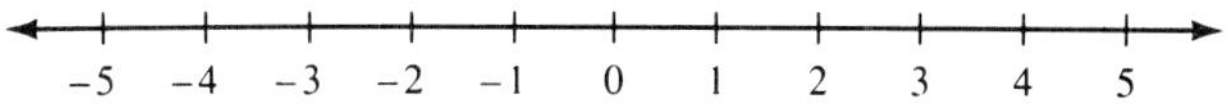

49. $\frac{3}{2}x - 1 < \frac{5}{4}x + 4$

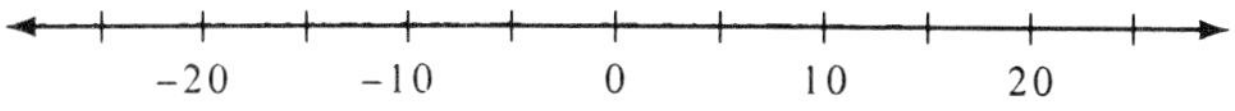

50. $0.2x - 0.3 > 0.7x - 1.8$

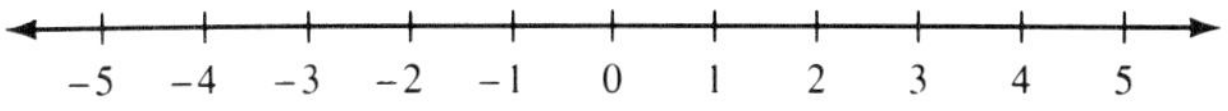

51. $-4 \leq 2x + 3 < 5$

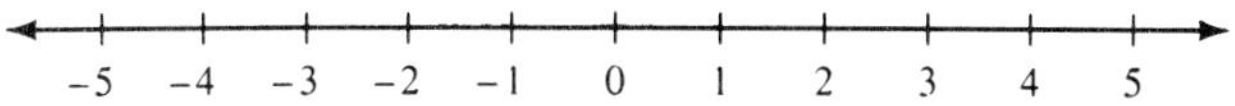

52. $-3 \leq 2x + 8 < 7$

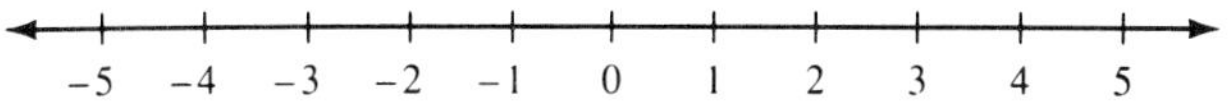

53. $-12 < 2x + 4x \leq 18$

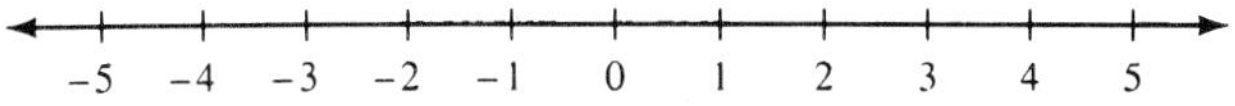

54. $-15 \leq 2x + 3x \leq 10$

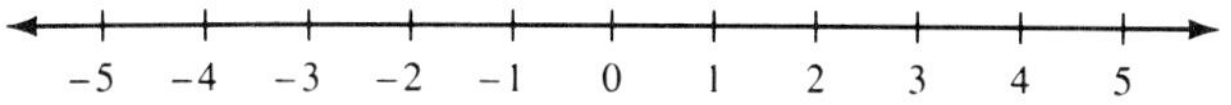

55. $3x - 4 < 2$ or $2x - 8 > 4$

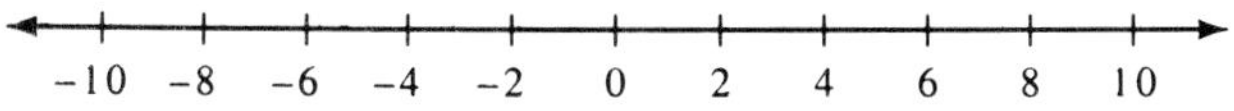

56. $5x + 10 \leq 15$ or $4x + 3 \geq 9$

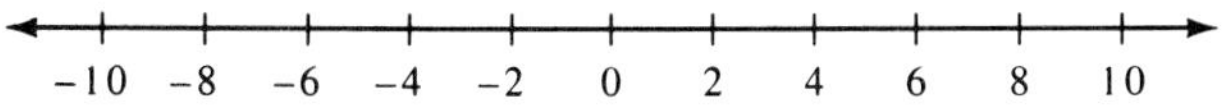

57. Tim needs an average of 90 to get an A in English. There are four tests and his first three test scores were 92, 78, and 94. What score on the last test does he need to get an A?

58. What amount must Tracy invest in her retirement account at 9% interest to earn at least $135 per year?

Section 2.7

Solve.

59. $|2x + 4| = 6$

60. $|3x - 1| = 8$

61. $|3x + 4| = -2$

62. $|5x - 7| = -8$

63. $\left|\frac{3}{4}x - 2\right| = 1$

64. $\left|x - \frac{3}{4}\right| = \frac{2}{9}$

65. $|2x - 5| > 3$

66. $|3x - 2| \leq 7$

67. $\left|\frac{3}{4}x + 5\right| < 3$

* **68.** $|x + 4| \geq x$

* **69.** $|x - 3| > x$

70. $|2.5x + 2| < 3$

71. $|x| < -3$

72. $|x| < -1$

73. $|x| > -5$

74. $|x| > -2$

* **75.** $|x + 6| = |x - 3|$

* **76.** $|z - 8| = |z + 7|$

COOPERATIVE LEARNING

1. Write a linear equation.

2. Is $-\frac{5}{6}$ a solution of $3x - \frac{5}{8} = -\frac{65}{24}$? Explain.

3. Write equations whose solutions are as follows.

a. $-\frac{23}{4}$ **b.** 59.6

4. Gil invested \$6000 for 1 year at 4% interest per year. He also invested \$5000 for 1 year. He earned \$415 total interest. At what rate did he invest the \$5000?

5. The number of different two-station telephone connections C that can be made between N different stations is given by the formula

$$2(C - N) = N(N - 3)$$

Solve this formula for C.

6. Write absolute value inequalities for the following graphs.

a.

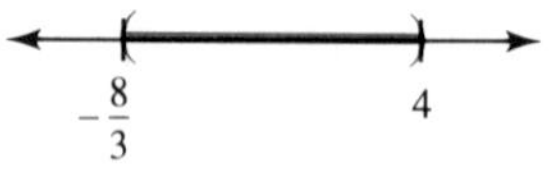

b.

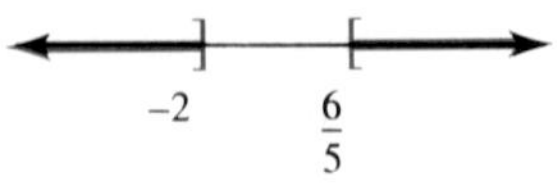

7. Sarah can spend \$1500 on her wedding reception. The caterer charges a fee of \$200 plus \$9.75 per person. What is the maximum number of people that Sarah can have at her wedding reception?

Chapter 2 Practice Test

NAME ____________

DATE ____________

CLASS ____________

Solve.

1. $y - 8 = 17$

2. $-\frac{2}{3}x = 4$

3. $2x - 7.4 = 6x + 3.4$

4. $6x - (2x - 4) = 4(2x + 1)$

5. The perimeter of a rectangular garden is 96 meters. If the length is twice the width, find the dimensions of the garden.

6. Find two consecutive even integers such that 6 times the smaller is 5 times the larger.

7. Laura invested \$12,000 for 1 year and earned \$910 interest. Some of the money was invested at 8% interest per year and the remainder at 7% interest per year. How much was invested at 8%?

8. Solve $C = \frac{5}{9}(F - 32)$ for F.

1. ____________

2. ____________

3. ____________

4. ____________

5. ____________

6. ____________

7. ____________

8. ____________

9. ______________________

9. Solve and graph $6x + 9 \leq 5x + 4$.

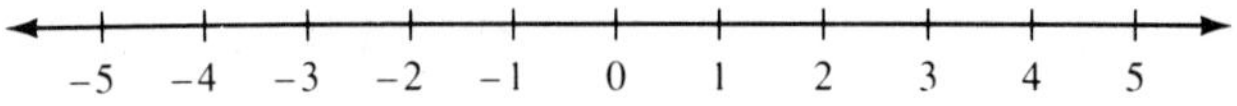

Solve.

10. ______________________

10. $-12y > 72$

11. ______________________

11. $-4(x - 4) \leq -6(x + 1)$

12. ______________________

12. $|3x - 1| = 13$

13. ______________________

13. $|5 - 3x| = -8$

14. ______________________

14. $|7x + 3| = |5x - 4|$

15. ______________________

15. Julie earned \$800 in April, \$1400 in May, and \$700 in June. What is the minimum amount that she must earn in July to be paid at least an average of \$1050 per month during this 4-month period?

Solve and graph.

16. ______________________

16. $|4x - 8| < 16$

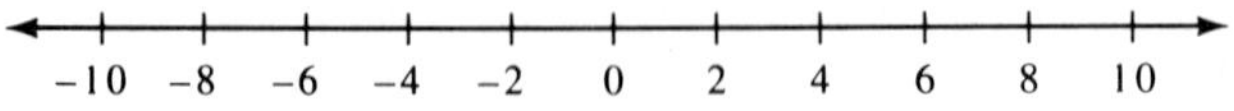

17. ______________________

17. $|2x - 3| \geq 4$

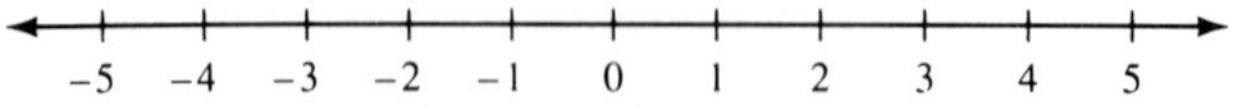

Chapters 1 and 2 Cumulative Review

NAME ______________________

DATE ______________________

CLASS ______________________

Find the following.

1. $\left|-\frac{3}{8}\right|$

2. $-|-6.4|$

3. Replace the ? with < or >: $-6 \ ? \ -8$

Add.

4. $3 + (-9)$

5. $(-7.8) + (-4.3)$

Subtract.

6. $8 - 11$

7. $\left(-\frac{3}{4}\right) - \left(-\frac{7}{5}\right)$

Multiply.

8. $(-6)(9)$

9. $(-3.4)(-7.8)$

10. $(-2)(-4)(-7)$

Divide.

11. $\frac{-35}{7}$

12. $\frac{-72}{-9}$

13. $\frac{-3}{5} \div \left(\frac{7}{15}\right)$

14. $\frac{0}{15}$

Find the reciprocal.

15. $\frac{-5}{9}$

16. 7

17. Evaluate when $x = -3$, $y = 4$: $2xy + 7x$.

18. Combine like terms: $4x - 9y - 7x - 8y$.

Simplify.

19. $3x - 2[2x - (3x - 4)]$

20. $\dfrac{5^2 - 3(2)}{4 + 2(5)}$

For Problems 21–23, assume that variables are nonzero real numbers.

Perform the operations indicated. Write with positive exponents.

21. $(3xy^2)(-5x^3y^{-5})$

22. $\dfrac{-64x^{-3}y^5}{-16x^{-2}y^4}$

23. Simplify and write with positive exponents:

$$\left(\frac{2x^2y^{-3}}{x^4y^{-2}}\right)^3$$

24. Write in scientific notation: 0.00354.

Solve.

25. $\dfrac{3x}{4} = \dfrac{3}{7}$

26. $3x - 2(4x - 1) = -3(x + 2)$

27. The perimeter of a rectangular flower bed is 36 feet. If the width is 4 feet less than the length, find the dimensions of the flower bed.

28. A collection of 25 dimes and quarters is worth $4.60. How many of each kind of coin is in the collection?

29. Solve $3x + 2y = 7$ for y.

Solve.

30. $3x + 4 \le 5x + 8$

31. $|2x + 4| = 16$

32. $|7x + 5| \ge -4$

33. $|x + 7| = |x - 9|$

34. Ken wants to earn $4440 per year in interest. What amount must he invest at 6% interest per year to earn at least $4440?

Solve and graph.

35. $|5x - 2| < 3$

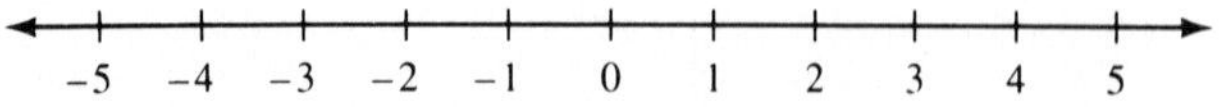

36. $|2x + 4| \ge 6$

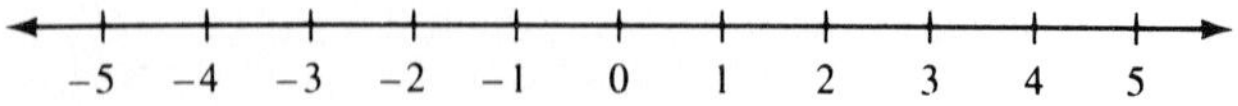

Graphing Linear Equations and Inequalities

CHAPTER 3

Pretest

1. Make a table of values for x and y. Then graph $y = 2x - 3$.

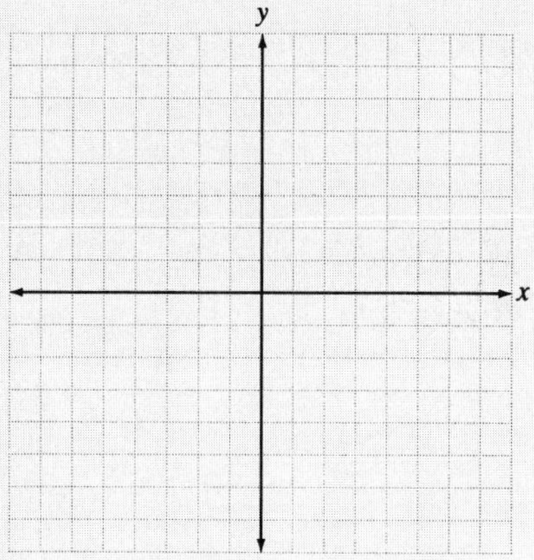

2. Graph using intercepts: $3x + 5y = 15$.

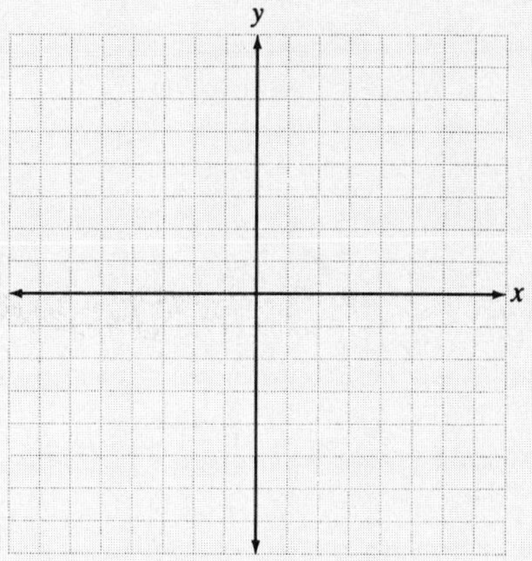

3. Find the slope of the line through the pair of points, if it exists: $(-7, -5)$ and $(-8, -9)$.

4. Find the slope and y-intercept and graph $3x - 7y = 14$.

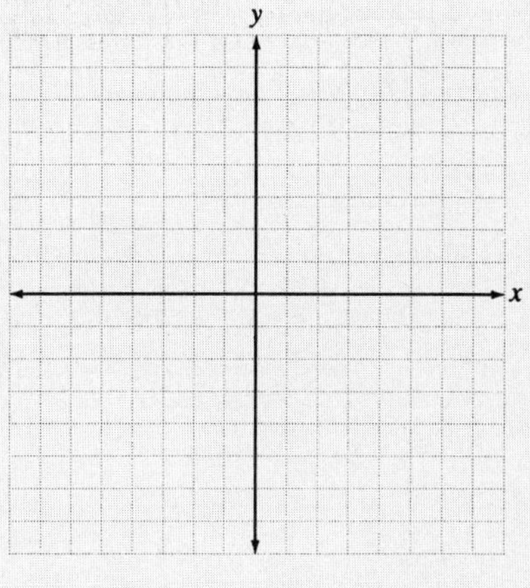

5. Find an equation for the line with the given slope and y-intercept: $m = -\frac{2}{3}$, $b = -4$.

6. Find an equation for the line through the points $(-5, -8)$ and $(-4, 3)$.

7. Find an equation for the line through the point $(-1, -7)$ and parallel to the line $4x - y = 12$.

8. Find an equation for the line through the point $(5, -2)$ and perpendicular to the line $5x - 3y = 22$.

9. For the Selden Company, the relationship between the number of units sold and the profit is linear. The profit is \$750 when 250 units are sold and \$3200 when 600 units are sold.
   a. Write the equation relating profit y to units sold x.
   b. Find the profit when 850 units are sold.

Graph.

10. $x < -2$

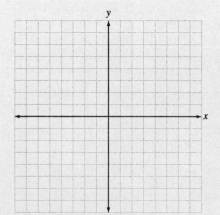

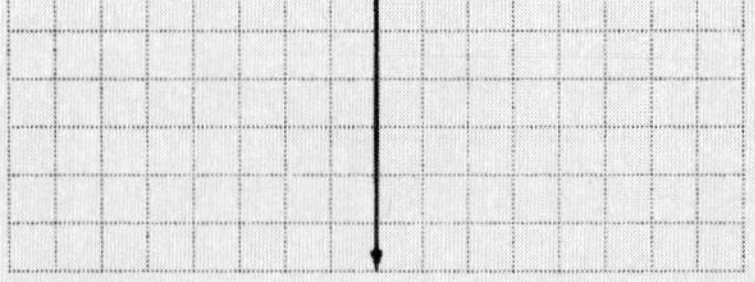

11. $3x + 2y \geq -4$

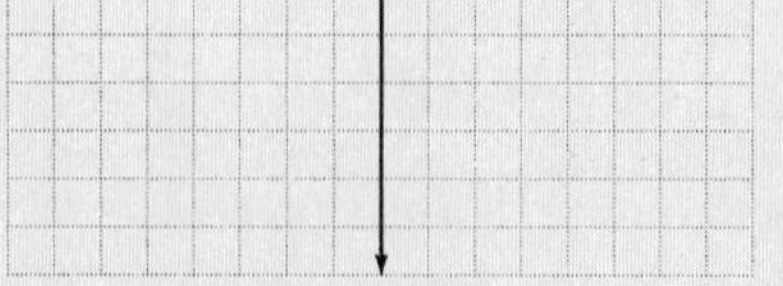

3.1 THE RECTANGULAR COORDINATE SYSTEM

OBJECTIVES

1. *Plot points*
2. *Determine in which quadrant points are located*
3. *Determine if given ordered pairs are solutions to an equation*
4. *Graph equations using ordered pairs*

In many of the applications of mathematics to other fields, we want to draw graphs of equations. For example, we might want a picture of the amount of goods in stock as related to the amount of sales. To draw a graph, we must be able to plot points, usually in the rectangular coordinate system. When two number lines, called *axes,* are drawn at right angles to each other, they form the ***rectangular*** (or ***Cartesian***) ***coordinate system.***

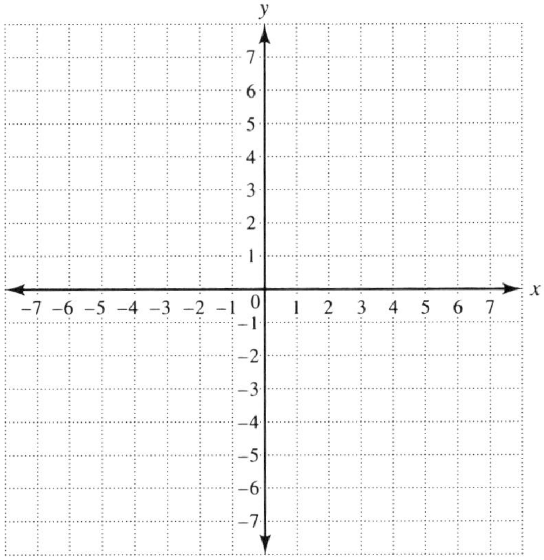

An ***ordered pair*** of real numbers is a pair of real numbers given in a definite order. This ordered pair corresponds to a point plotted on the rectangular coordinate system. The numbers are called ordered pairs because the order is important. The point corresponding to the ordered pair (6, 3) is placed in a different position than the point corresponding to the ordered pair (3, 6). The phrase "the point corresponding to the ordered pair (a, b)" is usually abbreviated "the point (a, b)."

The horizontal number line is called the *x-axis.* The first number in the ordered pair may be called the ***first coordinate,*** the ***x-coordinate,*** or the ***abscissa.*** The vertical number line is called the *y-axis.* The second number in the ordered pair may be called the ***second coordinate,*** the ***y-coordinate,*** or the ***ordinate.*** The point where the number lines cross is called the ***origin.*** It has coordinates (0, 0).

1 Plotting Points

To plot the point (a, b), we start at the origin and move a units to the right or left, (right if a is positive and left if a is negative). Then we move b units up or down (up if b is positive and down if b is negative). We plot a point at this location.

□ **Exercise 1** Plot the following points.

a. (3, 4)

b. (4, 3)

c. (−2, −3)

d. (−1, 4)

e. (3, −2)

f. (−2, 0)

g. (0, 3)

h. (0, −4)

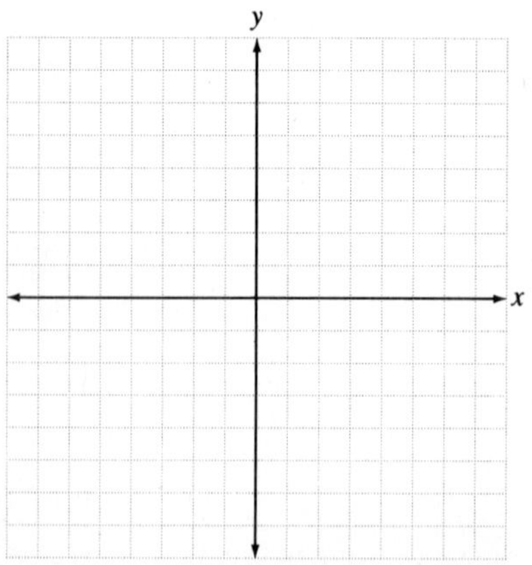

EXAMPLE 1

a. Plot the point (3, −4). Start at the origin. Move 3 units to the right along the x-axis, and then move 4 units down parallel to the y-axis.

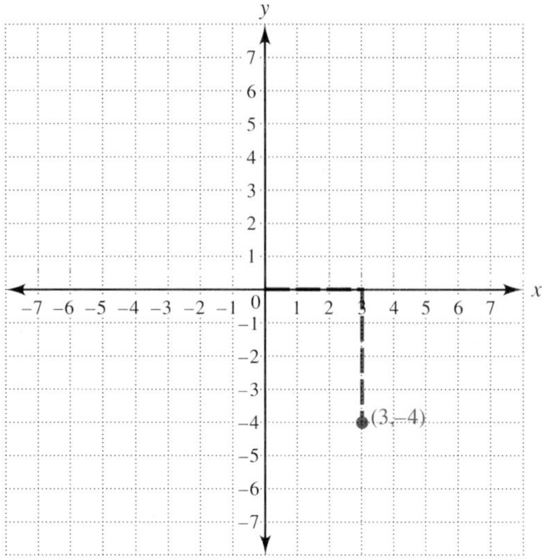

b. The points corresponding to (2, 3), (−1, −4), and (−3, 0) are shown.

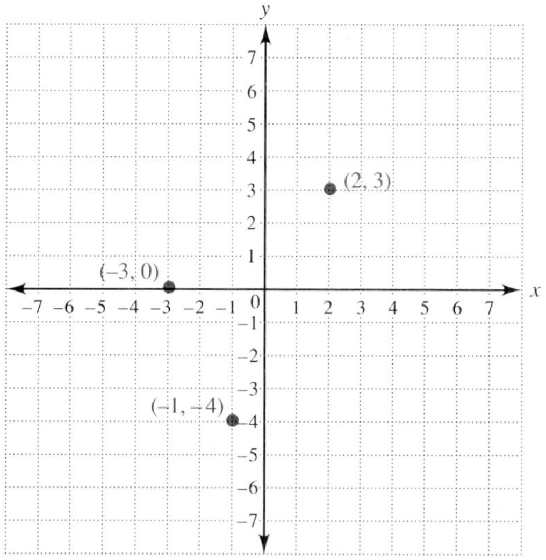

■

□ **DO EXERCISE 1.**

2 Quadrants

Quadrants are important in the study of trigonometry. The regions bordered by the coordinate axes are called the first, second, third, and fourth ***quadrants.***

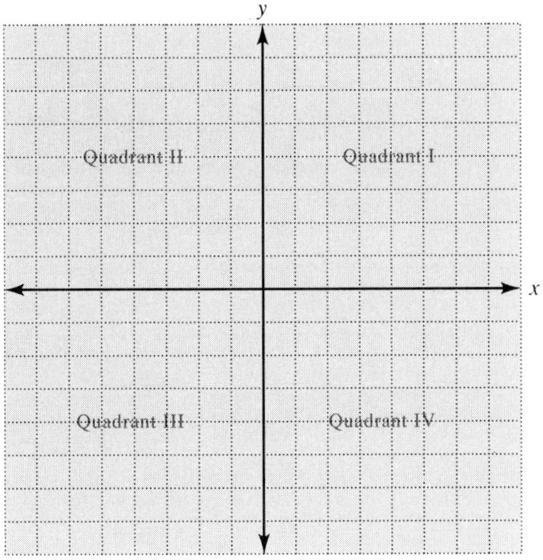

In quadrant I both coordinates of a point are positive. In quadrant II the x-coordinate of a point is negative and the y-coordinate is positive. The coordinates of a point are both negative in quadrant III. The x-coordinate is positive and the y-coordinate of a point is negative in quadrant IV. A point on an axis is not in any quadrant.

EXAMPLE 2 In which quadrant is the point $(-2, -7)$?

Since both the coordinates are negative, the point is in quadrant III. ■

□ **DO EXERCISE 2.**

□ **Exercise 2** In which quadrant are the following points located?

a. $(5, 8)$

b. $(-1, 3)$

c. $(1.5, -4)$

d. $(-6.3, -10.8)$

□ **Exercise 3** Are the following ordered pairs solutions to the equation $x - 3y = 8$?

a. $(-2, -2)$

b. $(17, 3)$

3 Solutions of Equations in Two Variables

The ***solutions of equations in two variables*** are *ordered pairs* of numbers. When the variables are x and y, the first number in the ordered pair is the solution for x and the second number in the ordered pair is the solution for y. When we substitute these solutions into the equation, we get a true equation. There are an infinite number of solutions.

EXAMPLE 3 Are the following ordered pairs solutions to the equation $x + 2y = 4$?

a. $(8, -2)$

$x + 2y$	$= 4$	
$8 + 2(-2)$	4	Substitute 8 for x and -2 for y.
$8 + (-4)$	4	
4	4	

The ordered pair $(8, -2)$ is a solution.

b. $(-6, 7)$

$x + 2y$	$= 4$
$-6 + 2(7)$	4
$-6 + 14$	4
8	4

The ordered pair $(-6, 7)$ is not a solution. ■

□ **DO EXERCISE 3.**

4 Graphing Equations

To graph an equation we make a drawing of its solutions. To find the solutions of an equation in two variables, we need to find ordered pairs that will make the equation true. To do this we *choose a value for one variable,* usually x, and find the value of the other variable. For example, in the equation $y = 2x - 3$, if we choose an x value of -1, the y value is found by substituting -1 in the equation.

$$y = 2x - 3$$

$$y = 2(-1) - 3 \qquad \text{Substitute } -1 \text{ for } x.$$

$$y = -2 - 3$$

$$y = -5$$

The ordered pair $(-1, -5)$ is a solution to the equation.

EXAMPLE 4 Graph $y = 2x - 3$.

Using the ordered pair solution that we found above, *choosing additional* x values of -2, 0, 1, and 3, and substituting these values in the equation to find the y values, we can make a table of x and y values as shown below. We can also list the ordered pairs. Notice that we have chosen several values for x, including some negative values.

We plot the points represented by the ordered pairs. The points appear to lie on a straight line. Since the number of solutions is infinite, we could plot enough points to form a solid straight line, so we sketch the line through the points that we have plotted.

x	y	**Ordered Pair**
-2	-7	$(-2, -7)$
-1	-5	$(-1, -5)$
0	-3	$(0, -3)$
1	-1	$(1, -1)$
3	3	$(3, 3)$

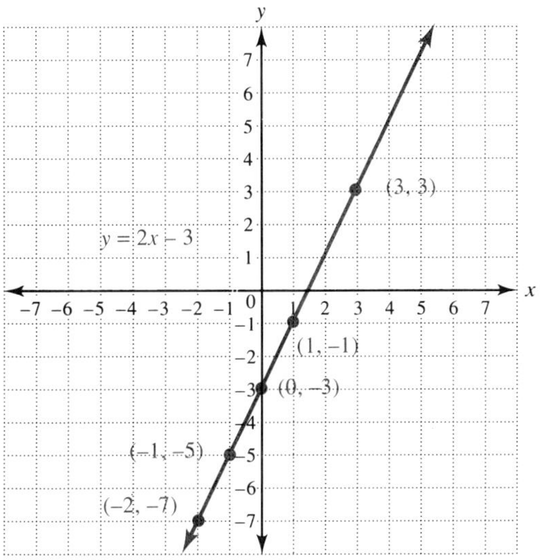

■

☐ **DO EXERCISE 4.**

☐ **Exercise 4** Graph.

a. $y = 3x + 1$

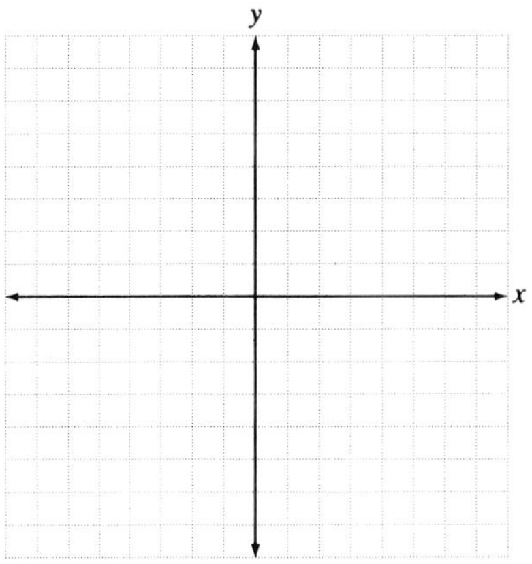

b. $y = 2x - 4$

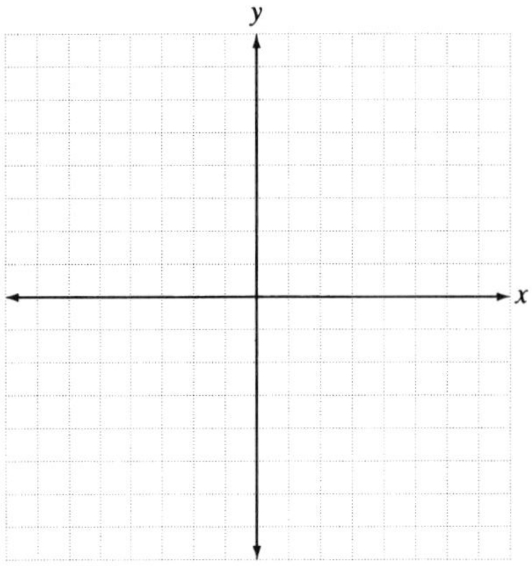

Answers to Exercises

1. 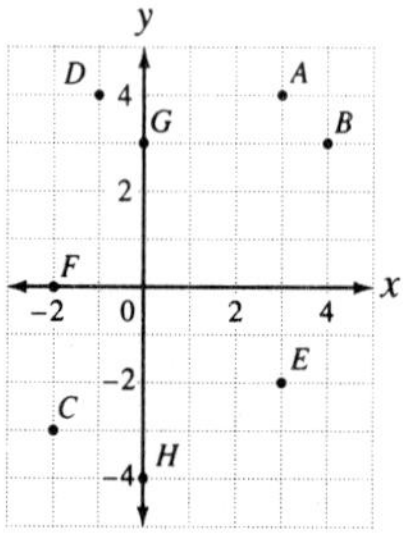

2. **a.** I **b.** II **c.** IV **d.** III

3. **a.** No **b.** Yes

4. **a.** 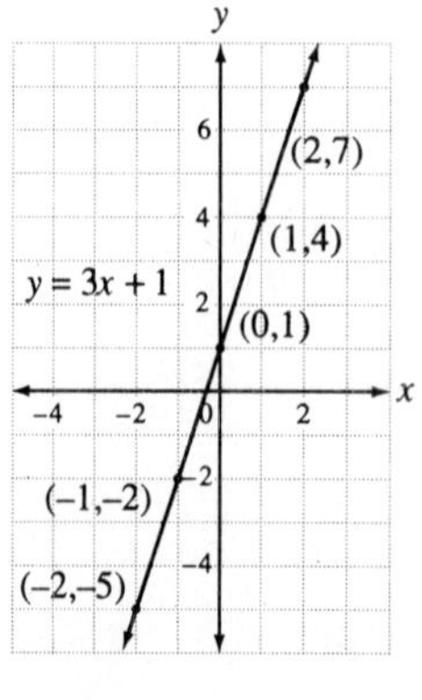

b. 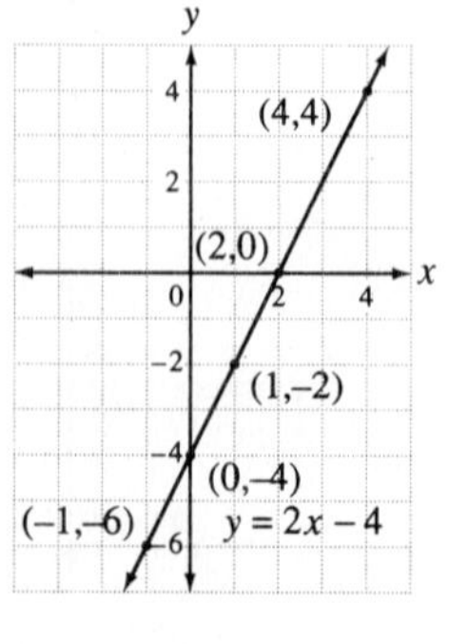

Problem Set 3.1

NAME ________________

DATE ________________

CLASS ________________

Plot the following points.

1. (2, 5)

2. (4, 2)

3. (−1, 3)

4. (2, −2)

5. (0, 5)

6. (0, −3)

7. (−3, −3)

8. (3, 0)

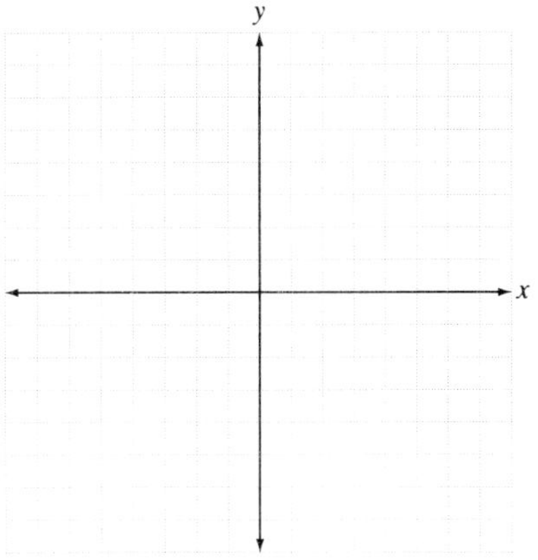

Plot the following points.

9. (1, 4)

10. (2, 3)

11. (−2, −2)

12. (−4, 0)

13. (0, 3)

14. (2, 0)

15. (−1, 1)

16. (1, −3)

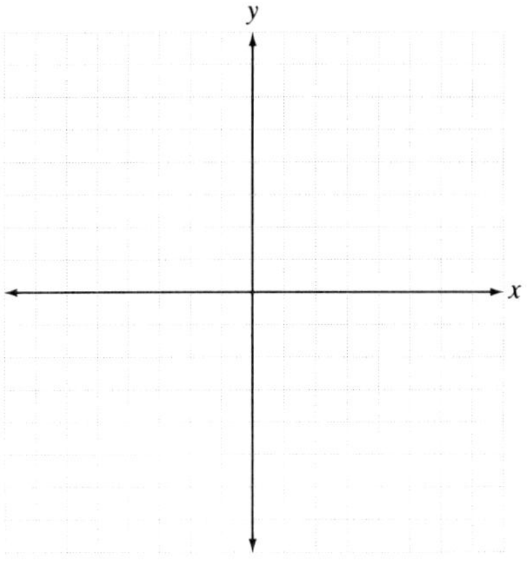

In which quadrant are the following points located?

17. (−2, 4) **18.** (5, 3) **19.** (−3, −3) **20.** (−4, 1)

Is the given ordered pair a solution of the equation?

21. (−2, −3); $2x - 3y = 5$

22. (5, −1); $3x + 4y = 9$

23. (4, 3); $3x - y = 2y + 6$

24. (9, 4); $4x - 2y = 5y + 8$

25. (5, −6); $3x + 8y = -33$

26. (7, 9); $8x - 7y = -7$

Make a table of values for x and y. Then graph.

27. $y = x + 2$

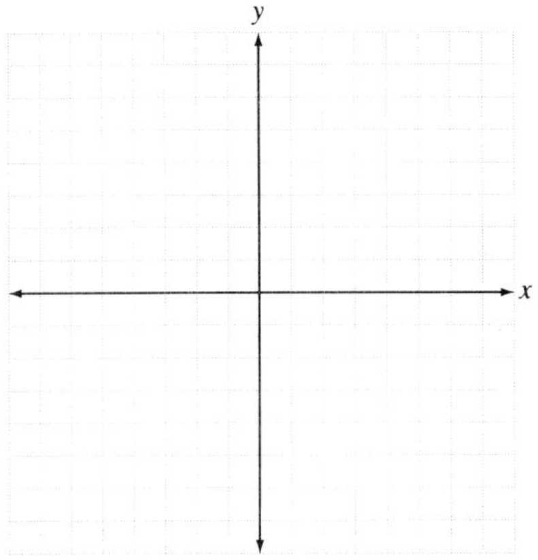

28. $y = x - 1$

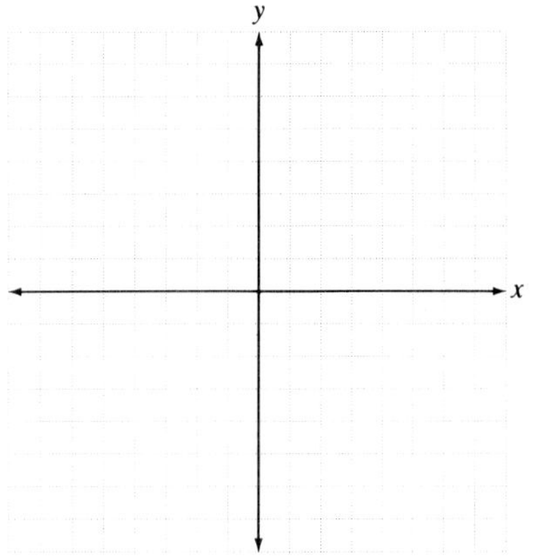

29. $y = -3x$

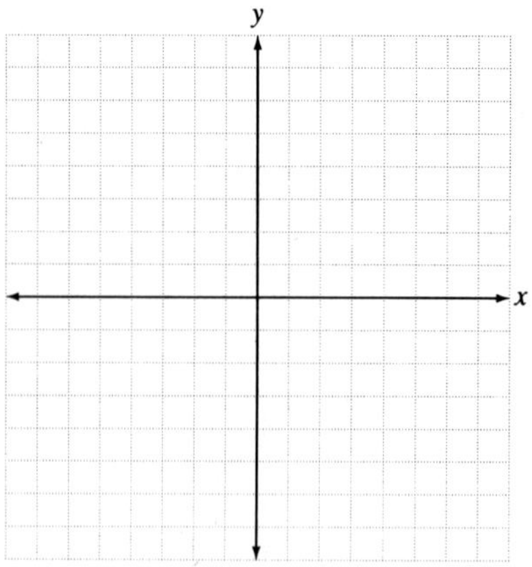

30. $y = 2x$

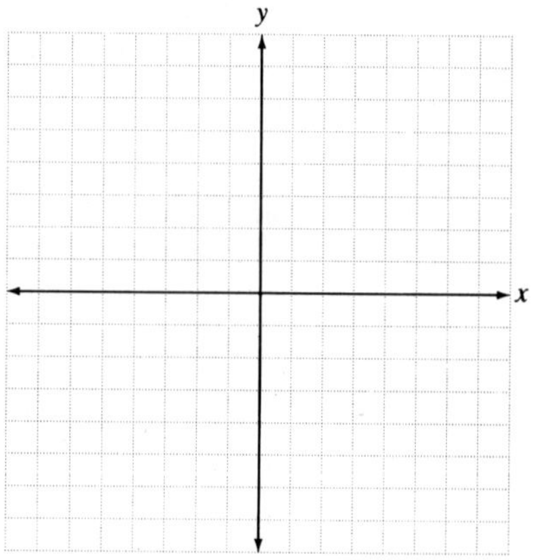

31. $y = 2x - 1$

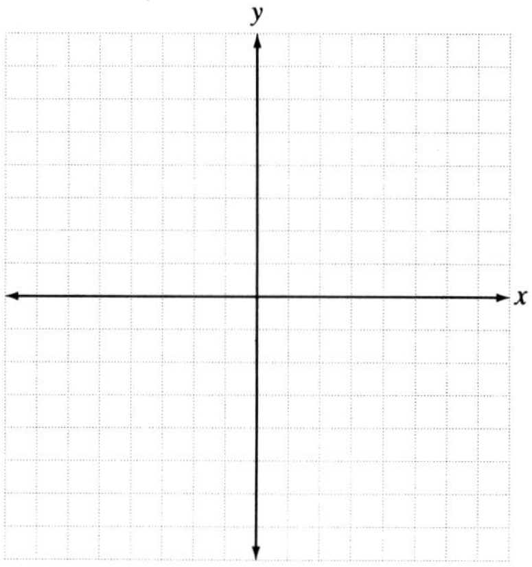

32. $y = -4x - 3$

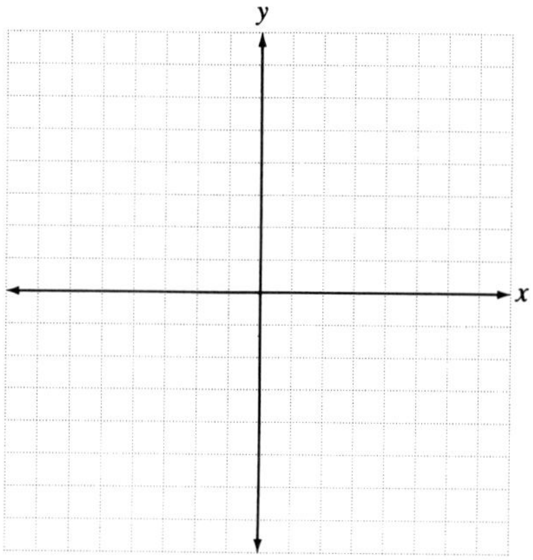

33. $y = 3x - 1$

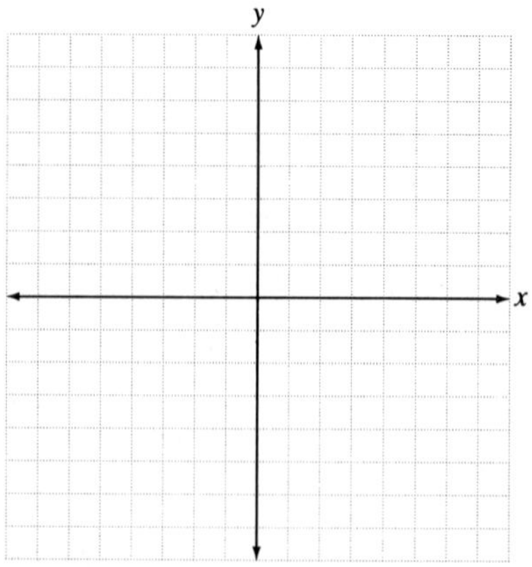

34. $y = 2x + 3$

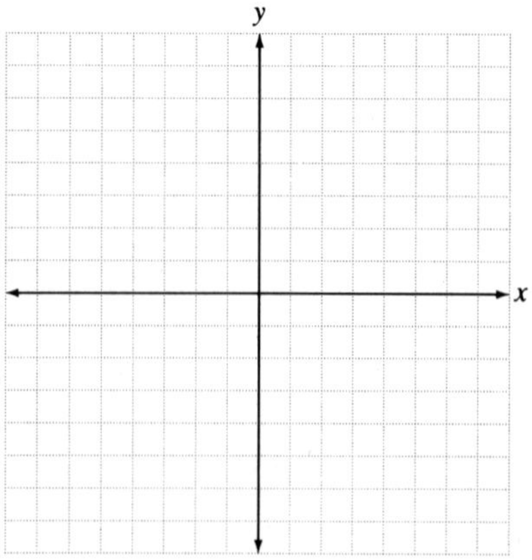

35. $y = -2x + 3$

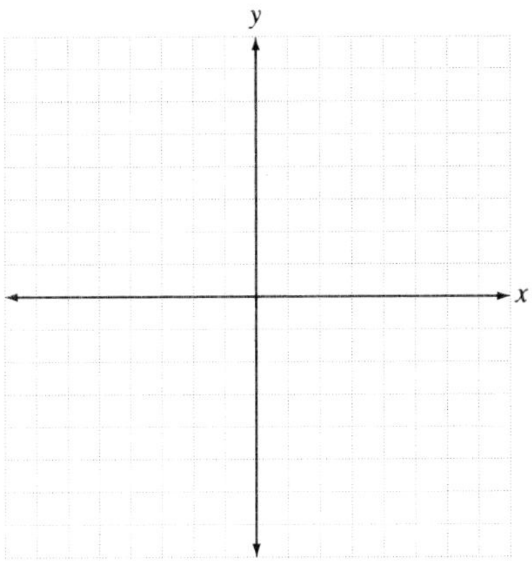

36. $y = -4x + 2$

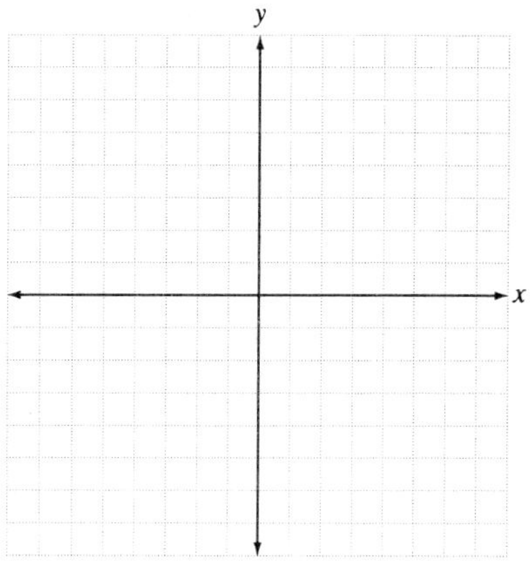

37. $y = 5x$

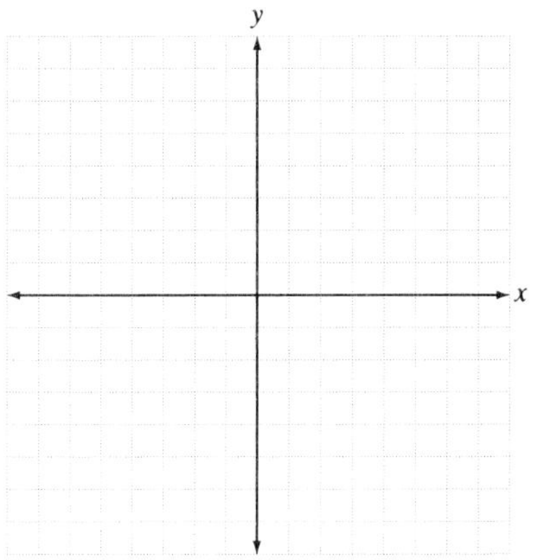

38. $y = -2x$

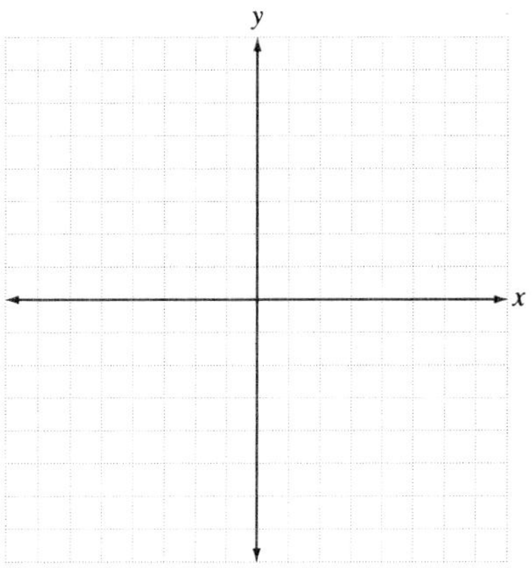

39. $y = -x - 1$

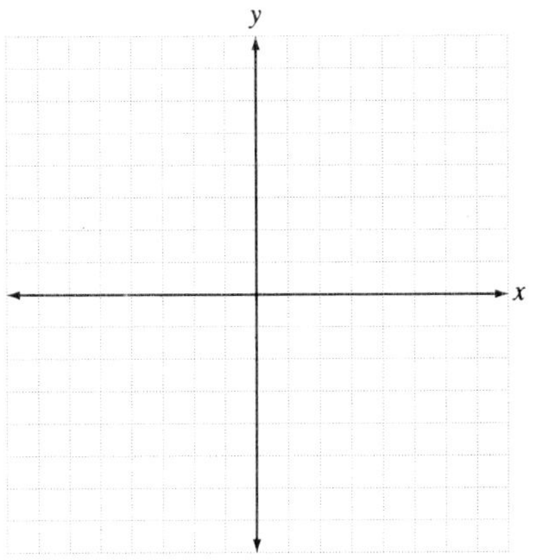

40. $y = -3x - 1$

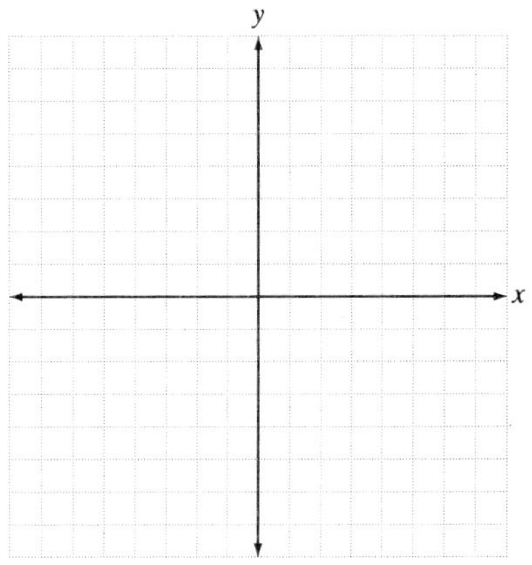

41. $y = -\frac{1}{3}x + 2$

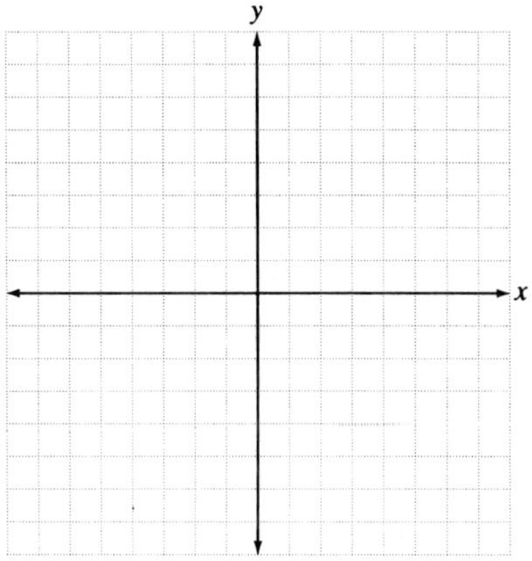

42. $y = \frac{1}{4}x - 3$

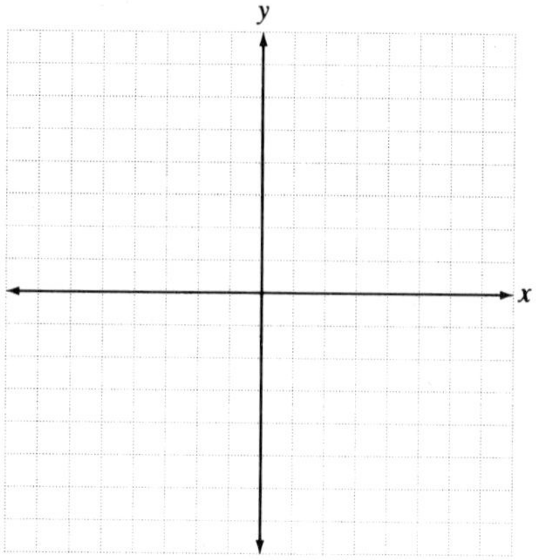

43. Graph $y = 3x$ and $y = \frac{1}{3}x$ on the same set of axes. How does the coefficient of x affect the slant of the line?

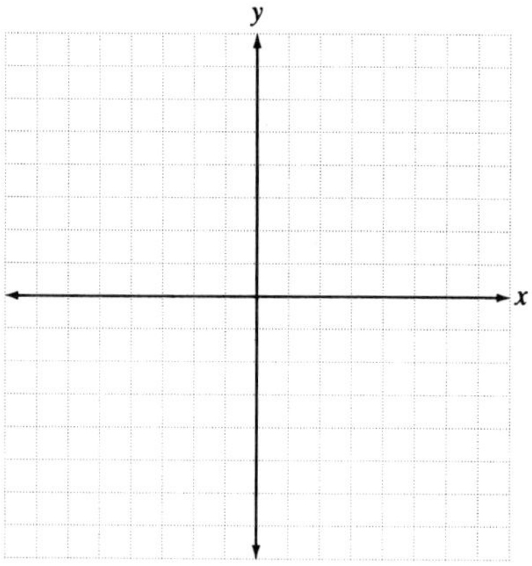

44. Graph $y = -4x$ and $y = -\frac{1}{4}x$ on the same set of axes. How do the coefficients of x affect the slant of these lines?

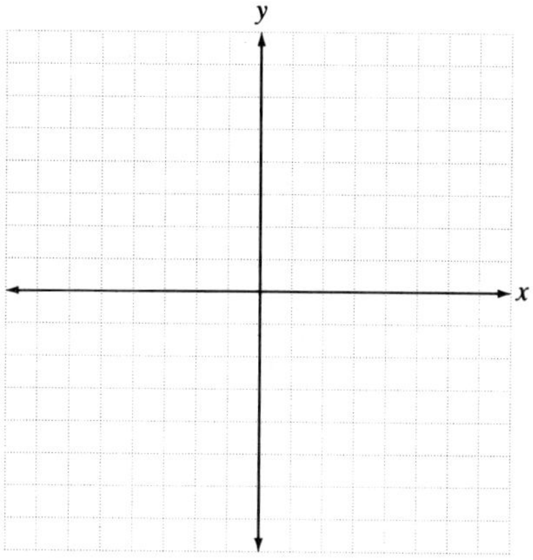

45. The pressure p in pounds per square inch (psi) on a diver at a depth d in ocean water is given by the formula

$$p = \frac{5}{11}d + 15$$

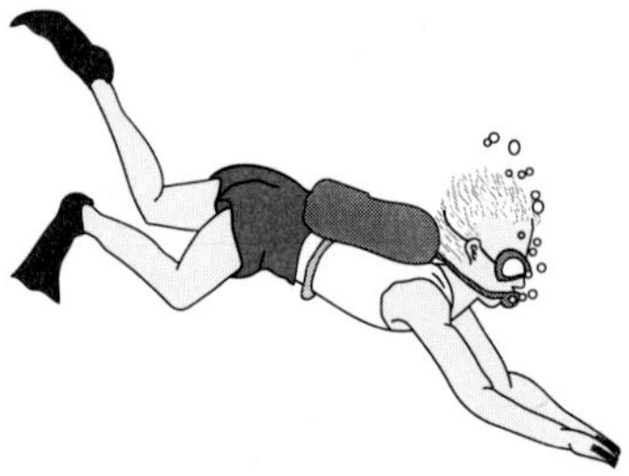

a. Find the pressures that correspond to depths of 11 feet, 33 feet, 55 feet, and 77 feet.
b. Graph the equation. Assume that d is the first coordinate and p is the second coordinate.

46. The book value V is related to the number of years t that an appliance is depreciated as shown in the equation

$$V = -50t + 600$$

a. Find the book values that correspond to 2, 4, 6, and 8 years.
b. Graph the equation. Assume that t is the first coordinate and V is the second coordinate.

47. Two number lines drawn at right angles to each other form the ____________ system.

48. The x-coordinate is the ____________ number in the ordered pair.

49. The point in the rectangular coordinate system where the number lines cross is called the ____________.

50. The regions bordered by the coordinate axes are called ____________.

51. The solutions of equations in two variables are ____________ ____________ of numbers.

52. To graph an equation, we make a drawing of its ____________.

Is the given ordered pair a solution of the equation?

53. $(-1.5, -2.5)$; $3.24x - 2.79y = 2.115$

54. $(4.3, -3.6)$; $2.37x + 1.29y = 5.547$

55. $(4.4, 7.8)$; $2.3x - 0.94y = 2.45y - 18.19$

56. $(-2.2, 0.7)$; $5.4x + 3.25y = 4.23x - 0.299$

* 57. Which of the equations has $(-\frac{3}{4}, \frac{1}{5})$ as a solution?

(a) $\dfrac{2}{3}x + 10y = \dfrac{3}{2}$

(b) $5x + 7y = -\dfrac{43}{20}$

(c) $3(4 - y) - \dfrac{3}{4} = 2(x + 1)$

(d) $0.16x = 0.2y - 0.16$

Think About It

Translate to an equation and graph the equation.

* 58. If 3 times the y-value is added to twice the x-value, the result is 6.

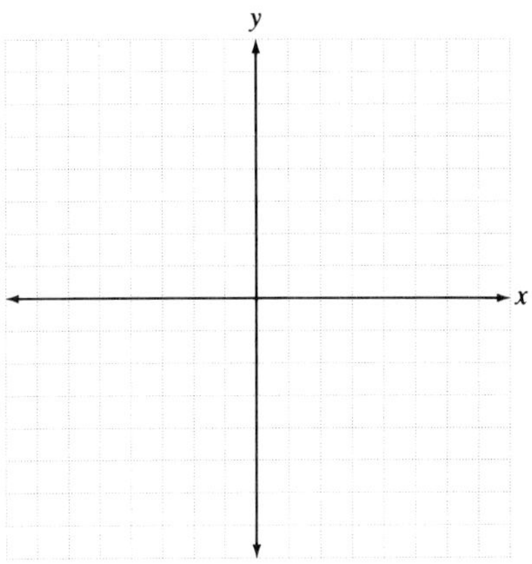

* 59. If 4 is added to the x-value, the result is twice the y-value.

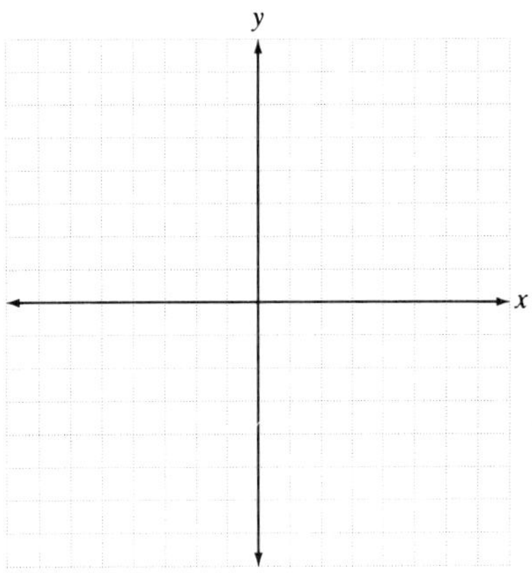

* **60.** If 5 is subtracted from the y-value, the result is 3 times the x-value.

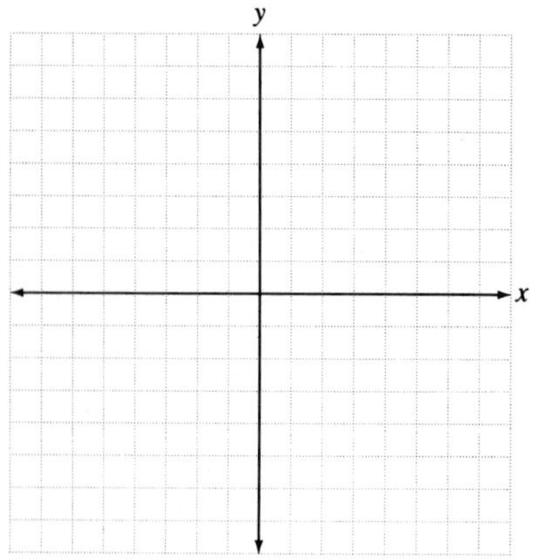

Checkup

The following problems provide a review of some of Section 2.2.

61. The width of a rectangle is 10 centimeters less than the length. If the perimeter is 128 centimeters, find the width of the rectangle.

62. The price of a house increased by 6%. If the new price is \$94,870, what was the original cost of the house?

63. If 12 is added to 4 times a number the result is 6 times the number. Find the number.

64. If 22 is subtracted from 3 times a number the result is -43. Find the number.

65. The sum of three consecutive integers is 216. Find the integers.

66. The sum of three consecutive even integers is 84. Find the integers.

67. What percent of 85 is 12.75?

68. 56.16 is what percent of 104?

69. A plumber cuts a 32-foot piece of tubing into three pieces. The second piece is 4 feet longer than the first piece. The third piece is the same length as the second piece. Find the length of each piece of tubing.

70. The price of a sofa was reduced 28%. If the sale price was \$540, what was the original price of the sofa?

3.2 GRAPHS OF LINEAR EQUATIONS

1 Linear Equations

Equations that can be written in the form

$$Ax + By = C$$

where A, B, and C are constants, A and B are not both zero, and x and y are variables are called ***linear equations*** in one or two variables. This form is called the ***standard form*** of a linear equation.

For an equation in two variables to be linear, the following must be true.

1. The exponent of each variable must be 1. Recall from Chapter 2 that to define a linear equation, we assume that the constant term does not contain a variable.
2. The equation may not contain a term where there is a product or quotient of two variables. The equations $xy = 5$ and $y/x + x = 6$ are *not* linear.

Linear equations in one variable were defined in Section 2.1.

The graph of a linear equation $Ax + By = C$ is a straight line.

EXAMPLE 1 Which equations are linear?

a. $3x + 2y = 7$ Linear; the exponents of the variables are understood to be 1.

b. $y = 4$ Linear; the exponent of y is 1.

c. $x^2 + 3y = 0$ Not linear; the exponent of x is not 1.

d. $y = \frac{1}{5}x + 3$ Linear. ■

□ **DO EXERCISE 1.**

OBJECTIVES

1 *Determine if an equation is linear*

2 *Graph linear equations in two variables using intercepts*

3 *Graph linear equations containing only one variable*

□ **Exercise 1** Which equations are linear?

a. $2x - y = 6$

b. $xy = 8$

c. $x = -2$

d. $x + 3y^2 - 2 = 5$

2 Graphing Linear Equations Using Intercepts

Many linear equations can be graphed more quickly using intercepts. The ***x-intercept*** of a graph of an equation is a point where it crosses the x-axis. The coordinates of the x-intercept are $(a, 0)$. The ***y-intercept*** is the point where the graph crosses the y-axis. The coordinates of the y-intercept are $(0, b)$.

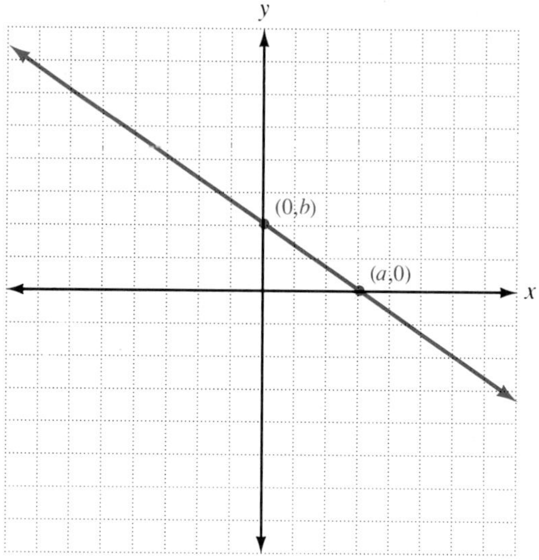

In the equation of a line, let $x = 0$ and solve for y to find the y-intercept $(0, b)$; let $y = 0$ and solve for x to find the x-intercept, $(a, 0)$.

Finding the intercepts gives us two points on the line. Since two points determine a line, this is enough points to draw the line. We should find a third point as a check.

EXAMPLE 2 Graph.

a. $2x + 3y = 6$

To find the y-intercept, let $x = 0$.

$$2x + 3y = 6$$

$$2(\mathbf{0}) + 3y = 6 \qquad \text{Substitute 0 for } x.$$

$$3y = 6$$

$$y = 2 \qquad \text{The } y\text{-intercept is } (0, 2).$$

To find the x-intercept, let $y = 0$.

$$2x + 3y = 6$$

$$2x + 3(\mathbf{0}) = 6 \qquad \text{Substitute 0 for } y.$$

$$2x = 6$$

$$x = 3 \qquad \text{The } x\text{-intercept is } (3, 0).$$

We have the following table of x and y values.

x	y
0	2
3	0

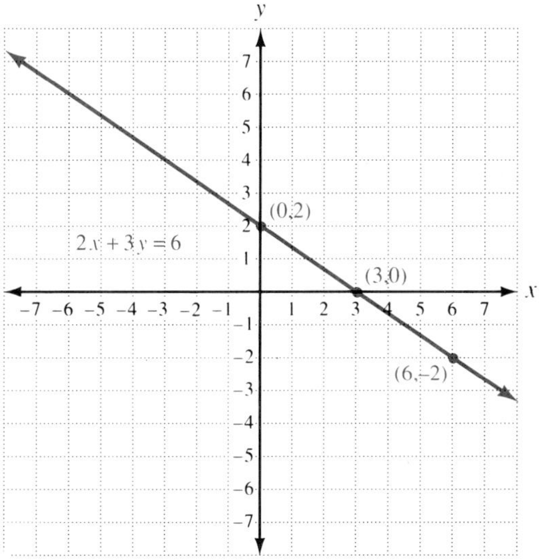

Plot the intercepts and draw the line. If we let $x = 6$ in the equation, then $y = -2$ and we can use the point $(6, -2)$ as a check point.

b. $2x - y = 2$

To find the y-intercept, we let $x = 0$ and solve for y. We found $y = -2$. To find the x-intercept, we let $y = 0$ and solve for x. We found $x = 1$. This gives us the following table of x and y values:

x	y
0	-2
1	0

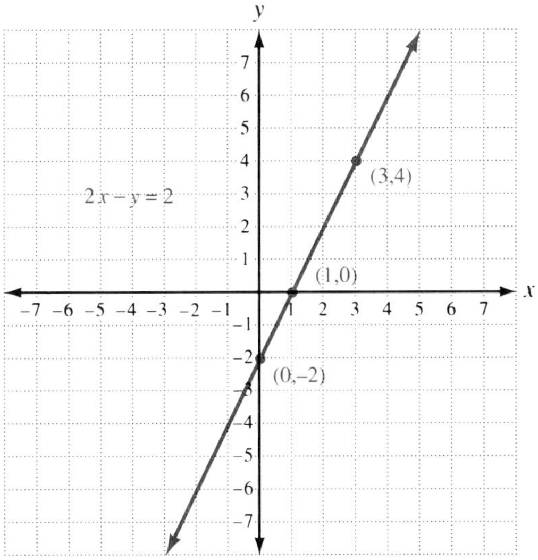

Plot these points and draw the line.

We use the point $(3, 4)$ as a check point since it is a solution of the equation. ■

□ **DO EXERCISE 2.**

3 Graphs of Linear Equations Containing Only One Variable

These equations are of the type $x = a$ or $y = b$, where a and b are constants. The graphs of these equations are parallel to an axis and have only one intercept.

□ **Exercise 2** Graph using intercepts.

a. $3x + 4y = 12$

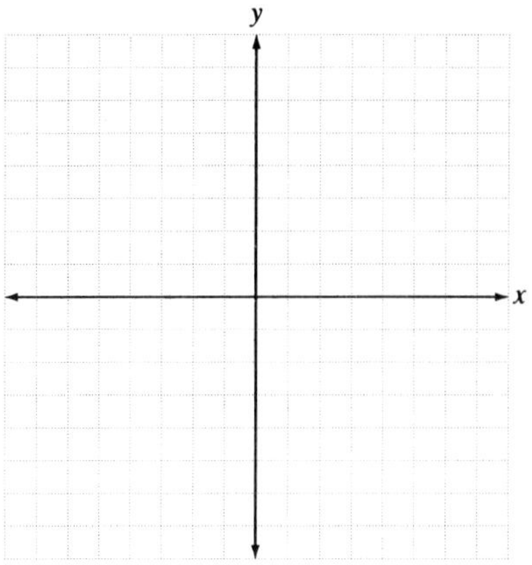

b. $x - 2y = 4$

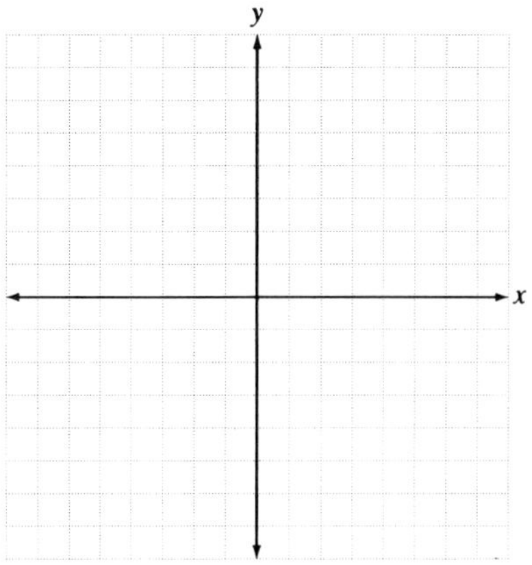

□ **Exercise 3** Graph.

a. $y = 3$

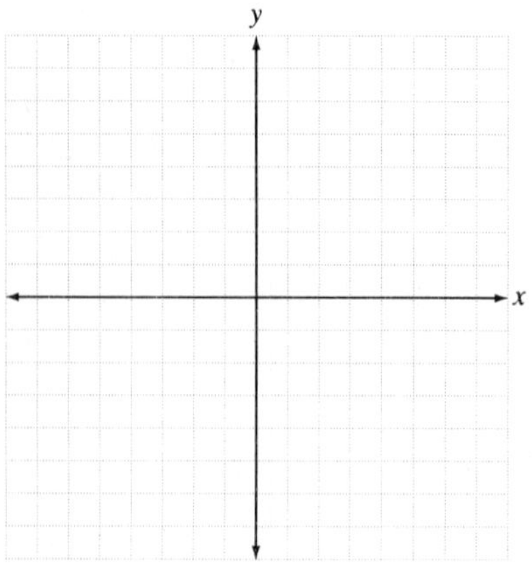

b. $x = -2$

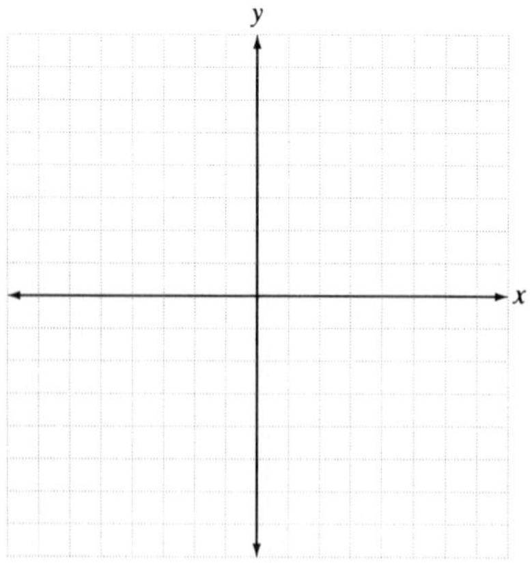

EXAMPLE 3 Graph.

a. $y = 4$

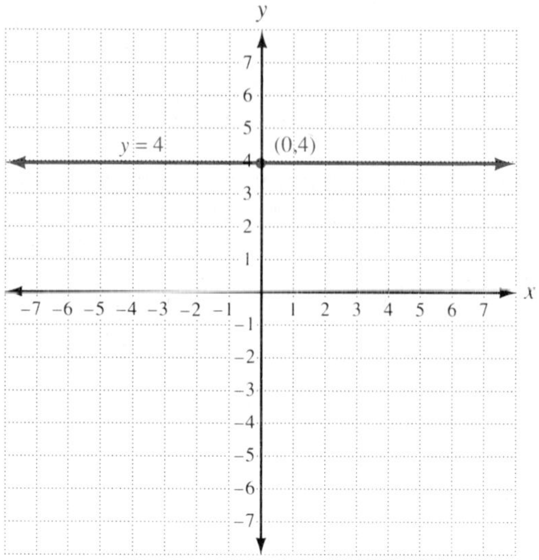

This is an equation of the form $y = b$. The equation shows that y is always 4 for any value of x. The solutions are all ordered pairs $(x, 4)$, including $(0, 4)$. The graph is a horizontal line parallel to the x-axis.

b. $x = -3$

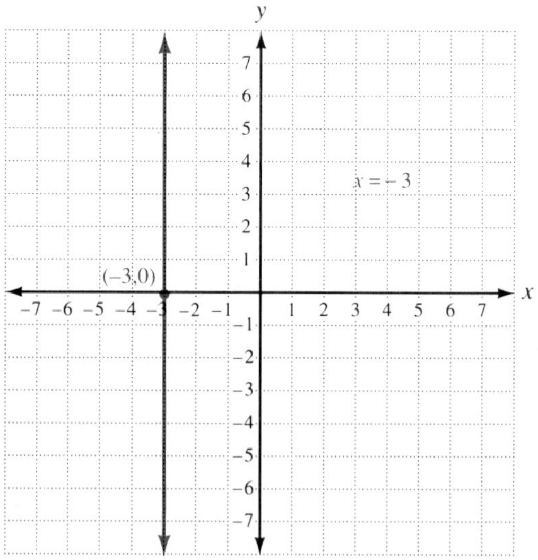

This is an equation of the form $x = a$. The x-coordinate is -3 for any value of y. The solutions are all ordered pairs $(-3, y)$, including $(-3, 0)$. The graph is a vertical line parallel to the y-axis. ■

> All graphs of equations of the form $y = b$ where b is a constant, are horizontal lines passing through the point $(0, b)$.
>
> All graphs of equations of the form $x = a$, where a is a constant, are vertical lines passing through the point $(a, 0)$.

□ **DO EXERCISE 3.**

DID YOU KNOW?

The genius Gottfried Leibniz, who lived from 1646 to 1716, gave us the words *coordinate, abscissa,* and *ordinate.* Leibniz earned his bachelor's degree by the time he was 17. He was refused his doctorate when he was 20 because the professors at the University of Leipzig envied his mental powers. Leibniz was so angry that he left Leipzig, never to live there again.

c. $3x - 12 = 0$
(Solve for x.)

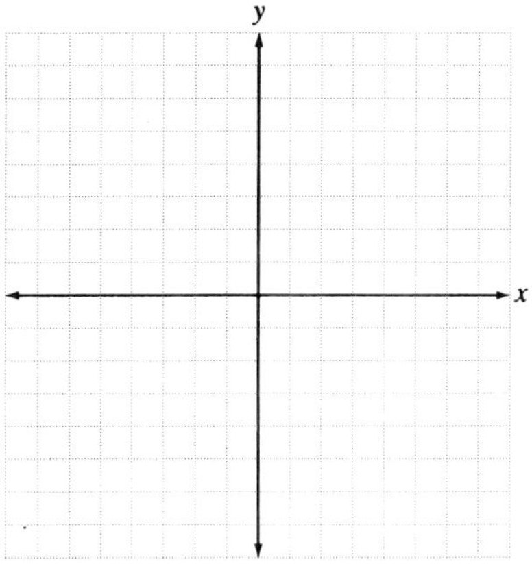

d. $4y + 8 = 0$
(Solve for y.)

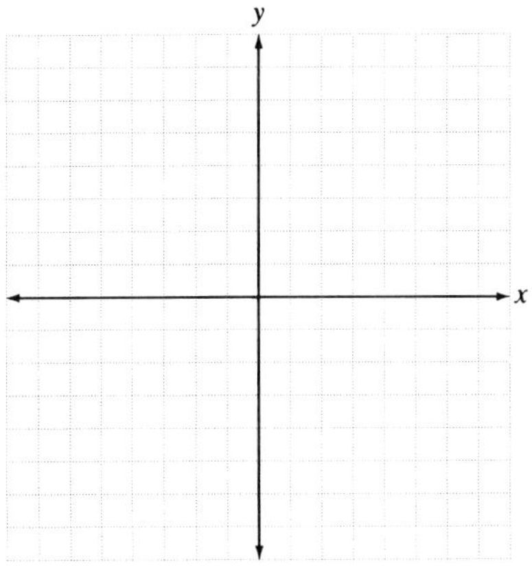

Answers to Exercises

1. a. Linear **b.** Not linear **c.** Linear **d.** Not linear

2. a.

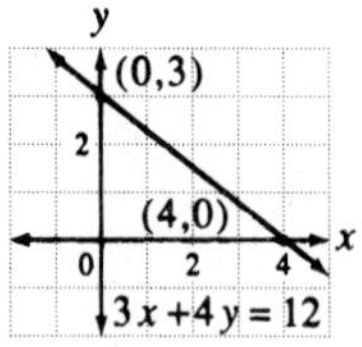

b.

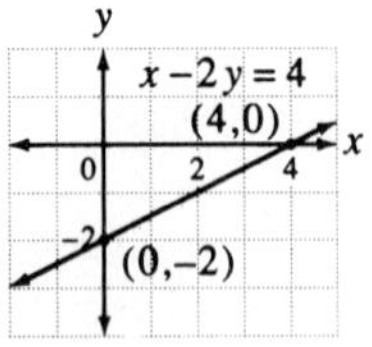

3. a.

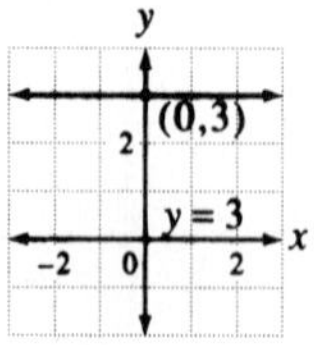

b.

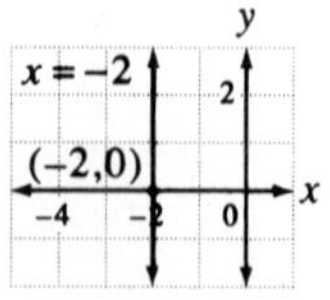

c.

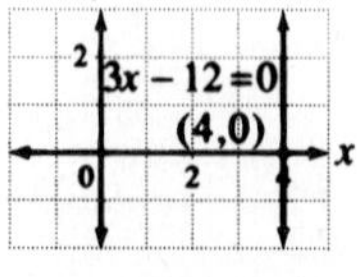

d.

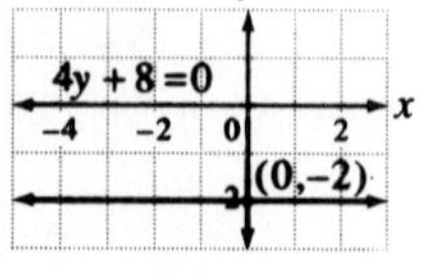

NAME

DATE

CLASS

Problem Set 3.2

Which equations are linear?

1. $3x - 4y - 8 = 0$

2. $2x + 3y = 7$

3. $x^2 + 2x - 3 = 0$

4. $-5y^2 - 3y = 7$

5. $5x + xy = 4$

6. $-yz = 8 + 3z$

7. $2x + 8 = 0$

8. $3y = -11$

9. $7 = 8x + 5y$

10. $3 - 2x = 4y$

Graph using intercepts.

11. $2x + 5y = 10$

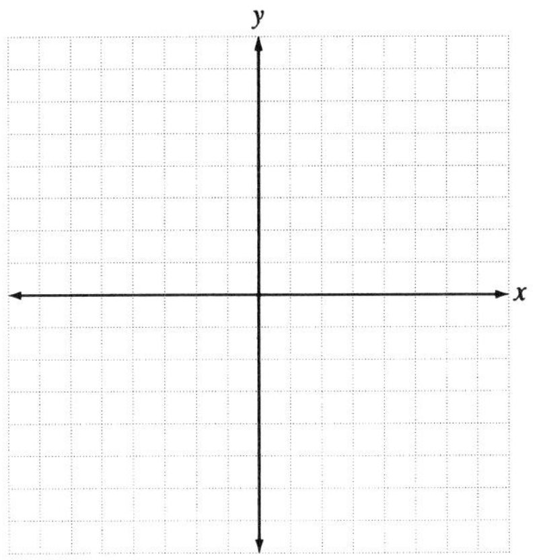

12. $3x - 2y = 12$

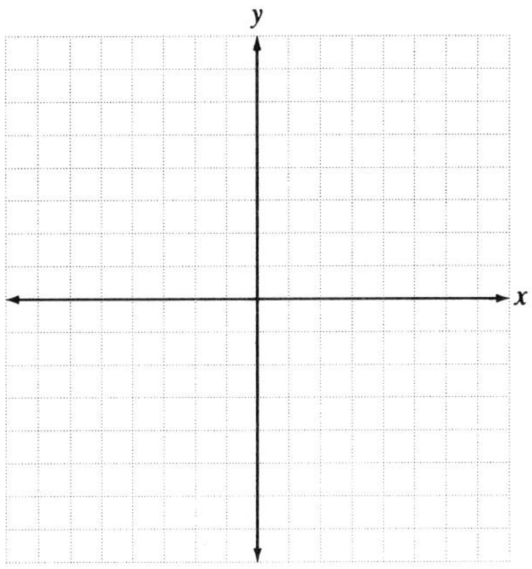

13. $7x - 2y = 21$

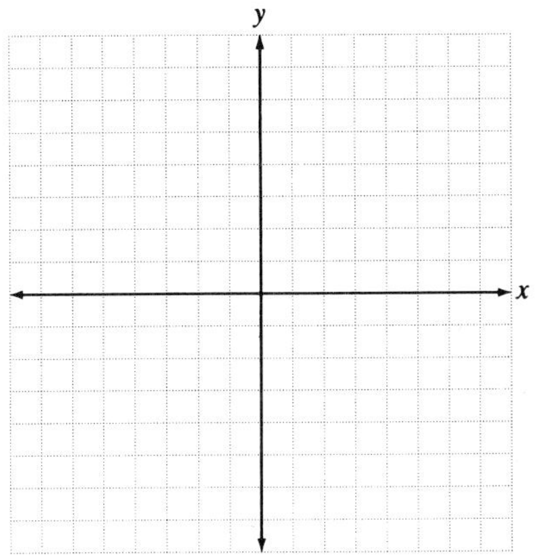

14. $5y + 3x = 15$

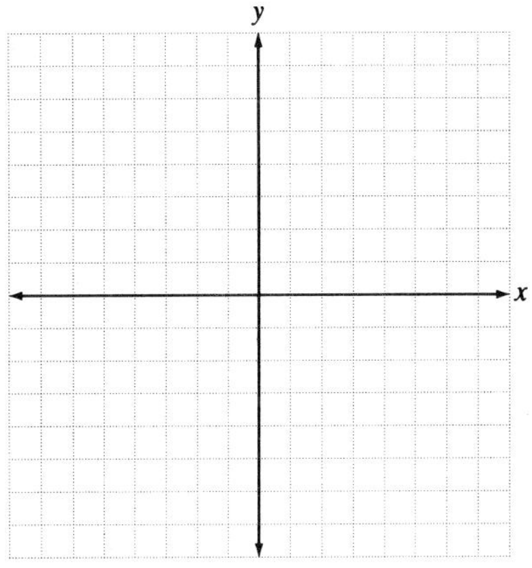

15. $x + 4y = 8$

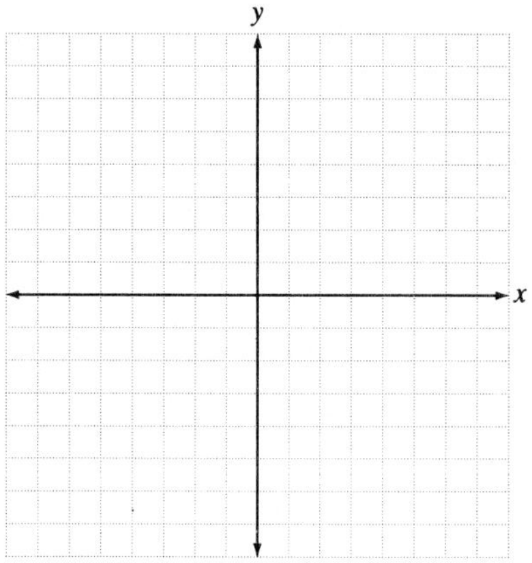

16. $3x - y = 6$

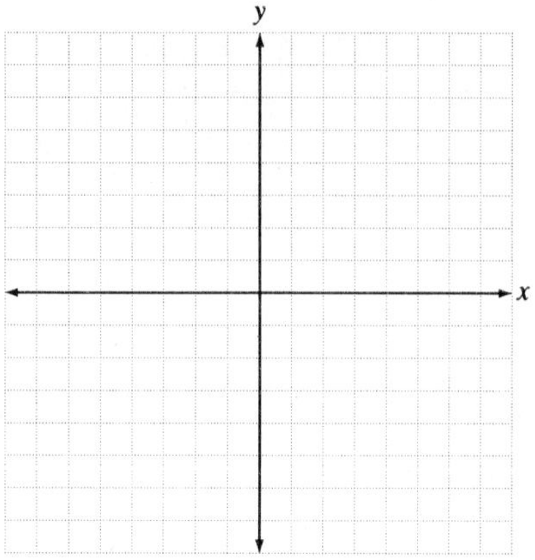

17. $4x = 3y - 12$

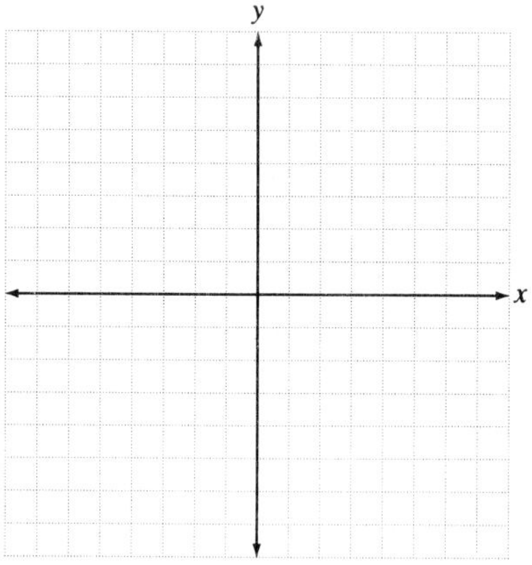

18. $5y = 4x - 20$

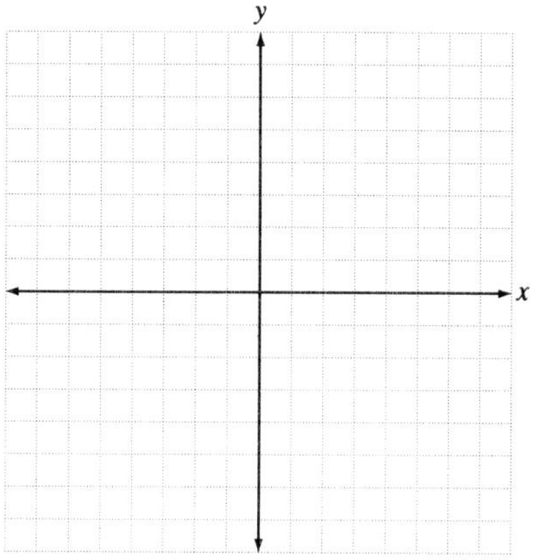

Graph.

19. $y = -1$

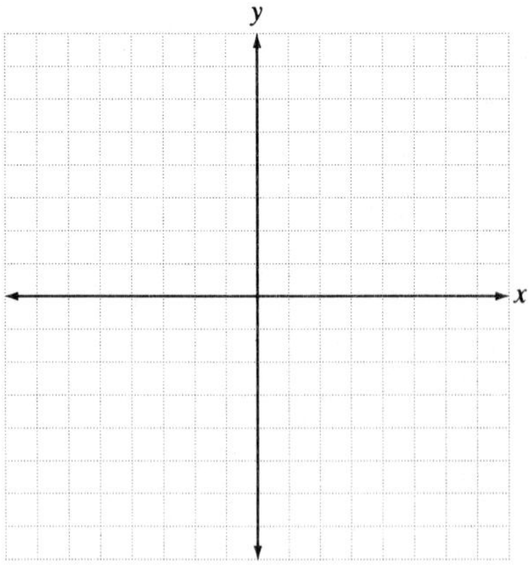

20. $x = 3$

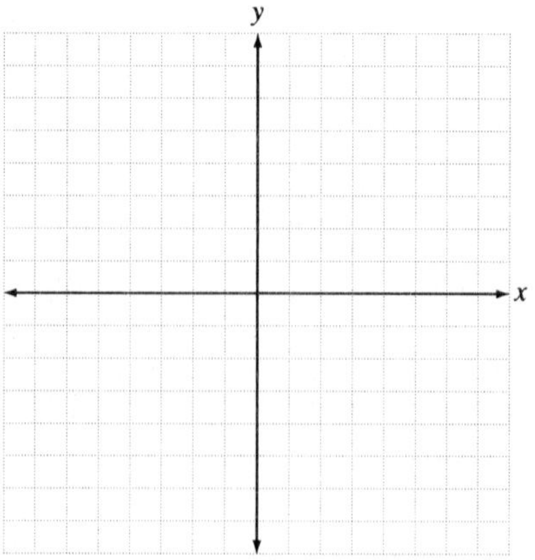

21. $2x + 4 = 0$

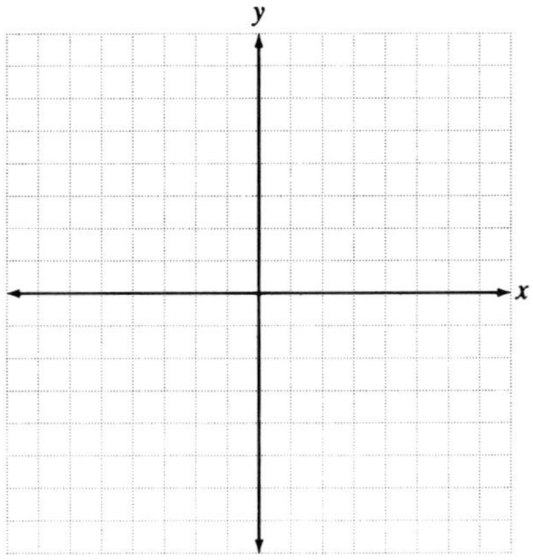

22. $3x - 15 = 0$

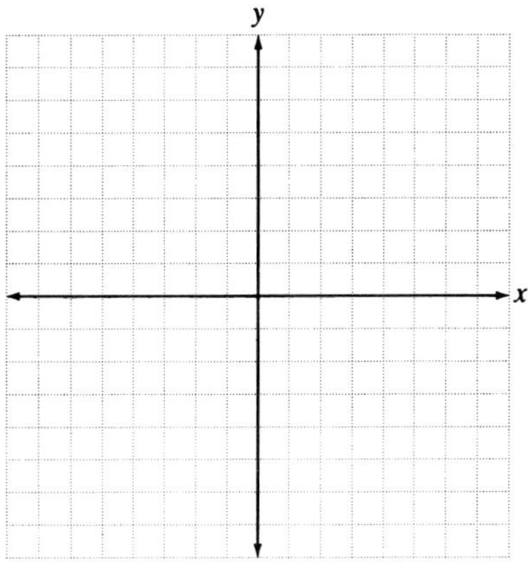

23. $6y - 18 = 0$

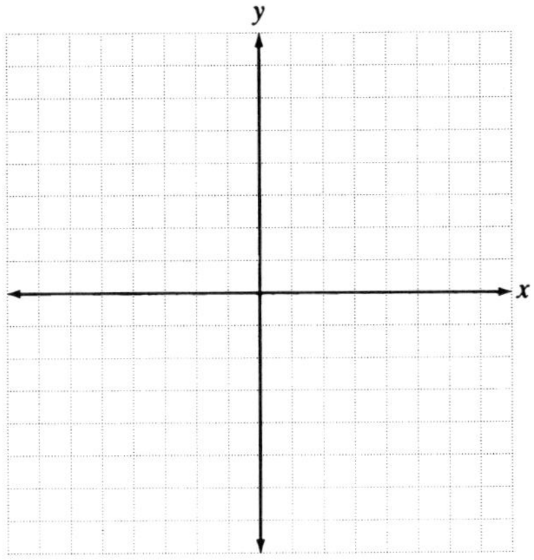

24. $5y + 20 = 0$

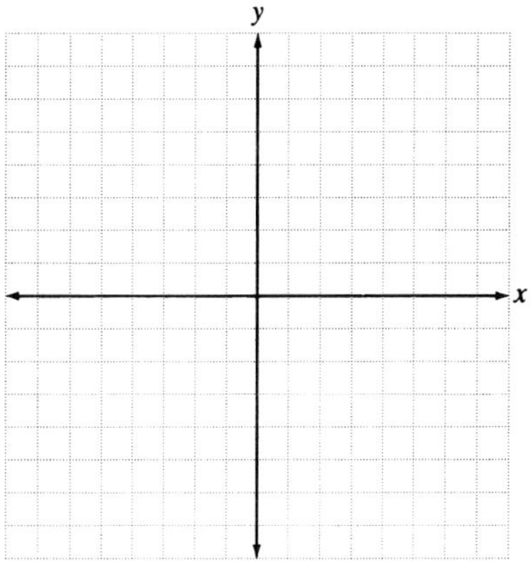

25. $x = \frac{5}{2}$

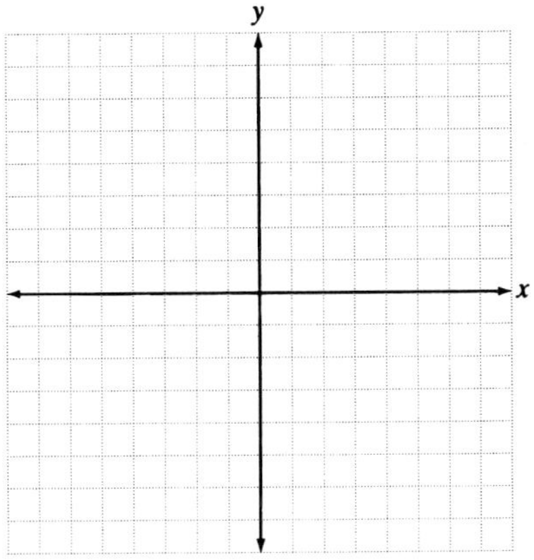

26. $y = -\frac{9}{2}$

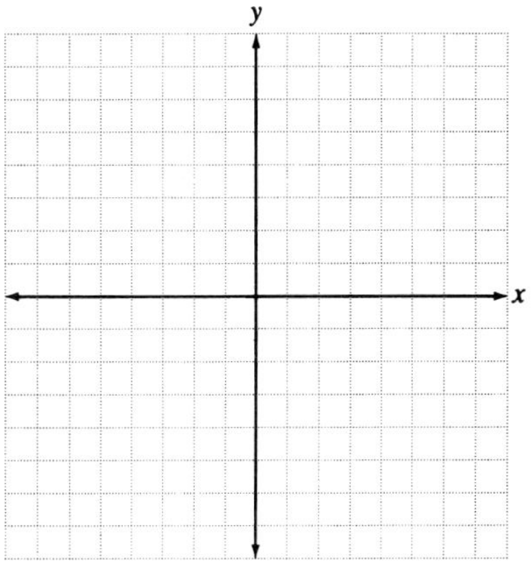

27. $x = 0$

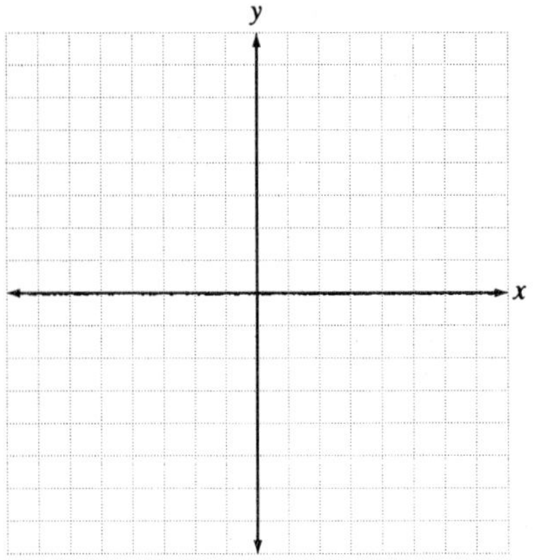

28. $y = 0$

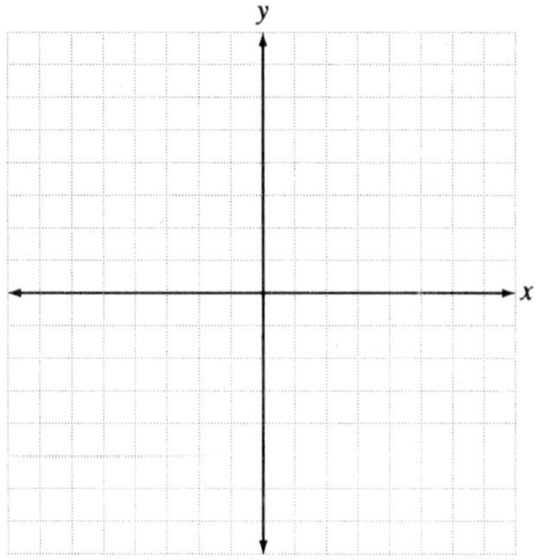

29. $5(x + 1) = -2(y + 1) - 3$

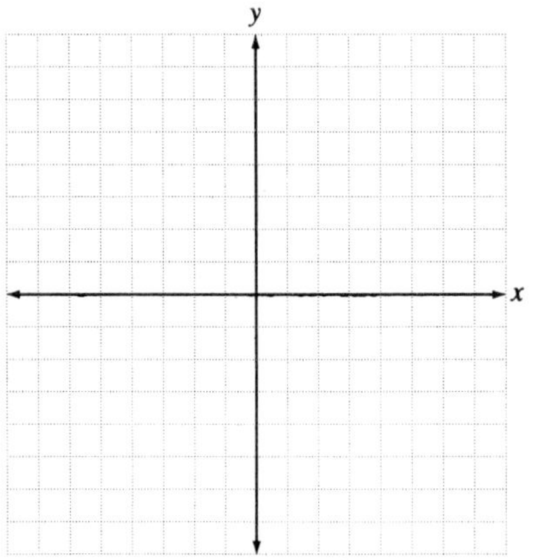

30. $3(x + 1) = 4(y + 1) + 11$

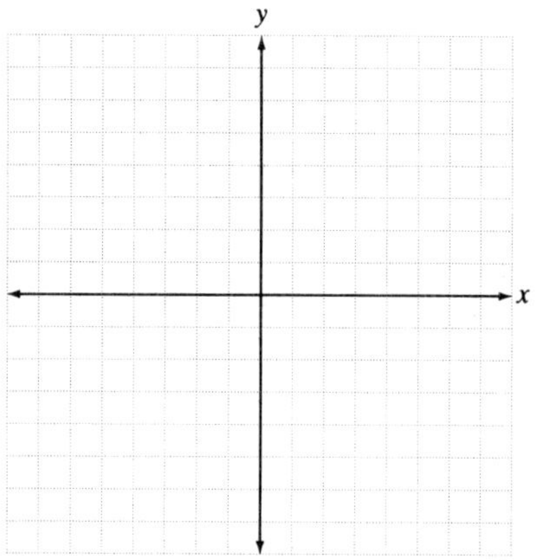

Which graphs are horizontal lines, which are vertical lines, and which are neither?

31. $x = 7$

32. $y = 3x$

33. $y = -2x + 5$

34. $y = -4$

35. $y = 8$

36. $x + 4y = 10$

37. $7x + 14 = 0$

38. $5x = 8y + 2$

39. $4y = 3x - 9$

40. $6y - 11 = 0$

41. The graph of a linear equation, $Ax + By = C$, is a ____________ ____________.

42. The point where a graph crosses the x-axis is called the ____________.

43. In the equation of a line we let x equal zero to find the ____________.

44. All graphs of equations of the form $x = a$ are ____________ lines.

Which equations are linear?

* **45.** $5(3x - 4y) = 6x + 10$

* **46.** $7x - 4 = 8x(x - 5)$

Graph.

* **47.** $y = x - 2 \quad \text{for } x \geq 0$

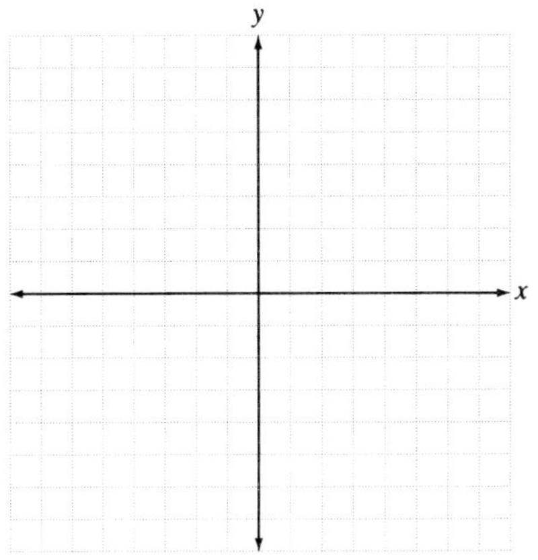

* **48.** $y = x + 3 \quad \text{for } x \leq 0$

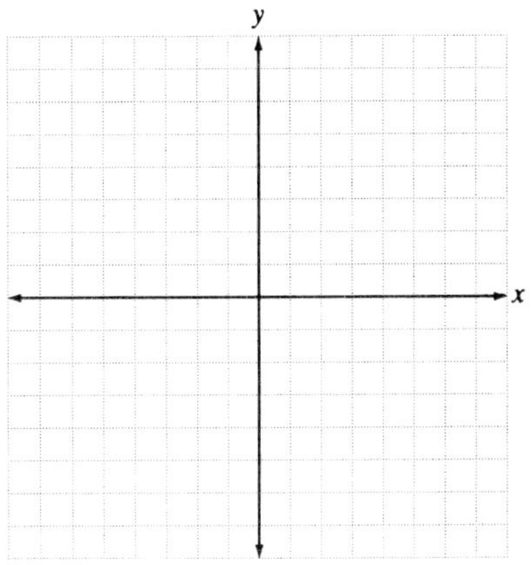

For the equation $y = |x|$, $y = x$ if $x \geq 0$ and $y = -x$ if $x < 0$. Graph the following.

* **49.** $y = |x|$

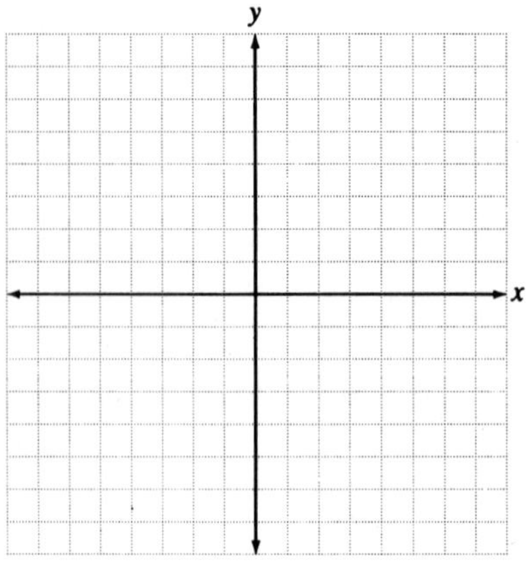

* **50.** $y = |x + 2|$

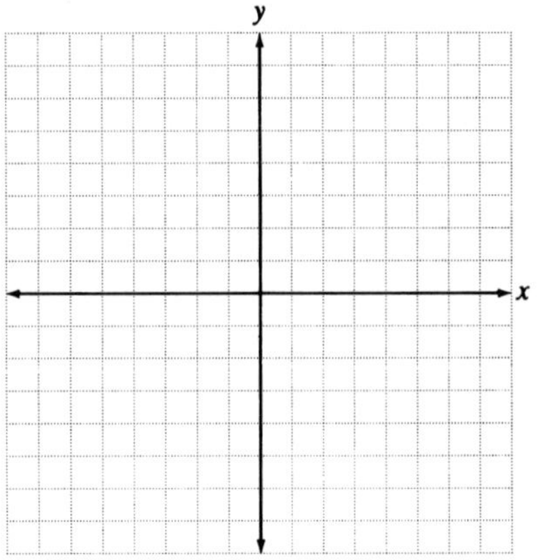

* **51.** $y = |x - 3|$

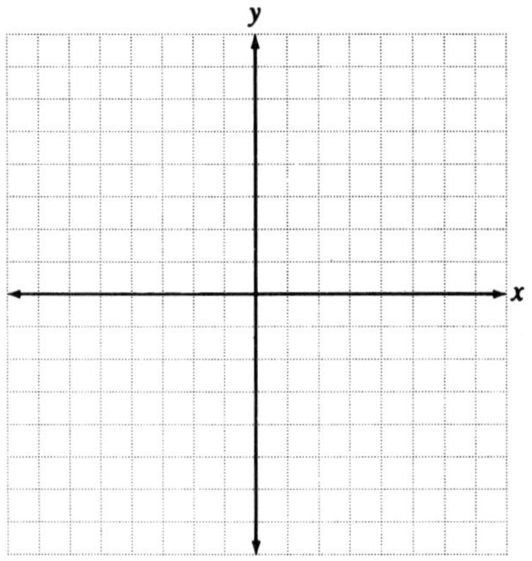

* **52.** $y = |x| - 1$

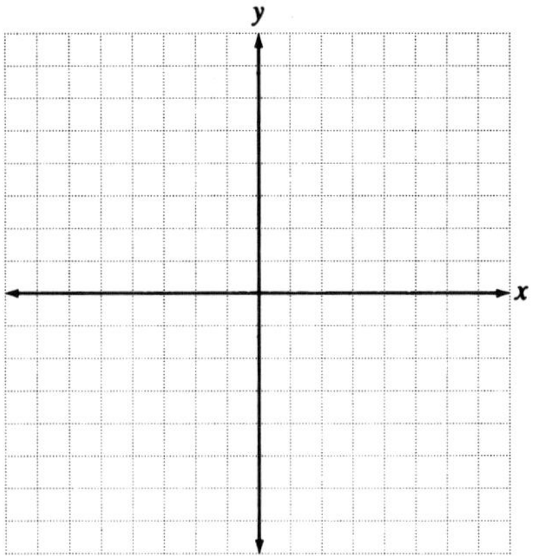

Checkup

The following problems provide a review of some of Section 1.3 and will help you with the next section.

Find two other names for the numbers by changing signs.

53. $\frac{-3}{5}$

54. $\frac{5}{-8}$

55. $-\frac{2}{3}$

56. $-\frac{9}{4}$

57. $-\frac{7}{9}$

58. $\frac{-5}{3}$

59. $\frac{11}{-4}$

60. $-\frac{14}{5}$

3.3 SLOPE

1 The Slope of a Line

It is helpful in mathematics and the applications of it to know the slant of a line. A number called the ***slope*** *of a line,* designated as m, tells us how the line slants. *Lines that slope upward to the right have positive slope. Lines that slope downward to the right have negative slope.* Geometrically, the slope of a line is the vertical change (difference of y-coordinates) divided by the horizontal change (difference of x-coordinates) between two points (x_1, y_1) and (x_2, y_2) on the line.

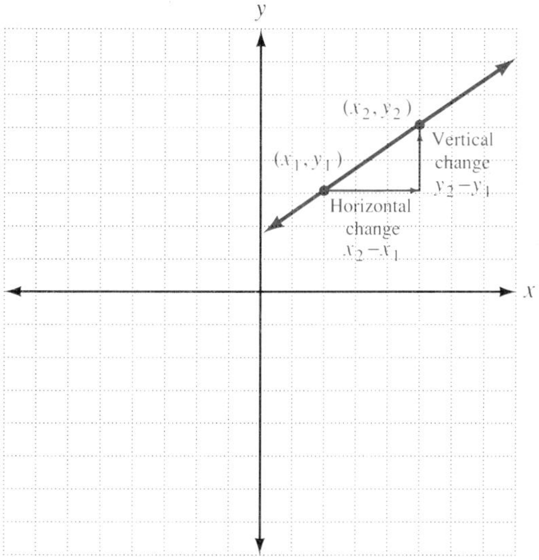

The vertical change is often called the ***rise*** and the horizontal change may be called the ***run.***

$$\text{Slope: } \frac{\text{vertical change}}{\text{horizontal change}} \quad \text{or} \quad \frac{\text{rise}}{\text{run}}$$

$$m = \frac{y_2 - y_1}{x_2 - x_1} = \frac{y_1 - y_2}{x_1 - x_2}$$

We may subtract the coordinates in either order, but we must be sure to subtract both the x- and y-coordinates in the same order. We may also choose any two points on a line to find its slope.

OBJECTIVES

1. *Find the slope of a line through a pair of points*
2. *Find the slopes of horizontal and vertical lines*
3. *Determine the slope and y-intercept from an equation of a line*
4. *Graph a linear equation using the slope and y-intercept*

□ **Exercise 1** Find the slope of the line through each pair of points.

a. $(-3, 2), (3, 9)$

b. $(6, 8), (1, 9)$

c. $(2, 6), (4, 1)$

d. $(5, 7), (-1, -2)$

EXAMPLE 1 Find the slope m of the line through each pair of points.

a. $(1, 2)$ and $(3, 5)$

Let $(x_2, y_2) = (3, 5)$ and $(x_1, y_1) = (1, 2)$:

$$m = \frac{y_2 - y_1}{x_2 - x_1} = \frac{5 - 2}{3 - 1} = \frac{3}{2}$$

The graph of the points and the line through them is shown here. Notice that the vertical change or rise is $+3$, and the horizontal change or run is $+2$, so geometrically we have the following:

$$\text{slope: } \frac{\text{vertical change}}{\text{horizontal change}} = \frac{3}{2}$$

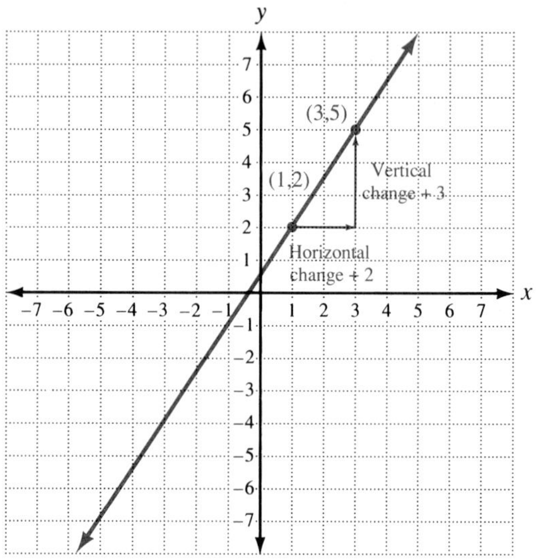

b. $(-3, 7), (5, 2)$

Let $(x_2, y_2) = (5, 2)$ and $(x_1, y_1) = (-3, 7)$:

$$m = \frac{y_2 - y_1}{x_2 - x_1} = \frac{2 - 7}{5 - (-3)} = \frac{-5}{8}$$

We could have subtracted in the opposite order and gotten the same result. Let $(x_2, y_2) = (-3, 7)$ and $(x_1, y_1) = (5, 2)$:

$$m = \frac{y_2 - y_1}{x_2 - x_1} = \frac{7 - 2}{-3 - 5} = \frac{5}{-8} = \frac{-5}{8}$$

Recall that $\frac{a}{-b} = \frac{-a}{b}$. ■

□ **DO EXERCISE 1.**

2 Slopes of Horizontal and Vertical Lines

We know that points on a horizontal line have the same second coordinate, and points on a vertical line have the same first coordinate. What are the slopes of horizontal and vertical lines?

EXAMPLE 2 Find the slope of the line.

a. $y = 3$

We know from Section 3.2 that this is a horizontal line. Two points on the line are (2, 3) and (4, 3). Let $(x_2, y_2) = (4, 3)$ and $(x_1, y_1) = (2, 3)$:

$$m = \frac{y_2 - y_1}{x_2 - x_1} = \frac{3 - 3}{4 - 2} = \frac{0}{2} = 0$$

The results of Example 2a are true for all horizontal lines.

b. $x = -4$

This is a vertical line. Two points on the line are (−4, −2) and (−4, 3). Let $(x_2, y_2) = (-4, 3)$ and $(x_1, y_1) = (-4, -2)$.

$$m = \frac{y_2 - y_1}{x_2 - x_1} = \frac{3 - (-2)}{-4 - (-4)} = \frac{5}{0}$$

Since division by zero is undefined, we say that the slope is undefined. ■

> All horizontal lines have a slope of 0.
>
> The slope of a vertical line is undefined.

□ DO EXERCISE 2.

3 Using the Equation of a Line to Find Its Slope and *y*-Intercept

The *y*-intercept (0, *b*) of an equation is often written "*b*." Consider the linear equation $y = 2x + 3$. Two points on the line are (1, 5) and (0, 3), and (0, 3) is the *y*-intercept. Let $(x_2, y_2) = (1, 5)$ and $(x_1, y_1) = (0, 3)$. The slope of the line is as follows:

$$m = \frac{y_2 - y_1}{x_2 - x_1} = \frac{5 - 3}{1 - 0} = \frac{2}{1} = 2$$

Notice that in the given equation, $y = 2x + 3$, the coefficient of x is 2, the same number as the slope. If we let $x = 0$ in the equation, we see that the *y*-intercept is 3.

> **Slope-Intercept Form of the Equation of a Line**
>
> The slope-intercept form of the equation of a line with slope m and *y*-intercept b is
>
> $$y = mx + b.$$
>
> (m: slope; b: *y*-intercept)

EXAMPLE 3 Write the equation $3y + 2x = 12$ in slope-intercept form and find the slope and the *y*-intercept.

$3y + 2x = 12$

$3y = -2x - 12$ Add $-2x$ to each side of the equation.

$y = \frac{-2}{3}x + 4$ Divide both sides of the equation by 3.

The slope is $-\frac{2}{3}$ and the *y*-intercept is 4. ■

□ DO EXERCISE 3.

□ **Exercise 2** Find the slope of each line.

a. $x = 3$

b. $y = -2$

c. $4y = 12$

d. $5x = 7$

e. $y + 8 = 0$

f. $2x - 3y = 5 - 3y$

□ **Exercise 3** Find the slope and the *y*-intercept.

a. $4y + 3x = 12$

b. $2y - 5x = 6$

□ **Exercise 4** Find the slope and y-intercept and graph.

a. $2x - 3y = 9$

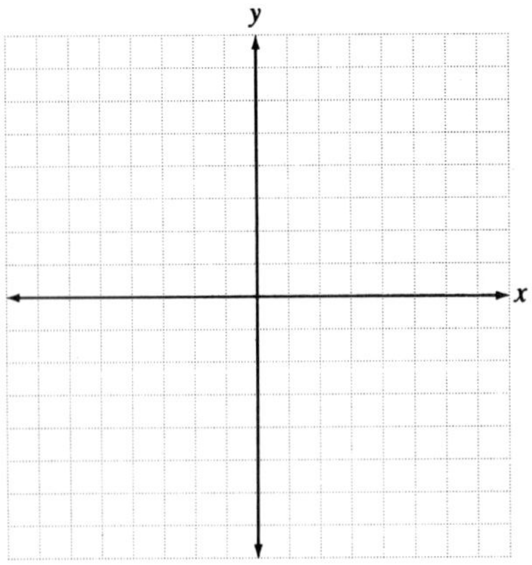

b. $3x + 4y = 8$

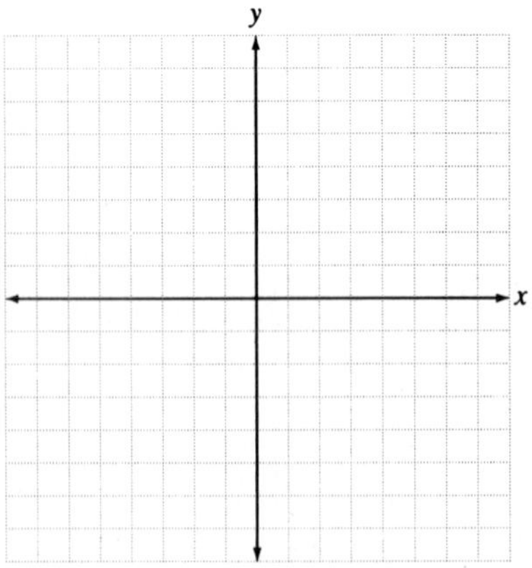

4 Graphing a Linear Equation Using the Slope and y-Intercept

If one or both of the intercepts of a line are fractions, it is often easier to graph the line using the slope and the y-intercept as shown in Example 4.

EXAMPLE 4 Graph.

a. $3y - 2x = 9$

1. Write the equation in the slope-intercept form, $y = mx + b$.

$$3y - 2x = 9$$

$$3y = 2x + 9 \quad \text{Add } 2x \text{ to each side.}$$

$$y = \frac{2}{3}x + 3 \quad \text{Divide both sides by 3.}$$

The slope is $\frac{2}{3}$ and the y-intercept is 3 or (0, 3).

2. Plot the y-intercept (0, 3).
3. Slope: $\dfrac{\text{vertical change}}{\text{horizontal change}} = \dfrac{2}{3}$

From the y-intercept (0, 3), move 2 units up in the y direction and then 3 units right in the x direction. Plot a point.

4. Draw the line through the two points.

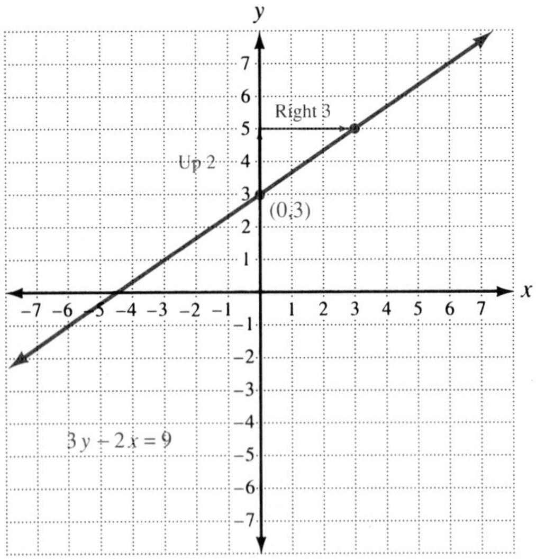

b. $y = -2x$

1. The slope is -2 and the y-intercept is 0 or (0, 0).
2. Plot the point (0, 0).
3. Write the slope with a denominator of 1 or -1.

$$-2 = \frac{-2}{1} = \frac{2}{-1}$$

We used the denominator of 1:

$$\text{slope: } \frac{\text{vertical change}}{\text{horizontal change}} = -2 = \frac{-2}{1}$$

From the point (0, 0) move down 2 units in the y direction and right 1 unit in the x direction. Plot a point.

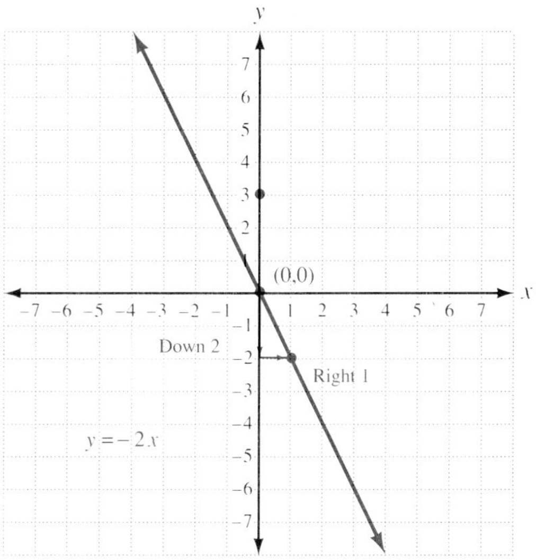

4. Draw the line through the points.

Since the points are close together, we could also use the equivalent value for slope, $2/-1$. From the y-intercept, move 2 units up in the y direction and 1 unit left in the x direction. Plot a point and draw the line between the three points.

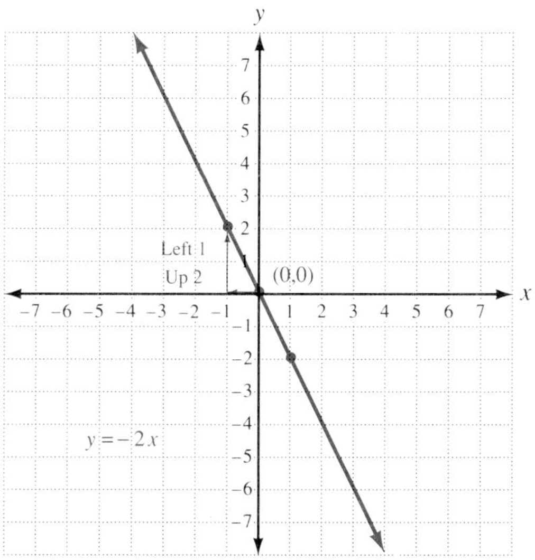

■

Notice that if we had tried to graph Example 4b using intercepts, we would have found that the x-intercept and the y-intercept are the same point (0, 0) and we would have had to find another point on the line. It is usually easier to graph equations of the form $y = kx$ using the slope and y-intercept.

□ **DO EXERCISE 4.**

There are many applications of slope in the real world.

c. $y = 4x$

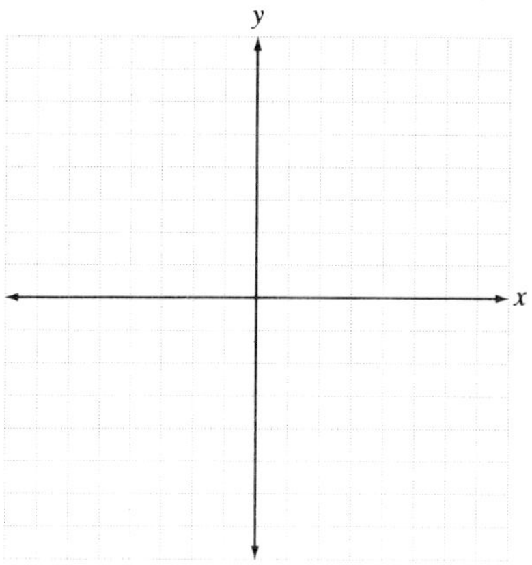

d. $3x + y = 0$

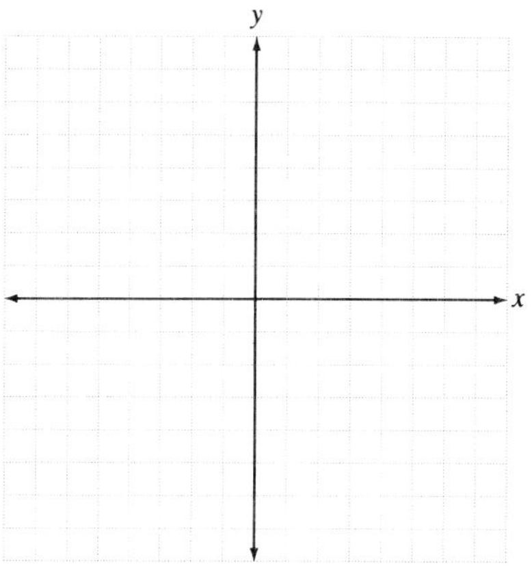

□ **Exercise 5** For every horizontal distance of 200 feet, a road rises 10 feet. What is the slope of the road?

EXAMPLE 5 A home in Anchorage, Alaska has a steep roof. The roof rises 12 feet for every 16 feet of run. What is the slope of the roof?

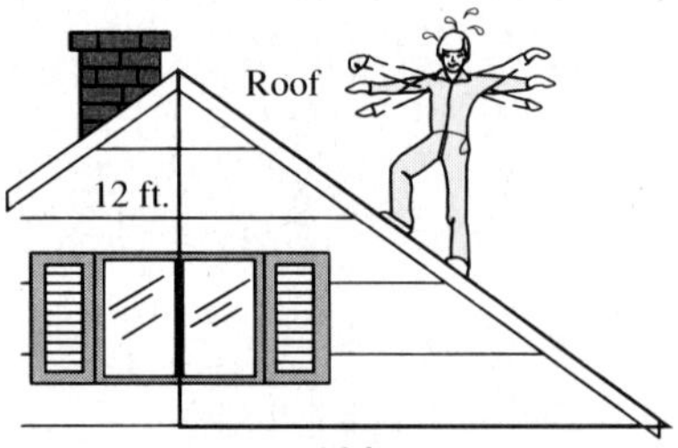

The slope is as follows.

$$m = \frac{12}{16} = \frac{3}{4}$$

The slope of the roof may also be called the *pitch*. ■

□ **DO EXERCISE 5.**

Graphing Linear Equations

1. Is the equation of the form $y = b$ or $x = a$? The graphs of each of these equations is a line parallel to an axis.
2. Determine by inspection if one or both of the intercepts will be fractions. If they are not, graph the line using intercepts unless the equation is of the form $y = kx$. If the points are too close together, choose another point farther from the intercepts.
3. If one or both of the intercepts are fractions or if the equation is of the form $y = kx$, it is often easier to graph the line using the slope and y-intercept.

GRAPHING CALCULATOR

Linear equations may be graphed with a graphing calculator. Since keystrokes for these calculators vary, general instructions for graphing an equation are as follows.

Graph $3x - 2y = 12$.

The calculator requires that the equation be in slope-intercept form.

$$3x - 2y = 12$$

$$-2y = -3x + 12$$

$$y = \frac{3}{2}x - 6 \quad \text{or} \quad y = \left(\frac{3}{2}\right)x - 6$$

1. Enter the equation in your calculator. The fraction $\frac{3}{2}$ must be entered in parentheses or the calculator will graph a different equation.
2. Set the graphics window. This should show the important aspects of the graph. Often, a convenient window is a left value of -10, a right value of 10, a lower value of -10, and an upper value of 10. This might be indicated by W: $[-10, 10]$ $[-10, 10]$, where the first set of brackets indicates the left and right values of x and the second set indicates the lower and upper values of y.
3. Plot the graph. We could *trace* the graph to estimate to the nearest whole number the x- and y-intercepts. For this equation, the x-intercept is $(4, 0)$ and the y-intercept is $(0, -6)$. We could also identify the coordinates of two other points and use the formula $m = \frac{y_2 - y_1}{x_2 - x_1}$ to show that the slope of the line is $\frac{3}{2}$.

Answers to Exercises

1. a. $\frac{7}{6}$ **b.** $-\frac{1}{5}$ **c.** $-\frac{5}{2}$ **d.** $\frac{3}{2}$

2. a. Undefined **b.** 0 **c.** 0 **d.** Undefined **e.** 0
f. Undefined

3. a. Slope: $-\frac{3}{4}$; y-intercept: 3 **b.** Slope: $\frac{5}{2}$; y-intercept: 3

4. a. Slope: $\frac{2}{3}$; y-intercept: -3 **b.** Slope: $-\frac{3}{4}$; y-intercept: 2

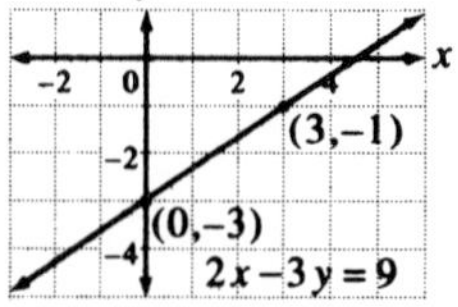

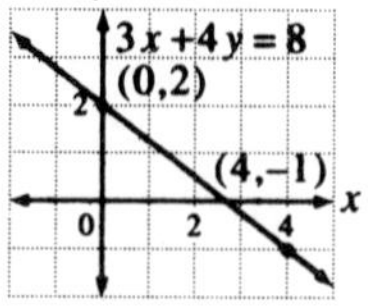

c. Slope: 4; y-intercept: 0 **d.** Slope: -3; y-intercept, 0

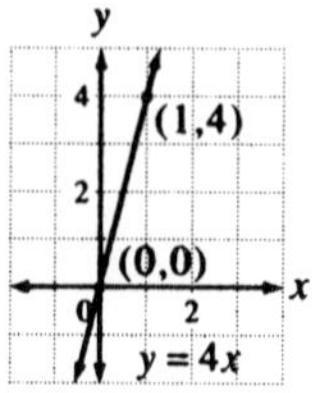

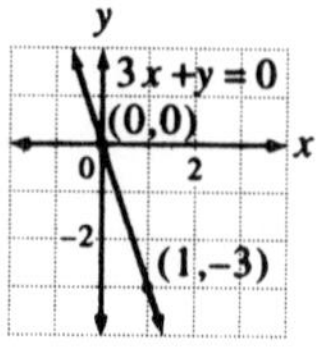

5. $\frac{1}{20}$

Problem Set 3.3

NAME

DATE

CLASS

Find the slope, if it exists, of the line through each pair of points.

1. $(3, -2), (2, 6)$

2. $(-3, 5), (2, 3)$

3. $(5, 2), (4, 3)$

4. $(8, 6), (9, 4)$

5. $(-3, -1), (-2, 4)$

6. $(7, -8), (-6, -4)$

7. $(-5, 4), (-5, 8)$

8. $(3, 9), (-3, 9)$

9. $(-7, 6), (3, -2)$

10. $(-8, -2), (4, 3)$

Find the slope and the y-intercept.

11. $y = \frac{2}{3}x + 4$

12. $y = -\frac{4}{3}x - 2$

13. $3x - 2y = 8$

14. $5x - 7y = 14$

15. $6x + 4y = 3$

16. $9x + 2y = 7$

17. $5y = x - 8$

18. $3y = 2x + 4$

19. $-4y = 5x - 2$

20. $-8y = -4x + 3$

Find the slope and y-intercept and graph.

21. $4x + 3y = 9$

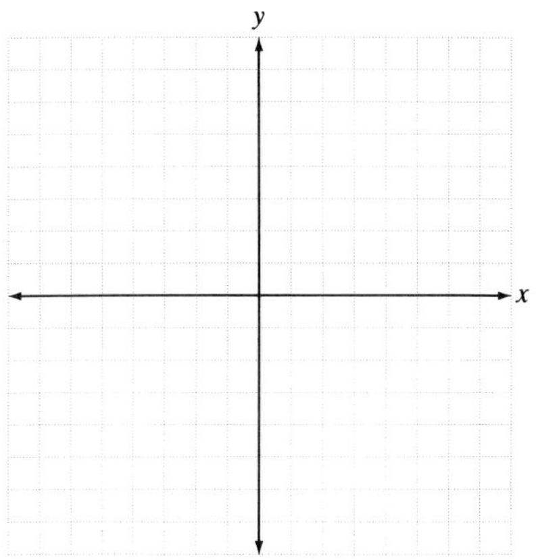

22. $3x + 2y = 4$

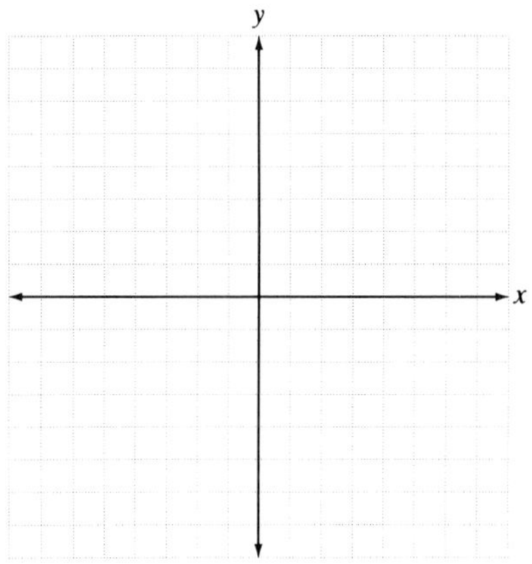

23. $2y - 5x = 2$

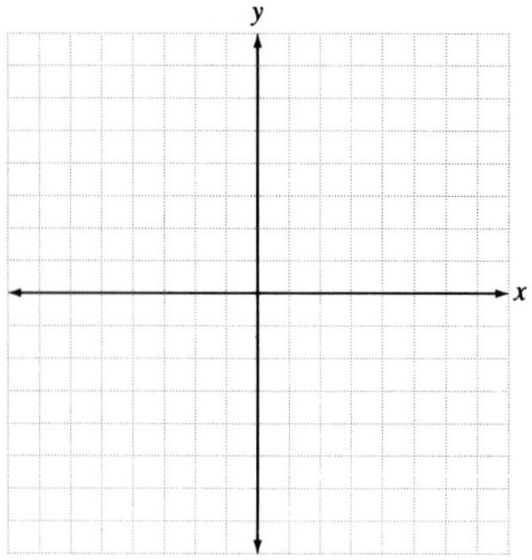

24. $5y - 2x = 10$

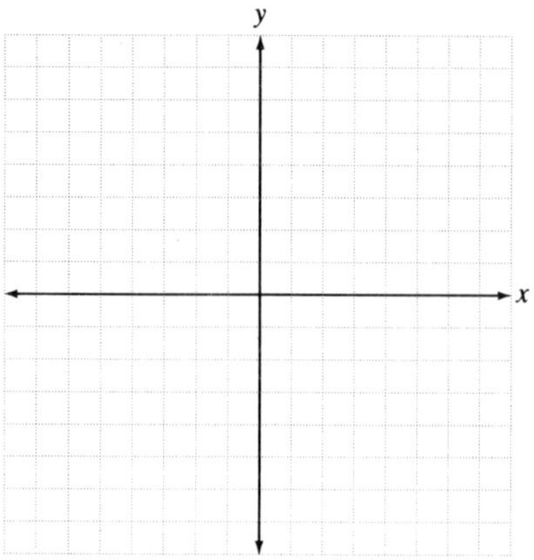

25. $6x + 5y = 30$

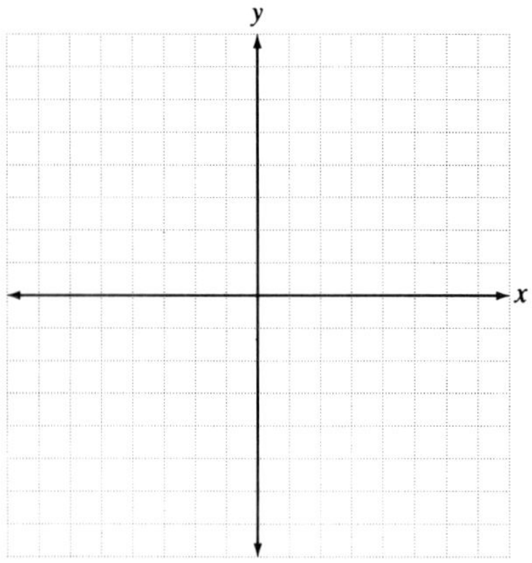

26. $7x - 2y = 8$

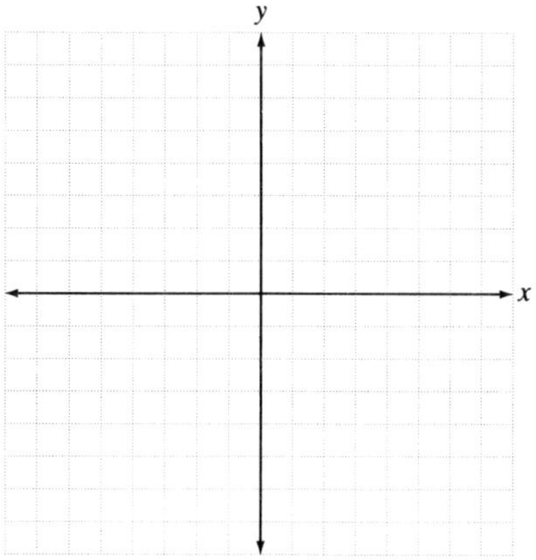

27. $3y = x + 6$

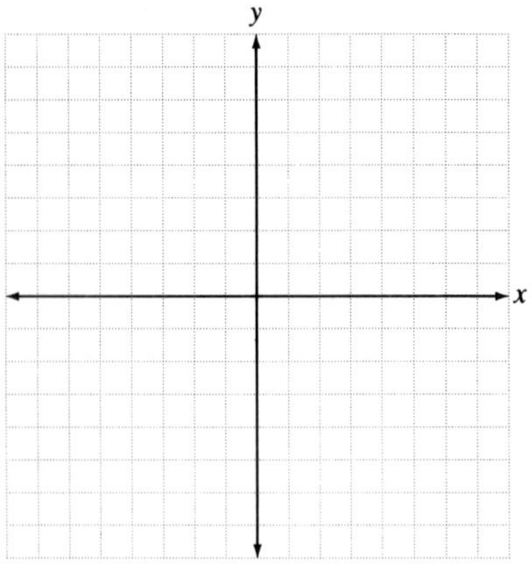

28. $-2y = 3x + 6$

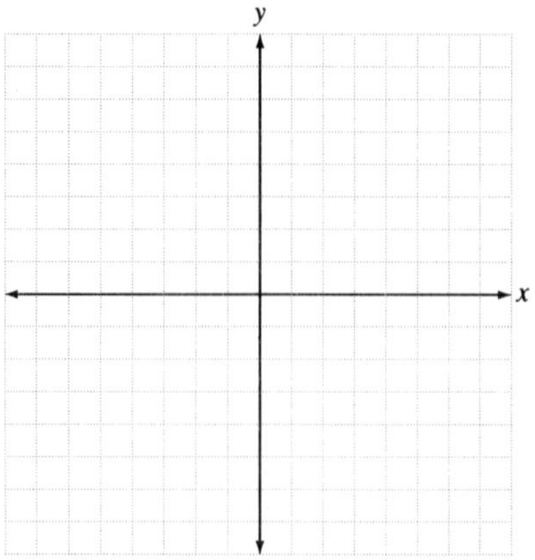

29. $-3x + y = 2$

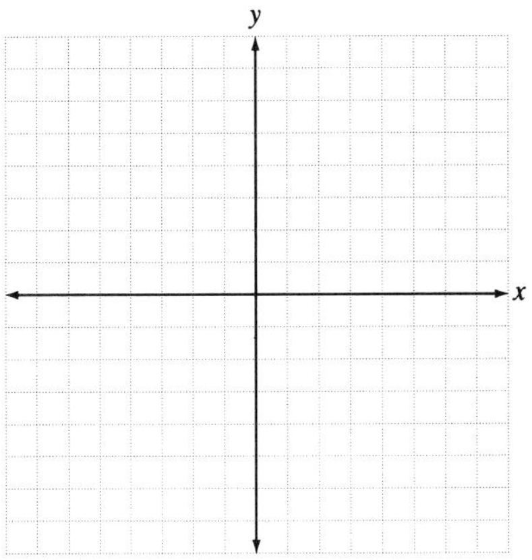

30. $4x - y = 3$

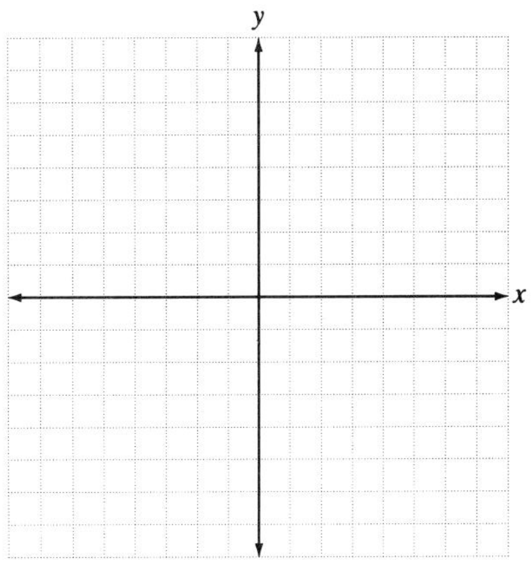

Graph. Use any method.

31. $y = x$

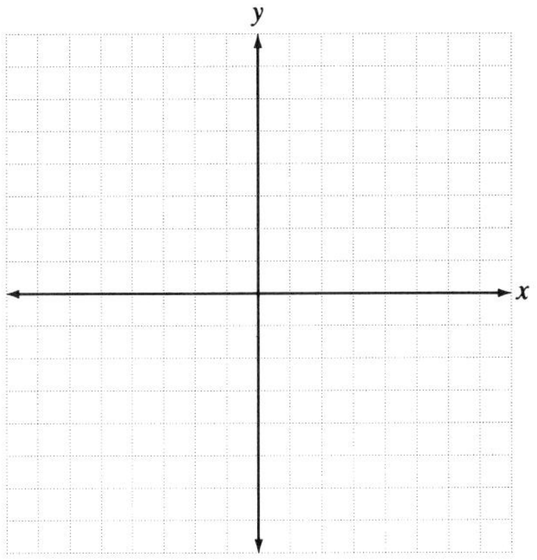

32. $y = -x$

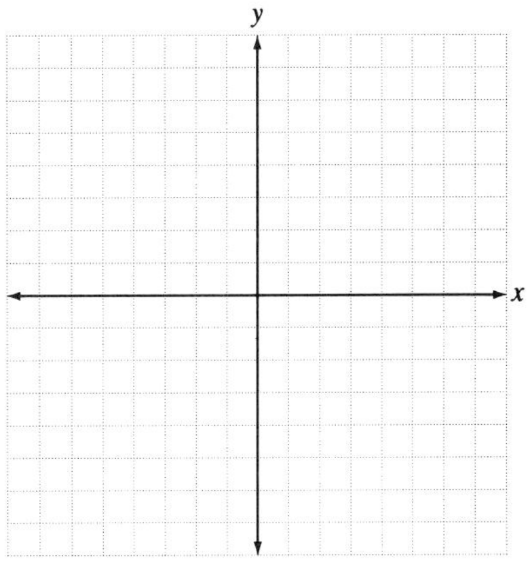

33. $2x + 3y = -6$

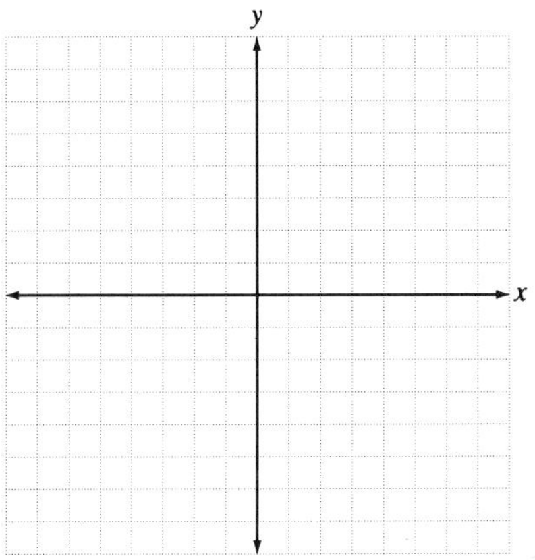

34. $3x - 7y = 21$

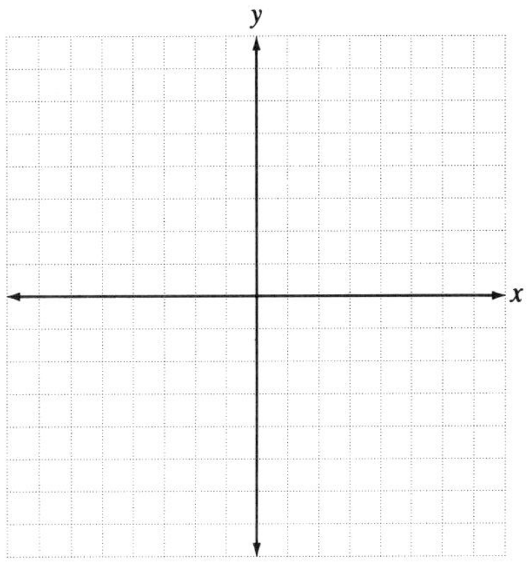

35. $2y = -5x - 8$

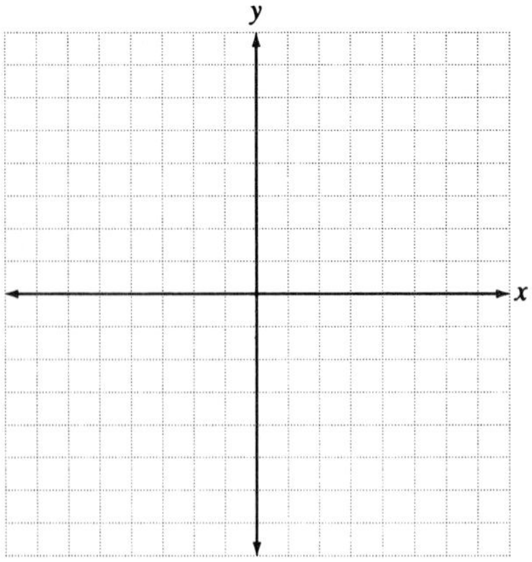

36. $4y = 3x + 4$

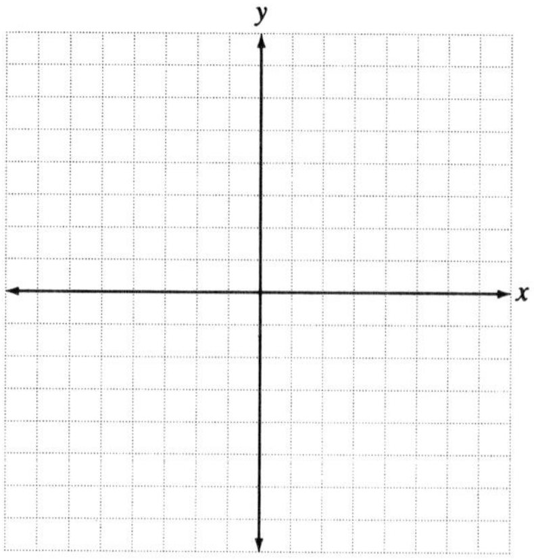

37. $y + 5x = 0$

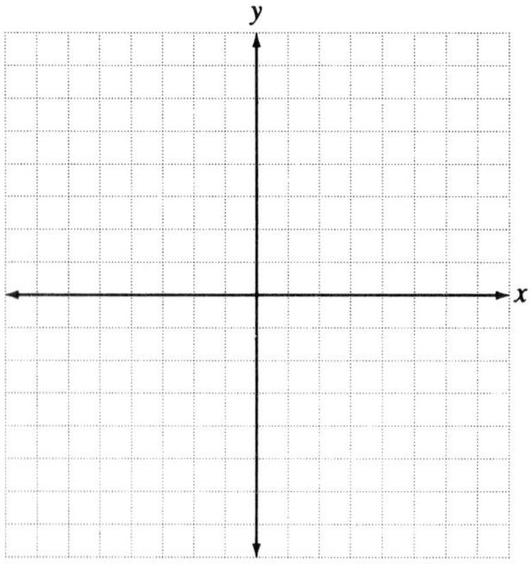

38. $y - 4x = 0$

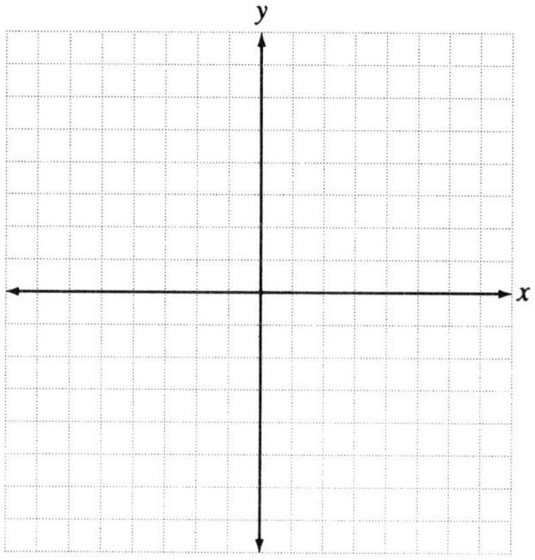

39. $x - 5y = 10$

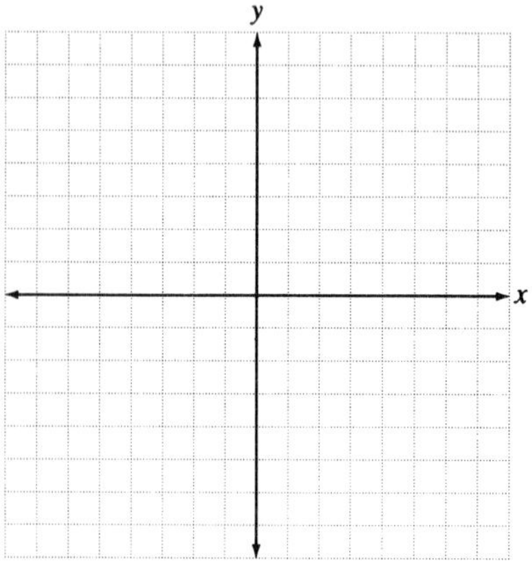

40. $x + 3y = -9$

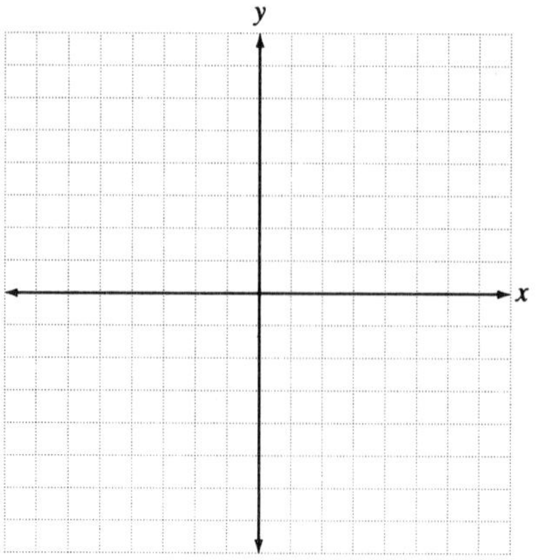

41. Line L has slope $\frac{3}{5}$. A horizontal change of 10 on the line will always give what vertical change?

42. A line has a slope of $-\frac{7}{4}$. A vertical change on the line of -14 will give what horizontal change?

43. A pile of sawdust is in the shape of a cone. If the slope of the side of the pile is $\frac{3}{4}$ and the diameter of the base of the pile is 16 feet, how high is the pile?

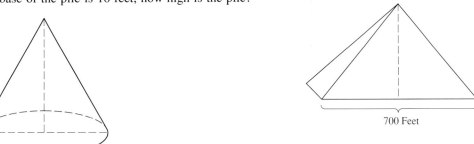

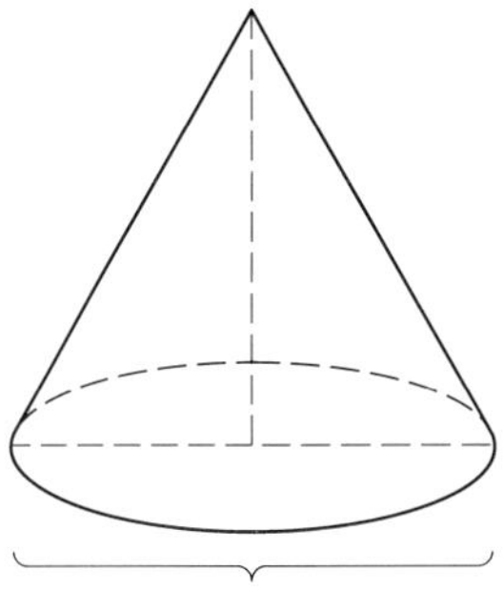

44. The slope of the sides of a pyramid is $\frac{13}{10}$. If the base of the pyramid is 700 feet, what is its height?

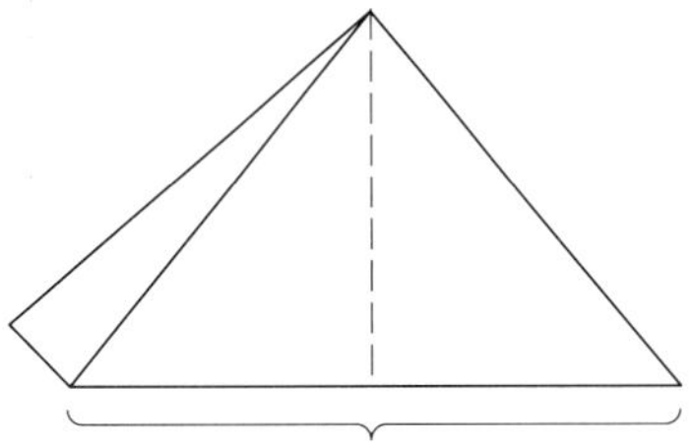

45. A river drops 20 feet vertically for every 250 feet that it falls horizontally. What is the slope of the river?

46. A home in Denver, Colorado has a steep roof. For every $2\frac{1}{2}$ feet of run it has $1\frac{1}{2}$ feet of rise. What is the pitch of the roof?

47. The slope of a road is 0.04. How many feet does it fall vertically for every 2350 feet of road?

48. The rate of climb or slope of a jet liner is 0.425. How far does it climb vertically for a horizontal distance of 8500 feet?

49. Lines that slope downward to the right have __________ slope.

50. All horizontal lines have a slope of __________.

51. The coordinates of the y-intercept of the line $y = mx + b$ are __________.

52. If an equation is of the form $y = kx$, it is often easier to graph the equation using the __________ and the __________.

Find the slope of the line through each pair of points.

53. (5.8, 6.7), (5.1, 3.2)

54. (−9.9, 2.4), (3.1, −4.1)

55. (8.2, −0.7), (−12.5, −7.6)

56. (6.2, −6.1), (6.5, −9.4)

* 57. $(6a, 2a), (-3a, 4a)$

* 58. $(7b, -4b), (8b, -2b)$

* 59. $(c, 4d), (7c, -3d)$

* 60. $(a + d, b - c), (a - d, 2b - c)$

Find the slope and y-intercept of the equation.

* 61. $\frac{1}{5}x = -\frac{5}{3}y - 2$

* 62. $\frac{7}{8}y = \frac{21}{5}x + 4$

* 63. Find the slope of the sides of a triangle with vertices (−4, 2), (1, 4), and (3, −1).

Think About It

* 64. Determine a so that the slope of the line through the pair of points has the given value.

$$(-2, a), (2, -2a); \qquad m = -\frac{7}{5}$$

Checkup

The following problems provide a review of some of Sections 2.1 and 1.3 and will help you with the next section.

Solve.

65. $y - 6 = \frac{1}{5}(y + 2)$

66. $x - 9 = \frac{1}{3}(x - 7)$

67. $y - 16 = -\frac{6}{5}(y + 6)$

68. $y - 2 = \frac{3}{8}(y + 3)$

Find the reciprocal.

69. $\frac{3}{8}$

70. $\frac{-9}{7}$

71. −2

72. 8

3.4 EQUATIONS OF STRAIGHT LINES

In the applications of mathematics we often have a set of data points that lie approximately on a straight line. Then we want to write a linear equation for this line. This is called *fitting a linear equation to the data points.* For example, the heights and weights of four people are shown on the following graph. A linear equation fits these data. We can write this equation by finding the slope of the line and one point on it or by finding two points on the line.

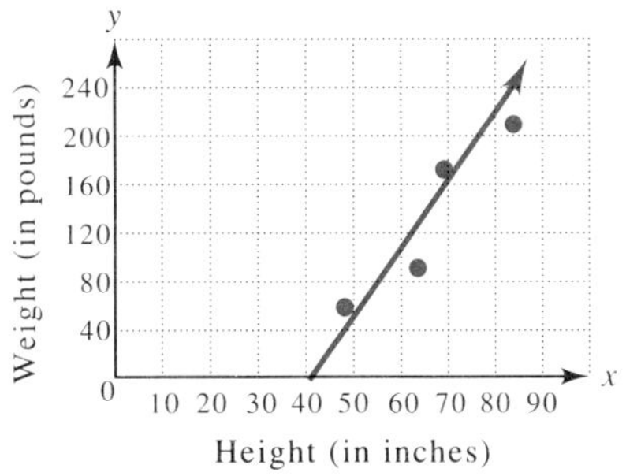

1 Using the Slope and y-Intercept to Write an Equation of a Line

Recall that the slope-intercept form of an equation, where m is the slope and b is the y-intercept, is

$$y = mx + b$$

If we know the slope and y-intercept of a line, we can write the equation of it.

EXAMPLE 1 Find an equation for the line with slope $-\frac{3}{4}$ and y-intercept -4.

$$m = -\frac{3}{4} \quad \text{and} \quad b = -4$$

$y = mx + b$ Slope-intercept form.

$y = -\frac{3}{4}x - 4$ Substitute for m and b. ■

□ **DO EXERCISE 1.**

2 Point-Slope Form of the Equation of a Line

Consider a line with slope m and passing through a given point (x_1, y_1). Any other point on the line has coordinates in the form (x, y), so the slope of the line is

$$\frac{y - y_1}{x - x_1} = m \qquad \text{for } x \neq x_1$$

Multiplying both sides of the equation by $x - x_1$ gives us the following:

Point-Slope Form of the Equation of a Line

$$y - y_1 = m(x - x_1)$$

We can use the point-slope form to write the equation of a line if we are given the *slope* and *one point* on the line or *two points* on the line.

OBJECTIVES

1 *Find an equation for a line with a given slope and y-intercept*

2 *Find an equation for a line through a given point with a given slope and an equation for a line through two points*

3 *Solve word problems using linear equations in two variables*

4 *Find an equation for a line through a given point and parallel to a given line*

5 *Find an equation for a line through a given point and perpendicular to a given line*

□ **Exercise 1** Find an equation for the lines with the following slopes and y-intercepts.

a. Slope: 3; y-intercept: -2

b. Slope: $-\frac{3}{4}$; y-intercept: $\frac{1}{8}$

c. Slope: $-\frac{2}{5}$; y-intercept: 0

d. Slope: 0; y-intercept: 6

EXAMPLE 2

a. Find an equation for the line with slope $\frac{1}{5}$ through the point (2, −3). Use the point-slope form.

$$y - y_1 = m(x - x_1)$$

$$m = \frac{1}{5} \quad \text{and} \quad (x_1, y_1) = (2, -3)$$

$$y - (-3) = \frac{1}{5}(x - 2) \qquad \text{Substitute for } m, x_1, \text{ and } y_1.$$

$$y + 3 = \frac{1}{5}(x - 2)$$

$$y + 3 = \frac{1}{5}x - \frac{2}{5} \qquad \text{Use a distributive law.}$$

$$y = \frac{1}{5}x - \frac{17}{5} \qquad \text{Add } -3 \text{ to both sides.}$$

b. Find an equation for the line through the points (−3, 2) and (4, −6). In order to use the point-slope form we must find the slope. We will let $(x_2, y_2) = (4, -6)$ and $(x_1, y_1) = (-3, 2)$.

$$m = \frac{y_2 - y_1}{x_2 - x_1} = \frac{-6 - 2}{4 - (-3)} = \frac{-8}{7}$$

$$y - y_1 = m(x - x_1) \qquad \text{Point-slope form.}$$

$$y - 2 = \frac{-8}{7}[x - (-3)] \qquad \text{Substitute for } m, x_1, \text{ and } y_1.$$

$$y - 2 = \frac{-8}{7}(x + 3)$$

$$y - 2 = -\frac{8}{7}x - \frac{24}{7} \qquad \text{Use a distributive law.}$$

$$y = -\frac{8}{7}x - \frac{10}{7} \qquad \text{Add 2 to each side.}$$

We would have gotten the same result if we had used $(x_2, y_2) = (-3, 2)$ and $(x_1, y_1) = (4, -6)$.

c. The equation for Example 2a could also be found using the slope-intercept form of the equation of a line as follows.

$$y = mx + b \qquad \text{Slope-intercept form.}$$

$$m = \frac{1}{5} \quad \text{and} \quad (x, y) = (2, -3)$$

$$-3 = \frac{1}{5}(2) + b \qquad \text{Substitute for } m, x, \text{ and } y.$$

$$-3 = \frac{2}{5} + b$$

$$-\frac{17}{5} = b \qquad \text{Add } -\frac{2}{5} \text{ to both sides.}$$

$$y = mx + b \qquad \text{Slope-intercept form.}$$

$$y = \frac{1}{5}x - \frac{17}{5} \qquad \text{Substitute for } m \text{ and } b.$$ ■

The slope of a vertical line is undefined. Therefore, vertical lines do not have point-slope equations.

□ DO EXERCISE 2.

3 Applied Problems

The following example shows how to fit a linear equation to data points.

EXAMPLE 3 Karen wants to build a doghouse. If she uses 48 square feet of plywood, the cost is \$21. If she uses 84 square feet of plywood, the cost is \$30. The relationship is linear. Write a linear equation that relates the total cost y to the number of square feet of plywood x used.

1. We are given two points that are solutions to our linear equation, (48, 21) and (84, 30). Since we are given two points, we use the point-slope form

$$y - y_1 = m(x - x_1)$$

2. Find the slope.

$$m = \frac{y_2 - y_1}{x_2 - x_1} = \frac{30 - 21}{84 - 48} = \frac{9}{36} = \frac{1}{4}$$

3. We may use either point for (x_1, y_1). We chose (48, 21).

$$y - y_1 = m(x - x_1)$$

$$y - 21 = \frac{1}{4}(x - 48)$$

$$y - 21 = \frac{1}{4}x - 12 \qquad \text{Use a distributive law.}$$

$$y = \frac{1}{4}x + 9$$

The equation is $y = \frac{1}{4}x + 9$. ■

Notice that the cost, y, depends on the number of square feet of plywood, x. Therefore, y is called the *dependent variable* and x is called the *independent variable*.

□ DO EXERCISE 3.

4 Parallel Lines

Vertical lines are parallel. Their slopes are undefined. Two distinct lines that are not vertical are parallel if they have the same slope. Similarly, if two lines are parallel, they have the same slope. Distinct lines with the same slope never intersect.

Two distinct lines that have the same slope are parallel.

□ **Exercise 2** Find equations for the following lines.

a. $m = 4$, through $(-2, 6)$

b. $m = -\frac{5}{8}$, through (3, 1)

c. Through $(-2, 3)$ and (4, 6)

d. Through $(-2, -4)$ and $(-3, 10)$

□ **Exercise 3** The relationship between the pressure in pounds per square inch (psi) and the depth in feet below the surface of the ocean is linear. If the pressure at 33 feet is 30 psi and the pressure at 110 feet is 65 psi, write an equation relating pressure y to depth d.

□ **Exercise 4** Find an equation for the line through the given point and parallel to the given line.

a. $(3, 5)$, $3x + 2y = 8$

b. $(4, -2)$, $5x - 7y = 10$

c. $(0, -3)$, $x - 5y = 3$

d. $(-4, 1)$, $2x + 9y = 8$

EXAMPLE 4 Find an equation for the line through the point $(-3, 5)$ and parallel to $4x + 3y = 7$.

1. Find the slope of the given line. To do this, write the equation in slope-intercept form, $y = mx + b$.

$$4x + 3y = 7$$

$$3y = -4x + 7 \qquad \text{Add } -4x \text{ to each side.}$$

$$y = -\frac{4}{3}x + \frac{7}{3} \qquad \text{Divide both sides by 3.}$$

The slope is the coefficient of x. Therefore, the slope of the given line and the parallel line is

$$m = -\frac{4}{3}$$

2. Use the point-slope form to write the equation of the line through the point $(-3, 5)$ with slope $-\frac{4}{3}$.

$$y - y_1 = m(x - x_1) \qquad \text{Point-slope form.}$$

$$y - 5 = -\frac{4}{3}[x - (-3)] \qquad \text{Substitute for } m, x, \text{ and } y_1.$$

$$y - 5 = -\frac{4}{3}(x + 3)$$

$$y - 5 = -\frac{4}{3}x - 4 \qquad \text{Use a distributive law.}$$

$$y = -\frac{4}{3}x + 1 \qquad \text{Add 5 to each side.}$$

The graph of the lines follows.

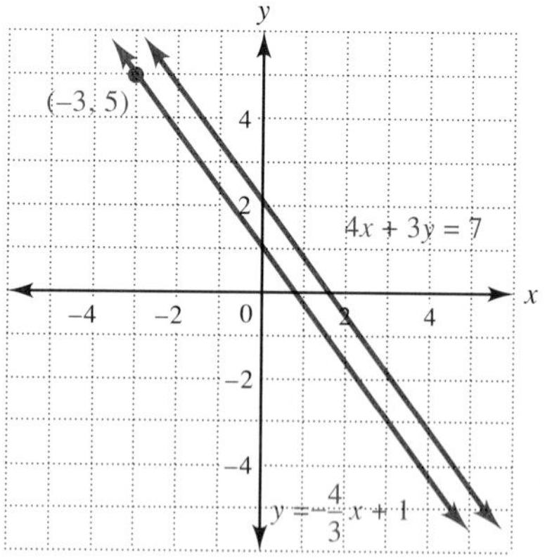

■

□ **DO EXERCISE 4.**

5 Perpendicular Lines

A vertical line (slope undefined) and a horizontal line (slope zero) are perpendicular. We can tell from the slope whether any two other lines are perpendicular. Consider a line L_1 with slope b/a. Then rotate the line 90° to get a new line L_2. The vertical change and the horizontal change are reversed for the new line and the vertical change is negative. Therefore, the slope of the perpendicular line is $-a/b$ (the negative reciprocal of b/a).

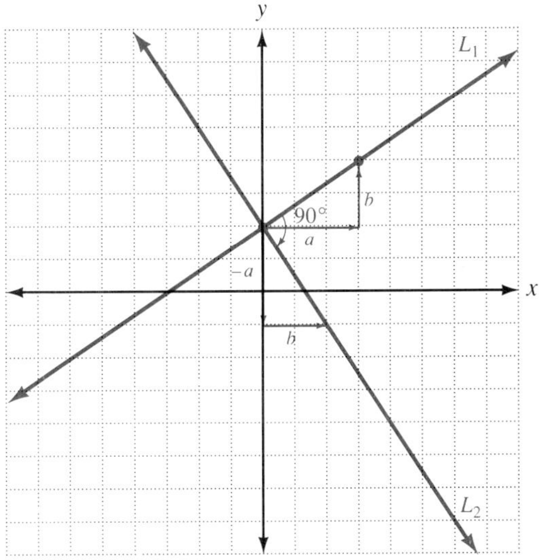

If the slope of one line is the negative reciprocal of the slope of another line, the lines are ***perpendicular.*** A line with undefined slope is perpendicular to a line with zero slope.

Notice that this also means that two lines are perpendicular if the product of their slopes is -1.

EXAMPLE 5 Find an equation for the line perpendicular to $3x + 4y = 7$ and passing through the point (2, 5).

1. Find the slope of the given line. To do this, write the equation in slope-intercept form, $y = mx + b$.

$$3x + 4y = 7$$

$$4y = -3x + 7 \qquad \text{Add } -3x \text{ to both sides.}$$

$$y = -\frac{3}{4}x + \frac{7}{4} \qquad \text{Divide by 4.}$$

The slope of the given line is the coefficient of x, which is $-\frac{3}{4}$. The slope of the perpendicular line is the negative reciprocal of $-\frac{3}{4}$, which is $-(4/-3)$ or $\frac{4}{3}$.

2. Write an equation for the perpendicular line that has a slope of $\frac{4}{3}$ and passes through the point (2, 5). Use the point-slope form.

$$y - y_1 = m(x - x_1) \qquad \text{Point-slope form.}$$

$$y - 5 = \frac{4}{3}(x - 2) \qquad \text{Substitute for } m, x, \text{ and } y_1.$$

$$y - 5 = \frac{4}{3}x - \frac{8}{3} \qquad \text{Use a distributive law.}$$

$$y = \frac{4}{3}x + \frac{7}{3} \qquad \text{Add 5 to each side.}$$

□ **Exercise 5** Find an equation for the line through the given point and perpendicular to the given line.

a. $(2, 7)$, $2x + 3y = 8$

b. $(-4, 5)$, $6x - 5y = 11$

c. $(3, -1)$, $3x + 4y = 9$

d. $(3, 6)$, $x + y = 7$

The graph is as follows:

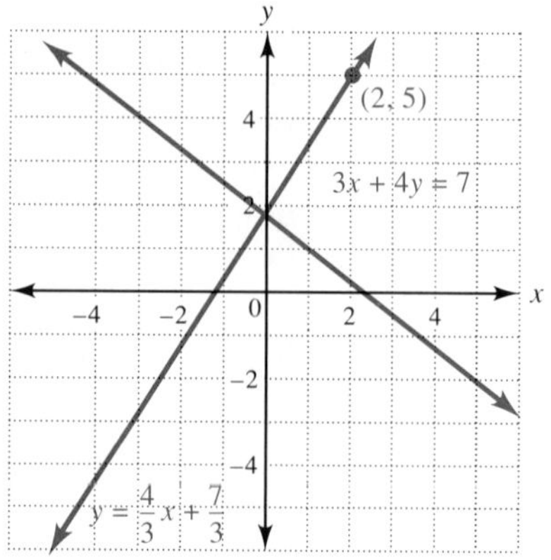

■

□ **DO EXERCISE 5.**

Some Forms of Linear Equations

$x = a$	Vertical line through the point $(a, 0)$. Slope is undefined.
$y = b$	Horizontal line through the point $(0, b)$. Slope is 0.
$y = mx + b$	Slope-intercept form with slope m, and y-intercept $(0, b)$.
$y - y_1 = m(x - x_1)$	Point-slope form with slope m; passes through the point (x_1, y_1).
$Ax + By = C$	Standard form where A, B, and C are constants and A and B are not both zero.

Answers to Exercises

1. a. $y = 3x - 2$ **b.** $y = -\frac{3}{4}x + \frac{1}{8}$ **c.** $y = -\frac{2}{5}x$ **d.** $y = 6$

2. a. $y = 4x + 14$ **b.** $y = -\frac{5}{8}x + \frac{23}{8}$ **c.** $y = \frac{1}{2}x + 4$
d. $y = -14x - 32$

3. $y = \frac{5}{11}d + \frac{165}{11}$

4. a. $y = -\frac{3}{2}x + \frac{19}{2}$ **b.** $y = \frac{5}{7}x - \frac{34}{7}$ **c.** $y = \frac{1}{5}x - 3$
d. $y = -\frac{2}{9}x + \frac{1}{9}$

5. a. $y = \frac{3}{2}x + 4$ **b.** $y = -\frac{5}{6}x + \frac{5}{3}$ **c.** $y = \frac{4}{3}x - 5$
d. $y = x + 3$

Problem Set 3.4

NAME ______________________

DATE ______________________

CLASS ______________________

Find an equation for the line with the given slope and y-intercept. Use the slope-intercept form.

1. $m = 4, b = 1$

2. $m = -3, b = 7$

3. $m = -\frac{5}{8}, b = \frac{1}{4}$

4. $m = -\frac{3}{5}, b = \frac{1}{10}$

5. $m = \frac{1}{3}, b = 5$

6. $m = \frac{5}{6}, b = -3$

7. $m = 0, b = 3$

8. $m = 0, b = -4$

9. $m = 3.4, b = -7.8$

10. $m = -2.1, b = -0.3$

Find an equation for the line with the given slope through the given point. Use the point-slope form.

11. $m = 3, (4, 1)$

12. $m = 5, (2, 3)$

13. $m = -2, (3, 8)$

14. $m = -4, (2, 7)$

15. $m = -5, (-2, 4)$

16. $m = -3, (-1, 6)$

17. $m = -\frac{3}{7}, (1, -5)$

18. $m = \frac{8}{5}, (-4, 2)$

19. $m = 0, (-3, -2)$

20. $m = 0, (-7, -9)$

Find an equation for the line through each pair of points.

21. $(5, 6), (7, 9)$

22. $(3, 5), (8, 6)$

23. $(3, -1), (4, -2)$

24. $(-6, 4), (-7, 9)$

25. $(-5, -2), (6, 6)$

26. $(5, 8), (-3, -4)$

27. $(-4, -4), (-8, -7)$

28. $(-6, -5), (-1, -1)$

29. $(-2, 5), (-3, 5)$

30. $(4, -7), (-3, -7)$

31. For the Big Company, the relationship between the number of units sold and the profit is linear. If the profit is \$500 when 300 units are sold and \$3500 when 900 units are sold, write an equation relating profit y to units sold x.

32. If the temperature x feet above the surface of the earth is $y°$ Celsius and the relationship between x and y is linear, find an equation that relates y to x given that the temperature on the surface of the earth is 20° Celsius and the temperature at 900 feet is 11° Celsius.

33. At a constant speed, the relationship between the distance an airplane travels and the time it travels is linear. If it travels 600 miles in 2 hours and 1350 miles in $4\frac{1}{2}$ hours, write an equation relating distance traveled d to time t.

34. If Bill sells \$260 worth of cleaning supplies, he earns \$65. If his sales are \$820, he earns \$205. The relationship between his sales and earnings is linear. Write an equation that relates his earnings E to his sales S.

35. The relationship between the value of a car and the number of years it has been depreciated for income tax purposes is linear. If the value of the car after 2 years was \$9600 and the value of the car after 4 years was \$7200, write an equation that relates value V to the number of years t that it has been depreciated.

36. For income tax purposes, Mrs. Redkin computes the depreciation of her rental house using straight-line depreciation. The value of the house after 4 years is \$64,000 and the value of the house after 7 years is \$52,000. Write an equation that relates value V to the number of years t that it has been depreciated.

37. At Sound Company, the cost of manufacturing 150 radios is \$5750 and the cost of manufacturing 300 radios is \$9500. If the relationship between the cost and the number of radios manufactured is linear, write an equation that relates the cost C to the number of radios r that are manufactured. Use of the equation to find the cost of manufacturing 400 radios.

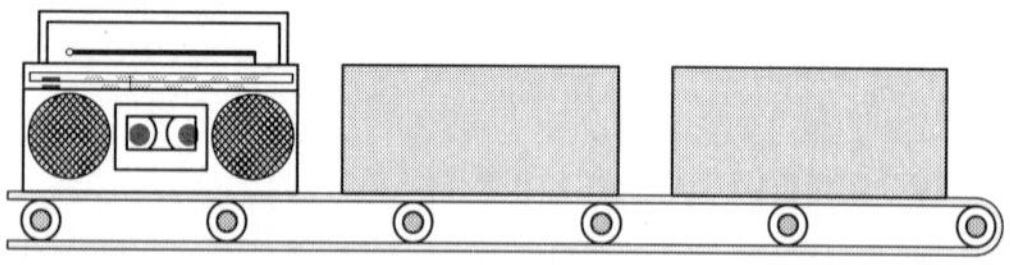

38. Samson rented a car for one day and drove it for 200 miles. The cost was \$45. Another day, he rented the same car and drove it 450 miles. The cost was \$82.50. The relationship between the cost of the car and the number of miles driven is linear. Write an equation that relates the cost C to the number of miles driven d. How much would it cost Samson to drive the car 300 miles?

The price of a product and the demand for that product may be expressed as a linear relationship. The graph of this relationship is called a demand curve.

39. At Shaw's Office Supply, if the cost of calculators is \$30, two calculators will be purchased. If the cost is \$10, four calculators will be purchased. Write an equation that relates the demand d to the cost of the calculator c. How many calculators will be purchased if the cost is \$40?

40. At Al's Tire Store, the demand for Goodo tires is 6 if the price is \$64 and 4 if the price is \$80. Write an equation that relates the demand d to the cost of the tire c. How many tires will be purchased if the cost is \$96?

Find an equation for the line through the given point and parallel to the given line.

41. $(2, 5)$, $4x - 3y = 9$

42. $(3, -2)$, $5x - 3y = 7$

43. $(0, -4)$, $x + 4y = 5$

44. $(-2, -1)$, $3x + y = 6$

45. $(-1, -4)$, $5x - y = 8$

46. $(-3, 2)$, $2x - y = 12$

47. $(-5, 4)$, $6x - 2y = 5$

48. $(-2, 1)$, $3x - 9y = 8$

Find an equation for the line through the given point and perpendicular to the given line.

49. $(3, 5)$, $3x + 2y = 7$

50. $(-3, 2)$, $5x - 4y = 8$

51. $(-1, -2)$, $6x - y = 9$

52. $(1, -4)$, $7x + y = 10$

53. $(-4, -2)$, $x - 5y = 20$

54. $(-6, -1)$, $-x + 3y = 15$

55. $(5, -6)$, $4x - y = 37$

56. $(-3, 4)$, $7x - y = 42$

57. Find an equation for the line that has an x-intercept of 3 and is parallel to the line $3x - 2y = 8$.

58. Find an equation for the line that has a y-intercept of -2 and is perpendicular to the line $3x + y = 7$.

59. Find an equation for the line through the point $(-2, 4)$ and perpendicular to the line through the points $(5, 7)$ and $(-1, 3)$.

60. Find an equation for the line through the point $(0, 8)$ and parallel to the line through the points $(-2, -4)$ and $(1, -5)$.

61. The ____________ form of the equation of a line may be used to write an equation of a line if we are given any two points on the line.

62. Two distinct lines that have the same slope are ____________.

63. If the slope of one line is the ____________ ____________ of the slope of another line, the lines are perpendicular.

Find an equation for the line through each pair of points.

64. $(-1.029, 7.184)$, $(3.004, -4.915)$

65. $(9.729, -6.138)$, $(-6.195, -2.157)$

Think About It

* **66.** A four-sided figure with opposite sides parallel is a parallelogram. Show that the points $(-4, -1)$, $(0, 2)$, $(-2, -1)$, and $(2, 2)$ are vertices of a parallelogram.

* **67.** Show that the points $(-2, 4)$, $(0, 8)$, and $(2, 2)$ are the vertices of a right triangle.

* **68.** Find a value of k so that the line $2x + ky + 2 = 0$ is parallel to $4x - 5y + 3 = 0$.

* **69.** Show that the equation $3x + 4y = 12$ can be written in the form $x/4 + y/3 = 1$. What are the x- and y-intercepts of this line?

Checkup

These problems provide a review of some of Sections 2.6 and 3.2 and will help you with the next section.

Solve.

70. $-3(x - 2) \leq -2x + 4$

71. $5x - 2 > 3(x + 8)$

72. $-4 < 3x + 5 \leq 11$

73. $3 \leq -2x + 7 \leq 5$

74. $-2x + 3 > 5$ or $3x > 15$

75. $4x \leq -8$ or $3x - 2 > 4$

Graph using intercepts.

76. $3x - y = 3$

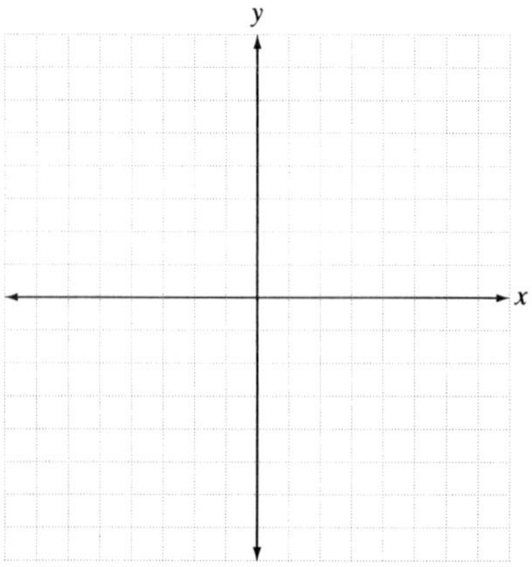

77. $y = x + 4$

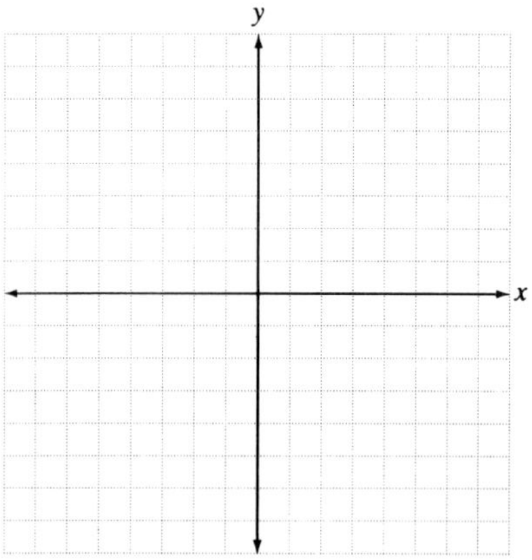

78. $3x - 2y = 6$

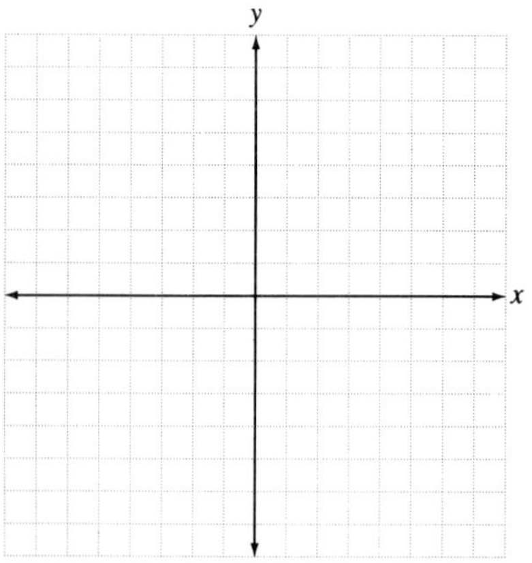

79. $3x + 5y = 15$

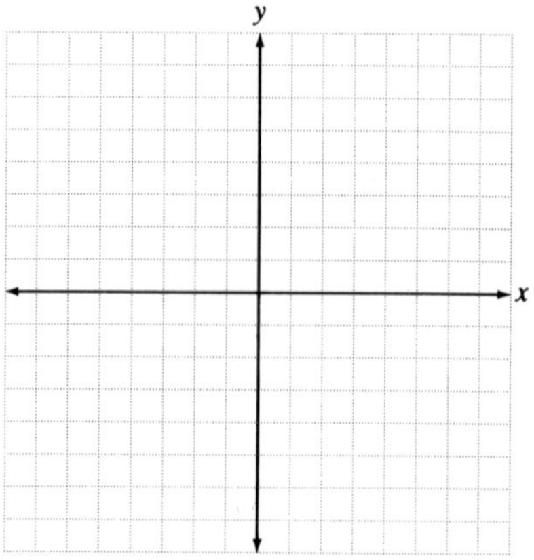

3.5 LINEAR INEQUALITIES IN TWO VARIABLES

Linear inequalities occur in the practical applications of mathematics, especially in the allocation of limited resources. A linear inequality is an expression of the form

$$Ax + By < C$$

where A and B are not both 0 and A, B, and C are real numbers. The symbol $<$ may be replaced with $>$, $\leq$, or $\geq$. Examples of linear inequalities are

$$3x + 4y < 7 \qquad 2y - 3x - 5 \geq 0 \qquad y \leq 4x + 6$$

1 Solutions of Linear Inequalities

There are many *solutions* to inequalities in two variables. The solutions are ordered pairs of real numbers.

EXAMPLE 1 Is the ordered pair $(-5, 3)$ a solution of $2x - 5y > 5$?

Since the ordered pair is $(-5, 3)$, $x = -5$ and $y = 3$.

$2x - 5y$	> 5
$2(-5) - 5(3)$	5
$-10 - 15$	5
-25	5

Since $-25 > 5$ is false, $(-5, 3)$ is not a solution. ■

□ DO EXERCISE 1.

2 Graphing Linear Inequalities in Two Variables

The following are steps for graphing a linear inequality.

1. Sketch the graph of the equation formed by replacing the inequality symbol with an equal sign. This is the boundary line. The boundary line will be *solid* if the inequality symbol is $\geq$ or $\leq$. This means that points on the boundary line are solutions. The boundary line will be *dashed* if the inequality symbol is $>$ or $<$. This means that points on the line are not solutions.
2. Choose a test point that is not on the line and substitute its coordinates into the *inequality*. The point $(0, 0)$ is an easy one to use if the boundary line does not pass through it. If the inequality is true for the chosen point, shade the side of the boundary line containing it. If the inequality is false for the point, shade the other side of the line. *All points* in the shaded region are solutions to the inequality.

OBJECTIVES

1 *Determine if a given ordered pair is a solution of an inequality*

2 *Graph linear inequalities in two variables*

3 *Graph linear inequalities in one variable on the rectangular coordinate system*

□ **Exercise 1** Is the ordered pair a solution of $3x - 4y < 7$?

a. $(2, 3)$

b. $(5, 1)$

EXAMPLE 2 Graph.

a. $3x - 4y < 12$.

1. Graph the line $3x - 4y = 12$. The easiest method of graphing the line is to use intercepts. The line is dashed because the inequality symbol is $<$.
2. Decide which side of the boundary line to shade by choosing a point not on the line and substituting its coordinates into the inequality to see if the inequality is true or false for this point. We chose (0, 0) since the boundary line does not pass through it.

$3x - 4y$	< 12	
$3(0) - 4(0)$	12	Substitute $x = 0$ and $y = 0$.
0	12	

3. Since $0 < 12$ is true, we shade the side of the line containing the point (0, 0). All points in the shaded region are solutions. Points on the line are *not* solutions. We cannot check all points in the

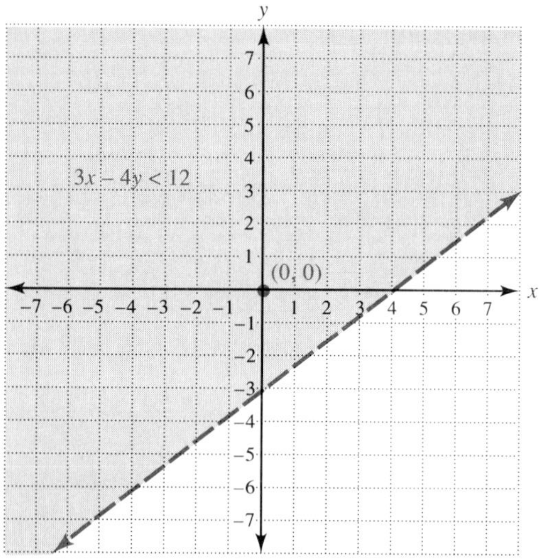

shaded region. We can do a partial check. Notice that the point (2, 2) is in the shaded region. It is a solution. The point (5, −2) is not in the shaded region. It is not a solution.

Let $x = 2$ and $y = 2$

$3x - 4y$	< 12
$3(2) - 4(2)$	12
$6 - 8$	12
-2	12

Since $-2 < 12$ is true, the point (2, 2) is a solution.

Let $x = 5$ and $y = -2$

$3x - 4y$	< 12
$3(5) - 4(-2)$	12
$15 + 8$	12
23	12

Since $23 < 12$ is false, the point (5, −2) is not a solution.

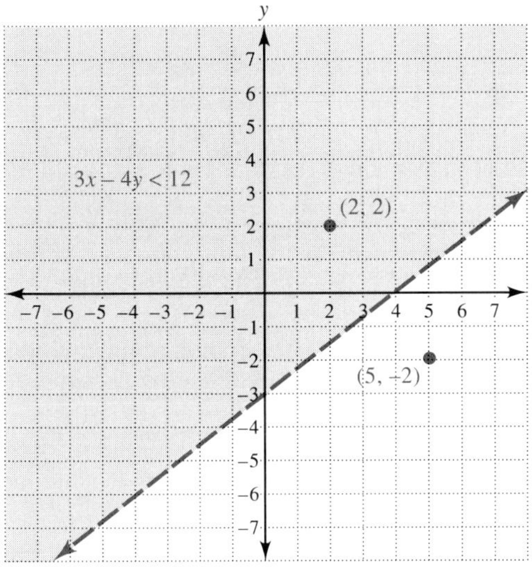

b. $y \geq 5x$

1. Graph the line $y = 5x$. The easiest method to use is the slope-intercept method shown in Section 3.3. The line is solid since the inequality symbol is $\geq$.
2. We cannot use the point (0, 0) as a test point since the boundary line passes through it. We chose the point (2, 0). The inequality is false for the point (2, 0) since, substituting in $y \geq 5x$, $0 \geq 5(2)$ or $0 \geq 10$ is false. We shade the side of the boundary line that does not contain the point (2, 0). All points in the shaded region and on the line are solutions.

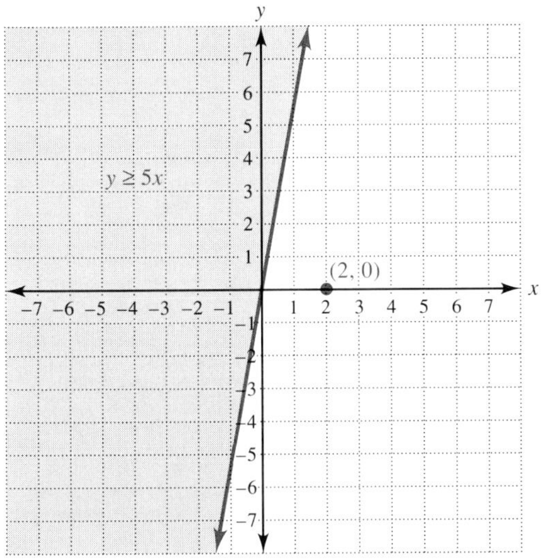

■

☐ **DO EXERCISE 2.**

☐ **Exercise 2** Graph.

a. $y < -4x$

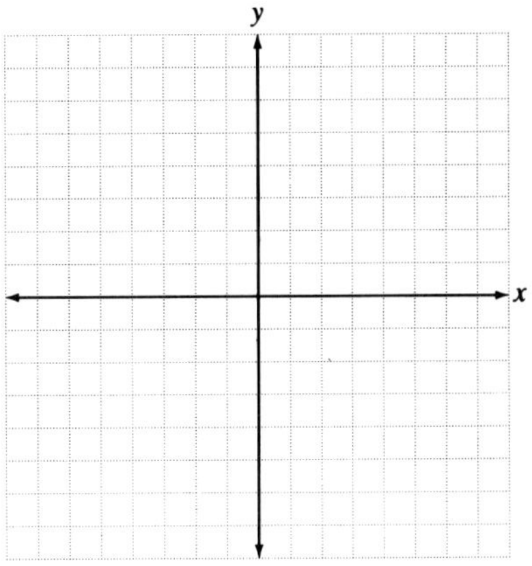

b. $2x - 5y \geq 10$

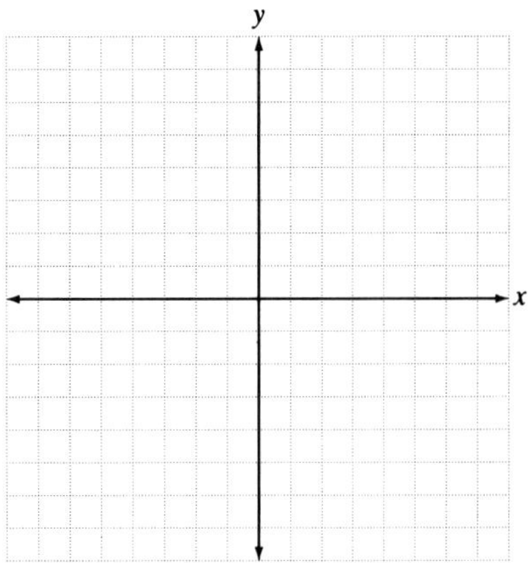

□ **Exercise 3** Graph.

a. $x \leq -2$

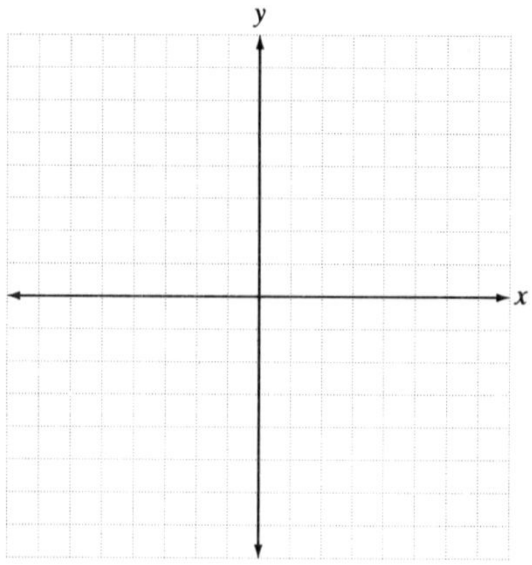

b. $y > 3$

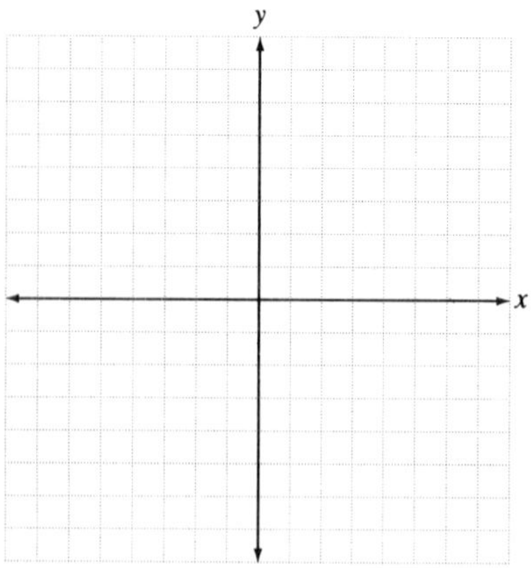

3 Graphing Linear Inequalities in One Variable

We may also graph linear inequalities in one variable on the rectangular coordinate system.

EXAMPLE 3 Graph $x > 2$.

1. Graph $x = 2$ using a dashed line.
2. We can check the point (0, 0). The inequality is false for the point (0, 0), since when we substitute $x = 0$ into $x > 2$, we get $0 > 2$, which is false. We shade the side of the boundary line that does not contain (0, 0). Notice that the shaded region consists of all points to the right of $x = 2$. All points to the right of $x = 2$ are solutions.

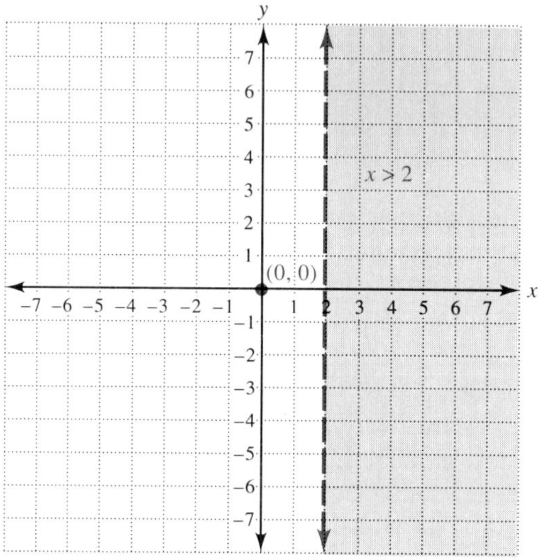

■

□ **DO EXERCISE 3.**

We find the intersection of two graphs by finding the region common to both graphs.

EXAMPLE 4 Graph $-3 \leq x < 4$.

1. Rewrite $-3 \leq x < 4$ as

$$x \geq -3 \quad \text{and} \quad x < 4$$

2. Graph $x = -3$. The graph of the inequality $x \geq -3$ is the line $x = -3$ and the shaded region to the right of the line $x = -3$.

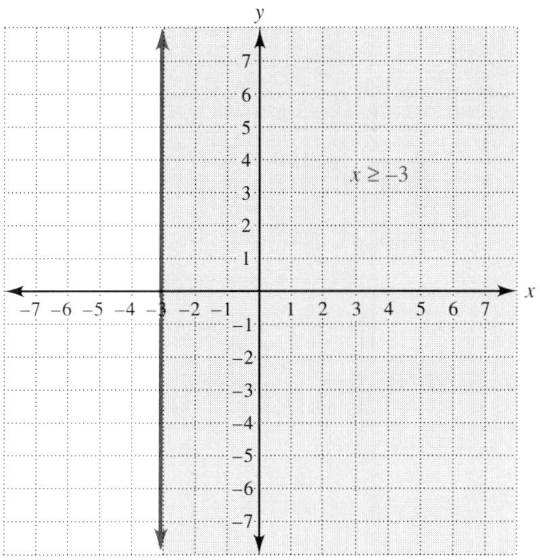

3. Graph $x < 4$. The graph of this inequality is the shaded region to the left of the line $x = 4$.

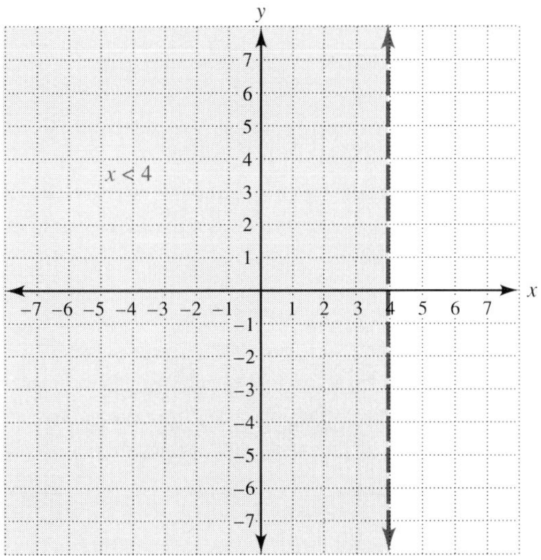

□ **Exercise 4** Graph.

a. $-2 < x \leq 3$

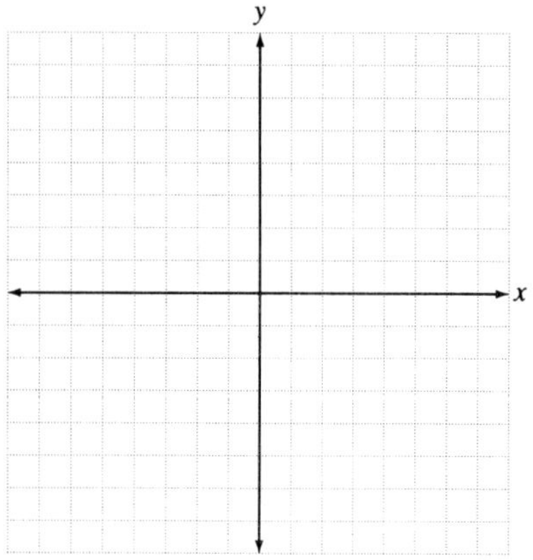

b. $-4 < y < 2$

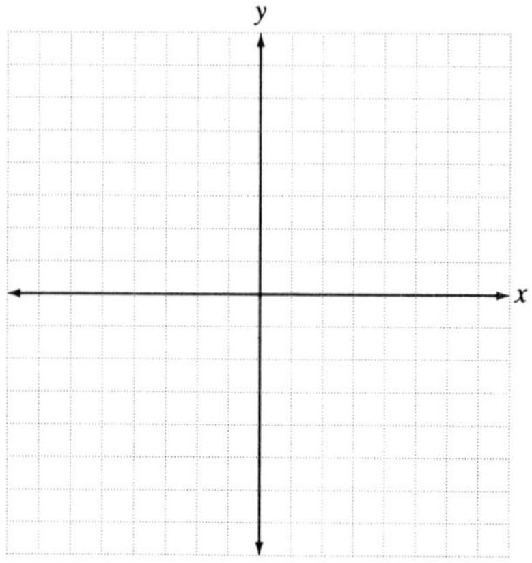

4. The solution is the region common to both of the preceding graphs. This is the intersection of the two graphs.

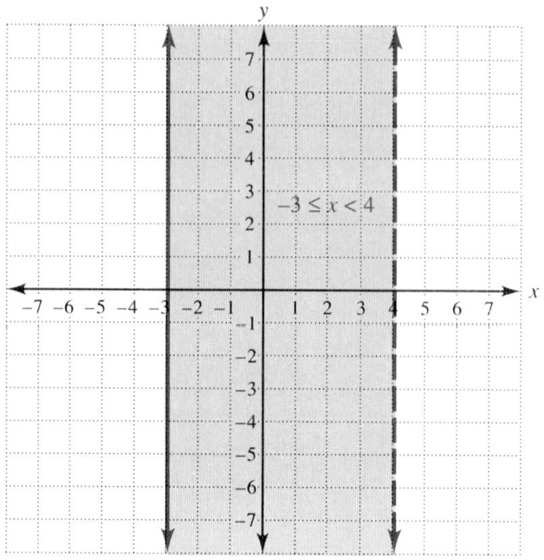

□ **DO EXERCISE 4.**

Answers to Exercises

1. a. Yes **b.** No

2. a.

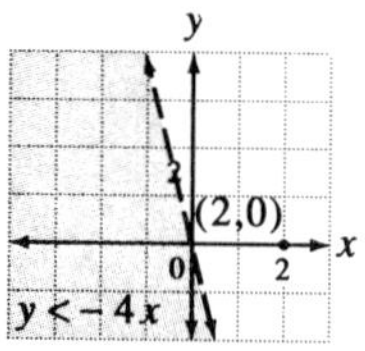

b.

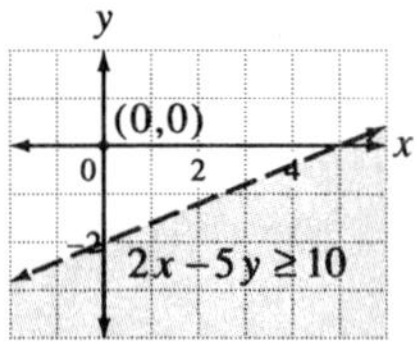

3. a.

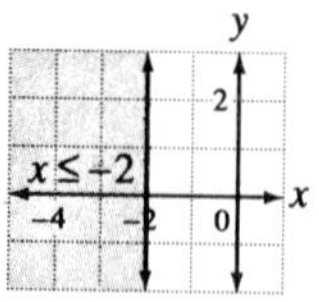

b.

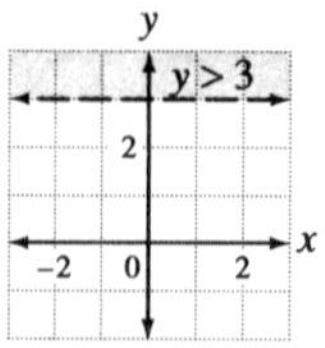

4. a.

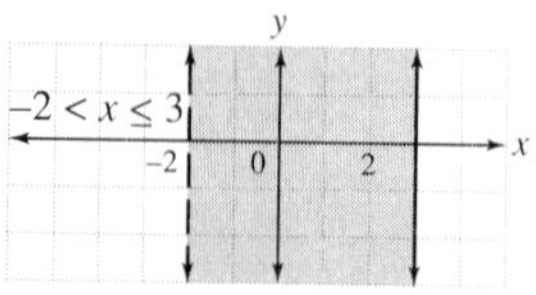

b.

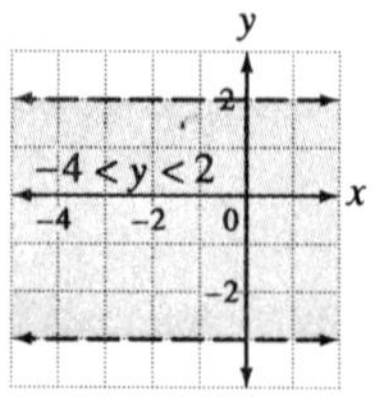

Problem Set 3.5

NAME ______________________

DATE ______________________

CLASS ______________________

Determine if the ordered pair is a solution of the inequality.

1. $(2, -3)$, $3x + 5y < 3$

2. $(3, -4)$, $4x + 2y < 3$

3. $(2, 5)$, $6x - 3y \geq -2$

4. $(6, 2)$, $2x - 4y \geq 2$

Graph on the rectangular coordinate system.

5. $x + y < 3$

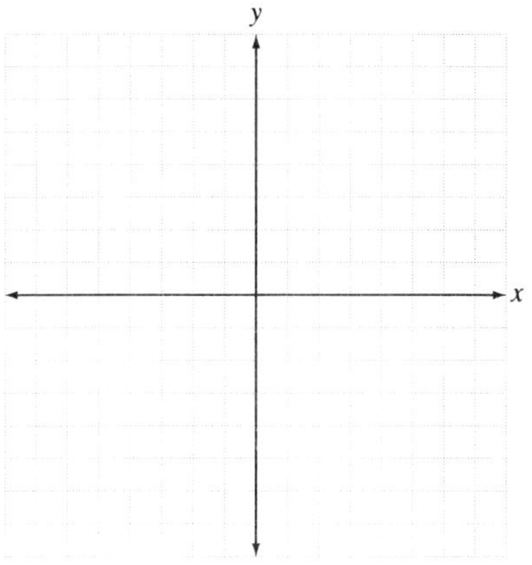

6. $x - y > 4$

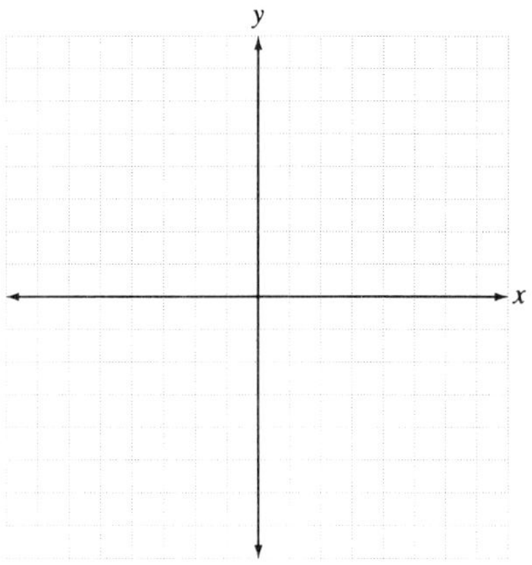

7. $y \geq x + 2$

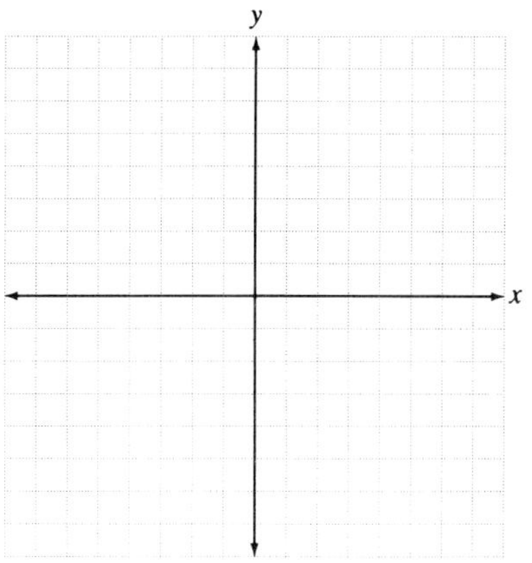

8. $y \leq x - 3$

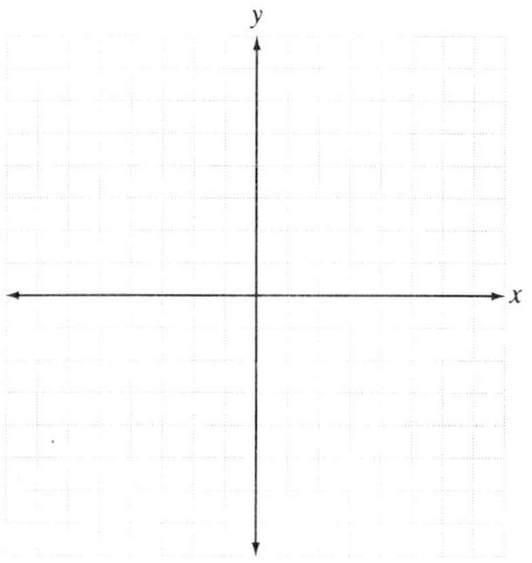

9. $3x - 5y < 15$

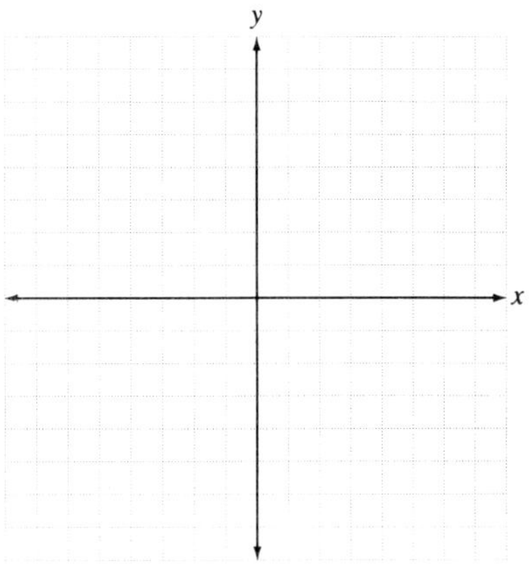

10. $3x + 2y > 6$

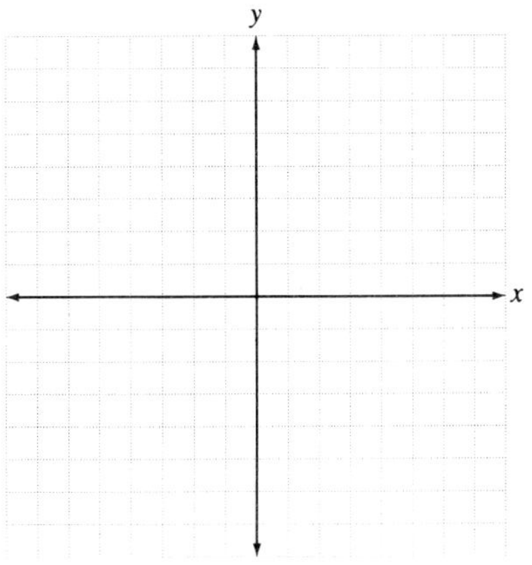

11. $5x - 2y \geq 10$

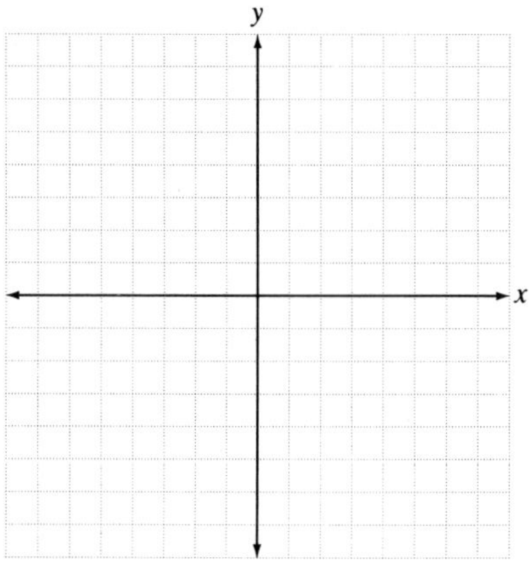

12. $3x + 4y \leq 12$

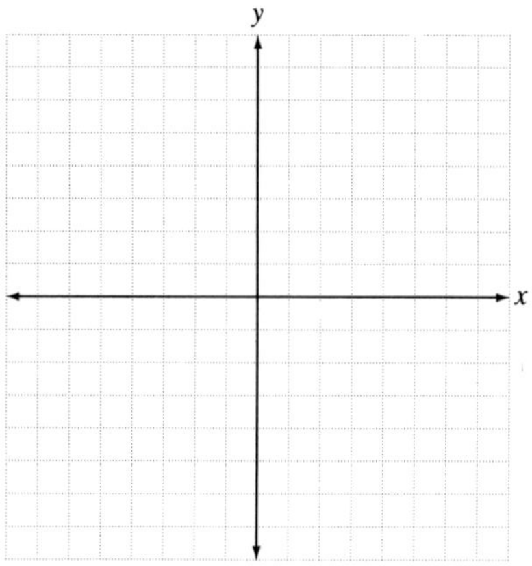

13. $2x + 3y < 9$

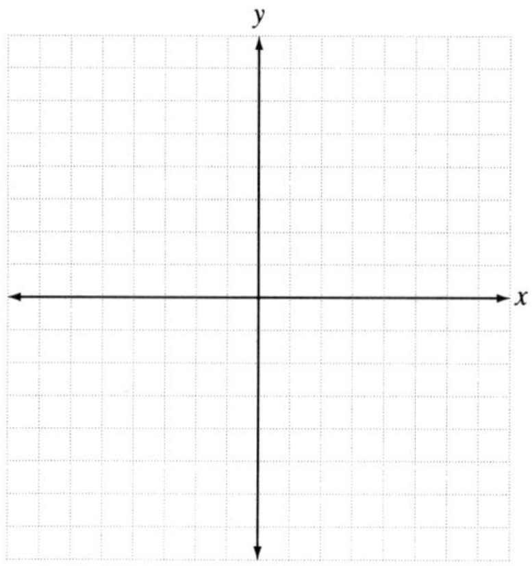

14. $4x - 5y \geq 10$

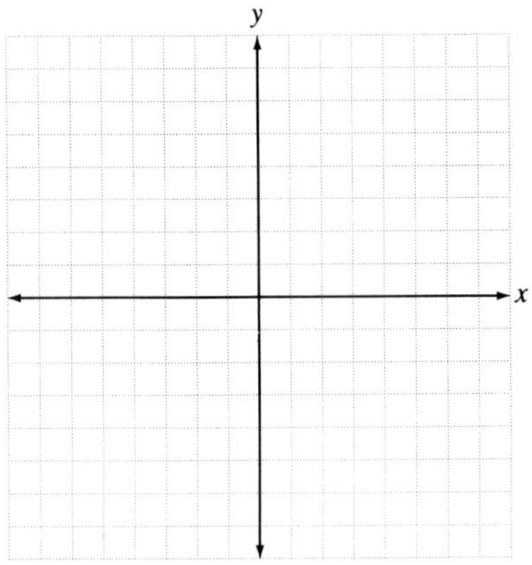

15. $y \geq 3x$

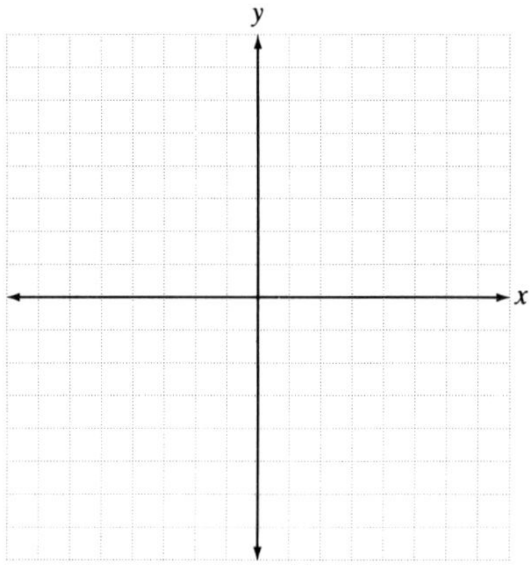

16. $y \leq -5x$

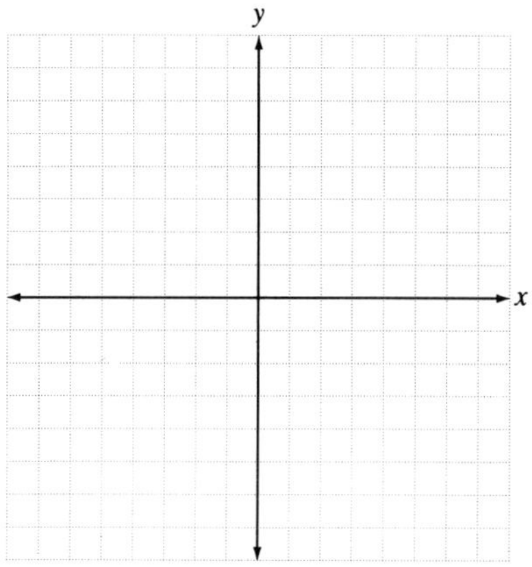

17. $x \geq -1$

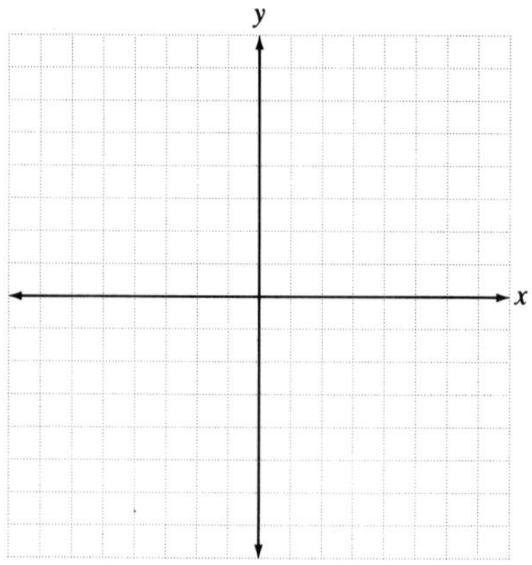

18. $y < 2$

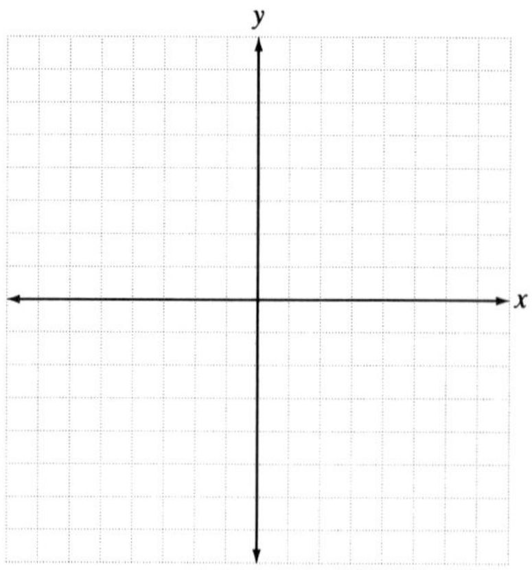

19. $-1 < x < 4$

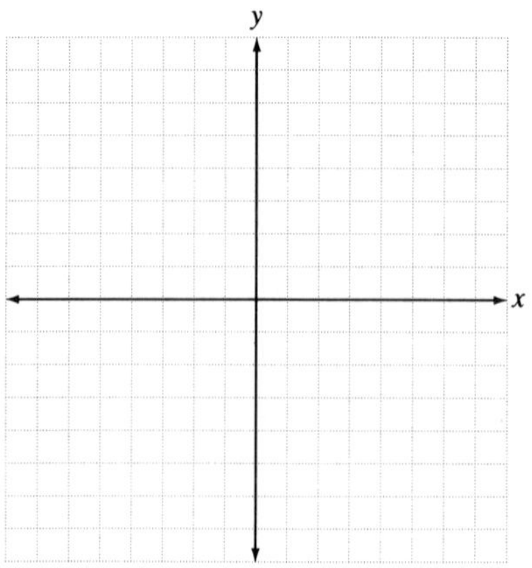

20. $-3 < y < 2$

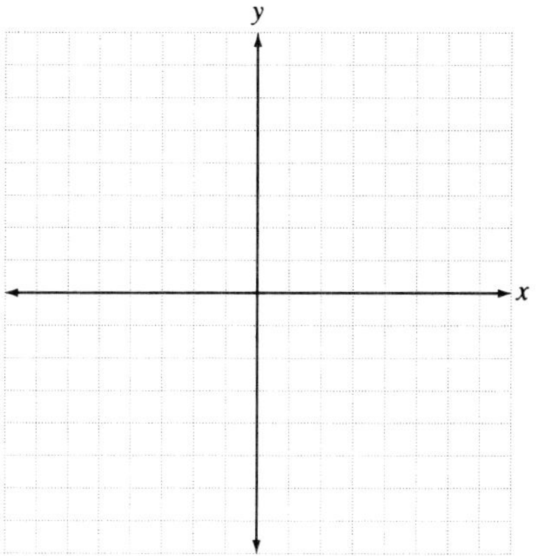

21. $y \geq 3$

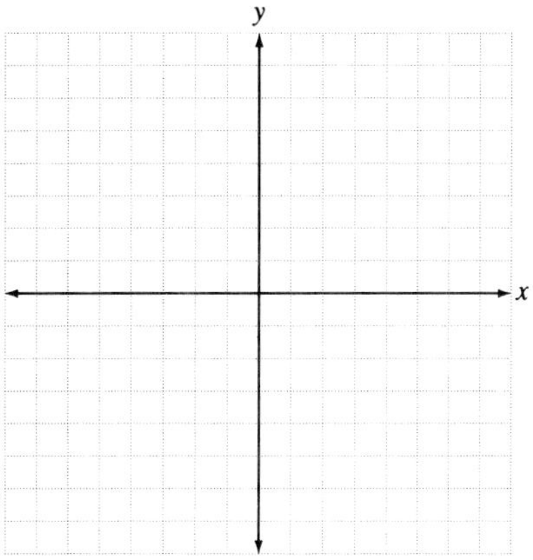

22. $x < 2$

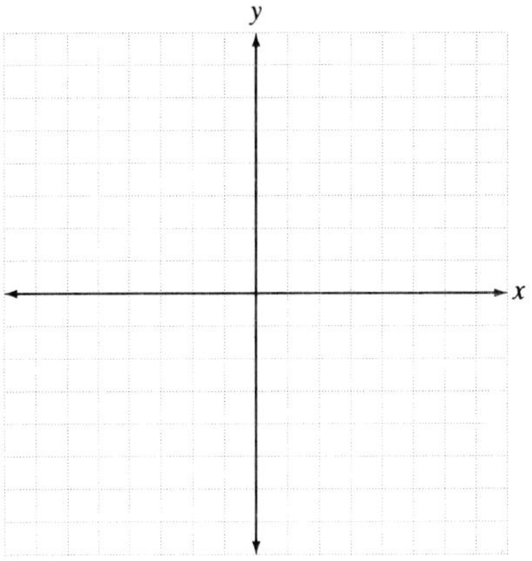

23. $1 \leq y < 4$

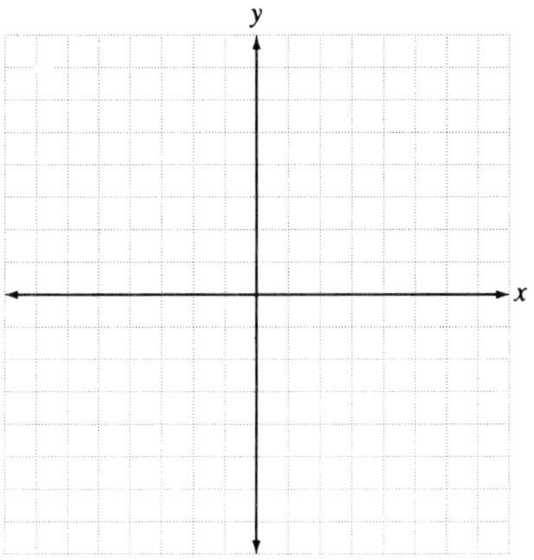

24. $-2 \leq x \leq 3$

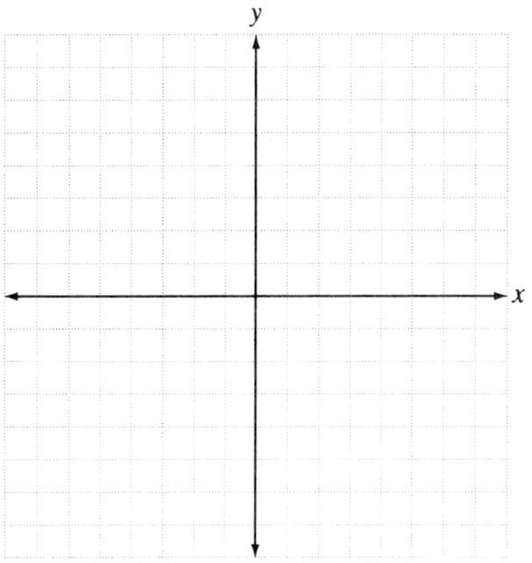

25. $5x + 5y > 2y + 15$

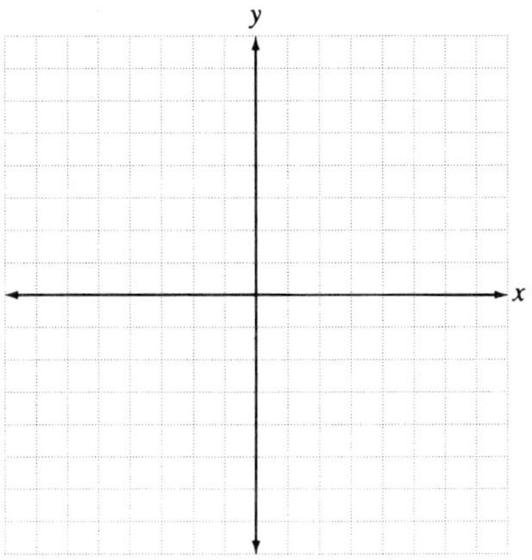

26. $x - y \leq 4 + y$

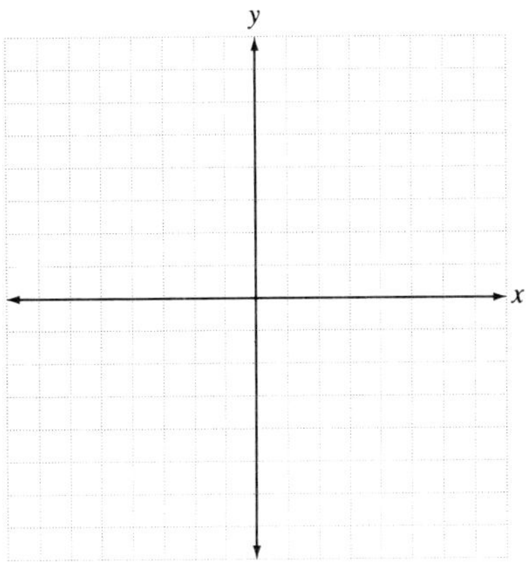

The solution of a system of two linear inequalities joined by the word "and" is all points that satisfy both inequalities. If we graph each inequality on the same axes, the solution is the intersection of the graphs of each inequality. Graph the solution of the following systems of inequalities.

27. $-x - y < 5$ and $x - y \leq 3$

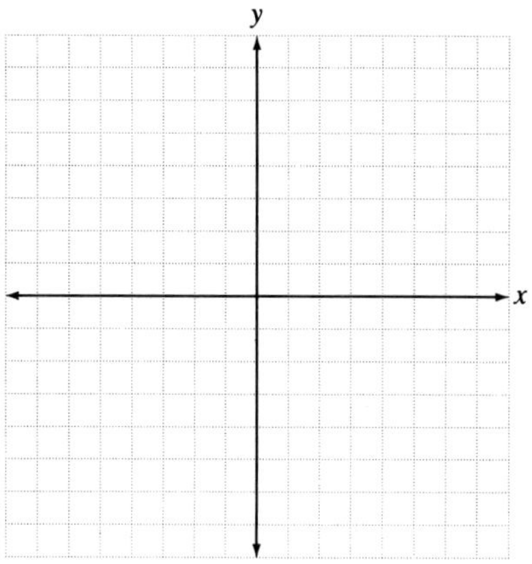

28. $3x - 2y < 4$ and $x + 2y \geq 4$

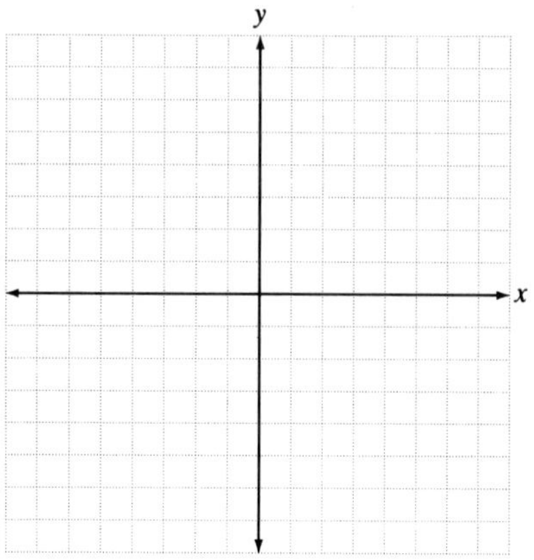

29. $5x - 2y + 10 \leq 0$ and $x \leq 2 - y$

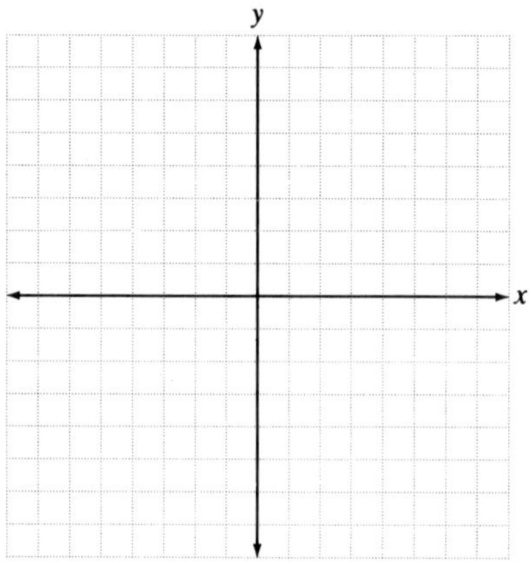

30. $x > -5 + y$ and $3x + 4y > 12$

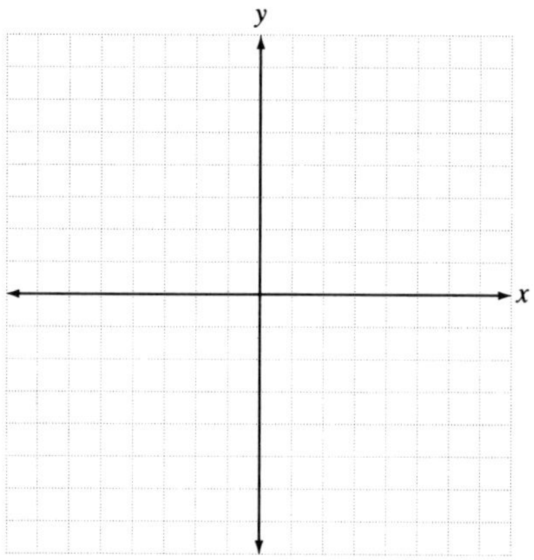

31. $x + y \leq 3$ and $x - y \leq 1$

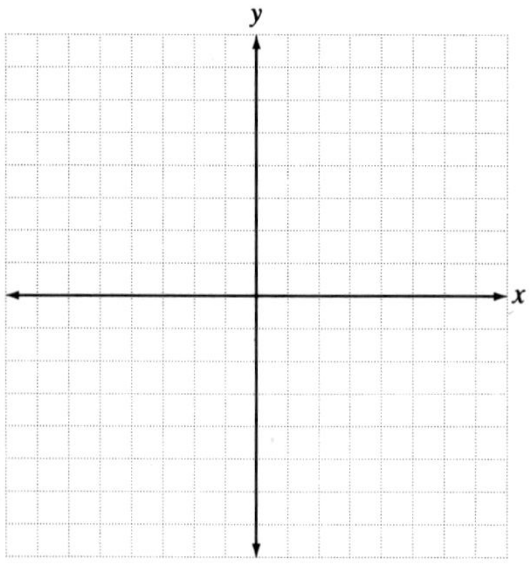

32. $x - y \leq 4$ and $x \geq -2$

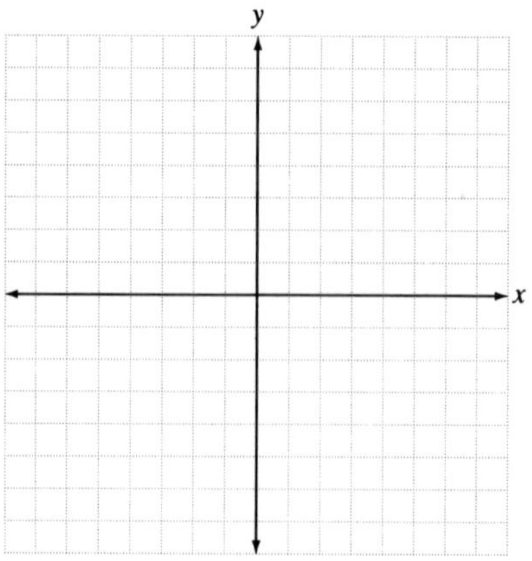

33. $2x + y > 5$ and $x + 2y > 4$

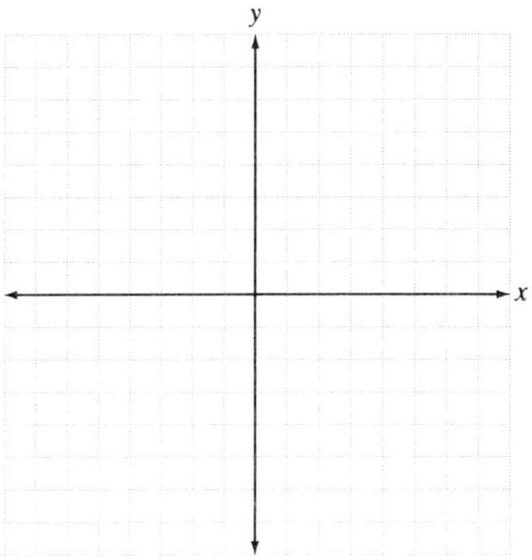

34. $3x + 2y < 6$ and $x - 2y > 2$

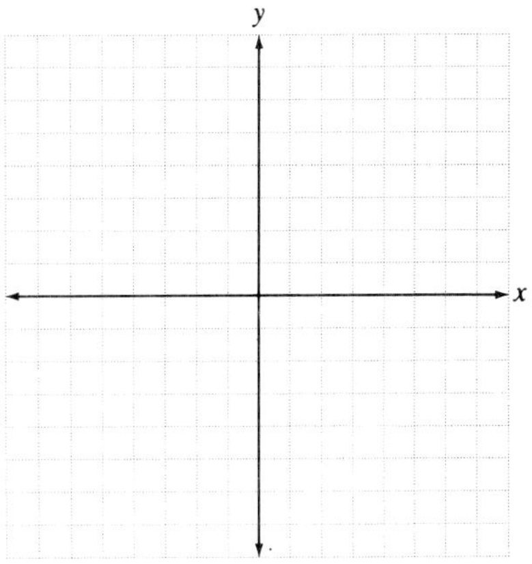

35. The solutions to linear inequalities in two variables are ____________ ____________ of numbers.

36. In the graph of a linear inequality in two variables, the boundary line will be ____________ if the inequality symbol is $<$ or $>$.

37. To decide which side of the boundary line to shade, choose a test point and substitute its coordinates into the ____________.

38. The point ____________ is an easy point to use as a test point if the boundary line does not pass through it.

Is the ordered pair a solution of the inequality?

39. $(2.2, 1.7)$; $2.34x - 4.19y > -1.675$

40. $(-4.2, 0.3)$; $1.27x + 9.84y > -2.905$

41. $(-3.3, -1.9)$; $0.78x - 3.2y \geq 3.506$

42. $(4.5, -2.7)$; $1.79x - 3.48y \leq 19.718$

* **43.** $\left(-\frac{3}{5}, \frac{4}{9}\right)$, $\frac{10}{9}x + \frac{3}{4}y \leq -\frac{1}{3}$

* **44.** $\left(-\frac{11}{8}, -\frac{5}{2}\right)$, $\frac{8}{7}x - \frac{2}{3}y > \frac{5}{21}$

Graph.

* **45.** $|x| > 3$

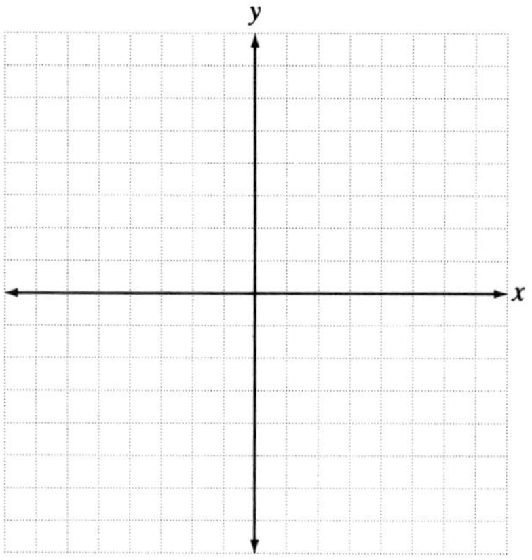

* **46.** $|y| \geq -1$

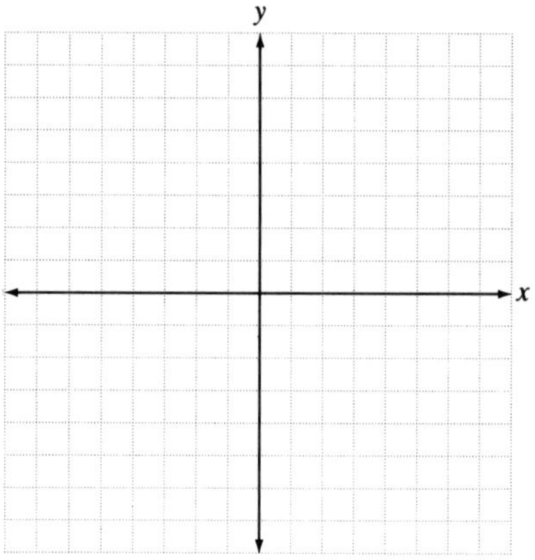

* **47.** $|x + 2| \leq 3$

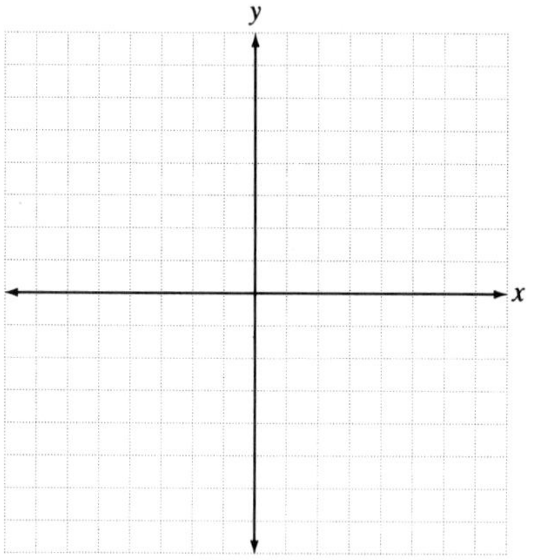

* **48.** $|y - 1| > 2$

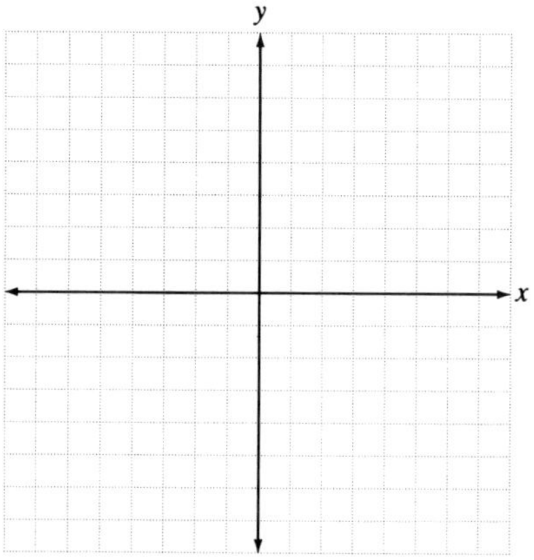

Checkup

The following problems provide a review of Section 2.7.

Solve.

49. $|x + 3| = 9$

50. $|2x - 3| = 7$

51. $|3x + 4| \leq 5$

52. $|2x - 7| > 4$

53. $|5x - 2| > 12$

54. $|x - 7| \leq 2$

55. $|3x - 2| = 7$

56. $|4x + 5| = 6$

57. $|9x - 1| < 4$

58. $|7x + 3| > 2$

59. $|3x + 8| \geq -2$

60. $|8x - 4| \leq -5$

Chapter 3 Summary

Section 3.1

The ***rectangular coordinate system*** is two number lines, called *axes*, drawn at right angles to each other. The system is also called the *Cartesian coordinate system*. The horizontal number line is the x-axis. The vertical number line is the y-axis.

An ***ordered pair of real numbers*** is a pair of real numbers given in a definite order, corresponding to a point on the rectangular coordinate system. The first number in the ordered pair may be called the ***first coordinate,*** the ***x-coordinate,*** or the ***abscissa.*** The second number in the ordered pair may be called the ***second coordinate,*** the ***y-coordinate,*** or the ***ordinate.*** The point where the number lines cross is called the ***origin.*** The origin has coordinates (0, 0).

The regions bordered by the coordinate axes are called the first, second, third, and fourth ***quadrants.***

The *solutions* of equations in *two variables* are ***ordered pairs*** of numbers.

Section 3.2

Equations that can be written in the form $Ax + By = C$ where A, B, and C are constants, and A and B are not both zero are called ***linear equations*** in one or two variables. This form is called the ***standard form*** of a linear equation.

For an equation in two variables to be linear, the following must be true.

1. The exponent of each variable must be 1.
2. The equation may not contain a term where there is a product or quotient of two variables.

The graph of a linear equation $Ax + By = C$ is a straight line.

The ***x-intercept*** of the graph of an equation is a point where it crosses the x-axis. The ***y-intercept*** is the point where the graph crosses the y-axis.

In the equation of a line, let $x = 0$ and solve for y to find the y-intercept $(0, b)$; let $y = 0$ and solve for x to find the x-intercept, $(a, 0)$.

All graphs of equations of the form $y = b$, where b is a constant, are horizontal lines passing through the point $(0, b)$.

All graphs of equations of the form $x = a$, where a is a constant, are vertical lines passing through the point $(a, 0)$.

Section 3.3

$$\text{Slope: } \frac{\text{vertical change}}{\text{horizontal change}} \quad \text{or} \quad \frac{\text{rise}}{\text{sun}}$$

$$m = \frac{y_2 - y_1}{x_2 - x_1} = \frac{y_1 - y_2}{x_1 - x_2}$$

Lines that slope *upward to the right* have *positive slope*. Lines that slope *downward to the right* have *negative slope*.

All horizontal lines have a slope of 0. The slope of a vertical line is undefined.

The slope-intercept form of the equation of a line with slope m and y-intercept b is

$$y = mx + b$$

Graphing linear equations

1. Is the equation of the form $y = b$ or $x = a$? The graphs of each of these equations is a line parallel to an axis.
2. Determine by inspection if one or both of the intercepts will be fractions. If they are not, graph the line using intercepts unless the equation is of the form $y = kx$. If the points are too close together, choose another point farther from the intercepts.
3. If one or both of the intercepts are fractions or if the equation is of the form $y = kx$, it is often easier to graph the line using the slope and y-intercept.

Section 3.4

Point-slope form of the equation of a line:

$$y - y_1 = m(x - x_1)$$

In the equation $y = mx + b$, y is called the ***dependent variable*** and x is called the ***independent variable.***

Two distinct lines that have the same slope are ***parallel.*** The slopes of vertical lines are undefined. Vertical lines are also parallel.

If the slope of one line is the negative reciprocal of the slope of another line, the lines are ***perpendicular.*** A line with undefined slope is perpendicular to a line with zero slope.

Some forms of linear equations

$x = a$	Vertical line through the point $(a, 0)$. Slope is undefined.
$y = b$	Horizontal line through the point $(0, b)$. Slope is 0.
$y = mx + b$	Slope-intercept form with slope m and y-intercept $(0, b)$.
$y - y_1 = m(x - x_1)$	Point-slope form with slope m; passes through the point (x_1, y_1).
$Ax + By = C$	Standard form where A, B, and C are constants and A and B are not both zero.

Section 3.5

A ***linear inequality*** is an expression of the form $Ax + By < C$, where A and B are not both 0 and A, B, and C are real numbers. The symbol $<$ may be replaced with $>$, $\leq$, or $\geq$.

The ***solutions*** to inequalities in two variables are ***ordered pairs*** of real numbers.

Graphing linear inequalities in two variables

1. Sketch the graph of the equation formed by replacing the inequality symbol with an equal sign. This is the boundary line. The boundary line will be *solid* if the inequality symbol is $\geq$ or $\leq$. This means that points on the boundary line are solutions. The boundary line will be *dashed* if the inequality symbol is $>$ or $<$. This means that points on the line are not solutions.
2. Choose a test point that is not on the line and substitute its coordinates into the *inequality*. The point $(0, 0)$ is an easy one to use if the boundary line does not pass through it. If the inequality is true for the chosen point, shade the side of the boundary line containing it. If the inequality is false for the point, shade the other side of the line. *All points* in the shaded region are solutions to the inequality.

Chapter 3 Additional Exercises (Optional)

NAME ______________________

DATE ______________________

CLASS ______________________

Section 3.1

In which quadrant are the following points located?

1. $(-5, 3)$ **2.** $(6, 7)$ **3.** $(0, 5)$ **4.** $(-3, 0)$

5. $(-3, -2)$ **6.** $(4, -8)$ **7.** $(-9, 1)$ **8.** $(-5, -3)$

Is the ordered pair $(\frac{1}{5}, \frac{1}{2})$ a solution of the equation?

9. $5x + 4y = 3$ **10.** $10x - 6y = -5$ * **11.** $\frac{5}{4}x - 3y = -\frac{3}{4}$ * **12.** $7x + 6y = \frac{22}{5}$

Make a table of values for x and y. Then graph.

13. $y = \frac{1}{2}x$

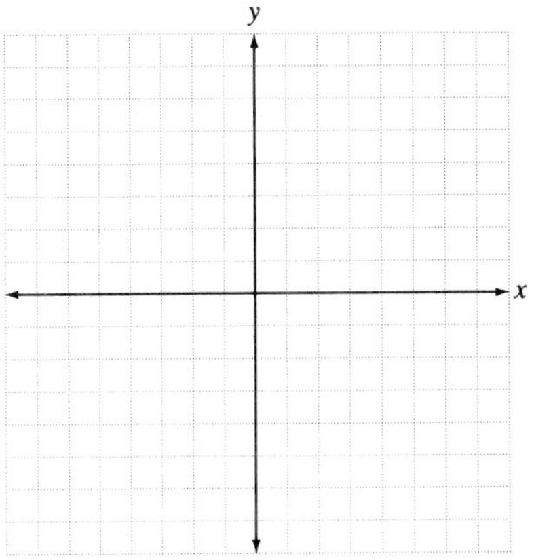

14. $y = -\frac{1}{3}x$

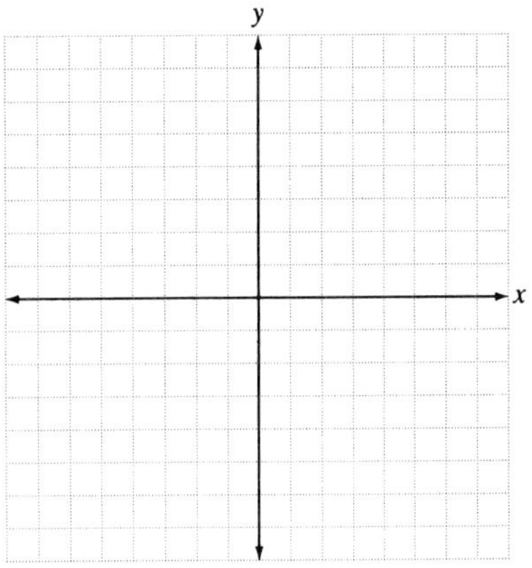

Section 3.2

Which equations are linear?

15. $x + 4y = 7$ **16.** $xy = 4$ **17.** $x^2 - 4 = 0$ **18.** $5x - 8y = 6x - 7$

19. $5 + xy = 7$ **20.** $7x = 4$ **21.** $x^3 + x = 2$ **22.** $3x - 5y = 8$

Graph using intercepts.

23. $x - 3y = 9$

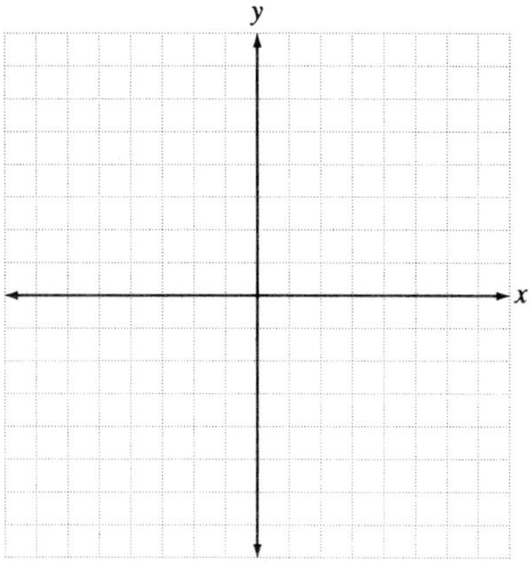

24. $4x + y = 4$

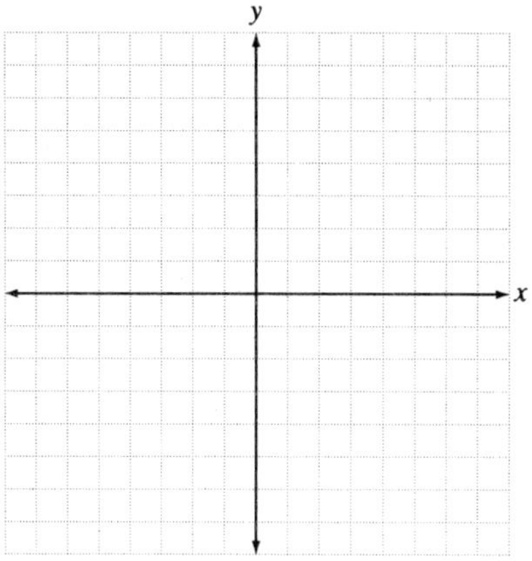

25. $3x + 2y = 12$

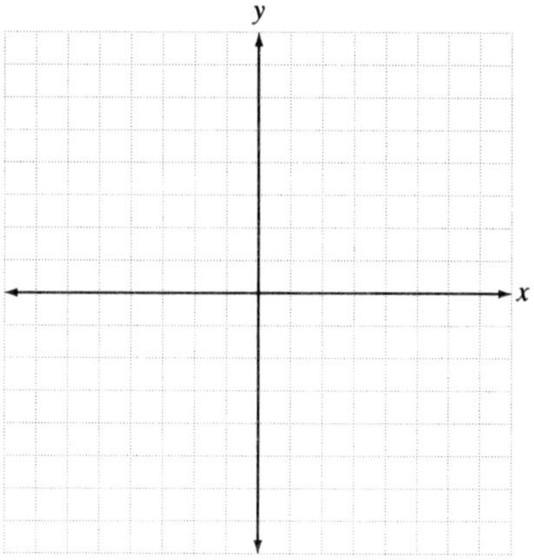

26. $2x - 5y = 10$

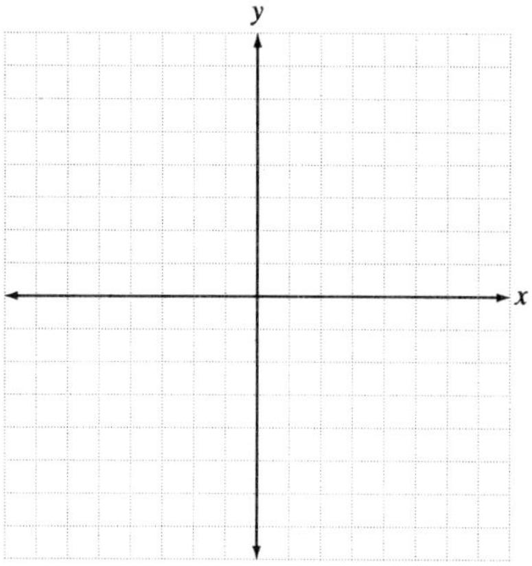

* **27.** $y - 1 = \frac{1}{3}(x + 6)$

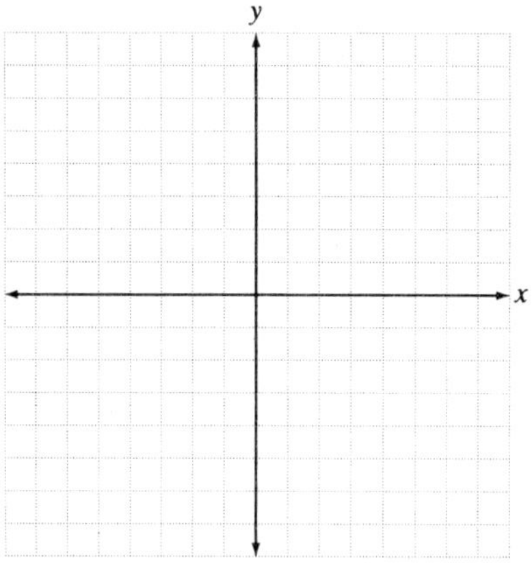

* **28.** $y - 2 = -3(x - 1)$

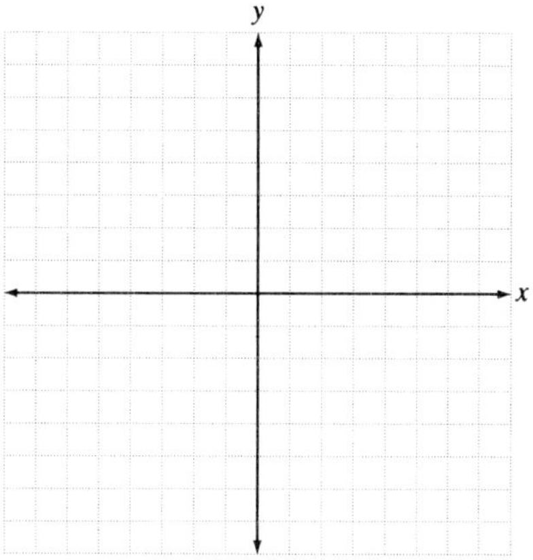

Section 3.3

Graph using the slope and y-intercept.

29. $4x - 3y = 9$

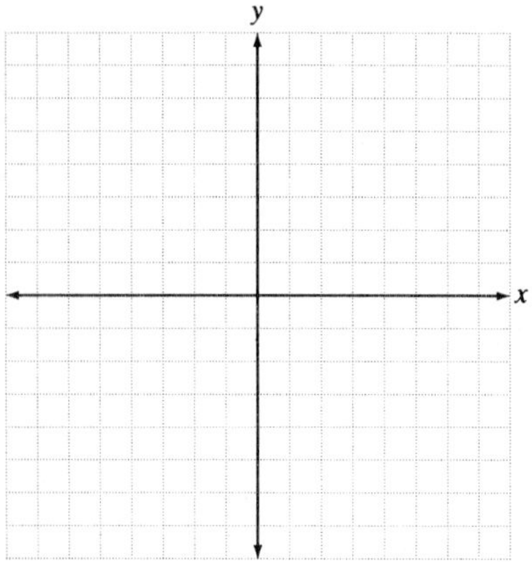

30. $2y + 5x = 2$

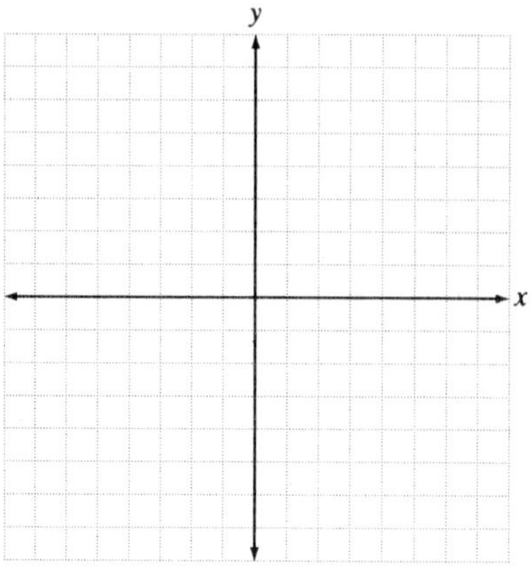

31. $7x + 2y = 8$

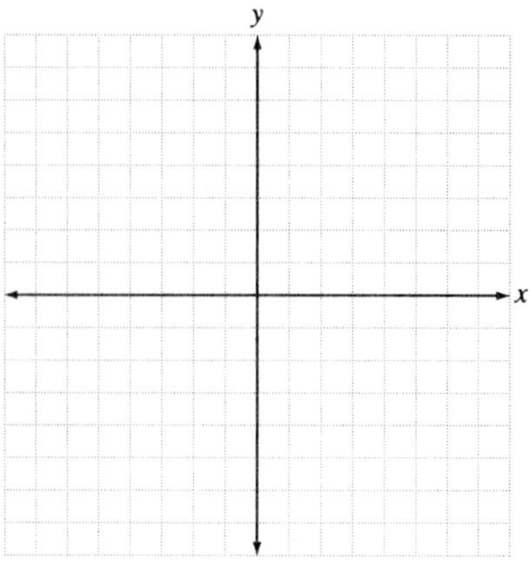

32. $-5y + 6x = 30$

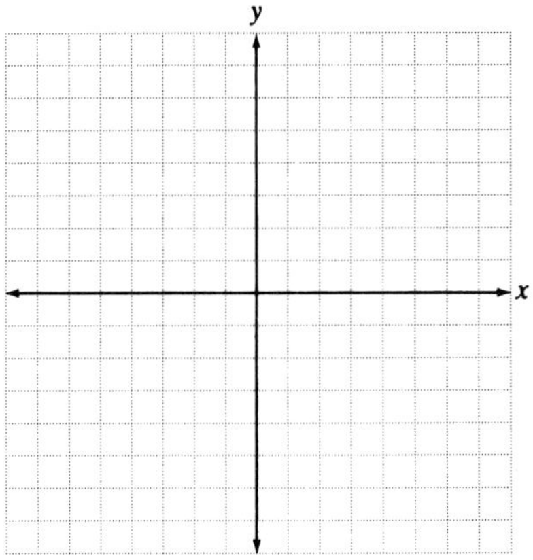

33. $3y - x = 9$

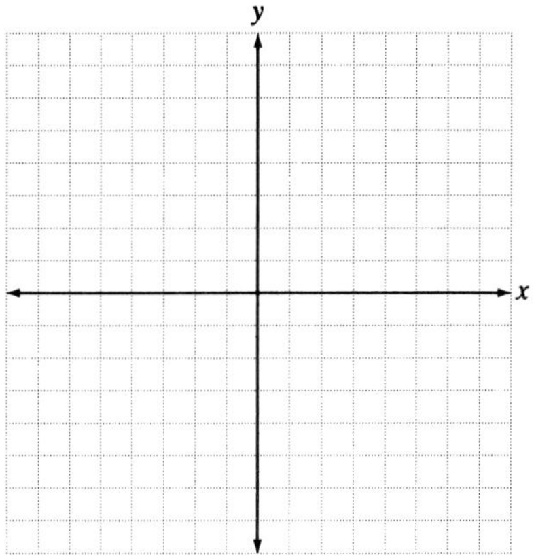

34. $3x + 2y = 6$

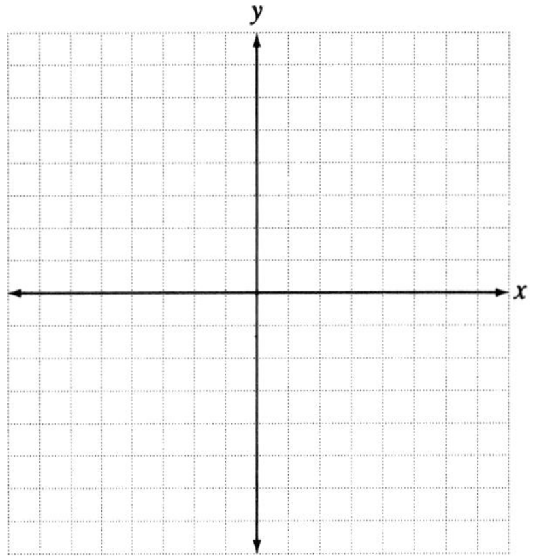

35. $2x - y = 3$

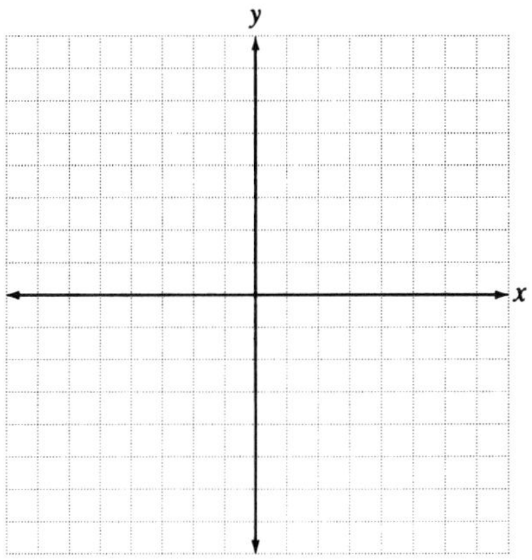

36. $y + 4x = 4$

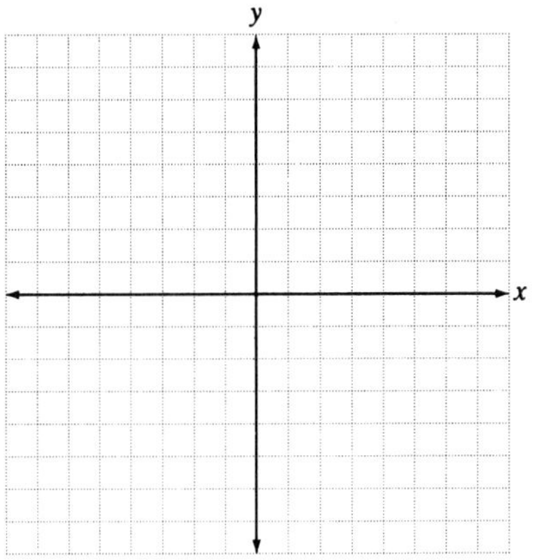

* **37.** $y - 2 = -\frac{3}{5}(x - 5)$

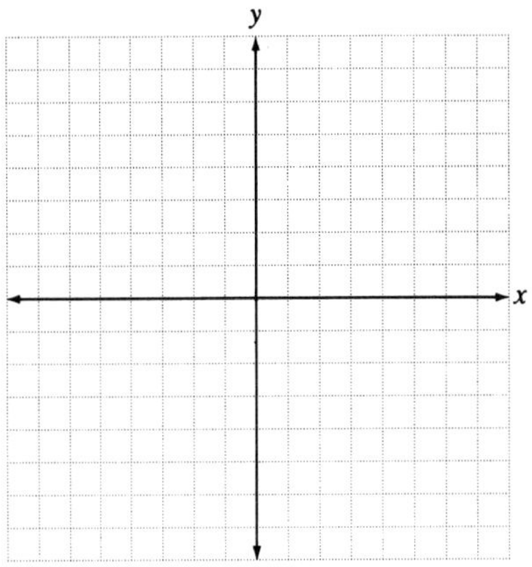

* **38.** $y - 3 = \frac{3}{4}(x - 8)$

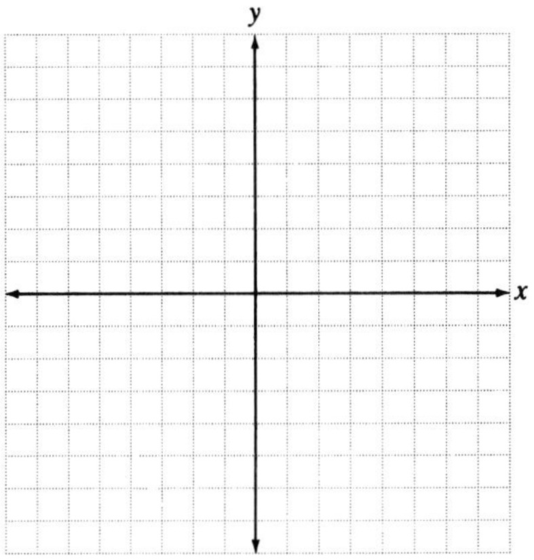

39. $y = \frac{1}{7}x$

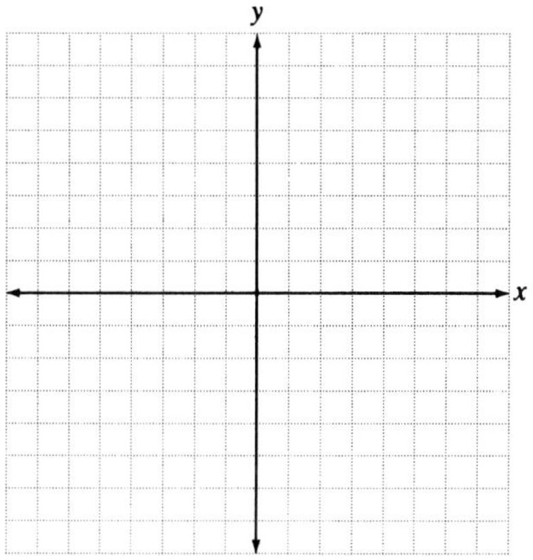

40. $y = -\frac{5}{2}x$

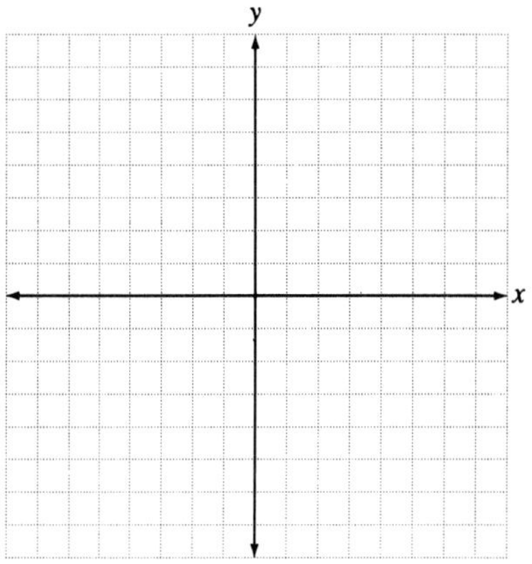

Section 3.4

Find an equation for the line with the given slope and y-intercept.

41. $m = 2, b = -3$

42. $m = -7, b = 5$

43. $m = \frac{2}{3}, b = \frac{1}{2}$

44. $m = -\frac{8}{7}, b = -1$

Find an equation for the line with the given slope through the given point.

* **45.** $m = 0.5, (0.3, 1.2)$

* **46.** $m = \frac{2}{3}, \left(\frac{3}{4}, \frac{5}{8}\right)$

47. $m = 0, \left(\frac{3}{7}, \frac{6}{5}\right)$

48. $m = 0, (3.5, 2.8)$

Find an equation for the line through each pair of points.

49. $(4, 6), (9, 7)$

50. $(-4, -1), (3, 3)$

51. $(-3, -3), (-7, -6)$

52. $(-1, 6), (-2, 6)$

* **53.** $(2.4, 6.6), (2.1, 4.5)$

* **54.** $(2a, b), (a, b)$

55. Robert wants to build a rabbit hutch. If he uses 68 square feet of plywood, the cost is \$26. If he uses 88 square feet of plywood, the cost is \$31. Write the linear equation that expresses the total cost c in terms of the number of square feet of plywood p used.

Consider the following lines:

(a) $3x + 6y = 7$ (b) $4y - 8x = 5$ (c) $y - 3 = \frac{2}{7}(x - 4)$ (d) $y - 5 = -\frac{1}{2}(x + 8)$

56. Which lines are parallel?

57. Which lines are perpendicular?

Section 3.5

58. Which of the following inequalities has $(-5, -2)$ as a solution?

(a) $2x - 5y > -3$ (b) $\frac{3}{10}x + \frac{5}{8}y > 2$ (c) $0.3x + 1.2y < -4$ (d) $-4x + 8y > 7$

Graph.

59. $3x - 4y > 12$

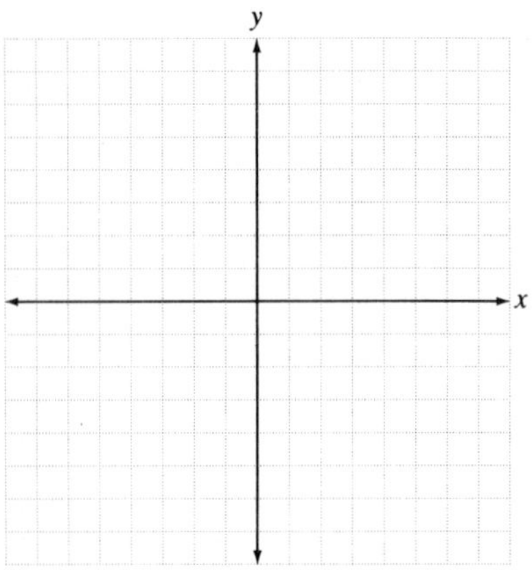

60. $3x + 5y < 15$

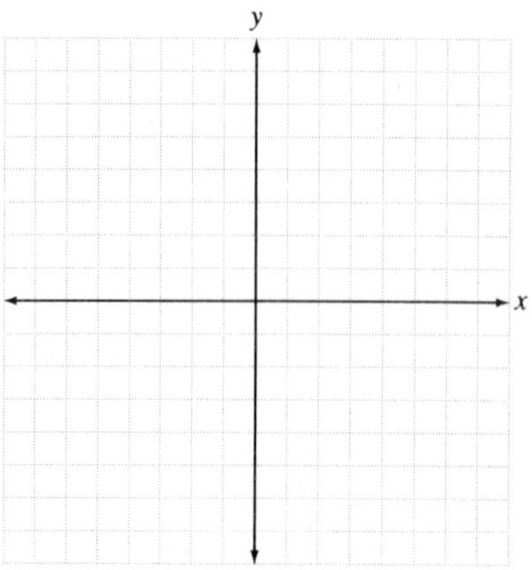

61. $y \geq \frac{3}{5}x + 1$

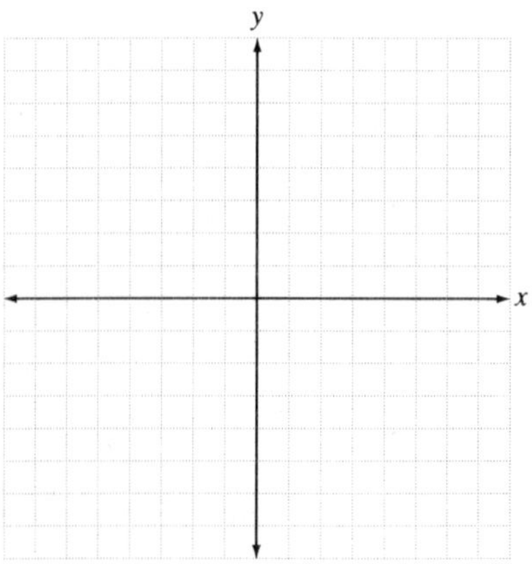

62. $y \leq -\frac{4}{3}x + 5$

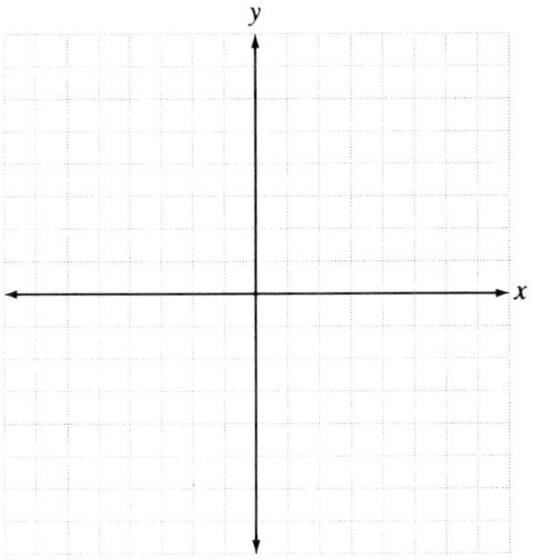

63. $y - \frac{1}{2}x \leq 0$

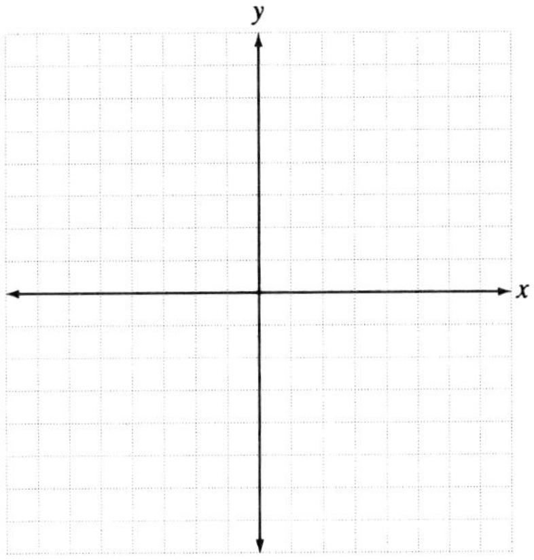

64. $y - \frac{3}{5}x > 0$

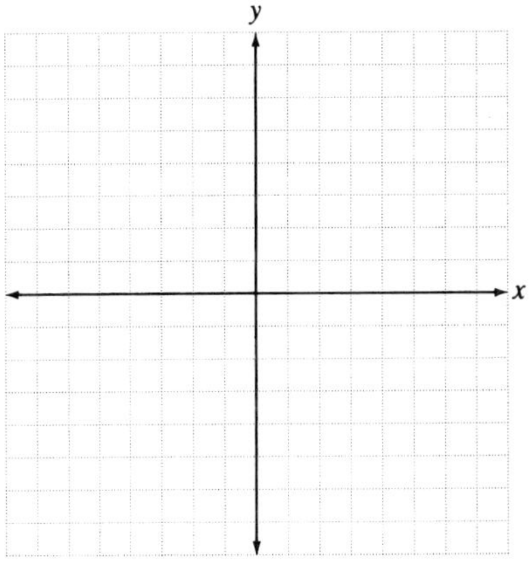

65. $-2 < x < 0$

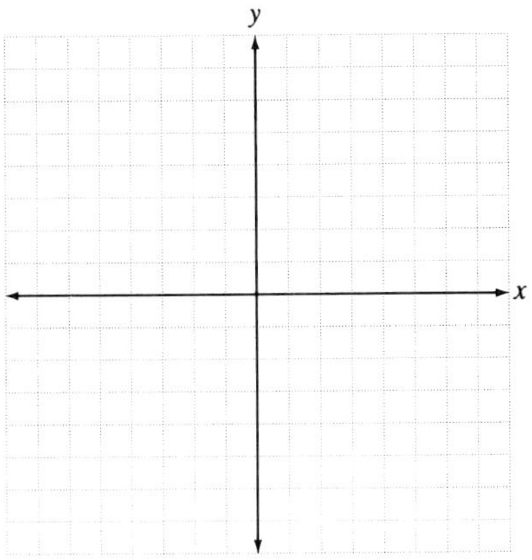

66. $-4 \leq y \leq 2$

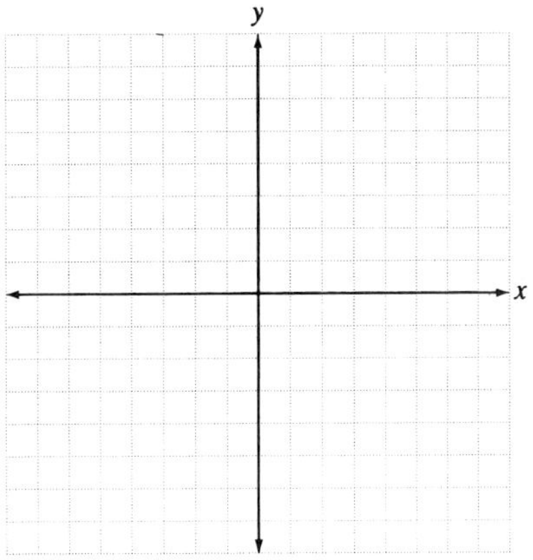

COOPERATIVE LEARNING

Write equations, in slope-intercept form, for the following graphs.

1.

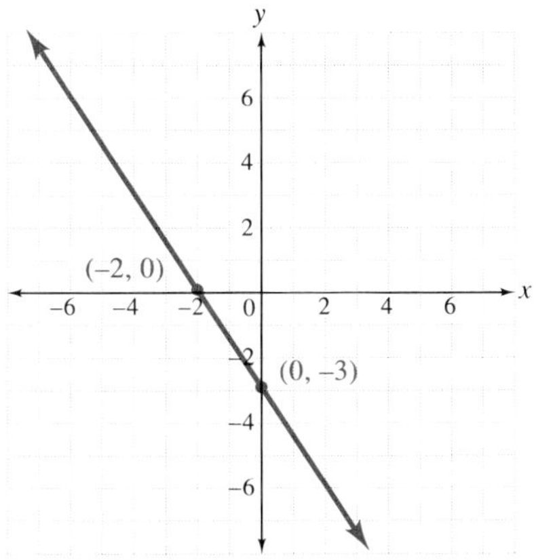

2.

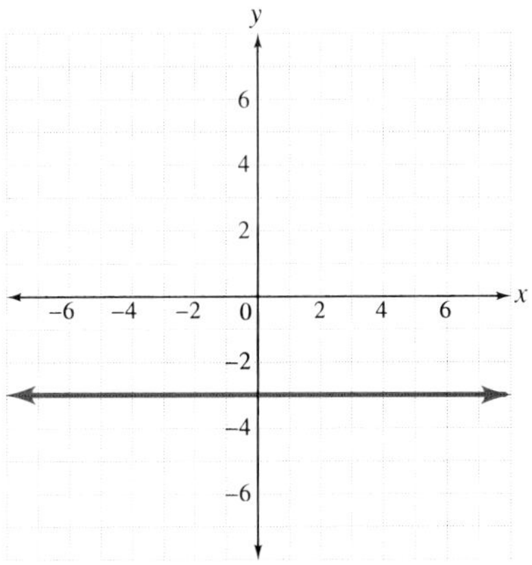

3.

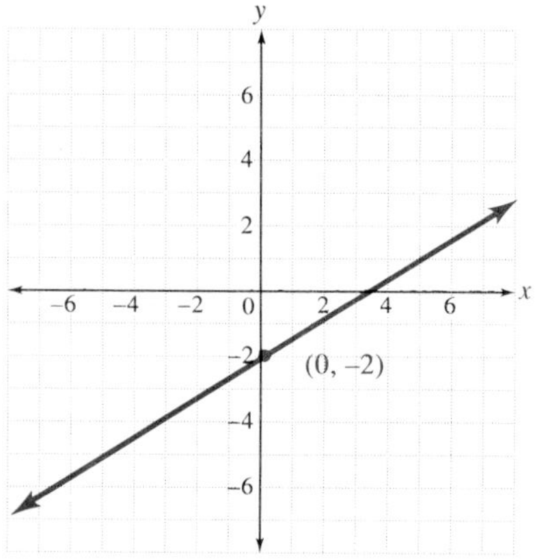

4.

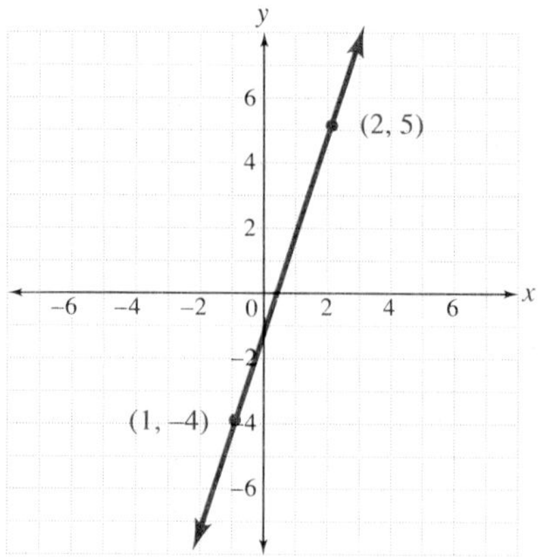

5. Draw a graph of the line through the point $(0, -2)$ with slope $-\frac{2}{3}$.

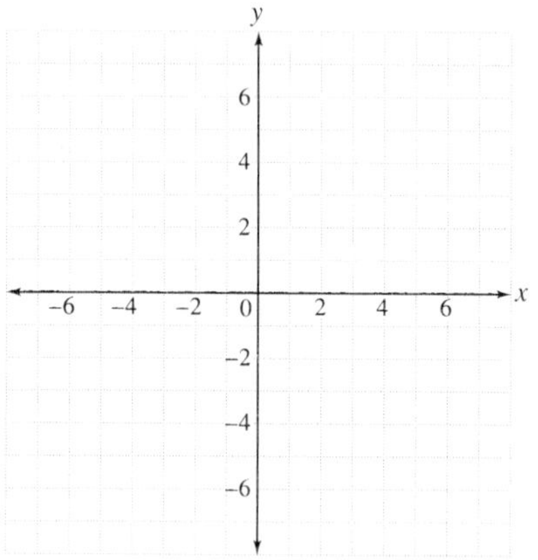

6. The graph of a line is shown below. Find an equation of the line passing through the point $(-2, -4)$ and

a. Parallel to the given line

b. Perpendicular to the given line

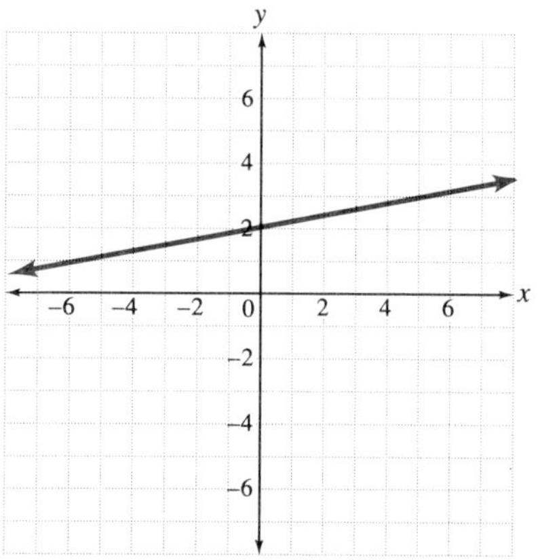

7. Write an inequality for the following graph.

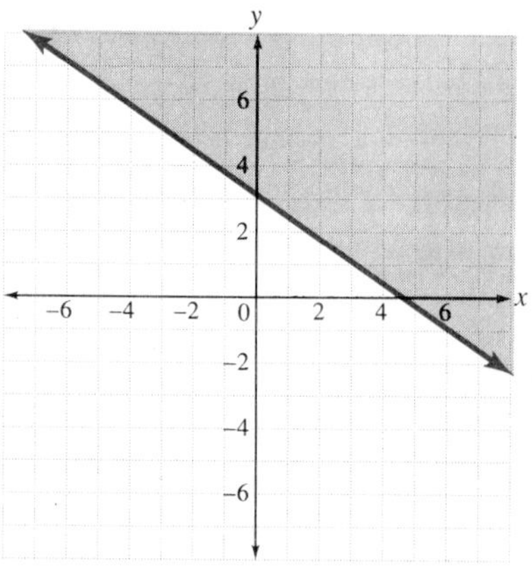

8. An equation that describes the value V of a rental home after t years is

$$V = -4000t + \$120{,}000$$

What was the cost of the home? What is the depreciation rate for tax purposes?

Chapter 3 Practice Test

NAME

DATE

CLASS

1. Make a table of values for x and y. Then graph $y = x$.

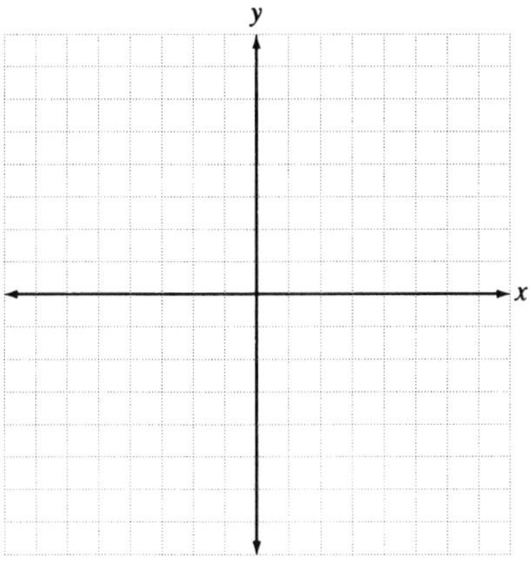

2. Graph using intercepts: $7y = 2x - 14$.

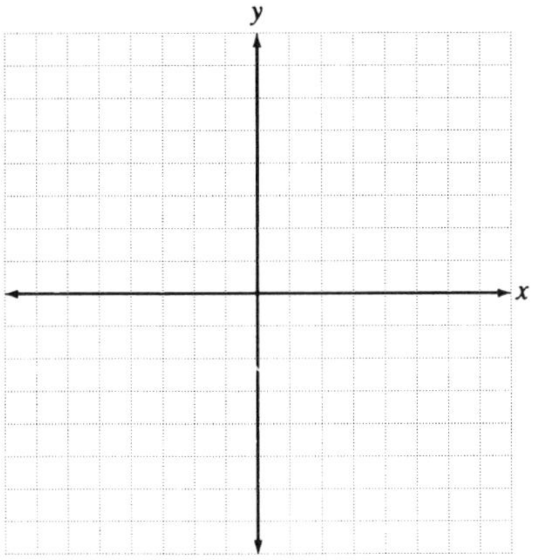

3. Find the slope of the line through the pair of points, if it exists: $(-4, -2)$ and $(-3, 5)$.

4. Find the slope and y-intercept and graph: $4y + 5x = 20$.

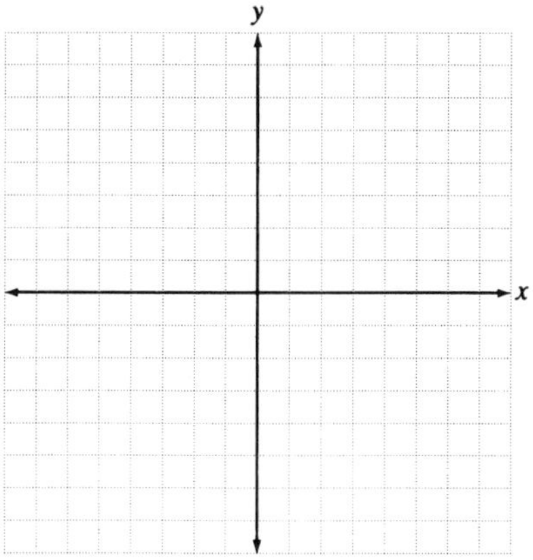

1. ______

2. ______

3. ______

4. ______

5. ______________________

5. Find an equation for the line with the given slope and y-intercept: $m = \frac{3}{5}$, $b = -2$.

6. ______________________

6. Find an equation for the line through the points $(-7, -4)$ and $(3, -2)$.

7. ______________________

7. Find an equation for the line through the point $(3, -2)$ and parallel to the line $2x + 5y = 4$.

8. ______________________

8. Find an equation for the line through the point $(1, 4)$ and perpendicular to the line $3x - y = 7$.

9. ______________________

9. At Sound Technology, the cost of manufacturing 75 compact discs is \$675 and the cost of manufacturing 200 compact discs is \$1050.

a. If the relationship between the cost and the number of compact discs manufactured is linear, write the equation that relates the cost C to the number of compact discs d that are manufactured.

b. Use the equation to find the cost of manufacturing 300 compact discs.

Graph.

10. ______________________

10. $y \geq -3$

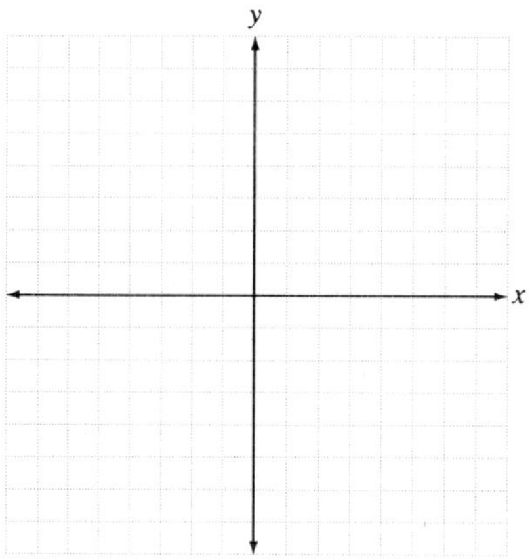

11. ______________________

11. $2x - 5y < 15$.

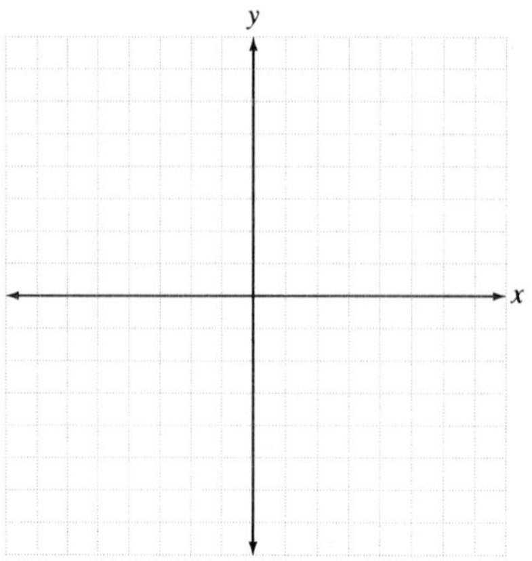

CHAPTER 4

Systems of Equations

Pretest

1. Solve by graphing. Then classify the system as consistent or inconsistent and as dependent or independent.

$4x + y = 2$

$y = 4x - 6$

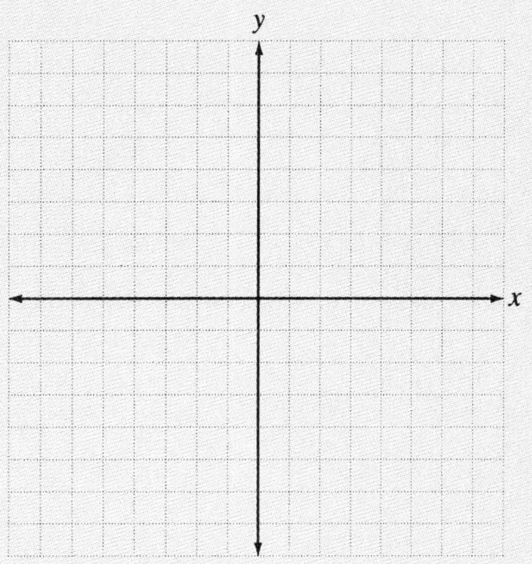

Solve.

2. $2x + 9y = -31$
$5x - 4y = 55$

3. $y = \frac{3}{2}x + 14$
$-3x + 2y = 28$

4. The perimeter of a rectangular house is 244 feet. If the width is 12 feet less than the length, what are the dimensions of the house?

5. Solution A is 7% alcohol and solution B is 15% alcohol. A medical technician wants to mix the two to get 18 liters of a 9.5% alcohol solution. How many liters of each should he use?

6. Solve:

$$\begin{aligned} 3x + 4y + 2z &= -13 \\ 2x - y + z &= -1 \\ x - 2y + 3z &= 0 \end{aligned}$$

7. Alan bought 15 pounds of apples, bananas, and grapes. He bought twice as many pounds of apples as bananas. If apples cost 80 cents per pound, bananas cost 50 cents per pound, and grapes cost 90 cents per pound and he spent $11.70, how many pounds of grapes did he buy?

Evaluate.

8. $\begin{vmatrix} -4 & 8 \\ -7 & -9 \end{vmatrix}$

9. $\begin{vmatrix} -3 & 9 & -7 \\ -4 & 2 & 6 \\ 5 & -3 & 2 \end{vmatrix}$

10. Solve using Cramer's rule:

$$\begin{aligned} 2x + 3y - 2z &= -23 \\ 3x - 4y - 3z &= -9 \\ 5x + 2y + 6z &= 36 \end{aligned}$$

4.1 SOLUTIONS OF LINEAR SYSTEMS BY GRAPHING

It is often easier to solve applied problems using systems of equations.

1 Solutions

A linear ***system of equations*** in two variables is composed of two or more linear equations. The solution of this system is all ordered pairs of real numbers that satisfy *all* the equations.

EXAMPLE 1 Determine whether the given ordered pair is a solution of the system

$$x + y = 9$$
$$4x - y = 21$$

a. (6, 3)

$x + y = 9$		$4x - y = 21$		
$6 + 3$	9	$4(6) - 3$	21	Replace x by 6 and y by 3.
9	9	$24 - 3$	21	
		21	21	

The ordered pair (6, 3) makes both equations true so it is a solution of the system.

b. (5, 4)

$x + y = 9$		$4x - y = 21$		
$5 + 4$	9	$4(5) - 4$	21	Replace x by 5 and y by 4.
9	9	$20 - 4$	21	
		16	21	

The pair (5, 4) does not make the second equation true, so (5, 4) is not a solution of the system since it does not satisfy *both* equations. ■

□ **DO EXERCISE 1.**

OBJECTIVES

1 *Determine if a given ordered pair is a solution of a system of equations*

2 *Solve a linear system of two equations in two variables by graphing*

□ **Exercise 1** Determine whether the ordered pair is a solution of the given system.

a. $3x + y = -7$ $\quad (-3, 2)$
$x + 4y = 5$

b. $8x - 5y = 10$ $\quad (5, 6)$
$6x + 7y = 62$

2 Solution by Graphing

Recall that the graph of a linear equation is a straight line and this graph is a picture of the solutions of the equation. There are three different situations that can occur when we graph a system of two linear equations in two variables.

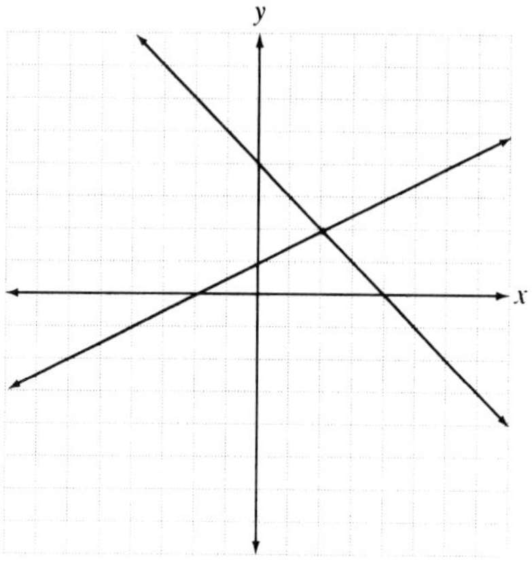

(a)

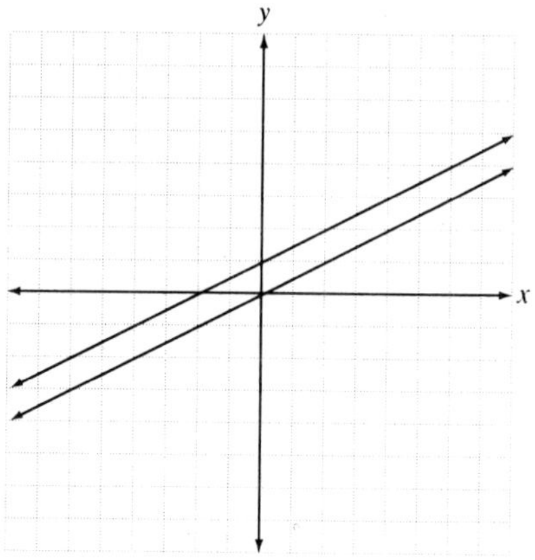

(b)

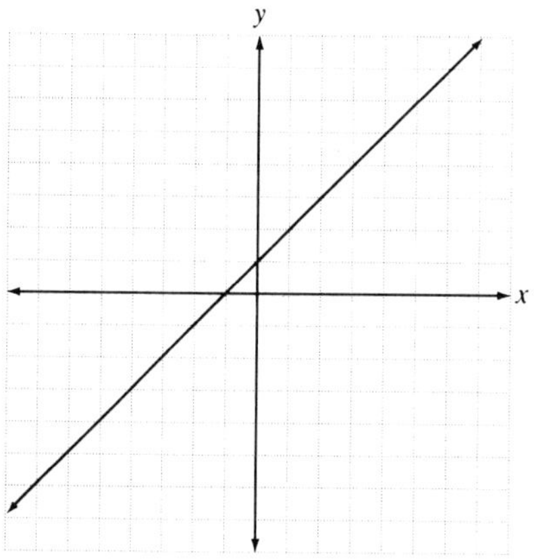

(c)

(a) *The graphs intersect at a single point.* The solution is the coordinates of the point of intersection. The system of equations is called ***consistent*** because it has a solution. This is the usual case.

(b) *The graphs are parallel lines.* There is no solution. The system of equations is ***inconsistent,*** because it has no solution.

(c) *The graphs are the same line.* There are infinitely many solutions. (Any solution of one equation is a solution of the other.) The system of equations is ***dependent*** and also consistent because there is at least one solution.

In cases (a) and (b) the systems are ***independent.***

> A system of equations is called ***consistent*** if it has a solution. It is called ***inconsistent*** if has no solution. A system of two equations in two variables is ***independent*** if it has exactly one solution or no solutions and ***dependent*** if it has infinitely many solutions.

EXAMPLE 2 Solve the system by graphing. Then classify the system as consistent or inconsistent and as dependent or independent.

$$x + y = 4$$

$$2x - y = 2$$

Graph each equation. The point of intersection appears to be (2, 2).

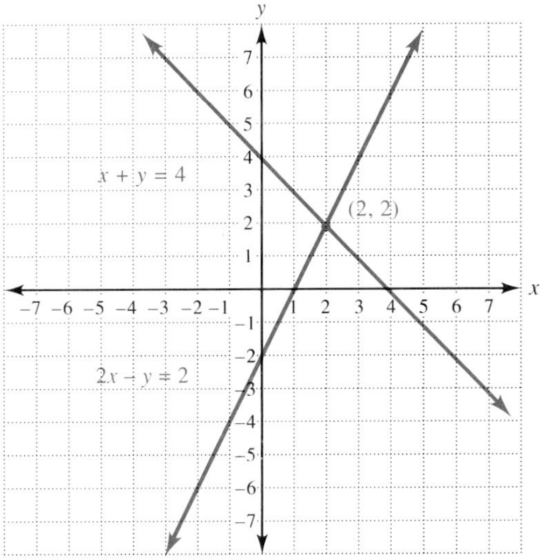

Replace x with 2 and y with 2 in each equation to check.

$x + y = 4$	
$2 + 2$	4
4	4

$2x - y = 2$	
$2(2) - 2$	2
$4 - 2$	2
2	2

The ordered pair (2, 2) checks, so it is the solution. Since the system has exactly one solution, it is consistent and independent. ■

It is difficult to read exact coordinates from a graph, so graphing the equations may give only approximate answers. To give us exact results, we will learn algebraic methods of solving linear systems of equations in the next section.

□ **DO EXERCISE 2.**

□ **Exercise 2** Solve by graphing. Then classify the system as consistent or inconsistent and as dependent or independent.

a. $2x - 3y = 12$
$-2x + y = -4$

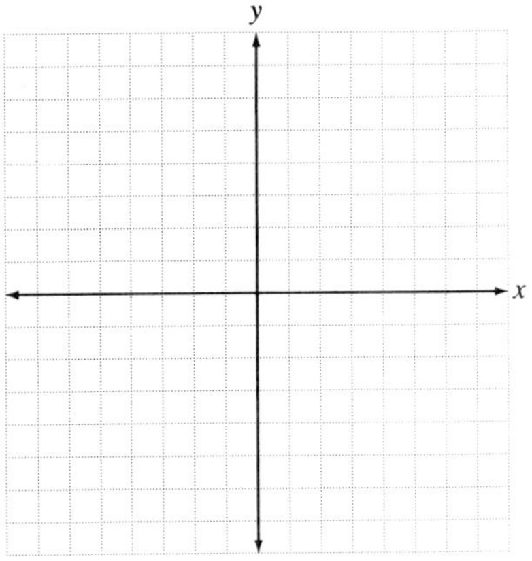

b. $x = 2y$
$3y + x = 5$

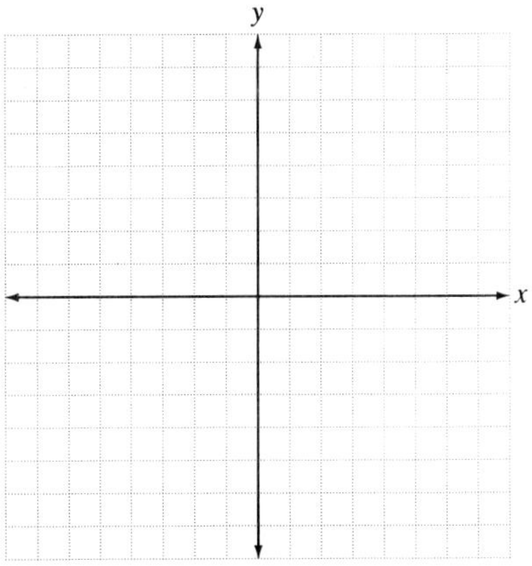

□ **Exercise 3** Solve by graphing. Then classify the system as consistent or inconsistent and as dependent or independent.

a. $y = 2x - 3$
$y = 2x + 1$

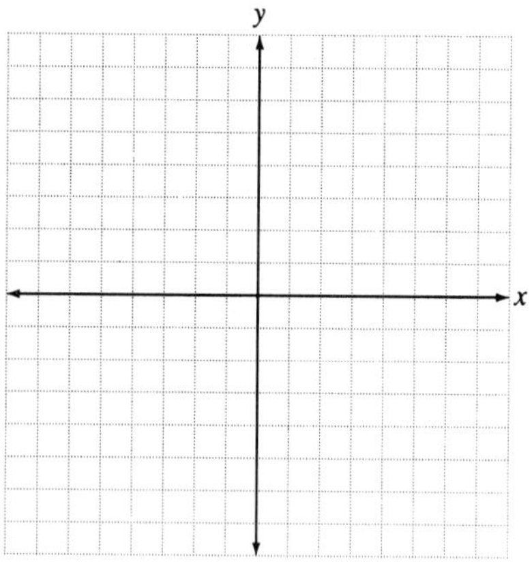

b. $3x + 4y = 12$
$-6x - 8y = -24$

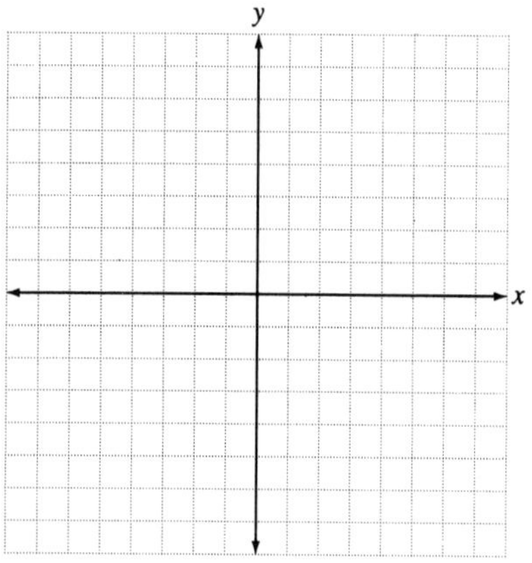

EXAMPLE 3 Solve the system by graphing. Then classify the system as consistent or inconsistent and as dependent or independent.

a. $y = 3x - 2$
$y = 3x - 4$

Graph each equation. The lines have the same slope and different y-intercepts, so they are parallel.

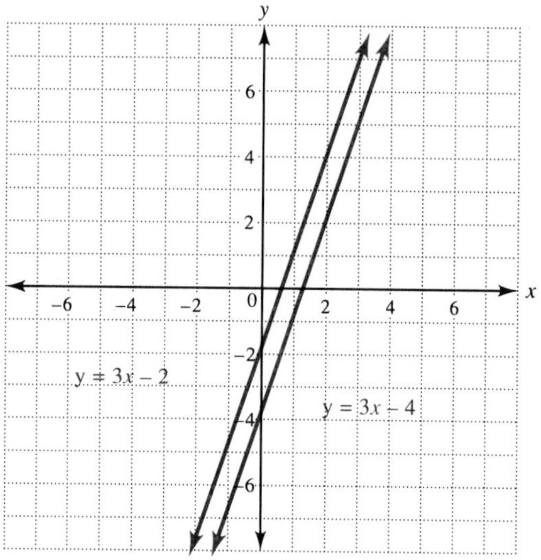

The system has no solution. The solution set is the empty set, ∅. The system is inconsistent and independent.

b. $y - 2x = 1$
$-4y + 8x = -4$

Graph each equation. The graphs are the same line.

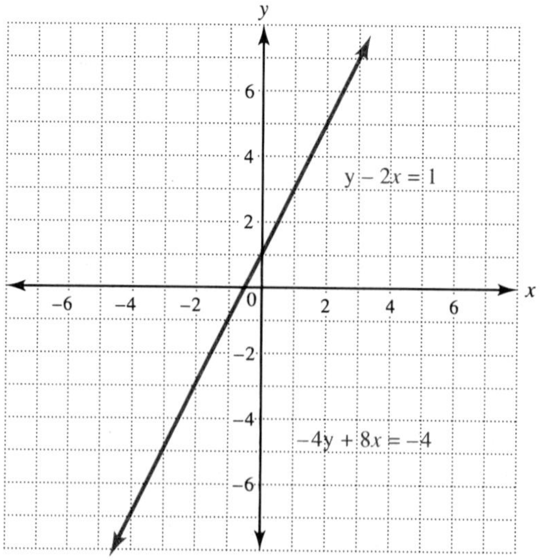

There are an infinite number of solutions that lie on this line, since any solution of one equation is a solution of the other. The system is consistent and dependent. ■

□ **DO EXERCISE 3.**

GRAPHING CALCULATOR

An approximate solution to a system of two linear equations in two variables may be obtained by graphing each equation on the same screen and tracing the graph to find the approximate solution. Remember from Section 3.3 that the calculator requires that the equations be in slope-intercept form.

DID YOU KNOW?

The Chinese emperor Shi Huang-ti ordered all the books in China, including all the mathematical works, burned in 213 B.C. He wished all recorded knowledge to begin with his reign. Because of this, very little is known about ancient Chinese mathematics.

Answers to Exercises

1. a. Yes **b.** No

2. a.

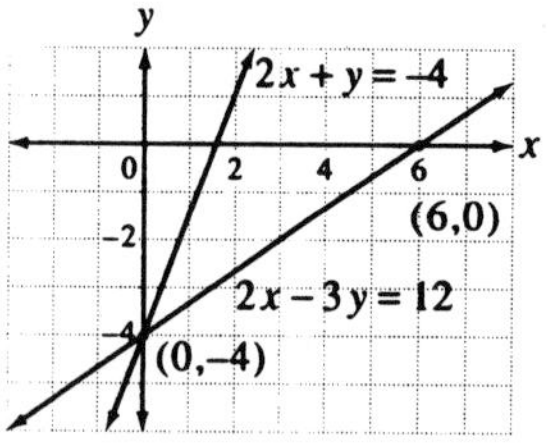

(0, −4); Consistent, independent

b.

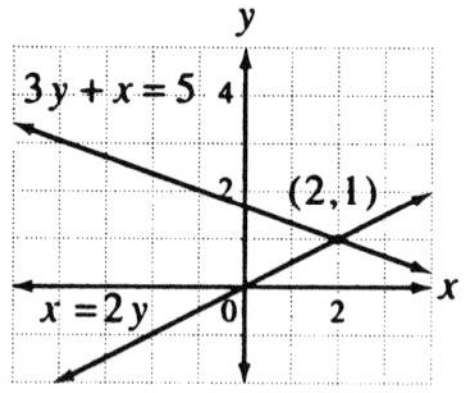

(2, 1); Consistent, independent

3. a.

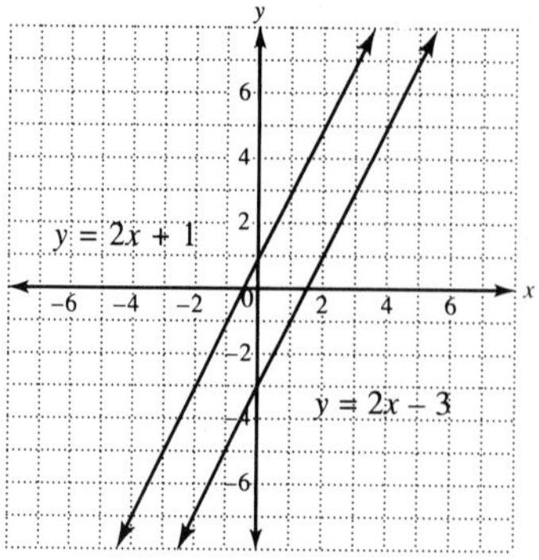

No solution; inconsistent, independent

b.

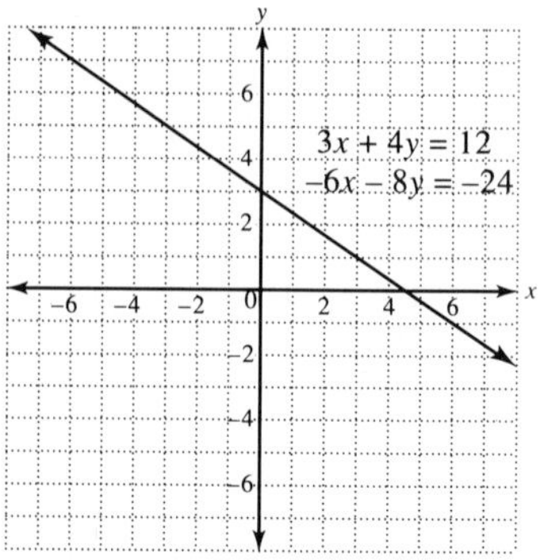

Infinite number of solutions; consistent, dependent

Problem Set 4.1

NAME ______________________

DATE ______________________

CLASS ______________________

Determine whether the ordered pair is a solution of the given system.

1. $3x + 7y = 13$ (2, 1)
$4x - y = 7$

2. $2x - y = 7$ (3, −1)
$x - 5y = 2$

3. $y = 2x + 6$ (−4, −2)
$5x - 6y = 12$

4. $y = 3x + 4$ (−3, −5)
$x - 4y = 17$

5. $2x - 5y = -14$ (3, 4)
$4x - y = 8$

6. $x + 2y = 7$ (−1, 4)
$3x - 4y = -19$

7. $8x - 5y = -15$ (0, 3)
$2x + 7y = 20$

8. $9x - 4y = 18$ (2, 0)
$6x + 7y = 15$

Solve by graphing. Then classify the system as consistent or inconsistent and as dependent or independent.

9. $x + y = 3$
$x - y = 1$

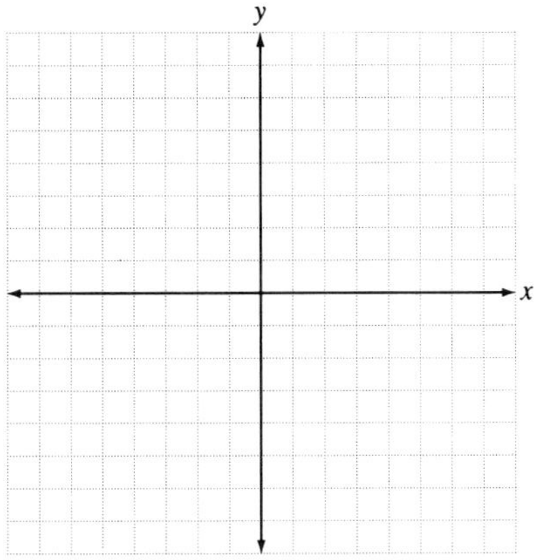

10. $x - y = 4$
$x + y = 6$

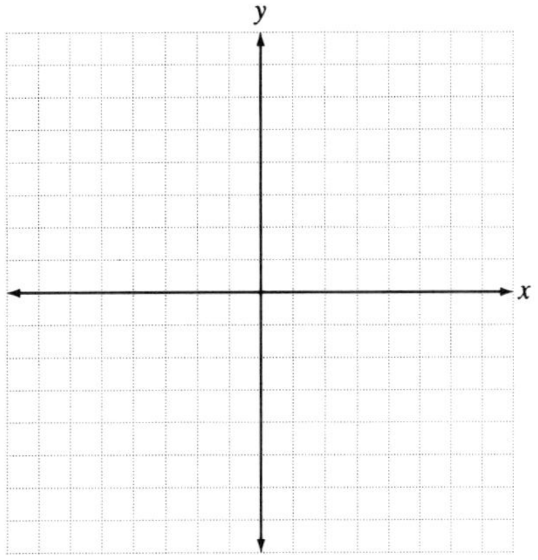

11. $3x - y = 6$
$2x - 3y = -3$

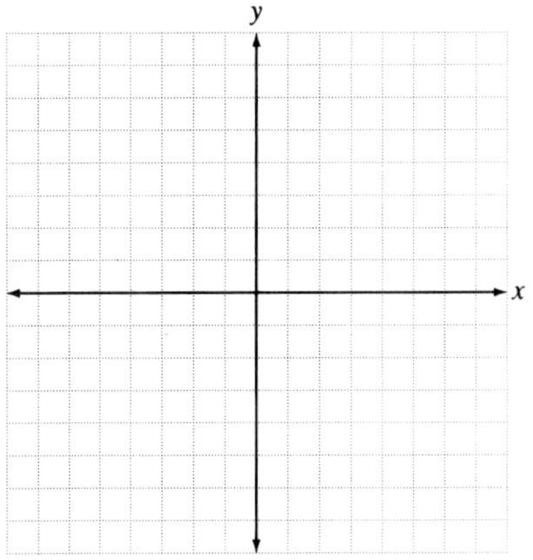

12. $2x - y = 7$
$x - 2y = 2$

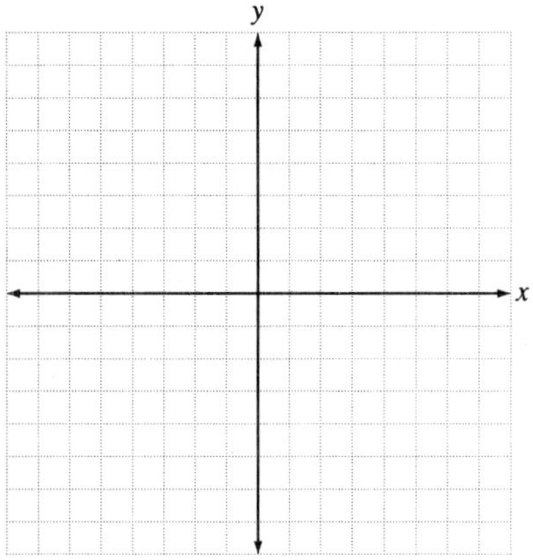

13. $y = 2x$
$x + y = -6$

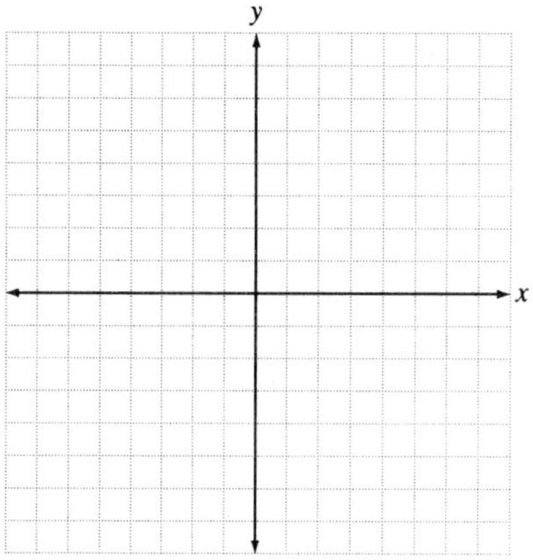

14. $y = -3x$
$x - y = -4$

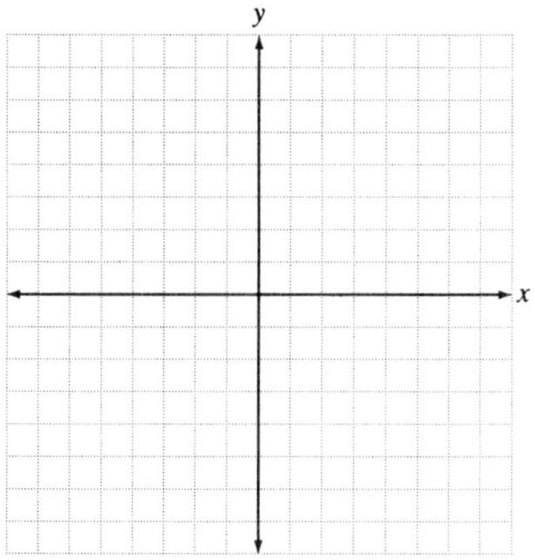

15. $x = y - 1$
$y = x + 1$

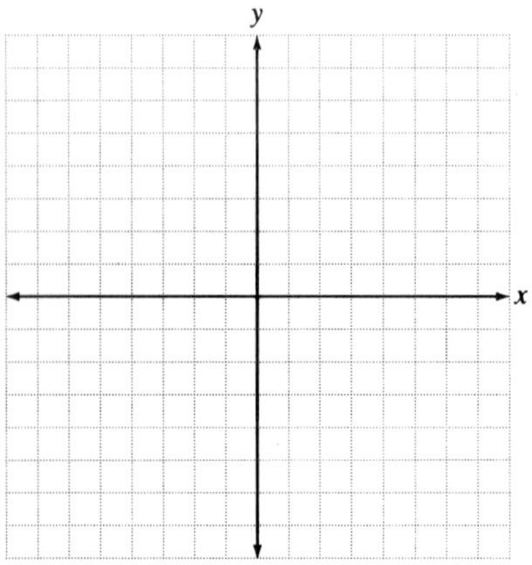

16. $2x = y + 9$
$y = 2x - 9$

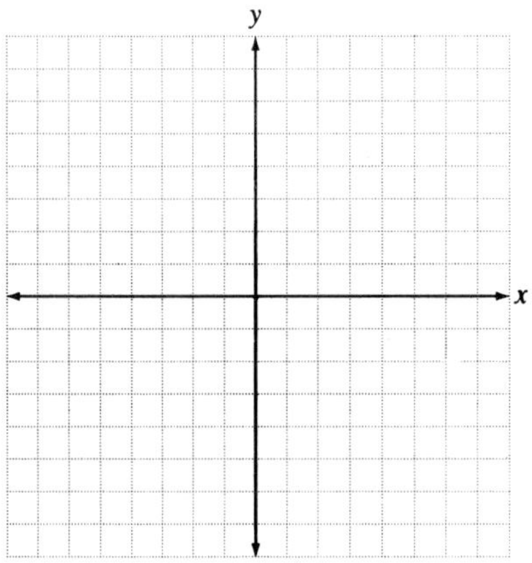

17. $y = 3x - 4$
$2x + y = 6$

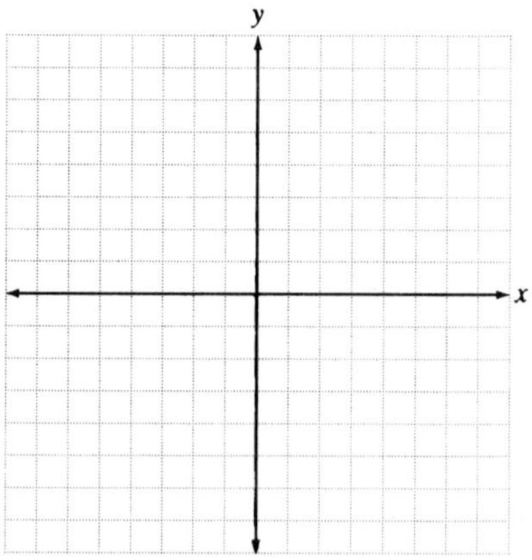

18. $4x + 3y = 21$
$x = 2y + 8$

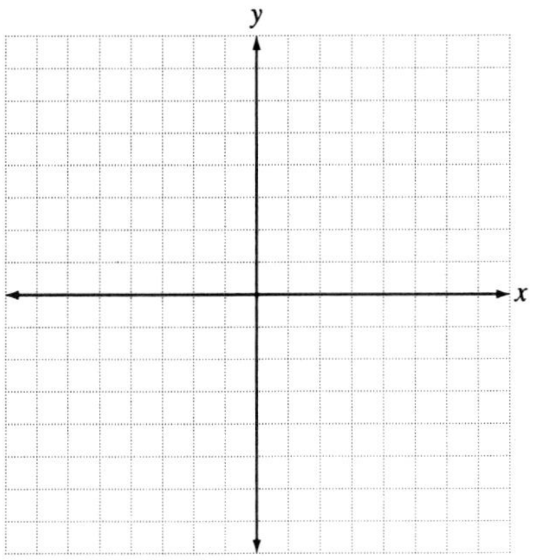

19. $5x - 2y = -4$
$8x - 4y = -4$

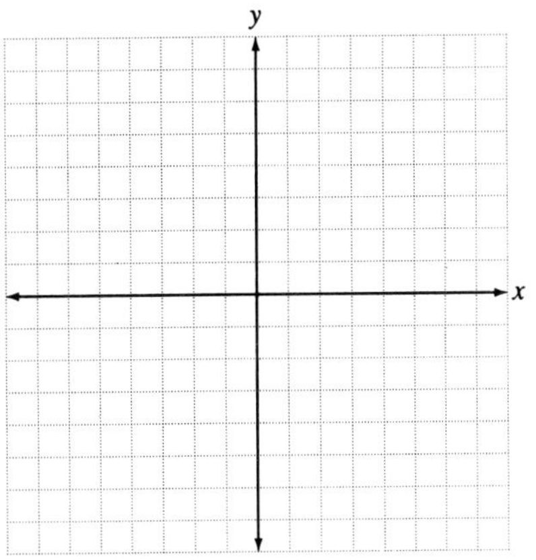

20. $3x - 2y = 4$
$2x + y = 6$

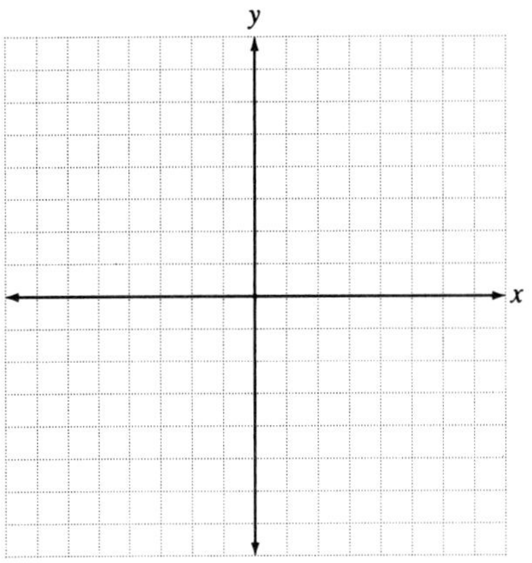

21. $y = \frac{2}{5}x + 2$
$x + y = 9$

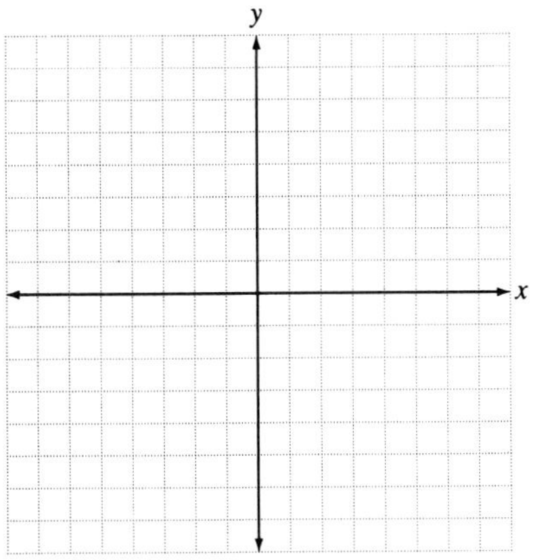

22. $y = \frac{3}{4}x + 5$
$x + 3y = 2$

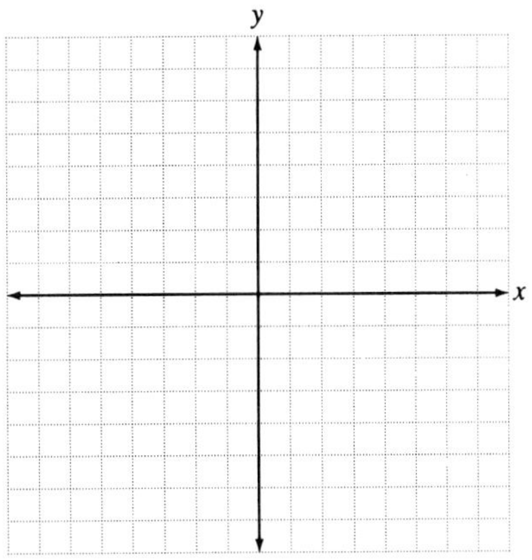

23. $y = -\frac{1}{4}x + 1$
$2y - x = -4$

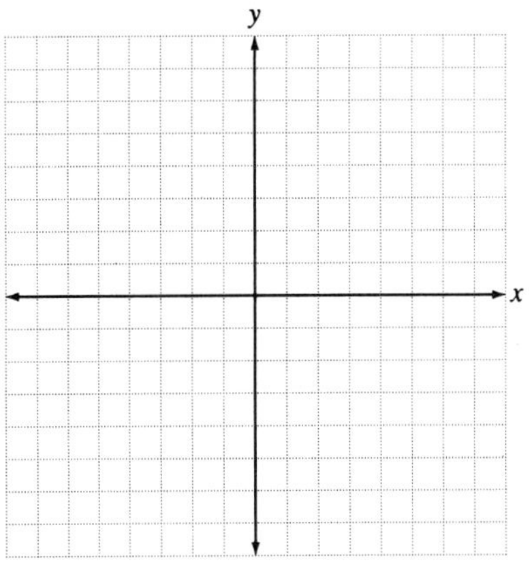

24. $y = -x - 1$

$\frac{4}{3}x - y = 8$

25. $y - 4x = -3$

$y - 4x = -1$

26. $y + 2x = 4$

$y + 2x = 1$

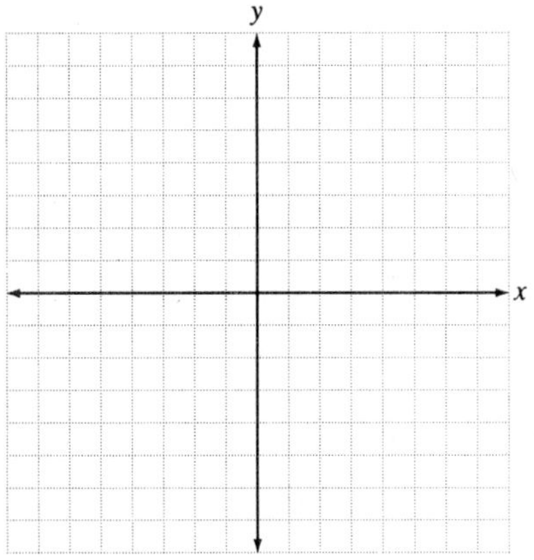

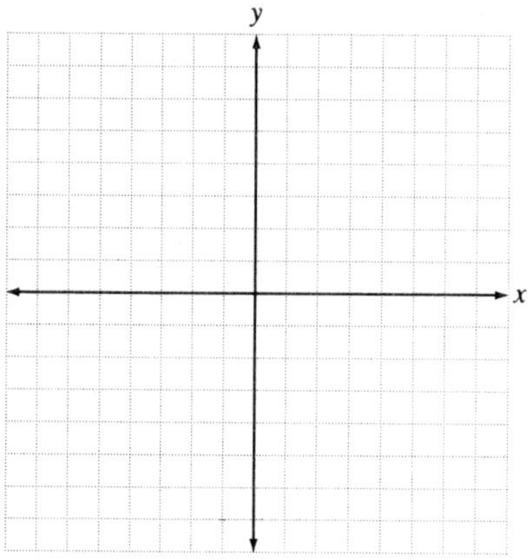

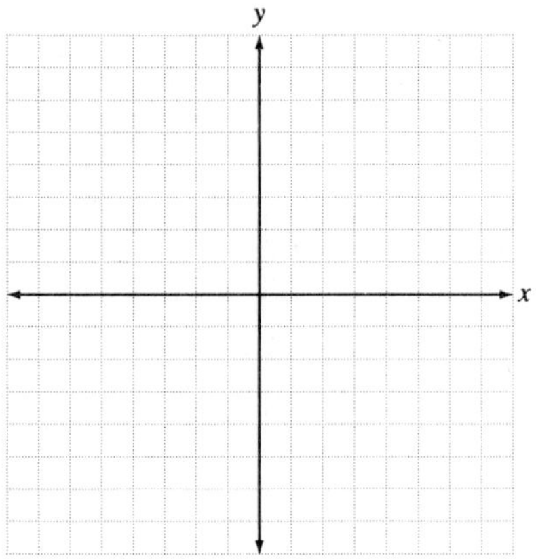

27. $x = -3$

$y = 2$

28. $x = 4$

$y = -1$

29. $2x - 3y = 6$

$-8x + 12y = -24$

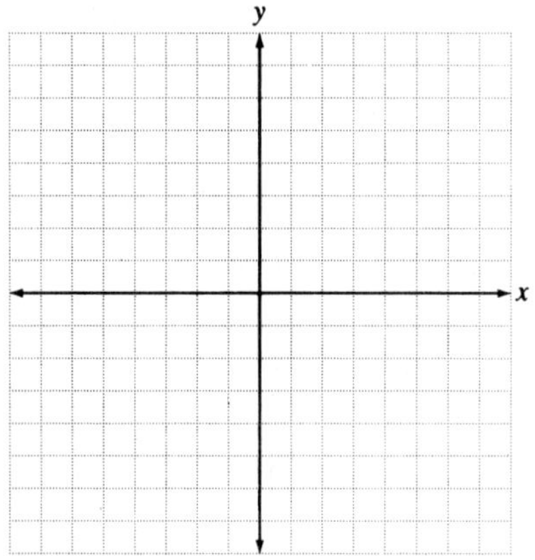

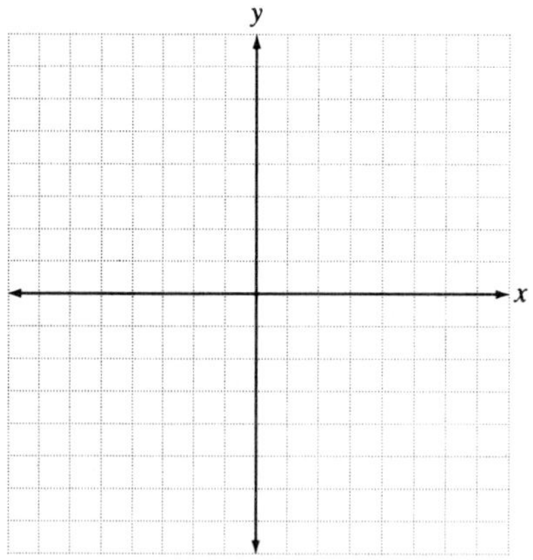

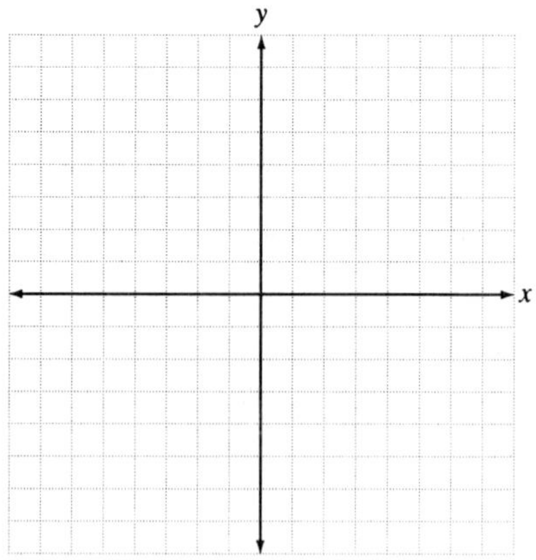

30. $\begin{aligned} 4x + y &= 3 \\ -12x - 3y &= -9 \end{aligned}$

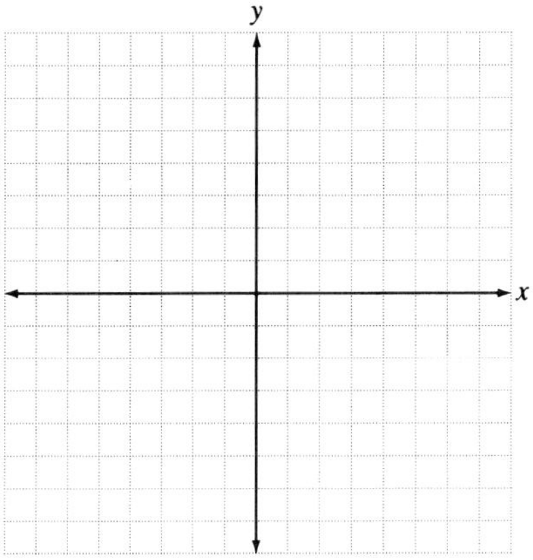

31. $\begin{aligned} y + 2x &= 4 \\ y + 2x &= -2 \end{aligned}$

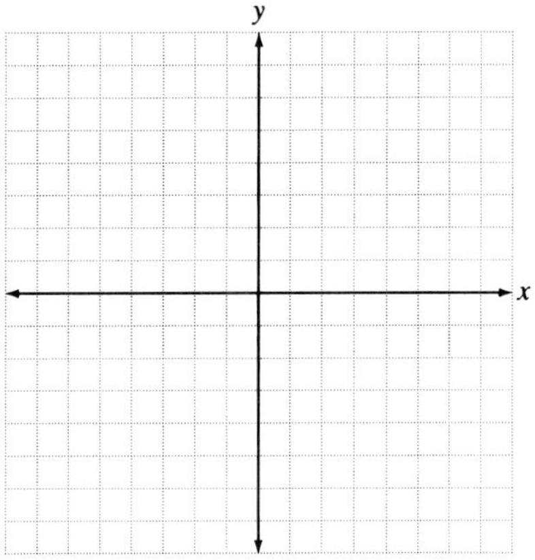

32. $\begin{aligned} y - 5x &= -1 \\ y - 5x &= -4 \end{aligned}$

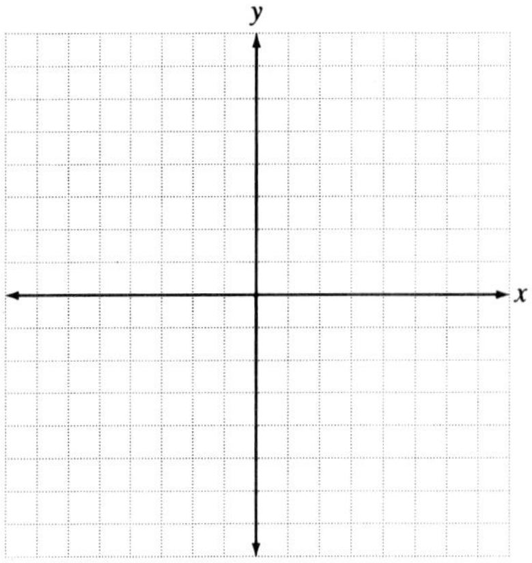

33. $\begin{aligned} x &= 0 \\ y &= 3 \end{aligned}$

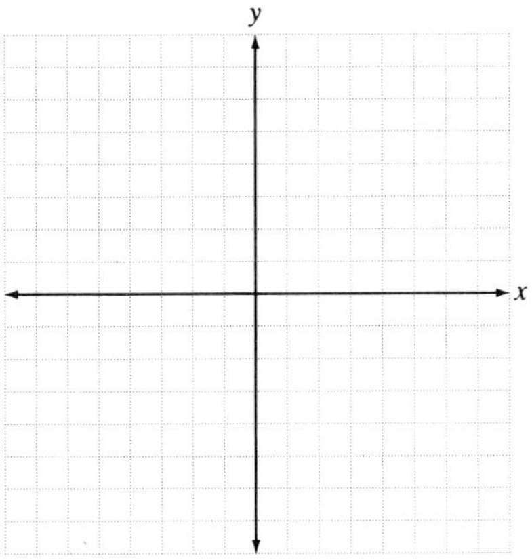

34. $\begin{aligned} x &= -1 \\ y &= 0 \end{aligned}$

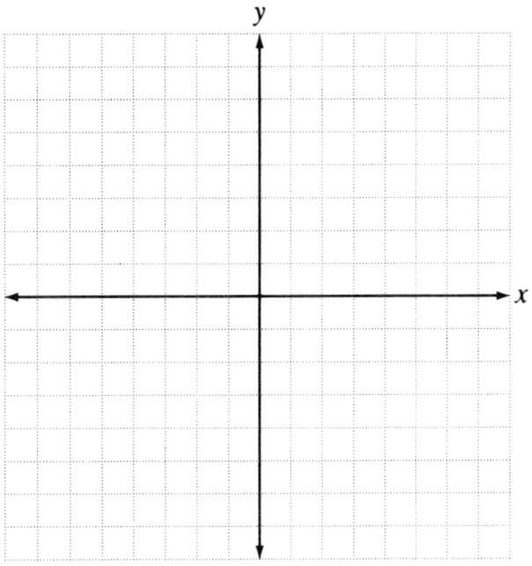

35. The solution of a system of two linear equations in two variables is all ____________ that satisfy both equations.

36. When the graphs of a system of two linear equations in two variables intersect, the system is called ____________.

37. If the graphs of the system are parallel lines, the system is ____________ and there is no solution.

38. When the graphs of the system of two linear equations in two variables are the same line, the system is ____________.

Is the ordered pair a solution of the given system?

39. $\begin{aligned} 2x + 3y &= -1 \\ x + 6y &= 2.2 \end{aligned}$ $(-1.4, 0.6)$

40. $\begin{aligned} 3x - y &= -1.23 \\ -5x + y &= 4.63 \end{aligned}$ $(2.2, 7.83)$

41. $\begin{aligned} 8x - 5y &= -8.4 \\ 2x + 7y &= 21 \end{aligned}$ $(0.7, 2.8)$

42. $\begin{aligned} 9x - 4y &= 18.2 \\ 6x + 7y &= 10.4 \end{aligned}$ $(2.2, -0.4)$

* **43.** $\begin{aligned} \frac{3}{5}x - \frac{3}{8}y &= \frac{39}{16} \\ \frac{1}{2}x + 5y &= -\frac{15}{8} \end{aligned}$ $\left(\frac{9}{4}, \frac{7}{10}\right)$

* **44.** $\begin{aligned} \frac{1}{4}x - 3y &= \frac{13}{8} \\ 6x + \frac{7}{4}y &= \frac{173}{12} \end{aligned}$ $\left(\frac{5}{2}, -\frac{1}{3}\right)$

Solve by graphing.

* **45.** $\begin{aligned} 2x + y &= 3 \\ 2x &= y + 7 \end{aligned}$

* **46.** $\begin{aligned} 4x + y &= 3 \\ y &= 2x \end{aligned}$

* **47.** $\begin{aligned} a &= b - 1 \\ 2a &= 3b \end{aligned}$

Think About It

* **48.** A line contains the points $(0, -4)$ and $(2, 2)$. Find three more points on the line.

Checkup

The following problems provide a review of some of Section 3.3.

Find the slope of the line through each pair of points.

49. $(5, 7), (3, -2)$

50. $(-1, -3), (-4, -5)$

51. $(6, -3), (7, -1)$

52. $(5, 3), (6, 3)$

53. $(-8, 2), (-8, -1)$

54. $(-7, 2), (-4, -1)$

55. $(-7, 4), (-9, 3)$

56. $(8, -2), (5, -6)$

57. $(8, 5), (-3, -7)$

58. $(-9, -3), (-7, -4)$

4.2 THE ELIMINATION AND SUBSTITUTION METHODS

OBJECTIVES

1. *Solve pairs of linear equations in two variables by the elimination method*
2. *Solve pairs of linear equations in two variables by the substitution method*

Solution by graphing allows us to see the geometric solution of a pair of equations. However, this method is slow and if the coordinates of the solution are fractions, it is difficult to read them from the graph.

1 The Elimination Method

The method most often used to solve pairs of equations is called the *elimination method.* We want to add the two equations so that one variable is eliminated and we have an equation in one variable that we can solve. This method may also be called the *addition method.*

EXAMPLE 1 Solve.

a. $3x + 2y = 1$

$4x - 2y = -8$

Add the two equations so that one variable is eliminated.

$$\begin{aligned} 3x + 2y &= 1 \\ 4x - 2y &= -8 \\ \hline 7x \qquad &= -7 \end{aligned} \qquad \text{Add.}$$

$$x = -1 \qquad \text{Divide both sides of the equation by 7.}$$

Remember that the solution must be an ordered pair. Now we must *solve for y.* Use either of the original equations and replace x by -1. Choose the equation that makes the arithmetic easier. We chose $3x + 2y = 1$.

$$3x + 2y = 1$$

$$3(\mathbf{-1}) + 2y = 1 \qquad \text{Substitute } -1 \text{ for } x.$$

$$-3 + 2y = 1$$

$$2y = 4 \qquad \text{Add 3 to both sides.}$$

$$y = 2$$

We found $x = -1$, $y = 2$, which gives the ordered pair $(-1, 2)$. Check this ordered pair by substituting -1 for x and 2 for y in *both* of the original equations.

$3x + 2y = 1$		$4x - 2y = -8$	
$3(-1) + 2(2)$	1	$4(-1) - 2(2)$	-8
$-3 + 4$	1	$-4 - 4$	-8
1	1	-8	-8

The ordered pair $(-1, 2)$ checks, so it is the solution.

Sometimes we must multiply one of the equations by a number before adding.

□ **Exercise 1** Solve.

a. $3x - y = 3$
$2x + y = 7$

b. $5x + 2y = -1$
$4x + y = -2$

c. $3x + 5y = -4$
$5x + 3y = -12$

d. $3x + 4y = 4$
$5x - 3y = 26$

b. $x + 2y = 1$ First equation.

$4x + 5y = 7$ Second equation.

If we add the two equations, we get $5x + 7y = 8$ and a variable is not eliminated. However, if we multiply both sides of the first equation by -4, we can eliminate the x variable when the equations are added.

$$\begin{aligned} -4x - 8y &= -4 && \text{Multiply the first equation by } -4. \\ 4x + 5y &= 7 && \text{Second equation.} \\ \hline -3y &= 3 && \text{Add.} \\ y &= -1 \end{aligned}$$

Now solve for x. Substitute -1 for y in one of the original equations. We chose $x + 2y = 1$.

$$\begin{aligned} x + 2y &= 1 \\ x + 2(\mathbf{-1}) &= 1 && \text{Substitute } -1 \text{ for } y. \\ x - 2 &= 1 \\ x &= 3 \end{aligned}$$

We found the pair $(3, -1)$. This checks in both original equations, so it is the solution.

Sometimes we must multiply both of the equations by a number before adding in order to eliminate a variable.

c. $2x + 3y = 12$ First equation.

$5x - 4y = 7$ Second equation.

If we add the two equations, we get $7x - y = 19$ and a variable is not eliminated. Multiply both sides of each equation by a number so that a variable disappears. One way to do this is to multiply the first equation by 4 and the second equation by 3.

$$\begin{aligned} 8x + 12y &= 48 && \text{First equation multiplied by 4 on both sides.} \\ 15x - 12y &= 21 && \text{Second equation multiplied by 3 on both sides.} \\ \hline 23x \qquad &= 69 && \text{Add.} \\ x &= 3 \end{aligned}$$

We chose to use the first equation to find y.

$$\begin{aligned} 2x + 3y &= 12 && \text{First equation.} \\ 2(\mathbf{3}) + 3y &= 12 && \text{Substitute 3 for } x. \\ 6 + 3y &= 12 \\ 3y &= 6 \\ y &= 2 \end{aligned}$$

The ordered pair $(3, 2)$ checks, so it is the solution.

We could also have multiplied the first equation by 5 and the second equation by -2, or the first equation by -5 and the second equation by 2, and eliminated x. The result would have been the same. ■

□ **DO EXERCISE 1.**

Occasionally, when we try to solve a system of two linear equations in two variables we find that the equations are dependent, and therefore they have infinitely many solutions.

EXAMPLE 2 Solve.

$$3x - y = 2 \qquad \text{First equation.}$$
$$6x - 2y = 4 \qquad \text{Second equation.}$$

We try to solve by multiplying the first equation by -2.

$$\begin{array}{rll} -6x + 2y = & -4 & \text{Multiply the first equation by } -2. \\ 6x - 2y = & 4 & \text{Second equation.} \\ \hline 0 = & 0 & \text{Add.} \end{array}$$

This is a true equation.

We could get the second equation by multiplying the first equation by 2. Hence the graph of the equations is the same line. The equations are dependent. When we try to solve a system of two linear equations in two variables and we get a true statement, we know that the system is *dependent* and there are an infinite number of ordered pairs that are solutions. The answer is: *infinite number of solutions.* ■

□ **DO EXERCISE 2.**

Some systems of equations are inconsistent, so they have no solution.

EXAMPLE 3 Solve.

$$x - 2y = 3 \qquad \text{First equation.}$$
$$-2x + 4y = 6 \qquad \text{Second equation.}$$

We try to solve by multiplying both sides of the first equation by 2.

$$\begin{array}{rll} 2x - 4y = & 6 & \text{Multiply the first equation by 2.} \\ -2x + 4y = & 6 & \text{Second equation.} \\ \hline 0 = & 12 & \text{Add.} \end{array}$$

We get a false statement: $0 = 12$. The graphs of these equations are parallel lines. There is no solution. The system is inconsistent. When we try to solve a system of two linear equations in two variables and we get a false statement, we know that the system is *inconsistent*. The answer is: *no solution.* ■

> If both variables are eliminated when a system of two linear equations in two variables is solved:
>
> 1. There is no solution if the resulting equation is false. The system is inconsistent and independent.
> 2. There are an infinite number of solutions if the resulting equation is true. The system is consistent and dependent.

To use the elimination method, it may be helpful to first write the equations in the form $Ax + By = C$. It is also usually easier to clear decimals or fractions before solving.

□ **Exercise 2** Solve.

a. $2x - 4y = 3$
$6x - 12y = 9$

b. $7x - 3y = 2$
$-14x + 6y = -4$

□ **Exercise 3** Solve.

a. $x - 5y = -2$
$-3x + 15y = -6$

b. $7x + 8y = -8$
$9x - 6y = 6$

Solving Linear Systems by Elimination

1. Write both equations in the form $Ax + By = C$.
2. Clear any decimals or fractions.
3. Multiply one or both equations by the appropriate numbers so that the sum of the coefficients of either variable is zero.
4. Eliminate a variable by adding the new equations and then solve for the remaining variable.
5. Substitute the result of step 4 in either of the original equations and solve for the other variable.

□ **DO EXERCISE 3.**

2 The Substitution Method

Sometimes in a linear system of equations one equation is already solved for one variable in terms of the other variable. In this case it is easier to use the substitution method to solve the system of equations. There are also other instances where we solve equations by substitution.

EXAMPLE 4 Solve.

a. $x + y = 5$ First equation.

$y = x + 3$ Second equation.

Notice that the second equation is solved for y in terms of x. We can substitute this expression for y in the first equation. When we substitute for y we replace y by $x + 3$. Be sure to use *parentheses* around the value that is substituted.

$x + y = 5$	First equation.
$x + (\mathbf{x + 3}) = 5$	Substitute $x + 3$, in parentheses, for y.
$2x + 3 = 5$	Remove parentheses and combine like terms.
$2x = 2$	Add -3 to both sides.
$x = 1$	Divide both sides by 2.

Use the original equation that is solved for y in terms of x to find y.

$y = x + 3$	Second equation.
$y = 1 + 3$	Substitute 1 for x.
$y = 4$	

Since $x = 1$ and $y = 4$ and these values check, the solution is $(1, 4)$.

b. $3x + y = -3$ First equation.

$x + 2y = 4$ Second equation.

To solve the system by substitution, we must solve one of the equations for one variable in terms of the other variable. If one of the equations has a variable with a coefficient of *one,* it is easier to solve for this variable. The coefficient of the y variable in the first equation is 1, so solve this equation for y. (We could also solve the second equation for x.)

$y = -3x - 3$ First equation solved for y.

Substitute $-3x - 3$ for y in the second equation.

$x + 2y = 4$ Second equation.

$x + 2(\mathbf{-3x - 3}) = 4$ Substitute $-3x - 3$ for y.

$x - 6x - 6 = 4$ Use a distributive law.

$-5x - 6 = 4$ Combine like terms.

$-5x = 10$ Add 6 to each side of the equation.

$x = -2$

To solve for y, substitute in the equation that was solved for y in terms of x.

$y = -3x - 3$ First equation solved for y.

$y = -3(-2) - 3$ Substitute -2 for x.

$y = 6 - 3$

$y = 3$

We found the ordered pair $(-2, 3)$ and this checks. The solution is $(-2, 3)$.

c. $x + \frac{y}{3} = -1$

$\frac{x}{4} + \frac{y}{2} = 1$

If there are fractions in an equation, multiply both sides of the equation by the least common denominator to clear the fractions.

$\mathbf{3}\left(x + \frac{y}{3}\right) = \mathbf{3}(-1)$ First equation multiplied on both sides by 3.

$\mathbf{4}\left(\frac{x}{4} + \frac{y}{2}\right) = \mathbf{4}(1)$ Second equation multiplied on both sides by 4.

The new system is

First equation: $3x + 3\left(\frac{y}{3}\right) = -3$

$3x + y = -3$

Second equation: $4\left(\frac{x}{4}\right) + 4\left(\frac{y}{2}\right) = 4$

$x + 2y = 4$

This system was solved in Example 4b. ■

□ **Exercise 4** Solve. Use the substitution method.

a. $x = 2y$
$2x - 5y = -2$

b. $3x + y = -8$
$4x - 2y = 6$

c. $\frac{x}{2} - \frac{y}{3} = 1$
$\frac{x}{10} + \frac{y}{5} = 1$

d. $x + \frac{y}{3} = -\frac{1}{3}$
$\frac{2x}{3} - y = -\frac{8}{3}$

Solving Linear Systems by Substitution

1. Clear any decimals or fractions.
2. Solve one of the equations for either variable. If one of the equations has a variable with a coefficient of 1 or -1, solve for this variable.
3. Substitute for that variable in the other equation.
4. Solve the equation from step 3.
5. Substitute the result from step 4 into the equation from step 2 to solve for the other variable.

□ **DO EXERCISE 4.**

DID YOU KNOW?

Problems involving two variables were found on papyri from ancient Egypt (1950 B.C.)

Answers to Exercises

1. **a.** (2, 3) **b.** $(-1, 2)$ **c.** $(-3, 1)$ **d.** $(4, -2)$
2. **a.** Infinite number of solutions **b.** Infinite number of solutions
3. **a.** No solution **b.** $(0, -1)$
4. **a.** (4, 2) **b.** $(-1, -5)$ **c.** (4, 3) **d.** $(-1, 2)$

Problem Set 4.2

NAME

DATE

CLASS

Solve using the elimination method.

1. $x + y = 4$
$3x - y = 8$

2. $x - 5y = -5$
$-x + 2y = -1$

3. $-x + 3y = -1$
$x - 4y = 2$

4. $4x - y = -5$
$2x + y = -7$

5. $7x - 3y = 1$
$2x + 3y = 8$

6. $3x + 4y = -1$
$5x - 4y = 9$

7. $3x + y = 15$
$2x - 5y = 10$

8. $x + 2y = 11$
$7x - 3y = -25$

9. $-7x - 2y = -3$
$14x + 4y = 6$

10. $3x + 4y = 2$
$6x + 8y = 4$

11. $5x + 2y = 19$
$2x - 5y = -4$

12. $7x + 3y = -26$
$5x - 4y = 6$

13. $8x + 3y = 1$
$7x - 2y = -13$

14. $3x + 4y = -7$
$4x - 3y = -26$

15. $5x - 3y = -4$
$10x - 6y = -9$

16. $2x + 7y = 4$
$6x + 21y = 8$

17. $\frac{x}{2} + \frac{y}{3} = -\frac{1}{3}$
$\frac{x}{2} + 2y = -7$

18. $\frac{x}{3} + \frac{y}{5} = 2$
$\frac{x}{3} - \frac{y}{2} = -\frac{1}{3}$

19. $\frac{x}{9} + \frac{y}{9} = 1$
$\frac{2x}{3} - \frac{y}{3} = 1$

20. $\frac{x}{6} - \frac{y}{3} = 1$
$-\frac{x}{4} + \frac{3y}{4} = -1$

Solve using the substitution method.

21. $4x + y = 3$
$y = 2x$

22. $x - 6y = 1$
$x = 3y$

23. $9x - 2y = 3$
$y = 3x - 6$

24. $x + 2y = 9$
$x = 3y - 3$

25. $5x - 2y = -20$
$3 - x = y$

26. $3x - 4y = -10$
$6 - y = x$

27. $5x + y = 8$
$3x - 4y = 14$

28. $-3x + y = 7$
$-4x + 6y = 14$

29. $5x - 4y = 9$
$x - 2y = 3$

30. $3x + 5y = 17$
$4x - y = -8$

31. $-3x + 2y = -1$
$6y = 9x + 7$

32. $y = \frac{2}{3}x + 4$
$3y - 2x = 12$

Solve by any method.

33. $2x - y = -2$
$5x + y = -12$

34. $7x + 3y = 4$
$6x - 3y = 9$

35. $x + 3y = 27$
$y = \frac{2}{3}x$

36. $8x - y = 8$
$x = \frac{1}{4}y$

37. $3x + 7y = 10$
$7x - 3y = 4$

38. $x + 4y = 14$
$2x = y + 1$

39. $-0.3x + 0.5y = 0.2$
$0.2x - 0.3y = 0.1$

40. $0.1x + 0.4y = 1.4$
$0.2x = 0.1y + 0.1$

41. $5p = 3q + 24$
$3p + 5q = 28$

42. $5x = 7y - 16$
$2x + 8y = 26$

43. $4y - 2x = 5$
$-16y = -8x - 20$

44. $3x - 7y = 4$
$-15x = -35y - 20$

45. $5x = 4y + 9$
$x = 2y - 3$

46. $6x = y - 9$
$y = -7x - 4$

47. $\frac{x}{6} + \frac{y}{9} = 1$
$y = \frac{3}{2}x$

48. $\frac{x}{3} + \frac{y}{5} = 2$
$\frac{x}{3} - \frac{y}{2} = -\frac{1}{3}$

49. We may begin to solve a linear system of two equations in two variables by ____________ the two equations so that one variable is eliminated.

50. Sometimes we must multiply ____________ or ____________ of the equations by a number before adding them together.

51. Two linear equations in two variables that are dependent have an ____________ ____________ of solutions.

52. Linear equations that are ____________ have no solution.

53. To solve a linear system of two equations in two variables by substitution, we must solve one of the equations for one ____________ in terms of the other variable.

54. If there are fractions in an equation, multiply both sides of the equation by the ____________ ____________ ____________ to clear the fractions.

Solve using the elimination method.

55. $4.2x + 3.1y = 0.9$
$6.3x - 6.2y = -31.2$

56. $5.4x - 2.3y = 20$
$10.8x + 6.9y = -6$

57. $1.5x + 0.25y = 4.7$
$0.75x - 0.5y = 1.85$

58. $0.45x + y = -0.01$
$-2.3x - 0.8y = -8.14$

Solve.

* **59.** $\frac{x}{8} - \frac{y}{10} = \frac{9}{2}$
$y = 5x$

* **60.** $\frac{1}{6}x + \frac{1}{9}y = \frac{4}{3}$
$\frac{1}{8}x + \frac{1}{6}y = \frac{3}{2}$

Solve for x and y in terms of a and b.

* **61.** $ax + by = 2$
$-ax + 2by = 1$

* **62.** $2ax - y = 6$
$y = 5ax$

Checkup

The following problems provide a review of some of Section 2.3 and will help you with the next section.

63. The sum of two numbers is 5. Three times the larger plus the smaller is 23. Find the smaller number.

64. The sum of two numbers is 3. Twice the smaller plus the larger is 1. Find the larger number.

65. Sarah has a total of 24 dimes and nickels. They are worth \$1.45. How many dimes does she have?

66. Dennis has three times as many quarters as nickels. If the value of the coins is \$6.40, how many nickels does he have?

67. Cheryl invested \$12,000 for 1 year and earned \$680 interest. If part of the money was invested at 6% interest per year and part at 5% interest per year, how much was invested at 5% interest?

68. A total of \$32,000 was invested in two certificates of deposit for 1 year. Part of the money was invested at 6% interest per year and part at 8% interest per year. If the money earned \$2320 interest, how much was invested at 6% interest?

69. The sum of Aaron and Heather's ages is 99 and the difference in their ages is 9 years. If Aaron is older, how old is Heather?

70. Jim is 4 times as old as Tami. The sum of their ages is 115. Find Tami's age.

71. Panos earned \$270 on a sum of money invested for 1 year. If the amount invested at 4% interest was \$2000 less than the amount invested at 3% interest, how much was invested at 4% interest?

72. Gloria invested some money at 6% interest per year and \$5500 more than that amount at 7% interest per year. She earned \$645 interest. How much did she invest at 7% interest?

4.3 APPLIED PROBLEMS

OBJECTIVE

[1] *Use two variables and two equations to solve word problems about perimeter, money, interest, mixtures, and angles*

[1] We solved applied problems in Chapter 2 using one variable because knowledge of how to work with one variable is necessary for more advanced mathematics. However, it is often easier to solve word problems using two variables. The steps in solving a word problem are shown in Section 2.2.

EXAMPLE 1 The perimeter of a rectangle is 210 feet. If the length is twice the width, what are the dimensions of the rectangle?

Variables Let L represent the length and W represent the width.

Drawing

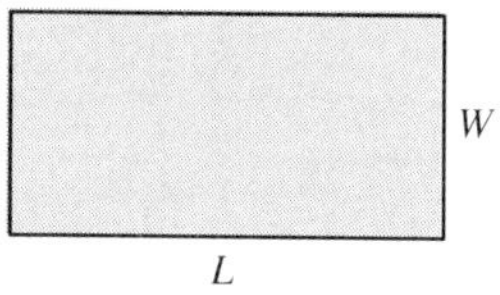

The formula for the perimeter of a rectangle is $2L + 2W = P$. The perimeter is 210.

$2L + 2W = P$ Perimeter formula.

$2L + 2W = 210$ First equation.

The length is twice the width.

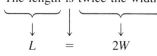

$L = 2W$ Second equation.

We now have the following equations.

Equations $2L + 2W = 210$ First equation.

$L = 2W$ Second equation.

Solve It is easy to use the substitution method to solve this system.

$2(2W) + 2W = 210$ Substitute $2W$ for L in the first equation.

$4W + 2W = 210$

$6W = 210$

$W = 35$

Substitute 35 for W in the second equation to find L.

$$L = 2W = 2(35) = 70$$

Check The perimeter of the rectangle is $2L + 2W = 2(70) + 2(35) = 210$. The length of 70 is twice the width of 35. The answer checks.
The dimensions of the rectangle are 35 feet and 70 feet. ■

☐ **DO EXERCISE 1.**

☐ **Exercise 1** Maria has 130 yards of fencing. She wants to build a rectangular fence with the width 25 yards less than the length. What should be the dimensions of the fence?

EXAMPLE 2 The Johnson family bought a total of 20 hamburgers and sundaes at Cindy's Restaurant for $15.40. If hamburgers cost 85 cents and sundaes cost 65 cents, how many of each did they buy?

□ **Exercise 2** The Greeting Card Shop sold a total of 20 Christmas cards and birthday cards. They received $21.80. If Christmas cards cost $1.25 and birthday cards cost 85 cents, how many of each did they sell?

Variables Let x represent the number of hamburgers and y represent the number of sundaes. They bought 20 hamburgers and sundaes.

$x + y = 20$ First equation.

	$\begin{pmatrix}\text{Cost of}\\ \text{each}\end{pmatrix}$	$\cdot$	$\begin{pmatrix}\text{number of}\\ \text{each}\end{pmatrix}$	$=$	$\begin{pmatrix}\text{total}\\ \text{value}\end{pmatrix}$
Hamburgers	0.85	$\cdot$	x	$=$	$0.85x$
Sundaes	0.65	$\cdot$	y	$=$	$0.65y$

$$\begin{pmatrix}\text{Total value}\\ \text{of hamburgers}\end{pmatrix} + \begin{pmatrix}\text{total value}\\ \text{of sundaes}\end{pmatrix} = \begin{pmatrix}\text{total value of}\\ \text{hamburgers and sundaes}\end{pmatrix}$$

$$0.85x + 0.65y = 15.40$$

Multiply both sides by 100 to clear the decimals.

$$85x + 65y = 1540 \quad \text{Second equation.}$$

We have the following equations.

Equations

$x + y = 20$ First equation.

$85x + 65y = 1540$ Second equation.

Solve

$-65x - 65y = -1300$ Multiply the first equation by -65.

$85x + 65y = 1540$ Second equation.

$20x = 240$ Add.

$x = 12$

Since $x + y = 20$ and $x = 12$, $y = 8$. We found that they bought 12 hamburgers and 8 sundaes.

Check The value of the hamburgers is $0.85(12) = \$10.20$. The value of the sundaes is $0.65(8)$ or $5.20. The total value is $\$10.20 + \$5.20 = \$15.40$. Since the number of hamburgers plus the number of sundaes is 12 plus 8, or 20, the results check in the original problem.

The Johnson family bought 12 hamburgers and 8 sundaes. ■

CALCULATOR

Example 2 may also be worked using a calculator.

Equations

$x + y = 20$ First equation.

$0.85x + 0.65y = 15.40$ Second equation.

Solve

$-0.65x - 0.65y = -13.00$ Multiply the first equation by -0.65.

$0.85x + 0.65y = 15.40$ Second equation.

$0.20x = 2.40$ Add.

$x = 12$

$x + y = 12$. Hence $y = 8$.

The Johnson family bought 12 hamburgers and 8 sundaes.

□ **DO EXERCISE 2.**

The following problems may also be done with a calculator.

EXAMPLE 3 The total interest earned on a sum of money for 1 year was \$228. If the amount invested at 7% interest per year was \$400 less than the amount invested at 9% interest per year, how much was invested at each rate?

Variables Let x represent the amount invested at 9% and y represent the amount invested at 7%.

One of the equations may be written as follows. The amount invested at 7% was \$400 less than the amount invested at 9%.

$$y = x - 400$$

Recall the simple interest formula, $Prt = I$.

P	$\cdot$	r	$\cdot$	t	$=$	I
x	$\cdot$	0.09	$\cdot$	1	$=$	$0.09x$
y	$\cdot$	0.07	$\cdot$	1	$=$	$0.07y$

Interest earned at 9%	+	interest earned at 7%	=	total interest
$0.09x$	+	$0.07y$	=	228

Multiply both sides by 100 to clear the decimals.

$$9x + 7y = 22{,}800 \quad \text{Second equation.}$$

Equations

$$y = x - 400 \quad \text{First equation.}$$

$$9x + 7y = 22{,}800 \quad \text{Second equation.}$$

Solve Use the substitution method to solve the equations.

$$9x + 7(\mathbf{x - 400}) = 22{,}800 \quad \text{Substitute for } y \text{ in the second equation.}$$

$$9x + 7x - 2800 = 22{,}800 \quad \text{Use a distributive law.}$$

$$16x - 2800 = 22{,}800$$

$$16x = 25{,}600$$

$$x = 1600$$

Since $y = x - 400$, we found that $y = 1200$. These results check.

\$1600 was invested at 9% and \$1200 was invested at 7%. ■

□ **DO EXERCISE 3.**

□ **Exercise 3** Richard invested \$1500 for 1 year and earned \$106 interest. If part of the money was invested at 6% interest per year and part of the money was invested at 8% interest per year, how much was invested at each rate?

□ **Exercise 4** Jon has a 25% alcohol solution and a 40% alcohol solution. He wants to make 100 grams of a solution that is 34% alcohol. How much of each solution should he use?

EXAMPLE 4 A chemist wants to make 60 liters of a 25% salt solution by mixing a 20% solution with a 35% solution. How many liters of the 20% solution and 35% solution should she use?

Variables Let x represent the liters of 20% solution and y represent the liters of 35% solution. The total number of liters of solution is 60, so

$$x + y = 60 \qquad \text{First equation.}$$

(Percent of salt in solution) ·	(liters of solution) =	(liters of pure salt)	
20	x	$0.20x$	Use these facts in the second equation.
35	y	$0.35y$	
25	60	$0.25(60)$	

The liters of pure salt in the solutions that are mixed must equal the liters of pure salt in the final solution.

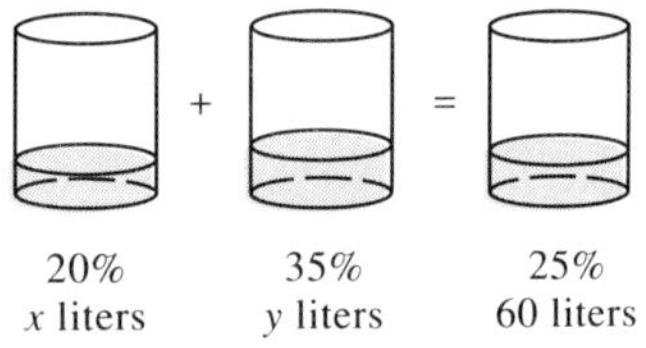

$$\begin{pmatrix}\text{Liters of}\\\text{pure salt}\\\text{in 20\%}\\\text{solution}\end{pmatrix} + \begin{pmatrix}\text{liters of}\\\text{pure salt}\\\text{in 35\%}\\\text{solution}\end{pmatrix} = \begin{pmatrix}\text{liters of}\\\text{pure salt}\\\text{in 25\%}\\\text{solution}\end{pmatrix}$$

$$0.20x + 0.35y = 0.25(60) \qquad \text{Second equation.}$$

Multiply both sides by 100 to clear the decimals.

$$20x + 35y = 25(60)$$

$$20x + 35y = 1500 \qquad \text{Second equation.}$$

We now have the following system of equations.

Equations

$$x + y = 60$$
$$20x + 35y = 1500$$

When we solved this system we found that $x = 40$ and $y = 20$.

Check If we add the 40 liters of 20% solution and the 20 liters of 35% solution, we get a total of 60 liters. The liters of pure salt in the 40 liters of 20% solution is $0.20(40) = 8$. The liters of pure salt in the 20 liters of 35% solution is $0.35(20) = 7$. The sum of the liters of salt in these two solutions, $8 + 7$, or 15, must equal the liters of pure salt in the 60 liters of the 25% final solution, $0.25(60)$, which is also 15. The answer checks.
The chemist should use 40 liters of the 20% solution and 20 liters of the 35% solution. ■

□ **DO EXERCISE 4.**

EXAMPLE 5 Two angles are complementary. One angle is 8° more than 4 times the other. Find the measures of the angles.

Variables Let x represent the measure of one angle and y represent the measure of the other angle.
Recall that two angles are complementary if the sum of their measures is 90°.

Drawing

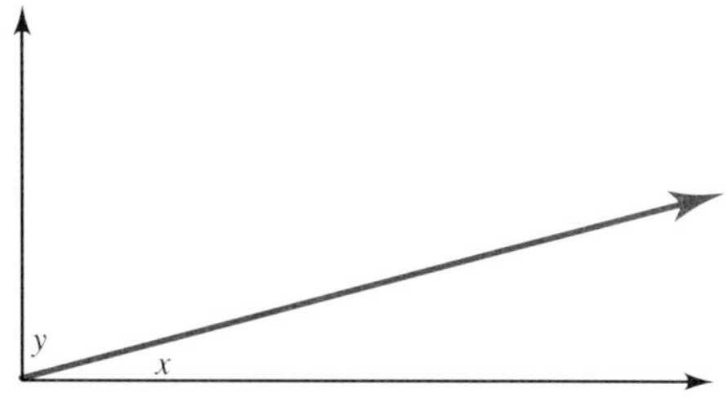

Complementary Angles

The sum of the measures of the angles is 90°.

$$x + y = 90 \qquad \text{First equation.}$$

One angle is 8 degrees more than 4 times the other.

$$y \quad = \quad 8 \quad + \quad 4x \qquad \text{Second equation.}$$

Equations $x + y = 90$ First equation.

$y = 8 + 4x$ Second equation.

Solve $x + (8 + 4x) = 90$ Substitute $8 + 4x$ for y in the first equation.

$$x + 8 + 4x = 90$$
$$5x + 8 = 90$$
$$5x = 82$$
$$x = 16.4$$
$$y = 8 + 4x = 8 + 4(16.4) = 73.6$$

Check The sum of the measures of the angles is 16.4° + 73.6° or 90°, so they are complementary. Eight more than 4 times the 16.4° angle is 73.6°, which is the measure of the second angle. The answer checks.

The measure of one angle is 16.4° and the measure of the other angle is 73.6°. ■

□ **DO EXERCISE 5.**

DID YOU KNOW?

According to records found on clay tablets, annual interest rates as high as 33% were charged around 2000 B.C. in Babylonia.

□ **Exercise 5** Two angles are complementary. One angle is 6 degrees less than twice the other angle. Find the measures of the angles.

Answers to Exercises

1. $W = L - 25$

$$\begin{aligned} 2L + 2W &= 130 \\ 2L + 2(L - 25) &= 130 \\ 2L + 2L - 50 &= 130 \\ 4L - 50 &= 130 \\ 4L &= 180 \\ L &= 45 \end{aligned}$$

$W = L - 25 = 45 - 25 = 20$

The length is 45 yards and the width is 20 yards.

2.

$$\begin{aligned} x + y &= 20 \\ 1.25x + 0.85y &= 21.80 \\ 125x + 85y &= 2180 && \text{Multiply by 100.} \\ -85x - 85y &= -1700 && \text{Multiply the first equation by } -85. \\ \hline 40x \qquad &= 480 \\ x &= 12 \end{aligned}$$

$x + y = 20,\ y = 20 - x = 20 - 12 = 8$

They sold 12 Christmas cards and 8 birthday cards.

3.

$$\begin{aligned} x + y &= 1500 \\ 0.06x + 0.08y &= 106 \\ 6x + 8y &= 10{,}600 \\ -6x - 6y &= -9000 \\ \hline 2y &= 1600 \\ y &= 800 \end{aligned}$$

$x + y = 1500,\ x = 1500 - y,\ x = 1500 - 800 = 700$

Richard invested \$800 at 8% and \$700 at 6%.

4.

$$\begin{aligned} x + y &= 100 \\ 0.25x + 0.40y &= 0.34(100) \\ 25x + 40y &= 3400 \\ -25x - 25y &= -2500 \\ \hline 15y &= 900 \\ y &= 60 \end{aligned}$$

$x + y = 100,\ x = 100 - y = 100 - 60 = 40$

Jon should use 60 grams of 40% solution and 40 grams of 25% solution.

5.

$$\begin{aligned} x + y &= 90 \\ y &= 2x - 6 \\ x + (2x - 6) &= 90 \\ x + 2x - 6 &= 90 \\ 3x - 6 &= 90 \\ 3x &= 96 \\ x &= 32 \end{aligned}$$

$y = 2x - 6 = 2(32) - 6 = 58$

The measures of the angles are 32 degrees and 58 degrees.

Problem Set 4.3

NAME

DATE

CLASS

1. The perimeter of a rectangle is 144 centimeters. If the length is 3 times the width, find the dimensions of the rectangle.

2. If the perimeter of a rectangle is 96 inches and the length is 5 times the width, what are the dimensions of the rectangle?

3. The perimeter of a rectangular house is 182 feet. If the width is 16 feet less than the length, what are the dimensions of the house?

4. Karen fenced a rectangular yard with 172 feet of fencing. If the length of the yard is 30 feet greater than the width, what are the dimensions of the fence?

5. The perimeter of a rectangle is 108 meters. If the length is 3 more than twice the width, what are the dimensions of the rectangle?

6. The length of a rectangle is 7 less than 3 times the width and the perimeter of the rectangle is 114 yards. Find the dimensions of the rectangle.

7. Kevin bought a total of 8 pounds of apples and oranges. Apples cost 65 cents a pound and oranges cost 59 cents a pound and Kevin spent $5.08. How many pounds of each did he buy?

8. A store sold a total of 20 pounds of candy. If the peanut brittle cost $6.50 per pound and the fudge cost $5.75 per pound and the store took in $119.50, how much of each type of candy was sold?

9. Tickets to an amusement park cost $16 for adults and $11 for children. The total receipts were $1500. If 4 times as many adult tickets as children's tickets were sold, how many of each type of ticket were sold?

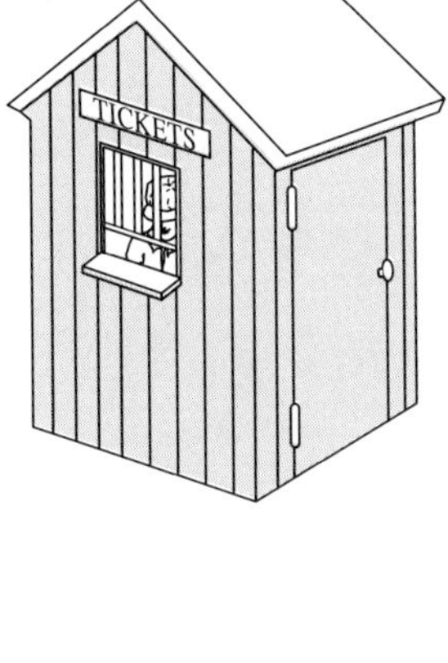

10. Jan deposited six more $5 bills than $10 bills in the bank. The bills were worth $210. How many of each denomination did she deposit?

11. A total of $1800 was invested for 1 year and the money earned $173 interest. If part of the money was invested at 9% interest per year and part at 10% interest per year, how much was invested at each rate?

12. A woman invested $15,000 and she earned an annual interest of $1350. She put some of the money in the bank at 6% interest per year and invested the remainder at 11% interest per year. How much did she invest at each rate?

13. The total interest earned on a sum of money for 1 year was $148. If $700 less was invested at 8% interest per year than was invested at 9% interest per year, how much was invested at each rate?

14. John invested a sum of money for 1 year and earned $920 interest. If $8000 more was invested at 7% interest per year than was invested at 5% interest per year, how much did he invest at each rate?

15. How much 15% alcohol solution and 40% alcohol solution should be mixed together to make 5 gallons of a 25% alcohol solution?

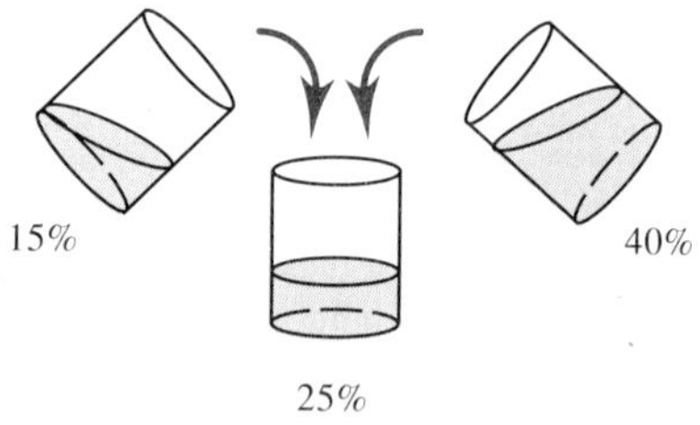

16. A 16% salt solution is to be mixed with a 9% salt solution to get 350 pounds of a 12% salt solution. How many pounds of each should be used?

17. A chemical technician combines a 20% acid solution with a 40% acid solution to obtain 12 liters of a 25% solution. How many liters of the 20% solution and how many liters of the 40% solution should he use?

18. Linda wants to mix a 15% alcohol solution with a 35% alcohol solution to get 100 milliliters of a 20% solution. How many milliliters of each solution should she use?

19. Susan bought three times as many candy bars as cans of pop and spent $5.49. If pop cost 48 cents a can and candy bars were 45 cents each, how many of each did she buy?

20. A store sold 3 more red sweaters than white sweaters. If red sweaters cost $25 and white sweaters cost $20 and the store received $435 for the sweaters, how many of each kind did they sell?

21. A total of $6000 is invested, part at 9% per year and part at 11% per year. If the interest for 1 year from the two investments is the same, how much is invested at each rate?

22. Ken invested $4000 for 1 year. Part of the money was invested at 8% per year and part of the money was invested at 12% per year. How much was invested at each rate if the interest from the two investments was the same?

23. How many liters of a 5% acid solution should be added to 20 liters of a 30% acid solution to obtain a 25% acid solution?

24. How many grams of an alloy containing 15% silver must be melted with 40 grams of an alloy containing 6% silver to obtain an alloy containing 10% silver?

25. The sum of two numbers is 8. Three times the first number minus twice the second number is 36. Find the numbers.

26. The sum of two numbers is -52. Twice the first number minus the second number is 31. Find the numbers.

27. The perimeter of a rectangle is 28 meters. If the length is 8 meters more than twice the width, find the length of the rectangle.

28. The length of a rectangle is 6 inches less than 5 times the width and the perimeter is 72 inches. Find the width of the rectangle.

29. Two angles are complementary. One angle is 4° less than 4 times the other angle. Find the measures of the angles.

30. Two angles are complementary. One angle is 7° more than twice the other angle. Find the measures of the angles.

31. At a banquet 275 dinners were served. Adult dinners cost $15 and children's dinners cost $8.50. How many of each dinner was served if the total receipts were $3800?

32. Karen's Bath Shop sold 58 towels. Bath towels cost $9.75 each and hand towels cost $6.50 each. If the total amount of money collected was $503.75, how many of each type of towel was sold?

33. Laura invested her savings. The amount invested at 7.1% interest per year was \$1000 more than twice the amount invested at 8.5% interest per year. How much did she invest at 7.1% interest if the total interest earned for 1 year was \$1433?

34. Juan invested in two certificates of deposit. One earned 7.5% interest per year and the other certificate earned 9.4% interest per year. The amount invested at 9.4% was \$2000 less than twice the amount invested at 7.5%. How much did he have invested at 7.5% interest if the total interest earned for 1 year was \$2705?

35. Solution A is 6% alcohol and solution B is 14% alcohol. A nurse wants to mix the two in order to get 20 milliliters of an 8.4% alcohol solution. How many milliliters of each should she use?

36. A biologist wants to make a 45.4% acid solution. How many liters of 25% acid solution should be mixed with 64 liters of a 70% acid solution to make the 45.4% solution?

Two angles are supplementary if the sum of their measures is 180°.

37. Two angles are supplementary. One angle is 24° more than twice the other. Find the measures of the angles.

38. Two angles are supplementary. One angle is 54° less than 4 times the other angle. Find the measures of the angles.

39. To find the perimeter of a rectangle, we add ____________ the length to ____________ the width.

40. The formula for simple interest is ____________.

41. If two salt solutions are mixed, the sum of the liters of pure salt in each solution must equal the liters of pure salt in the ____________ solution.

42. The perimeter of a rectangle is 165.42. If the length is 17.89 more than the width, what are the dimensions of the rectangle?

43. Karen bought a total of 13 pizzas and soft drinks for \$28.27. If pizzas cost \$4.39 and soft drinks cost \$0.79, how many of each did she buy?

44. The total interest earned on a sum of money for 1 year was \$365.75. If \$350 more was invested at 8.5% per year than was invested at 7.5% per year, how much was invested at each rate?

45. Jim wants to make 25.2 liters of a 35% salt solution by mixing a 15% solution with 55% solution. How many liters of 55% solution should he use? (Round your answer to the nearest tenth.)

Think About It

* **46.** Susan loaned \$20,000. Part of the money was loaned at 15% interest per year and the remainder at 18% interest per year. At the end of 1 year the money loaned at 18% interest earned \$1950 more than the money loaned at 15% interest. How much money was loaned at 18%?

* **47.** Mike bought 10 pounds of sugar and 2 pounds of flour for \$5.22. The next day he bought 5 pounds of sugar and 10 pounds of flour for \$5.40. How much did the flour cost per pound?

* **48.** A clerk has $3\frac{1}{2}$ times as many dimes as quarters. The value of the coins is \$25.20. How many coins of each kind does he have?

* **49.** A salesman earned an 8% commission and a 12% commission on two sales. The 12% commission was \$2000 less than twice the amount of the 8% commission. The sum of the two commissions was the same as a 10% commission on the total of both sales. What was the amount of each sale?

* **50.** Jim's automobile radiator contains 16 liters of antifreeze solution that is 60% antifreeze. How much of this solution should be drained and replaced with water to give a 45% antifreeze solution?

* **51.** A line passes through the points $(1, 2)$ and $(-1, 8)$. Another line passes through the point $(-1, 2)$ and has slope $\frac{1}{2}$. Where do the lines intersect?

* **52.** There were 26 A's given in a math class. If 20% of the men and 30% of the women in the class received A's and the number of men in the class doubled is the same as the number of women in the class, plus 20, how many women are in the class?

* **53.** A store sold women's suits at a discount. If the suits had been sold at their original prices, C. Fashon suits at \$180 and D. Stile suits for \$200, the total receipts would have been \$11,500. However, the suits were sold for \$150 and \$160, respectively, and the total receipts were \$9350. Find the number of suits of each type that were sold.

Checkup

The following problems provide a review of some of Section 3.4.

Find an equation for the line with the given slope and y-intercept.

54. $m = \frac{1}{3}, b = 2$

55. $m = 4, b = -\frac{1}{2}$

Find an equation for the line with the given slope through the given point.

56. $m = \frac{8}{7}, (-2, -3)$

57. $m = 3, (2.8, -0.3)$

58. $m = -\frac{5}{8}, (-4, 7)$

59. $m = -5, \left(\frac{3}{5}, -2\right)$

Find an equation for the line through each pair of points.

60. $(7, 2), (3, -1)$

61. $(-4, -5), (1, 3)$

62. $(-5, -6), (-8, -9)$

63. $(-2, 1), (-7, -4)$

4.4 SYSTEMS OF EQUATIONS IN THREE VARIABLES

OBJECTIVES

1 *Determine if a given ordered triple is a solution of a system of equations*

2 *Solve a system of three linear equations in three variables by the elimination method*

Sometimes, to solve applied problems, we use a system of equations in three variables.

1 Solutions

A *solution* of a system of equations in three variables is an *ordered triple* (x, y, z) that satisfies all three equations.

EXAMPLE 1 Is the ordered triple $(2, \frac{1}{2}, -4)$ a solution of the following system?

$$x + 2y + z = -1$$
$$3x - 4y + 2z = -4$$
$$2x + 4y - z = 10$$

Substitute 2 for x, $\frac{1}{2}$ for y and -4 for z in *each* equation.

$$\begin{array}{r|r} \multicolumn{2}{c}{x + 2y + z = -1} \\ \hline 2 + 2\left(\frac{1}{2}\right) + (-4) & -1 \\ 2 + 1 - 4 & -1 \\ -1 & -1 \end{array} \qquad \begin{array}{r|r} \multicolumn{2}{c}{3x - 4y + 2z = -4} \\ \hline 3(2) - 4\left(\frac{1}{2}\right) + 2(-4) & -4 \\ 6 - 2 - 8 & -4 \\ -4 & -4 \end{array}$$

$$\begin{array}{r|r} \multicolumn{2}{c}{2x + 4y - z = 10} \\ \hline 2(2) + 4\left(\frac{1}{2}\right) - (-4) & 10 \\ 4 + 2 + 4 & 10 \\ 10 & 10 \end{array}$$

The ordered triple $(2, \frac{1}{2}, -4)$ satisfies all three equations, so it is a solution. ■

□ **DO EXERCISE 1.**

□ **Exercise 1** Determine if the ordered triple is a solution to the following system.

$$2x - y + z = 8$$
$$3x + 2y - 4z = -8$$
$$x - 2y + 3z = 13$$

a. $(2, -1, 3)$

b. $(0, 2, -4)$

The definition of a linear equation may be enlarged to include equations with any finite number of terms with exponents of one on the variables, but no term may contain a product or quotient of variables.

Theoretically, we could solve a system of linear equations in three variables by graphing. The graph of a linear equation in three variables is a plane and we would need to know how to graph in three dimensions. We will not discuss this type of graphing, but we can see the graphs of the possible solution sets of a system of three linear equations in three variables.

(a) The three planes may intersect in a single point which is the solution of the system.

(b) The three planes may intersect in a line so that all points on the line are solutions to the system.

(c) The three planes may not intersect. Hence there is no solution to the system. There are several ways this can happen. One is shown below.

(d) The graph of the three planes may be the same plane, so all points on the plane are solutions.

The graphs of (a), (b), (c) and (d) are as follows.

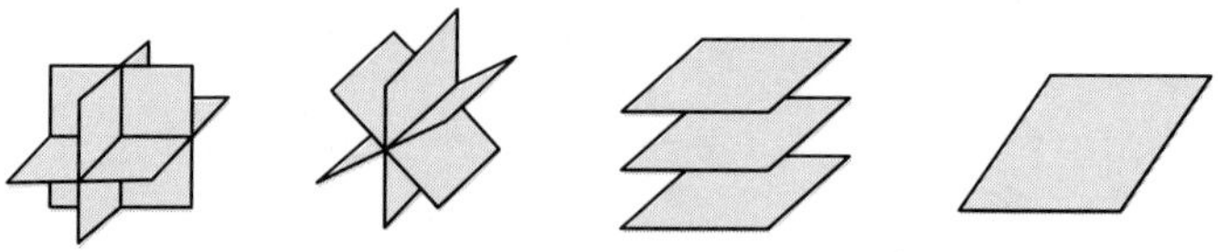

(a) (b) (c) (d)

2 Solving by the Elimination Method

The elimination method is easier to use than the substitution method to solve systems of equations in three variables. This is the method that we will demonstrate.

EXAMPLE 2 Solve

$$x + y + 2z = 0 \quad (1)$$
$$2x - 2y + z = 8 \quad (2)$$
$$3x + 2y + z = 2 \quad (3)$$

1. Find an equation in two variables by using *any* two of the three equations. We chose equations (2) and (3) and added to eliminate y.

$$\begin{array}{ll} 2x - 2y + z = 8 & (2) \\ \underline{3x + 2y + z = 2} & (3) \\ 5x \qquad + 2z = 10 & (4) \quad \text{Add.} \end{array}$$

2. Use a *different* pair of equations and eliminate the *same* variable that we eliminated in step 1. We chose equations (1) and (2). We multiplied equation (1) by 2 to eliminate y.

$$\begin{array}{ll} 2x + 2y + 4z = 0 & \text{Equation (1) multiplied by 2.} \\ \underline{2x - 2y + z = 8} & (2) \\ 4x \qquad + 5z = 8 & (5) \quad \text{Add.} \end{array}$$

3. Solve equations (4) and (5).

$$\begin{array}{ll} 5x + 2z = 10 & (4) \\ 4x + 5z = 8 & (5) \\ 25x + 10z = 50 & \text{Equation (4) multiplied by 5.} \\ \underline{-8x - 10z = -16} & \text{Equation (5) multiplied by } -2. \\ 17x \qquad = 34 & \\ x = 2 & \end{array}$$

Use either equation (4) or (5) to find z. Choose the equation that makes the arithmetic easier. We chose equation (4).

$$\begin{array}{ll} 5x + 2z = 10 & (4) \\ 5(2) + 2z = 10 & \text{Substitute 2 for } x. \\ 10 + 2z = 10 & \\ 2z = 0 & \\ z = 0 & \end{array}$$

4. Use any of the three original equations to find y. Again, choose the equation that makes the arithmetic easier. We choose equation (1).

$$x + y + 2z = 0 \quad (1)$$

$$2 + y + 2(0) = 0 \quad \text{Substitute 2 for } x \text{ and 0 for } z.$$

$$2 + y = 0$$

$$y = -2$$

The ordered triple $(2, -2, 0)$ checks in *all three* equations. The solution is $(2, -2, 0)$. ■

□ **DO EXERCISE 2.**

Sometimes one or more of the equations of a system has a missing variable, so one of the steps in the elimination is not necessary.

EXAMPLE 3 Solve

$$\begin{aligned} 2x \quad\quad + z &= 7 \quad (1) \\ y - z &= -2 \quad (2) \\ x + y \quad\quad &= 2 \quad (3) \end{aligned}$$

1. We chose to eliminate z by using equations (1) and (2) since there is no variable z in equation (3).

$$\begin{aligned} 2x \quad\quad + z &= 7 \quad (1) \\ y - z &= -2 \quad (2) \\ \hline 2x + y \quad\quad &= 5 \quad (4) \end{aligned}$$

2. Use equation (4) and equation (3) to solve for x and y. Multiply both sides of equation (3) by -1.

$$\begin{aligned} 2x + y &= 5 \quad (4) \\ -x - y &= -2 \quad \text{Equation (3) multiplied by } -1. \\ \hline x \quad\quad &= 3 \end{aligned}$$

We solved for y by substituting 3 for x in equation (3).

$$x + y = 2 \quad (3)$$

$$3 + y = 2$$

$$y = -1$$

3. Use one of the original equations containing the variable z to solve for it.

$$y - z = -2 \quad (2)$$

$$-1 - z = -2 \quad \text{Substitute } -1 \text{ for } y.$$

$$-z = -1$$

$$z = 1$$

The values that we found for x, y, and z check in *all three* equations. The solution of the system is $(3, -1, 1)$. ■

□ **DO EXERCISE 3.**

□ **Exercise 2** Solve.

a. $\begin{aligned} 3x + 2y + z &= 4 \\ 2x - 3y + 2z &= -7 \\ x + 4y - z &= 10 \end{aligned}$

b. $\begin{aligned} 2x + 3y - z &= 1 \\ x + 2y + 2z &= 5 \\ x - y + z &= 6 \end{aligned}$

□ **Exercise 3** Solve.

a. $\begin{aligned} 5y - 8z &= -19 \\ 5x \quad\quad - 8z &= 6 \\ 3x - 2y \quad\quad &= 12 \end{aligned}$

b. $\begin{aligned} 3x - 2y + z &= 4 \\ 6x \quad\quad - 5z &= 2 \\ 9x - 4y \quad\quad &= 9 \end{aligned}$

□ **Exercise 4** Solve.

$$x - 2y + 3z = 4$$
$$2x - y + z = 1$$
$$3x - 3y + 4z = 4$$

Linear systems in three variables may be inconsistent or dependent systems.

EXAMPLE 4 Solve.

$$x + 2y + z = 1 \qquad (1)$$
$$3x + 3y + z = 2 \qquad (2)$$
$$2x + y = 2 \qquad (3)$$

We eliminate z by multiplying equation (2) by -1 and adding it to equation (1).

$$\begin{array}{rl} -3x - 3y - z = -2 & \text{Equation (2) multiplied by } -1 \\ x + 2y + z = 1 & (1) \\ \hline -2x - y = -1 & (4) \end{array}$$

We add this result to equation (3).

$$\begin{array}{rl} -2x - y = -1 & (4) \\ 2x + y = 2 & (3) \\ \hline 0 = 1 & \end{array}$$

The result is a false statement. The system is inconsistent. There is no solution. The graph of this system would show at least two of the planes parallel to one another. ■

When we try to solve a linear system of three equations in three variables and we get a false statement, the system is *inconsistent*. There is *no solution* to the system.

□ **DO EXERCISE 4.**

Sometimes the graphs of the equations intersect in a line so that all points on the line are solutions to the system. When we try to solve the equations and the variable disappears and we get a true statement, this indicates that the system is dependent. For systems of three linear equations in three variables, however, the fact that a system is dependent does not always mean that it has infinite solutions. For example, the graph of two of the equations may be the same plane, but this plane may be parallel to the graph of the third plane, and there is no solution to the system.

When we try to solve a linear system of three equations in three variables and we get a true statement or we get two equations that are equivalent, the system is *dependent*. The system has *no unique solution*.

EXAMPLE 5 Solve.

$$2x + 3y - 4z = 1 \quad (1)$$
$$-4x - 6y + 8z = -2 \quad (2)$$
$$6x + 9y - 12z = 3 \quad (3)$$

If we multiply both sides of equation (1) by -2, we obtain equation (2). Also, if we multiply both sides of equation (1) by 3, we get equation (3). The equations are equivalent (have the same solutions). The graph of the equations is the same plane. They are dependent. In this case, there are an infinite number of ordered triples that are solutions. The system is *dependent* and has *no unique solution.* ■

□ **DO EXERCISE 5.**

To find the unique solution to a system of equations in three variables:

1. Use any two of the three equations to eliminate one variable. The result is an equation in two variables.
2. Eliminate the *same* variable from a different pair of equations. The result is an equation in the same two variables as the equation found in step 1.
3. Solve the equations found in steps 1 and 2 for the two variables.
4. Use any of the three original equations to find the value of the third variable.
5. Check your answer in all three original equations.

□ **Exercise 5** Solve.

$$-x + 3y - 2z = -2$$
$$3x - 9y + 6z = 6$$
$$-2x + 6y - 4z = -4$$

Answers to Exercises

1. **a.** Yes **b.** No
2. **a.** $(2, 1, -4)$ **b.** $(3, -1, 2)$
3. **a.** $\left(2, -3, \frac{1}{2}\right)$ **b.** $\left(\frac{1}{3}, -\frac{3}{2}, 0\right)$
4. No solution
5. Dependent, no unique solution

Problem Set 4.4

NAME

DATE

CLASS

1. Determine if (2, 0, 1) is a solution of the system.

$$\begin{aligned} x + 2y + z &= 3 \\ 2x - y + 2z &= 6 \\ 3x + y - z &= 5 \end{aligned}$$

2. Determine if (4, −5, 2) is a solution of the system.

$$\begin{aligned} 2x - y - 3z &= 4 \\ x + 2y - z &= -3 \\ 4x + 3y + 2z &= -5 \end{aligned}$$

3. Determine if $(\frac{1}{2}, \frac{4}{3}, -\frac{1}{2})$ is a solution of the system.

$$\begin{aligned} 2x + 3y - 2z &= 4 \\ x + 3y - 3z &= 4 \\ 3x - 6y + z &= -3 \end{aligned}$$

4. Determine if $(\frac{1}{2}, \frac{1}{3}, \frac{1}{4})$ is a solution of the system.

$$\begin{aligned} 2x - 3y &= 0 \\ 6y - 4z &= 1 \\ x + 2z &= 1 \end{aligned}$$

Solve.

5. $$\begin{aligned} x - y + z &= 2 \\ 2x + 3y + z &= 11 \\ 3x + y + z &= 8 \end{aligned}$$

6. $$\begin{aligned} 3x + 2y + z &= 4 \\ 4x - 2y + 3z &= 18 \\ x + y - 2z &= -4 \end{aligned}$$

7. $$\begin{aligned} x - 2y + 4z &= -3 \\ 3x + y - 2z &= 12 \\ 2x + y - 3z &= 11 \end{aligned}$$

8. $$\begin{aligned} x + y - z &= -2 \\ 2x - y + z &= 5 \\ x - 2y + 3z &= 4 \end{aligned}$$

9. $$\begin{aligned} x - y + z &= 6 \\ 2x + 3y + 2z &= 2 \\ 3x + 5y + 4z &= 4 \end{aligned}$$

10. $$\begin{aligned} 2x + 3y - 4z &= -4 \\ x - y + 2z &= 5 \\ 3x + 2y - 3z &= -2 \end{aligned}$$

11. $$\begin{aligned} 2x + 5y + 2z &= 9 \\ 4x - 7y - 3z &= 7 \\ 3x - 8y - 2z &= 9 \end{aligned}$$

12. $$\begin{aligned} x + 3y - 6z &= 7 \\ 2x - y + z &= 1 \\ x + 2y + 2z &= -1 \end{aligned}$$

13. $$\begin{aligned} x + y + z &= 0 \\ 2x + 3y + 2z &= -3 \\ -x + 2y - 3z &= -1 \end{aligned}$$

14. $3x - y + z = 8$
$x + 2y + 4z = 9$
$2x - y - z = 1$

15. $-2x + 4y - z = -8$
$x - 3y + 2z = 11$
$4x + y + z = 3$

16. $4x + 6y - z = 6$
$2x - y + z = 0$
$x + 3y - 2z = 5$

17. $x + 3y + 2z = 4$
$x - y + z = 6$
$2x + 6y + 4z = -2$

18. $3x - 6y + 3z = 4$
$-x + 2y - z = 7$
$5x - 8y + z = -3$

19. $x + y = 7$
$x - z = 6$
$y + 2z = -1$

20. $2x - z = -1$
$x + 3y = -3$
$4y + z = -11$

21. $x + y = 1$
$2x - z = 0$
$y + 2z = -2$

22. $-x + y = 1$
$y - z = 2$
$x + z = -2$

23. $3x - 4y = -1$
$x + 2z = 3$
$2y + 3z = 5$

24. $2x - 3z = -8$
$3x + y = -3$
$4y - z = -2$

25. $x + 2y - 4z = 3$
$5x - 2y + z = 7$
$-2x - 4y + 8z = -6$

26. $-4x + 8y + 4z = 12$
$3x - 7y + 9z = 8$
$x - 2y - z = -3$

27. $2x - 4y + z = 0$
$x - 3y - z = 0$
$3x - y + 2z = 0$

28. $2x + y + z = -2$
$2x - y + 3z = 6$
$3x - 5y + 4z = 7$

29. $2x + y + z = 3$
$3x - y + z = -2$
$4x - y + 2z = 0$

30. $x + y + z = 6$
$2x + 3y - z = 7$
$3x - y - z = 6$

31. $4x + 2y - 3z = 6$
$x - 4y + z = -4$
$-x + 2z = 2$

32. $2x - 3y + 2z = -1$
$x + 2y = 14$
$x - 3z = -5$

33. $2x - y + 3z = 0$
$x + 2y - z = 5$
$2y + z = 1$

34. $2x + y = 6$
$3y - 2z = -4$
$3x - 5z = -7$

35. $x + 3y + 2z = 4$
$3x + 9y + 6z = 5$
$x - y - z = 3$

36. $2x + 2y - 6z = 5$
$-x - y + 3z = 4$
$3x - y + z = 2$

37. A solution of a system of equations in three variables is an ____________ ____________.

38. The graph of a linear equation in three variables is a ____________.

39. Three planes may intersect in a ____________, ____________, or ____________.

40. If a linear system of three equations in three variables is inconsistent, the system has ____________ ____________.

41. There may be no solution to a linear system of equations in three variables even though it is ____________.

Solve. Round answers to the nearest hundredth.

42. $0.5x - 0.25z = -0.25$
$0.25x + 0.75y = -0.75$
$y + 0.25z = -2.75$

43. $1.5x - 2y = -0.5$
$0.5x + z = 1.5$
$y + 1.5z = 2.5$

44. $-1.34x + 0.82y = -2.38$
$1.23y - 0.34z = 4.98$
$1.45x + 0.68z = -2.86$

45. $8.2x + 1.34y = 4.88$
$2.05x - 3.8z = -7.3$
$4.15y + 7.6z = 0.47$

Solve each of the following systems of two equations in three variables in terms of x. The systems have an infinite number of solutions.

* **46.** $3x + 6y + 3z = 12$
$6x + 4y - 2z = -4$

* **47.** $x + 3y + 3z = 11$
$-4x - 4y - 3z = -10$

Checkup

The following problems provide a review of some of Section 3.5.

Determine if the ordered pair is a solution of the inequality.

48. $(3, 2)$, $4x - 3y > 8$ **49.** $(-4, 7)$, $2x + y < 1$ **50.** $(3, -1)$, $x + 5y \geq 2$ **51.** $(2, 4)$, $6x - 2y \leq 4$

Graph on graph paper.

52. $3x + y \leq 4$

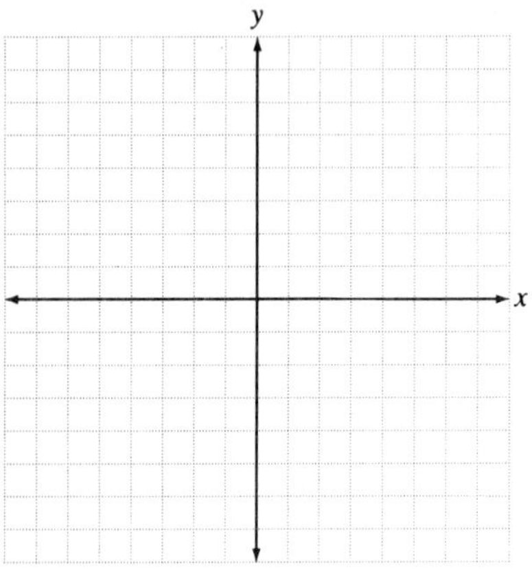

53. $2x + 5y < 10$

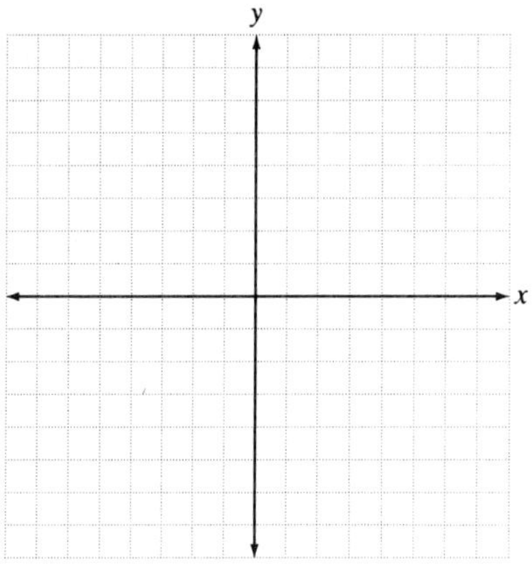

54. $3x - y > 3$

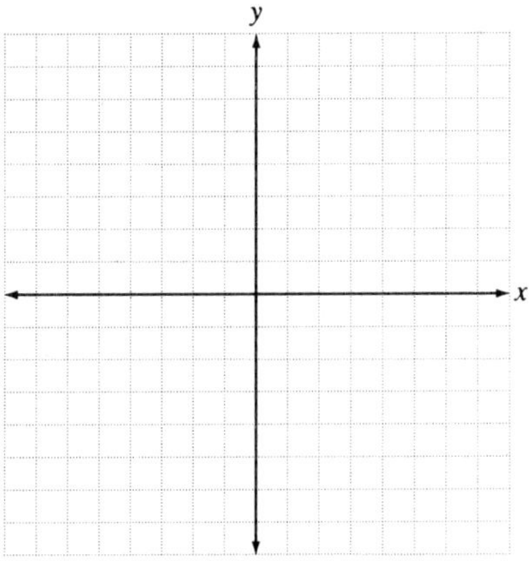

55. $3x - 2y \geq 9$

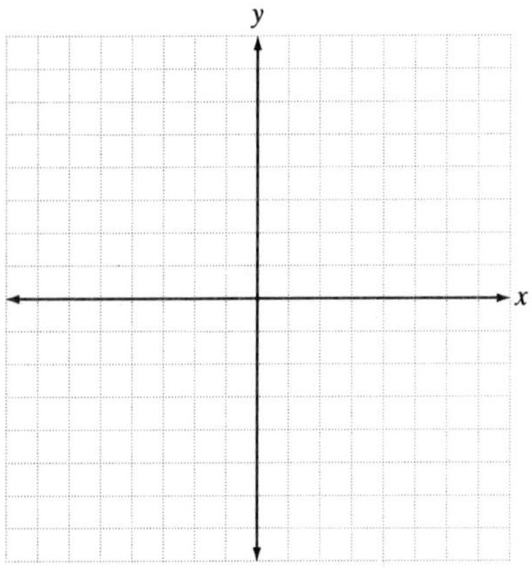

56. $5x + 6y < 12$

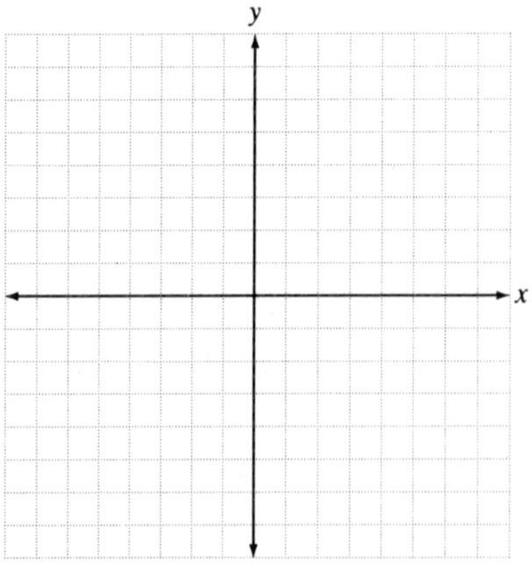

57. $4x - 5y \geq 20$

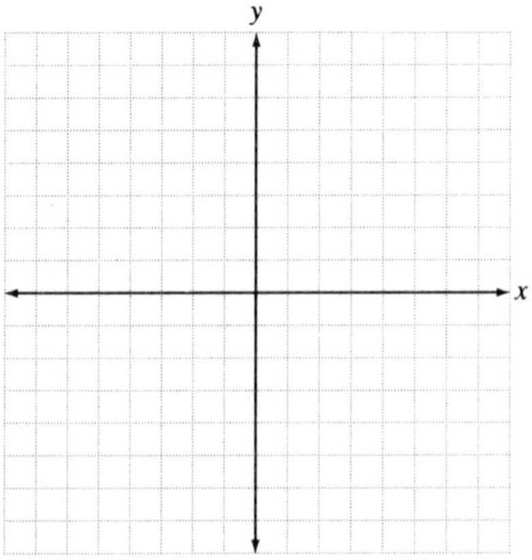

4.5 APPLIED PROBLEMS

[1] Systems of three or more equations are used to solve problems in science, business, engineering, and the social sciences. The method is similar to that used to solve applied problems with one or two variables.

EXAMPLE 1 The sum of three numbers is 5. Twice the first number, minus the second number, plus three times the third number, is -3. The first number minus twice the second number, minus the third number, is -5. Find the numbers.

Variables Let x represent the first number, y represent the second number, and z represent the third number.

$$\underbrace{\text{The sum of the three numbers}}_{x + y + z} \ \underbrace{\text{is}}_{=} \ \underbrace{5.}_{5}$$

$$\underbrace{\text{Twice the first number,}}_{2x} \ \underbrace{\text{minus}}_{-} \ \underbrace{\text{the second,}}_{y} \ \underbrace{\text{plus}}_{+} \ \underbrace{\text{three times the third,}}_{3z} \ \underbrace{\text{is}}_{=} \ \underbrace{-3.}_{-3}$$

$$\underbrace{\text{The first number,}}_{x} \ \underbrace{\text{minus}}_{-} \ \underbrace{\text{twice the second,}}_{2y} \ \underbrace{\text{minus}}_{-} \ \underbrace{\text{the third,}}_{z} \ \underbrace{\text{is}}_{=} \ \underbrace{-5.}_{-5}$$

Equations

$$\begin{aligned} x + y + z &= 5 \\ 2x - y + 3z &= -3 \\ x - 2y - z &= -5 \end{aligned}$$

Solve this system. Check in the original word problem to see that the solution is $(2, 4, -1)$. This ordered triple checks.

The first number is 2, the second number is 4, and the third number is -1. ■

□ **DO EXERCISE 1.**

OBJECTIVES

[1] *Use three variables and three equations to solve applied problems*

[2] *Use data points to write equations*

□ **Exercise 1** The sum of three numbers is 3. Twice the first number, minus the second number, plus the third number, is -4. The first number plus three times the second number, plus twice the third number, is 11. Find the numbers.

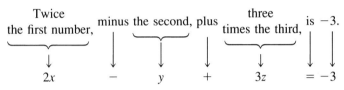

□ **Exercise 2** On three different days Anita sold a total of eight portraits. She sold twice as many on Tuesday as she sold on Monday. The number of portraits that she sold on Wednesday was the same as the number that she sold on Monday. How many portraits did she sell each day?

EXAMPLE 2 Keith picked blueberries for three days. He picked a total of 17 quarts. He picked 5 more quarts on Friday than he picked on Thursday. On Saturday he picked twice as many quarts as he picked on Thursday. How many quarts did he pick each day?

Variables Let x represent the number of quarts picked on Thursday
Let y represent the number of quarts picked on Friday
Let z represent the number of quarts picked on Saturday
He picked a total of 17 quarts. This translates to

$$x + y + z = 17. \qquad \text{First equation.}$$

The other statements may be translated as follows. The number of quarts that he picked on Friday was 5 more than the number that he picked on Thursday.

$$y = x + 5 \qquad \text{Second equation.}$$

The number of quarts that he picked on Saturday was twice as many as the number that he picked on Thursday.

$$z = 2x \qquad \text{Third equation}$$

Equations

$$\begin{aligned} x + y + z &= 17 \\ y &= x + 5 \\ z &= 2x \end{aligned} \qquad \text{or} \qquad \begin{aligned} x + y + z &= 17 \\ -x + y &= 5 \\ -2x + z &= 0 \end{aligned}$$

Verify that the solution to these equations is (3, 8, 6) and that the solution checks in the word problem.

Keith picked 3 quarts of blueberries on Thursday, 8 quarts on Friday, and 6 quarts on Saturday. ■

□ **DO EXERCISE 2.**

EXAMPLE 3 A bakery supplies bread to three stores, Shopway, Minimart and Saveco. The bakery produces 160 loaves of bread each day. They must send three times as many loaves to Shopway as they send to Minimart, and they send Saveco 80 loaves less than they send Shopway and Minimart together. How many loaves of bread should they send to each store to distribute the 160 loaves of bread to the three stores?

Variables Let x represent the number of loaves Shopway received
Let y represent the number of loaves Minimart received
Let z represent the number of loaves Saveco received

The bakery distributes 160 loaves to the three stores. This gives us the first equation.

$$x + y + z = 160 \qquad \text{First equation.}$$

They send three times as many loaves to Shopway as they send to Minimart. This translates to the following.

$$x = 3y \qquad \text{Second equation.}$$

The bakery sends Saveco 80 loaves less than they send to Shopway and Minimart together. The last equation is as follows:

$$z = x + y - 80 \qquad \text{Third equation.}$$

Equations

$$\begin{aligned} x + y + z &= 160 \\ x &= 3y \\ z &= x + y - 80 \end{aligned} \qquad \text{or} \qquad \begin{aligned} x + y + z &= 160 \\ x - 3y \quad &= 0 \\ -x - y + z &= -80 \end{aligned}$$

Solve this system. Check in the original word problem to see that the solution is (90, 30, 40). You may want to use your calculator. This ordered triple checks.

The bakery sends 90 loaves to Shopway, 30 loaves to Minimart, and 40 loaves to Saveco. ■

□ **DO EXERCISE 3.**

□ **Exercise 3** To meet his sales quota, Richard must sell 20 appliances. He must sell four more dishwashers than dryers and one more refrigerator than dryers. How many of each appliance must he sell?

□ **Exercise 4** Find a, b, and c so that the points $(1, -1)$, $(2, -1)$, and $(4, 5)$ lie on the graph of $y = ax^2 + bx + c$.

[2] The results of experiments are often data points which may be plotted on a rectangular coordinate system and a graph drawn through them. It is helpful to write the equation of this graph.

The coordinates of a point must satisfy an equation if the point is on the graph of the equation. If we know enough points on the graph and the general form of an equation, we can write a specific equation.

EXAMPLE 4 Suppose that the general formula of an equation is $y = ax^2 + bx + c$. The points $(2, 7)$, $(-4, 10)$, and $(0, 4)$ lie on the graph of the equation. Find a, b, and c and the equation.

Substitute the values for the three given points into the general formula.

$y = ax^2 + bx + c$	General formula.
$7 = a(2)^2 + b(2) + c$	Substitute 2 for x and 7 for y.
$10 = a(-4)^2 + b(-4) + c$	Substitute -4 for x and 10 for y.
$4 = a(0)^2 + b(0) + c$	Substitute 0 for x and 4 for y.

Simplify.

Equations	$7 = 4a + 2b + c$	or	$4a + 2b + c = 7$	(1)
	$10 = 16a - 4b + c$		$16a - 4b + c = 10$	(2)
	$4 = c$		$c = 4$	(3)

Since $c = 4$ we can substitute for it in equations (1) and (2).

$4a + 2b + 4 = 7$	Substitute 4 for c in equation (1).
$16a - 4b + 4 = 10$	Substitute 4 for c in equation (2).

We solve the equations and find the following values: $a = \frac{1}{2}$, $b = \frac{1}{2}$ and $c = 4$. This checks.

The equation is $y = \frac{1}{2}x^2 + \frac{1}{2}x + 4$. ■

□ **DO EXERCISE 4.**

Answers to Exercises

1. $-3, 2, 4$.

2. 2 on Monday, 4 on Tuesday, 2 on Wednesday.

3. 5 dryers, 9 dishwashers, 6 refrigerators

4. $a = 1$, $b = -3$, $c = 1$

Problem Set 4.5

NAME

DATE

CLASS

1. The sum of three numbers is -1. Five times the first, plus the second, minus the third, is 1. The first minus twice the second, minus 3 times the third, is 5. Find the numbers.

2. The sum of three numbers is 3. Twice the first, plus the second, minus 3 times the third, is -4. Three times the first, plus twice the second, minus the third, is 3. Find the numbers.

3. The sum of three numbers is 53. Five times the second is 3 less than 3 times the first. The third is 4 less than twice the second. Find the numbers.

4. The sum of three numbers is 62. The second is 5 more than twice the first. The third is 7 more than the second. Find the numbers.

5. The sum of the measures of the angles of a triangle is 180°. The second angle is 3 times as large as the first. The measure of the third angle equals the sum of the other two plus 20°. Find the measures of the three angles.

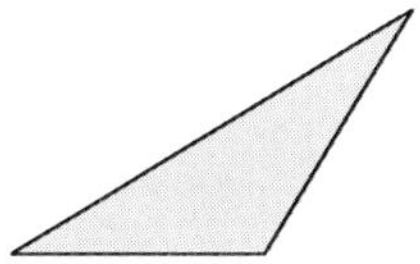

6. The sum of the measures of the angles of a triangle is 180°. The first angle is twice as large as the second angle. The measure of the third angle is 5° more than four times the measure of the second angle. Find the measures of the three angles.

7. In a triangle the measure of the second angle is 4 times the first plus 5°. The measure of the third angle is 5° less than the first. If the sum of the measures of the angles of a triangle is 180°, find the measures of the three angles.

8. In a triangle the third angle equals the sum of the measures of the other two angles. The third angle is also 3 times the first angle. If the sum of the angles of a triangle is 180°, find the measures of the three angles.

9. Kathy picked 19 quarts of raspberries. On Wednesday she picked 1 less quart than on Tuesday. She picked twice as many quarts on Thursday as she picked on Tuesday. How many quarts did she pick each day?

10. A bakery sold 30 pies. On Tuesday they sold 2 more than they sold on Monday. On Wednesday they sold 4 less than they sold on Tuesday. How many pies did they sell each day?

11. A manufacturer makes a total of 100 gallons of flat paint, semigloss, and enamel. The company makes 10 more gallons of semigloss than enamel. They also make 7 times as much flat paint as enamel. How much of each type of paint do they make?

12. If pumps A, B, and C are running at the same time, they can pump 1500 gallons per hour. When only pumps B and C are running, 950 gallons per hour can be pumped. If only A and C are running, 1250 gallons per hour can be pumped. What is the pumping capacity of each pump?

13. Find the values of a, b, and c so that the points $(1, -2)$, $(-2, -5)$, and $(3, 10)$ lie on the graph of the equation $y = ax^2 + bx + c$.

14. Find the values of a, b, and c so that the points $(-1, 3)$, $(0, 1)$, and $(2, -9)$ lie on the graph of the equation $y = ax^2 + bx + c$.

15. Suppose that the general formula of an equation is $y = ax^2 + bx + c$. The points $(-3, 1)$, $(0, -1)$, and $(3, 3)$ lie on the graph of the equation. Find a, b, and c and the equation.

16. Suppose that the general formula of an equation is $y = ax^2 + bx + c$. The points $(-2, -4)$, $(0, 1)$, and $(2, 2)$ lie on the graph of the equation. Find a, b, and c and the equation.

17. Kara bought a total of 9 pounds of apples, oranges, and bananas. She bought twice as many oranges as bananas. If apples cost 70 cents per pound, oranges cost 60 cents per pound, and bananas cost 40 cents per pound and she spent \$5.30, how many pounds of oranges did she buy?

18. Tickets for the theater cost \$15 for adults, \$12 for senior citizens, and \$6 for children. One day eight more senior citizens tickets were sold than adult tickets. Thirty-eight tickets were sold and the receipts were \$456. How many children's tickets were sold?

19. Carlos inherited \$60,000 from his aunt. He invested part of the money in a software company that produces a return of 12% per year, and divides the rest equally between a certificate of deposit at 5% per year and a savings account at 4% per year. How much is invested at each rate if his annual return on the investments is \$4950?

20. Tracy got three loans to buy rental properties. She borrowed a total of \$70,000. Some of the money was borrowed at 7% interest per year, and \$30,000 less than that amount was borrowed at 10% per year. The remainder was borrowed at 8% per year. How much was borrowed at each rate if the total simple interest was \$5400?

21. At Jim's car sales the cost of a new truck is \$14,745. The basic model with a radio and air conditioning is \$15,295. The basic model with just air conditioning and four-wheel drive is \$16,325. A basic model with radio and four-wheel drive is \$16,015. What is the cost of each of the three options?

22. Last year the Regan Company spent a total of \$22.2 million on salaries, overhead, and equipment. The total amount spent on overhead and equipment was \$7.8 million less than the amount spent on salaries. The amount spent on equipment was \$1.2 million more than the amount spent on overhead. How much was spent on overhead?

23. It is suggested that a person eat no more than 300 milligrams of cholesterol per day. If Ken drank 1 cup of whole milk, and ate 1 ounce of cheddar cheese and one 3-ounce serving of lean beef, he would get 140 milligrams of cholesterol. If he drank 3 cups of whole milk, and ate 2 ounces of cheddar cheese, he would take in 159 milligrams of cholesterol. If he drank 4 cups of whole milk and ate two servings of lean beef, he would get 286 milligrams of cholesterol. How much cholesterol is in each serving of 1 cup of whole milk, 1 ounce of cheddar cheese, and a 3-ounce serving of lean beef?

24. Susan, Maria, and Kanarat scored a total of 50 points in a basketball game. Maria got 12 more points than Susan. Kanarat scored 10 points less than Maria. How many points did each woman get?

25. Find the values of a, b, and c so that the points $(2, 12)$, $(-1, 9)$, and $(1, 5)$ lie on the graph of the equation $y = ax^2 + bx + c$.

26. Find the values of a, b, and c so that the points $(-2, -21)$, $(0, -5)$, and $(3, -11)$ lie on the graph of the equation $y = ax^2 + bx + c$.

27. Suppose that the general formula of an equation is $y = ax^2 + bx + c$. The points $(-3, -19)$, $(0, -4)$, and $(2, -4)$ lie on the graph of the equation. Find a, b, and c and the equation.

28. Suppose that the general formula of an equation is $y = ax^2 + bx + c$. The points $(-2, 9)$, $(1, -3)$, and $(4, 3)$ lie on the graph of the equation. Find a, b, and c and the equation.

29. Tickets for the theater cost \$15.75 for adults, \$11.50 for senior citizens, and \$5.50 for children. One day seven more senior citizens tickets and twice as many children's tickets were sold than adult tickets. The total receipts were \$1304.50. How many children's tickets were sold?

30. Gina bought a total of 7.8 pounds of nuts. She bought 3 more pounds of peanuts than pecans. If peanuts cost 69 cents per pound, pecans cost \$1.25 per pound, and filberts cost 75 cents per pound and she spent \$6.55, how many pounds of filberts did she buy?

* 31. A candy store fills an order for nuts. They combine 24 pounds of peanuts, 12 pounds of walnuts, and 10 pounds of cashews, and the total cost is \$180. The cost of peanuts is one-half the cost of the cashews, and the sum of the prices per pound of the nuts is \$13. What is the cost per pound of cashews?

* 32. Three salt solutions of strengths 30%, 40%, and 60% are to be mixed to give 1400 gallons of 50% solution. Four times as much 60% solution as 30% solution is used. How much of each solution should be used?

Checkup

The following problems provide a review of some of Section 1.5 and will help you with the next section.

Simplify.

33. $18 - 2(6)$

34. $15 + 2(-8)$

35. $8 \div 2(3)$

36. $9 \div 3(4)$

37. $\dfrac{5 + 2^2}{2 + 3(2)}$

38. $\dfrac{7 - 3^2}{-4}$

39. $\dfrac{2 + 3(4)^2}{5^2}$

40. $\dfrac{4(3)^2}{8(11) - 5(8)}$

41. $3(8) - 4(-7) - 6(-9)$

42. $5(-7) - 4(-19) + 5(21)$

43. $8^2 - 9^2 - 43$

44. $65 - 7^2 - 5^3$

45. $50 - 2^5 \div 4$

46. $4 \times 12^2 - 800$

4.6 DETERMINANTS AND CRAMER'S RULE

OBJECTIVES

1. *Evaluate second-order determinants*
2. *Evaluate third-order determinants*
3. *Use Cramer's rule to solve systems of linear equations in two or three variables*

In this section we show how linear systems of equations in two or three variables can be solved using determinants. Systems of equations can often be solved more easily in this way, especially if a calculator is used.

The following is a *determinant.*

$$\begin{vmatrix} a_1 & b_1 \\ a_2 & b_2 \end{vmatrix}$$

1 Second-Order Determinants

The determinant shown above is called a second-order determinant. It is defined as the following number:

$$\begin{vmatrix} a_1 & b_1 \\ a_2 & b_2 \end{vmatrix} = a_1b_2 - a_2b_1$$

EXAMPLE 1 Evaluate

$$\begin{vmatrix} 3 & -2 \\ 4 & 5 \end{vmatrix}$$

Here $a_1 = 3$, $b_1 = -2$, $a_2 = 4$, and $b_2 = 5$. Use the definition. Multiply and then subtract.

$$\begin{vmatrix} 3 & -2 \\ 4 & 5 \end{vmatrix} = 3(5) - 4(-2) = 15 + 8 = 23 \quad ■$$

□ **DO EXERCISE 1.**

□ **Exercise 1** Evaluate.

a. $\begin{vmatrix} 2 & -1 \\ 3 & 4 \end{vmatrix}$

b. $\begin{vmatrix} 5 & -2 \\ -4 & -1 \end{vmatrix}$

2 Third-Order Determinants

A third-order determinant is defined as follows:

$$\begin{vmatrix} a_1 & b_1 & c_1 \\ a_2 & b_2 & c_2 \\ a_3 & b_3 & c_3 \end{vmatrix} = (a_1b_2c_3 + b_1c_2a_3 + c_1a_2b_3) - (a_3b_2c_1 + b_3c_2a_1 + c_3a_2b_1)$$

This definition is complicated. One way to get the same result is to evaluate the third-order determinant using second-order determinants as follows:

$$\begin{vmatrix} a_1 & b_1 & c_1 \\ a_2 & b_2 & c_2 \\ a_3 & b_3 & c_3 \end{vmatrix} = \mathbf{a_1}\begin{vmatrix} b_2 & c_2 \\ b_3 & c_3 \end{vmatrix} - \mathbf{a_2}\begin{vmatrix} b_1 & c_1 \\ b_3 & c_3 \end{vmatrix} + \mathbf{a_3}\begin{vmatrix} b_1 & c_1 \\ b_2 & c_2 \end{vmatrix}$$

We can get the second-order determinants by crossing out the row and column in which the a appears.

For a_1: $\begin{vmatrix} \mathbf{a_1} & \mathbf{b_1} & \mathbf{c_1} \\ \mathbf{a_2} & b_2 & c_2 \\ \mathbf{a_3} & b_3 & c_3 \end{vmatrix}$ For a_2: $\begin{vmatrix} \mathbf{a_1} & b_1 & c_1 \\ \mathbf{a_2} & \mathbf{b_2} & \mathbf{c_2} \\ \mathbf{a_3} & b_3 & c_3 \end{vmatrix}$ For a_3: $\begin{vmatrix} \mathbf{a_1} & b_1 & c_1 \\ \mathbf{a_2} & b_2 & c_2 \\ \mathbf{a_3} & \mathbf{b_3} & \mathbf{c_3} \end{vmatrix}$

□ **Exercise 2** Evaluate.

a. $\begin{vmatrix} 3 & 1 & -1 \\ 4 & 5 & 1 \\ 2 & -1 & 3 \end{vmatrix}$

b. $\begin{vmatrix} -2 & 4 & 1 \\ -1 & 0 & 2 \\ 3 & 2 & -1 \end{vmatrix}$

EXAMPLE 2 Evaluate

$$\begin{vmatrix} 3 & 7 & 1 \\ -2 & 0 & -2 \\ -4 & 2 & 1 \end{vmatrix} = \mathbf{3}\begin{vmatrix} 0 & -2 \\ 2 & 1 \end{vmatrix} \mathbf{-(-2)}\begin{vmatrix} 7 & 1 \\ 2 & 1 \end{vmatrix} + \mathbf{(-4)}\begin{vmatrix} 7 & 1 \\ 0 & -2 \end{vmatrix}$$

Notice the minus sign.

$$= 3[0(1) - 2(-2)] - (-2)[7(1) - 2(1)]$$
$$+ (-4)[7(-2) - 0(1)]$$
$$= 3(4) + 2(5) - 4(-14)$$
$$= 12 + 10 + 56$$
$$= 78 \quad ■$$

□ **DO EXERCISE 2.**

3 Solution of Linear Systems of Equations by Determinants

A system of linear equations in two variables may be written as follows:

$$a_1x + b_1y = c_1$$
$$a_2x + b_2y = c_2$$

We designate the following three determinants as D, D_x, and D_y.

$D = \begin{vmatrix} a_1 & b_1 \\ a_2 & b_2 \end{vmatrix}$ — D is composed of the coefficients of x and y in the order that they are shown in the equations

$D_x = \begin{vmatrix} c_1 & b_1 \\ c_2 & b_2 \end{vmatrix}$ — Replacing the x-coefficients in D with the constants from the right side of the equations

$D_y = \begin{vmatrix} a_1 & c_1 \\ a_2 & c_2 \end{vmatrix}$ — Replacing the y-coefficients in D with the constants from the right side of the equations

We may solve a linear system of two equations in two variables by using the following, which is known as *Cramer's rule*.

$$x = \frac{D_x}{D} \quad \text{and} \quad y = \frac{D_y}{D} \qquad \text{if } D \neq 0$$

EXAMPLE 3 Use Cramer's rule to solve the system

$$2x - 3y = 0$$
$$-4x + 3y = -1$$

We find D, D_x, and D_y. Evaluate D first, since Cramer's rule does not work if $D = 0$, because division by zero is undefined.

$$D = \begin{vmatrix} 2 & -3 \\ -4 & 3 \end{vmatrix} = 2(3) - (-4)(-3) = 6 - 12 = -6$$

$$D_x = \begin{vmatrix} 0 & -3 \\ -1 & 3 \end{vmatrix} = 0(3) - (-1)(-3) = 0 - 3 = -3$$

$$D_y = \begin{vmatrix} 2 & 0 \\ -4 & -1 \end{vmatrix} = 2(-1) - (-4)0 = -2 - 0 = -2$$

Hence

$$x = \frac{D_x}{D} = \frac{-3}{-6} \text{ or } \frac{1}{2} \qquad \text{and} \qquad y = \frac{D_y}{D} = \frac{-2}{-6} = \frac{1}{3}$$

The solution is $(\frac{1}{2}, \frac{1}{3})$. ■

□ **DO EXERCISE 3.**

It is sometimes easier to use Cramer's rule to solve three linear equations in three variables than it is to use the addition method. To use Cramer's rule to solve three linear equations in three variables we again form a determinant, D, which is now made up of the coefficients of x, y, and z.

$$a_1x + b_1y + c_1z = d_1$$

$$a_2x + b_2y + c_2z = d_2$$

$$a_3x + b_3y + c_3z = d_3$$

$$D = \begin{vmatrix} a_1 & b_1 & c_1 \\ a_2 & b_2 & c_2 \\ a_3 & b_3 & c_3 \end{vmatrix}$$

The determinants D_x, D_y, and D_z are formed by replacing in turn the coefficients of x, y, and z with the constants on the right side of the equations as follows:

$$D_x = \begin{vmatrix} d_1 & b_1 & c_1 \\ d_2 & b_2 & c_2 \\ d_3 & b_3 & c_3 \end{vmatrix}$$

$$D_y = \begin{vmatrix} a_1 & d_1 & c_1 \\ a_2 & d_2 & c_2 \\ a_3 & d_3 & c_3 \end{vmatrix}$$

$$D_z = \begin{vmatrix} a_1 & b_1 & d_1 \\ a_2 & b_2 & d_2 \\ a_3 & b_3 & d_3 \end{vmatrix}$$

Then *Cramer's rule* for solving a system of three equations in three variables is

$$x = \frac{D_x}{D} \qquad y = \frac{D_y}{D} \qquad z = \frac{D_z}{D} \qquad \text{if } D \neq 0$$

EXAMPLE 4 Solve by Cramer's rule.

$$x - y + z = 6$$

$$2x + 3y + 2z = 2$$

$$3x + 5y + 4z = 4$$

We need to evaluate D, D_x, D_y, and D_z.

$$D = \begin{vmatrix} 1 & -1 & 1 \\ 2 & 3 & 2 \\ 3 & 5 & 4 \end{vmatrix} = \mathbf{1}\begin{vmatrix} 3 & 2 \\ 5 & 4 \end{vmatrix} \mathbf{-2}\begin{vmatrix} -1 & 1 \\ 5 & 4 \end{vmatrix} + \mathbf{3}\begin{vmatrix} -1 & 1 \\ 3 & 2 \end{vmatrix}$$

$$= 1(2) - 2(-9) + 3(-5)$$

$$= 5$$

□ **Exercise 3** Solve by Cramer's rule.

a. $9x - 2y = 3$
$3x - y = 6$

b. $x - 3y = -3$
$x + 2y = 9$

□ **Exercise 4** Solve by Cramer's rule.

a. $3x - 2y + 4z = 5$
$4x + y + z = 14$
$x - y - z = 1$

b. $3x + 2y - z = 4$
$3x - 2y + z = 5$
$4x - 5y - z = -1$

D_x, D_y, and D_z are evaluated by the same method.

$$D_x = \begin{vmatrix} 6 & -1 & 1 \\ 2 & 3 & 2 \\ 4 & 5 & 4 \end{vmatrix} = 10 \quad D_y = \begin{vmatrix} 1 & 6 & 1 \\ 2 & 2 & 2 \\ 3 & 4 & 4 \end{vmatrix} = -10$$

$$D_z = \begin{vmatrix} 1 & -1 & 6 \\ 2 & 3 & 2 \\ 3 & 5 & 4 \end{vmatrix} = 10$$

We use Cramer's rule to find the solution.

$$x = \frac{D_x}{D} = \frac{10}{5} = 2 \qquad y = \frac{D_y}{D} = -\frac{10}{5} = -2 \qquad z = \frac{D_z}{D} = \frac{10}{5} = 2$$

The solution is $(2, -2, 2)$. ■

Notice that if D is zero, we cannot divide by it. If D is zero and all of the other determinants are also zero, then the system is dependent. If D is zero and one or more of the other determinants is not zero, the system is inconsistent and there is no solution to it.

□ **DO EXERCISE 4.**

GRAPHING CALCULATOR

A rectangular array of real numbers is called a ***matrix.*** The matrix shown is a 2×2 matrix (read "2 by 2") because it has two rows and two columns.

$$\begin{bmatrix} a_{11} & a_{12} \\ a_{21} & a_{22} \end{bmatrix}$$

The first subscript on the variable indicates the row, and the second subscript indicates the column where the number is placed. Following are examples of a 3×3 matrix and a 3×1 matrix.

$$\underset{3 \times 3}{\begin{bmatrix} a_{11} & a_{12} & a_{13} \\ a_{21} & a_{22} & a_{23} \\ a_{31} & a_{32} & a_{33} \end{bmatrix}} \qquad \underset{3 \times 1}{\begin{bmatrix} a_{11} \\ a_{21} \\ a_{31} \end{bmatrix}}$$

Systems of equations may be solved with a graphing calculator using matrices. Since graphing calculators vary, consult your manual for instructions.

Answers to Exercises

1. a. 11 **b.** -13

2. a. 52 **b.** 26

3. a. $(-3, -15)$ **b.** $\left(\frac{21}{5}, \frac{12}{5}\right)$

4. a. $(3, 2, 0)$ **b.** $\left(\frac{3}{2}, \frac{13}{14}, \frac{33}{14}\right)$

Problem Set 4.6

NAME ______________________

DATE ______________________

CLASS ______________________

Evaluate.

1. $\begin{vmatrix} 2 & 6 \\ 3 & 1 \end{vmatrix}$

2. $\begin{vmatrix} 3 & 4 \\ 6 & -3 \end{vmatrix}$

3. $\begin{vmatrix} 4 & 5 \\ -1 & 2 \end{vmatrix}$

4. $\begin{vmatrix} -2 & 8 \\ 4 & 0 \end{vmatrix}$

5. $\begin{vmatrix} -1 & 7 \\ 0 & 0 \end{vmatrix}$

6. $\begin{vmatrix} 3 & 0 \\ 0 & 6 \end{vmatrix}$

7. $\begin{vmatrix} 6 & 8 \\ 9 & 7 \end{vmatrix}$

8. $\begin{vmatrix} 5 & 9 \\ 7 & 4 \end{vmatrix}$

9. $\begin{vmatrix} -3 & 8 \\ 7 & -5 \end{vmatrix}$

10. $\begin{vmatrix} 6 & -2 \\ -9 & 3 \end{vmatrix}$

11. $\begin{vmatrix} -4 & -7 \\ -8 & -5 \end{vmatrix}$

12. $\begin{vmatrix} -6 & -3 \\ -2 & -9 \end{vmatrix}$

13. $\begin{vmatrix} 2 & 0 & -1 \\ 3 & 4 & -2 \\ 1 & 0 & 5 \end{vmatrix}$

14. $\begin{vmatrix} 0 & 3 & 2 \\ 2 & 0 & -1 \\ 4 & 3 & 1 \end{vmatrix}$

15. $\begin{vmatrix} -1 & 0 & 4 \\ 3 & 4 & 1 \\ -2 & 0 & 5 \end{vmatrix}$

16. $\begin{vmatrix} -1 & 4 & 3 \\ -3 & 0 & 5 \\ 2 & 6 & 0 \end{vmatrix}$

17. $\begin{vmatrix} 4 & -5 & 0 \\ 2 & -2 & 0 \\ 8 & 9 & 0 \end{vmatrix}$

18. $\begin{vmatrix} 7 & 6 & -7 \\ 5 & 8 & 9 \\ 0 & 0 & 0 \end{vmatrix}$

19. $\begin{vmatrix} 5 & 0 & -5 \\ 1 & 4 & -2 \\ 3 & 2 & 0 \end{vmatrix}$

20. $\begin{vmatrix} 7 & 0 & -3 \\ -2 & 0 & 2 \\ 1 & 4 & 5 \end{vmatrix}$

21. $\begin{vmatrix} -3 & 7 & -5 \\ 4 & 6 & 8 \\ 7 & 6 & 11 \end{vmatrix}$

22. $\begin{vmatrix} 4 & 3 & -2 \\ -3 & 5 & 1 \\ -4 & 8 & 10 \end{vmatrix}$

Solve by Cramer's rule.

23. $$\begin{aligned} 2x - 3y &= 3 \\ 4x - 2y &= 10 \end{aligned}$$

24. $$\begin{aligned} 3x + y &= -2 \\ -3x + 2y &= -4 \end{aligned}$$

25. $$\begin{aligned} 2x - y &= 7 \\ 3x + 4y &= -6 \end{aligned}$$

26. $$\begin{aligned} 3x - 2y &= -8 \\ -2x + 3y &= 7 \end{aligned}$$

27. $$\begin{aligned} 2x - 3y &= 0 \\ -4x + 3y &= -1 \end{aligned}$$

28. $$\begin{aligned} -3x + 2y &= 0 \\ 3x - 4y &= -1 \end{aligned}$$

29. $$\begin{aligned} x - 3y &= 7 \\ 6x + 4y &= 9 \end{aligned}$$

30. $$\begin{aligned} x + 3y &= 5 \\ 2x - 11y &= 10 \end{aligned}$$

31. $$\begin{aligned} x + y - z &= 2 \\ -x + y + z &= 3 \\ x + y + z &= 4 \end{aligned}$$

32. $$\begin{aligned} -x - y + z &= 1 \\ x - y + z &= 3 \\ x + y - z &= 4 \end{aligned}$$

33. $$\begin{aligned} 3x - y + 2z &= 1 \\ x - y + 2z &= 3 \\ -2x + 3y + z &= 1 \end{aligned}$$

34. $$\begin{aligned} x - 2y + 3z &= 6 \\ 2x - y - z &= -3 \\ x + y + z &= 6 \end{aligned}$$

35. $$\begin{aligned} 2x - y + z &= 7 \\ 3x - 2y - z &= 4 \\ x - 3y - 2z &= 1 \end{aligned}$$

36. $$\begin{aligned} x + y + 3z &= 6 \\ 2x + 4y - z &= 3 \\ -x - y + 5z &= 2 \end{aligned}$$

37. $$\begin{aligned} x - 2y &= -4 \\ 3x + y &= -5 \\ 2x + z &= -1 \end{aligned}$$

38. $$\begin{aligned} 5x + 2z &= -20 \\ x + 3y &= -1 \\ z &= 1 \end{aligned}$$

39. $$\begin{aligned} 2x - 3z &= -1 \\ 2y + z &= 3 \\ 4x + y &= 5 \end{aligned}$$

40. $$\begin{aligned} x + 4y &= -2 \\ 3x - z &= 4 \\ y + 3z &= 5 \end{aligned}$$

41. $$\begin{aligned} 2x + 2y + z &= 6 \\ 4x - y + z &= 12 \\ -x + y - 2z &= -3 \end{aligned}$$

42. $$\begin{aligned} x - y + 6z &= 14 \\ 3x + 3y - z &= 10 \\ x + 9y + 2z &= 16 \end{aligned}$$

43. Ken deposited four more \$20 bills than \$5 bills in the bank. The bills were worth \$255. How many \$20 bills did he deposit?

44. The perimeter of a rectangle is 120 feet. If the length is 3 times the width, what is the length of the rectangle?

45. The sum of three numbers is 2. Twice the first, plus the second, plus twice the third, is -2. Three times the first, plus twice the second, plus twice the third, is 12. Find the numbers.

46. A bakery sold 37 cakes. On Friday they sold 3 more than they sold on Thursday. On Saturday they sold twice as many as they sold on Friday. How many cakes did they sell on Saturday?

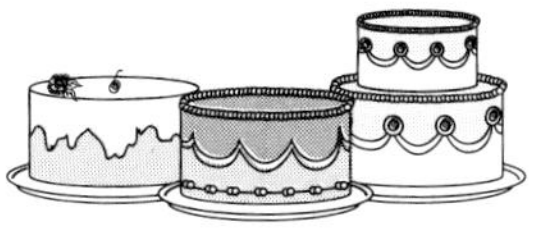

47. The sum of two numbers is 23. Twice the first plus 3 times the second is 54. Find the two numbers.

48. Two angles are complementary. One angle is 21° less than twice the other angle. Find the measures of the angles.

49. Dan, Megan, and Scott scored a total of 249 points on their art test. Dan's score was 15 points more than Megan's. Scott's was 9 more than Megan's. Find each student's score on the test.

50. In a triangle, the largest angle is 70° greater than the smallest angle. The third angle is 20° less than the largest angle. Find the measure of each angle.

51. The determinant $\begin{vmatrix} a_1 & b_1 \\ a_2 & b_2 \end{vmatrix}$ is called a ____________-____________ determinant.

52. A linear system of three equations in three variables may be solved using ____________ ____________ rule.

Evaluate. Round answers to the nearest thousandth.

53. $\begin{vmatrix} 9.71 & 2.86 \\ -6.25 & 4.97 \end{vmatrix}$

54. $\begin{vmatrix} -4.28 & -3.92 \\ 3.57 & 8.37 \end{vmatrix}$

55. $\begin{vmatrix} 7.5 & 0 & 6.8 \\ -3.4 & 5 & 0 \\ 6.3 & 2.1 & -1.1 \end{vmatrix}$

56. $\begin{vmatrix} 9.2 & 4.7 & 0 \\ 6.5 & 0 & -1.9 \\ 8 & 7.6 & -3.5 \end{vmatrix}$

Solve by Cramer's rule.

57. $6.72x + 3.49y = 2.97$
$2.59x - 1.57y = 9.89$

58. $5.42x - 3.85y = 13.12$
$7.43x + 6.2y = -4.97$

59. $3x \quad - 4.5z = -1.5$
$3y + 1.5z = 4.5$
$6x + 1.5y \quad = 7.5$

60. $2.5x + 10y \quad = -5$
$7.5x \quad - 2.5z = 10$
$2.5y + 7.5z = 12.5$

Evaluate.

* **61.** $\begin{vmatrix} 2 & 3p & 1 \\ 0 & 2p & 2 \\ 2 & 10p & 10 \end{vmatrix}$

* **62.** $\begin{vmatrix} u & 0 & v \\ x & y & z \\ 0 & 0 & 0 \end{vmatrix}$

Think About It

* **63.** Show that the slope-intercept form of a line can be written

$$\begin{vmatrix} y & x \\ m & 1 \end{vmatrix} = b$$

* **64.** Show that the following gives the formula for the surface area of a right circular cylinder, $S = 2\pi r^2 + 2\pi rh$.

$$\begin{vmatrix} \pi & 0 & \pi \\ r & r & 0 \\ -2h & r & r \end{vmatrix} = S$$

Checkup

The following problems provide a review of some of Section 1.5 and will help you with the next section.

Evaluate.

65. 2^3

66. 3^2

67. -3^2

68. -5^2

69. $(-3)^2$

70. $(-5)^2$

Write without exponents.

71. 3^0

72. $(-5)^0$

73. $(-7)^1$

74. 4^1

75. $(2y)^0$, $y \neq 0$

76. $-x^0$, $x \neq 0$

Chapter 4 Summary

Section 4.1

A ***linear system of equations*** in two variables is composed of two or more linear equations. The ***solution*** of this system is all ordered pairs of real numbers that satisfy all the equations.

There are three different situations that can occur when a system of two linear equations in two variables is graphed.

1. *The graphs intersect at a single point.* The solution is the coordinates of the point of intersection. The system of equations is called ***consistent*** because it has a solution.
2. *The graphs are parallel lines.* There is no solution. The system of equations is ***inconsistent,*** because it has no solution.
3. *The graphs are the same line.* There are infinitely many solutions. The system of equations is ***dependent*** and also consistent because there is at least one solution.

In cases 1 and 2 the systems are ***independent.***

A system of equations is called *consistent* if it has a solution. It is called *inconsistent* if it has no solution.

A system of two equations in two variables is *independent* if it has exactly one solution or no solutions and *dependent* if it has infinitely many solutions.

Section 4.2

Solving linear systems by elimination

1. Write both equations in the form $Ax + By = C$.
2. Clear any decimals or fractions.
3. Multiply one or both equations by the appropriate numbers so that the sum of the coefficients of either variable is zero.
4. Eliminate a variable by adding the new equations and then solve for the remaining variable.
5. Substitute the result of step 4 in either of the original equations and solve for the other variable.

If both variables are eliminated when a system of two linear equations in two variables is solved:

1. There is no solution if the resulting equation is false. The system is inconsistent and independent.
2. There are an infinite number of solutions if the resulting equation is true. The system is consistent and dependent.

Solving linear systems by substitution

1. Clear any decimals or fractions.
2. Solve one of the equations for either variable. If one of the equations has a variable with a coefficient of 1 or -1, solve for this variable.
3. Substitute for that variable in the other equation.
4. Solve the equation from step 3.
5. Substitute the result from step 4 into the equation from step 2 to solve for the other variable.

Section 4.3

Solving a word problem

1. Read the problem carefully. Decide what questions are to be answered.
2. Choose a variable to represent the unknown quantity. If it is helpful, make a drawing.
3. Write an equation. Sometimes a formula is needed that is not given. Some geometric formulas are given on the inside cover.
4. Solve the equation.
5. Determine if the questions have been answered and check the answer in the original word problem.

Interest formula:

$$I = Prt$$

where I is the simple interest, P is the principal, r is the interest rate, and t is the time in years.

Two angles are complementary if the sum of their measures is 90°.

Two angles are supplementary if the sum of their measures is 180°.

Section 4.4

A *solution* of a system of equations in three variables is an ***ordered triple*** (x, y, z) that satisfies all three equations.

The graph of a linear equation in three variables is a plane. The graphs of the possible solution sets of a system of three linear equations in three variables are as follows:

1. The three planes may intersect in a single point that is the solution of the system.
2. The three planes may intersect in a line so that all points on the line are solutions to the system.
3. The three planes may not intersect. Hence there is no solution to the system.
4. The graph of the three planes may be the same plane, so all points on the plane are solutions.

To find the unique solution to a system of equations in three variables:

1. Use any two of the three equations to eliminate one variable. The result is an equation in two variables.
2. Eliminate the *same* variable from a different pair of equations. The result is an equation in the same two variables as the equation found in step 1.
3. Solve the equations found in steps 1 and 2 for the two variables.
4. Use any of the three original equations to find the value of the third variable.
5. Check your answer in all three original equations.

Section 4.6

A ***second-order*** determinant is evaluated as follows:

$$\begin{vmatrix} a_1 & b_1 \\ a_2 & b_2 \end{vmatrix} = a_1b_2 - a_2b_1$$

A ***third-order*** determinant is evaluated as follows:

$$\begin{vmatrix} a_1 & b_1 & c_1 \\ a_2 & b_2 & c_2 \\ a_3 & b_3 & c_3 \end{vmatrix} = a_1\begin{vmatrix} b_2 & c_2 \\ b_3 & c_3 \end{vmatrix} - a_2\begin{vmatrix} b_1 & c_1 \\ b_3 & c_3 \end{vmatrix} + a_3\begin{vmatrix} b_1 & c_1 \\ b_2 & c_2 \end{vmatrix}$$

Cramer's rule for solving a linear system of two equations in two variables:

$$a_1x + b_1y = c_1$$

$$a_2x + b_2y = c_2$$

$$x = \frac{D_x}{D} \quad \text{and} \quad y = \frac{D_y}{D} \quad \text{if } D \neq 0$$

where

$$D = \begin{vmatrix} a_1 & b_1 \\ a_2 & b_2 \end{vmatrix} \quad D_x = \begin{vmatrix} c_1 & b_1 \\ c_2 & b_2 \end{vmatrix} \quad D_y = \begin{vmatrix} a_1 & c_1 \\ a_2 & c_2 \end{vmatrix}$$

Cramer's rule for solving a system of three equations in three variables.

$$a_1x + b_1y + c_1z = d_1$$

$$a_2x + b_2y + c_2z = d_2$$

$$a_3x + b_3y + c_3z = d_3$$

$$x = \frac{D_x}{D} \quad y = \frac{D_y}{D} \quad z = \frac{D_z}{D} \quad \text{if } D \neq 0$$

where

$$D = \begin{vmatrix} a_1 & b_1 & c_1 \\ a_2 & b_2 & c_2 \\ a_3 & b_3 & c_3 \end{vmatrix} \quad D_x = \begin{vmatrix} d_1 & b_1 & c_1 \\ d_2 & b_2 & c_2 \\ d_3 & b_3 & c_3 \end{vmatrix}$$

$$D_y = \begin{vmatrix} a_1 & d_1 & c_1 \\ a_2 & d_2 & c_2 \\ a_3 & d_3 & c_3 \end{vmatrix} \quad D_z = \begin{vmatrix} a_1 & b_1 & d_1 \\ a_2 & b_2 & d_2 \\ a_3 & b_3 & d_3 \end{vmatrix}$$

Chapter 4 Additional Exercises (Optional)

NAME ____________________

DATE ____________________

CLASS ____________________

Section 4.1

Solve by graphing. Then classify the system as consistent or inconsistent and as dependent or independent.

1. $\begin{aligned} 3x + y &= 15 \\ 2x - 5y &= 10 \end{aligned}$

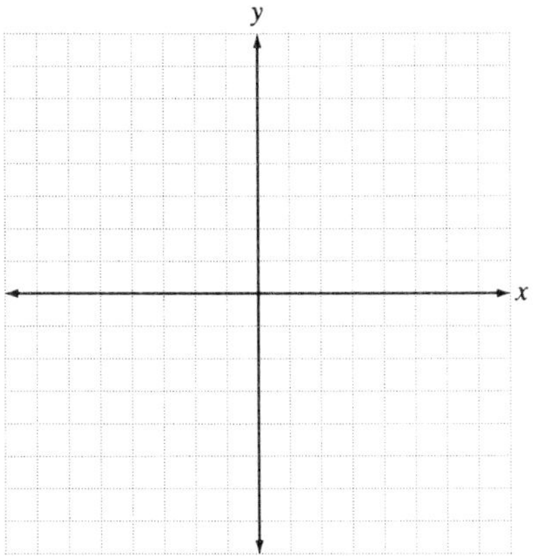

2. $\begin{aligned} x - 5y &= -5 \\ -x + 2y &= 2 \end{aligned}$

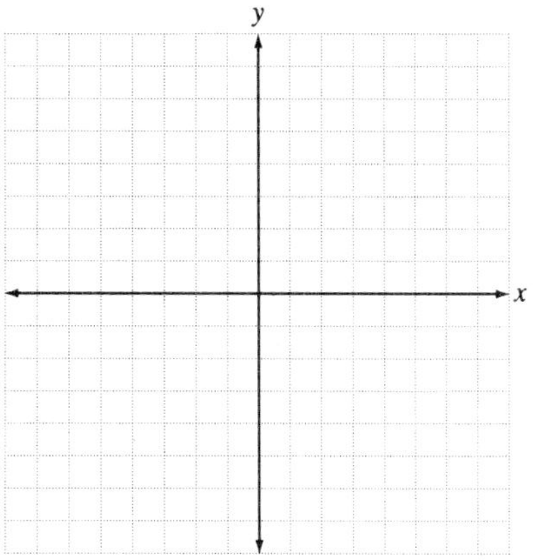

3. $\begin{aligned} 3x + 2y &= 10 \\ 4x - y &= 6 \end{aligned}$

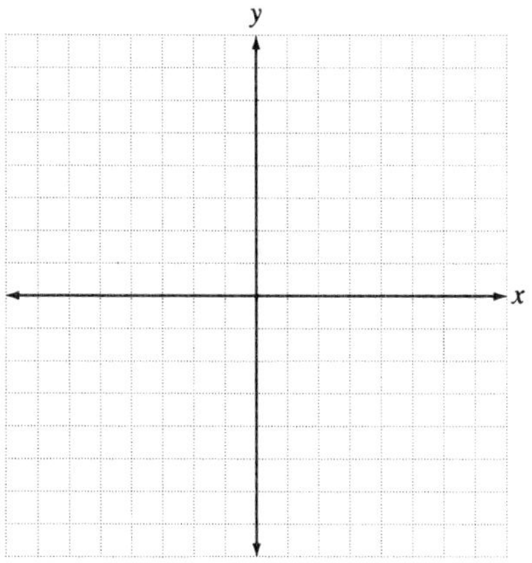

4. $\begin{aligned} 5x + y &= 3 \\ 3y + x &= 9 \end{aligned}$

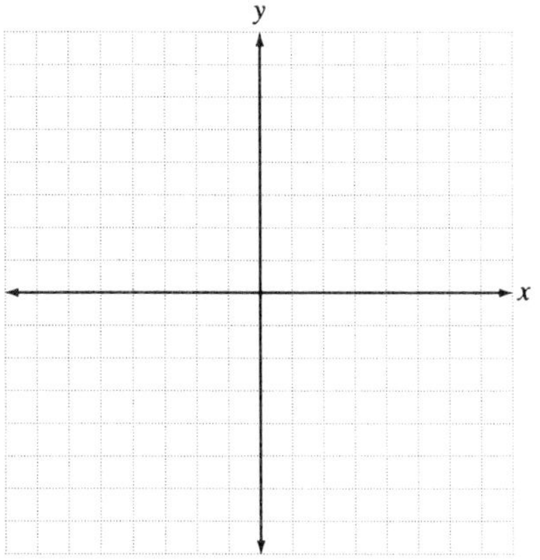

5. $\begin{aligned} x + y &= 5 \\ 2x - y &= 4 \end{aligned}$

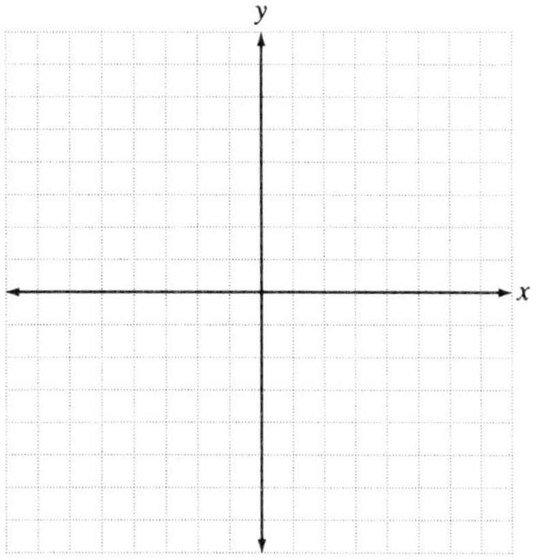

6. $\begin{aligned} 2x + 3y &= -6 \\ x - 3y &= 6 \end{aligned}$

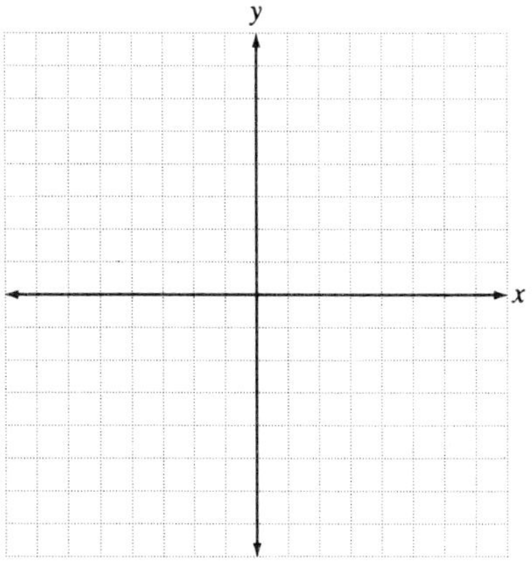

7. $3x - 2y = 4$
$3x + y = -2$

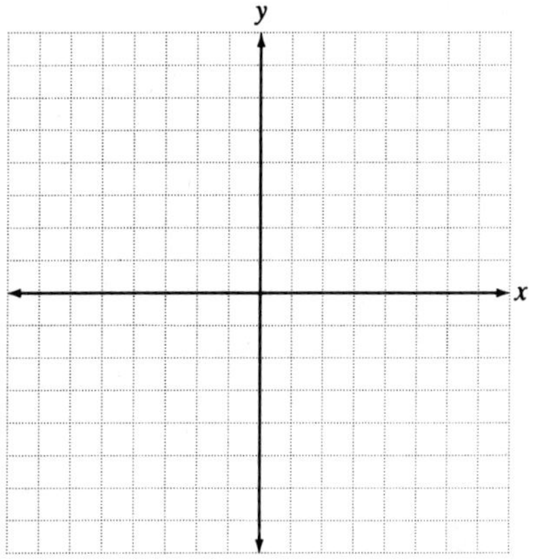

8. $2x - 3y = 3$
$2x + 2y = 8$

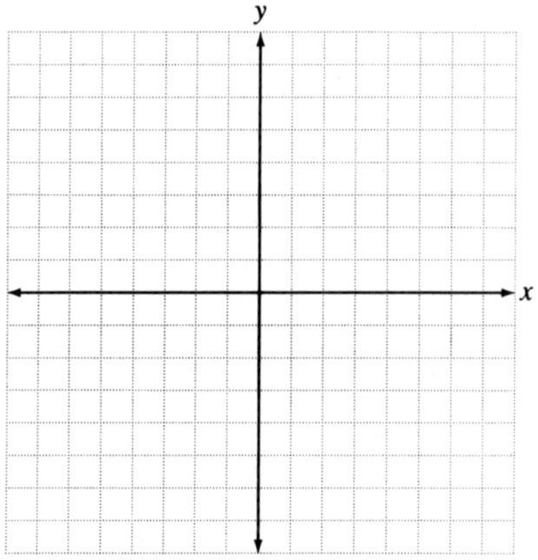

9. $3x - y = -7$
$2x + y = -3$

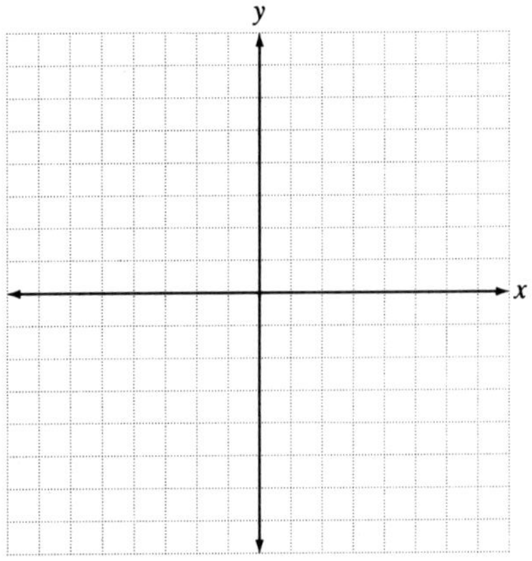

10. $2x - 3y = 12$
$-2x + y = -4$

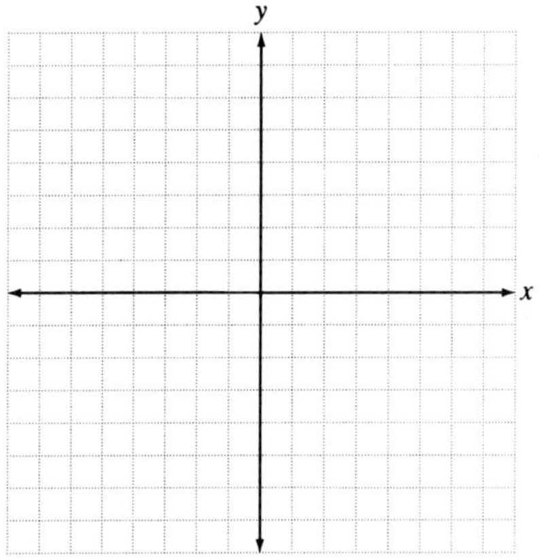

Section 4.2

Solve.

11. $-x + 3y = 2$
$x - 4y = -3$

12. $4x - y = -11$
$2x + y = -1$

13. $5x + 2y = -1$
$2x - 5y = 17$

14. $3x + 4y = -4$
$5x - 2y = 2$

15. $x + 2y = 1$
$y = x - 7$

16. $9x - 2y = 9$
$x = 3y + 1$

17. $x = 3y + 5$
$y = \frac{2}{3}x$

18. $3x = y - 2$
$x = \frac{1}{4}y$

19. $\frac{x}{2} + \frac{y}{3} = -\frac{1}{3}$
$\frac{x}{2} + 2y = -7$

20. $\frac{x}{3} + \frac{y}{5} = 2$
$\frac{x}{3} - \frac{y}{2} = -\frac{1}{3}$

21. $3.1x - 2.9y = -2.5$
$-4.3x + 5.8y = 8.8$

22. $5.2x + 3.1y = -12.5$
$-10.4x - 3.3y = 27.9$

Section 4.3

23. The perimeter of a rectangle is 7 feet more than three times the length. If the width is 2 feet less than the length, find the dimensions of the rectangle.

24. The length of a rectangle is 2 meters more than the width. The perimeter of the rectangle is 12 meters more than 3 times the width. Find the dimensions of the rectangle.

25. Carlos bought a total of 7 pounds of apples and bananas. Apples cost 49 cents a pound and bananas cost 35 cents a pound and Carlos spent \$3.15. How many pounds of apples did he buy?

* 26. If \$200 is removed from an investment, the new amount earns the same amount of interest at 9% per year as the original amount earned at 7% per year. Find the amount invested at 9%.

27. Melissa invested \$10,500. Part of the money was invested at 8% per year and part at 6% per year. If she earned interest of \$700, how much was invested at each rate?

28. Jim deposited eight more \$10 bills than \$5 bills in the bank. The bills were worth \$215. How many \$10 bills did he deposit?

* 29. How much water should be added to 10 gallons of a 45% salt solution to make a 30% solution? *Hint:* Water is a 0% salt solution.

* 30. How much pure acid (100% acid solution) should be added to 3 liters of a 40% acid solution to make a 65% solution?

Section 4.4

Solve.

31. $$\begin{aligned} 4x - y &= 2 \\ 3y + z &= 9 \\ x + 2z &= 7 \end{aligned}$$

32. $$\begin{aligned} x - 3y &= 6 \\ y + 2z &= 2 \\ 7x - 3y + 5z &= 16 \end{aligned}$$

33. $$\begin{aligned} x - y &= 0 \\ x + 2z &= 2 \\ 2y + 3z &= 3 \end{aligned}$$

34. $$\begin{aligned} 2x - 3z &= -9 \\ 3x + 3y &= -3 \\ 4y - z &= -7 \end{aligned}$$

35. $$\begin{aligned} 2x + 5y + 2z &= 1 \\ 4x - 7y - 3z &= 12 \\ 3x - 8y - 2z &= 12 \end{aligned}$$

36. $$\begin{aligned} x + 3y - 6z &= -4 \\ 2x - y + z &= -2 \\ x + 2y + 2z &= 3 \end{aligned}$$

37. $$\begin{aligned} x + y + z &= 1 \\ 2x + 3y + 2z &= 2 \\ -x + 2y - 3z &= 3 \end{aligned}$$

38. $$\begin{aligned} 2x + 3y - 4z &= 1 \\ x - y + 2z &= 3 \\ 3x + 2y - 3z &= 4 \end{aligned}$$

39. $$\begin{aligned} 2a + 5b - c &= 12 \\ a - b + 4c &= 10 \\ -8a - 20b + 4c &= 29 \end{aligned}$$

40. $$\begin{aligned} 2u - 3v + 4w &= 8 \\ 6u - 9v + 12w &= 24 \\ -4u + 6v - 8w &= -16 \end{aligned}$$

Section 4.5

41. Laura invested $11,000 at 6%, 8%, and 9% annual interest for 1 year. If twice as much was invested at 8% as at 6%, how much was invested at each rate? The total interest earned was $865.

42. The sum of three numbers is 6. Twice the first, plus 3 times the second, minus the third, is 7. Three times the first, minus the second, minus the third, is 6. Find the numbers.

43. A collection of 22 coins is worth $2.55. There are dimes, nickels, and quarters in the collection. If the number of dimes is one less than twice the number of quarters, how many of each type are in the collection?

44. The Anderson family bought a total of 18 hamburgers, packages of french fries, and milk shakes. They bought twice as many hamburgers as packages of french fries. If they purchased two fewer milk shakes than hamburgers, how many of each did they buy?

45. The sum of the measures of the angles of a triangle is 180°. The first angle is twice as large as the second angle. The third angle is 10° more than 7 times the first angle. Find the measures of the three angles.

46. Kristen's age is the sum of the ages of Ryan and Michael. Ryan's age is 4 less than twice Michael's age. The sum of all their ages is 64. Find Michael's age.

Section 4.6

47. $\begin{vmatrix} 3 & 4 \\ 7 & -1 \end{vmatrix}$

48. $\begin{vmatrix} 2 & 8 \\ -5 & 0 \end{vmatrix}$

49. $\begin{vmatrix} 7 & 9 \\ 0 & 2 \end{vmatrix}$

50. $\begin{vmatrix} -3 & 2 \\ 5 & 4 \end{vmatrix}$

51. $\begin{vmatrix} 0 & 4 & 1 \\ 3 & 0 & -2 \\ 2 & -5 & 1 \end{vmatrix}$

52. $\begin{vmatrix} 3 & -1 & 0 \\ 7 & 2 & 0 \\ 5 & -3 & 0 \end{vmatrix}$

53. $\begin{vmatrix} -3 & 0 & 4 \\ 2 & 1 & 0 \\ -1 & 3 & 2 \end{vmatrix}$

54. $\begin{vmatrix} -5 & 1 & 0 \\ -2 & 0 & 3 \\ 4 & 1 & 2 \end{vmatrix}$

Solve by Cramer's rule.

55. $\begin{aligned} 2x - 3y &= -6 \\ 4x - 2y &= 4 \end{aligned}$

56. $\begin{aligned} 2x - y &= -3 \\ 3x + 4y &= 1 \end{aligned}$

57. $\begin{aligned} 5x + 8y &= 1 \\ 3x + 7y &= 5 \end{aligned}$

58. $\begin{aligned} 4x - 3y &= 12 \\ 2x + 6y &= 16 \end{aligned}$

59. $\begin{aligned} 2x - 3y + 5z &= 27 \\ x + 2y - z &= -4 \\ 5x - y + 4z &= 27 \end{aligned}$

60. $\begin{aligned} 5x + y &= 10 \\ 3x + 2y + z &= -3 \\ -y - 2z &= 9 \end{aligned}$

61. $\begin{aligned} x - 2y &= 4 \\ 3x + y &= 5 \\ 2x + z &= 5 \end{aligned}$

62. $\begin{aligned} x + y + 3z &= -8 \\ 2x + 4y - z &= 5 \\ -x - y + 5z &= -16 \end{aligned}$

COOPERATIVE LEARNING

1. The solution of two equations by graphing is shown below. Write an equation for each line.

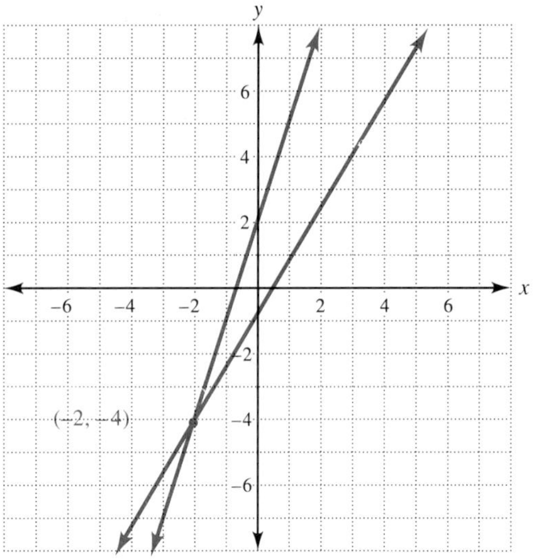

Write two equations whose solution is as follows.

2. $(-12, 15)$

3. $(-27, -54)$

4. A tank contains 40 liters of acid solution. How many liters of 65% solution should be drained from the tank and replaced with water to make a 43% solution?

5. Write three equations whose solution is $(9, -8, 27)$.

6. Garcia's Department Store had a sale on towels. Guest towels sold for \$1.95 each, hand towels sold for \$2.50 each, and bath towels sold for \$6.80 each. The total receipts from the sale of 128 towels were \$421.50. If two more guest towels were sold than hand towels, how many of each type were sold?

7. Write a second-order determinant whose value is -37.

Chapter 4 Practice Test	NAME ______ DATE ______ CLASS ______

1. Solve by graphing. Then classify the system as consistent or inconsistent and as dependent or independent. **1.** ______

$x + 3y = 6$

$y = \frac{2}{3}x - 1$

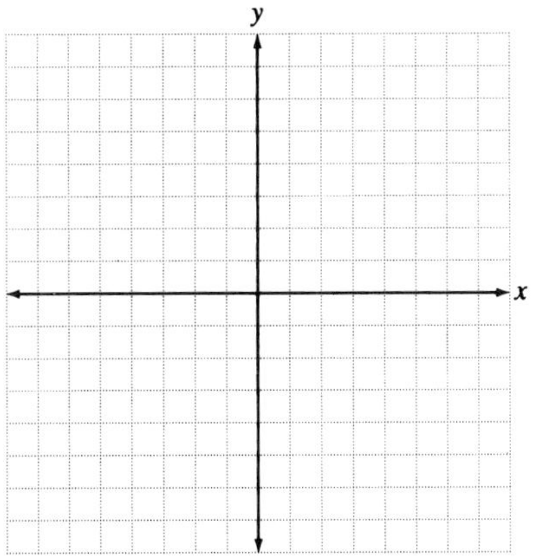

Solve.

2. $2x + 3y = 8$ **2.** ______
$x + y = 3$

3. $3x - 2y = -9$ **3.** ______
$y = -2x + 1$

4. Susan bought three more boxes of strawberries than blueberries. Strawberries cost 79 cents per box and blueberries cost 99 cents per box. If she spent $9.49, how many boxes of each type of berry did she buy? **4.** ______

5. ______

5. A 30% alcohol solution is to be mixed with an 80% alcohol solution to make 5 gallons of a 50% alcohol solution. How many gallons of each should be used?

6. ______

6. Solve:

$$\begin{aligned} x + 2y + z &= 1 \\ 3x - 2y - z &= 3 \\ x - y - 4z &= -6 \end{aligned}$$

7. ______

7. A store sold a total of 20 sweaters. They sold two fewer sweaters on Tuesday than they sold on Monday. On Wednesday they sold four fewer sweaters than the sum of those sold on Monday and Tuesday. How many sweaters did the store sell each day?

Evaluate.

8. ______

8. $\begin{vmatrix} 5 & 3 \\ 7 & 8 \end{vmatrix}$

9. ______

9. $\begin{vmatrix} 4 & 2 & -1 \\ -5 & 0 & 4 \\ 3 & 1 & -2 \end{vmatrix}$

10. ______

10. Solve using Cramer's rule. Show your work.

$$\begin{aligned} 4x - y - 3z &= 1 \\ 8x + y - z &= 5 \\ 2x + y + 2z &= 5 \end{aligned}$$

Chapters 3 and 4 Cumulative Review

NAME

DATE

CLASS

1. Graph using intercepts:
$x - 2y = 4$.

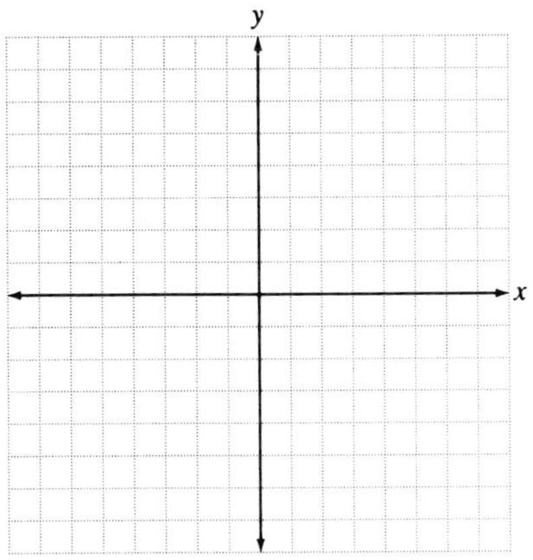

2. Find the slope of the line through the points $(7, -3)$ and $(9, -5)$.

3. Find the slope and y-intercept and graph:
$3x + 2y = 4$.

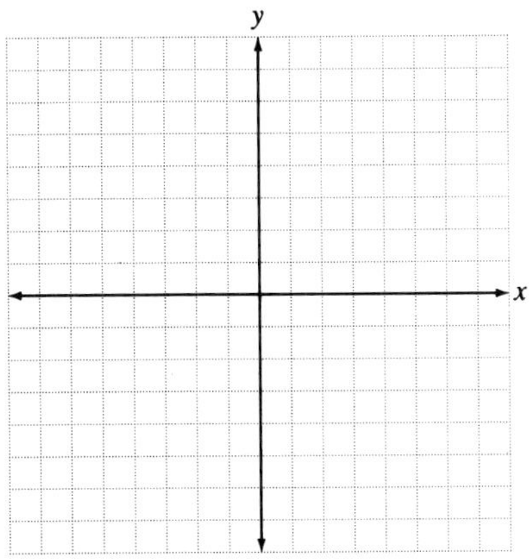

4. Find an equation for the line with slope $-\frac{4}{5}$ and y-intercept 2.

5. Find an equation for the line through the points $(-7, -2)$ and $(-3, 1)$.

6. Find an equation for the line through the point $(4, 5)$ and parallel to the line $3x + y = 7$.

7. Find an equation for the line through the point (3, 2) and perpendicular to the line $4x - y = 2$

8. Graph $2x + 3y \leq 9$.

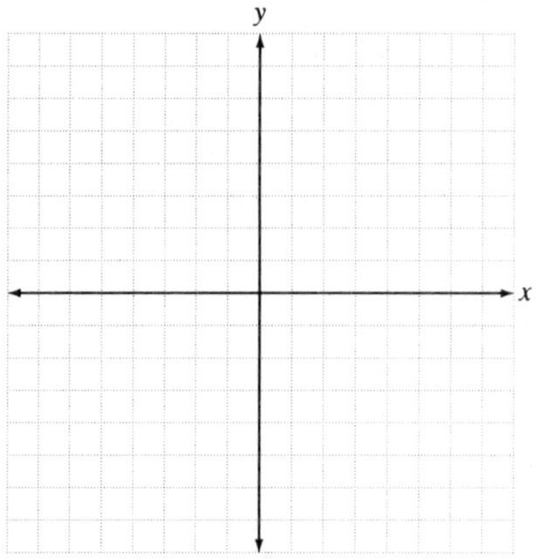

9. For income tax purposes, Mr. Slocum computes the depreciation of his automobile using straight-line depreciation. If the value of the car after 2 years is \$15,000 and the value of the car after 5 years is \$10,500, write the equation that relates value V to the number of years t that it is depreciated.

Solve.

10. $\begin{aligned} 3x - 2y &= 8 \\ 2x + 3y &= 14 \end{aligned}$

11. $\begin{aligned} 5x - 2y &= 3 \\ y &= 3x - 1 \end{aligned}$

12. A store sold a total of 24 pounds of candy. If chocolates cost \$6.25 per pound and caramels cost \$5.50 per pound and the receipts were \$144.75, how much of each type of candy was sold?

13. Susan invested \$2300 for 1 year and earned \$169 interest. If part of the money was invested at 7% interest per year and the rest at 8% interest per year, how much was invested at each rate?

14. Solve:

$$\begin{aligned} x + 3y - z &= 1 \\ 2x + 5y + z &= -3 \\ x - 2y + 3z &= -2 \end{aligned}$$

15. Evaluate:

$$\begin{vmatrix} 3 & 1 & 2 \\ 4 & 1 & -1 \\ 2 & 0 & -5 \end{vmatrix}$$

16. Solve using Cramer's rule. Show your work.

$$\begin{aligned} 2x - y + 3z &= 4 \\ x - 5y - 2z &= 1 \\ -4x - 2y + z &= 3 \end{aligned}$$

CHAPTER 5

Polynomials

Pretest

1. Evaluate $Q = 5x^2 - 7xy + x^3$ for $x = -2$ and $y = -3$.

2. Find the degree of the polynomial: $4x^3 + 6x - 9 + 8x^7 - 9x^5$.

3. Combine like terms: $9ab^2 - 6a^2b^2 + 4 - 10a^2b^2 - 15ab^2$.

4. Add: $9x^2 - 8xy + y^2$ and $8x^2 - 5xy - 7y^2$.

5. Subtract $5x^3 - 3x + 4$ from $7x^2 - 8x - 22$.

Multiply.

6. $(5x^2 - 7x - 3)(x - 4)$

7. $(3x - 7)(9x - 8)$

8. $(5y - 9)(5y + 9)$

9. $(7r - 6)^2$

Divide.

10. $(54p^3 - 42p^2 - 3) \div 6p^2$

11. $\dfrac{x^2 + 7x + 1}{x - 3}$

12. Divide using synthetic division: $(2x^3 - 1) \div (x - 2)$.

Factor.

13. $ax + bx - ay - by$

14. $2x^3 + 6x^2 - 56x$

15. $81x^2 - 126x + 49$

16. $100a^2b^2 - 36b^4$

17. $343x^3 - 27y^3$

Solve.

18. $15x^2 + 2 = 13x$

19. $p^2 = 19p$

20. The sum of two numbers is 9. The sum of their squares is 261. Find the numbers.

5.1 BASIC IDEAS ABOUT POLYNOMIALS

OBJECTIVES

1. *Evaluate polynomials for given values of variables*
2. *Find the degree of each term of a polynomial and the degree of the polynomial*
3. *Combine like terms*
4. *Write polynomials in descending powers of the variable*
5. *Identify trinomials, binomials, and monomials*

Operations with polynomials are very important in applications of mathematics to the real world.

> A polynomial in a variable or variables is a term or a finite sum of terms in which all variables have ***whole number exponents*** and no variables appear in denominators.

Remember that the whole numbers include zero.

Following are examples of polynomials:

$$4x \qquad 6x + 3 \qquad 4x^2 - 2x + 1 \qquad 5x^2y + 8y \qquad 9x^2y^2 \qquad 3$$

Notice that 3 is also a polynomial since it can be written $3x^0$, and that $5/x$ is not a polynomial since $5/x = 5x^{-1}$ and the exponent is not a whole number.

1 Evaluating Polynomials

A polynomial may be represented by a single letter such as $P = x^4 + 5x^2 - 2$. We evaluate a polynomial by substituting the given values for the variables and simplifying. Evaluating polynomials is important in the applications of mathematics. Recall the rules for order of operations from Section 1.5.

EXAMPLE 1

a. Evaluate $P = x^3 + 2x^2 - 3x$ for $x = -2$.

$$P = x^3 + 2x^2 - 3x = (\mathbf{-2})^3 + 2(\mathbf{-2})^2 - 3(\mathbf{-2}) \qquad \text{Substitute } -2 \text{ for } x.$$
$$= -8 + 2(4) - 3(-2)$$
$$= -8 + 8 + 6 = 6$$

b. Evaluate $P = 3xy + 4y^3 - 2x + 4$ for $x = 2$ and $y = -1$.

$$P = 3xy + 4y^3 - 2x + 4 = 3(2)(-1) + 4(-1)^3 - 2(2) + 4$$
$$= 3(2)(-1) + 4(-1) - 4 + 4$$
$$= -6 - 4 - 4 + 4 = -10$$

□ **Exercise 1** Evaluate.

a. $P = 5x^3 - 2x^2 + 8$ for $x = -1$

b. $P = 3y - 2x^2y + 4x$ for $x = -3$ and $y = 2$

c. If a car traveling at the rate of r miles per hour hits a tree, the force of the impact can be related to the force with which the car hits the ground if pushed off a building s feet high by the formula

$$s = 0.034r^2$$

If a car traveling at 40 miles per hour hits a tree, this is equivalent to pushing it from a building of what height?

$$s = 0.034r^2$$
$$= 0.034(40)^2 \quad \text{Substitute 40 for } r.$$
$$= 0.034(1600) = 54.4$$

The force of the impact is equivalent to pushing the car from a building 54.4 feet high. ■

□ **DO EXERCISE 1.**

Recall from Section 1.4 that in a term the number by which the variable is multiplied is called the *coefficient.* We usually do not write the term if the coefficient is zero. We say that a polynomial in one variable has a ***missing term*** if the coefficient of one of the variables is zero. In the polynomial $x^4 - 3x^2 + 4$, the x^3 and x terms are "missing."

2 Degree of a Polynomial

The ***degree of a term*** is the sum of the exponents on the variables.

> The ***degree of a polynomial*** is the same as the greatest degree of any of its terms.

Remember that a constant such as 4 can be written $4x^0$.

EXAMPLE 2 Find the degree of each term and the degree of the polynomial.

a. $8x^2 - 2x + 4$

The degrees of the terms are 2, 1 and 0, respectively. The degree of the polynomial is 2.

b. $4x + 7x^2y^3 - 4xy + 7$

The degrees of the terms are 1, 5, 2, and 0, respectively. The degree of the polynomial is 5. ■

□ **DO EXERCISE 2.**

□ **Exercise 2** Find the degree of each term and the degree of the polynomial.

a. $x^3 - 4x^2 + 8$

b. $7x^4y^2 - 3x + 4y^2 + 3xy$

The leading term of a polynomial is the term of greatest degree. Its coefficient is called the ***leading coefficient.***

3 Combining Like Terms

Like terms are terms that have exactly the same variables raised to exactly the same powers.

The following are examples of like terms.

$3x^2$ and $-5x^2$ $\quad$ $4x^2y$ and $7x^2y$ $\quad$ $2x$ and $-6x$

Notice that x^2y and xy^2 are *not* like terms.

EXAMPLE 3 Combine like terms.

a. $7x^2 - 3x + 4 - 6x^2 - 8 + 5x$

$7x^2 - 3x + 4 - 6x^2 - 8 + 5x$

$= 7x^2 - 6x^2 - 3x + 5x + 4 - 8$ — Rearrange terms using the commutative and associative laws.

$= 7x^2 + (-6x^2) + (-3x) + 5x + 4 + (-8)$ — Definition of subtraction.

$= x^2 + 2x - 4$

b. $5x^2y^3 + 3x - 3x^2y^3 - 2z$

$5x^2y^3 + 3x - 3x^2y^3 - 2z$

$= 5x^2y^3 - 3x^2y^3 + 3x - 2z$ — Rearrange terms using the commutative and associative laws.

$= 2x^2y^3 + 3x - 2z$ ■

□ **DO EXERCISE 3.**

□ **Exercise 3** Combine like terms.

a. $5x + 8x^2 - 7x + 3x^2$

b. $7xy^3 - 4x - 8xy^3 + 3x^3y$

□ **Exercise 4** Write the following polynomials in descending powers of the variable.

a. $-3 + 4y + 7y^3$

b. $-7x^9 + 4x^8 - 3x^{10}$

□ **Exercise 5** Identify as a trinomial, binomial, monomial, or none of these.

a. $8x + 3$

b. $7y^4 + 9y^2 - 3y + 4$

c. 15

d. $x^2 + 5x - 7$

4 Descending Order

The terms of polynomials in one variable are usually arranged so that the exponents on the variable decrease (descending order). The terms may also be arranged so that the exponents on the variable increase (ascending order).

EXAMPLE 4 Write the following polynomials in descending powers of the variable.

a. $x - 3x^2 + 5x^4 - 3 + 4x^3$

Write the polynomial as follows:

$$5x^4 + 4x^3 - 3x^2 + x - 3$$

b. $-3 + y - 8y^4 + 2y^2$

The polynomial is written as follows:

$$-8y^4 + 2y^2 + y - 3$$ ■

□ DO EXERCISE 4.

5 Special Polynomials

There are three polynomials that are given special names. ***Trinomials*** have exactly three terms, ***binomials*** have exactly two terms, and ***monomials*** have exactly one term.

EXAMPLE 5

Trinomials	*Binomials*	*Monomials*
$4x^2 - x + 3$	$x - 4$	$3x$
$x^5 - 3x^2 + x$	$y^2 + 4$	8
$x + 7 - x^3$	$9 - y^7$	$-3y^5$ ■

□ DO EXERCISE 5.

Answers to Exercises

1. a. 1 **b.** -42

2. a. The degree of each term is 3, 2, 0, respectively; the degree of the polynomial is 3 **b.** The degree of each term is 6, 1, 2, and 2, respectively; the degree of the polynomial is 6

3. a. $11x^2 - 2x$ **b.** $-xy^3 + 3x^3y - 4x$

4. a. $7y^3 + 4y - 3$ **b.** $-3x^{10} - 7x^9 + 4x^8$

5. a. Binomial **b.** None of these **c.** Monomial **d.** Trinomial

Problem Set 5.1

NAME

DATE

CLASS

Evaluate for $x = -1$ and $y = 2$.

1. $P = 4x^2 - 2x + 3$
2. $P = 5x^2 + 3x - 4$
3. $Q = -x^3 + 7x^2 - 2x$
4. $Q = -7x^3 - 3x^2 - 5x$
5. $P = x^2y + 5$
6. $Q = xy^2 - 3y$
7. $Q = 7xy - y^3 + 3x$
8. $P = 3y^2 - 2xy + y^3$

Find the degree of each polynomial.

9. $7x^2 + 3x + 4$
10. $y^3 - y^2 + 5$
11. $3x$
12. $4y$
13. -7
14. 10
15. $6x^5 - 3x^2 + x$
16. $7x^8 + 3x^5 + 4$
17. $8x + 9x^2y^2 - x^2y^3$
18. $7y - 3x^5y^2 - 2xy^4$

Combine like terms.

19. $8x^2 - 3x^2$
20. $7y - 3y$
21. $-7x^3y - 2x^3y$
22. $-8xy^4 - 3xy^4$
23. $-9x^2 + 6x^2$
24. $-5x^3 + 2x^3$
25. $3x - 5y - 2x + 3y$
26. $7a - 6y + 3a + 4y$

27. $-7p^2 + 3p^2 - 2p^2$

28. $3b^3 - 7b^3 + 4b^3$

29. $9x^2 - 2x + 3x^2$

30. $-3z^2 + 4z - 6z^2$

31. $6a^2 - 3 - 5 + 2a^2 + 8$

32. $-9 + 5r^2 - 6 + 3r^2 + 8$

33. $4a^2b - 6ab^2 + 2a^2b - 3ab^2$

34. $-3x^2y^2 + 4y^3 - 7x^2y^2 - 2y^3$

35. $3x^2y^2 + 2xy - 5x^2y^2 + 4xy$

36. $-7a^2b^2 + 3a - 2a^2b^2 - 4a$

Write each polynomial in descending powers of the variable.

37. $x^3 - 4x + 3x^2 - 1$

38. $x^2 - 8x + 7 - 3x^4$

39. $3 + y + 7y^4 - 9y^3$

40. $7y^4 - y + 3y^2 - 2y^5$

41. $13x^3 - 8x^4 - 6x + 9x^2 + 3$

42. $-7y^4 - 5y^2 + 18y^3 + 4y^5 - 8$

Identify as a trinomial, binomial, monomial, or none of these.

43. $3x + 4$

44. 7

45. $3a^2 + 4a - 2$

46. $6y^4 - 3$

47. $x^2 - 4$

48. $8x^3 - 3x + 2$

49. $5b$

50. 10

51. $5x^3 + 7x^2 - 8x + 3$

52. 43

53. $9x^2 - 5x - 12$

54. $6x^5 - 4x^3 + 6x - 5$

55. The cost of manufacturing n brushes is given by $C = \frac{1}{2}n^2 - 14n + 30$. Find the cost of manufacturing 40 brushes.

56. The cost of marketing n brushes is given by $C = n^2 - 30n + 60$. Find the cost of marketing 40 brushes.

57. The number of milk containers that a machine can fill in t hours is $N = 5t^2 - 3t + 500$. How many containers can it fill in 8 hours?

58. The number of toys a machine can produce in t hours is given by $N = 5t^2 - 4t + 20$. How many toys can the machine produce in 40 hours?

59. The volume of a sphere is given by the polynomial $V = \frac{4}{3}\pi r^3$. Find the volume of a sphere with a radius of 18 meters. Use 3.14 for π.

60. The surface area of a right circular cylinder is $S = 2\pi r^2 + 2\pi rh$. Use this formula to find the surface area of a can of chili with a radius of 1.5 inches and a height of 4.2 inches. Use 3.14 for π.

61. The total profit from manufacturing x calculators is $P = 0.5x^2 + x - 750$. Find the total profit from manufacturing 60 calculators.

62. If n compact disc players are manufactured, the total profit is $P = -1.2x^2 + 280x - 8000$. What is the total profit if 70 compact disc players are manufactured?

63. The formula for the volume of a cylindrical can is given by the polynomial $V = \pi r^2 h$, where r is the radius and h is the height of the can. Use 3.14 for π and evaluate this polynomial for $r = 2$ inches and $h = 7$ inches.

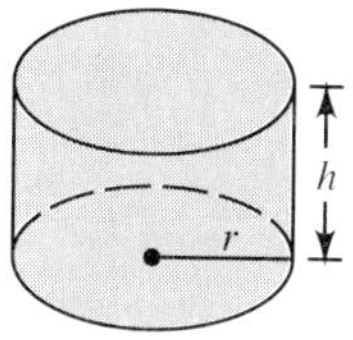

64. If an object is thrown straight up into the air with a velocity of 64 feet per second, its height h, above the ground t seconds later is $h = -16t^2 + 64t$. Find its height after 3 seconds.

65. A polynomial is a term or a finite sum of terms in which all variables have ____________ number exponents and no variables appear in denominators.

66. In the polynomial $3x^3 + 2x + 4$, the x^2 term is ____________.

67. The __________ of a term is the sum of the exponents on the variables.

68. Like terms are terms that have exactly the same __________ raised to exactly the same __________.

69. Binomials have exactly __________ terms.

Evaluate for $x = -3.95$ and $y = 4.68$. Round answers to the nearest hundredth.

70. $P = 3x^2y - 18.494$

71. $Q = 4xy^2 + 29.873$

72. $Q = 8xy - y^3 + 2x$

73. $P = 2y^2 - 3xy + x^3$

Combine like terms.

74. $7.348x - 5.9721y - 8.45x - 2.93y$

75. $9.8215x + 3.145y - 10.435x - 9.1y$

76. $7.3824 - 4.78x^2 - 6.14 + 13.9x^2$

77. $6.7a^2 - 3.9148 - 5.72 + 2.9389a^2 + 8.743$

* **78.** Evaluate for $y = -\frac{3}{2}$: $N = 12y^3 - 6y^2 + 5y + \frac{7}{8}$.

* **79.** Evaluate for $x = \frac{1}{3}$ and $y = -\frac{1}{5}$: $R = 4y^2 - 3xy + 4 - y$.

Checkup

The following problems provide a review of some of Section 1.4 and will help you with the next section.

Simplify.

80. $3x + 2 + (5x - 4)$

81. $7y - 3 + (2y - 8)$

82. $5a - 3 - (2a - 2)$

83. $-6x + 1 - (4x + 3)$

84. $3x + 4 + (5x - 9)$

85. $7y - 5 + (3y - 4)$

86. $8y - (7y + 6)$

87. $-9a - (3a - 4)$

88. $7x + 3 - (9x + 8)$

89. $-3y - 2 + (6y + 4)$

5.2 ADDITION AND SUBTRACTION OF POLYNOMIALS

OBJECTIVES

1 *Add polynomials*

2 *Subtract polynomials*

1 Addition

To add two polynomials, we combine like terms and arrange the terms in descending order.

Suppose the cost of manufacturing n toys is $C = \frac{1}{2}n^2 - 12n + 20$ and the cost of marketing n toys is $M = n^2 - 20n + 50$.

We can find the total cost of marketing and manufacturing the toys by adding the two polynomials.

$$\left(\frac{1}{2}n^2 - 12n + 20\right) + (n^2 - 20n + 50)$$

$$= \frac{1}{2}n^2 + n^2 - 12n - 20n + 20 + 50$$

$$= \frac{3}{2}n^2 - 32n + 70$$

The total cost of marketing and manufacturing n toys is $C = \frac{3}{2}n^2 - 32n + 70$.

Polynomials are usually added horizontally. However, they may be added vertically by placing like terms in columns and leaving spaces for the missing terms.

EXAMPLE 1

a. Add $-5x^3 - 7x + 3$ and $8x^3 + 4x^2 - 6$.

$$(-5x^3 - 7x + 3) + (8x^3 + 4x^2 - 6)$$

$$= -5x^3 + 8x^3 + 4x^2 - 7x + 3 - 6$$

$$= 3x^3 + 4x^2 - 7x - 3 \qquad \text{Combine like terms.}$$

b. Add vertically the two polynomials shown in part a.

$$\begin{array}{rrrr} -5x^3 & & -\,7x & +\,3 \\ 8x^3 & +\,4x^2 & & -\,6 \\ \hline 3x^3 & +\,4x^2 & -\,7x & -\,3 \end{array} \qquad \text{Add.}$$

c. Add $7x^2y + 3xy^2 - 2y$ and $2x^2y - 5xy^2 + 4y$.

$$(7x^2y + 3xy^2 - 2y) + (2x^2y - 5xy^2 + 4y)$$

$$= 7x^2y + 2x^2y + 3xy^2 - 5xy^2 - 2y + 4y$$

$$= 9x^2y - 2xy^2 + 2y \quad \blacksquare$$

□ **DO EXERCISE 1.**

□ **Exercise 1** Add. Use both horizontal and vertical addition.

a. $3y^3 + 5y^2 - 6y - 4$ and $-8y^3 - 2y^2 + 4y - 1$

b. $3x^3 + x - 7$ and $4x^3 + 5$

c. $8a^2b^2 - ab + 2$ and $-9a^2b^2 - 4ab - 7$

☐ **Exercise 2** Subtract.

a. $(3b^2 - 5) - (4b^2 + 2)$

b. $(8x^3 - 2x - 3) - (9x^3 + 4x - 6)$

c. $(7x^2y^2 + 3xy + 4y) - (-2x^2y^2 - 8xy - 3)$

☐ **Exercise 3** Subtract. Use vertical subtraction.

a. $(9x - 3) - (9x + 5)$

b. $(7y^2 - 3) - (7y^2 + 4y - 8)$

2 Subtraction

Recall that we subtract a real number from another real number by adding the opposite or additive inverse of the second real number to the first real number: $a - b = a + (-b)$. We use this same technique to subtract polynomials. To find the opposite of a polynomial, change the sign of every term in it.

EXAMPLE 2 Subtract.

a. $(-5x^2 - 3x + 4) - (2x^2 + 6x - 3)$

$= (-5x^2 - 3x + 4) + (-2x^2 - 6x + 3)$ Add the opposite of the second polynomial.

$= -5x^2 - 2x^2 - 3x - 6x + 4 + 3$

$= -7x^2 - 9x + 7$

b. $(4ab^2 - 3a^2b - ab) - (7ab^2 + 2a^2b - ab)$

$= (4ab^2 - 3a^2b - ab) + (-7ab^2 - 2a^2b + ab)$

$= 4ab^2 - 7ab^2 - 3a^2b - 2a^2b - ab + ab$

$= -3ab^2 - 5a^2b$ ■

☐ **DO EXERCISE 2.**

Vertical subtraction is used in long division of polynomials, which we will study later in this chapter. Vertical subtraction is similar to vertical addition except that we must remember to change the signs of the polynomial that is subtracted.

EXAMPLE 3 Subtract $(-3x^2 - 7x + 2) - (-4x^2 + 6x - 5)$.

Write the first polynomial above the second, placing like terms in columns.

$$\begin{array}{r} -3x^2 - 7x + 2 \\ \underline{-(-4x^2 + 6x - 5)} \end{array}$$

Change all the signs on the second polynomial and add.

$$\begin{array}{rl} -3x^2 - 7x + 2 & \\ \underline{4x^2 - 6x + 5} & \text{Change all the signs.} \\ x^2 - 13x + 7 & \text{Add.} \end{array}$$ ■

☐ **DO EXERCISE 3.**

Answers to Exercises

1. a. $-5y^3 + 3y^2 - 2y - 5$ **b.** $7x^3 + x - 2$ **c.** $-a^2b^2 - 5ab - 5$
2. a. $-b^2 - 7$ **b.** $-x^3 - 6x + 3$ **c.** $9x^2y^2 + 11xy + 4y + 3$
3. a. -8 **b.** $-4y + 5$

Problem Set 5.2

NAME

DATE

CLASS

Add vertically.

1. $5x - 8$ and
$-7x + 4$

2. $-6y + 7$ and
$-9y + 8$

3. $3x^2 - 2x + 5$ and
$-5x^2 - 6x - 3$

4. $-8x^2 - x - 1$ and
$-9x^2 - 4x + 1$

5. $6x^2y + 3x + 4$ and
$-2x^2y + 5x^2 - 8$

6. $3xy^2 - 2y^2 + 1$ and
$-5xy^2 - 3y + 5$

Add horizontally.

7. $7a + 3$ and
$9a - 4$

8. $b - 3$ and
$9b - 8$

9. $4x^2 + 3x + 2$ and
$3x^2 - 5x - 1$

10. $6x^2 + 5x - 3$ and
$4x^2 - 2x + 5$

11. $5a^4 + 3a^2 - 6$ and
$-7a^4 + 5a^2 - 8$

12. $-3b^3 - 5b + 9$ and
$2b^3 - 6b + 4$

13. $-3x^2 - 4x - 5$ and
$-6x^2 + 5x + 2$

14. $-5x^2 + 3x - 8$ and
$-2x^2 + 4x + 2$

15. $5x^2 - 3xy + y^2$ and
$-x^2 + 5xy + 9y^2$

16. $8a^2 + 5ab - 4b^2$ and
$3a^2 - 6ab + 9b^2$

17. $8x^2y - 5xy^2 - 7$ and
$-9x^2y - 3xy^2 - 8$

18. $-4a^2b - 6ab^2 + 5b$ and
$-3a^2b + 2ab^2 - 2b$

DID YOU KNOW?

In the Rhind papyrus (1650 B.C.), symbols are used for addition and subtraction. The symbol for addition shows a pair of legs walking from left to right, and the symbol for subtraction shows a pair of legs walking from right to left.

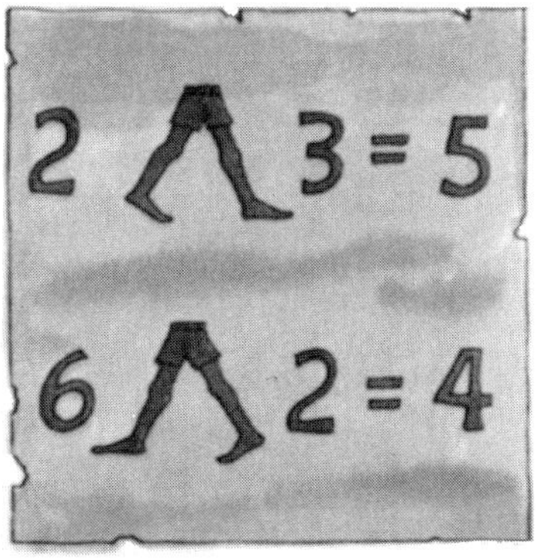

Subtract.

19. $(7y - 4) - (5y + 3)$

20. $(9x + 2) - (-3x - 5)$

21. $(3x - 2y + z) - (x - 4y - 3z)$

22. $(-5a + 4b - 3c) - (2a - 3b + 3c)$

23. $(x^2 - 6x + 4) - (4x^2 + 2x - 3)$

24. $(5x^2 + 3x + 2) - (7x^2 + 5x - 3)$

25. $(-4x^2 + 8x^3 - 3) - (6x^2 + 2x^3 + 1)$

26. $(9a^2 + 5a^3 - 4) - (8a^2 + 6a^3 + 8)$

27. $(x^2y^2 - 4xy + 4) - (3x^2y^2 + 2xy + 6)$

28. $(8y^2 - 7y + 5) - (11y^2 + 3y - 9)$

Subtract. Use vertical subtraction.

29. $(6x - 4) - (6x + 8)$

30. $(9x - 7) - (9x + 5)$

31. $(5x^2 - 7) - (4x^2 - 3x)$

32. $(8y^2 - 3) - (-9y^2 + 5)$

33. $(-9x^2 - 6x - 4) - (-3x^2 + 5x - 7)$

34. $(8x^3 - 6x^2 + 4) - (3x^3 - 2x^2 + 4)$

Add or subtract as indicated.

35. $(8x^2 + 2x) + (3x^2 - 5x) - (6x^2 + x)$

36. $(7y^2 - 3y) - (5y^2 + 2y) + 3y$

37. $(3a^2 + 4) - (5a^2 + a) + (2a - 3)$

38. $(-7z^2 - 3z) + (5z - 1) - (6z + 8)$

39. $(15x^2y - 8x^2y^2 - 5xy^2) - (12x^2y^2 - 21x^2y - 3xy^2)$

40. $(7pq - 18p^2q + 9pq^2) - (20pq^2 + 15pq + 4p^2q)$

41. $(9ab^2 - 6ab) + (8ab - 5a^2b) - (12ab^2 + 7a^2b)$

42. $(-8y^2z - 14yz) - (7y^2z + 6yz^2) + (19yz - 5yz^2)$

43. Subtract $9x^2 - 15x + 2$ from $7x^2 - 8x + 23$.

44. Subtract $12a^2 - 18a - 4$ from $-3a^2 - 5a - 17$.

45. Subtract $9y^2 - 14y + 7$ from $22y^2 + 2$.

46. Subtract $8ab^2 - 15ab$ from $9ab^2 - 8a^2b + 22ab$.

Total profit P is given by total revenue R minus total cost C as follows:

$$P = R - C$$

For Problems 47 and 48, find the total profit, P.

47. $R = 150n - 0.2n^2$, $C = 350 - 0.8n^2$

48. $R = 280n - 0.3n^2$, $C = 200 - 0.7n^2$

49. To add two polynomials, we combine ____________ ____________.

50. Polynomials may be added by placing like ____________ in columns.

51. To subtract one polynomial B, from another polynomial A, we add the ____________ of B to A.

52. Vertical subtraction is similar to vertical addition except that we must remember to ____________ ____________ ____________ of the polynomial that is subtracted.

Add or subtract as indicated.

53. $(8.43x^2 - 2.985) + (7.917x^2 - 9.496)$

54. $(-9.723a^2 + 14.73) - (8.34a^2 + 17.5)$

55. $(3.571a^3 + 4.987a) - (15.2a^3 + 0.94a^2) + (2.135a^2 - 9.095a)$

56. $(-7.8924z^5 - 3.802z^3) + (15.32z^3 - 0.9145z) + (6.35z^2 + 8.91z)$

* **57.** $[-(7x^2 - 8x + 5x^3) - (3x^2 + 4x + 6x^3)] + x^2$

* **58.** $(3y^4 - 5y^2 - 2) + [-(y^4 - y^2 + 2) - (y^4 + 3y^2 + 2)]$

* **59.** $6(5 - x) - 7(3 - 4x - x^3) + (2 - x - 3x^2)$

* **60.** $\frac{5}{7}(14x^2 - 21xy + 35y^2) - \frac{3}{8}(24x^2 - 72xy + 96y^2)$

Checkup

The following problems provide a review of some Sections 1.6 and 1.7 and will help you with the next section.

Multiply.

61. $x^4 \cdot x^2$

62. $x^3 \cdot x^5$

63. $x \cdot x$

64. $y^2 \cdot y$

65. $2x^2 \cdot 3x$

66. $3x^3 \cdot 2x$

67. $5x^2 \cdot (-3x^2)$

68. $-8x^3(-7x^2)$

Calculate. Write answers in scientific notation.

69. $(5.7 \times 10^{-5})(3.8 \times 10^{-4})$

70. $(4.26 \times 10^{-7})(9.6 \times 10^{2})$

71. $\dfrac{2.5 \times 10^{-3}}{7.5 \times 10^{-7}}$

72. $\dfrac{9.6 \times 10^{-4}}{3.2 \times 10^{5}}$

5.3 MULTIPLICATION OF POLYNOMIALS

OBJECTIVES

1. *Multiply monomials*
2. *Multiply a monomial and any polynomial*
3. *Multiply two polynomials*

1 Multiplying Monomials

Recall from Section 1.6 that to multiply exponential expressions with the same base, we keep the base and add the exponents. The rule is $a^m \cdot a^n = a^{m+n}$ for any integers m and n, and any nonzero real number a. We multiply two *monomials* by multiplying the coefficients and adding the exponents on the identical variables.

EXAMPLE 1 Multiply.

a. $(-2x)(6x^2) = -2(6)(x)(x^2) = -12x^3$ Recall that $x = x^1$.

b. $(3x^2y^2)(-5x^3y^5) = 3(-5)(x^2)(x^3)(y^2)(y^5)$

$= -15x^5y^7$ ■

□ DO EXERCISE 1.

□ **Exercise 1** Multiply.

a. $(4x)(7x^5)$

b. $(-8y^3)(-9y^6)$

c. $(-3a^4b^3)(7ab^6)$

d. $(7xy^2z^7)(-8x^3y^4z^2)$

2 Multiplying a Monomial and Any Polynomial

To multiply a monomial and any polynomial, multiply each term of the polynomial by the monomial. This process is based on the distributive laws.

EXAMPLE 2 Multiply.

a. $3x$ and $2x^2 - 4$

$3x(2x^2 - 4) = 3x(2x^2) - 3x(4)$ Use a distributive law.

$= 6x^3 - 12x$ Multiply the monomials.

b. $-4y^3$ and $5y^2 + 2y - 3$

$-4y^3(5y^2 + 2y - 3) = (-4y^3)(5y^2) + (-4y^3)(2y) - (-4y^3)(3)$

$= -20y^5 - 8y^4 + 12y^3$ ■

□ DO EXERCISE 2.

□ **Exercise 2** Multiply.

a. $5(2x - 3)$

b. $-4x^2(x + 3)$

c. $3y(y^2 - 4y + 7)$

d. $-y(5y^3 + 3y^2 - 1)$

□ **Exercise 3** Multiply.

a. $(2x^2 + 5)(x - 3)$

b. $(x - 4)(x + 6)$

c. $(a + 2)(a^2 - 3a + 4)$

d. $(y - 1)(y^3 - y + 5)$

□ **Exercise 4** Multiply

$2x^2 + 3x - 1$ and $4x^2 - 2x + 3$.

[3] Multiplying Two Polynomials

Two polynomials are multiplied by multiplying each term of one polynomial by each term of the other polynomial. We use the distributive laws more than once. Then combine like terms. The distributive laws may also be stated as $(b + c)a = ba + ca$ and $(b - c)a = ba - ca$.

EXAMPLE 3 Multiply.

a. $(4x^2 + 3)(\mathbf{x + 2})$

$= 4x^2(\mathbf{x + 2}) + 3(\mathbf{x + 2})$ Use a distributive law.

$= 4x^2(x) + 4x^2(2) + 3(x) + 3(2)$ Use a distributive law.

$= 4x^3 + 8x^2 + 3x + 6$ Multiply the monomials.

b. $(y + 3)(y^4 - 5y^3 + 2)$

$= y(\mathbf{y^4 - 5y^3 + 2}) + 3(\mathbf{y^4 - 5y^3 + 2})$

$= y(y^4) - y(5y^3) + y(2) + 3(y^4) - 3(5y^3) + 3(2)$

$= y^5 - 5y^4 + 2y + 3y^4 - 15y^3 + 6$

$= y^5 - 5y^4 + 3y^4 - 15y^3 + 2y + 6$

$= y^5 - 2y^4 - 15y^3 + 2y + 6$ Combine like terms. ■

□ **DO EXERCISE 3.**

Long multiplications of polynomials may be done by writing them vertically. The multiplication is then completed in a way that is similar to the multiplication of numbers, keeping like terms in columns.

EXAMPLE 4 Multiply $2m^2 + 3m - 4$ and $-3m^2 + m - 2$.

$$\begin{array}{rrrrrl}
 & & 2m^2 & +\;3m & -\;4 & \\
 & & -3m^2 & +\;\;m & -\;2 & \\
\hline
 & & -4m^2 & -\;6m & +\;8 & \text{Multiply by } -2. \\
 & 2m^3 & +\;3m^2 & -\;4m & & \text{Multiply by } m. \\
-6m^4 & -\;9m^3 & +\;12m^2 & & & \text{Multiply by } -3m^2. \\
\hline
-6m^4 & -\;7m^3 & +\;11m^2 & -\;10m & +\;8 & \text{Add.} \quad ■
\end{array}$$

□ **DO EXERCISE 4.**

Answers to Exercises

1. a. $28x^6$ **b.** $72y^9$ **c.** $-21a^5b^9$ **d.** $-56x^4y^6z^9$

2. a. $10x - 15$ **b.** $-4x^3 - 12x^2$ **c.** $3y^3 - 12y^2 + 21y$
d. $-5y^4 - 3y^3 + y$

3. a. $2x^3 - 6x^2 + 5x - 15$ **b.** $x^2 + 2x - 24$
c. $a^3 - a^2 - 2a + 8$ **d.** $y^4 - y^3 - y^2 + 6y - 5$

4. $8x^4 + 8x^3 - 4x^2 + 11x - 3$

Problem Set 5.3

NAME

DATE

CLASS

Multiply.

1. $(3x)(4x^3)$

2. $(4y)(-3y^2)$

3. $(-5x^2)(-2x^6)$

4. $(-7y^4)(8y^6)$

5. $(9x^2y^3)(8x^3y^5)$

6. $(7a^2b^3)(-9a^3b^5)$

7. $(-xyz^2)(-5x^3y^4z^2)$

8. $(x^2yz)(-6x^3y^2z)$

9. $-3x(2x - 4)$

10. $-5x(3x + 2)$

11. $4x^2(3x + 8)$

12. $3y^2(y - 2)$

13. $-5x^3(2x^2 - 4x)$

14. $7x^4(x^3 - 8x)$

15. $4x(x^2 - 7x + 2)$

16. $-3y(2y^2 - 8y + 1)$

17. $-2y^2(3y^2 - 2y + 8)$

18. $5x^3(x^2 - 3x + 7)$

19. $(x - 3)(x + 2)$

20. $(y + 4)(y + 3)$

21. $(x + y)(x + y)$

22. $(a - b)(a + b)$

23. $(4a - 3)(2a + 1)$

24. $(3x - 4)(4x + 5)$

25. $(2x^2 - 3)(x - 4)$

26. $(3y^3 - 2)(y^2 + 4)$

27. $(3a^2 - 2)(4a^2 + 3)$

28. $(5x^2 - 1)(2x^2 - 3)$

29. $(x + 2)(x^2 - 3x + 1)$

30. $(y - 1)(y^2 - 4y + 3)$

31. $(y - 2)(2y^2 - 3y + 4)$

32. $(2y + 1)(3y^2 + y - 1)$

33. $(y - 3)(y^4 - y + 3)$

34. $(x - 4)(x^3 - 3x + 1)$

35. $(a - b)(a^2 + ab + b^2)$

36. $(p + q)(p^2 - pq + p^2)$

37. $(x^2 + 2x - 1)(x^2 + 3x - 6)$

38. $(b^2 - 3b + 4)(b^2 + b - 1)$

39. $(3pq^2 - 4pq - 5q^2)(pq + 3q - p)$

40. $(7a^2 - 5a^2b + 3ab)(ab - b^2 - a^2)$

41. $\left(x - \frac{3}{4}\right)\left(x - \frac{3}{4}\right)$

42. $\left(y + \frac{5}{8}\right)\left(y + \frac{5}{8}\right)$

Multiply by writing vertically.

43. $x^2 + 2x - 3$ and $x^2 + x - 2$

44. $y^2 - 3y + 4$ and $y^2 + y - 1$

45. $3x^2 - 4x + 1$ and $2x^2 - x + 2$

46. $5x^2 - 2x + 3$ and $2x^2 + x - 1$

47. $-4x^2 - x - 3$ and $x^2 - 3x + 2$

48. $-3x^2 + 2x + 4$ and $-x^2 - x - 3$

49. Write the area A of this rectangle as a polynomial.

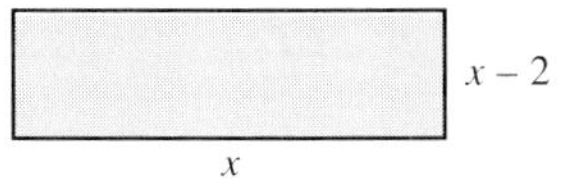

50. Write the area A of the following triangle as a polynomial.

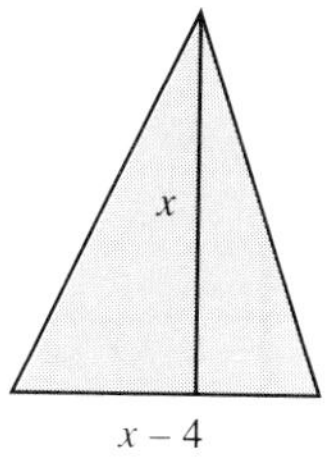

51. We multiply two monomials by multiplying the ___________ and ___________ the exponents on the identical variables.

52. To multiply a monomial and any polynomial, multiply each ___________ of the polynomial by the monomial.

53. Two polynomials that are not monomials may be multiplied by using the ___________ laws.

Multiply. Round answers to the nearest hundredth.

54. $5.35x^3(0.95x^2 - 13.22x + 7.48)$

55. $-3.785x(2.894x^2 - 18.093x + 1.08)$

56. $(7.89a - 2.34)(8.56a + 0.84)$

57. $(4.34x - 6.95y)(8.92x - 3.78y)$

Multiply.

* 58. $\left(5x - \frac{2}{5}\right)\left(3x - \frac{1}{7}\right)$

* 59. $\left(\frac{3}{4}x - \frac{1}{2}\right)\left(\frac{5}{8}x + \frac{1}{3}\right)$

* 60. $\left(y - \frac{4}{5}\right)\left(y^2 - 10y + \frac{3}{8}\right)$

* 61. $\left(x^3 - 6x - \frac{2}{3}\right)\left(x - \frac{1}{9}\right)$

Think About It

62. Write a polynomial that gives the area A of the trapezoid shown below.

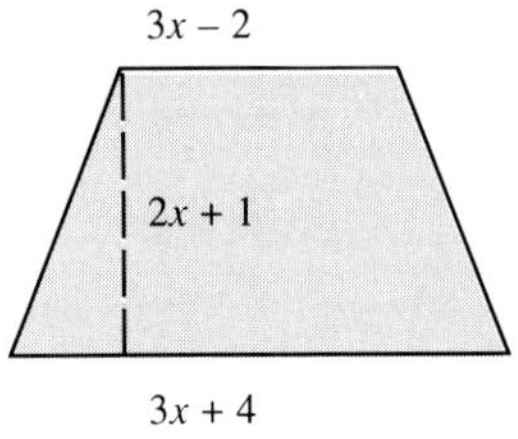

63. Write a polynomial that gives the volume V of the box.

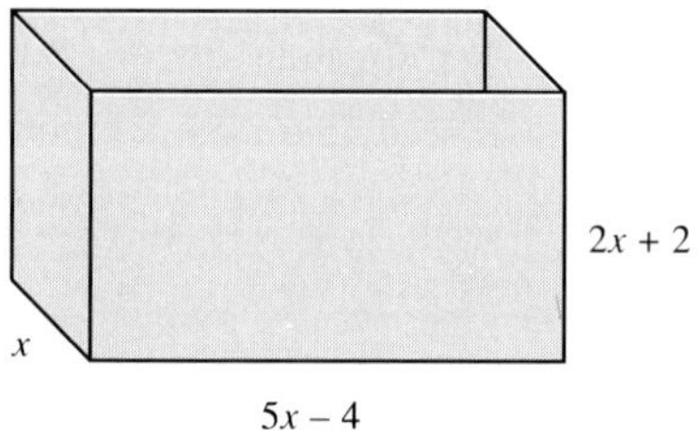

Checkup

The following problems provide a review of some of Section 4.2.

Solve.

64. $3x + 4y = 5$
$3x - 2y = -7$

65. $2x + 3y = 3$
$2x + 5y = 1$

66. $4x - 5y = -9$
$y = x + 1$

67. $3x - y = 2$
$y = 4x$

68. $2x + 3y = 1$
$5x - 2y = -7$

69. $4x - 3y = 5$
$2x + 5y = -4$

70. $-9x + 3y = 2$
$-3x + y = 7$

71. $6x + 3y = 1$
$y = -2x - 5$

72. $y = -3x + \frac{5}{2}$
$12x + 4y = 10$

73. $4y = 4x - 3$
$-8x + 8y = -6$

5.4 SPECIAL PRODUCTS OF BINOMIALS

There are special methods for multiplying a binomial by a binomial that make our work easier. Recall that when we multiply two polynomials the result is the product.

1 Multiplying a Binomial by a Binomial Using FOIL

To multiply two binomials we multiply the *f*irst terms, then the *o*uter terms, then the *i*nner terms, and finally the *l*ast terms. Then combine like terms. This method is called **FOIL:** for first, outer, inner, and last. The following diagram gives us a visual picture of the method.

$$\begin{array}{c} \quad\quad\quad\quad\quad\quad\quad\; F \quad\;\; O \quad\;\; I \quad\;\; L \\ (A + B)(C + D) = AC + AD + BC + BD \end{array}$$

EXAMPLE 1 Multiply.

a. $(x - 3)(x + 4)$

1. Multiply the **F**irst terms, $(x)(x) = x^2$.
2. Multiply the **O**uter terms, $(x)(4) = 4x$.
3. Multiply the **I**nner terms, $(-3)(x) = -3x$.
4. Multiply the **L**ast terms, $(-3)(4) = -12$.
5. Add the terms on the right. $x^2 + 4x - 3x - 12$
6. Combine like terms. $x^2 + x - 12$
7. $(x - 3)(x + 4) = x^2 + x - 12$

Or:

$$\begin{aligned} &\quad\quad\;\; \mathbf{F} \quad\; \mathbf{O} \quad\; \mathbf{I} \quad\; \mathbf{L} \\ (x - 3)(x + 4) &= x^2 + 4x - 3x - 12 \\ &= x^2 + x - 12 \end{aligned}$$

b. $(3x - 4)(2x - 5) = 6x^2 - 15x - 8x + 20$
$= 6x^2 - 23x + 20$

c. $(2x - 3y)(5x + 2y) = 10x^2 + 4xy - 15xy - 6y^2$
$= 10x^2 - 11xy - 6y^2$ ■

□ DO EXERCISE 1.

OBJECTIVES

1 *Multiply a binomial by a binomial using FOIL*

2 *Multiply a sum and a difference of the same two terms*

3 *Square a binomial*

□ **Exercise 1** Multiply.

a. $(x - 5)(x + 8)$

b. $(3a - 7)(4a + 1)$

c. $(4p - 3q)(2p + 3q)$

d. $(y^2 - 4)(2y^2 + 3)$

e. $(2x - 3)(2x + 3)$

f. $(5x - 2)(5x - 2)$

□ **Exercise 2** Multiply. Do not use FOIL.

a. $(x + 3)(x - 3)$

b. $(3y + 1)(3y - 1)$

c. $(x - y)(x + y)$

d. $(5p - 4q)(5p + 4q)$

e. $(x^2 + 3)(x^2 - 3)$

f. $(a^2 - 7b)(a^2 + 7b)$

2 Multiplying a Sum and a Difference of the Same Two Terms

This method is used when the binomials are identical except for the sign between the terms. By the FOIL method the product of $(A + B)(A - B)$ is as follows:

$$\begin{aligned} & \quad\;\;\; \textbf{F} \quad\; \textbf{O} \quad\;\; \textbf{I} \quad\;\; \textbf{L} \\ (A + B)(A - B) &= A^2 - AB + AB - B^2 \\ &= A^2 - B^2 \end{aligned}$$

$A^2 - B^2$ is called *the difference of two squares.*

From this result, we have the following rule.

> The product of the sum and difference of the same two terms is the square of the first term minus the square of the second term.
>
> $$(A + B)(A - B) = A^2 - B^2$$

EXAMPLE 2 Multiply.

a. $(x + 4)(x - 4) = x^2 - 4^2$
$= x^2 - 16$

b. $(5x + 2)(5x - 2) = (5x)^2 - 2^2$
$= 5^2x^2 - 2^2$ Recall that $(ab)^2 = a^2b^2$.
$= 25x^2 - 4$

c. $(3a - 4b)(3a + 4b) = (3a)^2 - (4b)^2$
$= 9a^2 - 16b^2$

d. $(x^2 - 2y)(x^2 + 2y) = (x^2)^2 - (2y)^2$
$= x^4 - 4y^2$ ■

□ **DO EXERCISE 2.**

3 Squaring a Binomial

When we square a binomial using FOIL, we get the following results:

$$(A + B)^2 = (A + B)(A + B)$$
$$= A^2 + AB + AB + B^2$$
$$= A^2 + 2AB + B^2$$

and

$$(A - B)^2 = (A - B)(A - B)$$
$$= A^2 - AB - AB + B^2$$
$$= A^2 - 2AB + B^2$$

Hence:

> The square of a binomial is the square of the first term, plus twice the product of the two terms, plus the square of the last term.
>
> $$(A + B)^2 = A^2 + 2AB + B^2$$
> $$(A - B)^2 = A^2 - 2AB + B^2$$

EXAMPLE 3 Multiply.

a. $(x - 4)^2 = x^2 - 2(x)(4) + 4^2$
$= x^2 - 8x + 16$

b. $(y + 5)^2 = y^2 + 2(y)(5) + 5^2$
$= y^2 + 10y + 25$

c. $(3p - 4)^2 = (3p)^2 - 2(3p)(4) + 4^2$
$= 9p^2 - 24p + 16$

d. $(2x + 3y)^2 = (2x)^2 + 2(2x)(3y) + (3y)^2$
$= 4x^2 + 12xy + 9y^2$

Caution: Notice that $(A + B)^2 \neq A^2 + B^2$ and $(A - B)^2 \neq A^2 - B^2$. Choose some numbers other than zero for A and B and you will see that this is the case. ■

□ DO EXERCISE 3.

Try to multiply polynomials mentally. Use the easiest method.

> **Special Products of Binomials**
>
> $$(A + B)(C + D) = AC + AD + BC + BD \qquad \text{(FOIL)}$$
> $$(A + B)(A - B) = A^2 - B^2$$
> $$(A + B)^2 = A^2 + 2AB + B^2$$
> $$(A - B)^2 = A^2 - 2AB + B^2$$

□ **Exercise 3** Multiply. Do not use FOIL.

a. $(x - 8)^2$

b. $(2x + 5)^2$

c. $(3x - 5)^2$

d. $(x + y)^2$

e. $(4x - 3y)^2$

f. $(2p + 6q)^2$

DID YOU KNOW?

Michael Stifel was one of the sixteenth-century mathematicians who made the symbols + and − popular in Europe. He predicted that the world would end October 3, 1533. Many people gave up all their worldly possessions and waited with Stifel on a hill on October third. Since the world did not end, he had himself locked up in prison to escape all the angry people later that day.

Answers to Exercises

1. **a.** $x^2 + 3x - 40$ **b.** $12a^2 - 25a - 7$ **c.** $8p^2 + 6pq - 9q^2$ **d.** $2y^4 - 5y^2 - 12$ **e.** $4x^2 - 9$ **f.** $25x^2 - 20x + 4$

2. **a.** $x^2 - 9$ **b.** $9y^2 - 1$ **c.** $x^2 - y^2$ **d.** $25p^2 - 16q^2$ **e.** $x^4 - 9$ **f.** $a^4 - 49b^2$

3. **a.** $x^2 - 16x + 64$ **b.** $4x^2 + 20x + 25$ **c.** $9x^2 - 30x + 25$ **d.** $x^2 + 2xy + y^2$ **e.** $16x^2 - 24xy + 9y^2$ **f.** $4p^2 + 24pq + 36q^2$

Problem Set 5.4

NAME

DATE

CLASS

Multiply. Use FOIL.

1. $(x - 8)(x - 2)$

2. $(y - 3)(y - 4)$

3. $(y + 7)(y + 8)$

4. $(x + 9)(x + 7)$

5. $(2x - 3)(3x + 4)$

6. $(4y - 1)(5y + 3)$

7. $(8a - 1)(2a - 3)$

8. $(6p - 5)(2p - 1)$

9. $(3x - 2y)(4x + 5y)$

10. $(7x + 3y)(2x - 3y)$

Multiply. Do not use FOIL. See Example 2.

11. $(x - 1)(x + 1)$

12. $(y - 2)(y + 2)$

13. $(x + 10)(x - 10)$

14. $(y + 9)(y - 9)$

15. $(3x - 4)(3x + 4)$

16. $(7x + 2)(7x - 2)$

17. $(4a + 3)(4a - 3)$

18. $(5b - 8)(5b + 8)$

19. $(8x - 9y)(8x + 9y)$

20. $(7x + 6y)(7x - 6y)$

Multiply. Do not use FOIL. See Example 3.

21. $(x + 3)^2$

22. $(x + 7)^2$

23. $(y - 5)^2$

24. $(y - 8)^2$

25. $(4x - 1)^2$

26. $(5x - 2)^2$

27. $(2s - 5)^2$

28. $(3r - 7)^2$

29. $(3y + 4z)^2$

30. $(5x - 2y)^2$

Multiply. Use the easiest method.

31. $(8x - 3)(8x + 3)$

32. $(3y + 8)(3y - 8)$

33. $(2x - 7)(3x + 4)$

34. $(5x - 9)(x + 1)$

35. $\left(x - \frac{1}{2}\right)\left(x - \frac{1}{3}\right)$

36. $\left(y - \frac{1}{5}\right)\left(y - \frac{1}{2}\right)$

37. $(2x - 7)^2$

38. $(6y + 5)^2$

39. $(3m - 8)(m + 1)$

40. $(7m - 2)(2m - 3)$

41. $(6 - x)(6 + x)$

42. $(9 + 2x)(9 - 2x)$

43. $\left(x - \frac{1}{2}\right)^2$

44. $\left(y - \frac{1}{3}\right)^2$

45. $(3x + 4y)(2x - y)$

46. $(6a - b)(3a + 2b)$

47. $(5p - 3q)(2p + 5q)$

48. $(8y - z)(3y + z)$

49. $(x^2 - 4)(x^2 + 4)$

50. $(y^2 - 8)(y^2 + 2)$

51. $\left(x - \frac{1}{6}\right)\left(x + \frac{1}{6}\right)$

52. $\left(x + \frac{1}{4}\right)\left(x - \frac{1}{4}\right)$

53. $(3y^2 - 2)(2y^2 + 1)$

54. $(6a^2 + 5)(3a^2 + 1)$

55. $(y^2 - 7)(y^2 + 7)$

56. $(3a^2 - b)(3a^2 + b)$

57. $(2a^2 - 3)^2$

58. $(5x^2 - 1)^2$

59. $(7 - 3x)(7 + 3x)$

60. $\left(\frac{1}{5} - 5x\right)^2$

61. $(a^2 + ac)(a^2 - ac)$

62. $(2x^2 + 5yz)(2x^2 - 5yz)$

63. $(x + 2)(x - 2)(x^2 + 4)$

64. $(y - 3)(y + 3)(y^2 + 9)$

65. $(2a - b)(2a + b)(4a^2 - b^2)$

66. $(5x + y)(5x - y)(25x^2 - y^2)$

67. The volume V of a cube with a side of length $x - 3$ is given by

$$V = (x - 3)^3$$

Find an equivalent formula for the volume of the cube.

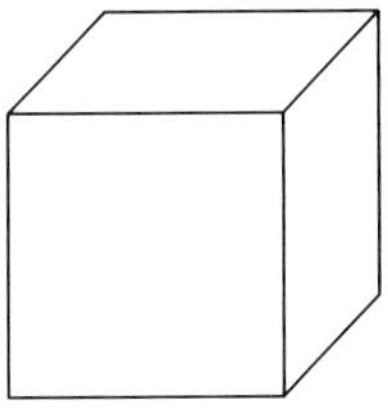

68. If P dollars is invested in a savings account at interest rate i, compounded annually for 3 years, the amount in the account after 3 years is

$$A = P\,(l + i)^3$$

Find an equivalent formula for A.

69. We may multiply two binomials by the FOIL method by multiplying the ____________ terms, outer terms, ____________ terms, and last terms and combining like terms.

70. The product of the sum and difference of two terms is the square of the first term ____________ the square of the second term.

71. The square of a binomial is the square of the first term, plus ____________ the product of the two terms, plus the square of the last term.

Multiply.

72. $(3.9x - 6.8)(3.9x + 6.8)$

73. $(4.7a - 10.8)(4.7a + 10.8)$

74. $(7.8a - 5.2b)^2$

75. $(9.8x + 3.7y)^2$

Assume that variables in exponents represent positive integers. Multiply.

* **76.** $\left(\frac{5}{3}x - \frac{1}{4}\right)\left(\frac{5}{3}x + \frac{1}{4}\right)$

* **77.** $\left(\frac{4}{5}y - \frac{3}{4}\right)^2$

* **78.** $\left(\frac{9}{8}x^n + \frac{3}{4}y^n\right)^2$

* **79.** $\left(\frac{7}{20}a^{2n} - \frac{3}{4}b^{2n}\right)\left(\frac{7}{20}a^{2n} + \frac{3}{4}b^{2n}\right)$

* **80.** $(3x^2 + 7xy)(-7xy + 3x^2)$

81. $(9a^2 + 6ab)(-6ab + 9a^2)$

Think About It

* **82.** A rectangular sheet of cardboard 24 inches long and 18 inches wide is used to make an open box by cutting squares from each corner and folding up the sides. If x represents a side of each square, write a polynomial to represent the volume of the box.

* **83.** Posters are to be made from sheets of material 50 centimeters wide and 90 centimeters long. If the margins at the top, bottom, and sides are each x centimeters, write a polynomial to represent the area of the printed part of the poster.

Checkup

The following problems provide a review of some of Section 4.3.

84. The perimeter of a rectangle is 24 meters. The length is 4 meters more than twice the width. Find the length of the rectangle.

85. Tickets to the movies cost $4 for adults and $2 for children. The total receipts were $124. If 8 more children's tickets were sold than adults' tickets, how many adults' tickets were sold?

86. Susan invested a sum of money for 1 year and earned $198 interest. If $600 more was invested at 9% interest per year than at 7% interest per year, how much was invested at 7%?

87. Juan wants to mix a 25% acid solution with a 65% acid solution to get 90 liters of a 40% solution. How many liters of 65% solution should he use?

88. Two angles are complementary. One angle is 4 times the sum of 5° and the measure of the other angle. Find the measures of the angles.

89. Two angles are complementary. One angle is 3 times the difference of the measure of the other angle and 6°. Find the measures of the angles.

90. McRonald's sold a total of 76 hamburgers and cheeseburgers. If hamburgers cost $0.65 and cheeseburgers cost $0.85 and the total amount of money collected was $53.80, how many cheeseburgers were sold?

91. The sum of two numbers is -85. Twice the first number minus the second number is -17. Find the numbers.

92. Two angles are supplementary. One angle is 24° less than 5 times the other. Find the measures of the angles.

93. Two angles are supplementary. One angle is 12° more than twice the other. Find the measures of the angles.

5.5 DIVISION OF POLYNOMIALS AND SYNTHETIC DIVISION

A knowledge of division of polynomials helps us to graph certain equations.

1 Division of a Polynomial by a Monomial

We may add or subtract fractions that have polynomials in the numerator and denominator by methods similar to the addition or subtraction of real numbers.

$$\frac{A+B}{C}=\frac{A}{C}+\frac{B}{C}$$ Since we can add the second expression to get the first expression, when A, B, and C are polynomials.

Also,

$$\frac{A-B}{C}=\frac{A}{C}-\frac{B}{C}$$

We may use the equalities above to divide a polynomial by a monomial.

> To divide a polynomial by a monomial, divide each term of the polynomial by the monomial.

Remember that the fraction c/d can be interpreted to mean c divided by d, where c and d are real numbers. Similarly, the polynomial A divided by B can be written A/B.

EXAMPLE 1

a. Divide $16x^2 + 12x - 4$ by 4.

$$\frac{16x^2+12x-4}{4}=\frac{16x^2}{4}+\frac{12x}{4}-\frac{4}{4}$$

$$=4x^2+3x-1$$

b. $$\frac{12x^3-9x^2+6x}{3x^2}=\frac{12x^3}{3x^2}-\frac{9x^2}{3x^2}+\frac{6x}{3x^2}$$

$$=4x-3+\frac{2}{x}$$ Recall that $\frac{a^m}{a^n}=a^{m-n}$.

Notice that the result is not a polynomial. The quotient of two polynomials is not necessarily a polynomial. ■

□ **DO EXERCISE 1.**

Caution: $\frac{B}{A+C}\neq\frac{B}{A}+\frac{B}{C}$

OBJECTIVES

1 *Divide a polynomial by a monomial*

2 *Divide a polynomial by a polynomial other than a monomial*

3 *Divide a polynomial by a binomial using synthetic division*

□ **Exercise 1** Divide.

a. $\frac{10x+30}{5}$

b. $\frac{24x^2-6x+12}{3x}$

c. $\frac{8y^3+10y^2-6y}{2y^2}$

d. $\frac{5z^4-14z^3-4z^2}{2z^2}$

2 Division of a Polynomial by a Polynomial Other Than a Monomial

The procedure that is used to divide a polynomial by a binomial or a trinomial is similar to the method that we used for long division in arithmetic. We will show division in arithmetic along with division of polynomials so that you can see the similarity. When we subtract one polynomial from another as part of the division process, we must be sure to change the signs of the expression that we are subtracting.

In the division problem $84 \div 7 = 12$, 84 is called the *dividend,* 7 is the *divisor,* and 12 is the *quotient.* Similar terms are used for polynomials.

EXAMPLE 2

a.

	Division in Arithmetic		Division in Algebra		
		$\frac{456}{6}$		$\frac{x^2 + 7x + 10}{x + 2}$	
Step 1	Divide 6 into 45. The whole-number part of the quotient is 7.	$\begin{array}{r} 7 \\ 6\overline{)456} \end{array}$	Divide x into x^2.	$\begin{array}{r} x \\ x + 2\overline{)x^2 + 7x + 10} \end{array}$	The quotient is x.
Step 2 / *Step 3*	Multiply 6(7) = / Subtract and bring down the 6.	$\begin{array}{r} 7 \\ 6\overline{)456} \\ \underline{42} \\ 36 \end{array}$	Multiply $x(x + 2)$ = / Subtract and bring down the next term, 10.	$\begin{array}{r} x \\ x + 2\overline{)x^2 + 7x + 10} \\ \underline{x^2 + 2x} \\ 5x + 10 \end{array}$	Remember to subtract!

The procedure begins to repeat at this point.

Step 4	Divide 6 into 36. The quotient is 6.	$\begin{array}{r} 76 \\ 6\overline{)456} \\ \underline{42} \\ 36 \end{array}$	Divide x into $5x$.	$\begin{array}{r} x + 5 \\ x + 2\overline{)x^2 + 7x + 10} \\ \underline{x^2 + 2x} \\ \mathbf{5x} + 10 \end{array}$	The quotient is 5.
Step 5	Multiply 6(6) = / Subtract	$\begin{array}{r} 76 \\ 6\overline{)456} \\ \underline{42} \\ 36 \\ \underline{36} \\ 0 \end{array}$	Multiply $5(x + 2)$ =	$\begin{array}{r} x + 5 \\ x + 2\overline{)x^2 + 7x + 10} \\ \underline{x^2 + 2x} \\ 5x + 10 \\ \underline{5x + 10} \\ 0 \end{array}$	

Hence $\frac{456}{6} = 76$ and $\frac{x^2 + 7x + 10}{x + 2} = x + 5$

b. Divide $6x^3 - 7x^2 - 3x + 1$ by $2x - 3$.

$$\begin{array}{r} 3x^2 + x \\ 2x - 3\overline{)6x^3 - 7x^2 - 3x + 1} \\ \underline{6x^3 - 9x^2} \\ 2x^2 - 3x \\ \underline{2x^2 - 3x} \\ + 1 \end{array}$$

Remember to subtract!

Subtract; the remainder is 1.

In the answer the remainder may be placed over the divisor. The answer is $3x^2 + x + 1/(2x - 3)$.

To check, multiply $3x^2 + x$ by the divisor and add the remainder to see if we obtain the dividend.

$$\underbrace{(3x^2 + x)\overbrace{(2x - 3)}^{\text{divisor}} + \overbrace{1}^{\text{remainder}} = 6x^3 - 9x^2 + 2x^2 - 3x + 1}$$

$$= \underbrace{6x^3 - 7x^2 - 3x + 1}_{\text{dividend}}$$

The result checks. ■

□ **DO EXERCISE 2.**

Caution: To divide polynomials, always arrange the polynomials in descending powers of the variable. If there are missing terms in the dividend, leave space for them or write them with zero coefficients.

EXAMPLE 3 Divide.

a. $\dfrac{x^3 + 1}{x + 1}$

We write the missing terms with zero coefficients.

$$\begin{array}{r|l l}
 & x^2 - x + 1 & \\
x + 1 & x^3 + 0x^2 + 0x + 1 & \\
 & \underline{x^3 + x^2} & \\
 & -x^2 + 0x & \text{Subtract and bring} \\
 & \underline{-x^2 - x} & \text{down the next term.} \\
 & x + 1 & \text{Subtract and bring} \\
 & \underline{x + 1} & \text{down the next term.}
\end{array}$$

The quotient is $x^2 - x + 1$.

b. Divide $x^3 + 3x^2 + 3$ by $x^2 - 2$.
We leave a space for the missing term.

$$\begin{array}{r|l l}
 & x + 3 & \\
x^2 - 2 & x^3 + 3x^2 \quad + 3 & \\
 & \underline{x^3 \quad - 2x} & \\
 & 3x^2 + 2x + 3 & \text{Subtract and bring} \\
 & \underline{3x^2 \quad - 6} & \text{down the next term.} \\
 & 2x + 9 & \text{Subtract.}
\end{array}$$

If the remainder is not zero, we continue to divide until the degree of the remainder is less than the degree of the divisor. In this case the degree of $2x + 9$ is less than the degree of $x^2 - 2$, so the result is

$$x + 3 + \frac{2x + 9}{x^2 - 2} \quad ■$$

□ **DO EXERCISE 3.**

□ **Exercise 2** Divide.

a. $\dfrac{x^2 + 8x + 15}{x + 3}$

b. $\dfrac{x^2 - 2x - 8}{x + 2}$

c. $\dfrac{6x^2 - 7x - 3}{2x - 3}$

d. $\dfrac{25y^3 + 15y^2 - 3y + 4}{5y + 2}$

□ **Exercise 3** Divide.

a. $\dfrac{x^3 - 1}{x - 1}$

b. $\dfrac{4y^4 - 7y^2 + y + 5}{2y^2 - y}$

3 Synthetic Division

When we divide a polynomial by a binomial of the form $x - k$, where the coefficient of the x term is 1 and k is a constant, we can use a shortcut. This easier procedure is called ***synthetic division.***

Notice the following. On the left we do the long division. On the right we show the division written without the variables.

$$
\begin{array}{r}
2x^2 + 3x + 4 \\
x - 4\overline{)2x^3 - 5x^2 - 8x - 18} \\
\underline{2x^3 - 8x^2} \qquad\qquad \\
3x^2 - 8x \qquad \\
\underline{3x^2 - 12x} \qquad \\
4x - 18 \\
\underline{4x - 16} \\
-2
\end{array}
\qquad\qquad
\begin{array}{rrrrr}
 & 2 & +3 & +4 & \\
1 - 4\overline{)} & 2 & -5 & -8 & -18 \\
 & \underline{2} & \underline{-8} & & \\
 & & 3 & -(8) & \\
 & & \underline{(3)} & \underline{-12} & \\
 & & & 4 & -(18) \\
 & & & \underline{(4)} & \underline{-16} \\
 & & & & -2
\end{array}
$$

The numbers in parentheses are repetitions of the numbers directly above them, so we may eliminate them and compress the problem. Then we have the following:

$$
\begin{array}{rrrrr}
 & 2 & +3 & +4 & \\
1 - 4\overline{)} & 2 & -5 & -8 & -18 \\
 & & -8 & -12 & -16 \\
\hline
 & 2 & 3 & 4 & -2
\end{array}
$$

We can eliminate the top line since it is the same as the first three terms of the bottom line. We can omit the one at the upper left. Also, we add in the second row instead of subtracting. To do this, we change the -4 at the upper left to $+4$. Now we have the following:

$$
\begin{array}{rrrrr}
4\overline{)} & 2 & -5 & -8 & -18 \\
 & & 8 & 12 & 16 \\
\hline
 & 2 & 3 & 4 & -2
\end{array}
$$

The numbers in the bottom row give the coefficients of the answer, which is always a polynomial of degree one less than the degree of the dividend. Since the dividend is of degree 3, the result is of degree 2. The answer is $2x^2 + 3x + 4 + -2/(x - 4)$.

The process is as follows.

EXAMPLE 4 Divide using synthetic division.

a. $\dfrac{x^2 - 9x + 20}{x - 5}$

1. $\begin{array}{r|rrr} 5 & 1 & -9 & 20 \\ & & & \\ \hline & 1 & & \end{array}$

Write the 5 from $x - 5$ changing the sign and write the coefficients of the dividend.

Bring down the first coefficient.

2. $\begin{array}{r|rrr} 5 & 1 & -9 & 20 \\ & & 5 & \\ \hline & 1 & -4 & \end{array}$

Multiply 1 by 5 to get 5; place 5 under and -9; add -9 and 5.

3. $\begin{array}{r|rrr} 5 & 1 & -9 & 20 \\ & & 5 & -20 \\ \hline & 1 & -4 & 0 \end{array}$

Multiply -4 by 5 to get -20; place -20 under 20; add -20 and 20.

The coefficients of the result appear in the bottom row. The result is of degree one less than the dividend.

The result is $x - 4$. There is no remainder.

b. $\dfrac{3x^3 - 2x + 21}{x + 2}$

Caution: Be sure to write zeros for missing terms in the dividend.

$\begin{array}{r|rrrr} -2 & 3 & 0 & -2 & 21 \\ & & -6 & 12 & -20 \\ \hline & 3 & -6 & 10 & 1 \end{array}$

Change the sign on the 2 from $x + 2$; write a zero for the x^2 term.

The result is $3x^2 - 6x + 10 + 1/(x + 2)$.

c. $\dfrac{x^4 - 1}{x + 1}$

$\begin{array}{r|rrrrr} -1 & 1 & 0 & 0 & 0 & -1 \\ & & -1 & 1 & -1 & 1 \\ \hline & 1 & -1 & 1 & -1 & 0 \end{array}$

Write zeros for the x^3, x^2, and x terms.

The quotient is $x^3 - x^2 + x - 1$. ■

□ **DO EXERCISE 4.**

□ **Exercise 4** Divide. Use synthetic division.

a. $\dfrac{x^2 - 5x - 14}{x - 7}$

b. $\dfrac{y^3 - 2y^2 + 5y - 4}{y + 2}$

c. $\dfrac{y^3 - 1}{y - 1}$

d. $\dfrac{3x^4 - 25x^2 - 18}{x - 3}$

Answers to Exercises

1. a. $2x + 6$ **b.** $8x - 2 + \frac{4}{x}$ **c.** $4y + 5 - \frac{3}{y}$ **d.** $\frac{5}{2}z^2 - 7z - 2$

2. a. $x + 5$ **b.** $x - 4$ **c.** $3x + 1$ **d.** $5y^2 + y - 1 + \frac{6}{5y + 2}$

3. a. $x^2 + x + 1$ **b.** $2y^2 + y - 3 + \frac{-2y + 5}{2y^2 - y}$

4. a. $x + 2$ **b.** $y^2 - 4y + 13 + \frac{-30}{y + 2}$ **c.** $y^2 + y + 1$

d. $3x^3 + 9x^2 + 2x + 6$

Problem Set 5.5

NAME

DATE

CLASS

Divide.

1. $\dfrac{6x^2 + 12x - 18}{6}$

2. $\dfrac{14y^2 - 21y + 42}{7}$

3. $\dfrac{9x^2 - 6x + 3}{3x}$

4. $\dfrac{16y^2 + 24y + 8}{8y}$

5. $\dfrac{8z^4 - 5z^3 + 10z}{5z^2}$

6. $\dfrac{16a^3 - 5a^2 + 12}{4a^2}$

7. $(8a^3b^4 + 32a^2b^3 - 56ab^2) \div 8ab$

8. $(81x^3y^4 - 45x^2y^3 + 63xy^4) \div 9xy^2$

Divide. Use long division.

9. $\dfrac{x^2 + 13x + 42}{x + 7}$

10. $\dfrac{x^2 + 7x - 18}{x - 2}$

11. $\dfrac{y^2 - 7y + 10}{y - 5}$

12. $\dfrac{y^2 - 8y + 16}{y - 4}$

13. $\dfrac{z^2 - 4z + 8}{z - 3}$

14. $\dfrac{r^2 - 7r - 12}{r - 2}$

15. $\dfrac{2x^2 + 13x + 21}{2x + 7}$

16. $\dfrac{10y^2 - 11y - 6}{5y + 2}$

17. $\dfrac{2z^3 + 7z^2 - 6}{2z + 3}$

18. $\dfrac{6p^3 - p^2 - 10}{3p + 4}$

19. $\dfrac{2a^4 - a^3 - 5a^2 + a - 6}{a^2 + 2}$

20. $\dfrac{3y^4 + 2y^3 - 11y^2 - 2y + 5}{y^2 - 2}$

21. $\dfrac{4x^4 + 6x^3 + 3x - 1}{2x^2 + 1}$

22. $\dfrac{15b^4 + 3b^3 + 4b^2 + 4}{3b^2 - 1}$

23. $(p^3 - 1) \div (p - 1)$

24. $(8a^3 + 1) \div (2a + 1)$

25. $(16x^4 + 1) \div (2x - 1)$

26. $(81b^4 - 1) \div (3b + 1)$

27. $(10b^3 + 6b^2 - 9b + 10) \div (5b - 2)$

28. $(12x^3 + 9x^2 - 10x + 21) \div (3x^2 - 2)$

Divide. Use synthetic division.

29. $\dfrac{x^2 - 9x + 8}{x - 1}$

30. $\dfrac{x^2 + 2x - 15}{x + 5}$

31. $\dfrac{5x^3 - 6x^2 + 3x + 14}{x + 1}$

32. $\dfrac{4y^3 - 3y^2 + 2y + 1}{y - 1}$

33. $\dfrac{x^3 + 2x^2 + 3x + 4}{x - 2}$

34. $\dfrac{a^3 - 2a^2 - 3a - 4}{a - 2}$

35. $\dfrac{3p^3 + 7p^2 - 4p + 3}{p + 3}$

36. $\dfrac{3y^3 + 7y^2 - 4y + 3}{y - 3}$

37. $\dfrac{x^6 + 2x^4 - 5x + 11}{x - 2}$

38. $\dfrac{6x^4 + 15x^3 + 28x + 6}{x + 3}$

39. $\dfrac{x^2 + x + 1}{x - 1}$

40. $\dfrac{x^2 - x + 1}{x + 1}$

41. $\dfrac{x^4 - 16}{x + 2}$

42. $\dfrac{y^4 + 81}{y - 3}$

43. $\dfrac{x^4 - 1}{x - 1}$

44. $\dfrac{x^4 + 1}{x - 1}$

45. $(6a^5 - 2a^3 - 4a^2 + 3a - 2) \div (a - 2)$

46. $(-4x^6 - 3x^5 - 3x^4 + 5x^3 - 6x^2 + 3x + 3) \div (x - 1)$

47. $(a^3 + 8) \div (a + 2)$

48. $(y^3 - 27) \div (y - 3)$

49. To divide a polynomial by a monomial, divide each ____________ of the polynomial by the monomial.

50. In the division problem $\frac{x^2 + 3x - 2}{x + 4}$, $x + 4$ is the ____________.

51. To divide a polynomial by a polynomial other than a monomial, use ____________ ____________.

52. If there are missing terms in the dividend, leave space for them or write them with ____________ coefficients.

53. An easier procedure for dividing a polynomial by a binomial of the form $x - k$ is called ____________ ____________.

Divide.

* **54.** $\frac{4p^3 - 18p^2 + 22p - 10}{2p^2 - 4p + 3}$

* **55.** $\frac{9x^4 + 6x^3 + x - 4}{3x^2 + x - 1}$

* **56.** $(a^2 + b^2 + 2ab - 6a - 6b + 9) \div (a + b - 3)$

* **57.** $(y^4 + 3y^3z + 2y^2z^2 + yz^3 - z^4) \div (y^2 + yz + z^2)$

* **58.** $(a^5 - b^5) \div (a - b)$

* **59.** $(x^7 + y^7) \div (x + y)$

* **60.** $(x^2 + \frac{7}{2}x + 3) \div (2x + 3)$

* **61.** $(3q^2 - \frac{23}{4}q - 5) \div (4q + 3)$

Checkup

The following problems provide a review of some of Section 5.3 and will help you with the next section.

Multiply.

62. $6(x - 3)$

63. $4(x + 8)$

64. $x^2(x^4 + x^2)$

65. $y^3(y^5 - y^2)$

66. $8x^2(x^3 - 4x^2)$

67. $3x^4(2x^2 + x^4)$

68. $6y^2(2y^3 - 3y^2 + 1)$

69. $5x^2(3x^2 + 2x - 3)$

70. $9xy^2(7x^2 - 5xy + 8xy^3)$

71. $8a^2b^2(6ab^2 - 7ab - 5b^4)$

5.6 FACTORING WHEN TERMS HAVE A COMMON FACTOR

One of the uses of factoring is to solve certain equations. Factoring reverses the process of multiplication. When we ***factor*** an expression we write it as a multiplication.

1 The Greatest Common Factor

Factoring Out the Greatest Common Factor

Factor out the largest number and the variable with the largest exponent that will divide evenly into each term without forming negative exponents. This factor is called the ***greatest common factor.***

Factoring out the greatest common factor is often a first step in factoring.

EXAMPLE 1 Factor out the greatest common factor.

a. $8x - 16$

The greatest common factor is 8 since it is the largest number that divides evenly into each term. Place it in front of a pair of parentheses.

$$\mathbf{8}(\quad)$$

In the parentheses place the quotient of the given expression and the number that we factored out.

$$\frac{8x - 16}{8} = x - 2 \qquad \text{The quotient is } x - 2.$$

The factored form is

$$8x - 16 = \mathbf{8}(x - 2)$$

b. $x^6 + x^4$

The variable with the largest exponent that divides into each term without forming negative exponents is x^4. Notice that this is the least exponent that appears on x.

$$\mathbf{x^4}(\quad) \qquad x^4 \text{ is the greatest common factor.}$$

The quotient of the given expression and x^4 is

$$\frac{x^6 + x^4}{x^4} = x^2 + 1 \qquad \text{Recall that } \frac{x^6}{x^4} = x^2 \text{ and } \frac{x^4}{x^4} = 1.$$

Hence $x^6 + x^4 = \mathbf{x^4}(x^2 + 1)$.

c. $16y^5 - 24y^4 + 40y^2$

The greatest common factor is $\mathbf{8y^2}$ since this is the largest number and the variable with the largest exponent that divides evenly into each term without creating negative exponents.

$$16y^5 - 24y^4 + 40y^2 = \mathbf{8y^2}(2y^3 - 3y^2 + 5)$$

since

$$\frac{16y^5 - 24y^4 + 40y^2}{8y^2} = 2y^3 - 3y^2 + 5 \quad ■$$

☐ **DO EXERCISE 1.**

OBJECTIVES

1 *Factor out the greatest common factor from a polynomial*

2 *Factor certain polynomials by grouping*

☐ **Exercise 1** Factor out the greatest common factor.

a. $7x + 21$

b. $y^5 - y^3$

c. $12x^2 + 20x^4 - 16x^3$

d. $36z^4 - 18z^2 + 27z^3$

e. $30x^2y - 12x^2y^3$

f. $15y^2z^2 + 10y^2z - 25yz^2$

□ **Exercise 2** Factor in two ways.

a. $-x^2 + 5x$

b. $-7p^2 + 3p$

□ **Exercise 3** Factor.

a. $a(x + y^2) - b(x + y^2)$

b. $y^2(m + n) + x^2(m + n)$

c. $(y + 2)(y - 5) + (y + 2)(2y + 7)$

d. $(x - 5)(x + 4) - (x - 5)(x - 3)$

Usually, we factor out a positive greatest common factor, but occasionally, there is a reason to factor out a negative greatest common factor.

EXAMPLE 2 Factor $-x^3 + 4x^2 - 2x$.

There are two ways to factor this expression. If we factor out x as the greatest common factor, we get

$$-x^3 + 4x^2 - 2x = x(-x^2 + 4x - 2).$$

We could factor out $-x$. Then

$$-x^3 + 4x^2 - 2x = -x(x^2 - 4x + 2) \quad ■$$

□ **DO EXERCISE 2.**

2 Factoring by Grouping

Sometimes the greatest common factor is a binomial.

EXAMPLE 3 Factor.

a. $a(m + n) + b(m + n)$

The greatest common factor is $m + n$. Factor it out.

$$a(\mathbf{m + n}) + b(\mathbf{m + n}) = (\mathbf{m + n})(a + b)$$

b. $(x - 4)(x + 3) + (x - 4)(3x + 2)$

The common factor is $x - 4$.

$$\begin{aligned}&(\mathbf{x - 4})(x + 3) + (\mathbf{x - 4})(3x + 2)\\ &= (\mathbf{x - 4})[(x + 3) + (3x + 2)]\\ &= (x - 4)[x + 3 + 3x + 2]\\ &= (x - 4)(4x + 5) \qquad \text{Combine like terms.}\end{aligned}$$

c. $$\begin{aligned}(\mathbf{a - 3})(4a + 5) - (\mathbf{a - 3})(2a - 3) &= (\mathbf{a - 3})[(4a + 5) - (2a - 3)]\\ &= (a - 3)[4a + 5 - 2a + 3]\\ &= (a - 3)(2a + 8)\\ &= (a - 3)(2)(a + 4)\\ &= 2(a - 3)(a + 4) \quad ■\end{aligned}$$

□ **DO EXERCISE 3.**

A polynomial with four terms can sometimes be factored by grouping the terms into groups of two that have common factors.

EXAMPLE 4 Factor.

a. $px - py + qx - qy = (px - py) + (qx - qy)$

$= p(\mathbf{x - y}) + q(\mathbf{x - y})$

$= (\mathbf{x - y})(p + q)$

b. $x^2 - 2x + 4x - 8 = (x^2 - 2x) + (4x - 8)$

$= x(\mathbf{x - 2}) + 4(\mathbf{x - 2})$

$= (\mathbf{x - 2})(x + 4)$

c. $y^2 + 3y + y + 3 = (y^2 + 3y) + (y + 3)$

$= y(y + 3) + 1(y + 3)$

$= (y + 3)(y + 1)$ Remember to keep the 1.

d. $ay + by - a - b$

Grouping terms as above gives

$(ay + by) + (-a - b)$

$= y(a + b) + (-a - b)$

There are no common factors here. We must group the terms as follows.

$ay + by - a - b$

$= (ay + by) - (a + b)$ Notice that $-(a + b) = -a - b$.

$= y(a + b) - 1(a + b)$

$= (a + b)(y - 1)$ ■

□ **DO EXERCISE 4.**

□ **Exercise 4** Factor.

a. $x + x^2 + y + xy$

b. $y^2 + 4y + 4y + 16$

c. $x^2 + 3x - x - 3$

d. $1 + x + y + xy$

DID YOU KNOW?

Diophantus (A.D. 250) is known as the father of algebra partially because he began using abbreviations to solve problems. Before Diophantus, in Greece, algebra problems were written out in words. Diophantus might write $6x^3 - 2x^2 + 5x$ as c6 u5 ↑ s2. Substitute c for variable cubed, u for variable, s for variable squared, and ↑ for minus.

Answers to Exercises

1. **a.** $7(x + 3)$ **b.** $y^3(y^2 - 1)$ **c.** $4x^2(3 + 5x^2 - 4x)$
d. $9z^2(4z^2 - 2 + 3z)$ **e.** $6x^2y(5 - 2y^2)$ **f.** $5yz(3yz + 2y - 5z)$

2. **a.** $x(-x + 5)$ or $-x(x - 5)$ **b.** $p(-7p + 3)$ or $-p(7p - 3)$

3. **a.** $(x + y^2)(a - b)$ **b.** $(m + n)(y^2 + x^2)$ **c.** $(y + 2)(3y + 2)$
d. $(x - 5)(7)$

4. **a.** $(1 + x)(x + y)$ **b.** $(y + 4)(y + 4)$ **c.** $(x + 3)(x - 1)$
d. $(1 + x)(1 + y)$

Problem Set 5.6

NAME ______________________

DATE ______________________

CLASS ______________________

Factor.

1. $12x + 24$
2. $6x - 15$
3. $9y - 18$
4. $8x + 32$
5. $x^2 + x^3$
6. $y^8 - 3y^6$
7. $3a^2 + 4a$
8. $7x^2 - 2x$
9. $9p^8 - 8p^4$
10. $3q^7 + 4q^3$
11. $20p^2 - 25p^4$
12. $14x^5 + 21x^3$
13. $24x + 18$
14. $35y - 28$
15. $15x^2 - 10x^4 + 25x^3$
16. $12y^3 + 18y^5 - 10y^7$
17. $32z^3 + 24z^5 - 16z^4$
18. $9p^3 - 18p^5 + 6p^2$
19. $15x^2y^3 - 60xy^4 + 45x^2y^2$
20. $48a^2b^3 + 24ab^2 - 36a^2b^2$

Factor out a factor with a negative coefficient.

21. $-4x + 20$
22. $-3x + 18$
23. $-5y - 30$
24. $-9y - 72$
25. $-3x^2 + 6x - 24$
26. $-2x^2 + 14x - 30$
27. $-5a^3 - 15a^2 - 35a$
28. $-6p^4 - 36p^3 - 12p$
29. $-x^3 - 6x + 5$
30. $-y^4 - 7y^2 - 9$

Factor by grouping.

31. $x(a^2 + b) - y(a^2 + b)$
32. $5(x - y) - a(x - y)$
33. $(x - 6)(x + 2) + (x - 6)(x + 7)$
34. $(x + 3)(2x - 3) - (x + 3)(x + 4)$
35. $ax + 3bx + ay + 3by$
36. $m + mn + n + n^2$
37. $x^2 - 6x + 3x - 18$

38. $y^2 - 5y + 2y - 10$

39. $x^2 + 4x + 3x + 12$

40. $y^2 + 2y + 4y + 8$

41. $x^2 - 8x - x + 8$

42. $p^2 - 2p - p + 2$

43. $y^2 - 3y - 4y + 12$

44. $x^2 + 2x - 3x - 6$

45. $a^7 + a^6 - a^4 - a^3$

46. $-x^5 - x^4 + x^3 + x^2$

47. $16x^3 - 24x^2 + 48x - 72$

48. $36a^5 - 24a^3 + 45a^2 - 30$

49. The sum of the first x natural numbers is given by the polynomial $S = \frac{1}{2}x^2 + \frac{1}{2}x$. Factor this polynomial and use it to find the sum of the first eight natural numbers.

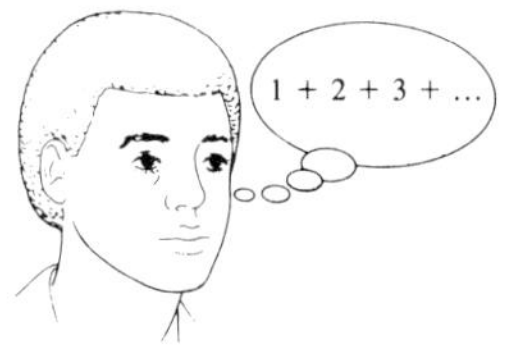

50. Use the polynomial from Problem 49 to find the sum of the first 12 natural numbers.

51. To factor an expression, write it as a ____________

52. The largest number and the variable with the largest exponent that will divide evenly into each term of a polynomial without forming negative exponents is called the ____________ ____________ ____________.

53. Sometimes it is necessary to factor out a ____________ greatest common factor.

54. A polynomial with four terms can sometimes be factored by grouping the terms into groups of two that have ____________ factors.

Factor.

* **55.** $(y - 2)^3 + (y - 2)^2 + 4(y - 2)$

* **56.** $p(p^2 - 4p + 1) - 3(p^2 - 4p + 1)$

* **57.** $16x^3 - 4x^2y^2 - 4xy + y^3$

* **58.** $40m^4 - 15m^2p - 8m^2p^3 + 3p^4$

Checkup

The following problems provide a review of some of Section 5.4 and will help you with the next section.

Multiply. Use FOIL.

59. $(x + 3)(x + 4)$

60. $(x - 3)(x - 5)$

61. $(x - 5)(x + 4)$

62. $(x + 1)(x - 3)$

63. $(3x + 2)(2x + 1)$

64. $(5x - 1)(2x - 3)$

65. $(6x - 1)(2x + 3)$

66. $(2x + 5)(4x - 1)$

5.7 FACTORING TRINOMIALS

OBJECTIVES

1. *Factor trinomials of the type $x^2 + bx + c$*
2. *Factor trinomials of the type $ax^2 + bx + c$, $a \neq 1$*

Many trinomials can be factored into the multiplication of two binomials. If the trinomial has a common factor it should be factored out first. Write the trinomial in descending order before you try to factor.

1 Factoring Trinomials of the Type $x^2 + bx + c$

Notice that the coefficient of x^2 is 1. We may factor trinomials of this type as follows. Recall that we may multiply two binomials using FOIL.

$$\begin{aligned} &\quad\ \ \textbf{F}\quad\ \textbf{O}\quad\ \textbf{I}\quad\ \textbf{L} \\ (x + 3)(x + 4) &= x^2 + 4x + 3x + 12 \\ &= x^2 + 7x + 12 \end{aligned}$$

In general,

$$\begin{aligned} &\quad\ \ \textbf{F}\quad\ \textbf{O}\quad\ \textbf{I}\quad\ \textbf{L} \\ (x + b)(x + a) &= x^2 + ax + bx + ab \\ &= x^2 + (a + b)x + ab \end{aligned}$$

We factor by reversing the procedure.

$$x^2 + (a + b)x + ab = (x + b)(x + a)$$

We need to find two numbers a and b whose sum is the coefficient of x and whose product is the last term.

EXAMPLE 1 Factor.

a. $x^2 + 5x + 6$

We need to find the correct numbers to put in the blanks.

$$x^2 + 5x + 6 = (x + \underline{\quad})(x + \underline{\quad})$$

We try to find pairs of integers whose product is +6 and whose sum is +5. We can find all the factors of 6 in a systematic way by starting with 1 as one of the factors, then −1, then 2, and so on. We use all the natural numbers that divide evenly into 6 until the factors begin to repeat.

Factors of +6	*Sum of Factors of +6*
1, 6	7
−1, −6	−7
2, 3	**5**
−2, −3	−5

The correct integers are 2 and 3 since their product is 6 and their sum is 5.

$$x^2 + 5x + 6 = (x + 2)(x + 3)$$

Note: It is also correct to write $(x + 3)(x + 2)$ since multiplication is commutative.

☐ **Exercise 1** Factor.

a. $x^2 + 6x + 8$

b. $x^2 + 3x - 10$

c. $x^2 - 6x - 27$

d. $y^2 - 7y + 12$

b. $x^2 + 4x - 21$

Factors of −21	*Sum of Factors of −21*
1, −21	−20
−1, 21	20
3, −7	−4
−3, 7	**4**

Notice that 2 does not divide evenly into 21, so 2 is not a factor of 21. The correct integers are −3 and 7 since their product is −21 and their sum is 4 (the coefficient of x).

$$x^2 + 4x - 21 = (x - 3)(x + 7)$$

We can check by multiplying $(x - 3)(x + 7)$.

c. $y^2 + 3y + 5$

We want to find two integers whose product is 5 and whose sum is 3. There are only two pairs of integers that are factors of 5. They are 1 and 5 and −1 and −5. Neither of these pairs add up to 3. This polynomial cannot be factored. It is called ***prime.*** ■

A polynomial that cannot be factored is called a ***prime polynomial.***

The following are steps to follow in factoring a trinomial when the coefficient of the x^2 term is 1. Be sure that the trinomial is written in descending order.

Factoring $x^2 + bx + c$

1. Factor the x^2 term.
2. Find all pairs of integers whose product equals the last term of the trinomial.
3. From the list of pairs of integers choose the pair whose sum equals the coefficient of the middle term.
4. Combine this pair of integers with the factors of the x^2 term.
5. If these integers do not exist, the polynomial is prime.

☐ **DO EXERCISE 1.**

2 Factoring Trinomials of the Type $ax^2 + bx + c$, $a \neq 1$

In this type of factoring the x^2 term must be factored and the last term must be factored. We use factors that give the correct middle term. First, we factor out the greatest common factor if there is one.

EXAMPLE 2 Factor.

a. $5x^2 + 11x + 2$

1. There is no common factor.
2. We need to find integers to fill in the blanks.

$$(\_x + \_)(\_x + \_)$$

We used plus signs since all the signs in the trinomial are plus. The first terms of each binomial have a product of $5x^2$, so they must be $5x$ and x. We use only the positive factors of the x^2 term since we usually want the first coefficients of the factors to be positive. Now we have the following:

$$(5x + \_\_)(x + \_\_)$$

3. We want to find two integers whose product is 2. They are 1 and 2 and -1 and -2. Since all the signs in the trinomial are plus, the integers are 1 and 2. We can use the 1 with the $5x$ or x. We try both combinations to see which arrangement gives the correct middle term when we multiply. Remember that the middle term is made up of the sum of the products of the inner and outer terms.

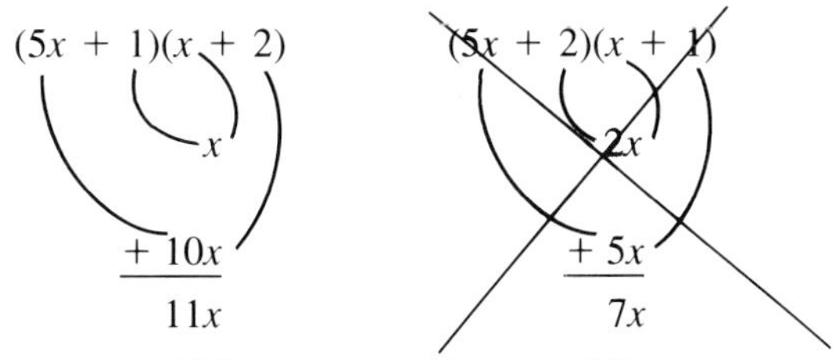

Correct middle term   Wrong middle term

Hence $5x^2 + 11x + 2 = (5x + 1)(x + 2)$.

b. $24x^2 + 10x - 4$.

1. There is a common factor of 2, so factor it out.

$$24x^2 + 10x - 4 = 2(12x^2 + 5x - 2)$$

2. Now try to factor $12x^2 + 5x - 2$.
The possible positive factors of 12 are 1 and 12, 2 and 6, and 3 and 4. We will try 3 and 4. If these do not work, we will try another pair.

$$12x^2 + 5x - 2 = (3x + \_\_)(4x + \_\_)$$

3. We have not decided what signs to use. The factors of -2 are 1 and -2 or -1 and 2. We try both combinations. We need to get a middle term of $5x$.

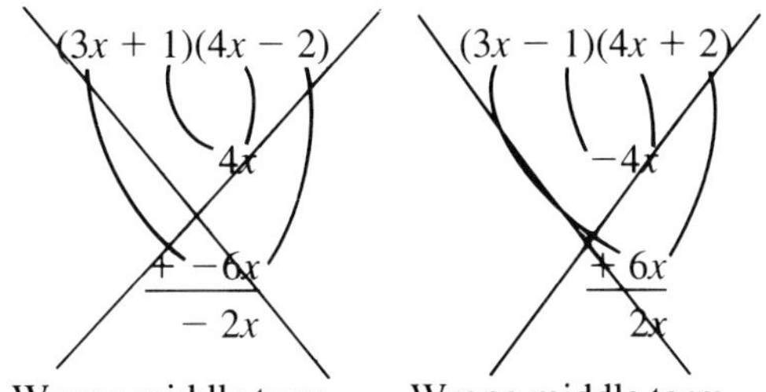

Wrong middle term   Wrong middle term

Before we try another pair of factors of 12, we try reversing the second pair of factors.

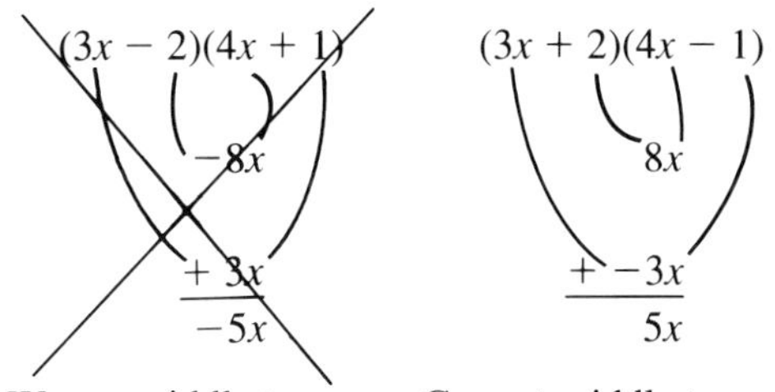

Wrong middle term   Correct middle term

Therefore, $24x^2 + 10x - 4 = 2(3x + 2)(4x - 1)$. ■

□ **Exercise 2** Factor.

a. $6x^2 + 8x + 2$

b. $2x^2 + 9x - 5$

c. $8x^2 - 16x + 6$

d. $24x^2 + 2x - 2$

The following method may be used to factor a trinomial when the coefficient of the x^2 term is other than 1.

> **Factoring $ax^2 + bx + c$, $a \neq 1$**
>
> 1. If there is a common factor, factor it out.
> 2. List all pairs of factors of the x^2 term.
> 3. List all pairs of integer factors of the last term.
> 4. Choose inner and outer terms by trying all combinations of these factors until the correct middle term is found.
> 5. If there are no such combinations, the trinomial is prime.

□ DO EXERCISE 2.

Alternative Method of Factoring $ax^2 + bx + c$, $a \neq 1$

In general,

$$(ax + b)(cx + d) = acx^2 + adx + bcx + bd$$

$$= acx^2 + (ad + bc)x + bd$$

We factor by reversing the procedure.

$$acx^2 + (ad + bc)x + bd = (ax + b)(cx + d)$$

To factor, we use a procedure similar to that of Example 1. We multiply *ac* times *bd* and factor the product in such a way that we can write the middle term as a sum. Then we factor the resulting polynomial.

EXAMPLE 3 Factor.

a. $2x^2 + 7x + 6$

1. There is no common factor.
2. Multiply the coefficient of x^2, 2, by the last term, 6.

$$2(6) = 12$$

3. Factor 12 so that the sum of the factors is 7 (the coefficient of the middle term).

Factors of +12	*Sum of Factors of +12*
1, 12	13
−1, −12	−13
−2, 6	8
2, −6	−8
3, 4	**7**

We may stop at this point since the correct factors are 3 and 4.

4. Write the middle term, $7x$, as a sum using the factors 3 and 4.

$$7x = \mathbf{3x + 4x}$$

5. Factor the polynomial by grouping it into groups of two terms.

$$2x^2 + 7x + 6 = 2x^2 + \mathbf{3x + 4x} + 6$$ Write the middle term as a sum.

$$= (2x^2 + 3x) + (4x + 6)$$

$$= x(2x + 3) + 2(2x + 3)$$

$$= (2x + 3)(x + 2)$$

b. $6x^3 - 57x^2 + 90x$

1. Factor out the common factor.

$$6x^3 - 57x^2 + 90x = 3x(2x^2 - 19x + 30)$$

2. Factor $2x^2 - 19x + 30$.

Multiply the coefficient of x^2, 2, by the last term, 30.

$$2(30) = 60$$

3. Factor 60 so that the sum of the factors is -19 (the coefficient of the middle term). We may omit the positive factors since the middle term is negative.

Factors of 60	*Sum of Factors of 60*
$-1, -60$	-61
$-2, -30$	-32
$-3, -20$	-23
$\mathbf{-4, -15}$	$\mathbf{-19}$

We stop at this point since the correct factors are -4 and -15.

4. Write the middle term, $-19x$, as a sum using the factors -4 and -15.

$$\mathbf{-19x = -4x - 15x}$$

5. Factor the polynomial by grouping it into groups of two terms.

$$2x^2 \mathbf{- 19x} + 30 = 2x^2 \mathbf{- 4x - 15x} + 30$$ Write the middle term as a sum.

$$= 2x(x - 2) - 15(x - 2)$$

$$= (x - 2)(2x - 15)$$

6. Write the factorization, *including the common factor.*

$$6x^3 - 57x^2 + 90x = 3x(x - 2)(2x - 15)$$ ■

□ **DO EXERCISE 3.**

□ **Exercise 3** Factor.

a. $3x^2 - 11x - 4$

b. $2x^2 + x - 15$

c. $8x^2 + 8x - 6$

d. $4y^2 + 37y + 9$

Answers to Exercises

1. a. $(x + 2)(x + 4)$ **b.** $(x + 5)(x - 2)$ **c.** $(x - 9)(x + 3)$
d. $(y - 3)(y - 4)$

2. a. $2(3x + 1)(x + 1)$ **b.** $(2x - 1)(x + 5)$ **c.** $2(2x - 1)(2x - 3)$
d. $2(3x + 1)(4x - 1)$

3. a. $(3x + 1)(x - 4)$ **b.** $(2x - 5)(x + 3)$ **c.** $2(2x + 3)(2x - 1)$
d. $(4y + 1)(y + 9)$

Problem Set 5.7

NAME ______

DATE ______

CLASS ______

Factor.

1. $x^2 + 6x + 9$

2. $x^2 + 10x + 25$

3. $y^2 + 4y - 12$

4. $y^2 - 3y - 10$

5. $x^2 - 10x + 16$

6. $x^2 - 10x + 21$

7. $x^2 - x - 20$

8. $x^2 + 2x - 24$

9. $2x^2 - 2x - 24$

10. $5x^2 + 10x - 15$

11. $x^3 + 4x^2 + 4x$

12. $y^3 - 6y^2 + 9y$

13. $2p^2 - 7p + 3$

14. $3q^2 + 5q - 2$

15. $5x^2 + x - 6$

16. $3x^2 + 14x + 8$

17. $12x^2 - 20x + 3$

18. $9x^2 + 17x - 2$

19. $24x^2 - x - 10$

20. $6x^2 + 19x - 7$

21. $2x^2 - 18x + 40$

22. $5x^2 + 5x - 30$

23. $x^2 + 9x + 8$

24. $y^2 + 4y - 5$

25. $6x^2 + 5x - 25$

26. $9x^2 - 3x - 6$

27. $16x^2 + 4x - 6$

28. $18x^2 - 21x + 3$

29. $4x^3 + 7x^2 - 15x$

30. $14y^3 - 9y^2 + y$

31. $9x^2 + 3x - 20$

32. $12x^2 - 8x - 15$

33. $8x^2 - 40x + 50$

34. $28x^2 - 18x + 2$

35. $35p^2 + 34p + 8$

36. $12z^2 - 7z + 1$

37. $6x^2 - 5x + 1$

38. $4x^2 - 4x + 1$

39. $x^4 + 6x^2 - 16$

40. $y^4 - 15y^2 + 14$

41. $p^4 - 8p^2 + 12$

42. $q^4 + 10q^2 + 9$

43. $12x^2 + 8xy - 15y^2$

44. $3a^2 + 7ab + 2b^2$

45. $18x^2 - 15x - 18$

46. $100b^2 - 90b + 20$

47. $35y^2 - 41y - 24$

48. $10x^2 + x - 3$

49. $8x^2 - 14xy - 39y^2$

50. $15p^2 - 22pq - 5q^2$

51. $8x^2 - 28x - 16$

52. $18y^2 - 6y - 24$

53. $9a^2 + 4 + 15a$

54. $6x^2 - 2 - x$

55. $-4x^2 + 24x - 35$

56. $-12y^2 + 17y - 6$

57. $40a^4 + 16a^2 - 12$

58. $24x^4 + 2x^2 - 15$

59. $3x^2 - 4 - x$

60. $6p^2 - 10 - 7p$

61. $x^2 - 6x - 16$

62. $y^2 - 3y - 40$

63. $3a^4 + 6a^3 - 72a^2$

64. $6p^3 + 12p^2 - 90p$

65. $3y^4 + 14y^2 + 8$

66. $8x^4 + 13x^2 - 6$

67. $35a^2 - 41ab - 24b^2$

68. $36x^2 + 30xy - 50y^2$

69. Jim has found that the profit P from his bakery is given by

$$P = x^2 - 40x + 400$$

where x is the number of loaves of bread sold each day. Write an equivalent expression for P in factored form.

70. Find a polynomial that can be factored as $-16b^2(b - 7)(b + 3)$.

71. To factor $x^2 + bx + c$, we must find pairs of integers whose products is c and whose ____________ is b.

72. A polynomial that cannot be factored is ____________.

73. A first step in factoring $ax^2 + bx + c$ is to factor out the ____________ ____________ factor if there is one.

74. A trinomial $ax^2 + bx + c$ without a common factor may be factored by listing combinations of factors of the coefficient of the x^2 term and the last term until we find the correct ____________ term.

Factor.

* **75.** $2a^3b^3 - 48a^2b^4 + 288ab^5$

* **76.** $6x^4y - 153x^3y + 867x^2y$

* **77.** $63p^2 + 214p - 240$

* **78.** $x^{2n+2} - 2x^{n+2} + x^2$

* **79.** $18x^{2n} - 36x^ny^n + 18y^{2n}$

* **80.** $72a^2 - 515ab - 350b^2$

* **81.** $6(x + 3)^2 + 13(x + 3) + 5$

* **82.** $10(m - 4)^2 - 9(m - 4) - 9$

* **83.** $6(p + k)^2 - (p + k) - 5$

* **84.** $30x^{7a} - 26x^{6a} - 40x^{5a}$

Checkup

The following problems provide a review of some of Section 5.4 and will help you with the next section.

Multiply

85. $(x - 5)(x + 5)$

86. $(y + 3)(y - 3)$

87. $(2y + 7)(2y - 7)$

88. $(3y - 4)(3y + 4)$

89. $(x - 7)^2$

90. $(x + 4)^2$

91. $(3x^2 + 2y)^2$

92. $(5x^2 - 4y)^2$

5.8 SPECIAL FACTORS OF POLYNOMIALS

OBJECTIVES

1. *Factor the difference of two squares*
2. *Factor a perfect square trinomial*
3. *Factor certain polynomials with four terms by grouping and using the factoring of a trinomial square*

Certain types of factoring may be done quickly by special methods.

1 Factoring the Difference of Two Squares

When we multiply a sum and a difference the result is the difference of two squares.

$$(A - B)(A + B) = A^2 - B^2$$

We reverse this procedure and factor the difference of two squares as follows:

$$\mathbf{A^2 - B^2 = (A - B)(A + B)}$$

Notice that both A^2 and B^2 are squares and there is a minus sign between them. *The sum of two squares $A^2 + B^2$ cannot be factored.*

The squares of the numbers from 1 to 10 are 1, 4, 9, 16, 25, 36, 49, 64, 81, and 100, respectively. The squares of x, x^2, x^3, x^4, and x^5 are x^2, x^4, x^6, x^8, and x^{10}. Since in the difference of two squares $A^2 - B^2$, both A^2 and B^2 are squares, they must factor into identical factors.

EXAMPLE 1 Factor.

a. $x^2 - 16$

$$A^2 - B^2 = (A - B)(A + B)$$

We need to fill in the following blanks.

$$x^2 - 16 = (\_\_ - \_\_)(\_\_ + \_\_)$$

The factors of x^2 and 16 must be identical.

$$x^2 - 16 = x^2 - 4^2 = (x - 4)(x + 4)$$

b. $9x^4 - 25y^2$

$$9x^4 - 25y^2 = (3x^2 - 5y)(3x^2 + 5y)$$

c. $8x^2 - 8y^2$

There is a common factor of 8. Factor it out *first*.

$\mathbf{8}x^2 - \mathbf{8}y^2 = \mathbf{8}(x^2 - y^2)$

$= \mathbf{8}(x - y)(x + y)$ Factor the difference of squares.

d. $16x^4 - 1$

$16x^4 - 1 = (4x^2 - 1)(4x^2 + 1)$

$= (2x - 1)(2x + 1)(4x^2 + 1)$ Factor $4x^2 - 1$ since it is the difference of squares. ■

□ DO EXERCISE 1.

□ **Exercise 1** Factor.

a. $x^2 - 9$

b. $4x^2 - 49y^4$

c. $5x^2 - 5y^6$

d. $9x^4 - 9$

2 Factoring a Perfect Square Trinomial

A ***perfect square trinomial*** is the square of a binomial. Recall that there is a special method for squaring a binomial.

$$(A + B)^2 = A^2 + 2AB + B^2$$

and

$$(A - B)^2 = A^2 - 2AB + B^2$$

We reverse this procedure and factor as follows.

$$A^2 + 2AB + B^2 = (A + B)^2$$
$$A^2 - 2AB + B^2 = (A - B)^2$$

Notice that the first and last terms of a perfect square trinomial must be squares. Since they are squares, they factor into identical factors. Also, twice the product of the first and last terms of the factored form must give the middle term of the trinomial.

EXAMPLE 2 Factor.

a. $x^2 - 12x + 36$

1. We want to fill in the following blanks. Since the sign on the middle term is negative, the signs between the factors must be minus.

$$(\_\_ - \_\_)(\_\_ - \_\_)$$

2. Factor the first and last terms of the trinomial into identical factors.

$$x^2 - 12x + 36 = x^2 - 12x + 6^2 = (x - 6)(x - 6) = (x - 6)^2$$

3. Check to see if this is correct by taking twice the product of the two terms.

$$2(x)(-6) = -12x$$

This is the middle term of the given trinomial. Hence

$$x^2 - 12x + 36 = (x - 6)^2.$$

b. $16y^2 + 40y + 25$

$$A^2 + 2AB + B^2 = (A + B)^2$$

The identical factors of $16y^2$ are $4y$ and $4y$, so A is $4y$. The identical factors of 25 are 5 and 5, so B is 5.

$$16y^2 + 40y + 25 = (4y)^2 + 40y + 5^2 = (4y + 5)^2$$

Check the middle term.

$$2(4y)(5) = 40y \qquad \text{This is correct.}$$

Therefore,

$$16y^2 + 40y + 25 = (4y + 5)^2$$

c. $5x^2 - 20x + 20$

This trinomial has a common factor. Factor it out first.

$$5x^2 - 20x + 20 = \mathbf{5}(x^2 - 4x + 4)$$

Now, try to factor $x^2 - 4x + 4$,

$$x^2 - 4x + 4 = (x - 2)(x - 2) = (x - 2)^2$$

Check the middle term.

$$2(x)(-2) = -4x \qquad \text{This is correct.}$$

Therefore,

$$5x^2 - 20x + 20 = \mathbf{5}(x - 2)^2$$

d. $x^2 + 10x + 9$

Does $x^2 + 10x + 9 = (x + 3)^2$? Check the middle term: $2(x)(3) = 6x$, which is *not* the middle term of the trinomial, $10x$. This is not a perfect square trinomial. However, it is factorable by the general method.

$$x^2 + 10x + 9 = (x + 1)(x + 9) \qquad ■$$

Remember that some trinomials are prime.

□ **DO EXERCISE 2.**

3 More Factoring by Grouping

Sometimes polynomials with four terms can be factored by grouping into a perfect square trinomial and another term. Then the perfect square trinomial is factored. Finally, the expression can be factored as the difference of two squares.

EXAMPLE 3 Factor.

a. $x^2 + 8x + 16 - y^2$

$$\begin{aligned} x^2 + 8x + 16 - y^2 &= (\mathbf{x^2 + 8x + 16}) - y^2 \\ &= (\mathbf{x + 4})^2 - y^2 && \text{Factor the trinomial.} \\ &= [(x + 4) - y][(x + 4) + y] && \text{Factor the difference of squares.} \\ &= (x + 4 - y)(x + 4 + y) \end{aligned}$$

b. $9 - x^2 - 4x - 4$

$$\begin{aligned} 9 - x^2 - 4x - 4 &= 9 - (\mathbf{x^2 + 4x + 4}) \\ &= 9 - (\mathbf{x + 2})^2 \\ &= [3 - (x + 2)][3 + (x + 2)] \\ &= [3 - x - 2][3 + x + 2] \\ &= (1 - x)(5 + x) \qquad ■ \end{aligned}$$

□ **DO EXERCISE 3.**

□ **Exercise 2** Factor.

a. $x^2 + 8x + 16$

b. $9y^2 - 30y + 25$

c. $16x^2 + 16x + 4$

d. $x^3 - 12x^2 + 36x$

e. $16x^2 + 72xy + 81y^2$

f. $9a^2 + 48ab + 64b^2$

□ **Exercise 3** Factor.

a. $x^2 - 6x + 9 - y^2$

b. $25 - x^2 - 12x - 36$

Special Factors of Polynomials

$$A^2 - B^2 = (A - B)(A + B)$$

$A^2 + B^2$ does not factor

$$A^2 + 2AB + B^2 = (A + B)^2$$

$$A^2 - 2AB + B^2 = (A - B)^2$$

Answers to Exercises

1. a. $(x - 3)(x + 3)$ **b.** $(2x - 7y^2)(2x + 7y^2)$
c. $5(x - y^3)(x + y^3)$ **d.** $9(x - 1)(x + 1)(x^2 + 1)$

2. a. $(x + 4)^2$ **b.** $(3y - 5)^2$ **c.** $4(2x + 1)^2$ **d.** $x(x - 6)^2$
e. $(4x + 9y)^2$ **f.** $(3a + 8b)^2$

3. a. $(x - 3 - y)(x - 3 + y)$ **b.** $(-1 - x)(11 + x)$

Problem Set 5.8

NAME ______________________

DATE ______________________

CLASS ______________________

Factor.

1. $x^2 - 25$

2. $y^2 - 100$

3. $49x^2 - 81$

4. $25a^2 - 36$

5. $p^2 - 64$

6. $x^2 - 4$

7. $x^2y^2 - 1$

8. $a^2b^2 - 9$

9. $9y^4 - 25y^2$

10. $16x^4 - 25x^2y^4$

11. $\frac{1}{36} - p^2$

12. $\frac{1}{81} - a^2$

13. $0.01x^2 - 0.25y^2$

14. $0.04a^2 - 0.16b^2$

15. $100y^2 - 4z^2$

16. $64a^2 - b^2$

17. $3x^2 - 3y^4$

18. $2x^2 - 32y^6$

19. $9x^3 - 36x$

20. $25a^4 - 49a^2$

21. $32x^4y - 8x^2y^3$

22. $16pq^4 - 36p^3q^2$

Factor.

23. $x^2 + 4x + 4$

24. $y^2 + 18y + 81$

25. $y^2 - 6y + 9$

26. $a^2 - 2a + 1$

27. $a^3 - 20a^2 + 100a$

28. $x^3 - 16x^2 + 64x$

29. $81x^2 + 18x + 1$

30. $4y^2 - 20y + 25$

31. $12x^2 - 36x + 27$

32. $18x^2 + 48x + 32$

33. $25x^2y^2 - 20xy + 4$

34. $9a^2b^2 + 24ab + 16$

35. $x^2 + 25 + 10x$

36. $y^2 + 25 - 10x$

37. $x^2 + 2xy + y^2$

38. $a^2 - 2ab + b^2$

39. $81a^2 - 72ab + 16b^2$

40. $49x^2 - 112xy + 64y^2$

41. $36 + 25a^2 - 60a$

42. $25 + q^2 - 10q$

43. $x^6 + 24x^3 + 144$

44. $y^6 - 16y^3 + 64$

45. $0.09a^2 - 0.24a + 0.16$

46. $0.25d^2 + 0.05d + 0.01$

47. $3y^4 - 30y^3 + 75y^2$

48. $4x^5 - 24x^4 + 36x^3$

49. $\frac{1}{16} - \frac{1}{2}y + y^2$

50. $\frac{1}{49} + \frac{2}{7}x + x^2$

51. $4a^8 + 20a^4 + 25$

52. $9p^8 + 24p^4 + 16$

Factor.

53. $x^2 + 2x + 1 - y^2$

54. $x^2 + 10x + 25 - y^2$

55. $a^2 - 16a + 64 - b^2$

56. $p^2 - 12p + 36 - q^2$

57. $x^2 + 2xy + y^2 - 100$

58. $a^2 - 2ab + b^2 - 144$

59. $16 - (x + y)^2$

60. $81 - (a + b)^2$

61. $a^2 + 4ab + 4b^2 - 16x^2$

62. $y^2 - 4y + 4 - 25z^2$

63. $100 - (p^2 + 2pq + q^2)$

64. $49 - (x^2 - 2xy + y^2)$

65. $64 - x^2 + 6x - 9$

66. $169 - a^2 + 10a - 25$

67. The product of the sum and difference of two terms is the difference of ____________ ____________.

68. In the binomial $A^2 - B^2$, since both A^2 and B^2 are squares, they must factor into ____________ factors.

69. The square of a binomial is a ____________ ____________ trinomial.

70. If the first and last terms of a trinomial are squares, we must check the ____________ ____________ to see if it is a perfect square trinomial.

Factor.

* 71. $x^{70} - 4x^{68}$

* 72. $x^{203} - 16x^{202} + 64x^{201}$

Checkup

The following problems provide a review of some of Section 4.4.

Solve.

73. $$\begin{aligned} 3x + 4y &= 6 \\ x + 2z &= 0 \\ 4y - 3z &= 9 \end{aligned}$$

74. $$\begin{aligned} 2x - 3z &= 2 \\ 3x + y &= -1 \\ 2y - z &= -8 \end{aligned}$$

75. $$\begin{aligned} x + 2y - 5z &= -8 \\ 3x - y + z &= 6 \\ x - 2y - 2z &= 3 \end{aligned}$$

76. $$\begin{aligned} 4x - y - 6z &= -6 \\ 2x - 6y + z &= -7 \\ x + 3y - 2z &= -1 \end{aligned}$$

77. $$\begin{aligned} x + 2y + 3z &= 4 \\ 2x + y + z &= 0 \\ 3x + y + 4z &= 2 \end{aligned}$$

78. $$\begin{aligned} 2x + 3y + z &= 3 \\ x + 4y + 2z &= 2 \\ x + y + 3z &= 7 \end{aligned}$$

79. $$\begin{aligned} x + y &= 9 \\ y + z &= 7 \\ x - z &= 2 \end{aligned}$$

80. $$\begin{aligned} 2x + y &= 2 \\ y + z &= 3 \\ 4x - z &= 0 \end{aligned}$$

81. $$\begin{aligned} 3x + 4y &= 15 \\ 2x - 5z &= -3 \\ 4y - 3z &= 9 \end{aligned}$$

82. $$\begin{aligned} 6x - 4y &= 2 \\ 3y + 3z &= 9 \\ 2x - 5z &= -8 \end{aligned}$$

83. $$\begin{aligned} x + y + z &= -1 \\ 2x - y + 5z &= 10 \\ 5x - y - z &= 19 \end{aligned}$$

84. $$\begin{aligned} 2x + 3y - z &= 20 \\ x - y + z &= -7 \\ 3x + 2y - z &= 16 \end{aligned}$$

5.9 SUMS OR DIFFERENCES OF TWO CUBES AND A GENERAL METHOD FOR FACTORING

OBJECTIVES

1. *Factor the sum or difference of two cubes*
2. *Factor polynomials using any of the methods shown previously*

1 Factoring the Sums or Differences of Two Cubes

We use special methods to factor the sum or difference of two cubes. Notice the results of the following multiplications.

$$\begin{aligned}(A + B)(A^2 - AB + B^2) &= A(A^2 - AB + B^2) + B(A^2 - AB + B^2)\\ &= A^3 - A^2B + AB^2 + A^2B - AB^2 + B^3\\ &= A^3 + B^3\end{aligned}$$

and

$$\begin{aligned}(A - B)(A^2 + AB + B^2) &= A(A^2 + AB + B^2) - B(A^2 + AB + B^2)\\ &= A^3 + A^2B + AB^2 - A^2B - AB^2 - B^3\\ &= A^3 - B^3\end{aligned}$$

We may reverse the equations above to factor the sum or difference of two cubes.

$$A^3 + B^3 = (A + B)(A^2 - AB + B^2)$$
$$A^3 - B^3 = (A - B)(A^2 + AB + B^2)$$

To get the last factor, we may think of the first factor and find the sum of the following: the square of the first term, the product of the two terms with the sign changed, and the square of the last term. The cubes of the natural numbers from 1 through 10 are 1, 8, 27, 64, 125, 216, 343, 512, 729, and 1000.

EXAMPLE 1 Factor.

a. $x^3 + 8$

$x^3 + 8 = x^3 + 2^3$

$$A^3 + B^3 = (A + B)(A^2 - AB + B^2)$$

Let $A = x$ and $B = 2$.

$$\begin{aligned}x^3 + \mathbf{8} = x^3 + \mathbf{2}^3 &= (x + 2)(x^2 - 2x + 2^2)\\ &= (x + 2)(x^2 - 2x + 4)\end{aligned}$$

b. $27x^3 + 64y^3$

$27x^3 + 64y^3 = (3x)^3 + (4y)^3$

$$A^3 + B^3 = (A + B)(A^2 - AB + B^2)$$

Let $A = 3x$ and $B = 4y$.

$$\begin{aligned}27x^3 + 64y^3 = (3x)^3 + (4y)^3 &= (3x + 4y)[(3x)^2 - (3x)(4y) + (4y)^2]\\ &= (3x + 4y)[9x^2 - 12xy + 16y^2]\end{aligned}$$

□ **Exercise 1** Factor.

a. $x^3 + 64$

b. $125x^3 + 216y^3$

c. $y^3 - 1$

d. $27p^3 - 1000q^3$

e. $128x^3 + 250$

f. $16y^4 - 2y$

c. $y^3 - 27$

$y^3 - 27 = y^3 - 3^3$

$$A^3 - B^3 = (A - B)(A^2 + AB + B^2)$$

Let $A = y$ and $B = 3$.

$$y^3 - \mathbf{27} = y^3 - \mathbf{3^3} = (y - 3)(y^2 + 3y + 3^2)$$
$$= (y - 3)(y^2 + 3y + 9)$$

Notice that $y^2 + 3y + 9$ does not factor. In general, $A^2 + AB + B^2$ and $A^2 - AB + B^2$ cannot be factored further.

d. $216x^3 - 125$

$216x^3 - 125 = (6x)^3 - 5^3$

$$A^3 - B^3 = (A - B)(A^2 + AB + B^2)$$

Let $A = 6x$ and $B = 5$.

$$216x^3 - 125 = (6x)^3 - 5^3 = (6x - 5)[(6x)^2 + (6x)(5) + 5^2]$$
$$= (6x - 5)(36x^2 + 30x + 25)$$

e. $128x^7 + 686xy^6$

There is a common factor of $2x$, so factor it out first.

$$128x^7 + 686xy^6 = 2x(64x^6 + 343y^6)$$
$$= 2x[(4x^2)^3 + (7y^2)^3]$$
$$= 2x(4x^2 + 7y^2)[(4x^2)^2 - (4x^2)(7y)^2) + (7y^2)^2]$$
$$= 2x(4x^2 + 7y^2)[16x^4 - 28x^2y^2 + 49y^4]$$ ■

The following is a summary of the rules for factoring sums or differences of squares or cubes.

Sum of two cubes:	$A^3 + B^3 = (A + B)(A^2 - AB + B^2)$
Sum of two squares:	$A^2 + B^2$ cannot be factored
Difference of two cubes:	$A^3 - B^3 = (A - B)(A^2 + AB + B^2)$
Difference of two squares:	$A^2 - B^2 = (A - B)(A + B)$

□ **DO EXERCISE 1.**

2

A General Method for Factoring

Following is a general method for factoring.

1. Factor out the greatest common factor.
2. If there are *two terms* to be factored: Try to factor the expression as the difference of squares. If it is not the difference of squares, try to factor the expression as the sum or difference of cubes.
   If there are *three terms:* Check to see if the expression is a perfect square trinomial. If it is not, try the general method for factoring a trinomial.

If there are *four terms:* Try to separate the expression into groups of two terms that have a common factor and then remove the common binomial factor. Next, try grouping into a difference of squares, one of which is a trinomial.

3. Factor completely. If a polynomial, other than a monomial, can be factored, you should factor it. Some polynomials are factored completely after the common factor is removed. Also, some polynomials are prime.

EXAMPLE 2 Factor.

a. $5x^4 + 7x^3 + 11x^2$

The greatest common factor is x^2. Factor it out.

$$5x^4 + 7x^3 + 11x^2 = x^2(5x^2 + 7x + 11)$$

The expression in parentheses, $5x^2 + 7x + 11$, cannot be factored.

b. $64x^2 - 25y^2$

There is no common factor. Since the expression has two terms, try factoring it as the difference of squares.

$$\begin{aligned} 64x^2 - 25y^2 &= (8x)^2 - (5y)^2 \\ &= (8x - 5y)(8x + 5y) \end{aligned}$$

We have factored completely. The binomial factors cannot be factored further.

c. $x^6 - 64$

There is no common factor. Try to factor as the difference of two squares.

$$\begin{aligned} x^6 - 64 &= (x^3)^2 - 8^2 \\ &= (x^3 - 8)(x^3 + 8) \qquad \text{Factor the difference of squares.} \end{aligned}$$

One of these factors is the difference of two cubes and the other is the sum of two cubes. Factor them.

$$\begin{aligned} (x^3 - 8)(x^3 + 8) &= (x^3 - 2^3)(x^3 + 2^3) \\ &= (x - 2)(x^2 + 2x + 2^2)(x + 2)(x^2 - 2x + 2^2) \\ &= (x - 2)(x^2 + 2x + 4)(x + 2)(x^2 - 2x + 4) \end{aligned}$$

Notice that we tried to factor as the difference of two squares first. If we had factored as the difference of two cubes, we would have gotten the following:

$$\begin{aligned} x^6 - 64 &= (x^2)^3 - 4^3 \\ &= (x^2 - 4)[(x^2)^2 + 4x^2 + 4^2] \\ &= (x - 2)(x + 2)(x^4 + 4x^2 + 16) \qquad \text{Not factored completely.} \end{aligned}$$

The expression $x^4 + 4x^2 + 16$ can be factored into two trinomials but this factoring is difficult.

Caution: *Always try to factor as the difference of two squares first.* If the expression is not the difference of squares, try to factor it as the difference of cubes.

□ **Exercise 2** Factor.

a. $7x^2 - 14x + 21$

b. $20y^2 + 40y$

c. $49x^2 - 4y^2$

d. $50a^2 - 32$

e. $8p^3 - q^3$

f. $128x^3 + 54$

g. $x^4 - 16$

h. $4x^3 + 12x^2 + 9x$

i. $6x^2 - x - 40$

j. $x^3 - xy^2 + x^2y - y^3$

k. $x^2 - 24x + 144 - y^2$

d. $4x^2 - 24x + 36$

Is there a common factor? Yes, 4 is a common factor. Factor it out.

$$4x^2 - 24x + 36 = \mathbf{4}(x^2 - 6x + 9)$$

The polynomial in parentheses has three terms. It is a perfect square trinomial. Factor it. Be sure to keep the common factor.

$$4(x^2 - 6x + 9) = 4(x - 3)^2$$

Hence

$$4x^2 - 24x + 36 = \mathbf{4}(x - 3)^2$$

e. $8x^2 - 10x - 3$

There is no common factor. Since 8 and 3 are not squares, the trinomial is not a perfect square trinomial. Factor it by the general method.

$$8x^2 - 10x - 3 = (2x - 3)(4x + 1)$$

f. $ab + 3b + 4a + 12$

There is no common factor. The polynomial has four terms. Factor by grouping.

$$ab + 3b + 4a + 12 = (ab + 3b) + (4a + 12)$$

$$= b(\mathbf{a+3}) + 4(\mathbf{a+3}) \quad \text{Factor each group.}$$

$$= (\mathbf{a+3})(b + 4) \quad \text{Factor out the binomial factor.}$$

g. $y^2 - 22y + 121 - z^2$

$$y^2 - 22y + 121 - z^2 = (\mathbf{y^2 - 22y + 121}) - z^2$$

$$= (\mathbf{y - 11})^2 - z^2 \quad \text{Factor the perfect square trinomial.}$$

$$= (y - 11 - z)(y - 11 + z) \quad \text{Factor the difference of two squares.}$$ ■

□ **DO EXERCISE 2.**

Answers to Exercises

1. a. $(x + 4)(x^2 - 4x + 16)$ **b.** $(5x + 6y)(25x^2 - 30xy + 36y^2)$ **c.** $(y - 1)(y^2 + y + 1)$ **d.** $(3p - 10q)(9p^2 + 30pq + 100q^2)$ **e.** $2(4x + 5)(16x^2 - 20x + 25)$ **f.** $2y(2y - 1)(4y^2 + 2y + 1)$

2. a. $7(x^2 - 2x + 3)$ **b.** $20y(y + 2)$ **c.** $(7x - 2y)(7x + 2y)$ **d.** $2(5a - 4)(5a + 4)$ **e.** $(2p - q)(4p^2 + 2pq + q^2)$ **f.** $2(4x + 3)(16x^2 - 12x + 9)$ **g.** $(x - 2)(x + 2)(x^2 + 4)$ **h.** $x(2x + 3)^2$ **i.** $(3x - 8)(2x + 5)$ **j.** $(x - y)(x + y)^2$ **k.** $(x - 12 - y)(x - 12 + y)$

Problem Set 5.9

NAME

DATE

CLASS

Factor completely.

1. $x^3 + 1$

2. $y^3 + 27$

3. $y^3 - 1$

4. $p^3 - 8$

5. $c^3 + 64$

6. $x^3 + 125$

7. $27a^3 - 1$

8. $8y^3 - 1$

9. $125 - 27x^3$

10. $64 - 27p^3$

11. $p^3 - q^3$

12. $x^3 - y^3$

13. $x^3 + \frac{1}{27}$

14. $a^3 + \frac{1}{64}$

15. $y^3 - 0.008$

16. $z^3 - 0.001$

17. $a^6 - b^6$

18. $t^6 - 64$

19. $y^3 - \frac{1}{125}$

20. $x^3 - \frac{1}{8}$

21. $8x^3 + 125$

22. $64y^3 + 1$

23. $y^3 - 8$

24. $x^3 - 343$

25. $343p^3 - 27q^3$

26. $216a^3 - 125b^3$

27. $9y^3 + 72$

28. $375x^3 + 3$

29. $512 - x^3$

30. $343 - 8y^3$

31. $250x^4 + 54x$

32. $2y^4 + 128y$

33. $16x^3 - 2000$

34. $15x^3 - 120$

35. $x^6 + y^6$

36. $8a^6 + 27b^6$

37. $125x^8 - x^2y^6$

38. $64x^6y - 27y^7$

Factor completely.

39. $ax^2 + 3ax + 4a$

40. $bx^3 + bx^2 + bc$

41. $5x^2 + 10x - 25$

42. $8x^2 + 2x - 6$

43. $x^2 - 100$

44. $y^2 - 64$

45. $8x^3 - 64y^3$

46. $125a^3 - b^3$

47. $6x^2 - x - 2$

48. $10x^2 - x - 3$

49. $18x^2 - 12x + 2$

50. $50x^2 - 40x + 8$

51. $xy + 5y + 5x + 25$

52. $ab - 3b + 7a - 21$

53. $7x^3 - 26x^2 - 8x$

54. $2y^3 + 8y^2 + 6y$

55. $64y^6 - 1$

56. $x^6 - 1$

57. $98x^3 - 8xy^2$

58. $50a^4 - 8a^2b^2$

59. $25x^2 - 20x + 4$

60. $64x^2 + 32x + 4$

61. $12x^2 - 2x + 30x - 5$

62. $15y^2 - 6y + 25y - 10$

63. $8x^3 - 22x^2 - 30x$

64. $108y^3 + 6y^2 - 2y$

65. $27x^4 + 1000xy^3$

66. $125a^4b + 8ab^4$

67. $9a^2 + 24ab + 16b^2$

68. $49p^2 - 28pq + 4q^2$

69. $x^3 + 512$

70. $8y^3 + 343$

71. $8x^2 - 6x - 20x + 15$

72. $18x^2 - 3x - 12x + 2$

73. $12x^2 - 48y^2$

74. $50a^2 - 98b^2$

75. $x^2 - 26x + 169 - y^2$

76. $a^2 + 30a + 225 - b^2$

77. $23ab + 15a^2b^2 - 28$

78. $-45xy + 42x^2y^2 + 12$

79. $(x - 3)(x + 4) + (x - 3)(x + 7)$

80. $(y - 8)(y - 2) - (y - 8)(y + 9)$

81. $9 - x^2 + 16x - 64$

82. $25 - a^2 - 18a - 81$

83. The first step in factoring is to factor out the ____________ ____________ ____________ if there is one.

84. After factoring out the greatest common factor, if there are two terms to be factored, first try to factor the expression as the ____________ ____________ ____________.

85. To factor a trinomial that does not have a common factor, first check to see if it is a ____________ ____________ ____________.

86. If there are four terms to be factored, first try to separate the expression into groups of ____________ ____________ which have a common factor.

Factor completely.

* **87.** $(x - y)^3 + (x + y)^3$

* **88.** $(p + 2)^3 - (p - 2)^3$

* **89.** $x^{3q} - 27$

* **90.** $8y^{3m} - 125$

* **91.** $x^4 - 11x^2 + 30$

* **92.** $a^6b^6 - c^6$

* **93.** $x^2 + y^2 - w^2 - z^2 - 2xy + 2wz$

* **94.** $x^{10} - x^2$

* **95.** $(3x + 5y)^2 - 4$

* **96.** $(2x + 5)^2 - 9$

* **97.** $9y^{2n} - 49$

* **98.** $(5x - 1)^2 - 6(5x - 1) + 9$

* **99.** $x^4y^4 + xy$

* **100.** $2m^3np^2 - 5m^2n^2p^2 - 12mn^3p^2$

* **101.** $(xz^2)^2 + 6xz^2 + 9$

* **102.** $1 - \frac{x^{36}}{64}$

Checkup

The following problems provide a review of some of Section 5.7 and will help you with the next section.

Factor.

103. $x^2 + 2x - 8$

104. $y^2 - 5y + 6$

105. $x^2 + 7x + 10$

106. $x^2 - 9x + 8$

107. $3x^2 + x - 10$

108. $6x^2 + 17x + 12$

109. $20x^2 - 19x + 3$

110. $7x^2 - 19x - 6$

5.10 SOLVING EQUATIONS BY FACTORING

OBJECTIVES

1 *Solve equations by factoring and using the zero product property*

2 *Solve applied problems by writing equations that can be solved by factoring*

1 Using the Zero Product Property

One of the uses of factoring is to solve some quadratic equations. These are equations where the greatest degree of any term is 2.

> A ***quadratic equation*** is an equation that can be written in the form
>
> $$ax^2 + bx + c = 0$$
>
> where a, b, and c are real numbers and $a \neq 0$. The form given is called ***standard form.***

To solve quadratic equations by factoring, we use a special property of zero called the ***zero product property.***

> **Zero Product Property**
>
> **If the product of two real numbers is zero, at least one of the numbers must be zero. That is, if $ab = 0$, either $a = 0$ or $b = 0$ or both.**

EXAMPLE 1

a. Solve $(x + 4)(3x - 2) = 0$.

The equation is already factored. Notice that if we multiplied out the left side, this would be quadratic equation. Since for given values of x, $x + 4$ is a number and $3x - 2$ is a number and their product is zero, then

$$x + 4 = 0 \quad \text{or} \quad 3x - 2 = 0$$

Solve these two equations.

$$\begin{aligned} x + 4 &= 0 &\quad \text{or} \quad 3x - 2 &= 0 \\ x &= -4 & 3x &= 2 \\ & & x &= \frac{2}{3} \end{aligned}$$

The solutions may be checked by substituting -4 and $\frac{2}{3}$ separately into the original equation. The numbers check.

The solutions are -4 and $\frac{2}{3}$.

b. Solve $2y^2 + 5y = 3$.

This is a quadratic equation. To use the zero product property we must get a zero on one side of the equation. The easiest way to do this is to add -3 to both sides of the equation. Then factor the other side. Set each factor equal to zero and solve.

$$2y^2 + 5y = 3$$

$$2y^2 + 5y - 3 = 0 \qquad \text{Add } -3 \text{ to each side.}$$

$$(2y - 1)(y + 3) = 0 \qquad \text{Factor.}$$

$$2y - 1 = 0 \quad \text{or} \quad y + 3 = 0 \qquad \text{Use the zero product property.}$$

$$2y = 1 \qquad y = -3$$

$$y = \frac{1}{2}$$

The solutions are $\frac{1}{2}$ and -3. ■

□ **Exercise 1** Solve.

a. $(x - 1)(x + 8) = 0$

b. $(3x - 4)(x - 5) = 0$

c. $x^2 + 6x + 8 = 0$

d. $10y^2 - y - 3 = 0$

□ **Exercise 2** Solve.

a. $x(x - 6) = 0$

b. $x^2 + 9x = 0$

c. $3x^2 = 6x$

d. $5y^2 = -10y$

To Solve an Equation Using the Zero Product Property:

1. Write the equation so that one side is equal to zero.
2. Factor the other side.
3. Use the zero product property to set each factor equal to zero.
4. Solve the resulting equations.

□ **DO EXERCISE 1.**

Caution: Do not divide both sides of an equation by a variable to try to solve it. It is a common error to do this, and doing so loses the solution of zero.

EXAMPLE 2 Solve $x^2 = 8x$.

Do *not* divide both sides of the equation by x. We add $-8x$ to both sides of the equation. Then factor and solve.

$$x^2 = 8x$$

$$x^2 - 8x = 0 \qquad \text{Add } -8x \text{ to both sides.}$$

$$x(x - 8) = 0 \qquad \text{Factor.}$$

$$x = 0 \quad \text{or} \quad x - 8 = 0 \qquad \text{Use the zero product property.}$$

$$x = 8$$

The solutions are 0 and 8. ■

□ **DO EXERCISE 2.**

2 Applied Problems

The zero product property may be used to solve applied problems.

EXAMPLE 3 The area of a rectangular garden is 18 square meters. Find the dimensions of the garden if the length is 3 meters greater than the width.

Variable If x represents the width, the length is $x + 3$.

Drawing

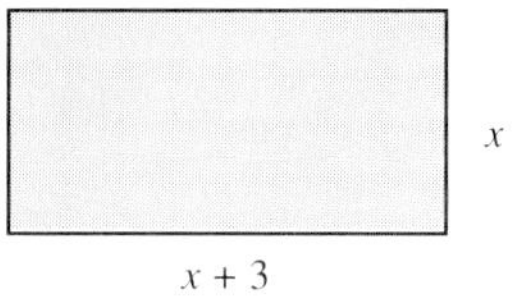

The formula for the area of a rectangle is $A = LW$, where A is the area, L is the length, and W is the width.

Equation $A = LW$

$18 = (x + 3)x$ Substitute $x + 3$ for L and x for W.

$18 = x^2 + 3x$ Multiply.

$0 = x^2 + 3x - 18$ Add -18 to each side.

$0 = (x - 3)(x + 6)$ Factor.

$x - 3 = 0$ or $x + 6 = 0$ Use the zero product property.

$x = 3$ $x = -6$

Check It would not make sense to have the width of the garden negative, so we reject -6 as a solution. If the width of the garden is 3 meters, then the length is 3 meters greater than the width or 6 meters. Then the area will be $3 \cdot 6$ or 18 square meters. The answer checks.

The garden has width 3 meters and length 6 meters. ■

□ **DO EXERCISE 3.**

□ **Exercise 3**

a. The width of a rectangle is 4 centimeters less than the length. If the area is 96 square centimeters, find the dimensions of the rectangle.

b. The square of a number minus the number is 42. Find the number.

Answers to Exercises

1. a. 1, −8 **b.** $\frac{4}{3}$, 5 **c.** −2, −4 **d.** $\frac{3}{5}$, $-\frac{1}{2}$

2. a. 0, 6 **b.** 0, −9 **c.** 0, 2 **d.** 0, −2

3. a.

$$x(x-4) = 96$$
$$x^2 - 4x = 96$$
$$x^2 - 4x - 96 = 0$$
$$(x+8)(x-12) = 0$$
$$x + 8 = 0 \quad \text{or} \quad x - 12 = 0$$
$$x = -8 \qquad\qquad x = 12$$
$$\text{reject} \qquad\qquad x - 4 = 8$$

The width is 8 cm and the length is 12 cm.

b.

$$x^2 - x = 42$$
$$x^2 - x - 42 = 0$$
$$(x-7)(x+6) = 0$$
$$x - 7 = 0 \quad \text{or} \quad x + 6 = 0$$
$$x = 7 \qquad\qquad x = -6$$

The number is 7 or −6.

Problem Set 5.10

NAME

DATE

CLASS

Solve using the zero product property.

1. $(x - 4)(x + 9) = 0$
2. $(y + 3)(y - 2) = 0$
3. $(2x - 1)(3x - 4) = 0$
4. $(5x + 2)(3x + 5) = 0$
5. $x(x - 5) = 0$
6. $y(y - 10) = 0$

Solve by factoring.

7. $x^2 + 5x - 14 = 0$
8. $x^2 + 5x - 36 = 0$
9. $y^2 - 17y + 72 = 0$
10. $y^2 - 15y + 56 = 0$
11. $x^2 + 11x + 30 = 0$
12. $x^2 + 12x + 32 = 0$
13. $x^2 + 3x = 0$
14. $x^2 + 7x = 0$
15. $2y^2 = 4y$
16. $5y^2 = 10y$
17. $4z^2 + 9z = 9$
18. $2x^2 - 11x = 40$
19. $6x^2 - x - 15 = 0$
20. $15x^2 - x - 2 = 0$
21. $x^2 = 16$
22. $y^2 = 36$
23. $0 = x^2 - 7x + 12$
24. $0 = x^2 - 11x + 24$
25. $x(x + 1) = 30$
26. $x(x - 4) = 21$

27. $-72 - x + x^2 = 0$

28. $56 - 15x + x^2 = 0$

29. $8p - p^2 = 0$

30. $9z - z^2 = 0$

31. $8x^2 - 41x + 5 = 0$

32. $4x^2 + 5x - 6 = 0$

33. $t^2 - 5t = 24$

34. $y^2 + 11y = -18$

35. $2x^3 + 8x^2 - 64x = 0$

36. $3y^3 - 24y^2 + 36y = 0$

37. $3p^3 = 48p$

38. $162p = 2p^3$

39. The area of a rectangular garden is 35 square meters. If the length is 2 meters greater than the width, find the dimensions of the garden.

40. The length of a rectangle is 8 feet more than the width. If the area of the rectangle is 105 square feet, find the length and width of the rectangle.

41. The square of a number plus twice the number is 80. Find the number.

42. Twice the square of an integer is 35 more than 9 times the integer. Find the integer.

43. The sum of the squares of two consecutive even integers is 100. Find the integers.

44. The sum of the squares of two consecutive odd integers is 130. Find the integers.

45. The area of a rectangular house is 176 square meters. The width is 5 meters less than the length. Find the width of the house.

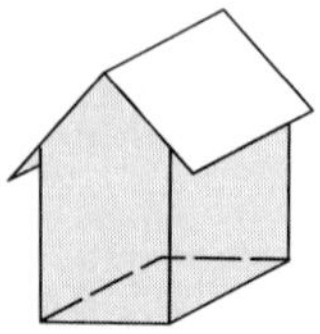

46. A rectangle has an area of 135 square centimeters. If the width is 6 centimeters less than the length, find the length of the rectangle.

47. The sum of two numbers is 8. The sum of their squares is 104. Find the numbers.

48. The sum of two numbers is 4. The sum of their squares is 106. Find the numbers.

49. The area of a rectangular parking area is 768 square meters. If the length is 3 times the width, find the dimensions of the parking area.

50. The area of a rectangle is 1764 square inches. If the length is 4 times the width, find the length of the rectangle.

51. Two numbers have a sum of 4 and a product of -96. Find the numbers.

52. Two numbers have a sum of -7 and a product of -60. Find the numbers.

53. If the product of two real numbers is zero, at least one of the numbers must be ____________.

54. If the greatest degree of any term of an equation is 2, the equation is a ____________ equation.

55. To use the zero product property, we must set one side of the equation equal to ____________.

56. Do not try to solve an equation by dividing both sides of the equation by a ____________.

Solve using the zero product property.

57. $(x + 9.2)(x + 3.4) = 0$

58. $(x - 8.75)(x - 1.38) = 0$

59. $(3x - 8.4)(2x + 4.7) = 0$

60. $(9x + 7.2)(2x + 3.2) = 0$

Solve.

* **61.** $x(x - 3) + x = 2(x + 4) - 3$

* **62.** $(x - 5)(x + 4) = (x + 7)(x + 2) + 6$

Think About It

* **63.** Find one solution of $x^3 - 27 = 0$.

* **64.** How much fencing is needed to enclose a flower garden containing 112 square feet if twice the length is 4 more than 3 times the width?

* **65.** A store is 100 feet long and 60 feet wide. If the area of the store is increased to 9600 square feet by adding rectangular sections of equal width to one side and one end, keeping the rectangular shape of the store, how wide are these strips?

* **66.** Suppose that the profit P for selling x handbags is given by

$$P = 54x - x^2 - 9$$

How many handbags must be sold to give a profit of $720?

* **67.** Two numbers have a sum of -6 and a product of -16. Find the numbers.

* **68.** A bakery makes rectangular cookies that have length 3 times the width. If the length is increased by 2 centimeters and the width by 1 centimeter, the new area is 44 square centimeters. What was the original length of the cookies?

Checkup

The following problems provide a review of some of Sections 1.3, 5.6, and 5.9 and will help you with the next section.

Find the reciprocal.

69. $\frac{3}{8}$

70. $\frac{11}{5}$

71. 9

72. 2

Factor completely.

73. $8y - 8$

74. $7y + 14$

75. $x^2 - 16$

76. $y^2 - 25$

77. $4x^2 - 12x + 9$

78. $9x^2 + 12x + 4$

79. $6x^2 + 13x - 5$

80. $6x^2 - 5x - 21$

Chapter 5 Summary

Section 5.1

A ***polynomial*** in a variable or variables is a term or a finite sum of terms in which all variables have ***whole-number exponents*** and no variables appear in denominators.

A polynomial in one variable has a ***missing term*** if the coefficient of one of the variables is zero.

The ***degree of a term*** is the sum of the exponents on the variables.

The ***degree of a polynomial*** is the same as the greatest degree of any of its terms.

The leading term of a polynomial is the term of greatest degree. Its coefficient is called the ***leading coefficient.***

Like terms are terms that have exactly the same variables raised to exactly the same powers.

If the terms of a polynomial in one variable are arranged so that the exponents on the variables decrease the polynomial is written in ***descending order.*** If the terms are arranged so that the exponents on the variables increase, the polynomial is written in ***ascending order.***

Trinomials have exactly three terms, ***binomials*** have exactly two terms, and ***monomials*** have exactly one term.

Section 5.2

To add two polynomials, combine like terms and arrange the terms in descending order.

To subtract one polynomial from another polynomial, add the opposite of the second polynomial to the first polynomial.

Section 5.3

Two monomials are multiplied by multiplying the coefficients and adding the exponents on the identical variables.

Two polynomials may be multiplied by multiplying each term of one polynomial by each term of the other polynomial. Then combine like terms.

Section 5.4

Special products of binomials

$$(A + B)(C + D) = AC + AD + BC + BD \qquad \text{(FOIL)}$$

$$(A + B)(A - B) = A^2 - B^2$$

$$(A + B)^2 = A^2 + 2AB + B^2$$

$$(A - B)^2 = A^2 - 2AB + B^2$$

Section 5.5

To divide a polynomial by a monomial, divide each term of the polynomial by the monomial.

To divide a polynomial by a binomial, or a trinomial, use a procedure similar to long division in arithmetic.

If $a \div b = c$, a is called the dividend, b is called the divisor, and c is the quotient.

If a polynomial is divided by a binomial of the form $x - k$, where k is a constant, the division may be done by an easier procedure called ***synthetic division.***

Section 5.6 To ***factor*** an expression means to write it as a multiplication.

To factor out the greatest common factor, factor out the largest number and the variable with the largest exponent that will divide evenly into each term without forming negative exponents.

Section 5.7 *Factoring* $x^2 + bx + c$

1. Factor the x^2 term.
2. Find all pairs of integers whose product equals the last term of the trinomial.
3. From the list of pairs of integers choose the pair whose sum equals the coefficient of the middle term.
4. Combine this pair of integers with the factors of the x^2 term.
5. If these integers do not exist, the polynomial is prime.

Factoring $ax^2 + bx + c, \quad a \neq 1$

1. If there is a common factor, factor it out.
2. List all pairs of factors of the x^2 term.
3. List all pairs of integer factors of the last term.
4. Choose inner and outer terms by trying all combinations of these factors until the correct middle term is found.
5. If there are no such combinations, the trinomial is prime.

Section 5.8 *Special factors of polynomials*

$$A^2 - B^2 = (A - B)(A + B)$$

$A^2 + B^2$ does not factor

$$A^2 + 2AB + B^2 = (A + B)^2$$

$$A^2 - 2AB + B^2 = (A - B)^2$$

Section 5.9

Sum of two cubes: $A^3 + B^3 = (A + B)(A^2 - AB + B^2)$

Sum of two squares: $A^2 + B^2$ cannot be factored

Difference of two cubes: $A^3 - B^3 = (A - B)(A^2 + AB + B^2)$

Difference of two squares: $A^2 - B^2 = (A - B)(A + B)$

A general method for factoring

1. Factor out the greatest common factor.
2. If there are *two terms* to be factored: Try to factor the expression as the difference of squares. If it is not the difference of squares, try to factor the expression as the sum or difference of cubes.

   If there are *three terms:* Check to see if the expression is a perfect square trinomial. If it is not, try the general method for factoring a trinomial.

   If there are *four terms:* Try to separate the expression into groups of two terms that have a common factor and then remove the common binomial factor. Next, try grouping into a difference of squares, one of which is a trinomial.
3. Factor completely. If a polynomial, other than a monomial, can be factored, factor it. Some polynomials are factored completely after the common factor is removed. Also, some polynomials are prime.

Section 5.10

A ***quadratic equation*** is an equation that can be written in the form

$$ax^2 + bx + c = 0$$

where a, b, and c are real numbers and $a \neq 0$. The form given is called standard form.

Zero product property: If the product of two real numbers is zero, at least one of the numbers must be zero. That is, if $ab = 0$, either $a = 0$ or $b = 0$ or both.

Solving an equation using the zero product property

1. Write the equation so that one side is equal to zero.
2. Factor the other side.
3. Use the zero product property to set each factor equal to zero.
4. Solve the resulting equations.

Chapter 5 Additional Exercises (Optional)

NAME ____________

DATE ____________

CLASS ____________

Section 5.1

Evaluate for $x = -3$ and $y = -1$.

1. $3x^2 - x + 2$

2. $2x^2 - 4x - 1$

3. $4y^3 + 2xy^2$

4. $3y^2 - 2xy + 2$

* **5.** $\frac{4}{3}x^2 + \frac{3}{5}y$

* **6.** $0.01x^3 - 0.04x^2 + 2.1x$

For each polynomial, (a) find the degree of the polynomial; (b) write in descending powers of the variable; and (c) identify as a trinomial, binomial, or none of these.

7. $8 + 3a^2 - 4a^3$

8. $6b^2 - 8b^5$

9. $7 + z^4$

10. $-3 + p^6 - p^8 + 4p^2$

Section 5.2

Add or subtract as indicated.

11. $(6x - 2y + z) + (-8x - 7y + 4z)$

12. $(-3a - 5b - 7c) - (6a - 8b - 3c)$

13. $(-4x^2 + 9x + 3) + (-7x^2 - 2x - 8)$

14. $(3x^2 - 2xy + 4y^2) - (5x^2 + 3xy - 6y^2)$

15. $(4x^3 - 2x^2 + 3) - (3x^2 + 2x - 5)$

16. $\left(\frac{3}{5}x^2 - \frac{1}{8}y^2\right) + \left(\frac{7}{3}x^2 + \frac{1}{2}y^2\right)$

17. $(0.05a^2 - 3.12ab) - (2.07a^2 + 1.45ab)$

Section 5.3

Multiply. Assume that variables in exponents represent positive integers.

18. $3a(-4a^5)$

19. $(-6b^3)(-3b^5)$

* **20.** $(x^{2n})x^{4n}$

* **21.** $y^{3n}(-y^{6n})$

* **22.** $(3x^{5n})x^{-2n}$

* **23.** $(4x^{-3n})(-2x^{7n})$

* **24.** $a^n(a^{2n} + 1)$

* **25.** $b^{2n}(b^n + b)$

26. $(3a - 2)(4a + 5)$

27. $(2x - 3)(6x - 1)$

* 28. $(x^n + y^n)(x^n + y^n)$

* 29. $(a^n + b^n)(a^n - b^n)$

30. $(x - 4)(2x^2 - x + 3)$

31. $(y - 1)(3y^2 + y - 2)$

Multiply by writing vertically.

32. $-2x^2 + 3x + 1$ and $3x^2 - x + 4$

33. $-x^2 + x - 3$ and $-4x^2 + 2x + 1$

34. $y^2 - y - 2$ and $3y^2 - 2y - 1$

Section 5.4

Multiply. Assume that variables in exponents represent positive integers.

35. $(3x - 4)(2x - 3)$

36. $(5x + 2)(2x + 3)$

37. $(5x - 7)(5x + 7)$

38. $(3x - 4)^2$

* 39. $(a^n + b^n)^2$

* 40. $(x^n - y^n)^2$

* 41. $(x^{2n} + y^{3n})(x^{2n} - y^{3n})$

* 42. $(a^{4n} - b^n)(a^{4n} + b^n)$

Section 5.5

Divide.

43. $\dfrac{7p^4 + 6p^2 - 3p}{3p^3}$

44. $\dfrac{8q^5 - 3q^3 - 2}{4q^3}$

45. $\dfrac{x^2 - 8x + 15}{x - 5}$

46. $\dfrac{8a^2 - 23a + 2}{a - 3}$

47. $\dfrac{3x^2 - 10x + 6}{x - 2}$

48. $\dfrac{x^2 - 8x + 7}{x - 3}$

49. $\dfrac{15x^2 + 11x - 17}{3x - 2}$

50. $\dfrac{6x^3 - 11x^2 + 11x - 2}{2x - 3}$

* 51. $\dfrac{2x^5 + x^3 - 2x - 3}{x^2 - 3x + 1}$

* 52. $\dfrac{9a^4 - 13a^3 + 13a^2 - 9a + 2}{a^2 - a + 2}$

Section 5.6

Factor out the greatest common factor.

53. $16x^2 + 96x^4 - 64x^3$

54. $24y^7 - 32y^3 + 48y^5$

55. $ax - 5bx + ay - 5by$

56. $3px + qx + 3py + qy$

57. $x^2 - 7x - x + 7$

58. $y^2 - 9y - y + 9$

59. $m^2 + n + m + mn$

* 60. $a^2 + b - a - ab$

61. $3(x - 2)^2 + 4(x - 2)$

62. $5(y + 1)^2 - 3(y + 1)$

Section 5.7

Factor completely. Assume that variables in exponents represent positive integers.

63. $24x^2 + 29x - 63$

64. $45a^2 - 103a + 56$

65. $35x^2 - 63x + 28$

66. $22x^5 + 55x^4 - 33x^3$

67. $42x^3 - 17x^2 - 15x$

68. $144y^2 - 26y - 30$

* 69. $18x^{2n} - 27x^n + 4$

* 70. $28y^{2n} - 33y^n + 9$

Section 5.8

Factor completely. Assume that variables in exponents represent positive integers.

71. $64x^2 - 81y^2$

72. $49x^2 - 9y^2$

* 73. $25x^{2n} - 16y^{2n}$

* 74. $100a^{2n} - 64b^{2n}$

75. $4x^2 - 24x + 36$

76. $9x^4 - 12x^3 + 4x^2$

* 77. $x^{3n} + 10x^{2n} + 25x^n$

* 78. $y^{5n} - 18y^{4n} + 81y^{3n}$

Section 5.9

Factor completely.

79. $64x^3 + 125$

80. $8y^3 - 1000$

81. $x^6 - y^6$

82. $27a^6 - 8b^6$

83. $x^3 - \dfrac{8}{125}$

84. $y^3 + \dfrac{1}{64}$

85. $686x^3 + 2000$

86. $375a^3 - 2187$

Section 5.10

Solve.

87. $x^2 - 10x + 21 = 0$

88. $x^2 - 3x - 10 = 0$

89. $3y^2 = 9y$

90. $4y^2 - 8y = 0$

91. $54a^2 = -21a + 20$

92. $72x^2 = 44x - 4$

93. The area of a rectangular rug is 154 square feet. If the length is 3 feet greater than the width, find the dimensions of the rug.

94. If 5 times the square of an integer is added to 11 times the integer, the result is 12. Find the integer.

95. The width of a rectangular picture is 6 inches less than the length. If the area of the picture is 280 square inches, find the dimensions of the picture.

96. The sum of two numbers is 10. The sum of their squares is 148. Find the numbers.

COOPERATIVE LEARNING

1. Write a polynomial with a degree of 9.

2. Subtract: $(75a^2 - 36a^2b^2 - 97b^2) - (104a^2 - 19a^2b^2 + 78b^2)$.

Multiply.

3. $(8y^2 - 7)(9y^3 - 16y^2 + 12)$

4. $(15x^2 - 32y)(15x^2 + 32y)$

5. The quotient of two polynomials is $4x + 7$. The dividend is $32x^2 + 44x - 21$. Find the divisor.

6. The product of two polynomials is $270x^4y^5 - 414x^5y^4 - 810x^6y^4$. One of the factors is $18x^3y^2$. Find the other factor.

7. The product of two polynomials is $ax^2 + bx^2 - ay^2 - by^2$. One of the factors is $a + b$. Find the other factor.

Find two polynomials whose product is the following.

8. $124x^2 + 263x - 36$

9. $81y^2 - 225y + 84$

Find second-degree equations with the following solutions.

10. $0, -\frac{11}{32}$

11. $\frac{3}{8}, -\frac{9}{7}$

12. Explain why the solutions of $(3x + 2)(7x - 8) = 14$ are not found from the two equations.

$$3x + 2 = 14 \quad \text{and} \quad 7x - 8 = 14$$

Chapter 5 Practice Test

NAME ______

DATE ______

CLASS ______

1. Evaluate $P = 4y^2 - 2xy + y^3$ for $x = 3$ and $y = -2$.

2. Find the degree of the polynomial: $4x^5 + 6x^7 - 8$.

3. Combine like terms: $3x^2 - 5x - 8x^2 - 2x$.

4. Add: $5a^2 - 3ab + b^2$ and $6a^2 - 5ab - 9b^2$.

5. Subtract: $(-3y^2 - 4y - 2) - (-5y^2 + 2y - 8)$.

Multiply.

6. $(3x^2 - 4x - 2)(x - 3)$

7. $(2x - 5)(3x - 4)$

8. $(3y - 8)(3y + 8)$

9. $(5p - 9)^2$

Divide.

10. $\dfrac{21a^3 - 14a^2 + 7}{7a^2}$

1. ______
2. ______
3. ______
4. ______
5. ______
6. ______
7. ______
8. ______
9. ______
10. ______

11. ______

11. $\dfrac{x^2 + 6x - 4}{x - 2}$

12. ______

12. Divide using synthetic division:

$$\frac{x^2 - 3x + 7}{x - 4}$$

Factor.

13. ______

13. $x^2 + 5x + 2x + 10$

14. ______

14. $30x^2 + 3x - 9$

15. ______

15. $4x^2 - 20x + 25$

16. ______

16. $49x^4 - 81x^2y^2$

17. ______

17. $64y^3 - 125z^3$

Solve.

18. ______

18. $2x^2 + 5x = 12$

19. ______

19. $y^2 = 15y$

20. ______

20. The area of a rectangular flower bed is 60 square meters. Find the dimensions of the flower bed if the width is 4 meters less than the length.

CHAPTER

6

Rational Expressions and Equations

Pretest

1. Multiply and simplify:

$$\frac{6x^2 + 11x - 10}{4x^2 - 25} \cdot \frac{6x - 15}{6x + 12}$$

2. Divide and simplify:

$$\frac{x^3 - 27}{3x - 6} \div \frac{x^2 + 3x + 9}{x^2 - 5x + 6}$$

Add or subtract and simplify.

3. $\dfrac{x}{x^2 - y^2} - \dfrac{y}{y^2 - x^2}$

4. $\dfrac{7}{3x - 2} + \dfrac{4}{x + 6} - \dfrac{2x - 8}{3x^2 + 16x - 12}$

5. Simplify:

$$\frac{\dfrac{5}{a} + \dfrac{4}{b}}{\dfrac{1}{a} - \dfrac{2}{b}}$$

6. Solve:

$$\frac{4}{3x - 2} - \frac{5}{2x + 3} = \frac{6}{6x^2 + 5x - 6}$$

7. Joel can prepare flower arrangements twice as fast as Carol. If they work together they can complete the arrangements in 6 hours. How long would it take Carol working alone to prepare the arrangements?

8. A canal has a current of 2 miles per hour. A canoe travels 3 miles upstream in the same amount of time it takes to go 7 miles downstream. What is the speed of the canoe in still water?

9. Solve $\frac{E}{e} = \frac{R + r}{r}$ for r.

10. The stopping distance of a car varies directly as the square of the speed. If a car traveling 30 miles per hour can stop in 50 feet, how many feet will it take to stop a car traveling 70 miles per hour? Round the answer to the nearest tenth.

11. The weight that a horizontal beam can support varies inversely as the length of the beam. If a 12-meter beam can support 800 kilograms, how many kilograms can a 5-meter beam support?

6.1 SIMPLIFYING, MULTIPLYING, AND DIVIDING

OBJECTIVES

1. *Simplify rational expressions*
2. *Multiply rational expressions*
3. *Divide rational expressions*

Recall that a rational number is a number that can be written as the quotient of two integers (with denominator not zero). The quotient of two polynomials is called a ***rational expression.*** Following are examples of rational expressions:

$$\frac{2}{3} \qquad \frac{4}{x-5} \qquad \frac{x^2+2x+3}{x-7} \qquad x-3$$

The last expression, $x - 3$, may be written $(x - 3)/1$. It is a special rational expression with a denominator of 1. The rules given in this chapter for multiplying and dividing rational expressions are used for rational expressions where at least one of the expressions has a denominator other than 1. Rules for multiplying and dividing rational expressions with denominators of 1 were discussed in Chapter 5.

1 Simplifying Rational Expressions

The methods for simplifying rational expressions are similar to the methods for simplifying fractions. We simplify fractions by factoring the numerator and denominator and eliminating a factor of "1." If a rational expression has the same numerator and denominator, it is the same as 1 (except that 0/0 is undefined).

$$\frac{-7}{-7} = 1 \qquad \frac{3x^2-2}{3x^2-2} = 1 \qquad \frac{y-4}{y-4} = 1$$

Fundamental Property of Rational Expressions

If A/B is a rational expression and C is any rational expression where $C \neq 0$, then

$$\frac{AC}{BC} = \frac{A}{B}$$

Notice that this is true since

$$\frac{AC}{BC} = \frac{A \cdot C}{B \cdot C} = \frac{A}{B} \cdot \frac{C}{C} = \frac{A}{B} \cdot 1 = \frac{A}{B} \qquad \text{Eliminate the factor of 1.}$$

Hence

$$\frac{(x-4)(x+2)}{(x+3)(x+2)} = \frac{x-4}{x+3}$$

Using the commutative property of multiplication, the fundamental property of rational expressions may also be stated as

$$\frac{AC}{BC} = \frac{CA}{BC} = \frac{AC}{CB} = \frac{A}{B}$$

For example,

$$\frac{(x+5)(x-4)}{(x-7)(x+5)} = \frac{x-4}{x-7}$$

We can make our work easier by using slashes to eliminate the common factors as follows:

$$\frac{\cancel{(x+5)}(x-4)}{(x-7)\cancel{(x+5)}} = \frac{x-4}{x-7}$$

□ **Exercise 1** Simplify by factoring and using the fundamental property of rational expressions.

a. $\dfrac{12x^2}{6x}$

b. $\dfrac{9}{18y}$

c. $\dfrac{2x+8}{x^2-16}$

d. $\dfrac{3y^2-y-10}{3y^2-7y-20}$

To simplify a rational expression, we factor the numerator and denominator. Then we use the fundamental property of rational expressions.

EXAMPLE 1 Simplify by factoring and using the fundamental property of rational expressions.

a. $\dfrac{5x^3}{15x}$

$$\frac{5x^3}{15x} = \frac{5 \cdot x \cdot x^2}{5 \cdot 3 \cdot x}$$ Factor; since there is an x in the denominator, we factor the numerator into $x \cdot x^2$.

$$= \frac{\cancel{5} \cdot \cancel{x} \cdot x^2}{\cancel{5} \cdot 3 \cdot \cancel{x}}$$

$$= \frac{x^2}{3}$$ Use the fundamental property.

b. $\dfrac{4}{8x+16}$

$$\frac{4}{8x+16} = \frac{\cancel{4}(1)}{\cancel{4}(2x+4)}$$ Factor.

$$= \frac{1}{2x+4}$$ Use the fundamental property.

Notice that the 1 remains in the numerator after using the fundamental property.

c. $\dfrac{2x^2+3x-2}{x^2-4}$

$$\frac{2x^2+3x-2}{x^2-4} = \frac{\cancel{(x+2)}(2x-1)}{\cancel{(x+2)}(x-2)}$$ Factor.

$$= \frac{2x-1}{x-2}$$ Use the fundamental property.

d. $\dfrac{2x^2-xy-3y^2}{2x^2-5xy+3y^2}$

$$\frac{2x^2-xy-3y^2}{2x^2-5xy+3y^2} = \frac{\cancel{(2x-3y)}(x+y)}{\cancel{(2x-3y)}(x-y)}$$

$$= \frac{x+y}{x-y}$$

Caution: The fundamental property of rational expressions applies only to *factors* of the numerator and denominator.

$$\frac{2x^2(3)}{2x^2} = 3 \quad \text{but} \quad \frac{2x^2-3}{2x^2} \neq -3$$ ■

□ **DO EXERCISE 1.**

Sometimes it appears that we cannot simplify a rational expression when we can simplify it by multiplying the numerator and denominator by −1. This method works when a factor in the numerator and denominator are the same except *the signs are exactly opposite.* We usually find the product in the denominator.

EXAMPLE 2 Simplify $(x - 4)/(4 - x)$ by multiplying the numerator and denominator by -1 and using the fundamental property of rational expressions.

$$\frac{x-4}{4-x} = \frac{(-1)(x-4)}{(-1)(4-x)}$$ Multiply numerator and denominator by -1.

$$= \frac{(-1)(x-4)}{-4+x}$$ Find the product in the denominator.

$$= \frac{(-1)(x-4)}{(x-4)}$$ Commutative property of addition.

$$= -1$$ Use the fundamental property.

Notice that $x - 4$ and $4 - x$ are opposites, since $4 - x = -x + 4$. The quotient of two nonzero expressions that are opposites is always -1. Hence we may shorten our work as follows.

$$\frac{x-4}{4-x} = -1 \quad ■$$

> The quotient of two nonzero expressions that are opposites is -1.

□ **DO EXERCISE 2.**

□ **Exercise 2** Simplify using opposites.

a. $\dfrac{y-5}{5-y}$

b. $\dfrac{x^2-4}{2-x}$

2 Multiplication of Rational Expressions

When we multiply two fractions, we multiply the numerators and multiply the denominators. Rational expressions are multiplied the same way.

> Rational expressions are multiplied by multiplying the numerators and multiplying the denominators.

When we multiply rational expressions, we *factor* the numerator and denominator of each expression, if possible. Then we indicate the multiplication and simplify.

EXAMPLE 3 Multiply.

a. $$\frac{3x^2}{4} \cdot \frac{8}{x^4} = \frac{3x^2}{4} \cdot \frac{4 \cdot 2}{x^2 \cdot x^2}$$ Factor; since there is a 4 in one of the denominators, we factor the 8 in one of the numerators into $4 \cdot 2$ so that we can eliminate factors of 4.

$$= \frac{3x^2 \cdot 4 \cdot 2}{4 \cdot x^2 \cdot x^2}$$ Multiply numerators and denominators.

$$= \frac{6}{x^2}$$ Use the fundamental property.

□ **Exercise 3** Multiply.

a. $\frac{10x}{9} \cdot \frac{3}{x^3}$

b. $\frac{7x+7}{x} \cdot \frac{3x^2}{14x+14}$

c. $\frac{x^2-2x-15}{x^2-4x+3} \cdot \frac{x-3}{x-5}$

d. $\frac{x^2-9}{3x-1} \cdot \frac{6x^2+x-1}{x^2-7x+12}$

b. $\frac{6x-6}{x} \cdot \frac{5x}{x^2-1} = \frac{6(x-1)}{x} \cdot \frac{5x}{(x-1)(x+1)}$ Factor.

$= \frac{6(x-1)(5x)}{x(x-1)(x+1)}$ Multiply numerators and denominators.

$= \frac{30}{x+1}$ Use the fundamental property.

c. $\frac{y^2+4y+4}{2y^2+5y+2} \cdot \frac{y^2-4y+3}{y-3}$

$= \frac{(y+2)(y+2)}{(2y+1)(y+2)} \cdot \frac{(y-1)(y-3)}{y-3}$ Factor.

$= \frac{(y+2)(y+2)(y-1)(y-3)}{(2y+1)(y+2)(y-3)}$

$= \frac{(y+2)(y-1)}{2y+1}$ Use the fundamental property.

It is often more useful to leave the answer in factored form. Integers, however, should be multiplied. ■

□ **DO EXERCISE 3.**

3 Division of Rational Expressions

We divide two fractions by multiplying by the reciprocal of the divisor. This is the same as inverting the divisor and multiplying. Division of rational expressions is done the same way.

> **To divide two rational expressions, multiply by the reciprocal of the divisor.**

Once we have inverted the divisor it may be necessary to factor the numerators and denominators of the rational expressions before completing the exercise.

EXAMPLE 4 Divide.

a. $\frac{x}{x+1} \div \frac{x}{x+4} = \frac{x}{x+1} \cdot \frac{\mathbf{x+4}}{\mathbf{x}}$ Multiply by the reciprocal of the divisor.

$= \frac{x(x+4)}{(x+1)x}$ Multiply numerators and denominators.

$= \frac{x+4}{x+1}$ Use the fundamental property.

b. $\dfrac{y^2 - 9}{2y + 6} \div \dfrac{y^2 + 2y - 15}{2y - 8}$

$= \dfrac{y^2 - 9}{2y + 6} \cdot \dfrac{\mathbf{2y - 8}}{\mathbf{y^2 + 2y - 15}}$ Multiply by the reciprocal of the divisor.

$= \dfrac{(y - 3)(y + 3)}{2(y + 3)} \cdot \dfrac{2(y - 4)}{(y - 3)(y + 5)}$ Factor.

$= \dfrac{\cancel{(y - 3)}\cancel{(y + 3)}(\cancel{2})(y - 4)}{\cancel{2}\cancel{(y + 3)}\cancel{(y - 3)}(y + 5)}$

$= \dfrac{y - 4}{y + 5}$ Use the fundamental property. ■

□ **DO EXERCISE 4.**

□ **Exercise 4** Divide.

a. $\dfrac{8x^2}{5} \div \dfrac{4x}{15}$

b. $\dfrac{3y + 6}{7} \div \dfrac{y^2 - 4}{14}$

c. $\dfrac{(a + 5)^2}{2a - 6} \div \dfrac{a^2 + 6a + 5}{a^2 - 5a + 6}$

d. $\dfrac{x^2 - 9}{3x^2 + 8x - 3} \div \dfrac{5x^2 - 14x - 3}{6x^2 + 13x - 5}$

Answers to Exercises

1. a. $2x$ b. $\frac{1}{2y}$ c. $\frac{2}{x-4}$ d. $\frac{y-2}{y-4}$

2. a. -1 b. $-(x + 2)$

3. a. $\frac{10}{3x^2}$ b. $\frac{3x}{2}$ c. $\frac{x+3}{x-1}$ d. $\frac{(x+3)(2x+1)}{x-4}$

4. a. $6x$ b. $\frac{6}{y-2}$ c. $\frac{(a+5)(a-2)}{2(a+1)}$ d. $\frac{2x+5}{5x+1}$

Problem Set 6.1

NAME

DATE

CLASS

Simplify by factoring and using the fundamental property of rational expressions.

1. $\dfrac{18x^4}{9x^2}$

2. $\dfrac{10y^3}{8y}$

3. $\dfrac{x}{5x^2}$

4. $\dfrac{y}{3y}$

5. $\dfrac{2x + 4}{2}$

6. $\dfrac{5x - 10}{5}$

7. $\dfrac{3x - 6}{4x - 8}$

8. $\dfrac{9y + 9}{7y + 7}$

9. $\dfrac{3x^2 + 6x}{x^2 + 2x}$

10. $\dfrac{5z^3 + 10z^2}{z + 2}$

11. $\dfrac{x^2 - 16}{2x + 8}$

12. $\dfrac{x^2 - 1}{3x - 3}$

13. $\dfrac{a^2 - 5a - 6}{a^2 - 8a + 12}$

14. $\dfrac{b^2 + 5b + 6}{b^2 + b - 6}$

15. $\dfrac{10x^2 - 3x - 1}{15x^2 + 8x + 1}$

16. $\dfrac{y^2 - 9y + 14}{y^2 - 3y - 28}$

17. $\dfrac{a^2 - 9}{4a^2 + 11a - 3}$

18. $\dfrac{p^2 - 4}{p^2 + 7p + 10}$

19. $\dfrac{x^3 + y^3}{x^2 - y^2}$

20. $\dfrac{a^2 - b^2}{a^3 - b^3}$

Simplify using opposites.

21. $\dfrac{x - 8}{8 - x}$

22. $\dfrac{x - 1}{1 - x}$

23. $\dfrac{9 - y}{y - 9}$

24. $\dfrac{4 - a}{a - 4}$

25. $\dfrac{25 - b^2}{b - 5}$

26. $\dfrac{49 - x^2}{x - 7}$

27. $\dfrac{x^2 - 9}{3 - x}$

28. $\dfrac{y^2 - 1}{1 - y}$

29. $\dfrac{y - x}{x - y}$

30. $\dfrac{x^2 - y^2}{y^2 - x^2}$

Multiply.

31. $\dfrac{15x^2}{2} \cdot \dfrac{6}{x^3}$

32. $\dfrac{21x}{3} \cdot \dfrac{8}{x^2}$

33. $\dfrac{2x}{8x+4} \cdot \dfrac{14x+7}{3}$

34. $\dfrac{6y-10}{5y} \cdot \dfrac{3}{9y-15}$

35. $\dfrac{x-1}{x-4} \cdot \dfrac{x^2-2x-8}{x^2+x-2}$

36. $\dfrac{x+3}{x+5} \cdot \dfrac{x^2+x-20}{x^2+x-6}$

37. $\dfrac{x^2-25}{x^2} \cdot \dfrac{3x^2+x}{x^2+6x+5}$

38. $\dfrac{y^2-5y-6}{y^2-8y+12} \cdot \dfrac{y^3-4y}{y^2}$

39. $\dfrac{6x^2+7x-3}{8x^2+10x-3} \cdot \dfrac{15x^2-16x+4}{9x^2-3x-2}$

40. $\dfrac{12x^2-7x+1}{4x^2-9x+2} \cdot \dfrac{6x^2-19x+3}{6x^2-13x+2}$

41. $\dfrac{p^3+27}{p^2-4} \cdot \dfrac{p^2-4p+4}{p^2-3p+9}$

42. $\dfrac{a^3-8}{a^2-9} \cdot \dfrac{a^2+6a+9}{a^2+2a+4}$

43. $\dfrac{12+4y-3x-xy}{24+8y-3x-xy} \cdot \dfrac{32+8y-4x-xy}{16-4y-4x+xy}$

44. $\dfrac{p^4-p^3+p^2-p}{2p^3+2p^2+p+1} \cdot \dfrac{2p^3-8p^2+p-4}{p^3-4p^2+p-4}$

Divide.

45. $\dfrac{8}{x^3} \div \dfrac{16}{x^3}$

46. $\dfrac{x}{21} \div \dfrac{x}{7}$

47. $\dfrac{4x+8}{5} \div \dfrac{x^2-4}{15}$

48. $\dfrac{3x+21}{x} \div \dfrac{x^2-49}{x^2}$

49. $\dfrac{(a-3)^2}{a^2-9} \div \dfrac{a-3}{4a^2+9a-9}$

50. $\dfrac{b^2-4}{b+2} \div \dfrac{3b^2-2b-8}{b-7}$

51. $\dfrac{x^3-4x}{x^2-16} \div \dfrac{x^2+5x-14}{x^2+4x-21}$

52. $\dfrac{y^2+5y}{y^2-25} \div \dfrac{y^2+8y+15}{y^2+y-20}$

53. $\dfrac{6x^2 + 5x - 6}{12x^2 - 11x + 2} \div \dfrac{4x^2 - 12x + 9}{8x^2 - 14x + 3}$

54. $\dfrac{8a^2 - 6a - 9}{6a^2 - 5a - 6} \div \dfrac{4a^2 + 11a + 6}{9a^2 - 4}$

55. $\dfrac{x^3 - 8}{x^3 + 8} \div \dfrac{x^2 - 4}{x^2 - 2x + 4}$

56. $\dfrac{64x^3 + 125}{64x^3 - 125} \div \dfrac{16x^2 - 25}{16x^2 + 20x + 25}$

57. $\dfrac{a^2 - b^2}{a^3 - b^3} \div \dfrac{a^2 + 2ab + b^2}{a^2 + ab + b^2}$

58. $\dfrac{4x^2 - 9y^2}{8x^3 - 27y^3} \div \dfrac{4x^2 + 12xy + 9y^2}{4x^2 + 6xy + 9y^2}$

59. $\dfrac{am - an + bm - bn}{am + an - bm - bn} \div \dfrac{am - an - 3bm + 3bn}{am + an - 3bm - 3bn}$

60. $\dfrac{x^3 + x + x^2 + 1}{x^3 + x^2 + xy^2 + y^2} \div \dfrac{x^3 + x + x^2y + y}{2x^2 + 2xy - xy^2 - y^3}$

61. A rational expression is the quotient of two ____________.

62. We simplify rational expressions by ____________ the numerator and denominator and using the fundamental property of rational expressions.

63. The fundamental property of rational expressions applies only to ____________ of the numerator and denominator.

64. Rational expressions are multiplied by multiplying the ____________ and multiplying the ____________.

65. To divide two rational expressions, multiply the first expression by the ____________ of the divisor.

Simplify.

66. $\dfrac{x - 17.4}{17.4 - x}$

67. $\dfrac{354.2 - y}{y - 354.2}$

Multiply or divide.

68. $\dfrac{75.2x}{x - 33.4} \cdot \dfrac{14.4x - 9.4}{x + 33.4}$

69. $\dfrac{x - 224.3}{19.8x} \cdot \dfrac{x + 224.3}{27.4x}$

70. $\dfrac{y - 53.7}{84.3y} \div \dfrac{98.2y}{y + 53.7}$

71. $\dfrac{79.4x - 3.2}{52x} \div \dfrac{15.6x + 8.3}{4.3x - 7.8}$

Simplify.

* **72.** $\dfrac{xz + xw - yz - yw}{xy + xz - y^2 - yz}$

* **73.** $\dfrac{x^3 - 4x^2 + 5x - 20}{4x^5 + 20x^3 - x^2 - 5}$

Multiply or divide as indicated.

* **74.** $\dfrac{x^2 + 4x + 4}{x - 2} \cdot \dfrac{x^2 - 5x + 6}{x + 2} \cdot \dfrac{3x - 12}{x - 3}$

* **75.** $\left[\dfrac{x + 1}{x^2 - 4} \cdot \dfrac{2x - 4}{x^2 - 1}\right] \div \dfrac{2x + 2}{x^2 - 4x + 4}$

* **76.** $\dfrac{6p^2 - 13p - 5}{p^2 + 4p} \cdot \dfrac{p^3 + 3p^2 - 4p}{2p - 5} \div \dfrac{3p^2 - 8p - 3}{p^2 - 5p + 6}$

* **77.** $\dfrac{7}{5x^2(x + 3)} \div \left[\dfrac{x^2 - 5x + 6}{8x^2} \cdot \dfrac{21}{5x^2 - 45}\right]$

* **78.** $\dfrac{3y^2 + 4y - 15}{8y^2 - 10y - 3} \div \left[\dfrac{2y^2 + 9y + 9}{8y^2 - 18y - 5} \div \dfrac{4y^2 - 9}{3y^2 + 10y - 25}\right]$

* **79.** $\dfrac{x^2y + xy^2}{xy - x^2} \div \left[\dfrac{x^2 - 2xy + y^2}{x^2 - y^2} \div \dfrac{xy - y^2}{x^2 + 2xy + y^2}\right]$

* **80.** $\dfrac{x^3 - y^3}{x^2 + 2xy + y^2} \div \left[\dfrac{x^2 + xy + y^2}{x^2 - y^2} \div \dfrac{x^2y + xy^2}{x^2 - 2xy + y^2}\right]$

* **81.** $\dfrac{a^2(2a + b) + 6a(2a + b) + 5(2a + b)}{3a^2(2a + b) - 2a(2a + b) - (2a + b)} \div \dfrac{a + 1}{a - 1}$

Checkup

The following problems provide a review of some of Section 5.2 and will help you with the next section.

Add or subtract as indicated.

82. $(3 + 2x) - (4x + 7)$

83. $(6 - 7x) + (9x - 3)$

84. $(5x^2 - 8x) + (6x^2 - 3)$

85. $(2x - 9) - (5x^2 + 8x)$

86. $(7x - 4) - (8x^2 - 3)$

87. $(x^2 - 3x) + (5 - 7x)$

88. $(3x^2 - 2x + 1) + (8x^2 - 3)$

89. $(7x + 8) - (5x^2 - 3x + 9)$

6.2 ADDITION AND SUBTRACTION

1 Addition and Subtraction of Rational Expressions with the Same Denominator

We add rational expressions in the same way that we add fractions.

> To add or subtract rational expressions with the same denominator, add or subtract the numerators and keep the common denominator. Simplify, if possible.

EXAMPLE 1 Add or subtract.

a. $\frac{5}{x} + \frac{x+4}{x} = \frac{5+x+4}{x} = \frac{9+x}{x}$ Add the numerators.

b. $\frac{x^2-3x}{x-2} + \frac{3x-4}{x-2} = \frac{x^2-3x+3x-4}{x-2}$

$= \frac{x^2-4}{x-2}$

$= \frac{(x-2)(x+2)}{x-2}$ Factor the numerator.

$= x+2$ Use the fundamental property of rational expressions.

c. $\frac{7x}{5x+5} - \frac{4x+3}{5x+5} = \frac{7x-(4x+3)}{5x+5}$ Use parentheses around the second numerator to subtract the entire numerator.

$= \frac{7x-4x-3}{5x+5}$

$= \frac{3x-3}{5x+5}$ ■

□ **DO EXERCISE 1.**

2 Least Common Denominator

To add fractions with different denominators, we need to find a common denominator. To make our work easier, we usually try to find the ***least common denominator,*** which is the smallest number that all denominators divide into with zero remainder. The procedure is similar for rational expressions.

> **Finding the Least Common Denominator**
>
> 1. Factor each denominator, factoring numbers into primes. Remember that natural numbers other than 1 are prime numbers if they are divisible only by themselves and 1. For example, the first few primes are 2, 3, 5, 7, and 11.
> 2. The least common denominator is found by using each factor the greatest number of times that it occurs in any one denominator.

OBJECTIVES

1 *Add or subtract rational expressions with the same denominator*

2 *Find the least common denominator for rational expressions*

3 *Multiply a rational expression by 1 to give an indicated denominator*

4 *Add or subtract rational expressions with different denominators*

□ **Exercise 1** Add or subtract.

a. $\frac{5}{y} + \frac{2-y}{y}$

b. $\frac{5x-1}{x+1} + \frac{x^2-5x}{x+1}$

c. $\frac{7}{x^2} - \frac{3x}{x^2}$

d. $\frac{3x}{x^2+x-2} - \frac{5x+4}{x^2+x-2}$

The least common denominator is often abbreviated "LCD."

EXAMPLE 2 Find the LCD.

a. $\frac{1}{8}$ and $\frac{7}{12}$

In the chart below, if a factor of the second number is the same as a factor of the first number, place this factor beneath the factor of the first number. For the LCD, list the numbers in each column once.

Prime Factorization

8	2	2	2	
12	2	2		3
LDC	2	2	2	3

List the numbers in each column once.

LCD $= 2 \cdot 2 \cdot 2 \cdot 3 = 24$

b. $\frac{5}{16x^2y}, \frac{7}{18xy^3}$

1. Factor the first denominator.

$$16x^2y = \mathbf{2 \cdot 2 \cdot 2 \cdot 2} \cdot \mathbf{x^2}y$$

2. Factor the second denominator.

$$18xy^3 = 2 \cdot \mathbf{3 \cdot 3} \cdot x\mathbf{y^3}$$

3. Each factor appears in the LCD the greatest number of times that it appears in any one denominator.

$$\text{LCD} = \mathbf{2 \cdot 2 \cdot 2 \cdot 2 \cdot 3 \cdot 3 \cdot x^2y^3} = 144x^2y^3$$

c. $\frac{x}{x^2 - 9}, \frac{2}{x - 3}$

$$x^2 - 9 = \mathbf{(x - 3)(x + 3)}$$ Factor the first denominator.

$x - 3$ does not factor.

$$\text{LCD} = \mathbf{(x - 3)(x + 3)}$$

Notice that $x - 3$ occurs once in each denominator, so it must appear once in the LCD. The factor $x + 3$ appears in one of the denominators, so it must be included in the LCD.

d. $\frac{x}{7}, \frac{4}{5x - 2}$

These denominators do not factor. The LCD is their product.

$$\text{LCD} = 7(5x - 2)$$

e. $\frac{4y + 2}{y^2 - y - 12}, \frac{y}{y^2 + y - 6}$

1. Factor the first denominator.

$$y^2 - y - 12 = (\mathbf{y + 3})(\mathbf{y - 4})$$

2. Factor the second denominator.

$$y^2 + y - 6 = (\mathbf{y - 2})(y + 3)$$

3. LCD = $(y + 3)(y - 4)(y - 2)$

Notice that we used all the factors of the first denominator; we did not repeat the $y + 3$ from the second denominator.

f. $\frac{5}{x^2 + 4x + 4}, \quad \frac{x}{3x + 6}, \quad \frac{2x - 1}{x^2 - 4}$

1. Factor the first denominator.

$$x^2 + 4x + 4 = (x + 2)(x + 2)$$

2. Factor the second denominator.

$$3x + 6 = 3(x + 2)$$

3. Factor the third denominator.

$$x^2 - 4 = (x - 2)(x + 2)$$

4. LCD = $3(x + 2)(x + 2)(x - 2)$ ■

□ **DO EXERCISE 2.**

3 Multiplication by 1

The numerator and denominator of a fraction may be multiplied by the same number to give an equivalent fraction. This is called multiplying by 1. In the same way, a rational expression may be multiplied by 1 to give an equivalent rational expression.

EXAMPLE 3 Multiply by 1 to give the rational expression the indicated denominator.

a. $\frac{3}{5x}$, denominator $20x^2y$

The denominator is $5x$, but we want a denominator of $20x^2y$. We multiply by $4xy/4xy$, which is 1.

$$\frac{3}{5x} = \frac{3}{5x} \cdot \frac{\mathbf{4xy}}{\mathbf{4xy}} = \frac{12xy}{20x^2y}$$

b. $\frac{7}{x + 5}$, denominator $x^2 + 8x + 15$

Since $x^2 + 8x + 15 = (x + 5)(x + 3)$, we multiply numerator and denominator by $x + 3$.

$$\frac{7}{x + 5} = \frac{7}{x + 5} \cdot \frac{\mathbf{x + 3}}{\mathbf{x + 3}} = \frac{7(x + 3)}{(x + 5)(x + 3)}$$

$$= \frac{7x + 21}{x^2 + 8x + 15}$$ ■

□ **DO EXERCISE 3.**

□ **Exercise 2** Find the LCD.

a. $\frac{7}{18}, \quad \frac{8}{15}$

b. $\frac{5}{24}, \quad \frac{9}{20}$

c. $\frac{3}{5x^3y}, \quad \frac{9}{10xy^4}$

d. $\frac{x}{x - 4}, \quad \frac{2}{x - 3}$

e. $\frac{x + 1}{x - 1}, \quad \frac{2x}{x^2 - 1}$

f. $\frac{y}{y^2 + 5y - 6}, \quad \frac{2y - 7}{y^2 + y - 2}$

g. $\frac{6}{x^2 + 2x + 1}, \quad \frac{3x}{4x^2 - 4x}, \quad \frac{5x + 2}{x^2 - 1}$

□ **Exercise 3** Multiply by 1 to give the fraction the indicated denominator.

a. $\frac{7}{3y} = \frac{\quad}{3x^2y}$

b. $\frac{8}{y - 2} = \frac{\quad}{y^2 - 4}$

4 Addition and Subtraction of Rational Expressions with Different Denominators

> To add or subtract rational expressions with different denominators find the least common denominator. Multiply each expression by one to give it the LCD. Then add or subtract the expressions, keeping the denominator. Simplify, if possible.

EXAMPLE 4 Add or subtract.

a. $\frac{5}{12} + \frac{7}{18}$

The LCD is $2 \cdot 2 \cdot 3 \cdot 3$, or 36. Multiply each rational number by 1 to give it the LCD and then add.

$$\frac{5}{12} + \frac{7}{18} = \frac{5}{12} \cdot \frac{\mathbf{3}}{\mathbf{3}} + \frac{7}{18} \cdot \frac{\mathbf{2}}{\mathbf{2}}$$

$$= \frac{15}{36} + \frac{14}{36} = \frac{29}{36}$$

b. $\frac{3x}{x+3} + \frac{4}{x-1}$

The LCD is $(x + 3)(x - 1)$.

$$\frac{3x}{x+3} + \frac{4}{x-1}$$

$$= \frac{3x}{x+3} \cdot \frac{\mathbf{x-1}}{\mathbf{x-1}} + \frac{4}{x-1} \cdot \frac{\mathbf{x+3}}{\mathbf{x+3}}$$ Multiply by 1.

$$= \frac{3x(x-1)}{(x+3)(x-1)} + \frac{4(x+3)}{(x-1)(x+3)}$$

$$= \frac{3x(x-1) + 4(x+3)}{(x+3)(x-1)}$$ Add numerators.

$$= \frac{3x^2 - 3x + 4x + 12}{(x+3)(x-1)}$$ Multiply in the numerator.

$$= \frac{3x^2 + x + 12}{(x+3)(x-1)}$$ Combine like terms.

It is not necessary to multiply out the denominator. The numerator cannot be factored to give any factors of the denominator, so the fraction cannot be simplified. In fact, the numerator of this fraction cannot be factored.

c. $\dfrac{5+x}{x-2} - \dfrac{3x}{x^2-4}$

$$\left.\begin{aligned} x-2 &= x-2 \\ x^2-4 &= (x-2)(x+2) \end{aligned}\right\} \text{ The LCD is } (x-2)(x+2).$$

$$\begin{aligned}
\frac{5+x}{x-2} - \frac{3x}{(x-2)(x+2)} &= \frac{5+x}{x-2}\cdot\frac{\mathbf{x+2}}{\mathbf{x+2}} - \frac{3x}{(x-2)(x+2)} \\
&= \frac{(5+x)(x+2)}{(x-2)(x+2)} - \frac{3x}{(x-2)(x+2)} \\
&= \frac{(5+x)(x+2) - 3x}{(x-2)(x+2)} \\
&= \frac{5x + 10 + x^2 + 2x - 3x}{(x-2)(x+2)} \qquad \text{Use FOIL.} \\
&= \frac{x^2 + 4x + 10}{(x-2)(x+2)} \qquad ■
\end{aligned}$$

□ **DO EXERCISE 4.**

> When one denominator is the opposite, or additive inverse, of the other, multiply one expression by $-1/-1$ to find a common denominator.

EXAMPLE 5 Add.

$\dfrac{3x}{x-4} + \dfrac{8x}{4-x}$

The denominators are opposites. We can make them the same by multiplying one of the rational expressions by $-1/-1$.

$$\begin{aligned}
\frac{3x}{x-4} + \frac{8x}{4-x} &= \frac{3x}{x-4} + \frac{8x}{4-x}\cdot\frac{\mathbf{-1}}{\mathbf{-1}} \qquad \text{Multiply by } \frac{-1}{-1}. \\
&= \frac{3x}{x-4} + \frac{-8x}{-4+x} \\
&= \frac{3x + (-8x)}{x-4} \\
&= \frac{-5x}{x-4} \qquad ■
\end{aligned}$$

□ **DO EXERCISE 5.**

□ **Exercise 4** Add or subtract.

a. $\dfrac{5}{18} + \dfrac{10}{27}$

b. $\dfrac{2y+3}{2y^2} - \dfrac{5}{6y}$

c. $\dfrac{3}{p+4} - \dfrac{p+2}{p-3}$

d. $\dfrac{3x-1}{x^2-25} + \dfrac{7-5x}{x+5}$

□ **Exercise 5** Add or subtract.

a. $\dfrac{m+3}{4} + \dfrac{2m}{-4}$

b. $\dfrac{x-6}{2-x} - \dfrac{3x}{x-2}$

□ **Exercise 6** Add or subtract.

a. $\dfrac{3y}{y^2 + y - 2} - \dfrac{4y}{y^2 + 2y - 3}$

b. $\dfrac{-2x}{x^2 + 3x - 4} + \dfrac{5x}{x^2 + 7x + 12}$

EXAMPLE 6 Subtract.

$$\frac{3y - 2}{y^2 - 5y + 4} - \frac{y + 2}{y^2 - 6y + 8}$$

$$= \frac{3y - 2}{(y - 1)(y - 4)} - \frac{y + 2}{(y - 2)(y - 4)}$$ Factor each denominator.

The LCD is $(y - 1)(y - 4)(y - 2)$.

$$= \frac{3y - 2}{(y - 1)(y - 4)} \cdot \frac{\mathbf{y - 2}}{\mathbf{y - 2}} - \frac{y + 2}{(y - 2)(y - 4)} \cdot \frac{\mathbf{y - 1}}{\mathbf{y - 1}}$$ Multiply by 1.

$$= \frac{(3y - 2)(y - 2)}{(y - 1)(y - 4)(y - 2)} - \frac{(y + 2)(y - 1)}{(y - 2)(y - 4)(y - 1)}$$

$$= \frac{3y^2 - 8y + 4}{(y - 1)(y - 4)(y - 2)} - \frac{y^2 + y - 2}{(y - 2)(y - 4)(y - 1)}$$

$$= \frac{3y^2 - 8y + 4 - (y^2 + y - 2)}{(y - 2)(y - 4)(y - 1)}$$ Do not forget the parentheses in the numerator.

$$= \frac{3y^2 - 8y + 4 - y^2 - y + 2}{(y - 2)(y - 4)(y - 1)}$$ Use a distributive law.

$$= \frac{2y^2 - 9y + 6}{(y - 2)(y - 4)(y - 1)}$$ Combine like terms. ■

□ **DO EXERCISE 6.**

When Do We Need a Common Denominator?

1. To add or subtract rational expressions, we need a common denominator.
2. To multiply or divide rational expressions, we do not need a common denominator.

Answers to Exercises

1. a. $\dfrac{7 - y}{y}$ **b.** $x - 1$ **c.** $\dfrac{7 - 3x}{x^2}$ **d.** $\dfrac{-2}{x - 1}$

2. a. 90 **b.** 120 **c.** $10x^3y^4$ **d.** $(x - 4)(x - 3)$
e. $(x - 1)(x + 1)$ **f.** $(y + 6)(y - 1)(y + 2)$
g. $4x(x + 1)(x + 1)(x - 1)$

3. a. $\dfrac{7x^2}{3x^2y}$ **b.** $\dfrac{8y + 16}{y^2 - 4}$

4. a. $\dfrac{35}{54}$ **b.** $\dfrac{y + 9}{6y^2}$ **c.** $\dfrac{-p^2 - 3p - 17}{(p + 4)(p - 3)}$ **d.** $\dfrac{-5x^2 + 35x - 36}{(x + 5)(x - 5)}$

5. a. $\dfrac{-m + 3}{4}$ **b.** $\dfrac{-4x + 6}{x - 2}$

6. a. $\dfrac{-y}{(y + 2)(y + 3)}$ **b.** $\dfrac{3x^2 - 11x}{(x + 4)(x - 1)(x + 3)}$

Problem Set 6.2

NAME ______________________

DATE ______________________

CLASS ______________________

Find the least common denominator.

1. $\frac{5}{8}, \frac{7}{12}$

2. $\frac{3}{16}, \frac{5}{24}$

3. $\frac{1}{18a^2}, \frac{5}{48a}$

4. $\frac{7}{30y^3}, \frac{3}{75y}$

5. $\frac{4}{15x^2y}, \frac{2}{25xy^4}$

6. $\frac{5}{14ab}, \frac{3}{42a^2b}$

7. $\frac{2}{x-3}, \frac{x}{x-5}$

8. $\frac{3y}{y-2}, \frac{5y}{y+4}$

9. $\frac{x+2}{x^2-9}, \frac{x}{3x+9}$

10. $\frac{x-3}{x^2-16}, \frac{3x}{2x-8}$

11. $\frac{5}{2x+3}, \frac{x}{x-5}, \frac{5}{x}$

12. $\frac{6x}{3x-1}, \frac{7}{3x+1}, \frac{x}{5}$

13. $\frac{y}{2y-4}, \frac{3}{y-2}$

14. $\frac{p}{3p+9}, \frac{2}{p^2-9}$

15. $\frac{7+z}{z^2-7z+12}, \frac{8}{z^2-4z+3}$

16. $\frac{9-q}{2q^2+7q-4}, \frac{3q}{q^2+6q+8}$

Multiply by 1 to give the rational expression the denominator indicated.

17. $\frac{7}{5x^2} = \frac{\quad}{10x^4}$

18. $\frac{3}{2xy} = \frac{\quad}{8x^2y^3}$

19. $\frac{5}{x-4} = \frac{\quad}{x^2-16}$

20. $\frac{3}{x+2} = \frac{\quad}{x^2-4}$

21. $\frac{7}{3z-4} = \frac{\quad}{6z-8}$

22. $\frac{4}{5z-2} = \frac{\quad}{10z-4}$

23. $\dfrac{3k}{k+1} = \dfrac{\quad}{k^2 + 6k + 5}$

24. $\dfrac{5p}{p-2} = \dfrac{\quad}{p^2 - 5p + 6}$

25. $\dfrac{8}{7-x} = \dfrac{\quad}{x-7}$

26. $\dfrac{x}{y-9} = \dfrac{\quad}{9-y}$

27. $\dfrac{3x+2}{x^2 - y^2} = \dfrac{\quad}{y^2 - x^2}$

28. $\dfrac{5x-4}{x^2-4} = \dfrac{\quad}{4-x^2}$

Add or subtract and simplify, if possible.

29. $\dfrac{6}{x} + \dfrac{3-x}{x}$

30. $\dfrac{5-3y}{y} + \dfrac{2y}{y}$

31. $\dfrac{8+6x}{x+4} - \dfrac{5x+4}{x+4}$

32. $\dfrac{6x-1}{x-3} - \dfrac{2+5x}{x-3}$

33. $\dfrac{3y-2}{7y} + \dfrac{3}{14y^2}$

34. $\dfrac{6k-1}{12k} - \dfrac{5}{18k^2}$

35. $\dfrac{8}{z-2} - \dfrac{z+3}{z+2}$

36. $\dfrac{5z+1}{z-4} - \dfrac{3}{z-3}$

37. $\dfrac{6}{3y-2} + \dfrac{7y}{6y-4}$

38. $\dfrac{3x}{4x-8} - \dfrac{2}{3x-6}$

39. $\dfrac{7x}{x^2-25} - \dfrac{3}{x+5}$

40. $\dfrac{8}{y-2} - \dfrac{3y}{y^2-4}$

41. $\dfrac{z+2}{3} + \dfrac{z}{-3}$

42. $\dfrac{x}{5} + \dfrac{x-7}{-5}$

43. $\dfrac{z}{z+3} - \dfrac{7}{z^2+6z+9}$

44. $\dfrac{3}{x-5} - \dfrac{6x}{x^2-7x+10}$

45. $\dfrac{3x-1}{x^2-6x+5} + \dfrac{8}{x^2-3x-10}$

46. $\dfrac{5y-2}{y^2-7y+12} + \dfrac{1}{y^2+y-12}$

47. $\dfrac{x}{x-5} - \dfrac{3x}{5-x}$

48. $\dfrac{5z}{7-x} - \dfrac{3z}{x-7}$

49. $\dfrac{6y}{x^2-y^2} + \dfrac{3y}{4x-4y}$

50. $\dfrac{-7a}{3a + 3b} + \dfrac{4a}{a^2 - b^2}$

51. $\dfrac{7}{x^2 - 16} - \dfrac{3x}{x^2 + x - 20}$

52. $\dfrac{5x}{x^2 - 6x + 9} - \dfrac{2}{x^2 - 9}$

53. $\dfrac{3x - 4}{x^2 - y^2} + \dfrac{2x + 1}{y^2 - x^2}$

54. $\dfrac{5a + 2b}{a^2 - b^2} + \dfrac{9a - 3b}{b^2 - a^2}$

55. $\dfrac{5y + 2}{2y^2 - 5y + 3} - \dfrac{8}{2y^2 - 7y + 6}$

56. $\dfrac{4x}{3x^2 + 5x - 2} - \dfrac{3x + 5}{3x^2 - 7x + 2}$

57. $\dfrac{2y - 1}{y^2 + 3y - 4} + \dfrac{3y - 2}{y^2 + 2y - 8}$

58. $\dfrac{3a - 2}{a^2 - 3a + 2} + \dfrac{5a - 1}{a^2 + a - 6}$

59. $\dfrac{7}{x} - \dfrac{9}{-x}$

60. $\dfrac{5}{-3y} - \dfrac{8}{3y}$

61. $\dfrac{3x - 7}{x^2 - 16} + \dfrac{5x - 2}{16 - x^2}$

62. $\dfrac{9x + 8}{12 - y^2} + \dfrac{6x - 15}{y^2 - 12}$

63. $\dfrac{3}{p - 1} + \dfrac{2}{p + 1} - \dfrac{3p}{p^2 - 1}$

64. $\dfrac{5x}{x - 5} - \dfrac{4}{x + 5} + \dfrac{3}{x^2 - 25}$

65. $\dfrac{2}{x - 3} - \dfrac{x}{x + 2} + \dfrac{x^2 - 4}{x^2 - x - 6}$

66. $\dfrac{3y}{y - 4} + \dfrac{2}{y + 1} - \dfrac{y^2 + 4}{y^2 - 3y - 4}$

67. $\dfrac{3}{(p - 4)^2} - \dfrac{3p}{p - 4} + 2$

68. $\dfrac{5a}{a + 3} - \dfrac{3}{(a + 3)^2} + 3$

69. $\dfrac{3p}{p^2 - q^2} - \dfrac{2}{p - q} + \dfrac{2}{q - p}$

70. $\dfrac{7}{5 - 2x} - \dfrac{4}{2x - 5} + \dfrac{x - 2}{2x^2 - 7x + 5}$

71. $\dfrac{4}{2t - 1} + \dfrac{1}{t + 2} - \dfrac{3t}{2t^2 + 3t - 2}$

72. $\dfrac{2c}{(c + d)^2} + \dfrac{5}{c + d} - \dfrac{4}{d + c}$

73. $\dfrac{3}{a + b} - \dfrac{2}{b - a} - \dfrac{a}{a^2 - b^2}$

74. $\dfrac{4}{x - y} + \dfrac{2}{y - x} - \dfrac{3y}{y^2 - x^2}$

75. To add or subtract rational expressions with the same denominator, add or subtract the numerators and keep the ____________ ____________.

76. The least common denominator is the smallest number that all denominators ____________ into with ____________ remainder.

77. A rational expression may be multiplied by ____________ to give an equivalent rational expression.

78. To add or subtract rational expressions with different denominators, first find the ____________ ____________ ____________.

Add or subtract and simplify, if possible.

* **79.** $\dfrac{2y^2 - y}{3y^2 - 12} - \dfrac{y - 2}{3y - 6} + \dfrac{4y^2}{4 - y^2}$

* **80.** $\dfrac{3x - 1}{3x^2 - x - 2} + \dfrac{x + 1}{x^2 + 2x - 3} - \dfrac{x - 4}{3x^2 + 11x + 6}$

* **81.** $5(x + 2)^{-1} - 5(x - 2)^{-1} - 25(x^2 - 4)^{-1}$

* **82.** $4(a^2 - 4)^{-1} + 2(a^2 - 4a + 4)^{-1} + 3(a^2 + 4a + 4)^{-1}$

* **83.** $\dfrac{a + b}{a^2 - ab - 2b^2} + \dfrac{a - b}{a^2 - 3ab + 2b^2}$

* **84.** $\dfrac{2x - 2y}{3x^2 - 2xy - y^2} - \dfrac{2x + y}{6x^2 - xy - y^2}$

Checkup

The following problems review some of Section 2.1 and will help you with the next section.

Solve.

85. $x - \dfrac{3}{4} = \dfrac{7}{8}$

86. $y + \dfrac{3}{5} = \dfrac{1}{10}$

87. $\dfrac{1}{2}(x - 4) = \dfrac{2}{3}x - 3$

88. $\dfrac{1}{6}(x + 5) = \dfrac{3}{7}x - 1$

89. $\dfrac{4}{5}(x + 2) = \dfrac{1}{4}(x + 12)$

90. $\dfrac{2}{9}(y - 4) = \dfrac{1}{5}(2y - 8)$

91. $\dfrac{1}{4}(4x + 2) - 3 = \dfrac{2}{3}(x - 4)$

92. $\dfrac{3}{16}(2x + 2) + 4 = \dfrac{1}{8}(4x - 4)$

6.3 COMPLEX FRACTIONS

OBJECTIVE

1 *Simplify complex fractions*

If the numerator or denominator of a fraction contains fractions, the fraction is called a ***complex fraction.*** The following are examples of complex fractions.

$$\frac{\frac{3}{y}+5}{7} \qquad \frac{3}{\frac{x}{5}+\frac{4x}{9}} \qquad \frac{\frac{2}{x+3}}{\frac{x-1}{x^2-9}}$$

A complex fraction is simplified by eliminating all fractions in both the numerator and the denominator.

1 There are two methods for simplifying complex fractions.

> **Method 1**
>
> Add or subtract all terms in the numerator and then all terms in the denominator to get a single fraction in both the numerator and denominator. Then divide the numerator by the denominator and simplify, if possible.

Remember that when we divide two fractions, we multiply the dividend by the reciprocal of the divisor.

EXAMPLE 1 Simplify by eliminating all fractions in the numerator and denominator of the complex fraction.

a. $\dfrac{\frac{2}{x}+3}{\frac{3}{8}}$

$$\frac{\frac{2}{x}+3}{\frac{3}{8}} = \frac{\frac{2}{x}+3\cdot\frac{x}{x}}{\frac{3}{8}}$$ Multiply by $\frac{x}{x}$ to get a common denominator.

$$= \frac{\frac{2+3x}{x}}{\frac{3}{8}}$$ Add in the numerator.

$$= \frac{2+3x}{x}\cdot\frac{8}{3}$$ Multiply by the reciprocal of the divisor.

$$= \frac{(2+3x)8}{3x}$$

$$= \frac{16+24x}{3x}$$

□ **Exercise 1** Simplify by eliminating all fractions in the numerator and denominator of the complex fraction.

a. $\dfrac{\dfrac{x+3}{x}}{\dfrac{2x+6}{x}}$

b. $\dfrac{x+\dfrac{7}{y}}{\dfrac{2}{3}}$

c. $\dfrac{\dfrac{3x}{2}-\dfrac{x}{7}}{x+\dfrac{x}{7}}$

d. $\dfrac{5+\dfrac{1}{y}}{5-\dfrac{1}{y}}$

e. $\dfrac{\dfrac{1}{a}+\dfrac{1}{b}}{\dfrac{1}{a}-\dfrac{1}{b}}$

b. $\dfrac{\dfrac{5}{x}-\dfrac{1}{4x}}{x+\dfrac{x}{4}}$

$\dfrac{\dfrac{5}{x}-\dfrac{1}{4x}}{x+\dfrac{x}{4}}=\dfrac{\dfrac{5}{x}\cdot\mathbf{\dfrac{4}{4}}-\dfrac{1}{4x}}{x\cdot\mathbf{\dfrac{4}{4}}+\dfrac{x}{4}}$ Multiply by $\dfrac{4}{4}$ to get common denominators in the numerator and denominator.

$=\dfrac{\dfrac{19}{4x}}{\dfrac{5x}{4}}$ Subtract in the numerator and add in the denominator.

$=\dfrac{19}{4x}\cdot\dfrac{4}{5x}$ Multiply by the reciprocal of the denominator.

$=\dfrac{19}{5x^2}$

c. $\dfrac{\dfrac{2}{a}-\dfrac{2}{b}}{\dfrac{1}{a^2}-\dfrac{1}{b^2}}$

$\dfrac{\dfrac{2}{a}-\dfrac{2}{b}}{\dfrac{1}{a^2}-\dfrac{1}{b^2}}=\dfrac{\dfrac{2}{a}\cdot\dfrac{b}{b}-\dfrac{2}{b}\cdot\dfrac{a}{a}}{\dfrac{1}{a^2}\cdot\dfrac{b^2}{b^2}-\dfrac{1}{b^2}\cdot\dfrac{a^2}{a^2}}$

$=\dfrac{\dfrac{2b-2a}{ab}}{\dfrac{b^2-a^2}{a^2b^2}}$ Subtract in the numerator and denominator.

$=\dfrac{2b-2a}{ab}\cdot\dfrac{a^2b^2}{b^2-a^2}$ Multiply by the reciprocal of the denominator.

$=\dfrac{2\cancel{(b-a)}}{\cancel{ab}}\cdot\dfrac{(\cancel{ab})(ab)}{(b+a)\cancel{(b-a)}}$ Factor.

$=\dfrac{2ab}{b+a}$ Use the fundamental property. ■

If there are common denominators in the fractions of both the numerator and denominator, the complex fraction can be simplified by dividing the numerator by the denominator.

□ **DO EXERCISE 1.**

There is another method for simplifying complex fractions.

Method 2

Multiply the numerator and denominator of the complex fraction by the least common denominator of the denominators. Then simplify the result.

EXAMPLE 2 Simplify by eliminating all fractions in the numerator and denominator of the complex fraction.

a. $\dfrac{\dfrac{5}{x} - \dfrac{1}{4x}}{x + \dfrac{x}{4}}$

The least common denominator of the denominators is $4x$, so multiply the numerator and denominator of the complex fraction by $4x$.

$$\frac{\frac{5}{x} - \frac{1}{4x}}{x + \frac{x}{4}} = \frac{\mathbf{4x}\left(\frac{5}{x} - \frac{1}{4x}\right)}{\mathbf{4x}\left(x + \frac{x}{4}\right)} \qquad \text{Multiply by } 4x.$$

$$= \frac{4x\left(\frac{5}{x}\right) - 4x\left(\frac{1}{4x}\right)}{4x(x) + 4x\left(\frac{x}{4}\right)} \qquad \text{Use a distributive law.}$$

$$= \frac{20 - 1}{4x^2 + x^2}$$

$$= \frac{19}{5x^2}$$

This is the same result that we obtained in Example 1b using Method 1.

b. $\dfrac{3a + \dfrac{5}{a - 2}}{4a - \dfrac{1}{a}}$

The least common denominator of the denominators is $a(a - 2)$.

$$\frac{3a + \frac{5}{a-2}}{4a - \frac{1}{a}} = \frac{\mathbf{a(a-2)}\left(3a + \frac{5}{a-2}\right)}{\mathbf{a(a-2)}\left(4a - \frac{1}{a}\right)} \qquad \text{Multiply by } a(a - 2).$$

$$= \frac{a(a-2)(3a) + a(a-2)\left(\frac{5}{a-2}\right)}{a(a-2)(4a) - a(a-2)\left(\frac{1}{a}\right)} \qquad \text{Use a distributive law.}$$

$$= \frac{3a^3 - 6a^2 + 5a}{4a^3 - 8a^2 - a + 2}$$

□ **Exercise 2** Simplify by eliminating all fractions in the numerator and denominator of the complex fraction.

a. $\dfrac{3 + \dfrac{4}{x}}{\dfrac{7}{5x} - 5}$

b. $\dfrac{\dfrac{2}{y} + \dfrac{1}{y - 2}}{\dfrac{2}{y} - \dfrac{1}{y - 2}}$

c. $\dfrac{\dfrac{1}{a} + \dfrac{1}{b}}{\dfrac{1}{a^2} + \dfrac{1}{b^2}}$

c. $\dfrac{\dfrac{2}{a} - \dfrac{2}{b}}{\dfrac{1}{a^2} - \dfrac{1}{b^2}}$

The least common denominator of the denominators is a^2b^2.

$$\frac{\dfrac{2}{a} - \dfrac{2}{b}}{\dfrac{1}{a^2} - \dfrac{1}{b^2}} = \frac{\mathbf{a^2b^2}\left(\dfrac{2}{a} - \dfrac{2}{b}\right)}{\mathbf{a^2b^2}\left(\dfrac{1}{a^2} - \dfrac{1}{b^2}\right)} \qquad \text{Multiply by } a^2b^2.$$

$$= \frac{a^2b^2\left(\dfrac{2}{a}\right) - a^2b^2\left(\dfrac{2}{b}\right)}{a^2b^2\left(\dfrac{1}{a^2}\right) - a^2b^2\left(\dfrac{1}{b^2}\right)} \qquad \text{Use a distributive law.}$$

$$= \frac{2ab^2 - 2a^2b}{b^2 - a^2}$$

$$= \frac{2ab\cancel{(b - a)}}{(b + a)\cancel{(b - a)}} \qquad \text{Factor.}$$

$$= \frac{2ab}{b + a}$$

This is the same result that we got for Example 1c using Method 1. ■

□ **DO EXERCISE 2.**

DID YOU KNOW?

The first well-known woman mathematician was Hypatia of Alexandria (370–415). She was a respected lecturer and was praised for her knowledge of philosophy and medicine. Unfortunately, she was skinned alive with oyster shells by a Christian mob, because they claimed she promoted paganism in her teachings.

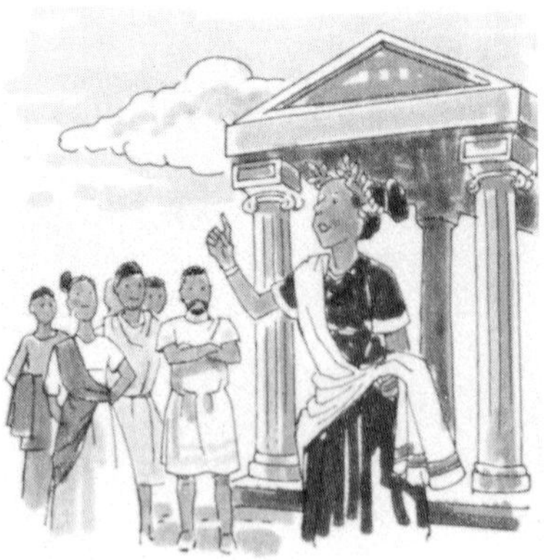

Answers to Exercises

1. a. $\dfrac{1}{2}$ **b.** $\dfrac{3xy + 21}{2y}$ **c.** $\dfrac{19}{16}$ **d.** $\dfrac{5y + 1}{5y - 1}$ **e.** $\dfrac{b + a}{b - a}$

2. a. $\dfrac{15x + 20}{7 - 25x}$ **b.** $\dfrac{3y - 4}{y - 4}$ **c.** $\dfrac{(b + a)ab}{b^2 + a^2}$

Problem Set 6.3

NAME ______________________

DATE ______________________

CLASS ______________________

Simplify by eliminating all fractions in the numerator and denominator of the complex fraction.

1. $\dfrac{\frac{5}{x} + 2}{\frac{3}{x}}$

2. $\dfrac{\frac{3}{2y} - 1}{\frac{7}{y}}$

3. $\dfrac{\frac{x - 3}{x}}{\frac{x + 3}{4x}}$

4. $\dfrac{\frac{a - b}{2}}{\frac{a + b}{8}}$

5. $\dfrac{\frac{1}{b} + 5}{\frac{1}{b} - 3}$

6. $\dfrac{\frac{1}{z} - 2}{\frac{1}{z} + 4}$

7. $\dfrac{\frac{x - 3}{x}}{\frac{x + 4}{x^2}}$

8. $\dfrac{\frac{y + 8}{10y}}{\frac{y - 8}{5y^2}}$

9. $\dfrac{a - \frac{1}{a}}{a + \frac{1}{a}}$

10. $\dfrac{b + \frac{1}{b}}{b - \frac{1}{b}}$

11. $\dfrac{\frac{p - 2}{p^2}}{\frac{p^2 - 4}{2p}}$

12. $\dfrac{\frac{4x^2 - 1}{3x^2}}{\frac{2x + 1}{6x}}$

13. $\dfrac{\frac{3}{2x} - \frac{5}{8x}}{1 + \frac{3}{8x}}$

14. $\dfrac{\frac{4x}{5} + 3}{\frac{2}{5} - \frac{3}{10x}}$

15. $\dfrac{3 + \frac{4}{x - 1}}{5 - \frac{1}{x}}$

16. $\dfrac{2 - \dfrac{3}{y}}{4 + \dfrac{2}{y - 2}}$

17. $\dfrac{\dfrac{1}{x} + \dfrac{1}{x - 1}}{\dfrac{1}{x} - \dfrac{1}{x - 1}}$

18. $\dfrac{\dfrac{3}{b} - \dfrac{4}{b - 2}}{\dfrac{3}{b} + \dfrac{5}{b - 2}}$

19. $\dfrac{\dfrac{2}{a} + \dfrac{3}{b}}{\dfrac{3}{a} - \dfrac{2}{b}}$

20. $\dfrac{\dfrac{3}{x} + \dfrac{5}{y}}{\dfrac{1}{x} - \dfrac{3}{y}}$

21. $\dfrac{\dfrac{x^2 - y^2}{xy}}{\dfrac{x + y}{x}}$

22. $\dfrac{\dfrac{a^2 - b^2}{ab}}{\dfrac{a + b}{a}}$

23. $\dfrac{\dfrac{3a}{b} - a}{\dfrac{b}{a} - b}$

24. $\dfrac{\dfrac{y}{x} + x}{\dfrac{4x}{y} + y}$

25. $\dfrac{\dfrac{1}{x} - \dfrac{4}{y}}{\dfrac{y^2 - 16x^2}{xy}}$

26. $\dfrac{\dfrac{2}{a} - \dfrac{3}{b}}{\dfrac{4b^2 - 9a^2}{2a}}$

27. $\dfrac{\dfrac{1}{2(x + h)} - \dfrac{1}{2x}}{h}$

28. $\dfrac{\dfrac{1}{5(x + h)} - \dfrac{1}{5x}}{h}$

29. $\dfrac{\dfrac{1}{x} + \dfrac{1}{y}}{\dfrac{1}{x^3} + \dfrac{1}{y^3}}$

30. $\dfrac{\dfrac{1}{x^2} - \dfrac{1}{y^2}}{\dfrac{1}{x^3} + \dfrac{1}{y^3}}$

31. In a complex fraction the ____________ or ____________ contains fractions.

32. A complex fraction is simplified by eliminating all ____________ in both the numerator and the denominator.

33. If there are common denominators in the numerator and denominator of a complex fraction, the fraction may be simplified by ____________ the numerator by the denominator.

34. One method of simplifying a complex fraction is to multiply the numerator and denominator by the ____________ ____________ ____________ and simplify the result.

Simplify.

* **35.** $\dfrac{\dfrac{3x-y}{3x+y}+\dfrac{3x+y}{3x-y}}{\dfrac{3x-y}{3x+y}-\dfrac{3x+y}{3x-y}}$

* **36.** $\dfrac{(a+b)^{-1}-(a-b)^{-1}}{4(a^2-b^2)^{-1}}$

* **37.** $\dfrac{\dfrac{x}{y}-2+\dfrac{y}{x}}{\dfrac{x^3+y^3}{xy^2+x^2y}}$

* **38.** $\dfrac{a+\dfrac{b}{a-b}}{b+\dfrac{a}{b-a}}$

* **39.** $4+\dfrac{1}{1+\dfrac{3a}{a-1}}$

* **40.** $\dfrac{y+2+\dfrac{3}{y+4}}{y+1-\dfrac{5}{y+3}}$

* **41.** $\dfrac{a-\dfrac{1}{a-\dfrac{a}{b}}}{a+\dfrac{1}{a-\dfrac{a}{b}}}$

* **42.** $1+\dfrac{1}{1+\dfrac{1}{1+\dfrac{1}{x+1}}}$

Think About It

Explain why the following are true.

43. $\dfrac{\dfrac{1}{(x+h)^2} - \dfrac{1}{x^2}}{h} \neq \dfrac{-2x - h}{x^2(x+h)}$

44. $\dfrac{\dfrac{1}{(x+h)^3} - \dfrac{1}{x^3}}{h} \neq \dfrac{-3x^2 - 3xh - h^2}{x(x+h)^3}$

Checkup

The following problems provide a review of some of Section 5.1 and will help you with the next section.

Evaluate for the given value of x.

45. $P = 3x - 2,\ x = \dfrac{2}{3}$

46. $P = 5x - 4,\ x = \dfrac{4}{5}$

47. $Q = x^2 + 2x - 3,\ x = -3$

48. $Q = y^2 + 6y + 8,\ y = -2$

49. $P = x^2 + 4,\ x = 1$

50. $P = y^2 + 5,\ y = 3$

Find the degree of the polynomial.

51. $3x^2 - 2x + 5$

52. $7x - 4$

53. $8x$

54. $x^2 - 6x$

6.4 SOLVING RATIONAL EQUATIONS

1 Undefined Rational Expressions

A ***rational equation*** is an equation with rational expressions. A rational expression is undefined for any values of the variable that make the denominator zero. Since division by zero is not allowed, the polynomial in the denominator of a rational expression cannot be equal to zero.

EXAMPLE 1 Find the values for which the rational expression is undefined.

a. $\dfrac{2}{3x - 6}$

To find the values of x that make the denominator 0, set the denominator equal to 0 and solve for x.

$$3x - 6 = 0$$
$$3x = 6$$
$$x = 2$$

The expression is undefined for 2.

b. $\dfrac{y - 2}{y^2 - 3y - 4}$

$$y^2 - 3y - 4 = 0 \quad \text{Set the denominator equal to zero.}$$
$$(y + 1)(y - 4) = 0 \quad \text{Factor.}$$
$$y + 1 = 0 \quad \text{or} \quad y - 4 = 0 \quad \text{Use the zero product property.}$$
$$y = -1 \qquad\qquad y = 4$$

The expression is undefined for -1 and 4.

c. $\dfrac{x}{x^2 + 3}$

Are there values of x that make the denominator 0?

$$\text{Let } x^2 + 3 = 0$$
$$x^2 = -3$$

There are no values of x such that squaring it will give -3 and make the denominator zero. The denominator, $x^2 + 3$, is always positive. The expression is defined for all real numbers. ■

□ **DO EXERCISE 1.**

2 Rational Equations

The easiest way to solve *rational equations* that contain fractions is the following.

> To solve a rational equation, first multiply both sides of the equation by the least common denominator to clear the fractions.

When we multiply both sides of an equation by the LCD, we eliminate the denominators and the result is an equation that we can solve.

OBJECTIVES

1 *Find values for which a rational expression is undefined*

2 *Solve rational equations*

3 *Differentiate between adding or subtracting rational expressions and solving rational equations*

□ **Exercise 1** Find the values for which the rational expression is undefined.

a. $\dfrac{7}{x}$

b. $\dfrac{y - 3}{4y + 2}$

c. $\dfrac{3}{y^2 - 7y + 6}$

d. $\dfrac{5}{x^2 + 2}$

□ **Exercise 2** Solve.

a. $\frac{3}{8}+\frac{3}{4}=\frac{x}{2}$

b. $\frac{x+1}{5}+\frac{x+2}{2}=4$

EXAMPLE 2 Solve $\frac{2x}{3}-\frac{x}{6}=-3$.

The LCD is 6. Multiply both sides of the equation by 6.

$$\mathbf{6}\left(\frac{2x}{3}-\frac{x}{6}\right)=\mathbf{6}(-3)$$

$$6\left(\frac{2x}{3}\right)-6\left(\frac{x}{6}\right)=6(-3)$$ Use a distributive law; notice that we multiply *each term* by the LCD.

$$4x-x=-18$$ Simplify.

$$3x=-18$$

$$x=-6$$

We check by substituting -6 in the original equation and find that -6 is the solution. ■

Notice that Example 2 is a *linear equation,* so we get the terms with variables on one side of the equation and all other terms on the other side and then solve.

□ **DO EXERCISE 2.**

When we solve rational equations containing a variable in the denominator by the method described above, we must check the answer. Numbers that appear to be solutions are not solutions if they make a denominator in the original equation zero.

> When both sides of an equation are multiplied by an expression containing a variable, we may not get an equivalent equation because the expression containing the variable may be zero. We must check the possible solutions in the given equation. Possible solutions that make a denominator zero in the given equation are called ***extraneous solutions.***

As a first step in solving rational equations we should determine what values of the variable will make any denominator zero.

EXAMPLE 3 Solve.

a. $\frac{-1}{3x}-\frac{3}{x}=\frac{10}{3}$

Notice that x cannot be 0 because this makes both of the denominators, x and $3x$, zero and we would have division by 0. The LCD is $3x$, so multiply both sides of the equation by $3x$.

$$\mathbf{3x}\left(\frac{-1}{3x}-\frac{3}{x}\right)=\mathbf{3x}\left(\frac{10}{3}\right)$$

$$3x\left(\frac{-1}{3x}\right)-3x\left(\frac{3}{x}\right)=3x\left(\frac{10}{3}\right)$$ Use a distributive law.

$$-1-9=10x$$ Simplify.

$$-10=10x$$

$$-1=x$$

Check $\dfrac{-1}{3x} - \dfrac{3}{x} = \dfrac{10}{3}$

$\dfrac{-1}{3(-1)} - \dfrac{3}{-1}$	$\dfrac{10}{3}$
$\dfrac{1}{3} - (-3)$	$\dfrac{10}{3}$
$\dfrac{1}{3} + \dfrac{9}{3}$	$\dfrac{10}{3}$
$\dfrac{10}{3}$	$\dfrac{10}{3}$

The solution is -1.

b. $\dfrac{x+1}{x-3} = \dfrac{4}{x-3} + 6$

Notice that 3 cannot be a solution because it would make the denominators zero. The LCD is $x - 3$, so multiply both sides of the equation by it.

$$(\boldsymbol{x-3})\left(\frac{x+1}{x-3}\right) = (\boldsymbol{x-3})\left(\frac{4}{x-3} + 6\right)$$

$$(x-3)\left(\frac{x+1}{x-3}\right) = (x-3)\left(\frac{4}{x-3}\right) + (x-3)6 \qquad \text{Multiply all terms by } x-3.$$

$$x + 1 = 4 + (x-3)6 \qquad \text{Simplify.}$$

$$x + 1 = 4 + 6x - 18 \qquad \text{Use a distributive law.}$$

$$x + 1 = 6x - 14 \qquad \text{Combine like terms.}$$

$$14 + 1 = 6x - x \qquad \text{Add 14 and } -x \text{ to both sides.}$$

$$15 = 5x$$

$$3 = x$$

Check $\dfrac{x+1}{x-3} = \dfrac{4}{x-3} + 6$

$\dfrac{3+1}{3-3}$	$\dfrac{4}{3-3} + 6$
$\dfrac{4}{0}$	$\dfrac{4}{0} + 6$

The number 3 is not a solution because it makes the denominators in the given equation zero. It is an extraneous solution. Remember that we decided at the beginning of the example that 3 could not be a solution. *There is no solution to this equation.* ■

□ **DO EXERCISE 3.**

Sometimes when we multiply both sides of an equation by the LCD, we get a second-degree equation. Then we set one side of the equation equal to zero and solve. Example 4a illustrates this technique.

□ **Exercise 3** Solve.

a. $\dfrac{3}{x} - \dfrac{7}{10} = \dfrac{8}{5x}$

b. $\dfrac{x-3}{x-4} = \dfrac{1}{x-4}$

□ **Exercise 4** Solve.

a. $2 + \frac{1}{y} = \frac{21}{y^2}$

b. $\frac{2}{x-3} - \frac{3}{x+3} = \frac{12}{x^2-9}$

EXAMPLE 4 Solve.

a. $y + \frac{8}{y} = 6$

Notice that 0 cannot be a solution. The LCD is y, so multiply both sides of the equation by it.

$$\boldsymbol{y}\left(y + \frac{8}{y}\right) = \boldsymbol{y}(6)$$

$$y(y) + y\left(\frac{8}{y}\right) = y(6) \qquad \text{Multiply each term by } y.$$

$$y^2 + 8 = 6y \qquad \text{Simplify.}$$

$$y^2 - 6y + 8 = 0 \qquad \text{Set one side equal to 0.}$$

$$(y-2)(y-4) = 0 \qquad \text{Factor.}$$

$$y - 2 = 0 \quad \text{or} \quad y - 4 = 0 \qquad \text{Use the zero product property.}$$

$$y = 2 \qquad\qquad y = 4$$

These values check. The solutions are 2 and 4. Notice that this equation has two solutions.

b. $\frac{2}{x-2} - \frac{1}{x+2} = \frac{1}{x^2-4}$

The numbers 2 and -2 cannot be solutions. The LCD is $(x - 2)(x + 2)$. We multiply both sides of the equation by $(x - 2)(x + 2)$.

$$\boldsymbol{(x-2)(x+2)}\left(\frac{2}{x-2} - \frac{1}{x+2}\right) = \boldsymbol{(x-2)(x+2)}\left(\frac{1}{x^2-4}\right)$$

$$(x-2)(x+2)\left(\frac{2}{x-2}\right) - (x-2)(x+2)\left(\frac{1}{x+2}\right)$$

$$= (x-2)(x+2)\left(\frac{1}{x^2-4}\right)$$

$$(x+2)2 - (x-2) = 1$$

$$2x + 4 - x + 2 = 1$$

$$x + 6 = 1$$

$$x = -5$$

Check that -5 is the solution. ■

□ **DO EXERCISE 4.**

3 Expressions or Equations?

It is important to understand the difference between adding or subtracting rational expressions and solving a rational equation.

> *To add or subtract rational expressions,* find the least common denominator. Multiply each expression by 1 to give it the LCD. Then add or subtract the expressions, *keeping the denominator.* Simplify, if possible.
>
> *To solve a rational equation,* find the LCD and multiply both sides of the equation by it. *This clears the denominators.* Then solve the equation.

EXAMPLE 5 Add or solve.

a. $\frac{-9}{x} + \frac{3}{4}$

This is *not* an equation. The LCD is $4x$. Multiply each expression by 1 to get the LCD in each denominator. Keep the LCD. Add the two expressions.

$$\frac{-9}{x} + \frac{3}{4} = \frac{-9}{x} \cdot \frac{\mathbf{4}}{\mathbf{4}} + \frac{3}{4} \cdot \frac{\boldsymbol{x}}{\boldsymbol{x}} \qquad \text{Multiply by 1.}$$

$$= \frac{-36}{4x} + \frac{3x}{4x} = \frac{-36 + 3x}{4x} \qquad \text{Add the numerators.}$$

The answer is $(-36 + 3x)/4x$.

b. $\frac{-9}{x} = \frac{3}{4}$

This is an equation. The LCD is $4x$. Multiply each side of the equation by $4x$ to clear denominators. Then solve the equation.

$$\mathbf{4x}\left(\frac{-9}{x}\right) = \mathbf{4x}\left(\frac{3}{4}\right) \qquad \text{Multiply each side by } 4x.$$

$$-36 = 3x \qquad \text{Simplify.}$$

$$-12 = x$$

The number -12 checks. It is the solution. ■

☐ **DO EXERCISE 5.**

☐ **Exercise 5** Add or subtract or solve.

a. $\frac{x-6}{x+2} = \frac{1}{5}$

b. $\frac{x-6}{x+2} - \frac{1}{5}$

DID YOU KNOW?

Our word *algebra* comes from the Arabic word *al-jabr*. *Al-jabr* can be translated as meaning restoration, and the word was used to indicate that terms subtracted from one side of an equation were restored on the other side of the equation. In Spain, *al-jabr* was translated to *algebrista* and meant "bonesetter" or "one who restored bones."

Answers to Exercises

1. a. 0 **b.** $-\frac{1}{2}$ **c.** 6, 1 **d.** None

2. a. $\frac{9}{4}$ **b.** 4

3. a. 2 **b.** No solution

4. a. $-\frac{7}{2}$, 3 **b.** No solution

5. a. 8 **b.** $\frac{4x - 32}{5(x + 2)}$

Problem Set 6.4

NAME ____________________

DATE ____________________

CLASS ____________________

Find the value(s) for which the rational expression is undefined.

1. $\dfrac{x-4}{x}$

2. $\dfrac{2-y}{y}$

3. $\dfrac{7}{3x-5}$

4. $\dfrac{5}{5x+4}$

5. $\dfrac{x-1}{x^2-8x+7}$

6. $\dfrac{y+3}{y^2+4y+3}$

7. $\dfrac{x-1}{x^2+8}$

8. $\dfrac{3}{y^2+4}$

9. $\dfrac{3x}{x+4}$

10. $\dfrac{5x-3}{x+8}$

11. $\dfrac{4y-1}{2y^2-9y+10}$

12. $\dfrac{8y}{6y^2+10y-4}$

Solve.

13. $\dfrac{x}{2}-\dfrac{x}{4}=4$

14. $\dfrac{y}{3}+\dfrac{y}{6}=3$

15. $\dfrac{2}{5}-\dfrac{2}{3}=\dfrac{x}{30}$

16. $\dfrac{5}{8}+\dfrac{1}{3}=\dfrac{z}{12}$

17. $\dfrac{2}{3}-\dfrac{5}{6}=\dfrac{1}{x}$

18. $\dfrac{5}{8}-\dfrac{1}{5}=\dfrac{1}{y}$

19. $\dfrac{x+6}{3}=\dfrac{x+8}{5}$

20. $\dfrac{x+1}{4}=\dfrac{2x+1}{5}$

21. $\dfrac{y+2}{3}+\dfrac{y+1}{5}=3$

22. $\dfrac{x-3}{2}-\dfrac{x-2}{5}=1$

23. $\dfrac{1}{x}+\dfrac{2}{3x}=\dfrac{1}{3}$

24. $\dfrac{2}{z}+\dfrac{3}{2z}=\dfrac{7}{6}$

25. $y+\dfrac{3}{y}=-4$

26. $x-\dfrac{5}{x}=-4$

27. $\dfrac{x-4}{x+2}=\dfrac{1}{4}$

28. $\dfrac{y-3}{y-5}=2$

29. $\dfrac{6}{y+6} = \dfrac{10}{y+8}$

30. $\dfrac{1}{3z-6} = \dfrac{3}{5z+1}$

31. $x + 2 = \dfrac{24}{x}$

32. $y + 8 = -\dfrac{15}{y}$

33. $\dfrac{z+2}{z+1} = \dfrac{1}{z+1} + 2$

34. $\dfrac{x+6}{x+3} = \dfrac{3}{x+3} + 2$

35. $\dfrac{x+1}{x+4} = \dfrac{-3}{x+4}$

36. $\dfrac{y-3}{y-2} = \dfrac{-1}{y-2}$

37. $\dfrac{7}{y} + 2 = \dfrac{4}{y^2}$

38. $\dfrac{13}{x} + 3 = \dfrac{30}{x^2}$

39. $\dfrac{9}{x-2} - \dfrac{8}{x-2} = \dfrac{1}{3}$

40. $\dfrac{3}{y-2} - \dfrac{1}{2} = \dfrac{2}{y-2}$

41. $\dfrac{2}{x-4} - \dfrac{6}{x+4} = \dfrac{8}{x^2-16}$

42. $\dfrac{3}{x-1} - \dfrac{10}{x+1} = \dfrac{-15}{x^2-1}$

43. $\dfrac{5}{y-4} - \dfrac{y+11}{y^2-5y+4} = \dfrac{3}{y-1}$

44. $\dfrac{7}{y-5} - \dfrac{48}{y^2-2y-15} = \dfrac{6}{y+3}$

45. $\dfrac{x}{2x-6} - \dfrac{x-2}{3x-9} = \dfrac{3}{x^2-6x+9}$

46. $\dfrac{2}{x+4} - \dfrac{1}{x-2} = \dfrac{-2x+1}{x^2+2x-8}$

47. $\dfrac{1}{y+4} = \dfrac{y+12}{y^2-16} + \dfrac{3}{y-4}$

48. $\dfrac{z-8}{z-3} - \dfrac{z}{z^2-9} = \dfrac{z+8}{z+3}$

49. $\dfrac{2}{x^2-7x+12} - \dfrac{1}{x^2-9} = \dfrac{4}{x^2-x-12}$

50. $\dfrac{1}{x^2+5x+4} + \dfrac{3}{x^2-1} = \dfrac{-1}{x^2+3x-4}$

Add or subtract or solve.

51. $\dfrac{1}{3} - \dfrac{5}{6} = \dfrac{1}{y}$

52. $\dfrac{x-5}{x+1} = \dfrac{3}{5}$

53. $\dfrac{1}{3} - \dfrac{5}{6} + \dfrac{1}{y}$

54. $\dfrac{x-5}{x+1} - \dfrac{3}{5}$

55. $\dfrac{5}{x+4} = \dfrac{3}{x-2}$

56. $z + \dfrac{6}{z} = -5$

57. $\dfrac{5}{x+4} - \dfrac{3}{x-2}$

58. $z + \dfrac{6}{z} + 5$

59. $\dfrac{5}{2} - \dfrac{1}{p-4} = \dfrac{-2}{2p-8}$

60. $\dfrac{4}{2y-6} - \dfrac{12}{4y+12} = \dfrac{12}{y^2-9}$

61. $\dfrac{5}{2} - \dfrac{1}{p-4} - \dfrac{-2}{2p-8}$

62. $\dfrac{4}{2y-6} - \dfrac{12}{4y+12} + \dfrac{12}{y^2-9}$

63. $\dfrac{22}{2b^2-9b-5} - \dfrac{3}{2b+1} = \dfrac{2}{b-5}$

64. $\dfrac{22}{2b^2-9b-5} - \dfrac{3}{2b+1} - \dfrac{2}{b-5}$

65. A rational equation is an equation with __________ expressions.

66. A rational expression is undefined for any values of the variable that make the denominator __________.

67. Any expression $x^2 + a$, where $a > 0$, is always greater than __________ for all values of x.

68. To solve a rational equation, first multiply both sides of the equation by the __________ __________ __________.

69. The equation $x + 3 = 5$ is a __________ equation.

70. Possible solutions that make a denominator zero in the given equation are called __________.

71. To add or subtract rational expressions, keep the __________.

Solve.

* **72.** $\frac{-16}{3x} = \frac{12}{3x + 1} + \frac{8}{x}$

* **73.** $\frac{8}{5 - a} + \frac{12}{a + 5} = \frac{-80}{a^2 - 25}$

* **74.** $\frac{(x - 2)^2 - (x + 2)(x - 2)}{(x - 2)^4} = 0$

* **75.** $\frac{1}{a^2 + 2a} + \frac{2}{a^2 + 4a + 4} - \frac{3}{a^2 - 2a} = 0$

Checkup

The following problems provide a review of some of Sections 5.5 and 1.3 and will help you with the next section.

Divide. Use long division.

76. $\frac{x^2 + x - 10}{x - 4}$

77. $\frac{x^2 + 8x + 7}{x + 3}$

78. $\frac{2x^2 - 11x - 7}{2x - 5}$

79. $\frac{6z^2 - 7z + 5}{3z - 2}$

80. $\frac{4x^4 + 4x^2 + 5}{2x^2 - 1}$

81. $\frac{3x^3 - 12x^2 - 2x + 9}{3x^2 - 2}$

Find the reciprocal.

82. 9

83. 4

84. $\frac{-7}{8}$

85. $\frac{11}{-5}$

6.5 APPLIED PROBLEMS

Rational equations are used to solve many applied problems. To help you solve these problems review the steps for solving word problems in Section 2.2.

OBJECTIVES

1 *Solve number problems*

2 *Solve work problems*

3 *Solve motion problems*

1 Number Problems

Our first example is a number problem that is included to give practice in working problems with rational equations.

Recall that the *reciprocal of a nonzero real number a is* $\frac{1}{a}$.

EXAMPLE 1 One number is three times another. The sum of their reciprocals is $\frac{2}{3}$. Find the numbers.

Variable Let x represent one number. Then $3x$ represents the other number. The reciprocals of the numbers are $1/x$ and $1/3x$.

Equation $\frac{1}{x} + \frac{1}{3x} = \frac{2}{3}$

Solve $\mathbf{3x}\left(\frac{1}{x} + \frac{1}{3x}\right) = \mathbf{3x}\left(\frac{2}{3}\right)$ Multiply both sides by the LCD of $3x$.

$$3x\left(\frac{1}{x}\right) + 3x\left(\frac{1}{3x}\right) = 3x\left(\frac{2}{3}\right)$$

$$3 + 1 = 2x \quad \text{Simplify.}$$

$$4 = 2x$$

$$2 = x$$

We found that $x = 2$, so $3x = 6$.

Check If we add the reciprocals of 2 and 6, we have $\frac{1}{2} + \frac{1}{6} = \frac{4}{6} = \frac{2}{3}$.

The answers check. The numbers are 2 and 6. ■

□ DO EXERCISE 1.

□ **Exercise 1** If the reciprocal of twice a number is subtracted from 3, the result is $\frac{5}{2}$. Find the number.

2 Work Problems

Rational equations are used to solve work problems.

The amount of work done, W, is the rate of work, r, times the amount of time, t, spent on the job.

$$W = rt$$

□ **Exercise 2** An electrician can complete a job in 12 hours. Another electrician can complete the job in 15 hours. How long would it take the two electricians to complete the job if they work together?

EXAMPLE 2 Andy can assemble an engine in 3 hours. His friend Jim can do the job in 4 hours. How long does it take them working together to assemble the engine?

Notice that it will take them less than 3 hours to do the job since Andy can assemble the engine alone in 3 hours.

Variable Let t represent the time it takes them to do the job together; then $\frac{1}{3}$ job per hour is the rate, r, of Andy since he can do the job in 3 hours and $\frac{1}{4}$ job per hour is the rate, r, of Jim since he can do the job in 4 hours

$$W = rt$$

$$\text{work of Andy} + \text{work of Jim} = \text{one job done together}$$

Equation

$$rt_{\text{Andy}} + rt_{\text{Jim}} = 1$$

$$\frac{1}{3}t + \frac{1}{4}t = 1$$

Solve The LCD is 12. Multiply both sides of the equation by 12.

$$\mathbf{12}\left(\frac{1}{3}t + \frac{1}{4}t\right) = \mathbf{12}(1)$$

$$12\left(\frac{1}{3}t\right) + 12\left(\frac{1}{4}t\right) = 12(1)$$

$$4t + 3t = 12$$

$$7t = 12$$

$$t = \frac{12}{7} \quad \text{or} \quad 1\frac{5}{7}$$

Check In $\frac{12}{7}$ hours, Andy does $\frac{1}{3}(\frac{12}{7}) = \frac{4}{7}$ of the job. Jim does $\frac{1}{4}(\frac{12}{7}) = \frac{3}{7}$ of the job. Together they complete $\frac{4}{7} + \frac{3}{7} = 1$ job. The answer checks.

Working together, it takes Andy and Jim $1\frac{5}{7}$ hours to assemble the engine. ■

□ **DO EXERCISE 2.**

EXAMPLE 3 At Sunsweet Factory, machine A can fill a certain number of cans with juice twice as fast as machine B. When they work together they can fill the cans in 9 hours. Find the time required by each machine working alone to fill the cans.

Variable Let x represent the number of hours that it takes machine A to fill the cans.
Then $2x$ represents the number of hours that it takes machine B to fill the cans.

$$\frac{1}{x} \text{ is the rate } r \text{ of machine A.}$$

$$\frac{1}{2x} \text{ is the rate } r \text{ of machine B.}$$

9 hours is the time that it takes the machines to do the job together.

$$W = rt$$

work of machine A + work of machine B = one job done together

Equation

$$rt_{\text{machine A}} + rt_{\text{machine B}} = 1$$

$$9\left(\frac{1}{x}\right) + 9\left(\frac{1}{2x}\right) = 1$$

$$\frac{9}{x} + \frac{9}{2x} = 1$$

Solve

$$\mathbf{2x}\left(\frac{9}{x} + \frac{9}{2x}\right) = \mathbf{2x}(1) \qquad \text{Multiply by } 2x.$$

$$2x\left(\frac{9}{x}\right) + 2x\left(\frac{9}{2x}\right) = 2x(1)$$

$$18 + 9 = 2x$$

$$27 = 2x$$

$$x = \frac{27}{2} \quad \text{or} \quad 13\frac{1}{2}$$

$$2x = 2\left(\frac{27}{2}\right) = 27$$

Check In 9 hours, machine A does $\frac{9}{13\frac{1}{2}}$ of the job. Machine B does $\frac{9}{27}$ of the job. Together they complete the following.

$$\frac{9}{13\frac{1}{2}} + \frac{9}{27} = \frac{9}{1} \cdot \frac{2}{27} + \frac{9}{27} = \frac{18}{27} + \frac{9}{27} = \frac{27}{27} = 1$$

They complete 1 job.

It takes machine A $13\frac{1}{2}$ hours working alone and machine B 27 hours working alone to fill the cans. ■

□ DO EXERCISE 3.

□ **Exercise 3** Factory A pollutes a river 3 times as fast as factory B. When they are both operating they produce a certain amount of pollution in 11 hours. What is the time it takes each factory to produce that amount of pollution alone?

3 Motion Problems

To work motion problems, we use the following formula:

$$\text{distance} = \text{rate} \times \text{time} \quad \text{or} \quad d = rt$$

We may solve this formula for r or t to get $r = d/t$ or $t = d/r$. Rate is often called "speed."

EXAMPLE 4 Ken drives to Boston 10 miles per hour (mph) faster than Laura. If Ken travels 220 miles in the same time that Laura travels 180 miles, find the speed that Ken drives.

Variable Let r represent the speed that Ken drives. Then $r - 10$ represents the speed that Laura drives.

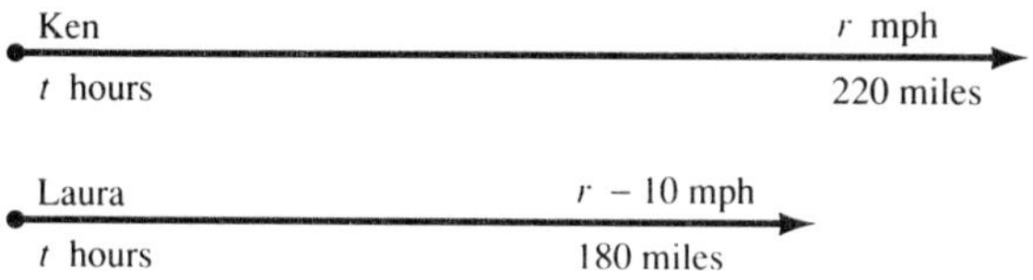

Since the time that they drive is the same, we use the equation $t = d/r$ and make a chart.

	(Distance in miles)	÷	(rate in miles per hour)	=	(time in hours)
Ken	220	÷	r	=	$\frac{220}{r}$
Laura	180	÷	$(r - 10)$	=	$\frac{180}{r - 10}$

Equation Since they travel for the same amount of time, we have the equation

$$\frac{220}{r} = \frac{180}{r - 10}$$

Solve The LCD is $r(r - 10)$. Multiply both sides of the equation by this term.

$$\mathbf{r(r-10)}\frac{220}{r} = \mathbf{r(r-10)}\frac{180}{r-10}$$

$$(r - 10)220 = r(180) \qquad \text{Simplify.}$$

$$220r - 2200 = 180r$$

$$40r = 2200$$

$$r = 55$$

We found Ken's speed to be 55 miles per hour.

Check If Ken's rate is 55 miles per hour, then Laura's rate is 55 − 10 or 45 miles per hour. The time Ken drives is d/r = 220 miles/55 mph = 4 hours and the time Laura drives is d/r = 180 miles/45 mph = 4 hours. They drive for the same time, so the answer checks.

Ken's speed is 55 miles per hour. ■

□ **DO EXERCISE 4.**

□ **Exercise 4** One train travels 5 miles per hour faster than a second train. If the slower train travels 210 miles in the same amount of time as the faster train travels 240 miles, what is the rate of the slower train?

EXAMPLE 5 The speed of a boat in still water is 15 miles per hour. If the boat can travel 1 mile upstream in the same amount of time that it travels 2 miles downstream, find the speed of the current.

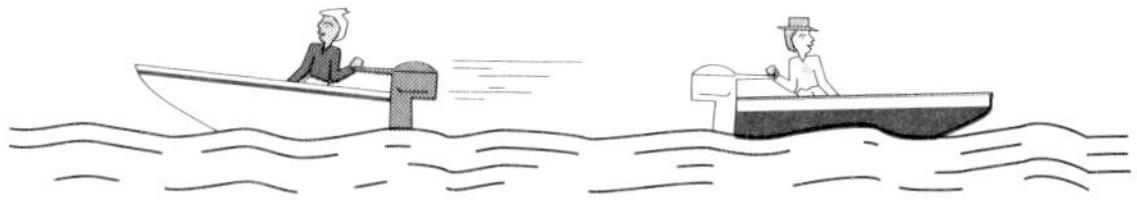

Variable Let x represent the speed of the current. The speed of the current decreases the rate of the boat upstream and increases the rate of the boat downstream. The rate upstream is $(15 - x)$ and the rate downstream is $(15 + x)$.

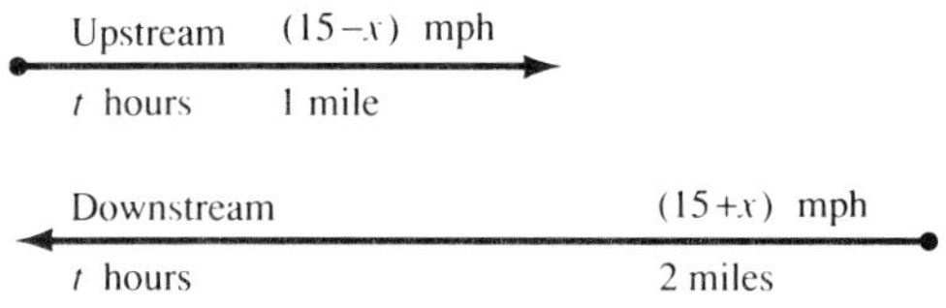

	(Distance in miles)	÷	(rate in miles per hour)	=	(time in hours)
Upstream	1	÷	$(15 - x)$	=	$\dfrac{1}{15 - x}$
Downstream	2	÷	$(15 + x)$	=	$\dfrac{2}{15 + x}$

□ **Exercise 5** A plane can fly 531 miles with the wind in the same amount of time it can fly 369 miles against the wind. The speed of the plane in still air is 200 miles per hour. Find the speed of the wind.

Equation Since the time to go upstream and downstream is the same, we have the following equation.

$$\frac{1}{15 - x} = \frac{2}{15 + x}$$

Solve The LCD is $(15 - x)(15 + x)$. Multiply both sides of the equation by that term.

$$\mathbf{(15 - x)(15 + x)}\left(\frac{1}{15 - x}\right) = \mathbf{(15 - x)(15 + x)}\left(\frac{2}{15 + x}\right)$$

$$15 + x = (15 - x)2$$

$$15 + x = 30 - 2x$$

$$3x = 15$$

$$x = 5$$

The answer checks. The speed of the current is 5 miles per hour. ■

□ **DO EXERCISE 5.**

Answers to Exercises

1. $$3 - \frac{1}{2x} = \frac{5}{2}$$

$$2x(3) - 2x\left(\frac{1}{2x}\right) = 2x\left(\frac{5}{2}\right)$$

$$6x - 1 = 5x$$

$$x = 1$$

The number is 1.

2. $$\frac{1}{12}t + \frac{1}{15}t = 1$$

$$60\left(\frac{1}{12}t\right) + 60\left(\frac{1}{15}t\right) = 60(1)$$

$$5t + 4t = 60$$

$$9t = 60$$

$$t = \frac{60}{9} = \frac{20}{3} \quad \text{or} \quad 6\frac{2}{3}$$

It would take them $6\frac{2}{3}$ hours.

3. $$\frac{11}{x} + \frac{11}{3x} = 1$$

$$3x\left(\frac{11}{x}\right) + 3x\left(\frac{11}{3x}\right) = 3x(1)$$

$$33 + 11 = 3x$$

$$44 = 3x$$

$$x = \frac{44}{3} \quad \text{or} \quad 14\frac{2}{3}$$

$$3x = 3\left(\frac{44}{3}\right) = 44$$

It takes factory A $14\frac{2}{3}$ hours and factory B 44 hours.

4. $$\frac{210}{r} = \frac{240}{r + 5}$$

$$r(r + 5)\frac{210}{r} = r(r + 5)\frac{240}{r + 5}$$

$$(r + 5)210 = r(240)$$

$$210r + 1050 = 240r$$

$$1050 = 30r$$

$$35 = r$$

The rate of the slower train is 35 miles per hour.

5. $$\frac{531}{200 + x} = \frac{369}{200 - x}$$

$$(200 + x)(200 - x)\frac{531}{200 + x} = (200 + x)(200 - x)\frac{369}{200 - x}$$

$$(200 - x)531 = (200 + x)369$$

$$106{,}200 - 531x = 73{,}800 + 369x$$

$$32{,}400 = 900x$$

$$36 = x$$

The speed of the wind is 36 miles per hour.

Problem Set 6.5

NAME

DATE

CLASS

1. When the reciprocal of 4 times a number is subtracted from 2, the result is the reciprocal of the number. Find the number.

2. If 5 is added to the reciprocal of 3 times a number, the result is the reciprocal of the number. Find the number.

3. What number must be added to both the numerator and denominator of $\frac{7}{9}$ to make the result equal $\frac{5}{6}$?

4. What number must be added to the numerator of $\frac{11}{15}$ to make the result equal to $\frac{7}{5}$?

5. If 14 is added to the reciprocal of a number, the result is the reciprocal of twice the number. What is the number?

6. One number is 3 times another number. The sum of their reciprocals is $\frac{1}{3}$. Find the numbers.

7. Juan can correct a stack of papers in 5 hours and Maria can correct them in 4 hours. How long does it take them to correct the papers together?

8. Karen can clean the house in 2 hours and Mark can clean it in 3 hours. How long will it take them to clean the house together?

9. A painter can paint a room in 4 hours and another painter can paint the room in 3 hours. How long does it take them to paint the room together?

10. A cold-water pipe can fill a swimming pool in 10 hours and a hot-water pipe can fill the pool in 15 hours. How long will it take to fill the pool if both pipes are left open?

11. Jeff can mow the lawn in $3\frac{1}{2}$ hours. Julie can mow the lawn in $4\frac{1}{4}$ hours. How long will it take them to mow the lawn if they work together?

12. Keith can wash and wax the car in $1\frac{1}{2}$ hours and Kevin can do the same job in 2 hours. How long will it take them to do the job together?

13. An inlet pipe can fill a barrel of vinegar in 8 hours and an outlet pipe can empty it in 12 hours. How long will it take to fill the barrel if both pipes are left open?

14. A swimming pool can be filled by an inlet pipe in 12 hours and emptied by an outlet pipe in 16 hours. How long will it take to fill the pool if both pipes are left open?

15. An electrician and her apprentice can wire a house in 8 hours. The electrician works twice as fast as her apprentice. How long would it take each person working alone to wire the house?

16. A carpenter works 3 times as fast as his apprentice. If they work together, they can construct a wall in 12 hours. How long would it take each person working alone to construct the wall?

17. Susan drives 20 miles per hour faster than Marilyn. If Susan travels 225 miles in the same time that Marilyn travels 125 miles, find the speed that Marilyn drives.

18. One train travels 50 kilometers per hour faster than a second train. If the slower train travels 225 kilometers in the same amount of time as the faster train travels 375 kilometers, what is the rate of the faster train?

19. The rate of a jet plane was 5 times the rate of a helicopter. If the helicopter travels 72 miles in 1 hour less time than the jet travels 810 miles, find the rate of the jet.

20. A car traveling to Denver travels twice as fast on the freeway as it does on a scenic route. If it travels 120 miles on the scenic route in 1 hour less time than it travels 300 miles on the freeway, find the speed of the car on the freeway.

21. Jenny drove her car 20 miles to the train station and then took the train 40 miles to work. If the train traveled 3 times as fast as her car and the total time of the trip was $1\frac{1}{9}$ hours, what was the speed of the train?

22. Ben drove 80 miles before running out of gas and walking 5 miles to the service station. He drove 12 times faster than he walked. The total time that he drove and walked was $3\frac{1}{4}$ hours. Find the rate at which he walked.

23. Wendy averages 8 miles per hour riding her bicycle to school. Returning home she averages 12 miles per hour. If she returns home in $\frac{1}{4}$ hour less time, how far is it from her home to school?

24. Dale traveled 120 miles in one direction. He made the return trip at double the speed and in 2 hours less time. Find his speed going to his destination.

25. A boat goes 12 miles per hour in still water. It can go 6 miles upstream in the same time as it can go 10 miles downstream. Find the speed of the current.

26. Tricia can fly her plane 150 miles with the wind in the same time that it takes her to fly 100 miles against the wind. Find the speed of her plane in still air if the speed of the wind is 30 miles per hour.

27. The speed of a boat in still water is 18 miles per hour. If it travels 7 miles downstream in the same amount of time that it travels 5 miles upstream, find the speed of the current.

28. A stream has a current of 2 miles per hour. Find the speed of John's boat if it goes 11 miles downstream in the same time as it goes 8 miles upstream.

29. A canal has a current of 3 miles per hour. A houseboat travels 22 miles downstream in the same amount of time it takes to go 16 miles upstream. What is the speed of the boat in still water?

30. John can fly his plane 123 miles with the wind in the same amount of time that he flies 87 miles against the wind. The speed of the plane in calm air is 175 miles per hour. Find the speed of the wind.

31. Mr. and Mrs. Ming can complete a job in 5 hours. If Mr. Ming works twice as fast as Mrs. Ming, how long does it take Mrs. Ming to complete the job alone?

32. Susan and John can complete a job in 3 hours. If Susan works 3 times as fast as John, how long would it take Susan to complete the job alone?

33. One person can clean a house in 5 hours. If a second person helps clean the house, the job can be done in 2 hours. How long would it take the second person to clean the house?

34. Binh can mow the lawn in 3 hours. If Binh and Mai work together, the job can be done in 2 hours. How long would it take Mai to mow the lawn alone?

35. For work problems, the amount of work done is the ____________ ____________ ____________ times the amount of time spent on the job.

36. For motion problems, the distance equals the ____________ times the ____________.

Think About It

* 37. Cheryl travels 10 miles per hour faster on her bicycle than her friend Janet jogs. Cheryl travels 20 miles in half as much time as Janet. What is Cheryl's speed?

* 38. Deepti is 5 years younger than her sister, and 8 years from now she will be four-fifths as old as her sister. How old is Deepti?

* 39. Joel rides his bicycle 3 miles to the bus stop at a rate of 8 miles per hour. He rides the bus, which travels 25 miles per hour, to work. If his total travel time is $1\frac{1}{2}$ hours, how far does he ride the bus?

* 40. The width of a rectangle is $\frac{1}{4}$ the length. If the perimeter is $\frac{5}{4}$ the length, plus 60, find the dimensions of the rectangle.

* 41. A boat has enough gasoline for 11 hours of travel. Its speed is 14 kilometers per hour. It travels upstream against a current of 10 kilometers per hour and returns downstream with the same current. How far can it travel?

* 42. Karen met her friend for lunch at a restaurant 15 miles from her home. She was gone from her home 2 hours and she spent 1 hour and 20 minutes at the restaurant. How fast did she drive going to and from the restaurant?

Checkup

The following problems provide a review of some of Section 2.4 and will help you with the next section.

Solve.

43. $E = IR$ for I

44. $PV = KT$ for V

45. $PS + PF = S$ for F

46. $d + dni = i$ for n

47. $PRT = A - P$ for A

48. $IR + InR = nE$ for E

49. $5x - 2y = 6$ for y

50. $7x - 8y = 12$ for y

51. $Q = \frac{1}{5}xy$ for x

52. $R = \frac{3}{4}ab$ for b

6.6 FORMULAS

1 Evaluating Formulas

Many formulas contain rational expressions. We studied some of these formulas in Section 2.4, but now that we know how to factor, we can solve more formulas for variables. First we learn how to evaluate a formula containing rational expressions for a variable.

EXAMPLE 1 The following diagram shows a section of an electronic circuit with two resistors, R_1 and R_2, connected in parallel. (R_1 and R_2 are symbols for different variables.)

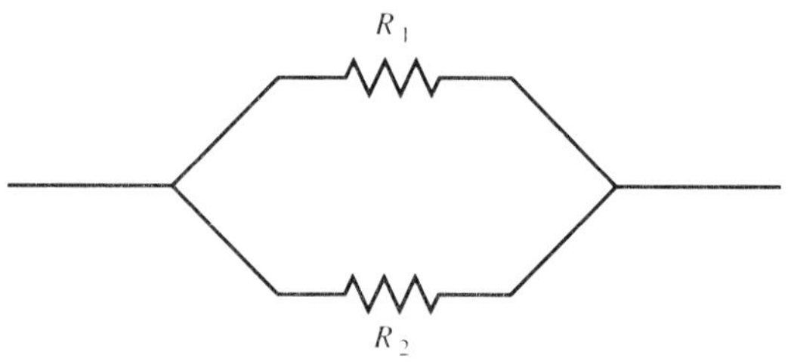

The equivalent resistance R of the two resistors R_1 and R_2 connected in parallel is given by the formula

$$\frac{1}{R} = \frac{1}{R_1} + \frac{1}{R_2}$$

Find R if $R_1 = 4$ and $R_2 = 5$.

In the formula, substitute 4 for R_1 and 5 for R_2.

$$\frac{1}{R} = \frac{1}{R_1} + \frac{1}{R_2}$$

$$\frac{1}{R} = \frac{1}{4} + \frac{1}{5} \qquad \text{Substitute } R_1 = 4 \text{ and } R_2 = 5.$$

Multiply both sides by the LCD, which is $20R$.

$$\mathbf{20R}\left(\frac{1}{R}\right) = \mathbf{20R}\left(\frac{1}{4} + \frac{1}{5}\right)$$

$$20R\left(\frac{1}{R}\right) = 20R\left(\frac{1}{4}\right) + 20R\left(\frac{1}{5}\right) \qquad \text{Use a distributive law.}$$

$$20 = 5R + 4R \qquad \text{Simplify.}$$

$$20 = 9R$$

$$\frac{20}{9} = R \quad ■$$

□ **DO EXERCISE 1.**

OBJECTIVES

1 *Evaluate formulas containing rational expressions for given values of the variables*

2 *Solve formulas with rational expressions for a specific variable*

□ **Exercise 1** Solve the formula $1/R = 1/R_1 + 1/R_2$ for the missing variable.

a. $R_1 = 6$ and $R_2 = 8$

b. $R = 10$ and $R_2 = 15$

□ **Exercise 2**

a. Solve $\frac{1}{R} = \frac{1}{R_1} + \frac{1}{R_2}$ for R_2.

b. Solve $R = \frac{gs}{g + s}$ for s.

2 Solving for a Specific Variable

In many cases we want to solve the formula for a specific variable. We do this in the same way that we have previously solved equations.

EXAMPLE 2 Solve $\frac{1}{R} = \frac{1}{R_1} + \frac{1}{R_2}$ for R_1.

Multiply both sides of the equation by the LCD, RR_1R_2.

$$\boldsymbol{RR_1R_2}\left(\frac{1}{R}\right) = \boldsymbol{RR_1R_2}\left(\frac{1}{R_1} + \frac{1}{R_2}\right)$$

$$RR_1R_2\left(\frac{1}{R}\right) = RR_1R_2\left(\frac{1}{R_1}\right) + RR_1R_2\left(\frac{1}{R_2}\right) \qquad \text{Use a distributive law.}$$

$$R_1R_2 = RR_2 + RR_1 \qquad \text{Simplify.}$$

To solve for R_1, we must get all the terms containing R_1 on one side of the equation.

$$R_1R_2 - RR_1 = RR_2 \qquad \text{Add } -RR_1 \text{ to each side.}$$

$$R_1(R_2 - R) = RR_2 \qquad \text{Factor out } R_1.$$

$$R_1 = \frac{RR_2}{R_2 - R} \qquad \text{Divide both sides by } R_2 - R. \quad \blacksquare$$

□ **DO EXERCISE 2.**

> To solve a formula containing a rational expression for a specific variable:
>
> 1. Multiply on both sides of the equation to clear fractions.
> 2. If necessary, multiply to remove parentheses.
> 3. Get all terms with the specific variable we are solving for alone on one side of the equation by using the addition property.
> 4. Factor out the specific variable, if necessary.
> 5. Isolate the specific variable that we are solving for on one side of the equation by using the multiplication property. Remember that the multiplication property allows us to divide both sides of an equation by the same number.

EXAMPLE 3 Solve.

a. $S = \frac{a - rL}{1 - r}$ for r.

$$(\mathbf{1 - r})S = (\mathbf{1 - r})\frac{a - rL}{1 - r} \qquad \text{Multiply both sides by } 1 - r.$$

$$S - rS = a - rL$$

$$S - a = rS - rL \qquad \text{Add } -a \text{ and } rS \text{ to both sides.}$$

$$S - a = r(S - L) \qquad \text{Factor out the unknown variable.}$$

$$\frac{S - a}{S - L} = r \qquad \text{Divide both sides by } S - L.$$

Caution: Notice that r is alone on one side of the equation and r must not appear on the other side of the equation if the formula is solved for r.

b. $F = \frac{Gm_1m_2}{d^2}$ for G

$$d^2F = d^2\frac{Gm_1m_2}{d^2} \qquad \text{Multiply both sides by } d^2.$$

$$d^2F = Gm_1m_2$$

$$\frac{d^2F}{m_1m_2} = G \qquad \text{Divide both sides by } m_1m_2.$$

This is a physics formula for the force between two masses. ■

☐ **DO EXERCISE 3.**

DID YOU KNOW?

The famous astronomer and mathematician Galileo invented the thermometer. He also built a telescope after he heard rumors of a new invention that could be used in astronomy. He discovered four moons around Jupiter using his new telescope.

☐ **Exercise 3** Solve for x.

a. $7y = \frac{5x}{x + 2}$

b. $-4a = \frac{-3 + x}{x - 2}$

Answers to Exercises

1. a. $\dfrac{24}{7}$ b. 30

2. a. $R_2 = \dfrac{RR_1}{R_1 - R}$ b. $\dfrac{Rg}{g - R} = s$

3. a. $x = \dfrac{-14y}{7y - 5}$ or $x = \dfrac{14y}{5 - 7y}$ b. $\dfrac{8a + 3}{1 + 4a} = x$

Problem Set 6.6

NAME ____________________

DATE ____________________

CLASS ____________________

Solve the following formulas for the variable whose value is not given.

1. $r = \frac{d}{t}$ (distance formula)
when $r = 40$ and $d = 170$

2. $\frac{P}{B} = R$ (percentage formula)
when $P = 12$ and $B = 100$

3. $\frac{1}{f} = \frac{1}{d_1} + \frac{1}{d_2}$ (focal length of a lens)
when $d_1 = 8$ and $d_2 = 12$

4. $\frac{P_1V_1}{T_1} = \frac{P_2V_2}{T_2}$ (gas law)
when $P_1 = 60$, $T_1 = 200$, $P_2 = 50$, $V_2 = 8$, and $T_2 = 250$

Solve.

5. $I = \frac{E}{R}$ for R

6. $V = \frac{KT}{P}$ for P

7. $A = \frac{h}{2}(b + c)$ for c

8. $A = \frac{h}{2}(b + c)$ for b

9. $\frac{1}{f} = \frac{1}{d_1} + \frac{1}{d_2}$ for d_1

10. $\frac{1}{f} = \frac{1}{d_1} + \frac{1}{d_2}$ for f

11. $P = \frac{S}{S + F}$ for S

12. $d = \frac{i}{1 + ni}$ for i

13. $\frac{1}{R} = \frac{1}{R_1} + \frac{1}{R_2}$ for R

14. $\frac{1}{f} = \frac{1}{d_1} + \frac{1}{d_2}$ for d_2

15. $A = \frac{p}{1 - rn}$ for n

16. $P = \frac{S}{S + F}$ for F

17. $T = \frac{A - P}{PR}$ for P

18. $I = \frac{nE}{R + nR}$ for R

19. $\frac{P_1V_1}{T_1} = \frac{P_2V_2}{T_2}$ for T_1

20. $\frac{P_1V_1}{T_1} = \frac{P_2V_2}{T_2}$ for T_2

21. $\frac{W_1}{W_2} = \frac{d_1}{d_2}$ for d_2

22. $\frac{W_1}{W_2} = \frac{d_1}{d_2}$ for W_2

23. $R = \frac{gs}{g + s}$ for g

24. $I = \frac{2V}{V + 2r}$ for V

25. $\frac{E}{e} = \frac{R + r}{r}$ for e

26. $\frac{E}{e} = \frac{R + r}{r}$ for r

27. $\frac{x}{6} + \frac{y}{8} = 1$ for y

28. $\frac{x}{7} - \frac{y}{5} = 1$ for y

29. $d = \frac{L - f}{n - 1}$ for f

30. $r = \frac{A - p}{Pt}$ for A

31. $g = \frac{2s - 2vt - 2x}{t^2}$ for v

32. $m = \frac{2E - 2P}{v^2}$ for P

33. In Problem 6 we solved the formula $V = (KT)/P$ for P. Use this new formula to find the pressure of a gas when the volume V is 100 liters, the temperature T is 30 kelvin, and K is 10.

34. In Problem 17 we solved the formula $T = (A - P)/PR$ for P. Use this new formula to find the principal borrowed if the amount A paid back was \$5800 for a loan for a time T of 5 years at an interest rate R of 9% per year.

35. To solve a formula for a variable, use the techniques for solving ____________.

36. If a formula contains rational expressions, the first step in solving it for a specific variable is to ____________ on both sides to clear fractions.

Solve the following formulas for the given variable. Round answers to the nearest tenth.

37. $r = \frac{d}{t}$ for t when $r = 63.4$ and $d = 327.8$

38. $\frac{P}{B} = R$ for P when $B = 100$ and $R = 0.085$

39. $\frac{1}{f} = \frac{1}{d_1} + \frac{1}{d_2}$ for f when $d_1 = 9.72$ and $d_2 = 14.4$

40. $\frac{P_1V_1}{T_1} = \frac{P_2V_2}{T_2}$ for V_1 when $P_1 = 70.8$, $T_1 = 180.3$, $P_2 = 47.3$, $V_2 = 9.8$, and $T_2 = 260.3$

Solve.

* **41.** $pv = k\left(1 + \frac{t}{m}\right)$ for t

* **42.** $F = f\left(\frac{v + v_0}{v - v_s}\right)$ for v

* **43.** $S = \frac{n}{2}[2a + (n - 1)d]$ for d

Think About It

* **44.** The formula $p = \frac{S}{S + F}$ gives the probability of the success of an event. Show that the formula $p = F^{-1}(S^{-1} + F^{-1})^{-1}$ is equivalent to the preceding formula.

Checkup

The following problems provide a review of some of Section 5.10.

Solve.

45. $4x^2 - 9x + 2 = 0$

46. $6x^2 + 7x - 3 = 0$

47. $x^2 + 5x = 0$

48. $x^2 - 9x = 0$

49. $x(x + 2) = 15$

50. $x(x + 5) = -6$

51. The area of a rectangular garden is 135 square feet. If the length is 6 feet more than the width, find the length of the garden.

52. The width of a rectangle is 6 centimeters less than the length. The area is 216 square centimeters. Find the width of the rectangle.

53. The sum of two numbers is 3. The sum of their squares is 89. Find the numbers.

54. The sum of two numbers is 8. The sum of their squares is 160. Find the numbers.

6.7 VARIATION

There are many examples of variation in the real world. For example, as the number of hours we work increases, our wages increase. This is an example of ***direct variation.***

1 Direct Variation

The ***ratio*** of two quantities is their quotient. For example, the ratio of 3 oranges to 5 oranges is $\frac{3}{5}$. A ***proportion*** is formed by setting two ratios equal to each other. The following are examples of proportions.

$$\frac{3}{4}=\frac{9}{12} \qquad \frac{14}{x}=\frac{2}{1} \qquad \frac{y}{8}=\frac{1}{4}$$

If quantities are directly proportional, then when one quantity *increases* from x_1 to x_2 the other quantity *increases* from y_1 to y_2, or when one quantity *decreases* from x_1 to x_2 the other quantity *decreases* from y_1 to y_2. This may be stated in symbols as

$$\frac{x_1}{x_2}=\frac{y_1}{y_2}$$

If quantities are directly proportional, this is called ***direct** variation.*

> y varies directly as x if there is some positive constant of variation, k, such that
>
> $$y = kx$$
>
> The number k is called the ***variation constant,*** or the ***constant of proportionality.***

There are two methods for solving direct variation problems. Method 1 is often used in arithmetic and introductory algebra. Be sure to learn *Method 2,* since it is used in more *advanced mathematics.*

EXAMPLE 1 The number of servings that can be obtained from a roast varies directly with the weight of the roast. If a 5-pound roast yields 11 servings, how many servings can be made from a 3-pound roast?

Method 1: To solve the problem, we write a proportion. Let y_2 represent the number of servings that can be obtained from a 3-pound roast. The ratios are formed by comparing like quantities. We compare pounds to pounds and servings to servings. If we let $x_1 = 5$ pounds, $y_1 = 11$ servings, and $x_2 = 3$ pounds, the equation is as follows.

OBJECTIVES

1 *Solve problems involving direct variation*

2 *Solve problems about inverse variation*

3 *Solve problems involving joint variation*

□ **Exercise 1**

a. If Mark works 70 hours, he earns \$280. How much would he earn if he worked 100 hours?

b. Hooke's law states that the force required to stretch a spring x inches beyond its natural length varies directly with x. If a 120-pound force stretches a spring 6 inches, how much force is required to stretch it 2 inches?

$$\frac{x_1}{x_2} = \frac{y_1}{y_2}$$

$$\frac{5 \text{ pounds}}{3 \text{ pounds}} = \frac{11 \text{ servings}}{y_2 \text{ servings}}$$

Solve the proportion.

$$\frac{5}{3} = \frac{11}{y_2}$$

Multiply both sides of the equation by the LCD, which is $3y_2$.

$$\mathbf{3y_2}\left(\frac{5}{3}\right) = \mathbf{3y_2}\left(\frac{11}{y_2}\right)$$

$$5y_2 = 33 \qquad \text{Simplify.}$$

$$y_2 = \frac{33}{5} \quad \text{or} \quad 6\frac{3}{5}$$

A 3-pound roast will make $6\frac{3}{5}$ servings.

Caution: The equation

$$\frac{5}{11} = \frac{3}{y_2}$$

also works when quantities are directly proportional. However, the type of comparison in Example 1 is necessary to work correctly the proportions shown in Example 3.

Method 2: We use the equation for direct variation, $y = kx$ and the given values of x and y to find the constant of variation. Then we write an equation of variation and use it to solve the problem. If y is the number of servings and x is the number of pounds of roast, then

$$y = kx \qquad \text{General equation.}$$

We know that a 5-pound roast will serve 11 people. Therefore, $x = 5$ and $y = 11$. We can find the constant of variation, k.

$$11 = k(5)$$

$$\frac{11}{5} = k$$

$$y = kx \qquad \text{General equation.}$$

$$y = \frac{11}{5}x$$ The equation of variation is found by substituting $k = \frac{11}{5}$ into the general equation.

For a 3-pound roast, $x = 3$. Find y, the number of servings.

$$y = \frac{11}{5}(\mathbf{3}) \qquad \text{Substitute 3 for } x.$$

$$y = \frac{33}{5} \quad \text{or} \quad 6\frac{3}{5}$$

A 3-pound roast yields $6\frac{3}{5}$ servings. ■

□ **DO EXERCISE 1.**

One variable may be proportional to the power of another variable.

> y varies directly as the nth power of x if there is some positive constant k such that $y = kx^n$.

☐ **Exercise 2** y varies directly as the cube of x and $y = 32$ when $x = 2$. Find y when $x = 3$.

EXAMPLE 2 The distance an object has fallen from rest toward the earth varies directly with the square of the time it has been falling. If an object falls 64 feet in 2 seconds, how far will it fall in 4 seconds?

If d is the distance and t is the time the object has been falling, then

$d = kt^2$ General equation.

We know that the object falls 64 feet in 2 seconds, so $d = 64$ and $t = 2$.

$64 = k(2)^2$

$64 = 4k$

$16 = k$

$d = kt^2$ General equation.

$d = 16t^2$ Substitute 16 for k in the general equation to get the equation of variation.

If the object falls 4 seconds, $t = 4$.

$d = 16(4)^2$ Substitute 4 for t.

$d = 256$

The object will fall 256 feet in 4 seconds. ■

☐ **DO EXERCISE 2.**

2 Inverse Variation

In ***inverse variation*** one quantity, x_1, *increases* to x_2 as the other quantity y_1 *decreases* to y_2. This may be shown as a proportion, as follows.

$$\frac{x_1}{x_2} = \frac{y_2}{y_1}$$

The quantities are inversely proportional. This may also be shown as follows:

> *y varies inversely as x* if there is a positive constant of variation k such that
>
> $$y = \frac{k}{x}$$
>
> Also, y varies inversely as the nth power of x if there is a positive constant of variation k such that
>
> $$y = \frac{k}{x^n}$$

There are also two methods for solving inverse variation problems. Be sure to learn *Method 2* since it is used in more *advanced mathematics*.

□ **Exercise 3**

a. y varies inversely as x and $y = 3$ when $x = 5$. Find y when $x = 3$.

b. p varies inversely as the cube of q. If p is 3 when q is 2, find p when q is 3.

EXAMPLE 3 y varies inversely as x. If y is 8 when x is 5, find y when x is 20.

Method 1: Write a proportion and solve it. Let $x_1 = 5$, $y_1 = 8$, and $x_2 = 20$. Find y_2. Since y varies inversely as x, the proportion is as follows:

$$\frac{x_1}{x_2} = \frac{y_2}{y_1}$$

$$\frac{5}{20} = \frac{y_2}{8}$$

$$\mathbf{40}\left(\frac{5}{20}\right) = \mathbf{40}\left(\frac{y_2}{8}\right) \qquad \text{Multiply both sides by the LCD of 40.}$$

$$10 = 5y_2 \qquad \text{Simplify.}$$

$$2 = y_2$$

Method 2: Write the general equation. Find the constant of variation using the given values of x and y. Then find the equation of variation and use it to solve the problem. Since y varies inversely as x, we use the equation $y = k/x$.

We are also given that $y = 8$ when $x = 5$.

$$y = \frac{k}{x} \qquad \text{General equation.}$$

$$8 = \frac{k}{5} \qquad \text{Substitute } y = 8 \text{ and } x = 5.$$

$$40 = k \qquad \text{Solve for } k\text{, the constant of variation.}$$

$$y = \frac{40}{x} \qquad \text{The equation of variation is found by substituting 40 into the general equation.}$$

When x is 20,

$$y = \frac{40}{20} \qquad \text{Substitute 20 for } x.$$

$$y = 2$$ ■

□ **DO EXERCISE 3.**

3 Joint Variation

If one variable varies as the product of other variables, we say that the first variable varies jointly as the others.

> y varies jointly as x and z if there is a positive constant of variation, k, such that
>
> $$y = kxz$$

EXAMPLE 4 Suppose that y varies jointly as x and z and $y = 70$ when $x = 7$ and $z = 5$. Find y when $x = 4$ and $z = 8$.

$$y = kxz \quad \text{General equation.}$$

We are given that $y = 70$ when $x = 7$ and $z = 5$. So

$$70 = k(7)(5)$$

$$2 = k$$

$$y = 2xz \quad \text{Substitute 2 for } k \text{ in the general equation to obtain the equation of variation.}$$

When $x = 4$ and $z = 8$,

$$y = 2(4)(8)$$

$$y = 64$$ ■

□ **DO EXERCISE 4.**

There are many combinations of direct and inverse variations.

EXAMPLE 5 The volume of a gas varies directly with its temperature and inversely with the pressure. If the volume of a gas is 10 cubic feet at a temperature of 100 kelvin and a pressure of 20 pounds per square inch, what is the volume of the gas at 450 kelvin when the pressure is 30 pounds per square inch?

We will use the equation for variation and the given values of the variables to solve for the constant of variation. Then we write the equation of variation and use it to solve the example.

Since the volume varies directly with temperature and inversely with pressure, we have the following equation:

$$V = \frac{kT}{P}$$

We are given that $V = 10$ when $T = 100$ and $P = 20$. Substitute these values in the equation.

$$10 = \frac{k(100)}{20}$$

$$\mathbf{20}(10) = \mathbf{20}\left(\frac{k(100)}{20}\right) \quad \text{Multiply by the LCD of 20.}$$

$$200 = 100k \quad \text{Simplify.}$$

$$2 = k \quad \text{Solve for the constant of variation.}$$

$$V = \frac{2T}{P} \quad \text{Equation of variation.}$$

For $T = 450$ and $P = 30$,

$$V = \frac{2(450)}{30}$$

$$V = 30$$

The volume is 30 cubic feet. ■

A calculator will often be helpful for doing variation problems.

□ **DO EXERCISE 5.**

□ **Exercise 4** If y varies jointly as x and the square of z, and y is 54 when x and z are 3, find y when $x = 2$ and $z = 4$.

□ **Exercise 5** If z varies jointly as x and the square of y and inversely as w and $z = \frac{3}{8}$ when $x = 2$, $y = 3$, and $w = 12$, find z when $x = 7$, $y = 2$, and $w = 10$.

Answers to Exercises

1. a. $400 b. 40 lb
2. 108
3. a. 5 b. $\frac{8}{9}$
4. 64
5. $\frac{7}{10}$

Problem Set 6.7

NAME

DATE

CLASS

Determine whether the equation represents direct, inverse, or joint variation.

1. $y = 5x$

2. $d = 7t^2$

3. $V = 9wh$

4. $y = 4xz$

5. $y = \frac{9}{x^3}$

6. $t = \frac{15}{r}$

Solve.

7. y varies directly as x and $x = 5$ when $y = 7$. Find y when $x = 3$.

8. y varies directly as x and $x = 9$ when $y = 2$. Find y when $x = 12$.

9. y varies directly as the square of x and $x = 2$ when $y = 6$. Find y when $x = 4$.

10. y varies directly as the cube of x and $x = 3$ when $y = 9$. Find y when $x = 2$.

11. y varies inversely as x and $x = 5$ when $y = 2$. Find y when $x = 4$.

12. p varies inversely as the square of q and $q = 5$ when $p = \frac{1}{5}$. Find p when $q = 3$.

13. y varies jointly as x and z^2 and $y = 100$ when $x = 4$ and $z = 5$. Find y when $x = 7$ and $z = 3$.

14. a varies jointly as b and c and inversely as d. If $a = 28$ when $b = 8$, $c = 7$, and $d = 4$, find a when $b = 3$, $c = 6$, and $d = 18$.

15. If Maria works 60 hours, she earns $195. How much does she earn if she works 35 hours? (This is an example of direct variation.)

16. The distance an object has fallen from rest varies directly as the square of the time it has been falling. If an object is dropped from a building 576 feet high and it hits the ground in 6 seconds, how far did it fall in the first 3 seconds?

17. The current in an electrical conductor varies inversely as the resistance of the conductor. If the current is $\frac{1}{4}$ ampere when the resistance is 120 ohms, what is the current when the resistance is 320 ohms?

18. The amount of time it takes a car to travel a certain distance is inversely proportional to the speed at which it travels. If a car travels a certain distance in 20 minutes at 40 miles per hour, how many minutes does it take to travel the same distance at 50 miles per hour?

19. The pressure exerted by a liquid at a given point varies directly as the depth of the point beneath the surface of the liquid. If a liquid exerts a pressure of 600 pounds per square foot at a depth of 10 feet, what is the pressure at 30 feet?

20. The length of a rectangle of a given area varies inversely with the width. What is the width of a rectangle of length 8 centimeters if it has the same area as a rectangle with length 7.5 centimeters and width 3 centimeters?

21. The resistance of a wire varies directly as its length and inversely with the square of its diameter. If 10 feet of wire with a diameter of 0.01 inch has a resistance of 1 ohm, what is the resistance of 30 feet of the same wire with a diameter of 0.02 inch?

22. The illumination produced by a light source varies inversely as the square of the distance from the light source. If 12.5 footcandles of illumination is produced at a distance of 2 meters from the source, find the illumination 7 meters from the source.

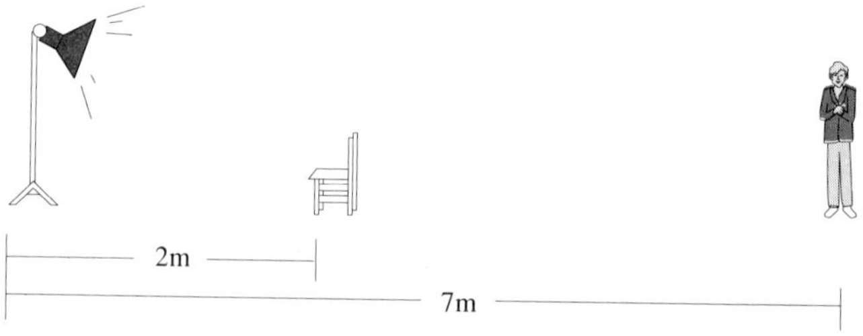

23. The strength of a rectangular beam varies jointly as its width and the square of its depth. If the strength of a beam 1 inch wide and 10 inches deep is 500 pounds, find the strength of a beam 2 inches wide and 4 inches deep.

24. The volume of a right circular cylinder varies jointly with the height and the square of the radius of its base. If the volume is 150 cubic meters when the height is 5.31 meters and the radius of the base is 3 meters, find the volume of a cylinder with radius 4 meters and height 12 meters.

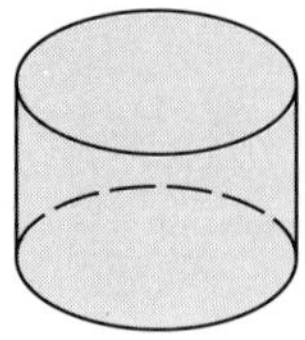

25. The number of kilograms of water in a human body varies directly as the total weight. A person weighing 84 kilograms contains 56 kilograms of water. How many kilograms of water does a person weighing 96 kilograms contain?

26. The weight of an object on Mars varies directly as its weight on Earth. A person weighing 155 pounds on Earth weighs 62 pounds on Mars. How much would a 125-pound person weigh on Mars?

27. The weight that a horizontal beam can support varies inversely as the length of the beam. If a 12-meter beam can support 800 kilograms, how many kilograms can an 8-meter beam support?

28. The wavelength of a radio wave varies inversely as its frequency. A wave with a frequency of 900 kilohertz has a length of 400 meters. What is the length of a wave with a frequency of 1400 kilohertz? Round your answer to the nearest tenth.

29. Two electrons repel each other with a force inversely proportional to the square of the distance between them. When the electrons are 5×10^{-8} meter apart, they repel each other with a force of 10^{-2} unit. With what force do they repel each other when they are 2.5×10^{-10} meter apart?

30. The stopping distance of a car varies directly as the square of the speed. If a car traveling 30 miles per hour can stop in 50 feet, how many feet will it take to stop a car traveling 90 miles per hour?

31. At Junflower Movers, the cost of shipping goods varies jointly as the distance shipped and the weight. If it costs \$351 to ship 4.5 tons of goods 12 miles, how much will it cost to ship 18 tons of goods 1234 miles?

32. The force of a wind blowing on a vertical surface varies jointly as the area of the surface and the square of the velocity. If a wind of 80 miles per hour exerts a force of 200 pounds on a surface of $\frac{1}{2}$ square foot, how much force will a wind of 40 miles per hour place on a surface of 2 square feet?

33. The ratio of two quantities is their ____________.

34. A ____________ is formed by setting two ratios equal to each other.

35. If there is some positive constant of variation k, such that $y = kx$, we say that y varies ____________ as x.

36. y varies ____________ as x if there is a positive constant of variation k, such that $y = k/x$.

37. If one variable varies as the product of other variables, we say that the first variable varies ____________ as the others.

Solve. Round your answers to the nearest hundredth.

38. y varies directly as x and $x = 3.65$ when $y = 7.38$. Find y when $x = 8.24$.

39. p varies inversely as q and $p = 9.24$ when $q = 6.29$. Find q when $p = 25.73$.

40. The illumination produced by a light source varies inversely as the square of the distance from the light source. If 8 footcandles of illumination is produced at a distance of 2.5 meters from the source, find the illumination 4.38 meters from the source.

41. The force F of the wind on a vertical surface varies jointly as the area A of the surface and the square of the wind velocity V. If the wind is blowing 20 miles per hour, the force on 1 square foot of surface is 1.8 pounds. Find the force of the wind on a 2-square foot surface when the wind velocity is 35.5 miles per hour.

Think About It

* **42.** The areas of two circles have the ratio of 25 to 4. What is the ratio of their radii?

* **43.** The force F with which two particles of mass m_1 and m_2 attract each other varies directly as the product of the masses and inversely with the square of the distance r between them. If one of the masses is doubled and the distance between the masses is also doubled, how is the force changed?

Checkup

The following problems provide a review of some of Section 1.5 and will help you with the next section.

Expand.

44. 8^2

45. 11^2

46. $\left(\frac{3}{5}\right)^2$

47. $\left(\frac{7}{9}\right)^2$

48. 2^3

49. 5^3

50. 2^5

51. 3^4

52. $(x^2)^2$

53. $(y^3)^4$

54. $(3y^2)^3$

55. $(2x^3)^4$

Chapter 6 Summary

Section 6.1

A ***rational expression*** is the quotient of two polynomials.

Fundamental property of rational expressions: If A/B is a rational expression and C is any rational expression where $C \neq 0$, then

$$\frac{AC}{BC} = \frac{A}{B}$$

The quotient of two nonzero expressions that are opposites is -1.

Rational expressions are multiplied by multiplying the numerators and multiplying the denominators.

To divide two rational expressions, multiply by the reciprocal of the divisor.

Section 6.2

To add or subtract rational expressions with the same denominator, add or subtract the numerators and keep the common denominator. Simplify, if possible.

The ***least common denominator*** is the smallest number that all denominators divide into with zero remainder.

Finding the least common denominator:

1. Factor each denominator, factoring numbers into primes. Remember that natural numbers other than 1 are prime numbers if they are divisible only by themselves and 1. The first few primes are 2, 3, 5, 7, and 11.
2. The least common denominator is found by using each factor the greatest number of times that it occurs in any one denominator.

To add or subtract rational expressions with different denominators, find the least common denominator. Multiply each expression by one to give it the LCD. Then add or subtract the expressions, keeping the denominator. Simplify, if possible.

When one denominator is the opposite, or additive inverse, of the other, multiply one expression by $-1/-1$ to find a common denominator.

When do we need a common denominator?

1. To add or subtract rational expressions, we need a common denominator.
2. To multiply or divide rational expressions, we do not need a common denominator.

Section 6.3

If the numerator or denominator of a fraction contains fractions, the fraction is called a ***complex fraction.***

There are two methods for simplifying complex fractions.

Method 1. Add or subtract all terms in the numerator and then all terms in the denominator to get a single fraction in both the numerator and denominator. Then divide the numerator by the denominator and simplify, if possible.

Method 2. Multiply the numerator and denominator of the complex fraction by the least common denominator of the denominators. Then simplify the result.

Section 6.4 A ***rational equation*** is an equation with rational expressions.

A rational expression is undefined for any values of the variable that make the denominator zero.

To solve a rational equation, first multiply both sides of the equation by the least common denominator to clear the fractions.

When both sides of an equation are multiplied by an expression containing a variable, we may not get an equivalent equation because the expression containing the variable may be zero. We must check the possible solutions in the given equation. Possible solutions that make a denominator zero in the given equation are called ***extraneous solutions.***

To *add or subtract rational expressions,* find the least common denominator. Multiply each expression by 1 to give it the LCD. Then add or subtract the expressions, keeping the denominator. Simplify, if possible.

To *solve a rational equation,* find the LCD and multiply both sides of the equation by it. This clears the denominators. Then solve the equation.

Section 6.5 The amount of work done, W, is the rate of work, r, times the amount of time, t, spent on the job.

$$W = rt$$

An equation of motion is

$$d = rt$$

where d is distance, r is rate, and t is time.

Section 6.6 *Solving a formula containing a rational expression for a specific variable*

1. Multiply on both sides of the equation to clear fractions.
2. If necessary, multiply to remove parentheses.
3. Get all terms with the specific variable we are solving for alone on one side of the equation by using the addition property.
4. Factor out the specific variable, if necessary.
5. Isolate the specific variable that we are solving for on one side of the equation by using the multiplication property. Remember that the multiplication property allows us to divide both sides of an equation by the same number.

Section 6.7 The ***ratio*** of two quantities is their quotient.

A ***proportion*** is formed by setting two ratios equal to each other.

y varies directly as x if there is some positive constant of variation k such that

$$y = kx$$

The number k is called the ***variation constant,*** or the ***constant of proportionality.***

y varies directly as the nth power of x if there is some positive constant k such that

$$y = kx^n$$

y varies inversely as x if there is a positive constant of variation k such that

$$y = \frac{k}{x}$$

Also, y varies inversely as the nth power of x if there is a positive constant of variation k such that

$$y = \frac{k}{x^n}$$

y varies jointly as x and z if there is a positive constant of variation k such that

$$y = kxz$$

Chapter 6 Additional Exercises (Optional)

NAME
DATE
CLASS

Section 6.1

Simplify.

1. $\frac{30x^2y^3}{25x^3y}$

2. $\frac{18m^5y^2}{36my^5}$

3. $\frac{9x^2 - 4}{12x + 8}$

4. $\frac{10x - 15}{4x^2 - 9}$

5. $\frac{a^2 + 3a - 10}{a^2 + 10a + 25}$

6. $\frac{2x^2 - 9x + 4}{6x^2 + 7x - 5}$

Multiply or divide.

* **7.** $\frac{-15p^9q}{10p^7q^4} \cdot \frac{12p^2q}{3p}$

* **8.** $\frac{16a^8b^3}{-14a^9b^5} \div \frac{8ab}{7a^3b}$

9. $\frac{12x^2 - 11xy + 2y^2}{3x^2 + 13xy - 10y^2} \div \frac{4x^2 + 3xy - y^2}{x + 5y}$

10. $\frac{15a^2 + 16ab + 4b^2}{6a^2 + ab - 2b^2} \cdot \frac{a - 3b}{5a^2 - 13ab - 6b^2}$

11. $\frac{x^3 - y^3}{y - x} \div \frac{x^2 + xy + y^2}{3x + y}$

12. $\frac{8a^3 + b^3}{b + 2a} \cdot \frac{-3}{4a^2 - 2ab + b^2}$

Section 6.2

Add or subtract and simplify, if possible.

13. $\frac{5}{x + 3} - \frac{2}{x + 3} + \frac{x}{x + 3}$

14. $\frac{8}{3z + 6} + \frac{4}{3z + 6} + \frac{6z}{3z + 6}$

15. $\frac{2}{16x^2y^2} + \frac{2x}{24xy^3} - \frac{3}{12xy^5}$

16. $\frac{5a}{21a^5b} - \frac{3}{35ab^4} + \frac{2a}{15a^7b^2}$

17. $\frac{x}{x - 5} - \frac{3}{x^2 - 2x - 15}$

18. $\frac{7}{y - 4} + \frac{y}{y^2 - 16}$

19. $\frac{3y - 1}{y^2 - 7y + 12} + \frac{2}{y^2 - 2y - 8}$

20. $\frac{3x - 2}{x^2 - 7x + 10} - \frac{4}{x^2 + 5x - 14}$

21. $\dfrac{4}{2y-3x} - \dfrac{3x}{5} + \dfrac{6x}{6x^2+17xy-14y^2}$

22. $\dfrac{80}{4p^2-9q^2} + \dfrac{7}{3q-2p} - \dfrac{9p}{8}$

23. $\dfrac{3x}{4x^2-20x+25} + \dfrac{x}{2x-5} - 4$

24. $\dfrac{5y}{3y-2} - \dfrac{y^2+3}{9y^2-12y+4} + 2$

Section 6.3

Simplify.

25. $\dfrac{\dfrac{x+4}{x}}{\dfrac{x-4}{5x}}$

26. $\dfrac{\dfrac{a-b}{6}}{\dfrac{a+b}{4}}$

27. $\dfrac{\dfrac{1}{x}+4}{\dfrac{2}{x}-3}$

28. $\dfrac{a+\dfrac{1}{a}}{a-\dfrac{1}{a}}$

29. $\dfrac{\dfrac{y-2}{y^2}}{\dfrac{y+2}{y^3}}$

30. $\dfrac{\dfrac{x+7}{8x}}{\dfrac{x-3}{2x^2}}$

31. $\dfrac{\dfrac{4}{3x}+\dfrac{2}{6x}}{2-\dfrac{5}{6x}}$

32. $\dfrac{\dfrac{6}{5x^2}-\dfrac{3}{x^3}}{5+\dfrac{1}{10x}}$

33. $\dfrac{\dfrac{3}{x}-\dfrac{4}{x-1}}{5+\dfrac{1}{x-1}}$

34. $\dfrac{\dfrac{7}{y+2}-\dfrac{3}{y-2}}{4-\dfrac{6}{y^2-4}}$

Section 6.4

Find the values (s) for which the rational expression is undefined.

35. $\dfrac{4}{x-3}$

36. $\dfrac{7y}{2y+8}$

* 37. $\dfrac{x+2}{6x^2+5x-4}$

* 38. $\dfrac{y-3}{4y^2-17y+15}$

39. $\dfrac{x^2+2x-3}{x^2-4}$

40. $\dfrac{y^2-3y+1}{y^2-1}$

Add or subtract or solve.

41. $\dfrac{x-2}{x-1} = \dfrac{3}{x-1} - 1$

42. $\dfrac{x-3}{x-4} + 1 = \dfrac{7}{x-4}$

43. $\dfrac{5}{y-4} - \dfrac{y+11}{y^2-5y+4} + \dfrac{3}{y-1}$

44. $\frac{2}{x+4} - \frac{1}{x-2} + \frac{-2x+1}{x^2+2x-8}$

45. $\frac{2}{x-1} + \frac{1}{x+1} = \frac{4x+1}{x^2-1}$

46. $\frac{3+y}{y^2-5y+4} - \frac{2}{y^2-4y} = \frac{1}{y}$

Section 6.5

47. If 5 is added to the reciprocal of a number, the result is twice the reciprocal of the number plus 2. What is the number?

48. One number is 5 times another number. The sum of their reciprocals is 12. Find the numbers.

49. Ken and Carol want to paint Carol's car. Ken can do it in 6 hours and Carol can do the job in 7 hours. How long will it take them to paint the car together?

50. Susan can wash the dishes in 50 minutes and Karl can do them in 30 minutes. How long does it take them to wash the dishes together?

51. The speed of a boat is 10 miles per hour. It can go 4 miles upstream in the same time as it can go 6 miles downstream. Find the speed of the current.

52. A stream has a current of 3 miles per hour. Find the speed of Carlos's boat if it goes 2 miles upstream in the same time that it goes 3 miles downstream.

Section 6.6

Solve.

53. $F = \frac{Gm_1m_2}{d^2}$ for G

54. $a = \frac{V_f - V_0}{t}$ for V_f

55. $A = \frac{Rr}{R+r}$ for R

56. $\frac{E}{e} = \frac{R+r}{r}$ for r

57. For the formula $P = S/(S+F)$, find S when $P = 0.4$ and $F = 6$.

58. For the formula $S = (a - rL)/(1 - r)$, find r when $S = 30$, $a = 2$, and $L = 16$.

Section 6.7

59. y varies directly as x and $x = 4$ when $y = 9$. Find y when $x = 5$.

60. y varies directly as the square of x and $x = 3$ when $y = 7$. Find y when $x = 4$.

61. y varies inversely as the square of x and $x = 6$ when $y = 3$. Find y when $x = 4$.

62. p varies inversely as q and $p = 8$ when $q = 3$. Find q when $p = 2$.

63. y varies jointly as x and z and inversely as w. If $y = 10$ when $x = 3$, $z = 2$, and $w = 5$, find y when $x = -2$, $z = 4$, and $w = 8$.

64. a varies jointly as b and the cube of c and $a = 18$ when $b = 2$ and $c = 3$. Find a when $b = 8$ and $c = -1$.

65. The volume of a gas varies directly with the temperature and inversely with the pressure. If the volume of a gas is 40 cubic feet at 200 kelvin and a pressure of 10 pounds per square inch, what is the volume of the gas at 300 kelvin when the pressure is 40 pounds per square inch?

66. The illumination produced by a light source on an object varies inversely as the square of the distance from the light source. What is the effect on the illumination of the object if the distance between it and the light source is doubled?

COOPERATIVE LEARNING

1. What is a rational expression?

2. What is the quotient of $74x^2 - 8$ and $8 - 74x^2$? Explain.

3. The product of two rational expressions is $(3x - 4)/(7x + 2)$. One of the rational expressions is $(18x^2 - 15x - 25)/(24x^2 - 22x - 35)$. Find the other rational expression.

4. The quotient of two rational expressions is $4y - 6$. The divisor is $(4y^2 + 6y + 9)/(10y - 4)$. Find the dividend.

5. What is wrong with the following?

a. $\dfrac{-15}{9 - x^2} + \dfrac{-23}{x^2 - 9} = -\dfrac{38}{9 - x^2}$

b. $\dfrac{5x - 3}{3x^2 + 7x - 6} - \dfrac{7x + 8}{9x^2 - 12x + 4} = \dfrac{-2x + 5}{(3x - 2)(x + 3)(3x - 2)}$

6. Marie can fly her plane 550 miles with the wind in the same amount of time that it takes her to fly a certain distance against the wind. The speed of her plane is 200 miles per hour and the speed of the wind is 25 miles per hour. How far can she fly against the wind?

Identify as direct or inverse variation.

7. $y = -32x$

8. $P = 75q$

Chapter 6 Practice Test

NAME ______

DATE ______

CLASS ______

1. Multiply and simplify:

$$\frac{2y^2 + 3y - 2}{4y^2 - 1} \cdot \frac{y + 3}{3y + 6}$$

1. ______

2. Divide and simplify:

$$\frac{x^3 - 1}{2x + 8} \div \frac{x^2 + x + 1}{x + 4}$$

2. ______

Add or subtract and simplify:

3. $\dfrac{a}{a^2 - b^2} + \dfrac{b}{b^2 - a^2}$

3. ______

4. $\dfrac{3}{x - 3} + \dfrac{x - 4}{x^2 - 5x + 6} - \dfrac{2}{x - 2}$

4. ______

5. Simplify:

$$\frac{2 + \dfrac{4}{x^2}}{2 - \dfrac{4}{x}}$$

5. ______

6. ______

6. Solve:

$$\frac{3}{x-3} - \frac{2}{x+3} = \frac{7}{x^2-9}$$

7. ______

7. Kevin can prepare a gourmet dinner in 3 hours and Julie can prepare it in 4 hours. How long does it take them to prepare the dinner together?

8. ______

8. Greg can fly his plane 300 miles with the wind in the same time that it takes him to fly 200 miles against the wind. Find the speed of his plane in still air if the speed of the wind is 25 miles per hour.

9. ______

9. Solve $I = \dfrac{nE}{R + nR}$ for n.

10. ______

10. The number of bottles a machine caps varies directly as the number of hours the machine is operating. If a machine caps 524 bottles in 2 hours, how many bottles does it cap in 1.5 hours?

11. ______

11. The weight of an object varies inversely as the square of its distance from the center of the earth. At sea level (6400 kilometers from the center) an object weighs 10 kilograms. Find its weight 300 kilometers above the earth.

Chapters 5 and 6 Cumulative Review

NAME

DATE

CLASS

1. Evaluate $Q = 5x^2 - 2xy - y^4$ for $x = -3$ and $y = -2$.

2. Find the degree of the polynomial: $3 + 5x^2 - x + 7$.

3. Add: $3a^2 - ab - 9b^2$ and $5a^2 + 7ab - 8b^2$.

4. Subtract: $(7x^2 - 8x - 5) - (9x^2 + 7x + 10)$.

Multiply.

5. $(3x - 2)(4x + 8)$

6. $(4y - 9)(4y + 9)$

7. $(x - 4)(2x^2 - x + 3)$

8. $(8p - 3q)^2$

Divide.

9. $\dfrac{6b^3 - 18b^2 - 3}{6b^2}$

10. $\dfrac{3x^2 - 7x - 15}{x - 4}$

Factor.

11. $18x^3 - 3x^2 - 6x$

12. $18x^2 - 60x + 50$

13. $x^3 - 4xy^2$

14. $27x^3 - 343y^3$

Solve.

15. $3x^2 = 7x + 40$

16. $4y^2 = 16y$

17. Divide and simplify:

$$\frac{8x^3 - 27}{x^2 - 9} \div \frac{4x^2 + 6x + 9}{4x^2 + 10x - 6}$$

18. Add and simplify:

$$\frac{x}{25x^2 + 20x + 4} + \frac{x - 3}{5x^2 - 13x - 6}$$

19. Subtract and simplify:

$$\frac{x + 3}{x - 7} - \frac{3x - 2}{2x^2 - 11x - 21}$$

20. Simplify:

$$\frac{\frac{3}{x - 4}}{\frac{2}{x} + 4}$$

21. Solve:

$$\frac{8}{(y - 3)^2} - \frac{3}{y - 3} = \frac{5}{4}$$

22. Greg can till the garden in 2 hours and Debra can till it in 3 hours. How long does it take them to till the garden together?

23. Jennifer can paddle her canoe 2 miles up the stream in the same time that she can paddle 5 miles down the stream. Find the speed of her canoe if the speed of the current is 2 miles per hour.

24. Solve $I = \dfrac{nE}{R + nR}$ for R.

25. The wages a person earns vary directly as the number of hours that he or she works. If Ken works 48 hours, he earns $271.20. How much does he earn if he works 35 hours?

26. The intensity of direct sunlight on a planet varies inversely as the square of the distance of the planet from the sun. The intensity of direct sunlight on Earth is 2 calories per square centimeter per minute. The planet Jupiter is 5.2 times farther than Earth from the sun. Find the intensity of direct sunlight on Jupiter.

* **27.** A computer can do a payroll in 12 hours. Another computer can do the payroll in 6 hours. It takes a third computer 9 hours to do the payroll. How long does it take to do the payroll if all three computers operate at the same time?

CHAPTER 7

Radical Expressions and Complex Numbers

Pretest

Find the root.

1. $\sqrt{64y^2}$

2. $\sqrt{81(x + 7)^2}$

3. $\sqrt[3]{-64x^3}$

4. $\sqrt[9]{(x - 5y)^9}$

For Problems 5 and 6, assume that all variables under radical signs represent positive numbers.

5. Multiply and simplify: $\sqrt[3]{2y^2} \cdot \sqrt[3]{4y^5}$.

6. Divide and simplify: $\dfrac{\sqrt{27x^2y^6}}{\sqrt{3y^2}}$.

7. Approximate to the nearest tenth: $\sqrt{175}$.

8. Subtract: $5\sqrt{98} - 8\sqrt{72}$.

9. Multiply and simplify: $(7\sqrt{2} - 3\sqrt{5})(4\sqrt{2} - 6\sqrt{5})$.

Rationalize the denominator.

10. $\dfrac{5}{3\sqrt{7}}$

11. $\dfrac{8\sqrt{2}-3}{2\sqrt{3}-9}$

12. Write with rational exponents: $\sqrt[4]{(xyz)^5}$.

13. Simplify and write with positive exponents: $\dfrac{7^{1/3}}{7^{2/5}}$.

14. Simplify: $\sqrt{x^5y^8}$. Assume that x and y are positive.

15. Solve: $3\sqrt{x+5} - 3\sqrt{x-3} = 6$.

16. Simplify: $\sqrt{-28}$.

17. Add: $(18 - 4i) + (-34 - 9i)$.

18. Multiply: $(8 - 3i)(9 - 7i)$.

19. $\sqrt{-2} \cdot \sqrt{-8}$

20. Divide: $\dfrac{4-3i}{7-9i}$.

21. The two equal sides of an isosceles right triangle are of length 21 meters. What is the length of the hypotenuse? Round the answer to the nearest thousandth.

7.1 RADICALS

1 Square Roots

Many of the formulas that describe the characteristics of objects in the real world involve roots. For example, the formula for the maximum speed that an automobile can travel around a curve of a given radius without skidding involves a square root.

When we square a number, we raise it to the second power. Hence the square of 3 is $3^2 = 3 \cdot 3 = 9$. We want to reverse the process. We want to find a number that when squared gives 9. One number is 3. It is called a ***square root*** of 9.

> The number c is a square root of b if $c^2 = b$.

There is a symbol for certain square roots.

> The symbol $\sqrt{\ }$, called a ***radical sign,*** is used to represent the nonnegative square root of a number. This is called the ***principal square root.***

We write $\sqrt{9} = 3$. If we want a negative number whose square is 9, we use $-\sqrt{\ }$. We write $-\sqrt{9} = -3$.

Negative numbers do not have square roots in the real number system. We cannot find any real number which squared gives a negative number. We say that square roots of negative numbers are not defined in the real number system. At the end of this chapter we define them in a new set of numbers called the *imaginary numbers.*

The number under the radical sign is called the ***radicand.*** The entire expression, including the radical sign and radicand, is called a ***radical.*** For example, $\sqrt{5}$, $\sqrt{x}$, $\sqrt{x^2 - 4}$ are radicals.

EXAMPLE 1 Find the following.

a. $\sqrt{36} = 6$ Since $6^2 = 36$.

b. $-\sqrt{36} = -6$

c. $\sqrt{0} = 0$

d. $\sqrt{\dfrac{16}{25}} = \dfrac{4}{5}$

e. $\sqrt{-25}$ Not defined; square roots of negative numbers are not defined in the real number system.

f. $\sqrt{625} = 25$ ■

If your calculator has a square root key, $\boxed{\sqrt{x}}$, you may want to use it for problems like Example 1f.

□ **DO EXERCISE 1.**

We may also find square roots of variable expressions. Consider the problem

$$\sqrt{x^2} = ?$$

OBJECTIVES

1 *Find principal square roots*

2 *Find higher roots*

□ **Exercise 1** Find the following.

a. $\sqrt{4}$ **b.** $\sqrt{49}$

c. $\sqrt{64}$ **d.** $\sqrt{16}$

e. $-\sqrt{16}$ **f.** $\sqrt{-16}$

g. $\sqrt{\dfrac{25}{9}}$ **h.** $-\sqrt{81}$

i. $\sqrt{121}$ **j.** $\sqrt{-100}$

□ **Exercise 2** Find the following.

a. $\sqrt{z^2}$

b. $\sqrt{(6y)^2}$

c. $\sqrt{(ab)^2}$

d. $\sqrt{a^2b^2}$

e. $\sqrt{(-15)^2}$

f. $\sqrt{(x+3)^2}$

g. $\sqrt{x^2-4x+4}$

h. $\sqrt{64(y+7)^2}$

The radical sign indicates that we must find the nonnegative square root, which means that x is not an acceptable answer because x may be negative. For example, $\sqrt{(-2)^2} = \sqrt{4} = 2 \neq -2$.

We use the absolute value sign to indicate the positive answer.

> For any real number x, $\sqrt{x^2} = |x|$.

To simplify these problems we also need to know the following about absolute value.

> The absolute value of a product is the product of the absolute values.
>
> $|x \cdot y| = |x| \cdot |y|$ for any numbers x and y

EXAMPLE 2 Find the following.

a. $\sqrt{(-7)^2} = |-7| = 7$

b. $\sqrt{(xy)^2} = |xy|$

c. $\sqrt{x^2+6x+9} = \sqrt{(x+3)^2} = |x+3|$

d. $\sqrt{(4y)^2} = |4y| = |4|\,|y| = 4|y|$ Since $|x \cdot y| = |x| \cdot |y|$. ■

□ **DO EXERCISE 2.**

2 Higher Roots

> The number c is an nth root of a if $c^n = a$.

There is also a symbol for principal nth roots.

> The symbol $\sqrt[n]{a}$ represents the principal nth root of a.

The number n is called the ***index.*** The index must be a positive integer greater than 1. We do not write the index when it is 2.

Odd Roots

A root is called odd when the index is an odd number. The ***cube root*** is a special name for the third root of a number. *Every number has exactly one odd real root.* Absolute value signs are not used on odd roots because if the number is positive, the root is positive, and if the number is negative, the root is negative.

EXAMPLE 3 Find the root.

a. $\sqrt[3]{8} = 2$ Since $2^3 = 8$.

b. $\sqrt[3]{-8} = -2$ Since $(-2)^3 = -8$.

c. $\sqrt[5]{243} = 3$ Since $3^5 = 243$.

d. $\sqrt[5]{-32} = -2$

e. $\sqrt[7]{x^7} = x$ Notice that an absolute value sign is not needed.

f. $\sqrt[9]{(x-4)^9} = x - 4$

g. $\sqrt[3]{x^6} = \sqrt[3]{(x^2)^3} = x^2$ ■

□ **DO EXERCISE 3.**

Even Roots

When the index, n, is an even number, the root is called **even.** Negative real numbers do not have real *even* nth roots. Positive real numbers have *both* a positive and a negative real even nth root. However, there is only *one* principal nth root. As with square roots, absolute value signs are sometimes necessary when we take even nth roots.

EXAMPLE 4 Find the root.

a. $\sqrt{100} = 10$ Remember that the index is 2.

b. $\sqrt[4]{81} = 3$

c. $-\sqrt[4]{81} = -3$

Notice that 3 and -3 are fourth roots of 81.

d. $\sqrt[4]{-16}$ Not defined.

e. $\sqrt[6]{(5x)^6} = 5|x|$ Since the root is even, we use an absolute value sign.

f. $\sqrt{(-3)^2} = |-3| = 3$

g. $\sqrt[8]{(y-3)^8} = |y - 3|$

h. $\sqrt{x^4} = \sqrt{(x^2)^2} = x^2$ Absolute value signs are not needed since x^2 is never negative.

i. $\sqrt[4]{256} = 4$ Since $4^4 = 256$.

A fourth root may be found by taking the square root twice. If your calculator has a root key, $\boxed{y^{1/x}}$ or $\boxed{\sqrt[x]{y}}$, it can be used to find a positive root. ■

> For any real number a:
>
> If *n is even,* $\sqrt[n]{a^n} = |a|$. We use absolute value when a contains a variable or is negative.
> If *n is odd,* $\sqrt[n]{a^n} = a$.

If the result is a variable raised to an even power, we may omit the absolute value sign.

□ **DO EXERCISE 4.**

□ **Exercise 3** Find the root.

a. $\sqrt[3]{27}$ **b.** $\sqrt[3]{-64}$

c. $\sqrt[5]{32y^5}$ **d.** $\sqrt[7]{(4x-3)^7}$

e. $\sqrt[3]{x^9}$ **f.** $\sqrt[5]{y^{10}}$

□ **Exercise 4** Find the root.

a. $\sqrt[4]{16}$ **b.** $-\sqrt[4]{16}$

c. $\sqrt[8]{x^8}$ **d.** $\sqrt[4]{-81}$

e. $\sqrt[6]{(x+3)^6}$ **f.** $\sqrt[8]{(2x-5)^8}$

g. $\sqrt{x^8}$ **h.** $\sqrt[4]{y^{20}}$

DID YOU KNOW?

The radical sign was first used in print in Christoff Rudolph's book, *Die Coss* in 1525. This symbol, $\sqrt{}$, is slightly different from $\sqrt{\ }$, the symbol we use now for a radical sign.

Answers to Exercises

1. a. 2 **b.** 7 **c.** 8 **d.** 4 **e.** -4 **f.** Not defined **g.** $\frac{5}{3}$
h. -9 **i.** 11 **j.** Not defined

2. a. $|z|$ **b.** $6|y|$ **c.** $|ab|$ **d.** $|ab|$ **e.** 15 **f.** $|x + 3|$
g. $|x - 2|$ **h.** $8|y + 7|$

3. a. 3 **b.** -4 **c.** $2y$ **d.** $4x - 3$ **e.** x^3 **f.** y^2

4. a. 2 **b.** -2 **c.** $|x|$ **d.** Not defined **e.** $|x + 3|$
f. $|2x - 5|$ **g.** x^4 **h.** $|y^5|$

Problem Set 7.1

NAME ______________________

DATE ______________________

CLASS ______________________

Find the following.

1. $\sqrt{9}$
2. $\sqrt{1}$
3. $-\sqrt{25}$
4. $-\sqrt{100}$
5. $\sqrt{\frac{4}{49}}$
6. $\sqrt{\frac{16}{25}}$
7. $-\sqrt{\frac{81}{64}}$
8. $-\sqrt{\frac{144}{121}}$
9. $\sqrt{0.16}$
10. $-\sqrt{0.04}$
11. $\sqrt{9y^2}$
12. $\sqrt{25x^2}$
13. $\sqrt{(-8b)^2}$
14. $\sqrt{(-5x)^2}$
15. $\sqrt{(p+2)^2}$
16. $\sqrt{(a-3)^2}$
17. $\sqrt{49(x+4)^2}$
18. $\sqrt{81(y-5)^2}$
19. $\sqrt{9x^2 - 12x + 4}$
20. $\sqrt{4x^2 - 20x + 25}$

Find each root, if it exists.

21. $\sqrt[3]{64}$
22. $\sqrt[3]{125}$
23. $\sqrt[3]{-1000x^3}$
24. $\sqrt[3]{-216y^3}$
25. $-\sqrt[3]{-27}$
26. $-\sqrt[3]{-125y^3}$
27. $\sqrt[5]{-32}$
28. $\sqrt[5]{-243}$
29. $-\sqrt[4]{625}$
30. $\sqrt[4]{256}$
31. $\sqrt[5]{p^5}$
32. $\sqrt[7]{q^7}$

33. $\sqrt[4]{x^4}$

34. $\sqrt[6]{y^6}$

35. $\sqrt[8]{(-4)^8}$

36. $\sqrt[10]{(-3)^{10}}$

37. $\sqrt[4]{-1}$

38. $\sqrt[4]{-625}$

39. $\sqrt[10]{(2a)^{10}}$

40. $\sqrt[12]{(5a)^{12}}$

41. $\sqrt[3]{(-4)^3}$

42. $\sqrt[5]{(-8)^5}$

43. $\sqrt[20]{(x+y)^{20}}$

44. $\sqrt[76]{(3p+2q)^{76}}$

45. $\sqrt[6]{-64}$

46. $\sqrt[4]{-256}$

47. $\sqrt[11]{(3x-4)^{11}}$

48. $\sqrt[15]{(2p+3)^{15}}$

49. $\sqrt[4]{16a^4}$

50. $-\sqrt[6]{64b^6}$

51. $\sqrt[3]{x^{12}}$

52. $\sqrt[5]{x^{15}}$

53. $\sqrt[4]{x^{12}}$

54. $\sqrt[6]{y^{12}}$

55. $\sqrt[7]{x^{14}}$

56. $\sqrt[8]{y^{24}}$

57. $\sqrt[4]{\frac{16}{81}}$

58. $\sqrt[5]{-\frac{1}{243}}$

59. $\sqrt[314]{(x+y)^{314}}$

60. $\sqrt[1432]{(a+b)^{1432}}$

61. $\sqrt[9]{(4ab)^9}$

62. $\sqrt[7]{(8xy)^7}$

63. The maximum speed v that a car can travel around a curve of radius r without skidding is given by the equation

$$v = \sqrt{\frac{5r}{2}}$$

where v is in miles per hour and r is measured in feet. What is the maximum speed a car can travel around a curve with a radius of 40 feet without skidding?

64. Television sets are often described by the length of the diagonal of their screens, as in a "25-inch television set." The length of the diagonal d may be calculated using the length a and the width b of the screen as follows:

$$d = \sqrt{a^2 + b^2}$$

What is the length of the diagonal of a television screen that has a length of 16 inches and a width of 12 inches?

65. The symbol $\sqrt{}$ is used to represent the ____________ square root of a number.

66. ____________ ____________ do not have square roots in the real number system.

67. In the expression $\sqrt[n]{a}$, n is the ____________.

68. Negative real numbers do not have real ____________ nth roots.

Find the root.

* **69.** $\sqrt{x^4 - 2x^2y^2 + y^4}$

* **70.** $\sqrt{a^4 - 6a^3b + 9a^2b^2}$

Checkup

The following problems provide a review of some of Section 4.5.

71. The sum of three numbers is 41. Four times the second is 4 more then twice the first. The third is 5 less than 3 times the second. Find the numbers.

72. The sum of the measures of the angles of a triangle is 180°. The first angle is 3 times as large as the second angle. The third angle is 6 degrees less than twice the second angle. Find the measures of the angles.

73. A fish market sold 45 pounds of halibut. On Tuesday they sold 3 more pounds than they sold on Monday. On Wednesday they sold 6 pounds less than they sold on Tuesday. How many pounds of halibut did they sell on Wednesday?

74. Suppose that the general formula of an equation is $y = ax^2 + bx + c$. The points $(0, 2)$, $(-1, 3)$, and $(1, 7)$ lie on the graph of the equation. Find the equation.

75. Stacie received a payment of \$40,000 when she retired. She invested part of the money in a mining company that produces a return of 12% per year and \$15,000 more than this amount in a certificate of deposit at 5% per year. She deposited the remainder in a savings account at 3% per year. How much is invested at each rate if her annual return on the investments is \$2600?

76. Panos borrowed money from three different lenders to expand his business. He borrowed a total of \$50,000. Some of the money was borrowed at 9% interest per year and an equal amount was borrowed at 10% interest per year. The remainder was borrowed at 7% interest per year. How much was borrowed at each rate if the total simple interest was \$4500?

77. Electron Company produces three different calculators, A, B, and C. Each calculator A requires 2 hours of electronics work, 2 hours of assembly time, and 1 hour of finishing time. Each calculator B requires 1 hour of electronics work, 3 hours of assembly time, and 1 hour of finishing time. Each calculator C requires 3 hours of electronics work, 2 hours of assembly time, and 2 hours of finishing time. If 100 hours can be spent on electronics, 100 hours on assembly, and 65 hours on finishing each week, how many of each calculator can be produced each week if all available time is used?

78. At Laura's car sales the cost of a new car is \$12,800. The basic model with a radio and air conditioning is \$13,450. The basic model with just air conditioning and automatic transmission is \$14,075. A basic model with radio and automatic transmission is \$13,825. What is the cost of each of the three options?

79. Find the values of a, b, and c so that the points $(-1, -9)$, $(0, -1)$, and $(2, -27)$ lie on the graph of the equation $y = ax^2 + bx + c$.

80. The sum of three numbers is -12. Twice the second is 5 less than the first. The third is twice the difference between the second and the first. Find the numbers.

7.2 SIMPLIFYING RADICALS

We may need to simplify radical expressions so that we can add or subtract them.

[1] Product Rule

Notice the following.

$$\sqrt{9} \cdot \sqrt{4} = 3 \cdot 2 = 6 \quad \text{and} \quad \sqrt{9 \cdot 4} = \sqrt{36} = 6$$

The example above suggests the following rule for nonnegative real numbers.

> **Product Rule for Radicals**
>
> If a and b are nonnegative real numbers, and n is a natural number greater than 1,
>
> $$\sqrt[n]{a} \cdot \sqrt[n]{b} = \sqrt[n]{ab}$$
>
> To multiply the radicals, we multiply the radicands when the indexes are the same.

Since many of our rules for radicals apply only to nonnegative real numbers, we often assume that all variables under the radical sign represent positive numbers.

EXAMPLE 1 Multiply. Assume that all variables under the radical sign represent positive numbers.

a. $\sqrt{2} \cdot \sqrt{5} = \sqrt{2 \cdot 5} = \sqrt{10}$

b. $\sqrt{3} \cdot \sqrt{xy} = \sqrt{3xy}$

c. $\sqrt[3]{6}\sqrt[3]{7} = \sqrt[3]{6 \cdot 7} = \sqrt[3]{42}$

d. $\sqrt[4]{\frac{5}{x}}\sqrt[4]{\frac{y}{3}} = \sqrt[4]{\frac{5}{x} \cdot \frac{y}{3}} = \sqrt[4]{\frac{5y}{3x}}$ Remember to write the index.

e. $\sqrt[6]{9}\,\sqrt[5]{7}$ This cannot be multiplied by the product rule for radicals since the radicals have different indexes. ■

□ **DO EXERCISE 1.**

[2] Quotient Rule

There is a quotient rule for radicals that is similar to the product rule.

> **Quotient Rule for Radicals**
>
> For any nonnegative number a, any positive number b, and any index n,
>
> $$\sqrt[n]{\frac{a}{b}} = \frac{\sqrt[n]{a}}{\sqrt[n]{b}}$$
>
> The radical of a quotient is the quotient of the radicals.

This rule allows us to simplify some radicals.

OBJECTIVES

[1] *Multiply radicals using the product rule*

[2] *Simplify radicals using the quotient rule*

[3] *Divide radicals, using the reverse of the quotient rule*

[4] *Simplify radicals by factoring and using the reverse of the product rule*

[5] *Identify rational and irrational numbers*

[6] *Approximate square roots*

□ **Exercise 1** Multiply. Assume that all variables under the radical sign represent positive numbers.

a. $\sqrt{5}\sqrt{11}$

b. $\sqrt{15x}\sqrt{2y}$

c. $\sqrt[5]{9}\sqrt[5]{8}$

d. $\sqrt[3]{7y^2}\sqrt[3]{3x^2}$

□ **Exercise 2** Simplify using the quotient rule for radicals. Assume that all variables under radical signs represent positive numbers.

a. $\sqrt{\dfrac{16}{49}}$

b. $\sqrt{\dfrac{81}{x^2}}$

c. $\sqrt[3]{\dfrac{27}{64y^6}}$

d. $\sqrt[5]{\dfrac{32}{y^5}}$

□ **Exercise 3** Divide and simplify, if possible, by finding roots. Assume that all variables under radical signs represent positive numbers.

a. $\dfrac{\sqrt{48}}{\sqrt{3}}$

b. $\dfrac{\sqrt[4]{35y^7}}{\sqrt[4]{y^5}}$

EXAMPLE 2 Simplify using the quotient rule for radicals. Assume that all variables under radical signs represent positive numbers.

a. $\sqrt{\dfrac{9}{25}} = \dfrac{\sqrt{9}}{\sqrt{25}} = \dfrac{3}{5}$

b. $\sqrt[3]{\dfrac{27}{125}} = \dfrac{\sqrt[3]{27}}{\sqrt[3]{125}} = \dfrac{3}{5}$

c. $\sqrt{\dfrac{36x^2}{y^4}} = \dfrac{\sqrt{36x^2}}{\sqrt{y^4}} = \dfrac{6x}{y^2}$

Absolute value is not needed in Example 2c because we assumed that all variables under radical signs represent positive numbers. ■

□ **DO EXERCISE 2.**

3 Dividing Radicals

If we reverse the quotient rule, we have a method for dividing radicals.

For any nonnegative number a, any positive number b, and any index n,

$$\frac{\sqrt[n]{a}}{\sqrt[n]{b}} = \sqrt[n]{\frac{a}{b}}$$

To divide radicals, we divide the radicands.

Sometimes we can simplify by finding roots after we have divided.

EXAMPLE 3 Divide and simplify if possible, by finding roots. Assume that all variables under radical signs represent positive numbers.

a. $\dfrac{\sqrt{32}}{\sqrt{8}} = \sqrt{\dfrac{32}{8}} = \sqrt{4} = 2$

b. $\dfrac{\sqrt[5]{70x^7}}{\sqrt[5]{2x^6}} = \sqrt[5]{\dfrac{70x^7}{2x^6}} = \sqrt[5]{35x}$ ■

□ **DO EXERCISE 3.**

4 Simplifying Radicals by Factoring

If we reverse the product rule, we have a statement that also helps us to simplify radicals, by factoring.

For any nonnegative real numbers a and b and any index n,

$$\sqrt[n]{a \cdot b} = \sqrt[n]{a} \cdot \sqrt[n]{b}$$

Whole numbers such as 1, 4, 9, 16, . . . that have integers for square roots are called ***perfect squares.*** A ***perfect cube*** is a whole number whose real cube root is an integer. Table 1 in the back of the book lists squares and cubes of numbers from 1 to 100.

> To simplify a radical by factoring, factor the radicand so that there are factors which are perfect nth powers, where n is the index. Find the nth root of those factors.

EXAMPLE 4 Simplify by factoring and finding roots. Assume that all variables under the radical sign are positive.

a. $\sqrt{28}$

Find the largest factor of 28 that is a perfect square. We may do this by looking in Table 1. The squares less than 28 are 25, 16, 9, 4, and 1. We start with the largest square and see if it divides evenly into 28. Then we try the others until we find one that does. Omit 1 since it divides all numbers evenly. Neither 25 nor 16 nor 9 divide evenly into 28. Four does, so it is the largest perfect square in 28. Write 28 as a product of 4 and 7. Then use the reverse of the product rule and simplify.

$$\sqrt{\mathbf{28}} = \sqrt{\mathbf{4 \cdot 7}} = \sqrt{4} \cdot \sqrt{7} = 2\sqrt{7}$$

b. $\sqrt[3]{54}$

Find the largest factor of 54 that is a perfect cube. In Table 1 we find that the cubes less than 54 are 27, 8, and 1. Twenty-seven divides 54 evenly, so it is the largest perfect cube in 54. Write 54 as a product of 27 and 2. Then use the reverse of the product rule and simplify.

$$\sqrt[3]{\mathbf{54}} = \sqrt[3]{\mathbf{27 \cdot 2}} = \sqrt[3]{27} \cdot \sqrt[3]{2} = 3\sqrt[3]{2}$$

c. $\sqrt{\mathbf{7x^2}} = \sqrt{\mathbf{x^2 \cdot 7}}$

$= \sqrt{x^2} \cdot \sqrt{7}$

$= x \cdot \sqrt{7}$ or $x\sqrt{7}$

d. $\sqrt{\mathbf{300x^7y^8}} = \sqrt{\mathbf{(100x^6y^8)(3x)}}$ Notice that x^6 and y^8 are perfect squares since $x^6 = (x^3)^2$ and $y^8 = (y^4)^2$.

$= \sqrt{100x^6y^8}\sqrt{3x}$

$= \sqrt{100}\sqrt{x^6}\sqrt{y^8}\sqrt{3x}$

$= \sqrt{100}\sqrt{(x^3)^2}\sqrt{(y^4)^2}\sqrt{3x}$

$= 10x^3y^4\sqrt{3x}$

e. $\sqrt[3]{8x^6\mathbf{y^4}} = \sqrt[3]{(8x^6\mathbf{y^3})(\mathbf{y})}$ Notice that x^6 is a perfect cube since $x^6 = (x^2)^3$.

$= \sqrt[3]{8x^6y^3}\sqrt[3]{y}$

$= \sqrt[3]{8}\sqrt[3]{x^6}\sqrt[3]{y^3}\sqrt[3]{y}$

$= \sqrt[3]{8}\sqrt[3]{(x^2)^3}\sqrt[3]{y^3}\sqrt[3]{y}$

$= 2x^2y\sqrt[3]{y}$ ■

□ **DO EXERCISE 4.**

Sometimes we can simplify after we have multiplied.

□ **Exercise 4** Simplify by factoring and finding roots. Assume that all variables under the radical sign are positive.

a. $\sqrt{108}$

b. $\sqrt[3]{32}$

c. $\sqrt{700}$

d. $\sqrt{15x^2}$

e. $\sqrt{72x^4y^5}$

f. $\sqrt[3]{40x^3y^7}$

g. $\sqrt{162}$

h. $\sqrt[4]{64a^7b^9}$

□ **Exercise 5** Multiply and simplify by factoring and finding roots. Assume that all variables under the radical sign are positive.

a. $\sqrt{5}\sqrt{15}$

b. $\sqrt{6y}\sqrt{15y}$

c. $\sqrt[3]{9x}\sqrt[3]{3x^2}$

d. $\sqrt{5x}\sqrt{10y}$

□ **Exercise 6** Are the following numbers rational or irrational?

a. $\sqrt{16}$

b. $\sqrt{5}$

c. 8.15

d. 0.454545 . . .
(decimal number repeats)

EXAMPLE 5 Multiply and simplify by factoring and finding roots. Assume that all variables under the radical sign are positive.

a. $\sqrt{30}\sqrt{5} = \sqrt{\mathbf{150}} = \sqrt{\mathbf{25 \cdot 6}} = 5\sqrt{6}$

b. $4\sqrt[3]{2} \cdot 2\sqrt[3]{32} = 8\sqrt[3]{2 \cdot 32}$

$= 8\sqrt[3]{64}$

$= 8(4) = 32$

c. $\sqrt[3]{6}\sqrt[3]{12y^3} = \sqrt[3]{6 \cdot 12y^3}$

$= \sqrt[3]{72y^3}$

$= \sqrt[3]{8y^3 \cdot 9}$

$= 2y\sqrt[3]{9}$ ■

□ **DO EXERCISE 5.**

5 Rational and Irrational Numbers

Rational numbers are numbers that can be written in the form a/b, where a and b are integers and b is not zero. When we write a rational number as a decimal, the decimal either repeats or terminates.

Most positive integers are not perfect squares. Hence their square roots are not rational numbers. These roots are called *irrational numbers.* A fraction will have rational square roots if the fraction is a ratio of two perfect squares. For example, $\sqrt{\frac{1}{4}}$ is a rational number and $\sqrt{\frac{1}{3}}$ is an irrational number.

Irrational numbers are:

1. Numbers that *cannot* be written in the form a/b, where a and b are integers and b is not zero.
2. Decimals that do not repeat and do not terminate.
3. Real numbers that are not rational, such as $\sqrt{7}$, since 7 is not a perfect square.

There are irrational numbers that are not found by taking square roots. The constant π is one of these numbers.

EXAMPLE 6 Are the following numbers rational or irrational?

a. $\sqrt{2} = 1.4142136\ldots$ — Irrational; the decimal does not repeat.

b. $\pi = 3.14159265\ldots$ — Irrational; the decimal does not repeat.

c. $\frac{3}{4} = 0.75$ — Rational; the decimal terminates.

d. $\frac{2}{11} = 0.181818\ldots = 0.\overline{18}$ — Rational; this is a repeating decimal. ■

□ **DO EXERCISE 6.**

The set of real numbers is composed of the rational numbers and the irrational numbers.

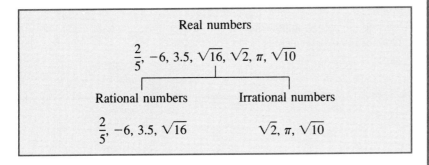

Real numbers	
$\frac{2}{5}$, −6, 3.5, $\sqrt{16}$, $\sqrt{2}$, π, $\sqrt{10}$	
Rational numbers	**Irrational numbers**
$\frac{2}{5}$, −6, 3.5, $\sqrt{16}$	$\sqrt{2}$, π, $\sqrt{10}$

6 Approximating Square Roots

We know that since 5 is not a perfect square, its square root is not a rational number. However, we can approximate the irrational number that is its square root. We do this by using the $\boxed{\sqrt{x}}$ key on our calculators or referring to Table 1 in the Appendix. The roots in the table are rounded, so they are approximate. We use ≈ as the symbol for *approximately equal* to.

If a number is not listed in the table we may be able to factor it into numbers that are listed there and find the square roots of the factors. Then find the product of the square roots. Answers may vary because different calculators give different numbers of digits in their readouts. Also, if we use Table 1, the result may vary from the answer found with a calculator.

EXAMPLE 7 Approximate to the nearest tenth.

a. $\sqrt{175}$

Using a calculator we might get

$$\sqrt{175} \approx 13.228757 \approx 13.2$$

Using factoring and Table 1, we find the following:

$$\begin{aligned}\sqrt{175} &= \sqrt{25 \cdot 7}\\ &= \sqrt{25} \cdot \sqrt{7}\\ &= 5\sqrt{7}\\ &\approx 5(2.646) \qquad \text{From Table 1, } \sqrt{7} \approx 2.646.\\ &\approx 13.230 \approx 13.2\end{aligned}$$

b. $\sqrt{155}$

Using a calculator, we get

$$\sqrt{155} \approx 12.4499 \approx 12.4$$

Using factoring and Table 1, we find that there are no factors of 155 that are perfect squares, so we factor it into two numbers whose square roots are shown in Table 1.

$$\begin{aligned}\sqrt{155} &= \sqrt{31 \cdot 5}\\ &= \sqrt{31} \cdot \sqrt{5}\\ &\approx (5.568)(2.236)\\ &\approx 12.5 \qquad \text{Round to the nearest tenth.}\end{aligned}$$

■

□ **DO EXERCISE 7.**

□ **Exercise 7** Approximate to the nearest tenth.

a. $\sqrt{250}$

b. $\sqrt{105}$

GRAPHING CALCULATOR

Higher roots may be approximated with a graphing calculator. See your calculator manual for details.

Answers to Exercises

1. a. $\sqrt{55}$ **b.** $\sqrt{30xy}$ **c.** $\sqrt[5]{72}$ **d.** $\sqrt[3]{21x^2y^2}$

2. a. $\frac{4}{7}$ **b.** $\frac{9}{x}$ **c.** $\frac{3}{4y^2}$ **d.** $\frac{2}{y}$

3. a. 4 **b.** $\sqrt[4]{35y^2}$

4. a. $6\sqrt{3}$ **b.** $2\sqrt[3]{4}$ **c.** $10\sqrt{7}$ **d.** $x\sqrt{15}$ **e.** $6x^2y^2\sqrt{2y}$
f. $2xy^2\sqrt[3]{5y}$ **g.** $9\sqrt{2}$ **h.** $2ab^2\sqrt[4]{4a^3b}$

5. a. $5\sqrt{3}$ **b.** $3y\sqrt{10}$ **c.** $3x$ **d.** $5\sqrt{2xy}$

6. a. Rational **b.** Irrational **c.** Rational **d.** Rational

7. a. 15.8 **b.** 10.2

Problem Set 7.2

NAME ______________________

DATE ______________________

CLASS ______________________

Assume that all variables under radical signs represent positive numbers.

Multiply.

1. $\sqrt{5}\sqrt{3}$

2. $\sqrt{3}\sqrt{7}$

3. $\sqrt{7x}\sqrt{11y}$

4. $\sqrt{2a}\sqrt{5b}$

5. $\sqrt[3]{4}\sqrt[3]{5}$

6. $\sqrt[3]{6}\sqrt[3]{3}$

7. $\sqrt[4]{9}\sqrt[4]{7}$

8. $\sqrt[4]{8}\sqrt[4]{7}$

9. $\sqrt[3]{9}\sqrt[4]{3}$

10. $\sqrt[6]{8}\sqrt[5]{15}$

11. $\sqrt[5]{9t^3}\sqrt[5]{3t}$

12. $\sqrt[6]{8y^2}\sqrt[6]{12y}$

Simplify using the quotient rule for radicals.

13. $\sqrt{\dfrac{25}{9}}$

14. $\sqrt{\dfrac{36}{49}}$

15. $\sqrt{\dfrac{7}{4}}$

16. $\sqrt{\dfrac{3}{25}}$

17. $\sqrt[3]{\dfrac{1}{8}}$

18. $\sqrt[3]{\dfrac{1}{125}}$

19. $\sqrt[3]{\dfrac{27}{8}}$

20. $\sqrt[3]{\dfrac{216}{125}}$

21. $\sqrt{\dfrac{x^6}{64}}$

22. $\sqrt{\dfrac{100}{y^4}}$

23. $\sqrt[5]{\dfrac{y^{10}}{32}}$

24. $\sqrt[5]{\dfrac{243}{x^5}}$

25. $\sqrt{\dfrac{64a^3}{b^4}}$

26. $\sqrt{\dfrac{25y^5}{x^8}}$

27. $\sqrt[3]{\dfrac{27a^5}{64b^3}}$

28. $\sqrt[3]{\dfrac{8x^7}{125y^9}}$

29. $\sqrt[4]{\dfrac{a^8b^7}{c^8}}$

30. $\sqrt[4]{\dfrac{x^9y^{11}}{z^{12}}}$

Divide and simplify, if possible, by finding roots.

31. $\dfrac{\sqrt{125}}{\sqrt{5}}$

32. $\dfrac{\sqrt{108}}{\sqrt{3}}$

33. $\dfrac{\sqrt{5x}}{\sqrt{x}}$

34. $\dfrac{\sqrt{35y}}{\sqrt{5y}}$

35. $\dfrac{\sqrt[3]{80}}{\sqrt[3]{10}}$

36. $\dfrac{\sqrt[3]{320}}{\sqrt[3]{5}}$

37. $\dfrac{\sqrt[5]{y^7}}{\sqrt[5]{y^3}}$

38. $\dfrac{\sqrt[7]{y^9}}{\sqrt[7]{y^5}}$

39. $\dfrac{\sqrt[4]{162x^7y^{11}}}{\sqrt[4]{2x^3y^3}}$

40. $\dfrac{\sqrt[5]{32a^7b^{24}}}{\sqrt[5]{ab^4}}$

41. $\dfrac{\sqrt{(x+y)^3}}{\sqrt{x+y}}$

42. $\dfrac{\sqrt{(a-b)^3}}{\sqrt{a-b}}$

Simplify by factoring and finding roots.

43. $\sqrt{12}$

44. $\sqrt{27}$

45. $\sqrt{24}$

46. $\sqrt{72}$

47. $-\sqrt{90}$

48. $-\sqrt{160}$

49. $\sqrt[3]{72}$

50. $\sqrt[3]{256}$

51. $\sqrt{76x^8}$

52. $\sqrt{63y^6}$

53. $\sqrt[3]{54x^7}$

54. $\sqrt[3]{96y^5}$

55. $\sqrt[4]{96}$

56. $\sqrt[4]{32}$

57. $\sqrt[4]{162x^6y^8}$

58. $\sqrt[4]{243a^4b^{12}}$

59. $\sqrt[3]{(a-b)^4}$

60. $\sqrt[3]{(c+d)^5}$

61. $\sqrt[3]{-32p^4q^7}$

62. $\sqrt[3]{-375x^6y^2}$

63. $\sqrt[5]{64x^6y^{10}}$

64. $\sqrt[5]{243p^{11}q^{18}}$

Multiply and simplify by factoring and finding roots.

65. $\sqrt{8}\sqrt{10}$

66. $\sqrt{5}\sqrt{20}$

67. $\sqrt[3]{4}\sqrt[3]{10}$

68. $\sqrt[3]{10}\sqrt[3]{25}$

69. $\sqrt{35}\sqrt{20}$

70. $\sqrt{24}\sqrt{50}$

71. $\sqrt{3x^2}\sqrt{5y}$

72. $\sqrt{7a^3}\sqrt{6b^2}$

73. $\sqrt[3]{16y}\sqrt[3]{3y^2}$

74. $\sqrt[3]{24x^2}\sqrt[3]{x^4}$

75. $\sqrt{3x}\sqrt{27y}$

76. $\sqrt{2a}\sqrt{50b}$

77. $\sqrt{8a^2b}\sqrt{12a^3b^2}$

78. $\sqrt{15x^3y}\sqrt{5x^2y^2}$

Are the following numbers rational or irrational?

79. $\sqrt{25}$

80. $\frac{3}{8}$

81. $7.\overline{24}$

82. $\sqrt{10}$

83. 3.91

84. 7.1415936 . . . (does not repeat)

Approximate to the nearest tenth.

85. $\sqrt{134}$

86. $\sqrt{190}$

87. $\frac{\sqrt{360}}{4}$

88. $\frac{\sqrt{720}}{8}$

89. $\frac{12 + \sqrt{320}}{5}$

90. $\frac{9 - \sqrt{470}}{8}$

91. $\frac{15 - \sqrt{240}}{2}$

92. $\frac{32 + \sqrt{530}}{9}$

93. To multiply radicals, we multiply the radicands when the ____________ are the same.

94. The radical of a quotient is the ____________ of the radicals.

95. To divide radicals, we divide the ____________ if the indexes are the same.

96. To begin to simplify a radical by factoring, factor the radicand so that there are factors which are perfect ____________ powers, where n is the index.

97. Rational numbers are numbers that can be written in the form a/b, where a and b are ____________ and b is not zero.

98. The set of real numbers is composed of the rational numbers and the ____________ numbers.

Use a calculator to approximate the following square roots. Round answers to the nearest tenth.

99. $\sqrt{5}$

100. $\sqrt{11}$

101. $\sqrt{94}$

102. $\sqrt{135}$

103. $-\sqrt{168}$

104. $-\sqrt{47}$

105. $\sqrt{378}$

106. $-\sqrt{734}$

107. $\frac{\sqrt{360}}{9}$

108. $\frac{\sqrt{750}}{5}$

109. $\frac{\sqrt{269}}{7}$

110. $\frac{\sqrt{982}}{3}$

Assume that all variables under radical signs represent positive numbers.

Divide and simplify.

* **111.** $\dfrac{\sqrt{23{,}328}}{\sqrt{8}}$

* **112.** $\dfrac{\sqrt[3]{4374}}{\sqrt[3]{6}}$

* **113.** $\dfrac{\sqrt{9}}{\sqrt{x^2 + 2x + 1}}$

* **114.** $\dfrac{\sqrt{x^4 - 2x^2 + 1}}{\sqrt{x^2 - 8xy + 16y^2}}$

Multiply and simplify.

* **115.** $3\sqrt{a^{-1}b^8}\sqrt{16x^4y^3}\sqrt{a^9b^{-2}}\sqrt{4x^{-2}y^{-2}}$

* **116.** $(\sqrt[4]{x^2})^4$

* **117.** $\sqrt[3]{\dfrac{4}{289}}\sqrt[3]{\dfrac{432}{17}}$

* **118.** Kepler's third law of planetary motion is $T = \sqrt{d^3}$, where T is the time in years that it takes a planet to revolve around the sun and d is the average distance in astronomical units between the planet and the sun. Find T when $d = 4$.

Checkup

The following problems provide a review of some of Sections 5.2 and 6.2 and will help you with the next section.

Add or subtract as indicated.

119. $(6a - 7) + (5a + 4)$

120. $4x - 3 - (8x + 2)$

121. $\dfrac{3}{y} + \dfrac{5 - y}{y}$

122. $\dfrac{7x}{x + 1} - \dfrac{2x + 4}{x + 1}$

123. $\dfrac{4}{x - 3} - \dfrac{3x}{x^2 - 9}$

124. $\dfrac{2}{y + 4} + \dfrac{3y}{(y + 4)^2}$

125. $\dfrac{2x - 1}{x^2 - 4x + 3} + \dfrac{3x - 2}{x^2 + 3x - 4}$

126. $\dfrac{3a + 1}{a^2 + 3a - 10} - \dfrac{2a - 1}{a^2 + 7a + 10}$

7.3 ADDITION AND SUBTRACTION OF RADICAL EXPRESSIONS

OBJECTIVE

1 *Add or subtract radical expressions*

1 Recall that the sum of 6 and x is $6 + x$. Similarly, the sum of 6 and $\sqrt{5}$ is $6 + \sqrt{5}$. Also, the sum of $5x$ and $2x$ is $7x$ and the sum of $5\sqrt{3}$ and $2\sqrt{3}$ is $7\sqrt{3}$. We can combine terms containing variables if the variables are identical and raised to the same power. Similarly, we can combine radical expressions if the radicals have the same index and radicand.

EXAMPLE 1 Add or subtract.

a. $7\sqrt{\mathbf{5}} + 3\sqrt{\mathbf{5}} = 10\sqrt{\mathbf{5}}$

b. $4\sqrt[3]{\mathbf{7}} + x\sqrt[3]{2} - 6\sqrt[3]{\mathbf{7}} = 4\sqrt[3]{\mathbf{7}} - 6\sqrt[3]{\mathbf{7}} + x\sqrt[3]{2}$ Use a commutative law.

$= -2\sqrt[3]{\mathbf{7}} + x\sqrt[3]{2}$ Combine like terms.

c. $8\sqrt[3]{3x} - \sqrt[3]{3x} + 4\sqrt[5]{3x} = 7\sqrt[3]{3x} + 4\sqrt[5]{3x}$ Notice that $\sqrt[3]{3x} = 1\sqrt[3]{3x}$. ■

Caution: If the radicals in the radical expressions do not have the same index, the radical expressions cannot be combined.

□ **DO EXERCISE 1.**

□ **Exercise 1** Add or subtract. Assume that variables under radical signs represent positive numbers.

a. $4\sqrt{7} + 8\sqrt{7}$

b. $9\sqrt[4]{2x} - \sqrt[4]{2x} - 7\sqrt[6]{2x}$

Sometimes it appears that we cannot add or subtract radical expressions because the radicands are different. It is often possible to make the radicands the same by simplifying radicals.

EXAMPLE 2 Add or subtract.

a. $3\sqrt{\mathbf{24}} + 5\sqrt{6} = 3\sqrt{\mathbf{4 \cdot 6}} + 5\sqrt{6}$ Factor 24.

$= 3\sqrt{4} \cdot \sqrt{6} + 5\sqrt{6}$ Use the product rule.

$= 3 \cdot 2\sqrt{6} + 5\sqrt{6}$ Simplify $\sqrt{4}$.

$= 6\sqrt{6} + 5\sqrt{6}$ Multiply.

$= 11\sqrt{6}$ Combine like terms.

b. $4\sqrt[3]{\mathbf{32x^4}} - 7x\sqrt[3]{4x} = 4\sqrt[3]{\mathbf{8x^3 \cdot 4x}} - 7x\sqrt[3]{4x}$ Factor $\sqrt[3]{32x^4}$.

$= 4\sqrt[3]{8x^3} \cdot \sqrt[3]{4x} - 7x\sqrt[3]{4x}$ Use the product rule.

$= 4 \cdot 2x \cdot \sqrt[3]{4x} - 7x\sqrt[3]{4x}$ Find $\sqrt[3]{8x^3}$.

$= 8x\sqrt[3]{4x} - 7x\sqrt[3]{4x}$ Multiply.

$= x\sqrt[3]{4x}$

c. $3\sqrt{2} - 4\sqrt{5}$ Cannot be simplified. ■

□ **DO EXERCISE 2.**

□ **Exercise 2** Add or subtract.

a. $3\sqrt{11} - 5\sqrt{44}$

b. $2\sqrt{8} - 6\sqrt{50} + 3\sqrt{200}$

c. $4\sqrt[3]{54} - 2\sqrt[3]{16}$

d. $\sqrt[3]{y^5} + 3y^2 - \sqrt[3]{27y^6}$

Answers to Exercises

1. a. $12\sqrt{7}$ **b.** $8\sqrt[4]{2x} - 7\sqrt[6]{2x}$

2. a. $-7\sqrt{11}$ **b.** $4\sqrt{2}$ **c.** $8\sqrt[3]{2}$ **d.** $y\sqrt[3]{y^2}$

Problem Set 7.3

NAME ______________________

DATE ______________________

CLASS ______________________

Add or subtract. Assume that all variables under radical signs represent positive numbers.

1. $4\sqrt{3} + 7\sqrt{3}$

2. $5\sqrt{7} + 8\sqrt{7}$

3. $3\sqrt[3]{6} - 5\sqrt[3]{6}$

4. $8\sqrt[3]{9} - 7\sqrt[3]{9}$

5. $8\sqrt[3]{3x} + 9\sqrt[3]{3x}$

6. $2\sqrt[3]{4x} - 7\sqrt[3]{4x}$

7. $9\sqrt{5} - 6\sqrt{5} - 4\sqrt{5}$

8. $3\sqrt{2} - 7\sqrt{2} + 8\sqrt{2}$

9. $3\sqrt[4]{5} - \sqrt{6} + 2\sqrt[4]{5} - \sqrt{6}$

10. $4\sqrt[5]{9} - \sqrt{3} - 8\sqrt[5]{9} + 7\sqrt{3}$

11. $3\sqrt{32} - 2\sqrt{2}$

12. $8\sqrt{3} + 4\sqrt{75}$

13. $4\sqrt{12} - 7\sqrt{27}$

14. $3\sqrt{80} + 6\sqrt{45}$

15. $11\sqrt{72} + 2\sqrt{98}$

16. $12\sqrt{45} - 6\sqrt{125}$

17. $\sqrt[3]{54} - 4\sqrt[3]{16}$

18. $6\sqrt[3]{81} - 2\sqrt[3]{24}$

19. $3\sqrt{40} + 5\sqrt{90} - 3\sqrt{160}$

20. $4\sqrt{128} - \sqrt{18} - 5\sqrt{32}$

21. $3\sqrt{2x} - 2\sqrt{72x}$

22. $4\sqrt{50y} - 3\sqrt{18y}$

23. $4\sqrt{72x^2} + 2\sqrt{32x^2}$

24. $5\sqrt{27y^2} - 3\sqrt{108y^2}$

25. $3\sqrt[3]{27x} - 4\sqrt[3]{8x}$

26. $7\sqrt{3x^3} - 8x\sqrt{3x}$

27. $\sqrt[3]{a^2b} + \sqrt[3]{8a^2b}$

28. $2\sqrt[4]{xy} + \sqrt[4]{81xy}$

29. $4\sqrt{18} + \sqrt{32} - 5\sqrt{50}$

30. $3\sqrt{8} - 4\sqrt{72} + 5\sqrt{50}$

31. $4\sqrt{72m^2} + 3\sqrt{32m^2} - 5\sqrt{18m^2}$

32. $8\sqrt{27q^2} - 6\sqrt{108q^2} - \sqrt{48q^2}$

33. $5x\sqrt[3]{xy^2} - 3\sqrt[3]{8x^4y^2}$

34. $7x^2\sqrt[3]{5x} + 3x\sqrt[3]{40x^4}$

35. $3\sqrt[4]{32} - 5\sqrt[4]{162}$

36. $2\sqrt[4]{512} - 5\sqrt[4]{32}$

37. $\sqrt{12a - 12} + \sqrt{3a - 3}$

38. $\sqrt{7x + 7} + \sqrt{28x + 28}$

39. $\sqrt{x^3 - x^2} - \sqrt{16x - 16}$

40. $\sqrt{25x + 25} + \sqrt{x^3 + x^2}$

41. $4\sqrt[3]{32} + \sqrt[3]{108} - 3\sqrt[3]{256}$

42. $5\sqrt[3]{8y} - 2\sqrt[3]{27y} + 5\sqrt[3]{64y}$

43. We can combine radical expressions if the radicals have the same ___________ and radicand.

44. It is often possible to make two radicands the same by ___________ one or both radicals.

Assume that all variables under radical signs represent positive numbers. Add or subtract.

* **45.** $x\sqrt[3]{27x} - 11\sqrt[3]{x^4} + \dfrac{5\sqrt[3]{x^7}}{x}$

* **46.** $\sqrt{72x^2y} - \sqrt{1200xy^2} + x\sqrt{648y}$

* **47.** $\sqrt{9x - 9} - \sqrt{4x - 4} + \sqrt{25x - 25}$

* **48.** $\dfrac{3}{4}y\sqrt{12x^2y} - \dfrac{5}{2}x\sqrt{75y^3} + \dfrac{2}{3}\sqrt{18x^2y}$

* **49.** $\dfrac{5}{7}b\sqrt{54a^2b} + \dfrac{8}{3}a\sqrt{24b^3} - \dfrac{1}{6}\sqrt{96a^2b}$

* **50.** Use a pair of numbers to show that

$$\sqrt{a + b} \neq \sqrt{a} + \sqrt{b}$$

Checkup

The following problems provide a review of some of Sections 5.3, 5.4, and 6.1 and some will help you with the next section.

Multiply.

51. $2x(3x - 7)$

52. $5y(6y + 3)$

53. $(4x - 5)(2x + 9)$

54. $(7x - 3)(2x - 1)$

55. $(3x - 4)(3x + 4)$

56. $(x + 5y)(x - 5y)$

57. $(x - 2y)^2$

58. $(4x + 3)^2$

Divide.

59. $\dfrac{7}{x^3} \div \dfrac{14}{x^5}$

60. $\dfrac{6}{y^2} \div \dfrac{8}{3y^4}$

61. $\dfrac{3x + 6}{4} \div \dfrac{x^2 - 4}{20}$

62. $\dfrac{(x + y)^2}{x} \div \dfrac{x + y}{x^3}$

63. $\dfrac{x^2 + 6x}{x^2 - 36} \div \dfrac{3x^2 + 13x - 10}{3x^2 - 20x + 12}$

64. $\dfrac{y^2 - 9}{6y^2 + 7y - 3} \div \dfrac{(y + 3)^2}{2y^2 + 7y + 6}$

7.4 MULTIPLICATION OF RADICAL EXPRESSIONS

We learned how to multiply radical expressions with one term in Section 7.2. Radical expressions with more than one term are multiplied by the methods used for multiplying polynomials.

1 Multiplying Radical Expressions Using the Distributive Laws

EXAMPLE 1 Multiply. Assume that all variables under radical signs represent positive numbers.

a. $\sqrt{2}(x - \sqrt{3}) = \sqrt{2} \cdot x - \sqrt{2} \cdot \sqrt{3}$ Use a distributive law.

$= x\sqrt{2} - \sqrt{6}$ Use the product rule.

b. $\sqrt[3]{x}(\sqrt[3]{x^5} + \sqrt[3]{4}) = \sqrt[3]{x} \cdot \sqrt[3]{x^5} + \sqrt[3]{x} \cdot \sqrt[3]{4}$ Use a distributive law.

$= \sqrt[3]{x^6} + \sqrt[3]{4x}$ Use the product rule.

$= x^2 + \sqrt[3]{4x}$ Simplify $\sqrt[3]{x^6}$. ■

□ **DO EXERCISE 1.**

2 Special Products

To multiply radical expressions with the same index we multiply each term of one expression by each term of the other expression. Many of the radical expressions that we will work with are binomial expressions. We will demonstrate the methods used to multiply two binomials containing radicals. They are the same as the methods used for multiplying binomials without radicals.

EXAMPLE 2 Multiply. Assume that all variables under radical signs represent positive numbers.

a. $(2 - \sqrt{6})(5 - \sqrt{3})$

We may use the FOIL method.

F O I L

$(2 - \sqrt{6})(5 - \sqrt{3}) = 2 \cdot 5 - 2 \cdot \sqrt{3} - 5 \cdot \sqrt{6} + \sqrt{6} \cdot \sqrt{3}$

$= 10 - 2\sqrt{3} - 5\sqrt{6} + \sqrt{18}$

$= 10 - 2\sqrt{3} - 5\sqrt{6} + 3\sqrt{2}$ Simplify $\sqrt{18}$.

b. $(\sqrt{5} + \sqrt{2})(\sqrt{3} - \sqrt{2})$

$(\sqrt{5} + \sqrt{2})(\sqrt{3} - \sqrt{2}) = \sqrt{5} \cdot \sqrt{3} - \sqrt{5} \cdot \sqrt{2} + \sqrt{2} \cdot \sqrt{3} - \sqrt{2} \cdot \sqrt{2}$

$= \sqrt{15} - \sqrt{10} + \sqrt{6} - \sqrt{4}$

$= \sqrt{15} - \sqrt{10} + \sqrt{6} - 2$ Simplify $\sqrt{4}$.

Notice that $\sqrt{2} \cdot \sqrt{2} = 2$. ■

□ **DO EXERCISE 2.**

OBJECTIVES

1 *Multiply radical expressions using the distributive laws*

2 *Multiply binomial radical expressions using the rules for special products of binomials*

□ **Exercise 1** Multiply. Assume that all variables under radical signs represent positive numbers.

a. $\sqrt{6}(3 + \sqrt{2})$

b. $\sqrt[3]{y}(\sqrt[3]{y^2} - \sqrt[3]{5})$

□ **Exercise 2** Multiply. Assume that all variables under radical signs represent positive numbers.

a. $(3 - \sqrt{5})(4 + \sqrt{2})$

b. $(\sqrt{6} + \sqrt{2})(\sqrt{3} - \sqrt{7})$

c. $(\sqrt{x} + 3\sqrt{2})(\sqrt{y} - 4\sqrt{2})$

□ **Exercise 3** Multiply. Assume that all variables under radical signs represent positive numbers.

a. $(5\sqrt{2} - b)(5\sqrt{2} + b)$

b. $(\sqrt{a} + \sqrt{b})(\sqrt{a} - \sqrt{b})$

c. $(3\sqrt{7} - 4)^2$

EXAMPLE 3 Multiply. Assume that all variables under radical signs represent positive numbers.

a. $(\sqrt{7} + \sqrt{3})(\sqrt{7} - \sqrt{3})$

This is the product of a sum and a difference. Recall that $(A + B)(A - B) = A^2 - B^2$.

$$\begin{aligned}(\sqrt{7} + \sqrt{3})(\sqrt{7} - \sqrt{3}) &= (\sqrt{7})^2 - (\sqrt{3})^2 \\ &= 7 - 3 \\ &= 4\end{aligned}$$

b. $(\sqrt{x} + \sqrt{y})(\sqrt{x} - \sqrt{y})$

$$\begin{aligned}(\sqrt{x} + \sqrt{y})(\sqrt{x} - \sqrt{y}) &= (\sqrt{x})^2 - (\sqrt{y})^2 \\ &= x - y\end{aligned}$$

c. $(\sqrt{5} - x)^2$

We may use the rule for squaring a binomial, $(A - B)^2 = A^2 - 2AB + B^2$.

$$\begin{aligned}(\sqrt{5} - x)^2 &= (\sqrt{5})^2 - 2x\sqrt{5} + x^2 \\ &= 5 - 2x\sqrt{5} + x^2\end{aligned}$$ ■

□ **DO EXERCISE 3.**

Answers to Exercises

1. a. $3\sqrt{6} + 2\sqrt{3}$ **b.** $y - \sqrt[3]{5y}$

2. a. $12 + 3\sqrt{2} - 4\sqrt{5} - \sqrt{10}$ **b.** $3\sqrt{2} - \sqrt{42} + \sqrt{6} - \sqrt{14}$
c. $\sqrt{xy} - 4\sqrt{2x} + 3\sqrt{2y} - 24$

3. a. $50 - b^2$ **b.** $a - b$ **c.** $79 - 24\sqrt{7}$

Problem Set 7.4

NAME

DATE

CLASS

Multiply. Assume that all variables under radical signs represent positive numbers.

1. $\sqrt{7}(4 - \sqrt{7})$ **2.** $\sqrt{5}(9 + \sqrt{5})$ **3.** $\sqrt{3}(\sqrt{5} + \sqrt{2})$ **4.** $\sqrt{2}(\sqrt{6} - \sqrt{5})$

5. $\sqrt{2}(3\sqrt{7} - 2\sqrt{6})$ **6.** $\sqrt{3}(5\sqrt{8} - 4\sqrt{3})$ **7.** $\sqrt[3]{4}(\sqrt[3]{3} - 6\sqrt[3]{2})$ **8.** $\sqrt[3]{9}(2\sqrt[3]{6} + 3\sqrt[3]{5})$

9. $\sqrt[4]{2}(\sqrt[4]{8} + 3\sqrt[4]{5})$ **10.** $\sqrt[4]{9}(\sqrt[4]{9} + 2\sqrt[4]{7})$ **11.** $\sqrt[3]{b}(\sqrt[3]{3b^2} - \sqrt[3]{24b^2})$ **12.** $\sqrt[3]{a}(\sqrt[3]{2a^2} + \sqrt[3]{54a^2})$

13. $(5 + \sqrt{3})(5 - \sqrt{3})$ **14.** $(7 + \sqrt{6})(7 - \sqrt{6})$ **15.** $(\sqrt{2} - 8)(\sqrt{2} + 8)$ **16.** $(\sqrt{3} - 9)(\sqrt{3} + 9)$

17. $(3 - \sqrt{2})(3 + \sqrt{2})$ **18.** $(\sqrt{5} - 4)(\sqrt{5} + 4)$ **19.** $(2\sqrt{6} - \sqrt{8})(2\sqrt{6} + \sqrt{8})$

20. $(7\sqrt{11} - \sqrt{12})(7\sqrt{11} + \sqrt{12})$ **21.** $(\sqrt{x} + \sqrt{z})(\sqrt{x} - \sqrt{z})$ **22.** $(\sqrt{p} + \sqrt{q})(\sqrt{p} - \sqrt{q})$

23. $(\sqrt{5} + 3)(\sqrt{5} - 2)$ **24.** $(\sqrt{7} + 6)(\sqrt{7} - 2)$ **25.** $(\sqrt{3} + 5)(4\sqrt{3} + 1)$

26. $(\sqrt{6} - 8)(3\sqrt{6} - 2)$ **27.** $(3\sqrt{3} - 4\sqrt{5})(2\sqrt{3} + 2\sqrt{5})$ **28.** $(6\sqrt{2} + 3\sqrt{7})(2\sqrt{2} - 4\sqrt{7})$

29. $(\sqrt{x} - \sqrt{3})(\sqrt{x} - \sqrt{2})$ **30.** $(5 + \sqrt{a})(1 + \sqrt{a})$ **31.** $(\sqrt[3]{4} - 2\sqrt[3]{3})(\sqrt[3]{4} - 5\sqrt[3]{3})$

32. $(\sqrt[4]{7} + 2\sqrt[4]{6})(3\sqrt[4]{7} - \sqrt[4]{6})$ **33.** $(\sqrt{3} - x)^2$

34. $(\sqrt{5} + x)^2$ **35.** $(7 + \sqrt{2})^2$ **36.** $(10 - \sqrt{5})^2$

37. $(2\sqrt{3} - a)^2$ **38.** $(4\sqrt{2} + b)^2$ **39.** $(\sqrt{a} + \sqrt{5})(\sqrt{a} + \sqrt{2})$

40. $(\sqrt{3} - \sqrt{b})(\sqrt{7} - \sqrt{b})$ **41.** $(\sqrt[5]{8} + \sqrt[5]{2})(\sqrt[5]{6} - \sqrt[5]{16})$ **42.** $(\sqrt[4]{7} - \sqrt[4]{3})(\sqrt[4]{12} - \sqrt[4]{5})$

43. The FOIL method may be used to multiply two two __________ containing radicals.

44. The product of a sum and a difference containing radicals may be multiplied using the rules for __________.

Assume that all variables under radical signs represent positive numbers.

Multiply and simplify.

* **45.** $(3 - \sqrt{x - 4})^2$

* **46.** $(7 + \sqrt{x + 2})^2$

* **47.** $(\sqrt[3]{x} - \sqrt[3]{y})(\sqrt[3]{x} + \sqrt[3]{y})$

* **48.** $\sqrt[4]{5}(3 - \sqrt[4]{125})(3 + \sqrt[4]{125})$

* **49.** $(3\sqrt{2} - 4\sqrt{5})(2\sqrt{3} + 6\sqrt{7})$

* **50.** $(\sqrt[3]{x} - 2\sqrt[3]{y})^2$

* **51.** $(x + \sqrt{y - 3})^2$

* **52.** $(\sqrt[3]{a} - \sqrt[3]{b})(\sqrt[3]{a^2} - \sqrt[3]{a}\sqrt[3]{b} + \sqrt[3]{b^2})$

* **53.** $(\sqrt{5} + 2\sqrt{6})(\sqrt{5} - 2\sqrt{6})$

* **54.** $(\sqrt{2} + \sqrt{3} + \sqrt{5})^4$

Think About It

* **55.** Use the fact that $(A - B)(A^2 + AB + B^2) = A^3 - B^3$ to show that

$$(\sqrt{9} - \sqrt{4})(\sqrt{81} + \sqrt{36} + \sqrt{16}) = 19$$

* **56.** Choose the incorrect addition and explain why it is incorrect.
(a) $3\sqrt{20} + 5\sqrt{45} = 21\sqrt{5}$
(b) $7\sqrt{8} + 3\sqrt{32} = 26\sqrt{2}$
(c) $4\sqrt{28} + 2\sqrt{63} = 12\sqrt{7}$

Checkup

The following problems provide a review of some of Section 7.2 and will help you with the next section. Assume that all variables under radical signs represent positive numbers.

Multiply and simplify.

57. $\sqrt{7}\sqrt{7}$

58. $\sqrt{3}\sqrt{3}$

59. $\sqrt{5y}\sqrt{5y}$

60. $\sqrt{2x}\sqrt{2x}$

61. $\sqrt[3]{5}\sqrt[3]{25}$

62. $\sqrt[3]{2}\sqrt[3]{4}$

63. $\sqrt[3]{x}\sqrt[3]{9x^2}$

64. $\sqrt[3]{4y}\sqrt[3]{16y}$

65. $\sqrt{2}\sqrt{12}$

66. $\sqrt{5}\sqrt{18}$

67. $\sqrt[3]{3a^2}\sqrt[3]{18a^5}$

68. $\sqrt[3]{4x}\sqrt[3]{6x^4}$

7.5 RATIONALIZING NUMERATORS AND DENOMINATORS

To simplify calculations with radical expressions, it is common practice to write them without radicals in the denominator. One process of removing the radicals from the denominator is called ***rationalizing the denominator.***

1 Rationalizing Monomial Denominators

To rationalize a monomial denominator we multiply by 1 to make the radicand in the denominator a perfect nth power, where n is the index of the radical. Then take the nth root of the radicand in the denominator.

EXAMPLE 1 Rationalize the denominator.

a. $\dfrac{2}{\sqrt{5}}$

$$\frac{2}{\sqrt{5}} = \frac{2}{\sqrt{5}} \cdot \frac{\sqrt{5}}{\sqrt{5}}$$ Multiply by 1, $\sqrt{5}/\sqrt{5}$, to make the denominator a perfect square.

$$= \frac{2\sqrt{5}}{\sqrt{5^2}} = \frac{2\sqrt{5}}{\sqrt{25}}$$ The radicand in the denominator is a perfect square.

$$= \frac{2\sqrt{5}}{5}$$

We have eliminated the radical from the denominator.

b. $\dfrac{\sqrt[3]{4}}{\sqrt[3]{3}}$

$$\frac{\sqrt[3]{4}}{\sqrt[3]{3}} = \frac{\sqrt[3]{4}}{\sqrt[3]{3}} \cdot \frac{\sqrt[3]{9}}{\sqrt[3]{9}}$$ Multiply by 1 to get the smallest possible perfect cube in the denominator.

$$= \frac{\sqrt[3]{36}}{\sqrt[3]{27}}$$ The radicand of 27 in the denominator is a perfect cube.

$$= \frac{\sqrt[3]{36}}{3}$$

Caution: Do not multiply by a 1 of $\sqrt[3]{3}/\sqrt[3]{3}$. This gives $\sqrt[3]{9}$ in the denominator, which is not a perfect cube. ■

□ **DO EXERCISE 1.**

If the expression involves quotients under a radical sign, we simplify by using the quotient rule for radicals and then rationalizing the denominator.

OBJECTIVES

1 *Rationalize monomial denominators and numerators*

2 *Rationalize binomial denominators and numerators*

□ **Exercise 1** Rationalize the denominator.

a. $\dfrac{\sqrt{7}}{\sqrt{3}}$

b. $\dfrac{\sqrt{11}}{\sqrt{6}}$

c. $\dfrac{\sqrt{1}}{\sqrt[3]{4}}$

d. $\dfrac{\sqrt[3]{9}}{\sqrt[3]{2}}$

☐ **Exercise 2** Simplify by rationalizing the denominator. Assume that all variables under radical signs represent positive numbers.

a. $\sqrt{\dfrac{5}{6}}$

b. $\sqrt{\dfrac{11a}{5b}}$

c. $\sqrt[3]{\dfrac{15}{9}}$

d. $\sqrt[3]{\dfrac{a}{3x^2}}$

☐ **Exercise 3** Rationalize the numerator. Assume that all variables under the radical sign represent positive numbers.

a. $\dfrac{\sqrt{10}}{\sqrt{7}}$

b. $\dfrac{\sqrt[3]{3a}}{\sqrt[3]{5b}}$

EXAMPLE 2 Simplify by rationalizing the denominator. Assume that all variables under radical signs represent positive numbers.

a. $\sqrt{\dfrac{3x}{7y}}$

$$\sqrt{\frac{3x}{7y}} = \frac{\sqrt{3x}}{\sqrt{7y}}$$ Use the quotient rule for radicals.

$$= \frac{\sqrt{3x}}{\sqrt{7y}} \cdot \frac{\sqrt{\mathbf{7y}}}{\sqrt{\mathbf{7y}}}$$ Multiply by 1.

$$= \frac{\sqrt{21xy}}{\sqrt{49y^2}}$$

$$= \frac{\sqrt{21xy}}{7y}$$ Notice that we may not divide 21 by 7 since 21 is under the radical sign and 7 is not.

b. $\sqrt[3]{\dfrac{5}{16x^2}}$

$$\sqrt[3]{\frac{5}{16x^2}} = \frac{\sqrt[3]{5}}{\sqrt[3]{16x^2}}$$ Use the quotient rule for radicals.

$$= \frac{\sqrt[3]{5}}{\sqrt[6]{16x^2}} \cdot \frac{\sqrt[3]{\mathbf{4x}}}{\sqrt[3]{\mathbf{4x}}}$$ Multiply by 1 to get the smallest possible perfect cube in the denominator.

$$= \frac{\sqrt[3]{20x}}{\sqrt[3]{64x^3}}$$

$$= \frac{\sqrt[3]{20x}}{4x}$$ ■

☐ **DO EXERCISE 2.**

Rationalizing Monomial Numerators

Occasionally, in calculus, we want to ***rationalize a monomial numerator.*** The procedure is similar to that for rationalizing a monomial denominator.

EXAMPLE 3 Rationalize the numerator of $\sqrt[3]{2x}/\sqrt[3]{9}$. Assume that all variables under radical signs represent positive numbers.

We want a perfect cube in the numerator. Since $2x \cdot 4x^2 = 8x^3$ is a perfect cube, the "1" that we multiply by is $\sqrt[3]{4x^2}/\sqrt[3]{4x^2}$.

$$\frac{\sqrt[3]{2x}}{\sqrt[3]{9}} = \frac{\sqrt[3]{2x}}{\sqrt[3]{9}} \cdot \frac{\sqrt[3]{\mathbf{4x^2}}}{\sqrt[3]{\mathbf{4x^2}}}$$ Multiply by 1.

$$= \frac{\sqrt[3]{8x^3}}{\sqrt[3]{36x^2}}$$

$$= \frac{2x}{\sqrt[3]{36x^2}}$$ ■

☐ **DO EXERCISE 3.**

2 Rationalizing a Binomial Denominator

The binomials $A + B$ and $A - B$ are called ***conjugates.*** Recall that $(A + B)(A - B) = A^2 - B^2$. We use this fact to rationalize a fraction containing a binomial denominator. We multiply the fraction by a "1" that has the *conjugate of the denominator* as its numerator and denominator.

EXAMPLE 4 Rationalize the denominator.

a. $\dfrac{5}{\sqrt{3}-\sqrt{2}} = \dfrac{5}{\sqrt{3}-\sqrt{2}} \cdot \dfrac{\sqrt{3}+\sqrt{2}}{\sqrt{3}+\sqrt{2}}$ Multiply by 1, $\dfrac{\sqrt{3}+\sqrt{2}}{\sqrt{3}+\sqrt{2}}$.

$= \dfrac{5(\sqrt{3}+\sqrt{2})}{(\sqrt{3}-\sqrt{2})(\sqrt{3}+\sqrt{2})}$

$= \dfrac{5(\sqrt{3}+\sqrt{2})}{(\sqrt{3})^2-(\sqrt{2})^2}$ Since $(A - B)(A + B) = A^2 - B^2$.

$= \dfrac{5(\sqrt{3}+\sqrt{2})}{3-2}$

$= \dfrac{5(\sqrt{3}+\sqrt{2})}{1}$

$= 5\sqrt{3} + 5\sqrt{2}$

b. $\dfrac{2+\sqrt{7}}{x+\sqrt{6}} = \dfrac{2+\sqrt{7}}{x+\sqrt{6}} \cdot \dfrac{x-\sqrt{6}}{x-\sqrt{6}}$ Multiply by 1.

$= \dfrac{(2+\sqrt{7})(x-\sqrt{6})}{(x+\sqrt{6})(x-\sqrt{6})}$

$= \dfrac{2x - 2\sqrt{6} + x\sqrt{7} - \sqrt{42}}{x^2 - (\sqrt{6})^2}$ Use FOIL in the numerator.

$= \dfrac{2x - 2\sqrt{6} + x\sqrt{7} - \sqrt{42}}{x^2 - 6}$ ■

□ **DO EXERCISE 4.**

□ **Exercise 4** Rationalize the denominator.

a. $\dfrac{\sqrt{3}+5}{\sqrt{7}+\sqrt{2}}$

b. $\dfrac{8}{\sqrt{5}-x}$

□ **Exercise 5** Rationalize the numerator.

a. $\dfrac{4-\sqrt{3}}{7}$

b. $\dfrac{\sqrt{7}+\sqrt{2}}{6-\sqrt{5}}$

Rationalizing Binomial Numerators

When a fraction has a radical in a binomial numerator, we rationalize the numerator by a method similar to the procedure for rationalizing a fraction with a binomial denominator. Multiply the fraction by a "1" that has the ***conjugate of the numerator*** as its numerator and denominator.

EXAMPLE 5 Rationalize the numerator of $\dfrac{\sqrt{2}+\sqrt{3}}{5-\sqrt{7}}$.

$$\frac{\sqrt{2}+\sqrt{3}}{5-\sqrt{7}} = \frac{\sqrt{2}+\sqrt{3}}{5-\sqrt{7}} \cdot \frac{\sqrt{2}-\sqrt{3}}{\sqrt{2}-\sqrt{3}} \qquad \text{Multiply by 1.}$$

$$= \frac{(\sqrt{2}+\sqrt{3})(\sqrt{2}-\sqrt{3})}{(5-\sqrt{7})(\sqrt{2}-\sqrt{3})}$$

$$= \frac{(\sqrt{2})^2-(\sqrt{3})^2}{5\sqrt{2}-5\sqrt{3}-\sqrt{14}+\sqrt{21}}$$

$$= \frac{2-3}{5\sqrt{2}-5\sqrt{3}-\sqrt{14}+\sqrt{21}}$$

$$= \frac{-1}{5\sqrt{2}-5\sqrt{3}-\sqrt{14}+\sqrt{21}}$$ ■

□ **DO EXERCISE 5.**

> **Rationalizing a Denominator**
>
> To rationalize a *monomial* denominator, multiply by a "1" to make the radicand in the denominator a perfect *n*th power, where *n* is the index of the radical.
>
> To rationalize a *binomial* denominator, multiply by a "1" that has the conjugate of the denominator as its numerator and denominator.

Answers to Exercises

1. a. $\dfrac{\sqrt{21}}{3}$ **b.** $\dfrac{\sqrt{66}}{6}$ **c.** $\dfrac{\sqrt[3]{2}}{2}$ **d.** $\dfrac{\sqrt[3]{36}}{2}$

2. a. $\dfrac{\sqrt{30}}{6}$ **b.** $\dfrac{\sqrt{55ab}}{5b}$ **c.** $\dfrac{\sqrt[3]{45}}{3}$ **d.** $\dfrac{\sqrt[3]{9ax}}{3x}$

3. a. $\dfrac{10}{\sqrt{70}}$ **b.** $\dfrac{3a}{\sqrt[3]{45a^2b}}$

4. a. $\dfrac{\sqrt{21}-\sqrt{6}+5\sqrt{7}-5\sqrt{2}}{5}$ **b.** $\dfrac{8\sqrt{5}+8x}{5-x^2}$

5. a. $\dfrac{13}{28+7\sqrt{3}}$ **b.** $\dfrac{5}{6\sqrt{7}-6\sqrt{2}-\sqrt{35}+\sqrt{10}}$

Problem Set 7.5

NAME

DATE

CLASS

Assume that all variables under the radical signs are positive.

Rationalize the denominator.

1. $\dfrac{4}{\sqrt{3}}$ **2.** $\dfrac{9}{\sqrt{7}}$ **3.** $\dfrac{8}{\sqrt{10}}$ **4.** $\dfrac{9}{\sqrt{15}}$

5. $\dfrac{\sqrt{5}}{\sqrt{7}}$ **6.** $\dfrac{\sqrt{11}}{\sqrt{2}}$ **7.** $\dfrac{2\sqrt{3}}{3\sqrt{5}}$ **8.** $\dfrac{3\sqrt{5}}{5\sqrt{11}}$

9. $\dfrac{1}{\sqrt[3]{4}}$ **10.** $\dfrac{12}{\sqrt[3]{2}}$ **11.** $\dfrac{5}{\sqrt[3]{4x}}$ **12.** $\dfrac{14}{\sqrt[3]{3y}}$

13. $\dfrac{-3}{7+\sqrt{6}}$ **14.** $\dfrac{-9}{8-\sqrt{5}}$ **15.** $\dfrac{\sqrt{3}}{\sqrt{7}-\sqrt{10}}$ **16.** $\dfrac{\sqrt{11}}{\sqrt{2}+\sqrt{3}}$

17. $\dfrac{\sqrt{3}+\sqrt{5}}{\sqrt{6}+\sqrt{2}}$ **18.** $\dfrac{\sqrt{7}-\sqrt{2}}{\sqrt{11}-\sqrt{3}}$ **19.** $\dfrac{8+\sqrt{5}}{x+\sqrt{10}}$ **20.** $\dfrac{9-\sqrt{2}}{y-\sqrt{7}}$

21. $\dfrac{2\sqrt{3}-\sqrt{2}}{5\sqrt{2}-\sqrt{6}}$ **22.** $\dfrac{8\sqrt{2}-\sqrt{7}}{9\sqrt{3}+\sqrt{7}}$ **23.** $\dfrac{3\sqrt{5}-2\sqrt{6}}{2\sqrt{3}+3\sqrt{7}}$ **24.** $\dfrac{2\sqrt{11}+4\sqrt{3}}{5\sqrt{2}+3\sqrt{3}}$

25. $\dfrac{\sqrt{a}-\sqrt{b}}{\sqrt{a}+\sqrt{b}}$ **26.** $\dfrac{\sqrt{x}+\sqrt{y}}{\sqrt{x}-\sqrt{y}}$ **27.** $\dfrac{\sqrt{p}+3\sqrt{q}}{3\sqrt{p}-\sqrt{q}}$ **28.** $\dfrac{\sqrt{a}-4b}{\sqrt{a}-3b}$

Simplify by rationalizing the denominator.

29. $\sqrt{\frac{7}{5}}$

30. $\sqrt{\frac{7}{2}}$

31. $\sqrt{\frac{4a}{3b}}$

32. $\sqrt{\frac{15x}{6y}}$

33. $\sqrt[3]{\frac{7}{25}}$

34. $\sqrt[3]{\frac{9}{10}}$

35. $\sqrt[3]{\frac{5}{9x}}$

36. $\sqrt[3]{\frac{7}{16y^2}}$

37. $\sqrt[4]{\frac{1}{8x^6y^3}}$

38. $\sqrt[5]{\frac{1}{8x^6y^3}}$

39. $\sqrt[5]{\frac{3x}{8x^8y^7}}$

40. $\sqrt[4]{\frac{12a}{27a^7y^5}}$

Rationalize the numerator.

41. $\frac{\sqrt{11}}{9}$

42. $\frac{\sqrt{15}}{2}$

43. $\frac{\sqrt{5}}{\sqrt{7}}$

44. $\frac{\sqrt{10}}{\sqrt{3}}$

45. $\frac{\sqrt{6}}{\sqrt{2}}$

46. $\frac{\sqrt{10}}{\sqrt{5}}$

47. $\frac{\sqrt{2a}}{\sqrt{5}}$

48. $\frac{\sqrt{6x}}{3}$

49. $\frac{\sqrt[3]{5}}{8}$

50. $\frac{\sqrt[3]{16}}{3}$

51. $\frac{\sqrt[3]{25a^2}}{\sqrt[3]{3b}}$

52. $\frac{\sqrt[3]{4b}}{\sqrt[3]{2a}}$

53. $\frac{\sqrt{xy}}{2}$

54. $\frac{\sqrt{ab}}{7}$

55. $\frac{\sqrt{2} - 9}{7}$

56. $\frac{\sqrt{3} + 4}{2}$

57. $\frac{9 + \sqrt{3}}{9 - \sqrt{2}}$

58. $\frac{8 - \sqrt{5}}{7 - \sqrt{3}}$

59. $\frac{\sqrt{3} + \sqrt{7}}{\sqrt{3} + \sqrt{6}}$

60. $\frac{\sqrt{5} - \sqrt{2}}{\sqrt{7} + \sqrt{3}}$

61. $\frac{3\sqrt{2} - 7\sqrt{3}}{5\sqrt{2} - 2\sqrt{5}}$

62. $\frac{6\sqrt{3} + 8\sqrt{2}}{5\sqrt{5} + 7\sqrt{6}}$

63. $\frac{\sqrt{3} - \sqrt{x}}{\sqrt{3} + \sqrt{x}}$

64. $\frac{\sqrt{5} + \sqrt{b}}{\sqrt{5} - \sqrt{b}}$

65. $\frac{a\sqrt{b} - c}{\sqrt{b} - c}$

66. $\frac{y + x\sqrt{z}}{x + \sqrt{z}}$

67. The first step in rationalizing a monomial denominator is to multiply by ____________ to make the radicand in the denominator a perfect nth power, where n is the index of the radical.

68. To rationalize a binomial denominator, multiply the fraction by a 1 that has the ____________ of the denominator as its numerator and denominator.

Assume that all variables under radical signs represent positive numbers.

Rationalize the denominator.

* 69. $\frac{\sqrt{x} - 3\sqrt{y}}{2\sqrt{x} + \sqrt{y}}$

* 70. $\frac{\sqrt{a} + \sqrt{a-2}}{\sqrt{a} - \sqrt{a-2}}$

* 71. $\frac{\sqrt{x} + \sqrt{5} - \sqrt{3}}{\sqrt{7}}$

* 72. $\frac{2\sqrt{x+y}}{2 + \sqrt{x+y}}$

Rationalize the numerator.

* 73. $\frac{5 - \sqrt{5-x}}{4}$

* 74. $\frac{\sqrt{11}}{\sqrt{3} + \sqrt{x} - \sqrt{7}}$

Perform the operations indicated and simplify.

* **75.** $\sqrt{x} + \dfrac{3}{\sqrt{x}}$

* **76.** $\dfrac{5}{\sqrt{3x}} - \dfrac{4\sqrt{3x}}{x} + \sqrt{\dfrac{3}{x}}$

* **77.** $\sqrt{\dfrac{2}{5}} - \dfrac{25}{\sqrt{10}} - \dfrac{7\sqrt{10}}{5}$

* **78.** $2\sqrt[3]{36} - \dfrac{10}{\sqrt[3]{6}} + 5\sqrt[3]{\dfrac{1}{6}}$

Checkup

The following problems provide a review of some of Sections 1.6 and 6.3 and will help you with the next section. Assume that all variables represent nonzero real numbers.

Multiply and write with positive exponents.

79. $2^3 \cdot 2^5$

80. $3^4 \cdot 3^7$

81. $x^{-3} \cdot x^5$

82. $y^{-2} \cdot y^{-4}$

Divide and write positive exponents.

83. $\dfrac{8^{-3}}{8^{-2}}$

84. $\dfrac{7^{-4}}{7^{-5}}$

85. $\dfrac{x^2y}{x^5y^{-3}}$

86. $\dfrac{y^7z^{-2}}{y^3z^4}$

Raise to powers. Write with positive exponents.

87. $(3x^2y^{-2})^3$

88. $(2x^{-3}y^2)^{-2}$

Simplify.

89. $\dfrac{\dfrac{x-2}{x^3}}{\dfrac{x^2-4}{2x^2}}$

90. $\dfrac{\dfrac{x-3}{4x}}{\dfrac{x+2}{x^2}}$

91. $\dfrac{2 + \dfrac{5}{x-1}}{3 - \dfrac{1}{x}}$

92. $\dfrac{\dfrac{2}{3x} + \dfrac{1}{6x}}{5 - \dfrac{5}{6x}}$

7.6 RATIONAL EXPONENTS

1 Using Rational Exponents

We want to define expressions like $3^{1/4}$ and $2^{1/2}$ so that the rules for exponents hold. Recall that for integer exponents, $a^m \cdot a^n = a^{m+n}$. What meaning should we give to $2^{1/2}$? If the rules for exponents are to hold, then

$$2^{1/2} \cdot 2^{1/2} = 2^{1/2 + 1/2} = 2^1 = 2$$

We know that $\sqrt{2} \cdot \sqrt{2} = 2$. Using the rules of exponents, we found that $2^{1/2} \cdot 2^{1/2} = 2$. Hence we define $2^{1/2} = \sqrt{2}$. In general, we define $a^{1/n}$ as the principal nth root of a.

> $$a^{1/n} = \sqrt[n]{a}$$
>
> for any nonnegative real number a and any index n.

We assume that the *base* is *nonnegative* when we use *rational exponents.*

EXAMPLE 1 Write without rational exponents.

a. $8^{1/3} = \sqrt[3]{8} = 2$ **b.** $25^{1/2} = \sqrt{25} = 5$

c. $a^{1/5} = \sqrt[5]{a}$ **d.** $3^{1/4} = \sqrt[4]{3}$ ■

□ **DO EXERCISE 1.**

EXAMPLE 2 Write with rational exponents.

a. $\sqrt{5} = 5^{1/2}$ **b.** $\sqrt[4]{7x^2y} = (7x^2y)^{1/4}$ ■

□ **DO EXERCISE 2.**

How should we define $8^{2/3}$ for our rules of exponents to hold? We must have

$$8^{2/3} = (8^{1/3})^2 \quad \text{or} \quad (8^2)^{1/3}, \qquad \text{so } 8^{2/3} = (\sqrt[3]{8})^2 \quad \text{or} \quad \sqrt[3]{8^2}$$

Generalizing, we define $a^{m/n}$ as follows.

> $$a^{m/n} = (\sqrt[n]{a})^m \quad \text{or} \quad a^{m/n} = \sqrt[n]{a^m}$$
>
> for any nonnegative real number a and any natural numbers m and n ($n \neq 1$).

EXAMPLE 3 Write without rational exponents.

a. $64^{2/3} = (\sqrt[3]{64})^2 = 4^2 = 16$

Using the alternative definition,

$$64^{2/3} = \sqrt[3]{64^2} = \sqrt[3]{4096} = 16$$

Usually, the definition $a^{m/n} = (\sqrt[n]{a})^m$ is easier to use when a is a specific number.

b. $100^{3/2} = (\sqrt{100})^3 = 10^3 = 1000$

c. $x^{15/4} = \sqrt[4]{x^{15}} = \sqrt[4]{x^{12}x^3} = x^3\sqrt[4]{x^3}$ ■

OBJECTIVES

1. *Write expressions with and without rational exponents*
2. *Use the definition of negative rational exponents to write expressions with positive rational exponents*
3. *Use the rules for exponents with rational exponents*
4. *Simplify radicals using rational exponents*

□ **Exercise 1** Write without rational exponents.

a. $9^{1/2}$ **b.** $16^{1/4}$

c. $x^{1/3}$ **d.** $(abc)^{1/6}$

e. $6^{1/5}$ **f.** $8^{1/7}$

□ **Exercise 2** Write with rational exponents.

a. $\sqrt[3]{7}$ **b.** $\sqrt{3ab}$

□ **Exercise 3** Write without rational exponents.

a. $27^{2/3}$ **b.** $36^{3/2}$

c. $x^{4/5}$ **d.** $y^{5/3}$

e. $y^{8/3}$ **f.** $x^{9/4}$

□ **Exercise 4** Write with rational exponents.

a. $\sqrt[5]{8^3}$ **b.** $\sqrt[4]{(7x)^3}$

□ **Exercise 5** Write with positive exponents. Simplify numbers, if possible, by writing as radicals and evaluating.

a. $36^{-3/2}$ **b.** $(8xy)^{-7/9}$

Notice that the denominator of the rational exponent is the index of the radical and the numerator is the power.

□ **DO EXERCISE 3.**

EXAMPLE 4 Write with rational exponents.

a. $\sqrt[3]{7^4} = 7^{4/3}$ **b.** $\sqrt[5]{(9ab)^2} = (9ab)^{2/5}$ ■

□ **DO EXERCISE 4.**

2 Negative Rational Exponents

We define negative rational exponents in a way that is identical to the definition of negative integer exponents.

> $$a^{-m/n} = \frac{1}{a^{m/n}}$$
>
> for any rational number m/n and any positive real number a.

For the rest of this section, we will assume that *bases* are *positive*.

EXAMPLE 5 Write with positive exponents. Simplify numbers, if possible, by writing as radicals and evaluating.

a. $8^{-2/3} = \dfrac{1}{8^{2/3}}$ Write with a positive exponent.

$$= \frac{1}{(\sqrt[3]{8})^2}$$

$$= \frac{1}{2^2}$$

$$= \frac{1}{4}$$

b. $(3ab)^{-3/5} = \dfrac{1}{(3ab)^{3/5}}$ ■

□ **DO EXERCISE 5.**

3 Rules for Rational Exponents

The following rules for integer exponents also hold for rational number exponents.

> For any positive real number a and any rational number exponents m and n,
>
> **1.** $a^m \cdot a^n = a^{m+n}$
>
> **2.** $\dfrac{a^m}{a^n} = a^{m-n}$
>
> **3.** $(a^m)^n = a^{mn}$

EXAMPLE 6 Simplify using the rules for rational exponents. Write answers with positive exponents.

a. $5^{3/2} \cdot 5^{1/2} = 5^{3/2+1/2} = 5^{4/2} = 5^2 = 25$

b. $\dfrac{3^{2/3}}{3^{5/3}} = 3^{2/3-5/3} = 3^{-1} = \dfrac{1}{3}$

c. $(7^{2/3})^5 = 7^{(2/3)(5)} = 7^{10/3}$

d. $\dfrac{(x^{1/4})^{4/5}}{x} = \dfrac{x^{1/5}}{x^1} = x^{1/5-1} = x^{-4/5} = \dfrac{1}{x^{4/5}}$ ■

□ **DO EXERCISE 6.**

4 Simplifying Radicals Using Rational Exponents

We can simplify many expressions involving radicals by converting the radical expressions to expressions with rational number exponents. Then if the exponent is not an integer, we simplify using the rules for exponents, if necessary, and convert the answer back to radical form.

EXAMPLE 7 Simplify by converting to a rational exponent. If the exponent is not an integer, use the rules for exponents, if necessary, and convert back to radical form.

a. $\sqrt{x^6} = x^{6/2} = x^3$

b. $\sqrt[3]{x^{14}} = x^{14/3}$ Convert to an exponential expression.

$= x^{4\,2/3}$ Write 14/3 as a mixed number.

$= x^{4+2/3}$

$= x^4 \cdot x^{2/3}$ Use the first rule for exponents.

$= x^4\sqrt[3]{x^2}$ Convert to radical notation.

c. $\sqrt[4]{9} = 9^{1/4}$ Convert to exponential notation.

$= (3^2)^{1/4}$ Write 9 as 3^2.

$= 3^{1/2}$ Use the third rule for exponents.

$= \sqrt{3}$ Convert to radical notation.

d. $\sqrt[6]{p^2q^4} = (p^2q^4)^{1/6}$ Convert to exponential notation.

$= p^{2/6} \cdot q^{4/6}$ Use the third rule for exponents.

$= p^{1/3} \cdot q^{2/3}$ Simplify.

$= (pq^2)^{1/3}$ Use the third rule for exponents.

$= \sqrt[3]{pq^2}$ Convert to radical notation. ■

□ **DO EXERCISE 7.**

We can use the properties of rational exponents to write a single radical expression for a product, quotient or term containing more than one radical.

EXAMPLE 8 Use rational exponents to write a single radical expression.

a. $\sqrt[3]{x^2} \cdot \sqrt{x} = x^{2/3} \cdot x^{1/2}$ Convert to exponential notation.

$= x^{4/6} \cdot x^{3/6}$ Write with a common denominator.

$= x^{7/6}$ Use the first rule for exponents.

$= \sqrt[6]{x^7}$ Convert to radical notation.

$= x\sqrt[6]{x}$

□ **Exercise 6** Simplify using the rules for rational exponents. Write answers with positive exponents.

a. $6^{1/3} \cdot 6^{3/4}$

b. $\dfrac{8^{1/6}}{8^{5/6}}$

c. $(9^{1/2})^{1/4}$

d. $y^{3/8} \cdot y^{5/8}$

□ **Exercise 7** Simplify by converting to a rational exponent. If the exponent is not an integer, use the rules for exponents, if necessary, and convert back to radical form.

a. $\sqrt{y^8}$ **b.** $\sqrt[7]{x^7}$

c. $\sqrt[4]{x^{11}}$ **d.** $\sqrt{x^7y^9}$

e. $\sqrt[6]{27}$ **f.** $\sqrt[4]{a^4b^8}$

g. $\sqrt[5]{x^5y^{15}}$ **h.** $\sqrt[8]{x^4y^2}$

□ **Exercise 8** Use rational exponents to write a single radical expression.

a. $\sqrt[3]{x^2} \cdot \sqrt[4]{x}$

b. $\dfrac{\sqrt[4]{x^3}}{\sqrt{x}}$

c. $\sqrt[4]{\sqrt{z}}$

b. $\dfrac{\sqrt{x^3}}{\sqrt[4]{x}} = \dfrac{x^{3/2}}{x^{1/4}}$ Convert to exponential notation.

$= \dfrac{x^{6/4}}{x^{1/4}}$ Write with a common denominator.

$= x^{5/4}$ Use the second rule for exponents.

$= \sqrt[4]{x^5}$ Convert to radical notation.

$= x\sqrt[4]{x}$

c. $\sqrt[3]{\sqrt{b}} = \sqrt[3]{b^{1/2}}$

$= (b^{1/2})^{1/3}$

$= b^{1/6}$ Use the third rule for exponents.

$= \sqrt[6]{b}$ Convert to radical notation. ■

□ **DO EXERCISE 8.**

Methods of Simplifying Radical Expressions

1. *Factor the radicand.* Remove all factors that are perfect powers. That is, the radicand should contain no factor to a power greater than or equal to the index.
2. *Rationalize the denominator.* Radical expressions are usually said to be simplified if the denominator does not contain a radical.
3. *Simplify using rational exponents.* Convert to exponential notation. If the exponent is not an integer, use the rules for exponents, if necessary, and then convert back to a radical expression. Use this method only for positive radicands, since we have only defined rules for rational exponents for positive radicands in this book.

DID YOU KNOW?

Isaac Newton (1642–1727) was probably the first mathematician to use rational numbers as exponents. Before Newton, only integers were used for exponents. For instance, $x^{3/2}$ would be written $\sqrt{x^3}$, which is much more cumbersome.

Answers to Exercises

1. a. 3 **b.** 2 **c.** $\sqrt[3]{x}$ **d.** $\sqrt[6]{abc}$ **e.** $\sqrt[5]{6}$ **f.** $\sqrt[7]{8}$

2. a. $7^{1/3}$ **b.** $(3ab)^{1/2}$

3. a. 9 **b.** 216 **c.** $\sqrt[5]{x^4}$ **d.** $y\sqrt[3]{y^2}$ **e.** $y^2\sqrt[3]{y^2}$ **f.** $x^2\sqrt[4]{x}$

4. a. $8^{3/5}$ **b.** $(7x)^{3/4}$

5. a. $\dfrac{1}{216}$ **b.** $\dfrac{1}{(8xy)^{7/9}}$

6. a. $6^{13/12}$ **b.** $\dfrac{1}{8^{2/3}}$ **c.** $9^{1/8}$ **d.** y

7. a. y^4 **b.** x **c.** $x^2\sqrt[4]{x^3}$ **d.** $x^3y^4\sqrt{xy}$ **e.** $\sqrt{3}$ **f.** ab^2 **g.** xy^3 **h.** $\sqrt[4]{x^2y}$

8. a. $\sqrt[12]{x^{11}}$ **b.** $\sqrt[4]{x}$ **c.** $\sqrt[8]{z}$

Problem Set 7.6

NAME ______________________

DATE ______________________

CLASS ______________________

Assume that all bases are positive.

Write without rational exponents.

1. $y^{1/3}$ **2.** $x^{1/6}$ **3.** $25^{1/2}$ **4.** $27^{1/3}$

5. $81^{1/4}$ **6.** $121^{1/2}$ **7.** $64^{3/2}$ **8.** $32^{2/5}$

9. $125^{4/3}$ **10.** $49^{3/2}$ **11.** $32^{4/5}$ **12.** $16^{5/4}$

13. $(x^2yz)^{1/3}$ **14.** $(pq^2r)^{1/5}$ **15.** $(7xy)^{4/9}$ **16.** $(6abc)^{10/3}$

Write with rational exponents.

17. $\sqrt{a}$ **18.** $\sqrt[3]{b}$ **19.** $\sqrt[4]{15}$ **20.** $\sqrt[6]{11}$

21. $\sqrt[4]{7^5}$ **22.** $\sqrt[5]{9^3}$ **23.** $\sqrt[3]{5^7}$ **24.** $\sqrt{3^9}$

25. $\sqrt{(xyz)^9}$ **26.** $\sqrt[3]{(a^2bc)^4}$ **27.** $\sqrt[9]{(pq)^8}$ **28.** $\sqrt[7]{(rs)^6}$

Write with positive exponents.

29. $y^{-1/5}$ **30.** $x^{-3/4}$ **31.** $3^{-2/5}$ **32.** $7^{-8/5}$

33. $\dfrac{1}{x^{-2/3}}$ **34.** $\dfrac{1}{y^{-3/5}}$ **35.** $\dfrac{3}{a^{-4/3}}$ **36.** $\dfrac{5}{b^{-1/2}}$

Simplify using the rules for rational exponents. Write answers with positive exponents.

37. $7^{1/2} \cdot 7^{3/2}$ **38.** $2^{4/3} \cdot 2^{2/3}$ **39.** $8^{3/5} \cdot 8^{1/10}$ **40.** $5^{1/3} \cdot 5^{1/4}$

41. $\dfrac{3^{5/6}}{3^{1/6}}$ **42.** $\dfrac{4^{7/8}}{4^{3/8}}$ **43.** $\dfrac{6^{2/3}}{6^{1/4}}$ **44.** $\dfrac{9^{7/6}}{9^{1/2}}$

45. $(5^{1/2})^{2/3}$ **46.** $(2^{7/8})^{4/3}$ **47.** $x^{3/8} \cdot x^{1/8}$ **48.** $y^{5/4} \cdot y^{7/4}$

49. $\dfrac{a^{1/5} \cdot a^{1/2}}{a^{3/10}}$ **50.** $\dfrac{b^{1/3} \cdot b^{1/4}}{b^{5/12}}$ **51.** $\dfrac{(x^{1/2})^{-2}}{x^2}$ **52.** $\dfrac{(y^{-4/5})^{10}}{y^6}$

Simplify by converting to rational exponents. If the exponent is not an integer, use the rules for exponents, if necessary, and convert back to radical form.

53. $\sqrt[8]{x^4}$ **54.** $\sqrt[8]{y^2}$ **55.** $\sqrt{2^6}$ **56.** $\sqrt{3^8}$

57. $\sqrt[6]{7^3}$ **58.** $\sqrt[10]{2^5}$ **59.** $\sqrt[8]{x^{17}}$ **60.** $\sqrt[5]{3^{12}}$

61. $\sqrt[3]{x^8}$

62. $\sqrt[4]{y^9}$

63. $\sqrt[6]{8}$

64. $\sqrt[8]{16}$

65. $\sqrt{x^6y^4}$

66. $\sqrt{a^2b^8}$

67. $\sqrt[8]{x^6y^2}$

68. $\sqrt[9]{x^3y^6}$

69. $\sqrt[6]{64p^6q^{18}}$

70. $\sqrt[4]{81x^{12}y^8}$

71. $\dfrac{\sqrt[3]{a^5}}{\sqrt[3]{a^4}}$

72. $\dfrac{\sqrt[3]{x^8}}{\sqrt[3]{x^4}}$

Use rational exponents to write a single radical expression.

73. $\sqrt[5]{x^4} \cdot \sqrt{x}$

74. $\sqrt[4]{x^3} \cdot \sqrt[3]{x}$

75. $\sqrt[3]{2} \cdot \sqrt{2}$

76. $\sqrt[5]{4} \cdot \sqrt[3]{4}$

77. $\dfrac{\sqrt{x^5}}{\sqrt[3]{x^2}}$

78. $\dfrac{\sqrt[4]{z}}{\sqrt[8]{z}}$

79. $\dfrac{\sqrt{5}}{\sqrt[4]{5}}$

80. $\dfrac{\sqrt[3]{7}}{\sqrt[4]{7}}$

81. $\sqrt[5]{\sqrt{x}}$

82. $\sqrt{\sqrt[3]{n}}$

83. $\sqrt{\sqrt[4]{27}}$

84. $\sqrt[4]{\sqrt{32}}$

85. We define $a^{1/n} = \sqrt[n]{a}$ for any __________ real number a and any index n.

86. To write a variable with a rational exponent in radical form, use the __________ of the fraction as the index of the radical.

Use a calculator with a y^x key to approximate each expression. Round answers to the nearest hundredth.

87. $36^{0.5}$

88. $81^{0.25}$

89. $49^{1.5}$

90. $16^{1.25}$

91. $32^{0.6}$

92. $16^{0.75}$

93. $76.34^{0.375}$

94. $0.8^{4.29}$

Assume that all bases are positive. Simplify. Write answers with positive exponents.

* **95.** $\left(\dfrac{a^4b^{5/2}}{a^{-2}b^{3/2}}\right)^{1/6}$

* **96.** $\left[\dfrac{(x^{2/3}y^{3/4})(x^{2/5}y^{-3/2})}{x^2y^{-1/4}}\right]^{-1}$

* **97.** $\left(\dfrac{16c^{-8}}{d^{12}}\right)^{-1/4} \cdot \left(\dfrac{-c^{-3/2}}{d^{1/2}}\right)^2$

* **98.** $\left(\dfrac{x^0y^{1/2}}{z^6}\right)^{1/2} \cdot \left(\dfrac{x^{-3/7}y^{3/4}}{z^{-6}}\right)^{-1/3}$

99. $\sqrt[3]{\sqrt{x}}$

100. $\sqrt[4]{\sqrt[3]{x}}$

* **101.** $(\sqrt[4]{9}\sqrt[5]{x})^{20}$

* **102.** $\sqrt[12]{9x^2 + 24x + 16}$

Checkup

The following problems provide a review of some of Sections 2.1 and 6.4 and will help you with the next section.

Solve.

103. $3y - 5y + 2 = y - 8$

104. $8x - 12 + 3x = 5x - 4$

105. $\dfrac{5}{x-4} - \dfrac{2}{x-4} = 3$

106. $\dfrac{2}{x+3} + \dfrac{4}{x+3} = \dfrac{3}{2}$

107. $\dfrac{x}{x-2} = \dfrac{x+3}{x+2} - \dfrac{x}{x^2-4}$

108. $x + \dfrac{6}{x} = 5$

109. $\dfrac{15}{2} - \dfrac{3}{p-4} = \dfrac{-6}{2p-8}$

110. $\dfrac{8}{2x-6} - \dfrac{24}{4x+12} = \dfrac{24}{x^2-9}$

7.7 RADICAL EQUATIONS

OBJECTIVE

[1] *Solve radical equations*

[1] In our study of business and the physical sciences, we often find equations with radicals. A ***radical equation*** contains at least one radical with a variable in the radicand. To solve these equations we use a new rule. The rule says that if both sides of an equation are raised to the same power, we get another true equation.

> **Power Rule for Equations**
>
> If
>
> $$a = b$$
>
> then
>
> $$a^n = b^n$$
>
> for any natural number n.

The power rule shows that all solutions of the original equation are also solutions of the new equation. It *does not* say that all solutions to the new equation are solutions to the original equation. Therefore, *solutions to the new equation must be checked in the original equation.* Solutions that do not satisfy the original equation are *extraneous.* To solve equations containing radicals, we isolate a radical on one side of the equal sign.

EXAMPLE 1 Solve.

a. $\sqrt{x} + 2 = 5$

$\sqrt{x} = 3$ Add -2 to both sides to isolate the radical.

$(\sqrt{x})^2 = 3^2$ Use the power rule to square both sides.

$x = 9$

Check $\sqrt{x} + 2 = 5$

$\sqrt{x} + 2$	5
$\sqrt{9} + 2$	5
$3 + 2$	5
5	5

The solution is 9.

b. $\sqrt{x} = -7$

Notice that this equation has no solution since the principal square root of a number cannot be a negative number. We will try to solve it and see what happens.

$\sqrt{x} = -7$

$(\sqrt{x})^2 = (-7)^2$ Square both sides.

$x = 49$

Check $\sqrt{x} = -7$

$\sqrt{x}$	-7
$\sqrt{49}$	-7
7	-7

The number 49 does not check. It is an extraneous solution. The equation has no real number solution. ■

□ **DO EXERCISE 1.**

□ **Exercise 1** Solve.

a. $\sqrt{x} + 3 = 9$

b. $\sqrt{y} = -5$

□ **Exercise 2** Solve.

a. $\sqrt{x-9} = 3$

b. $x + 1 = \sqrt{x+3}$

Following is a general method for solving equations with radicals.

1. Isolate one radical on one side of the equal sign.
2. Raise each side of the equation to the power that is the same as the index of the radical.
3. If the equation still contains a radical, repeat steps 1 and 2.
4. Solve the equation.
5. Possible solutions must be checked.

EXAMPLE 2 Solve $x - 7 = \sqrt{x-5}$.

$$x - 7 = \sqrt{x-5}$$

$$(x-7)^2 = (\sqrt{x-5})^2 \quad \text{Square both sides.}$$

$$x^2 - 14x + 49 = x - 5 \quad \text{Recall that } (A-B)^2 = A^2 - 2AB + B^2.$$

$$x^2 - 15x + 54 = 0 \quad \text{Add } -x \text{ and 5 to both sides.}$$

$$(x-6)(x-9) = 0 \quad \text{Factor.}$$

$$x - 6 = 0 \quad \text{or} \quad x - 9 = 0 \quad \text{Use the zero product property.}$$

$$x = 6 \qquad x = 9$$

Check If $x = 6$: If $x = 9$:

$x - 7$	$\sqrt{x-5}$
$6 - 7$	$\sqrt{6-5}$
-1	$\sqrt{1}$
-1	1

$x - 7$	$\sqrt{x-5}$
$9 - 7$	$\sqrt{9-5}$
2	$\sqrt{4}$
2	2

The number 6 does not check. It is an extraneous solution. The number 9 checks. The solution is 9. ■

□ **DO EXERCISE 2.**

The following equations contain two terms with radicals. To solve them, we square both sides of the equation twice. If necessary, first we isolate one radical on one side of the equation.

EXAMPLE 3 Solve.

a. $\sqrt{5y + 9} = 3 + \sqrt{y}$

$\sqrt{5y+9} = 3 + \sqrt{y}$	
$(\sqrt{5y+9})^2 = (3 + \sqrt{y})^2$	Square both sides.
$5y + 9 = 3^2 + 2 \cdot 3\sqrt{y} + (\sqrt{y})^2$	Recall that $(A + B)^2 = A^2 + 2AB + B^2$.
$5y + 9 = 9 + 6\sqrt{y} + y$	
$4y = 6\sqrt{y}$	Add -9 and $-y$ to isolate the radical.
$2y = 3\sqrt{y}$	Divide both sides by 2.
$(2y)^2 = (3\sqrt{y})^2$	Square both sides.
$4y^2 = 9y$	
$4y^2 - 9y = 0$	Add $-9y$ to both sides.
$y(4y - 9) = 0$	Factor.
$y = 0$ or $4y - 9 = 0$	Use the zero product property.
$4y = 9$	
$y = \dfrac{9}{4}$	

Both 0 and $\frac{9}{4}$ check. The solutions are 0 and $\frac{9}{4}$.

b. $\sqrt{x + 6} - \sqrt{x - 2} = 2$

$\sqrt{x+6} - \sqrt{x-2} = 2$	
$\sqrt{x+6} = 2 + \sqrt{x-2}$	Add $\sqrt{x-2}$ to both sides to isolate one radical.
$(\sqrt{x+6})^2 = (2 + \sqrt{x-2})^2$	Square both sides.
$(\sqrt{x+6})^2 = 2^2 + 2 \cdot 2\sqrt{x-2} + (\sqrt{x-2})^2$	Recall that $(A + B)^2 = A^2 + 2AB + B^2$.
$x + 6 = 4 + 4\sqrt{x-2} + (x - 2)$	
$x + 6 = 2 + 4\sqrt{x-2} + x$	Combine like terms.
$4 = 4\sqrt{x-2}$	Add $-x$ and -2 to isolate the radical.
$1 = \sqrt{x-2}$	Divide both sides by 4.
$(1)^2 = (\sqrt{x-2})^2$	Square both sides.
$1 = x - 2$	
$3 = x$	

The number 3 checks. It is the solution. ■

□ **DO EXERCISE 3.**

□ **Exercise 3** Solve.

a. $\sqrt{y + 1} = 1 + \sqrt{y - 4}$

b. $\sqrt{4x - 3} - \sqrt{2x - 5} = 2$

□ **Exercise 4** Solve.

a. $\sqrt[3]{6y+9} + 8 = 5$

b. $\sqrt[4]{z+11} = \sqrt[4]{2z+6}$

The power rule may be used to solve radical equations containing radicals with an index greater than 2.

EXAMPLE 4 Solve $\sqrt[4]{y+8} = \sqrt[4]{2y}$.

$$\sqrt[4]{y+8} = \sqrt[4]{2y}$$

$$(\sqrt[4]{y+8})^4 = (\sqrt[4]{2y})^4 \qquad \text{Raise both sides to the fourth power.}$$

$$y + 8 = 2y$$

$$8 = y \qquad \text{Add } -y \text{ to both sides.}$$

Check $\sqrt[4]{y+8} = \sqrt[4]{2y}$

$\sqrt[4]{8+8}$	$\sqrt[4]{2(8)}$
$\sqrt[4]{16}$	$\sqrt[4]{16}$
2	2

The solution is 8. ■

□ **DO EXERCISE 4.**

Answers to Exercises

1. a. 36 **b.** No solution

2. a. 18 **b.** 1

3. a. 8 **b.** 7, 3

4. a. −6 **b.** 5

Problem Set 7.7

NAME ______________________

DATE ______________________

CLASS ______________________

Solve.

1. $\sqrt{x} = 4$

2. $\sqrt{y} = 7$

3. $\sqrt{x} + 3 = 5$

4. $\sqrt{x} - 2 = 4$

5. $\sqrt{x - 2} = 3$

6. $\sqrt{y + 1} = 7$

7. $\sqrt{7x - 3} = 5$

8. $\sqrt{3x + 1} = 9$

9. $\sqrt{3x + 1} - 2 = 2$

10. $\sqrt{5y + 1} + 3 = 8$

11. $\sqrt{x - 3} = -2$

12. $\sqrt{y + 7} = -5$

13. $\sqrt{x - 2} = 4$

14. $\sqrt{4y + 1} = 9$

15. $1 - \sqrt{4x + 3} = -5$

16. $7 - \sqrt{3y + 1} = -1$

17. $y + 3\sqrt{y - 2} = 12$

18. $x + 2\sqrt{x + 1} = 7$

19. $7 = \dfrac{1}{\sqrt{x}}$

20. $\dfrac{1}{\sqrt{y}} = 5$

21. $\sqrt{x + 1} = \sqrt{2x - 3}$

22. $\sqrt{2y + 3} = \sqrt{3y - 5}$

23. $\sqrt{5x - 2} = \sqrt{4x + 3}$

24. $\sqrt{8x - 4} = \sqrt{7x + 2}$

25. $\sqrt{x - 9} + \sqrt{x} = 1$

26. $\sqrt{x - 5} + \sqrt{x} = 5$

27. $\sqrt{x + 8} = \sqrt{x - 4} + 2$

28. $\sqrt{x+5} = \sqrt{x-3} + 2$

29. $\sqrt{x+2} - \sqrt{x-3} = 1$

30. $\sqrt{y+2} + \sqrt{y-1} = 3$

31. $\sqrt[3]{3y+6} = 1$

32. $\sqrt[3]{2x+4} = 2$

33. $\sqrt{x-5} - \sqrt{x-8} = 3$

34. $\sqrt{x-3} - \sqrt{x-3} = 4$

35. $\sqrt[4]{x-1} = 2$

36. $\sqrt[3]{2y+5} = \sqrt[3]{6y+1}$

37. $\sqrt{4x+16} = 4 + \sqrt{x}$

38. $\sqrt{2x-1} - 2 = \sqrt{x-4}$

39. $\sqrt{x} - \sqrt{x+9} + 1 = 0$

40. $\sqrt{x} - \sqrt{x+24} + 2 = 0$

41. $\sqrt{x+1} + \sqrt{x} - 2 = 0$

42. $\sqrt{2x+4} + \sqrt{2x} - 3 = 0$

43. $\sqrt[4]{x^2+2x+8} - 2 = 0$

44. $\sqrt[4]{y^2+y-1} - 1 = 0$

45. Possible solutions to a radical equation that do not satisfy the equation are called ____________.

46. To solve an equation containing two terms with radicals, we ____________ both sides of the equation twice.

Solve.

Round answers to the nearest tenth.

47. $\sqrt{x} + 3.8 = 9.5$

48. $\sqrt{y} - 10.5 = 4.8$

49. $\sqrt{x+9.4} - \sqrt{x-3.8} = 1.5$

50. $\sqrt{y+1.6} = 1.2 + \sqrt{y-3.7}$

Think About It

51. The area A of a triangle can be found from the lengths of the sides a, b, and c by the following formula where $s = (a + b + c)/2$.

$$A = \sqrt{s(s - a)(s - b)(s - c)}$$

Find the area when the sides of the triangle are 52.8 centimeters, 49.7 centimeters and 41.5 centimeters.

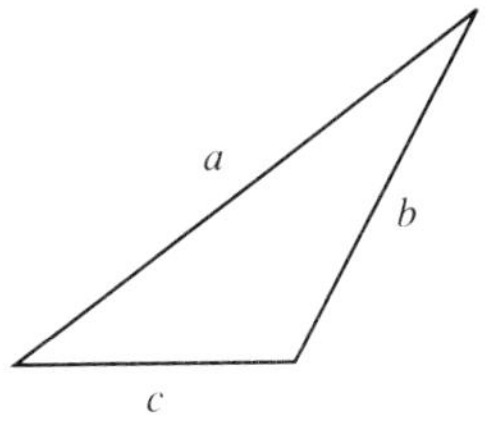

52. The population P of a culture of bacteria at time t is given by the equation

$$P = \sqrt[4]{100t^5 + 10t^2 + 9}$$

Find the population P when $t = 3.4$ hours.

Solve.

* **53.** $\sqrt[3]{x^2 + 2x + 1} = \sqrt[3]{x^2 + 5}$

* **54.** $\sqrt[4]{p^2 - 4p} = \sqrt[4]{5}$

* **55.** $\sqrt{\frac{y}{3}} = \frac{1}{2}$

* **56.** $\sqrt{x + 5} - \sqrt{x - 3} = 2$

* **57.** $\sqrt[3]{x^2 - 3x - 2} - 2 = 0$

* **58.** $\sqrt{\sqrt{x} + 9} = \sqrt{x - 3}$

* **59.** $2\sqrt{x} = \sqrt{4x + 65} - 5$

* **60.** $\sqrt{x - 4} \cdot \sqrt{x + 4} - 2\sqrt{5} = 0$

* **61.** $\sqrt{x} - \sqrt{x - 8} = \frac{2}{\sqrt{x - 8}}$

* **62.** $\sqrt{\sqrt{x^2 + 15x - 19}} = 3$

* **63.** $\dfrac{\sqrt{1+y}+\sqrt{1-y}}{\sqrt{1+y}-\sqrt{1-y}} = 2$

* **64.** $\sqrt{5x + 2\sqrt{x^2 - x + 7}} - 4 = 0$

Checkup

The following problems provide a review of some of Sections 6.6 and 7.5 and some will help you with the next section.

Solve.

65. $x = \dfrac{y}{y+z}$ for y

66. $a = \dfrac{b}{1 - bc}$ for b

67. $\dfrac{1}{x} = \dfrac{1}{y} + \dfrac{1}{z}$ for z

68. $\dfrac{1}{x} = \dfrac{1}{y} + \dfrac{1}{z}$ for x

Rationalize the denominator.

69. $\dfrac{5}{\sqrt{2}}$

70. $\dfrac{7}{\sqrt{6}}$

71. $\dfrac{3}{\sqrt[3]{2}}$

72. $\dfrac{8}{\sqrt[3]{4}}$

73. $\dfrac{6}{5 - \sqrt{3}}$

74. $\dfrac{-8}{3 - \sqrt{7}}$

75. $\dfrac{3 + \sqrt{2}}{5 + \sqrt{2}}$

76. $\dfrac{4 - \sqrt{3}}{7 + \sqrt{6}}$

7.8 APPLICATIONS

1 There are many problems that involve powers and roots. Some of these problems involve the Pythagorean theorem, which is a relationship between the lengths of the sides of a right triangle. A ***right triangle*** has a 90° angle.

Pythagorean Theorem

In a right triangle, the sum of the squares of the lengths of the two shorter sides, *a* and *b*, is equal to the square of the length of the longest side, *c*.

$$a^2 + b^2 = c^2$$

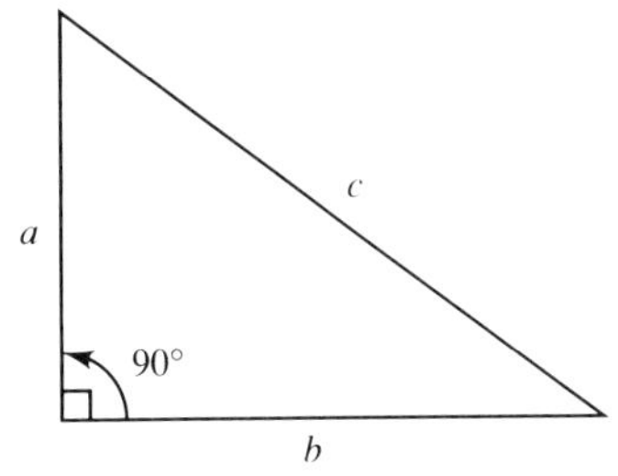

The longest side is called the *hypotenuse* and the shorter sides are called the *legs*.

From this formula $c^2 = a^2 + b^2$, the length of the hypotenuse is given by $c = \sqrt{a^2 + b^2}$.

If the sides of a triangle have lengths a, b, and c and $a^2 + b^2 = c^2$, the triangle is a right triangle.

A ***Pythagorean triple*** is a set of three numbers a, b, c for which $a^2 + b^2 = c^2$. The set of 3, 4, 5 is a Pythagorean triple.

EXAMPLE 1 Jim wants to stake out a rectangular region for a foundation for his home. He measures lengths of 50 feet and widths of 30 feet. He wants to be certain that the angles are right angles. He finds the lengths of each diagonal and adjusts the stakes so that the diagonals are the correct length to form right triangles. What are the lengths of the diagonals?

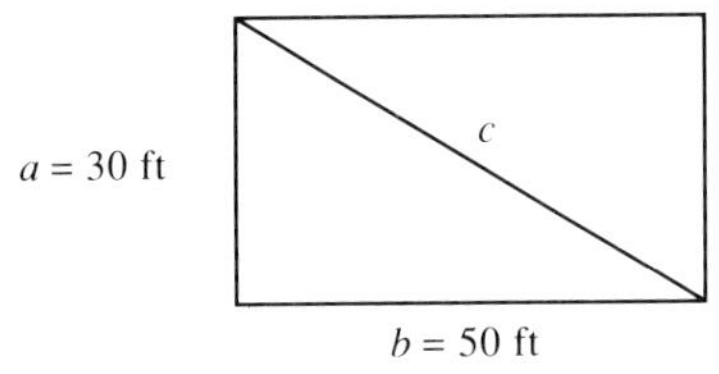

We use the Pythagorean theorem.

$$a^2 + b^2 = c^2$$

$$\mathbf{30}^2 + \mathbf{50}^2 = c^2$$

$$900 + 2500 = c^2$$

$$3400 = c^2$$

$$\sqrt{3400} = c$$

$$c \approx 58.3 \qquad \text{Use a calculator.}$$

The length of each diagonal is approximately 58.3 feet.

OBJECTIVE

1 *Find the length of a side of a right triangle, given the lengths of the other two sides and solve applied problems using powers and roots*

Answers to Exercises

1. 50 ft

2. **a.** 13 ft **b.** 6.708 m

3. 16 ft

Problem Set 7.8

NAME ______________________

DATE ______________________

CLASS ______________________

Find the length of the side of the right triangle that is not given (c is the length of the hypotenuse). If necessary, use a calculator or the table of powers and roots to find an approximation to the nearest thousandth.

1. $a = 8$ feet, $b = 6$ feet

2. $a = 3$ centimeters, $b = 4$ centimeters

3. $b = 6$ meters, $c = 10$ meters

4. $a = 12$ centimeters, $c = 20$ centimeters

5. $a = 4$ feet, $b = 5$ feet

6. $a = 6$ centimeters, $b = 7$ centimeters

7. $a = 4$ meters, $c = 8$ meters

8. $b = 3$ centimeters, $c = 10$ centimeters

9. $a = 10$ feet, $b = 8$ feet

10. $b = 5$ meters, $c = 15$ meters

11. $a = 1$ centimeter, $c = \sqrt{2}$ centimeters

12. $a = 3$ inches, $b = \sqrt{7}$ inches

13. In clear weather, the miles M that a person can see the view at an altitude A in miles is given by $M = 1.22\sqrt{A}$. How far can she see at an altitude of 9 miles?

14. The optimal inventory level I for many stores is given by $I = k\sqrt{S}$, where k is a constant and S is the amount of sales. Find the optimal inventory of washing machines if k is 5 and sales for 1 month were 36.

15. A 20-inch television set has a screen diagonal of 20 inches. If the height of the screen is 12 inches, what is the width of the screen?

16. The length of the diagonal of a 25-inch television screen is 25 inches. If the width of the screen is 20 inches, what is the height of the screen?

17. The two equal sides of an isosceles right triangle are of length 16 centimeters. What is the length of the hypotenuse? Round the answer to the nearest hundredth.

18. The lengths of the sides of an equilateral triangle are as shown. What is the height of the triangle? Round the answer to the nearest hundredth.

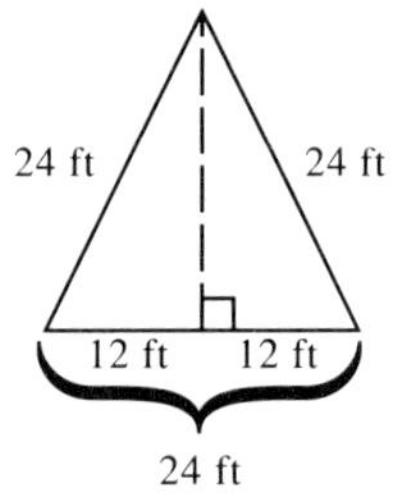

19. Parsons Pavers want to stake out a rectangular parking lot with a length of 120 feet and a width of 90 feet. What should be the lengths of the diagonals to insure that the parking lot is a rectangle?

20. Carpenters stabilize wall frames with a diagonal brace as shown below. If the bottom of the brace is 8 feet from the wall and the brace is 12 feet long, how far up the wall should it be nailed? Round the answer to the nearest tenth.

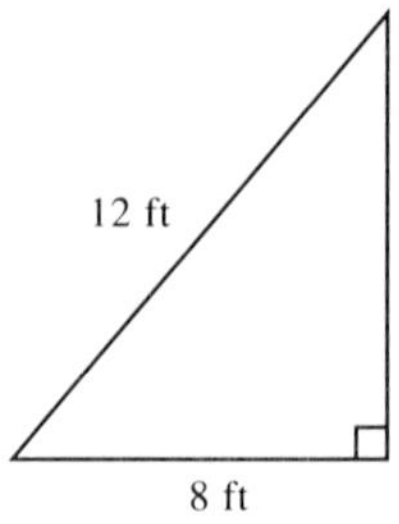

21. The length L of certain animals is related to their surface area S by the equation

$$L = \sqrt{\frac{5S}{2}}$$

Find the surface area of an animal with a length of 50 centimeters.

22. In the study of accidents on some road surfaces, the formula

$$S = 5.5\sqrt{0.8D}$$

may be used to determine the speed S in miles per hour of a car that leaves skid marks of D feet. What is the length of the skid marks left by a vehicle traveling 44 miles per hour?

Use the following formula for Problems 23 and 24. The interest rate r, compounded annually, needed for P dollars to grow to A dollars at the end of 2 years is given by

$$r = \sqrt{\frac{A}{P}} - 1$$

23. Find the value of A if \$8000 is invested at 6% interest for 2 years, compounded annually.

24. Find the value of A if \$8000 is invested at 10% interest for 2 years, compounded annually.

25. The period of a pendulum P in seconds depends on its length L in feet, and is given by

$$P = 2\pi\sqrt{\frac{L}{32}}$$

Find the length of a pendulum with a period of 4 seconds.

26. The pressure of the stream from a fire hydrant may be given by the formula

$$G = 26.8d^2\sqrt{p}$$

where G is the discharge in gallons per minute, d is the diameter of the outlet in inches, and p represents the pressure in pounds per square inch (psi). What is the pressure of water from an outlet that is 3.5 inches in diameter and discharges 600 gallons per minute? Round your answer to the nearest tenth.

Find the length of the side of the right triangle that is not given (c is the length of the hypotenuse). Round answers to the nearest tenth.

27. $a = 16.4$ feet, $b = 12.3$ feet

28. $a = 25.2$ meters, $b = 37.4$ meters

29. $b = 84.3$ inches, $c = 125.2$ inches

30. $c = 242.3$ feet, $a = 73.5$ feet

Find the length of the side of the right triangle that is not given.

* **31.** $a = 70$ feet, $b = 50$ feet

* **32.** $a = 34$ inches, $c = 61$ inches

* **33.** $a = x$ centimeters, $b = 3$ centimeters

* **34.** $b = \frac{x}{2}$ meters, $c = 5$ meters

* **35.** $a = 4$ feet, $b = x$ feet

Checkup

The following problems provide a review of some of Section 6.7.

Solve.

36. y varies directly as x and $x = 4$ when $y = 6$. Find y when $x = 3$.

37. y varies inversely as x and $x = 2$ when $y = 10$. Find y when $x = 5$.

38. The strength of a rectangular beam varies jointly as its width and the square of its depth. If the strength of a beam 5 inches wide and 4 inches deep is 400 pounds, find the strength of a beam 4 inches wide and 8 inches deep.

39. The illumination produced by a light source varies inversely as the square of the distance from the light source. If 3.125 footcandles of illumination is produced at a distance of 4 meters from the source, find the illumination 6 meters from the source.

40. The wavelength of a radio wave varies inversely as its frequency. A wave with a frequency of 900 kilohertz has a length of 400 meters. What is the length of a wave with a frequency of 1200 kilohertz?

41. The stopping distance of a car varies directly as the square of the speed. If a car traveling 30 miles per hour can stop in 50 feet, how many feet will it take to stop a car traveling 70 miles per hour?

7.9 THE COMPLEX NUMBERS

OBJECTIVES

1. *Simplify numbers of the form* $\sqrt{-b}, b > 0$
2. *Add or subtract complex numbers*
3. *Multiply complex numbers*
4. *Divide complex numbers*

1 Imaginary Numbers

In the physical sciences and engineering it is necessary to solve equations like $x^2 = -9$. The equation $x^2 = -9$ has no real number solutions since a solution must be a number whose square is -9. To solve such equations a new system was invented so that negative numbers would have square roots. We define the number i, whose square is -1.

> $i^2 = -1$ or $i = \sqrt{-1}$
>
> An ***imaginary number*** is a number of the form $a + bi$, where b is a *nonzero* real number and a is a real number.

In electronics, j may be used instead of i, that is, $j = \sqrt{-1}$. The product rule for radicals, $\sqrt[n]{ab} = \sqrt[n]{a} \cdot \sqrt[n]{b}$, is also true if a and b are not both negative. Hence for any positive real number b,

$$\sqrt{-b} = \sqrt{-1 \cdot b} = \sqrt{-1}\sqrt{b} = i\sqrt{b}$$

> For any positive number b, $\sqrt{-b} = i\sqrt{b}$

EXAMPLE 1 Simplify using the fact that $\sqrt{-b} = i\sqrt{b}$.

a. $\sqrt{-6} = i\sqrt{6}$

This may also be written as $\sqrt{6}i$. Notice that i is not under the radical sign.

b. $\sqrt{-64} = i\sqrt{64} = 8i$

Since $i = \sqrt{-1}$, Example b may also be worked as follows:

$$\begin{aligned} \sqrt{-64} &= \sqrt{-1 \cdot 64} \\ &= \sqrt{-1}\sqrt{64} \quad \text{Use the product rule.} \\ &= i\sqrt{64} \\ &= i \cdot 8 \\ &= 8i \end{aligned}$$ ■

☐ **DO EXERCISE 1.**

☐ **Exercise 1** Simplify using the fact that $\sqrt{-b} = i\sqrt{b}$.

a. $\sqrt{-2}$

b. $\sqrt{-7}$

c. $\sqrt{-25}$

d. $-\sqrt{-100}$

The real numbers and the imaginary numbers are subsets of a set called the ***complex numbers.*** A complex number may be a real number or an imaginary number.

> A complex number is of the form $a + bi$, where a and b are real numbers.

Notice that b can be zero. The following chart shows the relationship of the real, imaginary, and complex numbers.

Complex numbers
$3, \; 5i, \; 7 + 2i$

Real numbers	Imaginary numbers
$2, \; \frac{3}{8}, \; 0, \; \sqrt{5}$	$4i, \; -3i, \; 7 + 12i$

□ **Exercise 2** Add or subtract.

a. $(5 + 2i) + (3 - 7i)$

b. $(-1 - 4i) + (-3 - 6i)$

c. $(9 + 8i) - (4 + 9i)$

d. $(2 - i) - (6 - 4i)$

□ **Exercise 3** Multiply.

a. $4i(-6 + 3i)$

b. $-7i(3 - 2i)$

c. $(1 + 4i)(1 - 7i)$

d. $(5 - 4i)(2 - i)$

□ **Exercise 4** Multiply.

a. $\sqrt{-5} \cdot \sqrt{-5}$

b. $\sqrt{-3} \cdot \sqrt{-7}$

c. $\sqrt{-3} \cdot \sqrt{-8}$

d. $\sqrt{-3} \cdot \sqrt{-6}$

The commutative, associative, and distributive laws hold for complex numbers.

2 Addition and Subtraction

Since the laws described above hold for complex numbers, we can add and subtract them in the way that we add and subtract real numbers. However, we combine the real parts and we combine the imaginary parts.

EXAMPLE 2 Add or subtract.

a. $(3 + 4i) + (7 + 2i) = (3 + 7) + (4i + 2i)$ — Combine the real parts and the imaginary parts.

$= 10 + 6i$

b. $(8 + 3i) - (9 - 5i) = (8 + 3i) + (-9 + 5i)$ — Add the opposite.

$= -1 + 8i$ ■

□ **DO EXERCISE 2.**

3 Multiplication

Many of the procedures for multiplication of complex numbers are similar to the methods used for the multiplication of real numbers. Remember that $i^2 = -1$.

EXAMPLE 3 Multiply.

a. $(-2i)(4 - 3i) = (-2i)(4) - (-2i)(3i)$ — Use a distributive law.

$= -8i + 6i^2$

$= -8i + 6(-1)$ — Since $i^2 = -1$.

$= -8i - 6$

$= -6 - 8i$ — Write in $a + bi$ form.

b.

F O I L

$(2 + 5i)(1 - 4i) = 2 - 8i + 5i - 20i^2$ — Use FOIL.

$= 2 - 8i + 5i - 20(-1)$ — Since $i^2 = -1$.

$= 2 - 8i + 5i + 20$

$= 22 - 3i$ ■

□ **DO EXERCISE 3.**

The property $\sqrt{a}\sqrt{b} = \sqrt{ab}$ *does not hold when a and b are* both *negative.* Therefore, we change numbers of the form $\sqrt{-b}$ to $i\sqrt{b}$ before multiplying. For example,

$$\sqrt{-3} \cdot \sqrt{-5} = i\sqrt{3} \cdot i\sqrt{5} = i^2\sqrt{15} = -\sqrt{15}$$

but $\sqrt{-3} \cdot \sqrt{-5} = \sqrt{(-3)(-5)} = \sqrt{15}$ is *wrong*.

EXAMPLE 4 Multiply.

a. $\sqrt{-2} \cdot \sqrt{-7} = i\sqrt{2} \cdot i\sqrt{7} = i^2\sqrt{14} = (-1)\sqrt{14} = -\sqrt{14}$

b. $\sqrt{-3} \cdot \sqrt{-12} = i\sqrt{3} \cdot i\sqrt{12} = i^2\sqrt{36} = (-1)6 = -6$ ■

□ **DO EXERCISE 4.**

4 Division

Some complex numbers may be divided by *removing* the *imaginary terms* from the *denominator.* This division of complex numbers may be done by a procedure similar to rationalizing denominators. We multiply the fraction containing the complex numbers by 1. What expression shall we use for 1?

> **The complex numbers $a + bi$ and $a - bi$ are complex conjugates.**

To divide some complex numbers, we multiply the numerator and the denominator by the conjugate of the denominator. This gives a real number in the denominator.

Notice the following.

$$(A + Bi)(A - Bi) = A^2 - ABi + ABi - B^2i^2$$
$$= A^2 - B^2(-1)$$
$$= A^2 + B^2$$

> $$(A + Bi)(A - Bi) = A^2 + B^2$$

EXAMPLE 5 Divide and write in $a + bi$ form.

a. $\dfrac{4 - 5i}{2 + 4i}$

$$\frac{4 - 5i}{2 + 4i} = \frac{4 - 5i}{2 + 4i} \cdot \frac{\mathbf{2 - 4i}}{\mathbf{2 - 4i}} \qquad \text{Multiply by 1.}$$

$$= \frac{(4 - 5i)(2 - 4i)}{(2 + 4i)(2 - 4i)}$$

$$= \frac{8 - 16i - 10i + 20i^2}{(2 + 4i)(2 - 4i)} \qquad \text{Use FOIL.}$$

$$= \frac{8 - 16i - 10i + 20i^2}{2^2 + 4^2} \qquad \text{Since } (A + Bi)(A - Bi) = A^2 + B^2.$$

$$= \frac{-12 - 26i}{20} = \frac{2(-6 - 13i)}{20}$$

$$= \frac{-6 - 13i}{10} = \frac{-6}{10} - \frac{13}{10}i = \frac{-3}{5} - \frac{13}{10}i$$

b. $\dfrac{3 + i}{i}$ The conjugate of i is $-i$. Multiply numerator and denominator by $-i$.

$$\frac{3 + i}{i} = \frac{3 + i}{i} \cdot \frac{\mathbf{-i}}{\mathbf{-i}}$$

$$= \frac{(3 + i)(-i)}{-i^2}$$

$$= \frac{-3i - i^2}{-i^2}$$

$$= \frac{-3i - (-1)}{-(-1)}$$

$$= \frac{-3i + 1}{1}$$

$$= 1 - 3i \quad ■$$

□ **DO EXERCISE 5.**

□ **Exercise 5** Divide.

a. $\dfrac{1 + i}{4 + i}$

b. $\dfrac{5 + 2i}{3 - 4i}$

c. $\dfrac{7}{2 - 3i}$

d. $\dfrac{3 - 4i}{i}$

□ **Exercise 6** Find each of the following powers of i.

a. i^{32}

b. i^{49}

c. i^{75}

d. i^{66}

The fact that $i^2 = -1$ can be used to find higher powers of i.

$$i^3 = i^2 \cdot i = (-1)i = -i$$

$$i^4 = (i^2)^2 = (-1)^2 = 1$$

$$i^5 = i^4 \cdot i = (i^2)^2 \cdot i = (-1)^2 \cdot i = i$$

$$i^6 = (i^2)^3 = (-1)^3 = -1$$

A few powers of i are as follows:

$i^1 = i$	$i^5 = i$	$i^9 = i$
$i^2 = -1$	$i^6 = -1$	$i^{10} = -1$
$i^3 = -i$	$i^7 = -i$	$i^{11} = -i$
$i^4 = 1$	$i^8 = 1$	$i^{12} = 1$

As these examples suggest, the powers of i cycle through the values i, -1, $-i$, and 1.

Recall the following:

$(-1)^n = 1$	if n is even
$(-1)^n = -1$	if n is odd

EXAMPLE 6 Find each of the following powers of i.

a. $i^{24} = (i)^{12} = (-1)^{12} = 1$

b. $i^{45} = i^{44} \cdot i = (i^2)^{22} \cdot i = (-1)^{22} \cdot i = 1 \cdot i = i$

c. $i^{78} = (i^2)^{39} = (-1)^{39} = -1$

d. $i^{59} = i^{58} \cdot i = (i^2)^{29} \cdot i = (-1)^{29} \cdot i = -1 \cdot i = -i$ ■

□ **DO EXERCISE 6.**

Answers to Exercises

1. a. $i\sqrt{2}$ **b.** $i\sqrt{7}$ **c.** $5i$ **d.** $-10i$

2. a. $8 - 5i$ **b.** $-4 - 10i$ **c.** $5 - i$ **d.** $-4 + 3i$

3. a. $-12 - 24i$ **b.** $-14 - 21i$ **c.** $29 - 3i$ **d.** $6 - 13i$

4. a. -5 **b.** $-\sqrt{21}$ **c.** $-2\sqrt{6}$ **d.** $-3\sqrt{2}$

5. a. $\frac{5}{17} + \frac{3}{17}i$ **b.** $\frac{7}{25} + \frac{26}{25}i$ **c.** $\frac{14}{13} + \frac{21}{13}i$ **d.** $-4 - 3i$

6. a. 1 **b.** i **c.** $-i$ **d.** -1

Problem Set 7.9

NAME ____________________

DATE ____________________

CLASS ____________________

Simplify using the fact that $\sqrt{-b} = i\sqrt{b}$.

1. $\sqrt{-5}$

2. $\sqrt{-11}$

3. $\sqrt{-9}$

4. $\sqrt{-4}$

5. $-\sqrt{-18}$

6. $-\sqrt{-28}$

Add or subtract. Write answers in the form $a + bi$.

7. $(8 + 3i) + (2 + 4i)$

8. $(6 + 5i) + (4 - i)$

9. $(7 - 8i) + (5 - 2i)$

10. $(3 - 6i) + (-9 - 4i)$

11. $(-4 + i) + (-7 - 3i)$

12. $(-6 - i) + (-2 + 8i)$

13. $(6 - i) + (-6 + i)$

14. $(-8 + 2i) + (8 - 2i)$

15. $(4 - i) - (3 - 5i)$

16. $(9 - 5i) - (6 - 2i)$

17. $(3 - 2i) - (7 + 3i)$

18. $(5 - 9i) - (6 - 8i)$

19. $(4 + 8i) - (-3 - i)$

20. $(9 + 2i) - (3 + 5i)$

21. $(2 + 2i) - (3 + 6i)$

22. $(5 - 3i) - (2 - 8i)$

23. $(-2 - 3i) - (4 - i)$

24. $(-3 - 4i) - (-5 + 6i)$

Multiply.

25. $4i(6 - 2i)$

26. $2i(5 - i)$

27. $-3i(7 + 4i)$

28. $-i(5 + 3i)$

29. $(5 + 3i)(2 + i)$

30. $(7 + 5i)(2 + 3i)$

31. $(2 - 3i)(4 + 5i)$

32. $(5 - 6i)(7 + 3i)$

33. $(4 - 2i)(5 - 7i)$ **34.** $(6 - 5i)(8 - 2i)$ **35.** $(-3 + i)(5 - 7i)$ **36.** $(-4 - 3i)(6 + 7i)$

37. $(3 + 2i)(3 - 2i)$ **38.** $(5 - 4i)(5 + 4i)$ **39.** $(2 + i)(2 - i)$ **40.** $(7 - 3i)(7 + 3i)$

41. $(6 + 2i)^2$ **42.** $(5 + i)^2$ **43.** $(3 - 4i)^2$ **44.** $(1 - 8i)^2$

45. $(9 - 3i)(7 + 4i)$ **46.** $(8 + 5i)(9 + 2i)$ **47.** $(5 + 2i)(5 - 2i)$ **48.** $(7 - i)(7 + i)$

49. $\sqrt{-3}\sqrt{-5}$ **50.** $\sqrt{-2}\sqrt{-3}$ **51.** $\sqrt{-7}\sqrt{-7}$

52. $\sqrt{-8}\sqrt{-8}$ **53.** $\sqrt{-2}\sqrt{-6}$ **54.** $\sqrt{-3}\sqrt{-6}$

Divide.

55. $\frac{3}{4 + i}$ **56.** $\frac{-2}{3 - i}$ **57.** $\frac{4 + i}{6 - i}$ **58.** $\frac{7 - 2i}{3 + 4i}$

59. $\frac{1 - 5i}{3 - 5i}$ **60.** $\frac{4 + 3i}{2 + 7i}$ **61.** $\frac{9 - 2i}{3i}$ **62.** $\frac{8 - 5i}{2i}$

63. $\frac{9i}{5 - 4i}$ **64.** $\frac{3i}{7 + 2i}$ **65.** $\frac{5 + 6i}{9i}$ **66.** $\frac{8 - 2i}{i}$

Find the following powers of i.

67. i^{11} **68.** i^{15} **69.** i^{53} **70.** i^{81}

71. i^{14} **72.** i^{22} **73.** i^{67} **74.** i^{43}

75. i^9

76. i^6

77. $(-i)^5$

78. $(-i)^8$

79. An ____________ number is a number of the form $a + bi$, where b is a nonzero real number.

80. A complex number is of the form $a + bi$, where a and b are ____________ numbers.

81. To add or subtract complex numbers, we combine the real parts and we combine the ____________ parts.

82. The property $\sqrt{a}\sqrt{b} = \sqrt{ab}$ does not hold when a and b are both ____________.

83. Some complex numbers may be divided by removing the ____________ terms from the denominator.

Write answers in the form $a + bi$.

Add or subtract.

84. $(-8.6345 + 2i) + (7.3421 - 6i)$

85. $(6.7849 - 5i) - (9.3490 - 8i)$

86. $(7.53 - 2.67i) - (4.8 + 3.934i)$

87. $(8.269 + 3.457i) + (5.2 - 6.87i)$

Multiply. Round answers to the nearest hundredth.

88. $5.374i(7.352 - 4.854i)$

89. $-4.85i(6.832 + 8.32i)$

90. $(5.56 - 6.87i)(3.48 + 7.82i)$

91. $(6.28 - 3.47i)(8.29 - 5.77i)$

92. $(5.25 - 4.78i)(5.25 + 4.78i)$

93. $(9.89 - 7.83i)(9.89 + 7.83i)$

Think About It

Simplify.

* **94.** $\frac{1}{5}(-25 - \sqrt{-625})$

* **95.** i^{150}

* **96.** $(3 + i^{10})(3 - i^{10})$

* **97.** $\dfrac{i^4 + i^7 + i^3}{(1 - i)^2 (1 + i)^2}$

* **98.** i^{170}

* **99.** $\dfrac{3 - \sqrt{-16}}{2 + \sqrt{-9}}$

* **100.** $5\sqrt{-100} - 8\sqrt{-289}$

* **101.** i^{-41}

* **102.** $\dfrac{7}{1 + \dfrac{2}{i}}$

* **103.** $(7 - 4i)^{-2}$

* **104.** i^{-58}

* **105.** Find the value of $3x^2 + 2x + 2$ for $x = \dfrac{-3 - \sqrt{7}i}{6}$.

Checkup

The following problems provide a review of some of Section 7.1 and some will help you with the next section.

Find the following.

106. $\sqrt{81}$

107. $\sqrt{100}$

108. $\sqrt{\dfrac{9}{16}}$

109. $\sqrt{\dfrac{4}{25}}$

110. $\sqrt{\dfrac{144}{49}}$

111. $\sqrt{\dfrac{121}{64}}$

112. $\sqrt[4]{81}$

113. $\sqrt[6]{64}$

114. $\sqrt[3]{y^{15}}$

115. $\sqrt[5]{x^{20}}$

Chapter 7 Summary

Section 7.1

The number c is a ***square root*** of b if $c^2 = b$.

The symbol $\sqrt{}$, called a ***radical sign,*** is used to represent the nonnegative square root of a number. This is called the ***principal square root.***

The number under the radical sign is called the ***radicand.*** The entire expression, including the radical sign and radicand, is called a ***radical.***

For any real number x, $\sqrt{x^2} = |x|$.

The absolute value of a product is the product of the absolute values.

$$|x \cdot y| = |x| \cdot |y| \qquad \text{for any numbers } x \text{ and } y$$

The number c is an nth root of a if $c^n = a$.

The symbol $\sqrt[n]{a}$ represents the principal nth root of a. The number n is called the ***index.*** The index must be a positive integer greater than 1. The index is not written when it is 2.

A root is called ***odd*** when the index is an odd number. The ***cube root*** is the third root of a number. Every number has exactly one odd real root.

If the index is an even number, the root is called ***even.*** Negative real numbers do not have real even nth roots. Positive real numbers have both a positive and a negative real even nth root. There is only one principal nth root.

For any real number a:

$$\text{If } n \text{ is } even, \qquad \sqrt[n]{a^n} = |a|.$$

$$\text{If } n \text{ is } odd, \qquad \sqrt[n]{a^n} = a.$$

Section 7.2

Product rule for radicals: If a and b are nonnegative real numbers, and n is a natural number greater than 1,

$$\sqrt[n]{a} \cdot \sqrt[n]{b} = \sqrt[n]{ab}$$

Quotient rule for radicals: For any nonnegative number a, any positive number b, and any index n,

$$\sqrt[n]{\frac{a}{b}} = \frac{\sqrt[n]{a}}{\sqrt[n]{b}}$$

The quotient rule may also be written as follows. For any nonnegative number a, any positive number b, and any index n,

$$\frac{\sqrt[n]{a}}{\sqrt[n]{b}} = \sqrt[n]{\frac{a}{b}}$$

The product rule is also as follows:

$$\sqrt[n]{a \cdot b} = \sqrt[n]{a} \cdot \sqrt[n]{b}$$

Whole numbers that have integers for square roots are called ***perfect squares.***

A ***perfect cube*** is a whole number whose real cube root is an integer.

To simplify a radical by factoring, factor the radicand so that there are factors which are perfect nth powers, where n is the index. Find the nth root of those factors.

Rational numbers are numbers that can be written in the form a/b, where a and b are integers and b is not zero. If a rational number is written as a decimal, the decimal either repeats or terminates.

Irrational numbers are:

1. Numbers that cannot be written in the form a/b, where a and b are integers and b is not zero
2. Decimals that do not repeat and do not terminate
3. Real numbers that are not rational, such as $\sqrt{7}$.

The set of real numbers is composed of the rational numbers and the irrational numbers.

Section 7.3

Radical expressions can be combined if the radicals have the same index and radicand.

Section 7.4

Radical expressions with more than one term are multiplied by the methods used for multiplying polynomials.

Section 7.5

One process of removing the radicals from the denominator is called ***rationalizing the denominator.***

To ***rationalize a numerator,*** methods are used to remove radicals from the numerator.

The binomials $A + B$ and $A - B$ are called ***conjugates.***

Rationalizing a denominator

1. To rationalize a *monomial* denominator, multiply by a "1" to make the radicand in the denominator a perfect nth power, where n is the index of the radical.
2. To rationalize a *binomial* denominator, multiply by a "1" that has the conjugate of the denominator as its numerator and denominator.

Section 7.6

$$a^{1/n} = \sqrt[n]{a}$$

for any nonnegative real number a and any index n.

We assume that the *base is nonnegative* when we use rational exponents.

$$a^{m/n} = \sqrt[n]{a^m} \qquad \text{or} \qquad a^{m/n} = (\sqrt[n]{a})^m$$

for any nonnegative real number a and any natural numbers m and n $(n \neq 1)$.

$$a^{-m/n} = \frac{1}{a^{m/n}}$$

for any rational number m/n and any positive real number a.

For any positive real number a and any rational number exponents m and n,

1. $a^m \cdot a^n = a^{m+n}$
2. $\dfrac{a^m}{a^n} = a^{m-n}$
3. $(a^m)^n = a^{mn}$

Methods of simplifying radical expressions

1. Factor the radicand. The radicand should contain no factor to a power greater than or equal to the index.
2. Rationalize the denominator.
3. Simplify using rational exponents.

Section 7.7

A radical equation contains at least one radical with a variable in the radicand.

Power rule for equations:

$$\text{If} \quad a = b$$
$$\text{then} \quad a^n = b^n$$

for any natural number n.

Solving equations with radicals:

1. Isolate one radical on one side of the equal sign.
2. Raise each side of the equation to the power that is the same as the index of the radical.
3. If the equation still contains a radical, repeat steps 1 and 2.
4. Solve the equation.
5. Possible solutions must be checked.

Section 7.8

Pythagorean theorem: In a right triangle, the sum of the squares of the lengths of the two shorter sides, a and b, is equal to the square of the length of the longest side, c.

$$a^2 + b^2 = c^2$$

The longest side is called the ***hypotenuse*** and the shorter sides are called the ***legs.***

Section 7.9

$$i^2 = -1 \qquad \text{or} \qquad i = \sqrt{-1}$$

An ***imaginary number*** is a number of the form $a + bi$, where b is a nonzero real number and a is a real number.

For any positive number b,

$$\sqrt{-b} = i\sqrt{b}$$

A ***complex number*** is of the form $a + bi$, where a and b are real numbers.

The complex numbers are composed of the real numbers and the imaginary numbers.

The complex numbers $a + bi$ and $a - bi$ are complex conjugates.

$$(-1)^n = 1 \qquad \text{if } n \text{ is even}$$
$$(-1)^n = -1 \qquad \text{if } n \text{ is odd}$$

Chapter 7 Additional Exercises (Optional)

NAME ____________________

DATE ____________________

CLASS ____________________

Section 7.1

Find each root, if it exists. Assume that variables under the radical sign represent any real number.

1. $\sqrt{100y^2}$ **2.** $\sqrt{36x^2}$ **3.** $\sqrt{25(x-4)^2}$ **4.** $\sqrt{16(y+2)^2}$

5. $\sqrt{16x^2 + 40x + 25}$ **6.** $\sqrt{4y^2 - 28y + 49}$ **7.** $-\sqrt[3]{27x^{15}}$ **8.** $-\sqrt[5]{32y^{10}}$

9. $\sqrt[4]{-81}$ **10.** $\sqrt[6]{-1}$ **11.** $\sqrt[6]{64x^{18}}$

12. $\sqrt[4]{16x^{12}}$ **13.** $\sqrt[4]{x^8}$ **14.** $\sqrt[6]{y^{24}}$

Section 7.2

Assume that all variables under radical signs represent positive numbers.

Multiply and simplify, if possible.

15. $\sqrt[3]{5}\sqrt[3]{3}$ **16.** $\sqrt[4]{12} \cdot \sqrt[4]{2}$ **17.** $\sqrt{2}\sqrt{50}$

18. $\sqrt{3}\sqrt{8}$ **19.** $\sqrt{3y}\sqrt{21y}$ **20.** $\sqrt{8y^3}\sqrt{5y}$

Divide and simplify.

21. $\sqrt[4]{\dfrac{80x^{10}}{5x^2}}$ **22.** $\sqrt[3]{\dfrac{81y^{11}}{3y^8}}$ **23.** $\sqrt{\dfrac{150x^5y^3}{6xy}}$

Section 7.3

Combine like terms. Assume that all variables under radical signs represent nonnegative numbers.

24. $5\sqrt[3]{7} - 8\sqrt[3]{7}$

25. $-3\sqrt[4]{5} + 9\sqrt[4]{5}$

26. $6\sqrt{48} + 3\sqrt{27}$

27. $8\sqrt{20} - 6\sqrt{80}$

28. $\sqrt{108} - \sqrt{192} + \sqrt{75}$

29. $\sqrt[3]{54} + \sqrt[3]{432} - \sqrt[3]{128}$

30. $\sqrt[3]{8x^4y^2} - \sqrt[3]{x^7y^8} - \sqrt[3]{27x^4y^2}$

31. $\sqrt{9x^8y^3} + \sqrt{25x^6y^9} - \sqrt{4x^8y^3}$

Section 7.4

Multiply and simplify. Assume that all variables under radical signs represent nonnegative numbers.

32. $\sqrt{6}(3 - \sqrt{6})$

33. $\sqrt{5}(\sqrt{5} - 4)$

34. $\sqrt[3]{3}(\sqrt[3]{4} - \sqrt[3]{9})$

35. $\sqrt[4]{2}(\sqrt[4]{8} + \sqrt[4]{7})$

36. $(\sqrt{7} + \sqrt{2})(\sqrt{7} - \sqrt{2})$

37. $(\sqrt{10} - \sqrt{5})(\sqrt{10} + \sqrt{5})$

38. $(3\sqrt{2} - \sqrt{6})^2$

39. $(5\sqrt{3} - \sqrt{7})^2$

40. $(\sqrt{11} - 4\sqrt{5})(\sqrt{11} + 2\sqrt{5})$

41. $(3\sqrt{6} + \sqrt{13})(4\sqrt{6} - \sqrt{13})$

* **42.** $(3 - \sqrt{y+2})^2$

* **43.** $(\sqrt{x+3} - 4)^2$

Section 7.5

Assume that all variables under radical signs represent positive numbers.

Rationalize the denominator.

44. $\dfrac{\sqrt{5}}{\sqrt{7}}$

45. $\dfrac{\sqrt{6}}{\sqrt{3}}$

46. $\dfrac{\sqrt[3]{5}}{\sqrt[3]{9}}$

47. $\dfrac{\sqrt[3]{7}}{\sqrt[3]{4}}$

48. $\frac{\sqrt[4]{3}}{\sqrt[4]{8}}$

49. $\frac{\sqrt[5]{2}}{\sqrt[5]{16}}$

50. $\frac{\sqrt{x} - \sqrt{5}}{\sqrt{x} + \sqrt{7}}$

51. $\frac{\sqrt{y} - \sqrt{3}}{\sqrt{y} + \sqrt{5}}$

* 52. $\frac{3}{\sqrt{x+2} - \sqrt{x}}$

* 53. $\frac{-2}{\sqrt{x+3} - \sqrt{x}}$

Simplify.

54. $\sqrt{\frac{5}{y}}$

55. $\sqrt{\frac{3}{y^3}}$

56. $\sqrt[3]{\frac{15}{x^4}}$

57. $\sqrt[3]{\frac{3}{x}}$

Rationalize the numerator.

58. $\frac{9 + \sqrt{2}}{9 - \sqrt{3}}$

59. $\frac{\sqrt{5} + 3}{7}$

* 60. $\frac{\sqrt{x+h-1} - \sqrt{x-1}}{h}$

* 61. $\frac{\sqrt{x+h+3} - \sqrt{x+3}}{h}$

Section 7.6

Simplify. Write answers with positive exponents.

* 62. $-\left(\frac{16}{49}\right)^{3/2}$

* 63. $\left(\frac{81}{16}\right)^{3/4}$

* 64. $\left(\frac{27}{8}\right)^{-2/3}$

* 65. $\left(\frac{36}{81}\right)^{-3/2}$

66. $5^{-1/2} \cdot 5^{-2/3}$

67. $x^{1/8} \cdot x^{-3/4}$

68. $\frac{x^{1/5} \cdot x^{3/10}}{x^{2/5}}$

69. $\frac{a^{1/3}}{a^{5/6} \cdot a^{-1/2}}$

Multiply or divide. Simplify by converting to rational exponents and converting back to radical form. Assume that all variables under radical signs represent positive numbers.

* 70. $\sqrt[3]{2x^4}\sqrt[3]{4x^6}$

* 71. $\sqrt[4]{9y^7}\sqrt[4]{9y^{12}}$

72. $\frac{\sqrt[4]{96x^5}}{\sqrt[4]{6x^{-4}}}$

73. $\frac{\sqrt{75y^{-2}}}{\sqrt{3y^9}}$

Section 7.7

Solve.

74. $\sqrt{x} - 2 = 3$

75. $\sqrt{y} - 6 = 4$

76. $\sqrt{5x - 1} = 3$

77. $\sqrt{4y + 1} = 5$

78. $\sqrt{x - 6} - \sqrt{x + 9} - 3 = 0$

79. $\sqrt{z + 15} - \sqrt{2z + 7} - 1 = 0$

80. $\sqrt{x^2 - 6x + 9} = x + 7$

81. $\sqrt{y^2 + 3y + 7} = y + 2$

* **82.** $\sqrt{7 + x} = 3\sqrt{7 + x}$

* **83.** $\sqrt{5 - 2x} = 4\sqrt{5 - 2x}$

84. $\sqrt[4]{x + 5} - \sqrt[4]{2x - 6} = 0$

85. $\sqrt[3]{y^2 + 3y + 12} - \sqrt[3]{y^2} = 0$

Section 7.8

Find the length of the side of the right triangle that is not given (c is the length of the hypotenuse). If necessary, use a calculator to find an approximation to the nearest thousandth.

86. $a = 9$ meters, $b = 15$ meters

87. $a = 12$ centimeters, $b = 4$ centimeters

88. $b = 8$ feet, $c = 17$ feet

89. $a = 3$ inches, $c = 11$ inches

90. $a = 3$ meters, $b = \sqrt{15}$ meters

91. $b = 4$ feet, $c = \sqrt{27}$ feet

92. The two equal sides of an isosceles right triangle are of length 15 feet. What is the length of the hypotenuse? Round the answer to the nearest tenth.

93. Gem Construction wants to stake out an office building with a length of 150 feet and a width of 110 feet. What should be the lengths of the diagonals to ensure that the building is a rectangle? Round the answer to the nearest tenth.

94. In clear weather, the miles M that a person can see the view at an altitude of A miles is given by

$$M = 1.22\sqrt{A}$$

What is the altitude of a person that can see for 4.22 miles?

95. The length of the diagonal d of a box with length L, width W, and height H is given by

$$d = \sqrt{L^2 + W^2 + H^2}$$

What is the height of a box that is 4 inches long, 3 inches wide, and has a diagonal that measures $5\sqrt{2}$ inches?

Section 7.9

Multiply.

96. $(3 + 2i)(5 - 4i)$

97. $(7 - i)(6 - 4i)$

98. $\sqrt{-5}\sqrt{-2}$

99. $\sqrt{-7}\sqrt{-9}$

Identify as real or imaginary numbers.

100. $\sqrt{5}$

101. $\sqrt{7}$

102. $6 + 2i$

103. $5 - 3i$

104. $\sqrt[3]{-8}$

105. $\sqrt{-15}$

106. $-4i$

107. $15i$

Simplify.

108. i^{23}

109. i^{12}

* **110.** i^{-28}

* **111.** i^{-17}

* **112.** $\frac{3}{1 - 2i} + \frac{4}{3 + 4i}$

* **113.** $\frac{5}{1 - i} - \frac{2}{3 + 5i}$

* **114.** $\frac{4 + 3i}{2i} - \frac{7 - i}{4i}$

* **115.** $\frac{6 + 2i}{3i} + \frac{5 - i}{i}$

COOPERATIVE LEARNING

Find a number that has the following principal square root.

1. $8|y|$

2. $4|x + 5|$

Find a number that has the following cube root.

3. $-7x^3$

4. $-4y^5$

What are two binomials that have the following product?

5. $1 + \sqrt{3}$

6. $42 - 22\sqrt{5}$

7. $-10 - 3\sqrt{6}$

8. Cynthia has a square yard with a diagonal of length 75 feet. What are the lengths of the sides of the yard? If fencing costs $3.25 per foot, what is the cost of fencing the yard?

Find two binomials whose product is the following.

9. $31 + i$

10. $33 - 27i$

Find two imaginary numbers whose product is the following.

11. $-\sqrt{14}$

12. $-\sqrt{33}$

Chapter 7 Practice Test

NAME ____________________

DATE ____________________

CLASS ____________________

Find the root.

1. $\sqrt{81x^2}$

2. $\sqrt{25(x + 3)^2}$

3. $\sqrt[3]{-27y^3}$

4. $\sqrt[8]{(x - 3y)^8}$

For Problems 5 and 6, assume that all variables under radical signs represent positive numbers.

5. Multiply and simplify: $\sqrt[3]{3y^2}\sqrt[3]{9y^3}$.

6. Divide and simplify: $\dfrac{\sqrt{2x^2y^3}}{\sqrt{2y^3}}$.

7. Approximate to the nearest tenth: $\sqrt{170}$.

8. Subtract: $5\sqrt{72} - 3\sqrt{18}$.

1. ____________________

2. ____________________

3. ____________________

4. ____________________

5. ____________________

6. ____________________

7. ____________________

8. ____________________

9. ______________________

10. ______________________

11. ______________________

12. ______________________

13. ______________________

14. ______________________

15. ______________________

16. ______________________

17. ______________________

18. ______________________

19. ______________________

20. ______________________

21. ______________________

9. Multiply and simplify: $(3\sqrt{2} - 4\sqrt{3})(5\sqrt{2} + \sqrt{3})$.

10. Rationalize the denominator: $\frac{6}{2\sqrt{3}}$.

11. Rationalize the denominator: $\frac{3\sqrt{3} - 5}{2\sqrt{2} + 6}$.

12. Write with rational exponents: $\sqrt[3]{(ab^2c)^4}$.

13. Simplify and write with positive exponents: $\frac{6^{1/4}}{6^{2/3}}$.

14. Simplify: $\sqrt{x^4y^6}$. Assume that x and y are positive.

15. Solve: $\sqrt{x + 5} - \sqrt{x - 3} = 2$.

16. Simplify: $\sqrt{-24}$.

17. Add: $(3 - 5i) + (-9 - 7i)$.

18. Multiply: $(6 - i)(7 + 3i)$.

19. $\sqrt{-3} \cdot \sqrt{-6}$

20. Divide: $\frac{3 - 2i}{5 - 6i}$.

21. The length of a room is 15 feet and the width is 12 feet. What is the length of the diagonal of the room? Round the answer to the nearest tenth.

CHAPTER

8

Quadratic Equations

Pretest

1. Solve: $(5x + 2)^2 = 8$.

2. The hypotenuse of a right triangle is $\sqrt{89}$ meters. One leg is 3 meters shorter than the other. Find the length of the shortest leg.

3. Solve by completing the square: $3y^2 + 6y + 5 = 0$.

4. Solve by the quadratic formula: $2x^2 - x - 7 = 0$.

5. Solve. Approximate solutions to the nearest tenth: $x^2 - 4x + 2 = 0$.

6. Predict the number and type of solutions of $4x^2 - 12x + 9 = 0$.

7. Find a quadratic equation with solutions of $\sqrt{3}$ and $2\sqrt{3}$.

8. Solve for r. Assume that variables represent positive numbers:

$$V = \pi (r^2 + R^2)h$$

9. The speed of the current in a stream is 2 miles per hour. A boat travels 12 miles upstream and back downstream in 6 hours. What is the speed of the boat in still water? Round the answer to the nearest tenth.

10. Solve: $(x^2 + 2)^2 - 3(x^2 + 2) - 10 = 0$.

Graph.

11. $y = \frac{1}{2}x^2 - 3$

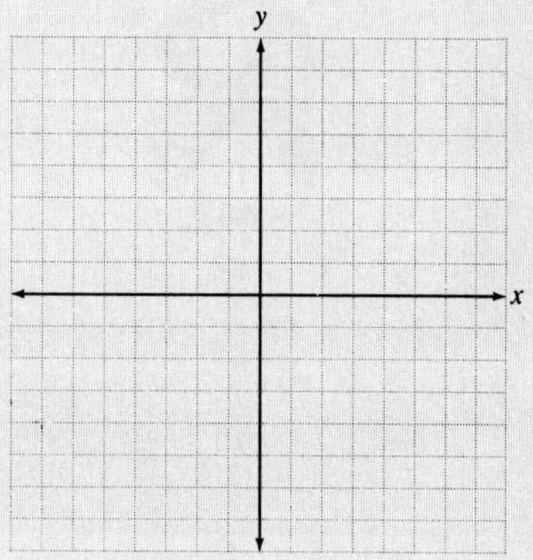

12. $y = x^2 - 6x + 8$

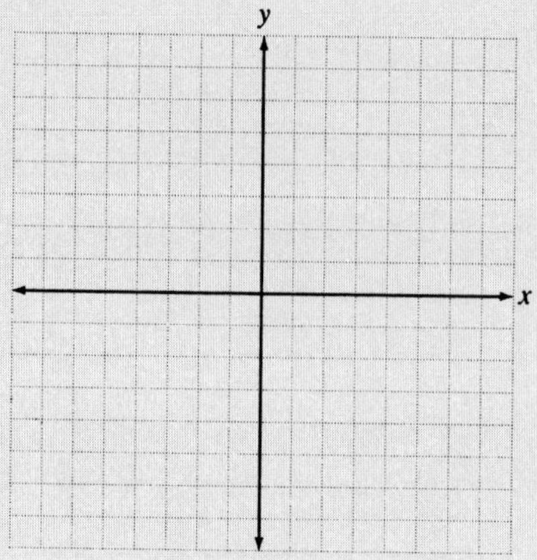

13. $x = y^2 + 5y + 6$

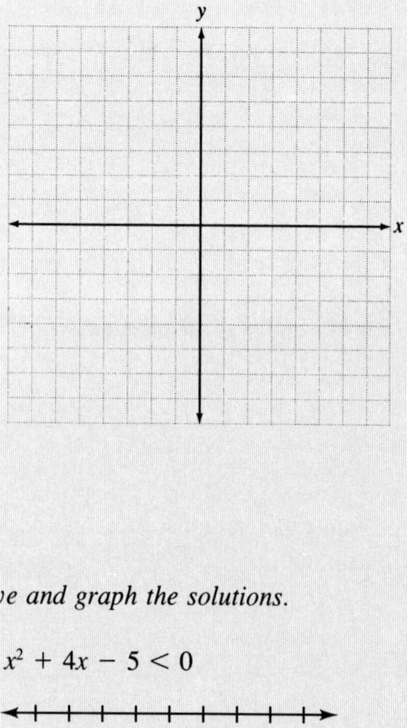

14. The demand D for a new calculator is estimated by the quadratic equation $D = -\frac{1}{4}x^2 + 18x + 750$, where x is the week of sales. Find the week of highest demand and the number of calculators needed that week to meet the demand.

Solve and graph the solutions.

15. $x^2 + 4x - 5 < 0$

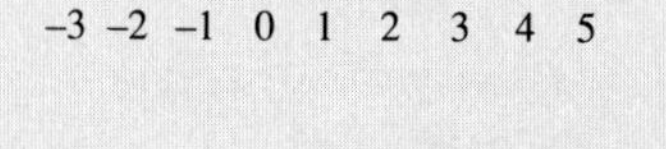

16. $3x^2 \geq 6x$

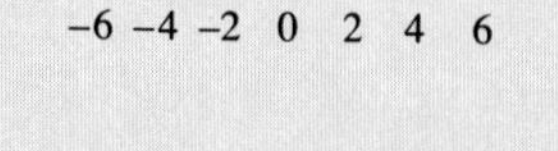

17. Solve: $\frac{4x + 4}{x - 5} \leq 8$.

8.1 QUADRATIC EQUATIONS AND SOLVING USING THE SQUARE ROOT PROPERTY

Second-degree (quadratic) equations are often used to solve applied problems. For example, the height in feet after t seconds of an object thrown upward with an initial velocity of 32 feet per second is $h = 32t - 16t^2$.

A ***quadratic equation*** is any second-degree equation that can be written in the form

$$ax^2 + bx + c = 0$$

where a, b, and c are real numbers and $a \neq 0$. This form is called the *standard form of a quadratic equation.*

Following are examples of quadratic equations:

$$3x^2 - 2x + 4 = 0 \qquad y^2 - 3 = 0 \qquad x^2 + 5x = 0$$

Quadratic equations have exactly *two solutions.* Sometimes they are identical. The solutions may also be two different real numbers or two different imaginary numbers.

1 Solving by Taking the Square Root

The easiest way to solve equations of the type $x^2 - c = 0$ or $x^2 = c$ is to take the principal square roots of both sides of the equation.

If $x^2 = c$, then $x = \sqrt{c}$ or $x = -\sqrt{c}$, since $(\sqrt{c})^2 = c$ and $(-\sqrt{c})^2 = c$. This gives us the square root property.

Square Root Property

If x and c are complex numbers and if $x^2 = c$, then $x = \sqrt{c}$ or $x = -\sqrt{c}$.

Caution: If $c \neq 0$, using the square root property always gives two square roots, one positive and one negative.

EXAMPLE 1 Solve.

a. $x^2 = 7$

$x = \sqrt{7}$ or $x = -\sqrt{7}$. Use the square root property.

Check For $x = \sqrt{7}$ For $x = -\sqrt{7}$

$x^2 = 7$			$x^2 = 7$	
$(\sqrt{7})^2$	7		$(-\sqrt{7})^2$	7
7	7		7	7

The solutions are $\sqrt{7}$ and $-\sqrt{7}$. This is often written as $\pm\sqrt{7}$.

OBJECTIVES

1 *Solve quadratic equations by taking the square root of both sides of the equation*

2 *Solve quadratic equations by factoring*

3 *Solve applied problems using quadratic equations*

□ **Exercise 1** Solve.

a. $2x^2 = 6$

b. $-3x^2 + 4 = 0$

c. $-4x^2 + 7 = 0$

d. $5x^2 + 15 = 0$

b. $-3x^2 + 5 = 0$

$-3x^2 = -5$ Add -5 to each side.

$x^2 = \frac{5}{3}$ Divide by -3.

$x = \sqrt{\frac{5}{3}}$ or $x = -\sqrt{\frac{5}{3}}$ Use the square root property.

We rationalize the denominator.

$x = \sqrt{\frac{5}{3}} = \frac{\sqrt{5}}{\sqrt{3}}$ or $x = -\sqrt{\frac{5}{3}} = -\frac{\sqrt{5}}{\sqrt{3}}$ Use the quotient rule for radicals.

$x = \frac{\sqrt{5}}{\sqrt{3}} \cdot \frac{\sqrt{3}}{\sqrt{3}}$ $x = -\frac{\sqrt{5}}{\sqrt{3}} \cdot \frac{\sqrt{3}}{\sqrt{3}}$ Rationalize the denominators.

$x = \frac{\sqrt{15}}{3}$ $x = \frac{-\sqrt{15}}{3}$

Check that the solutions are $\frac{\sqrt{15}}{3}$ and $\frac{-\sqrt{15}}{3}$.

c. $6x^2 + 12 = 0$

$6x^2 = -12$

$x^2 = -2$

$x = \sqrt{-2}$ or $x = -\sqrt{-2}$ Use the square root property.

$x = i\sqrt{2}$ $x = -i\sqrt{2}$

Check For $x = i\sqrt{2}$ For $x = -i\sqrt{2}$

$6x^2 + 12 = 0$	
$6(i\sqrt{2})^2 + 12$	0
$6(-2) + 12$	0
0	0

$6x^2 + 12 = 0$	
$6(-i\sqrt{2})^2 + 12$	0
$6(-2) + 12$	0
0	0

The solutions are $i\sqrt{2}$ and $-i\sqrt{2}$. The solutions are imaginary numbers. ■

□ **DO EXERCISE 1.**

We can solve more complicated equations by taking the principal square roots of both sides of the equations.

EXAMPLE 2 Solve.

a. $(x-3)^2 = 16$

$x - 3 = \sqrt{16}$ or $x - 3 = -\sqrt{16}$ Use the square root property.

$x - 3 = 4$ $\quad$ $x - 3 = -4$

$x = 7$ $\quad$ $x = -1$

Check For $x = 7$ $\quad$ For $x = -1$

$(x-3)^2$	16
$(7-3)^2$	16
4^2	16
16	16

$(x-3)^2$	16
$(-1-3)^2$	16
$(-4)^2$	16
16	16

The solutions are 7 and -1.

b. $(2y-3)^2 = -7$

$2y - 3 = \sqrt{-7}$ or $2y - 3 = -\sqrt{-7}$ Use the square root property.

$2y - 3 = i\sqrt{7}$ $\quad$ $2y - 3 = -i\sqrt{7}$ Since $\sqrt{-7} = i\sqrt{7}$.

$2y = 3 + i\sqrt{7}$ $\quad$ $2y = 3 - i\sqrt{7}$

$y = \dfrac{3 + i\sqrt{7}}{2}$ $\quad$ $y = \dfrac{3 - i\sqrt{7}}{2}$

$y = \dfrac{3}{2} + \dfrac{\sqrt{7}}{2}i$ $\quad$ $y = \dfrac{3}{2} - \dfrac{\sqrt{7}}{2}i$ Divide each term by 2.

These answers check. The solutions are $\dfrac{3}{2} + \dfrac{\sqrt{7}}{2}i$ and $\dfrac{3}{2} - \dfrac{\sqrt{7}}{2}i$. ■

□ **DO EXERCISE 2.**

2 Solution by Factoring

We solved quadratic equations in Section 5.10 by factoring and applying the *zero product property*. The zero product property also holds for complex numbers. Since solution by factoring is often the easiest method of solving a quadratic equation, we review the procedure.

To Solve an Equation by Factoring:

1. Write the equation in standard form.
2. Factor.
3. Use the zero product property.
4. Solve the resulting equations.

□ **Exercise 2** Solve.

a. $(x-5)^2 = 4$

b. $(y+2)^2 = -9$

c. $(3y+4)^2 = 12$

d. $(4x-5)^2 = -16$

□ **Exercise 3** Solve.

a. $3x^2 + 2x - 8 = 0$

b. $2x^2 = 5x$

EXAMPLE 3 Solve.

a. $2x^2 - 7x - 4 = 0$

$(2x + 1)(x - 4) = 0$ Factor.

$2x + 1 = 0$ or $x - 4 = 0$ Use the zero product property.

$2x = -1$ $\qquad x = 4$

$x = -\frac{1}{2}$

Check that the solutions are $-\frac{1}{2}$ and 4.

b. $2x^2 = 3x$

$2x^2 - 3x = 0$ Write the equation in standard form.

$x(2x - 3) = 0$ Factor.

$x = 0$ or $2x - 3 = 0$ Use the zero product property.

$2x = 3$

$x = \frac{3}{2}$

Check For $x = 0$ For $x = \frac{3}{2}$

$2x^2$	$3x$
$2(0)^2$	$3(0)$
$2(0)$	$3(0)$
0	0

$2x^2$	$3x$
$2\left(\frac{3}{2}\right)^2$	$3\left(\frac{3}{2}\right)$
$2\left(\frac{9}{4}\right)$	$3\left(\frac{3}{2}\right)$
$\frac{9}{2}$	$\frac{9}{2}$

Caution: Notice that we may not solve this equation by dividing both sides of the equation by x. If we do that, we lose a solution. ■

□ **DO EXERCISE 3.**

3 Applied Problems

Recall the *Pythagorean theorem* from Section 7.8. It states that in any right triangle, the square of the longest side c (the *hypotenuse*) is the sum of the squares of the other two sides a and b (the legs):

$$c^2 = a^2 + b^2$$

EXAMPLE 4 If one leg of a right triangle is 1 centimeter longer than the other leg and the length of the hypotenuse is 5 centimeters, what are the lengths of the legs?

Variable Let x represent the length of one leg. Then $x + 1$ represents the length of the other leg.

Drawing

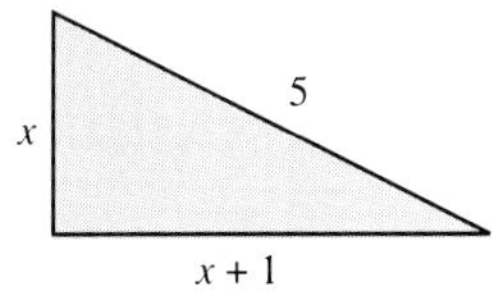

Equation $5^2 = x^2 + (\mathbf{x + 1})^2$ Use the Pythagorean theorem.

Solve $25 = x^2 + \mathbf{x^2 + 2x + 1}$ Square $x + 1$.

$25 = 2x^2 + 2x + 1$ Combine like terms.

$0 = 2x^2 + 2x - 24$ Write in standard form.

$0 = x^2 + x - 12$ Divide both sides by 2 to make factoring easier.

$0 = (x - 3)(x + 4)$ Factor.

$x - 3 = 0$ or $x + 4 = 0$ Use the zero product property.

$x = 3$ $\qquad$ $x = -4$

The number -4 cannot be a solution because the length of the side of a triangle cannot be negative.

If $x = 3$, $x + 1 = 4$. The numbers 3 and 4 check. The lengths of the sides of the triangle are 3 centimeters and 4 centimeters. ■

□ DO EXERCISE 4.

□ **Exercise 4** The hypotenuse of a right triangle is 10 feet long. Find the lengths of the legs if one leg is 2 feet longer than the other.

DID YOU KNOW?

Pythagoras, a Greek, founded the Pythagorean school about 540 B.C. It was a religious sect. He lectured on arithmetic, music, geometry, and astronomy. The Pythagoreans studied mathematics as a religious duty. Pythagoras considered beans sacred and would not eat them.

Answers to Exercises

1. a. $\sqrt{3}, -\sqrt{3}$ **b.** $\frac{2\sqrt{3}}{3}, \frac{-2\sqrt{3}}{3}$ **c.** $\frac{\sqrt{7}}{2}, \frac{-\sqrt{7}}{2}$
d. $i\sqrt{3}, -i\sqrt{3}$

2. a. 7, 3 **b.** $-2 + 3i, -2 - 3i$ **c.** $\frac{-4 + 2\sqrt{3}}{3}, \frac{-4 - 2\sqrt{3}}{3}$
d. $\frac{5}{4} + i, \frac{5}{4} - i$

3. a. $\frac{4}{3}, -2$ **b.** $0, \frac{5}{2}$

4. 6 feet, 8 feet

Problem Set 8.1

NAME ______

DATE ______

CLASS ______

Solve.

1. $x^2 = 5$

2. $y^2 = 11$

3. $2x^2 - 16 = 0$

4. $3x^2 - 36 = 0$

5. $7x^2 - 6 = 0$

6. $3x^2 - 5 = 0$

7. $(x - 4)^2 = 25$

8. $(y - 3)^2 = 1$

9. $(2x + 3)^2 = 3$

10. $(4x + 1)^2 = 7$

11. $(5x + 2)^2 = 12$

12. $(3x - 4)^2 = 18$

13. $(x - 3)^2 = -1$

14. $(y - 2)^2 = -3$

15. $(2y - 6)^2 = -2$

16. $(3x - 5)^2 = -5$

17. $x^2 - x - 6 = 0$

18. $y^2 - 3y - 10 = 0$

19. $8x^2 + 2x - 1 = 0$

20. $6x^2 + 7x - 10 = 0$

21. $3x^2 - 7x + 4 = 0$

22. $12x^2 - 20x + 3 = 0$

23. $16x^2 + 20x - 14 = 0$

24. $9x^2 + 52x - 12 = 0$

25. $10x^2 + 7x = 0$

26. $15x^2 + 8x = 0$

27. $3x^2 = 7x$

28. $9x^2 = 8x$

29. $5x = 6x^2$

30. $10x = 3x^2$

31. $16x^2 + 9 = 0$

32. $4x^2 + 25 = 0$

33. $(3x + 4)^2 = 8$

34. $(5x - 2)^2 = 24$

35. $(7x - 3)^2 = 0$

36. $(9y + 5)^2 = 0$

37. $(4p - 3)^2 = -5$

38. $(8z + 2)^2 = -4$

39. $x^2 + 4x + 4 = 9$

40. $y^2 + 10y + 25 = 81$

41. $9p^2 - 12p + 4 = 100$

42. $16x^2 - 40x + 25 = 49$

43. If one leg of a right triangle is 3 meters longer than the other leg and the length of the hypotenuse is 15 meters, find the lengths of the legs.

44. If one leg of a right triangle is 7 centimeters longer than the other leg and the length of the hypotenuse is 13 centimeters, find the lengths of the legs.

45. The hypotenuse of a right triangle is 100 feet long. Find the lengths of the legs if one leg is 20 feet shorter than the other.

46. The hypotenuse of a right triangle is 150 yards long. Find the lengths of the legs if one leg is 30 yards shorter than the other.

47. The area of a square garden is 24 square meters. Find the length of a side of the garden. Give your answer to the nearest tenth.

48. Find the length of the side of a square if its area is 54 square inches.

49. The area of a rectangular house is 4000 square feet. If the length is 30 feet more than the width, find the width of the house.

50. A rectangle has an area of 112 square centimeters. If the width is 6 centimeters less than the length, find the width of the rectangle.

51. The square of a number plus three times the number is 54. Find the number.

52. Twice the square of an integer is 8 more than six times the integer. Find the integer.

53. The sum of the squares of two consecutive positive odd integers is 34. Find the integers.

54. The sum of the squares of two consecutive negative odd integers is 130. Find the integers.

55. Any equation that can be written in the form $ax^2 + bx + c = 0$, where a, b, and c are real numbers and $a \neq 0$, is a ____________ equation.

56. Quadratic equations have exactly ____________ solutions.

57. Equations of the type $x^2 = c$ may be solved by taking the ____________ roots of both sides of the equation.

58. In a right triangle, the longest side is called the ____________.

59. The square of the longest side in a right triangle is equal to the sum of the ____________ of the other two sides.

Solve. Round answers to the nearest thousandth.

60. $x^2 = 10$

61. $y^2 = 38$

62. $7x^2 - 12 = 0$

63. $8x^2 - 36 = 0$

64. $(3x + 7.82)^2 = 45$

65. $(5x - 2.973)^2 = 19$

Solve.

* 66. $\left(x - \frac{1}{4}\right)^2 = \frac{25}{4}$

* 67. $\left(z + \frac{1}{3}\right)^2 = \frac{9}{64}$

* 68. $3x^3 - 4x^2 - 15x = 0$

* 69. $x(8x^2 - 14x + 3)(2x - 15) = 0$

Think About It

* **70.** How much should the radius of a circle of radius 6 be increased to double the area?

* **71.** The length of a rectangle is 6 more than the width. If the perimeter is 2 times the width, plus 72, find the width of the rectangle.

Checkup

The following problems provide a review of some of Sections 6.2 and 7.4 and some will help you with the next section.

Add or subtract.

72. $3 + \frac{7}{y}$

73. $2 - \frac{4}{x}$

74. $x - \frac{3}{2}$

75. $y + \frac{7}{4}$

Multiply.

76. $\sqrt{3}(2 - \sqrt{7})$

77. $\sqrt{5}(\sqrt{3} + \sqrt{5})$

78. $(3\sqrt{6} + \sqrt{5})(3\sqrt{6} - \sqrt{5})$

79. $(\sqrt{7} - \sqrt{2})(\sqrt{7} + \sqrt{2})$

80. $(4 + \sqrt[3]{6})(2 - 5\sqrt[3]{6})$

81. $(2\sqrt[3]{7} - 1)(4\sqrt[3]{7} - 5)$

82. $(\sqrt{5} - x)^2$

83. $(\sqrt{10} + y)^2$

8.2 SOLVING BY COMPLETING THE SQUARE

OBJECTIVE

[1] *Solve quadratic equations by completing the square*

[1] If a quadratic equation can be factored, the easiest method of solving it is by factoring. However, many quadratic equations cannot be factored. We can solve them by a method called ***completing the square.*** This method is not usually as easy to use as the quadratic formula that we will study in the next section. However, the method of completing the square is used to derive the quadratic formula and for other purposes in mathematics, so we will discuss the procedure.

Recall that a *trinomial* may be the square of a *binomial.*

$$(x + 4)^2 = x^2 + 8x + 16$$

Notice that the last term, 16, is the square of one-half the coefficient of x.

$$\frac{1}{2}(8) = 4, 4^2 = 16$$

We use this fact to find the last term (complete the square) of an expression.

EXAMPLE 1 Complete the square.

a. $x^2 + 6x$

$\frac{1}{2}(6) = 3$ Find one-half the coefficient of x.

$3^2 = \mathbf{9}$ Square one-half the coefficient.

Add this quantity, **9,** to the given expression.

$$x^2 + 6x + \mathbf{9} = (x + 3)^2$$

The square has been completed.

b. $x^2 - 3x$

$\frac{1}{2}(-3) = \frac{-3}{2}$ Find one-half the coefficient of x.

$\left(\frac{-3}{2}\right)^2 = \mathbf{\frac{9}{4}}$ Square one-half the coefficient.

Add $\frac{9}{4}$ to the original expression.

$$x^2 - 3x + \mathbf{\frac{9}{4}} = \left(x - \frac{3}{2}\right)^2 \quad ■$$

□ **DO EXERCISE 1.**

To solve an equation, we complete the square on one side of the equation. Then we can solve the equation using the square root property. To begin the process of completing the square, all terms with variables must be on one side of the equation and the constant term must be on the other side.

□ **Exercise 1** Complete the square.

a. $x^2 + 10x$

b. $y^2 - 4y$

c. $x^2 - x$

d. $y^2 + 5y$

□ **Exercise 2** Solve by completing the square.

a. $x^2 + 8x - 9 = 0$

b. $x^2 - 4x - 16 = 0$

c. $x^2 - x + 3 = 0$

d. $x^2 + 7x - 1 = 0$

EXAMPLE 2 Solve.

a. $x^2 + 8x - 5 = 0$

$x^2 + 8x = 5$ Isolate all terms with variables on one side and the constant on the other side.

$\frac{1}{2}(8) = 4$ Find one-half the coefficient of x.

4^2 Square one-half of the coefficient.

Add 4^2 to both sides of the equation.

$$x^2 + 8x + \mathbf{4^2} = 5 + \mathbf{4^2}$$

$$x^2 + 8x + 4^2 = 5 + 16 \quad \text{Square 4 on the right side.}$$

$$(x + 4)^2 = 21 \quad \text{Factor.}$$

Notice that the numbers connected by the arrows are always the same, so the factorization is easy. Solve the equation using the square root property.

$$x + 4 = \sqrt{21} \quad \text{or} \quad x + 4 = -\sqrt{21}$$

$$x = -4 + \sqrt{21} \qquad x = -4 - \sqrt{21}$$

The solutions are $-4 + \sqrt{21}$ and $-4 - \sqrt{21}$.

b. $x^2 + 5x + 7 = 0$

$x^2 + 5x = -7$ Add -7 to both sides.

$\frac{1}{2}(5) = \frac{5}{2}$ Find one-half the coefficient of x.

$x^2 + 5x + \left(\frac{\mathbf{5}}{\mathbf{2}}\right)^2 = -7 + \left(\frac{\mathbf{5}}{\mathbf{2}}\right)^2$ Add $\left(\frac{5}{2}\right)^2$ to both sides.

$x^2 + 5x + \left(\frac{5}{2}\right)^2 = -7 + \frac{25}{4}$ Square $\frac{5}{2}$ on the right side.

$x^2 + 5x + \left(\frac{5}{2}\right)^2 = \frac{-28 + 25}{4}$ Find a common denominator.

$$\left(x + \frac{5}{2}\right)^2 = \frac{-3}{4}$$

$$x + \frac{5}{2} = \sqrt{\frac{-3}{4}} \quad \text{or} \quad x + \frac{5}{2} = -\sqrt{\frac{-3}{4}}$$

$$x + \frac{5}{2} = i\frac{\sqrt{3}}{2} \qquad x + \frac{5}{2} = -i\frac{\sqrt{3}}{2}$$

$$x = -\frac{5}{2} + \frac{\sqrt{3}}{2}i \qquad x = -\frac{5}{2} - \frac{\sqrt{3}}{2}i$$

The solutions are $-\frac{5}{2} + \frac{\sqrt{3}}{2}i$ and $-\frac{5}{2} - \frac{\sqrt{3}}{2}i$. ■

□ **DO EXERCISE 2.**

The preceding method of completing the square should not be used unless the coefficient of x^2 is 1. If the coefficient of x^2 is not 1, we divide each side of the equation by this coefficient.

EXAMPLE 3 Solve $2x^2 - 8x - 5 = 0$.

$2x^2 - 8x - 5 = 0$

$x^2 - 4x - \frac{5}{2} = 0$ Divide each side of the equation by 2.

$x^2 - 4x = \frac{5}{2}$ Add $\frac{5}{2}$ to each side.

$\frac{1}{2}(-4) = -2$ Find one-half the coefficient of x.

$x^2 - 4x + (\mathbf{-2})^2 = \frac{5}{2} + (\mathbf{-2})^2$ Add $(-2)^2$ to each side.

$x^2 - 4x + (-2)^2 = \frac{5}{2} + 4$ Square -2 on the right side.

$(x - 2)^2 = \frac{13}{2}$

$x - 2 = \sqrt{\frac{13}{2}}$ or $x - 2 = -\sqrt{\frac{13}{2}}$

$x = 2 + \sqrt{\frac{13}{2}}$ $x = 2 - \sqrt{\frac{13}{2}}$

$x = 2 + \frac{\sqrt{26}}{2}$ $x = 2 - \frac{\sqrt{26}}{2}$ Rationalize the denominator.

$x = \frac{4 + \sqrt{26}}{2}$ $x = \frac{4 - \sqrt{26}}{2}$ Find a common denominator.

The solutions are $\frac{4 + \sqrt{26}}{2}$ and $\frac{4 - \sqrt{26}}{2}$. ■

□ **DO EXERCISE 3.**

Solving a Quadratic Equation by Completing the Square

1. If the coefficient of x^2 is 1, go to step 2. If the coefficient of x^2 is not 1, divide each side of the equation by this coefficient.
2. Be sure that terms containing variables are on one side of the equation and constant terms are on the other side.
3. Take half the coefficient of the x term and square it.
4. Add the square to both sides. The side containing the variables is now a perfect square trinomial.
5. Factor the perfect square trinomial.
6. Solve the equation by using the square root property and solving the resulting equations.

□ **Exercise 3** Solve.

a. $3x^2 + 6x - 5 = 0$

b. $2x^2 - 4x + 7 = 0$

Answers to Exercises

1. a. $x^2 + 10x + 25$ **b.** $y^2 - 4y + 4$ **c.** $x^2 - x + \frac{1}{4}$

d. $y^2 + 5y + \frac{25}{4}$

2. a. $-9, 1$ **b.** $2 + 2\sqrt{5}, 2 - 2\sqrt{5}$ **c.** $\frac{1}{2} + \frac{\sqrt{11}}{2}i, \frac{1}{2} - \frac{\sqrt{11}}{2}i$

d. $\frac{-7 + \sqrt{53}}{2}, \frac{-7 - \sqrt{53}}{2}$

3. a. $\frac{-3 + 2\sqrt{6}}{3}, \frac{-3 - 2\sqrt{6}}{3}$ **b.** $1 + \frac{\sqrt{10}}{2}i, 1 - \frac{\sqrt{10}}{2}i$

Problem Set 8.2

NAME

DATE

CLASS

Complete the square.

1. $x^2 + 14x$

2. $y^2 - 18y$

3. $y^2 - \frac{3}{4}y$

4. $x^2 + \frac{5}{3}x$

5. $x^2 - \frac{2}{5}x$

6. $y^2 + \frac{2}{3}y$

Solve by completing the square.

7. $x^2 + 2x - 3 = 0$

8. $y^2 + 8y - 2 = 0$

9. $x^2 - 6x + 4 = 0$

10. $x^2 - 10x + 3 = 0$

11. $x^2 - 5x + 7 = 0$

12. $x^2 - 7x + 8 = 0$

13. $x^2 - 2x + 5 = 0$

14. $y^2 - 4y + 9 = 0$

15. $3x^2 + 9x - 4 = 0$

16. $4x^2 - 8x + 3 = 0$

17. $x^2 + 6x = -7$

18. $y^2 + 8y = -14$

19. $2x^2 + 5x - 1 = 0$

20. $3x^2 - 2x - 4 = 0$

21. $x^2 - 2x + 9 = 0$

22. $y^2 + 6y + 10 = 0$

23. $x^2 = 6x + 13$

24. $x^2 = -2x + 2$

25. $3x^2 - 8x - 3 = 0$

26. $25y^2 + 10y - 3 = 0$

27. $x^2 - 8x - 33 = 0$

28. $p^2 - 4p - 60 = 0$

29. $m^2 + 7m = -9$

30. $y^2 - 3y = -1$

31. $z^2 - z + 5 = 0$

32. $x^2 + 5x = -7$

33. $2x^2 - 12x = 32$

34. $3x^2 - 6x = 24$

35. $3p^2 + 6p - 4 = 0$

36. $4m^2 - 8m + 2 = 0$

37. $2x^2 + 3x = -6$

38. $5x^2 + x = -1$

39. $x^2 - \frac{3}{2}x - 1 = 0$

40. $x^2 + \frac{3}{2}x + \frac{1}{2} = 0$

41. If a quadratic equation can be factored, the easiest method of solving it is by ___________.

42. To begin the process of solving an equation by completing the square, all terms with variables must be on one side of the equation and the ___________ term on the other side.

43. If the coefficient of x^2 is not 1, ___________ each side of the equation by this coefficient in order to solve it by completing the square.

Solve by completing the square.

* **44.** $x^3 + 5x^2 + x = 0$

* **45.** $y^3 - 27 = 0$

* **46.** $x^2 - 4ax = 0$

* **47.** $x^2 + ax + b = 0$

Checkup

The following problems provide a review of some of Sections 7.2 and 7.3 and will help you with the next section.

Simplify.

48. $\sqrt{32}$

49. $\sqrt{24}$

50. $\sqrt{48}$

51. $\sqrt{40}$

Add or subtract.

52. $2\sqrt[3]{6} + 5\sqrt[3]{6}$

53. $8\sqrt[3]{3} - 9\sqrt[3]{3}$

54. $6\sqrt{2} - 4\sqrt{50}$

55. $4\sqrt{5} - 3\sqrt{45}$

56. $3\sqrt{27} + 8\sqrt{108} - \sqrt{12}$

57. $2\sqrt{8} - 3\sqrt{18} + \sqrt{72}$

8.3 THE QUADRATIC FORMULA

OBJECTIVES

1. *Solve quadratic equations using the quadratic formula*
2. *Approximate solutions to quadratic equations*
3. *Solve work problems using quadratic equations*

1 Solving Equations Using the Quadratic Formula

The method of completing the square may be used to solve any quadratic equation. However, the method is often difficult to use. The quadratic formula may also be used to solve any quadratic equation. We use the method of completing the square to derive the quadratic formula by solving the general quadratic equation $ax^2 + bx + c = 0$, where a, b, and c are real numbers and $a \neq 0$. We assume that $a > 0$ for the derivation.

$ax^2 + bx + c = 0$ General quadratic equation.

$x^2 + \frac{b}{a}x + \frac{c}{a} = 0$ Divide both sides by a.

$x^2 + \frac{b}{a}x = -\frac{c}{a}$ Add $-\frac{c}{a}$ to both sides.

$\frac{1}{2}\left(\frac{b}{a}\right) = \frac{b}{2a}$ Find $\frac{1}{2}$ the coefficient of x.

Add $\left(\frac{b}{2a}\right)^2$ to both sides of the equation.

$$x^2 + \frac{b}{a}x + \left(\mathbf{\frac{b}{2a}}\right)^2 = -\frac{c}{a} + \left(\mathbf{\frac{b}{2a}}\right)^2$$

$$\left(x + \frac{b}{2a}\right)^2 = -\frac{c}{a} + \frac{b^2}{4a^2}$$

$\left(x + \frac{b}{2a}\right)^2 = \frac{-c(4a) + b^2}{4a^2}$ Find a common denominator.

$\left(x + \frac{b}{2a}\right)^2 = \frac{b^2 - 4ac}{4a^2}$ Use a commutative law.

$x + \frac{b}{2a} = \pm\sqrt{\frac{b^2 - 4ac}{4a^2}}$ Use the square root property.

$$x + \frac{b}{2a} = \pm\frac{\sqrt{b^2 - 4ac}}{2a}$$

$$x = \frac{-b \pm \sqrt{b^2 - 4ac}}{2a}$$

It can be shown that if $a < 0$, the same solutions are obtained.

> **The Quadratic Formula**
>
> The solutions of $ax^2 + bx + c = 0$, $a \neq 0$, are
>
> $$x = \frac{-b \pm \sqrt{b^2 - 4ac}}{2a}$$

□ **Exercise 1** Solve using the quadratic formula.

a. $x^2 + x = 1$

b. $6x^2 - 5x = 4$

□ **Exercise 2** Solve using the quadratic formula.

a. $x^2 - 2x + 5 = 0$

b. $2x^2 + x + 2 = 0$

EXAMPLE 1 Solve $2x^2 + 3x = 1$.

$$2x^2 + 3x = 1$$

$$2x^2 + 3x - 1 = 0 \qquad \text{Write in standard form.}$$

$$ax^2 + bx + c = 0 \qquad \text{General quadratic equation.}$$

$$a = 2, \quad b = 3, \quad \text{and} \quad c = -1$$

$$x = \frac{-b \pm \sqrt{b^2 - 4ac}}{2a} \qquad \text{Quadratic formula.}$$

$$x = \frac{-3 \pm \sqrt{3^2 - 4(2)(-1)}}{2(2)} \qquad \text{Substitute for } a, b, \text{ and } c.$$

$$x = \frac{-3 \pm \sqrt{9 + 8}}{4}$$

$$x = \frac{-3 \pm \sqrt{17}}{4}$$

The solutions are $(-3 + \sqrt{17})/4$ and $(-3 - \sqrt{17})/4$. ■

□ **DO EXERCISE 1.**

Remember that the solutions to some quadratic equations are not real numbers.

EXAMPLE 2 Solve $3x^2 - 4x + 4 = 0$

$$3x^2 - 4x + 4 = 0$$

$$ax^2 + bx + c = 0 \qquad \text{General quadratic equation.}$$

$$a = 3, \quad b = -4, \quad \text{and} \quad c = 4$$

$$x = \frac{-b \pm \sqrt{b^2 - 4ac}}{2a} \qquad \text{Quadratic formula.}$$

$$x = \frac{-(-4) \pm \sqrt{(-4)^2 - 4(3)(4)}}{2(3)} \qquad \text{Substitute for } a, b, \text{ and } c.$$

$$x = \frac{4 \pm \sqrt{16 - 48}}{6}$$

$$x = \frac{4 \pm \sqrt{-32}}{6}$$

$$x = \frac{4 \pm 4i\sqrt{2}}{6} \qquad \text{Simplify.}$$

$$x = \frac{4(1 \pm i\sqrt{2})}{6} \qquad \text{Factor.}$$

$$x = \frac{2(1 \pm i\sqrt{2})}{3} = \frac{2 \pm 2i\sqrt{2}}{3}$$

The solutions are $\frac{2}{3} + \left(2 \pm 2i\frac{\sqrt{2}}{3}\right)i$ and $\frac{2}{3} - \left(2\frac{\sqrt{2}}{3}\right)i$. ■

□ **DO EXERCISE 2.**

> **Solving a Quadratic Equation**
>
> 1. Check to see if it is of the form $ax^2 = c$ or $(x + k)^2 = d$. If it is either of these forms, use the square root property.
> 2. If it is not of the form of step 1, write it in standard form $ax^2 + bx + c = 0$.
> 3. Try to solve by factoring.
> 4. If it cannot be solved by factoring, use the quadratic formula.

2 Approximating Solutions

Solutions to quadratic equations may be approximated using a calculator or the square root table in Appendix B.

EXAMPLE 3 Solve $x^2 + 5x - 1 = 0$.

$$a = 1, \quad b = 5, \quad \text{and} \quad c = -1$$

$$x = \frac{-b \pm \sqrt{b^2 - 4ac}}{2a}$$

$$x = \frac{-5 \pm \sqrt{5^2 - 4(1)(-1)}}{2(1)}$$

$$x = \frac{-5 \pm \sqrt{25 + 4}}{2}$$

$$x = \frac{-5 \pm \sqrt{29}}{2}$$

$$x = \frac{-5 + \sqrt{29}}{2} \quad \text{or} \quad x = \frac{-5 - \sqrt{29}}{2}$$

Using a calculator, we find that $\sqrt{29} \approx 5.385$.

$$x \approx \frac{-5 + 5.385}{2} \quad \text{or} \quad x \approx \frac{-5 - 5.385}{2}$$

$$x \approx \frac{0.385}{2} \qquad x \approx \frac{-10.385}{2}$$

$$x \approx 0.1925 \qquad x \approx -5.1925$$

$$x \approx 0.2 \qquad x \approx -5.2 \qquad \text{Round to the nearest tenth.}$$ ■

Remember that the answer using Table 1 may vary from the answer using a calculator.

□ **DO EXERCISE 3.**

3 Work Problems

We solved some work problems in Section 6.5. Sometimes it is necessary to solve a quadratic equation to complete the solution of these problems.

EXAMPLE 4 It takes two computers 3 hours to run a certain job. It would take computer A 1 hour less time than computer B to run the job alone. How long would it take computer B to do the job alone?

□ **Exercise 3** Solve and round solutions to the nearest tenth.

a. $x^2 - 4x - 9 = 0$

b. $2x^2 - 14x - 19 = 0$

□ **Exercise 4** It takes two people 4 hours to paint a room. If each worked alone, the slower painter would take 1 hour longer than the faster painter. How long would it take the faster painter to do the job?

Variable Let x represent the time it takes computer B to do the job. Then $x - 1$ represents the time it takes computer A to do the job. $1/x$ job per hour is the rate, r, of computer B. $1/(x - 1)$ job per hour is the rate, r, of computer A. 3 hours is the time it takes to do the job together.

$$W = rt \qquad \text{where } W \text{ is work, } r \text{ is rate, and } t \text{ is time}$$

Work of computer B + work of computer A = one job done together

$$rt_{\text{computer B}} + rt_{\text{computer A}} = 1$$

Equation $\frac{1}{x}(3) + \frac{1}{x - 1}(3) = 1$

Solve $\frac{3}{x} + \frac{3}{x - 1} = 1$

$$x(x - 1)\frac{3}{x} + x(x - 1)\frac{3}{x - 1} = x(x - 1)$$ Multiply both sides by the LCD.

$$3(x - 1) + 3x = x(x - 1)$$

$$3x - 3 + 3x = x^2 - x$$

$$6x - 3 = x^2 - x$$

$$0 = x^2 - 7x + 3$$ Write in standard form.

Since the equation does not factor, we use the quadratic formula. Here $a = 1$, $b = -7$, and $c = 3$.

$$x = \frac{-b \pm \sqrt{b^2 - 4ac}}{2a}$$

$$x = \frac{-(-7) \pm \sqrt{(-7)^2 - 4(1)(3)}}{2(1)}$$

$$x = \frac{7 \pm \sqrt{49 - 12}}{2}$$

$$x = \frac{7 \pm \sqrt{37}}{2}$$

Using a calculator or the table of powers and roots, we find that $\sqrt{37} \approx 6.083$.

$$x \approx \frac{7 \pm 6.083}{2}$$

If we use the plus sign, we get $x \approx 6.5$. The minus sign gives $x \approx 0.5$. (We rounded to the nearest tenth.) Since it must take computer B longer to do the job alone than it takes the computers to do the job together, only the solution 6.5 makes sense.

It takes computer B approximately 6.5 hours to do the job alone. ■

□ **DO EXERCISE 4.**

Answers to Exercises

1. a. $\frac{-1 + \sqrt{5}}{2}, \frac{-1 - \sqrt{5}}{2}$ **b.** $\frac{4}{3}, -\frac{1}{2}$

2. a. $1 + 2i, 1 - 2i$ **b.** $-\frac{1}{4} + \frac{\sqrt{15}}{4}i, -\frac{1}{4} - \frac{\sqrt{15}}{4}i$

3. a. 5.6, −1.6 **b.** 8.2, −1.2

4. 7.5 hours

Problem Set 8.3

NAME ______________________

DATE ______________________

CLASS ______________________

Solve.

1. $x^2 + 7x + 2 = 0$

2. $x^2 - 7x + 2 = 0$

3. $4y^2 - 3y - 2 = 0$

4. $3y^2 + 9y + 4 = 0$

5. $x^2 + x + 3 = 0$

6. $y^2 + y + 5 = 0$

7. $2x^2 + 3x = -2$

8. $3y^2 + 5y = -5$

9. $x^2 + 8 = -x$

10. $p^2 + 5 = 4p$

11. $3q^2 + 2q - 1 = 0$

12. $5x^2 - 3x - 4 = 0$

13. $x^2 = 2x - 2$

14. $5x^2 = 8x + 3$

15. $2x + x(x - 3) = 4$

16. $4x + x(x - 3) = 5$

17. $(x - 1)^2 + (x + 2)^2 = 0$

18. $(x - 3)^2 + (x - 2)^2 = 0$

19. $15x^2 + 3x = 0$

20. $25p^2 + 10p = 0$

21. $3q(q - 1) - 5q(q - 2) = 1$

22. $4x(x + 1) - 6x(x + 2) = 3$

DID YOU KNOW?

By 2000 B.C., the Babylonians were solving problems using written instructions that we can now translate into the quadratic formula.

23. $x^2 + 7 = 3x$

24. $y^2 + 8 = 2y$

25. $x + \frac{1}{x} = \frac{5}{6}$

26. $\frac{2}{x} + \frac{x}{2} = \frac{11}{3}$

27. $\frac{1}{y} + \frac{1}{y+2} = \frac{1}{5}$

28. $\frac{1}{z} + \frac{1}{z+1} = \frac{1}{4}$

29. $x^3 - 1 = 0$

30. $x^3 + 27 = 0$

Use a calculator or Table 1 in Appendix B to approximate the solutions to the nearest tenth.

31. $p^2 - 8p = 3$

32. $z^2 - 4z = 9$

33. $7x^2 = 4 + 5x$

34. $8x^2 - 3 = 7x$

35. $x^2 + 5x - 5 = 0$

36. $x^2 - 7x + 3 = 0$

37. $3x^2 + 2x - 2 = 0$

38. $3y^2 - 3y - 2 = 0$

39. Terry can weed the garden in 2 hours less time than Andrew. If it takes them 8 hours to weed the garden together, how long does it take Terry to weed the garden alone? Give your answer to the nearest tenth.

40. Susan and Brian can do a job in 10 hours. If it takes Susan 4 hours longer to do the job, how long does it take each of them working alone? Give your answers to the nearest tenth.

41. Janice can clean a house in 2 hours less time than Amy can. If it takes them 6 hours to clean the house together, how long does it take Janice to clean the house alone? Give your answer to the nearest tenth.

42. Tim and Gary can mow the lawn in 3 hours. If it takes Gary 1 hour longer than Tim to mow the lawn, how long does it take Tim to do the job? Give your answers to the nearest tenth.

43. The ____________ formula may be used to solve any quadratic equation.

44. To solve a quadratic equation, first check to see if it is of the form ____________ or ____________.

Solve. Round answers to the nearest hundredth.

45. $3.4x^2 - 0.7x - 2.8 = 0$

46. $5.9y^2 + 7.34y - 2.2 = 0$

47. $2.3x^2 + 6.97x = -1.8$

48. $3.5x^2 - 4.2x = 0.83$

Solve.

* 49. $(x + 9)(x - 3) = -37$

* 50. $x + 4 - \frac{5}{x} = \frac{x + 4}{x}$

* 51. $im^2 + 2m - i = 0$

* 52. $\sqrt{2}p^2 - 4p + \sqrt{2} = 0$

* 53. $ix^2 - x - 2 = 0$

Think About It

* **54.** A group of students decides to buy a house for \$80,000. Then three more students join the group, which decreases the cost to each student by \$6000. How many people were in the original group?

Checkup

The following problems provide a review of some of Section 5.10 and will help you with the next section.

Solve.

55. $x^2 + 6x + 9 = 0$

56. $x^2 - 4x + 4 = 0$

57. $x^2 + 3x = 10$

58. $x^2 + 15 = 8x$

59. $2x^2 - x = 3$

60. $4x^2 + 13x + 3 = 0$

61. $3x^2 + 10x - 8 = 0$

62. $15x^2 + 29x - 14 = 0$

63. $8b^2 + 38b = -9$

64. $21m^2 - 31m = -4$

8.4 SOLUTIONS OF QUADRATIC EQUATIONS

1 The Discriminant

The discriminant allows us to determine the nature of a quadratic equation without solving it. Consider the following equations and the solutions to them.

(a) $x^2 + x - 1 = 0$

$a = 1, b = 1, c = -1$

$$x = \frac{-b \pm \sqrt{b^2 - 4ac}}{2a}$$

$$x = \frac{-1 \pm \sqrt{1^2 - 4(1)(-1)}}{2(1)}$$

$$x = \frac{-1 \pm \sqrt{1 + 4}}{2}$$

$$x = \frac{-1 \pm \sqrt{5}}{2}$$

Notice that the number under the radical sign is positive. The solutions are real numbers.

(b) $x^2 + x + 1 = 0$

$a = 1, b = 1, c = 1$

$$x = \frac{-b \pm \sqrt{b^2 - 4ac}}{2a}$$

$$x = \frac{-1 \pm \sqrt{1^2 - 4(1)(1)}}{2(1)}$$

$$x = \frac{-1 \pm \sqrt{1 - 4}}{2}$$

$$x = \frac{-1 \pm \sqrt{-3}}{2}$$

The number under the radical sign is negative. The solutions are not real numbers. They are imaginary.

Also notice that if the number under the radical sign (the radicand) is zero, there is only one solution. It will be a real number, called a *double root,* since quadratic equations have two solutions.

> The expression under the radical sign, $b^2 - 4ac$, is called the ***discriminant.***

We can use the discriminant to decide what type of solution a quadratic equation will have. Sometimes if the solutions are not real, we may not want to solve the equation.

Discriminant **($b^2 - 4ac$)**	***Type of Solution***
Positive	Two different real numbers
Negative	Two different imaginary numbers
Zero	One real number (a double root)

If the discriminant is positive and not a perfect square, the real number solutions are irrational. If the discriminant is positive and a perfect square, the solutions are rational and the equation can be factored.

OBJECTIVES

1 *Predict the number and type of solutions of a quadratic equation using the discriminant.*

2 *Write a quadratic equation with given solutions*

□ **Exercise 1** Predict the number and type of solutions.

a. $2x^2 - 3x + 4 = 0$

b. $5x^2 + 3x - 2 = 0$

c. $x^2 + 6x + 2 = 0$

d. $4x^2 - 12x + 9 = 0$

EXAMPLE 1

a. Predict the number and type of solutions of $x^2 + 5x + 2 = 0$.

$$a = 1, \quad b = 5, \quad \text{and} \quad c = 2$$

Find the discriminant.

$$b^2 - 4ac = 5^2 - 4(1)(2)$$
$$= 25 - 8 = 17$$

The discriminant is positive, so there are two real number solutions.

b. Predict the number and type of solutions of $6x^2 + 3x + 4 = 0$.

$$a = 6, \quad b = 3, \quad \text{and} \quad c = 4$$

Find the discriminant.

$$b^2 - 4ac = 3^2 - 4(6)(4)$$
$$= 9 - 96 = -87$$

The discriminant is negative. There are two imaginary number solutions.

c. Predict the number and type of solutions of $16x^2 - 8x + 1 = 0$.

$$a = 16, \quad b = -8, \quad \text{and} \quad c = 1$$

Find the discriminant.

$$b^2 - 4ac = (-8)^2 - 4(16)(1)$$
$$= 64 - 64 = 0$$

Since the discriminant is zero, there is one real number solution. ■

Use a calculator to help you find the discriminant.

□ **DO EXERCISE 1.**

2 Equations from Solutions

> Given two numbers r_1 and r_2, then
> $$(x - r_1)(x - r_2) = 0$$
> is an equation with solutions r_1 and r_2 by the zero product property.

Notice that this is true since if

$$(x - r_1)(x - r_2) = 0$$

$x - r_1 = 0$ or $x - r_2 = 0$ Zero product property.

$x = r_1$ $x = r_2$

EXAMPLE 2 Find quadratic equations with the following solutions.

a. 3, -4

$[x - 3][x - (-4)] = 0$ Since 3 and -4 are solutions.

$(x - 3)(x + 4) = 0$

$x^2 + 4x - 3x - 12 = 0$ Use FOIL.

$x^2 + x - 12 = 0$

b. $\sqrt{11}, -\sqrt{11}$

$(x - \sqrt{11})(x + \sqrt{11}) = 0$ Since $\sqrt{11}$ and $-\sqrt{11}$ are solutions.

$x^2 - 11 = 0$ Recall that $(A - B)(A + B) = A^2 - B^2$.

c. $1 + i, 1 - i$

$[x - (1 + i)][x - (1 - i)] = 0$ Since $1 + i$ and $1 - i$ are solutions.

$x^2 - x(1 - i) - x(1 + i) + (1 + i)(1 - i) = 0$ Use FOIL.

$x^2 - x + ix - x - ix + 1 - i^2 = 0$ Multiply.

$x^2 - 2x + 1 - i^2 = 0$ Combine like terms.

$x^2 - 2x + 1 - (-1) = 0$ Since $i^2 = -1$.

$x^2 - 2x + 2 = 0$ ■

□ **DO EXERCISE 2.**

□ **Exercise 2** Find quadratic equations with the following solutions.

a. 7, -3

b. $\sqrt{5}, -\sqrt{5}$

c. $4i, -4i$

d. 6 (it is a double root)

DID YOU KNOW?

Thomas Harriot (1560–1621) was the first to use a dot (·) for multiplication. Harriot was also a well-known astronomer who viewed the moons of Jupiter at about the same time as Galileo saw them.

Answers to Exercises

1. a. Two imaginary **b.** Two real **c.** Two real **d.** One real

2. a. $x^2 - 4x - 21 = 0$ **b.** $x^2 - 5 = 0$ **c.** $x^2 + 16 = 0$
d. $x^2 - 12x + 36 = 0$

Problem Set 8.4

NAME ____________________

DATE ____________________

CLASS ____________________

Predict the number and type of solutions.

1. $x^2 + 6x + 1 = 0$

2. $x^2 - 3x + 1 = 0$

3. $3x^2 + 2x + 2 = 0$

4. $5x^2 - x + 4 = 0$

5. $4x^2 - 3x - 3 = 0$

6. $2x^2 + 4x - 3 = 0$

7. $4x^2 - 20x + 25 = 0$

8. $9x^2 + 12x + 4 = 0$

9. $25x^2 + 8x + 1 = 0$

10. $9x^2 + 6x + 4 = 0$

11. $5y^2 + 3y + 4 = 0$

12. $8y^2 - 2y + 3 = 0$

13. $4x^2 + 12x + 9 = 0$

14. $9x^2 - 24x + 16 = 0$

15. $3p^2 + 8p = -2$

16. $7q^2 + 3q = 4$

17. $9x^2 + 3x + 2 = 0$

18. $8x^2 + 3x - 2 = 0$

19. $4m^2 - 5m = -1$

20. $3y^2 - 7y = -2$

21. $x^2 = \frac{1}{2}x + \frac{3}{4}$

22. $p^2 + \frac{7}{3} = 5p$

23. $5y^2 - 3\sqrt{2}y + 4 = 0$

24. $4x^2 + 4\sqrt{2}x - 3 = 0$

25. $\frac{1}{3}x^2 - \frac{2}{3}x + \frac{4}{3} = 0$

26. $\frac{2}{5}y^2 + \frac{7}{10}y - \frac{1}{5} = 0$

Find quadratic equations with the following solutions.

27. $-5, 2$

28. $-3, 3$

29. $2i, -2i$

30. $7i, -7i$

31. 4

32. -8

33. $\sqrt{10}, -\sqrt{10}$

34. $\sqrt{3}, -\sqrt{3}$

35. $\frac{1}{2}, -\frac{3}{4}$

36. $-\frac{1}{3}, \frac{3}{5}$

37. $1 - 3i, 1 + 3i$

38. $1 - 2i, 1 + 2i$

39. $\frac{a}{3}, \frac{b}{3}$

40. $\frac{c}{5}, \frac{d}{10}$

41. In the quadratic formula, the radicand is called the ____________.

42. If the discriminant is negative, the solutions of an equation are two different ____________ numbers.

43. A discriminant of zero indicates that the equation has a ____________ root.

44. If we are given two numbers r_1 and r_2, then $(x - r_1)(x - r_2) = 0$ is an equation with ____________ r_1 and r_2.

Predict the number and type of solutions.

45. $4.8x^2 - 3.25x + 7.9 = 0$

46. $3.9x^2 + 4.7x - 5.24 = 0$

47. $25.4x^2 + 10.4x + 1 = 0$

48. $2.14x^2 - 4.2x = 1.83$

Find quadratic equations with the following solutions.

* 49. $\dfrac{4 - \sqrt{5}}{3}, \dfrac{4 + \sqrt{5}}{3}$

* 50. $\dfrac{\sqrt{3}i}{7}, \dfrac{-\sqrt{3}i}{7}$

Think About It

Find a so that the following equations have a double root.

* 51. $ax^2 + 3x + 6 = 0$

* 52. $ax^2 - 2x + 10 = 0$

Determine which equations have rational solutions.

* 53. $x^2 - 15x + 54 = 0$

* 54. $7y^2 - y = 35$

Checkup

The following problems provide a review of some of Sections 2.4 and 6.6 and will help you with the next section.

Solve.

55. $V = LWH$ for H

56. $V = \dfrac{1}{3}\pi r^2 h$ for h

57. $R = \dfrac{C - S}{n}$ for n

58. $V = \dfrac{KT}{P}$ for T

59. $I = \dfrac{nE}{R + nR}$ for n

60. $A = \dfrac{h}{2}(b + c)$ for b

61. $\dfrac{E}{e} = \dfrac{R + r}{r}$ for R

62. $\dfrac{E}{e} = \dfrac{R + r}{r}$ for E

63. $\dfrac{1}{f} = \dfrac{1}{d_1} + \dfrac{1}{d_2}$ for d_2

64. $\dfrac{1}{f} = \dfrac{1}{d_1} + \dfrac{1}{d_2}$ for f

8.5 FORMULAS AND APPLICATIONS

1 Formulas

Recall that to make computations easier, we may want to solve a formula for a specific variable. Any of the techniques that we have used to solve equations can be used to solve formulas for one of the variables. We may use the methods of squaring both sides of an equation or taking the square root of both sides of an equation. Sometimes we use the quadratic formula to solve for the variable.

OBJECTIVES

1 *Solve a formula for a variable using the techniques for solving equations*

2 *Solve additional word problems using quadratic equations*

EXAMPLE 1

a. A formula for the velocity V of sound in air at a given temperature T may be given by the formula $V = k\sqrt{T}$. Solve the formula for T.

$$V = k\sqrt{T}$$

$$V^2 = (k\sqrt{T})^2 \qquad \text{Square both sides.}$$

$$V^2 = k^2T$$

$$\frac{V^2}{k^2} = T \qquad \text{Divide by } k^2.$$

b. Solve the Pythagorean theorem, $c^2 = a^2 + b^2$, for a.

$$c^2 = a^2 + b^2$$

$$c^2 - b^2 = a^2 \qquad \text{Add } -b^2 \text{ to each side.}$$

$$\pm\sqrt{c^2 - b^2} = a \qquad \text{Use the square root properly.}$$

$$a = \sqrt{c^2 - b^2} \quad \text{or} \quad a = -\sqrt{c^2 - b^2}$$

Since a is the length of the side of a triangle, we reject the negative value.

$$a = \sqrt{c^2 - b^2}$$

c. The height s (in feet) at time t (in seconds) of a projectile launched upward from the ground at a velocity of v feet per second is given by the equation $s = vt - 16t^2$. Solve the equation for t.

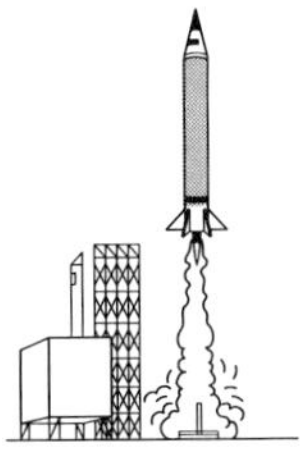

$$s = vt - 16t^2$$

$$16t^2 - vt + s = 0 \qquad \text{Write the equation in standard form.}$$

Use the quadratic formula. $a = 16$, $b = -v$, and $c = s$.

$$t = \frac{-(-v) \pm \sqrt{(-v)^2 - 4(16)(s)}}{2(16)}$$

$$= \frac{v \pm \sqrt{v^2 - 64s}}{32}$$

The solutions are

$$t = \frac{v + \sqrt{v^2 - 64s}}{32} \quad \text{and} \quad t = \frac{v - \sqrt{v^2 - 64s}}{32}$$ ■

□ DO EXERCISE 1.

□ **Exercise 1** Assume that variables represent positive numbers.

a. Solve $F = \dfrac{k}{\sqrt{d}}$ for d.

b. Solve $A = 2\pi r^2$ for r.

c. Solve $2t^2 - 5t - s = 0$ for t.

□ **Exercise 2** How many seconds will it take for the projectile in Example 2 to get 34 feet above the ground?

□ **Exercise 3** Jan and Sharon bought a 9- by 12-foot rug for their room. The area of the room is 154 square feet and there is an even strip of hardwood surrounding the rug. How wide is the strip of hardwood?

2 Applications

The following example shows how quadratic equations are used to solve motion problems.

EXAMPLE 2 A projectile is launched upward from the ground at a velocity of 50 feet per second. In how many seconds will the object be 36 feet above the ground?

From Example 1c, the equation of motion is

$$s = vt - 16t^2$$

We know that $s = 36$ and $v = 50$, so we substitute these values into the equation.

$$36 = 50t - 16t^2$$

$16t^2 - 50t + 36 = 0$ Write in standard form.

$8t^2 - 25t + 18 = 0$ Divide both sides by 2.

$(8t - 9)(t - 2) = 0$ Factor.

$8t - 9 = 0$ or $t - 2 = 0$ Use the zero product property.

$8t = 9$ $\qquad t = 2$

$t = \frac{9}{8}$

The number $\frac{9}{8}$ checks. It will take $1\frac{1}{8}$ seconds for the projectile to be 36 feet above the ground. ■

□ **DO EXERCISE 2.**

EXAMPLE 3. A rectangular flower garden that is 14 feet by 20 feet has a path of uniform width around it. If the area of the garden and path is 352 square feet, what is the width of the path?

Variable Let x represent the width of the path.

Drawing

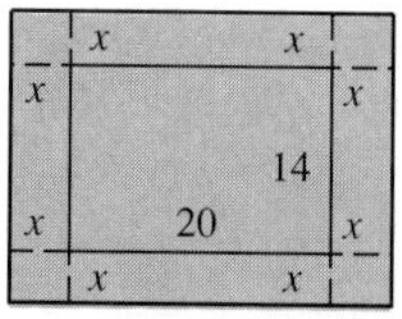

Since the area of a rectangle is the length times the width, we have the following.

Equation $(14 + 2x)(20 + 2x) = 352$

Solve $280 + 68x + 4x^2 = 352$

$4x^2 + 68x - 72 = 0$ Write in standard form.

$x^2 + 17x - 18 = 0$ Divide both sides by 4.

$(x + 18)(x - 1) = 0$ Factor.

$x + 18 = 0$ or $x - 1 = 0$ Use the zero product property.

$x = -18$ $\qquad x = 1$

The width of the path cannot be -18 feet, so the solution is 1 foot, since this number checks. ■

□ **DO EXERCISE 3.**

Recall that we worked some motion problems in Section 6.5. It may be necessary to use a quadratic equation to solve a motion problem.

EXAMPLE 4 The speed of the current in a stream is 3 miles per hour. A boat travels 18 miles upstream and back downstream in 8 hours. What is the speed of the boat in still water?

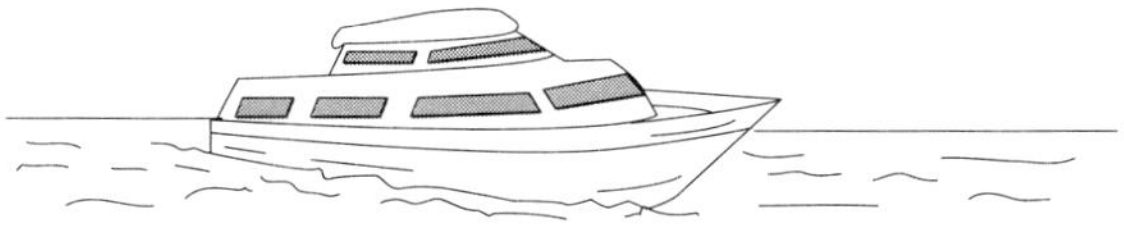

Variable Let r represent the speed of the boat in still water.

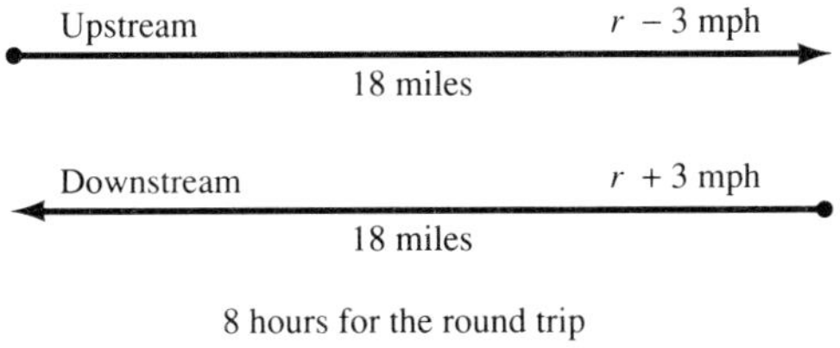

The following chart helps us to organize the given information.

	(Distance in miles)	÷	(rate in miles per hour)	=	(time in hours)
Upstream	18	÷	$(r-3)$	=	$\frac{18}{r-3}$
Downstream	18	÷	$(r+3)$	=	$\frac{18}{r+3}$

Equation Since the time to go upstream plus the time to go downstream is 8 hours, we have the following equation:

$$\frac{18}{r-3}+\frac{18}{r+3}=8$$

Solve The LCD is $(r-3)(r+3)$. Multiply both sides of the equation by the LCD.

$$\mathbf{(r-3)(r+3)}\left(\frac{18}{r-3}+\frac{18}{r+3}\right)=\mathbf{(r-3)(r+3)}8$$

$$(r-3)(r+3)\frac{18}{r-3}+(r-3)(r+3)\frac{18}{r+3}=(r-3)(r+3)8 \quad \text{Use a distributive law.}$$

$$(r+3)18+(r-3)18=(r-3)(r+3)8 \quad \text{Simplify.}$$

$$18r+54+18r-54=8r^2-72$$

$$36r=8r^2-72$$

$$0=8r^2-36r-72$$

$$0=2r^2-9r-18 \quad \text{Divide by 4.}$$

$$0=(2r+3)(r-6) \quad \text{Factor.}$$

$$r=-\frac{3}{2} \quad \text{or} \quad r=6$$

The speed of the boat cannot be $-\frac{3}{2}$, so the solution is 6 miles per hour since this number checks. ■

□ **DO EXERCISE 4.**

□ **Exercise 4** Juan's boat can go 10 miles per hour in still water. It takes Juan 4 hours to go 15 miles upstream and return. Find the speed of the current.

Answers to Exercises

1 a. $d = \frac{k^2}{F^2}$ **b.** $r = \sqrt{\frac{A}{2\pi}}$ **c.** $t = \frac{5 + \sqrt{25 + 8s}}{4}$

2. 1

3. $(9 + 2x)(12 + 2x) = 154$

$$108 + 42x + 4x^2 = 154$$
$$4x^2 + 42x - 46 = 0$$
$$2x^2 + 21x - 23 = 0$$
$$(2x + 23)(x - 1) = 0$$

$x = -\frac{23}{2}$ (reject), $x = 1$

The strip of hardwood is 1 foot wide.

4. $\frac{15}{10 - x} + \frac{15}{10 + x} = 4$

$$(10 - x)(10 + x)\frac{15}{10 - x} + (10 - x)(10 + x)\frac{15}{10 + x} = 4(10 - x)(10 + x)$$
$$(10 + x)15 + (10 - x)15 = 4(10 - x)(10 + x)$$
$$150 + 15x + 150 - 15x = 400 - 4x^2$$
$$300 = 400 - 4x^2$$
$$4x^2 = 100$$
$$x^2 = 25$$

$x = 5$ or $x = -5$ (reject)

The speed of the current is 5 miles per hour.

Problem Set 8.5

NAME

DATE

CLASS

Solve each equation for the variable indicated. Assume that all variables represent positive numbers.

1. $D = \sqrt{ch}$ for h

2. $A = \sqrt{\dfrac{w_1}{w_2}}$ for w_2

3. $p = \sqrt{\dfrac{kL}{g}}$ for g
(a pendulum formula)

4. $W = \sqrt{\dfrac{1}{LC}}$ for C
(an electricity formula)

5. $c^2 = a^2 + b^2$ for b
(Pythagorean formula in two dimensions)

6. $a^2 + b^2 + c^2 = d^2$ for a
(Pythagorean formula in three dimensions)

7. $V = \dfrac{1}{3}\pi r^2 h$ for r
(volume of a cone)

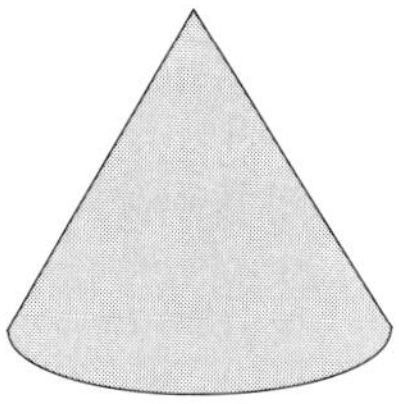

8. $A = 6s^2$ for s
(area of a cube)

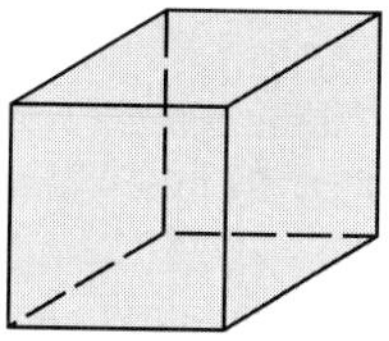

9. $s = kwd^2$ for d
(strength of a rectangular beam)

10. $d = kt^2$ for t
(distance an object has fallen)

11. $L = \dfrac{kd^4}{h^2}$ for h

12. $F = \dfrac{kA}{V^2}$ for V

13. $g(P_1 - P_2) = \dfrac{kLV}{D^2}$ for D

14. $p(N + p) = \dfrac{kA^2}{R^2}$ for R

15. $At^2 + Bt + C = 0$ for t

16. $A = \pi r^2 + \pi rs$ for r

17. $R = \frac{k}{d^2}$ for d

18. $I = \frac{ks}{d^2}$ for d

19. $S = 2\pi rh + \pi r^2$ for r

20. $V = \pi(r^2 + R^2)h$ for R

Solve the following word problems.

21. The position s (in feet) at time t (in seconds) of an object moving in a straight line is given by the equation

$$s = 2t^2 - 5t$$

How many seconds will it take the object to move 25 feet?

22. At what time is the object in Problem 17 at a position of zero?

23. A rectangular pool 30 by 40 feet has a strip of concrete of uniform width around it. If the total area of the pool and the concrete is 1496 square feet, find the width of the strip of concrete.

24. A rectangular garden is 15 by 30 feet. There is a path of uniform width around the garden. If the area of the path is 196 square feet, find the width of the path.

25. A car traveled 300 miles to a city at a certain speed. It made the return trip at a speed 10 miles per hour slower. The total time for the trip was 11 hours. What was the speed of the car going to the city?

26. A car traveled 60 miles at a certain speed. Then it traveled 50 miles at a speed 10 miles per hour slower. If the total time for the trip was 2 hours, what was the speed on the first part of the trip?

27. Laura rides her motorcycle 20 miles per hour faster than Kevin rides his motorcycle. If Laura goes 60 miles in $\frac{1}{4}$ hour less time than Kevin, find Kevin's average speed.

28. An airplane travels 5600 kilometers at a certain speed. Another airplane travels 4000 kilometers at a speed 50 kilometers per hour faster than the first plane, in 6 hours less time. What is the speed of the slower plane?

29. A boat travels 6 miles per hour in still water. It travels 8 miles upstream and back again in 8 hours. What is the speed of the current?

30. The speed of the current of a river is 5 miles per hour. It takes twice as long for a boat to go 20 miles upstream as it does for it to go downstream. Find the speed of the boat in still water.

31. The revenue gained from selling x television sets is given by $R = 100x^2 - 500x + 600$. How many television sets does the Telemark need to sell for a revenue of \$600?

32. The cost of manufacturing x teddy bears is $C = 2x^2 - 1600x + 5000$. How many teddy bears can be produced at a cost of \$5000?

33. A teacher has determined that the number of students in her trigonometry class has increased over a 5-year period at a rate r given by the equation $5r^2 + 14r - 3 = 0$. Find r.

34. The rate of reduction in taxes over a 3-year period in California is given by the equation $12r^2 + 5r - 2 = 0$. Find r.

35. The outside of a picture frame measures 8 inches by 14 inches; 40 square inches of picture is visible. Find the width of the frame.

36. A sheet of computer paper measures 9 inches by 11 inches. There are 63 square inches of typing on the sheet. Find the width of the uniform margins.

37. The speed of the current in a canal is 4 miles per hour. It takes twice as long for a houseboat to go 18 miles upstream as it does for it to go downstream. Find the speed of the boat in still water.

38. A canoe travels 4 miles per hour in still water. It travels 6 miles upstream and back again in 4 hours. What is the speed of the current?

39. An airplane travels 360 miles per hour in still air. It travels 400 miles to Denver and back again in $2\frac{1}{4}$ hours. What is the speed of the wind?

40. The speed of the wind is 50 miles per hour. An airplane travels 1000 miles against the wind to Chicago and back again in $5\frac{1}{2}$ hours. What is the speed of the airplane?

41. A formula can be solved for one of the variables using any of the techniques that are used to solve ____________.

42. It does not make sense for the length of the side of a triangle to be less than or equal to ____________.

Round answers to the nearest tenth.

43. The position s (in feet) at time t (in seconds) of an object moving in a straight line is given by the equation

$$s = 2t^2 - 5t$$

How many seconds will it take the object to move 8.9 feet?

44. A rectangular garden is 15.8 feet by 25.5 feet. There is a path of uniform width around the garden. If the area of the path is 182.9 square feet, find the width of the path.

Solve each equation for the variable indicated. Assume that all variables represent positive numbers.

* **45.** $x^2 - (3a - b)x = 3ab$ for x

* **46.** $\frac{p}{p-q} + \frac{q-p}{p} = 2$ for q

* **47.** $D = \sqrt{\frac{kLV}{g(P_1 - P_2)}}$ for P_1

* **48.** $V = \frac{4}{3}\pi r^3$ for r

Think About It

* **49.** The volume of a rectangular solid is 320 cubic centimeters. If the length is 10 centimeters and the surface area is 304 square centimeters, find the other dimensions. (*Hint:* Use the geometric formulas on the inside front cover.)

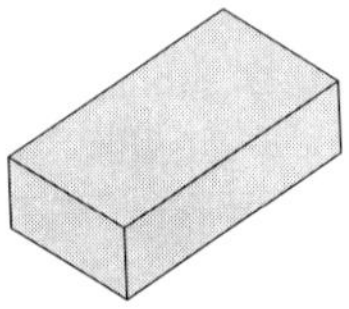

* **50.** A rectangular garden is 3 times as long as it is wide. If the width is increased by 20 feet, the area is doubled. Find the area of the garden.

Checkup

The following problems provide a review of some of Sections 7.6 and 8.1 and some will help you with the next section.

Simplify. Write answers with positive exponents. Assume that all bases are positive.

51. $x^{1/3} \cdot x^{3/5}$

52. $y^{2/3} \cdot y^{-5/6}$

53. $\frac{6^{3/4}}{6^{7/8}}$

54. $\frac{x^{1/5}}{x^{-3/8}}$

55. $\frac{x^{1/2} \cdot x^{2/3}}{x^{5/6}}$

56. $\frac{y^{1/4} \cdot y^{-1/8}}{y^{-3/16}}$

57. $(2^{-3/4})^{4/5}$

58. $(x^{1/3})^{6/5}$

Solve.

59. $x^2 = 5$

60. $x^2 = 15$

8.6 EQUATIONS REDUCIBLE TO QUADRATIC EQUATIONS

OBJECTIVE

1 *Solve equations that are reducible to quadratic equations*

1 Sometimes equations that are not quadratic can be solved as quadratic equations by making a substitution. These equations are said to be reducible to quadratic equations.

EXAMPLE 1 Solve $(x^2 + 1)^2 - (x^2 + 1) - 6 = 0$.

Let $u = x^2 + 1$.

Substitute u for $x^2 + 1$ in the equation.

$$(x^2 + 1)^2 - (x^2 + 1) - 6 = 0$$

$$u^2 - u - 6 = 0 \qquad \text{Substitute } u \text{ for } x^2 + 1.$$

$$(u - 3)(u + 2) = 0 \qquad \text{Factor.}$$

$$u - 3 = 0 \quad \text{or} \quad u + 2 = 0 \qquad \text{Use the zero product property.}$$

$$u = 3 \qquad\qquad u = -2$$

Caution: This is *not* the solution. We must substitute $x^2 + 1$ for u and solve the resulting equations.

$$x^2 + 1 = 3 \quad \text{or} \quad x^2 + 1 = -2 \qquad \text{Substitute } x^2 + 1 \text{ for } u.$$

$$x^2 = 2 \qquad\qquad x^2 = -3$$

$$x = \pm\sqrt{2} \qquad\qquad x = \pm\sqrt{-3}$$

$$x = \pm i\sqrt{3}$$

These numbers check. The solutions are $\sqrt{2}$, $-\sqrt{2}$, $\sqrt{3}i$, and $-\sqrt{3}i$. ■

□ **DO EXERCISE 1.**

□ **Exercise 1** Solve:

$$(x^2 - 1)^2 - 3(x^2 - 1) - 40 = 0$$

We solved equations containing radicals in Chapter 7. The following is another technique for solving some equations with radicals.

If we square both sides of an equation to solve the problem, we must be sure to check the answer because we may get possible solutions that are not solutions to the original equation.

EXAMPLE 2 Solve $x + 3\sqrt{x} - 10 = 0$.

Let $u = \sqrt{x}$. Then $u^2 = x$. We substitute u^2 for x and u for $\sqrt{x}$.

$$x + 3\sqrt{x} - 10 = 0$$

$$u^2 + 3u - 10 = 0 \qquad \text{Substitute for } x \text{ and } \sqrt{x}.$$

$$(u + 5)(u - 2) = 0$$

$$u = -5 \quad \text{or} \quad u = 2$$

Replace u with $\sqrt{x}$ and solve these equations.

$$\sqrt{x} = -5 \quad \text{or} \quad \sqrt{x} = 2$$

The first equation does not have a real number solution since the principal square root of a number cannot be negative. Squaring the second equation gives $x = 4$. This checks. The solution is 4. ■

□ **DO EXERCISE 2.**

□ **Exercise 2** Solve:

$$x - 5\sqrt{x} + 6 = 0$$

□ **Exercise 3** Solve:

$x^4 - 3x^2 - 10 = 0.$

EXAMPLE 3 Solve $x^4 - 7x^2 + 6 = 0$.

Rewrite the equation as

$$(x^2)^2 - 7x^2 + 6 = 0.$$

Let $u = x^2$.

$$u^2 - 7u + 6 = 0 \qquad \text{Substitute } u \text{ for } x^2.$$
$$(u - 6)(u - 1) = 0$$
$$u - 6 = 0 \quad \text{or} \quad u - 1 = 0$$
$$u = 6 \qquad\qquad u = 1$$

Substitute x^2 for u.

$$x^2 = 6 \qquad\qquad x^2 = 1$$
$$x = \pm\sqrt{6} \qquad\qquad x = \pm\sqrt{1}$$
$$x = \pm 1$$

These numbers check. The solutions are $\sqrt{6}, -\sqrt{6}, 1$, and -1. ■

Notice that this fourth-degree equation has four solutions. In general, an nth-degree equation has n or fewer distinct solutions.

□ **DO EXERCISE 3.**

□ **Exercise 4** Solve:

$x^{2/3} - 2x^{1/3} - 3 = 0.$

EXAMPLE 4 Solve $p^{2/3} + 6p^{1/3} + 8 = 0$.

Let $u = p^{1/3}$. Then $u^2 = p^{2/3}$. Substitute these values into the equation.

$$p^{2/3} + 6p^{1/3} + 8 = 0$$
$$u^2 + 6u + 8 = 0$$
$$(u + 4)(u + 2) = 0$$
$$u = -4 \quad \text{or} \quad u = -2$$

Substitute $p^{1/3}$ for u.

$$p^{1/3} = -4 \quad \text{or} \quad p^{1/3} = -2$$
$$(p^{1/3})^3 = (-4)^3 \qquad (p^{1/3})^3 = (-2)^3 \qquad \text{Cube both sides of the equation.}$$
$$p = -64 \qquad\qquad p = -8$$

These numbers check. The solutions are -64 and -8. ■

□ **DO EXERCISE 4.**

Answers to Exercises

1. $3, -3, 2i, -2i$
2. 4, 9
3. $\sqrt{5}, -\sqrt{5}, \sqrt{2}i, -\sqrt{2}i$
4. $27, -1$

Problem Set 8.6

NAME

DATE

CLASS

Solve.

1. $x^4 + 3x^2 - 4 = 0$

2. $x^4 - 7x^2 + 12 = 0$

3. $x - 4\sqrt{x} + 3 = 0$

4. $y + 2\sqrt{y} - 24 = 0$

5. $(x^2 - 3)^2 - (x^2 - 3) - 2 = 0$

6. $(x^2 - 5)^2 + (x^2 - 5) - 2 = 0$

7. $x - 2\sqrt{x} - 15 = 0$

8. $5x - 7\sqrt{x} + 2 = 0$

9. $x^4 - 5x^2 - 14 = 0$

10. $p^4 - 3p^2 - 10 = 0$

11. $(x^2 - 6x)^2 - 2(x^2 - 6x) - 35 = 0$

12. $(x^2 - x)^2 - 14(x^2 - x) + 24 = 0$

13. $y^{2/3} + 3y^{1/3} - 4 = 0$

14. $3x^{2/3} - 5x^{1/3} + 2 = 0$

15. $2x^{2/5} - x^{1/5} = 6$

16. $x^{4/3} - 5x^{2/3} = -4$

17. $x^{-2} - 6x^{-1} + 8 = 0$

18. $3x^{-2} + 14x^{-1} - 5 = 0$

19. $(2 + \sqrt{x})^2 - 4(2 + \sqrt{x}) - 5 = 0$

20. $(1 + \sqrt{x})^2 + 5(1 + \sqrt{x}) + 6 = 0$

21. $3x^{-2} - 7x^{-1} + 4 = 0$

22. $x^{-2} - 7x^{-1} - 8 = 0$

23. $8x^4 - 22x^2 = -15$

24. $6x^4 - 32x^2 = -10$

25. $\left(x - \frac{1}{2}\right)^2 + 5\left(x - \frac{1}{2}\right) - 4 = 0$
(Use the quadratic formula.)

26. $\left(x + \frac{1}{4}\right)^2 - 4\left(x + \frac{1}{4}\right) + 2 = 0$
(Use the quadratic formula.)

27. $\left(\frac{x+2}{x-2}\right)^2 - 2\left(\frac{x+2}{x-2}\right) - 8 = 0$

28. $\left(\frac{x+3}{x-5}\right)^2 + \left(\frac{x+3}{x-5}\right) - 42 = 0$

29. $\left(\frac{x+6}{x-1}\right)^2 - 6\left(\frac{x+6}{x-1}\right) - 16 = 0$

30. $3\left(\frac{x+5}{x-2}\right)^2 + 2\left(\frac{x+5}{x-2}\right) - 8 = 0$

31. Equations that are not quadratic can sometimes be solved as quadratic equations by making a ____________.

32. An nth-degree equation has ____________ or fewer distinct solutions.

Solve.

* **33.** $x^6 - 1 = 0$

* **34.** $9y^3 - 10y^{3/2} + 1 = 0$

* **35.** $\left(x - \frac{8}{x}\right)^2 - \left(x - \frac{8}{x}\right) = 42$

* **36.** $\frac{x^2+1}{x} + \frac{4x}{x^2+1} = 4$

Checkup

The following problems provide a review of some of Section 3.1 and will help you with the next section.

Make a table of values for x and y. Then graph.

37. $y = 3x + 6$

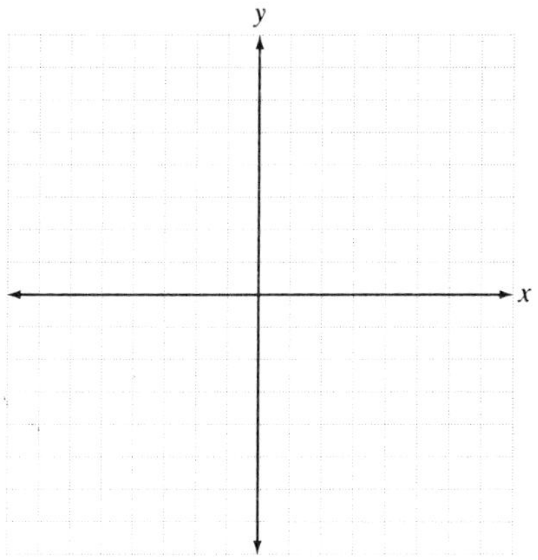

38. $y = 2x - 4$

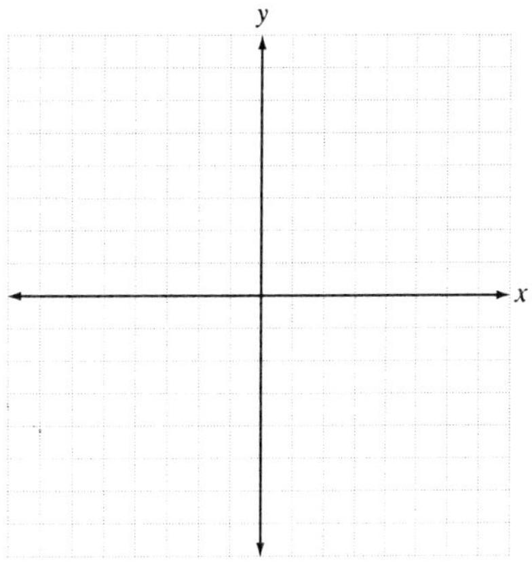

39. $y = -2x$

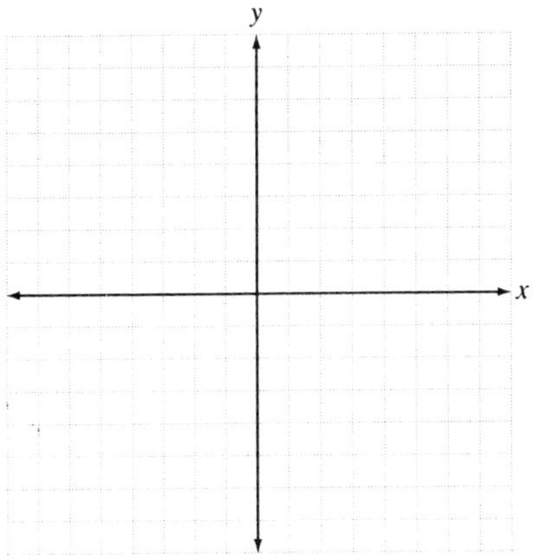

40. $y = 3x$

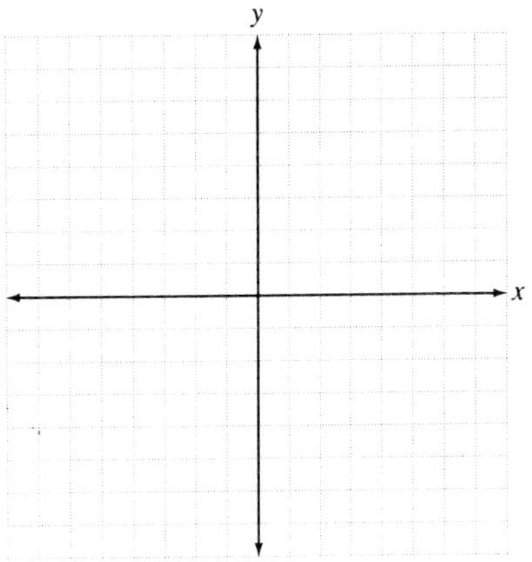

41. $y = -2x - 6$

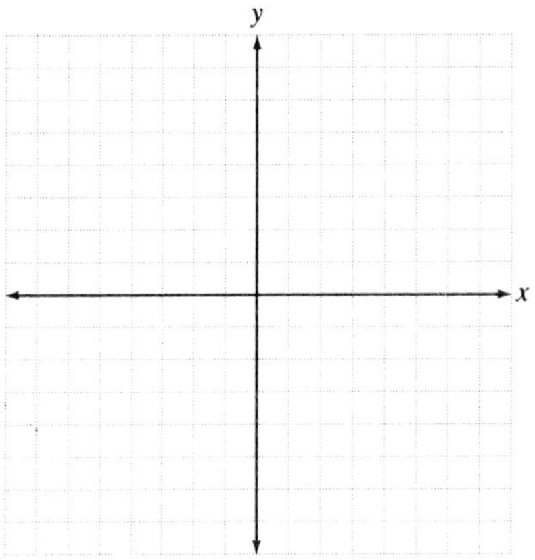

42. $y = -3x + 3$

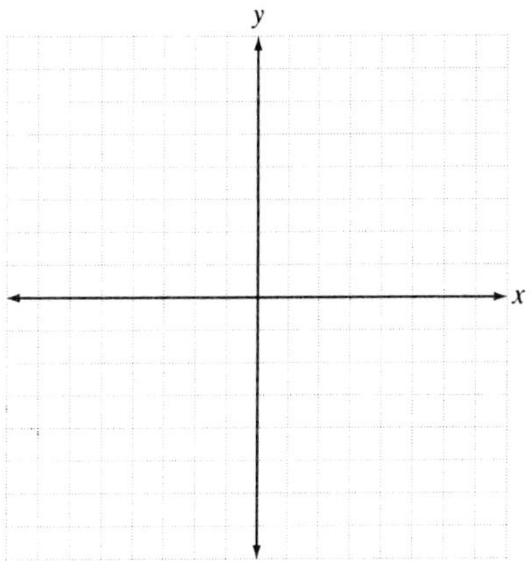

43. $y = -\frac{3x}{4} + 2$

(*Hint:* Use multiples of 4 for x.)

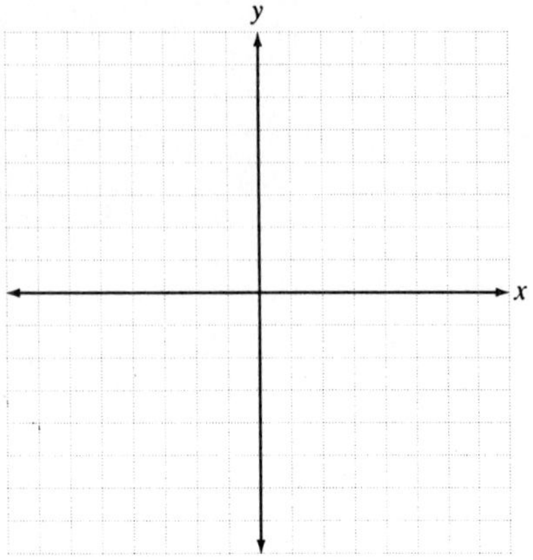

44. $y = \frac{4x}{5} + 1$

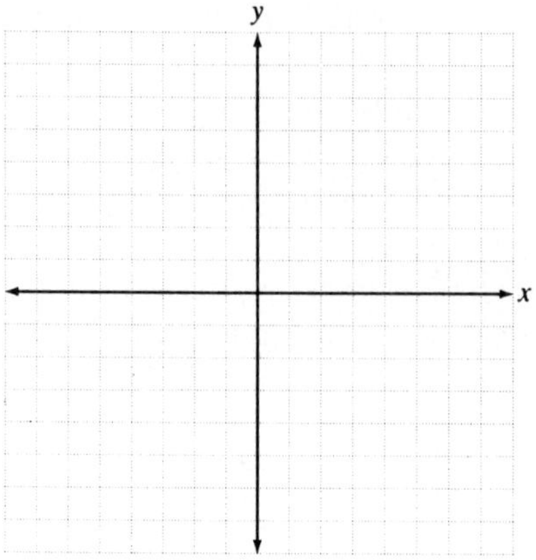

45. $y = \frac{2x}{3} - 3$

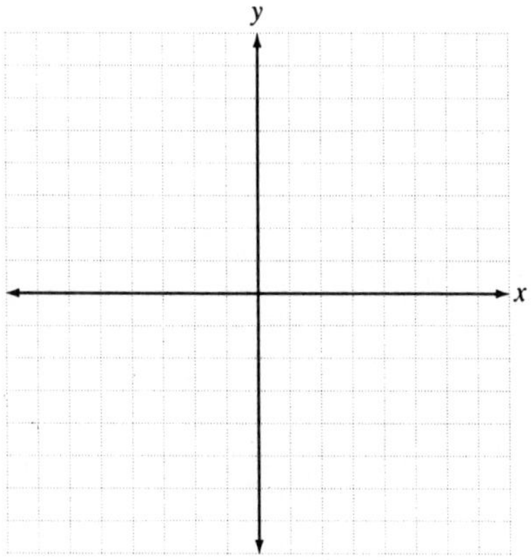

46. $y = \frac{5x}{2} - 2$

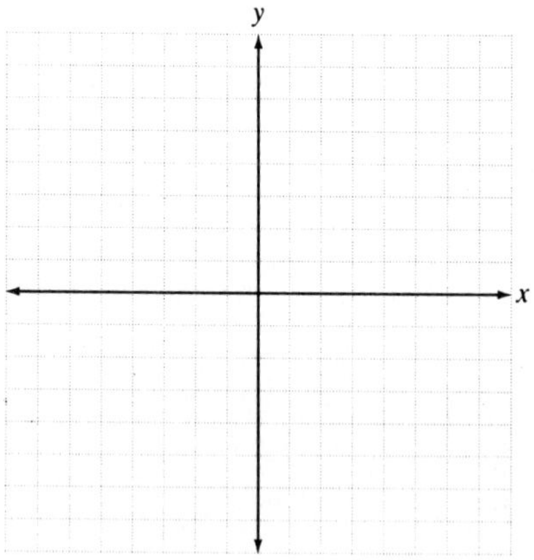

8.7 PARABOLAS

The equation

$$y = ax^2 + bx + c \qquad \text{with } a \neq 0$$

is a quadratic equation with two variables if a, b, and c are real-number constants. The *solutions* of the equation will be *ordered pairs* of numbers. We may graph a quadratic equation with two variables by finding ordered pairs of numbers that satisfy the equation and graphing them. This method is often the easiest way to graph equations of the form $y = ax^2 + c$ ($b = 0$). When $b = 0$; we use an x value of 0 to give us the highest or lowest point on the graph. This point is called the ***vertex.***

OBJECTIVES

1. *Graph equations of the form* $y = ax^2 + c$
2. *Graph equations of the form* $y = ax^2 + bx + c$

1 Graphing Equations of the Form $y = ax^2 + c$

EXAMPLE 1 Graph of $y = x^2 - 2$.

Find ordered pairs that satisfy the equation. We choose at least five values for x, *including* 0 and two negative values. (We must choose some negative values to draw a complete graph.) Let $x = -2$.

$y = x^2 - 2$

$y = (\mathbf{-2})^2 - 2$

$y = 4 - 2 = 2$

x	y	Ordered Pair
−2	2	(−2, 2)
−1	−1	(−1, −1)
0	−2	(0, −2) Vertex
1	−1	(1, −1)
2	2	(2, 2)

We also chose values for x of −1, 0, 1, and 2 and found the corresponding y values in the table. Plot these points and draw a smooth curve through them. (If we had chosen other values for x, the graphs of the resulting ordered pairs would lie on this curve.) The graph is shown below.

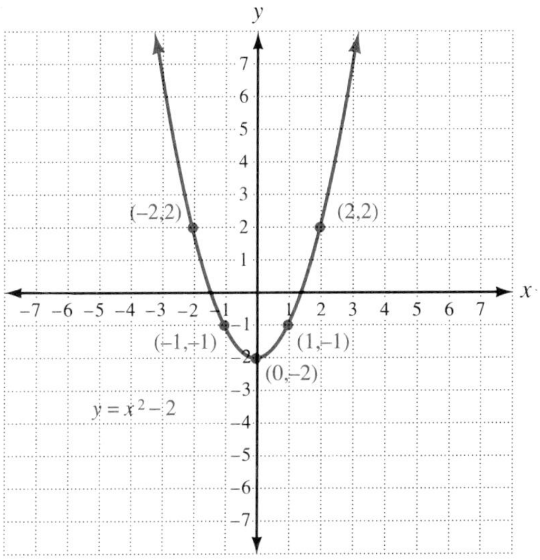

□ **Exercise 1** Graph $y = 2x^2$.

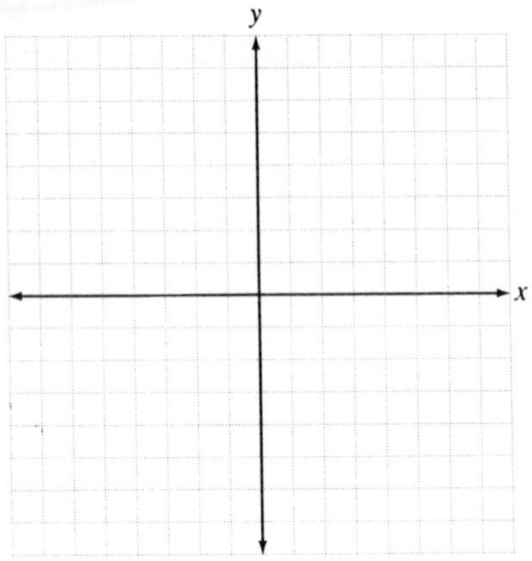

Notice that the preceding graph is a U-shaped curve. Graphs of quadratic equations with two variables are always U-shaped. The U may be inverted. These curves are called *parabolas*. ■

□ **DO EXERCISE 1.**

2 Graphing Equations of the Form $y = ax^2 + bx + c$

When a quadratic equation with two variables has an x term, one method of graphing it is to use the vertex and the intercepts.

Intercepts

To find the y-intercept, let $x = 0$. Recall that this is the point where the graph crosses the y-axis.

To find the x-intercept, let $y = 0$. This is the point or points where the graph crosses the x-axis.

Vertex

> The x-coordinate of the vertex is $x = -\dfrac{b}{2a}$.

To see why this is true we complete the square on the general quadratic equation in two variables. For purposes of the demonstration we want to isolate y on one side of the equation. So instead of dividing both sides of the equation by the coefficient of x^2, we factor it out of the x^2 and x terms.

$$y = ax^2 + bx + c \qquad \text{General quadratic equation.}$$

$$y = a\left(x^2 + \frac{bx}{a}\right) + c \qquad \text{Factor out } a\text{, the coefficient of } x^2.$$

We want to add $[\frac{1}{2}(b/a)]^2 = (b/(2a))^2$ inside the parentheses to complete the square. This means that we are adding $a(b/(2a))^2$ and we must subtract $a(b/(2a))^2$.

$$y = a\left[x^2 + \frac{bx}{a} + \left(\frac{\mathbf{b}}{\mathbf{2a}}\right)^2\right] + c - a\left(\frac{\mathbf{b}}{\mathbf{2a}}\right)^2 \qquad \text{Complete the square.}$$

$$y = a\left(x + \frac{b}{2a}\right)^2 + c - \frac{b^2}{4a} \qquad \text{Factor and simplify.}$$

$$y = a\left(x + \frac{b}{2a}\right)^2 + \frac{4ac - b^2}{4a} \qquad \text{Find a common denominator.}$$

The graph of the quadratic equation with two variables will be at its highest or lowest point (the vertex) when $x = -b/(2a)$ since this will make the first term on the right zero and the y value will be as small as possible when $a > 0$ (a minimum) or as large as possible when $a < 0$ (a maximum). If a is *positive*, the parabola opens *upward*. If a is *negative*, the parabola opens *downward*.

The line $x = -b/(2a)$ through the x-coordinate of the vertex is called the *axis of symmetry*. If the graph is folded on this line, the two halves of the parabola will match. To find the y-coordinate of the vertex, we substitute the x-coordinate of the vertex into the given equation.

EXAMPLE 2 Sketch the graph of $y = x^2 - 2x - 3$.

1. Find the y-intercept. Let $x = 0$ and solve for y.

$$y = x^2 - 2x - 3$$

$$y = \mathbf{0}^2 - 2(\mathbf{0}) - 3 = -3 \qquad \text{Substitute 0 for } x.$$

The y-intercept is $(0, -3)$.

2. Find the x-intercepts. Let $y = 0$ and solve for x.

$$y = x^2 - 2x - 3$$

$$\mathbf{0} = x^2 - 2x - 3 \qquad \text{Substitute 0 for } y.$$

$$0 = (x - 3)(x + 1) \qquad \text{Factor.}$$

$$x = 3 \quad \text{or} \quad x = -1$$

The coordinates of the x-intercepts are $(3, 0)$ and $(-1, 0)$.

3. Find the vertex. Notice that in the given equation, $y = x^2 - 2x - 3$, $a = 1$, and $b = -2$. The x-coordinate of the vertex is

$$x = -\frac{b}{2a} = -\frac{-2}{2(1)} = \frac{2}{2} = 1$$

The y-coordinate of the vertex is found by substituting this value for x into the given equation.

$$y = x^2 - 2x - 3$$

$$y = 1^2 - 2(1) - 3 = -4$$

The coordinates of the vertex are $(1, -4)$.

4. Plot these points and draw a parabola through them. Notice that the vertex is the lowest point on the graph and the parabola opens upward. The line $x = 1$ is the axis of symmetry.

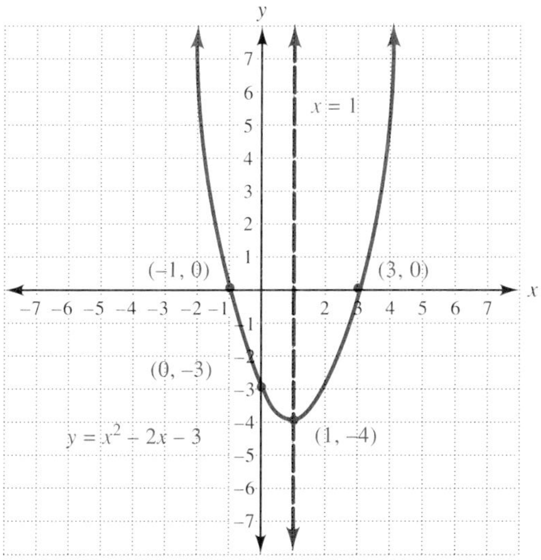

■

□ **DO EXERCISE 2.**

□ **Exercise 2** Sketch the graph of $y = x^2 - 6x + 5$.

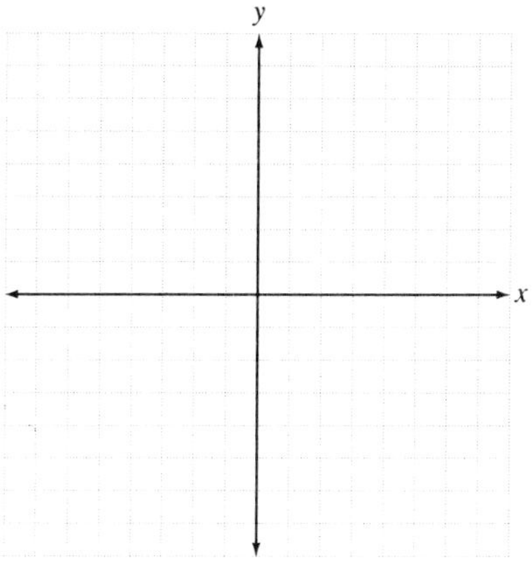

Sometimes when we let y equal zero to find the x-intercepts, the resulting equation does not factor. Then we evaluate the discriminant. If the discriminant is positive, we can use the quadratic formula to find the x-intercepts. If the discriminant is negative, the equation does not have x-intercepts.

EXAMPLE 3 Sketch the graph of $y = -x^2 + 4x - 2$.

1. Find the y-intercept by letting x equal zero.

$$y = -x^2 + 4x - 2$$
$$y = -\mathbf{0}^2 + 4(\mathbf{0}) - 2 = -2$$

The coordinates of the y-intercept are $(0, -2)$.

2. Find the x-intercepts, if they exist. Let $y = 0$.

$$y = -x^2 + 4x - 2$$
$$\mathbf{0} = -x^2 + 4x - 2$$

The equation does not factor, so we evaluate the discriminant. Notice that $a = -1$, $b = 4$, and $c = -2$. The discriminant is

$$b^2 - 4ac = 4^2 - 4(-1)(-2)$$
$$= 16 - 8 = 8$$

The discriminant is positive, so we use the quadratic formula to find the x-intercepts.

$$x = \frac{-b \pm \sqrt{b^2 - 4ac}}{2a} \qquad \text{Quadratic formula}$$

$$x = \frac{-4 \pm \sqrt{4^2 - 4(-1)(-2)}}{2(-1)}$$

$$x = \frac{-4 \pm \sqrt{16 - 8}}{-2} = \frac{-4 \pm \sqrt{8}}{-2} = \frac{-4 \pm 2\sqrt{2}}{-2}$$

$$= \frac{-2(2 \pm \sqrt{2})}{-2} = 2 \pm \sqrt{2}$$

Using a calculator, we find that $\sqrt{2} \approx 1.4$ (rounded to the nearest tenth). Hence

$$x \approx 2 + 1.4 \qquad \text{or} \qquad x \approx 2 - 1.4$$
$$x \approx 3.4 \qquad\qquad\qquad x \approx 0.6$$

The coordinates of the x-intercepts are approximately $(3.4, 0)$ and $(0.6, 0)$.

3. Find the vertex. Since $a = -1$ and $b = 4$, the x-coordinate of the vertex is as follows.

$$x = -\frac{b}{2a} = -\frac{4}{2(-1)} = 2$$

The y-coordinate of the vertex is found by substituting 2 for x in the given equation.

$$y = -x^2 + 4x - 2$$
$$y = -2^2 + 4(2) - 2 = -4 + 8 - 2 = 2$$

The coordinates of the vertex are $(2, 2)$.

4. Plot these points and draw a parabola through them. Notice that the parabola opens downward and the vertex is the highest point on the graph. The line $x = 2$ is the axis of symmetry.

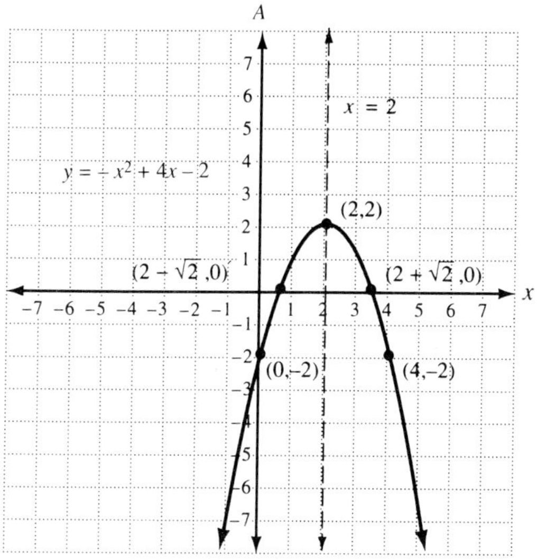

☐ **DO EXERCISE 3.**

If the discriminant of a quadratic equation with two variables is negative, this tells us that the solutions of the equation are not real numbers. So there are no x-intercepts. To graph the equation, we find the vertex and the y-intercept and plot an additional point.

☐ **Exercise 3** Sketch the graph of $y = -x^2 + 2x + 4$.

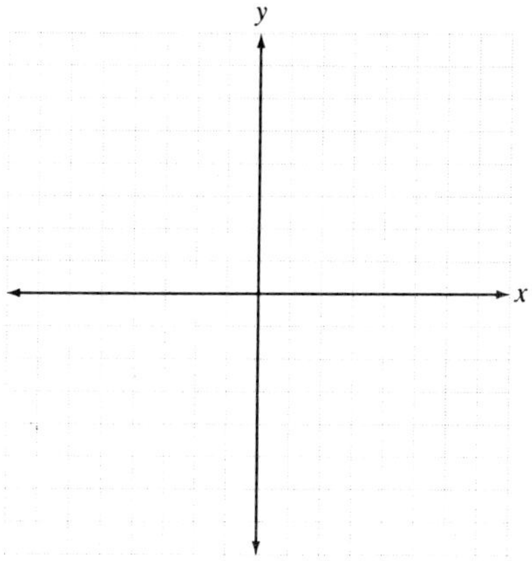

☐ **Exercise 4** Sketch the graph of $y = -2x^2 + 6x - 5$.

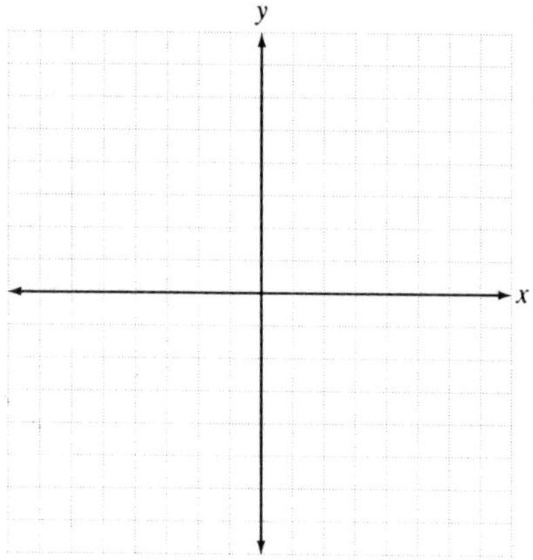

EXAMPLE 4 Sketch the graph of $y = x^2 - 4x + 5$.

1. Find the y-intercept by letting $x = 0$.

$$y = x^2 - 4x + 5$$

$$y = \mathbf{0}^2 - 4(\mathbf{0}) + 5 = 5$$

The y-intercept is $(0, 5)$.

2. We try to find the x-intercepts by letting $y = 0$.

$$\mathbf{0} = x^2 - 4x + 5$$

Before we use the quadratic formula, we evaluate the discriminant to see if the equation has x-intercepts. Notice that $a = 1$, $b = -4$, and $c = 5$. The discriminant is

$$b^2 - 4ac = (-4)^2 - 4(1)(5) = 16 - 20 = -4$$

Since the discriminant is negative, the equation has no x-intercepts.

3. Find the vertex.

$$x = -\frac{b}{2a} = -\frac{(-4)}{2(1)} = 2$$

To find the y-coordinate, let $x = 2$.

$$y = x^2 - 4x + 5$$

$$y = 2^2 - 4(2) + 5 = 1$$

The coordinates of the vertex are $(2, 1)$. The axis of symmetry is $x = 2$.

4. Since we only have two points, choose an additional value for x and find the corresponding y value. We chose $x = 4$.

$$y = x^2 - 4x + 5$$

$$y = 4^2 - 4(4) + 5 = 16 - 16 + 5 = 5$$

The coordinates of another point are $(4, 5)$. Note that we could use symmetry to eliminate this calculation. The point $(4, 5)$ can be found from the mirror image of $(0, 5)$ across the axis of symmetry, $x = 2$.

5. Plot these points and draw a parabola through them.

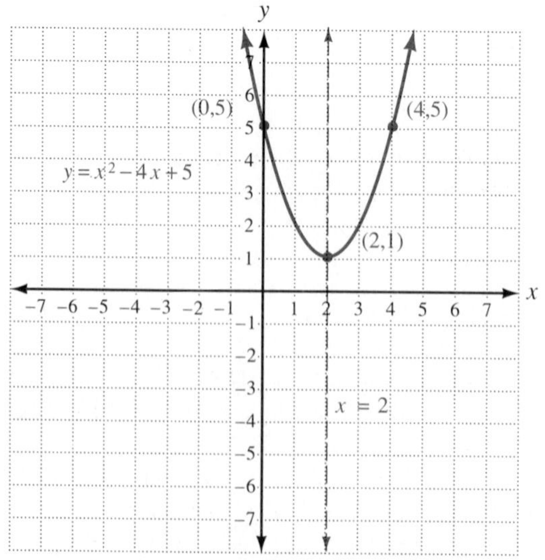

■

☐ **DO EXERCISE 4.**

GRAPHING CALCULATOR

Quadratic equations with two variables may be graphed with a graphing calculator. Since keystrokes for these calculators vary, general instructions for graphing an equation are as follows. Graph $y = x^2 - 2x - 3$.

1. Enter the equation in your calculator. The calculator requires that the equation be solved for y.
2. Set the graphics window. An appropriate window for this equation is W: [−10, 10] [−10, 10].
3. Plot the graph.

We could *trace* the graph to estimate to the nearest whole number the x- and y-intercepts and the vertex. This graph was shown in Example 2.

Graphing Equations of the Form $y = ax^2 + bx + c$

1. Find the y-intercept by letting $x = 0$.
2. Find the x-intercepts by letting $y = 0$. If the equation does not factor, evaluate the discriminant to see if there are x-intercepts. If there are no x-intercepts, find an additional point.
3. Find the vertex. The x-coordinate is $-b/(2a)$. Use this value to find the y-coordinate in the given equation. If the vertex and a point from step 2 are the same, find an additional point.
4. Plot all of the points above and draw a parabola through them. In the equation, if a is positive, the parabola opens upward. If a is negative, the parabola opens downward. For a more accurate graph, plot at least two more points.

Answers to Exercises

1.

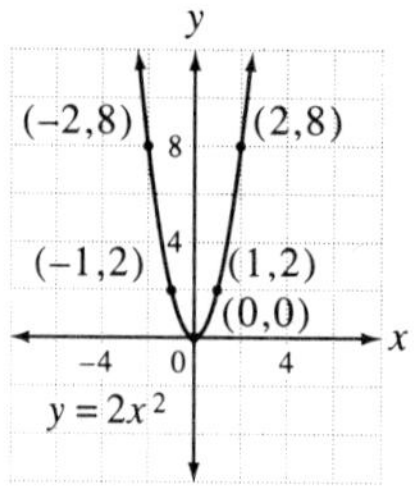

2.

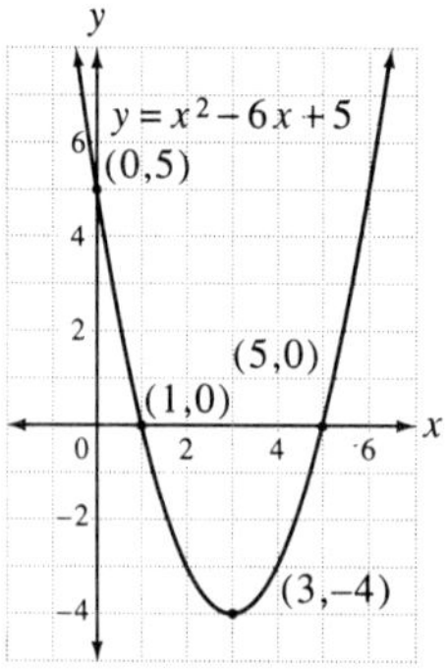

3.

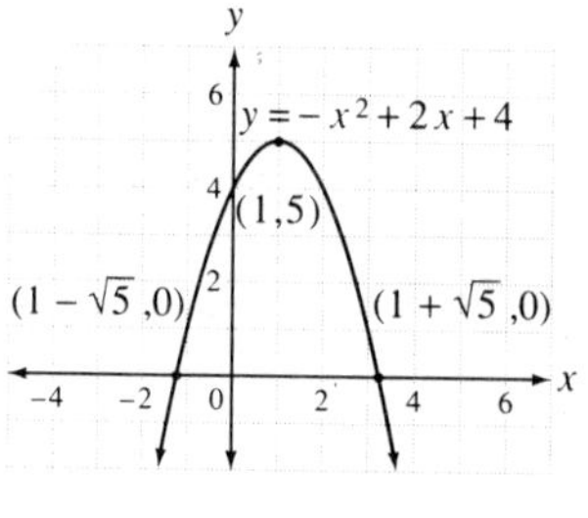

4.

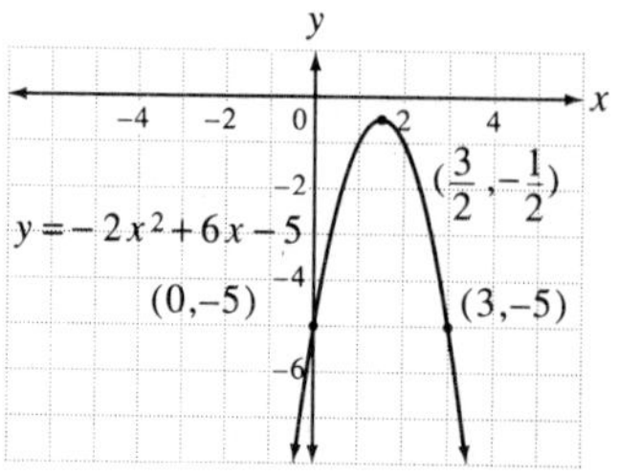

Problem Set 8.7

NAME

DATE

CLASS

Graph, using $x = 0$ to find the vertex and plotting four additional points.

1. $y = x^2 - 4$

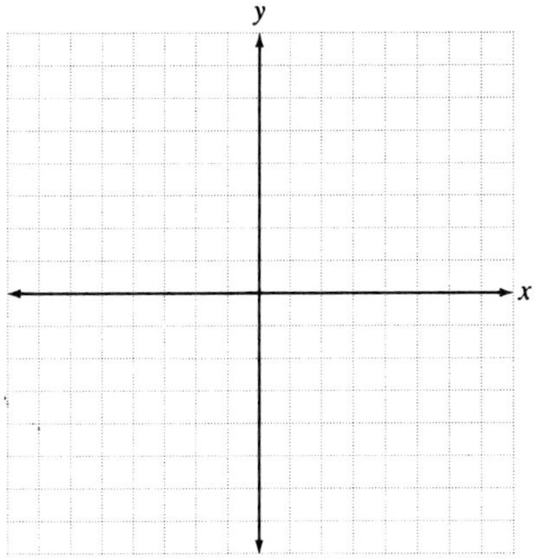

2. $y = x^2 + 3$

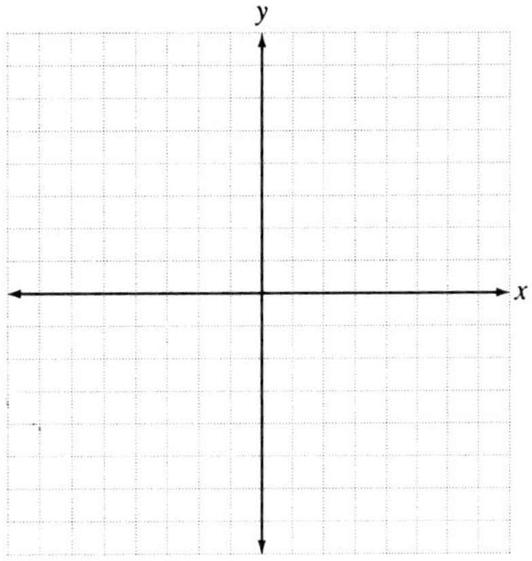

3. $y = x^2$

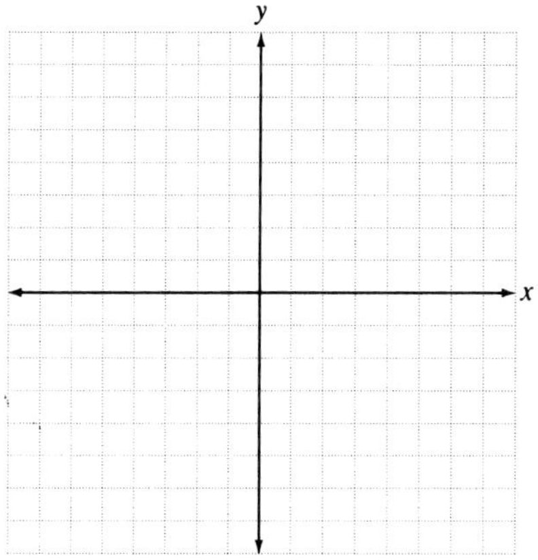

4. $y = -x^2$

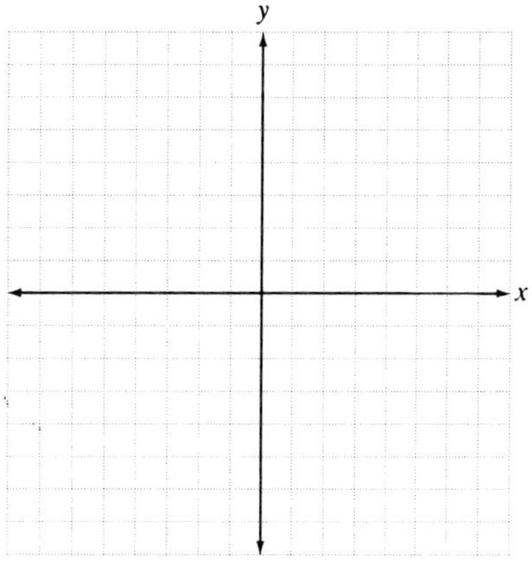

5. $y = -2x^2 + 6$

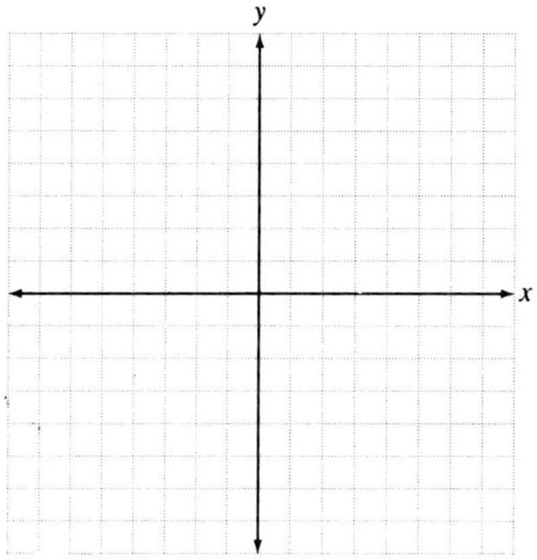

6. $y = 3x^2 - 4$

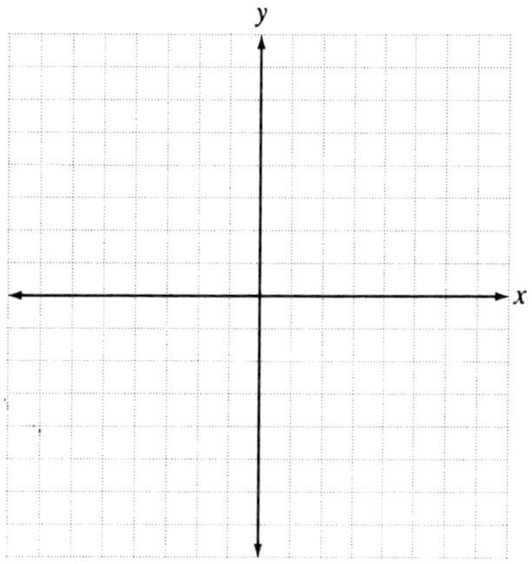

7. $y = 3x^2$

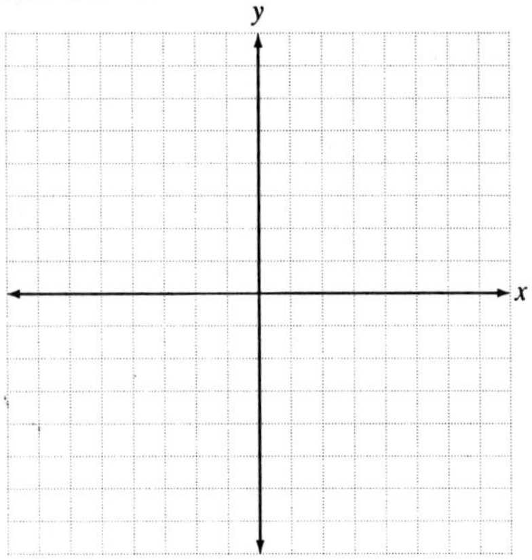

8. $y = -2x^2$

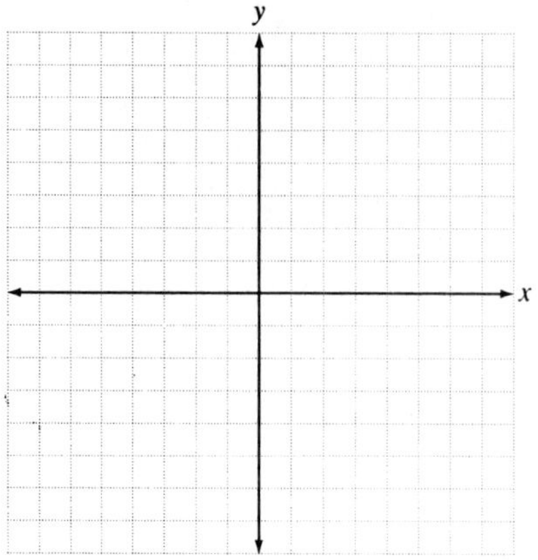

9. $y = 3 - x^2$

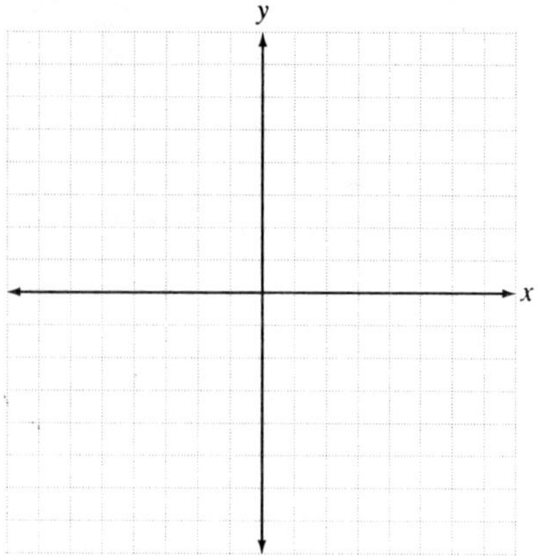

10. $y = -2 - x^2$

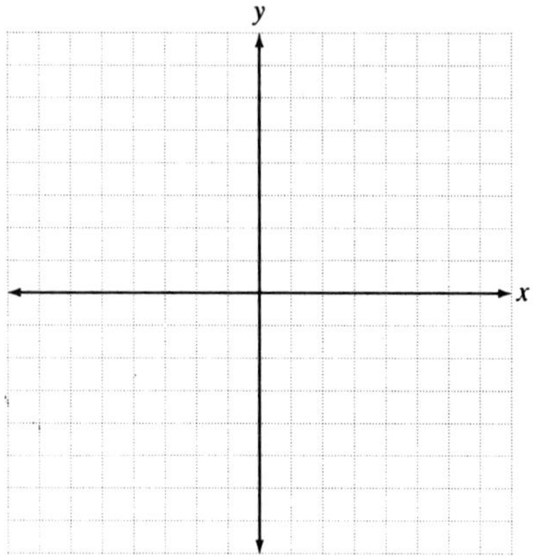

11. $y = -5x^2$

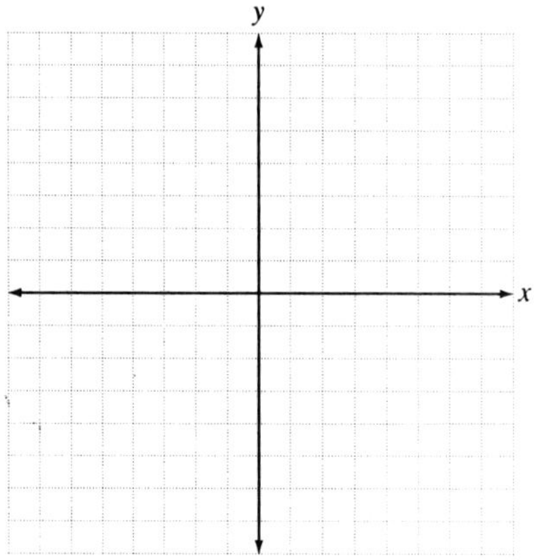

12. $y = -4x^2$

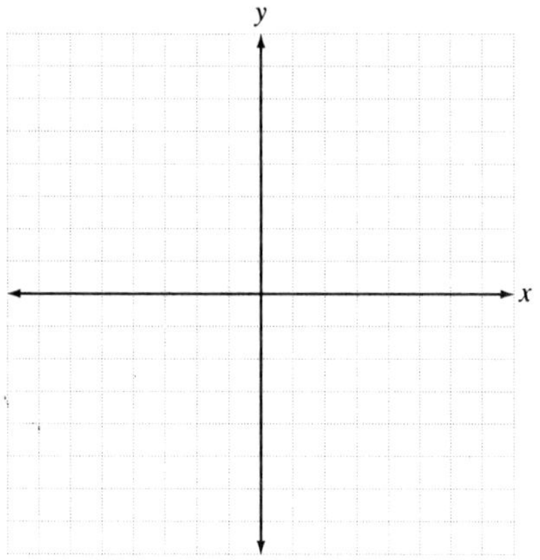

13. $y = \frac{1}{2}x^2$

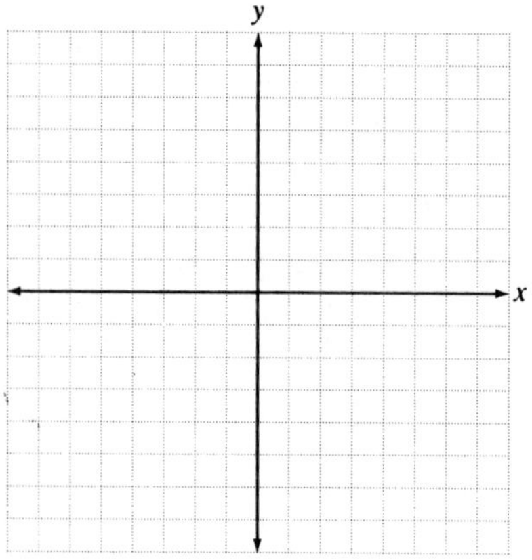

14. $y = \frac{1}{4}x^2$

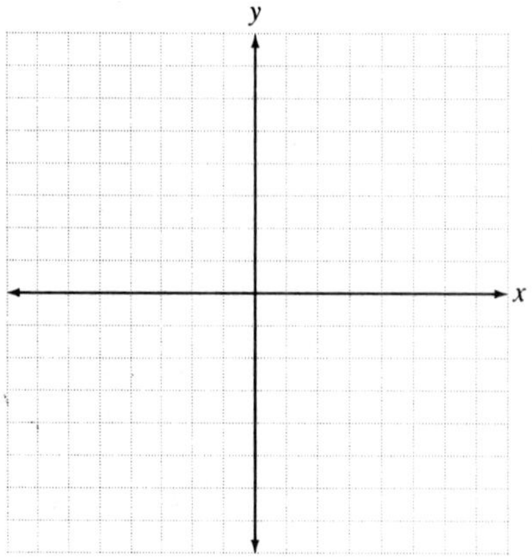

Graph, using the intercepts and the vertex. If the x-intercepts do not exist, plot an additional point.

15. $y = x^2 + 4x - 5$

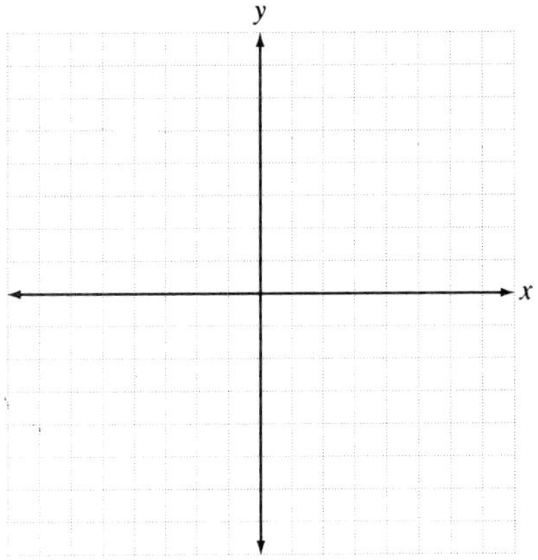

16. $y = x^2 - 4x - 5$

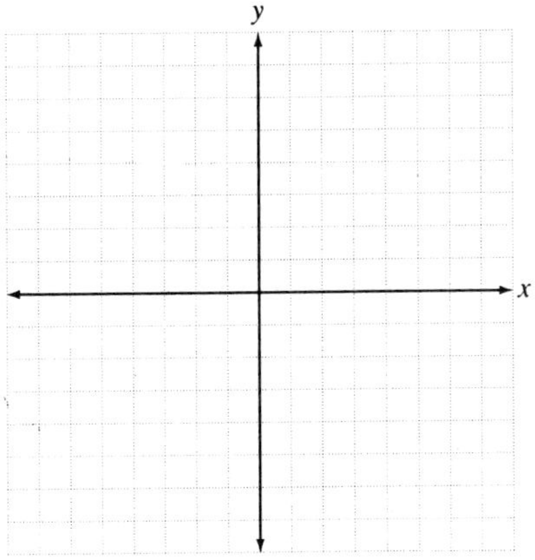

17. $y = -x^2 - 2x + 3$

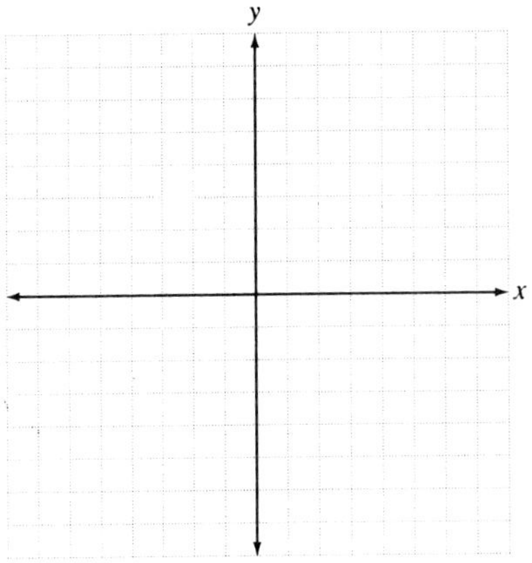

18. $y = -x^2 + 2x + 3$

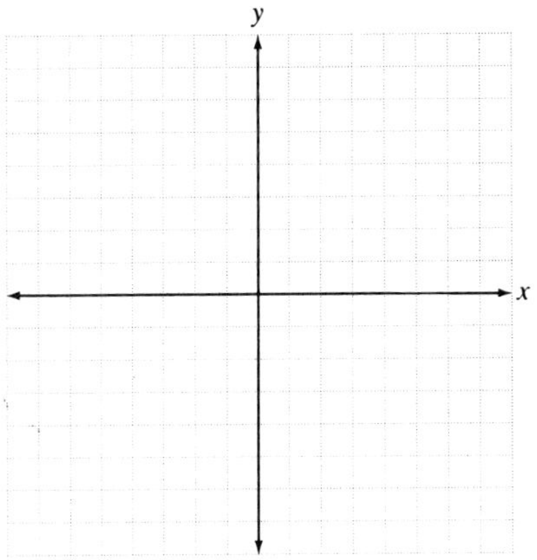

19. $y = 2x^2 - 4x - 4$

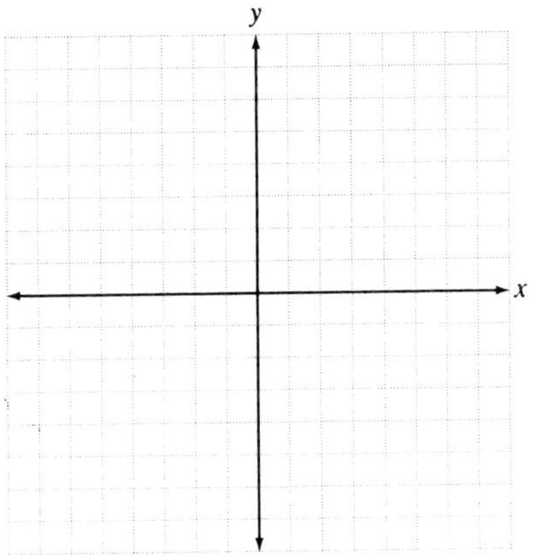

20. $y = 2x^2 + 8x + 5$

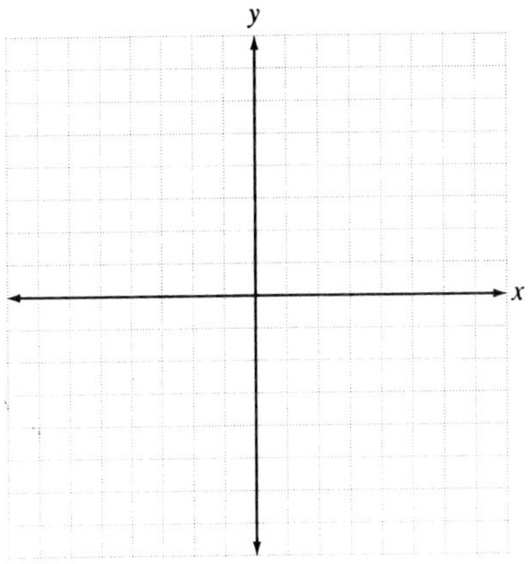

21. $y = x^2 + 4x + 3$

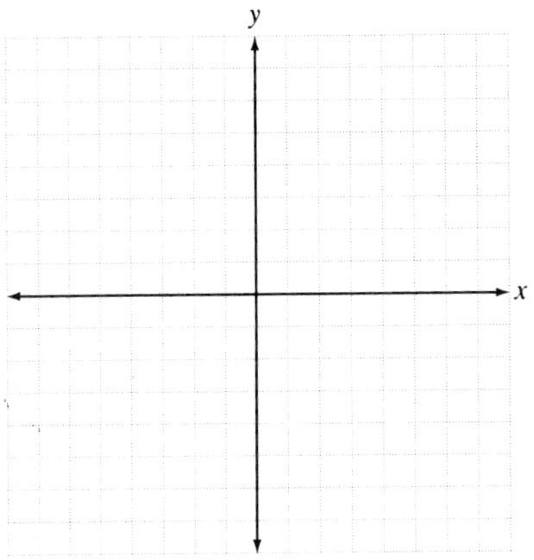

22. $y = x^2 + 6x + 3$

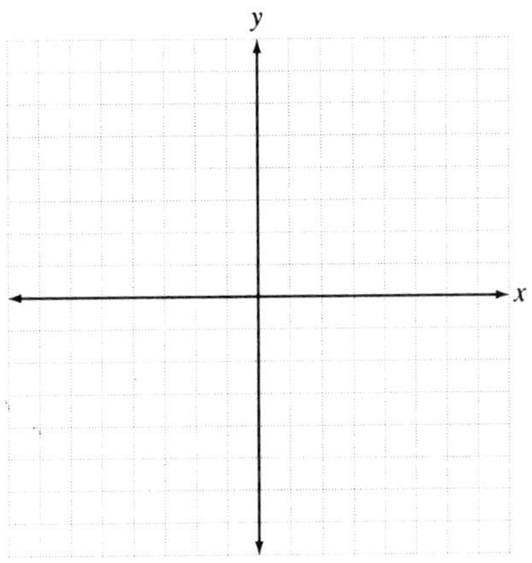

23. $y = x^2 - x + 4$

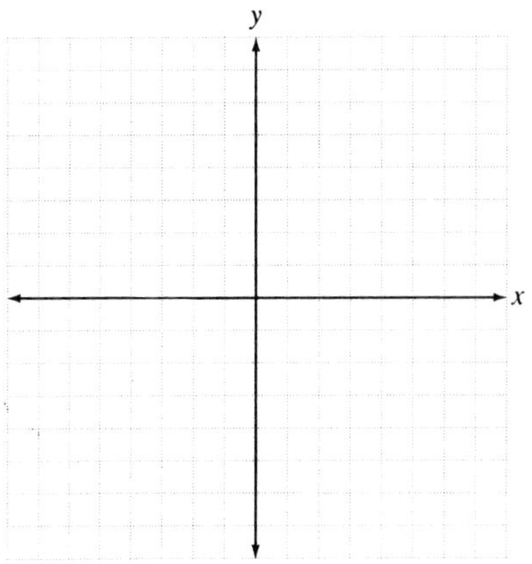

24. $y = -x^2 + 2x - 5$

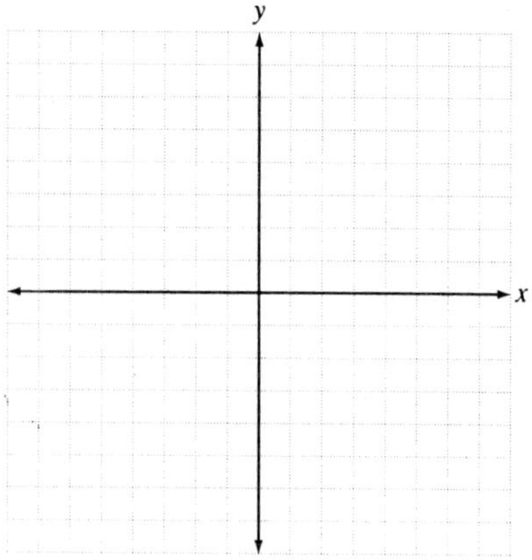

25. $y = 3x^2 + 4x + 2$

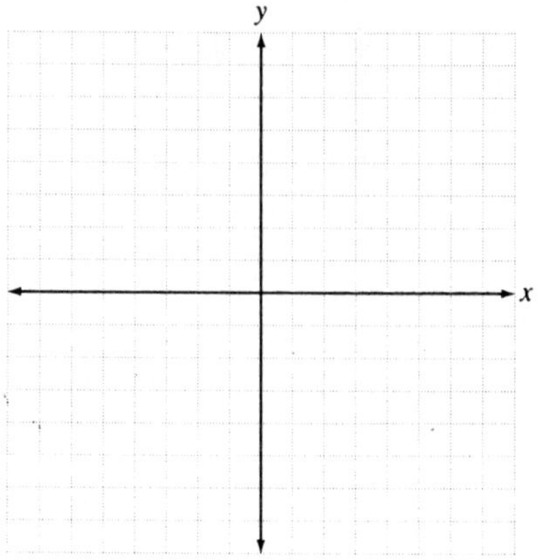

26. $y = 2x^2 + 4x + 3$

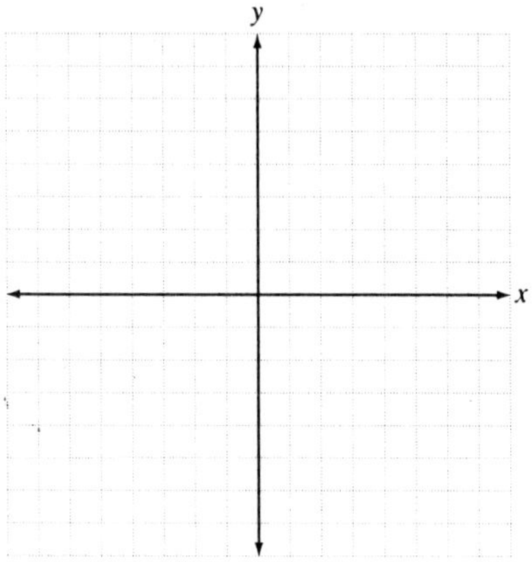

27. $y = -2x^2 + 2x + 1$

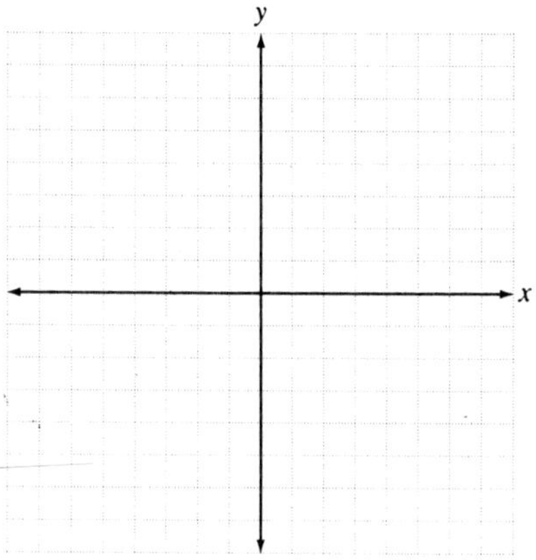

28. $y = -2x^2 - 2x + 3$

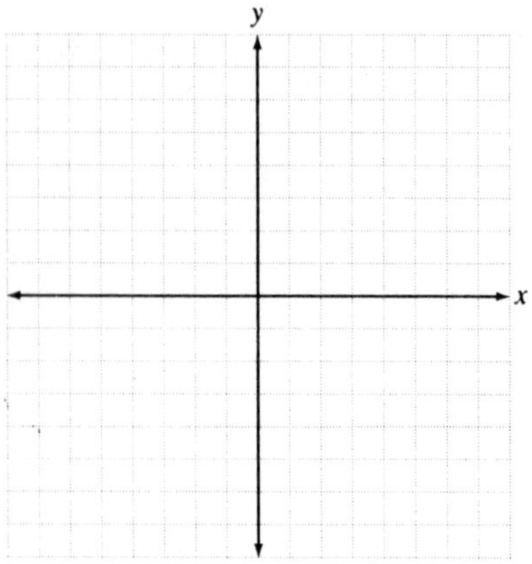

29. $y = 3x^2 - 9x + 8$

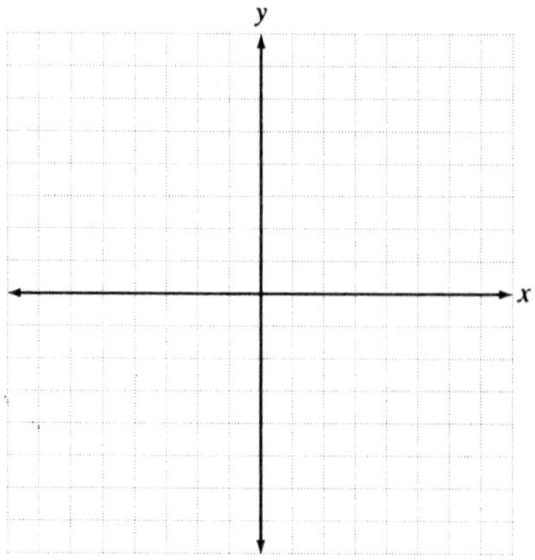

30. $y = -5x^2 - 10x + 2$

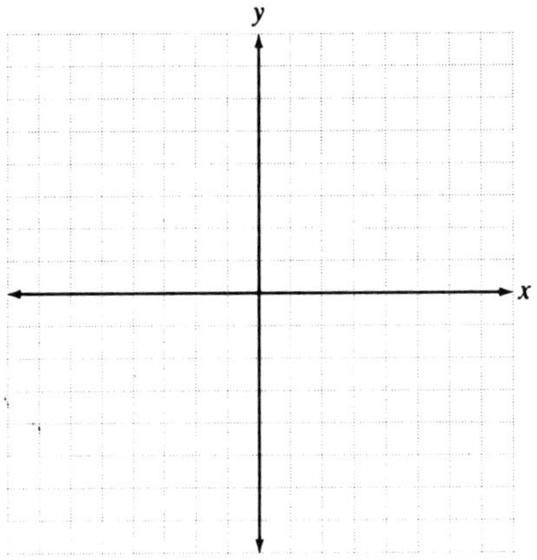

31. The solutions of a quadratic equation with two variables are ______________ ______________ of numbers.

32. The highest or lowest point on the graph of $y = ax^2 + bx + c$ is called the ______________.

33. The graph of a quadratic equation with two variables is a ______________.

34. In the graph of $y = ax^2 + bx + c$ if a is ______________, the graph opens downward.

35. For the graph of $y = ax^2 + bx + c$, the line $x = -b/(2a)$ through the x-coordinate of the vertex is called the ______________ ______________ ______________.

36. If the discriminant of $ax^2 + bx + c = 0$ for the equation $y = ax^2 + bx + c$ is ______________, the graph of the equation does not have x-intercepts.

Graph.

37. $y = 1.86x^2 + 3.72x - 3$

38. $y = -2.72x^2 + 5.44x + 3$

* **39.** $y = -3(x + 1)^2 + 2$

* **40.** $y = \frac{1}{3}(x + 3)^2 - 1$

Checkup

The following problems provide a review of some of Sections 6.5 and 7.8.

41. What number must be added to both the numerator and denominator of $\frac{3}{5}$ to make the result equal to $\frac{3}{4}$?

42. What number must be added to the denominator of $\frac{8}{5}$ to make the result equal to $\frac{4}{7}$?

43. Binh drives 25 miles per hour faster than Mai. If Binh travels 260 miles in the same time that Mai travels 160 miles, find the speed that Mai drives.

44. Beth averages 10 miles per hour riding her bicycle to school. Returning home she averages 8 miles per hour. If she gets to school in $\frac{1}{8}$ hour less time, how long does it take her to return from school?

45. A boat goes 15 miles per hour in still water. It can go 8 miles upstream in the same time that it can go 12 miles downstream. Find the speed of the current.

46. Gary can fly his plane 120 miles with the wind in the same time that it takes him to fly 90 miles against the wind. Find the speed of his plane in still air if the speed of the wind is 20 miles per hour.

If necessary, use a calculator to find an approximation to the nearest thousandth.

Find the length of the side of the right triangle that is not given (c is the length of the hypotenuse).

47. $a = 15$ meters, $b = 12$ meters

48. $b = 11$ feet, $c = 24$ feet

49. The two equal sides of an isosceles right triangle are of length 18 centimeters. What is the length of the hypotenuse?

50. Angela wants to stake out a rectangular foundation with a length of 50 feet and a width of 40 feet. What should be the lengths of the diagonals to ensure that the foundation is a rectangle?

51. The interest rate r, compounded annually, needed for P dollars to grow to A dollars at the end of 2 years is given by

$$r = \sqrt{\frac{A}{P}} - 1$$

Find the value of A if \$6000 is invested at 8% interest for 2 years, compounded annually.

52. The period P of a pendulum in seconds depends on its length L in feet, and is given by

$$P = 2\pi\sqrt{\frac{L}{32}}$$

Find the length of a pendulum with a period of 6 seconds.

8.8 APPLICATIONS AND MORE PARABOLAS

OBJECTIVES

1. *Use quadratic equations to solve maximum and minimum problems*
2. *Graph equations of the form* $x = ay^2 + c$
3. *Graph equations of the form* $x = ay^2 + by + c$

1 Applications

We know that the vertex of the graph of the quadratic equation $y = ax^2 + bx + c$, $a \neq 0$, is the highest or the lowest point on the graph. This gives us a maximum or minimum value for y. Many applied problems involve finding maximum or minimum values.

EXAMPLE 1 A farmer wants to fence a rectangular plot of land against a river. He has 80 feet of fencing. What is the maximum area that he can enclose?

Variable Let x represent the width of the plot. Then $80 - 2x$ represents the length of the plot since the side along the river is not fenced. (See the drawing.)

Drawing

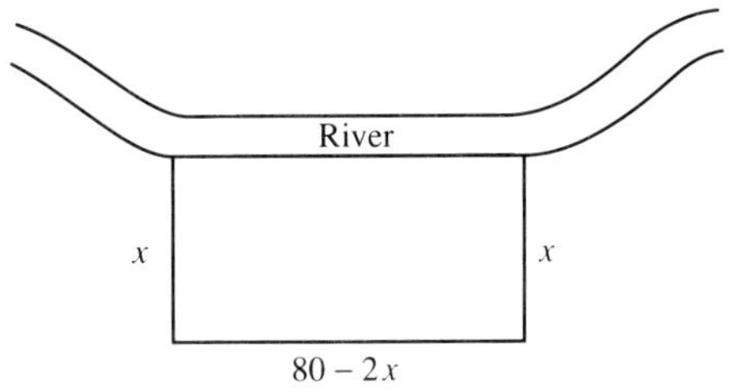

We want to find the maximum area that can be enclosed.

Equation $A = LW$

Solve $A = \mathbf{(80 - 2x)x}$ Substitute for L and W.

$A = 80x - 2x^2$

$A = -2x^2 + 80x$

The area is a maximum at the vertex of the parabola, so we find the x-coordinate of the vertex.

$$x = -\frac{b}{2a} = -\frac{80}{2(-2)} = \frac{-80}{-4} = 20$$

The plot will have maximum area when the width of the plot is 20 feet. Substitute this value for x into the equation to find the maximum area.

$$A = -2x^2 + 80x$$

$$A = -2(20)^2 + 80(20)$$

$$A = -800 + 1600 = 800$$

The maximum area is 800 square feet. The graph of the parabola follows.

☐ **Exercise 1** What is the maximum product of two numbers whose sum is 40? (*Hint:* Let x represent one number and $40 - x$ represent the other number.)

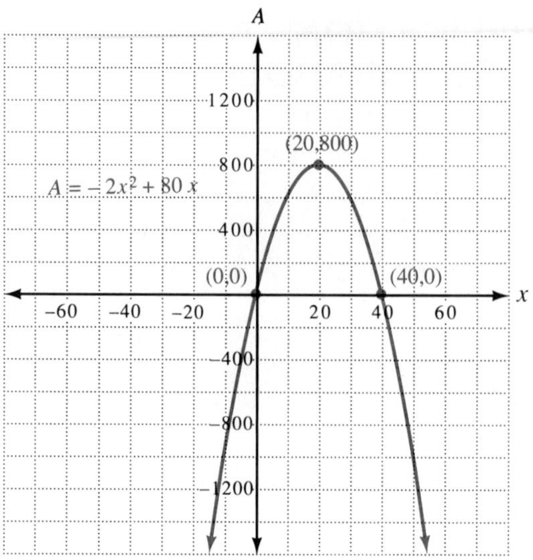

■

☐ DO EXERCISE 1.

Other Parabolas

There are also quadratic equations with two variables of the form

$$x = ay^2 + by + c \qquad a \neq 0$$

and a, b, and c are real numbers. The graph of this equation is also a parabola. However, these parabolas *open* to the *right or* to the *left* instead of upward or downward. They open to the right if a is positive and to the left if a is negative. The vertex will be the point with the smallest or largest x-value of any point on the parabola.

2 Graphing Parabolas of the Form $x = ay^2 + c$

One of the easiest methods of graphing these equations when they do not have a y term is to plot at least five points that are not too close together. This gives a good picture of the curve. We *choose values for y* and find the corresponding x values. Be sure to choose at least two negative values and zero for y. (Choosing zero for y gives us the vertex of parabolas of this form.)

EXAMPLE 2 Graph $x = y^2$. Let $y = -2$.

$x = y^2$

$x = (\mathbf{-2})^2$

$x = 4$

x	y	Ordered Pair
4	−2	(4, −2)
1	−1	(1, −1)
0	0	(0, 0)
1	1	(1, 1)
4	2	(4, 2)

We also chose values for y of −1, 0, 1, and 2 and found the corresponding x values shown in the table. Plot these points and draw a parabola through them. The graph follows.

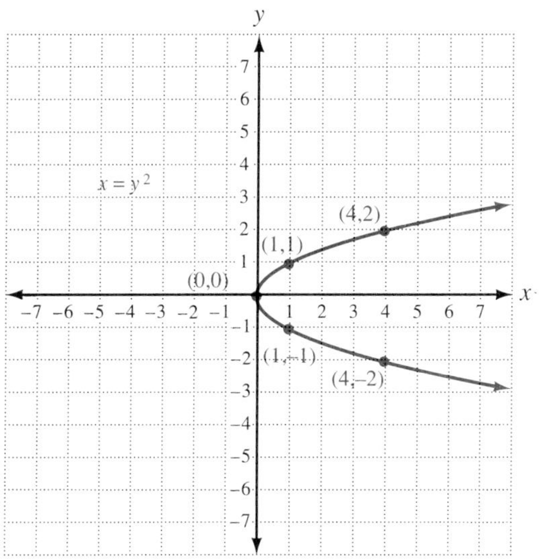

□ DO EXERCISE 2.

3 Graphing Equations of the Form $x = ay^2 + by + c$

When the quadratic equation with two variables has a y term, one method of graphing it is to use the vertex and the intercepts. In this case, the y value of the vertex is given by

$$y = -\frac{b}{2a}$$

The x- and y-intercepts are found in the usual manner. The line $y = -b/(2a)$ through the y-coordinate of the vertex is the axis of symmetry.

EXAMPLE 3 Sketch the graph of $x = -y^2 + 2y + 3$.

1. Find the x-intercept by letting $y = 0$ and solving for x.

$$x = -y^2 + 2y + 3$$

$$x = -\mathbf{0}^2 + 2(\mathbf{0}) + 3 \qquad \text{Substitute 0 for } y.$$

$$x = 3$$

The x-intercept is $(3, 0)$.

2. Find the y-intercepts. Let $x = 0$ and solve for y.

$$x = -y^2 + 2y + 3$$

$$\mathbf{0} = -y^2 + 2y + 3 \qquad \text{Substitute 0 for } x.$$

$$0 = y^2 - 2y - 3 \qquad \text{Multiply both sides of the equation by } -1.$$

$$0 = (y - 3)(y + 1) \qquad \text{Factor.}$$

$$y = 3 \qquad \text{or} \qquad y = -1$$

The coordinates of the y-intercepts are $(0, 3)$ and $(0, -1)$.

3. Find the vertex. Notice that in the given equation,

$$x = -y^2 + 2y + 3,\ a = -1, \text{ and } b = 2$$

The y-coordinate of the vertex is

$$y = -\frac{b}{2a} = -\frac{2}{2(-1)} = 1$$

□ **Exercise 2** Graph.

a. $x = -y^2$

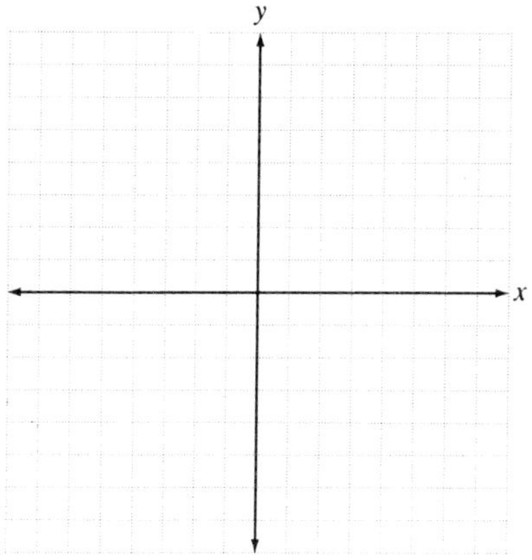

b. $x = 2y^2 - 3$

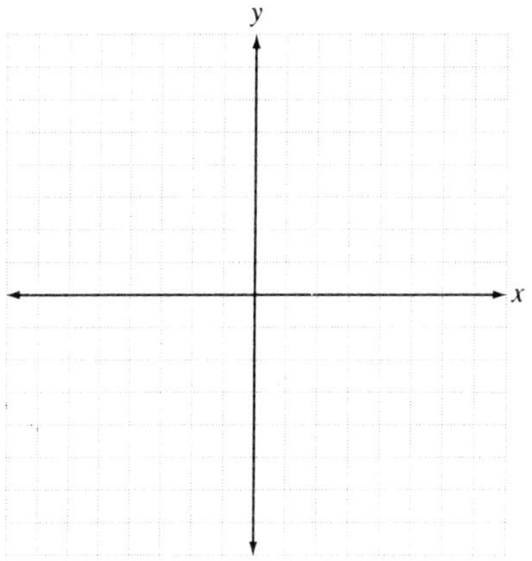

□ **Exercise 3** Sketch the graph of $x = y^2 + 6y + 5$.

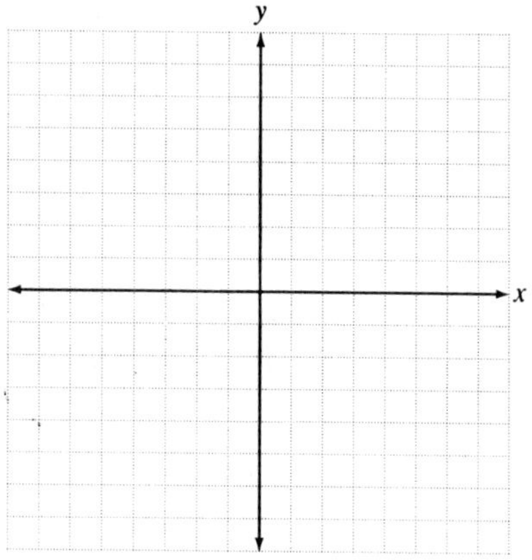

The x-coordinate of the vertex is found by substituting this value for y into the given equation.

$$x = -y^2 + 2y + 3$$

$$x = -1^2 + 2(1) + 3 = 4$$

The coordinates of the vertex are (4, 1).

4. Plot these points and draw a parabola through them. The axis of symmetry is $y = 1$.

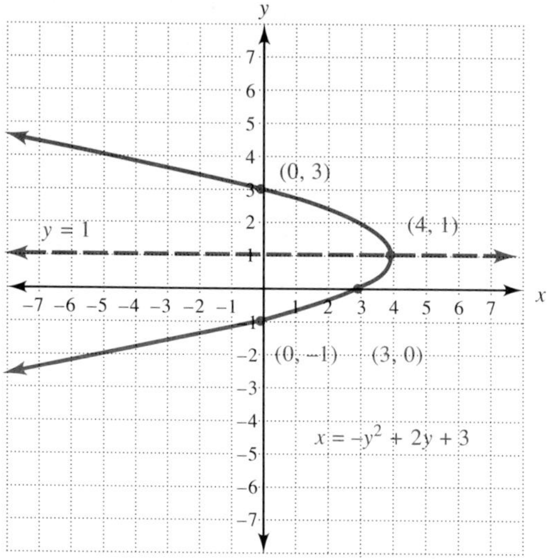

■

□ **DO EXERCISE 3.**

When we try to find the y-intercepts, the quadratic equation may not factor. If it does not factor, evaluate the discriminant to see if the graph of the equation has y-intercepts. If it does, use the quadratic formula to find them. If there are no y-intercepts, find the x-intercept, vertex, and an additional point and sketch a parabola through them.

Answers to Exercises

1. 400

2. a.

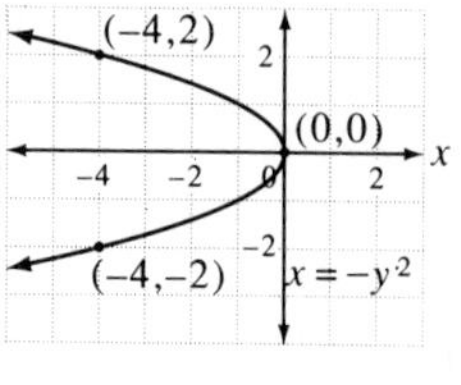

b.

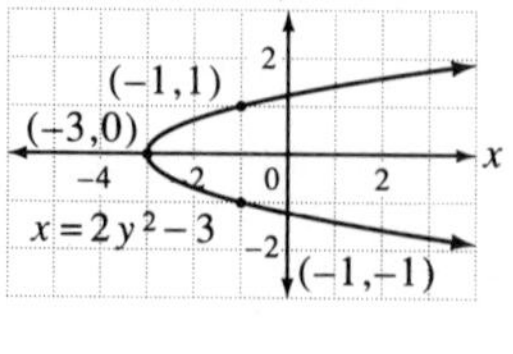

3.

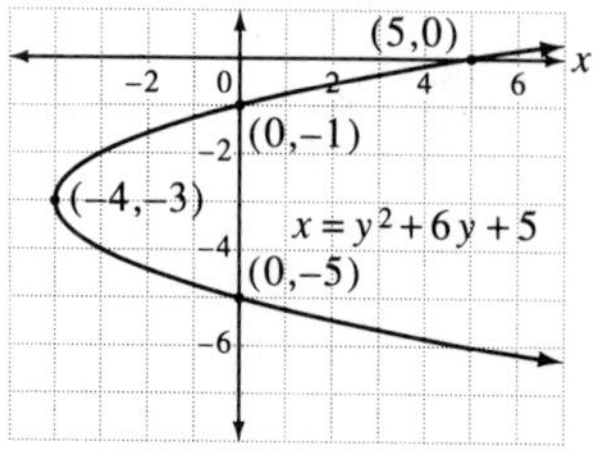

Problem Set 8.8

NAME ____________________

DATE ____________________

CLASS ____________________

Solve.

1. The height of an object thrown upward with an initial velocity of 64 feet per second after t seconds is given by

$$h = 64t - 16t^2$$

Find the maximum height attained by the object.

2. Jean owns a clothing store. She found that her profits on shirts are given by

$$P = -x^2 + 12x + 22$$

where P represents profit and x is the number of shirts sold daily. What is the maximum profit on shirts that Jean can make daily?

3. What is the maximum product of two numbers whose sum is 32?

4. Find the maximum product of two numbers whose sum is 26.

5. A farmer has 84 feet of fence to enclose a rectangular area. What is the maximum area that he can fence? What are the dimensions of the rectangle?

6. A rectangular room is to have a perimeter of 42 feet. What is the maximum area possible for the room? What are the dimensions of the room for this area?

7. What is the minimum product of two numbers whose difference is 3? What are the numbers?

8. What is the minimum product of two numbers whose difference is 5? What are the numbers?

9. The number of cars C that Hert Motor Company can build with n robots is given by $C = -\frac{1}{4}n^2 + 20n$. Find the maximum number of cars that can be built and the number of robots needed to build them.

10. The number of radios R that Sound, Inc. can produce with n employees is given by $R = -\frac{1}{3}n^2 + 15n$. Find the maximum number of radios that can be produced and the number of employees needed to produce them.

11. The demand D for a new compact disc is estimated to be a quadratic equation defined by $D = -\frac{1}{2}x^2 + 36x + 1500$, where x is the week of sales. Find the week of highest demand and the number of discs needed that week to meet the demand.

12. The demand curve of a manufacturer of jeans is $y = -\frac{1}{2}x^2 + 4x + 2000$, where y is the demand and x is the week of sales. Find the week of highest demand and the number of pairs of jeans needed to meet the demand.

Graph.

13. $x = 2y^2$

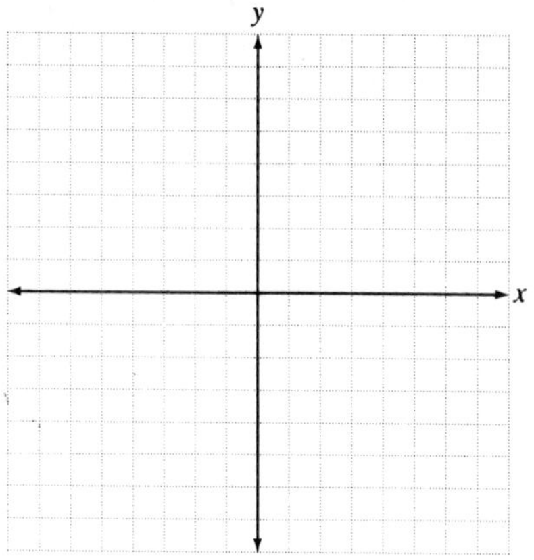

14. $x = -2y^2$

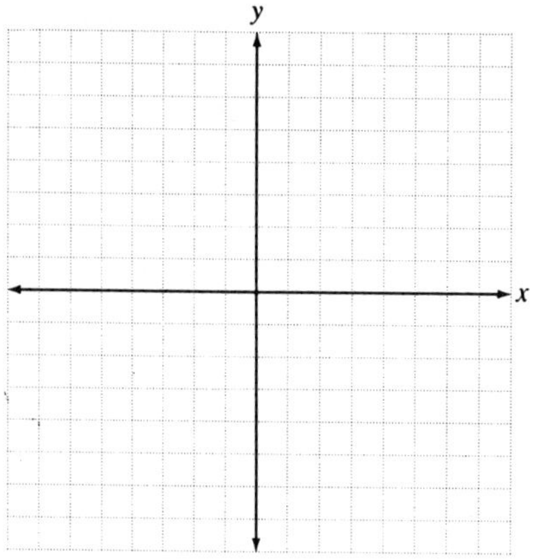

15. $x = -y^2 + 1$

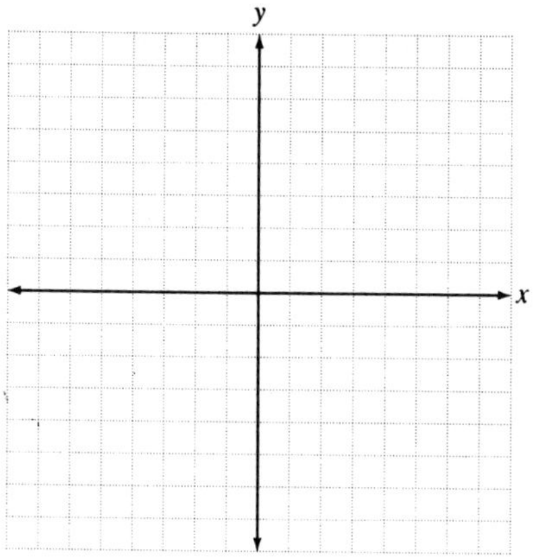

16. $x = y^2 - 2$

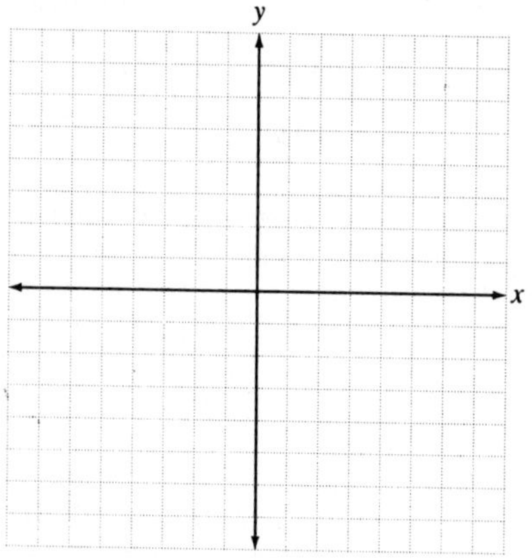

17. $x = 3y^2 - 4$

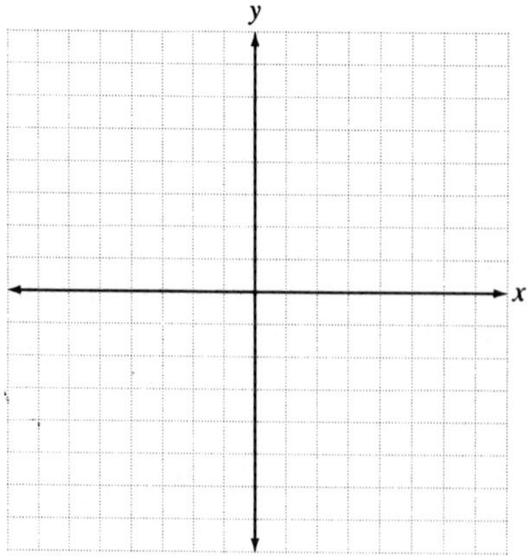

18. $x = -2y^2 + 3$

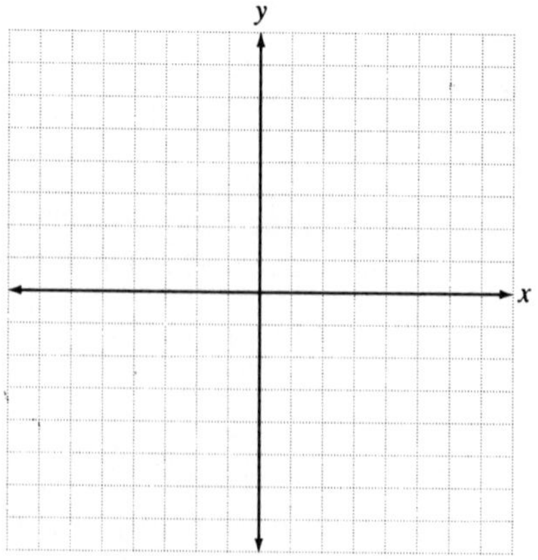

19. $x = y^2 - 4y + 3$

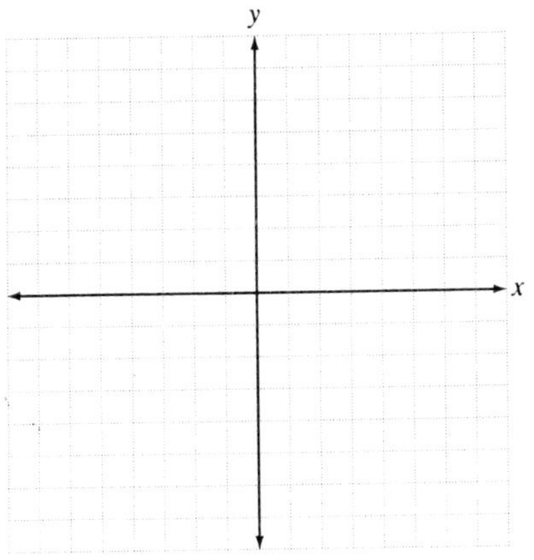

20. $x = -y^2 + 4y + 5$

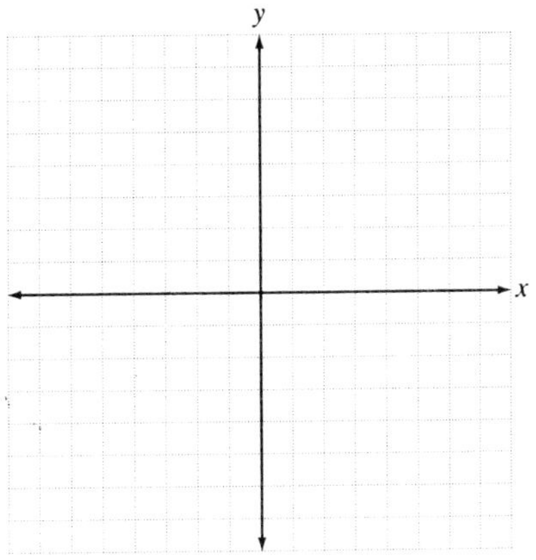

21. $x = y^2 + y - 2$

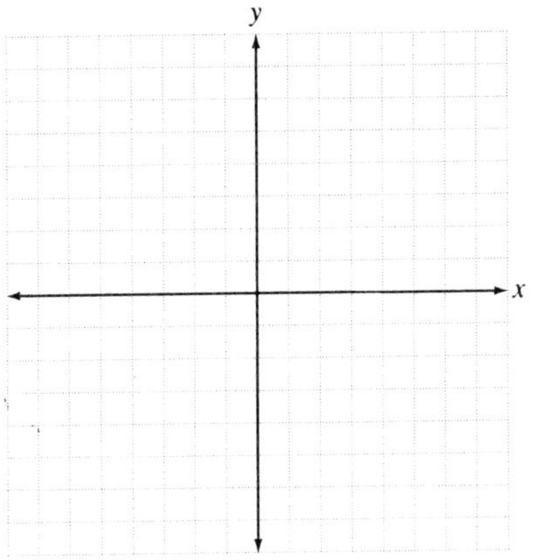

22. $x = y^2 - y - 6$

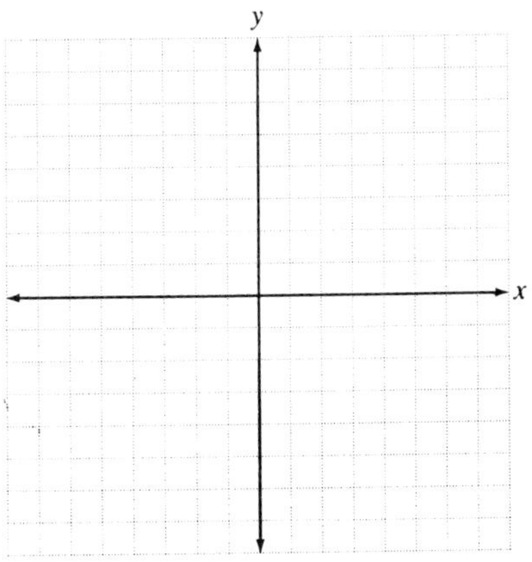

23. $x = 2y^2 + 3y - 2$

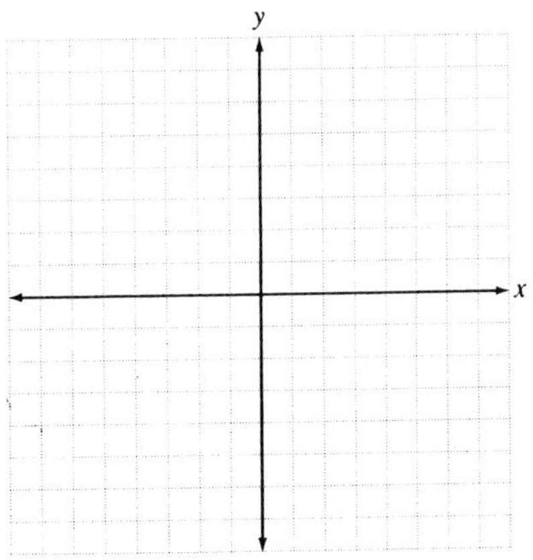

24. $x = 2y^2 + 5y - 3$

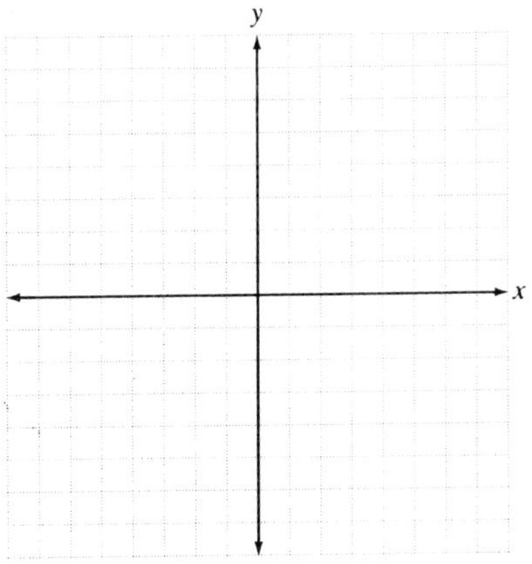

25. $x = y^2 + 4y - 1$

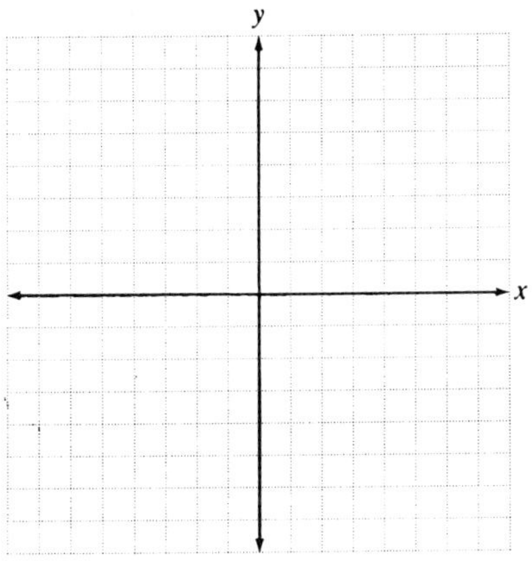

26. $x = y^2 + 2y + 6$

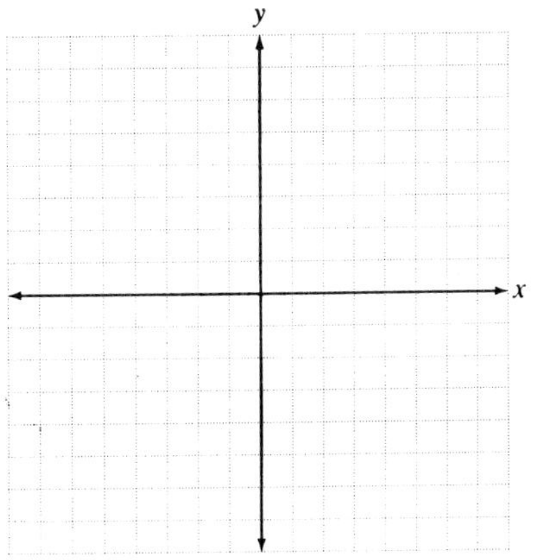

27. $x = -3y^2$

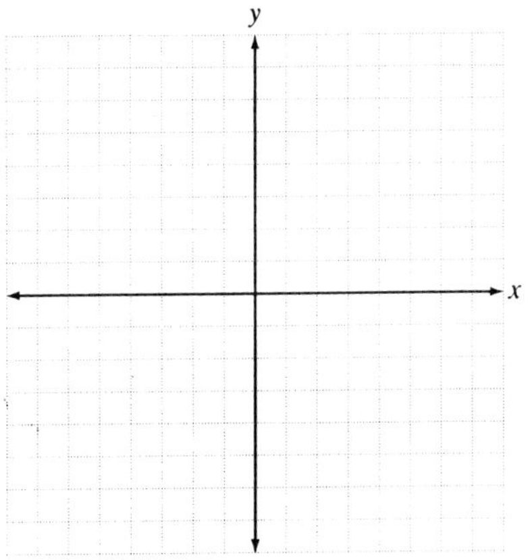

28. $x = 5y^2$

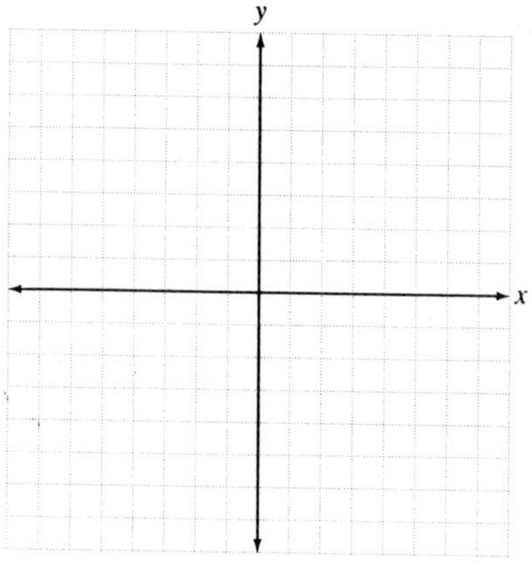

29. $x = \frac{1}{4}y^2$

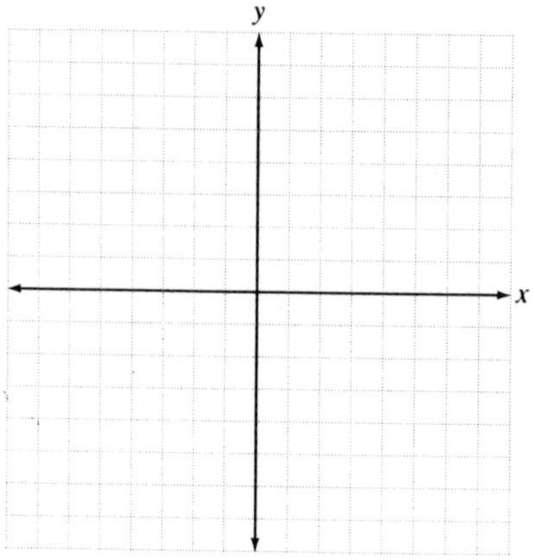

30. $x = -\frac{1}{2}y^2$

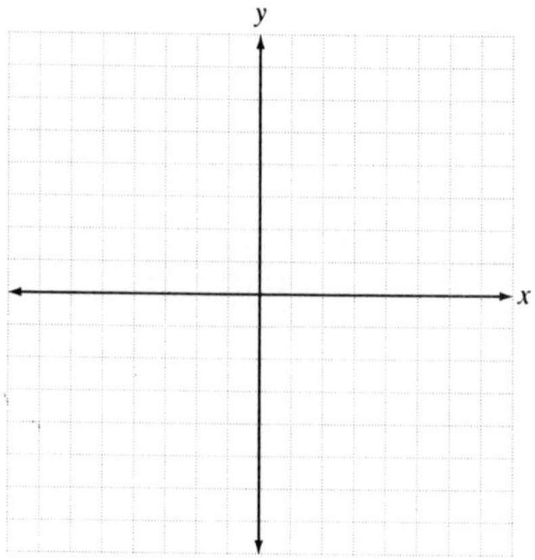

31. $x = \frac{1}{2}y^2 - 3$

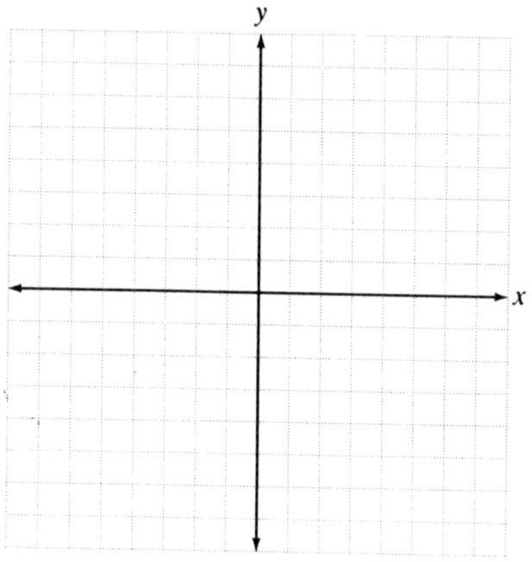

32. $x = -\frac{1}{4}y^2 + 4$

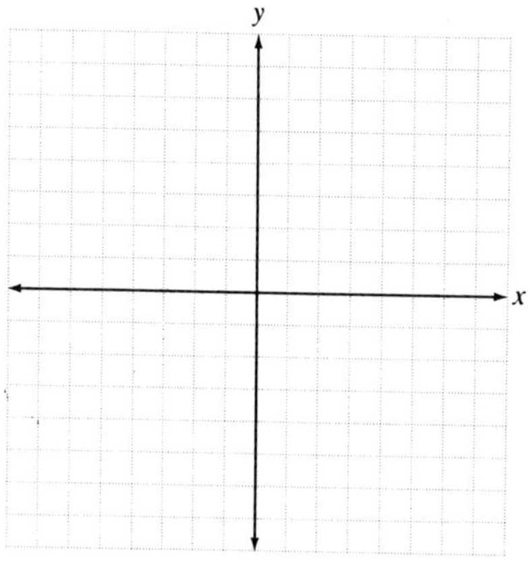

33. $x = y^2 - 6y$

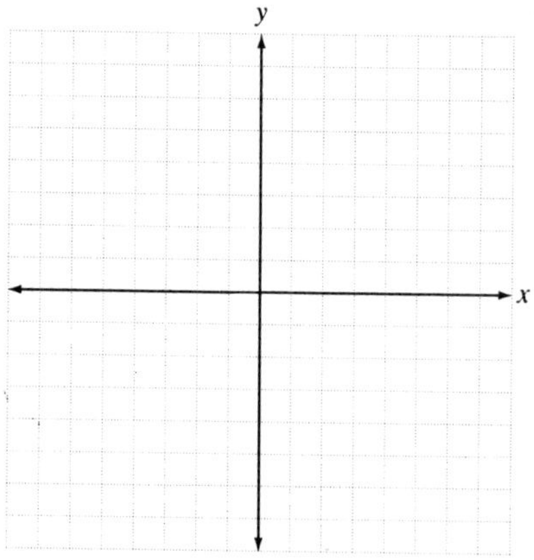

34. $x = -y^2 - 2y$

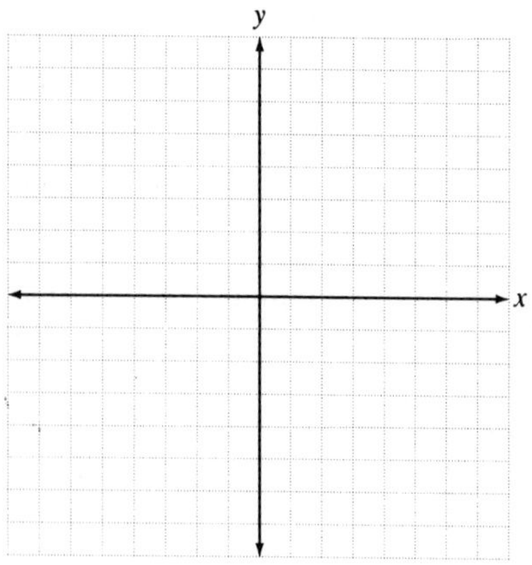

35. $x = -y^2 - 8y - 9$

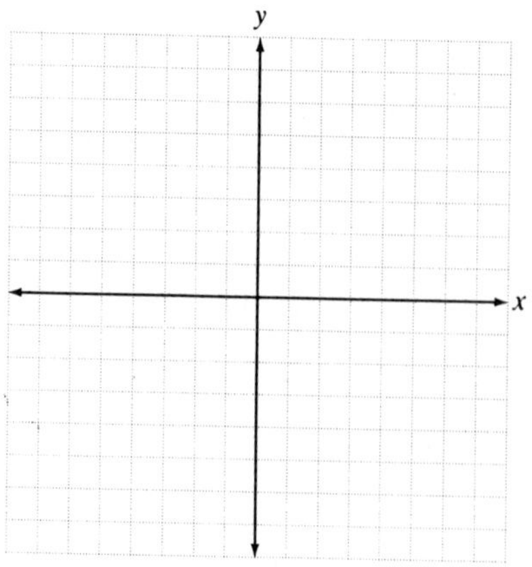

36. $x = y^2 - 6y + 1$

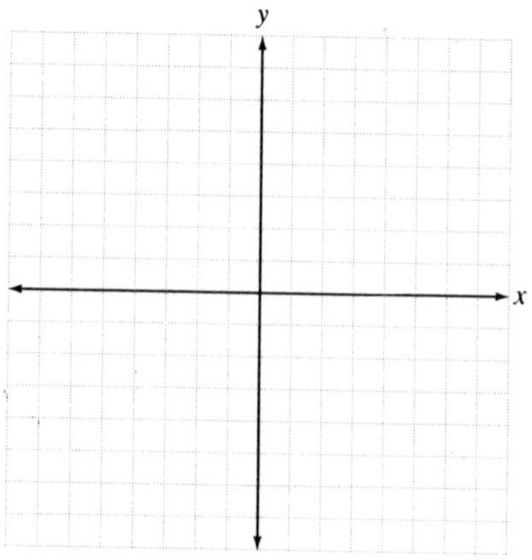

37. The graph of $x = ay^2 + by + c$, $a \neq 0$, opens to the ______ if a is positive.

38. To graph an equation of the form $x = ay^2 + c$, $a \neq 0$, choose at least five values for y, including two negative values and ______.

39. Graphs of equations of the form $x = ay^2 + by + c$, $a \neq 0$, are ______.

40. The profit on berries at Jim's Berry Farm is given by

$$P = -0.015x^2 + 9.33x$$

where P represents profit and x is the number of quarts of berries sold. What is the maximum profit on berries that Jim can make daily?

41. A rancher wants to fence a rectangular plot of land against a river. He has 125.5 feet of fencing. What is the maximum area that he can enclose? Round your answer to the nearest tenth.

Graph.

* 42. $x = (y - 2)^2 - 1$

* 43. $3x = -(y + 1)^2$

* 44. $4x = -2(y - 3)^2 + 3$

Think About It

* 45. Find the maximum amount that a number can be greater than its square.

* 46. The sum of the base and the height of a triangle is 28 inches. For what height is the area a maximum?

* 47. The length and width of a rectangle have a sum of 44 centimeters. For what width is the area a maximum? Find the maximum area.

Checkup

The following problems provide a review of some of Section 7.9.

Add or subtract.

48. $(3 + 5i) + (6 - 8i)$

49. $(2 - 4i) + (-3 - 5i)$

50. $(6 + i) - (7 - 5i)$

51. $(3 + 8i) - (-2 + 9i)$

Multiply.

52. $(5 + 2i)(6 - 7i)$

53. $(3 - 4i)(2 - i)$

54. $(8 - 9i)(8 + 9i)$

55. $(6 + 7i)(6 - 7i)$

56. $\sqrt{-5}\sqrt{-7}$

57. $\sqrt{-2}\sqrt{-8}$

Divide.

58. $\dfrac{7}{3 + i}$

59. $\dfrac{-4}{7 - i}$

8.9 NONLINEAR INEQUALITIES

OBJECTIVES

1 *Solve quadratic inequalities*

2 *Solve rational inequalities*

1 Quadratic Inequalities

Quadratic inequalities in one variable are of the form

$$ax^2 + bx + c > 0$$

where a, b, and c are real numbers and $a \neq 0$. The $>$ symbol may be replaced by $<$, $\leq$, or $\geq$.

Consider the inequality $x^2 + 2x - 8 > 0$. Let $y = x^2 + 2x - 8$. We can graph $y = x^2 + 2x - 8$ as follows:

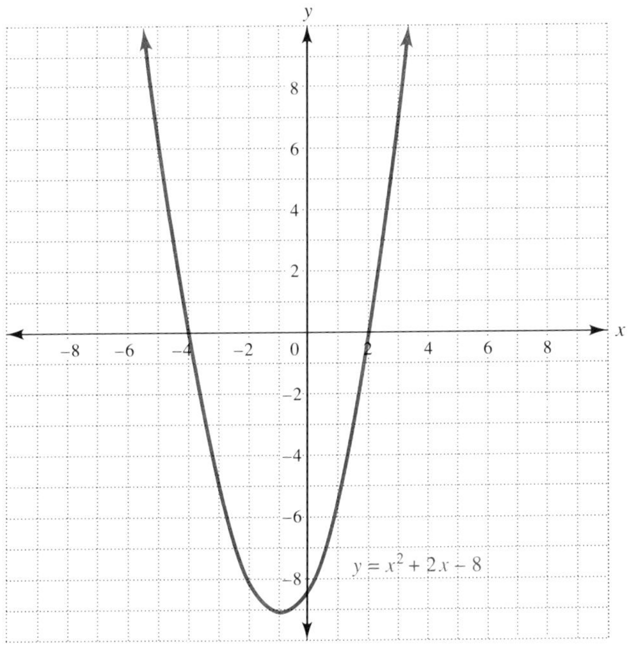

Values of y will be positive to the left and right of the x-intercepts. This means that $x^2 + 2x - 8 > 0$ for $x < -4$ or $x > 2$. The solution set is $\{x \mid x < -4 \text{ or } x > 2\}$.

We can solve the inequality without drawing the graph as shown in Example 1a (Method 1). An additional method of solving the inequality will be shown in Example 3.

Recall that on the number line, numbers to the right of zero are positive and numbers to the left of zero are negative.

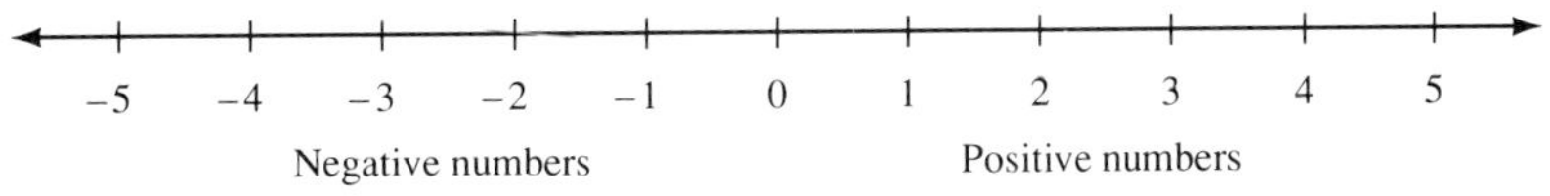

EXAMPLE 1

a. Solve $x^2 + 2x - 8 > 0$.

1. Factor the left side.

$$(x + 4)(x - 2) > 0$$

2. We want the product of $x + 4$ and $x - 2$ to be greater than zero (positive). The only way that the product can be positive is if both

factors are negative or both factors are positive. We check to see for what values of x these factors are positive (greater than zero).

$$x + 4 > 0 \qquad x - 2 > 0$$
$$x > -4 \qquad x > 2$$

The factor $x + 4$ is positive for all values of $x > -4$. The factor will be negative for all values of $x < -4$. Similarly, the factor $x - 2$ is positive for all values of $x > 2$ and it will be negative for all values of $x < 2$. We show this information and the product on a number line.

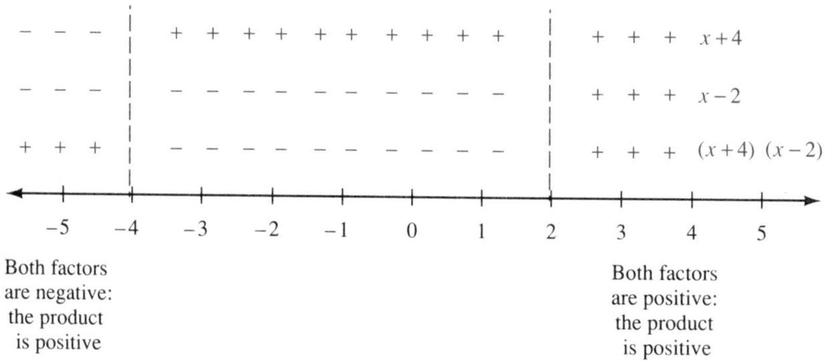

3. As you can see from the diagram, the solution to $x^2 + 2x - 8 > 0$ is $\{x \mid x < -4 \text{ or } x > 2\}$. The solution may be graphed as follows:

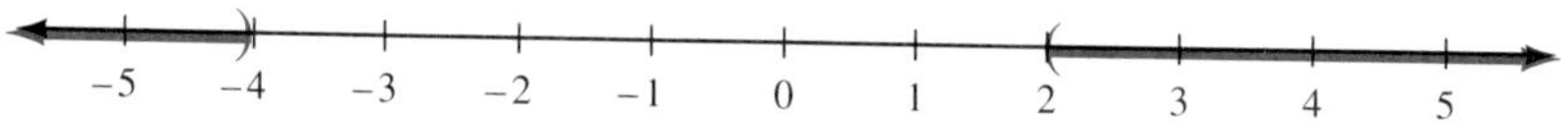

b. Solve $x^2 \leq 3x$.

1. Set one side of the inequality equal to zero.

$$x^2 - 3x \leq 0 \qquad \text{Add } -3x \text{ to each side.}$$

2. Factor the left side.

$$x(x - 3) \leq 0$$

The product $x(x - 3)$ is negative or zero. The factors must be opposite in sign to give a negative product.

3. Determine what values of x make each factor negative (less than zero). We could also determine which values of x make each factor positive.

$$x < 0 \qquad x - 3 < 0$$
$$x < 3$$

These facts and the result are shown in the following diagram.

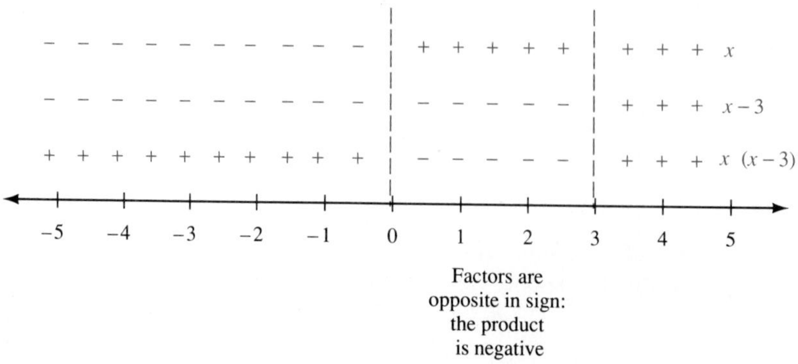

4. The solution set is $\{x \mid 0 \leq x \leq 3\}$. The graph is shown below.

■

Solving a Quadratic Inequality: Method 1

1. Set one side of the inequality equal to zero.
2. Factor the other side.
3. Find which values of the variable make each factor positive and which values make each factor negative. Draw a diagram showing this information and the product of the factors.
4. Find the solution set from the diagram.

□ **DO EXERCISE 1.**

There are special quadratic inequalities that may be solved by inspection.

EXAMPLE 2

a. Solve $x^2 - 4x + 4 \geq 0$.
Factor the left side.

$$(x - 2)^2 \geq 0$$

Since $(x - 2)^2$ is always positive or zero for any value of x, the solution set is $\{x \mid x \text{ is a } \textit{real number}\}$. The graph of the solutions includes the entire number line.

b. Solve $(3y - 4)^2 < 0$.
Since $(3y - 4)^2$ is never less than zero for any value of y, there is *no solution* to this inequality. The solution set is $\varnothing$. ■

□ **DO EXERCISE 2.**

There is another method for solving quadratic inequalities.

□ **Exercise 1** Solve and graph each solution.

a. $x^2 + 3x - 4 \geq 0$

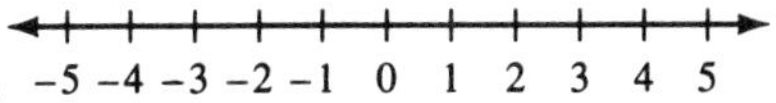

b. $x^2 < -2x$

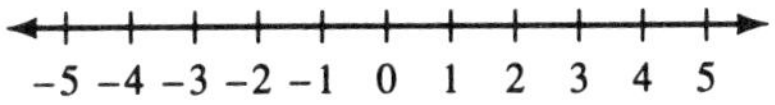

□ **Exercise 2** Solve.

a. $(2x + 1)^2 \geq 0$

b. $x^2 + 6x + 9 < 0$

□ **Exercise 3** Solve $x^2 + 4x - 12 \geq 0$.

EXAMPLE 3 Solve $x^2 + 2x - 8 > 0$.

Solve the quadratic equation

$$x^2 + 2x - 8 = 0$$

The equation can be factored.

$$(x + 4)(x - 2) = 0$$

$$x + 4 = 0 \quad \text{or} \quad x - 2 = 0$$

$$x = -4 \qquad\qquad x = 2$$

The numbers -4 and 2 divide the number line into three regions as shown.

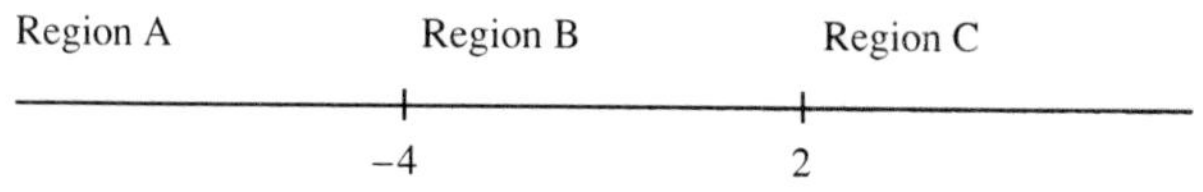

If one number in a region satisfies the inequality, all numbers in the region will satisfy the inequality. We chose $\mathbf{-5}$ from region A. Does it satisfy the inequality?

$$x^2 + 2x - 8 > 0$$

$$(\mathbf{-5})^2 + 2(\mathbf{-5}) - 8 > 0 \quad ? \qquad \text{Substitute } -5 \text{ for } x.$$

$$25 - 10 - 8 > 0 \quad ?$$

$$7 > 0 \quad \text{True}$$

Since -5 from region A satisfies the inequality, all numbers in region A satisfy the inequality.

Next, choose a test number from region B. We chose *0*.

$$x^2 + 2x - 8 > 0$$

$$(\mathbf{0})^2 + 2(\mathbf{0}) - 8 > 0 \quad ? \qquad \text{Substitute 0 for } x.$$

$$-8 > 0 \quad \text{False}$$

The numbers in region B are not solutions.

Finally, choose a test number from region C. Verify that 3 satisfies the inequality, hence all numbers in region C satisfy the inequality.

The solution set is $\{x \mid x < -4 \text{ or } x > 2\}$. ■

□ **DO EXERCISE 3.**

Solving a Quadratic Inequality: Method 2

1. Write the inequality as an equation and solve the equation.
2. Divide a number line into regions by writing the numbers found in step 1 on the number line.
3. Find the intervals that make the inequality true by substituting a number from each region into the inequality. All numbers in those regions that make the inequality true are in the solution set.
4. Write set-builder notation for the solution set.

2 Rational Inequalities

When an inequality contains a rational expression, it is called a ***rational inequality.*** We may adapt either of our two methods for solving quadratic inequalities to solving rational inequalities. We show the adaptation of Method 2.

EXAMPLE 4 Solve $\dfrac{x+1}{x-5} \leq 2$.

Write the related equation and solve it.

$$\frac{x+1}{x-5} = 2$$

$$(\boldsymbol{x-5})\frac{x+1}{x-5} = (\boldsymbol{x-5})2 \qquad \text{Multiply by the LCD.}$$

$$x + 1 = 2x - 10$$

$$11 = x$$

For rational inequalities, we also need to determine those values for which the inequality is undefined. These are the values that make the denominator 0. We set the denominator equal to 0 and solve.

$$x - 5 = 0$$

$$x = 5$$

These two numbers, 5 and 11, divide a number line into three regions.

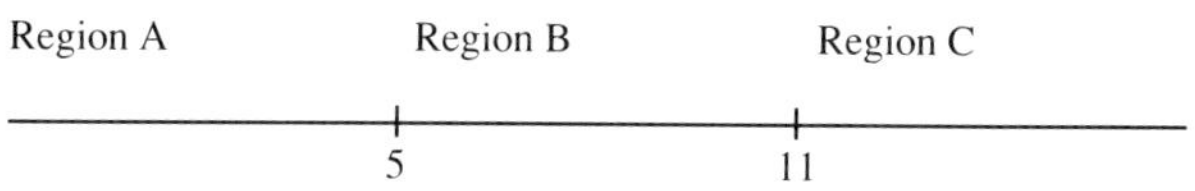

We test one number from each region to see if it satisfies the original inequality. We chose 0 from region A.

$$\frac{x+1}{x-5} \leq 2$$

$$\frac{0+1}{0-5} \leq 2 \quad ? \qquad \text{Substitute 0 for } x.$$

$$-\frac{1}{5} \leq 2 \quad \text{True}$$

All numbers in region A are solutions of the inequality. However, 5 is not a solution since the inequality is undefined for this value.

Testing one number from the other two regions shows that the solution set is $\{x \mid x < 5 \text{ or } x \geq 11\}$. ■

□ **DO EXERCISE 4.**

□ **Exercise 4** Solve $\dfrac{x-1}{x-2} > 3$.

Solving a Rational Inequality

1. Write the inequality as an equation and solve the equation.
2. Set the denominator equal to zero and solve that equation.
3. Use the numbers found in steps 1 and 2 to divide a number line into regions.
4. Substitute a number from each region into the inequality to determine the intervals that satisfy the inequality.
5. Write set-builder notation for the solution set, excluding any values that make the denominator equal to zero.

GRAPHING CALCULATOR

Approximate solutions to quadratic inequalities may be found using a graphing calculator. Solve $x^2 \leq 3x$.

1. Enter $y = x^2$ and $y = 3x$ into the calculator.
2. Set the graphics window. We need to set a window that includes the intersection of the two graphs. An appropriate window is W: $[-10, 10]$ $[-5, 15]$.
3. Plot the graphs. Trace either graph to identify the points of intersection. The part of the graph of $y = x^2$ that lies below the graph of $y = 3x$ is contained approximately between the x values of 0 and 3. The solution set is $\{x \mid 0 \leq x \leq 3\}$. This is the same solution that we found in Example 1b.

Approximate solutions to rational inequalities can also be obtained with a graphing calculator. Solve $\dfrac{x+1}{x-5} \leq 2$.

1. Enter $y = \dfrac{x+1}{x-5}$ and $y = 2$ into the calculator.
2. Set the graphics window. An appropriate window for these two graphs is W: $[-5, 15]$ $[-10, 10]$.
3. Plot the graphs. Trace either graph to see that the graph of $y = \dfrac{x+1}{x-5}$ lies below the graph of $y = 2$ for x values less than 5 or greater than or equal to 11. The solution set is $\{x \mid x < 5$ or $x \geq 11\}$. This is the solution we found for Example 4.

Answers to Exercises

1. a. $\{x \mid x \leq -4 \quad \text{or} \quad x \geq 1\}$ **b.** $\{x \mid -2 < x < 0\}$

2. a. $\{x \mid x \text{ is a real number}\}$ **b.** $\varnothing$

3. $\{x \mid x \leq -6 \quad \text{or} \quad x \geq 2\}$

4. $\left\{x \mid 2 < x < \dfrac{5}{2}\right\}$

Problem Set 8.9

NAME
DATE
CLASS

Solve and graph each solution, if it exists.

1. $(x + 4)(x - 3) > 0$

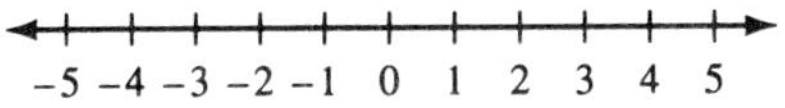

2. $(x - 3)(x + 2) < 0$

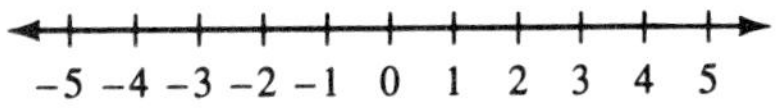

3. $(x + 1)(x + 5) \leq 0$

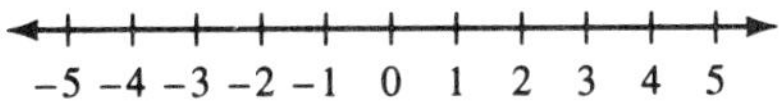

4. $(x - 2)(x - 4) \geq 0$

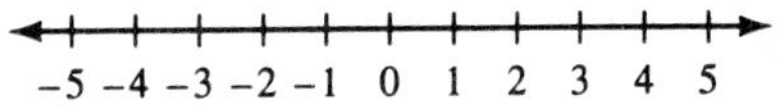

5. $x^2 - 2x - 24 < 0$

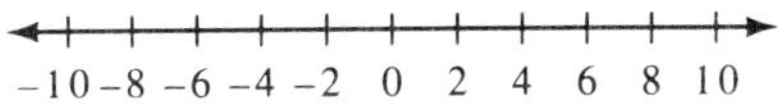

6. $x^2 + 3x - 10 > 0$

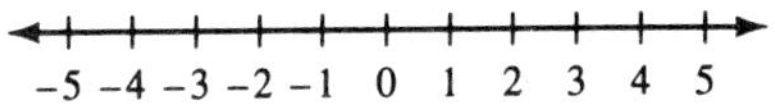

7. $x(x + 4) \geq 0$

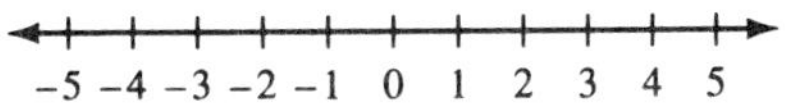

8. $x(x - 3) \leq 0$

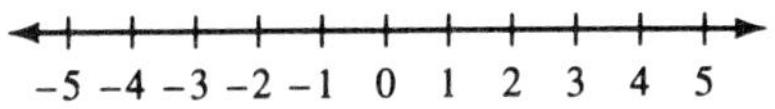

9. $x^2 > x$

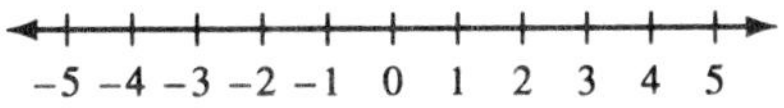

10. $x^2 < -5x$

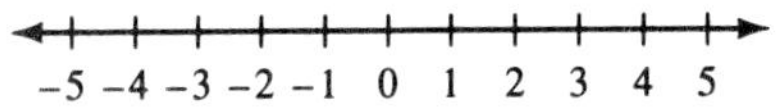

11. $3x^2 + 10x \le 8$

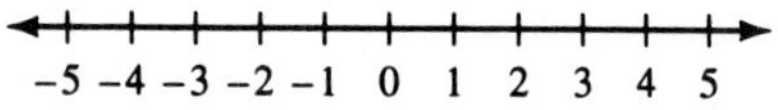

12. $9x^2 + 3x \ge 2$

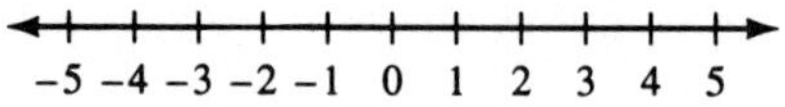

13. $4x^2 - 12x + 9 \ge 0$

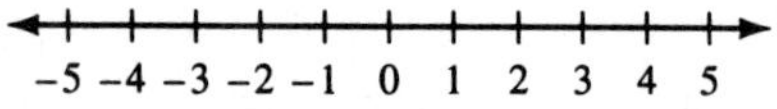

14. $9y^2 - 24y + 16 < 0$

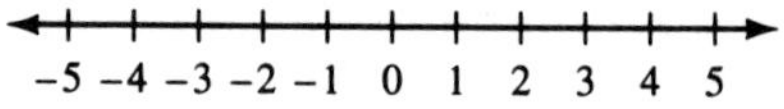

15. $x^2 - 16 < 0$

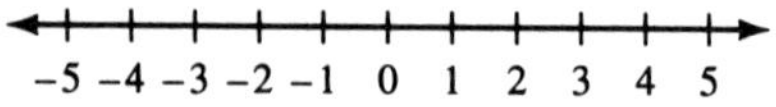

16. $x^2 - 4 > 0$

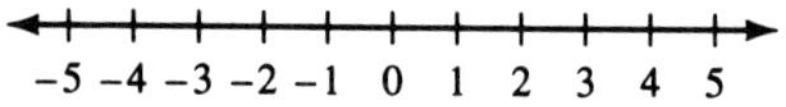

17. $4x^2 + 7x + 3 \ge 0$

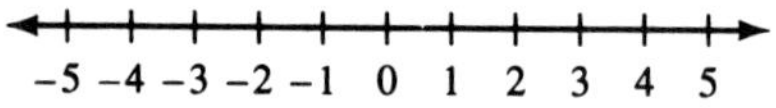

18. $3x^2 + 16x + 5 \le 0$

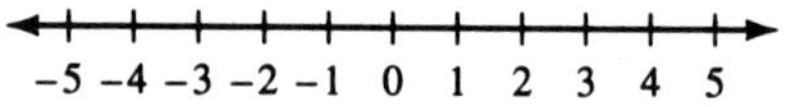

19. $(x - 1)(x + 2)(x - 3) > 0$

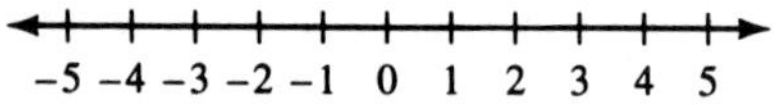

20. $(x - 4)(x + 5)(x - 1) < 0$

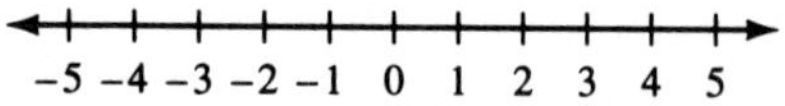

Solve.

21. $\frac{8}{x-2} \ge 2$

22. $\frac{15}{x-1} \le 1$

23. $\frac{1}{x-3} < 0$

24. $\frac{1}{x+6} > 0$

25. $\frac{x+1}{x-5} > 0$

26. $\frac{x+1}{x-3} > 0$

27. $\frac{1-q}{4-q} \geq 0$

28. $\frac{5-3p}{4p+3} \leq 0$

29. $\frac{3x}{x-2} > 1$

30. $\frac{2y}{y+1} < 1$

31. $\frac{1}{x} < 2$

32. $\frac{1}{x} \geq 4$

33. $\frac{3p}{p+6} - \frac{8}{p+6} > 0$

34. $\frac{5m}{m+1} - \frac{3}{m+1} > 0$

35. $\frac{x}{(x-1)(x+2)} \geq 0$

36. $\frac{y-2}{(y+1)(y-1)} \leq 0$

37. On the number line, numbers to the left of zero are ____________.

38. If a value of x is chosen for the expression $(x - 3)(x + 2)$ so that both factors are negative, the product is ____________.

39. In the expression $(x - 4)(x - 1)$, if a value of x is chosen so that one factor is negative and the other is ____________, the product is negative.

40. Consider the inequality $(2x - 3)^2 < 0$. Since $(2x - 3)^2$ is never less than zero, there is ____________ to this inequality.

Solve.

* **41.** $x^2 - 5x \leq 3$

* **42.** $5x^2 + 3x > 4$

* **43.** At Goodware Tire Company, the cost C of producing n tires in 1 day is given by $C = n^2 - 100n + 5400$. If the company wants to keep the cost under \$3000, how many tires can they produce?

* **44.** The velocity v of a particle in feet per second is given by $v = 2t^2 - 19t + 42$, where t is the time in seconds. When is the velocity negative?

Checkup

The following problems provide a review of some of Section 7.7.

Solve.

45. $\sqrt{5x - 9} = 6$

46. $\sqrt{3x + 4} = 4$

47. $\sqrt{x + 4} = \sqrt{3x - 4}$

48. $\sqrt{2x - 1} = \sqrt{x + 4}$

49. $\sqrt{3x - 5} = -2$

50. $\sqrt{6x + 8} = -7$

51. $\sqrt{x - 3} + \sqrt{x + 2} = 5$

52. $\sqrt{y - 2} + \sqrt{y + 5} = 7$

53. $\sqrt[3]{2y + 4} = 2$

54. $\sqrt[3]{5x + 6} = 1$

Chapter 8 Summary

Section 8.1

A ***quadratic equation*** is an equation that can be written in the form $ax^2 + bx + c = 0$, where a, b, and c are real numbers and $a \neq 0$. This form is called the *standard form of a quadratic equation.*

Square root property: If x and c are complex numbers and if $x^2 = c$, then $x = \sqrt{c}$ or $x = -\sqrt{c}$.

Solving an equation by factoring

1. Write the equation in standard form.
2. Factor.
3. Use the zero product property.
4. Solve the resulting equations.

Section 8.2

Solving a quadratic equation by completing the square

1. If the coefficient of x^2 is 1, go to step 2. If the coefficient of x^2 is not 1, divide each side of the equation by this coefficient.
2. Be sure that terms containing variables are on one side of the equation and constant terms are on the other side.
3. Take half the coefficient of the x term and square it.
4. Add the square to both sides. The side containing the variables is now a perfect square trinomial.
5. Factor the perfect square trinomial.
6. Solve the equation by using the square root property and solving the resulting equations.

Section 8.3

Quadratic formula: The solutions of $ax^2 + bx + c$ $(a \neq 0)$ are

$$x = \frac{-b \pm \sqrt{b^2 - 4ac}}{2a}$$

Solving a quadratic equation

1. Check to see if it is of the form $ax^2 = c$ or $(x + k)^2 = d$. If it is either of these forms, use the square root property.
2. If it is not of the form of step 1, write it in standard form $ax^2 + bx + c = 0$.
3. Try to solve by factoring.
4. If it cannot be solved by factoring, use the quadratic formula.

Work formula: $W = rt$, where W is work, r is rate, and t is time.

Section 8.4

In the quadratic formula, the expression under the radical sign, $b^2 - 4ac$, is called the ***discriminant.***

Discriminant	***Type of Solution***
Positive	Two different real numbers
Negative	Two different imaginary numbers
Zero	One real number (a double root)

Given two numbers r_1 and r_2, then $(x - r_1)(x - r_2) = 0$ is an equation with solutions r_1 and r_2 by the zero product property.

Section 8.7

The equation $y = ax^2 + bx + c$ ($a \neq 0$ and a, b, and c are constants) is a quadratic equation with two variables. The solutions of the equation are ordered pairs of numbers. The highest or lowest point on the graph of these equations is called the ***vertex.*** Graphs of quadratic equations in two variables are called ***parabolas.***

The x-coordinate of the vertex of the graph of a quadratic equation

$$y = ax^2 + bx + c \text{ is } x = -\frac{b}{2a}$$

The line $x = -b/(2a)$ is called the ***axis of symmetry.***

If a is positive in the equation $y = ax^2 + bx + c$, the parabola opens upward. If a is negative, the parabola opens downward.

Graphing equations of the form $y = ax^2 + bx + c$

1. Find the y-intercept by letting $x = 0$.
2. Find the x-intercepts by letting $y = 0$. If the equation does not factor, evaluate the discriminant to see if there are x-intercepts. If there are no x-intercepts, find an additional point.
3. Find the vertex. The x-coordinate is $-b/(2a)$. Use this value to find the y-coordinate in the given equation. If the vertex and a point from step 2 are the same, find an additional point.
4. Plot all of the points above and draw a parabola through them. In the equation, if a is positive, the parabola opens upward. If a is negative, the parabola opens downward. For a more accurate graph, plot at least two more points.

Section 8.8

There are also quadratic equations in two variables of the form $x = ay^2 + by + c$ ($a \neq 0$ and a, b, and c are constants). The graph of this equation is also a parabola. These parabolas open to the right if a is positive and to the left if a is negative.

Section 8.9

Quadratic inequalities in one variable are of the form $ax^2 + bx + c > 0$, where a, b, and c are real numbers and $a \neq 0$. The $>$ symbol may be replaced by $<$, $\leq$, or $\geq$.

Solving a quadratic inequality (Method 1)

1. Set one side of the inequality equal to zero.
2. Factor the other side.
3. Find which values of the variable make each factor positive and which values make each factor negative. Draw a diagram showing this information and the product of the factors.
4. Find the solution set from the diagram.

Solving a quadratic inequality (Method 2)

1. Write the inequality as an equation and solve the equation.
2. Divide a number line into regions by writing the numbers found in step 1 on the number line.
3. Find the intervals that make the inequality true by substituting a number from each region into the inequality. All numbers in those regions that make the inequality true are in the solution set.
4. Write set-builder notation for the solution set.

If an inequality contains a rational expression, it is called a ***rational inequality.***

Solving a rational inequality

1. Write the inequality as an equation and solve the equation.
2. Set the denominator equal to zero and solve that equation.
3. Use the numbers found in steps 1 and 2 to divide a number line into regions.
4. Substitute a number from each region into the inequality to determine the intervals that satisfy the inequality.
5. Write set-builder notation for the solution set, excluding any values that make the denominator equal to zero.

Chapter 8 Additional Exercises (Optional)

NAME

DATE

CLASS

Section 8.1

Solve.

1. $9x^2 + 7 = 0$

2. $17x^2 + 3 = 0$

3. $(x - 3)^2 = 16$

4. $(y + 4)^2 = 8$

5. $4x^2 - 12x + 9 = -7$

6. $16x^2 - 8x + 1 = 5$

7. $72x^2 - 5x - 25 = 0$

8. $20y^2 + 64y - 21 = 0$

9. $27x = 36x^2$

10. $24x^2 = 28x$

Section 8.2

Solve by completing the square.

11. $x^2 + 4x - 2 = 0$

12. $y^2 + 6y + 3 = 0$

13. $3x^2 + 6x + 2 = 0$

14. $2x^2 + 8x - 5 = 0$

15. $64y^2 + 32y - 5 = 0$

16. $36x^2 + 27x + 2 = 0$

17. $15y^2 - 30y + 4 = 0$

18. $11x^2 - 22x - 1 = 0$

Section 8.3

Solve by the quadratic formula. Assume that variables under the radical sign represent positive numbers.

19. $2x^2 - 3x + 4 = 0$

20. $3x^2 + 2x - 5 = 0$

21. $x^2 + 7 = -3x$

22. $x^2 - 2 = 5x$

23. $2p^2 + 3p = 4$

24. $4k^2 - 8k = -1$

* **25.** $L^2x^2 + 3Lx + 2 = 0$

* **26.** $3A^2y^2 - Ay - 5 = 0$

Section 8.4

Predict the number and type of solutions.

27. $5x^2 - 2x = 4$

28. $4x^2 + 3x = -5$

* **29.** $\frac{1}{2}p^2 + \frac{3}{4}p + 2 = 0$

* **30.** $\frac{9}{5}m^2 - \frac{3}{5}m + 5 = 0$

* **31.** $\frac{9}{16}x^2 - \frac{3}{2}x + 1 = 0$

* **32.** $k^2 + \frac{8}{3}k + \frac{1}{3} = 0$

Find quadratic equations with the following solutions.

33. $-3, 5$

34. $4, -4$

* **35.** $\frac{3 - \sqrt{2}}{4}, \frac{3 + \sqrt{2}}{4}$

* **36.** $\frac{1 - \sqrt{5}}{2}, \frac{1 + \sqrt{5}}{2}$

Section 8.5

Solve for the variable indicated. Assume that all variables represent positive numbers.

37. $z = \sqrt{\frac{xy}{2}}$ for y

38. $d = \frac{a}{\sqrt{c}}$ for c

39. $k = \frac{rF}{wV^2}$ for V

40. $V = (r^2 + R^2)h$ for R

41. $N = \frac{S^2 - 3S}{2}$ for S

42. $A = 2\pi r^2 + 2\pi rh$ for r

43. The number of diagonals d, of a polygon of n sides is given by $d = (n/2)(n - 3)$. How many sides does a polygon with 54 diagonals have?

44. The formula

$$S = \frac{n}{2}(n + 1)$$

gives the sum of the first n natural numbers 1, 2, 3,. . . . How many consecutive natural numbers, starting with 1, are needed to get a sum of 55?

Section 8.6

Solve.

45. $x^4 + 6x^2 + 8 = 0$

46. $x^4 - 9x^2 + 8 = 0$

47. $x - 4\sqrt{x} - 12 = 0$

48. $x - 3\sqrt{x} - 10 = 0$

49. $x^{-2} + x^{-1} = 6$

50. $y^{-2} + 4y^{-1} = 5$

51. $6p^{-2} + 11p^{-1} + 4 = 0$

52. $6x^{-2} - 11x^{-1} + 3 = 0$

Section 8.7

Graph on graph paper.

53. $y = 3x^2 + 1$

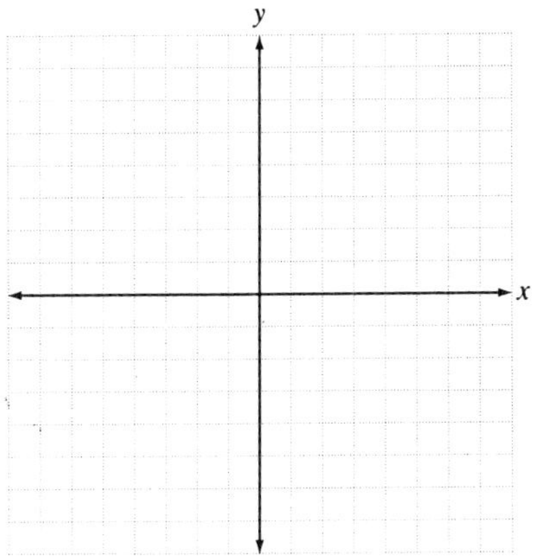

54. $y = -3x^2 + 4$

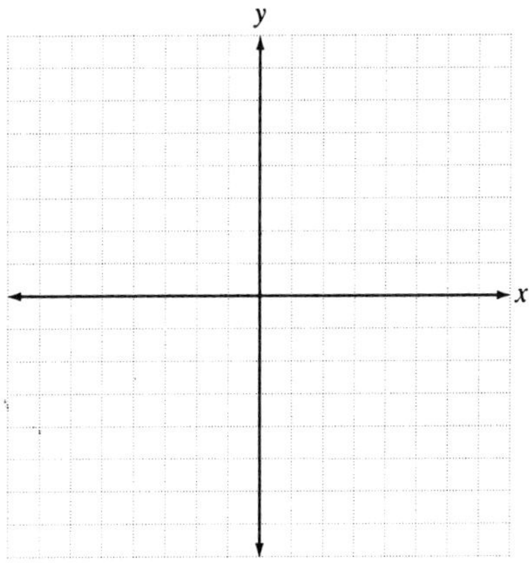

55. $y = x^2 - 2x - 8$

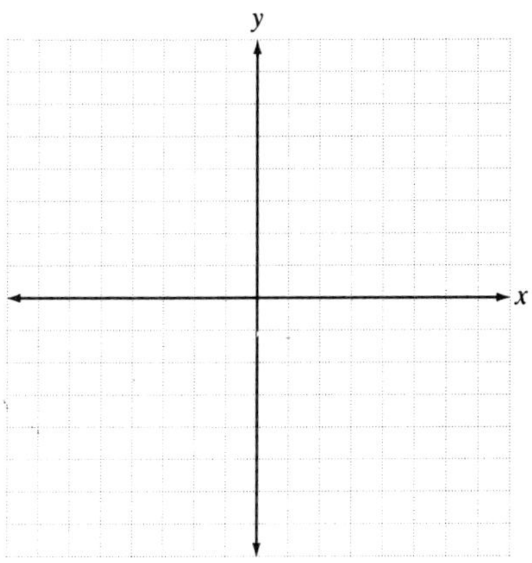

56. $y = x^2 - 8x + 15$

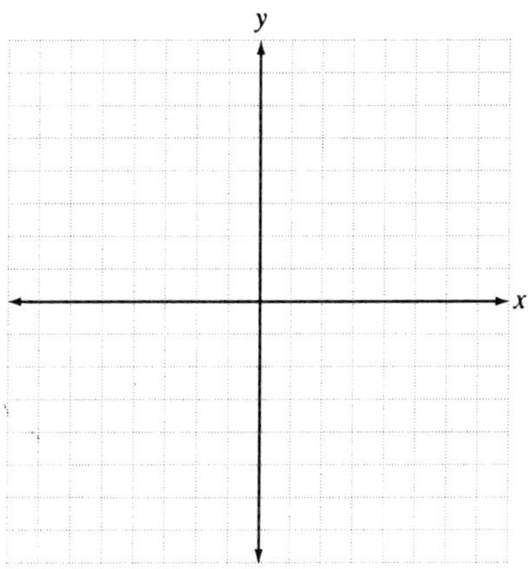

57. $y = -3x^2 + 4x - 2$

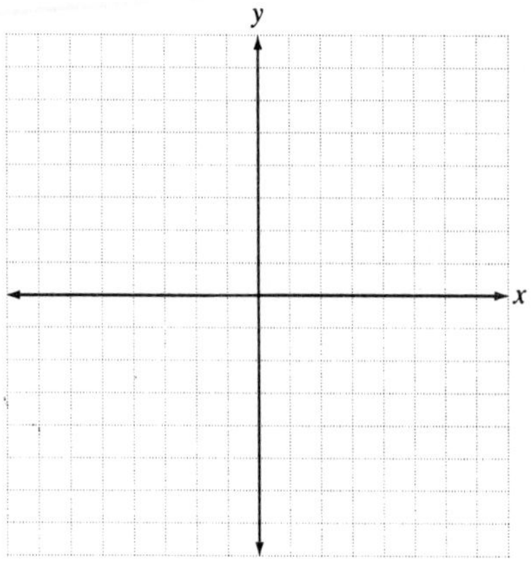

58. $y = -6x^2 + 12x - 3$

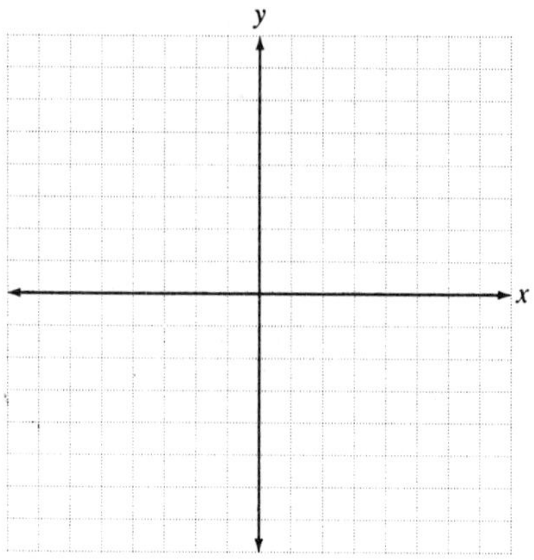

Section 8.8

59. Laurie owns a muffin shop. She has found that the cost of operating her shop is given by

$$C = 3x^2 - 240x + 5000$$

where x is the number of muffins sold each day. What is the number of muffins that Laurie must sell each day to have the lowest operating cost?

60. A projectile is fired upward. Its height h in feet, after t seconds is given by

$$h = -16t^2 + 192t$$

Find the maximum height attained by the object.

Graph.

61. $x = -y^2 + 2$

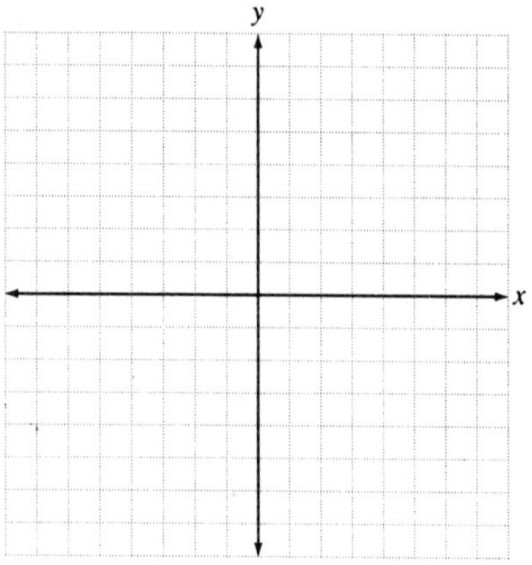

62. $x = y^2 - 3$

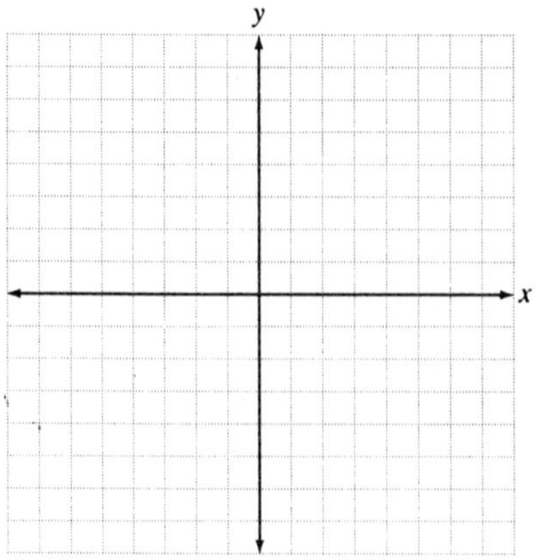

63. $x = 3y^2 + 4y + 2$

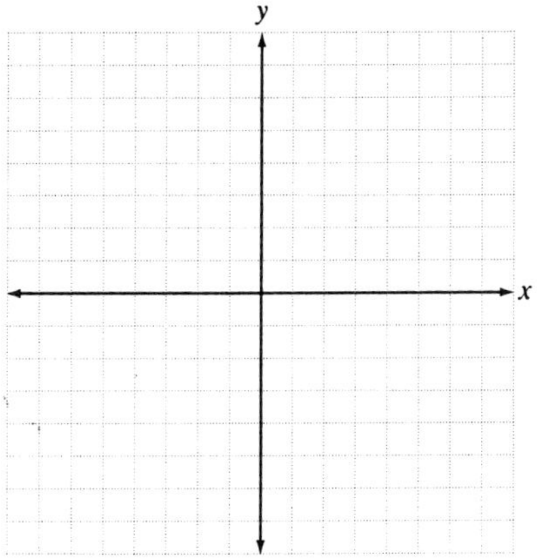

64. $x = -2y^2 + 4y + 1$

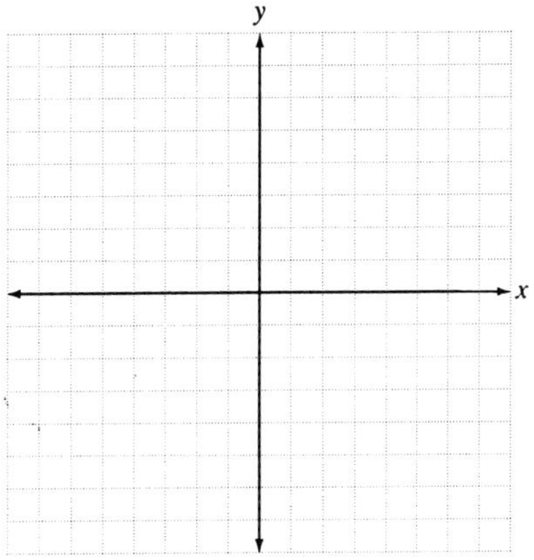

65. $x = 2y^2 - 5y - 1$

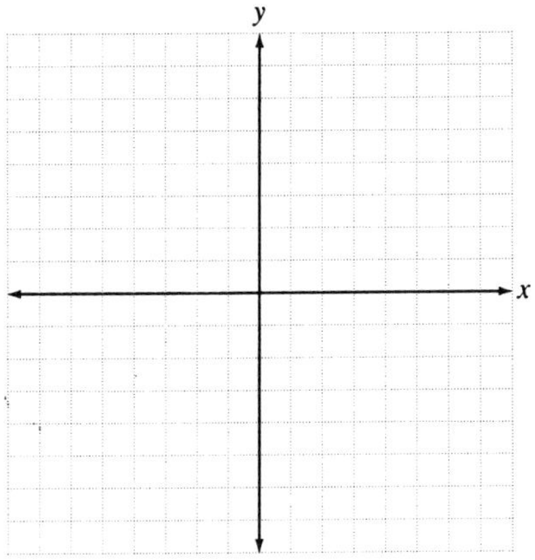

66. $x = 3y^2 + 6y + 3$

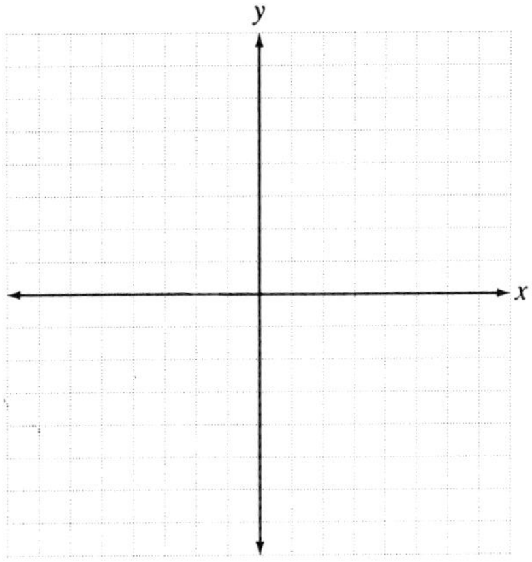

Section 8.9

Solve and graph each solution, if it exists.

67. $(x - 3)(x + 2) < 0$

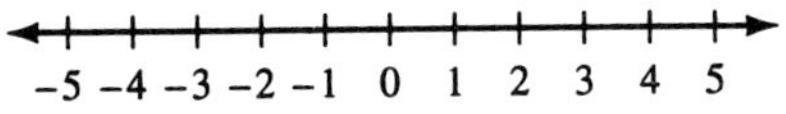

68. $(x - 5)(x - 1) \geq 0$

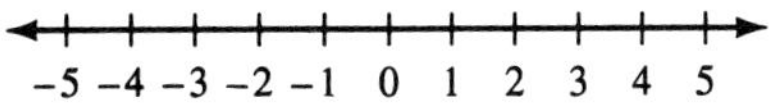

69. $3x^2 > 9x$

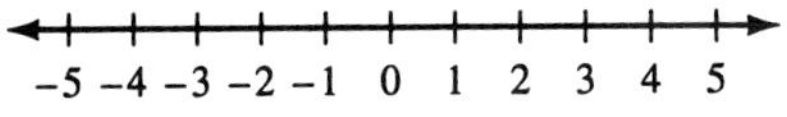

70. $2x^2 \leq 4x$

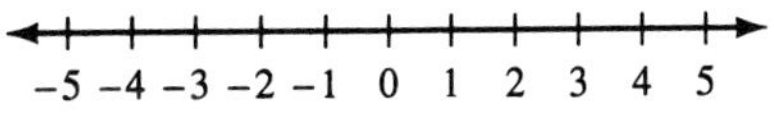

71. $9x^2 - 24x + 16 < 0$

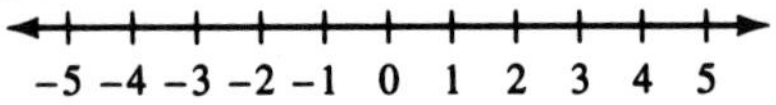

72. $25x^2 + 20x + 4 \geq 0$

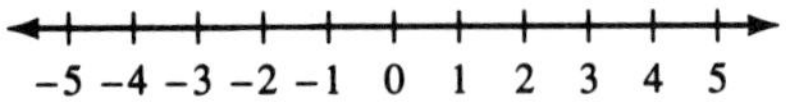

73. $(y - 1)(2y + 3)(y + 4) \geq 0$

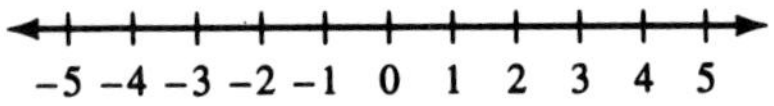

74. $(x - 3)(2x - 5)(x + 2) < 0$

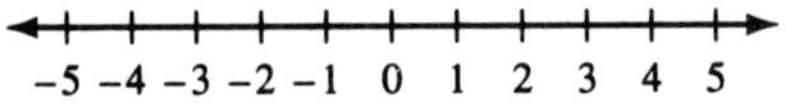

* **75.** $\dfrac{2x}{x + 1} < 2$

* **76.** $\dfrac{3y}{y + 2} > 3$

* **77.** $\dfrac{17}{p + 3} \geq \dfrac{-8}{p - 2}$

* **78.** $\dfrac{5}{b - 1} \leq \dfrac{1}{b + 1}$

COOPERATIVE LEARNING

Solve by the easiest method.

1. $3x^2 + 5x = 2$

2. $4y^2 - 3y = 8$

3. $(5x - 3)^2 = 15$

4. $9x^2 = 14x + 8$

5. $2y^2 = 6y + 3$

6. $7y^2 = -8$

7. Near New York City, a car traveled 55 miles per hour. Later, in a less populated area, the car traveled 65 miles per hour. At the faster rate it went 40 miles more than twice as far as it did at the slower rate. How far did it travel at each rate?

Find a quadratic equation for the following graphs.

8.

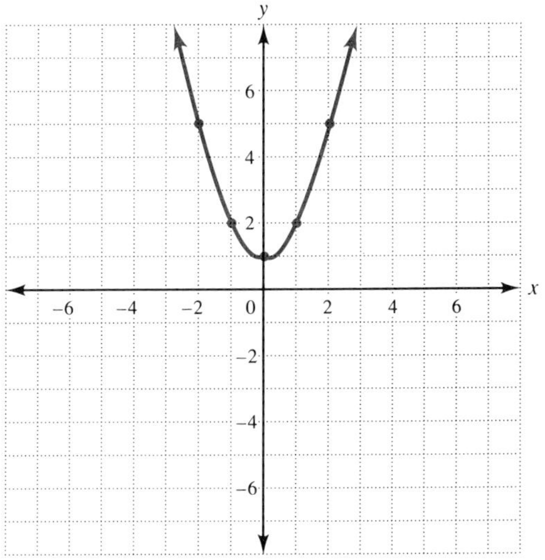

9.

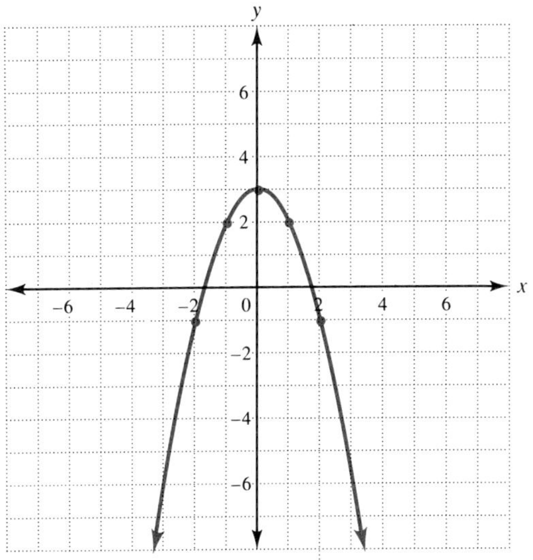

10.

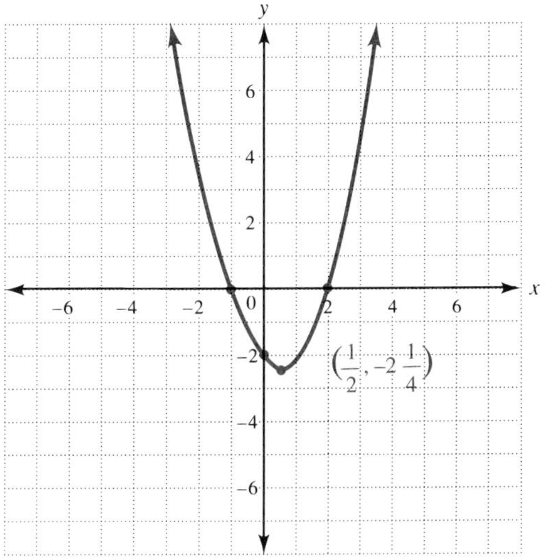

11.

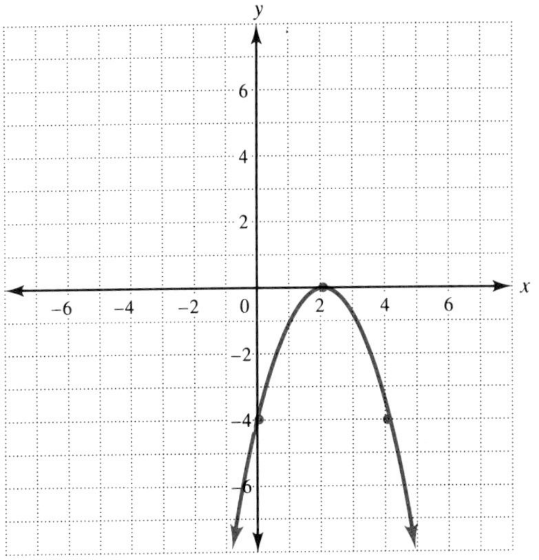

12.

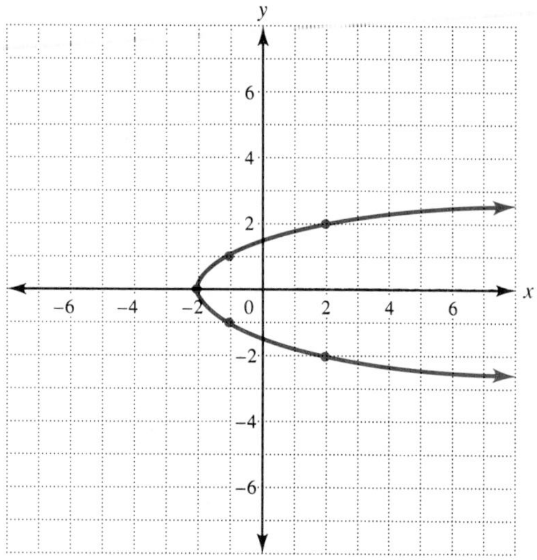

13.

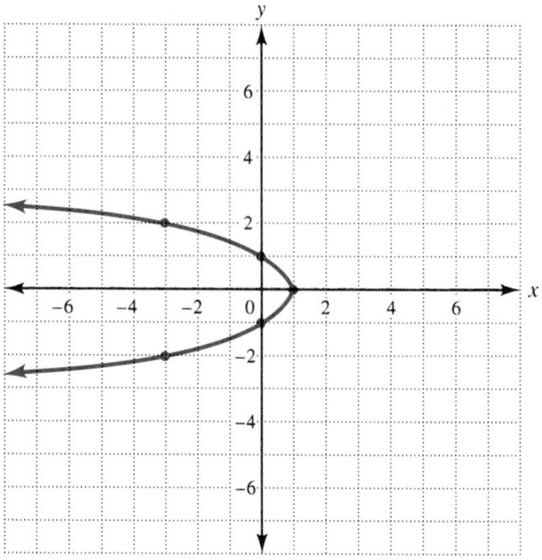

14. The revenue from selling n rings is given by $R = 10n^2 - 70n + 80$. The cost of manufacturing n rings is $C = 7n^2 - 10n - 70$. Profit P is given by $P = R - C$. How many rings must be sold to gain a profit of \$150?

Chapter 8 Practice Test

NAME ______________________

DATE ______________________

CLASS ______________________

1. Solve: $(3x + 1)^2 = -7$. **1.** ______

2. The hypotenuse of a right triangle is 13 centimeters long. One leg is 7 centimeters longer than the other. Find the length of the longest leg. **2.** ______

3. Solve by completing the square: $y^2 + 6y - 5 = 0$. **3.** ______

4. Solve by the quadratic formula: $x^2 + 3x + 3 = 0$. **4.** ______

5. Solve and approximate solutions to the nearest tenth: $2x^2 + 6x - 3 = 0$. **5.** ______

6. Predict the number and type of solutions of $3x^2 + 6x + 2 = 0$. **6.** ______

7. Find a quadratic equation with solutions of 5 and -2. **7.** ______

8. Solve for r (assume that variables represent positive numbers): $S = 4\pi r^2$. **8.** ______

9. A rectangular picture, 8 by 10 inches, is surrounded by a frame of uniform width. If the area of the picture and frame is 120 square inches, how wide is the frame? **9.** ______

10. Solve: $x^4 + 6x^2 - 7 = 0$. **10.** ______

Graph.

11. $y = 3x^2 - 4$ **11.** ______

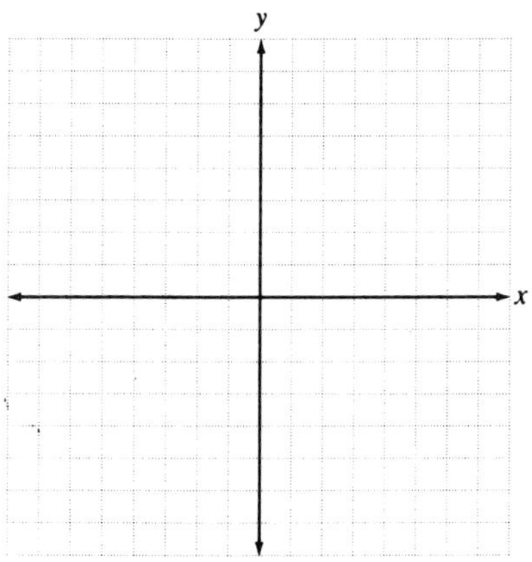

12. ____________

12. $y = x^2 - 3x - 4$

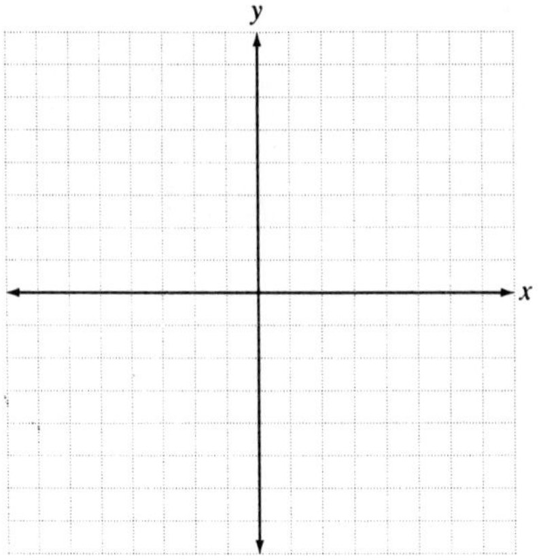

13. ____________

13. $x = -2y^2 + 4$

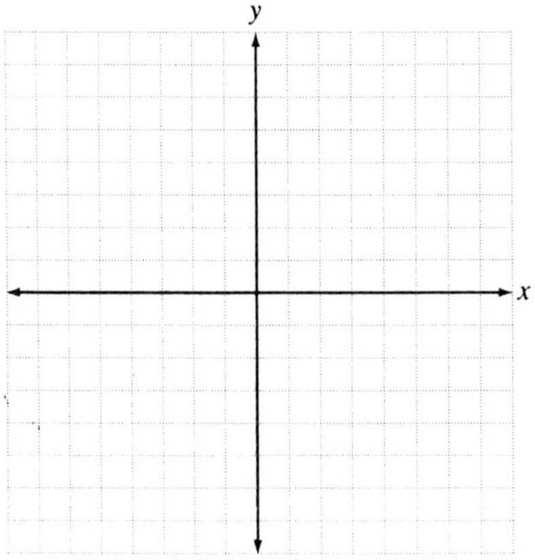

14. ____________

14. What is the maximum product of two numbers whose sum is 12?

Solve and graph the solutions.

15. ____________

15. $x^2 - x - 2 \geq 0$

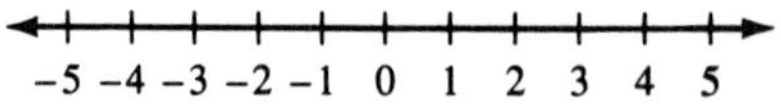

16. ____________

16. $x^2 < x$

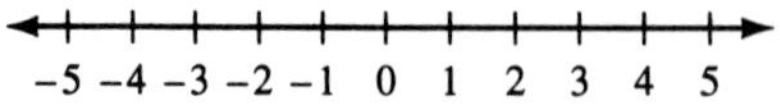

17. ____________

17. Solve: $\dfrac{-4}{x + 5} \geq 4.$

Chapters 7 and 8 Cumulative Review

NAME ______________________

DATE ______________________

CLASS ______________________

Find the root.

1. $\sqrt{49x^2}$

2. $\sqrt[3]{-64y^6}$

For Problems 3 and 4, assume that variables under radical signs represent positive numbers.

3. Multiply and simplify: $\sqrt{2x^3}\sqrt{32x}$.

4. Divide and simplify:

$$\frac{\sqrt[3]{81y^5}}{\sqrt[3]{3y^2}}$$

5. Add: $4\sqrt{48} + 3\sqrt{108}$.

6. Multiply and simplify: $(7 - 5\sqrt{2})(7 + 3\sqrt{2})$.

7. Rationalize the denominator:

$$\frac{x - \sqrt{2}}{x + \sqrt{5}}$$

8. Write without fractional exponents: $(a^2b)^{2/3}$.

9. Simplify and write with positive exponents (assume that $x > 0$):

$$\frac{(x^{-1/3})^9}{x^5}$$

10. Solve: $\sqrt{y - 4} + \sqrt{y + 8} = 6$.

11. Simplify: $\sqrt{-32}$.

12. Subtract: $(2 - 6i) - (4 + 3i)$.

Multiply.

13. $(5 - 2i)(3 - 4i)$

14. $\sqrt{-5} \cdot \sqrt{-6}$

15. Divide:

$$\frac{4 + i}{3i}$$

16. In clear weather, the miles M that a person can see the view at an altitude A miles is given by $M = 1.22\sqrt{A}$. How far can Shaw see at an altitude of 8 miles? Round the answer to the nearest tenth.

Solve.

17. $3x^2 = 48$

18. $2x^2 - 3x = 6$

19. Predict the number and type of solutions of $4x^2 - 3x + 6 = 0$.

20. Solve for L (assume that variables represent positive numbers):

$$W = \sqrt{\frac{1}{LC}}$$

21. A rectangular piece of metal has a width that is 4 inches less than the length. A square piece with sides of 2 inches is cut from each corner and the metal is turned up to form a box with volume 192 cubic inches. Find the dimensions of the original piece of metal.

Graph.

22. $y = 2x^2 - 5$

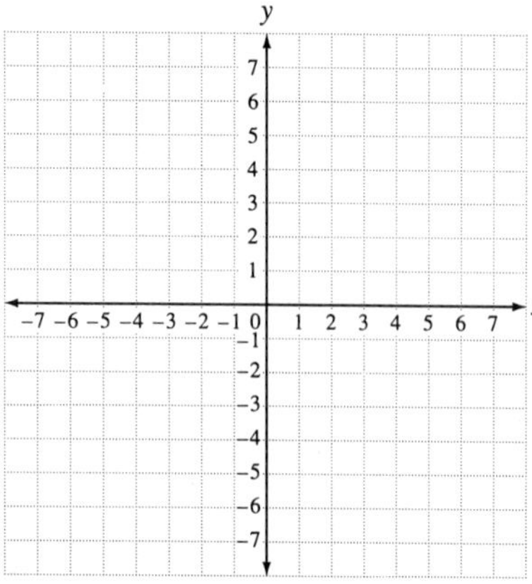

23. $y = -x^2 - x + 6$

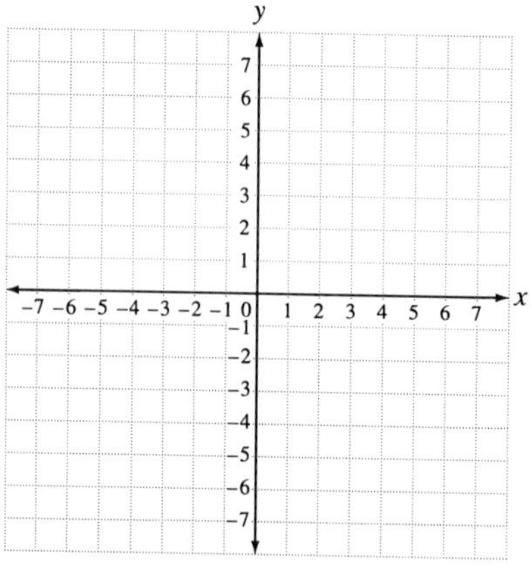

24. $x = 3y^2 - 2$

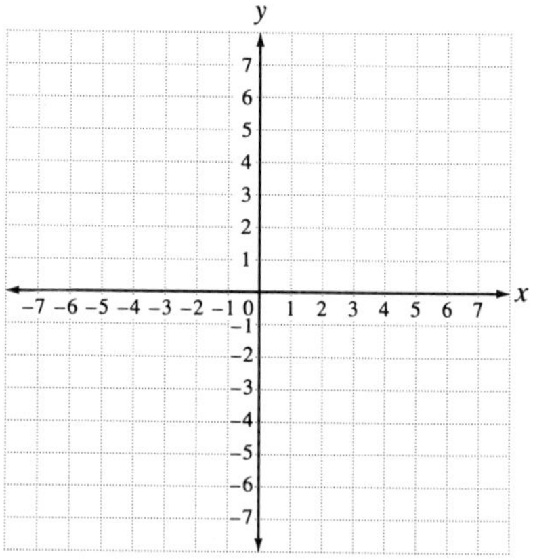

Solve and graph the solutions.

25. $x^2 + 2x - 15 < 0$

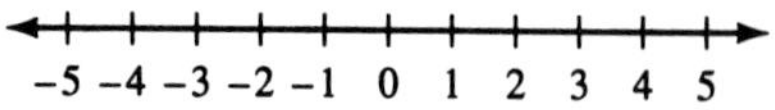

26. $x^2 - 3x \geq 0$

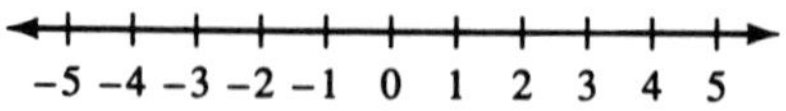

27. Solve: $\dfrac{2}{b+1} \leq \dfrac{3}{b+2}$.

CHAPTER 9

More Second-Degree Equations

Pretest

1. Find the distance between $(-3, -5)$ and $(-6, -12)$.

2. Find the midpoint of the line segment with endpoints $(\frac{1}{2}, \frac{3}{8})$ and $(\frac{5}{6}, \frac{7}{5})$.

3. Write an equation for the circle with center $(-5, -3)$ and radius $3\sqrt{6}$.

4. Find the center and radius of the circle:

$$x^2 + y^2 - 10x - 14y = -10$$

Graph.

5. $\dfrac{x^2}{\frac{9}{4}} + \dfrac{y^2}{\frac{25}{4}} = 1.$

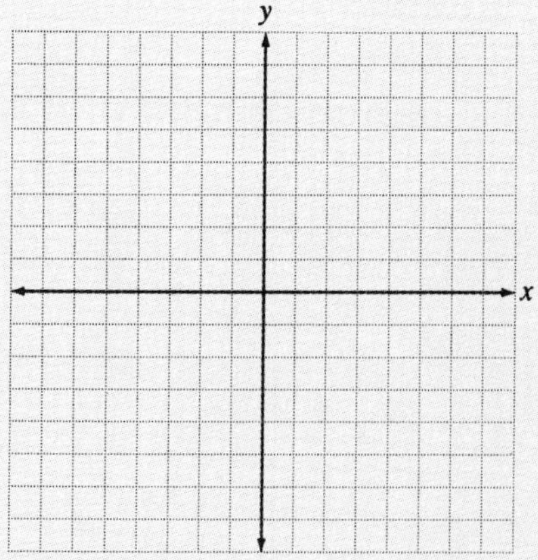

6. $x = \dfrac{1}{2}(y - 3)^2 - 2.$

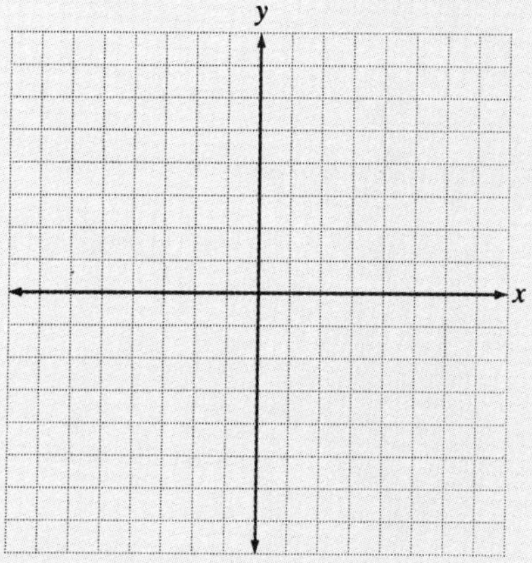

7. $\frac{x^2}{16} - \frac{y^2}{25} = 1$

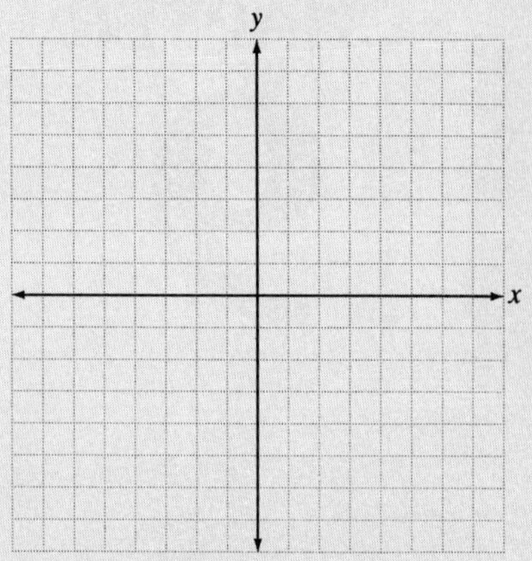

8. Identify the conic section:

$$x^2 + y^2 - 6x + 16y = -24$$

Solve.

9. $2L + 2W = 30$
$LW = 56$

10. $3x^2 - 4y^2 = 32$
$4x^2 + 5y^2 = 84$

11. The area of a rectangular garden is 154 square meters and the length of a diagonal is $\sqrt{317}$ meters. Find the dimensions of the garden.

9.1 THE DISTANCE AND MIDPOINT FORMULAS

In Chapter 8 we discussed quadratic equations. They are a special type of second-degree equation. In this chapter, we consider other kinds of second-degree equations. First, we study the distance and midpoint formulas.

OBJECTIVES

1 *Find the distance between a pair of points*

2 *Find the midpoint of a line segment*

1 The Distance Formula

It is easy to find the distance between two points on a horizontal or vertical line. First we consider a horizontal line. Points on a horizontal line have the same second coordinate. Consider the points (2, 3) and (6, 3).

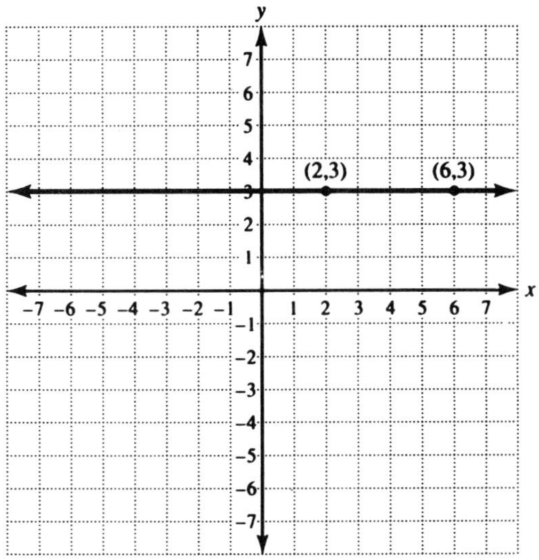

We find the distance d between the points on a horizontal line by subtracting their *first* coordinates in a way that yields a positive result. If we use absolute value, we may subtract either way. (The distance is never negative.) The distance between the points (2, 3) and (6, 3) is

$$d = |6 - 2| = |2 - 6| = 4$$

In general, the distance d between two points on a horizontal line is

$$d = |x_2 - x_1|$$

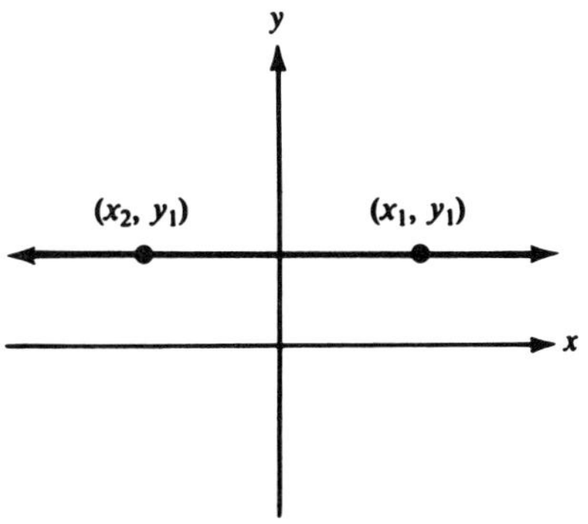

A similar formula can be developed for a vertical line. To find the distance d between two points on a vertical line, we subtract their *second* coordinates and find the absolute value.

$$d = |y_2 - y_1|$$

☐ **Exercise 1** Find the distance between each pair of points.

a. $(3, 5)$ and $(-4, 5)$

EXAMPLE 1 Find the distance between each pair of points.

a. $(5, 2)$ and $(-3, 2)$

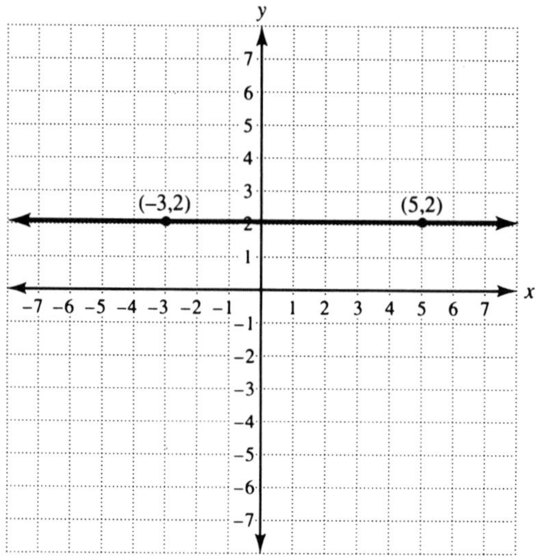

These are points on a horizontal line.

$$d = |5 - (-3)| = |8| = 8 \qquad \text{Subtract first coordinates.}$$

b. $(7, -2)$ and $(7, 8)$

b. $(4, -1)$ and $(4, 6)$

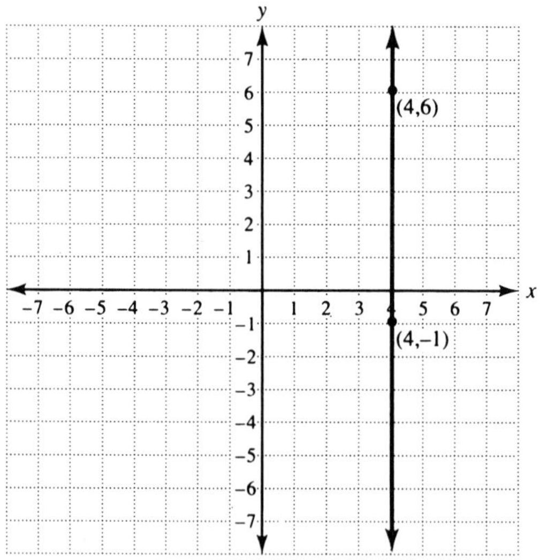

These are points on a vertical line.

$$d = |-1 - 6| = |-7| = 7 \qquad \text{Subtract second coordinates.}$$ ■

☐ **DO EXERCISE 1.**

We want to find a general formula for the distance d between any two points (x_1, y_1) and (x_2, y_2) on a line. We form a right triangle by drawing a vertical line through the point (x_2, y_2) and a horizontal line through the point (x_1, y_1). The point where the lines intersect is (x_2, y_1). Use the Pythagorean theorem (Section 7.8) to find the distance between the points. Since (x_2, y_1) and (x_1, y_1) are points on a horizontal line and (x_2, y_2) and (x_2, y_1) are points on a vertical line, the lengths of the legs are $|x_2 - x_1|$ and $|y_2 - y_1|$.

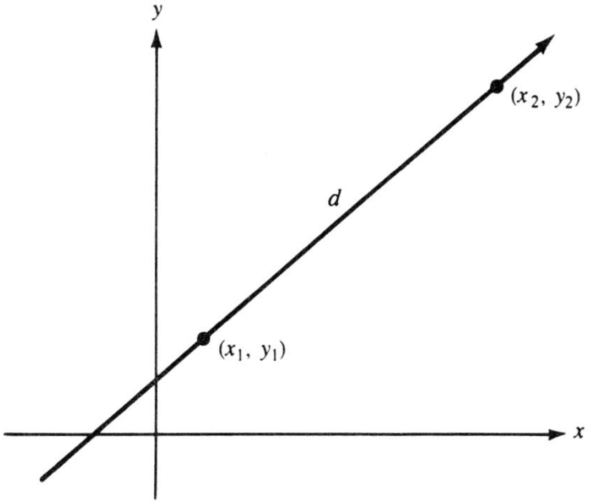

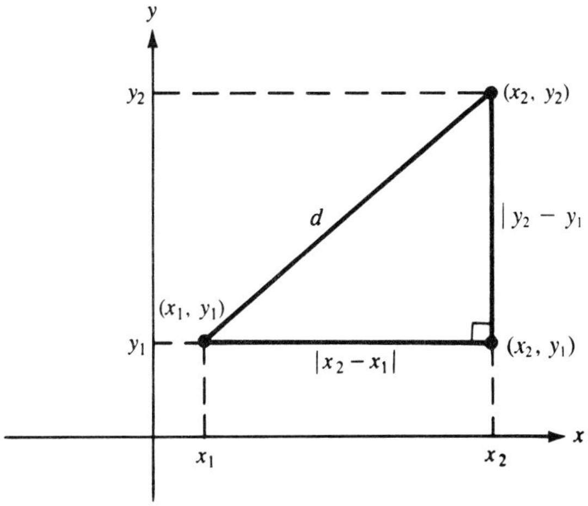

□ **Exercise 2** Find the distance between each pair of points. If necessary, use a calculator to approximate to the nearest thousandth.

a. (5, −4) and (−2, 1)

b. (6, 3) and (−4, 5)

Since squares of numbers are not negative,

$$|x_2 - x_1|^2 = (x_2 - x_1)^2$$

and

$$|y_2 - y_1|^2 = (y_2 - y_1)^2$$

Hence

$$d^2 = (x_2 - x_1)^2 + (y_2 - y_1)^2. \qquad \text{Use the Pythagorean theorem.}$$

The distance is found by taking the square root of both sides of the equation. Since distance is never negative, we use the principal square root.

The Distance Formula

The distance d between any two points (x_2, y_2) and (x_1, y_1) is

$$d = \sqrt{(x_2 - x_1)^2 + (y_2 - y_1)^2}$$

The formula may also be used to find the distance between two points on a horizontal or a vertical line.

EXAMPLE 2 Find the distance between (−2, 5) and (7, 3).

Let $(x_2, y_2) = (7, 3)$ and $(x_1, y_1) = (-2, 5)$.

$$d = \sqrt{(x_2 - x_1)^2 + (y_2 - y_1)^2}$$

$$d = \sqrt{[7 - (-2)]^2 + (3 - 5)^2}$$

$$d = \sqrt{9^2 + (-2)^2}$$

$$d = \sqrt{85} \approx 9.220 \quad ■$$

□ DO EXERCISE 2.

□ **Exercise 3** Find the midpoint of the line segment with the given endpoints.

a. (5, −2), (4, −6)

b. (−8, −3), (−4, −2)

2 The Midpoint Formula

A line segment is part of a line and has two endpoints.

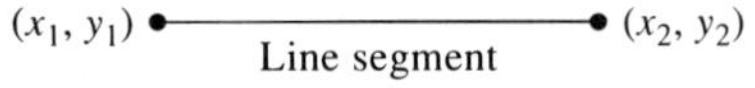

If the coordinates of the endpoints of a line segment are known, the distance formula can be used to derive a formula for the midpoints of the segment. The formula is as follows.

> The coordinates of the midpoint of a line segment whose endpoints are (x_1, y_1) and (x_2, y_2) are
> $$\left(\frac{x_1 + x_2}{2}, \frac{y_1 + y_2}{2}\right).$$

EXAMPLE 3 Find the midpoint of the line segment with endpoints (−3, −4) and (6, −7).

Let $(x_1, y_1) = (-3, -4)$ and

$(x_2, y_2) = (6, -7)$

$$\left(\frac{x_1 + x_2}{2}, \frac{y_1 + y_2}{2}\right) \quad \text{Midpoint formula.}$$

$$\left(\frac{-3 + 6}{2}, \frac{-4 + (-7)}{2}\right) \quad \text{Substitute for } (x_1, y_1) \text{ and } (x_2, y_2).$$

$$= \left(\frac{3}{2}, \frac{-11}{2}\right)$$

The midpoint is $\left(\frac{3}{2}, \frac{-11}{2}\right)$. ■

□ **DO EXERCISE 3.**

DID YOU KNOW?

Archaeologists found a clay tablet, Plimpton 322, that proves that the ancient Babylonians were using the Pythagorean theorem more than a thousand years before Pythagoras was born. However, Pythagoras was the first to prove the theorem in the general case.

Answers to Exercises

1. a. 7 **b.** 10

2. a. 8.602 **b.** 10.198

3. a. $\left(\frac{9}{2}, -4\right)$ **b.** $\left(-6, -\frac{5}{2}\right)$

Problem Set 9.1

NAME ____________________

DATE ____________________

CLASS ____________________

Find the distance between each pair of points. If necessary, use a calculator to find an approximation to the nearest thousandth.

1. (8, 6) and (7, 1)

2. (1, 5) and (3, 2)

3. (5, 2) and (5, −4)

4. (−3, 4) and (−5, 4)

5. (0, −9) and (5, −3)

6. (3, 0) and (−2, −1)

7. (−3, 5) and (6, 4)

8. (1, −5) and (6, 3)

9. $\left(\frac{5}{6}, \frac{3}{8}\right)$ and $\left(\frac{7}{6}, \frac{1}{8}\right)$

10. $\left(\frac{3}{2}, \frac{5}{7}\right)$ and $\left(\frac{9}{2}, \frac{6}{7}\right)$

11. (−15, 24) and (−32, 48)

12. (−72, −18) and (16, 23)

13. $(0, \sqrt{3})$ and $(5, 3\sqrt{3})$

14. $(\sqrt{7}, 0)$ and $(0, \sqrt{2})$

15. (100, −320) and (300, 460)

16. (620, −200) and (−380, 100)

Find the midpoint of the line segment with the given endpoints.

17. (−7, 4) and (−3, 5)

18. (9, 4) and (−2, −5)

19. (6, 3) and (−5, −8)

20. (−4, −3) and (11, −7)

21. (−9, −4) and (15, 6)

22. (8, −1) and (−9, 2)

23. (−4.2, −3.6) and (−5.2, 6.1)

24. (9.3, −4.7) and (8.1, −2.6)

25. $\left(-\frac{1}{8}, \frac{3}{5}\right)$ and $\left(-\frac{1}{4}, \frac{3}{2}\right)$

26. $\left(-\frac{3}{8}, \frac{3}{7}\right)$ and $\left(-\frac{1}{6}, \frac{9}{14}\right)$

27. (c, d) and $(0, 0)$

28. $(0, 0)$ and (p, q)

If the sides of a triangle have lengths a, b, and c and $a^2 + b^2 = c^2$, the triangle is a right triangle. Determine whether or not the following are vertices of a right triangle.

29. $(0, 6)$, $(9, -6)$, $(-3, 0)$

30. $(5, 6)$, $(0, -3)$, $(9, 2)$

31. In a right triangle, the sum of the squares of the lengths of the ____________ is equal to the square of the length of the hypotenuse.

32. The distance between two points on a horizontal line may be found by taking the absolute value of the difference of the ____________ coordinates of the points.

33. The distance between two points on a line is never ____________.

Find the distance between each pair of points. Round answers to the nearest tenth.

34. $(9.4, 5.7)$ and $(6.8, 1.3)$

35. $(-3.8, 4.2)$ and $(-5.5, 4.6)$

Think About It

Find the perimeter of the triangle determined by each set of points.

* 36. $(1, 3)$, $(-3, -5)$, $(1, -2)$

* 37. $(-4, -2)$, $(6, 4)$, $(1, 10)$

If three points lie on the same line, the sum of the lengths of the two shorter segments equals the length of the longest segment. Determine if the following points lie on the same line.

* 38. $(-1, 1)$, $(-3, -5)$, $(2, 10)$

* 39. $(-1, -7)$, $(2, -1)$, $(4, 5)$

Checkup

The following problems provide a review of some of Section 8.2 and will help you with the next section.

Solve by completing the square.

40. $x^2 + 4x - 5 = 0$

41. $x^2 + 2x - 4 = 0$

42. $x^2 + 3x + 4 = 0$

43. $x^2 + 2x + 8 = 0$

44. $x^2 - x - 2 = 0$

45. $x^2 + 5x + 1 = 0$

46. $3x^2 + 9x - 2 = 0$

47. $2x^2 + 6x + 3 = 0$

9.2 CIRCLES AND ELLIPSES

1 Circles

All points on a ***circle*** are a distance r from the center (h, k). If (x, y) is any point on the circle, by the distance formula the distance r between the center (h, k) and (x, y) is given by

$$r = \sqrt{(x - h)^2 + (y - k)^2}$$

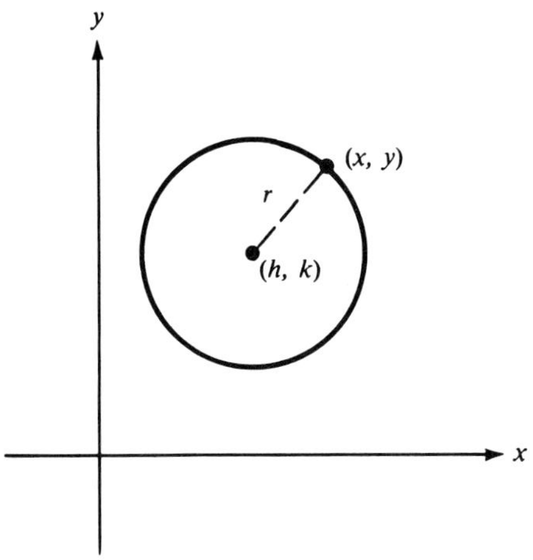

If we square both sides of the equation, we obtain

$$r^2 = (x - h)^2 + (y - k)^2$$

> The equation of a circle with center at (h, k) and radius r is
>
> $$(x - h)^2 + (y - k)^2 = r^2 \qquad \text{Standard form.}$$

Notice that if the center of the circle is at the origin, h and k are zero.

> The equation of a circle with *center* at the *origin* and radius r is
>
> $$x^2 + y^2 = r^2$$

OBJECTIVES

1 *Find an equation for a circle with a given center and radius, and given an equation of a circle, find its center and radius*

2 *Graph ellipses*

☐ **Exercise 1** Find an equation for each circle.

a. Center at the origin and radius $\sqrt{5}$

b. Center at $(-5, -7)$ and radius 6

EXAMPLE 1 Find an equation for each circle.

a. Center at (3, 4) and radius 5

$(x - h)^2 + (y - k)^2 = r^2$ Standard form.

$(x - 3)^2 + (y - 4)^2 = 5^2$ Substitute for h, k, and r.

$(x - 3)^2 + (y - 4)^2 = 25$

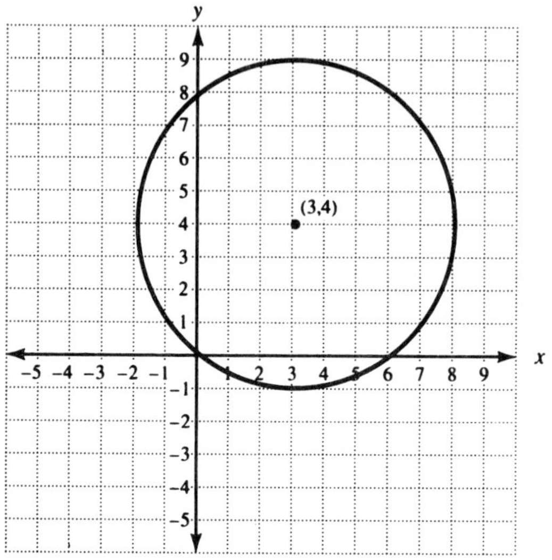

b. Center at $(-2, 3)$ and radius 7

$(x - h)^2 + (y - k)^2 = r^2$ Standard form.

$[x - (-2)]^2 + (y - 3)^2 = 7^2$ Substitute for h, k, and r.

$(x + 2)^2 + (y - 3)^2 = 49$

c. Center at the origin and radius $\sqrt{3}$

$x^2 + y^2 = r^2$ Equation of a circle with center at the origin.

$x^2 + y^2 = (\sqrt{3})^2$ Substitute for r.

$x^2 + y^2 = 3$ ■

☐ **DO EXERCISE 1.**

We may also find the center and radius of a circle if we are given the equation of it.

EXAMPLE 2 Find the center and radius and sketch the graph of the following circles.

a. $x^2 + y^2 = 25$

This is the equation of a circle with center at the origin,

$$x^2 + y^2 = r^2$$

Hence

$$x^2 + y^2 = 25$$

$$x^2 + y^2 = 5^2$$

The center is at (0, 0) and the radius is 5. The graph is shown below.

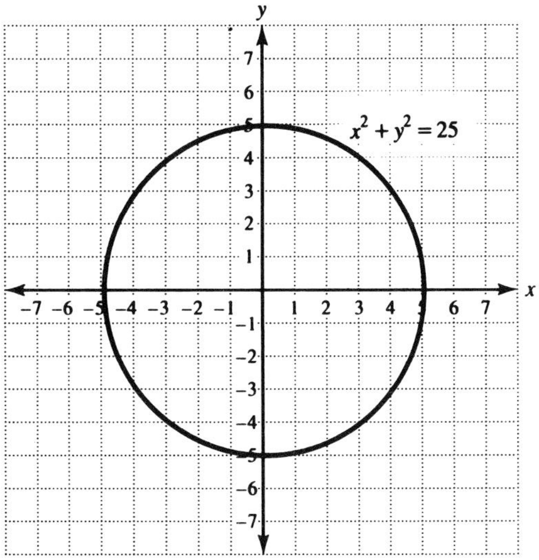

b. $(x - 1)^2 + (y + 2)^2 = 9$

Write the equation in the standard form.

$(x - h)^2 + (y - k)^2 = r^2$ Standard form.

$(x - 1)^2 + [y - (-2)]^2 = 3^2$ Write 2 as $-(-2)$ and 9 as 3^2.

The center is $(1, -2)$ and the radius is 3. The graph is as follows.

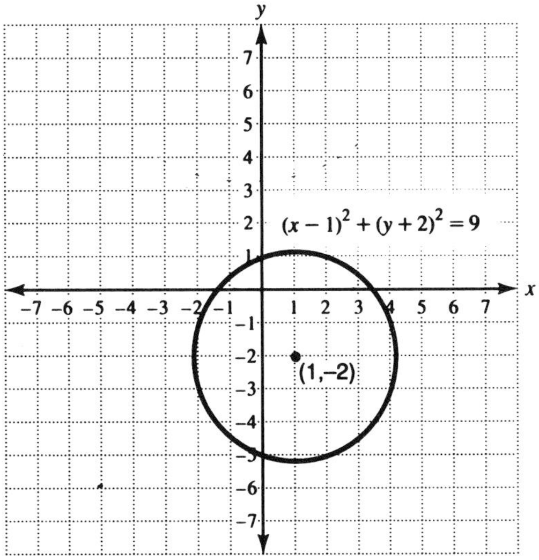

□ **DO EXERCISE 2.**

There are some equations that may not be in the standard form of a circle but they can be written in the standard form by completing the square.

□ **Exercise 2** Find the center and radius and sketch the graph of the following circles.

a. $x^2 + y^2 = 1$

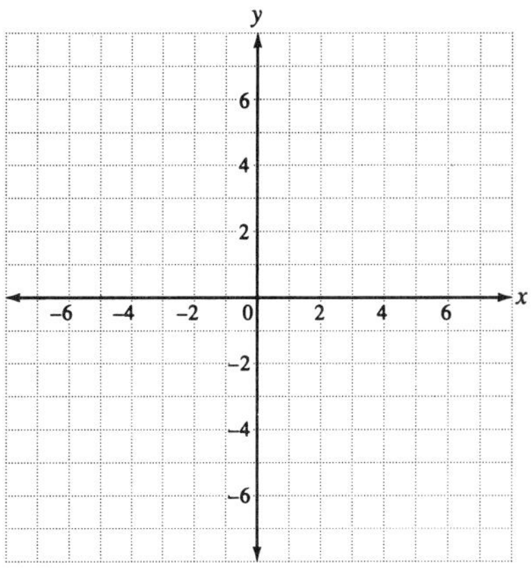

b. $(x + 3)^2 + (y + 2)^2 = 16$

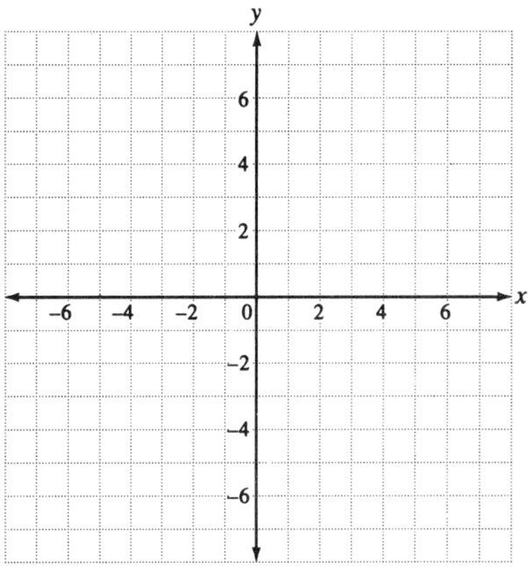

□ **Exercise 3** Find the center and radius of the following circles.

a. $x^2 + y^2 + 6x - 10y = 2$

b. $x^2 + y^2 - 6x - y = -\frac{1}{4}$

EXAMPLE 3 Find the center and radius of the following circle.

$$x^2 + y^2 - 4x + 2y = 11$$

$$(x^2 - 4x) + (y^2 + 2y) = 11 \qquad \text{Group terms.}$$

$$[x^2 - 4x + (-2)^2] + (y^2 + 2y + 1^2) = 11 + (-2)^2 + 1^2 \qquad \text{Complete the square.}$$

$$(x - 2)^2 + (y + 1)^2 = 16$$

$$(x - 2)^2 + [y - (-1)]^2 = 4^2$$

The center of the circle is $(2, -1)$ and the radius is 4. ■

□ **DO EXERCISE 3.**

2 Ellipses

The orbit of the earth around the sun is an ellipse. An ***ellipse*** is the set of all points in a plane such that the sum of the distances from two fixed points F_1 and F_2 is a constant.

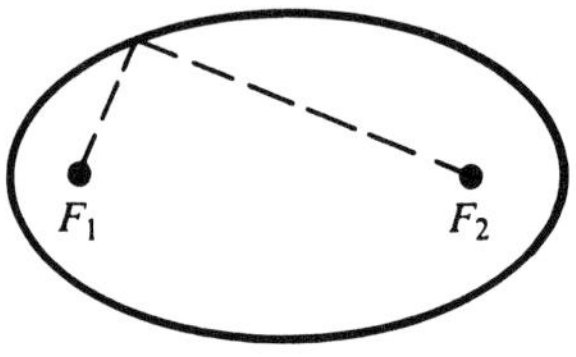

The equation of an ellipse with center at the origin is

$$\frac{x^2}{a^2} + \frac{y^2}{b^2} = 1 \qquad a, b > 0, \quad a \neq b. \qquad \text{Standard form.}$$

Where will the ellipse cross the y-axis? Let $x = 0$ to find the y-intercepts.

$$\frac{x^2}{a^2} + \frac{y^2}{b^2} = 1 \qquad \text{Standard form.}$$

$$\frac{0^2}{a^2} + \frac{y^2}{b^2} = 1 \qquad \text{Substitute 0 for } x.$$

$$\frac{y^2}{b^2} = 1 \qquad \text{Since } \frac{0^2}{a^2} \text{ is zero.}$$

$$y^2 = b^2 \qquad \text{Multiply each side by } b^2.$$

$$y = \pm b$$

The y-intercepts are $(0, b)$ and $(0, -b)$.

What are the x-intercepts? Let $y = 0$ in the standard form to find them.

$$\frac{x^2}{a^2} + \frac{y^2}{b^2} = 1 \qquad \text{Standard form.}$$

$$\frac{x^2}{a^2} + \frac{0^2}{b^2} = 1 \qquad \text{Substitute 0 for } y.$$

$$\frac{x^2}{a^2} = 1$$

$$x^2 = a^2 \qquad \text{Multiply each side by } a^2.$$

$$x = \pm a$$

The x-intercepts are $(a, 0)$ and $(-a, 0)$.

The following ellipse has intercepts as shown.

$$\frac{x^2}{a^2} + \frac{y^2}{b^2} = 1 \qquad \text{Standard form.}$$

The x-intercepts are $(a, 0)$ and $(-a, 0)$. The y-intercepts are $(0, b)$ and $(0, -b)$.

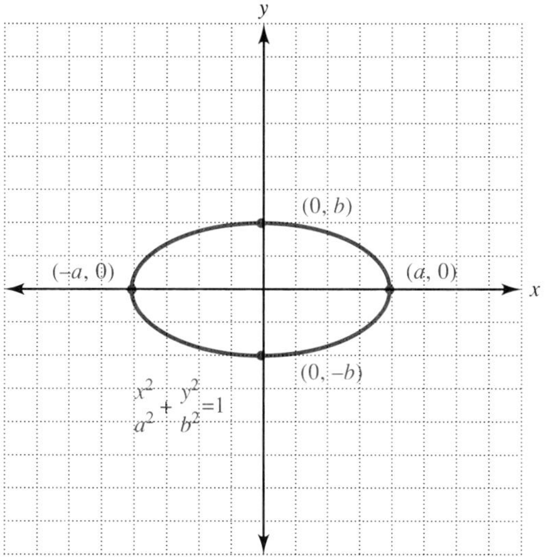

Sketch the graph of the equation.

a. $\dfrac{x^2}{9} + \dfrac{y^2}{16} = 1$

$\dfrac{x^2}{a^2} + \dfrac{y^2}{b^2} = 1$ Standard form.

Therefore,

$a^2 = 9$	$b^2 = 16$
$a = \pm 3$	$b = \pm 4$
The x-intercepts are $(3, 0)$ and $(-3, 0)$.	The y-intercepts are $(0, 4)$ and $(0, -4)$.

Plot the points and sketch an ellipse through them.

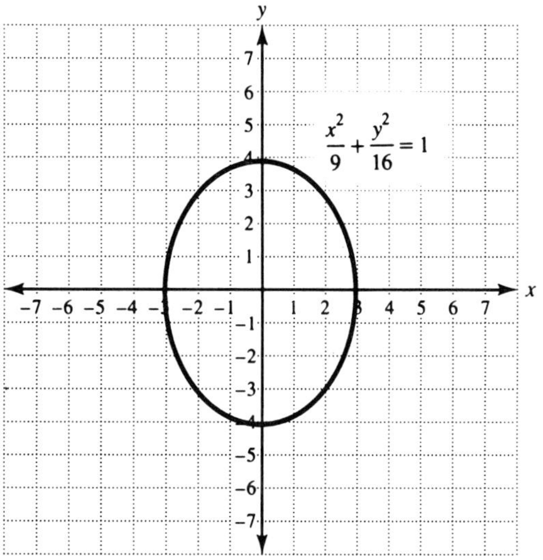

□ **Exercise 4** Sketch the graph of the following.

a. $\frac{x^2}{25} + \frac{y^2}{4} = 1$

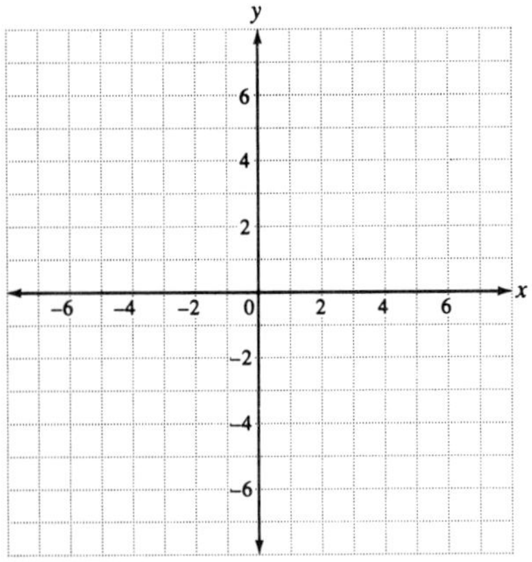

b. $4x^2 + y^2 = 100$

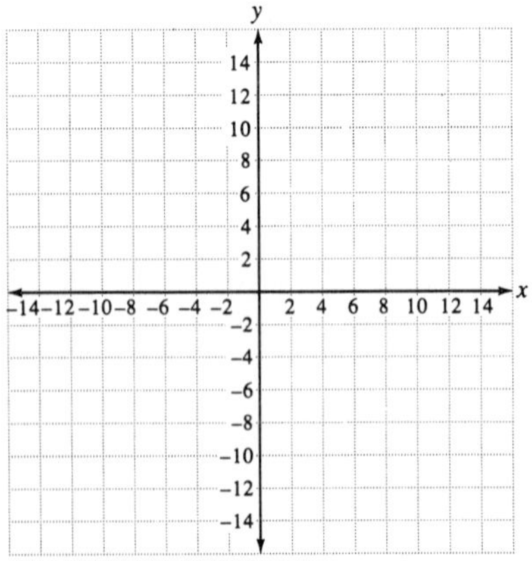

b. $4x^2 + 9y^2 = 36$

If we divide both sides of the equation by 36, the equation will be in standard form and we can find the intercepts.

$$\frac{x^2}{a^2} + \frac{y^2}{b^2} = 1 \qquad \text{Standard form.}$$

$$4x^2 + 9y^2 = 36$$

$$\frac{4x^2 + 9y^2}{36} = \frac{36}{36} \qquad \text{Divide both sides by 36.}$$

$$\frac{4x^2}{36} + \frac{9y^2}{36} = 1$$

$$\frac{x^2}{9} + \frac{y^2}{4} = 1 \qquad \text{Simplify.}$$

$a^2 = 9$ $b^2 = 4$

$a = \pm 3$ $b = \pm 2$

The x-intercepts are $(3, 0)$ and $(-3, 0)$. The y-intercepts are $(0, 2)$ and $(0, -2)$.

Plot the intercepts and sketch an ellipse through them.

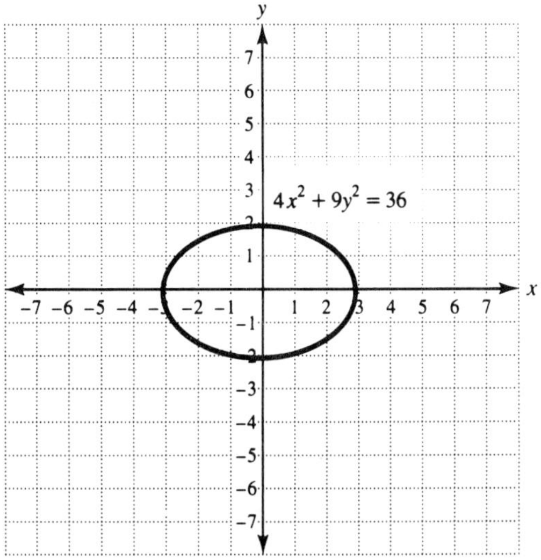

■

□ **DO EXERCISE 4.**

GRAPHING CALCULATOR

Circles and ellipses may also be graphed with the graphing calculator.

a. Graph $x^2 + y^2 = 25$. The calculator requires that the equation be solved for y.

$$y^2 = 25 - x^2$$
$$y = \pm\sqrt{25 - x^2}$$

1. Enter $y = \sqrt{25 - x^2}$ and $y = -\sqrt{25 - x^2}$ into your calculator.
2. Set the graphics window. An appropriate window is W: [−10, 10] [−10, 10].
3. Plot the graph. This is the graph that is shown in Example 2a.

b. Graph $\frac{x^2}{9} + \frac{y^2}{16} = 1$. The calculator requires that the equation be solved for y.

$$144\left(\frac{x^2}{9} + \frac{y^2}{16}\right) = 144(1)$$
$$16x^2 + 9y^2 = 144$$
$$9y^2 = 144 - 16x^2$$
$$y^2 = \frac{144 - 16x^2}{9}$$
$$y = \pm\sqrt{\frac{144 - 16x^2}{9}}$$

1. Enter $y = \sqrt{\frac{144 - 16x^2}{9}}$ and $y = -\sqrt{\frac{144 - 16x^2}{9}}$ into your calculator.
2. Set the graphics window. An appropriate window is W: [−10, 10] [−10, 10].
3. Plot the graph. This is the graph that is shown in Example 4.

Answers to Exercises

1. **a.** $x^2 + y^2 = 5$ **b.** $(x + 5)^2 + (y + 7)^2 = 36$

2. **a.** Center: (0, 0); radius: 1 **b.** Center: $(-3, -2)$; radius: 4

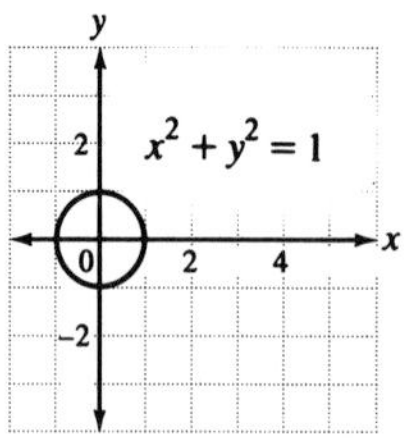

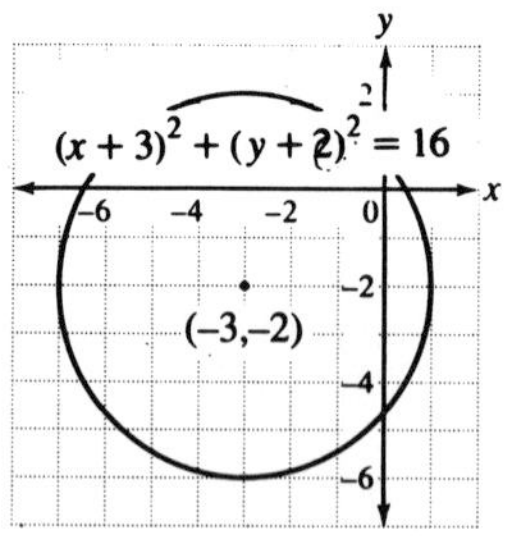

3. **a.** Center: $(-3, 5)$; radius: 6 **b.** Center: $\left(3, \frac{1}{2}\right)$; radius: 3

4. **a.**

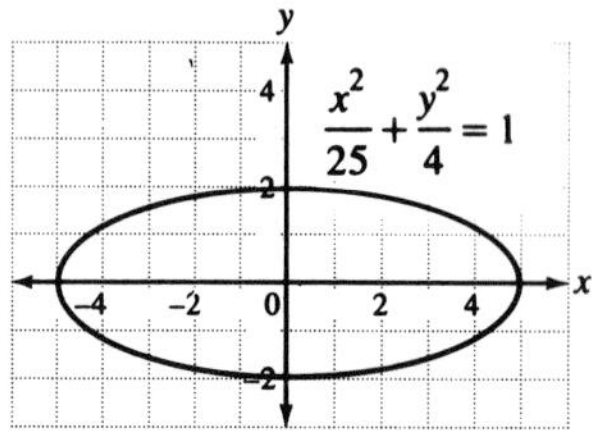

b.

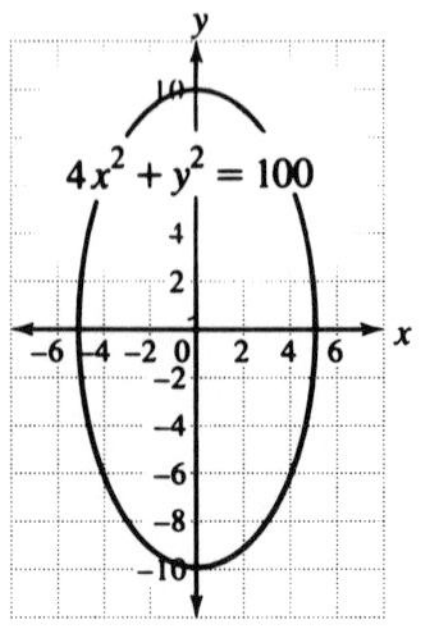

Problem Set 9.2

NAME

DATE

CLASS

Find an equation for each circle.

1. Center $(-3, 2)$, radius 2

2. Center $(-1, 5)$, radius 4

3. Center $(0, 0)$, radius $\sqrt{3}$

4. Center $(0, 0)$, radius $3\sqrt{2}$

5. Center $(-4, -6)$, radius 5

6. Center $(-2, -8)$, radius 7

7. Center $(5, -1)$, radius $\sqrt{6}$

8. Center $(3, -9)$, radius $\sqrt{5}$

Find the center and radius and graph the following circles.

9. $x^2 + y^2 = 25$

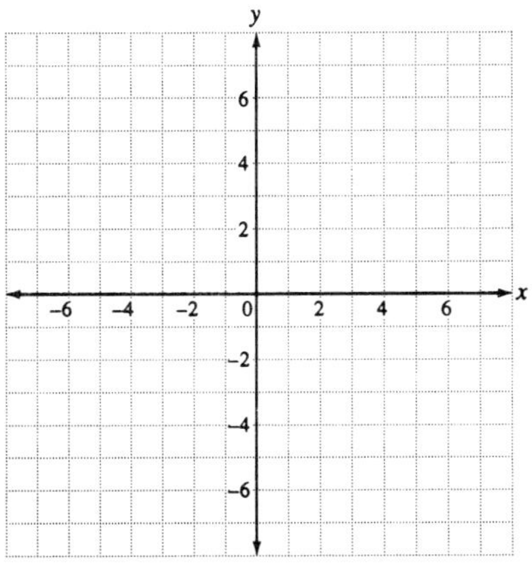

10. $x^2 + y^2 = 9$

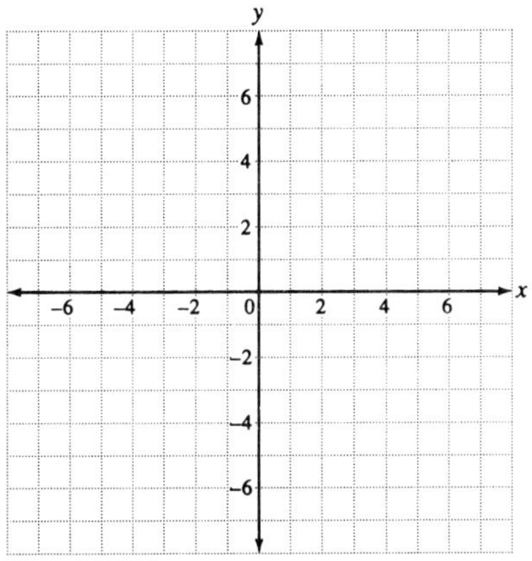

11. $(x - 2)^2 + (y - 3)^2 = 4$

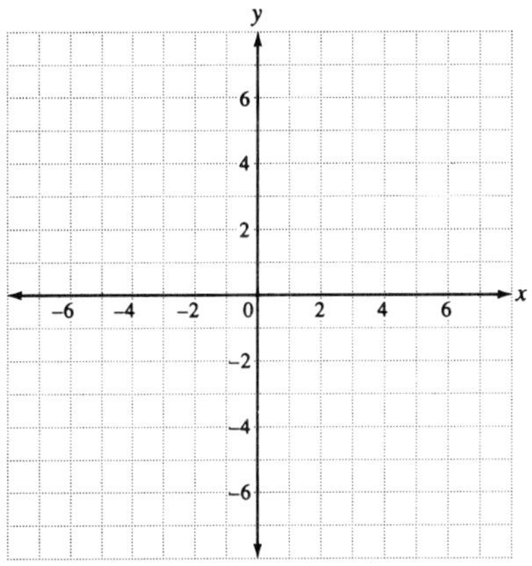

12. $(x - 5)^2 + (y - 1)^2 = 36$

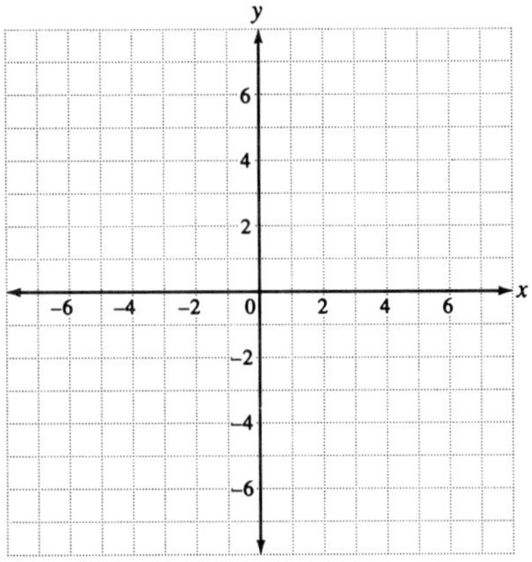

13. $(x + 2)^2 + (y - 4)^2 = 8$

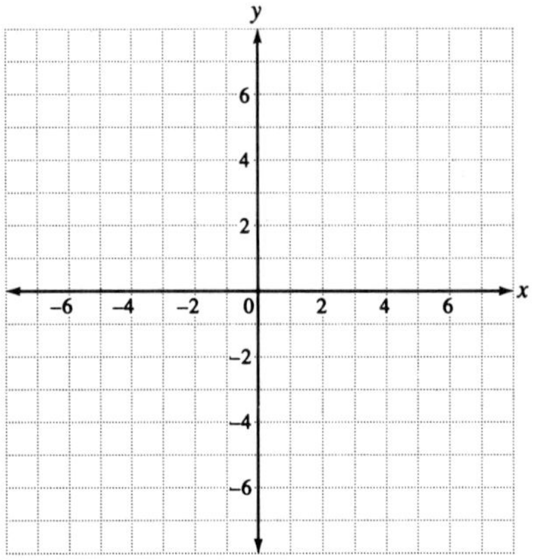

14. $(x - 4)^2 + (y + 2)^2 = 7$

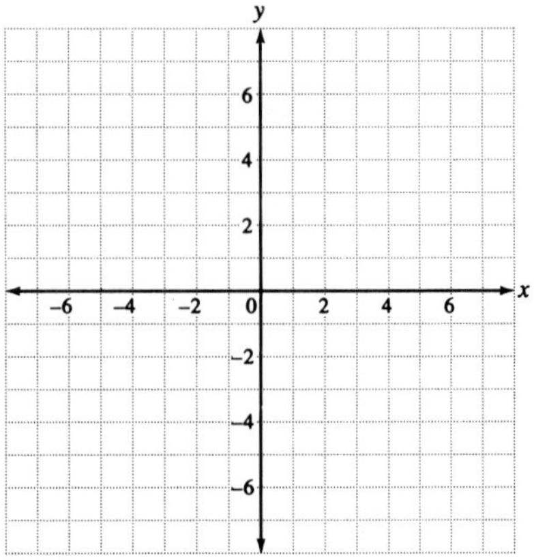

15. $x^2 = 5 - (y + 3)^2$

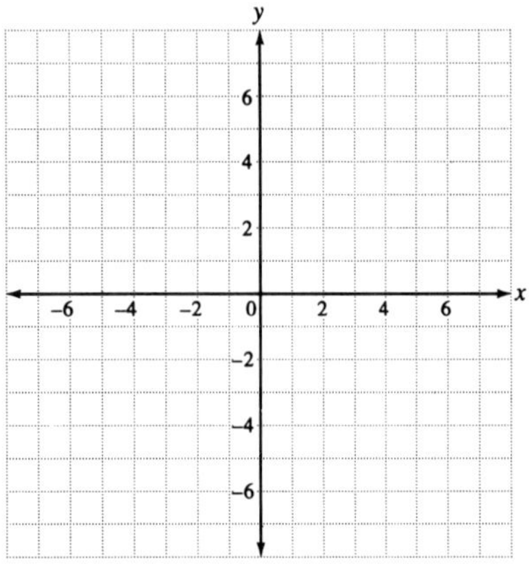

16. $y^2 = 16 - (x - 1)^2$

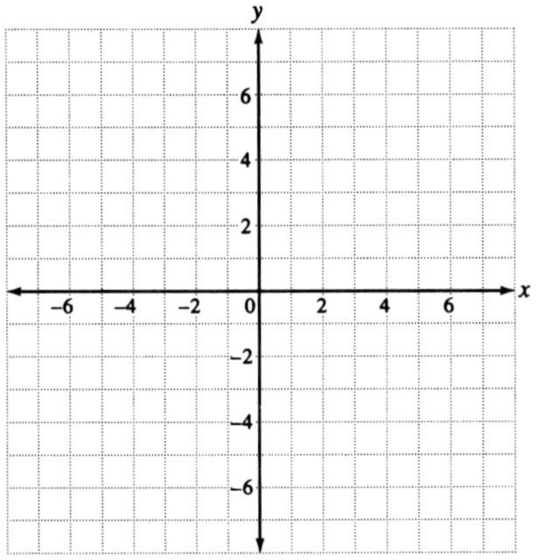

Find the center and radius of the following circles.

17. $x^2 + y^2 + 4x + 12y = -36$

18. $x^2 + y^2 + 2x + 6y = 12$

19. $x^2 + y^2 - 8x - 2y = 7$

20. $x^2 + y^2 - 10x + 4y = -9$

21. $x^2 + y^2 - 8x - 5 = 0$

22. $x^2 + y^2 + 6y - 9 = 0$

23. $x^2 + y^2 + 4x - 8y + 11 = 0$

24. $x^2 + y^2 + 8x + 10y + 5 = 0$

Graph the following.

25. $\frac{x^2}{4} + \frac{y^2}{16} = 1$

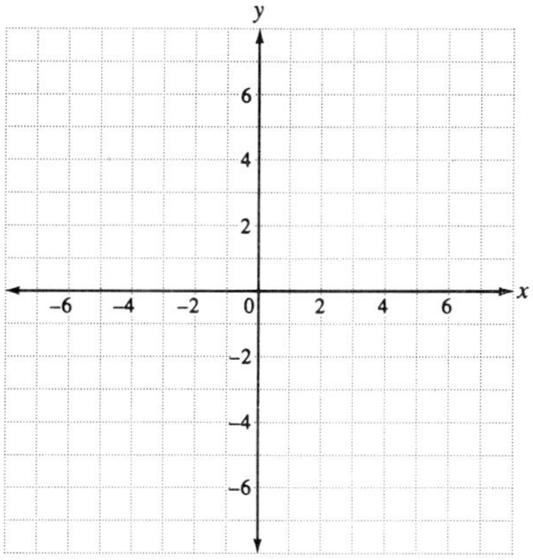

26. $\frac{x^2}{9} + \frac{y^2}{25} = 1$

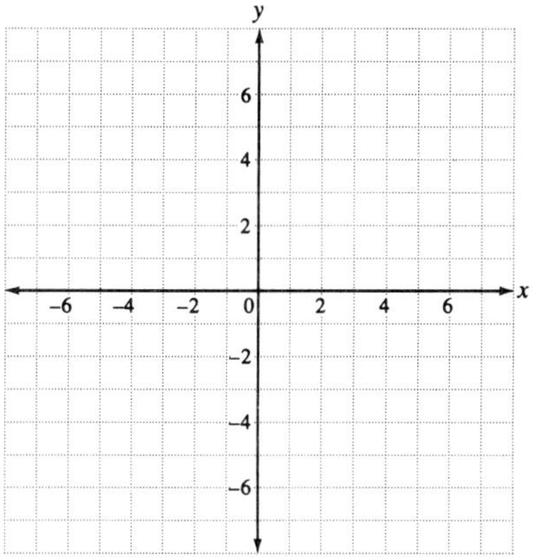

27. $\frac{x^2}{36} + \frac{y^2}{1} = 1$

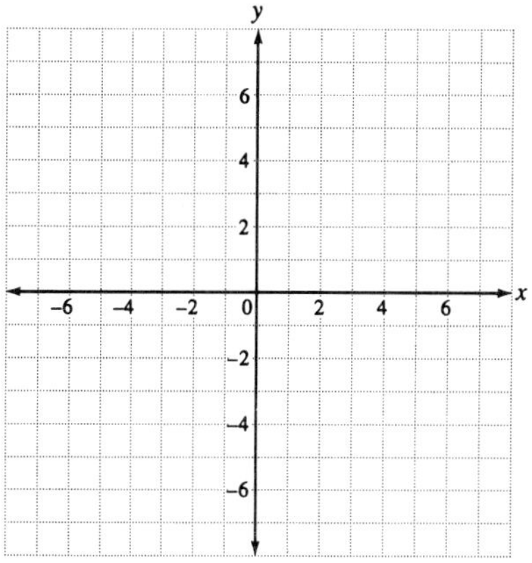

28. $\frac{x^2}{9} + \frac{y^2}{1} = 1$

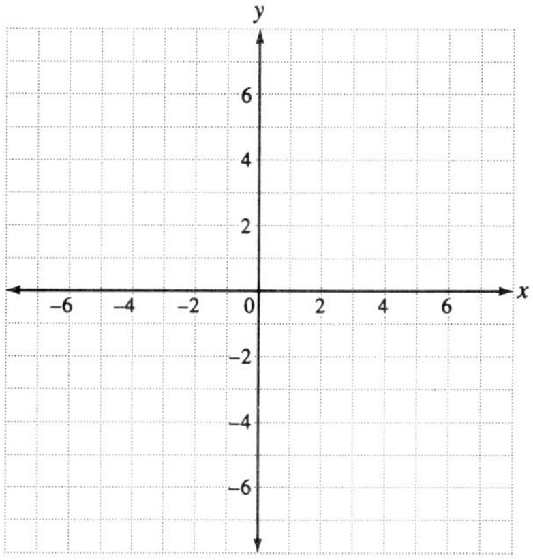

29. $25x^2 + y^2 = 25$

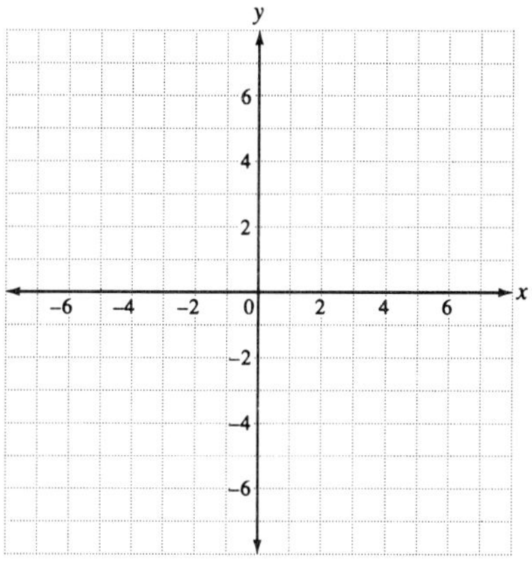

30. $x^2 + 36y^2 = 36$

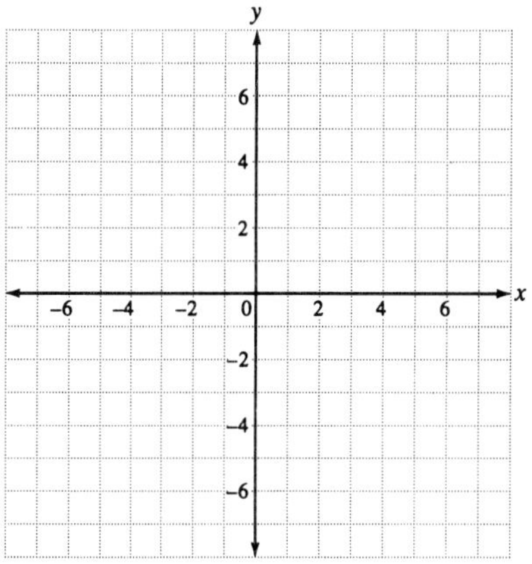

31. $\frac{x^2}{4} + \frac{y^2}{9} = 1$

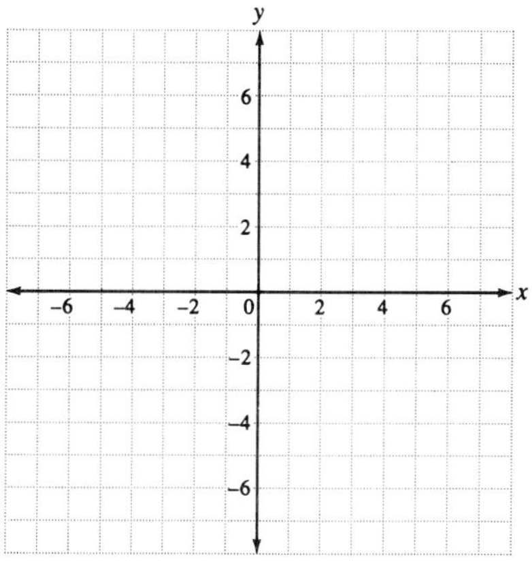

32. $\frac{x^2}{16} + \frac{y^2}{25} = 1$

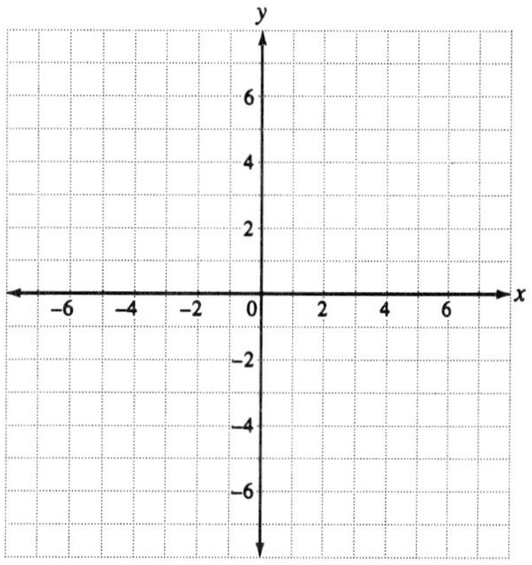

33. $x^2 + 4y^2 = 36$

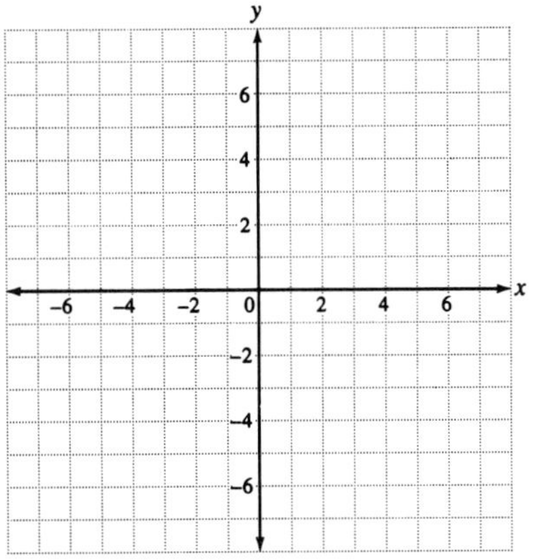

34. $x^2 + 4y^2 = 16$

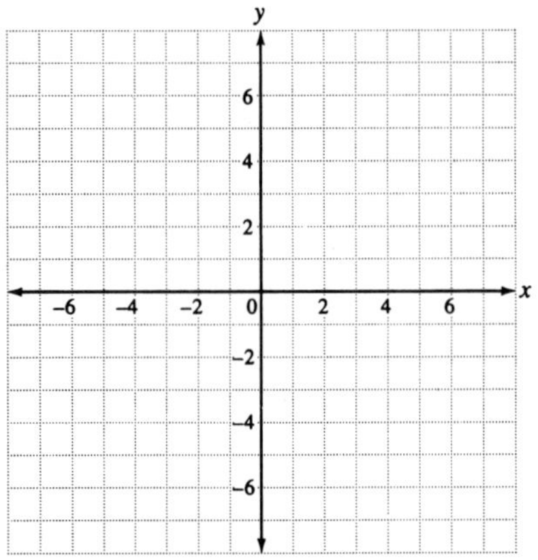

35. $\dfrac{x^2}{9/4} + \dfrac{y^2}{25/16} = 1$

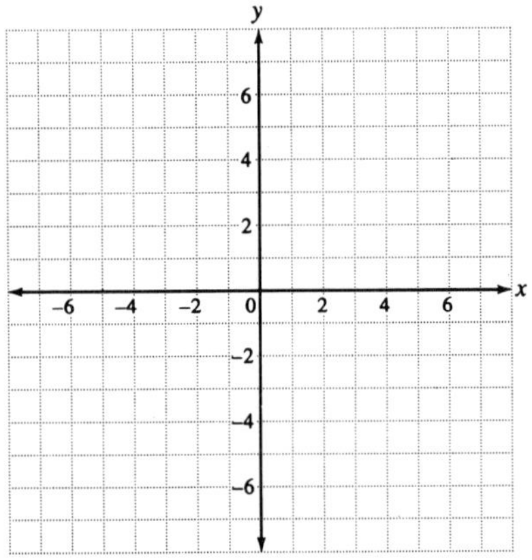

36. $\dfrac{x^2}{64/9} + \dfrac{y^2}{36/49} = 1$

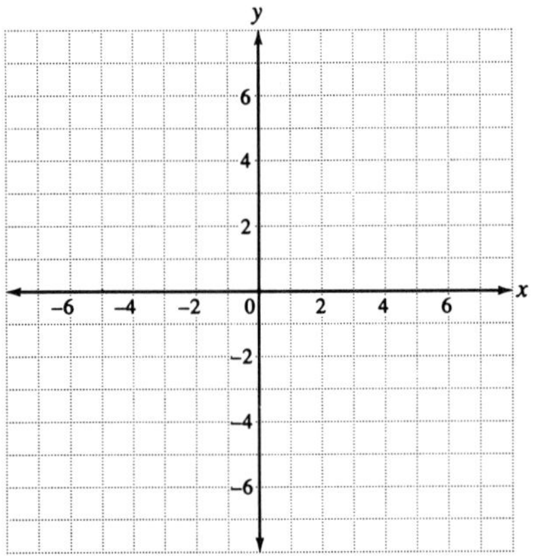

37. The equation $x^2 + y^2 = r^2$ is the equation of a circle with center at the ____________ and radius r.

38. Some equations that are not in the standard form of a circle may be written in standard form by ____________ ____________.

39. An ellipse is the set of all points in a plane such that the sum of the distances from two fixed points is a ____________.

40. The graph of $x^2/a^2 + y^2/b^2 = 1$ crosses the ____________ at $(a, 0)$ and $(-a, 0)$.

Think About It

Find equations of the following circles.

* **41.** Center (0, 0); passing through the point $\left(\dfrac{\sqrt{35}}{2}, \dfrac{1}{2}\right)$

* **42.** Center (2, 3); passing through the point $(-3, -1)$

Find the center and radius of the following circles.

* **43.** $2x^2 + 2y^2 + 3x + 5y = -4$

* **44.** $5x^2 + 5y^2 - 4x + 2y = 3$

* **45.** $3x^2 + 3y^2 - 3x - 2y = -\dfrac{7}{12}$

* **46.** A sign is made in the form of an ellipse whose equation is $64x^2 + 49y^2 = 3136$. How long (in feet) is the sign?

See the Additional Exercises to learn how to graph an ellipse with center at (h, k).

Checkup

The following problems provide a review of some of Sections 8.7 and 8.8 and will help you with the next section.

Graph.

47. $y = x^2 - 5x + 6$

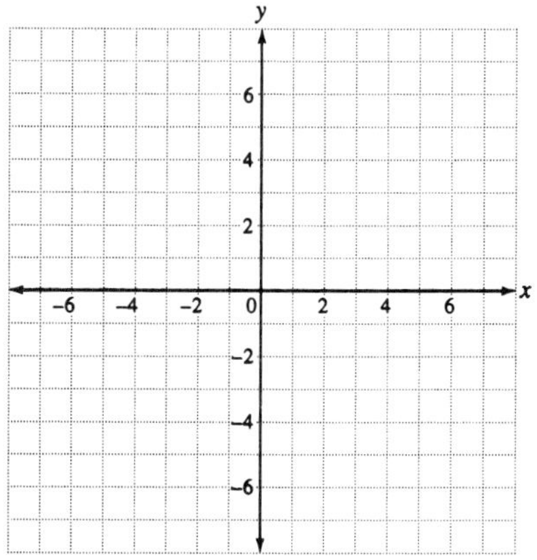

48. $y = x^2 + 5x + 4$

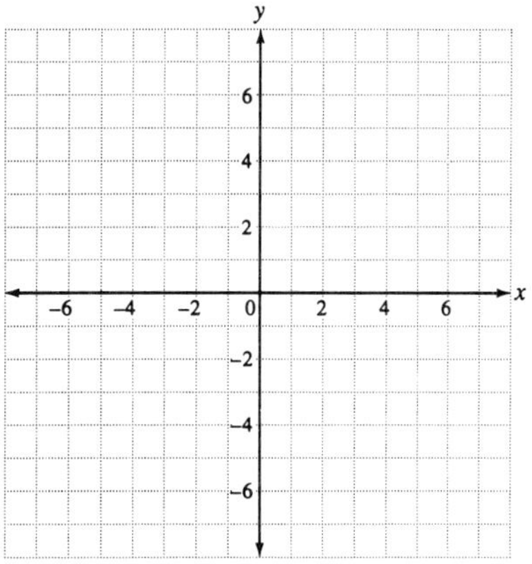

49. $y = x^2 - 6x + 8$

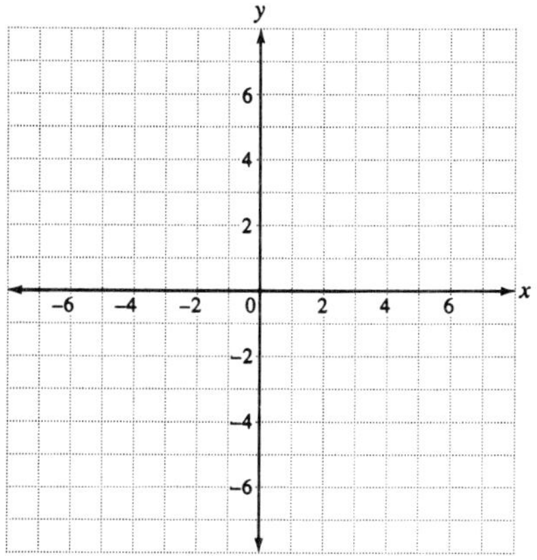

50. $y = x^2 + 6x + 8$

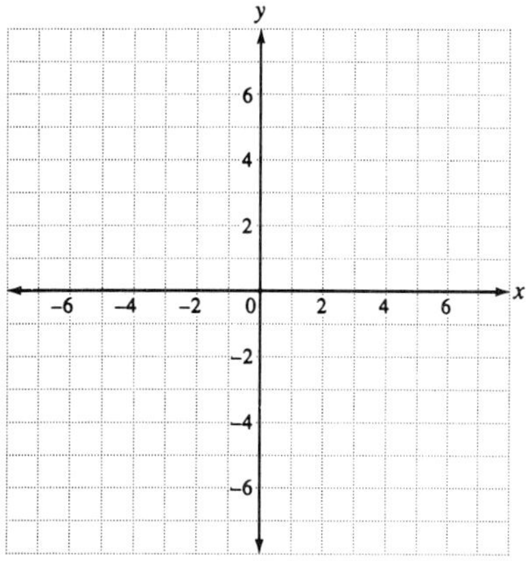

51. $y = -x^2 - 4x - 3$

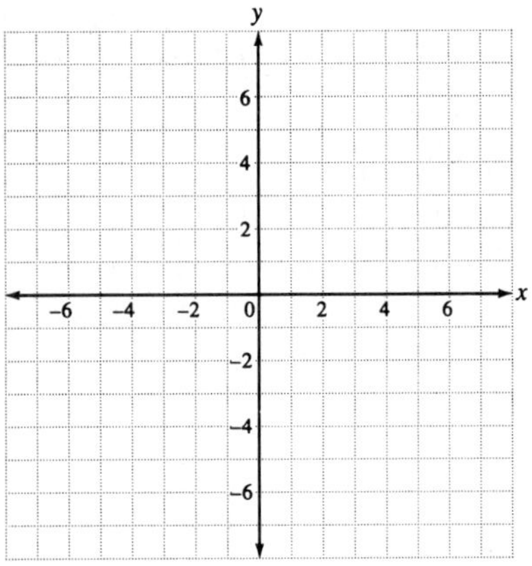

52. $y = -x^2 + 4x + 5$

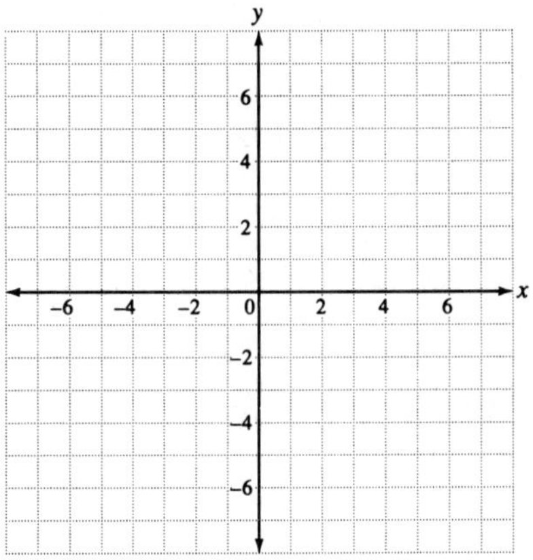

53. $x = y^2 - 4y - 5$

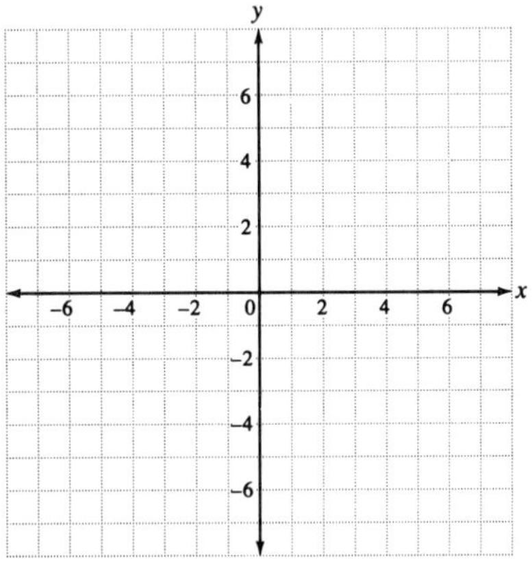

54. $x = -y^2 - y + 2$

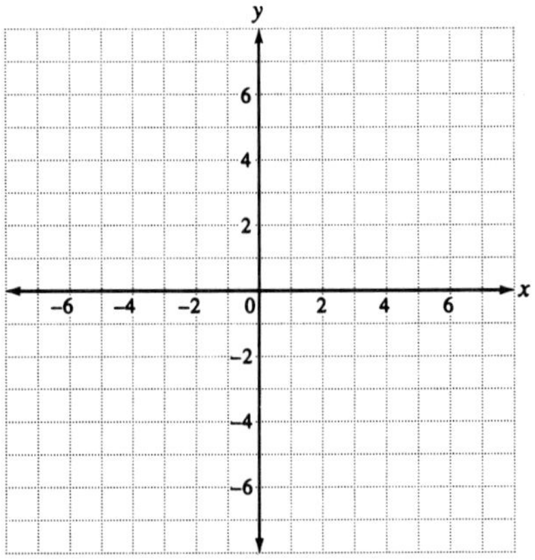

55. $y = x^2 - 3$

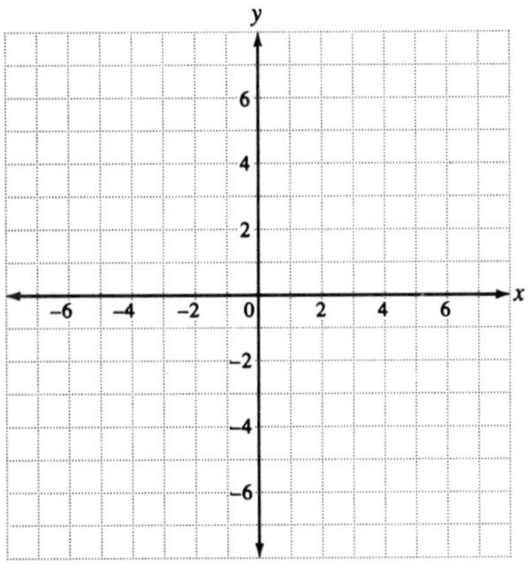

56. $y = -x^2 + 4$

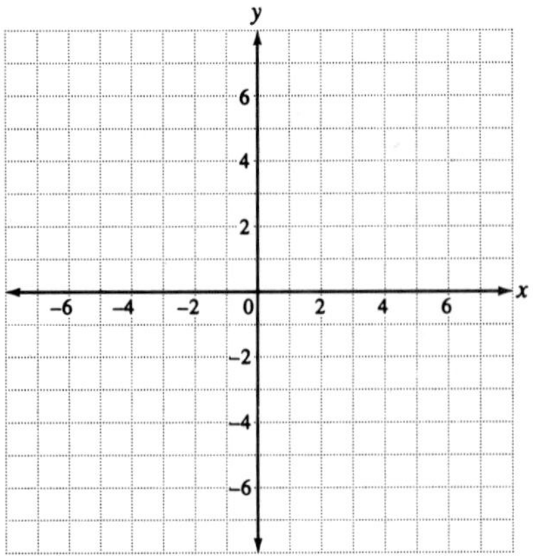

9.3 PARABOLAS AND HYPERBOLAS

1 Parabolas

There are many applications of ***parabolas.*** If we disregard air resistance, objects thrown into the air travel in parabolic paths. Cross sections of radar dishes and automobile headlights are also parabolas.

We will consider the graphs of $y = x^2$ and $y = 2(x - 1)^2 - 3$. We make a table of values for each equation.

$y = x^2$

x	y
−2	4
−1	1
0	0
1	1
2	4

$y = 2(x - 1)^2 - 3$

x	y
−2	15
−1	5
0	−1
1	−3
2	−5
3	5

The graphs are as follows:

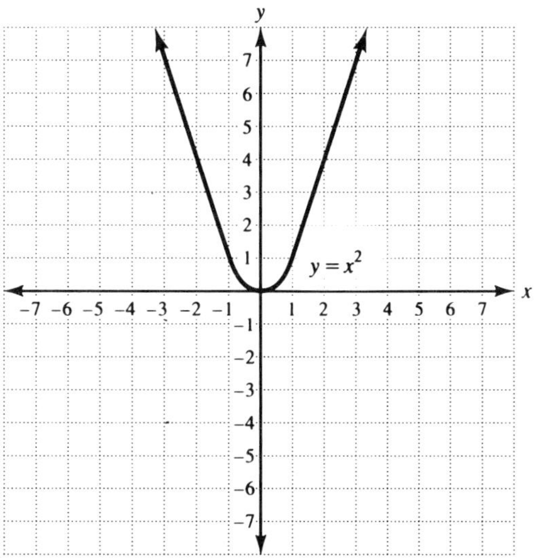

OBJECTIVES

1 *Graph parabolas*

2 *Graph hyperbolas*

3 *Identify conic sections from equations*

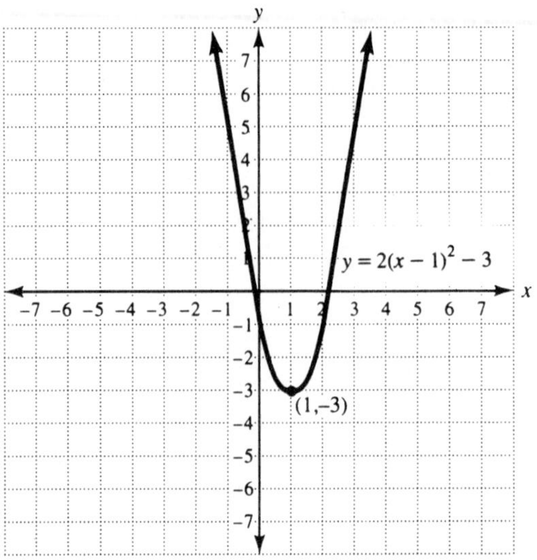

Notice that the graph of $y = 2(x - 1)^2 - 3$ is like the graph of $y = x^2$ except that it is thinner and shifted 1 unit to the right and 3 units down. The vertex of $y = 2(x - 1)^2 - 3$ is not at (0, 0). It is at (1, −3). The equation $y = 2(x - 1)^2 - 3$ is in the form $y = a(x - h)^2 + k$, and as we shall see, we can find the vertex from the equation. Recall from Section 8.7 that since $a > 0$, the parabola opens upward.

We can get a similar result for parabolas of the form $x = a(y - k)^2 + h$. If $a > 0$, the parabola opens to the right. If $a < 0$, the parabola opens to the left. When equations of parabolas are written in the following form, it is easy to find the vertex.

Graphs of the following equations are parabolas.

$y = a(x - h)^2 + k$,	vertex is at (h, k),	$a > 0$ opens upward, $a < 0$ opens downward
$x = a(y - k)^2 + h$,	vertex is at (h, k),	$a > 0$ opens to the right $a < 0$ opens to the left

EXAMPLE 1 Sketch the graph of the following.

a. $y = -3(x - 1)^2 - 2$

$y = a(x - h)^2 + k$ General equation, vertex (h, k).

The vertex is at $(1, -2)$. The graph opens downward since $a < 0$. Plot a few ordered pairs as needed to complete the graph.

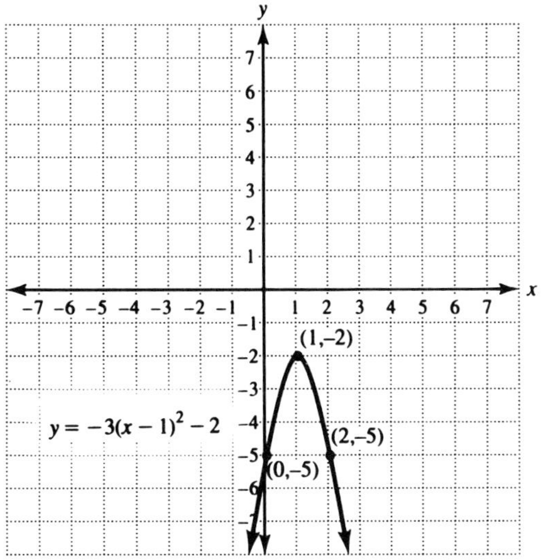

b. $x = (y + 2)^2 - 4$

$x = a(y - k)^2 + h$ General equation, vertex (h, k).

$x = [y - (-2)]^2 - 4$ Write 2 as $-(-2)$.

The vertex is at $(-4, -2)$. The parabola opens to the right since a is 1. It also passes through the points $(0, 0)$ and $(0, -4)$.

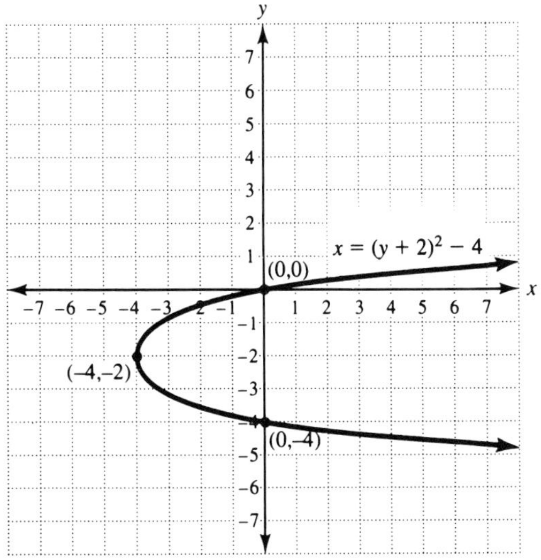

☐ **DO EXERCISE 1.**

☐ **Exercise 1** Sketch the graph.

a. $y = (x + 4)^2 + 3$

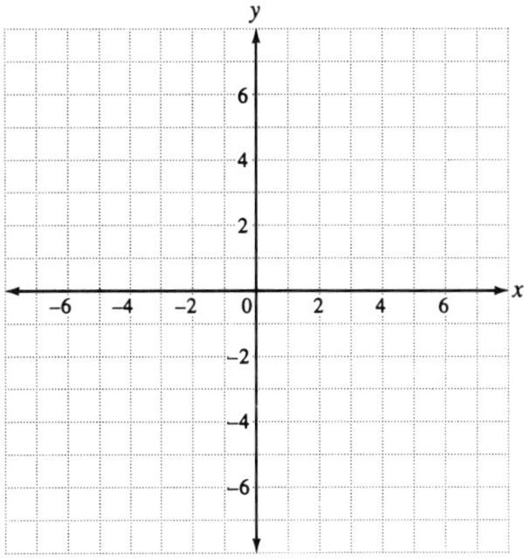

b. $x = -2(y - 3)^2 - 1$

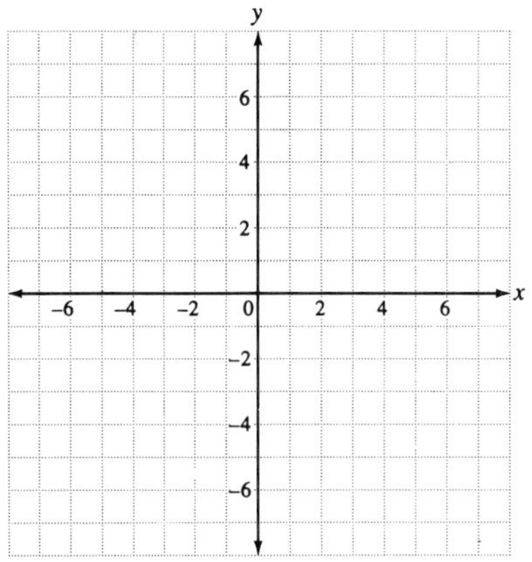

2 Hyperbolas

Some comets that come in contact with Earth's gravitational field travel in hyperbolic paths.

A ***hyperbola*** is the set of all points in a plane such that the absolute value of the difference of the distance of these points from two fixed points F_1 and F_2 is constant.

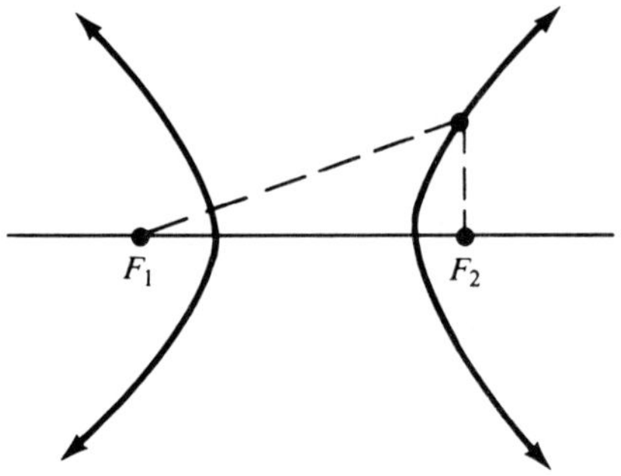

It can be shown using this definition and the distance formula that hyperbolas with centers at the origin have equations of the following standard forms.

> Graphs of the following equations are hyperbolas with centers at the origin.
>
> $\frac{x^2}{a^2} - \frac{y^2}{b^2} = 1$ The x-intercepts are $(a, 0)$ and $(-a, 0)$. There are no y-intercepts.
>
> $\frac{y^2}{b^2} - \frac{x^2}{a^2} = 1$ The y-intercepts are $(0, b)$ and $(0, -b)$. There are no x-intercepts.

The graph of the first equation has no y-intercepts because if we let $x = 0$,

$$\frac{0^2}{a^2} - \frac{y^2}{b^2} = 1$$

$$-y^2 = b^2$$

$$y^2 = -b^2$$

This equation has no real number solutions. We can, however, use b^2 to help us draw the graph. The four points (a, b), $(a, -b)$, $(-a, b)$, and $(-a, -b)$ may be used to sketch a rectangle. The lines through the opposite corners of the rectangle are called ***asymptotes.*** The graph will approach these lines.

EXAMPLE 2 Sketch the graph.

a. $\frac{x^2}{9} - \frac{y^2}{4} = 1$

$$\frac{x^2}{a^2} - \frac{y^2}{b^2} = 1 \qquad \text{General equation.}$$

$$a^2 = 9 \qquad \text{and} \qquad b^2 = 4$$

$$a = \pm 3 \qquad\qquad b = \pm 2$$

The x-intercepts are $(3, 0)$ and $(-3, 0)$. Plot these points. Plot the points $(3, 2)$, $(3, -2)$, $(-3, 2)$, and $(-3, -2)$. Sketch a rectangle through them.

Then draw the lines through the corners of the rectangle. Sketch the graph through the intercepts and approaching the asymptotes.

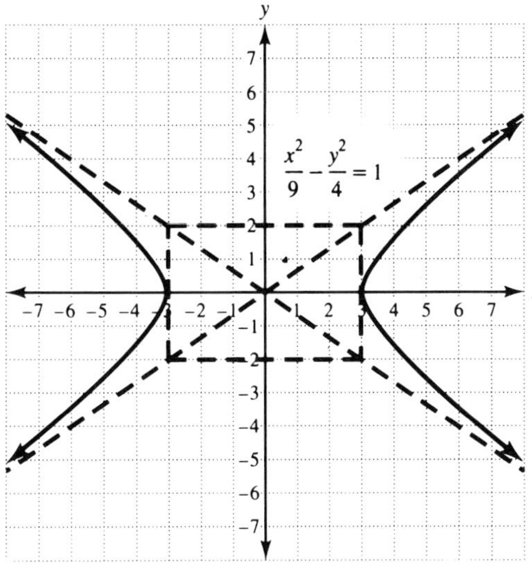

b. $\frac{y^2}{16} - \frac{x^2}{25} = 1$

$$\frac{y^2}{b^2} - \frac{x^2}{a^2} = 1$$

$$b^2 = 16 \qquad \text{and} \qquad a^2 = 25$$

$$b = \pm 4 \qquad\qquad a = \pm 5$$

The y-intercepts are $(0, 4)$ and $(0, -4)$. Plot these points. The equation has no x-intercepts. The corners of the rectangle are $(5, 4)$, $(5, -4)$, $(-5, 4)$, and $(-5, -4)$. Plot these points and sketch a rectangle through them. Then draw the lines through the corners of the rectangle to form the asymptotes. Sketch the curve through the intercepts and approaching the asymptotes.

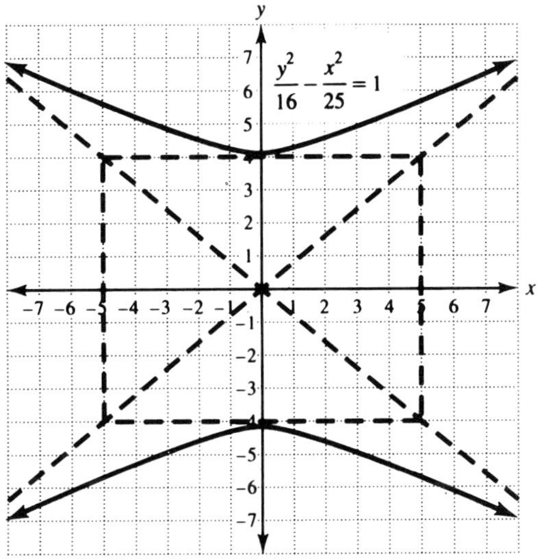

■

□ **DO EXERCISE 2.**

□ **Exercise 2** Sketch the graph.

a. $\frac{x^2}{9} - \frac{y^2}{16} = 1$

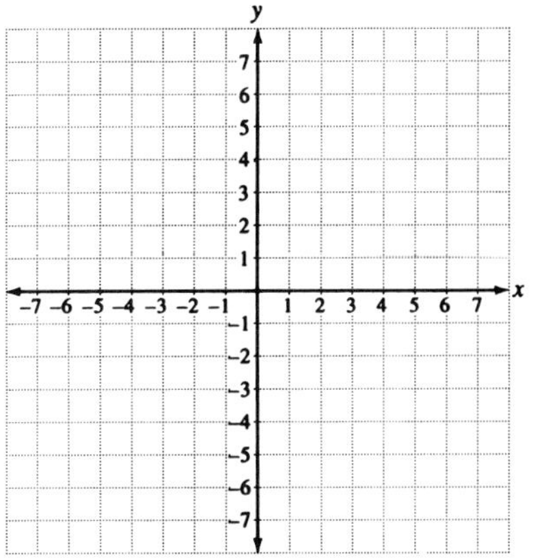

b. $\frac{y^2}{4} - \frac{x^2}{36} = 1$

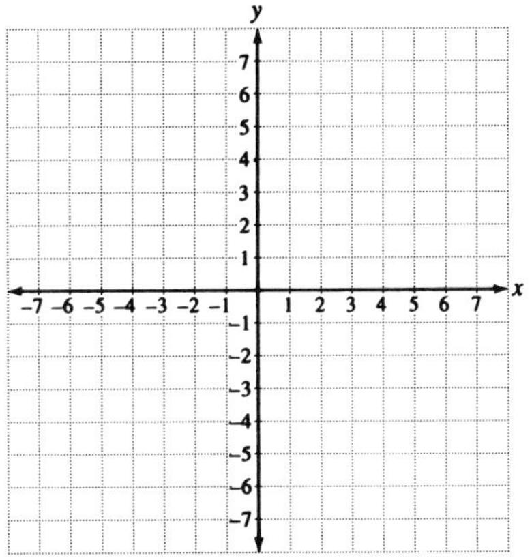

□ **Exercise 3** Identify the conic section.

a. $x^2 - 9y^2 = 36$

b. $4y^2 - x = 0$

c. $5x^2 = 20 - 4y^2$

d. $(x - 3)^2 = 49 - y^2$

3 Identifying Conic Sections

Circles, ellipses, parabolas, and hyperbolas are called ***conic sections*** because they can be formed by slicing a cone with a plane as shown.

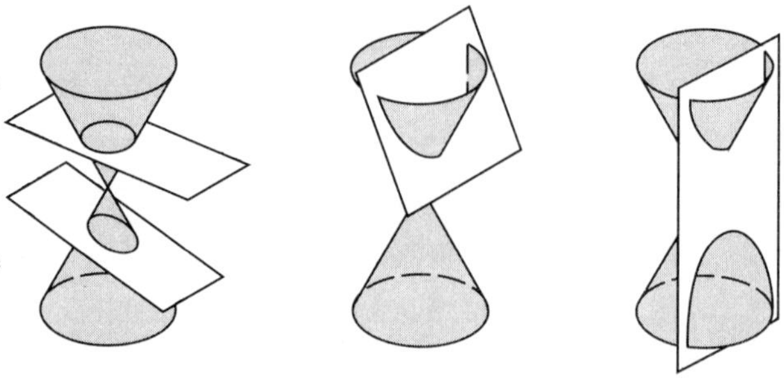

Equation	Conic Section	
$(x - h)^2 + (y - k)^2 = r^2$	Circle	Center at $(h, k,)$ radius r
$\frac{x^2}{a^2} + \frac{y^2}{b^2} = 1$	Ellipse	x-intercepts at $(a, 0)$ and $(-a, 0)$; y-intercepts at $(0, b)$ and $(0, -b)$
$y = a(x - h)^2 + k$	Parabola	Vertex at (h, k); if $a > 0$, opens upward; if $a < 0$, opens downward
$x = a(y - k)^2 + h$	Parabola	Vertex at (h, k); if $a > 0$, opens right; if $a < 0$, opens left
$\frac{x^2}{a^2} - \frac{y^2}{b^2} = 1$	Hyperbola	x-intercepts are $(a, 0)$ and $(-a, 0)$
$\frac{y^2}{b^2} - \frac{x^2}{a^2} = 1$	Hyperbola	y-intercepts are $(0, b)$ and $(0, -b)$

We should be able to determine when the graph of a second-degree equation is a circle, ellipse, parabola, or hyperbola. Sometimes we must write an equivalent equation to recognize it.

EXAMPLE 3 Identify the conic section.

$4x^2 = 16 - 4y^2$

$4x^2 + 4y^2 = 16$ Add $4y^2$ to each side.

$x^2 + y^2 = 4$ Divide both sides by 4.

This is the equation of a circle with center at the origin and radius 2. ■

□ **DO EXERCISE 3.**

GRAPHING CALCULATOR

From Section 8.7, we know that equations of parabolas that open upward or downward may be graphed with the graphing calculator. The calculator may also be used to graph hyperbolas.

Graph $\frac{x^2}{9} - \frac{y^2}{4} = 1$. The calculator requires that the equation be solved for y.

$$36\left(\frac{x^2}{9} - \frac{y^2}{4}\right) = 36(1)$$

$$4x^2 - 9y^2 = 36$$

$$-9y^2 = -4x^2 + 36$$

$$y^2 = \frac{4x^2 - 36}{9}$$

$$y = \pm\sqrt{\frac{4x^2 - 36}{9}}$$

1. Enter $y = \sqrt{\frac{4x^2 - 36}{9}}$ and $y = -\sqrt{\frac{4x^2 - 36}{9}}$ into your calculator.
2. Set the graphics window. An appropriate window is W: [−10, 10] [−10, 10].
3. Plot the graph. This is the graph that is shown in Example 2a.

DID YOU KNOW?

Euclid, famous for his book, *Elements of Geometry,* lectured at the University of Alexandria about 300 B.C. One of Euclid's students complained that there was no reason to learn mathematics. Euclid responded by having one of his slaves give the student a penny, so that he would gain from what he learned!

Answers to Exercises

1. a.

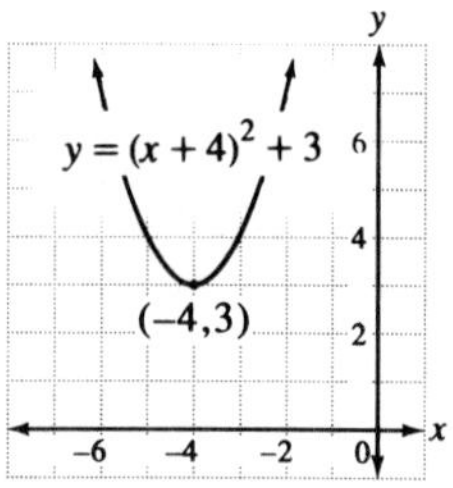

b.

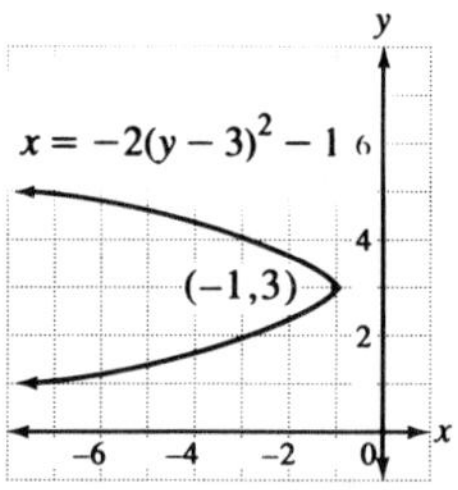

2. a.

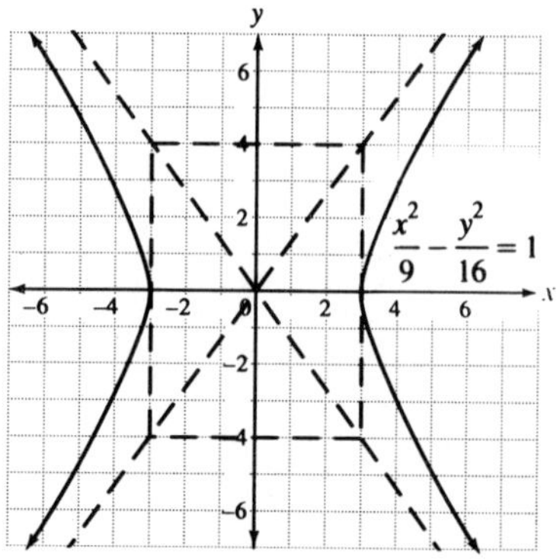

b.

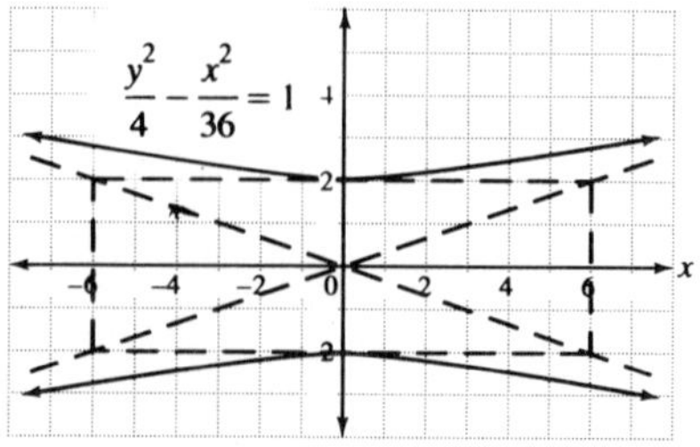

3. a. Hyperbola **b.** Parabola **c.** Ellipse **d.** Circle

Problem Set 9.3

NAME

DATE

CLASS

Graph.

1. $y = -x^2 + 2$

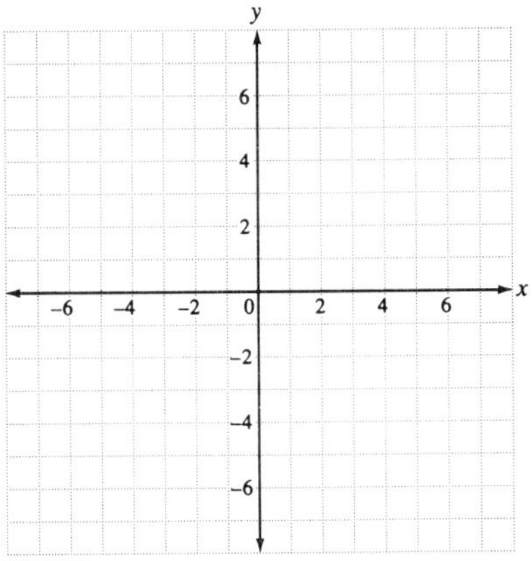

2. $x = y^2 - 4$

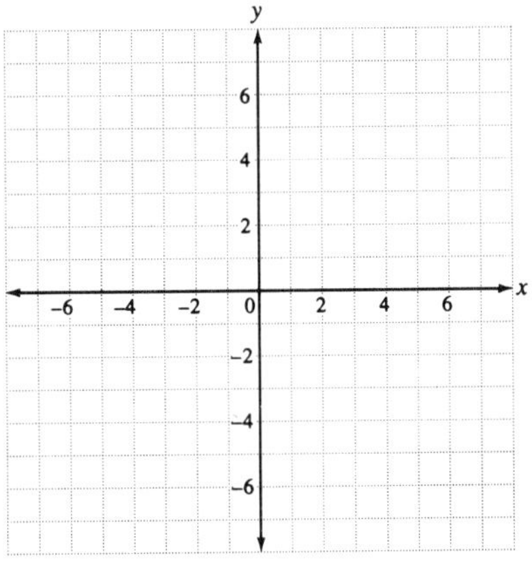

3. $x = (y - 3)^2$

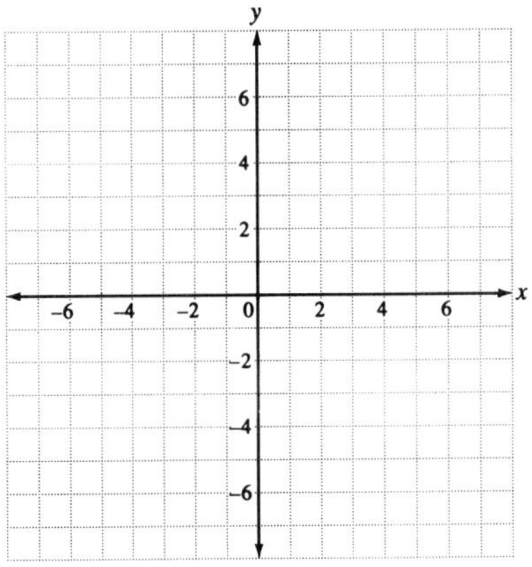

4. $y = (x - 2)^2$

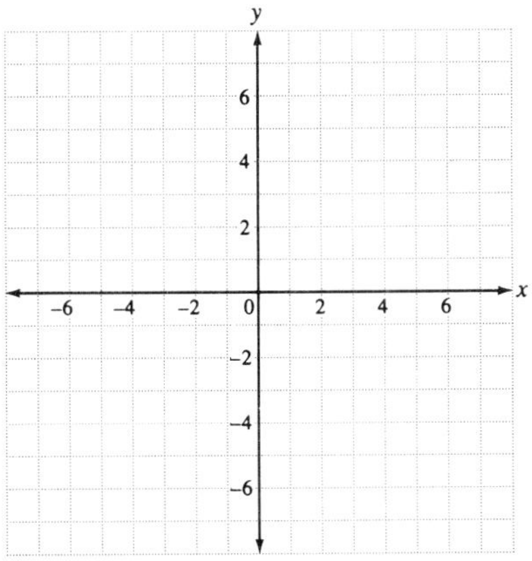

5. $y = (x - 3)^2 + 4$

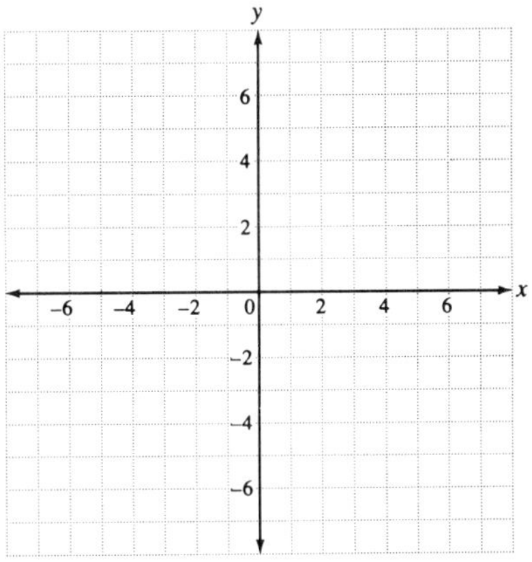

6. $x = (y - 1)^2 - 3$

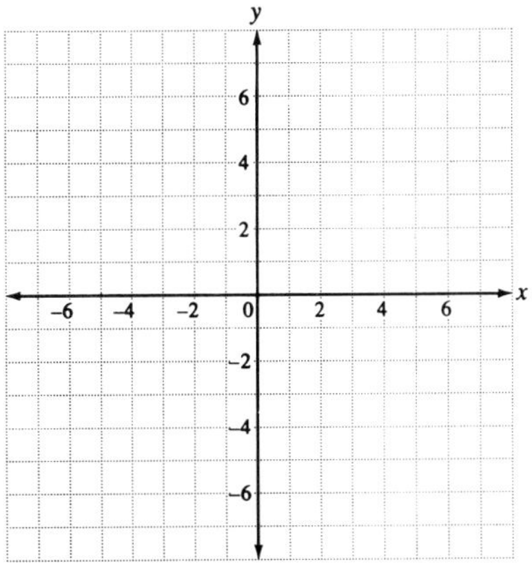

7. $x = 3(y + 1)^2 - 2$

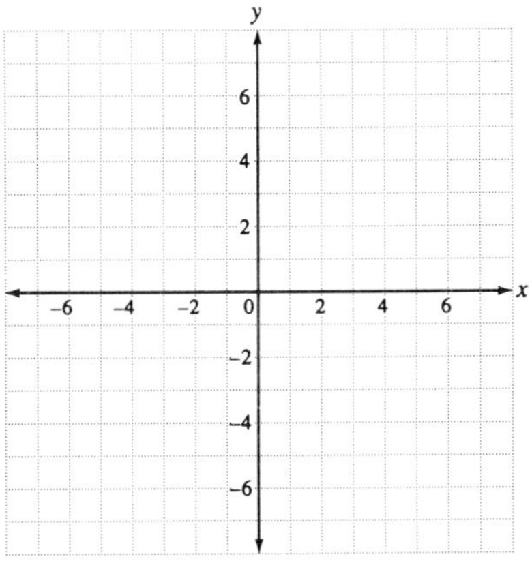

8. $y = -2(x + 3)^2 + 1$

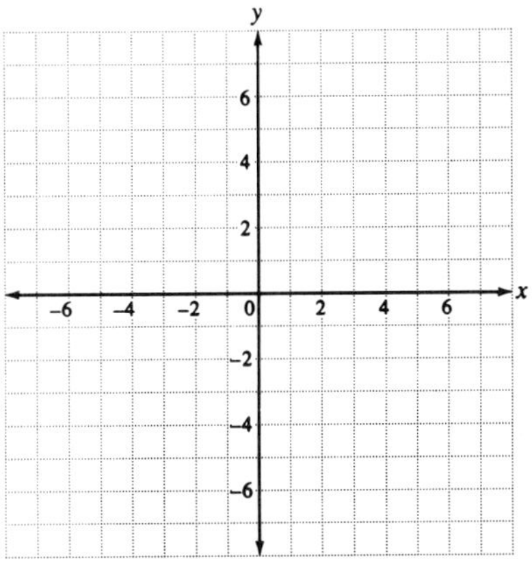

9. $y = -4(x + 3)^2 + 5$

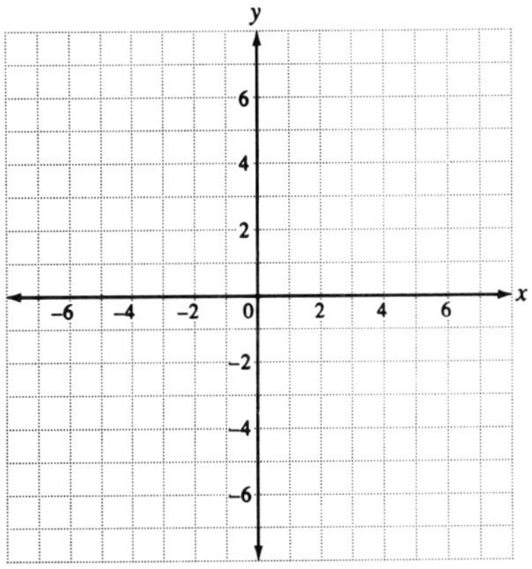

10. $x = 3(y + 1)^2 - 3$

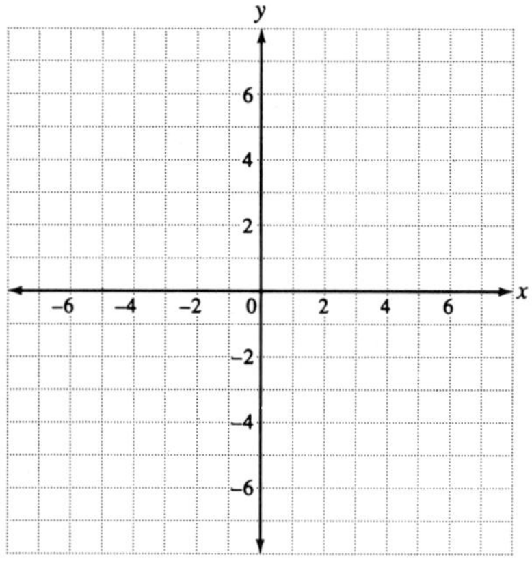

11. $y = 2(x + 1)^2$

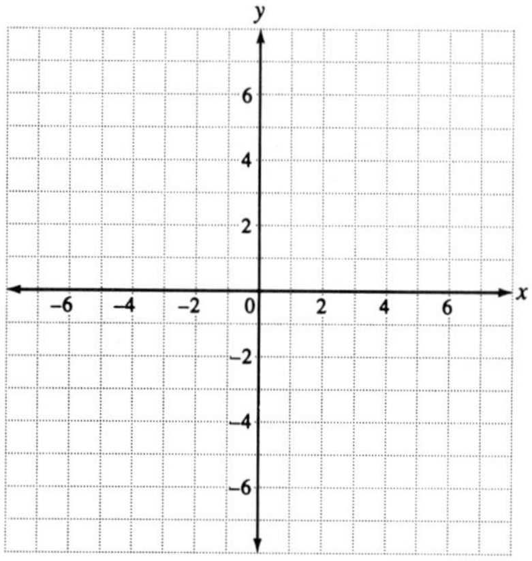

12. $y = -3(x + 1)^2$

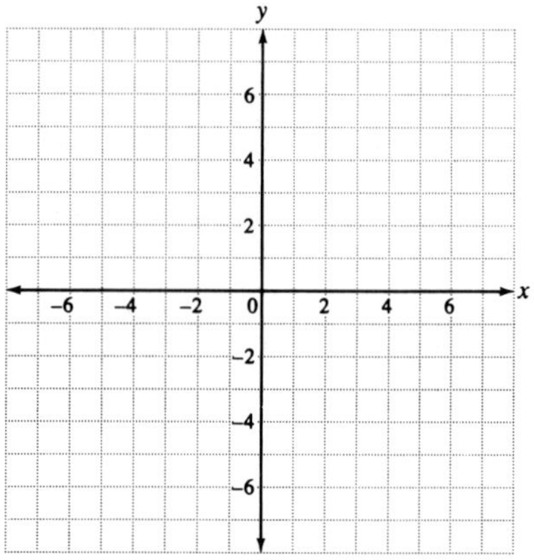

13. $x = \frac{3}{4}(y + 2)^2 - 2$

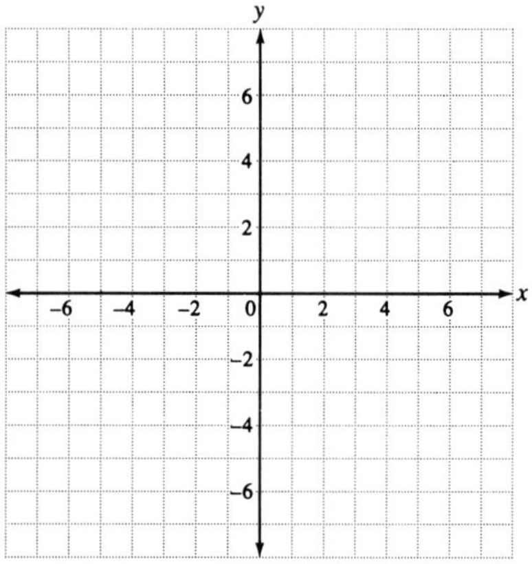

14. $x = \frac{3}{2}(y - 3)^2 + 1$

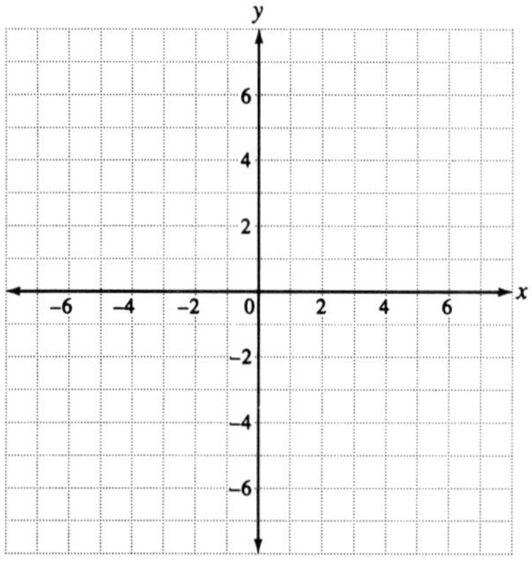

15. $y = \frac{5}{4}(x - 1)^2 - 3$

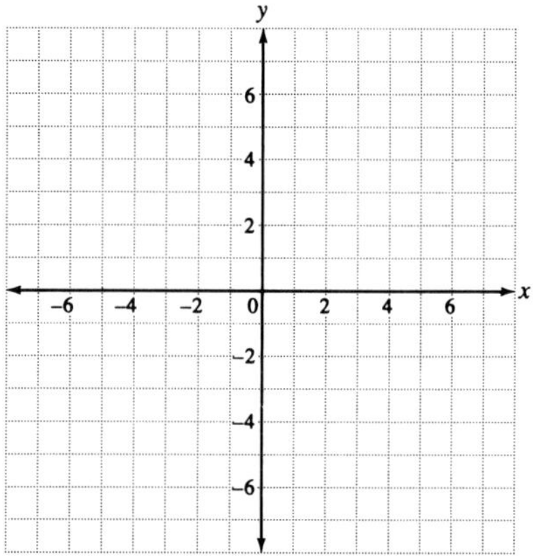

16. $y = \frac{2}{3}(x + 1)^2 - 2$

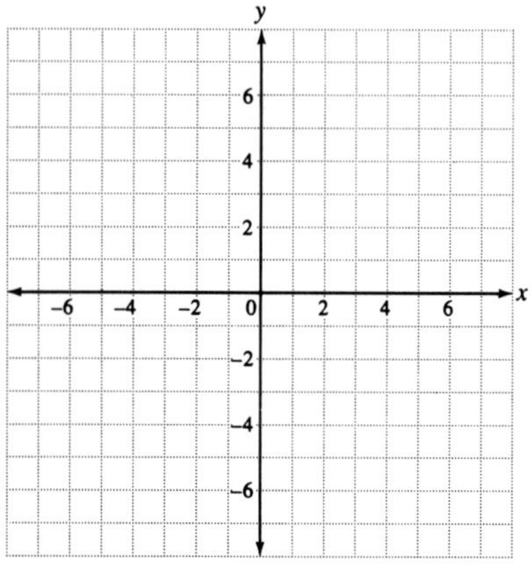

17. $\frac{x^2}{4} - \frac{y^2}{25} = 1$

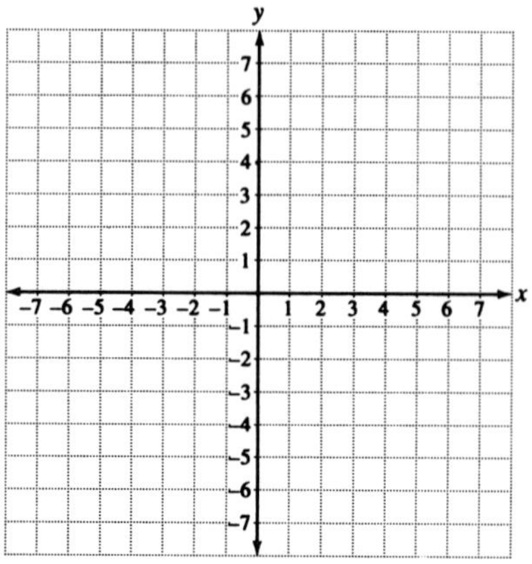

18. $\frac{y^2}{16} - \frac{x^2}{49} - 1$

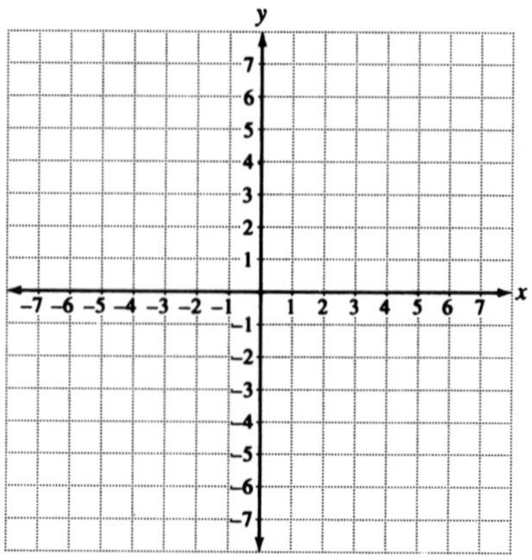

19. $\dfrac{y^2}{9} - \dfrac{x^2}{9} = 1$

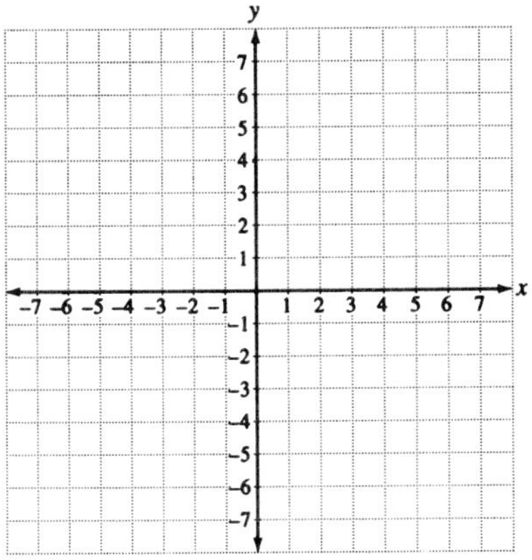

20. $\dfrac{x^2}{4} - \dfrac{y^2}{4} = 1$

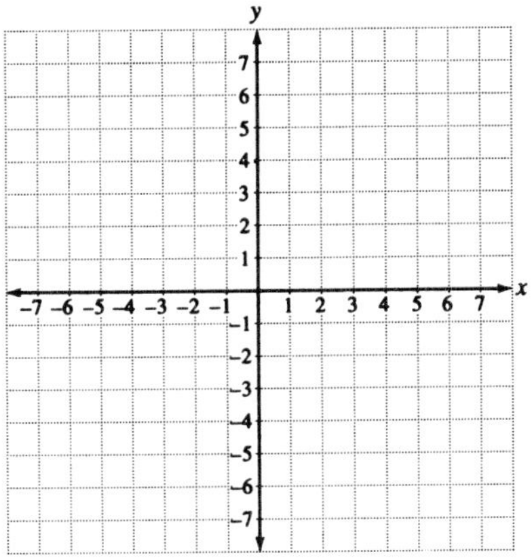

21. $\dfrac{x^2}{36} - y^2 = 1$

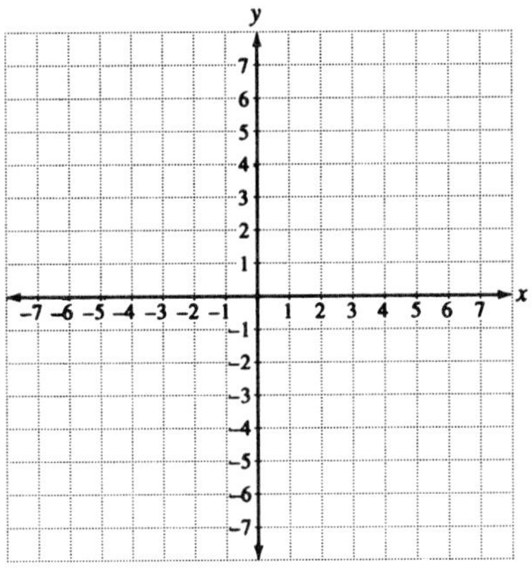

22. $y^2 - \dfrac{x^2}{9} = 1$

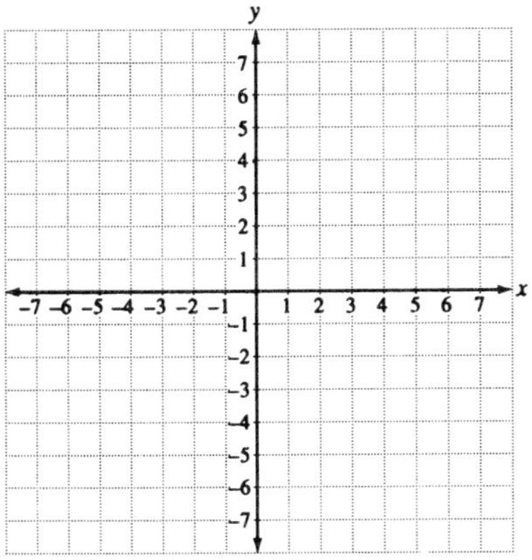

23. $\dfrac{y^2}{25} - \dfrac{x^2}{25} = 1$

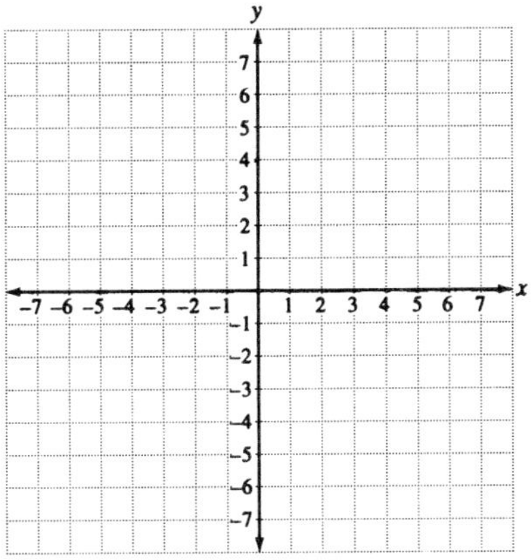

24. $\dfrac{x^2}{49} - \dfrac{y^2}{64} = 1$

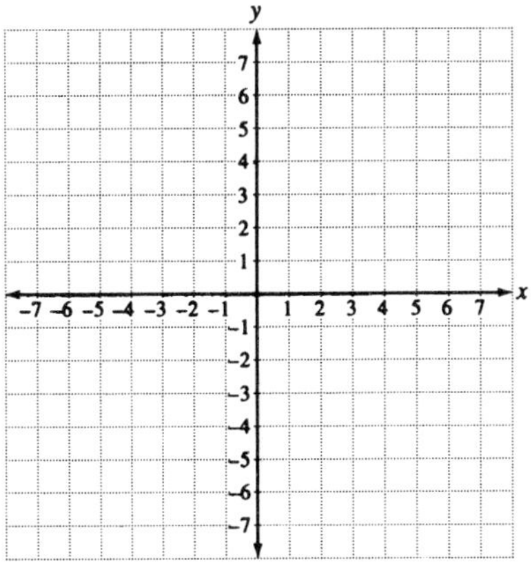

25. $\dfrac{x^2}{49} - \dfrac{y^2}{36} = 1$

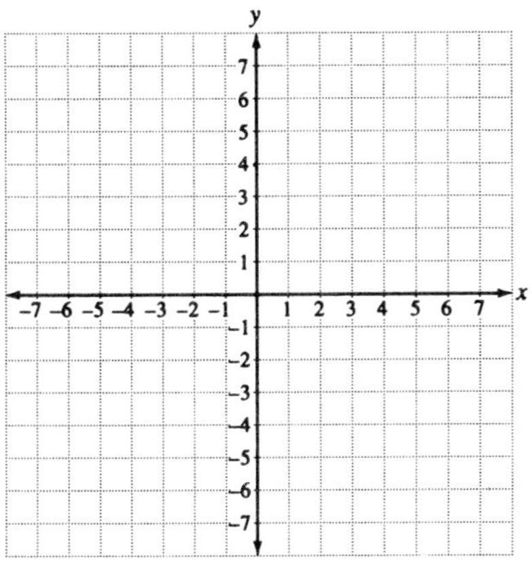

26. $\dfrac{y^2}{144} - \dfrac{x^2}{100} = 1$

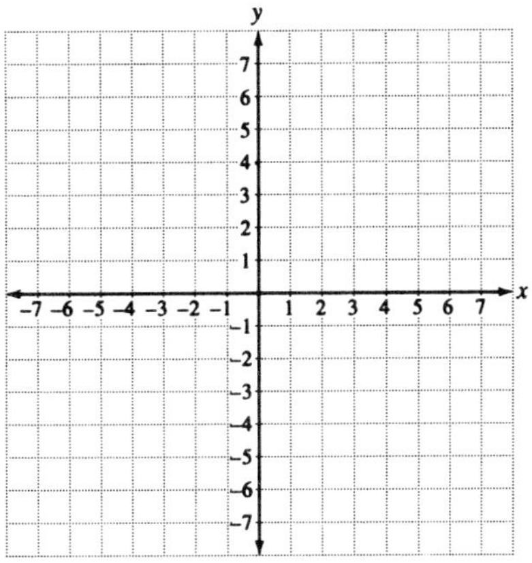

27. $\dfrac{y^2}{64} - \dfrac{x^2}{81} = 1$

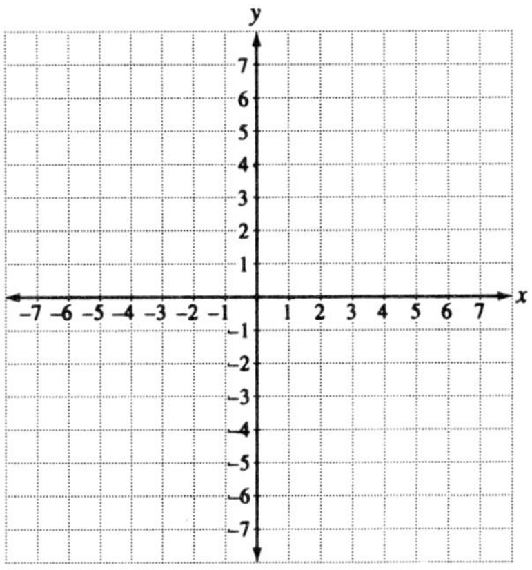

28. $\dfrac{x^2}{36} - \dfrac{y^2}{36} = 1$

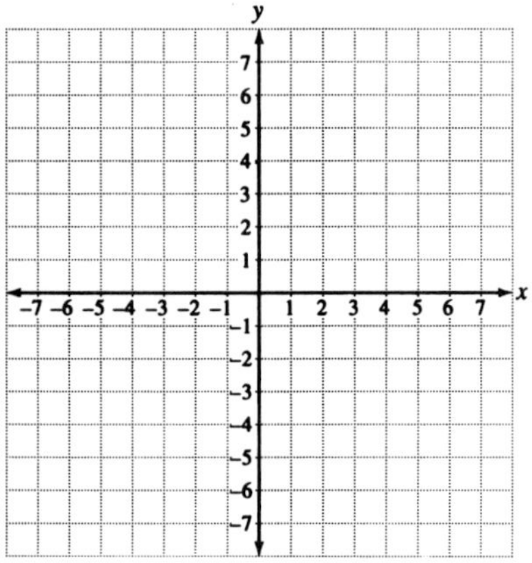

29. $\dfrac{x^2}{9/16} - \dfrac{y^2}{16/9} = 1$

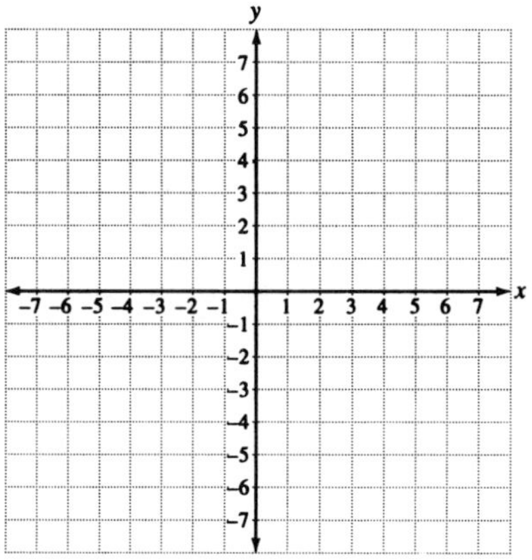

30. $\dfrac{y^2}{81/2} - \dfrac{x^2}{49/4} = 1$

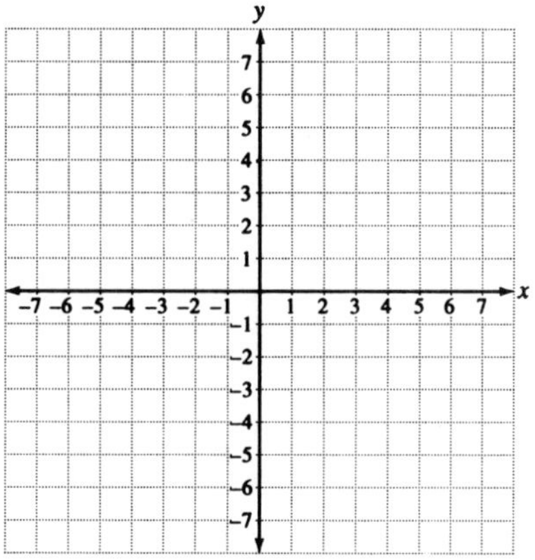

Identify the conic section.

31. $4x^2 - 9y^2 = 36$

32. $3x^2 + y = 0$

33. $(x - 3)^2 = 25 - y^2$

34. $9x^2 = 144 - 16y^2$

35. $(y - 4)^2 = 2x$

36. $x^2 + y^2 = 34$

37. $25x^2 = 4y^2 + 100$

38. $25x^2 = 25y^2 + 25$

39. $3y = (x - 2)^2 + 6$

40. $(x - 2)^2 = 27 - (y + 4)^2$

41. The graph of $y = a(x - h)^2 + k$ opens ____________ if a is positive.

42. The ____________ of $x = a(y - k)^2 + h$ is at (h, k).

43. The graph of $x = a(y - k)^2 + h$ opens to the left if a is ____________.

44. The graph of the equation $x^2/a^2 - y^2/b^2 = 1$ has ____________ of $(a, 0)$ and $(-a, 0)$.

45. Circles, ellipses, parabolas, and hyperbolas are called ____________ sections.

Find the vertex of the following parabolas by completing the square. Graph each parabola.

* **46.** $y = 3x^2 + 6x + 2$

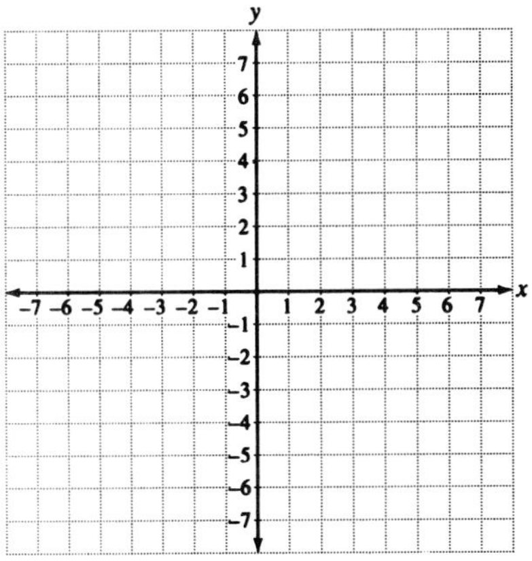

* **47.** $y = \frac{1}{2}x^2 - 3x + 1$

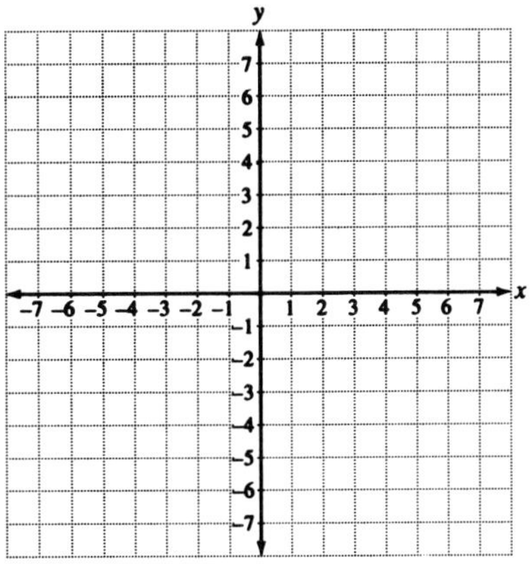

* **48.** $x = \frac{2}{3}y^2 - 2y - 1$

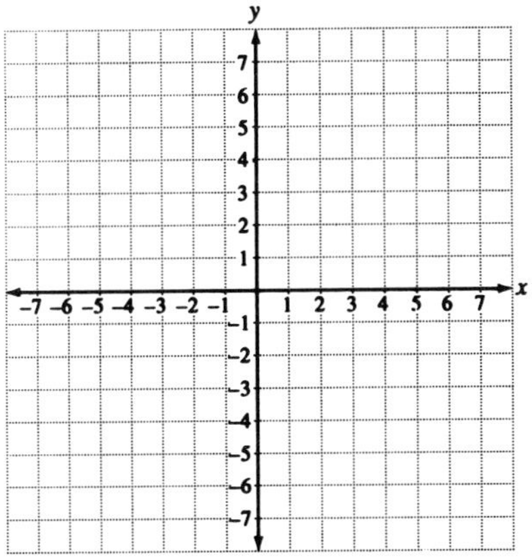

* **49.** $x = \frac{5}{3}y^2 + 10y + 11$

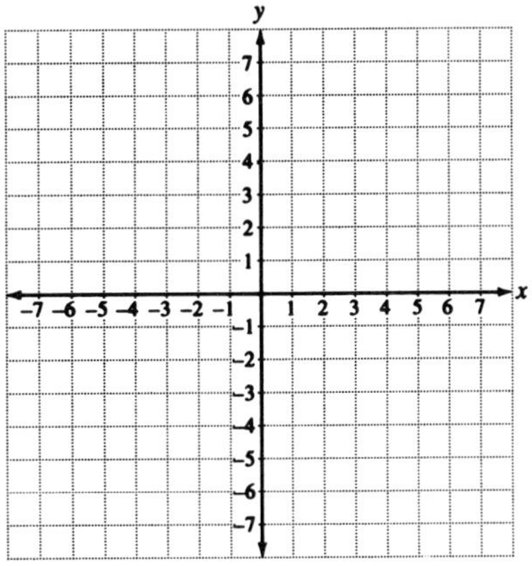

Think About It

Graph by plotting points and identify the conic section.

* **50.** $xy = 1$

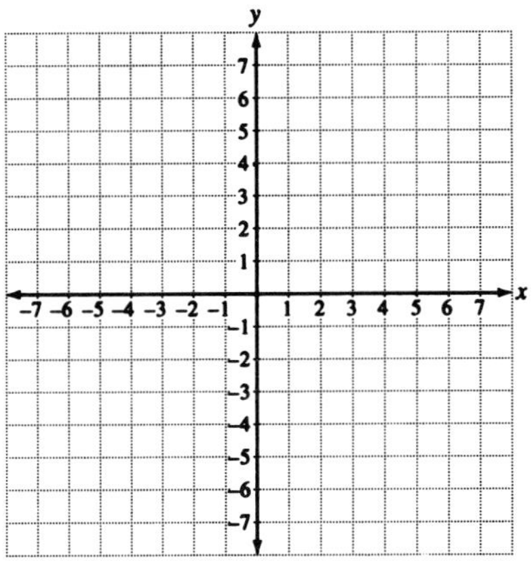

* **51.** $xy = -1$

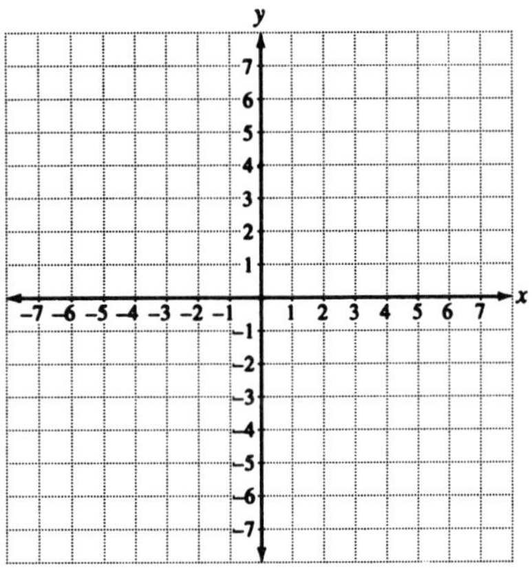

Graph.

* **52.** $\dfrac{x^2}{\frac{9}{16}} - \dfrac{y^2}{\frac{25}{4}} = 1$

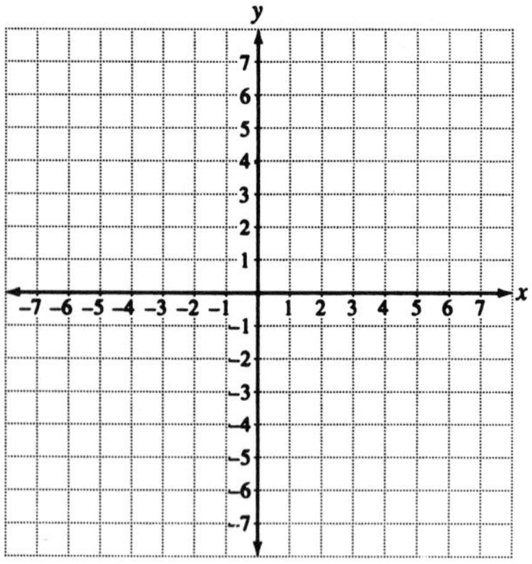

* **53.** $\dfrac{y^2}{6.25} - \dfrac{x^2}{0.81} = 1$

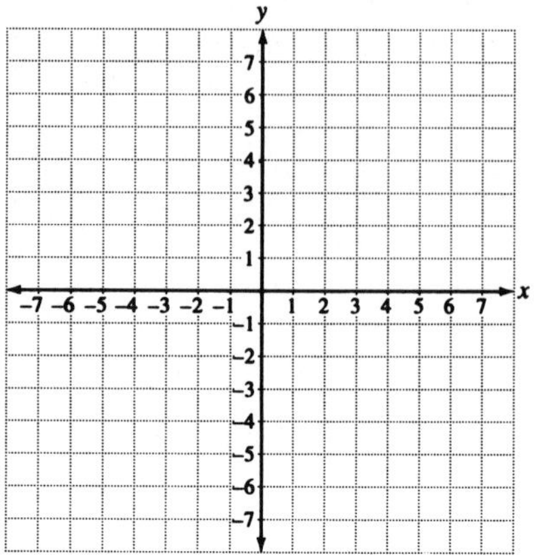

See the additional exercises to graph a hyperbola with center at (h, k).

Checkup

The following problems provide a review of some of Section 4.2 and will help you with the next section.

Solve by substitution.

54. $3x + y = 5$
$x = -2y$

55. $3x + 2y = 9$
$y = 3x$

56. $x - 4y = 5$
$y = x - 1$

57. $2x - y = 1$
$y = x - 3$

58. $4x + y = 7$
$6x - 5y = -9$

59. $2x + y = -2$
$5x + 3y = -3$

60. $2x - 7y = 1$
$x - 5y = 2$

61. $2x - 5y = -8$
$x + 6y = -4$

62. $3x - 4y = 7$
$x + 2y = -11$

63. $4x - y = 27$
$2x + 2y = -4$

9.4 SYSTEMS OF EQUATIONS

1 Solving Systems

Systems of equations may be required to solve applied problems in which one or more of the equations is nonlinear. Each system of two equations in this section contains one second-degree equation and one first-degree equation. Usually, the easiest method of solving these equations is by substitution.

EXAMPLE 1 Solve.

a. $x^2 + y^2 = 9 \quad (1)$

$x + 2y = 3 \quad (2)$

Solve the linear equation for x since x has a coefficient of 1.

$$x = 3 - 2y \qquad (3)$$

Substitute this value for x in equation (1) and solve it for y.

$$\begin{aligned} x^2 + y^2 &= 9 && (1) \\ (\mathbf{3 - 2y})^2 + y^2 &= 9 && \text{Substitute for } x. \\ 9 - 12y + 4y^2 + y^2 &= 9 && \text{Recall that } (A - B)^2 = A^2 - 2AB + B^2. \\ 9 - 12y + 5y^2 &= 9 \\ -12y + 5y^2 &= 0 \\ y(-12 + 5y) &= 0 && \text{Factor.} \end{aligned}$$

$$y = 0 \quad \text{or} \quad -12 + 5y = 0 \qquad \text{Use the zero product property.}$$

$$5y = 12$$

$$y = \frac{12}{5}$$

Substitute these y values into equation (3) and solve for x. [If we used equation (1) to solve for x, we would get some additional values for x that do not check.]

If $y = 0$:	If $y = \frac{12}{5}$:	
$x = 3 - 2y$	$x = 3 - 2y$	Equation (3).
$x = 3 - 2(0)$	$x = 3 - 2\left(\frac{12}{5}\right)$	
$x = 3$	$x = -\frac{9}{5}$	

These values check, so the solutions are $(3, 0)$ and $(-\frac{9}{5}, \frac{12}{5})$. The solutions are shown on the following graph of the system.

OBJECTIVES

1. *Solve systems of equations containing one second-degree equation and one first-degree equation*
2. *Solve word problems using systems of first- and second-degree equations*

□ **Exercise 1** Solve.

a. $x^2 + y^2 = 25$
$y - x = 1$

b. $xy = -10$
$x + 3y = 7$

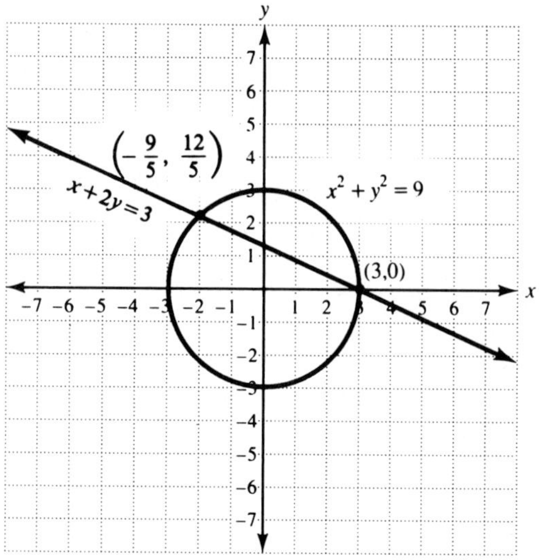

b. $xy = -4$ (1)

$2x + y = -2$ (2)

Notice that $xy = -4$ is a second-degree equation since the sum of the exponents on the only term containing variables is 2. Solve the linear equation for y.

$$y = -2x - 2 \quad (3)$$

Substitute this value for y into equation (1).

$$xy = -4 \quad (1)$$

$$x(\mathbf{-2x - 2}) = -4 \quad \text{Substitute for } y.$$

$$-2x^2 - 2x = -4$$

$$x^2 + x = 2 \quad \text{Divide both sides by } -2.$$

$$x^2 + x - 2 = 0$$

$$(x + 2)(x - 1) = 0 \quad \text{Factor.}$$

$$x + 2 = 0 \quad \text{or} \quad x - 1 = 0 \quad \text{Use the zero product property.}$$

$$x = -2 \qquad x = 1$$

Substitute these values into equation (3) and find y.

$$\text{If } x = -2, y = 2 \qquad \text{If } x = 1, y = -4$$

The ordered pairs $(-2, 2)$ and $(1, -4)$ check. They are solutions. ■

□ **DO EXERCISE 1.**

2 Applied Problems

Systems of first- and second-degree equations may be used to solve word problems.

EXAMPLE 2 The sum of the squares of two numbers is 34. The sum of the two numbers is 8. Find the numbers.

Let x represent one number and y represent the other number. The sum of the squares of the two numbers is 34, so we have the following equation.

$$x^2 + y^2 = 34 \qquad (1)$$

The sum of the two numbers is 8. This gives us another equation.

$$x + y = 8 \qquad (2)$$

Solve equation (2) for x or y. We solve for y.

$$y = 8 - x$$

Substitute this value for y into equation (1).

$$x^2 + y^2 = 34 \qquad (1)$$

$$x^2 + (\mathbf{8 - x})^2 = 34 \qquad \text{Substitute for } y.$$

$$x^2 + 64 - 16x + x^2 = 34$$

$$2x^2 - 16x + 64 = 34$$

$$2x^2 - 16x + 30 = 0$$

$$x^2 - 8x + 15 = 0 \qquad \text{Divide by 2.}$$

$$(x - 5)(x - 3) = 0 \qquad \text{Factor.}$$

$$x - 5 = 0 \quad \text{or} \quad x - 3 = 0 \qquad \text{Use the zero product property.}$$

$$x = 5 \qquad\qquad x = 3$$

When $x = 5$, $y = 3$ and when $x = 3$, $y = 5$. These values check and they give the same result. The numbers are 3 and 5. ■

□ **DO EXERCISE 2.**

□ **Exercise 2** The sum of the squares of two numbers is 193. The larger number minus the smaller number is 5. Find the numbers.

Answers to Exercises

1. a. $(-4, -3), (3, 4)$ **b.** $\left(-3, \frac{10}{3}\right), (10, -1)$

2. 7 and 12

Problem Set 9.4

NAME ______

DATE ______

CLASS ______

Solve.

1. $x^2 + y^2 = 4$
$x - 2y = 4$

2. $x^2 + y^2 = 1$
$x + 2y = 1$

3. $xy = 2$
$x + 3y = -5$

4. $xy = 4$
$x + y = 5$

5. $x^2 + 3y^2 = 3$
$x = 3y$

6. $3x^2 + 4y^2 = 171$
$y = 2x$

7. $y = x^2 + 4x + 4$
$y - x = 2$

8. $y = x^2 + 6x + 9$
$x + y = 3$

9. $4x^2 + 9y^2 = 36$
$3y + 2x = 6$

10. $9x^2 + 4y^2 = 36$
$3x + 2y = 6$

11. $y = x^2$
$4x = -y - 3$

12. $y^2 = x$
$3y = x - 4$

13. $x^2 - xy + 3y^2 = 5$
$x - y = 2$

14. $2y^2 + xy = 5$
$x + 4y = 7$

15. $xy = -12$
$x + 3y = 5$

16. $xy = -6$
$x + y = -1$

17. The positive difference of the squares of two numbers is 32. The sum of the two numbers is 16. Find the numbers.

18. The positive difference of the squares of two numbers is 108. The larger number minus the smaller number is 6. Find the numbers.

19. The area of a rectangle is 108 square feet and the perimeter is 42 feet. Find the dimensions of the rectangle.

20. Find the dimensions of a rectangle if the perimeter is 44 centimeters and the area is 105 square centimeters.

21. A rectangular flower garden is enclosed by 44 feet of fencing. The area of the garden is 112 square feet. What are the dimensions of the garden?

22. The area of a rectangular picture is 154 square inches. If the perimeter is 50 inches, find the dimensions of the picture.

23. The cost C of producing x mixers at Lux Corporation is given by $C = 8x$. The revenue R received from the sale of x mixers if given by $R = \frac{1}{8}x^2$. Find the break-even point where cost equals revenue.

24. Toys International has found that the price p (in dollars) of its teddy bears is related to the supply x (in thousands) by the equation $px = 8$. The price is related to the demand x (in thousands) for the teddy bears by the equation $p = 10x + 2$. The equilibrium price is the value of p where demand equals supply. What is the equilibrium price and the supply/demand at that price?

25. The position of a ball thrown in the air at night is given by $y = -x^2 + 5$. A beam of light is shining on the line $x - y + 3 = 0$. At what points will the light shine on the ball?

26. An object is moving in an elliptical path given by $4x^2 + y^2 = 4$. A light is shining on the line $y = x - 2$. At what points will the light shine on the object?

27. A system of one second-degree equation and one first-degree equation is usually solved by ____________.

Solve. Round answers to the nearest hundredth.

28. $x = 3.5y$
$x^2 + 3.75y^2 = 4.29$

29. $y = 2.4x$
$3.46x^2 + 4y^2 = 165.54$

30. $xy = 2.7$
$x + 1.8y = 4.53$

31. $xy = -5.9$
$1.35x + 2.5y = 9.63$

* 32. The diagonal of a rectangle is 50 meters and the perimeter of the rectangle is 140 meters. Find the dimensions of the rectangle.

* 33. A number divided by a second number is 4. Their product divided by the difference of the first number and the second number is $\frac{8}{3}$. Find the numbers.

* 34. If a number containing two digits is divided by the sum of the digits, the quotient is 5. The product of the digits is 20. Find the number. (*Hint:* The number may be written $10t + u$ where t is the tens digit and u is the ones digit.)

Checkup

The following problems provide a review of some of Sections 4.2 and 8.5 and will help you with the next section.

Solve by the elimination method.

35. $x + 4y = -22$
$4x + y = -13$

36. $x + 2y = 5$
$3x - y = 8$

37. $4x + y = 1$
$x - 2y = -11$

38. $x + 2y = 7$
$3x + y = 1$

39. $2x + y = -7$
$x + 2y = -8$

Solve for the missing variable.

40. $W = \sqrt{\dfrac{1}{LC}}$ for L

41. $A = \sqrt{\dfrac{W_1}{W_2}}$ for W_1

42. Dr. and Mrs. Minh bought an 11- by 14-foot rug for their living room. The area of the room is 340 square feet and there is an even strip of hardwood surrounding the rug. How wide is the strip of hardwood?

43. A rectangular pool 20 by 30 feet has a strip of concrete of uniform width around it. If the total area of the pool and the concrete is 816 square feet, what is the width of the strip of concrete?

44. A boat travels 8 miles per hour in still water. It travels 10 miles upstream and back again in $1\frac{2}{3}$ hours. What is the speed of the current?

45. The speed of the current of a river is 3 miles per hour. It takes twice as long for a boat to go 15 miles upstream as it does for it to go downstream. Find the speed of the boat in still water.

9.5 NONLINEAR SYSTEMS OF EQUATIONS

OBJECTIVES

1 *Solve certain systems of two second-degree equations*

2 *Solve applied problems using two second-degree equations*

1 Solving Nonlinear Systems

Some systems of two second-degree equations may be solved by either the elimination or the substitution method. In each case we use the easier method.

EXAMPLE 1

a. Solve:

$$x^2 + 4y^2 = 25 \qquad (1)$$
$$4x^2 + y^2 = 25 \qquad (2)$$

It is easier to solve this system by the elimination method. We multiply both sides of equation (2) by -4 and add the result to equation (1).

$$\begin{aligned} -16x^2 - 4y^2 &= -100 && \text{Multiply both sides by } -4. \\ x^2 + 4y^2 &= 25 && (1) \\ \hline -15x^2 \qquad &= -75 && \text{Add.} \\ x^2 &= 5 \\ x &= \pm\sqrt{5} \end{aligned}$$

To solve for y, we substitute these values for x into one of the original equations.

For $x = \sqrt{5}$:

$$\begin{aligned} x^2 + 4y^2 &= 25 && \text{Use equation (1).} \\ (\sqrt{5})^2 + 4y^2 &= 25 \\ 5 + 4y^2 &= 25 \\ 4y^2 &= 20 \\ y^2 &= 5 \\ y &= \pm\sqrt{5} \end{aligned}$$

For $x = -\sqrt{5}$:

$$\begin{aligned} x^2 + 4y^2 &= 25 \\ (-\sqrt{5})^2 + 4y^2 &= 25 \\ 5 + 4y^2 &= 25 \\ 4y^2 &= 20 \\ y^2 &= 5 \\ y &= \pm\sqrt{5} \end{aligned}$$

These values check. The solutions are $(\sqrt{5}, \sqrt{5})$, $(\sqrt{5}, -\sqrt{5})$, $(-\sqrt{5}, \sqrt{5})$, and $(-\sqrt{5}, -\sqrt{5})$. The graph is as follows:

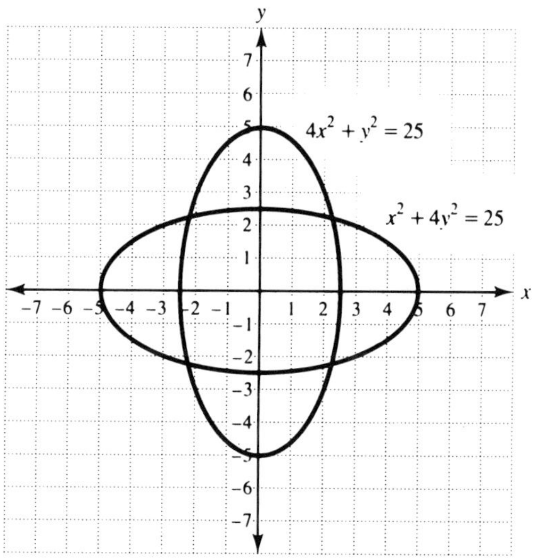

□ **Exercise 1** Solve.

a. $2x^2 + 5y^2 = 22$
$3x^2 - y^2 = -1$

b. $2y = x^2 - 2$
$y = x^2 - 3$

b. Solve:

$$x^2 - 3y = -5 \qquad (1)$$

$$y = x^2 - 1 \qquad (2)$$

We may solve this system by substitution. Substitute the value for y in equation (2) into equation (1).

$$\begin{aligned} x^2 - 3y &= -5 && (1) \\ x^2 - 3(\mathbf{x^2 - 1}) &= -5 && \text{Substitute } x^2 - 1 \text{ for } y. \\ x^2 - 3x^2 + 3 &= -5 && \text{Use a distributive law.} \\ -2x^2 + 3 &= -5 \\ -2x^2 &= -8 \\ x^2 &= 4 \\ x &= \pm 2 \end{aligned}$$

Substituting these values for x in equation (2) gives $y = 3$. The ordered pairs $(2, 3)$ and $(-2, 3)$ check. They are the solutions. ■

□ **DO EXERCISE 1.**

2 Word Problems

Systems of two second-degree equations may also be used to solve word problems.

EXAMPLE 2 The area of a sign in the shape of a right triangle is 7.5 square feet and the hypotenuse is $\sqrt{34}$ feet. Find the lengths of the other two sides.

Drawing

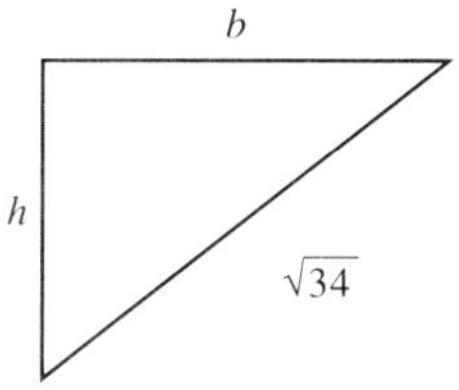

Equations

$A = \frac{1}{2}bh$ Area of a triangle.

$7.5 = \frac{1}{2}bh$ Substitute 7.5 for A.

$b = \frac{15}{h}$ (1)

$b^2 + h^2 = 34$ (2) Pythagorean theorem

Solve

$\left(\frac{15}{h}\right)^2 + h^2 = 34$

$\frac{225}{h^2} + h^2 = 34$

$225 + h^4 = 34h^2$ Multiply both sides by h^2.

$h^4 - 34h^2 + 225 = 0$

$u^2 - 34u + 225 = 0$ Let $u = h^2$.

$u - 9 = 0$ or $u - 25 = 0$

$u = 9$ $u = 25$

Substitute h^2 for u and solve the equations.

$h^2 = 9$ or $h^2 = 25$

$h = \pm 3$ $h = \pm 5$

The height of a triangle cannot be negative, so we reject $h = -3$ and $h = -5$. Since $b = 15/h$, if $h = 3$, $b = 5$. If $h = 5$, $b = 3$. These solutions check. The lengths of the other two sides are 3 feet and 5 feet. ■

□ **DO EXERCISE 2.**

□ **Exercise 2** The area of a right triangle is 10 square meters and the length of the hypotenuse is $\sqrt{41}$ meters. Find the lengths of the two legs.

Answers to Exercises

1. **a.** (1, 2), (1, −2), (−1, 2), (−1, −2) **b.** (2, 1), (−2, 1)

2.
$$\frac{1}{2}bh = 10$$
$$b = \frac{20}{h}$$
$$b^2 + h^2 = 41$$
$$\frac{20^2}{h^2} + h^2 = 41$$
$$\frac{400}{h^2} + h^2 = 41$$
$$400 + h^4 = 41h^2$$
$$h^4 - 41h^2 + 400 = 0$$
$$(h^2 - 25)(h^2 - 16) = 0$$
$$h^2 - 25 = 0 \quad \text{or} \quad h^2 - 16 = 0$$
$$h^2 = 25 \qquad h^2 = 16$$
$$h = \pm 5 \qquad h = \pm 4$$

Reject $h = -5$ and $h = -4$

When $h = 5$, $b = 4$

When $h = 4$, $b = 5$

The lengths of the two legs are 5 meters and 4 meters.

Problem Set 9.5

NAME

DATE

CLASS

Solve.

1. $x^2 - 25y^2 = 20$
$2x^2 + 25y^2 = 88$

2. $x^2 - 2y^2 = 1$
$x^2 + 4y^2 = 25$

3. $2x^2 + 3y^2 = 6$
$x^2 + 3y^2 = 3$

4. $6x^2 + y^2 = 9$
$3x^2 + 4y^2 = 36$

5. $2x^2 + 4x = y$
$y = x^2 - 3x - 10$

6. $x^2 + y^2 = 9$
$y^2 = x^2 - 9$

7. $2x^2 = y$
$y = 8x^2 - 2x - 4$

8. $x^2 + y^2 = 25$
$y^2 = x + 5$

9. $x^2 + y^2 = 9$
$y = x^2 - 3$

10. $x^2 + y^2 = 1$
$y = x^2 - 1$

11. $x^2 + y^2 = 10$
$xy = 3$

12. $x^2 + y^2 = 8$
$xy = 4$

13. $x^2 + y^2 = 17$
$xy = 4$

14. $x^2 + y^2 = 13$
$xy = 6$

15. $y = 3x^2 + x - 4$
$y = 3x^2 - 8x + 5$

16. $y = 2x^2 + 3x + 1$
$y = 2x^2 + 9x + 7$

17. $2ab + 3b^2 = 7$
$3ab - 2b^2 = 4$

18. $pq - q^2 = 2$
$2pq - 3p^2 = 0$

19. The sum of the squares of two numbers is 58. The positive difference of their squares is 40. Find the numbers.

20. The product of two numbers is 24. The sum of their squares is 73. Find the numbers.

21. The area of a rectangle is 3 square meters and the length of a diagonal is $\sqrt{10}$. Find the dimensions of the rectangle.

22. The area of a rectangle is 10 square inches and the length of a diagonal is $\sqrt{29}$. Find the dimensions of the rectangle.

23. The sum of the squares of the length and width of a garden is 346 square feet. The width times the length is 165 square feet. Find the dimensions of the garden.

24. A house contains two square rooms. Find the length of each room if the sum of their areas is 400 square feet and the positive difference of their areas is 112 square feet.

25. Jim wants to cut a piece of plywood in the shape of a right triangle that has an area of 8 square feet and a hypotenuse of length $2\sqrt{5}$ feet. What are the lengths of the other sides of the piece of plywood?

26. Tara wants to make a right rectangular countertop that has an area of 6 square meters and a diagonal of length $\sqrt{13}$ meters. Find the dimensions of the countertop.

27. The product of the lengths of the legs of a right triangle is 168 square centimeters. The hypotenuse has length $2\sqrt{85}$ centimeters. What are the lengths of the legs? (*Hint:* You may want to use the quadratic formula and a calculator.)

28. The hypotenuse of a right triangle has length $\sqrt{337}$ inches. The product of the lengths of the legs is 144 square inches. Find the lengths of the legs.

29. The sum of the squares of two numbers is 89. The positive difference of their squares is 39. What are the two numbers?

30. The product of two numbers is 42. The positive difference of their squares is 13. Find the numbers.

31. Systems of two second-degree equations may be solved by either the ____________ or the substitution method.

Solve. Round answers to the nearest hundredth.

32. $0.34x^2 + 7.29y^2 = 13.86$
$3.91x^2 - 14.58y^2 = 22.35$

33. $5.78x^2 - 4.28y^2 = 25.46$
$2.89x^2 - 6.31y^2 = -42.57$

34. $2.3x^2 = y$
$y = 8.7x^2 - 2.3x - 4.8$

35. $3.74x^2 + y^2 = 25.78$
$y^2 = 1.27x + 4.89$

36. $xy = 6.72$
$x^2 + y^2 = 14.65$ (Use the quadratic formula.)

Think About It

* 37. A rectangular garden has an area of 1536 square feet. If the length and width are each decreased by 5 feet, the area is 415 square feet less than the original area. Find the dimensions of the garden.

* 38. A guy wire to a pole is attached to a stake 6 feet from the pole. If the wire were 7 feet longer, it would reach to a stake 15 feet from the pole. Find the height of the pole and the length of the wire.

* 39. A side opposite the right angle formed by the perpendicular bisector of the base of a triangle is 10 centimeters. The area of the triangle is 48 square centimeters. Find the height and the base of the triangle. (*Hint:* The perpendicular bisector divides the base into two equal lengths.)

Checkup

The following problems provide a review of some of Sections 6.4 and 8.3.

Find the value(s) for which the rational expression is undefined.

40. $\frac{-2}{x - 3}$

41. $\frac{4}{x + 2}$

42. $\frac{x}{3x + 4}$

43. $\frac{4x}{2x - 1}$

44. $\frac{5}{x^2 + 2x - 3}$

45. $\frac{-3}{x^2 - 4x - 5}$

Solve using the quadratic formula.

46. $x^2 + 3x - 2 = 0$

47. $y^2 - 5y + 2 = 0$

48. $3x^2 - 2x - 1 = 0$

49. $4v^2 + v - 3 = 0$

Chapter 9 Summary

Section 9.1

Distance formula: The distance d between any two points (x_2, y_2) and (x_1, y_1) is

$$d = \sqrt{(x_2 - x_1)^2 + (y_2 - y_1)^2}$$

A ***line segment*** is part of a line and has two endpoints.

The coordinates of the ***midpoint*** of a line segment whose endpoints are (x_1, y_1) and (x_2, y_2) are

$$\left(\frac{x_1 + x_2}{2}, \frac{y_1 + y_2}{2}\right)$$

Section 9.2

The equation of a circle with center at (h, k) and radius r is

$$(x - h)^2 + (y - k)^2 = r^2 \qquad \text{Standard form.}$$

The equation of a circle with center at the origin and radius r is

$$x^2 + y^2 = r^2$$

The equation of an ellipse with center at the origin is

$$\frac{x^2}{a^2} + \frac{y^2}{b^2} = 1 \qquad a, b < 0, \quad a \neq b \qquad \text{Standard form.}$$

The following ellipse has intercepts as shown.

$$\frac{x^2}{a^2} + \frac{y^2}{b^2} = 1 \qquad \text{Standard form.}$$

The x-intercepts are $(a, 0)$ and $(-a, 0)$. The y-intercepts are $(0, b)$ and $(0, -b)$.

Section 9.3

Graphs of the following equations are parabolas.

$y = a(x - h)^2 + k$	vertex is at (h, k)	$a > 0$ opens upward, $a < 0$ opens downward.
$x = a(y - k)^2 + h$	vertex is at (h, k)	$a > 0$ opens to the right, $a < 0$ opens to the left.

Graphs of the following equations are hyperbolas with centers at the origin.

$\frac{x^2}{a^2} - \frac{y^2}{b^2} = 1$ The x-intercepts are $(a, 0)$ and $(-a, 0)$. There are no y-intercepts.

$\frac{y^2}{b^2} - \frac{x^2}{a^2} = 1$ The y-intercepts are $(0, b)$ and $(0, -b)$. There are no x-intercepts.

Equation	Conic Section	
$(x - h)^2 + (y - k)^2 = r^2$	Circle	Center at (h, k), radius r
$\frac{x^2}{a^2} + \frac{y^2}{b^2} = 1$	Ellipse	x-intercepts at $(a, 0)$ and $(-a, 0)$; y-intercepts at $(0, b)$ and $(0, -b)$
$y = a(x - h)^2 + k$	Parabola	Vertex at (h, k); if $a > 0$, opens upward; if $a < 0$, opens downward
$x = a(y - k)^2 + h$	Parabola	Vertex at (h, k); if $a > 0$, opens right; if $a < 0$, opens left
$\frac{x^2}{a^2} - \frac{y^2}{b^2} = 1$	Hyperbola	x-intercepts are $(a, 0)$ and $(-a, 0)$
$\frac{y^2}{b^2} - \frac{x^2}{a^2} = 1$	Hyperbola	y-intercepts are $(0, b)$ and $(0, -b)$

Section 9.4 Usually, the easiest method of solving a system of one second-degree equation and one first-degree equation is by substitution.

Section 9.5 Some systems of two second-degree equations may be solved by either the elimination or the substitution method. In each case, use the easier method.

Chapter 9 Additional Exercises (Optional)

NAME

DATE

CLASS

Section 9.1

Find the distance between the points.

1. $(7, 5)$ and $(3, 2)$

2. $(8, 4)$ and $(6, 7)$

3. $(-3, 5)$ and $(-2, 9)$

4. $(5, -6)$ and $(-1, -2)$

5. $(-4, -1)$ and $(-8, -7)$

6. $(-3, 4)$ and $(-7, -1)$

7. $(-0.2, 4.3), (-0.4, 5.1)$

8. $\left(\frac{5}{8}, \frac{1}{2}\right), \left(\frac{1}{4}, \frac{3}{4}\right)$

* **9.** $(\sqrt{5}, 0), (0, \sqrt{3})$

* **10.** $(3\sqrt{2}, 2 - \sqrt{5}), (\sqrt{2}, 3 - \sqrt{5})$

* **11.** $(-6a, 4b)$ $(a, -3b)$

* **12.** $(p, 2q), (p, 5q)$

Find the midpoint of the line segment with the given endpoints.

13. $(-8, -9)$ and $(-4, -2)$

14. $(5, -3)$ and $(-4, 7)$

15. $(-3.5, 4.7)$ and $(-2.8, 3)$

16. $(8.9, -6.5)$ and $(-3.2, 4.1)$

17. $\left(\frac{5}{6}, -\frac{7}{4}\right)$ and $\left(\frac{3}{8}, -\frac{5}{12}\right)$

18. $\left(-\frac{5}{9}, -\frac{11}{3}\right)$ and $\left(\frac{3}{4}, -\frac{10}{9}\right)$

Determine if the following are vertices of a right triangle.

19. $(2, 3), (-2, -2), (3, -6)$

20. $(1, 6), (2, 1), (-4, 1)$

21. $(-5, 2), (-3, 4), (-9, 6)$

22. $(3, 1), (4, 5), (7, -8)$

Section 9.2

Find an equation for each circle.

23. Center: $(-2, -3)$; radius: $\sqrt{7}$

24. Center: $(3, -4)$; radius: $\sqrt{5}$

25. Center $(0, 0)$, passing through the point $(1, 2)$

26. Center $(3, -4)$, passing through the point $(0, 0)$

Find the center and radius of the following circles.

27. $(x - 4)^2 + (y + 3)^2 = 9$

28. $(x - 3)^2 + (y - 5)^2 = 16$

29. $x^2 + y^2 + 6x - 2y = 8$

30. $x^2 + y^2 - 10x + 4y = 4$

Graph.

31. $\frac{x^2}{16} + \frac{y^2}{25} = 1$

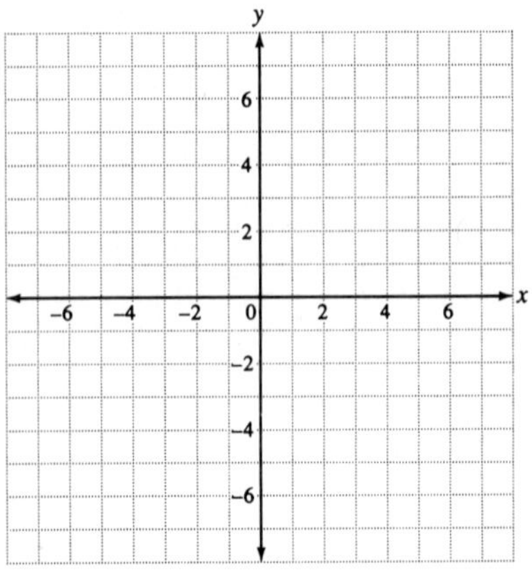

32. $\frac{x^2}{9} + \frac{y^2}{4} = 1$

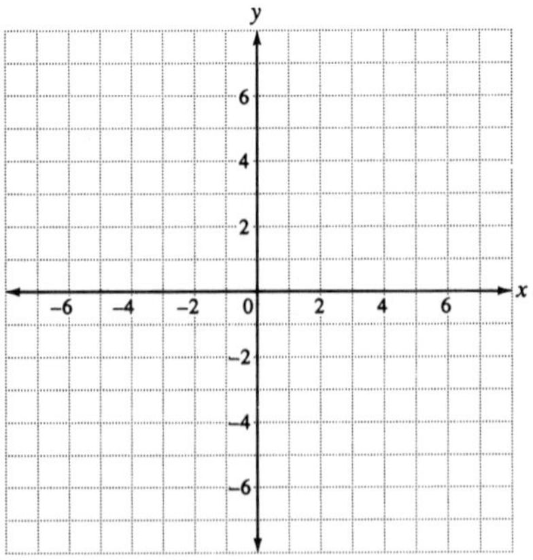

The equation of an ellipse with center at the origin is

$$\frac{x^2}{a^2} + \frac{y^2}{b^2} = 1 \qquad \text{Standard form.}$$

The x-intercepts are $(a, 0)$ and $(-a, 0)$. The y-intercepts are $(0, b)$ and $(0, -b)$. These intercepts are often called ***vertices.***

The equation of an ellipse with center at (h, k) is

$$\frac{(x - h)^2}{a^2} + \frac{(y - k)^2}{b^2} = 1 \qquad \text{Standard form.}$$

The vertices are $(a + h, k)$, $(-a + h, k)$, $(h, b + k)$, and $(h, -b + k)$. When the center of an ellipse is not the origin the vertices are not the x- and y-intercepts.

Find the center and vertices of the following ellipses. Graph each ellipse.

* **33.** $\frac{(x - 3)^2}{16} + \frac{(y - 2)^2}{4} = 1$

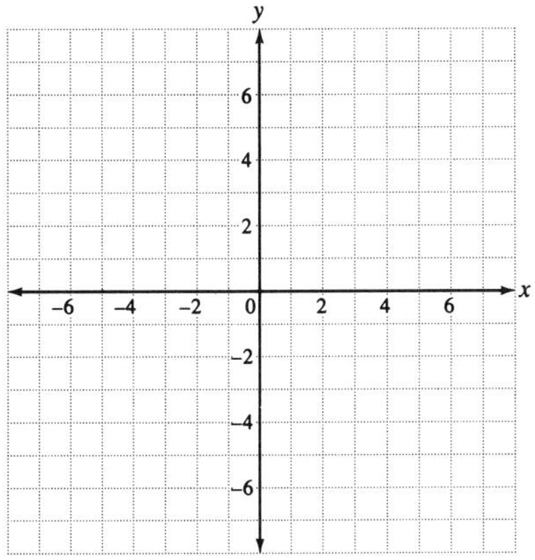

* **34.** $\frac{(x - 1)^2}{9} + \frac{(y + 2)^2}{4} = 1$

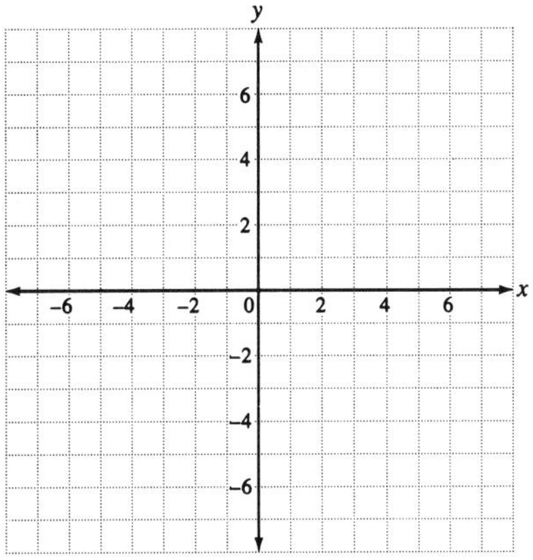

* **35.** $4(x - 2)^2 + 9(y - 1)^2 = 36$

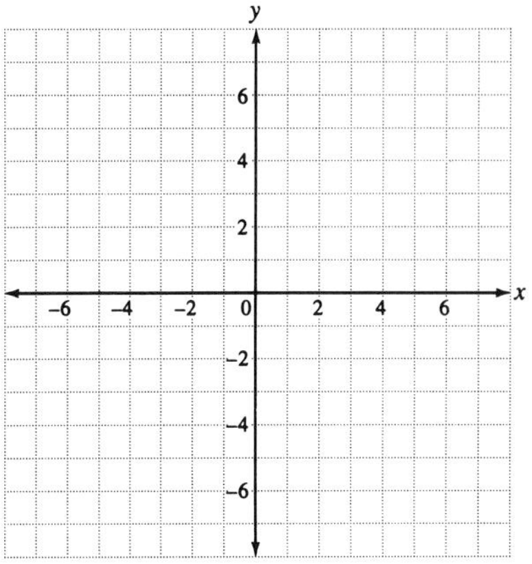

* **36.** $25(x + 1)^2 + 9(y - 3)^2 = 225$

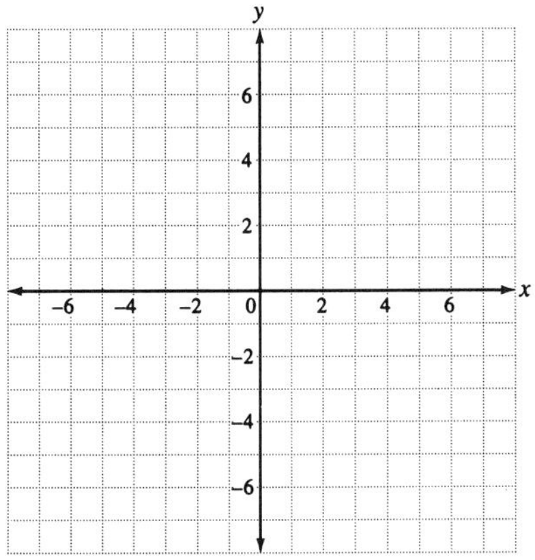

* **37.** $x^2 + 4y^2 - 2x - 16y + 13 = 0$

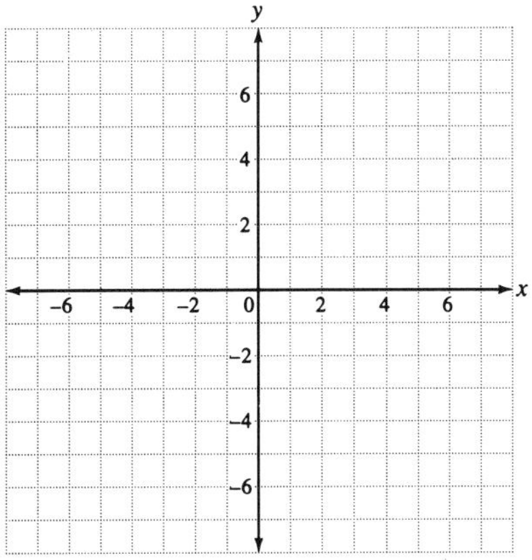

* **38.** $x^2 + 4y^2 + 2x - 8y + 1 = 0$

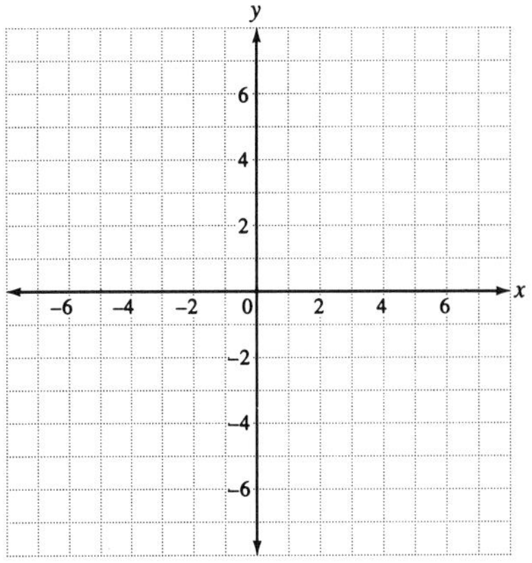

Section 9.3

Find the vertex of the following parabolas by completing the square. Graph each parabola.

* **39.** $y = x^2 - 4x - 1$

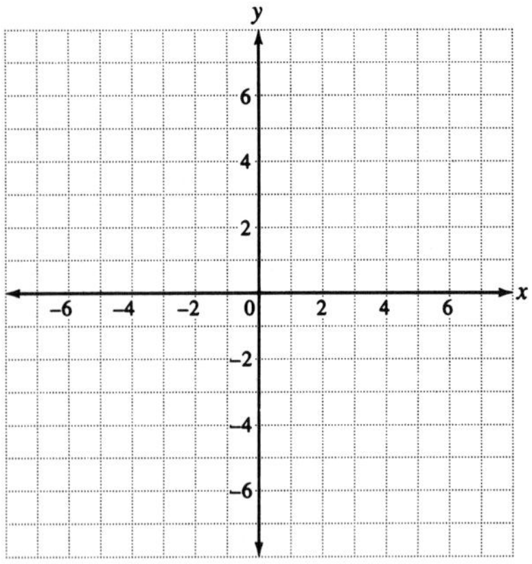

* **40.** $x = y^2 + 3y + 2$

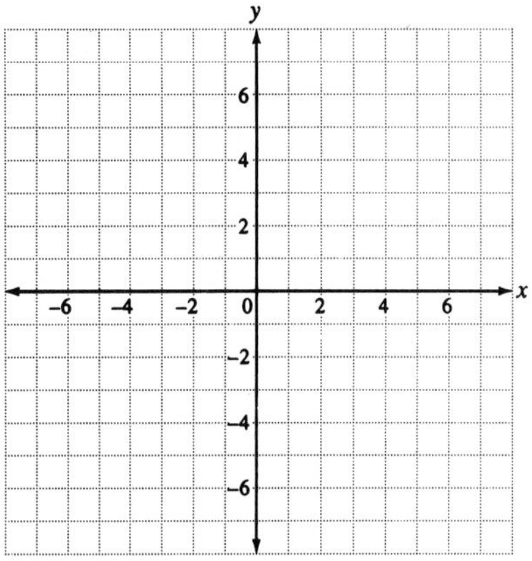

* **41.** $x = 2y^2 - 8y + 1$

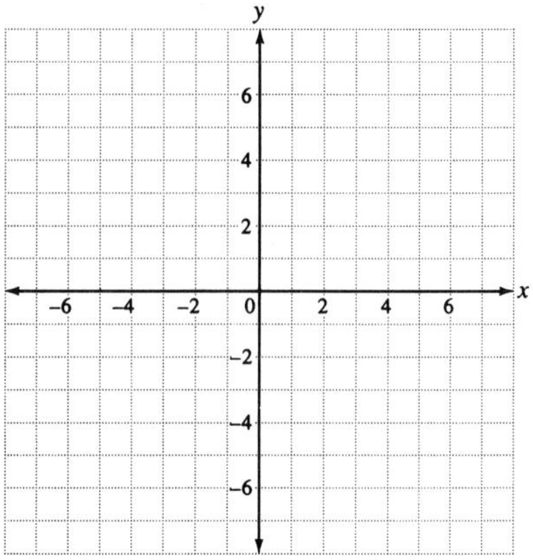

* **42.** $y = 3x^2 + 6x - 2$

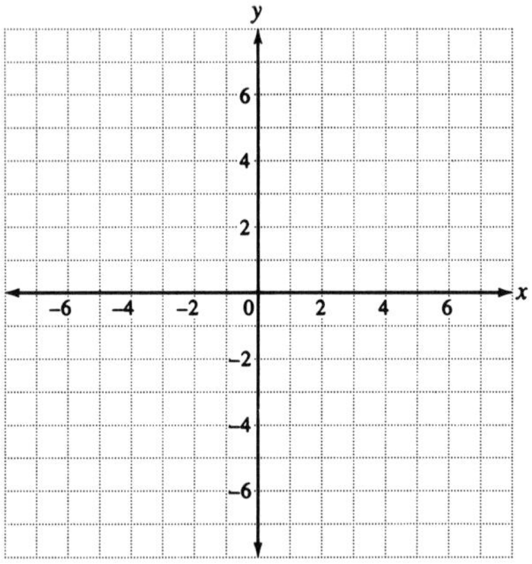

43. $y = 2x^2 - 4x + 5$

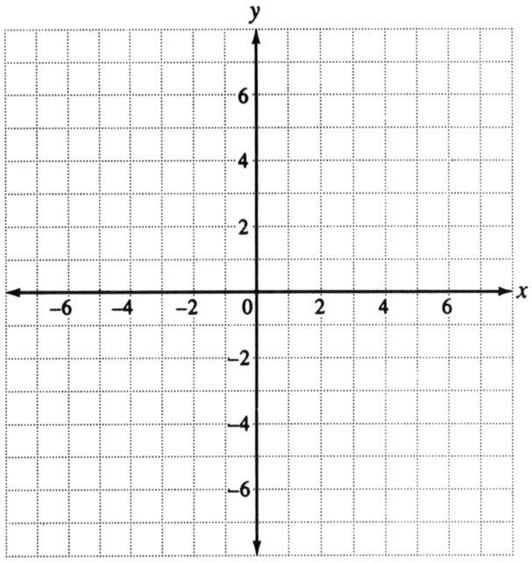

44. $y = 2x^2 - 8x + 7$

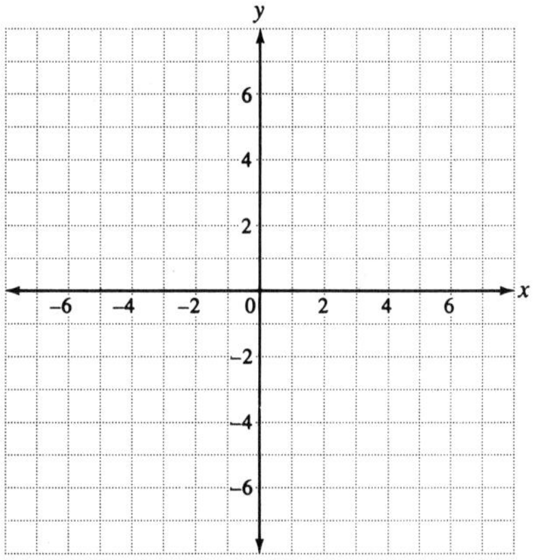

Graph.

45. $\frac{x^2}{16} - \frac{y^2}{4} = 1$

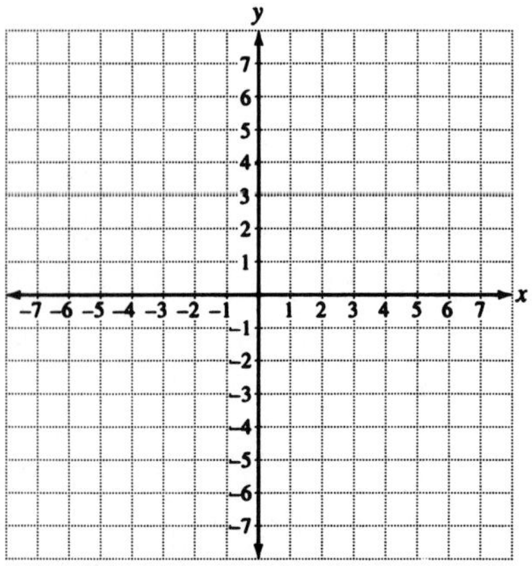

46. $\frac{y^2}{25} - \frac{x^2}{49} = 1$

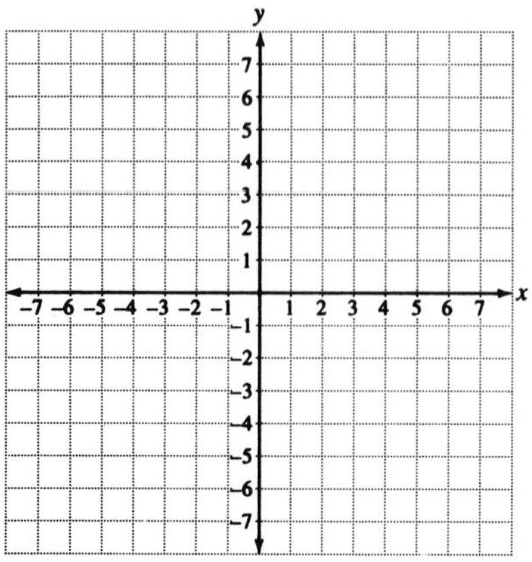

The equation of a hyperbola with center at the origin is

$$\frac{x^2}{a^2} - \frac{y^2}{b^2} = 1 \quad \text{or} \quad \frac{y^2}{b^2} - \frac{x^2}{a^2} = 1 \quad \text{Standard forms.}$$

The x-intercepts of the first equation are $(a, 0)$ and $(-a, 0)$. The y-intercepts of the second equation are $(0, b)$ and $(0, -b)$. These intercepts are often called ***vertices.***

The equation of a hyperbola with center at (h, k) is

$$\frac{(x-h)^2}{a^2} - \frac{(y-k)^2}{b^2} = 1 \quad \text{or} \quad \frac{(y-k)^2}{b^2} - \frac{(x-h)^2}{a^2} = 1 \quad \text{Standard forms.}$$

The vertices of the first equation are $(a + h, k)$ and $(-a + h, k)$. The vertices of the second equation are $(h, b + k)$ and $(h, -b + k)$. The corners of the rectangle that determines the asymptotes are $(a + h, b + k)$, $(a + h, -b + k)$, $(-a + h, b + k)$, and $(-a + h, -b + k)$.

Find the center, vertices, and the corners of the rectangle that determines the asymptotes for each hyperbola. Graph the hyperbola.

* **47.** $\frac{(y-1)^2}{4} - \frac{(x-1)^2}{1} = 1$

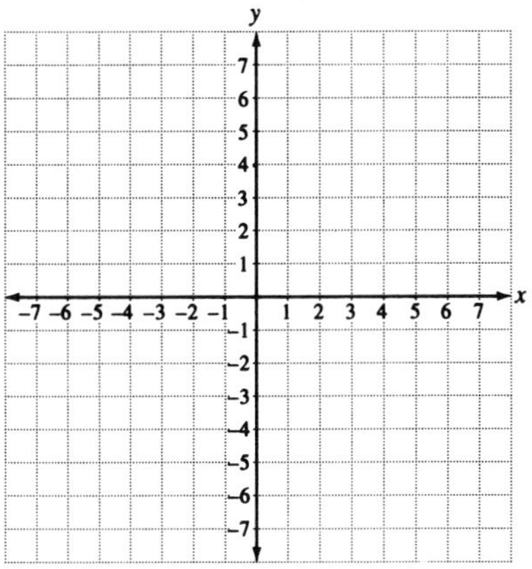

* **48.** $\frac{(x-2)^2}{16} - \frac{(y-1)^2}{25} = 1$

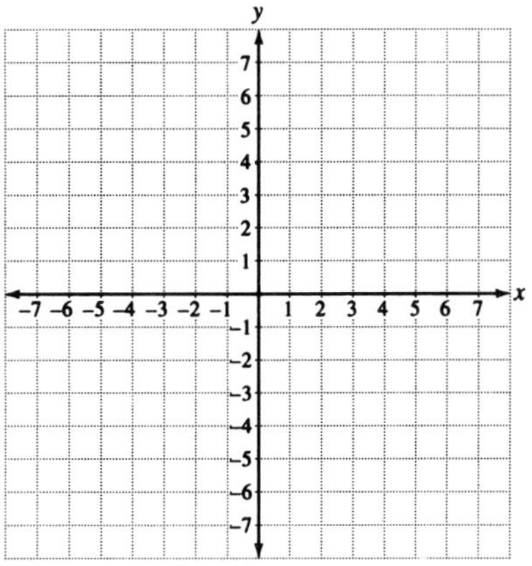

* **49.** $\frac{(x-5)^2}{25} - \frac{(y-4)^2}{4} = 1$

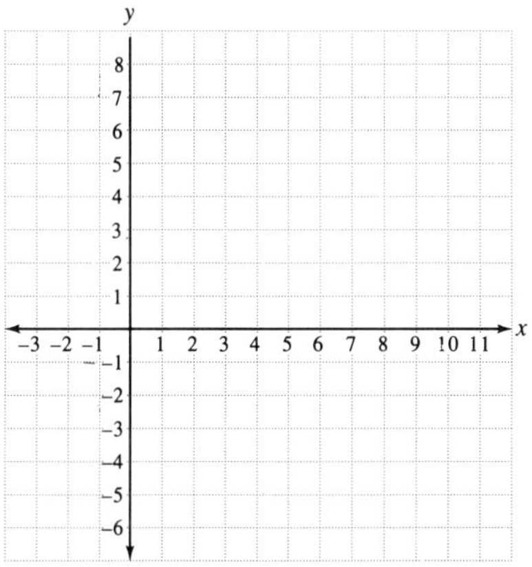

* **50.** $\frac{(y-2)^2}{9} - \frac{(x-1)^2}{16} = 1$

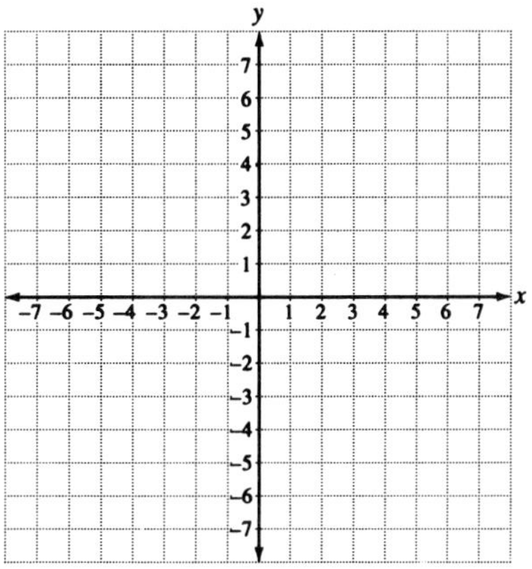

Section 9.4

Solve.

51. $x^2 - y^2 = 24$
$x - 2y = 3$

52. $x^2 - 2y = 0$
$x + 2y = 6$

53. $xy = 6$
$x + 2y = 7$

54. $xy = -4$
$2x + y = 2$

55. $y = 2x$
$x^2 + 2y^2 = 9$

56. $x = 3y$
$2x^2 + 3y^2 = 21$

57. $y - x = 4$
$y = x^2 - 2x + 6$

58. $x + y = 4$
$y = x^2 - 4x - 6$

59. A template in the shape of a rectangle has an area of 96 square centimeters and its perimeter is 40 centimeters. Find the length and width of the template.

Section 9.5

Solve.

60. $x^2 - 16y^2 = 4$
$3x^2 + 16y^2 = 12$

61. $4x^2 + 3y^2 = 27$
$4x^2 + 2y^2 = 18$

62. $2x^2 + 3y^2 = 20$
$x^2 + 3y^2 = 16$

63. $3y^2 = x$
$x = y^2 - y + 1$

64. $y = 2x^2 + 2x - 1$
$y = x^2 + 2x$

65. $x^2 + y^2 = 13$
$y = x^2 - 1$

66. $x^2 + y^2 = 5$
$xy = 2$

67. $x^2 + 4y^2 = 20$
$xy = 4$

68. A rectangular piece of metal has an area of 256 square inches. Four squares with sides 2 inches long are cut from each corner. The edges are turned up to form an open box. If the volume of the box is 224 cubic inches, what are the dimensions of the piece of metal?

COOPERATIVE LEARNING

1. Find the distance between $(6a, 4a)$ and $(-3a, -8a)$.

2. Find the midpoint of the line segment with endpoints $(6\sqrt{2}, 5\sqrt{12})$ and $(\sqrt{8}, 2\sqrt{27})$.

Find an equation for each of the following.

3.

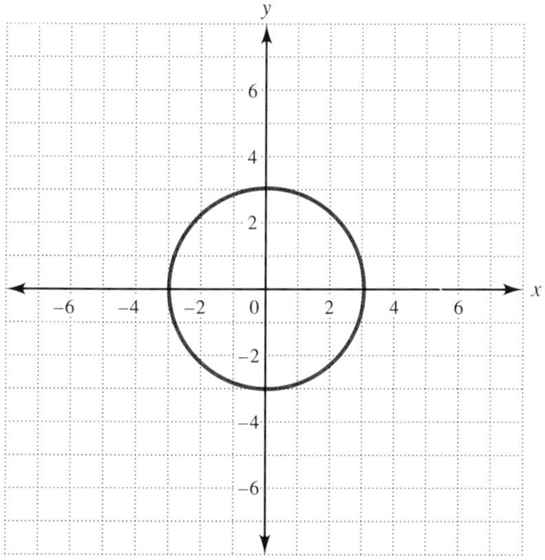

4.

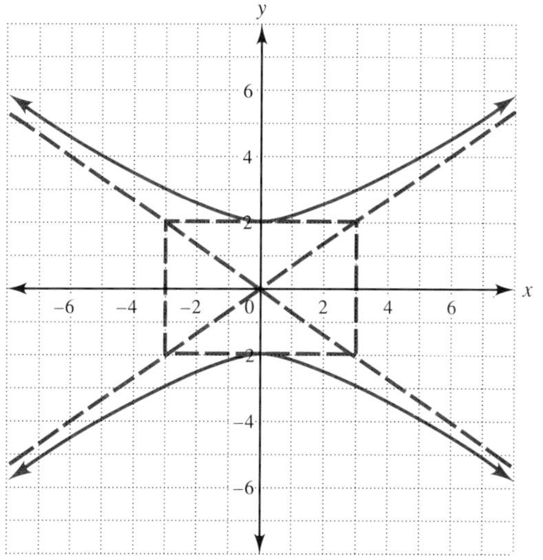

5.

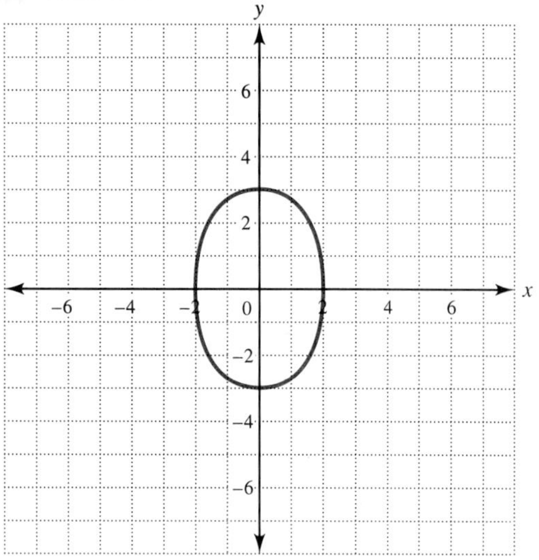

6. A steel plate is in the shape of an ellipse. If the horizontal length of the plate is 10 centimeters and the vertical length of the plate is 8 centimeters, write an equation to describe the plate.

7. Solve the equation $x^2 + y^2 = 25$ for y. What is the equation for the top half of the circle? What is the equation for the bottom half of the circle?

8. The plaza at Downtown University has opposite sides that are hyperbolic. One set of sides is described by the equation $x^2 - y^2 = 900$. The other pair is described by the equation $2y^2 - x^2 = 1225$. Draw the graph of the plaza.

9. A conic section passes through the points $(4, 0)$, $(-1, \sqrt{15})$, $(2, 2\sqrt{3})$, $(-2, -2\sqrt{3})$, $(1, -\sqrt{15})$, and $(-4, 0)$. Write an equation for the conic section.

Chapter 9 Practice Test

NAME ______________________

DATE ______________________

CLASS ______________________

1. Find the distance between (5, 7) and (−2, 9).

2. Find the midpoint of the line segment with endpoints (5.4, −9.2) and (−6, 4.1).

3. Write an equation for the circle with center (4, −2) and radius $\sqrt{7}$.

4. Find the center and radius of the circle $x^2 + y^2 + 4x - 8y = 5$.

Graph.

5. $\dfrac{x^2}{25} + \dfrac{y^2}{9} = 1$

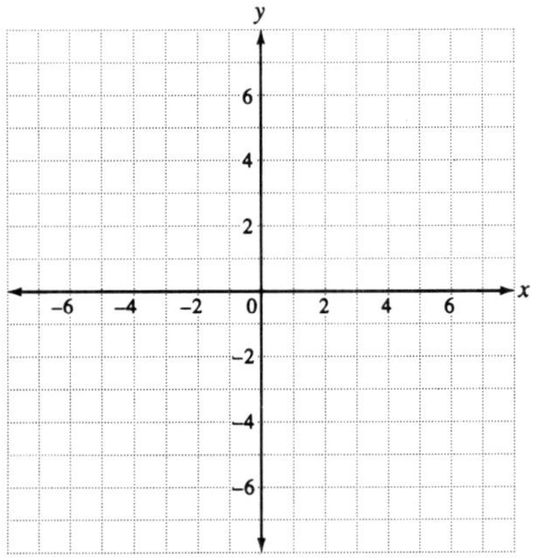

1. ______________________

2. ______________________

3. ______________________

4. ______________________

5. ______________________

6. ____________________

6. $x = 2(y - 1)^2 + 3$.

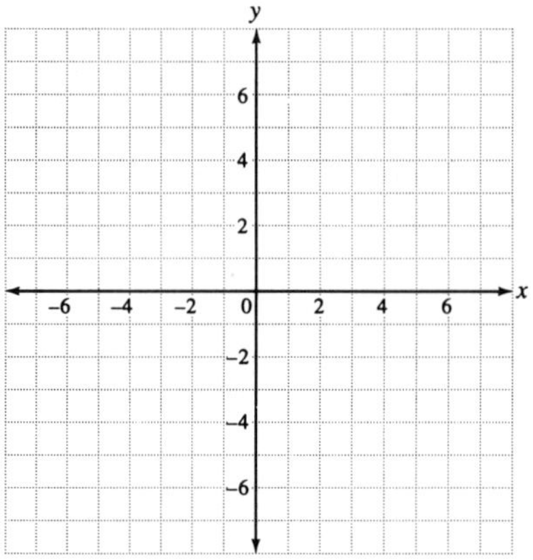

7. ____________________

7. $\frac{x^2}{4} - \frac{y^2}{25} = 1$

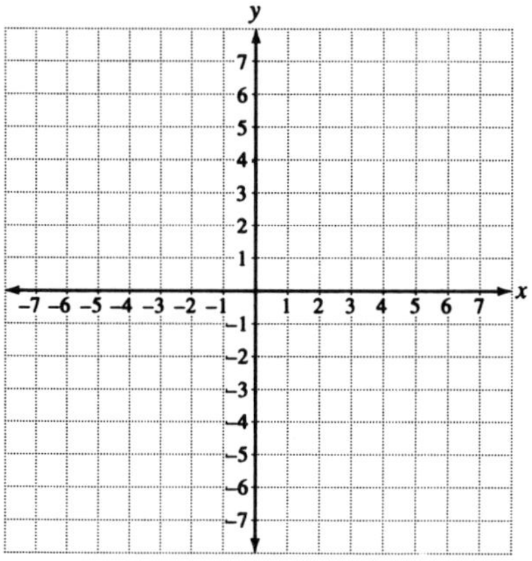

8. ____________________

8. Identify the conic section: $7x^2 + 4y^2 = 28$.

Solve.

9. ____________________

9. $x^2 + y^2 = 16$

$x + 2y = 8$

10. ____________________

10. $5x^2 - 2y^2 = -13$

$3x^2 + 4y^2 = 39$

11. ____________________

11. The area of a rectangular stamp is 20 square centimeters and the length of a diagonal is $\sqrt{41}$ centimeters. Find the dimensions of the stamp.

CHAPTER

10

Relations and Functions

Pretest

Consider the set $\{(-4, 0), (-3, -11), (5, 8), (6, -12), (-7, -3)\}$.

1. Is the set a function?

2. Find the domain.

3. Find the range.

4. Find the domain of the function: $y = \dfrac{8}{x - 4}$.

5. Use the vertical line test to decide if the following is a function:

$$\frac{x^2}{16} - \frac{y^2}{25} = 1$$

6. If $f(x) = 3x^2 - 7x - 2$, find $f(-3)$.

7. If $f(x) = 3x^2 - 4x + 8$ and $g(x) = x - 2$, find $f[g(x)]$.

Let $f(x) = 4x^2 + 17x - 20$ *and* $g(x) = 4x - 3$.

8. Find $g - f$.

9. Find $\frac{f}{g}$.

10. Identify as a linear function, identity function, constant function, absolute value function, quadratic function, or power function.

$$f(x) = -4x + 8$$

11. Determine whether $F(x) = x^2 - 7x + 12$ is one-to-one.

12. Find the inverse of the function: $f(x) = 2x^2 - 3$, $x \geq 0$.

10.1 BASIC DEFINITIONS

Relations

There are many applications of relations in the real world. The wages that a person earns are "related" to the number of hours that he or she works.

Recall from Section 1.1 that we define a ***set*** to be a collection of objects. The objects in the set are called ***members*** of the set. We use braces { } to enclose the members of the set.

> Any set of ordered pairs is a ***relation.***

The ordered pairs may be listed or we may be given an equation that shows how to find the ordered pairs.

The set $\{(1, 3), (2, 5), (1, 7), (3, -8)\}$ is a relation. The set of ordered pairs given by $W = 5h$ is a relation. In this equation we could define h as the number of hours worked. Then if the person earned \$5 per hour, W would be the wages earned.

1 Functions

A function is a special type of relation that is used extensively in mathematics.

> A ***function*** is a relation in which no two ordered pairs have the same first coordinates and different second coordinates.

EXAMPLE 1 Are the following functions?

a. $\{(1, 5), (-2, 4), (0, 8), (6, 7)\}$

The set $\{(1, 5), (-2, 4), (0, 8), (6, 7)\}$ is a function since no two ordered pairs have the same first coordinates.

b. $\{(3, 7), (4, 5), (4, -2), (9, 0)\}$

The set $\{(3, 7), (4, 5), (4, -2), (9, 0)\}$ is *not* a function since two ordered pairs have the same first coordinate, which is 4, and different second coordinates.

c. $y = x + 3$

We may obtain an infinite number of ordered pairs from this equation. However, for each different value of x that we use, we will get a different value for y. Therefore, no two ordered pairs have the same first coordinates and different second coordinates.

The relation $y = x + 3$ is a function.

d. $y^2 = x$

If we solve this equation for y, we get

$$y = \pm\sqrt{x}$$

For each value of x except 0, we get two values for y. For example, we could get the ordered pairs (4, 2) and (4, −2). Two ordered pairs have the same first coordinates and different second coordinates.

This is *not* a function. ■

□ DO EXERCISE 1.

OBJECTIVES

1. *Identify functions*
2. *Find the domains and ranges of functions defined by a set of ordered pairs*
3. *Find the domains of functions defined by equations*
4. *Determine if equations are functions by the vertical line test*

□ **Exercise 1** Are the following functions?

a. $\{(3, -3), (-5, 7), (6, 0), (3, 4)\}$

b. $\{(8, 7), (-1, 4), (-3, 4), (0, 8)\}$

c. $y = 2x - 3$

d. $y^2 = x - 9$

□ **Exercise 2** Find the domain and range.

a. $\{(0, 7), (-2, 5), (-1, 3), (8, 7)\}$

b. $\{(3, -9), (4, 8), (2, 5), (6, -1)\}$

A function can also be expressed as a correspondence or mapping from one set to another. For example, the equation $W = 5h$, where W is wages and h is the number of hours worked, is also a function. The mapping is as follows.

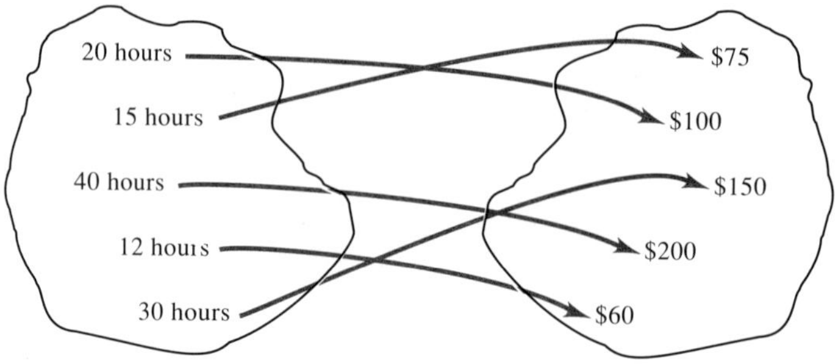

> A function is a correspondence or mapping that assigns to each member of some set exactly one member of another set.

2 Domain and Range

> The set of all first coordinates of the ordered pairs of a function is called the ***domain*** of the function. The set of all second coordinates of the ordered pairs is called the ***range.***

EXAMPLE 2 Find the domain and the range of the function $\{(1, 5), (-7, 2), (4, 5), (0, 3)\}$.

The domain is $\{1, -7, 4, 0\}$. The range is $\{5, 2, 3\}$.

Caution: Do not repeat the 5. (We do not repeat members of a set.) ■

□ DO EXERCISE 2.

3 Domains of Functions Defined by Equations

Generally, it is not possible to list completely all the members of the domain and range of a function whose members are given by an equation. The number of ordered pairs that can be obtained from an equation is usually infinite. We can, however, find the domain and range of these functions. We will consider only the domain.

> The domain of a function given in terms of an equation is the set of all real numbers that can be substituted for x in the equation.

We cannot substitute values of x in an equation that will give a zero denominator. We also cannot use values of x that will give the square root of a negative number since we do not want imaginary numbers.

EXAMPLE 3 Find the domain of the following.

a. $y = x - 2$

We can substitute any value for x in this equation. The domain is *all real numbers*.

b. $y = \dfrac{3}{x-5}$

We cannot substitute 5 for x in this equation or we will get a zero denominator. The domain is all real numbers except 5.

c. $y = \sqrt{3x-6}$

The number under the radical sign must be greater than or equal to zero.

$$3x - 6 \geq 0$$
$$3x \geq 6$$
$$x \geq 2$$

The domain is $\{x \mid x \geq 2\}$. ■

□ **DO EXERCISE 3.**

4 Vertical Line Test for Functions

A relation is a function if no two ordered pairs have the same first coordinates and different second coordinates. If a relation does have two ordered pairs with the same first coordinates and different second coordinates, they will lie on a vertical line. Therefore, if we graph an equation and a vertical line intersects the graph in more than one point, the relation given by the equation is not a function.

If a vertical line intersects the graph of an equation in more than one point, the relation given by the equation is not a function.

EXAMPLE 4 Use the vertical line test to decide if the following are functions.

a. $y = x^2$

The graph is as shown. Notice that the vertical line cuts the graph in only one point. The relation is a function.

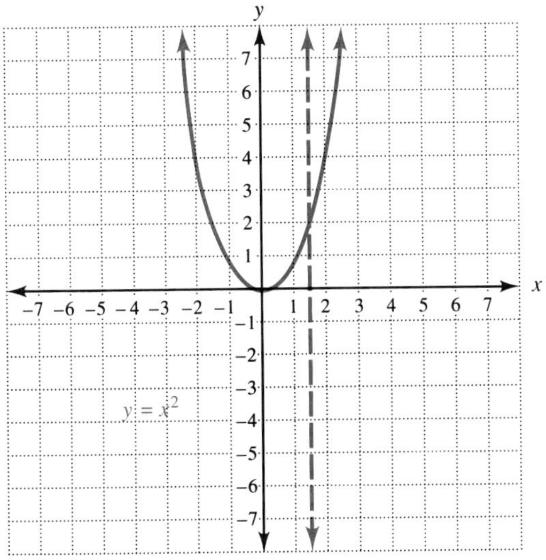

□ **Exercise 3** Find the domain of the following functions.

a. $y = x + 8$

b. $y = \dfrac{1}{x+7}$

c. $y = \sqrt{x-8}$

d. $y = x^2$

□ **Exercise 4** Use the vertical line test to decide if the following are functions.

a. $y = x - 2$

b. $x^2 + y^2 = 16$

b. $y^2 = x$

On the graph the vertical line intersects the graph in more than one point. The relation is not a function.

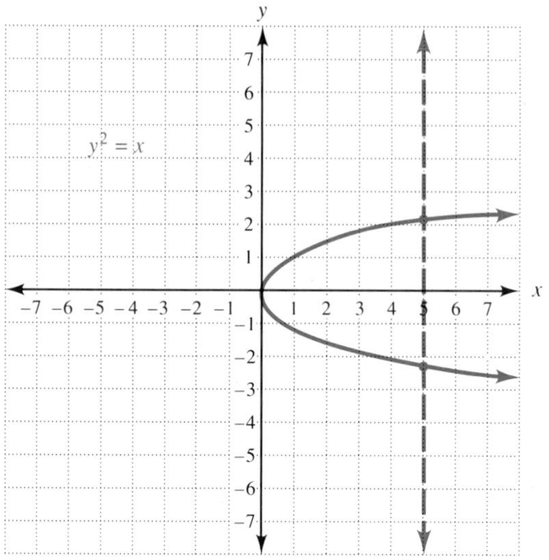

■

□ **DO EXERCISE 4.**

DID YOU KNOW?

Lejeune Dirichlet (1805–1859) defined function as we use it today. He was a student of the great mathematician Gauss, and once rescued a bit of manuscript Gauss was using to light his pipe.

Answers to Exercises

1. a. No **b.** Yes **c.** Yes **d.** No

2. a. Domain: $\{0, -2, -1, 8\}$; Range: $\{7, 5, 3\}$
b. Domain: $\{3, 4, 2, 6\}$; Range: $\{-9, 8, 5, -1\}$

3. a. All real numbers **b.** All real numbers except -7
c. $\{x|x \geq 8\}$ **d.** All real numbers

4. a. Function **b.** Not a function

Problem Set 10.1

NAME

DATE

CLASS

Are the following functions? If so, give the domain and range of each.

1. $\{(1, 5), (3, 7), (-1, 8)\}$

2. $\{(0, 7), (-3, -6), (5, 7)\}$

3. $\{(3, 4), (7, 1), (7, -5)\}$

4. $\{(-6, 8), (0, 2), (-6, 9)\}$

5. $\{(-5, 2), (3, 2), (-4, -7)\}$

6. $\{(1, 4), (-1, 5), (3, 8)\}$

7. $\{(4, 12), (-5, 3), (-8, -2), (4, 7)\}$

8. $\{(-3, 5), (6, -8), (-5, -7), (6, 0)\}$

9. $\{(3, -8), (0, 10), (4, 7), (-5, 3), (-9, -8)\}$

10. $\{(5, 0), (-3, 4), (2, 11), (-6, -7), (-1, 4)\}$

11. $\{(5, 12), (-3, 8), (27, 4), (15, -32), (7, 4)\}$

12. $\{(35, 3), (-7, 16), (8, -2), (-7, 29), (5, 88)\}$

13. $\{(300, -200), (400, 600), (-100, -450), (325, -650)\}$

14. $\{(-750, -230), (-280, -500), (-280, 475), (0, 895)\}$

Are the following functions? If so, give the domain.

15. $y = x - 4$

16. $y = x + 7$

17. $y = x^2 + 2$

18. $y = x^2 - 8$

19. $x = y^2 + 4$

20. $x = y^2 - 3$

21. $y = \frac{1}{x - 3}$

22. $y = \frac{3}{x + 8}$

23. $x^2 + y^2 = 9$

24. $x^2 - y^2 = 25$

25. $y = \frac{5}{3x - 4}$

26. $y = \frac{1}{2x + 5}$

27. $y = \sqrt{x}$

28. $y = \sqrt{x + 5}$

29. $y = \sqrt{3x - 1}$

30. $y = \sqrt{4x - 5}$

Use the vertical line test to decide if the following are functions.

31. $x + y = 4$

32. $y + 2x = 3$

33. $x^2 + y^2 = 9$

34. $x^2 + y^2 = 25$

35. $y = -(x - 2)^2 + 3$

36. $y = (x + 4)^2 - 1$

37. $4x^2 - 9y^2 = 36$

38. $4x^2 + 9y^2 = 36$

39. $y = x^2 + 3x - 1$

40. $y = -x^2 - 2x + 2$

41. If a person earns \$6 per hour, her wages W are given by the equation $W = 6h$, where h is hours worked. Is this a function?

42. The distance d an object has fallen from rest toward the earth is given by the equation $d = 16t^2$, where t is the time the object has been falling. If $t > 0$, is this a function?

43. Any set of ordered pairs is a ____________.

44. A function is a relation in which no two ordered pairs have the same ____________ coordinates and different second coordinates.

45. The set of all second coordinates of the ordered pairs of a function is called the ____________ of the function.

46. The ____________ of a function given in terms of an equation is the set of all real numbers that can be substituted for x in the equation.

47. If a vertical line intersects the graph of an equation in only one point, the equation is a ____________.

Think About It

The range of a function given in terms of an equation is the set of all real numbers that can be found for y, where y is a function of x. Find the range of the following.

* **48.** $y = x + 7$

* **49.** $y = \dfrac{1}{x}$

* **50.** $y = \dfrac{1}{x - 4}$

* **51.** $y = \sqrt{x}$

* **52.** $y = \sqrt{3x + 6}$

* **53.** $x^2 + y^2 = 16$
(*Hint:* Draw the graph.)

* **54.** $y = |x|$

* **55.** For the Best Company, the relationship between profit (in hundreds of dollars) and units sold is given by $P = -x^2 + 26x - 160$, where P is profit in dollars and x is the number of units sold.
a. Draw the graph of the equation.
b. Is this equation a function?
c. For what number of units is the profit a maximum?

Checkup

The following problems provide a review of some of Section 8.9.

Solve.

56. $x^2 - 3x - 10 \leq 0$

57. $x^2 - 7x - 8 \geq 0$

58. $2x^2 - 7x - 15 > 0$

59. $3x^2 - 14x + 8 < 0$

60. $y^2 > 2y$

61. $y^2 < -3y$

62. $\frac{1}{x - 4} < 0$

63. $\frac{1}{x + 3} > 0$

64. $\frac{x - 2}{x + 5} \geq 0$

65. $\frac{x + 1}{x + 4} \leq 0$

10.2 FUNCTION NOTATION AND COMPOSITE FUNCTIONS

1 Function Notation

Function notation gives us another way of finding y when y is a function of x. Instead of writing $y = x^2 + 2$, we write $f(x) = x^2 + 2$. This is called *function notation.* The symbol $f(x)$ is read "f of x." Notice that $f(x)$ and y mean the same thing when we are discussing functions. We may also use symbols such as $g(x)$, $h(x)$, and $F(x)$ for functions.

The equations $y = x^2 + 2$ and $f(x) = x^2 + 2$ are the same functions. If we replace x with 3, we get the same result.

	$y = x^2 + 2$	$f(x) = x^2 + 2$
Let $x = \mathbf{3}$		$f(\mathbf{3}) = \mathbf{3}^2 + 2$
	$y = \mathbf{3}^2 + 2$	$f(3) = 11$
	$y = 11$	

EXAMPLE 1 Find the following if $f(x) = 3x - 4$.

a. $f(\mathbf{2}) = 3(\mathbf{2}) - 4$ Substitute 2 for x.

$= 6 - 4 = 2$

b. $f(\mathbf{-5}) = 3(\mathbf{-5}) - 4$ Substitute -5 for x.

$= -15 - 4 = 4 = -19$

c. $f(\mathbf{0}) = 3(\mathbf{0}) - 4$ Substitute 0 for x.

$= 0 - 4 = -4$ ■

Caution: $f(x)$ does not mean f times x, but represents the y value for the indicated x value.

□ **DO EXERCISE 1.**

EXAMPLE 2 Find the following if $f(x) = 2x^2 + 4$.

a. $f(\mathbf{3}) = 2(\mathbf{3})^2 + 4 = 2(9) + 4 = 18 + 4 = 22$

b. $f(\mathbf{-1}) = 2(\mathbf{-1})^2 + 4 = 2(1) + 4 = 2 + 4 = 6$

c. $f(\boldsymbol{a}) = 2(\boldsymbol{a})^2 + 4$ Substitute a for x.

$= 2a^2 + 4$

d. $f(\boldsymbol{x + h}) = 2(\boldsymbol{x + h})^2 + 4$ Substitute $x + h$ for x.

$= 2[x^2 + 2hx + h^2] + 4$

$= 2x^2 + 4hx + 2h^2 + 4$ ■

□ **DO EXERCISE 2.**

The following expression is important in the calculus:

$$\frac{f(x + h) - f(x)}{h}$$

OBJECTIVES

1 *Use function notation*

2 *Write composite functions, given two functions*

□ **Exercise 1** Find the following if $f(x) = 5x - 7$.

a. $f(3)$

b. $f(-4)$

□ **Exercise 2** Find the following if $f(x) = x^2 + 3x - 1$.

a. $f(2)$

b. $f(-2)$

c. $f(b)$

d. $f(x + h)$

□ **Exercise 3** Find $(f(x + h) - f(x))/h$ for the following.

a. $f(x) = 2x + 3$

b. $f(x) = 3x^2 - 1$

EXAMPLE 3 Find $(f(x + h) - f(x))/h$ for the following.

a. $f(x) = 3x - 4$

First, find $f(x + h)$.

$$f(\boldsymbol{x} + \boldsymbol{h}) = 3(\boldsymbol{x} + \boldsymbol{h}) - 4 = 3x + 3h - 4$$

Then

$$\frac{f(x + h) - f(x)}{h} = \frac{(3x + 3h - 4) - (3x - 4)}{h} \qquad \text{Substitute for } f(x + h) \text{ and } f(x).$$

$$= \frac{3x + 3h - 4 - 3x + 4}{h}$$

$$= \frac{3h}{h} = 3$$

b. $f(x) = x^2 - 9$

Find $f(x + h)$.

$$f(\boldsymbol{x} + \boldsymbol{h}) = (\boldsymbol{x} + \boldsymbol{h})^2 - 9 = x^2 + 2hx + h^2 - 9$$

Then

$$\frac{f(x + h) - f(x)}{h} = \frac{(x^2 + 2hx + h^2 - 9) - (x^2 - 9)}{h} \qquad \text{Substitute for } f(x + h) \text{ and } f(x).$$

$$= \frac{x^2 + 2hx + h^2 - 9 - x^2 + 9}{h}$$

$$= \frac{h(2x + h)}{h} = 2x + h \qquad ■$$

□ **DO EXERCISE 3.**

2 Composite Functions

We may want to replace the variable in a function by another function. When we do this, we form a ***composite function.*** There are real-world examples of composite functions. The volume of gasoline in an upright cylindrical tank is a function of the radius and the height of gasoline in the tank, but if gasoline is being pumped into the tank at a constant rate, the height is a function of the time that gasoline has been pumped into the tank. The new composite function is $f[g(x)]$ or $f \circ g(x)$. The following diagram illustrates a composite function.

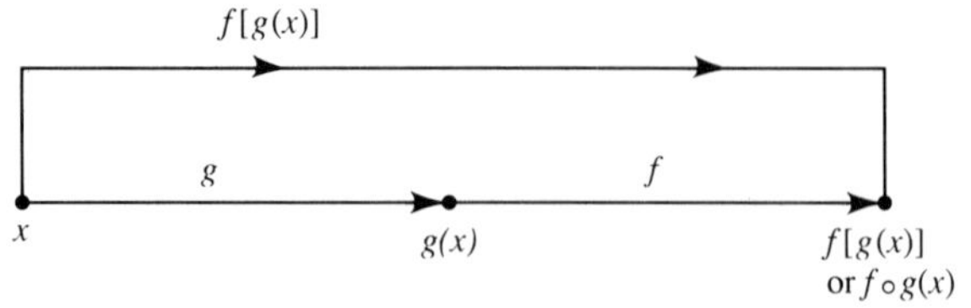

EXAMPLE 4 If $f(x) = 3x - 2$ and $g(x) = x^2 + 4$, find the following.

a. $f[g(x)]$

We replace x on the right side of the equation $f(x) = 3x - 2$ by $x^2 + 4$.

$$f(x) = 3x - 2$$

$$f[g(x)] = 3(\mathbf{x^2 + 4}) - 2 \qquad \text{Substitute } x^2 + 4 \text{ for } x \text{ on the right.}$$

$$= 3x^2 + 12 - 2$$

$$= 3x^2 + 10$$

b. $g[f(x)]$

Replace x on the right side of $g(x) = x^2 + 4$ by $3x - 2$.

$$g(x) = x^2 + 4$$

$$g[f(x)] = (\mathbf{3x - 2})^2 + 4 \qquad \text{Substitute } 3x - 2 \text{ for } x \text{ on the right.}$$

$$= 9x^2 - 12x + 4 + 4$$

$$= 9x^2 - 12x + 8 \quad \blacksquare$$

The expressions $f[g(x)]$ and $g[f(x)]$ are both called *composite functions* and $f[g(x)]$ is called the *composition* of f with g.

□ **DO EXERCISE 4.**

□ **Exercise 4** If $f(x) = 2x - 5$ and $g(x) = 4x^2 - 1$, find the following.

a. $f[g(x)]$

b. $g[f(x)]$

Answers to Exercises

1. a. 8 **b.** −27

2. a. 9 **b.** −3 **c.** $b^2 + 3b - 1$

d. $x^2 + 2hx + h^2 + 3x + 3h - 1$

3. a. 2 **b.** $3(2x + h)$

4. a. $8x^2 - 7$ **b.** $16x^2 - 80x + 99$

Problem Set 10.2

NAME
DATE
CLASS

If $f(x) = 2x - 5$ and $g(x) = 3x^2 - 2$, find the following.

1. $f(3)$ **2.** $f(7)$ **3.** $g(2)$ **4.** $g(1)$

5. $f(-4)$ **6.** $f(-8)$ **7.** $g(-2)$ **8.** $g(-5)$

9. $f(0)$ **10.** $g(-3)$ **11.** $g(a)$ **12.** $f(b)$

13. $g(x + h)$ **14.** $f(x + h)$

If $P(x) = x^4 + 2x^3 - 3x^2 - x + 1$, find the following.

15. $P(2)$ **16.** $P(-2)$ **17.** $P(-1)$ **18.** $P(0)$

Find $(f(x + h) - f(x))/h$ for the following functions.

19. $f(x) = 4x - 1$ **20.** $f(x) = 5x + 2$ **21.** $f(x) = x^2 + 2$ **22.** $f(x) = x^2 - 5$

23. $f(x) = 2x^2 - 4$ **24.** $f(x) = 4x^2 + 1$ **25.** $f(x) = x^2 + 3x$ **26.** $f(x) = x^2 - 2x$

For the following functions, find $f[g(x)]$ and $g[f(x)]$.

27. $f(x) = 2x - 1$
$g(x) = x^2 + 3$

28. $f(x) = x^2 - 5$
$g(x) = 3x + 2$

29. $f(x) = 3x^2 - 1$
$g(x) = x - 4$

30. $f(x) = x + 7$
$g(x) = 2x^2 + 3$

31. $f(x) = x^2 + 3x - 8$
$g(x) = x - 2$

32. $g(x) = x + 5$
$f(x) = x^2 - 2x + 7$

33. $f(x) = 3x^2 + 1$
$g(x) = \frac{1}{x}$

34. $f(x) = 5x^2 - 2$
$g(x) = \frac{4}{x}$

35. $f(x) = \frac{5}{x^2}$
$g(x) = x - 4$

36. $f(x) = \frac{2}{x^2}$
$g(x) = x^2 + 8$

37. The function $V(t) = -800t + 14{,}000$ gives the book value V after t years of an automobile. Find $V(6)$.

38. The cost $C(x)$ of oranges is a function of the number of pounds x that are purchased. If oranges are \$0.59 per pound, then $C(x) = 0.59x$. Find $C(4)$.

39. The amount paid for gasoline, $f(x)$, is a function of the number of gallons, x, purchased. If the price of gasoline is 90 cents per gallon, then $f(x) = 0.9x$. What does $f(10)$ represent in this problem?

40. The postage rate for a letter is 29 cents per ounce and 20 cents for each additional ounce. If x is the number of additional ounces and $f(x)$ is the cost of mailing the letter, then $f(x) = 0.20x + 0.29$. What is $f(3)$?

41. The symbol $f(x)$ is read "f ________ x."

42. When a variable in a function is replaced by another function, a ________ function is formed.

If $f(x) = 3x^2 - 7.1x + 5.28$, find the following.

43. $f(4.9)$ **44.** $f(-3.7)$ **45.** $f(-0.6)$ **46.** $f(0.8)$

Find $\dfrac{f(x+h) - f(x)}{h}$ for the following functions.

* **47.** $f(x) = \dfrac{1}{x}$

* **48.** $f(x) = x + \dfrac{1}{x}$

* **49.** $f(x) = \sqrt{x}$ (*Hint:* Rationalize the numerator.)

* **50.** $f(x) = \sqrt{2x + 1}$

* **51.** $f(x) = \sqrt{x - 3}$

* **52.** Let $f(x) = x^2$ and $g(x) = \sqrt{x}$. Are $f[g(x)]$ and $g[f(x)]$ the same function? Explain.

Checkup

The following problems provide a review of some of Section 8.4.

Predict the number and type of solutions.

53. $3x^2 + 6x + 5 = 0$ **54.** $5x^2 + 2x + 3 = 0$ **55.** $4x^2 - 20x = -25$ **56.** $4x^2 - 12x = -9$

57. $2x^2 + 3x + 1 = 0$ **58.** $3x^2 + 7x + 2 = 0$ **59.** $x^2 + x + 3 = 0$ **60.** $x^2 + 10x + 8 = 0$

Find quadratic equations with the following solutions.

61. $\sqrt{5}, -\sqrt{5}$ **62.** $\sqrt{17}, -\sqrt{17}$ **63.** $\dfrac{5}{8}, \dfrac{16}{3}$ **64.** $\dfrac{7}{5}, \dfrac{10}{21}$

10.3 ALGEBRA OF FUNCTIONS

OBJECTIVE

[1] *Add, subtract, multiply, and divide functions*

[1] We know how to add, subtract, multiply, and divide polynomials. The algebra of functions gives us a shorter way of writing these problems.

If f and g are two functions with a common domain, we define the following functions.

> The function $f + g$ is the sum of the functions f and g.
>
> $$(f + g)(x) = f(x) + g(x)$$
>
> The function $f - g$ is the function f minus the function g.
>
> $$(f - g)(x) = f(x) - g(x)$$
>
> The function fg is the product of the functions f and g.
>
> $$(fg)(x) = f(x) \cdot g(x)$$
>
> The function f/g is the function f divided by the function g, where $g(x) \neq 0$.
>
> $$\left(\frac{f}{g}\right)(x) = \frac{f(x)}{g(x)} \quad \text{if } g(x) \neq 0$$

Notice that if $f(x) = 2x^2 - 7x + 3$ and $g(x) = x - 3$, we can find $f + g$, $f - g$, fg, and f/g. We could also find $g - f$ and g/f. This notation is more convenient than writing the addition, subtraction, multiplication, and division problems separately.

EXAMPLE 1 Let $f(x) = 2x^2 - 7x + 3$ and $g(x) = x - 3$.

a. Find $f + g$.

$$\begin{aligned}(f + g)(x) &= f(x) + g(x)\\ &= (2x^2 - 7x + 3) + (x - 3)\\ &= 2x^2 - 7x + 3 + x - 3\\ &= 2x^2 - 6x\end{aligned}$$

b. Find $f - g$.

$$\begin{aligned}(f - g)(x) &= f(x) - g(x)\\ &= (2x^2 - 7x + 3) - (x - 3)\\ &= 2x^2 - 7x + 3 - x + 3\\ &= 2x^2 - 8x + 6\end{aligned}$$

c. Find fg.

$$\begin{aligned}(fg)(x) &= f(x) \cdot g(x)\\ &= (2x^2 - 7x + 3)(x - 3)\\ &= (2x^2 - 7x + 3)x - (2x^2 - 7x + 3)(3) && \text{Use a distributive law.}\\ &= (2x^3 - 7x^2 + 3x) - (6x^2 - 21x + 9) && \text{Use a distributive law.}\\ &= 2x^3 - 7x^2 + 3x - 6x^2 + 21x - 9\\ &= 2x^3 - 13x^2 + 24x - 9\end{aligned}$$

□ **Exercise 1** If $f(x) = 3x^2 - 10x - 8$ and $g(x) = 3x + 2$, find the following.

a. $f + g$

b. $f - g$

c. fg

d. $\frac{f}{g}$

□ **Exercise 2** If $f(x) = x^2 + 7x - 3$, $g(x) = x - 2$, and $h(x) = -4x^2 - 8x + 9$, find the following.

a. $f + h$

b. $h - f$

c. $\frac{f}{g}$

d. $\frac{h}{g}$

d. Find $\frac{f}{g}$.

$$\left(\frac{f}{g}\right)(x) = \frac{f(x)}{g(x)}$$

$$= \frac{2x^2 - 7x + 3}{x - 3}$$

$$= \frac{(x - 3)(2x - 1)}{x - 3} \quad \text{Factor.}$$

$$= 2x - 1$$

Notice that the quotient is undefined for 3. ■

□ **DO EXERCISE 1.**

EXAMPLE 2 If $f(x) = x - 5$, $g(x) = x^2 - 6x - 5$, and $h(x) = 3x^2 + 2x - 6$, find the following.

a. $g + h$

$$(g + h)(x) = (x^2 - 6x - 5) + (3x^2 + 2x - 6)$$
$$= x^2 - 6x - 5 + 3x^2 + 2x - 6$$
$$= 4x^2 - 4x - 11$$

b. $h - g$

$$(h - g)(x) = (3x^2 + 2x - 6) - (x^2 - 6x - 5)$$
$$= 3x^2 + 2x - 6 - x^2 + 6x + 5$$
$$= 2x^2 + 8x - 1$$

c. $\frac{g}{f}$

$$\left(\frac{g}{f}\right)(x) = \frac{g(x)}{f(x)}$$

$$= \frac{x^2 - 6x - 5}{x - 5}$$

Since $x^2 - 6x - 5$ does not factor, we use long division.

$$\begin{array}{r} x - 1 \\ x - 5\overline{)x^2 - 6x - 5} \\ \underline{x^2 - 5x} \\ -x - 5 \\ \underline{-x + 5} \\ -10 \end{array}$$

$$\left(\frac{g}{f}\right)(x) = x - 1 + \frac{-10}{x - 5}$$

Notice that the quotient is undefined for 5. ■

□ **DO EXERCISE 2.**

Answers to Exercises

1. a. $3x^2 - 7x - 6$ **b.** $3x^2 - 13x - 10$ **c.** $9x^3 - 24x^2 - 44x - 16$ **d.** $x - 4$

2. a. $-3x^2 - x + 6$ **b.** $-5x^2 - 15x + 12$ **c.** $x + 9 + \frac{15}{x - 2}$ **d.** $-4x - 16 + \frac{-23}{x - 2}$

Problem Set 10.3

NAME ______________________

DATE ______________________

CLASS ______________________

If $f(x) = 2x^2 - 7x + 3$ *and* $g(x) = 2x - 1$*, find the following.*

1. $f + g$

2. $f - g$

3. $g - f$

4. $g + f$

5. fg

6. gg

7. $\frac{f}{g}$

8. $\frac{g}{f}$

9. ff

10. gf

If $f(x) = x^2 - 6x + 4$, $g(x) = 2x^2 + 7x - 3$, $h(x) = x - 4$, *and* $p(x) = 3x + 7$*, find the following.*

11. $f - g$

12. $g - f$

13. $f + g$

14. $g + p$

15. fh

16. gp

17. hp

18. hh

19. $\frac{f}{h}$

20. $\frac{g}{h}$

21. pp

22. ph

23. $p - h$

24. $h - p$

25. $\frac{h}{p}$

26. $\frac{p}{h}$

27. $f - h$

28. $f - p$

29. fp

30. gh

31. $\frac{g}{f}$

32. $\frac{f}{g}$

33. ff

34. gg

35. Data Stor Company finds that the revenue function for selling x boxes of disks per day is $R(x) = 12x - 0.06x^2$ and the cost function is $C(x) = 200 + x$. Find the profit function $P(x)$ if $P(x) = R(x) - C(x)$.

36. Miniword Company sells word processing programs for computers. The revenue function for selling x programs per day is $R(x) = 35x - 0.1x^2$ and the cost function is $C(x) = 3x + 100$. Find the profit function $P(x)$ if $P(x) = R(x) - C(x)$.

37. The function $f - g$ is the function f ____________ the function g.

38. The function fg is the ____________ of the functions f and g.

If $f(x) = 3.792x^2 + 8.34x - 7.89$, $g(x) = -6.85x^2 - 9.2x + 4.1$, $h(x) = 4.3x - 7.1$, $p(x) = 5.7x + 9.3$, find the following.

39. $f + g$

40. $f - g$

41. hp

42. pp

43. $g - f$

44. $h - g$

If $f(x) = 3 + x^2$ and $g(x) = 2x^2 - 2x + 4$, find the following.

* **45.** $4f(x) + 3g(x)$

* **46.** $f(a) - f(b)$

* **47.** $x \cdot f(x) - g(x)$

* **48.** $g(a + b)$

* **49.** $f(a - b)$

Think About It

* **50.** Find the values of x for which $f(x) = g(x)$.

Checkup

The following problems provide a review of some of Sections 8.7 and 9.3 and will help you with the next section.

Graph.

51. $y = -x^2 - 4x - 3$

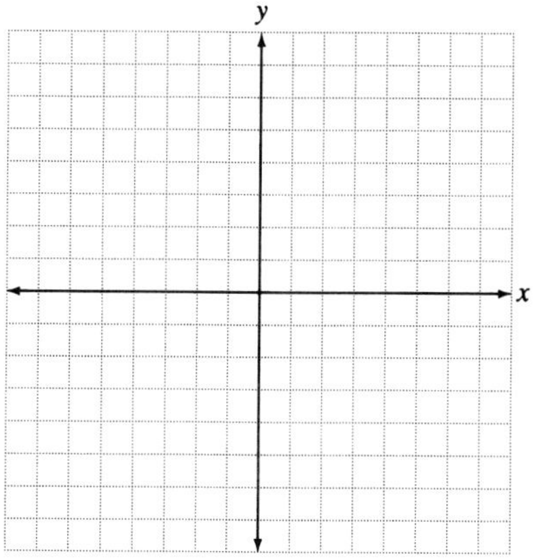

52. $y = x^2 + 5x + 6$

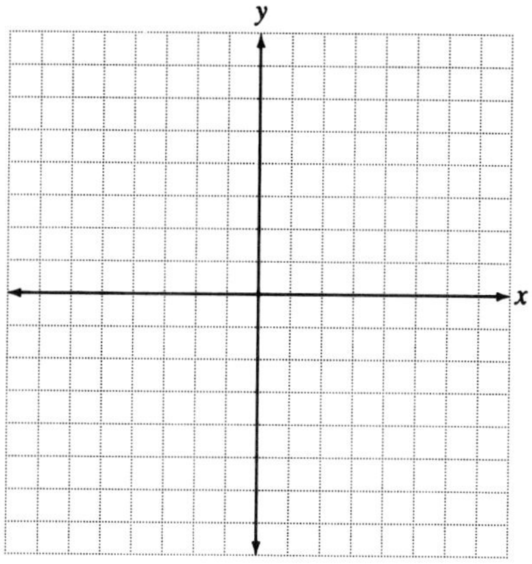

53. $y = 2x^2 - 1$

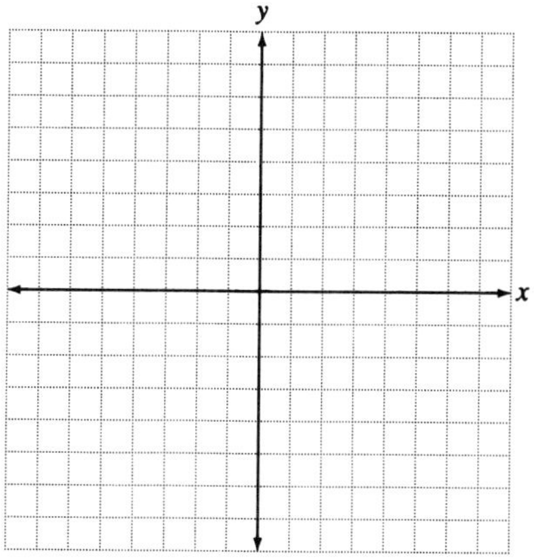

54. $y = 3x^2 + 1$

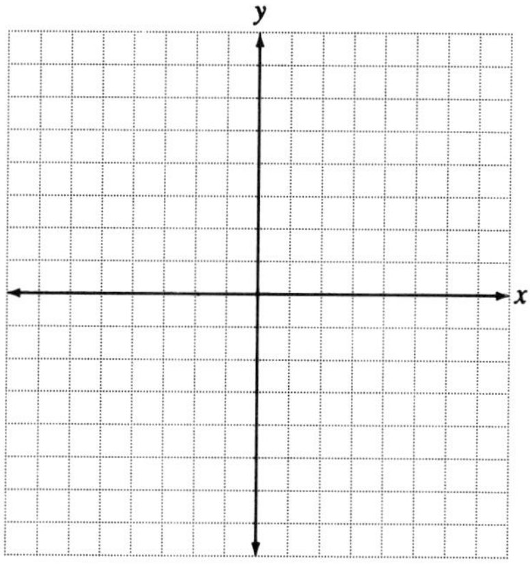

55. $y = (x + 2)^2 - 3$

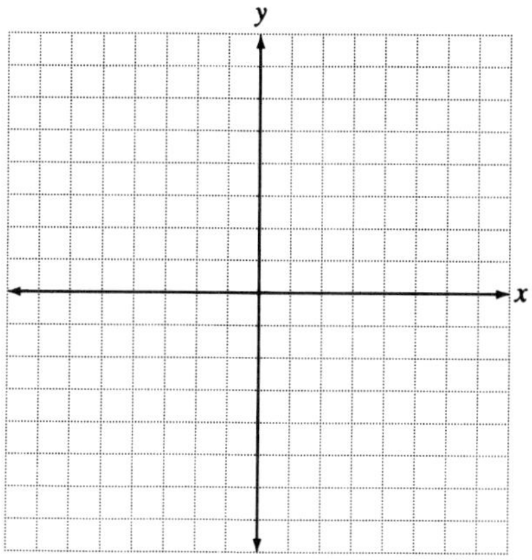

56. $y = (x - 1)^2 + 2$

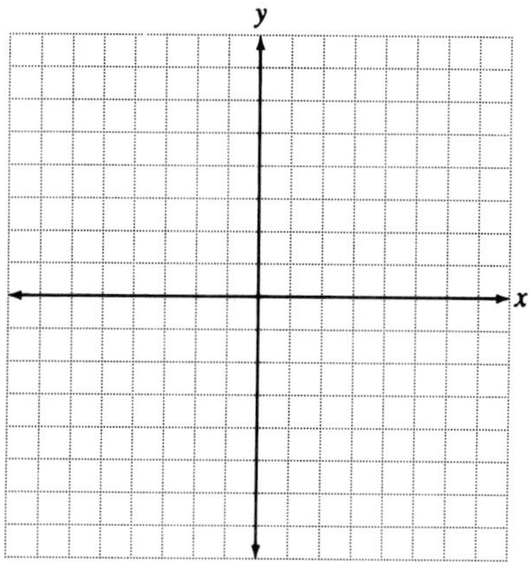

57. $\dfrac{x^2}{16} - \dfrac{y^2}{9} = 1$

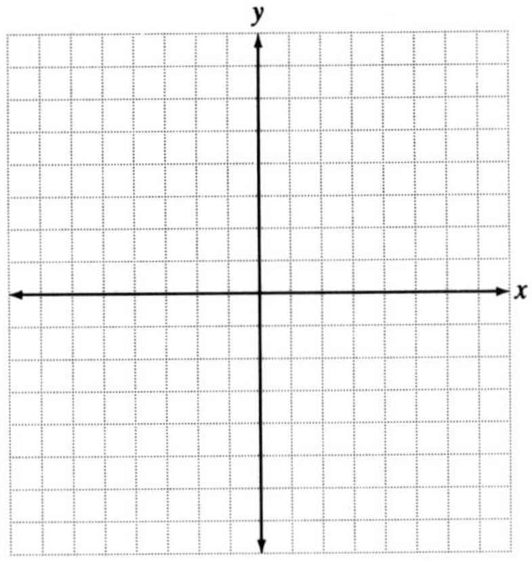

58. $\dfrac{y^2}{25} - \dfrac{x^2}{4} = 1$

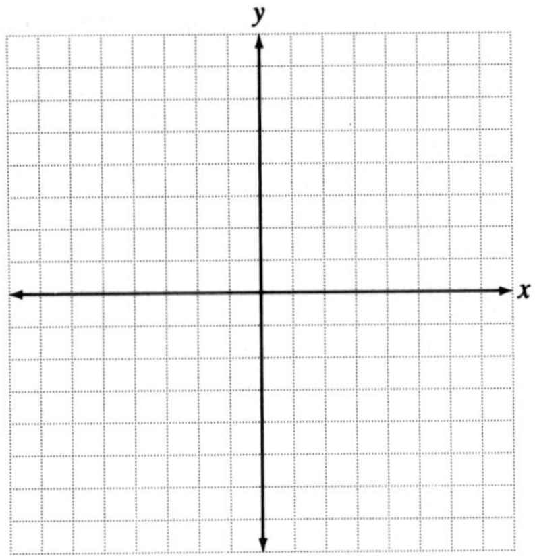

59. $\dfrac{x^2}{9} + y^2 = 1$

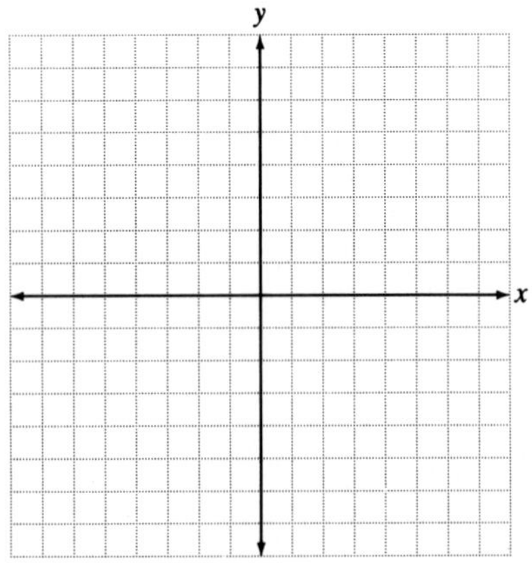

60. $x^2 + \dfrac{y^2}{49} = 1$

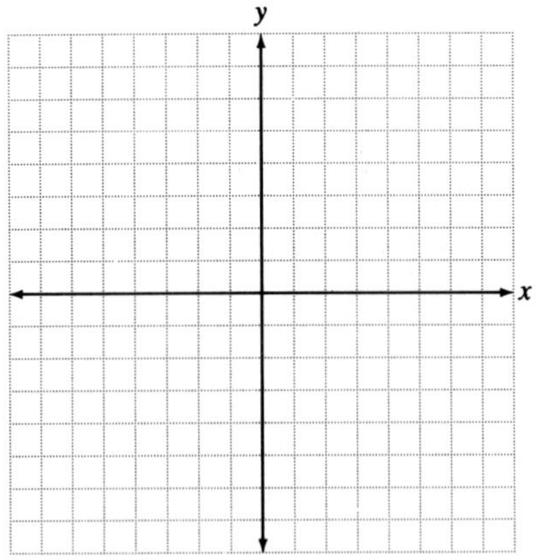

10.4 CLASSIFICATION OF FUNCTIONS

OBJECTIVE

1 *Identify and graph linear functions, identity functions, constant functions, absolute value functions, quadratic functions, and power functions*

1 There are many types of functions. We have already worked with most of the following kinds of functions. They are often given special names.

Linear Functions

A ***linear function*** is any function of the form

$$f(x) = mx + b$$

where m and b are real numbers.

The graph of any linear function is a straight line.

EXAMPLE 1 Graph the linear function $f(x) = 2x + 1$.

Recall that the equations $f(x) = 2x + 1$ and $y = 2x + 1$ are equivalent. Therefore, the graphs of the two equations are the same line. Graph the line using the slope and the y-intercept. The slope is 2 and the y-intercept is 1. The graph is shown below. The y-axis is usually labeled y instead of $f(x)$. Notice that this is a function because a vertical line would not intersect the graph in more than one point.

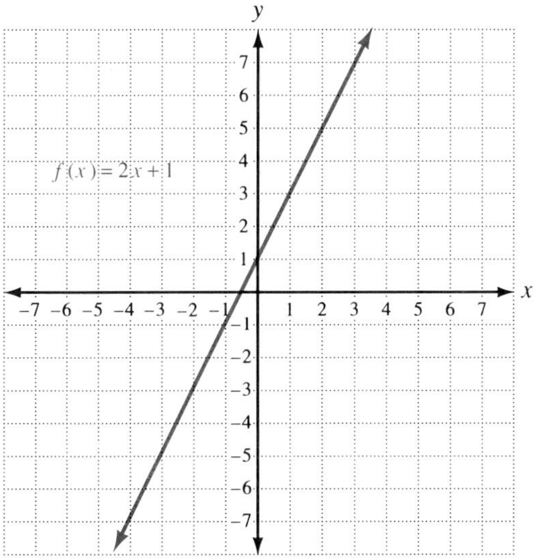

■

□ **DO EXERCISE 1.**

□ **Exercise 1** Graph the linear function $f(x) = x - 4$.

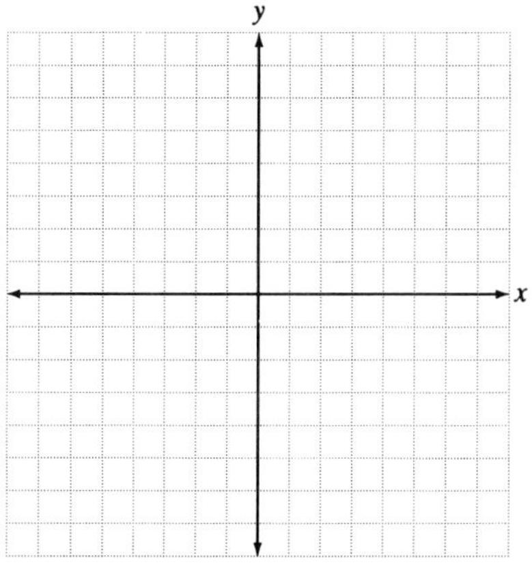

The identity function and constant functions are special cases of linear functions.

Identity Function

The ***identity function*** is a function of the form

$$f(x) = x$$

□ **Exercise 2** Graph the constant function $f(x) = 2$.

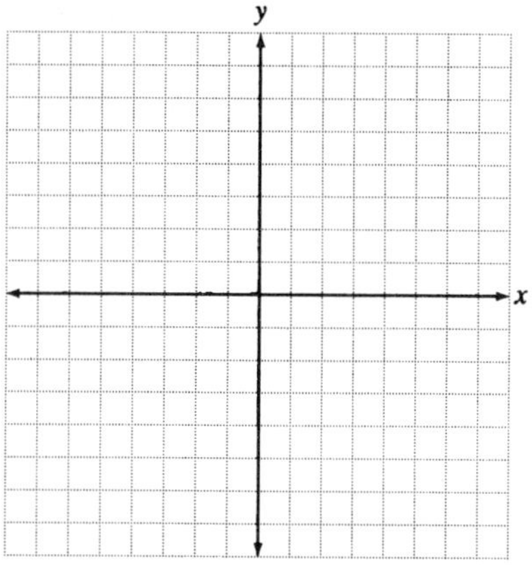

Notice that this is a linear function where $m = 1$ and $b = 0$. The graph of the identity function is as follows. It is the same as the graph of $y = x$.

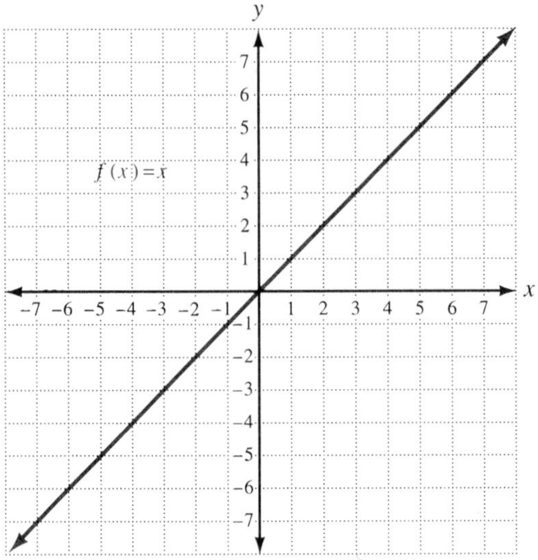

Constant Functions

> A ***constant function*** is any function of the form
> $$f(x) = b$$
> where b is a constant.

This is a linear function where $m = 0$. The *graph* of a constant function is a *horizontal line*.

EXAMPLE 2 Graph the constant function $f(x) = -3$.

Since $y = f(x)$, we graph $y = -3$.

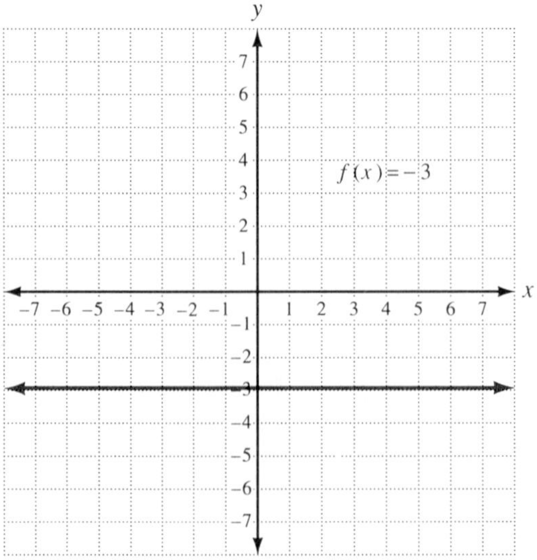

■

□ **DO EXERCISE 2.**

Graphs of vertical lines are not functions. The graph of an equation of the form $x = a$ is not a function.

Absolute Value Functions

An ***absolute value function*** is a function of the form

$$f(x) = |ax + b| + c$$

where a, b, and c are real numbers and $a \neq 0$.

Graphs of absolute value functions are V-shaped. The V may be inverted.

EXAMPLE 3 Graph the absolute value function $f(x) = |x + 3|$.

We graph the function by choosing some values for x. We include some negative values and zero.

x	y	Ordered Pair
-5	2	$(-5, 2)$
-4	1	$(-4, 1)$
-3	0	$(-3, 0)$
-2	1	$(-2, 1)$
0	3	$(0, 3)$
1	4	$(1, 4)$
2	5	$(2, 5)$

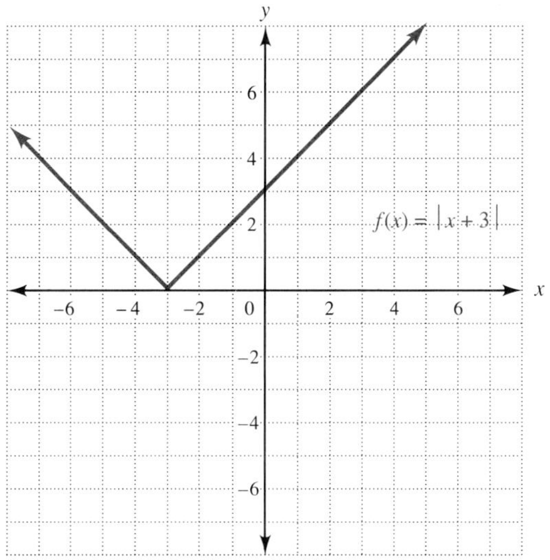

■

□ **DO EXERCISE 3.**

□ **Exercise 3** Graph the absolute value function $f(x) = |x - 2|$.

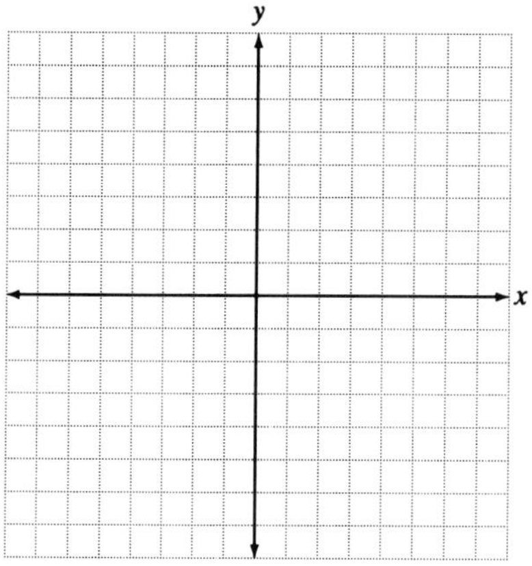

□ **Exercise 4** Graph the quadratic function $f(x) = x^2 - 4x + 4$.

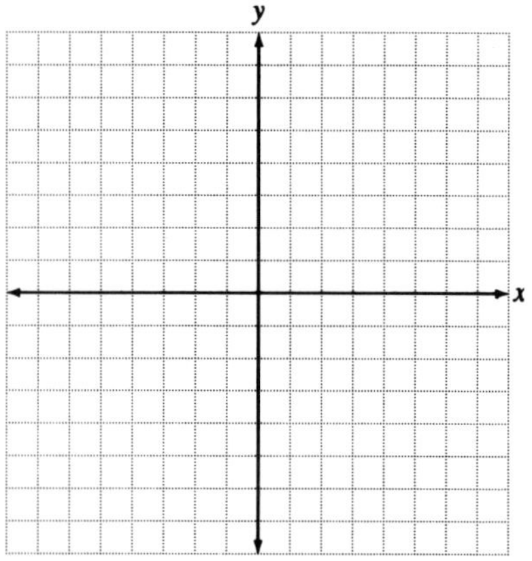

Quadratic Functions

A ***quadratic function*** is any function of the form

$$f(x) = ax^2 + bx + c$$

where a, b, and c are real numbers and $a \neq 0$.

The graph of a quadratic function is a parabola. We studied graphs of quadratic equations in Chapter 8.

Caution: Not all quadratic equations in two variables are functions. Equations of the form $x = ay^2 + by + c$, $a \neq 0$, are *not* functions. For example, the equation $x = y^2$ in Example 4b of Section 10.1 is not a function.

EXAMPLE 4 Graph the quadratic function $f(x) = -x^2 + 4x - 3$.

We graph the equation $y = -x^2 + 4x - 3$. Using the methods of Section 8.7, we find that the y-intercept is $(0, -3)$, the x-intercepts are $(1, 0)$ and $(3, 0)$, and the vertex is $(2, 1)$. The graph is as shown.

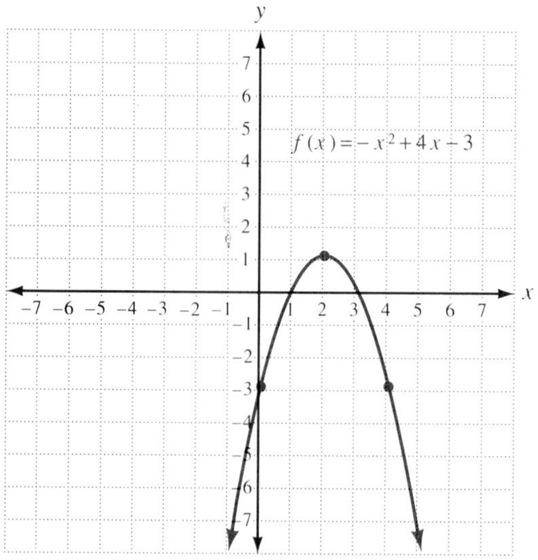

■

□ **DO EXERCISE 4.**

Power Functions

Power functions are of the form

$$f(x) = x^n$$

where n is a fixed positive integer.

The graph of $f(x) = x^2$ is one example of a power function. We know that this graph is a parabola. In Example 5 we show the graph of another power function.

EXAMPLE 5 Graph the power function $f(x) = x^3$.

We graph the function by choosing some values for x. We include some negative values and zero.

x	$f(x) = y$	Ordered Pair
-2	-8	$(-2, -8)$
-1	-1	$(-1, -1)$
$-\frac{1}{2}$	$-\frac{1}{8}$	$\left(-\frac{1}{2}, -\frac{1}{8}\right)$
0	0	$(0, 0)$
$\frac{1}{2}$	$\frac{1}{8}$	$\left(\frac{1}{2}, \frac{1}{8}\right)$
1	1	$(1, 1)$
2	8	$(2, 8)$

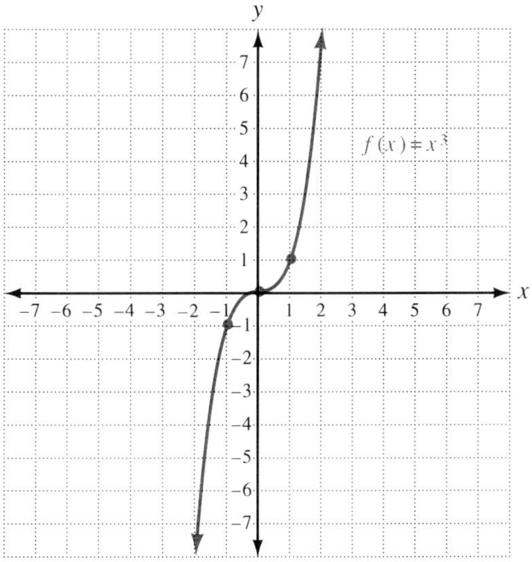

□ **DO EXERCISE 5.**

There are many other types of functions. We will study exponential and logarithmic functions in Chapter 11.

□ **Exercise 5** Graph the power function $f(x) = x^4$.

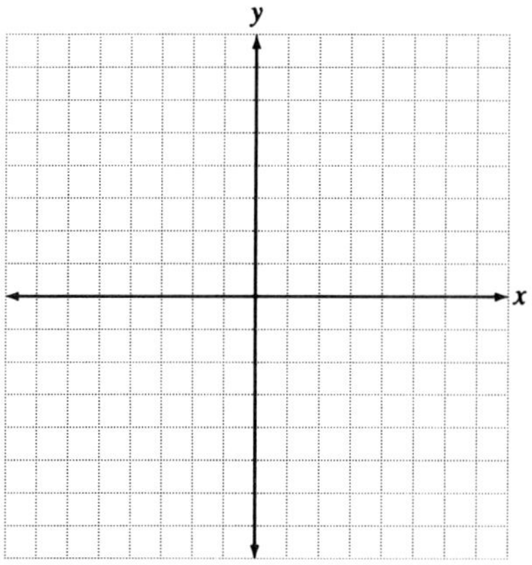

DID YOU KNOW?

René Descartes broke his lifelong habit of staying in bed until noon to teach Queen Christina of Sweden philosophy one winter. She liked her lessons at 5 A.M. with the windows open. Descartes died after about two months of these lessons.

Answers to Exercises

1.

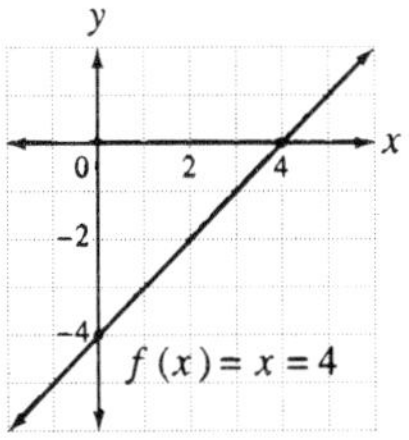

2.

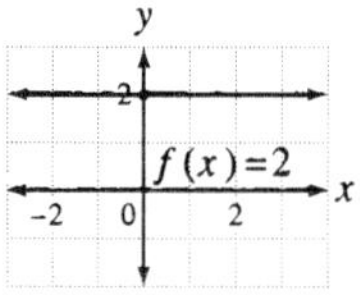

3.

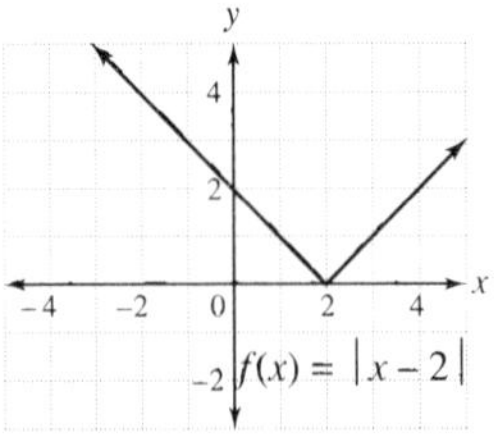

4.

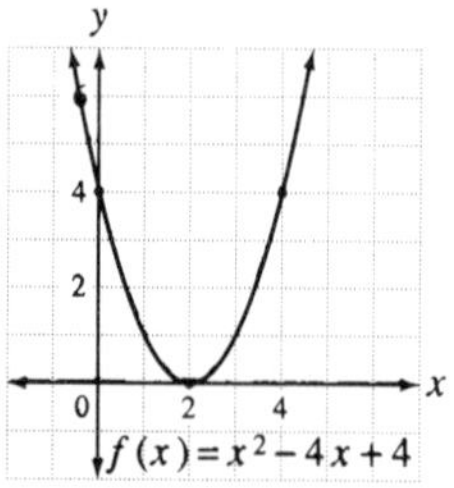

5.

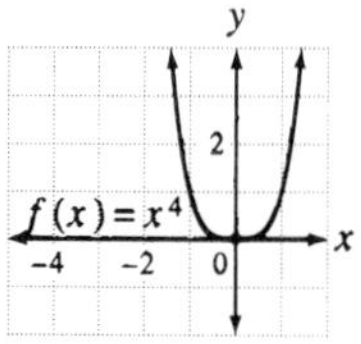

Problem Set 10.4

NAME

DATE

CLASS

Identify as a linear function, identity function, constant function, absolute value function, quadratic function, or power function. Sketch the graph of each.

1. $f(x) = 4$

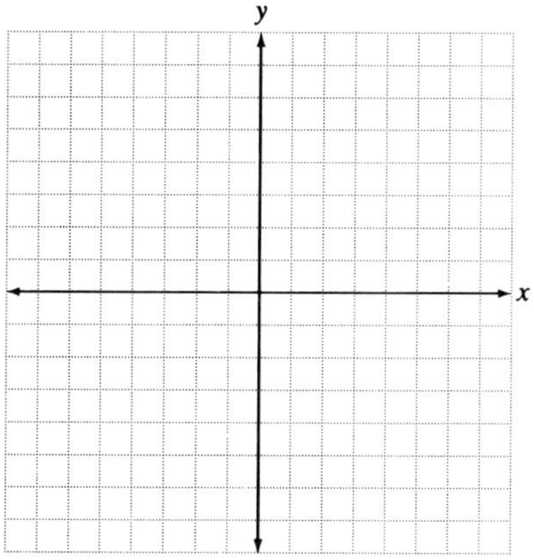

2. $f(x) = -1$

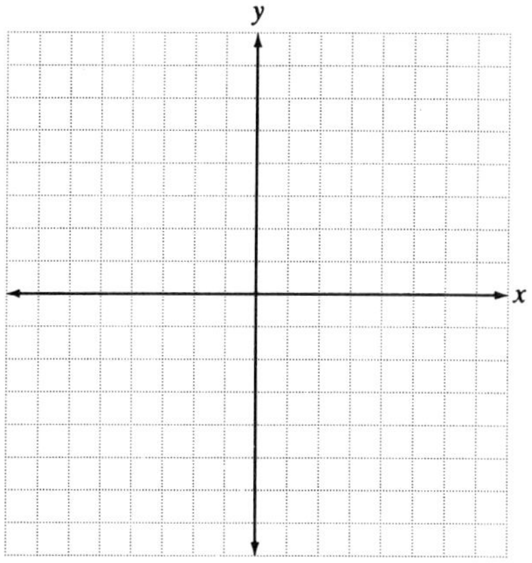

3. $f(x) = x$

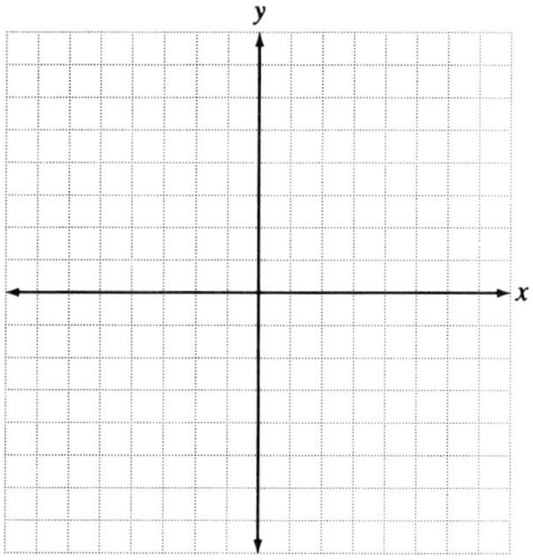

4. $g(x) = 0$

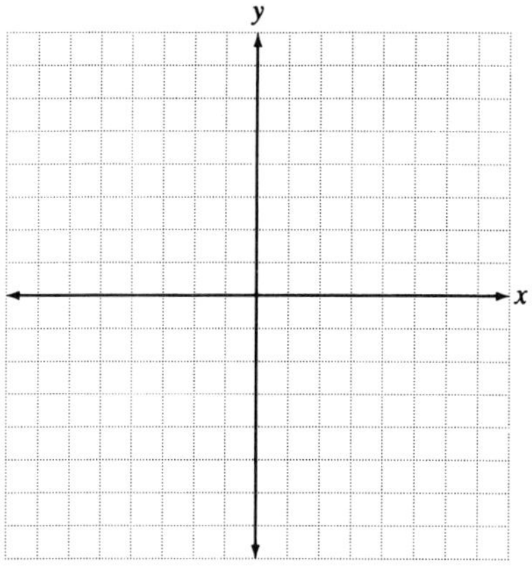

5. $f(x) = x^2 - 4$

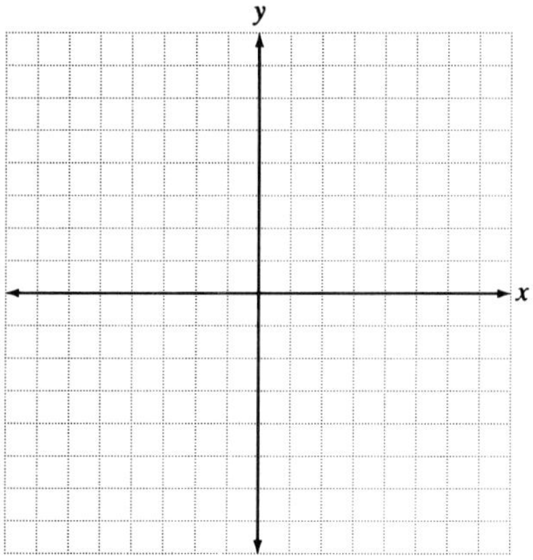

6. $g(x) = 3x^2 + 2$

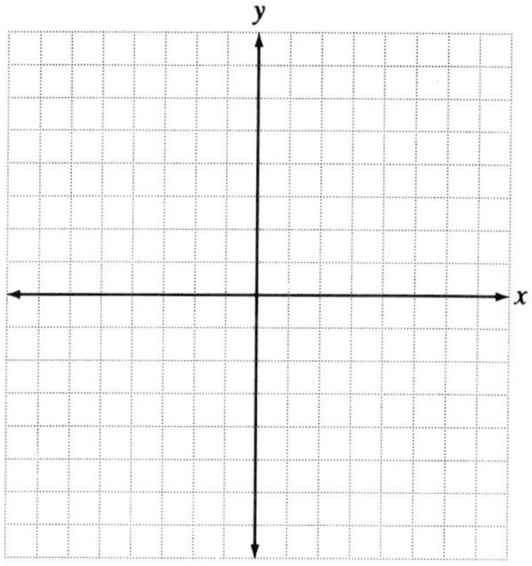

7. $f(x) = 2x - 2$

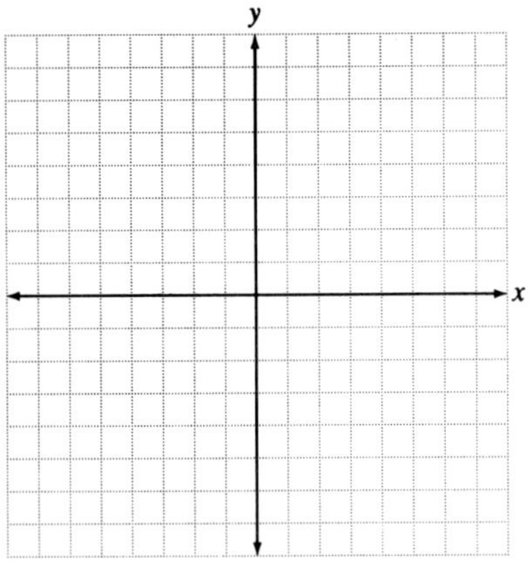

8. $f(x) = 5x + 1$

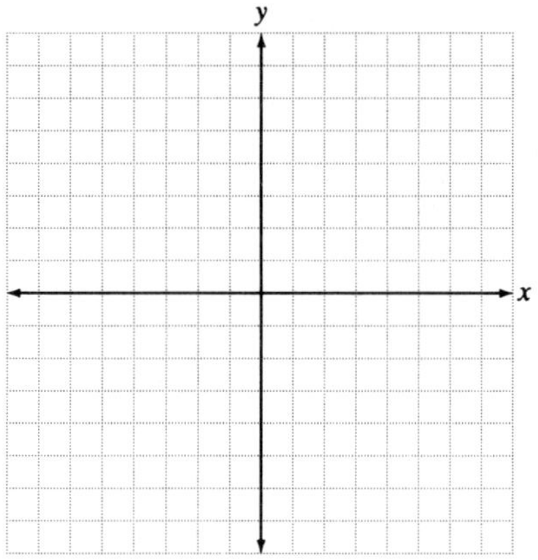

9. $g(x) = (x - 2)^2 + 3$

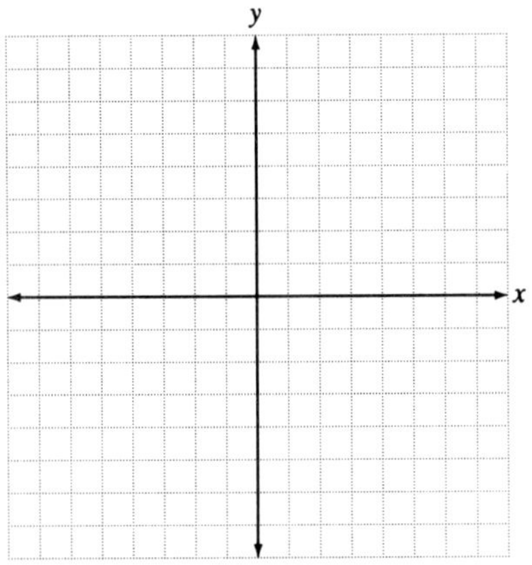

10. $f(x) = (x + 1)^2 - 2$

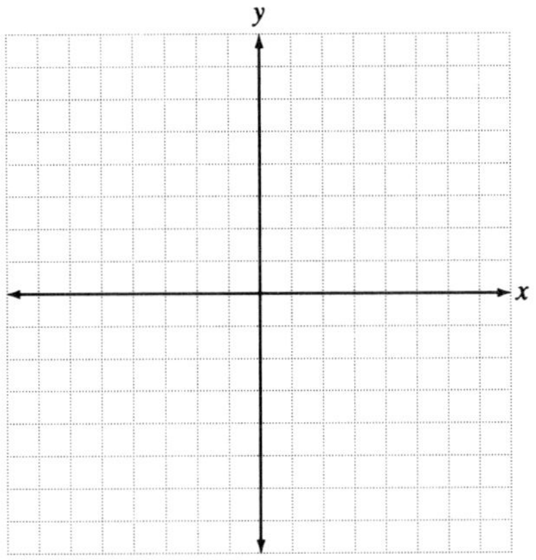

11. $f(x) = 4x$

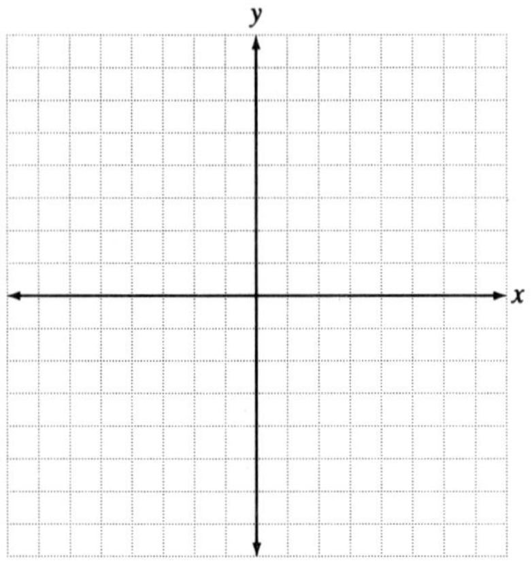

12. $h(x) = -3x$

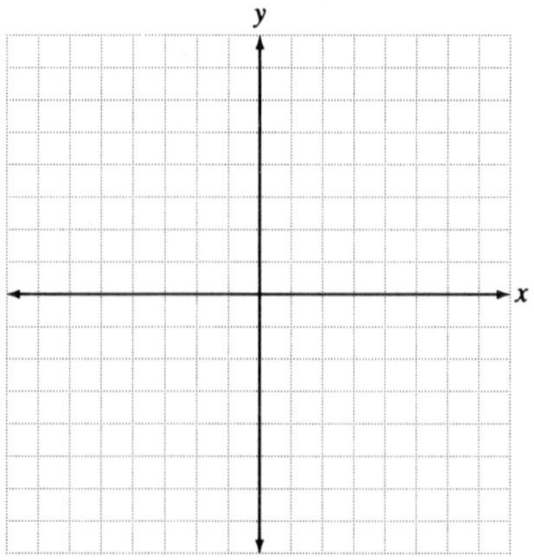

13. $f(x) = x^5$

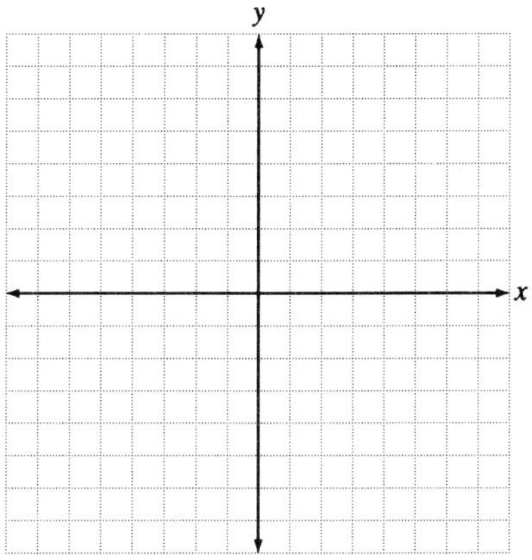

14. $f(x) = x^2$

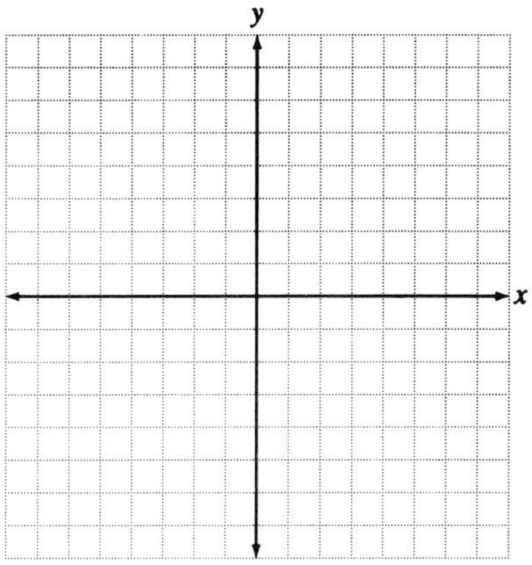

15. $f(x) = -3x - 2$

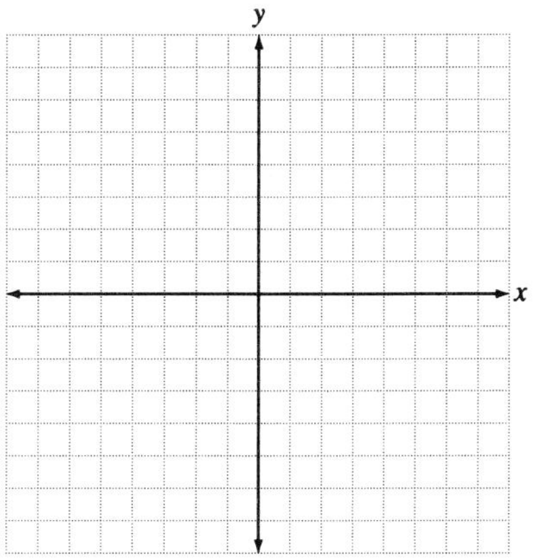

16. $f(x) = -x - 4$

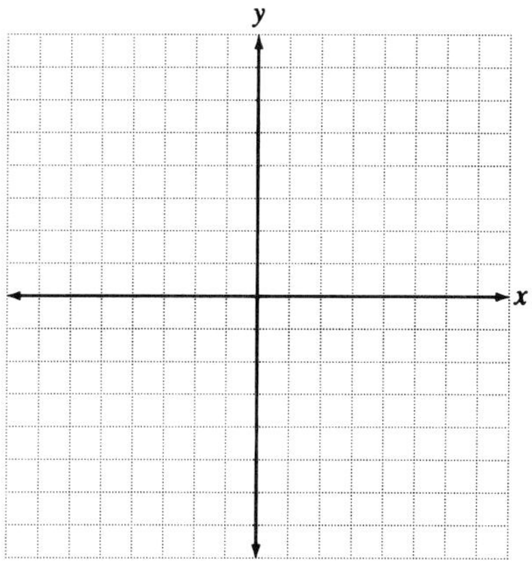

17. $f(x) = |x + 4|$

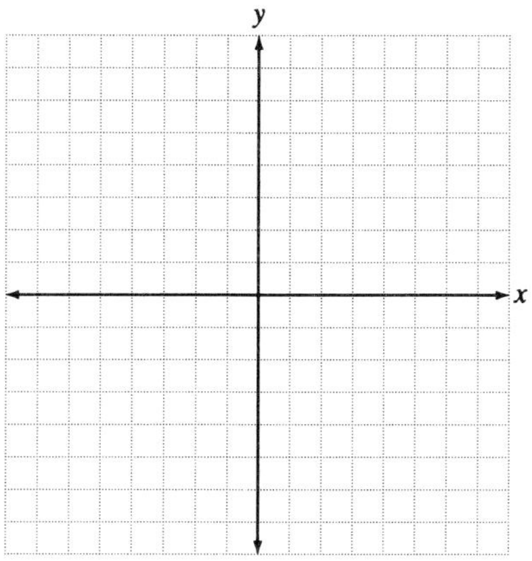

18. $f(x) = |x - 3|$

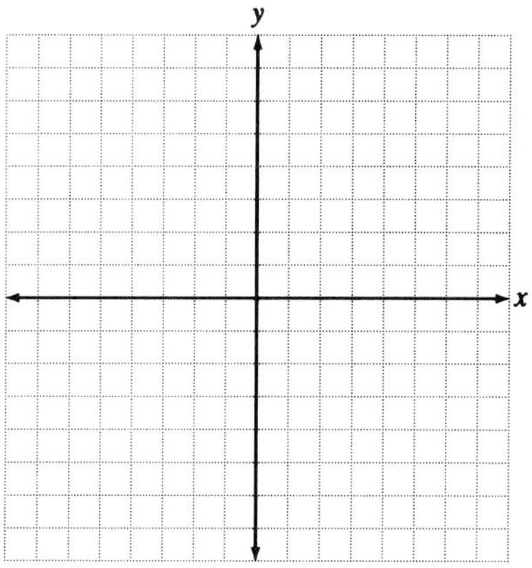

19. $f(x) = 4 - x^2$

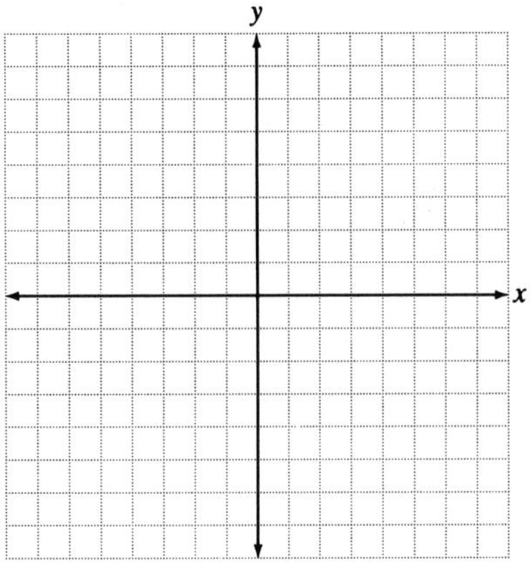

20. $f(x) = 2 - x^2$

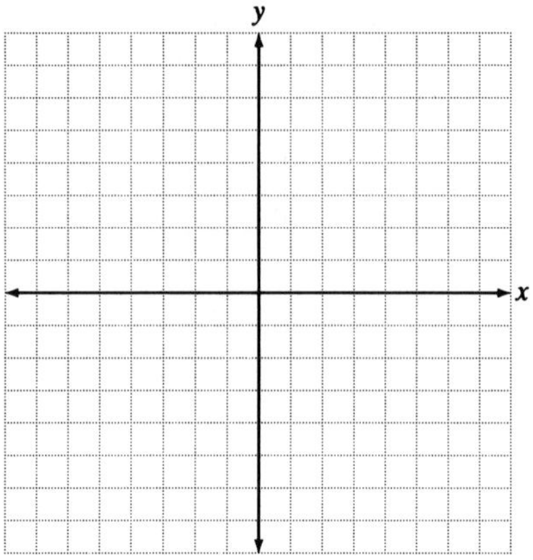

21. $f(x) = |x|$

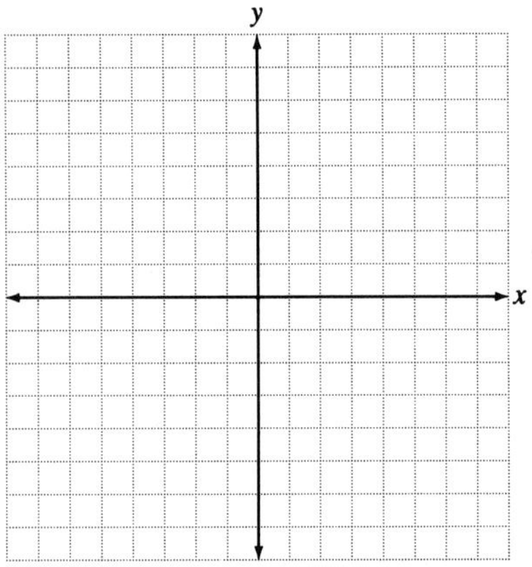

22. $f(x) = |x| - 1$

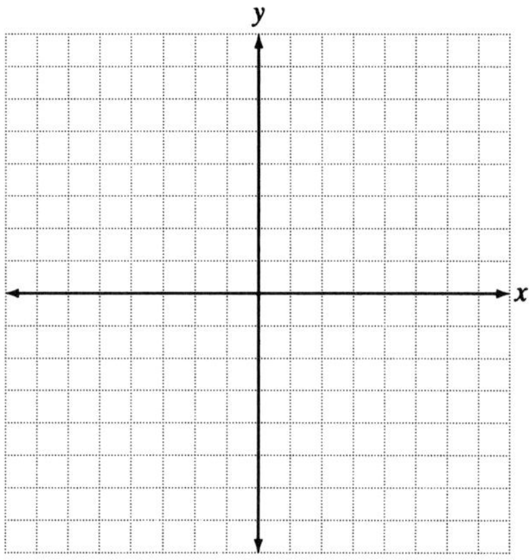

23. $f(x) = x^2 + 2x - 8$

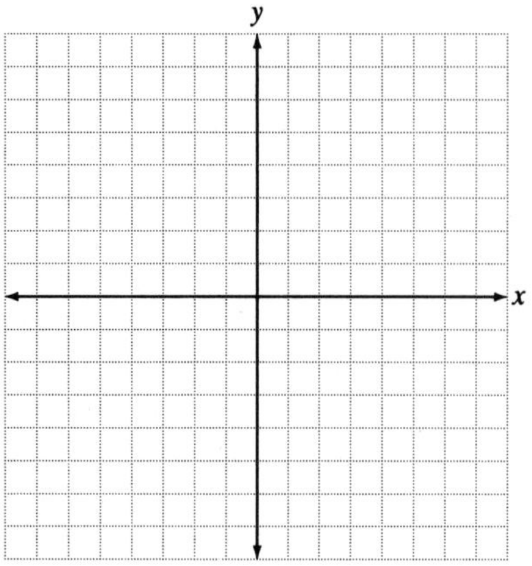

24. $f(x) = x^2 + 4x + 4$

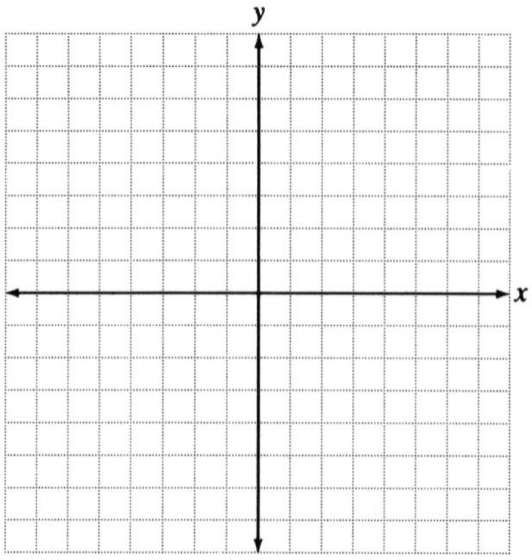

25. $f(x) = 5 - |x|$

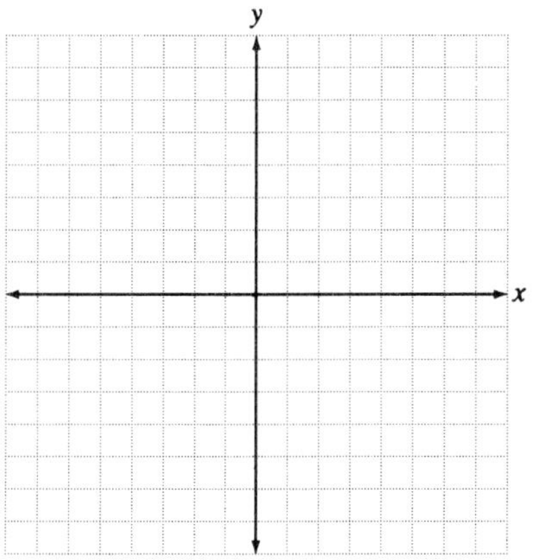

26. $f(x) = -2 + |x|$

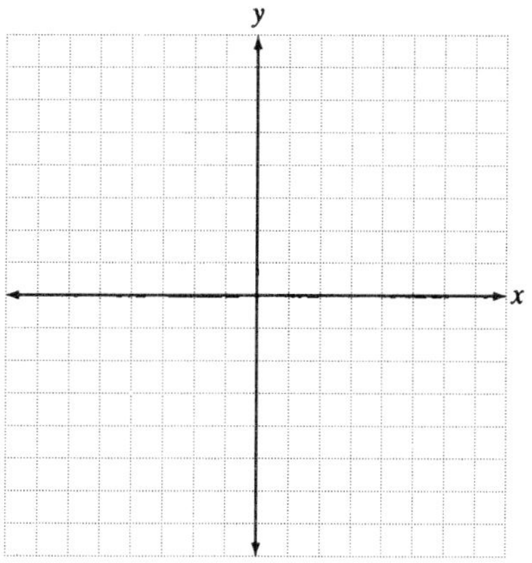

27. The function $S(r) = 4\pi r^2$ gives the surface area of a ball of radius r. Find the surface area of a ball with a radius of 6 inches. Use 3.14 for π.

28. The function $V(s) = s^3$ gives the volume of a cube with sides s. Find the volume of a cube with a side of 8 centimeters.

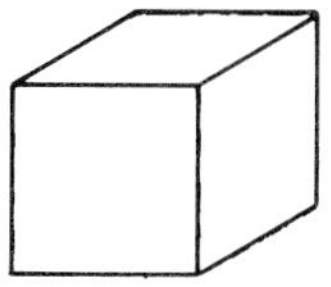

29. The graph of a linear function is a ______________ ______________.

30. The graph of a constant function is a ______________ line.

31. The function $f(x) = ax^2 + bx + c$, where a, b, and c are real numbers and $a \neq 0$, is a ______________ function.

32. The function $A = 6s^2$ gives the surface area of a cube with side s. Find the surface area of a cube with side 6.84 meters.

33. The function $V = \frac{4}{3}\pi r^3$ gives the volume of a sphere of radius r. Find the volume of a sphere of radius 19.3 centimeters. Use 3.14 for π.

Think About It

Find a linear function f such that each condition is satisfied.

* **34.** $f(-3) = 4$ and $f(2) = 4$

* **35.** $f(-7) = -7$ and $f(5) = 5$

* **36.** $f(0) = 3$ and $f(-2) = -1$

* **37.** $f(1) = -2$ and $f(3) = 8$

* **38.** $f(-2) = -12$ and $f(5) = 9$

* **39.** Find a linear function f such that $f(1) = 1$ and the graph of f is perpendicular to the graph of $f(x) = \frac{1}{3}x - 2$.

Checkup

The following problems provide a review of some of Sections 9.1 and 9.2.

Find the distance between each pair of points. If necessary, find an approximation to the nearest thousandth.

40. $(3, 4)$ and $(-1, -2)$

41. $(-2, 5)$ and $(1, 3)$

42. $(-6, -8)$ and $(-4, -7)$

43. $(-9, -5)$ and $(-4, 1)$

44. $(12, 20)$ and $(6, 18)$

45. $(-35, -8)$ and $(-40, -2)$

Graph.

46. $(x - 2)^2 + (y + 3)^2 = 9$

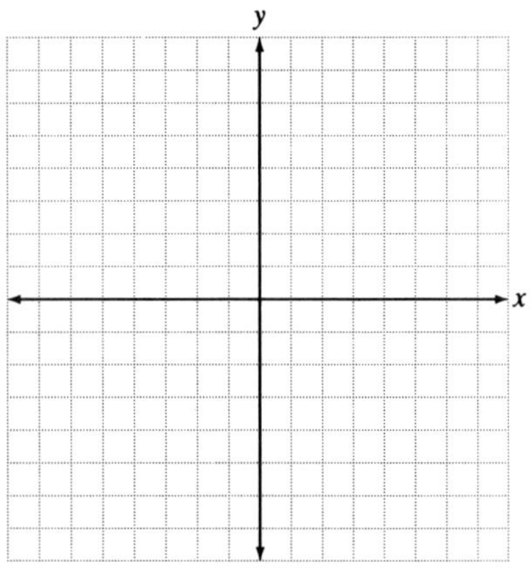

47. $(x + 4)^2 + (y - 1)^2 = 4$

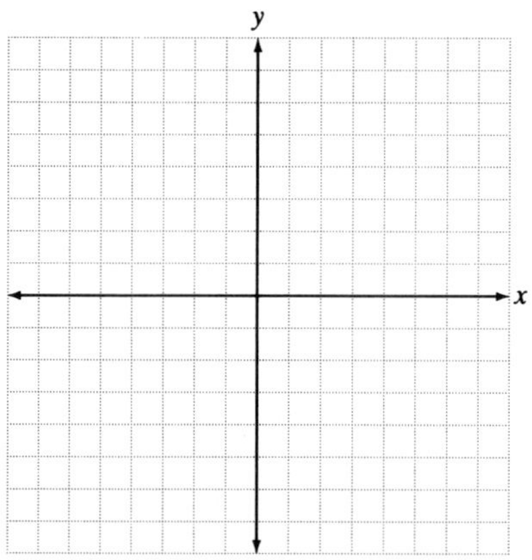

48. $\frac{x^2}{9} + \frac{y^2}{25} = 1$

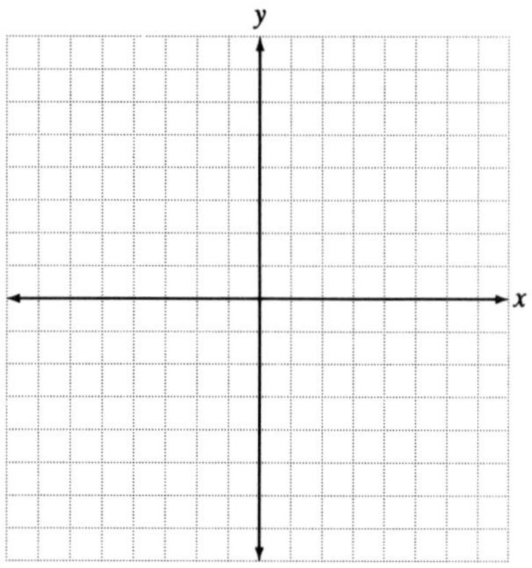

49. $\frac{x^2}{36} + \frac{y^2}{4} = 1$

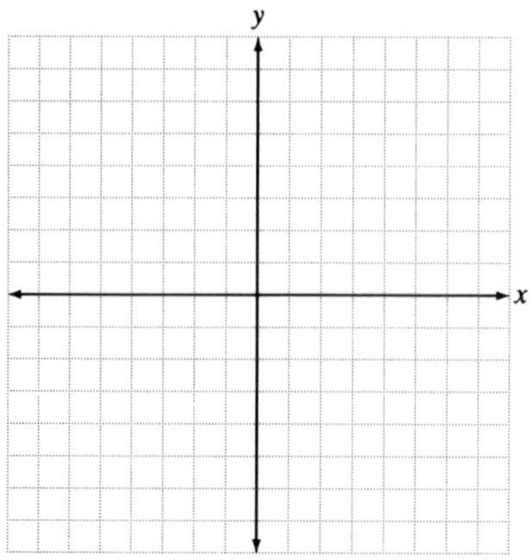

10.5 INVERSE FUNCTIONS

OBJECTIVES

1. *Determine if a function is one-to-one*
2. *Use the horizontal line test to determine if a function is one-to-one*
3. *Find and graph inverses of functions*

1 One-to-one Functions

Consider the function

$$f = \{(1, 3), (2, 4), (3, 5), (4, 6)\}$$

The inverse relation g of f is obtained by interchanging the x-coordinate and the y-coordinate in each ordered pair in f.

$$g = \{(3, 1), (4, 2), (5, 3), (6, 4)\}$$

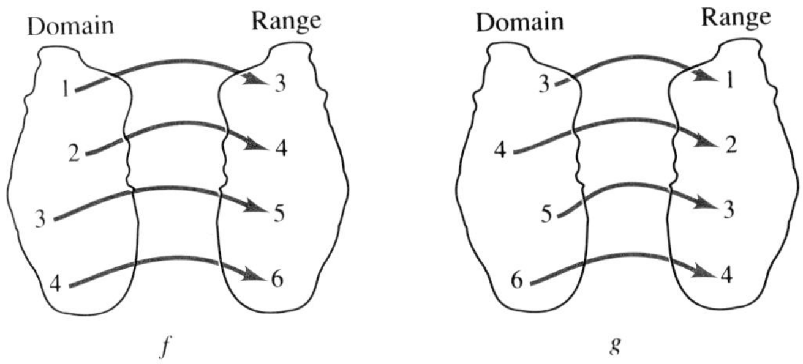

Notice that g is also a function since no two different ordered pairs have the same first coordinates. The inverse relation of a function is not always a function.

If the inverse relation is a *function,* it is called ***f*-inverse,** which is written f^{-1}.

> The ***inverse function*** of a function f, written f^{-1}, is the set of all ordered pairs of the form (y, x), where (x, y) is a member of f.

Caution: The -1 in f^{-1} is *not an exponent.* It is part of the symbol for f-inverse.

The domain of f is the range of f^{-1} and the range of f is the domain of f^{-1}. This is helpful in graphing. If we make a table of values for x and $f(x)$, we can get a table of values for x and $f^{-1}(x)$ by interchanging the x and $f(x)$ values.

For the inverse relation of a function f to be a function, f must be one-to-one. In a one-to-one function, each x value corresponds to only one y value and each y value corresponds to exactly one x value.

> A function f is one-to-one if each y value corresponds to exactly one x value. That is,
>
> $$\text{if } a \neq b, \qquad \text{then } f(a) \neq f(b).$$
>
> The inverse of a function is a function only when the function is one-to-one.

□ **Exercise 1** Determine if the function is one-to-one. If it is, find the inverse function.

a. $f = \{(-4, 8), (7, 0), (-6, -3), (5, 2)\}$

b. $f = \{(7, -3), (4, -8), (0, 5), (-6, 3), (2, -8)\}$

EXAMPLE 1 Determine if the function is one-to-one. If it is, find the inverse function.

a. $f = \{(3, 5), (-2, 4), (0, -8), (-3, -6), (5, 2)\}$

Every y value corresponds to only one x value. (No two different ordered pairs have the same second coordinates.) Therefore, f is one-to-one. The inverse function is found by interchanging the numbers in each ordered pair.

$$f^{-1} = \{(5, 3), (4, -2), (-8, 0), (-6, -3), (2, 5)\}$$

b. $f = \{(-3, 8), (0, -4), (-7, -6), (5, 8), (6, 4)\}$

Each y value does not correspond to only one x value. There are two ordered pairs, $(-3, \mathbf{8})$ and $(5, \mathbf{8})$ that have the same second coordinates. f is not one-to-one. ■

□ **DO EXERCISE 1.**

2 Horizontal Line Test

If a function is given in the form of an equation, it may be helpful to graph the function to see if it is one-to-one and therefore has an inverse. The function is one-to-one if no two x values have the same y value. This means that no horizontal line will intersect the graph of the function more than once.

> **Horizontal-Line Test**
>
> A function is one-to-one and has an inverse that is a function if there is no horizontal line that intersects the graph of the function more than once.

EXAMPLE 2 Graph the function and use the horizontal line test to determine whether it is one-to-one.

a. $f(x) = x - 3$

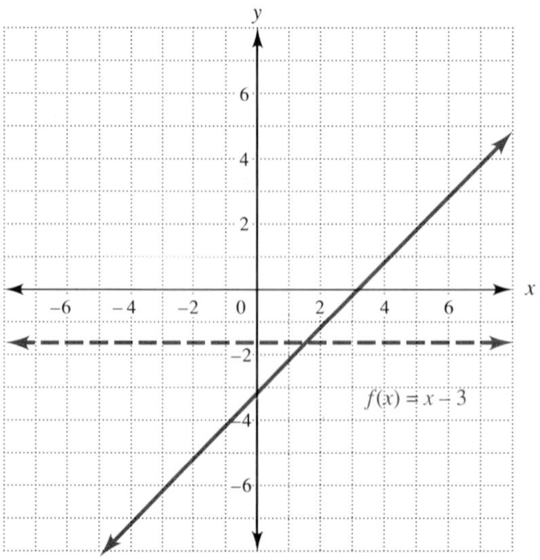

No horizontal line crosses the graph more than once. The function is one-to-one.

b. $f(x) = x^2 - 1$

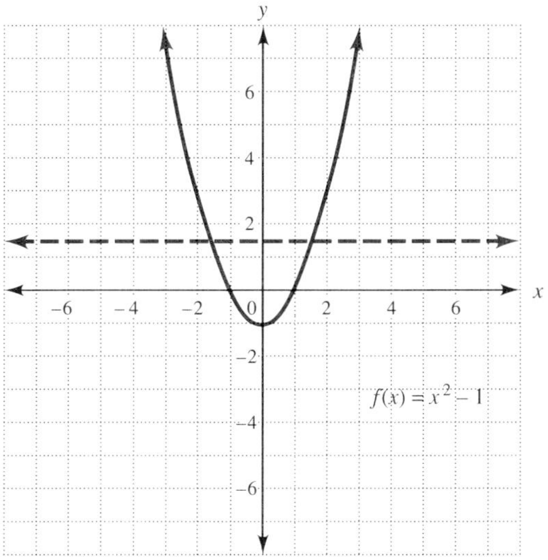

A horizontal line crosses the graph more than once. The function is not one-to-one. ■

□ DO EXERCISE 2.

3 Finding and Graphing Inverses of Functions

Usually, a function is given in the form of an equation instead of as a set of ordered pairs. How do we find the inverse relation? Since the inverse relation of a function is obtained by interchanging the x-coordinate and y-coordinate in each ordered pair of the function, we obtain the equation of the inverse relation by interchanging x and y in the equation of the function. Then we solve the equation for y.

EXAMPLE 3 Find the inverse of the function $f(x) = 2x + 3$ and graph the function and its inverse on the same axes.

1. $f(x) = 2x + 3$

This function is one-to-one. Hence its inverse is a function.
Let $y = f(x)$.

$$y = 2x + 3$$

2. Interchange x and y.

$$\mathbf{x} = 2\mathbf{y} + 3$$

3. Solve the equation for y.

$$x - 3 = 2y$$

$$\frac{x-3}{2} = y$$

4. Since this is a function, we let $y = f^{-1}(x)$.

$$y = f^{-1}(x) = \frac{x-3}{2}$$

□ **Exercise 2** Graph the function and use the horizontal line test to determine whether it is one-to-one.

a. $f(x) = 2x - 4$

b. $f(x) = -x^2 + 3$

□ **Exercise 3** Find the inverse of the function $f(x) = 5 - 3x$. Graph the function and its inverse on the same axes.

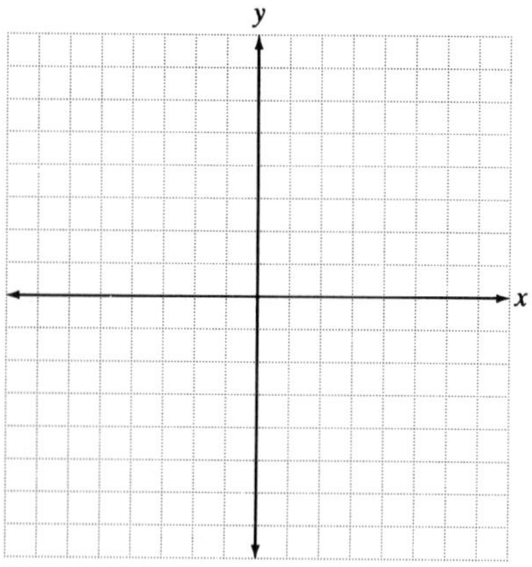

5. If we graph the function $f(x) = 2x + 3$, by using ordered pairs, we get a set of ordered pairs that belong to f-inverse by interchanging the x-coordinates and the y-coordinates.

x	$f(x)$	Ordered Pair
-4	-5	$(-4, -5)$
0	3	$(0, 3)$
1	5	$(1, 5)$

x	$f^{-1}(x)$	Ordered Pair
-5	-4	$(-5, -4)$
3	0	$(3, 0)$
5	1	$(5, 1)$

The graph is as shown. Notice that since f^{-1} is obtained from f by interchanging x and y, the graph of f^{-1} is a reflection of f across the line $y = x$.

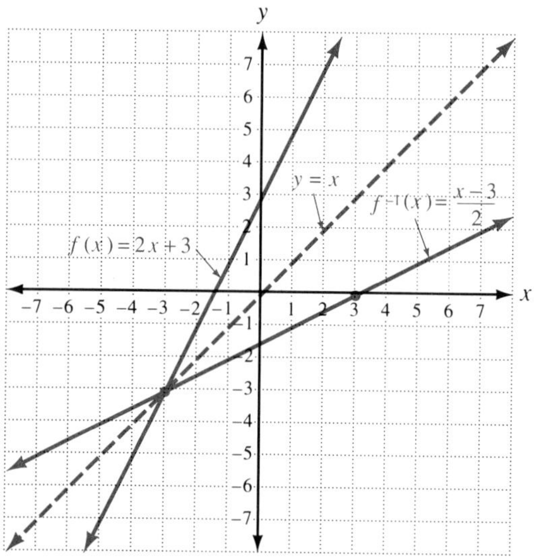

■

□ **DO EXERCISE 3.**

> The equation for the inverse of a one-to-one function may be obtained by the following steps.
>
> **1.** Let $y = f(x)$.
> **2.** Interchange x and y.
> **3.** Solve for y.
> **4.** Let $f^{-1}(x) = y$.

EXAMPLE 4 Find the inverse of the function $f(x) = x^3$ and graph the function and its inverse on the same axes.

This function is one-to-one.

1. Let $y = f(x)$

   $y = x^3$
2. $\mathbf{x = y^3}$ Interchange x and y.
3. $\sqrt[3]{x} = y$ Solve for y.
4. $f^{-1}(x) = \sqrt[3]{x}$ Replace y with $f^{-1}(x)$ since $y = \sqrt[3]{x}$ is a function.

We graphed $f(x) = x^3$ in Section 10.4. A set of ordered pairs that are solutions to this equation are

$$(-2, -8), (-1, -1), \left(-\frac{1}{2}, -\frac{1}{8}\right), (0, 0), \left(\frac{1}{2}, \frac{1}{8}\right), (1, 1), \text{ and } (2, 8)$$

Therefore, some ordered pairs that are solutions to f^{-1} are

$$(-8, -2), (-1, -1), \left(-\frac{1}{8}, -\frac{1}{2}\right), (0, 0), \left(\frac{1}{8}, \frac{1}{2}\right), (1, 1), \text{ and } (8, 2)$$

The graph is as shown.

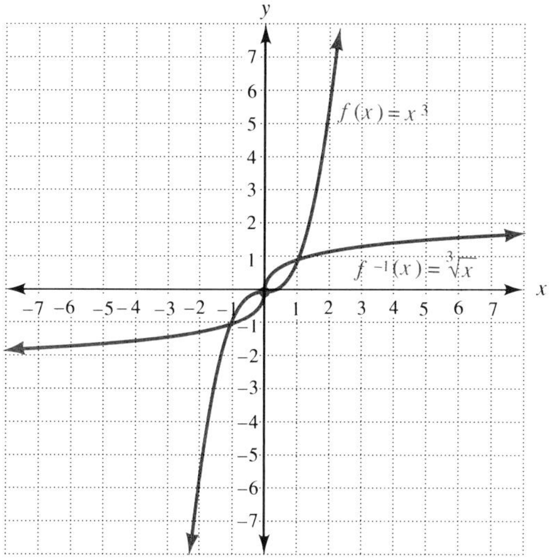

☐ **DO EXERCISE 4.**

Remember that sometimes the inverse relation of a function is not a function. We can often restrict the domain of the function to form a new function whose inverse relation is a function.

☐ **Exercise 4** Find the inverse of $f(x) = x^3 + 2$ and graph the function and its inverse on the same axes.

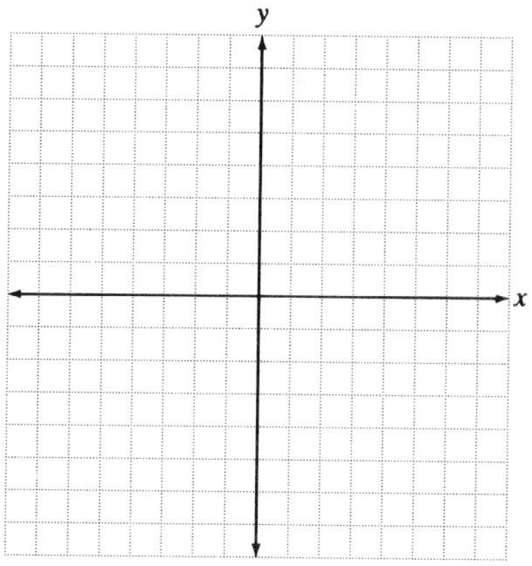

EXAMPLE 5 Find the inverse of $f(x) = x^2 - 3$ and graph the function and its inverse on the same axes.

1.	$y = x^2 - 3$	Replace $f(x)$ with y.
2.	$x = y^2 - 3$	Interchange x and y.
3.	$x + 3 = y^2$	
4.	$y = \pm\sqrt{x + 3}$	Solve for y.

Some ordered pairs that belong to $f(x)$ are

$$(-3, 6), \quad (-2, 1), \quad (0, -3) \quad (2, 1), \quad \text{and} \quad (3, 6)$$

Therefore, some ordered pairs that belong to the inverse relation are

$$(6, -3), \quad (1, -2), \quad (-3, 0) \quad (1, 2), \quad \text{and} \quad (6, 3)$$

We graph $f(x)$ and the inverse relation as follows. Notice that the inverse relation is not a function since a vertical line intersects the graph in more than one point. Also, a horizontal line would intersect the graph of the function in more than one point.

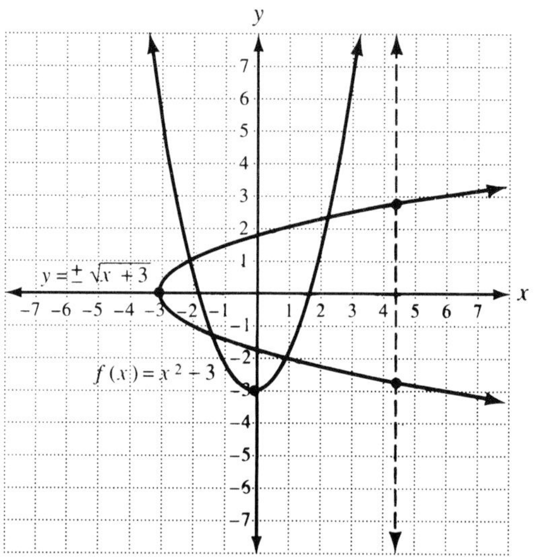

We may restrict the domain of $f(x)$ to $x \geq 0$ so that the inverse relation is a function. Now $f^{-1}(x) = \sqrt{x + 3}$. We omit the ordered pairs $(-3, 6)$ and $(-2, 1)$ that belong to $f(x)$ and use the ordered pairs $(0, -3)$, $(2, 1)$, and $(3, 6)$ to graph the function. Therefore, we use the ordered pairs $(-3, 0)$, $(1, 2)$, and $(6, 3)$ to graph $f^{-1}(x)$.

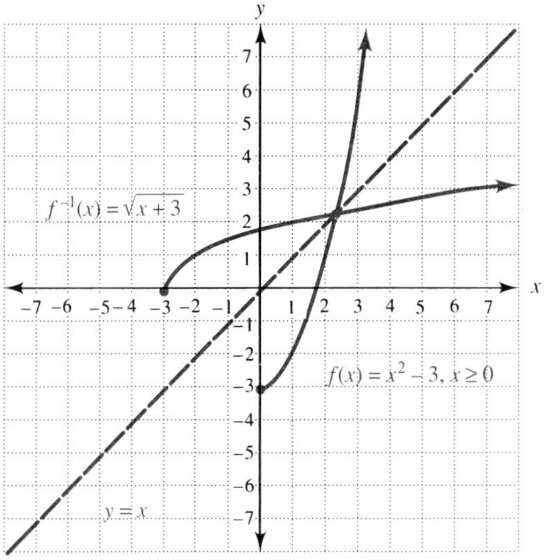

□ **DO EXERCISE 5.**

□ **Exercise 5** Find the inverse of $f(x) = x^2 + 2$, $x \geq 0$, and graph the function and its inverse on the same axes.

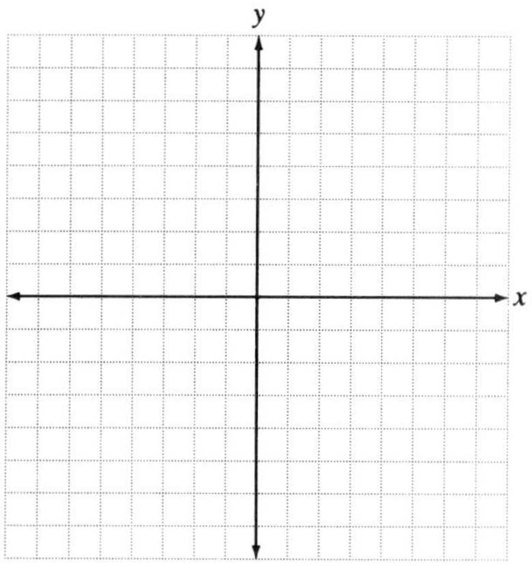

GRAPHING CALCULATOR

A function and its inverse may be graphed with a graphing calculator. Graph $f(x) = x^3$ and $f^{-1}(x) = \sqrt[3]{x}$ on the same axes.

1. Enter $y = x^3$ and $y = \sqrt[3]{x}$ into your calculator.
2. An appropriate window is W: $[-10, 10]$ $[-10, 10]$.
3. Plot the graph. This is the graph that is shown in Example 4.

Answers to Exercises

1. a. Yes; $f^{-1} = \{(8, -4), (0, 7), (-3, -6), (2, 5)\}$ **b.** No

2. a. Yes **b.** No

3. $f^{-1}(x) = \dfrac{5 - x}{3}$

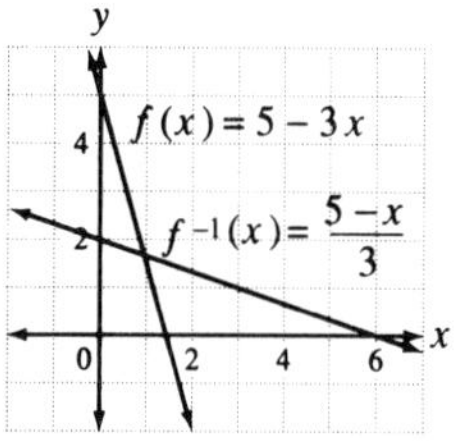

4. $f^{-1}(x) = \sqrt[3]{x - 2}$

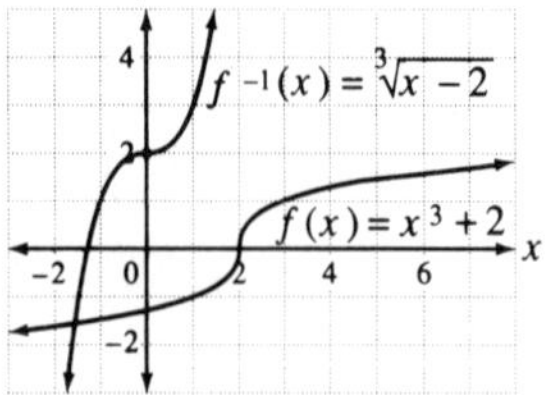

5. $f^{-1}(x) = \sqrt{x - 2}$

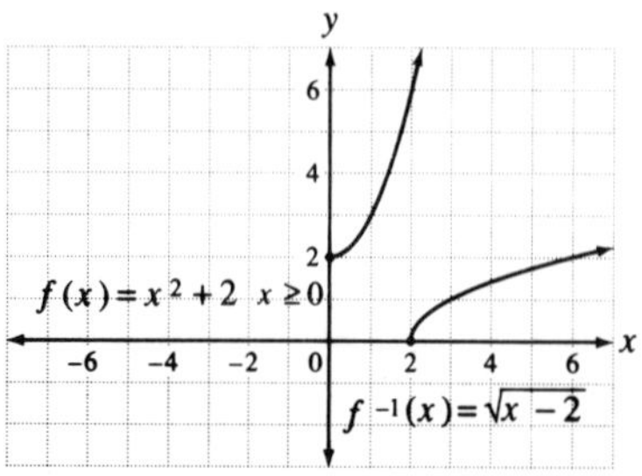

Problem Set 10.5

NAME

DATE

CLASS

Determine if the function is one-to-one. If it is, find the inverse function.

1. $f = \{(7, 8), (-5, 4), (3, 0), (-7, -2)\}$

2. $f = \{(-6, 2), (5, 9), (0, 0), (-3, 1)\}$

3. $g = \{(4, -9), (-8, 15), (-7, 3), (12, 6), (-4, 3)\}$

4. $g = \{(-11, 5), (-4, 3), (-15, 2), (8, 4), (7, -3)\}$

Graph the function and use the horizontal line test to determine whether it is one-to-one.

5. $f(x) = 5x - 2$

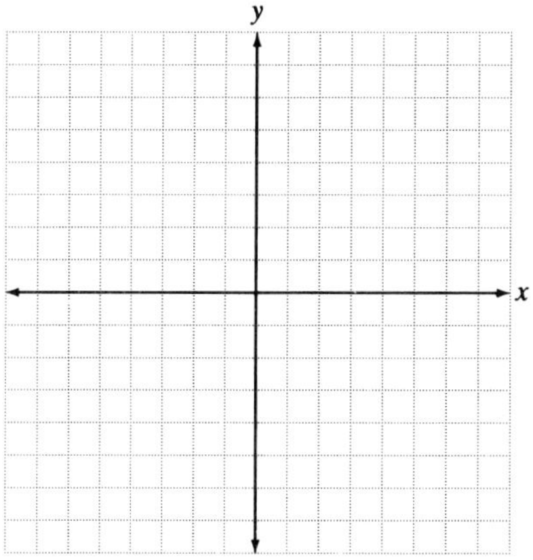

6. $f(x) = 2 - 4x$

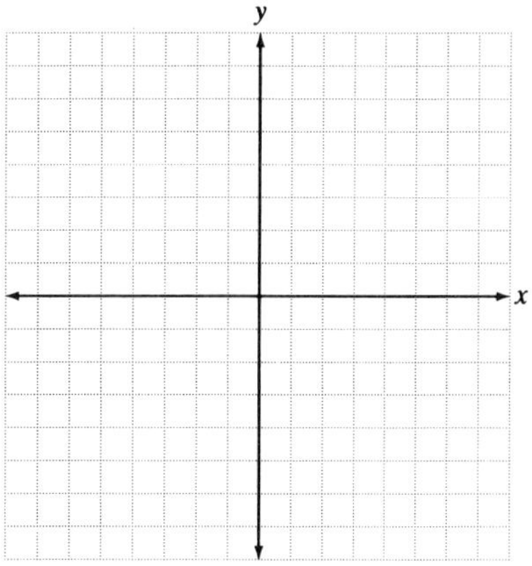

7. $f(x) = -x^2 + 4$

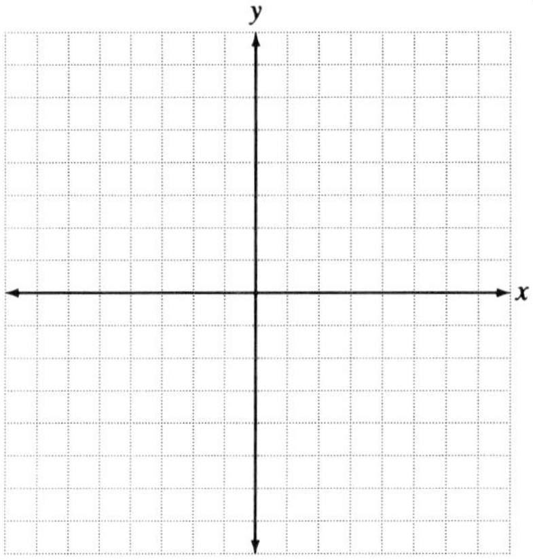

8. $f(x) = x^2 - 3$

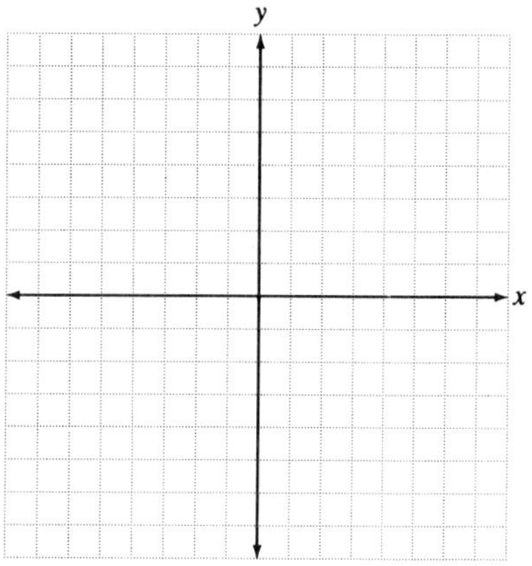

9. $g(x) = x^3 - 3$

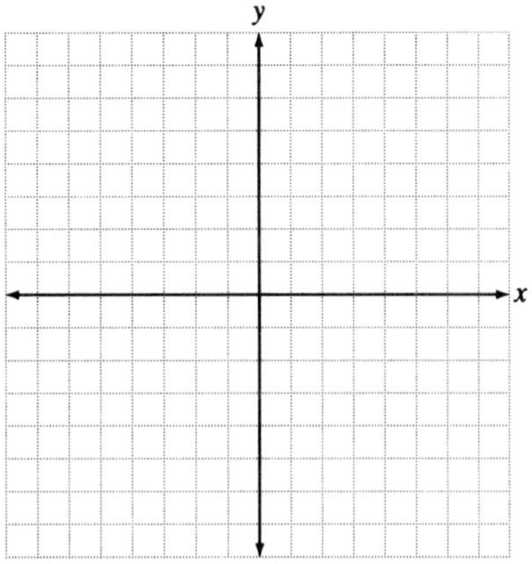

10. $g(x) = -x^3 + 1$

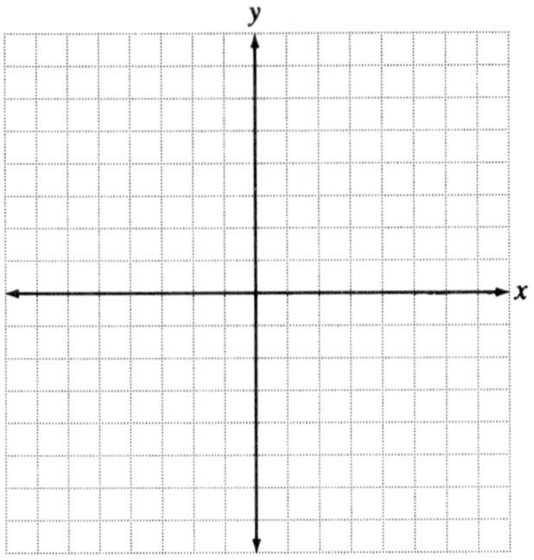

11. $h(x) = |x| - 4$

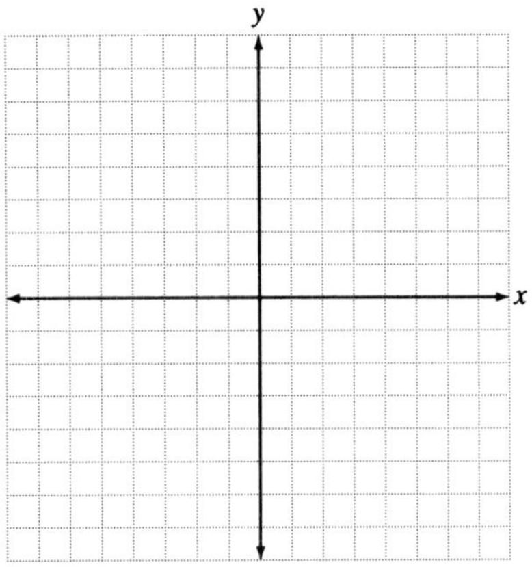

12. $h(x) = 5 - |x|$

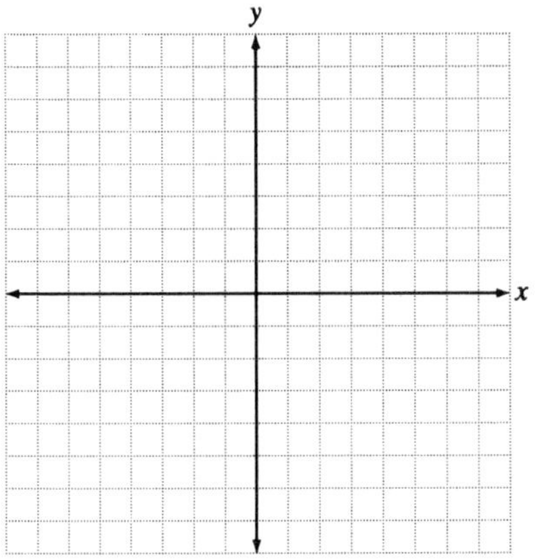

13. $F(x) = |x - 2|$

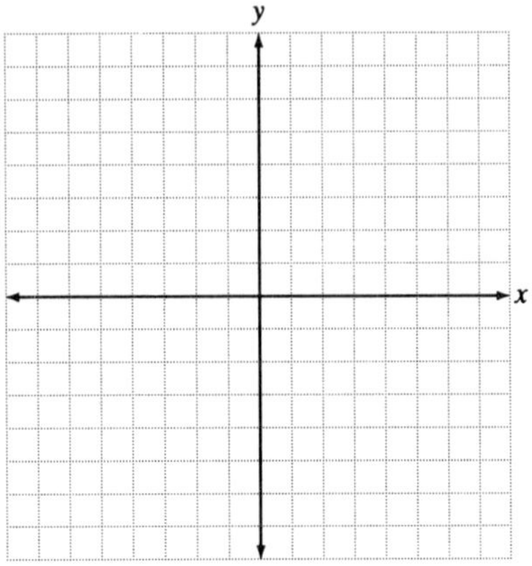

14. $F(x) = |x + 4|$

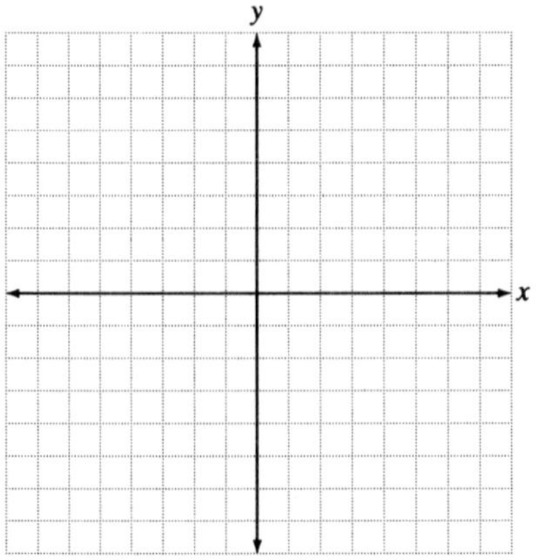

Find the inverse of the function.

15. $f(x) = 5x$

16. $f(x) = -2x$

17. $g(x) = 6 - 5x$

18. $g(x) = 3x + 7$

19. $f(x) = \dfrac{4x - 2}{5}$

20. $f(x) = \dfrac{2x - 8}{7}$

21. $H(x) = 2x^3 - 3$

22. $G(x) = -x^3 + 4$

23. $f(x) = \sqrt[3]{x - 3}$

24. $f(x) = \sqrt[3]{x + 2}$

25. $F(x) = 3x^2 - 1,\ x \geq 0$

26. $F(x) = -3x^2 + 2,\ x \geq 0$

For each of the functions, find the inverse of the function and graph the function and its inverse on the same axes.

27. $f(x) = 3x + 2$

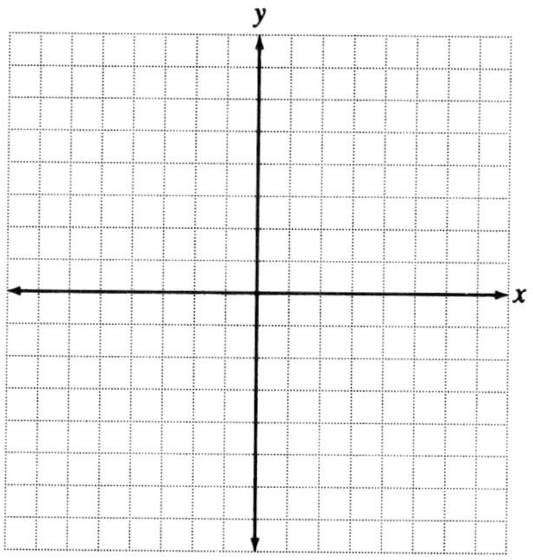

28. $f(x) = 2x - 1$

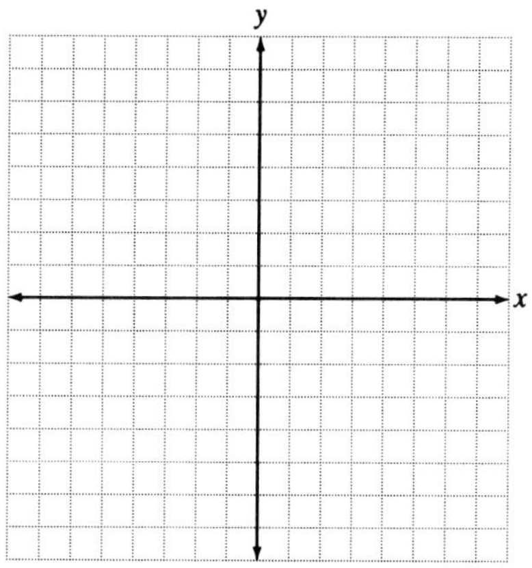

29. $f(x) = 4 - x$

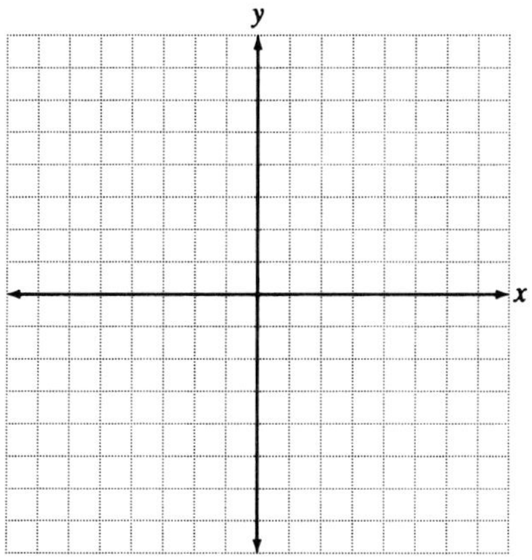

30. $f(x) = 2 - x$

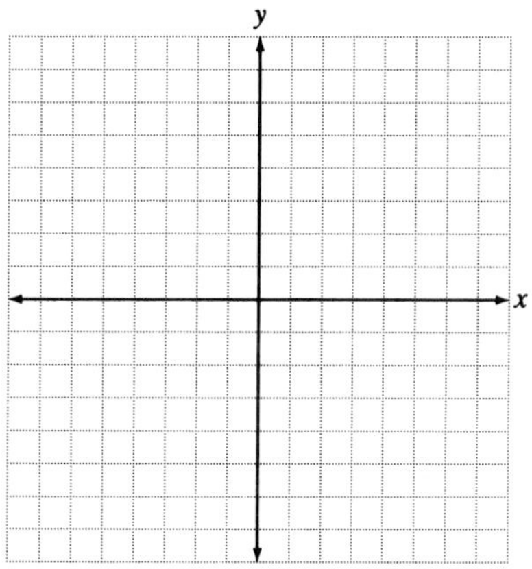

31. $f(x) = x^3 - 1$

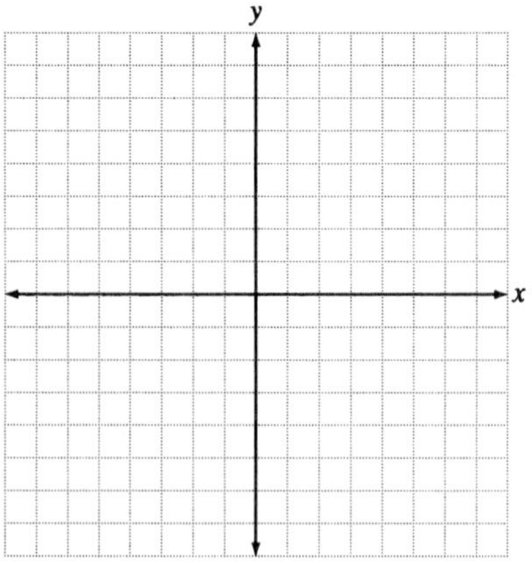

32. $f(x) = x^3 + 1$

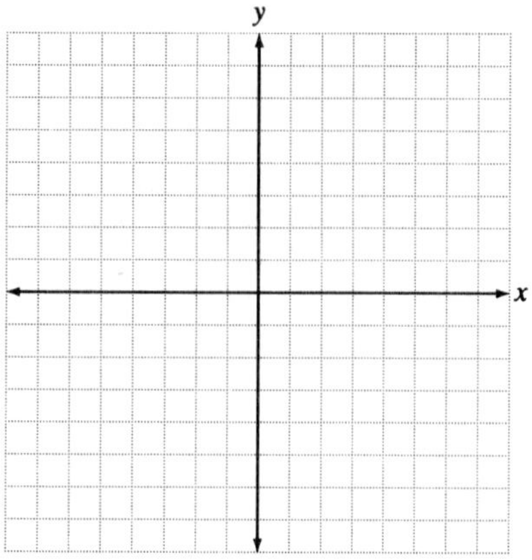

33. $f(x) = 2 - 3x$

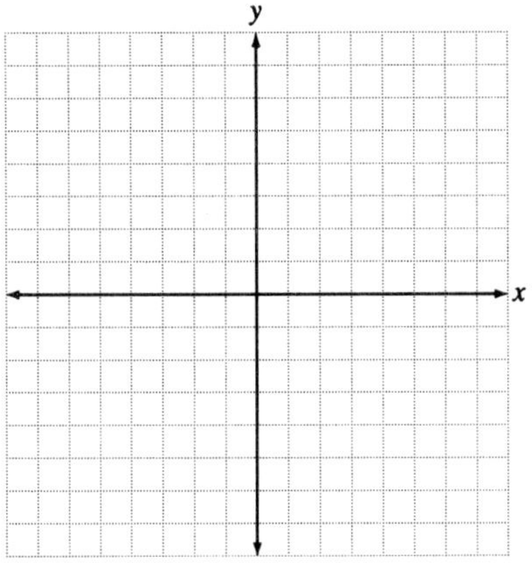

34. $f(x) = 3 - 4x$

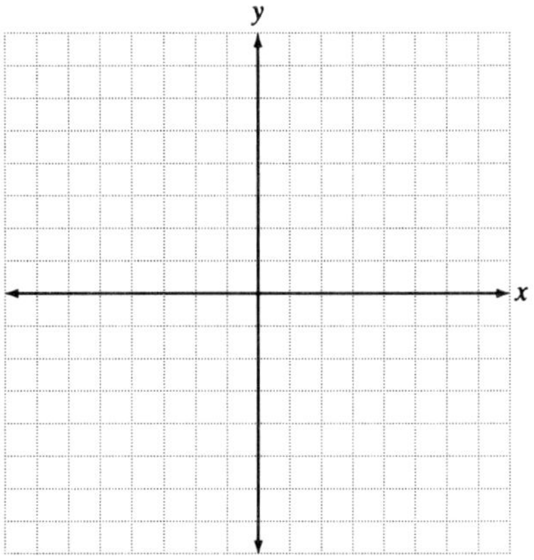

35. $f(x) = x^3 - 2$

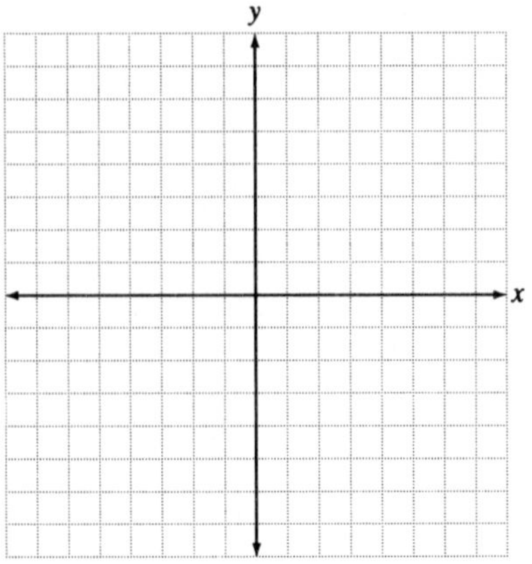

36. $f(x) = x^3 + 3$

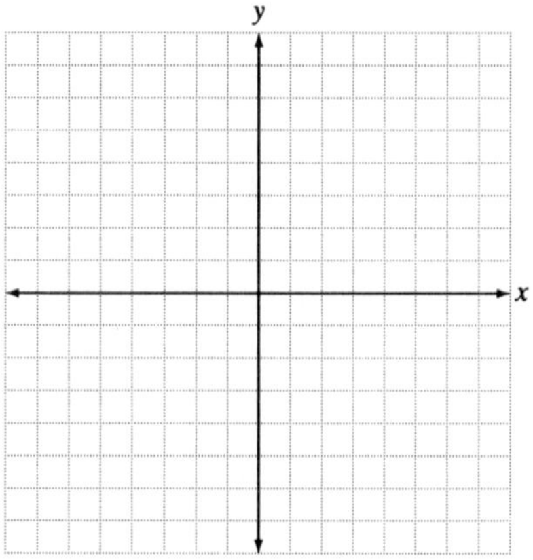

37. $f(x) = x^2 + 1, x \geq 0$

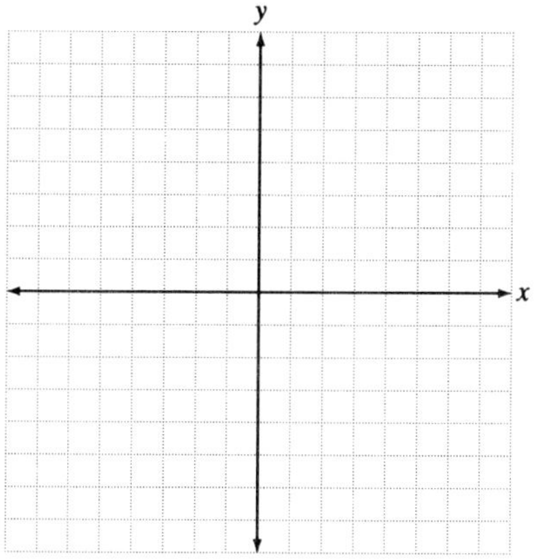

38. $f(x) = -x^2 + 2, x \geq 0$

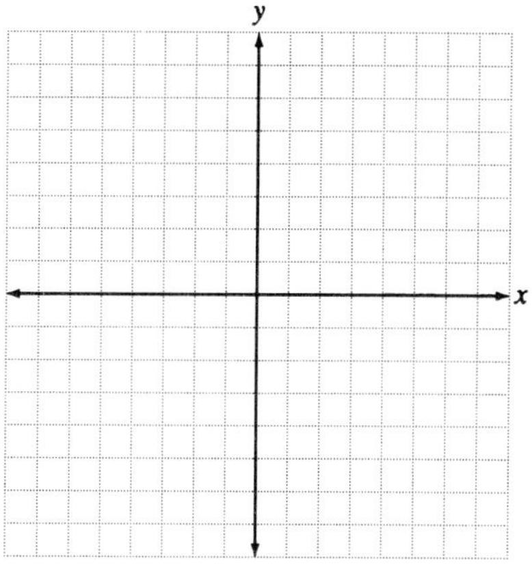

39. $f(x) = -2x^2 + 1, x \geq 0$

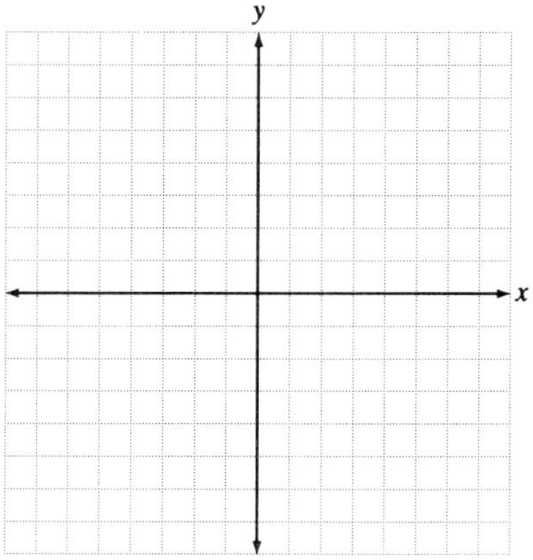

40. $f(x) = 2x^2 - 1, x \geq 0$

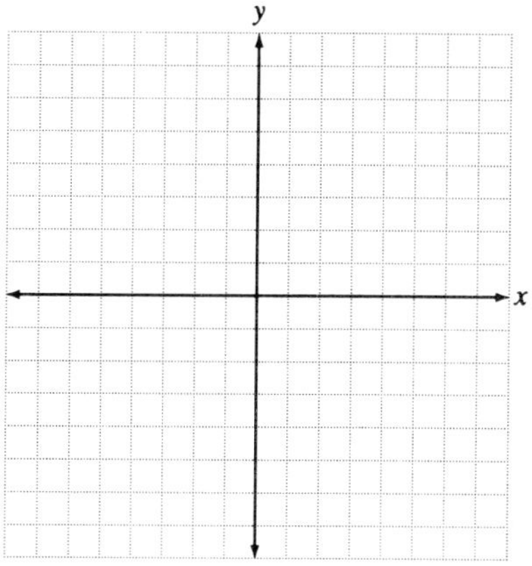

41. The domain of f is the ____________ of f^{-1}.

42. The equation of the inverse relation is found by ____________ x and y in the equation of the function.

43. Sometimes it is necessary to restrict the ____________ of a function in order for its inverse relation to be a function.

Find $f[f^{-1}(x)]$ and $f^{-1}[f(x)]$ for the following functions.

* **44.** $f(x) = 2x - 4, f^{-1}(x) = \dfrac{x + 4}{2}$

* **45.** $f(x) = x^3 - 4, f^{-1}(x) = \sqrt[3]{x + 4}$

Think About It

For the following functions, find a restriction on the domain so that the inverses of the functions are functions.

* **46.** $f(x) = |x|$

* **47.** $f(x) = (x + 1)^2 + 2$

Checkup

The following problems provide a review of some of Sections 9.4 and 9.5.

Solve.

48. $x^2 + y^2 = 5$
$x - 3y = 7$

49. $x^2 + y^2 = 9$
$x + 2y = -3$

50. $y = x^2 - 2x + 3$
$y - x = 3$

51. $y = x^2 - 4x + 2$
$2x + y = 2$

52. $x^2 + 4y^2 = 13$
$2x^2 - 4y^2 = 14$

53. $x^2 - 2y^2 = 2$
$x^2 + 5y^2 = 9$

54. $y = x^2 - 2$
$y = x^2 - x - 1$

55. $4x^2 = y$
$y = x^2 - 2x + 5$

Chapter 10 Summary

Section 10.1

Any set of ordered pairs is a ***relation.***

A ***function*** is a relation in which no two ordered pairs have the same first coordinates and different second coordinates.

A ***function*** is also a correspondence or mapping that assigns to each member of some set exactly one member of another set.

The set of all first coordinates of the ordered pairs of a function is called the ***domain*** of the function. The set of all second coordinates of the ordered pairs is called the ***range.***

The domain of a function given in terms of an equation is the set of all real numbers that can be substituted for x in the equation.

If a vertical line intersects the graph of an equation in more than one point, the relation given by the equation is not a function.

Section 10.2

The equation $f(x) = x^2 + 2$ is written in function notation.

A ***composite function*** is formed by replacing the variable in a function with another function.

Section 10.3

The function $f + g$ is the sum of the functions f and g.

$$(f + g)(x) = f(x) + g(x)$$

The function $f - g$ is the function f minus the function g.

$$(f - g)(x) = f(x) - g(x)$$

The function fg is the product of the functions f and g.

$$(fg)(x) = f(x) \cdot g(x)$$

The function f/g is the function f divided by the function g, where $g(x) \neq 0$.

$$\left(\frac{f}{g}\right)(x) = \frac{f(x)}{g(x)} \qquad \text{if } g(x) \neq 0$$

Section 10.4

A ***linear function*** is any function of the form

$$f(x) = mx + b$$

where m and b are real numbers.

The ***identity function*** is a function of the form

$$f(x) = x$$

A ***constant function*** is any function of the form

$$f(x) = b$$

where b is a constant.

An ***absolute value function*** is a function of the form

$$f(x) = |ax + b| + c$$

where a, b, and c are real numbers and $a \neq 0$.

A ***quadratic function*** is any function of the form

$$f(x) = ax^2 + bx + c$$

where a, b, and c are real numbers and $a \neq 0$.

Power functions are of the form

$$f(x) = x^n$$

where n is a fixed positive integer.

Section 10.5

The inverse function of a function f, written f^{-1}, is the set of all ordered pairs of the form (y, x), where (x, y) is a member of f.

A function f is one-to-one if each y-value corresponds to exactly one x-value. That is, if $a \neq b$, then $f(a) \neq f(b)$. The inverse of a function is a function only when the function is one-to-one.

Horizontal-line test: A function is one-to-one and has an inverse that is a function if there is no horizontal line that intersects the graph of the function more than once.

The equation for the inverse of a one-to-one function may be obtained by the following steps.

1. Let $y = f(x)$.
2. Interchange x and y.
3. Solve for y.
4. Let $f^{-1}(x) = y$.

Chapter 10 Additional Exercises (Optional)

NAME

DATE

CLASS

Section 10.1

Are the following functions? If so, give the domain.

1. $\{(-3, 4), (-1, 2), (0, 4), (2, 5)\}$

2. $\{(2, -8), (3, 7), (2, 5)\}$

3. $y = x - 5$

4. $y = \dfrac{2}{x + 5}$

5. $x = y^2 + 3$

6. $y = \sqrt{x + 2}$

7. $y = x^3 + 2$

8. $y = x^4 - 3$

* **9.** $y < 3x + 4$

* **10.** $y \geq 2x$

* **11.** $|y| = x$

* **12.** $y = |x|$

Which of the following graphs represent functions?

13.

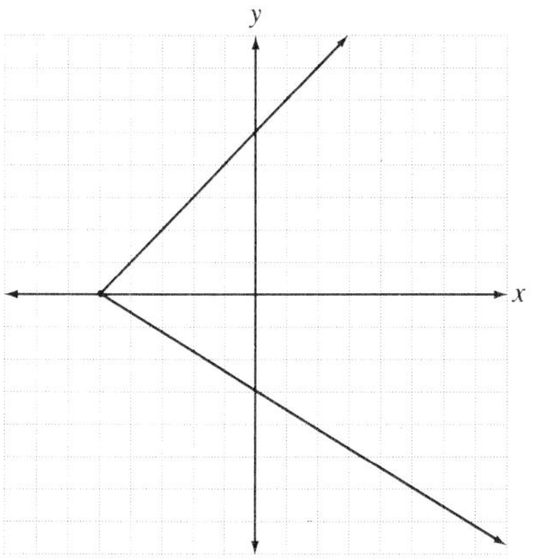

14.

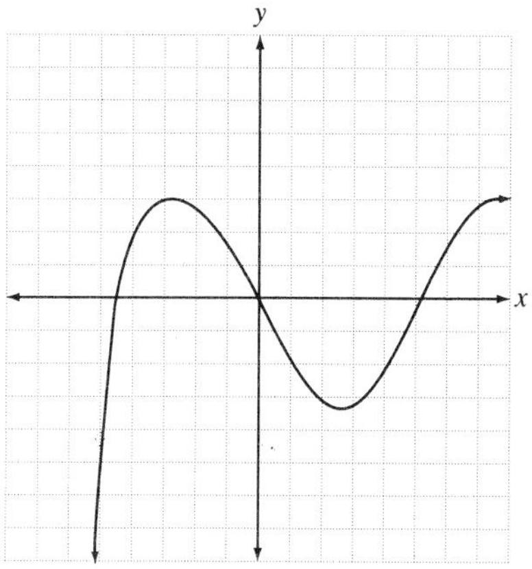

Section 10.2

If $f(x) = x^3$ *and* $g(x) = 2x + 3$, *find the following.*

15. $g(-3)$

16. $f(2)$

17. $f(-2)$

18. $g(-6)$

19. $f\left(\frac{1}{4}\right)$

20. $g(0.3)$

* **21.** $f(\sqrt{2})$

* **22.** $f(-\sqrt{5})$

23. $f[g(2)]$

24. $g[g(0)]$

25. $g[f(2)]$

26. $f[f(1)]$

27. $f[g(x)]$

28. $g[f(x)]$

Section 10.3

For the following problems, find (a) $f + g$, *(b)* $f - g$, *(c)* fg, *and (d)* $\frac{f}{g}$.

29. $f(x) = x - 7,\ g(x) = x - 3$

30. $f(x) = -x^3,\ g(x) = -x$

31. $f(x) = x^2,\ g(x) = -x$

32. $f(x) = x^2 - 1,\ g(x) = x + 1$

33. $f(x) = 4x^2 - 25,\ g(x) = 2x - 5$

34. $f(x) = 3x + 4,\ g(x) = 3x - 4$

35. $f(x) = \frac{3}{x + 4},\ g(x) = \frac{7}{2x - 1}$

36. $f(x) = \frac{1}{3 - x},\ g(x) = \frac{5}{x - 3}$

37. $f(x) = 6x^2 + 7x - 3,\ g(x) = 2x + 3$

38. $f(x) = x^2 - 10x + 9,\ g(x) = x - 9$

Section 10.4

Identify as a linear function, identity function, constant function, absolute value function, quadratic function, or power function.

39. $f(x) = x + 4$

40. $f(x) = x$

41. $f(x) = 5$

42. $f(x) = 3x^2 - 2x + 4$

43. $f(x) = 3x - 7$

44. $f(x) = x^3$

45. $f(x) = x^2 + x - 6$

46. $f(x) = -2$

47. $f(x) = \frac{2x}{3} - 3$

48. $f(x) = (x + 1)^2 + 3$

49. $f(x) = |x| + 4$

50. $f(x) = |x - 7|$

Section 10.5

Determine if the function is one-to-one. If it is, find the inverse function.

51. $f = \{(0, -5), (-6, -8), (5, 9), (-7, 2), (3, 8)\}$

52. $g = \{(-3, 4), (0, 0), (-15, 9), (3, -22), (5, 50)\}$

53. $f(x) = x^2 - 5$

54. $g(x) = x^2 - 8x + 7$

Find the inverse of the following functions.

55. $f(x) = 4x - 1$

56. $f(x) = 6 - 2x$

57. $f(x) = x^3 - 4$

58. $f(x) = x^3 + 5$

59. $f(x) = -x^2 + 4, x \geq 0$

60. $f(x) = -3x^2 - 2, x \geq 0$

Recall that the graph of f^{-1} is a reflection of the graph of f across the line $y = x$. Graph the inverse of each function on the same axes as the function.

61.

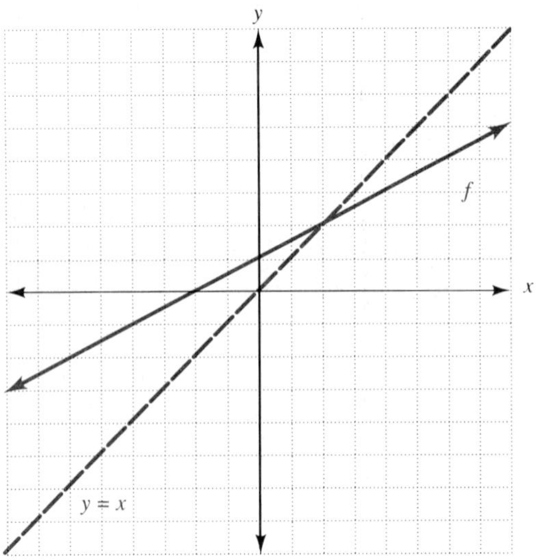

62.

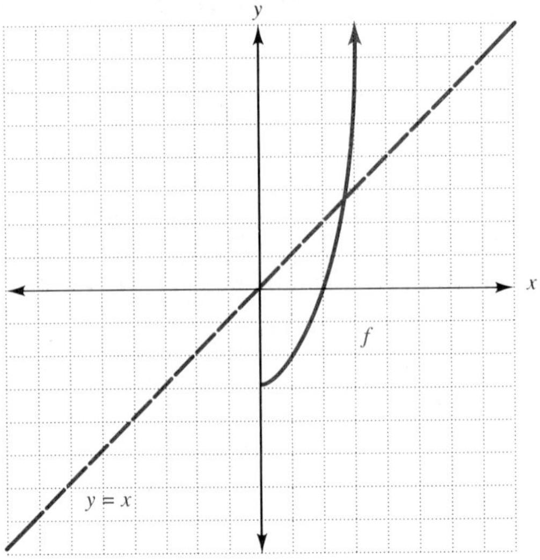

COOPERATIVE LEARNING

1. Change the set $\{(4, 2), (-3, 0), (-4, -8), (-3, 7), (0, -6)\}$ to make it a function.

2. Write a function whose domain is all real numbers except 8. (Answers may vary.)

3. What number is not contained in the range of $y = \dfrac{3}{x - 2}$?

4. Find $g(x)$ if $f[g(x)] = \sqrt{x^2 - 3}$ and $f(x) = \sqrt{x - 3}$.

5. Find $f(x)$ if $f[g(x)] = x$ and $g(x) = \dfrac{1}{x}$.

6. What are $f(x)$ and $g(x)$ if $fg(x) = 51x^2 - 22x - 8$?

7. What is $f(x)$ if $\dfrac{f(x)}{g(x)} = 6x^2 - 23x + 20$ and $g(x) = 3x - 4$?

8. The inverse of a function is $y = x - 4$. Find the function.

Chapter 10 Practice Test

NAME

DATE

CLASS

Consider the set $\{(1, 3), (-2, 5), (4, 5), (0, 6)\}$.

1. Is the set a function?

1. ______

2. Find the domain.

2. ______

3. Find the range.

3. ______

4. Find the domain of the function: $y = \sqrt{x - 3}$.

4. ______

5. Use the vertical line test to decide if the following is a function:

$$\frac{x^2}{4} + \frac{y^2}{9} = 1$$

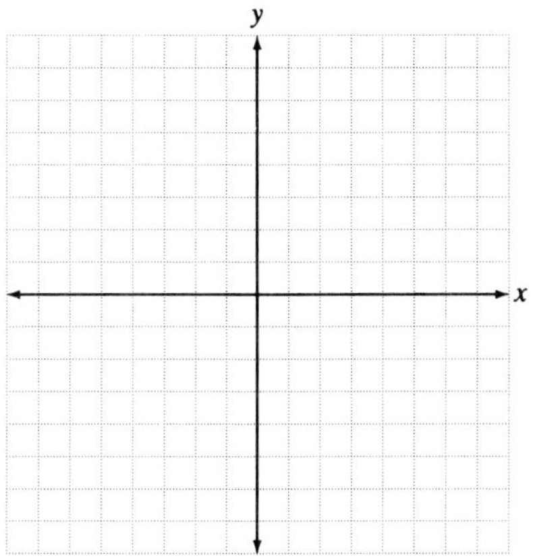

5. ______

6. ______________________

6. If $f(x) = x^2 - 5x + 3$, find $f(-2)$.

7. ______________________

7. If $f(x) = 2x^2 + 3$ and $g(x) = 3x - 1$, find $f[g(x)]$.

Let $f(x) = 3x^2 - x + 7$ and $g(x) = 2x - 5$.

8. ______________________

8. Find $g - f$.

9. ______________________

9. Find fg.

10. ______________________

10. Identify as a linear function, identity function, constant function, absolute value function, quadratic function, or power function: $f(x) = x^2 + 3x - 4$.

11. ______________________

11. Determine whether $h(x) = |x - 1|$ is one-to-one.

12. ______________________

12. Find the inverse of the function: $f(x) = x^3 - 7$.

Chapters 9 and 10 Cumulative Review

NAME ____________________

DATE ____________________

CLASS ____________________

1. Find the distance between the points $(4, -1)$ and $(-2, -5)$.

2. Find the midpoint of the line segment with endpoints $(\frac{3}{8}, \frac{1}{4}), (\frac{5}{16}, \frac{7}{6})$.

3. Find an equation for the circle with center $(-3, 5)$ and radius 4.

Graph.

4. $\frac{x^2}{16} + \frac{y^2}{4} = 1$

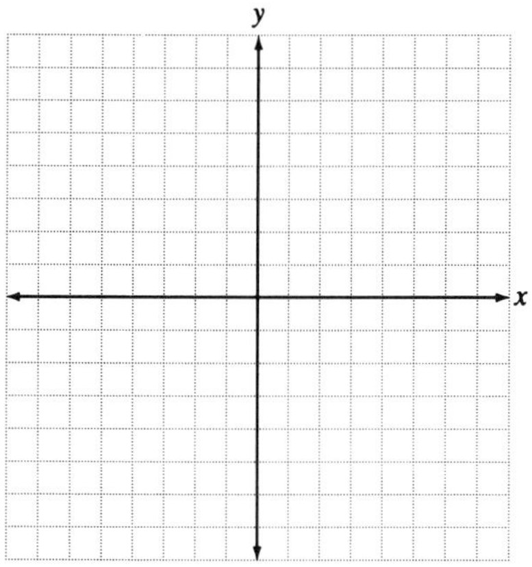

5. $y = 3(x - 2)^2 - 1$

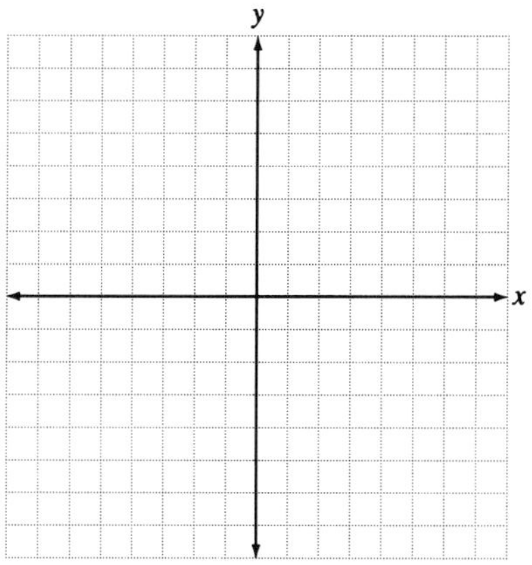

6. $\frac{x^2}{25} - \frac{y^2}{9} = 1$

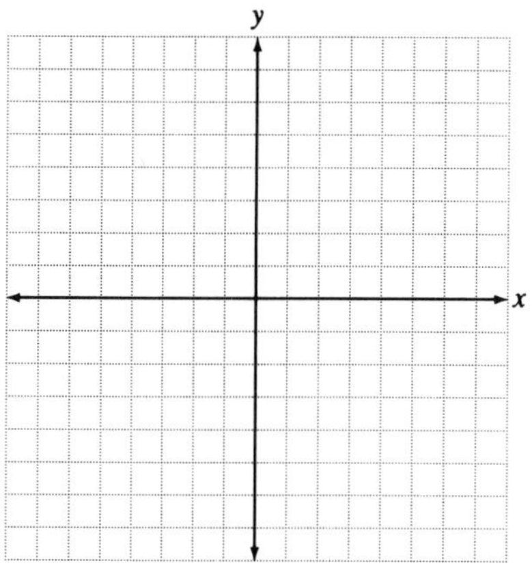

7. Identify the conic section:

$$\frac{x^2}{4} + \frac{y^2}{4} = 1$$

8. A rectangular floor has an area of 240 square feet and its perimeter is 64 feet. Find the dimensions of the floor.

9. Solve:

$$4x^2 + 3y^2 = 19$$
$$2x^2 + 5y^2 = 13$$

Consider the set $\{(-1, 4), (2, 5), (-3, 4)\}$.

10. Is the set a function?

11. Find the domain.

12. Find the range.

13. Find the domain of the function:

$$y = \frac{3}{x - 8}$$

14. Use the vertical line test to decide if the following is a function:

$$y = x^2 - 4$$

15. If $f(x) = 3x^2 + 4x - 7$, find $f(-3)$.

16. If $f(x) = x + 4$ and $g(x) = x^2 - 3$, find $g[f(x)]$.

Let $f(x) = 2x^2 - 11x + 12$ *and* $g(x) = x - 4$.

17. Find $f + g$.

18. Find $\frac{f}{g}$

19. Identify as a linear function, identity function, constant function, quadratic function, or power function.

$$f(x) = 3x + 4$$

20. Find the inverse of the function.

$$f(x) = 7x - 2$$

21. Determine whether $h(x) = |3x - 4|$ is one-to-one.

CHAPTER

11

Exponential and Logarithmic Functions

Pretest

Graph.

1. $y = 2^{3-x}$

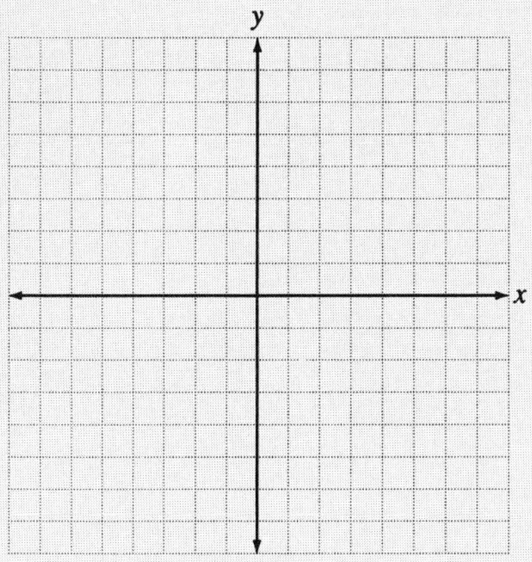

2. $x = \left(\frac{1}{2}\right)^y$

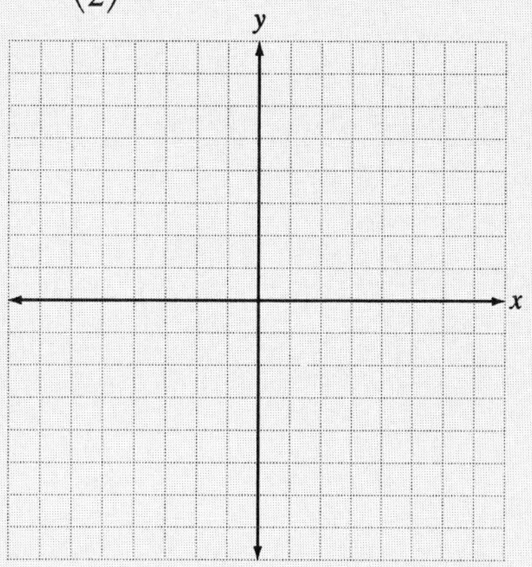

3. Change to an exponential equation: $\log_e 52 = 3.9512$.

4. Change to a logarithmic equation: $p^q = y$.

5. Graph: $y = \log_2 x$.

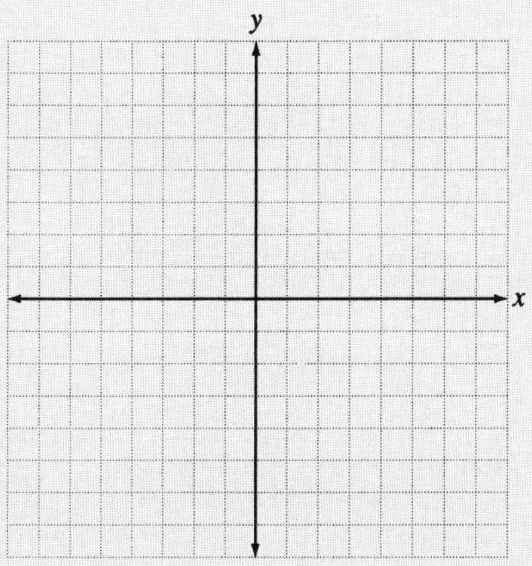

6. Write as a single logarithm: $\frac{1}{2}\log_b V + \frac{2}{3}\log_b W$.

7. Write in terms of logarithms of x, y, p, and q: $\log_b \sqrt{\frac{x^2y^6}{p^4q^8}}$.

Find the following. Use a calculator.

8. log 0.049

9. ln 8.2

10. The common antilogarithm of 3.718

11. The natural antilogarithm of -4.28

Find the following.

12. $\log_a a$

13. $\log_p 1$

14. $\log_b b^{21}$

15. Use the change-of-base formula to find $\log_5 46$.

Solve.

16. $7^{2x-1} = 343$

17. $5^{3x-4} = 22$

18. $\log_2(9x^2 - 1) - \log_2(3x - 1) = 5$

19. How many years will it take \$8500 invested at 6.5% interest compounded annually to yield \$14,517.46?

20. The amount of turkey, in pounds per person per year, eaten t years after 1937 is given by the equation $N = 2.3(3)^{0.033t}$. What is the consumption of turkey per person in 1997?

11.1 EXPONENTIAL FUNCTIONS

OBJECTIVES

1. *Graph exponential functions*
2. *Graph inverses of exponential functions*

Exponential functions have many applications in the real world. For example, an exponential function can be used to give the number of bacteria in a culture after t days. We study applied problems in Section 11.6.

In Section 7.6 we saw how to evaluate a^x for any positive real number a and any rational values of x. It can be shown that a^x also exists for irrational values of x such as $\sqrt{2}$. Then we can define an exponential function.

An ***exponential function*** with base a is a function of the form

$$f(x) = a^x \qquad \text{where } a > 0 \text{ and } a \neq 1$$

1 Graphing Exponential Functions

To graph these functions, we find ordered pairs by choosing positive and negative values and zero for the variable and finding the corresponding values of the function.

EXAMPLE 1 Graph $f(x) = 2^x$.

	x	$f(x)$ or y	Ordered Pair
$f(-3) = 2^{-3} = \frac{1}{2^3} = \frac{1}{8}$	-3	$\frac{1}{8}$	$\left(-3, \frac{1}{8}\right)$
$f(-2) = 2^{-2} = \frac{1}{2^2} = \frac{1}{4}$	-2	$\frac{1}{4}$	$\left(-2, \frac{1}{4}\right)$
$f(-1) = 2^{-1} = \frac{1}{2^1} = \frac{1}{2}$	-1	$\frac{1}{2}$	$\left(-1, \frac{1}{2}\right)$
$f(0) = 2^0 = 1$	0	1	(0, 1)
$f(1) = 2^1 = 2$	1	2	(1, 2)
$f(2) = 2^2 = 4$	2	4	(2, 4)
$f(3) = 2^3 = 8$	3	8	(3, 8)

We graph these ordered pairs and connect them with a smooth curve.

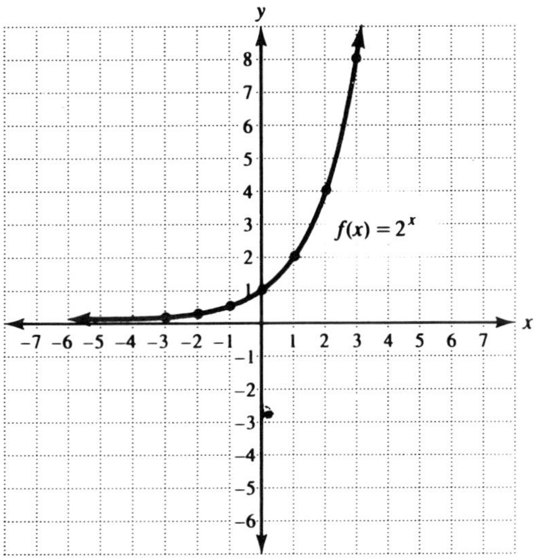

■

□ **Exercise 1** Graph $f(x) = 4^x$.

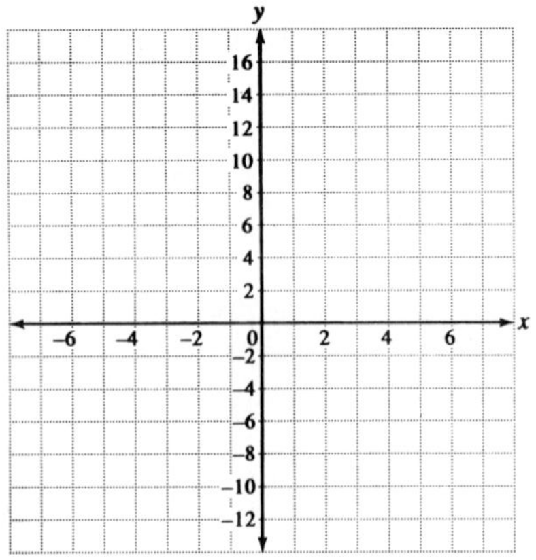

Notice that as x increases, y increases. This always happens when a is greater than 1.

> When $a > 1$, the function values $f(x) = a^x$ increase as x increases.

The graph of $f(x) = 2^x$ is typical of the graph of a function of the form $f(x) = a^x$ when $a > 1$. The greater the value of a, the steeper the curve.

□ **DO EXERCISE 1.**

EXAMPLE 2 Graph $f(x) = \left(\frac{1}{2}\right)^x$.

We choose some values for x, including some negative values, and compute the function values to find ordered pairs that belong to the function.

	Ordered Pair
$f(-3) = \left(\frac{1}{2}\right)^{-3} = \dfrac{1}{\left(\frac{1}{2}\right)^3} = \dfrac{1}{\frac{1}{8}} = 8$	$(-3, 8)$
$f(-2) = \left(\frac{1}{2}\right)^{-2} = \dfrac{1}{\left(\frac{1}{2}\right)^2} = \dfrac{1}{\frac{1}{4}} = 4$	$(-2, 4)$
$f(-1) = \left(\frac{1}{2}\right)^{-1} = \dfrac{1}{\left(\frac{1}{2}\right)^1} = \dfrac{1}{\frac{1}{2}} = 2$	$(-1, 2)$
$f(0) = \left(\frac{1}{2}\right)^0 = 1$	$(0, 1)$
$f(1) = \left(\frac{1}{2}\right)^1 = \frac{1}{2}$	$\left(1, \frac{1}{2}\right)$
$f(2) = \left(\frac{1}{2}\right)^2 = \frac{1}{4}$	$\left(2, \frac{1}{4}\right)$
$f(3) = \left(\frac{1}{2}\right)^3 = \frac{1}{8}$	$\left(3, \frac{1}{8}\right)$

We graph the ordered pairs and connect them with a smooth curve.

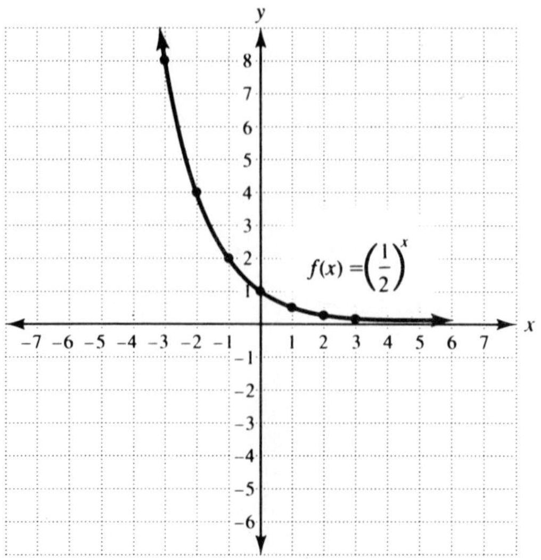

■

Notice that as x increases, the y values decrease. This always happens when a is less than 1. Remember that the exponential function is defined only for $a > 0$ and $a \neq 1$.

> When $0 < a < 1$, the function values $f(x) = a^x$ decrease as x increases.

The graph of $f(x) = (\frac{1}{2})^x$ is typical of the graph of a function of the form $f(x) = a^x$ when $0 < a < 1$.

Notice also that the graphs in both Examples 1 and 2 go through the point (0, 1).

> The graph of $f(x) = a^x$ passes through the point (0, 1).

The graphs in both examples also approach the x-axis without crossing it. We say that the x-axis is an ***asymptote.***

☐ **DO EXERCISE 2.**

☐ **Exercise 2** Graph $f(x) = \left(\frac{1}{3}\right)^x$.

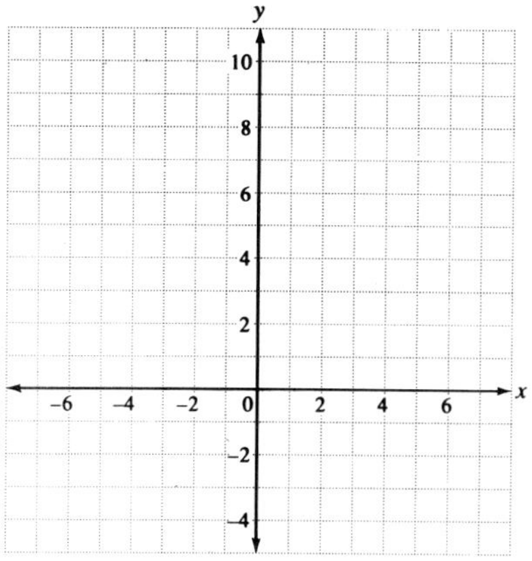

We will substitute y for $f(x)$ in the rest of this section, since y is easier to use in graphing.

EXAMPLE 3 Graph $y = 4^{x+1}$.

Some ordered pairs that satisfy the equation are

$$\left(-3, \frac{1}{16}\right), \quad \left(-2, \frac{1}{4}\right), \quad (-1, 1), \quad (0, 4), \quad \text{and} \quad (1, 16)$$

Plot these ordered pairs and connect them with a smooth curve. The graph looks like the graph of $y = 2^x$ except that it is shifted up and is steeper. Notice that this curve does not pass through (0, 1).

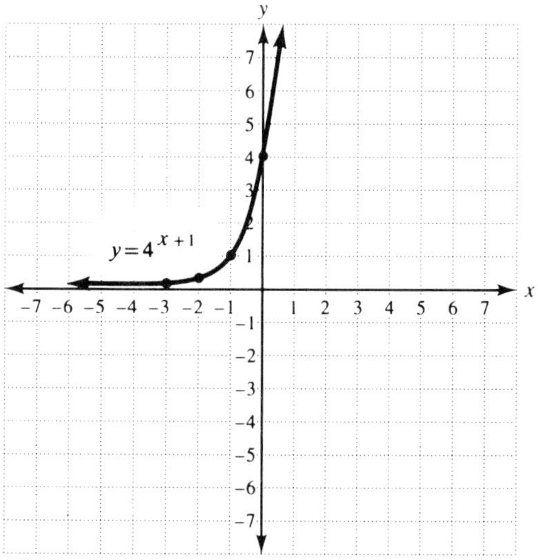

■

☐ **DO EXERCISE 3.**

☐ **Exercise 3** Graph $y = 3^{x-1}$.

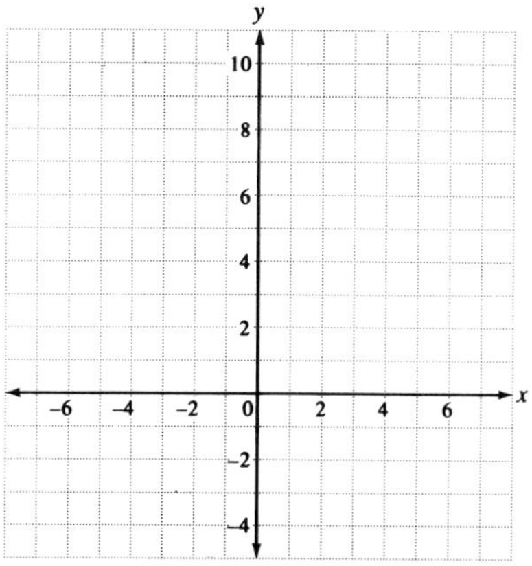

2 Inverses of Exponential Functions

We can find the inverses of exponential functions by interchanging x and y and solving for y. If we have a set of ordered pairs that belong to a function, we can get a set of ordered pairs that belong to the inverse by interchanging the x- and y-coordinates. This is an easy way to graph the inverse.

EXAMPLE 4 Graph $x = 3^y$.

Notice that $x = 3^y$ is the inverse of $y = 3^x$.
A set of ordered pairs that satisfy $y = 3^x$ is

$$\left\{\left(-3, \frac{1}{27}\right), \left(-2, \frac{1}{9}\right), \left(-1, \frac{1}{3}\right), (0, 1), (1, 3), (2, 9)\right\}$$

Therefore, a set of ordered pairs that satisfy $x = 3^y$ is

$$\left\{\left(\frac{1}{27}, -3\right), \left(\frac{1}{9}, -2\right), \left(\frac{1}{3}, -1\right), (1, 0), (3, 1), (9, 2)\right\}$$

We graph these ordered pairs and connect them with a smooth curve. Notice that the y-axis is an asymptote for this curve. The curve passes through $(1, 0)$ since it is the graph of the inverse of a function whose graph passes through $(0, 1)$.

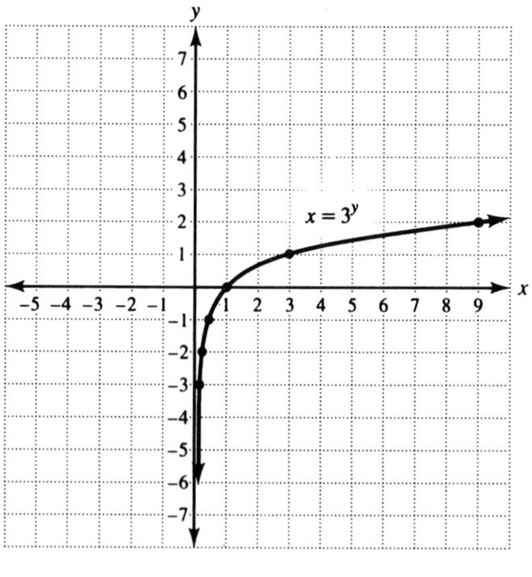

■

The graph of $x = 3^y$ is a reflection of the graph of $y = 3^x$ across the line $y = x$.

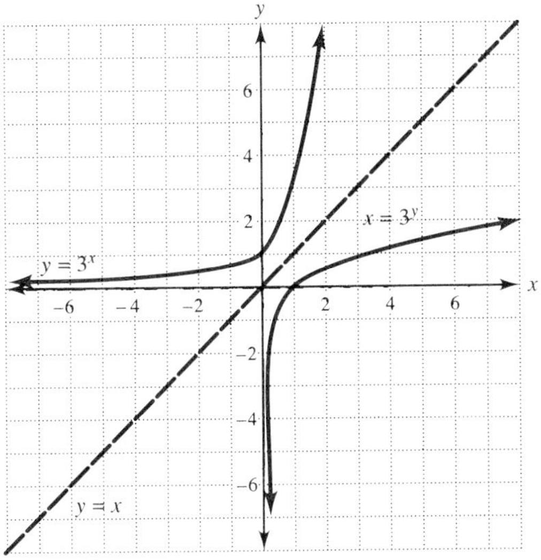

□ **DO EXERCISE 4.**

□ **Exercise 4** Graph $x = 2^y$.

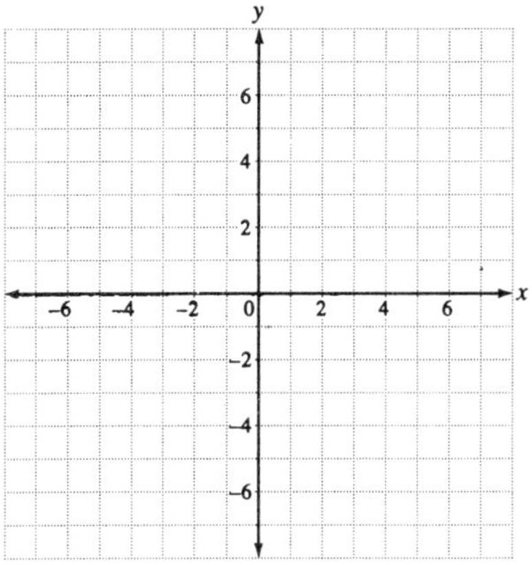

GRAPHING CALCULATOR

Exponential functions may also be graphed with a graphing calculator. Graph $y = 4^{x+1}$.

1. Enter $y = 4^{(x+1)}$ into your calculator. You must use the parentheses.
2. An appropriate window is W: $[-5, 10]$ $[-1, 20]$.
3. Plot the graph. This is the graph that is shown in Example 3.

We will graph equations like $x = 3^y$ with the graphing calculator in Section 11.4.

Answers to Exercises

1.

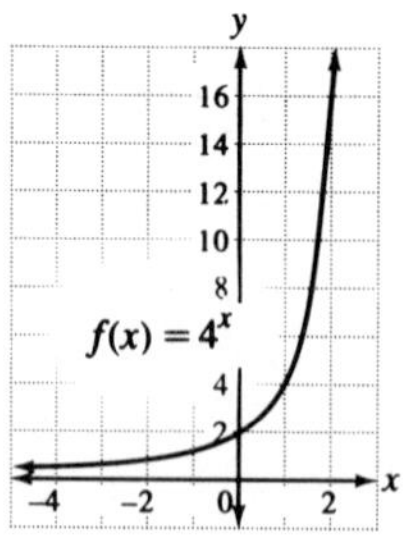

2.

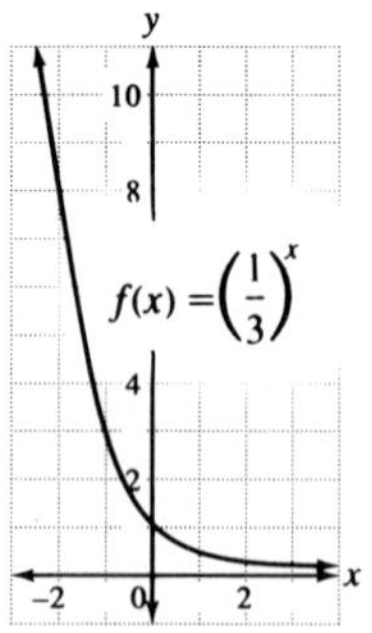

3.

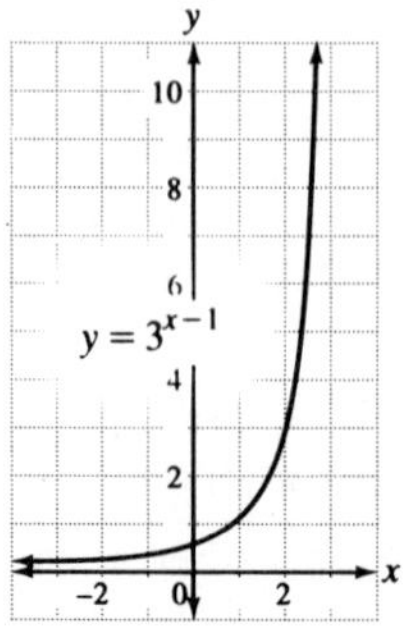

4.

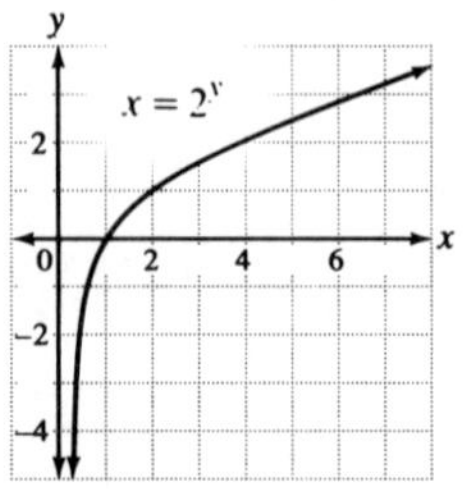

Problem Set 11.1

NAME ____________________

DATE ____________________

CLASS ____________________

Graph.

1. $y = 3^x$

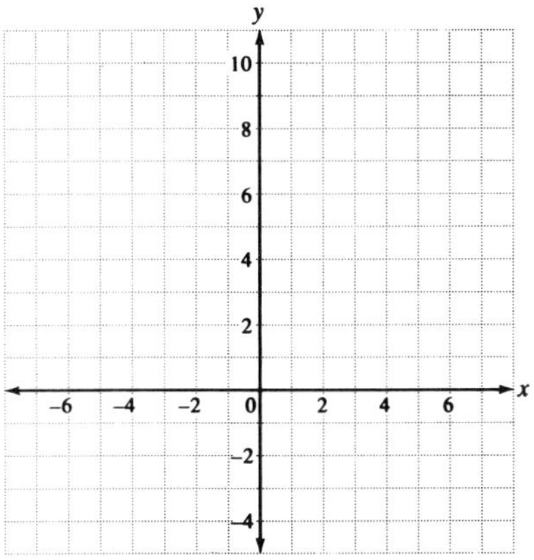

2. $y = 5^x$

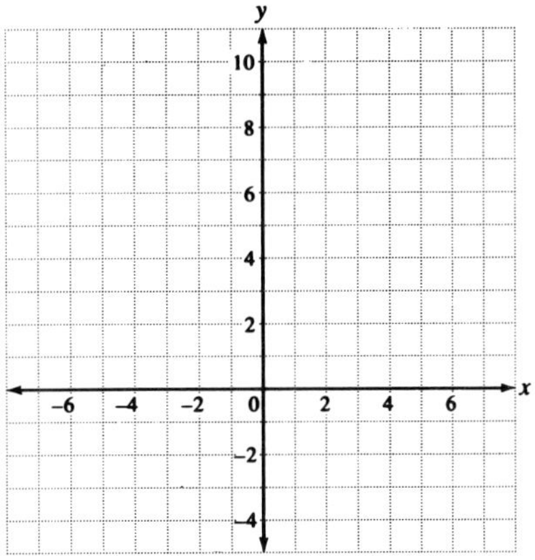

3. $y = 10^x$

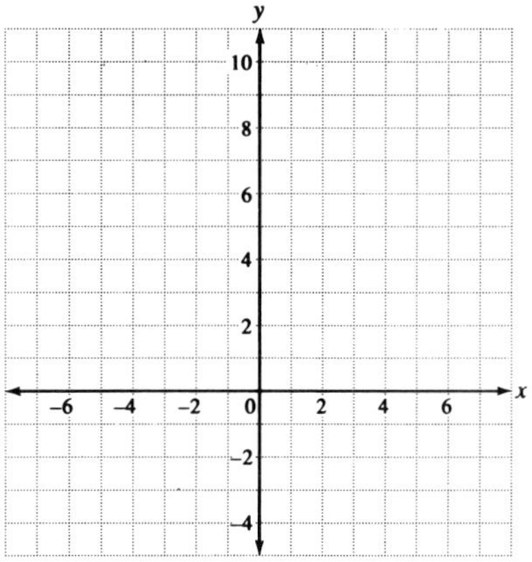

4. $y = 6^x$

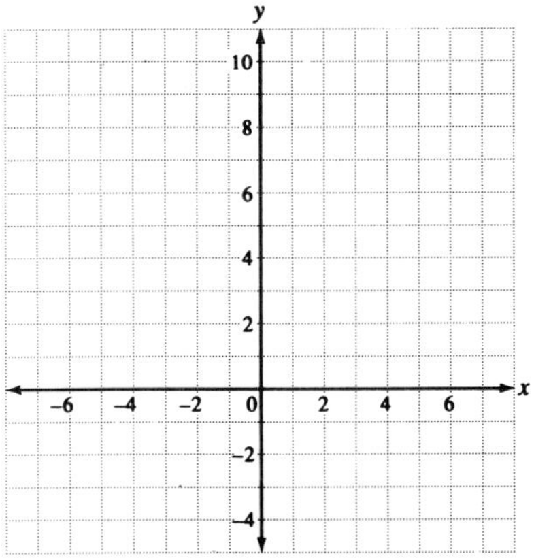

5. $y = \left(\frac{1}{4}\right)^x$

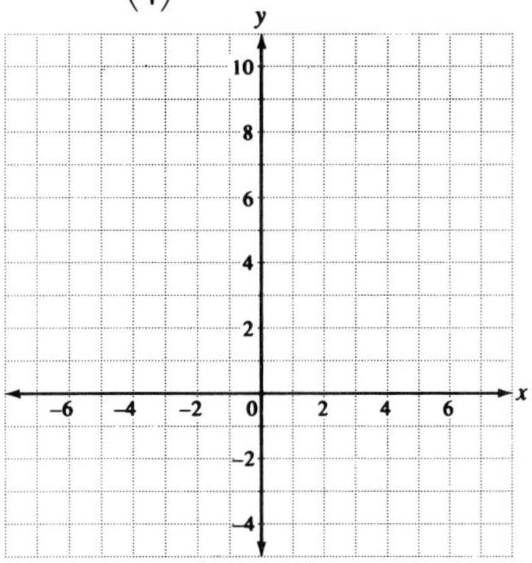

6. $y = \left(\frac{1}{2}\right)^x$

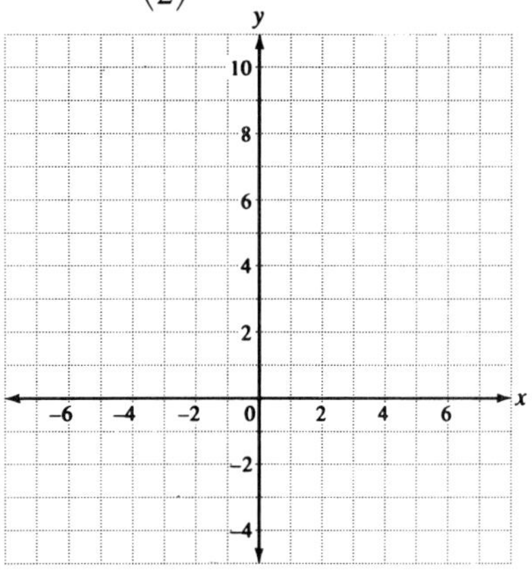

7. $y = \left(\frac{1}{10}\right)^x$

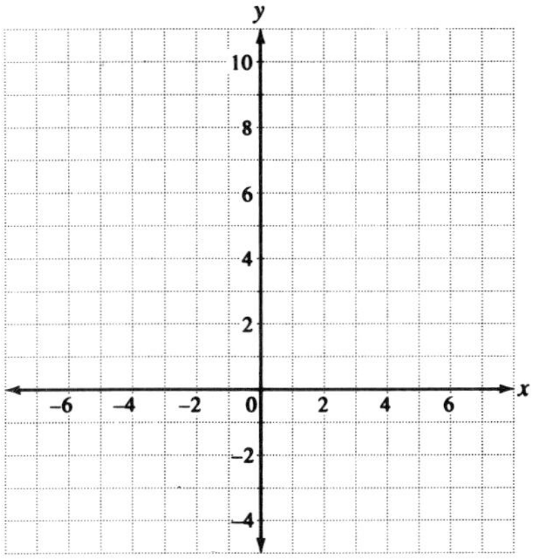

8. $y = \left(\frac{3}{4}\right)^x$

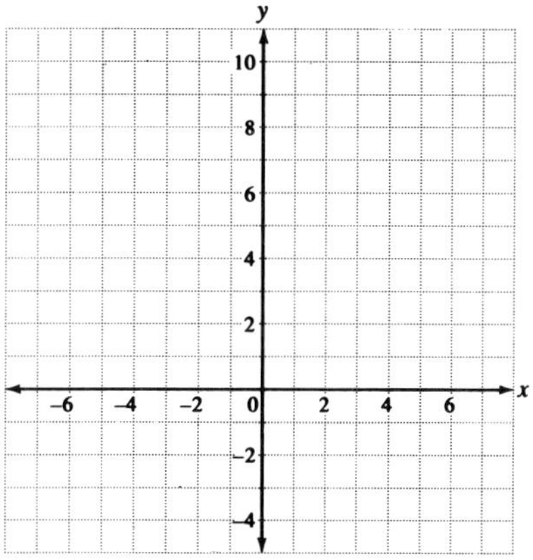

9. $y = 2^{x+1}$

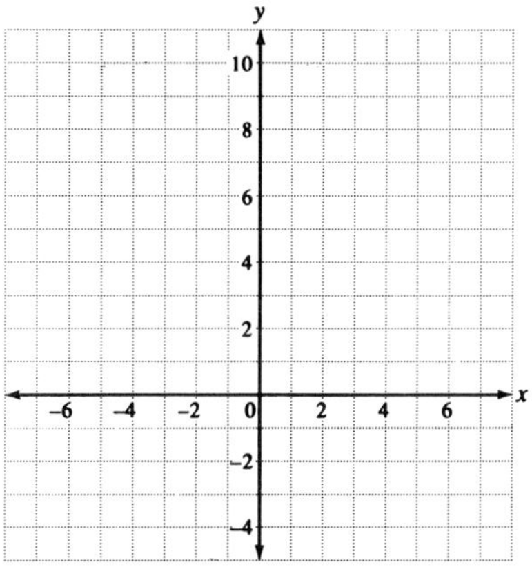

10. $y = 2^{x-1}$

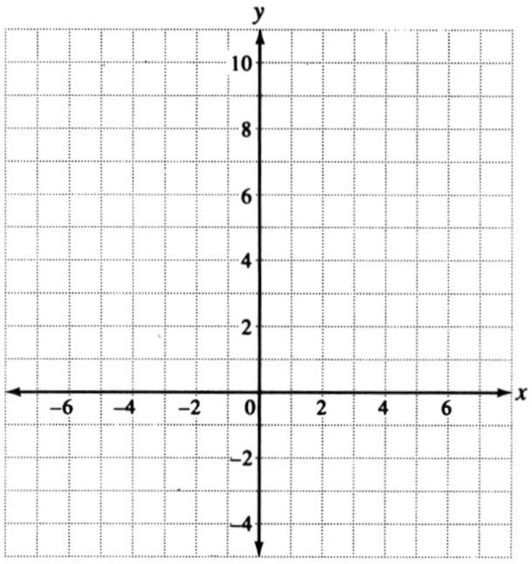

11. $y = 3^{x+1}$

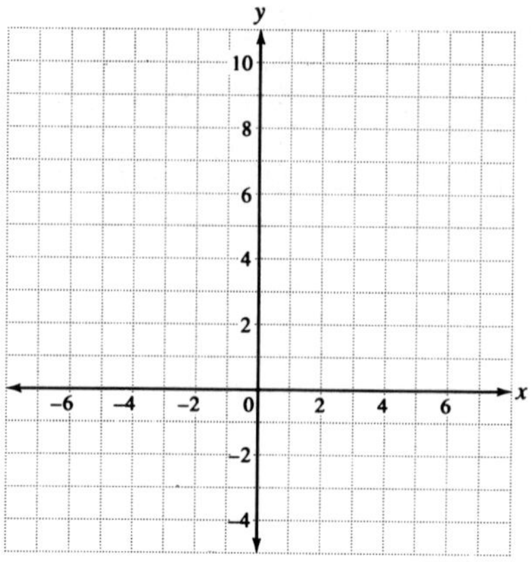

12. $y = 4^{x-1}$

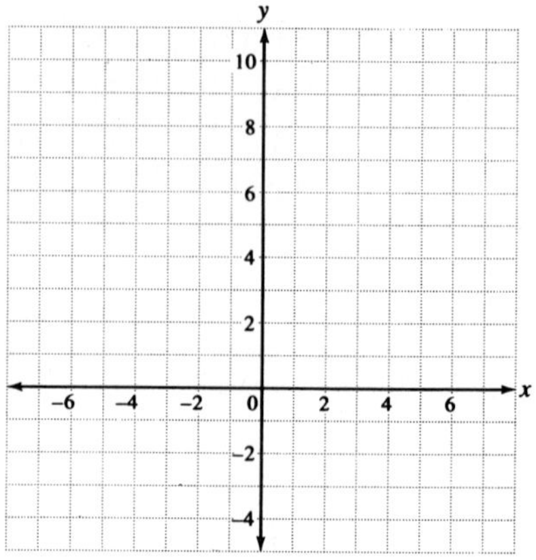

13. $y = 3^{x-1}$

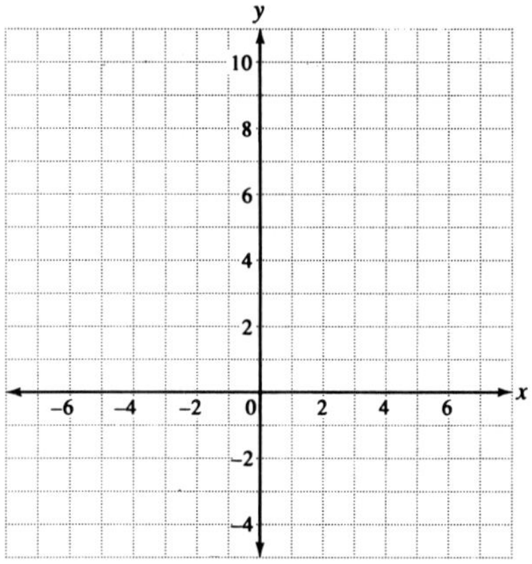

14. $y = 5^{x+2}$

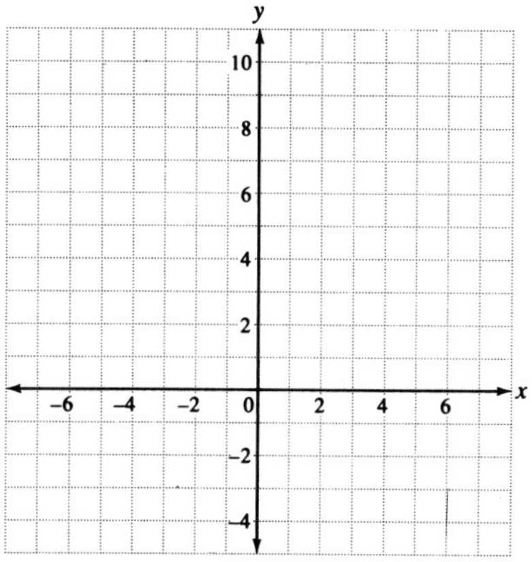

15. $y = 2^{x} + 2$

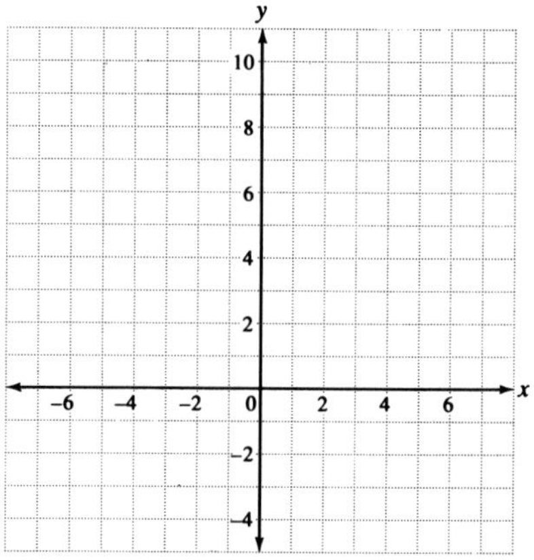

16. $y = 2^{x} - 4$

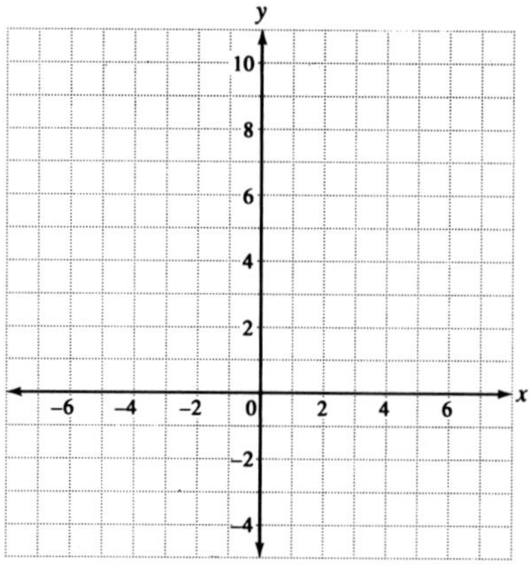

17. $y = 2^{3-x}$

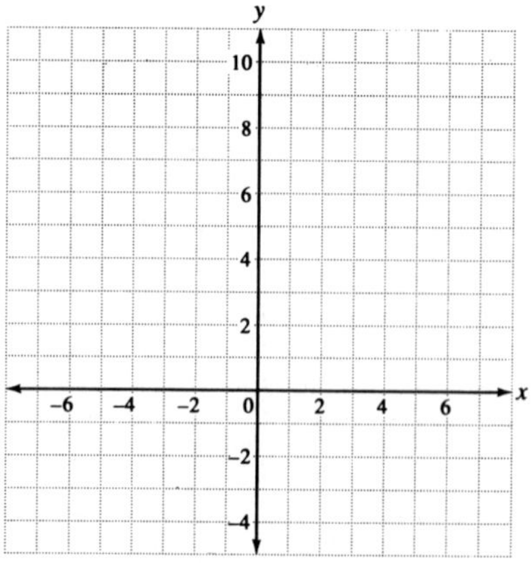

18. $y = 3^{3x-1}$

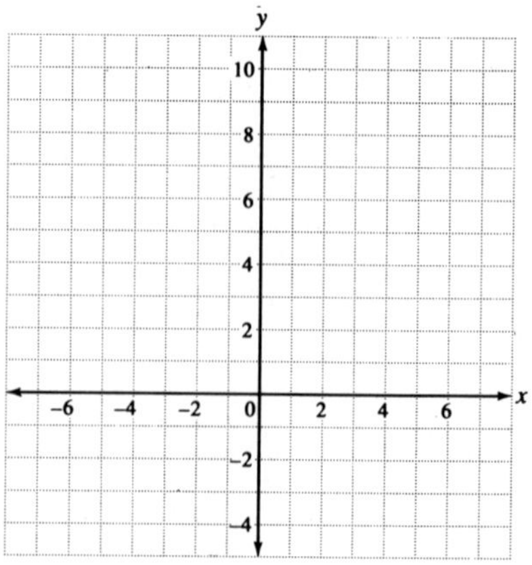

19. $x = 4^y$

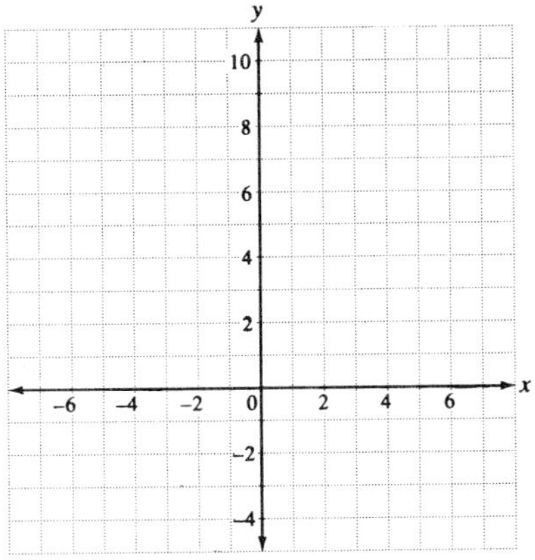

20. $x = 3^y$

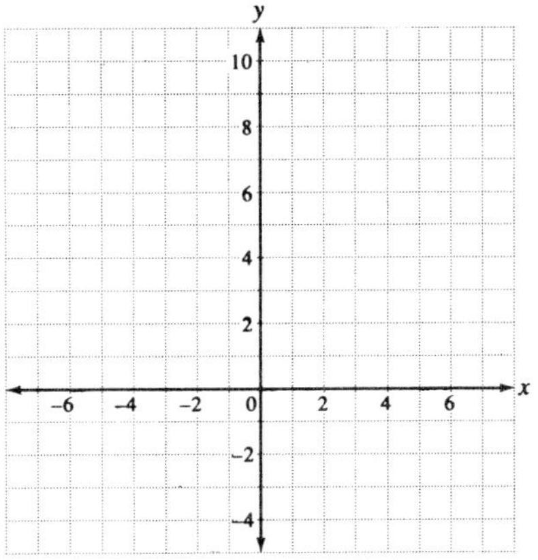

21. $x = \left(\frac{1}{2}\right)^y$

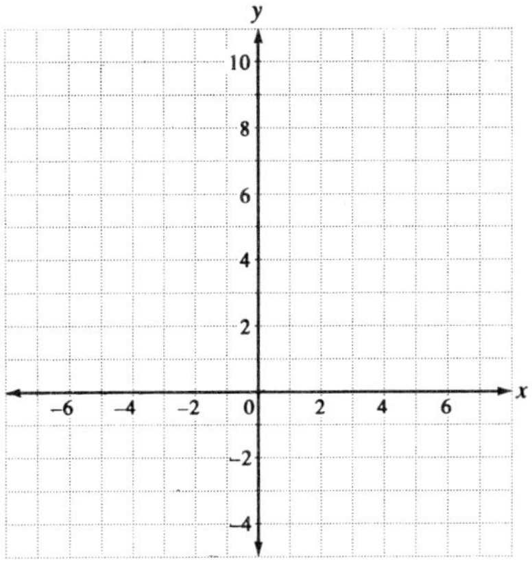

22. $x = \left(\frac{1}{4}\right)^y$

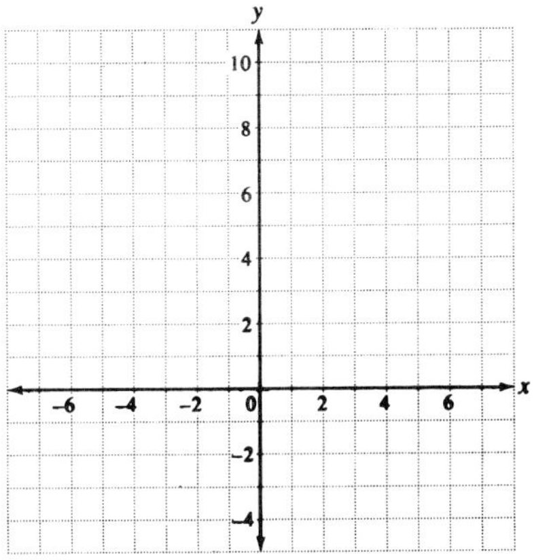

23. $x = 5^y$

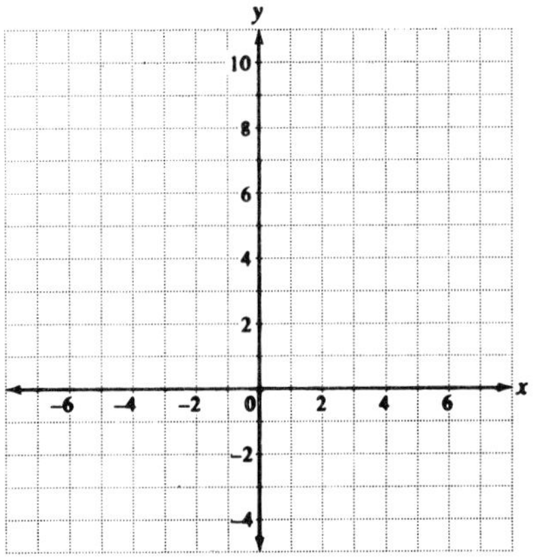

24. $x = \left(\frac{2}{3}\right)^y$

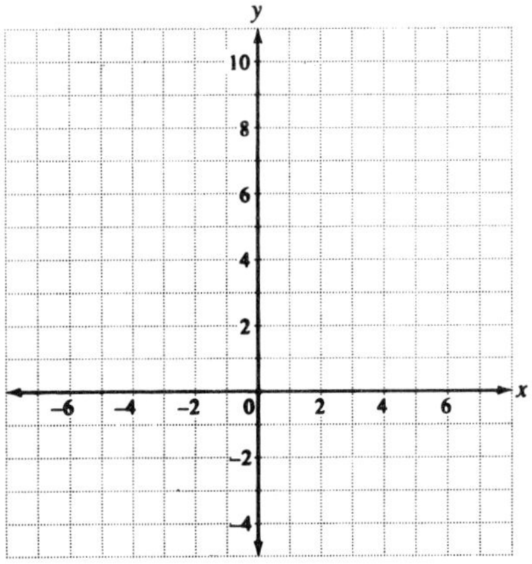

Graph both equations using the same set of axes.

25. $y = 4^x, \quad x = 4^y$

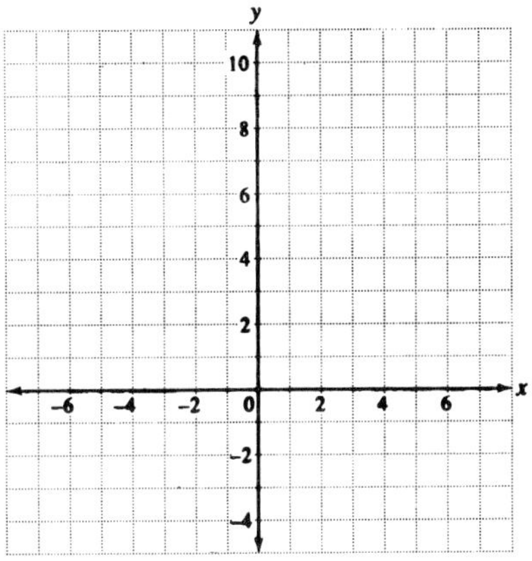

26. $y = 2^x, \quad x = 2^y$

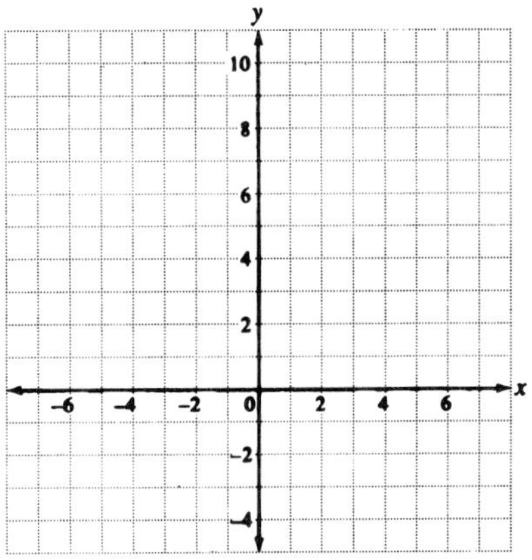

27. $y = \left(\frac{1}{3}\right)^x, \quad x = \left(\frac{1}{3}\right)^y$

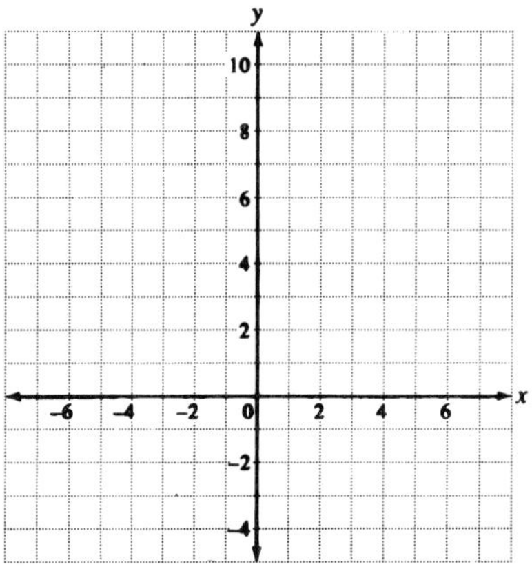

28. $y = \left(\frac{3}{4}\right)^x, \quad x = \left(\frac{3}{4}\right)^y$

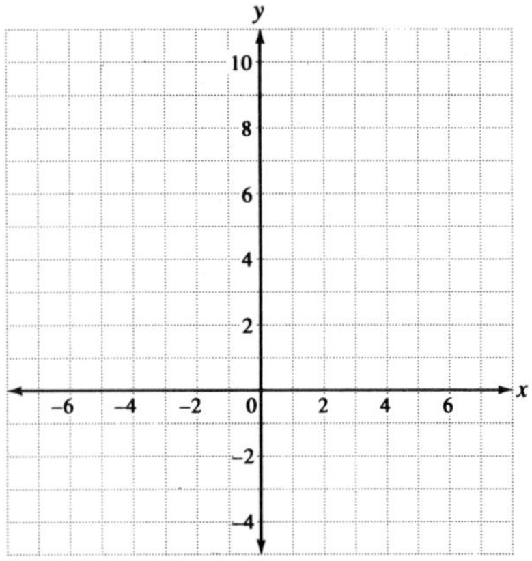

29. When $a > 1$, the function values $f(x) = a^x$ ____________ as x increases.

30. The x-axis is an ____________ for functions of the form $f(x) = a^x$, $a > 0$ and $a \neq 1$.

Graph.

* 31. $y = -3^x$

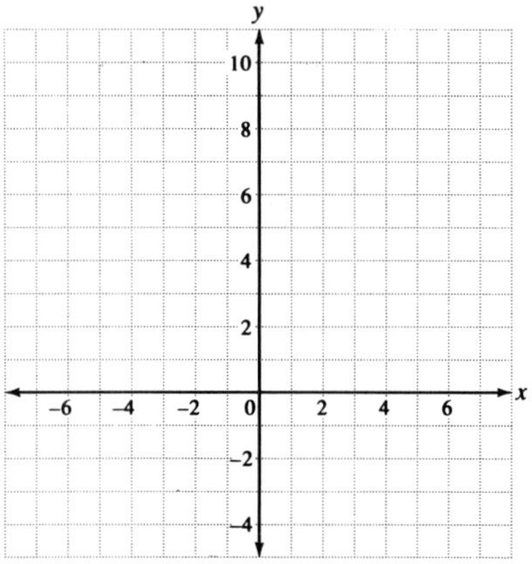

* 32. $y = -\left(\frac{1}{2}\right)^x$

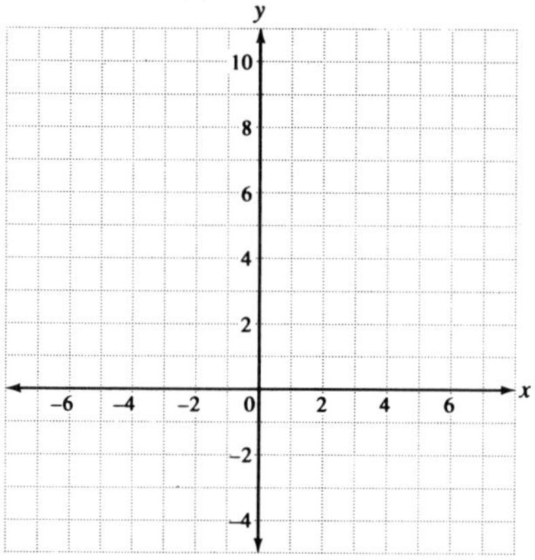

Think About It

Determine the base of the exponential function $f(x) = a^x$ if the graph of $f(x)$ contains the following points.

* 33. (4, 81)

* 34. $\left(\frac{1}{3}, 4\right)$

* 35. $\left(3, \frac{1}{27}\right)$

* 36. Let f be an exponential function with base a. Show that $f(u + v) = f(u) \cdot f(v)$ for any real numbers u and v.

Checkup

The following problems provide a review of some of Section 10.1.

Are the following functions? If so, give the domain.

37. $y = x + 4$

38. $y = 2x - 3$

39. $y = \frac{3}{x - 2}$

40. $y = \frac{7}{x + 8}$

41. $x = y^2 - 2$

42. $x^2 + y^2 = 4$

43. $y = \sqrt{2x - 5}$

44. $y = \sqrt{x + 3}$

45. $\frac{x^2}{81} - \frac{y^2}{144} = 1$

46. $\frac{x^2}{49} + \frac{y^2}{64} = 1$

11.2 LOGARITHMS

Logarithms were once used to do complicated calculations. Since we now have calculators, we seldom use logarithms to do computations. However, they have many other applications in advanced mathematics and science.

Logarithms are exponents. Consider some whole-number powers of 2. If $2^y = x$, then $y = \log_2 x$. We say that y is the logarithm of x base 2. Since $2^0 = 1$, $2^1 = 2$, $2^2 = 4$, $2^3 = 8$, and $2^4 = 16$,

$0 = \log_2 1$ Zero is the logarithm of 1 base 2.

$1 = \log_2 2$

$2 = \log_2 4$

$3 = \log_2 8$

$4 = \log_2 16$

OBJECTIVES

1. *Write logarithmic equations as exponential equations*
2. *Write exponential equations as logarithmic equations*
3. *Solve certain equations containing logarithms*
4. *Graph equations of the form $y = \log_a x$*

1 Logarithmic Functions

In Section 11.1 we graphed $x = 3^y$ and noticed that it is the inverse of $y = 3^x$. In general, the inverse of the exponential function $y = a^x$ is $x = a^y$. However, we usually want to write $x = a^y$ so that y is a function of x. We generalize from the previous notation for base 2 to give the following definition.

> $y = \log_a x$ is equivalent to $x = a^y$ $\quad a > 0, \quad a \neq 1, \quad x > 0$

The equation $y = \log_a x$ is read "y is the logarithm of x, base a."

> The ***logarithmic function*** is a function of the form
>
> $y = \log_a x \quad a > 0, \quad a \neq 1, \quad x > 0$

EXAMPLE 1 Write each logarithmic equation as an exponential equation.

a. $3 = \log_2 8$

This is equivalent to

$$8 = 2^3$$

Notice that the ***base*** *does not change* and the ***logarithm*** is the *exponent.*

b. $2 = \log_{10} 100$

This is equivalent to

$$100 = 10^2$$

c. $y = \log_3 4$

We use the definition to write the exponential equation.

$4 = 3^y$ The base does not change and the logarithm is the exponent.

d. $-2 = \log_x 5$

Using the definition, we have

$$5 = x^{-2}$$ ■

□ **DO EXERCISE 1.**

□ **Exercise 1** Write each logarithmic equation as an exponential equation.

a. $4 = \log_2 16$

b. $2 = \log_{10} 100$

c. $y = \log_5 7$

d. $-1 = \log_x 9$

2 Converting from Exponential to Logarithmic Equations

We may also use the definition of logarithms to convert from exponential to logarithmic equations.

EXAMPLE 2 Write each exponential equation as a logarithmic equation.

a. $9 = 3^2$

Since $x = a^y$ is equivalent to $y = \log_a x$,

$9 = 3^2$ is equivalent to $2 = \log_3 9$.

b. $1 = 6^0$

This is equivalent to

$$0 = \log_6 1$$

c. $7 = 4^x$

Using the definition, we obtain

$$x = \log_4 7$$

d. $y^{-2} = 5$

Since this is an equation, it may also be written $5 = y^{-2}$. Using the definition gives us

$$-2 = \log_y 5 \quad ■$$

□ DO EXERCISE 2.

3 Solving Equations

Some equations containing logarithms may be solved by changing them to exponential equations.

EXAMPLE 3 Solve.

a. $\log_5 x = -2$

This may also be written $-2 = \log_5 x$. We write the equation as an exponential equation.

$$x = 5^{-2}$$

$$x = \frac{1}{25}$$

b. $\log_x 81 = 2$

$x^2 = 81$ Write the equivalent exponential equation.

$x = 9$ or $x = -9$

Check For $x = 9$:

$\log_x 81$	$= 2$
$\log_9 81$	2
$\log_9 9^2$	2
2	2

For $x = -9$:

The base of a logarithm must be positive. Therefore, $\log_{-9} 81$ is not defined and -9 is not a solution.

The solution is 9. ■

□ DO EXERCISE 3.

□ **Exercise 2** Write each exponential equation as a logarithmic equation.

a. $25 = 5^2$

b. $1000 = 10^3$

c. $6 = 3^x$

d. $y^{-3} = 8$

□ **Exercise 3** Solve.

a. $\log_{1/2} x = -3$

b. $\log_x 16 = 2$

4 Graphing Equations of the Form $y = \log_a x$

We graph equations of the form $y = \log_a x$ by recalling that this equation is equivalent to $x = a^y$. The inverse of $x = a^y$ and therefore of $y = \log_a x$ is $y = a^x$.

EXAMPLE 4 Graph $y = \log_2 x$.

$x = 2^y$ Write the equivalent equation.

The equation of the inverse is

$$y = 2^x$$

Some ordered pairs that satisfy $y = 2^x$ are

$$\left(-3, \frac{1}{8}\right), \quad \left(-2, \frac{1}{4}\right), \quad \left(-1, \frac{1}{2}\right), \quad (0, 1), \quad (1, 2), \quad (3, 8)$$

Therefore, some ordered pairs that satisfy $y = \log_2 x$ are

$$\left(\frac{1}{8}, -3\right), \quad \left(\frac{1}{4}, -2\right), \quad \left(\frac{1}{2}, -1\right), \quad (1, 0), \quad (2, 1), \quad (8, 3)$$

We plot these points and connect them with a smooth curve. We also show the graph of $y = 2^x$. Notice that the graph of $y = \log_2 x$ is a reflection of the graph of $y = 2^x$ across the line $y = x$.

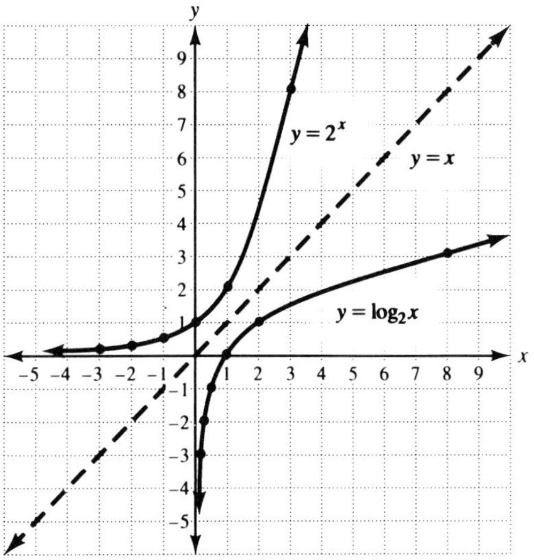

■

□ **DO EXERCISE 4.**

□ **Exercise 4** Graph $y = \log_5 x$.

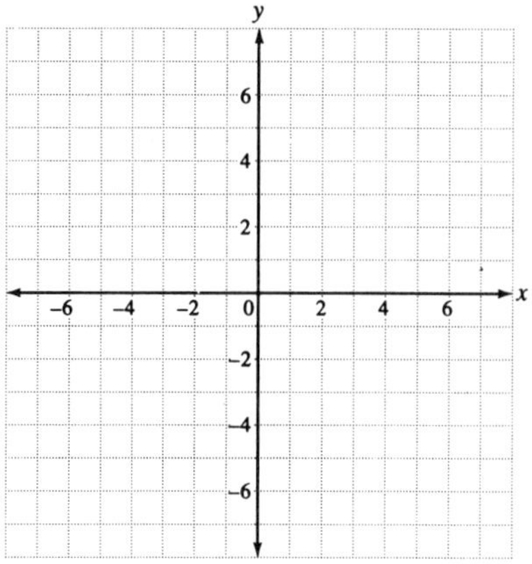

DID YOU KNOW?

John Napier published a book in 1614 explaining his invention of logarithms. Since he studied mathematics as a hobby, he spent more than 20 years putting together the book that explained his theory. Napier also predicted the invention of the submarine and accompanied his writings with drawings.

Answers to Exercises

1. **a.** $16 = 2^4$ **b.** $100 = 10^2$ **c.** $7 = 5^y$ **d.** $9 = x^{-1}$

2. **a.** $2 = \log_5 25$ **b.** $3 = \log_{10} 1000$ **c.** $x = \log_3 6$
d. $-3 = \log_y 8$

3. **a.** 8 **b.** 4

4.

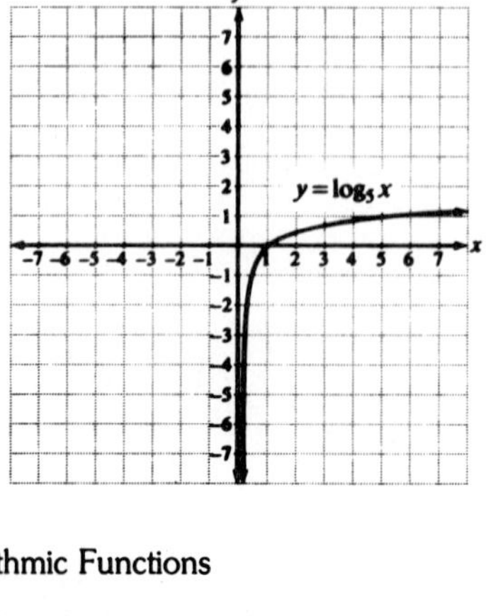

Problem Set 11.2

NAME ______________________

DATE ______________________

CLASS ______________________

Change to exponential equations.

1. $2 = \log_2 4$

2. $5 = \log_2 32$

3. $2 = \log_7 49$

4. $3 = \log_3 27$

5. $0 = \log_8 1$

6. $0 = \log_{10} 1$

7. $\log_2 \frac{1}{2} = -1$

8. $\log_{10} \frac{1}{100} = -2$

9. $1 = \log_6 6$

10. $1 = \log_{10} 10$

11. $\log_{10} 9 = 0.9542$

12. $\log_{10} 2 = 0.3010$

13. $\log_e 15 = 2.7081$

14. $\log_e 30 = 3.4012$

15. $\log_b R = a$

16. $\log_p S = q$

Change to logarithmic equations.

17. $64 = 8^2$

18. $36 = 6^2$

19. $125 = 5^3$

20. $1000 = 10^3$

21. $2^0 = 1$

22. $9^0 = 1$

23. $3^{-2} = \frac{1}{9}$

24. $5^{-2} = \frac{1}{25}$

25. $8^1 = 8$

26. $7^1 = 7$

27. $10^{0.6990} = 5$

28. $10^{0.4771} = 3$

29. $e^3 = 20.0855$

30. $e^{-2} = 0.1353$

31. $r^t = x$

32. $a^b = y$

Solve.

33. $\log_3 x = 4$

34. $\log_2 x = 5$

35. $\log_7 x = 2$

36. $\log_4 x = 3$

37. $\log_{1/2} x = -4$

38. $\log_{1/3} x = -2$

39. $\log_x 25 = 2$

40. $\log_x 8 = 3$

41. $\log_x 17 = 1$

42. $\log_x 37 = 1$

43. $\log_5 x = 0$

44. $\log_7 = 0$

45. $\log_4 x = \frac{1}{2}$

46. $\log_{27} x = \frac{1}{3}$

47. $\log_{32} x = \frac{1}{5}$

48. $\log_{25} x = \frac{1}{2}$

Graph.

49. $y = \log_3 x$

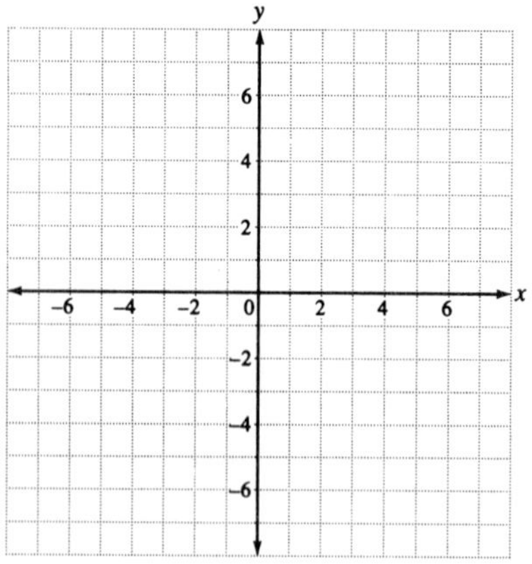

50. $y = \log_4 x$

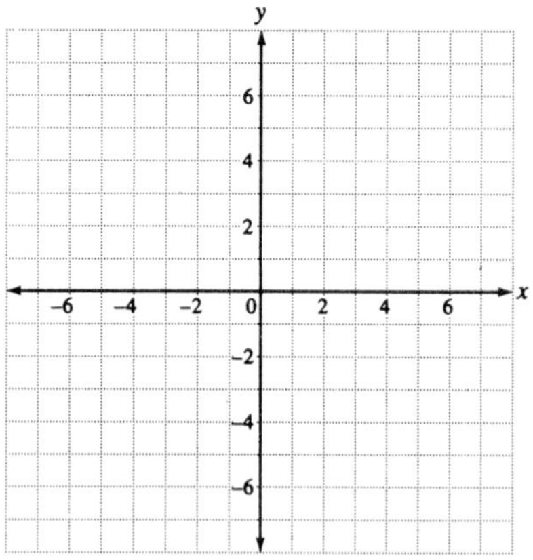

51. $y = \log_{1/2} x$

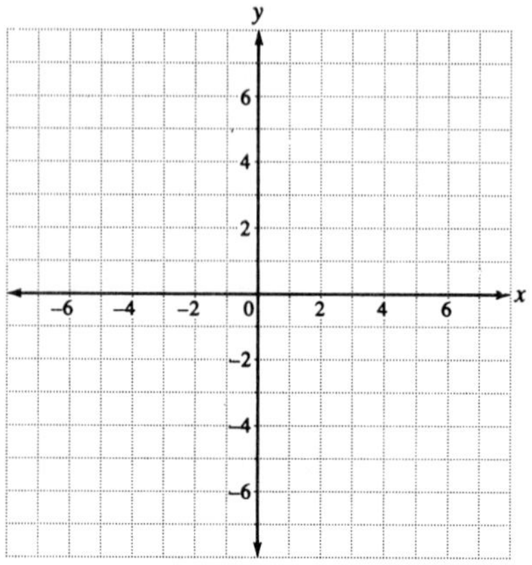

52. $y = \log_{1/3} x$

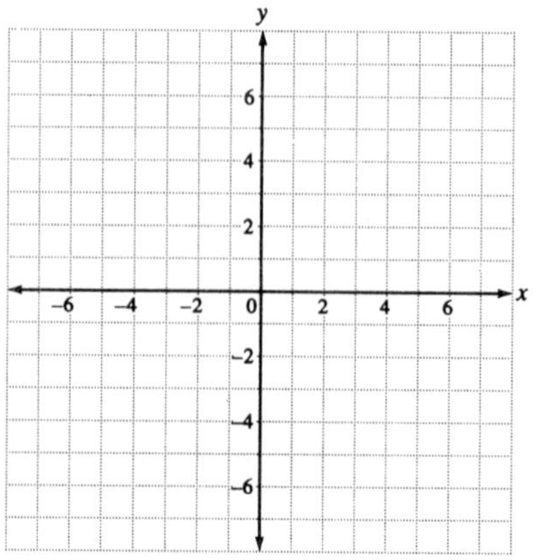

Graph both functions using the same set of axes.

53. $y = 6^x$, $y = \log_6 x$

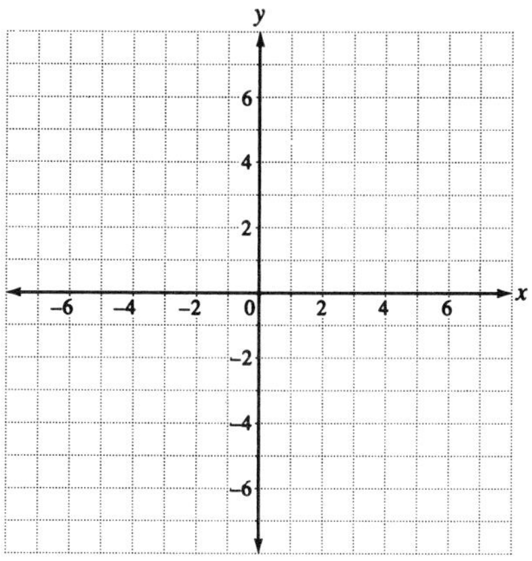

54. $y = 10^x$, $y = \log_{10} x$

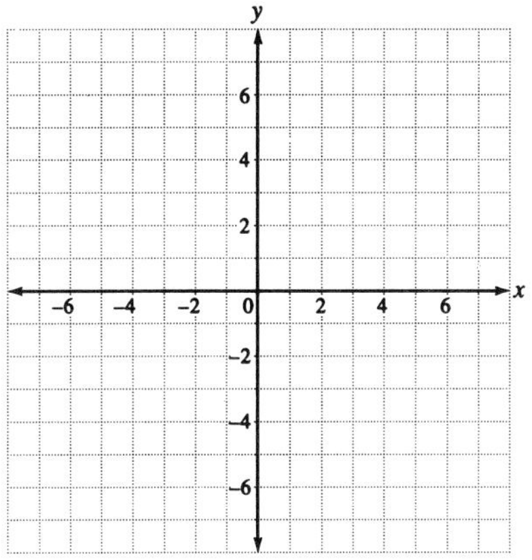

55. Logarithms are ____________.

56. The ____________ in an exponential equation and its equivalent logarithmic equation is the same.

57. One method of graphing equations of the form $y = \log_a x$ is to find ordered pairs that satisfy the inverse $y = a^x$ and ____________ the x-coordinates and the y-coordinates.

Graph.

* **58.** $y = -\log_2 x$

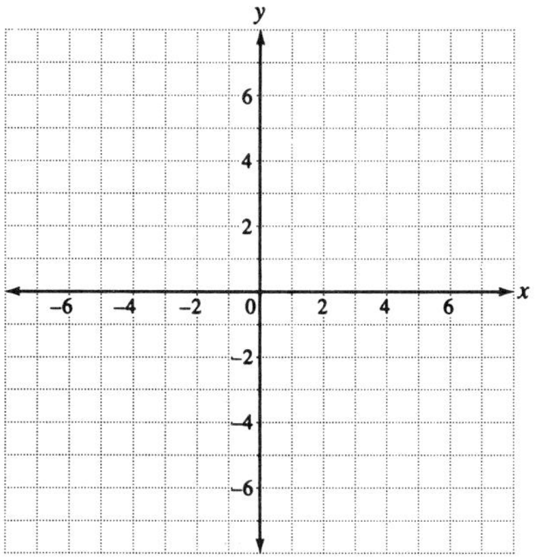

* **59.** $y = -\log_{1/4} x$

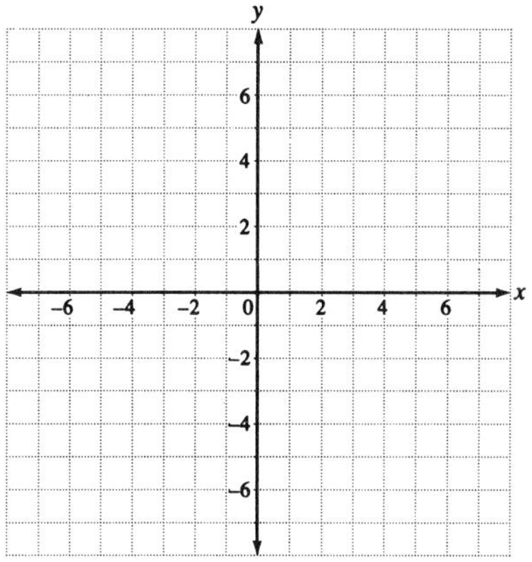

* **60.** $y = -\log_2 (x + 1)$

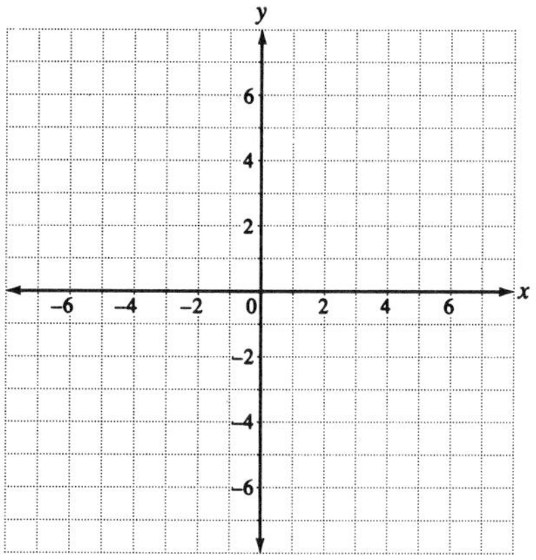

* **61.** $y = \log_3(x - 2)$

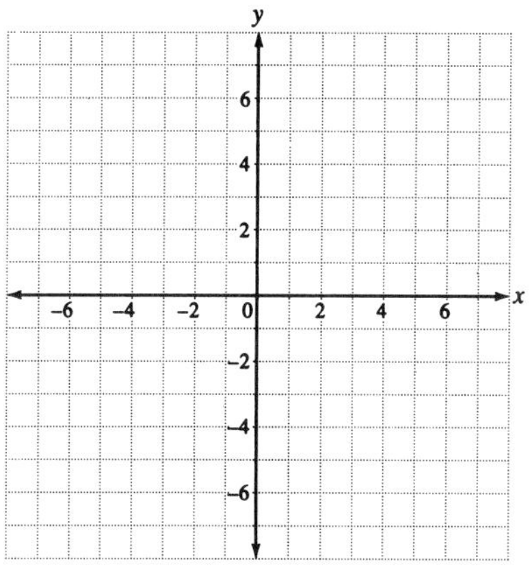

Think About It

Determine the base of the logarithmic function $y = \log_a x$ if the graph contains the following points.

* **62.** $(4, 2)$

* **63.** $(125, 3)$

* **64.** $\left(\frac{1}{4}, -1\right)$

* **65.** $(8, -3)$

* **66.** $(c^3, 3)$

Solve.

* **67.** $\log_4 x = -\frac{1}{2}$

* **68.** $\log_x 64 = -2$

* **69.** $y = \log_{3/2} \frac{9}{4}$

* **70.** $y = \log_{1/3} \frac{1}{27}$

* **71.** $\log_7 (2x - 1) = -1$

Checkup

The following problems provide a review of some of Section 10.2.

If $f(x) = -x^2 + 3x + 4$, find the following.

72. $f(2)$

73. $f(-3)$

74. $f(0)$

75. $f(5)$

For the following functions, find $f[g(x)]$.

76. $f(x) = 3x + 4$
$g(x) = x^2 - 2$

77. $f(x) = -2x + 5$
$g(x) = x^2 + 8$

78. $f(x) = x^2 + 7$
$g(x) = 2x + 1$

79. $f(x) = 2x^2 - 3$
$g(x) = 3x - 1$

11.3 PROPERTIES OF LOGARITHMS

[1] Finding $\log_a a$ and $\log_a 1$

We know that for any positive real number a, $a^1 = a$ and $a^0 = 1$. For example, $2^1 = 2$ and $2^0 = 1$. Converting $a^1 = a$ and $a^0 = 1$ to logarithmic equations gives us the following ***two properties of logarithms:***

> For any positive real number a, $a \neq 1$,
>
> $$\log_a a = 1 \qquad \text{and} \qquad \log_a 1 = 0$$

EXAMPLE 1 Find the following.

a. $\log_3 3$

$$\log_3 3 = 1$$

b. $\log_{10} 10$

$$\log_{10} 10 = 1$$

c. $\log_7 1$

$$\log_7 1 = 0$$

d. $\log_{10} 1$

$$\log_{10} 1 = 0$$ ■

□ **DO EXERCISE 1.**

[2] Logarithms of Products

Consider the following table of logarithms for base 2:

Base 2						
Number	$2^0 = 1$	$2^1 = 2$	$2^2 = 4$	$2^3 = 8$	$2^4 = 16$	$2^2 = 32$
Logarithm	0	1	2	3	4	5

Notice that $\log_2(4 \cdot 8) = \log_2 32 = 5$.

Also, $\log_2 4 + \log_2 8 = 2 + 3 = 5$.

Hence $\log_2(4 \cdot 8) = \log_2 4 + \log_2 8$.

We generalize the result above to give the rule for logarithms of products.

> **The Product Rule**
>
> For all positive numbers x, y, and a, $a \neq 1$,
>
> $$\log_a xy = \log_a x + \log_a y$$

We may prove this rule as follows.

Proof: Let $\log_a x = m$ and $\log_a y = n$

$a^m = x$ and $a^n = y$ Change to exponential notation.

Then

$$xy = a^m \cdot a^n = a^{m+n}$$

OBJECTIVES

[1] *Find $\log_a a$ and $\log_a 1$*

[2] *Write logarithms of products as sums of logarithms, and conversely*

[3] *Write logarithms of quotients as differences of logarithms, and conversely*

[4] *Express logarithms of powers as a product*

[5] *Express logarithms of products, quotients, and powers as individual logarithms, and conversely*

[6] *Find $\log_a a^k$.*

□ **Exercise 1** Find the following.

a. $\log_5 5$

b. $\log_9 1$

□ **Exercise 2** Rewrite the following, using the product rule for logarithms.

a. $\log_4 (3 \cdot 9)$

b. $\log_6 5 + \log_6 8$

c. $\log_2 2x$

d. $\log_a MN$

Now change to logarithmic form.

$$\log_a xy = \log_a a^{m+n} = m + n$$

Recall that $m = \log_a x$ and $n = \log_a y$. Then

$$\log_a xy = \log_a x + \log_a y$$

which proves the result.

EXAMPLE 2 Rewrite the following, using the product rule for logarithms.

a. $\log_3 (7 \cdot 2)$

$$\log_3(7 \cdot 2) = \log_3 7 + \log_3 2 \qquad \text{Use the product rule.}$$

b. $\log_2 8 + \log_2 9$

$$\log_2 8 + \log_2 9 = \log_2(8 \cdot 9) \qquad \text{Use the product rule (in reverse).}$$
$$= \log_2 72$$

c. $\log_5 5x,\ x > 0$

$$\log_5 5x = \log_5 5 + \log_5 x$$
$$= 1 + \log_5 x \qquad \text{Recall that } \log_5 5 = 1.$$

d. $\log_7 x^2$

$$\log_7 x^2 = \log_7(x \cdot x)$$
$$= \log_7 x + \log_7 x$$
$$= 2 \log_7 x \quad ■$$

□ **DO EXERCISE 2.**

3 Logarithms of Quotients

If we again consider the table of logarithms for base 2 at the beginning of this section, we will see the following.

$$\log_2 \frac{16}{2} = \log_2 8 = 3$$

Also, $\log_2 16 - \log_2 2 = 4 - 1 = 3$.

Hence $\log_2 \frac{16}{2} = \log_2 16 - \log_2 2$.

We can generalize the example above.

> **The Quotient Rule**
>
> For all positive numbers x, y, and a, $a \neq 1$,
>
> $$\log_a \frac{x}{y} = \log_a x - \log_a y$$

The proof of the rule for logarithms of quotients is similar to the proof of the rule for logarithms of products.

EXAMPLE 3 Rewrite the following using the quotient rule for logarithms.

a. $\log_{10} \frac{5}{9}$

$$\log_{10} \frac{5}{9} = \log_{10} 5 - \log_{10} 9 \qquad \text{Use the quotient rule.}$$

b. $\log_a 3 - \log_a y$

$$\log_a 3 - \log_a y = \log_a \frac{3}{y} \qquad \text{Use the quotient rule.}$$

Caution: $\dfrac{\log_a x}{\log_a y} \neq \log_a x - \log_a y.$ ■

□ **DO EXERCISE 3.**

4 Logarithms of Powers

If we again consider some logarithms for base 2, we see the following.

$$\log_2 2^3 = \log_2 8 = 3$$

$$\begin{aligned}\log_2 2^3 &= \log_2(2 \cdot 2 \cdot 2)\\ &= \log_2 2 + \log_2 2 + \log_2 2\\ &= 3(\log_2 2) = 3\end{aligned}$$

Hence $\log_2 2^3 = 3(\log_2 2)$.

We can generalize to give the following result.

> **The Power Rule**
>
> **If p is any real number and x and a are positive numbers, $a \neq 1$,**
>
> $$\log_a x^p = p(\log_a x)$$

The proof is similar to the proof of the rule for logarithms of products.

EXAMPLE 4 Use the power rule to write the following as a product.

a. $\log_5 7^2$

$$\log_5 7^2 = 2(\log_5 7) \qquad \text{Use the power rule.}$$

b. $\log_a \sqrt{3}$

$$\log_a \sqrt{3} = \log_a 3^{1/2} \qquad \text{Use exponential notation.}$$

$$= \frac{1}{2}\log_a 3 \qquad \text{Use the power rule.}$$ ■

□ **DO EXERCISE 4.**

5 Using the Properties Together

We use the properties of logarithms to write expressions in different forms in advanced mathematics.

EXAMPLE 5

a. Write as logarithms in terms of m and n.

$$\log_a \sqrt[3]{\frac{m^2}{n}} = \log_a \left(\frac{m^2}{n}\right)^{1/3} \qquad \text{Use exponential notation.}$$

$$= \frac{1}{3}\log_a \frac{m^2}{n} \qquad \text{Use the power rule.}$$

$$= \frac{1}{3}(\log_a m^2 - \log_a n) \qquad \text{Use the quotient rule.}$$

$$= \frac{1}{3}(2\log_a m - \log_a n) \qquad \text{Use the power rule.}$$

□ **Exercise 3** Rewrite the following using the quotient rule for logarithms.

a. $\log_5 \dfrac{29}{6}$

b. $\log_b M - \log_b N$

□ **Exercise 4** Use the power rule to write the following as a product.

a. $\log_6 5^2$

b. $\log_b \sqrt[3]{7}$

c. $\log_3 y^5$

d. $\log_2 3^{-4}$

□ **Exercise 5**

a. Write in terms of logarithms of x and y: $\log_a \sqrt[5]{x^3/y^2}$.

b. Write as a single logarithm:

$$3 \log_a x - \log_a x + \frac{1}{3} \log_a y.$$

□ **Exercise 6** Simplify.

a. $\log_5 5^7$

b. $\log_3 3^{-9}$

c. $\log_e e^6$

d. $\log_p p^q$

b. Write as a single logarithm.

$3 \log_b m + \log_b m - 5 \log_b n$

$= \log_b m^3 + \log_b m - \log_b n^5$	Use the power rule.
$= \log_b(m^3 \cdot m) - \log_b n^5$	Use the product rule; notice that the bases must be the same.
$= \log_b m^4 - \log_b n^5$	
$= \log_b \dfrac{m^4}{n^5}$	Use the quotient rule. ■

□ **DO EXERCISE 5.**

6 Logarithm of the Base to a Power

We can simplify $\log_4 4^3$ as follows.

$\log_4 4^3 = 3(\log_4 4)$	Use the power rule.
$= 3(1)$	Since $\log_4 4 = 1$.
$= 3$	

We generalize to the following rule, which is used in advanced mathematics.

> If p is any real number and a is a positive number, $a \neq 1$,
>
> $$\log_a a^p = p$$

The rule may be proved as follows.

Proof:	$\log_a a^p = p(\log_a a)$	Use the power rule.
	$= p(1)$	Since $\log_a a = 1$.
	$= p$	

EXAMPLE 6 Simplify.

a. $\log_2 2^8 = 8$

b. $\log_e e^{-3} = -3$

c. $\log_b b^c = c$ ■

□ **DO EXERCISE 6.**

Answers to Exercises

1. a. 1 **b.** 0

2. a. $\log_4 3 + \log_4 9$ **b.** $\log_6 40$ **c.** $1 + \log_2 x$ **d.** $\log_a M + \log_a N$

3. a. $\log_5 29 - \log_5 6$ **b.** $\log_b \dfrac{M}{N}$

4. a. $2 \log_6 5$ **b.** $\dfrac{1}{3} \log_b 7$ **c.** $5 \log_3 y$ **d.** $-4 \log_2 3$

5. a. $\dfrac{1}{5}(3 \log_a x - 2 \log_a y)$ **b.** $\log_a x^2\sqrt[3]{y}$

6. a. 7 **b.** −9 **c.** 6 **d.** q

Problem Set 11.3

NAME ______________________

DATE ______________________

CLASS ______________________

Find the following.

1. $\log_7 7$
2. $\log_2 2$
3. $\log_4 1$
4. $\log_8 1$
5. $\log_9 9$
6. $\log_5 1$

Write as a sum or difference of logarithms.

7. $\log_4(27 \cdot 5)$
8. $\log_5(32 \cdot 9)$
9. $\log_6 6x$
10. $\log_7 7x$
11. $\log_a PQ$
12. $\log_b 3R$
13. $\log_6 \dfrac{7}{15}$
14. $\log_5 \dfrac{24}{9}$
15. $\log_c \dfrac{32}{V}$
16. $\log_a \dfrac{W}{25}$

Write as a product.

17. $\log_8 7^2$
18. $\log_5 3^4$
19. $\log_a x^5$
20. $\log_a y^3$
21. $\log_a 4^{-3}$
22. $\log_a 7^{-4}$
23. $\log_3 \sqrt[3]{4}$
24. $\log_7 \sqrt[4]{5}$

Write as a single logarithm.

25. $\log_2 7 + \log_2 9$
26. $\log_8 12 + \log_8 11$
27. $\log_a 45 + \log_a 8$
28. $\log_b 17 + \log_b 6$
29. $\log_c M + \log_c y$
30. $\log_b a + \log_b c$
31. $\log_a 25 - \log_a 5$
32. $\log_b 64 - \log_b 16$
33. $2 \log_b 5 - 3 \log_b 4$
34. $5 \log_b 6 - 4 \log_b 7$
35. $5 \log_b M + \log_b N$
36. $3 \log_b x - 2 \log_b y$
37. $2 \log_a x - \dfrac{1}{5} \log_a y$
38. $\dfrac{1}{3} \log_b V + 4 \log_b W$

Write in terms of logarithms.

39. $\log_b x^3y^2$

40. $\log_a xy^4$

41. $\log_a \sqrt[4]{\frac{x}{y^3}}$

42. $\log_b \sqrt[3]{\frac{x^2}{y^2}}$

43. $\log_b \frac{pq^3}{r^2}$

44. $\log_b \frac{x^2y^7}{u^5v^4}$

45. $\log_a \sqrt[5]{\frac{x^9y^{11}}{b^2c^3}}$

46. $\log_a \sqrt{\frac{p^2q^5}{r^8s^{12}}}$

Simplify.

47. $\log_2 2^4$

48. $\log_7 7^3$

49. $\log_3 3^{-5}$

50. $\log_4 4^{-2}$

51. $\log_e e^8$

52. $\log_e e^{-5}$

53. $\log_b b^{-t}$

54. $\log_m m^c$

55. The logarithm of 1 to any defined base is ____________.

56. The logarithm of a for any defined base a is ____________.

57. To find the logarithm of a product, ____________ the logarithms of the factors.

58. The logarithm of a power of x is the power ____________ the logarithm of x.

True or false?

* **59.** $(\log_5 7)(\log_5 4) = \log_5(7 + 4)$

* **60.** $\frac{\log_{10} 8}{\log_{10} 2} = \log_{10} 4$

* **61.** $\log_2(\log_6 72 - \log_6 12) = 0$

* **62.** $\frac{\log_8 8}{\log_{64} 8} = 2$

Write as a single logarithm.

* **63.** $3 \log_a 2x^3 - 2 \log_a 2x^2y + 3 \log_a 3x$

* **64.** $\frac{1}{3}\log_a 27x^3y^2z + \frac{1}{2}\log_a 9x^2z^4 - \frac{2}{3}\log_a y^2z$

Checkup

The following problems provide a review of some of Sections 1.7 and 10.3.

Change to scientific notation.

65. 4250

66. 374

67. 0.0024

68. 0.018

If $f(x) = 2x - 3$ and $g(x) = 4x^2 - 5x + 7$, find the following.

69. $f + g$

70. $f - g$

71. fg

72. $\frac{f}{g}$

73. $\frac{g}{f}$

74. gg

11.4 FINDING LOGARITHMIC FUNCTION VALUES ON A CALCULATOR

1 Common Logarithms

Base 10 logarithms are called ***common logarithms.*** We usually do not write the base. Hence

$$\log_{10} x = \log x$$

Logarithms were used for many years to do complicated calculations. Now we do most of our calculations with calculators and computers. However, working with logarithms helps us to understand the properties of logarithmic functions. Recall that logarithms are exponents. If we make a table of logarithms for numbers base 10 using whole-number exponents, we see that many numbers are missing.

Base 10				
Number	$10^0 = 1$	$10^1 = 10$	$10^2 = 100$	$10^3 = 1000$
Logarithm	0	1	2	3

Methods have been developed for finding the logarithms of numbers that are not whole-number powers of 10. These logarithms are shown in the *table of common logarithms* in Appendix B. The logarithms in the table have been rounded, so they are approximate.

Today we use a calculator with a log key to find common logarithms. This may be a second function key on some calculators. We find the common logarithm of 3.86 as follows.

ENTER	DISPLAY
[3.86] [log]	0.5866 (rounded to four decimal places)

$$\log 3.86 \approx 0.5866$$

EXAMPLE 1 Use a calculator to find the following.

a. $\log 9.28$
$\log 9.28 \approx 0.9675$

b. $\log 3740$
$\log 3740 \approx 3.5729$

c. $\log 0.0463$
$\log 0.0463 \approx -1.3344$ (rounded to four decimal places)

The common logarithm of a number between 0 and 1 is always negative because the logarithm is the exponent of 10 that gives the number.

$$10^{-1.3344} \approx 0.463$$

d. $\log 0.000245$
$\log 0.000245 \approx -3.6108$ ■

□ **DO EXERCISE 1.**

OBJECTIVES

1 *Find logarithms and antilogarithms, base 10, using a calculator*

2 *Use base 10 logarithms in an application*

3 *Find logarithms and antilogarithms, base e, using a calculator*

4 *Use base e logarithms in an application*

5 *Use the change of base rule*

□ **Exercise 1** Use a calculator to find the following. Round the answer to four decimal places.

a. $\log 7.34$

b. $\log 35{,}400$

c. $\log 768{,}000$

d. $\log 0.0325$

e. $\log 0.00984$

□ **Exercise 2** Use a calculator to find the common antilogarithm.

a. 4.7042

b. −2.2573

c. 0.014

d. −0.483

□ **Exercise 3** Find the pH of lye, which has a hydronium ion concentration of 3.2×10^{-14}.

Suppose we know that $\log x = a$ and we want to find x. We are looking for a number x that has a logarithm of a. This is called finding the ***antilogarithm*** of a. "Antilogarithm" is abbreviated "antilog."

We may use a calculator to find common logarithms by using the *log* key. To find *antilogs,* we usually use the 10^x key. We can do this because

$$\text{if } \log_{10} y = x$$

$$\text{then } y = 10^x$$

If your calculator does not have a 10^x key, follow the directions in the manual for finding antilogarithms.

We use a calculator to find the common antilogarithm of 2.9047 as follows. The 10^x key may be a second function on some calculators.

$$\text{If} \quad \log x = 2.9047$$

$$\text{then} \quad x = 10^{2.9047}$$

ENTER	DISPLAY
[2.9047] [10^x]	802.9713 (rounded to four decimal places)

The common antilogarithm of 2.9047 is approximately 802.9713.

EXAMPLE 2 Find the common antilogarithm.

a. 3.6031
$10^{3.6031} \approx 4009.59$ (rounded to two decimal places).

b. −1.2197
$10^{-1.2197} \approx 0.0603$ (rounded to four decimal places).

c. −4.2
$10^{-4.2} \approx 0.000063$ (rounded to six decimal places). ■

□ **DO EXERCISE 2.**

2 An Application of Common Logarithms

In chemistry, the acidity or alkalinity of a solution may be described by pH, where pH is defined as follows.

$$pH = -\log[H_3O^+]$$

where $[H_3O^+]$ is the hydronium ion concentration in moles per liter.

EXAMPLE 3 Find the pH of milk that has a hydronium ion concentration of 4×10^{-7}.

$$\begin{aligned} pH &= -\log[H_3O^+] \\ &= -\log(4 \times 10^{-7}) \\ &= -(\log 4 + \log 10^{-7}) && \text{Use the product rule.} \\ &\approx -(0.6021 - 7) && \log 10^{-7} = -7. \\ &\approx 6.3979 \approx 6.4 \end{aligned}$$

The pH of the milk is 6.4. ■

□ **DO EXERCISE 3.**

3 Natural Logarithms

Logarithms used in the calculus and many other applications are base e. The number e is an irrational number.

$$e \approx 2.7182818$$

Logarithms to the base e are called ***natural logarithms.*** The base e logarithm is usually abbreviated as ln.

$$\log_e x = \ln x$$

We use the ln key on our calculators to find natural logarithms. The natural logarithm of 45 may be found as follows.

ENTER	DISPLAY
45 ln	3.8067 (rounded to four decimal places)

$$\ln 45 \approx 3.8067$$

EXAMPLE 4 Use a calculator to find the following. Round the answer to four decimal places.

a. ln 3
$\ln 3 \approx 1.0986$

b. ln 0.048
$\ln 0.048 \approx -3.0366$

c. ln 0.00032
$\ln 0.00032 \approx -8.0472$ ■

□ **DO EXERCISE 4.**

To find antilogarithms base e, we use the e^x key, since if $\ln y = x$, then $y = e^x$.

We find the natural antilogarithm of 3.2845 as follows.

ENTER	DISPLAY
3.2845 e^x	26.6956

The natural antilogarithm of 3.2845 is approximately 26.6956.

EXAMPLE 5 Use a calculator to find the natural antilogarithm.

a. 1.234
$e^{1.234} \approx 3.4349$ (rounded to four decimal places).

b. -4.3287
$e^{-4.3287} \approx 0.013$ (rounded to three decimal places).

c. -0.84
$e^{-0.84} \approx 0.4317$ (rounded to four decimal places). ■

□ **DO EXERCISE 5.**

□ **Exercise 4** Use a calculator to find the following. Round the answer to four decimal places.

a. ln 205.3

b. ln 70,320

c. ln 0.537

d. ln 0.04

□ **Exercise 5** Find the natural antilogarithm.

a. 2.8413

b. 0.03

c. -5.2

d. -3.9875

□ **Exercise 6** The number of years that it will take Pleasant View to reach a population of 12,200 is estimated to be

$$t = \frac{\ln 12{,}200 - \ln 10{,}000}{0.04}$$

In how many years will the population of Pleasant View reach 12,200?

4 An Application of Natural Logarithms

EXAMPLE 6 Suppose that the estimated number of years for a country to reach a population of 264 million is given by

$$t = \frac{\ln 264 - \ln 234}{0.008}$$

How many years does it take the population to reach 264 million?

$$t \approx \frac{5.5759 - 5.4553}{0.008}$$

$$\approx 15.1$$

It will take approximately 15.1 years for the population to reach 264 million. ■

□ **DO EXERCISE 6.**

5 The Change-of-Base Rule

A calculator can be used to approximate the values of common logarithms and natural logarithms. However, sometimes we want to find a logarithm to some other base. We can use the following rule.

Change-of-Base Rule

For any logarithmic bases a and b and any positive number x,

$$\log_a x = \frac{\log_b x}{\log_b a}$$

We may prove the rule as follows.

Let $\log_a x = m$

$a^m = x$	Write as an exponential equation.
$\log_b (a^m) = \log_b x$	Take the logarithm base b on both sides.
$m \log_b a = \log_b x$	Use the power rule.
$m = \dfrac{\log_b x}{\log_b a}$	Divide both sides by $\log_b a$.
$\log_a x = \dfrac{\log_b x}{\log_b a}$	Substitute $\log_a x$ for m.

EXAMPLE 7 Find each logarithm using a calculator.

a. $\log_4 15$

Use natural logarithms and the change-of-base rule.

$$\log_4 15 = \frac{\ln 15}{\ln 4}$$

$$\approx \frac{2.7081}{1.3863} \approx 1.9534$$

b. $\log_2 127$

Use common logarithms and the change-of-base rule.

$$\log_2 127 = \frac{\log 127}{\log 2}$$

$$\approx \frac{2.1038}{0.3010} \approx 6.9887$$ ■

□ **DO EXERCISE 7.**

□ **Exercise 7** Use common logarithms or natural logarithms to find each of the following to four decimal places.

a. $\log_3 8$

b. $\log_8 53.4$

GRAPHING CALCULATOR

Equations such as $y = \log_2 x$, which is equivalent to $x = 2^y$, may be graphed with the graphing calculator using the change-of-base rule. Graph $y = \log_2 x$.

$$\log_2 x = \frac{\ln x}{\ln 2}$$

1. Enter $y = \ln x/\ln 2$ into your calculator.
2. An appropriate window is W: $[-10, 10]$ $[-10, 10]$.
3. Plot the graph. This is the graph that is shown in Example 4, Section 11.2.

DID YOU KNOW?

The mathematician Pierre de Laplace said that logarithms saved astronomers so much time on calculations that it doubled the length of their lives. Logarithms enabled them to change multiplication and division operations to addition and subtraction.

Answers to Exercises

1. **a.** 0.8657 **b.** 4.5490 **c.** 5.8854 **d.** −1.4881 **e.** −2.007

2. **a.** 50,605.77 **b.** 0.00553 **c.** 1.033 **d.** 0.3289

3. 13.5

4. **a.** 5.3245 **b.** 11.1608 **c.** −0.6218 **d.** −3.2189

5. **a.** 17.138 **b.** 1.0305 **c.** 0.0055 **d.** 0.0185

6. 5 years

7. **a.** 1.8928 **b.** 1.9129

Problem Set 11.4

NAME ______________________

DATE ______________________

CLASS ______________________

Use a calculator to find the following. Round the answer to four decimal places.

1. log 7.18 **2.** log 5.29 **3.** log 2.39 **4.** log 9.16

5. log 250 **6.** log 360 **7.** log 71.3 **8.** log 29,300

9. log 713,000 **10.** log 2,850 **11.** log 0.0348 **12.** log 0.00752

13. log 0.345 **14.** log 0.0007 **15.** log 0.293 **16.** log 0.0458

17. log 0.03 **18.** log 0.08

Use a calculator to find the common antilogarithm.

19. 2.5366 **20.** 5.8768 **21.** 1.6020 **22.** 4.8542

23. −3.9914 **24.** −1.2233 **25.** −0.0985 **26.** −2.3401

27. −1.6853 **28.** −4.0885

Use a calculator to find the following. Round the answer to four decimal places.

29. ln 2 **30.** ln 4 **31.** ln 10 **32.** ln 18

33. ln 35 **34.** ln 60 **35.** ln 0.09 **36.** ln 0.37

37. ln 370.2 **38.** ln 50,420 **39.** ln 0.0046 **40.** ln 0.0005

Use a calculator to find the natural antilogarithm.

41. 5.6132 **42.** 2.5879 **43.** −3.1084 **44.** −1.0375

45. 0.06 **46.** 0.007 **47.** 10.3 **48.** 12.2

49. −4.08 **50.** −3.954 **51.** 1.3478 **52.** 3.5792

Use common logarithms or natural logarithms to find each of the following to four decimal places.

53. $\log_3 20$

54. $\log_5 37$

55. $\log_6 18$

56. $\log_2 54$

57. $\log_{50} 24$

58. $\log_{30} 12$

59. $\log_{0.6} 6$

60. $\log_{0.2} 4$

61. $\log_2 7.38$

62. $\log_8 4.57$

63. $\log_{12} 43.2$

64. $\log_{15} 53.8$

65. Find the pH of a solution for which $[H_3O^+] = 3.7 \times 10^{-2}$. (See Example 3.)

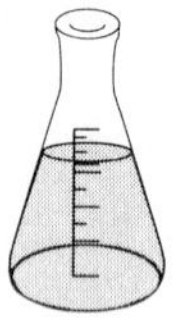

66. Suppose that the number of days that it takes to produce 638 flies is given by

$$t = \frac{\ln 638 - \ln 350}{0.02}$$

Find the time it takes to produce 638 flies.

67. Base 10 logarithms are called ___________ logarithms.

68. A base 10 logarithm, $\log_{10} x$, is usually written without the ___________.

69. Finding a number that has a given logarithm is called finding the ___________.

Simplify.

* **70.** $\dfrac{\log_2 9}{\log_2 7}$

* **71.** $\dfrac{\log_4 33}{\log_4 21}$

* **72.** $\dfrac{\log_3 17}{\log_3 25}$

* **73.** Find the concentration of $[H_3O^+]$ for a solution with a pH of 2.9.

Checkup

The following problems provide a review of some of Section 10.4.

Identify as a linear function, constant function, quadratic function, absolute value function, or power function.

74. $f(x) = x^2 - 2$

75. $f(x) = x^5$

76. $f(x) = 3x + 7$

77. $f(x) = -4$

78. $f(x) = |x - 7|$

79. $f(x) = x$

80. $f(x) = x^8$

81. $f(x) = 3x^2 - 4x + 2$

82. $f(x) = 7x$

83. $f(x) = |x| - 15$

11.5 EXPONENTIAL AND LOGARITHMIC EQUATIONS

OBJECTIVES

1 *Solve exponential equations*

2 *Solve logarithmic equations*

Exponential and logarithmic equations are used to solve problems in business and science. ***Exponential equations*** are equations with a *variable in at least one of the exponents.* ***Logarithmic equations*** are equations involving the logarithm of a variable. To solve these equations, we will use the following properties.

> For any real numbers x, y, and a, $a > 0$, $a \neq 1$,
>
> **1.** If $a^x = a^y$, then $x = y$.
>
> **2.** If $x = y$, and $x > 0$, $y > 0$, then $\log_a x = \log_a y$.
>
> **3.** If $\log_a x = \log_a y$, and $x > 0$, $y > 0$, then $x = y$.

1 Exponential Equations

If we can make the bases the same on both sides of an equation, one of the easiest methods of solving exponential equations is to use property 1.

EXAMPLE 1 Solve $4^{3x-1} = 16$.

$$4^{3x-1} = 16$$

Notice that $16 = 4^2$. Therefore,

$$4^{3x-1} = 4^2$$

$$3x - 1 = 2 \qquad \text{Use property 1.}$$

$$3x = 3$$

$$x = 1$$

This answer checks. The solution is 1. ■

□ **DO EXERCISE 1.**

□ **Exercise 1** Solve.

a. $2^{2x-5} = 8$

b. $4^x = 32$

It is often not possible to make the bases the same on both sides of an exponential equation. Then we use property 2.

EXAMPLE 2 Solve $4^x = 15$.

$$4^x = 15$$

$$\log 4^x = \log 15 \qquad \text{Use property 2.}$$

$$x \log 4 = \log 15 \qquad \text{Use the power rule.}$$

$$x = \frac{\log 15}{\log 4}$$

This is the exact solution. We can get an approximate solution by finding the logarithms with a calculator. Using a calculator gives us

$$x \approx \frac{1.1761}{0.6021}$$

$$x \approx 1.9533$$

Caution: $\dfrac{\log 15}{\log 4} \neq \log 15 - \log 4$. ■

□ **DO EXERCISE 2.**

□ **Exercise 2** Solve.

a. $6^x = 12$

b. $7^{3x} = 5$

If the base is e, we take the logarithm base e on both sides of the equation.

☐ **Exercise 3** Solve.

a. $e^{0.5t} = 30$

b. $e^{3t} = 7$

☐ **Exercise 4** Solve.

a. $\log_2 (3x - 1) = 3$

b. $\log_7 (2x - 7) = 2$

☐ **Exercise 5** Solve.

a. $\log_3 (x + 1) + \log_3 (x + 3) = 1$

b. $\log_4 x + \log_4 (6x + 10) = 1$

EXAMPLE 3 Solve $e^{0.04t} = 750$.

$$e^{0.04t} = 750$$

$$\ln e^{0.04t} = \ln 750 \qquad \text{Use property 2.}$$

$$0.04t = \ln 750 \qquad \text{Since } \log_a a^p = p.$$

$$t = \frac{\ln 750}{0.04} \approx 165.5 \quad ■$$

☐ **DO EXERCISE 3.**

2 Logarithmic Equations

Some logarithmic equations may be solved by using the definition of logarithms to change the equations to exponential form. The solution must give a logarithm of a number greater than zero in the original problem since logarithms are defined only for positive numbers.

EXAMPLE 4 Solve $\log_4 (3x - 2) = 2$.

$$\log_4 (3x - 2) = 2$$

$$3x - 2 = 4^2 \qquad \text{Change to exponential form.}$$

$$3x - 2 = 16$$

$$3x = 18$$

$$x = 6$$

In this case, $3x - 2$ must be positive. The expression is positive for $x - 6$ and this number checks. The solution is 6. ■

☐ **DO EXERCISE 4.**

EXAMPLE 5 Solve $\log_3 x + \log_3 (x - 6) = 3$.

$$\log_3 x + \log_3 (x - 6) = 3$$

$$\log_3 x(x - 6) = 3 \qquad \text{Use the product rule.}$$

$$x(x - 6) = 3^3 \qquad \text{Change to exponential form.}$$

$$x^2 - 6x = 27$$

$$x^2 - 6x - 27 = 0 \qquad \text{Factor.}$$

$$(x - 9)(x + 3) = 0$$

$$x = 9 \quad \text{or} \quad x = -3 \qquad \text{Use the zero product property.}$$

Both x and $x - 6$ must be positive. This is not true for $x = -3$. The answer 9 checks. The solution is 9. ■

☐ **DO EXERCISE 5.**

Answers to Exercises

1. a. 4 **b.** $\frac{5}{2}$

2. a. 1.3868 **b.** 0.2757

3. a. 6.8024 **b.** 0.6486

4. a. 3 **b.** 28

5. a. 0 **b.** $\frac{1}{3}$

Problem Set 11.5

NAME
DATE
CLASS

Solve.

1. $4^x = 64$

2. $5^x = 25$

3. $3^{x+4} = 9$

4. $2^{x-1} = 16$

5. $3^{2x+1} = 27$

6. $4^{2x-1} = 4$

7. $9^x = 81$

8. $25^x = 125$

9. $7^x = 18$

10. $5^x = 21$

11. $3^x = 14$

12. $6^x = 15$

13. $8^{2x} = 7$

14. $4^{3x} = 10$

15. $5^{2x+1} = 14$

16. $7^{3x-1} = 12$

17. $6^x = 36$

18. $2^x = 16$

19. $2^{x^2} \cdot 2^{-x} = 4$

20. $3^{x^2} \cdot 3^{-4x} = \frac{1}{27}$

21. $e^t = 25$

22. $e^t = 200$

23. $e^{-x} = 0.04$

24. $e^{-x} = 0.3$

25. $e^{-0.5t} = 8$

26. $e^{-0.07t} = 0.09$

27. $2^{3x} \cdot 5^{4x} = 17$

28. $4^{x+2} = 3^{x-5}$

29. $\log_4(x + 9) = 2$

30. $\log_5(x - 3) = 1$

31. $\log_2(2x + 4) = 3$

32. $\log_3(3x - 6) = 3$

33. $\log_2 x + \log_2(x - 7) = 3$

34. $\log x + \log(x + 9) = 1$

35. $\log x + \log(3x - 7) = 1$

36. $\log x + \log(x - 9) = 1$

37. $\log_2(x^2 - 9) - \log_2(x + 3) = 2$

38. $\log_4(x^2 - 2x) - \log_4 x = 2$

39. $\ln x = 4$

40. $\ln x = -2$

41. $\log_2(5 + 2x) - \log_2(4 - x) = 3$

42. $\log_3(3x - 2) - \log_2(6 - x) = 2$

43. Exponential equations have a __________ in at least one of the exponents.

44. In an exponential equation of the form $a^x = a^y$, $a > 0$, since the bases are the same, the exponents are __________.

45. The solution of a logarithmic equation of the form $\log_a x = c$ must be a number __________ than zero.

Solve.

* 46. $2^{3x^2-2x} = 32$

* 47. $3^{w^2+4w} = \frac{1}{27}$

* 48. $\left(\frac{1}{3}\right)^{2x+7} = 81$

* 49. $4^{3x-2} = \frac{1}{64}$

* 50. $\log_2|2x - 3| = 4$

* 51. $\log_2(\log_6 6) = x$

* 52. $\log_2(3x) - \log_2(7x - 1) = -1$

* 53. $\log x + \log(2x - 7) = \log 4$

* 54. $\log_5 t - \log_5(3 + t) = \log_5 2 - \log_5(t + 1)$

* 55. $\log_3 x - \log_3(4 + 4x) = \log_3 1 - \log_3(x + 1)$

Checkup

The following problems provide a review of some of Section 10.5.

Find the inverse of the following functions.

56. $f(x) = 2x - 3$

57. $f(x) = -3x + 7$

58. $f(x) = x^3 - 4$

59. $f(x) = x^3 + 5$

60. $f(x) = x^2 + 5, x \geq 0$

61. $f(x) = -x^2 + 4, x \geq 0$

Determine if the function is one-to-one.

62. $f = \{(5, 9), (-3, 8), (0, 4), (-6, 2), (-2, 2)\}$

63. $g = \{(2, -7), (0, 0), (3, -8), (-4, -4), (6, 12)\}$

64. $f(x) = -x^2 + 2x - 3$

65. $f(x) = |x - 7|$

11.6 APPLIED PROBLEMS

OBJECTIVE

1 *Solve applied problems using exponential and logarithmic equations*

1 Exponential and logarithmic equations are used to solve many problems in the real world. We consider examples from the sciences and business.

If N_0 is the number of bacteria present initially and b is the rate of increase per day, the number N of bacteria in certain cultures after t days is given by

$$N = N_0 \cdot b^t$$

EXAMPLE 1 Suppose that 5000 bacteria are present initially in a culture and the bacteria double in population every day. What is the number of bacteria present after 3.5 days?

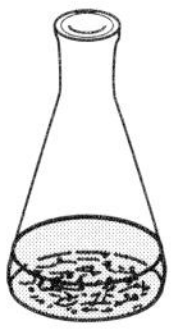

We use the formula

$$N = N_0 \cdot b^t$$

$N_0 = 5000$, $b = 2$, and $t = 3.5$. Therefore,

$N = 5000(2)^{3.5}$

$N \approx 5000(11.3137)$ Find $2^{3.5}$ with the $\boxed{y^x}$ key on a calculator.

$N \approx 56{,}569$

The number of bacteria present after 3 days is approximately 56,569.

Recall from Section 1.5 that $2^{3.5}$ can be found on a calculator as follows.

ENTER	DISPLAY
$\boxed{2}\ \boxed{y^x}\ \boxed{3.5}\ \boxed{=}$	11.313708

It is not necessary to round 11.313708 to four decimal places. There may be a small variation in answers if we round. Answers to exercises will be found without rounding. ■

□ **DO EXERCISE 1.**

□ **Exercise 1** A colony of bacteria triple in population every day and the number of bacteria present initially is 8300. What is the number of bacteria present after 4.5 days?

If an amount of money P is invested at an interest rate r compounded annually, the total amount of money A in the account after t years is given by the formula

$$A = P(1 + r)^t$$

□ **Exercise 2** Suppose that \$4000 is invested at 7% interest compounded annually. How much money will be in the account after 30 years?

□ **Exercise 3** How long does it take \$4000 to double if it is invested in an account that pays 9% interest per year compounded annually?

EXAMPLE 2 If \$2000 is invested at 9% interest, compounded annually, how much money will be in the account after 25 years?

$$A = P(1 + r)^t$$

$$P = 2000, \quad r = 0.09, \quad \text{and} \quad t = 25$$

$$A = 2000(1 + 0.09)^{25}$$

$$A = 2000(1.09)^{25}$$

$$A \approx 2000(8.6231) \qquad \text{Find } (1.09)^{25} \text{ with a calculator.}$$

$$A \approx 17{,}246$$

There will be approximately \$17,246 in the account after 25 years! ■

□ **DO EXERCISE 2.**

EXAMPLE 3 If \$6000 invested at 8% interest compounded annually yields \$10,600, how many years was the money invested?

$$A = P(1 + r)^t$$

$$A = 10{,}600, \quad P = 6000, \quad \text{and} \quad r = 0.08$$

$$10{,}600 = 6000(1 + 0.08)^t$$

$$10{,}600 = 6000(1.08)^t$$

$$\log 10{,}600 = \log[6000(1.08)^t] \qquad \text{Use Property 2.}$$

$$\log 10{,}600 = \log 6000 + t \log 1.08 \qquad \text{Use the product rule and the power rule.}$$

$$\frac{\log 10{,}600 - \log 6000}{\log 1.08} = t$$

$$\frac{4.0253 - 3.7782}{0.0334} \approx t \qquad \text{Find the logarithms using a calculator.}$$

$$7.4 \approx t$$

The money was invested approximately 7.4 years. ■

□ **DO EXERCISE 3.**

Exponential equations may be used to model the decay of a radioactive substance.

EXAMPLE 4 All living plants and animals contain a radioactive form of carbon called carbon 14. After a plant or animal dies, the decay of carbon 14 is given by the equation

$$y = y_0 e^{-0.00012t}$$

where y_0 is the original amount of carbon 14 present, t is the time in years, and y is the amount of carbon 14 present after t years. How old is an animal bone that originally contained 100 milligrams of carbon 14 and now contains 70 milligrams?

$$y = y_0e^{-0.00012t}$$

$$\mathbf{70} = \mathbf{100}e^{-0.00012t} \qquad \text{Substitute for } y \text{ and } y_0.$$

$$0.7 = e^{-0.00012t} \qquad \text{Divide by 100.}$$

$$\ln 0.7 = \ln e^{-0.00012t} \qquad \text{Use property 2.}$$

$$\ln 0.7 = -0.00012t(\ln e) \qquad \text{Use the power rule.}$$

$$\ln 0.7 = -0.00012t \qquad \text{Since } \ln e = 1.$$

$$\frac{\ln 0.7}{-0.00012} = t$$

$$t \approx \frac{-0.3567}{-0.00012} \approx 2972$$

The bone is approximately 2972 years old. ■

□ DO EXERCISE 4.

□ **Exercise 4** What is the age of a bone fragment that originally contained 100 milligrams of carbon 14 and now contains 65 milligrams?

Interest may also be compounded continuously according to the formula

$$A = Pe^{rt}$$

where P is the amount invested at an interest rate r and A is the amount of money in the account after t years.

EXAMPLE 5 Jane deposits \$620 in a savings account that pays 6 percent interest compounded continuously. How much money will she have in the account after 8 years?

$$A = Pe^{rt}$$

$$A = 620e^{0.06(8)} \qquad \text{Substitute for } P, r, \text{ and } t.$$

$$= 620e^{0.48}$$

$$\approx 620(1.6161) \qquad \text{Find } e^{0.48} \text{ with a calculator.}$$

$$\approx 1001.97$$

Jane has approximately \$1001.97 in the account after 8 years. ■

□ DO EXERCISE 5.

□ **Exercise 5** How much money will Panos have in an account after 4 years if he deposits \$3000 and the account pays 8 percent interest compounded continuously?

We are often interested in the loudness of sounds of things, such as construction equipment or jet engines. The loudness L of a sound in decibels is given by the formula

$$L = 10 \log \frac{I}{I_0}$$

where I_0 is lowest sound that can be heard by a human being and I is the intensity of a certain sound.

☐ **Exercise 6** Find the loudness of a sound if the intensity of this sound is 49,900 times the lowest sound a human being can hear.

EXAMPLE 6 Find the loudness of a sound if I is 30,000 times I_0.

$$L = 10 \log \frac{I}{I_0}$$

$$L = 10 \log \frac{30{,}000 I_0}{I_0}$$

$$L = 10 \log 30{,}000$$

$$L \approx 10(4.4771)$$

$$L \approx 44.8$$

The loudness of the sound is approximately 44.8 decibels. ■

☐ **DO EXERCISE 6.**

Answers to Exercises

1. 1,164.458

2. $30,449.02

3. 8 years

4. 3590 years

5. $4131.38

6. 47 decibels

Problem Set 11.6

NAME

DATE

CLASS

Solve.

For Problems 1–4, use the formula $N = N_0 \cdot b^t$, where N_0 is the number of bacteria present initially, b is the rate of increase per day, and N is the number of bacteria present after c days.

1. The number of bacteria initially present in a sample is 2000. If the bacteria double in population every day, what is the number of bacteria present after 6.2 days?

2. A colony of bacteria triple in population every day and the number of bacteria present initially is 1500. What is the number of bacteria present after 2.3 days?

3. The number of bacteria initially present in a sample is 5450. In how many days will the bacteria count be 115,000 if the population doubles every day?

4. In how many days will the bacteria count be 35,500 if the population triples every day and the number of bacteria initially present is 2050?

For Problems 5–14, use the formula $A = P(1 + r)^t$, where P is the amount invested at an interest rate r compounded annually, and A is the amount in the account after t years.

5. If \$3500 is invested at 8% interest compounded annually, how much money will be in the account after 15 years?

6. Suppose that \$9000 is invested at 6% interest compounded annually. How much money will be in the account after 20 years?

7. How much money is in an account after 25 years if \$1500 was originally deposited at an interest rate of 7% compounded annually?

8. If \$6000 is invested at 10% interest compounded annually, how much money will be in the account after 9 years?

9. Suppose \$3800 is invested at 9% interest compounded annually. How many years after the money was invested will the account yield \$22,000?

10. If \$4500 invested at 7% interest compounded annually yields \$15,600, how many years was the money invested?

11. How long does it take \$2500 to double if it is invested in an account that pays 6% interest per year compounded annually?

12. If \$6000 is invested in an account that earns 8% interest compounded annually, how long does it take the money in the account to triple?

13. If \$3500 invested at 8.5% interest compounded annually yields \$5851.61, how many years was the money invested?

14. Janet invested \$2000 at 9.5% interest compounded annually. In how many years will the account yield \$12,283.22?

Radioactive strontium decays according to the formula $y = y_0e^{-0.0239t}$, where y_0 is the original amount of radioactive strontium, t is the time in years, and y is the amount of radioactive strontium present after t years. Use this formula for Problems 15–18.

15. How many years will it take 6 grams of radioactive strontium to decay to 3 grams?

16. In how many years will a 5-gram sample of radioactive strontium decay to 1 gram?

17. If the original amount of radioactive strontium was 8 grams, how many grams will be present after 30 years?

18. After 60 years, a sample that originally contained 7 grams of radioactive strontium contains how many grams?

Use the formula $A = Pe^{rt}$, where P is the amount invested at an interest rate r, compounded continuously, and A is the amount of money in the account after t years for Problems 19–22.

19. Sean deposited \$700 in a certificate of deposit that paid 6% interest compounded continuously. How much money will he have in the account after 9 years?

20. Stacie deposited \$4000 in a money market fund that paid 4% interest compounded continuously. How much money will be in her account after 12 years?

21. If \$5000 is invested at 9% interest compounded continuously, how many years will it take for the account to yield \$47,438.68?

22. How long does it take \$10,000 to double if it is invested at 7% interest compounded continuously?

For Problems 23 and 24, use the formula $L = 10 \log I/I_0$, where L is the loudness of a given sound, I_0 is the lowest sound that can be heard by a human being, and I is the intensity of a given sound.

23. Find the loudness of a given sound if I is 25,000 times I_0.

24. Find the loudness of a sound if the intensity of this sound is 37,600 times the lowest sound a human being can hear.

For Problems 25 and 26, use the formula $A = P(1 + r)^t$, where P is the amount invested at an interest rate r compounded annually and A is the total amount in the account after t years.

* **25.** What is the interest rate if \$4000 compounded annually for 20 years grows to \$15,478.74?

* **26.** What interest rate is Bryan's money earning if \$2500 compounded annually for 12 years yields \$4002.58?

Checkup

The following problems provide a review of some of Section 8.6.

Solve.

27. $x^4 - 5x^2 + 6 = 0$

28. $y^4 + 6y^2 + 8 = 0$

29. $x + 2\sqrt{x} - 3 = 0$

30. $x - 7\sqrt{x} + 10 = 0$

31. $(x^2 - 2)^2 + 7(x^2 - 2) + 10 = 0$

32. $(x^2 - 1)^2 - 7(x^2 - 1) + 12 = 0$

33. $x^{-2} - 2x^{-1} - 15 = 0$

34. $x^{-2} - 5x^{-1} - 14 = 0$

35. $\left(\dfrac{x+3}{x-3}\right)^2 - 5\left(\dfrac{x+3}{x-3}\right) + 6 = 0$

36. $\left(\dfrac{x+2}{x-4}\right)^2 + 7\left(\dfrac{x+2}{x-4}\right) + 12 = 0$

Chapter 11 Summary

Section 11.1

An exponential function with base a is a function of the form

$$f(x) = a^x \qquad \text{where } a > 0 \text{ and } a \neq 1$$

When $a > 1$, $f(x) = a^x$ increases as x increases. When $0 < a < 1$, $f(x) = a^x$ decreases as x increases.

The graph of $f(x) = a^x$ passes through the point (0, 1).

Section 11.2

$y = \log_a x$ is equivalent to $x = a^y$, $a > 0$, $a \neq 1$, $x > 0$.

The logarithmic function is a function of the form

$$y = \log_a x, \qquad a > 0, \quad a \neq 1, \quad x > 0$$

Section 11.3

For any positive real number a, $a \neq 1$,

$$\log_a a = 1 \qquad \text{and} \qquad \log_a 1 = 0$$

Product rule: For all positive numbers x, y, and a, $a \neq 1$,

$$\log_a xy = \log_a x + \log_a y$$

Quotient rule: For all positive numbers x, y, and a, $a \neq 1$,

$$\log_a \frac{x}{y} = \log_a x - \log_a y$$

Power rule: If p is any real number and x and a are positive numbers, $a \neq 1$,

$$\log_a x^p = p(\log_a x)$$

Logarithm of the base to a power: If p is any real number and a is a positive number, $a \neq 1$,

$$\log_a a^p = p$$

Section 11.4

Base 10 logarithms are called ***common logarithms.***

$$\log_{10} x = \log x$$

Finding a number x that has a logarithm of a is called finding the ***antilogarithm*** of a. Antilogarithm is abbreviated antilog.

The number e is an irrational number.

$$e \approx 2.7182818$$

Logarithms to the base e are called ***natural logarithms.*** The base e logarithm is usually abbreviated ln.

$$\log_e x = \ln x$$

Change-of-base rule: For any logarithmic bases a and b and any positive number x,

$$\log_a x = \frac{\log_b x}{\log_b a}$$

Section 11.5 ***Exponential equations*** are equations with a variable in at least one of the exponents.

Logarithmic equations are equations involving the logarithm of a variable.

For any real numbers x, y, and a, $a > 0$, $a \neq 1$,

1. If $a^x = a^y$, then $x = y$.
2. If $x = y$, and $x > 0$, $y > 0$, then $\log_a x = \log_a y$.
3. If $\log_a x = \log_a y$, and $x > 0$, $y > 0$, then $x = y$.

Chapter 11 Additional Exercises (Optional)

NAME ______________________

DATE ______________________

CLASS ______________________

Section 11.1

Graph.

1. $y = 2^x$

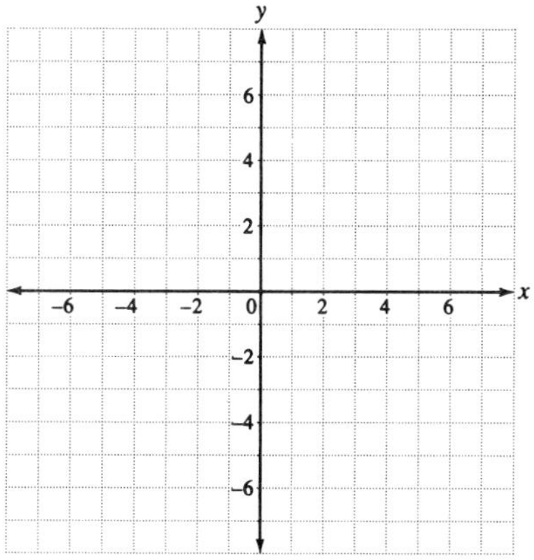

2. $y = \left(\frac{1}{4}\right)^x$

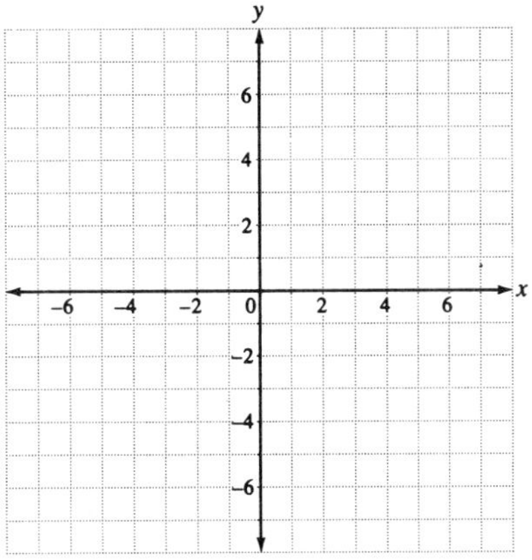

3. $y = 5^{x-2}$

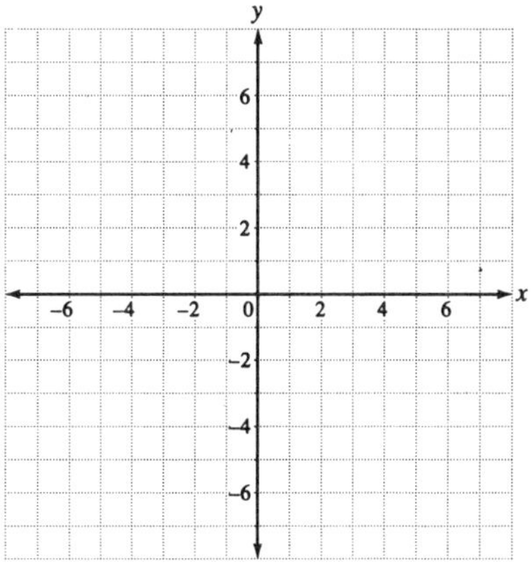

4. $y = \left(\frac{1}{3}\right)^{-x}$

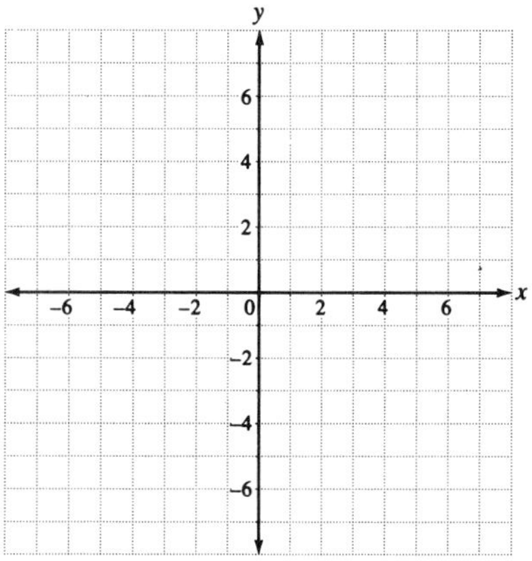

5. $y = -2^x$

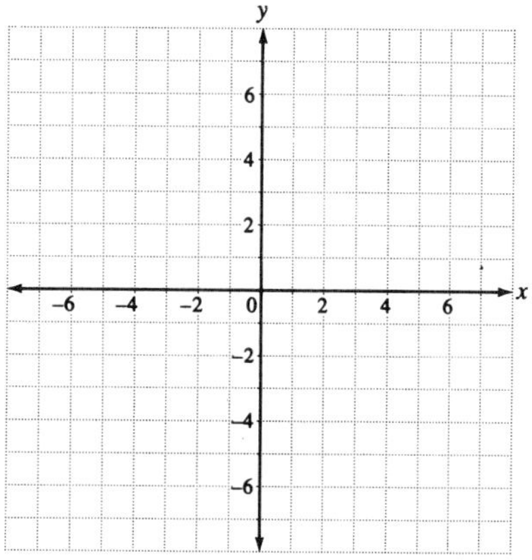

6. $y = 2^x + 1$

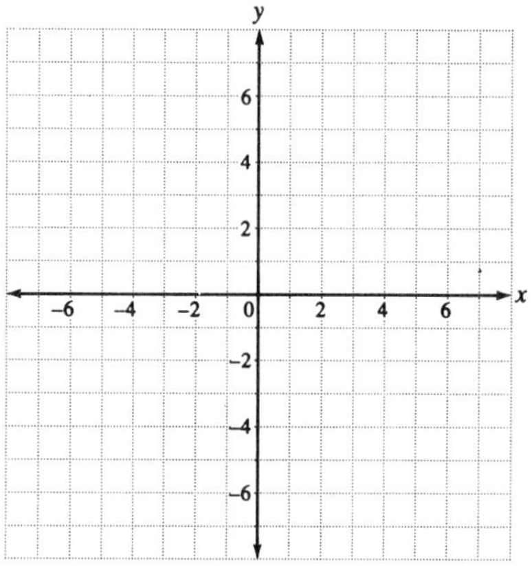

7. $y = 3^x - 2$

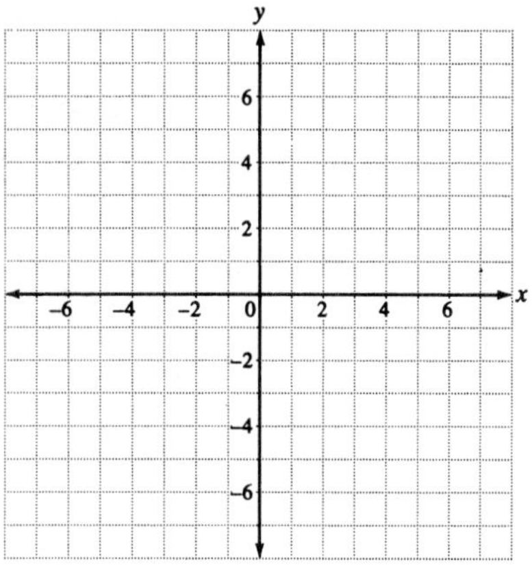

8. $y = 2^{-x} - 1$

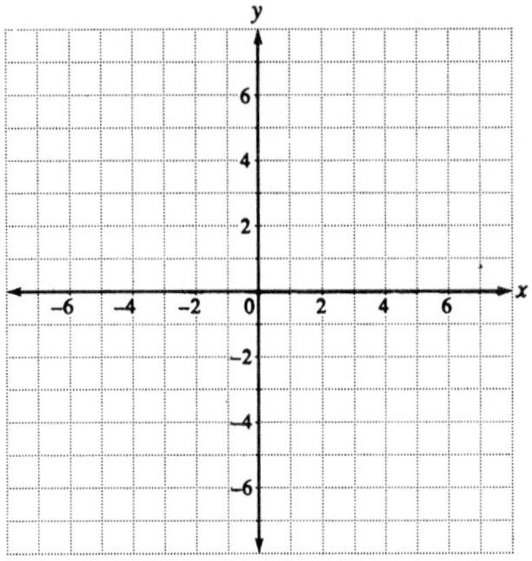

9. $y = 4 - 3^x$

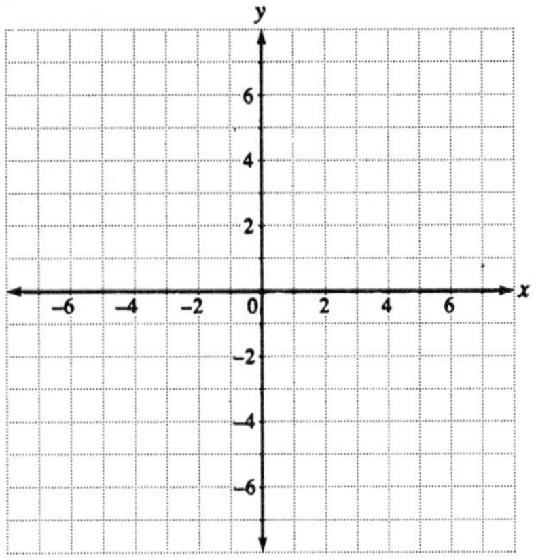

Section 11.2

Change to exponential equations.

10. $1 = \log_7 7$

11. $0 = \log_{10} 1$

12. $6 = \log_2 64$

13. $\log_3 \frac{1}{9} = -2$

14. $\log_{10} 5 = 0.6990$

15. $\log_{10} 3 = 0.4771$

Change to logarithmic equations.

16. $25 = 5^x$

17. $y = 3^5$

18. $36 = a^2$

19. $32 = b^5$

20. $10^{0.5478} = 3.53$

21. $10^{2.9425} = 876$

Solve.

22. $\log_{64} x = \frac{2}{3}$

23. $\log_{49} x = \frac{1}{2}$

24. $\log_{32} x = \frac{1}{5}$

25. $\log_{10} x = -4$

26. $\log_{1/2} x = -3$

27. $\log_{1/3} x = -4$

Section 11.3

Find the following.

28. $\log_6 1$

29. $\log_{10} 1$

30. $\log_4 4$

31. $\log_{10} 10$

32. $\log_2 1$

33. $\log_6 6$

Write as a sum or difference of logarithms, or as a product.

34. $\log_2 x(x - 1)$

35. $\log_5 \dfrac{x - 1}{x + 1}$

36. $\log_b \sqrt{3x - 1}$

37. $\log_a(x - 2)^4$

38. $\log_a \dfrac{x^8}{y^5}$

39. $\log_7(3x + 2)4y$

Write as a single logarithm.

40. $\log_3 \dfrac{3}{8} + \log_3 \dfrac{40}{9}$

41. $\log_5 \dfrac{7}{3} - \log_5 \dfrac{35}{18}$

42. $2 \log_c(a - b) - \log_c(3a - 3b)$

43. $\dfrac{1}{2} \log_b x^4 + 3 \log_b y^2$

44. $\log_a \dfrac{a^2}{\sqrt{x}} - \log_a \dfrac{\sqrt{x}}{a}$

45. $\dfrac{1}{2} \log_b(3x - 1) + 3 \log_b(x + 4)$

Section 11.4

Use a calculator to find the following. Round the answer to four decimal places.

46. log 327

47. log 5280

48. log 0.0342

49. log 0.0078

50. ln 15

51. ln 320

52. ln 0.6

53. ln 0.003

Find the common antilogarithm.

54. 2.9143

55. 4.7427

56. −4.5735

57. −0.9318

Find the natural antilogarithm.

58. 3.8142

59. 0.05

60. −2.9

61. −4.103

Use common logarithms or natural logarithms to find each of the following to four decimal places.

62. $\log_3 29$

63. $\log_2 15$

64. $\log_{40} 12$

65. $\log_{20} 38$

Section 11.5

Solve.

66. $3^x = 81$

67. $10^x = 1000$

68. $5^x = 14$

69. $9^x = 23$

* **70.** $a^{x^2} = a^x$

* **71.** $b^{x^2} = b^{x+6}$

72. $2^{x+1} = 3^x$

73. $5^{x-2} = 4^x$

74. $\log_4(x + 3) = -1$

75. $\log(x^2 - 9) = 0$

76. $\log_5 \sqrt{x^2 + 16} = 1$

77. $\log_3(x + 1)^2 = 4$

78. $\log_4 3x = \log_4 5 + \log_4 3$

79. $\log_5 2x = \log_5 2 - \log_5 8$

80. $\log_{1/2}(9x^2 - 1) - \log_{1/2}(3x + 1) = 1$

81. $\log_3(x - 1) + \log_3(x - 2) = \log_3 6$

Section 11.6

For Problems 82 and 83, use the formula $N = N_0 \cdot b^t$, where N is the number of bacteria present after t days, N_0 is the number present initially, and b is the rate of increase per day.

82. The number of bacteria initially present in a sample is 3000. If the bacteria double in population every day, what is the number of bacteria present after 3.4 days?

83. In how many days will the bacteria count be 481,000 if the population doubles every day and the number of bacteria initially present is 1520?

84. If \$5000 is invested in an account that pays 7% interest compounded annually, how much will be in the account after 6 years? Use the formula $A = P(1 + r)^t$, where P is the amount invested at an interest rate r and A is the amount in the account after t years.

Use the following formula for Problems 77 and 78. If the principal, P, invested in an account with interest rate, r, is compounded continuously, the balance, S, in the account after t years is

$$S = Pe^{rt}$$

85. If \$5000 is invested in an account that pays 7% interest compounded continuously, find the amount in the account after 6 years.

86. In how many years will an amount of money in an account triple if the interest rate is 6% compounded continuously?

COOPERATIVE LEARNING

Write an exponential equation for the following graphs.

1.

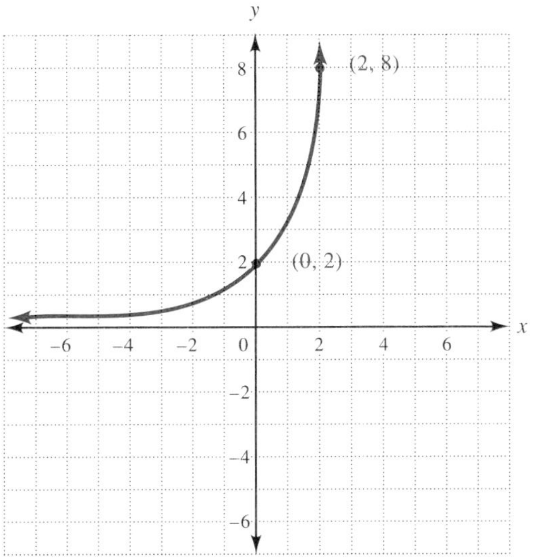

2.

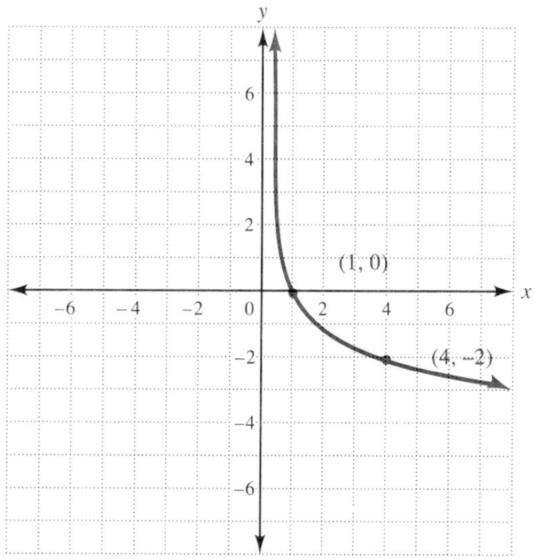

3. Find a logarithmic equation for the function represented by the data.

x	1	2	3	4	5
y	0	0.43	0.68	0.86	1

4. Given that log 10 = 1, log 100 = 2, and log 1000 = 3, write the following as a sum of logarithms.

a. $\log(10x)$ **b.** $\log(100x)$ **c.** $\log(1000x)$

5. Write a logarithmic equation for the following graph.

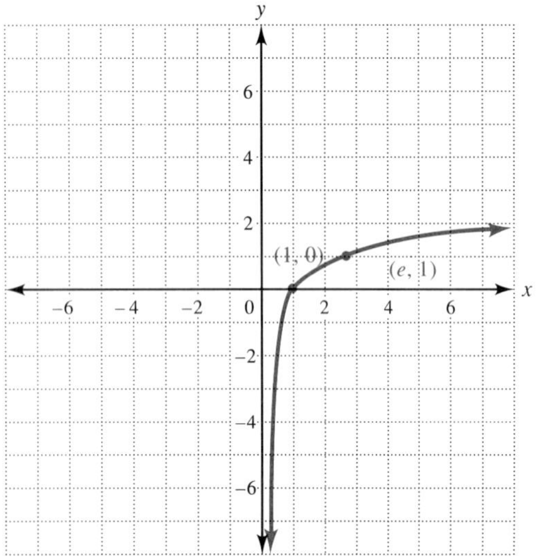

Solve for t using natural logarithms.

6. $B = ce^t$

7. $be^{ct} = e^{at}$, where $c \neq a$

8. Water is being filtered so that the quantity of pollutant, P (measured in milligrams per liter) is decreasing according to the equation $P = P_0e^{-kt}$, where P_0 is the initial amount of pollutant, t is time in hours, and k is a constant. If 10% of the pollution is removed in the first 5 hours:

a. What percentage of the pollution is left after 12 hours?
b. How long will it take for the pollution to be reduced by 60%?
c. Plot a graph of pollution versus time.
d. Explain why the quantity of pollutant might decrease in this way.

Chapter 11 Practice Test

NAME

DATE

CLASS

Graph.

1. $y = 3^{x-1}$

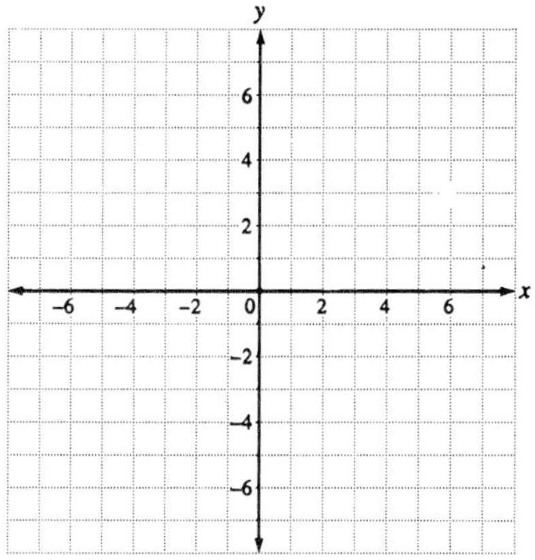

1. ______

2. $x = \left(\frac{1}{4}\right)^y$

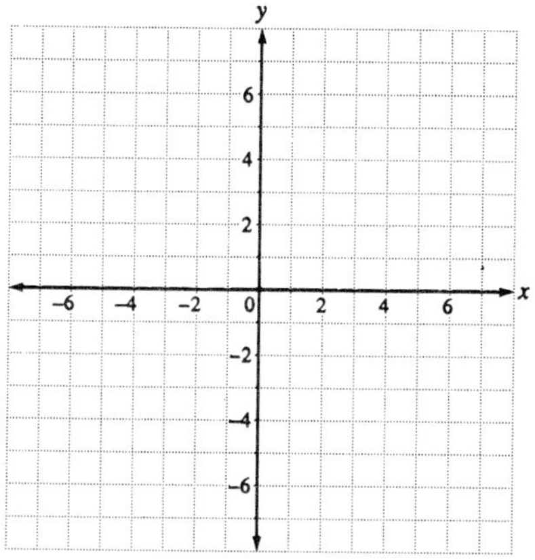

2. ______

3. Change to an exponential equation: $\log_{10} 6.7 = 0.8261$.

3. ______

4. ______________________

4. Change to a logarithmic equation: $4^{-2} = \frac{1}{16}$.

5. ______________________

5. Graph $y = \log_3 x$.

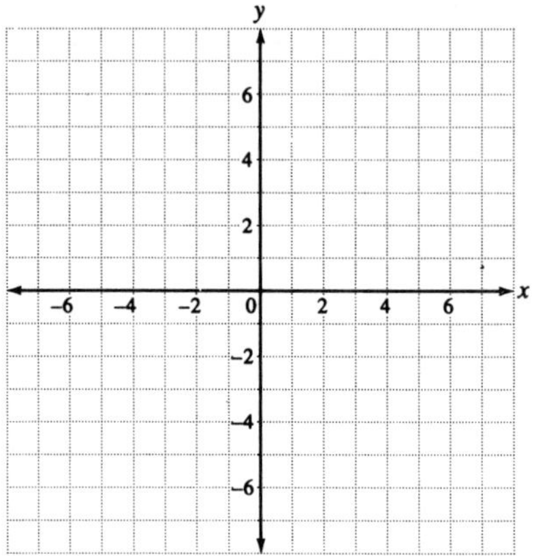

6. ______________________

6. Write as a single logarithm: $6 \log_b M - 4 \log_b N$.

7. ______________________

7. Write in terms of logarithms of x and y: $\log_a \sqrt[3]{x^2 y}$.

Find the following. Use a calculator.

8. ______________________

8. log 137

9. The common antilogarithm of 2.9069

9. ______

10. ln 0.0485

10. ______

11. The natural antilogarithm of -3.45

11. ______

Find the following.

12. $\log_b b$

12. ______

13. $\log_p 1$

13. ______

14. $\log_a a^{14}$

14. ______

15. $\log_9 25$ (Use the change-of-base formula.)

15. ______

16. ____________

17. ____________

18. ____________

19. ____________

20. ____________

Solve.

16. $3^{2x-2} = 243$

17. $6^{3x-1} = 15$

18. $\log x + \log(x + 3) = 1$

19. How long does it take \$3500 to double if it is invested in an account at 8% interest per year compounded annually?

20. What is the age of a bone fragment that originally contained 100 milligrams of carbon 14 and now contains 60 milligrams?

Final Examination

NAME

DATE

CLASS

Chapter 1

1. Evaluate when $x = 4$, $y = -3$, and $z = -1$: $-3xy + 4yz$.

1. ______

2. Simplify: $3 - 2[2x - 4(2x + 1)]$.

2. ______

3. Simplify and write with positive exponents. Assume that variables represent nonzero real numbers:

$$\left(\frac{8xy^4}{4x^3y^{-2}}\right)^2$$

3. ______

Chapter 2

4. Solve: $9 - 3(2x - 1) = 3(4x - 2)$.

4. ______

5. The perimeter of a rectangular flower bed is 105 feet. If the length is 8 feet more than the width, find the dimensions of the flower bed.

5. ______

6. Solve: $A = \frac{h}{2}(b + c)$ for c.

6. ______

7. Solve: $|4x + 2| = 10$.

7. ______

8.

8. Solve and graph $|2x - 3| < 5$.

9.

9. The owner of a pizza parlor rents the store for \$1200 per month plus 8% of the total sales during the month. The owner wishes to earn a minimum of \$3000 per month from the store. Find the minimum sales that will enable the owner to achieve her goal.

Chapter 3

10.

10. Graph $3y = 2x + 9$.

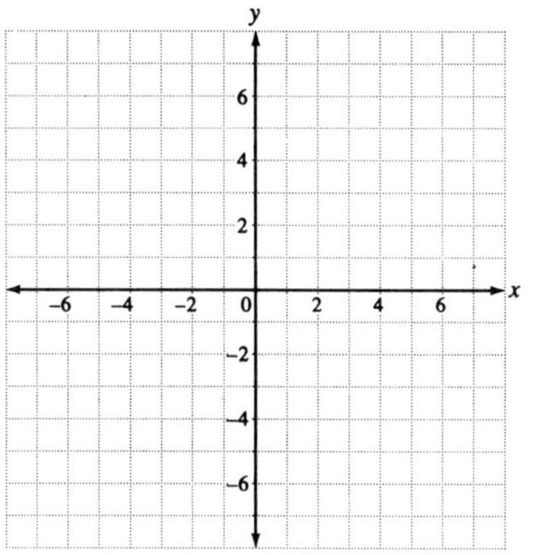

11.

11. Find an equation for the line through the points $(-2, 3)$ and $(6, -1)$.

12.

12. Find an equation for the line through the point $(2, -3)$ and perpendicular to the line $x - 3y = 4$.

13.

13. At Tech Corp., the cost of manufacturing 120 calculators is \$4500 and the cost of manufacturing 200 calculators is \$6500. If the relationship between the cost and the number of calculators manufactured is linear, write the equation that relates the cost C to the number of calculators n that are manufactured. Use the equation to find the cost of manufacturing 350 calculators.

14. Graph $3x - 2y < 12$.

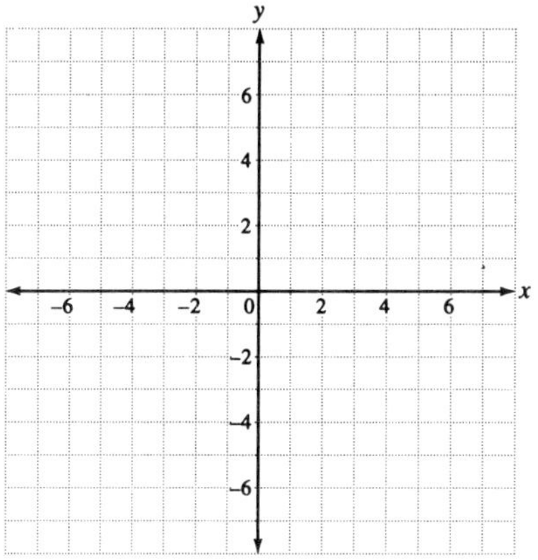

14. ____________

Chapter 4

15. Solve:

$$3x + 2y = 7$$
$$4x - 5y = -6$$

15. ____________

16. How much 25% alcohol solution should be mixed with 60% alcohol solution to make 8 gallons of a 40% solution?

16. ____________

17. Solve:

$$2x - 3y + 2z = -10$$
$$5x + 2y + 3z = -2$$
$$4x + 6y + 5z = 2$$

17. ____________

18. Evaluate:

$$\begin{vmatrix} 2 & 1 & 3 \\ -1 & 0 & 4 \\ 2 & 4 & 3 \end{vmatrix}$$

18. ____________

19. ____________

20. ____________

21. ____________

22. ____________

23. ____________

24. ____________

25. ____________

26. ____________

27. ____________

Chapter 5

19. Find the degree of the polynomial: $3x^4 + 6x^5 - 3$.

20. Add: $4x^2 + 6xy - y^2$ and $7x^2 - 9xy - 8y^2$.

21. Subtract: $(8x^3 - 3x^2 + 7) - (5x^2 + 3x - 4)$.

Multiply.

22. $(2y^2 - 3y + 4)(y - 2)$

23. $(3x + 8)(2x - 3)$

24. Use synthetic division to divide:

$$\frac{x^2 + 5x - 8}{x - 4}$$

Factor.

25. $12x^2 - 14x - 6$

26. $16x^4 - y^4$

27. $8y^3 - 125z^3$

Chapter 6

28. Divide and simplify:

$$\frac{x^2 - 16}{3x + 15} \div \frac{x + 4}{2x^2 + 7x - 15}$$

28. ______

29. Perform the operations and simplify:

$$\frac{2}{x + 4} + \frac{5}{x^2 + 2x - 8} - \frac{3}{x - 2}$$

29. ______

30. Simplify:

$$\frac{\frac{4}{3x} - \frac{1}{6x}}{1 + \frac{5}{6x}}$$

30. ______

31. Solve:

$$\frac{7}{x - 2} - \frac{5}{x + 2} = 6$$

31. ______

32. Solve $A = \dfrac{P}{1 - rn}$ for r.

32. ______

33. The maximum load a cylindrical column of circular cross section can support varies directly as the fourth power of the diameter and inversely as the square of the height. If an 8-foot column 2 feet in diameter holds 12 tons, how much load will a column 6 feet high and 3 feet in diameter hold?

33. ______

34. A plumber works twice as fast as his apprentice. If they work together, they can wire a house in 15 hours. How long would it take the apprentice working alone to wire the house?

34. ______

35. ____________________

35. The rate of a jet plane was four times the rate of a helicopter. If the helicopter travels 84 miles in 1 hour less time than the jet travels 720 miles, find the rate of the jet.

Chapter 7

Assume that variables under radical signs represent positive numbers.

36. ____________________

36. Multiply and simplify: $\sqrt[3]{4x^2y}\sqrt[3]{16x^2y^2}$.

37. ____________________

37. Subtract: $3\sqrt{24} - 5\sqrt{96}$.

38. ____________________

38. Rationalize the denominator:

$$\frac{\sqrt{3} + x}{\sqrt{5} - x}$$

39. ____________________

39. Solve: $\sqrt{x - 1} = 5 - \sqrt{x + 4}$.

40. ____________________

40. Divide:

$$\frac{4 - 3i}{2 + 5i}$$

41. ____________________

41. The two equal sides of an isosceles right triangle are of length 38 meters. What is the length of the hypotenuse? Round the answer to the nearest tenth.

Chapter 8

Solve.

42. ____________________

42. $x^2 + 5x = 2$

43. Solve: $x^4 - 4x^2 - 5 = 0$.

43. ______

44. Graph $y = x^2 + x - 6$.

44. ______

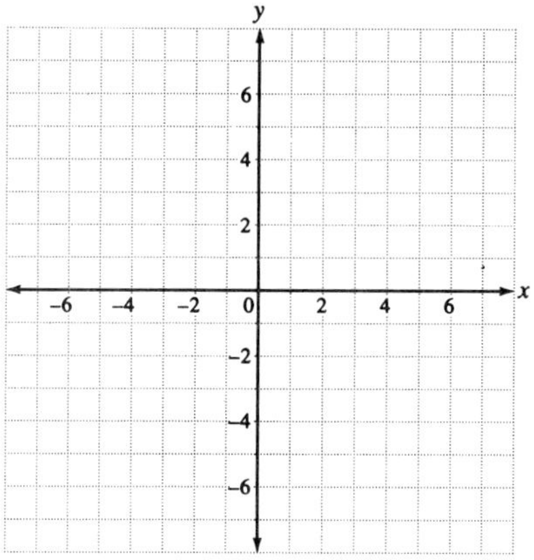

45. What is the minimum product of two numbers whose difference is 4?

45. ______

46. Solve and graph $x^2 + 3x - 18 < 0$.

46. ______

Chapter 9

47. Find the distance between $(-3, 2)$ and $(-4, 5)$.

47. ______

48. Find an equation for the circle with center at $(-1, 4)$ and radius $\sqrt{5}$.

48. ______

49. Graph $\frac{x^2}{9} - \frac{y^2}{4} = 1$.

49. ______

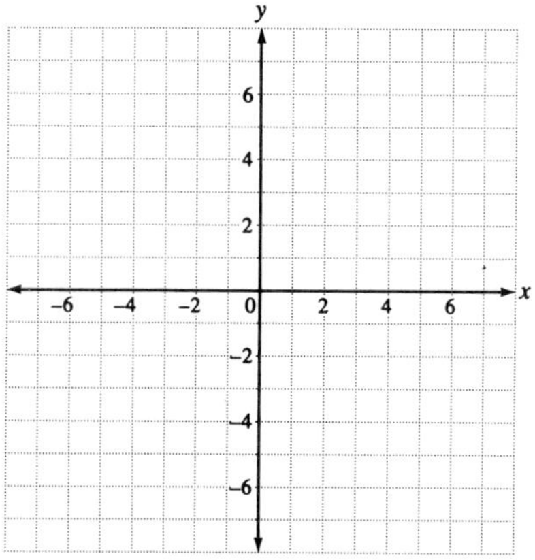

50. ______________________

50. Solve:

$$3x^2 + 4y^2 = 16$$
$$4x^2 - 3y^2 = 13$$

Chapter 10

51. ______________________

51. Find the domain:

$$y = \frac{5}{x - 3}$$

52. ______________________

52. If $f(x) = x^2 - 4x - 7$, find $f(-3)$.

Chapter 11

53. ______________________

53. Graph $y = 2^{x+2}$.

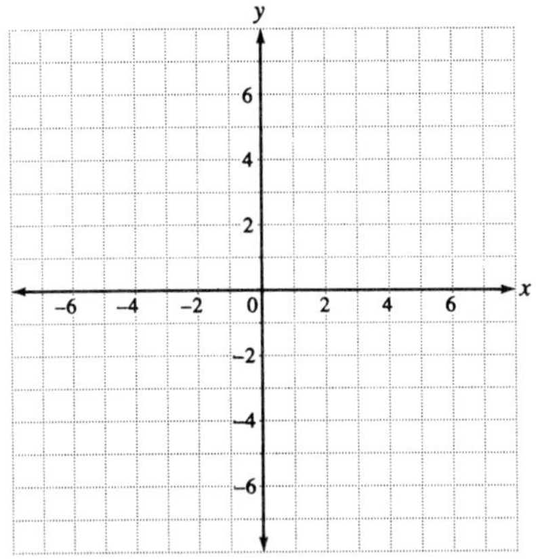

54. ______________________

54. Change to a logarithmic equation: $5^{-2} = \frac{1}{25}$.

Solve.

55. ______________________

55. $3^{x-1} = 7$

56. ______________________

56. Katera deposited $4500 in a money market fund that paid 5% interest compounded continuously. How much money will be in the account after 9 years?

Appendices

APPENDIX A: MORE ON SETS

1 Elements of a Set

Recall from Section 1.1 that a set is a collection of objects. The objects in the set are called ***elements of the set.*** We use braces, { }, to enclose the elements of the set. The elements of the set may be listed in any order.

Sets in which the number of elements can be counted are called ***finite sets.*** We cannot determine the number of elements in some sets. These sets are called ***infinite sets.*** We list some elements of the set and indicate the rest of the elements with three dots. For example, the set of counting or natural numbers is

$$\{1, 2, 3, 4, \ldots\}.$$

Capital letters are usually used to name sets and elements of the set are denoted by lowercase letters.

> The symbol $\in$ is used to indicate membership in a set.

EXAMPLE 1 Determine whether the following are true or false.

a. $3 \in \{1, 3, 5, 7, \ldots\}$
True

b. $d \in \{a, c, e\}$
False ■

□ DO EXERCISE 1.

2 Subsets

When all the elements of a set A are also elements of a set B, we say that A is a ***subset*** of B, written $A \subset B$.

> $A \subset B$ means that A is a subset of B.

OBJECTIVES

1 *Determine if an object is an element of a set*

2 *Determine whether a set is a subset of another set*

3 *Find all of the subsets of a given set*

□ **Exercise 1** Determine whether each statement is true or false for the set

$$R = \{u, v, w, x, z\}.$$

a. $u \in R$

b. $y \in R$

c. $t \in R$

d. $x \in R$

□ **Exercise 2** Decide if the following are true or false.

a. $\{d, e, f\} \subset \{b, c, d, e, f\}$

b. $\{2, 4, 6\} \subset \{4, 6, 8, 10\}$

c. The set of all cars is a subset of the set of all Chevrolets.

□ **Exercise 3** List all the subsets of the following.

a. $\{2, 3\}$

b. $\{x, y, z\}$

EXAMPLE 2 Decide if the following are true or false.

a. $\{a, b, c\} \subset \{a, b, c, d, e\}$
True

b. $\{4, 5, 6\} \subset \{2, 4, 6, 8\}$
False

c. The set of all German Shepherds is a subset of the set of all dogs.
True ■

□ **DO EXERCISE 2.**

3 Finding Subsets

Every set is a subset of itself. A set with no elements is called the ***empty set,*** denoted by $\varnothing$. The empty set is defined to be a subset of every set.

> The empty set, $\varnothing$, has no elements.

The empty set may also be written as { }.

EXAMPLE 3 List all the subsets of the following.

a. $\{a, b\}$

$\{a, b\}$	Every set is a subset of itself.
$\{a\}, \{b\}$	There are two subsets with one element.
$\varnothing$	The empty set is a subset of every set.

Notice that there are 2^2 or 4 subsets.

b. $\{2, 4, 6\}$

The subsets are as follows.

$\{2, 4, 6\}$	
$\{2\}, \{4\}, \{6\}$	
$\{2, 4\}, \{2, 6\}, \{4, 6\}$	There are three subsets with two elements.
$\varnothing$	

Notice that there are 2^3 or 8 subsets. ■

We can generalize to the following:

> For any nonnegative integer n, a set with n elements has 2^n subsets.

□ **DO EXERCISE 3.**

Answers to Exercises

1. a. True **b.** False **c.** False **d.** True

2. a. True **b.** False **c.** False

3. a. $\{2, 3\}, \{2\}, \{3\}, \varnothing$
b. $\{x, y, z\}, \{x\}, \{y\}, \{z\}, \{x, y\}, \{x, z\}, \{y, z\}, \varnothing$

Problem Set Appendix A

NAME ______________________

DATE ______________________

CLASS ______________________

Determine if the following are true or false.

1. $4 \in \{2, 4, 6, 8\}$

2. $0 \in \{2, 4, 6, 8\}$

3. $q \in \{l, o, p\}$

4. $2 \in \{1, 2, 3, 4, \ldots\}$

5. $\{3, 7, 11\} \subset \{3, 5, 7, 9, 11\}$

6. $\{8, 12\} \subset \{6, 12, 18\}$

7. $\{a, j, q\} \subset \{a, j, p, t\}$

8. $\{u, x\} \subset \{u, v, x, y\}$

9. $\{6, 12, 17\} \subset \{2, 6, 15, 20\}$

10. $\{8, 9\} \subset \{8, 19\}$

11. $a \in \{b, a, d\}$

12. $z \in \{q, z, p, r\}$

13. $\{8, 11, 5\} \subset \{3, 8, 5, 14, 11\}$

14. $\{6, 9, 22, 4\} \subset \{4, 6, 35, 10\}$

List all the subsets of the following.

15. $\{5, 7\}$

16. $\{p, q\}$

17. $\{a, b, c\}$

18. $\{2, 4, 6\}$

19. $\{3, 8, 9, 5\}$

20. $\{a, b, c, d\}$

21. $\{u, v, w, x\}$

22. $\{0, 4, 8, 3\}$

Let $A = \{1, 2, 3, 4\}$, $B = \{2, 4, 6, 8\}$, *and* $C = \{5, 6, 7, 8\}$. *Find the following. (Hint: See Section 2.5.)*

* **23.** $(A \cup B) \cap (A \cup C)$

* **24.** $(B \cap C) \cup (A \cap C)$

* **25.** $A \cup [(A \cap B) \cap (A \cup C)]$

* **26.** $B \cup [(A \cup B) \cup (B \cap C)]$

APPENDIX B: TABLES

Table 1 Powers and Roots

n	n^2	$\sqrt{n}$	n^3	$\sqrt[3]{n}$
1	1	1.000	1	1.000
2	4	1.414	8	1.260
3	9	1.732	27	1.442
4	16	2.000	64	1.587
5	25	2.236	125	1.710
6	36	2.449	216	1.817
7	49	2.646	343	1.913
8	64	2.828	512	2.000
9	81	3.000	729	2.080
10	100	3.162	1,000	2.154
11	121	3.317	1,331	2.224
12	144	3.464	1,728	2.289
13	169	3.606	2,197	2.351
14	196	3.742	2,744	2.410
15	225	3.873	3,375	2.466
16	256	4.000	4,096	2.520
17	289	4.123	4,913	2.571
18	324	4.243	5,832	2.621
19	361	4.359	6,859	2.668
20	400	4.472	8,000	2.714
21	441	4.583	9,261	2.759
22	484	4.690	10,648	2.802
23	529	4.796	12,167	2.844
24	576	4.899	13,824	2.884
25	625	5.000	15,625	2.924
26	676	5.099	17,576	2.962
27	729	5.196	19,683	3.000
28	784	5.292	21,952	3.037
29	841	5.385	24,389	3.072
30	900	5.477	27,000	3.107
31	961	5.568	29,791	3.141
32	1,024	5.657	32,768	3.175
33	1,089	5.745	35,937	3.208
34	1,156	5.831	39,304	3.240
35	1,225	5.916	42,875	3.271
36	1,296	6.000	46,656	3.302
37	1,369	6.083	50,653	3.332
38	1,444	6.164	54,872	3.362
39	1,521	6.245	59,319	3.391
40	1,600	6.325	64,000	3.420
41	1,681	6.403	68,921	3.448
42	1,764	6.481	74,088	3.476
43	1,849	6.557	79,507	3.503
44	1,936	6.633	85,184	3.530
45	2,025	6.708	91,125	3.557
46	2,116	6.782	97,336	3.583
47	2,209	6.856	103,823	3.609
48	2,304	6.928	110,592	3.634
49	2,401	7.000	117,649	3.659
50	2,500	7.071	125,000	3.684

n	n^2	$\sqrt{n}$	n^3	$\sqrt[3]{n}$
51	2,601	7.141	132,651	3.708
52	2,704	7.211	140,608	3.733
53	2,809	7.280	148,877	3.756
54	2,916	7.348	157,464	3.780
55	3,025	7.416	166,375	3.803
56	3,136	7.483	175,616	3.826
57	3,249	7.550	185,193	3.849
58	3,364	7.616	195,112	3.871
59	3,481	7.681	205,379	3.893
60	3,600	7.746	216,000	3.915
61	3,721	7.810	226,981	3.936
62	3,844	7.874	238,328	3.958
63	3,969	7.937	250,047	3.979
64	4,096	8.000	262,144	4.000
65	4,225	8.062	274,625	4.021
66	4,356	8.124	287,496	4.041
67	4,489	8.185	300,763	4.062
68	4,624	8.246	314,432	4.082
69	4,761	8.307	328,509	4.102
70	4,900	8.367	343,000	4.121
71	5,041	8.426	357,911	4.141
72	5,184	8.485	373,248	4.160
73	5,329	8.544	389,017	4.179
74	5,476	8.602	405,224	4.198
75	5,625	8.660	421,875	4.217
76	5,776	8.718	438,976	4.236
77	5,929	8.775	456,533	4.254
78	6,084	8.832	474,552	4.273
79	6,241	8.888	493,039	4.291
80	6,400	8.944	512,000	4.309
81	6,561	9.000	531,441	4.327
82	6,724	9.055	551,368	4.344
83	6,889	9.110	571,787	4.362
84	7,056	9.165	592,704	4.380
85	7,225	9.220	614,125	4.397
86	7,396	9.274	636,056	4.414
87	7,569	9.327	658,503	4.431
88	7,744	9.381	681,472	4.448
89	7,921	9.434	704,969	4.465
90	8,100	9.487	729,000	4.481
91	8,281	9.539	753,571	4.498
92	8,464	9.592	778,688	4.514
93	8,649	9.644	804,357	4.531
94	8,836	9.695	830,584	4.547
95	9,025	9.747	857,375	4.563
96	9,216	9.798	884,736	4.579
97	9,409	9.849	912,673	4.595
98	9,604	9.899	941,192	4.610
99	9,801	9.950	970,299	4.626
100	10,000	10.000	1,000,000	4.642

Table 2 Common Logarithms

x	0	1	2	3	4	5	6	7	8	9
1.0	.0000	.0043	.0086	.0128	.0170	.0212	.0253	.0294	.0334	.0374
1.1	.0414	.0453	.0492	.0531	.0569	.0607	.0645	.0682	.0719	.0755
1.2	.0792	.0828	.0864	.0899	.0934	.0969	.1004	.1038	.1072	.1106
1.3	.1139	.1173	.1206	.1239	.1271	.1303	.1335	.1367	.1399	.1430
1.4	.1461	.1492	.1523	.1553	.1584	.1614	.1644	.1673	.1703	.1732
1.5	.1761	.1790	.1818	.1847	.1875	.1903	.1931	.1959	.1987	.2014
1.6	.2041	.2068	.2095	.2122	.2148	.2175	.2201	.2227	.2253	.2279
1.7	.2304	.2330	.2355	.2380	.2405	.2430	.2455	.2480	.2504	.2529
1.8	.2553	.2577	.2601	.2625	.2648	.2672	.2695	.2718	.2742	.2765
1.9	.2788	.2810	.2833	.2856	.2878	.2900	.2923	.2945	.2967	.2989
2.0	.3010	.3032	.3054	.3075	.3096	.3118	.3139	.3160	.3181	.3201
2.1	.3222	.3243	.3263	.3284	.3304	.3324	.3345	.3365	.3385	.3404
2.2	.3424	.3444	.3464	.3483	.3502	.3522	.3541	.3560	.3579	.3598
2.3	.3617	.3636	.3655	.3674	.3692	.3711	.3729	.3747	.3766	.3784
2.4	.3802	.3820	.3838	.3856	.3874	.3892	.3909	.3927	.3945	.3962
2.5	.3979	.3997	.4014	.4031	.4048	.4065	.4082	.4099	.4116	.4133
2.6	.4150	.4166	.4183	.4200	.4216	.4232	.4249	.4265	.4281	.4298
2.7	.4314	.4330	.4346	.4362	.4378	.4393	.4409	.4425	.4440	.4456
2.8	.4472	.4487	.4502	.4518	.4533	.4548	.4564	.4579	.4594	.4609
2.9	.4624	.4639	.4654	.4669	.4683	.4698	.4713	.4728	.4742	.4757
3.0	.4771	.4786	.4800	.4814	.4829	.4843	.4857	.4871	.4886	.4900
3.1	.4914	.4928	.4942	.4955	.4969	.4983	.4997	.5011	.5024	.5038
3.2	.5051	.5065	.5079	.5092	.5105	.5119	.5132	.5145	.5159	.5172
3.3	.5185	.5198	.5211	.5224	.5237	.5250	.5263	.5276	.5289	.5302
3.4	.5315	.5328	.5340	.5353	.5366	.5378	.5391	.5403	.5416	.5428
3.5	.5441	.5453	.5465	.5478	.5490	.5502	.5514	.5527	.5539	.5551
3.6	.5663	.5575	.5587	.5599	.5611	.5623	.5635	.5647	.5658	.5670
3.7	.5682	.5694	.5705	.5717	.5729	.5740	.5752	.5763	.5775	.5786
3.8	.5798	.5809	.5821	.5832	.5843	.5855	.5866	.5877	.5888	.5899
3.9	.5911	.5922	.5933	.5944	.5955	.5966	.5977	.5988	.5999	.6010
4.0	.6021	.6031	.6042	.6053	.6064	.6075	.6085	.6096	.6107	.6117
4.1	.6128	.6138	.6149	.6160	.6170	.6180	.6191	.6201	.6212	.6222
4.2	.6232	.6243	.6253	.6263	.6274	.6284	.6294	.6304	.6314	.6325
4.3	.6335	.6345	.6355	.6365	.6375	.6385	.6395	.6405	.6415	.6425
4.4	.6435	.6444	.6454	.6464	.6474	.6484	.6493	.6503	.6513	.6522
4.5	.6532	.6542	.6551	.6561	.6571	.6580	.6590	.6599	.6609	.6618
4.6	.6628	.6637	.6646	.6656	.6665	.6675	.6684	.6693	.6702	.6712
4.7	.6721	.6730	.6739	.6749	.6758	.6767	.6776	.6785	.6794	.6803
4.8	.6812	.6821	.6830	.6839	.6848	.6857	.6866	.6875	.6884	.6893
4.9	.6902	.6911	.6920	.6928	.6937	.6946	.6955	.6964	.6972	.6981
5.0	.6990	.6998	.7007	.7016	.7024	.7033	.7042	.7050	.7059	.7067
5.1	.7076	.7084	.7093	.7101	.7110	.7118	.7126	.7135	.7143	.7152
5.2	.7160	.7168	.7177	.7185	.7193	.7202	.7210	.7218	.7226	.7235
5.3	.7243	.7251	.7259	.7267	.7275	.7284	.7292	.7300	.7308	.7316
5.4	.7324	.7332	.7340	.7348	.7356	.7364	.7372	.7380	.7388	.7396

x	0	1	2	3	4	5	6	7	8	9
5.5	.7404	.7412	.7419	.7427	.7435	.7443	.7451	.7459	.7466	.7474
5.6	.7482	.7490	.7497	.7505	.7513	.7520	.7528	.7536	.7543	.7551
5.7	.7559	.7566	.7574	.7582	.7589	.7597	.7604	.7612	.7619	.7627
5.8	.7634	.7642	.7649	.7657	.7664	.7672	.7679	.7686	.7694	.7701
5.9	.7709	.7716	.7723	.7731	.7738	.7745	.7752	.7760	.7767	.7774
6.0	.7782	.7789	.7796	.7803	.7810	.7818	.7825	.7832	.7839	.7846
6.1	.7853	.7860	.7868	.7875	.7882	.7889	.7896	.7903	.7910	.7917
6.2	.7924	.7931	.7938	.7945	.7952	.7959	.7966	.7973	.7980	.7987
6.3	.7993	.8000	.8007	.8014	.8021	.8028	.8035	.8041	.8048	.8055
6.4	.8062	.8069	.8075	.8082	.8089	.8096	.8102	.8109	.8116	.8122
6.5	.8129	.8136	.8142	.8149	.8156	.8162	.8169	.8176	.8182	.8189
6.6	.8195	.8202	.8209	.8215	.8222	.8228	.8235	.8241	.8248	.8254
6.7	.8261	.8267	.8274	.8280	.8287	.8293	.8299	.8306	.8312	.8319
6.8	.8325	.8331	.8338	.8344	.8351	.8357	.8363	.8370	.8376	.8382
6.9	.8388	.8395	.8401	.8407	.8414	.8420	.8426	.8432	.8439	.8445
7.0	.8451	.8457	.8463	.8470	.8476	.8482	.8488	.8494	.8500	.8506
7.1	.8513	.8519	.8525	.8531	.8537	.8543	.8549	.8555	.8561	.8567
7.2	.8573	.8579	.8585	.8591	.8597	.8603	.8609	.8615	.8621	.8627
7.3	.8633	.8639	.8645	.8651	.8657	.8663	.8669	.8675	.8681	.8686
7.4	.8692	.8698	.8704	.8710	.8716	.8722	.8727	.8733	.8739	.8745
7.5	.8751	.8756	.8762	.8768	.8774	.8779	.8785	.8791	.8797	.8802
7.6	.8808	.8814	.8820	.8825	.8831	.8837	.8842	.8848	.8854	.8859
7.7	.8865	.8871	.8876	.8882	.8887	.8893	.8899	.8904	.8910	.8915
7.8	.8921	.8927	.8932	.8938	.8943	.8949	.8954	.8960	.8965	.8971
7.9	.8976	.8982	.8987	.8993	.8998	.9004	.9009	.9015	.9020	.9025
8.0	.9031	.9036	.9042	.9047	.9053	.9058	.9063	.9069	.9074	.9079
8.1	.9085	.9090	.9096	.9101	.9106	.9112	.9117	.9122	.9128	.9133
8.2	.9138	.9143	.9149	.9154	.9159	.9165	.9170	.9175	.9180	.9186
8.3	.9191	.9196	.9201	.9206	.9212	.9217	.9222	.9227	.9232	.9238
8.4	.9243	.9248	.9253	.9258	.9263	.9269	.9274	.9279	.9284	.9289
8.5	.9294	.9299	.9304	.9309	.9315	.9320	.9325	.9330	.9335	.9340
8.6	.9345	.9350	.9355	.9360	.9365	.9370	.9375	.9380	.9385	.9390
8.7	.9395	.9400	.9405	.9410	.9415	.9420	.9425	.9430	.9435	.9440
8.8	.9445	.9450	.9455	.9460	.9465	.9469	.9474	.9479	.9484	.9489
8.9	.9494	.9499	.9504	.9509	.9513	.9518	.9523	.9528	.9533	.9538
9.0	.9542	.9547	.9552	.9557	.9562	.9566	.9571	.9576	.9581	.9586
9.1	.9590	.9595	.9600	.9605	.9609	.9614	.9619	.9624	.9628	.9633
9.2	.9638	.9643	.9647	.9652	.9657	.9661	.9666	.9671	.9675	.9680
9.3	.9685	.9689	.9694	.9699	.9703	.9708	.9713	.9717	.9722	.9727
9.4	.9731	.9736	.9741	.9745	.9750	.9754	.9759	.9763	.9768	.9773
9.5	.9777	.9782	.9786	.9791	.9795	.9800	.9805	.9809	.9814	.9818
9.6	.9823	.9827	.9832	.9836	.9841	.9845	.9850	.9854	.9859	.9863
9.7	.9868	.9872	.9877	.9881	.9886	.9890	.9894	.9899	.9903	.9908
9.8	.9912	.9917	.9921	.9926	.9930	.9934	.9939	.9943	.9948	.9952
9.9	.9956	.9961	.9965	.9969	.9974	.9978	.9983	.9987	.9991	.9996

Bibliography

BELL, E. T., *The Development of Mathematics*. New York: McGraw-Hill, 1945.

———, *Men of Mathematics*. New York: Simon & Schuster, 1986.

BOYER, CARL B., *A History of Mathematics*. New York: Wiley and Sons, Inc., 1968.

BURTON, DAVID M., *The History of Mathematics: An Introduction*. Newton, Mass.: Allyn and Bacon, 1985.

DANZIG, TOBIAS, *Number, The Language of Science*. New York: Free Press, 1967.

EVES, HOWARD, *In Mathematical Circles*. Boston: Prindle, Weber & Schmidt, 1969.

———, *An Introduction to the History of Mathematics*, 5th ed. New York: Saunders College Publishing, 1983.

———, *Mathematical Circles Squared*. Boston: Prindle, Weber & Schmidt, 1972.

———, *Mathematical Circles Revisited*. Boston: Prindle, Weber & Schmidt, 1971.

———, and CARROLL V. NEWSOM, *An Introduction to the Foundations and Fundamental Concepts of Mathematics*. New York: Holt, Rinehart and Winston, 1958.

GAMOW, GEORGE, *One Two Three . . . Infinity*. New York: Bantam Books, 1971.

Historical Topics for the Mathematics Classroom (Thirty-first Yearbook). Washington, D.C.: National Council of Teachers of Mathematics, 1969.

HOGBEN, LANCELOT, *Mathematics in the Making*. New York: Doubleday & Company, 1960.

HOOPER, ALFRED, *Makers of Mathematics*. London: Faber & Faber, 1948.

KLINE, MORRIS, *Mathematics and the Physical World*. New York: Thomas Y. Crowell, 1959.

———, *Mathematics in Western Culture*. New York: Oxford University Press, 1953.

———, *Mathematics for the Liberal Arts*. Reading, Mass.: Addison-Wesley, 1967.

LARRIVEE, JULES A., "A History of Computers I," *Mathematics Teacher* LI, October 1958, pp. 469–473.

SAWYER, W. W., *The Search for Pattern*. New York: Penguin Books, 1970.

———, *Scientific American*, June 1975, p. 49.

SCOTT, J. F., *A History of Mathematics*. London: Taylor & Francis, 1960.

SMITH, DAVID E., *History of Mathematics* (2 vols.). New York: Dover, 1923.

SMITH, KARL J., *The Nature of Modern Mathematics*, 3rd ed. Belmont, Calif.: Wadsworth, 1980.

WAMPLER, J. F., "The Concept of Function," *Mathematics Teacher* LIII, November 1960, pp. 581–583.

WEST, BEVERLY HENDERSON, et al., *The Prentice-Hall Encyclopedia of Mathematics*. Englewood Cliffs, N.J.: Prentice Hall, 1982.

Answers

CHAPTER 1

Chapter 1 Pretest

1. 17.4 **2.** $\frac{11}{8}$ **3.** > **4.** -17 **5.** -5.8 **6.** $-\frac{1}{21}$ **7.** -37 **8.** 8.7 **9.** $-\frac{47}{24}$ or $-1\frac{23}{24}$ **10.** -74.88 **11.** -1 **12.** -168 **13.** 8 **14.** 8 **15.** -9 **16.** $-\frac{40}{147}$ **17.** $-\frac{4}{3}$ or $-1\frac{1}{3}$ **18.** 0 **19.** $\dfrac{8}{-7}$ **20.** $\frac{1}{12}$ **21.** 6 **22.** $-12x + 32$ **23.** $35y - 45$ **24.** $-22x - 2y$ **25.** $-3a - 19b$ **26.** $6a - 8$ **27.** $-2x + 11$ **28.** $46x - 75$ **29.** -4 **30.** $\dfrac{27x^9}{y^9}$ **31.** $-\dfrac{3x^6}{2y^4}$ **32.** $\dfrac{x^{12}}{-27y^{15}}$ **33.** $\dfrac{81x^{44}}{y^{24}}$ **34.** 7.54×10^7 **35.** 0.0000052

Problem Set 1.1

1. 3 **3.** $-2, 3, 0$ **5.** 3, 0, 4 **7.** $-8, 3, -\frac{5}{4}, 6.2, 0, 5\frac{3}{7}, 4, 9.1$ **9.** {A, p, r, i, l} **11.** {0, 2, 4, 6, 8, 10} **13.** $\{\ldots, -4, -2, 0, 2, 4, \ldots\}$ **15.** $\{x \mid x$ is a whole number less than or equal to 6$\}$ **17.** $\{x \mid x$ is a natural number$\}$ **19.** $\{x \mid x$ is a real number less than $-4\}$ **21.** 9 **23.** 2.1 **25.** $\frac{3}{4}$ **27.** -7 **29.** $-\frac{9}{5}$ **31.** 18 **33.** > **35.** > **37.** < **39.** > **41.** < **43.** < **45.** > **47.** > **49.** < **51.** -7 **53.** 9.4 **55.** $\frac{1}{5}$ **57.** -15 **59.** Natural **61.** Absolute value **63.** True **65.** False **67.** True **69.** $\dfrac{-9}{5}, -1.15, 0.09, 0.235, \frac{5}{8}, \frac{2}{3}, 0.74, \frac{87}{100}$

Problem Set 1.2

1. 15 **3.** -12 **5.** 2 **7.** 0 **9.** -7 **11.** -10 **13.** 5.3 **15.** -3.8 **17.** $\frac{5}{12}$ **19.** 0 **21.** $-\frac{5}{6}$ **23.** -9.0 **25.** -10 **27.** -1 **29.** 11 **31.** 35 **33.** -1.1 **35.** -21.4 **37.** $-\frac{9}{14}$ **39.** $\frac{1}{4}$ **41.** -1.3 **43.** -106 **45.** -63 **47.** $-\frac{25}{72}$ **49.** $\frac{26}{15}$ or $1\frac{11}{15}$ **51.** 13°F **53.** 13,962 ft **55.** Commutative law of addition **57.** Associative law of addition **59.** Commutative law of addition **61.** Commutative law of addition **63.** Commutative law of addition **65.** -96 **67.** Subtract, larger **69.** Commutative **71.** -150.112 **73.** -28.732 **75.** 141.211 **77.** -158.05 **79.** 35.75 **81.** -47.436 **83.** -8.053 **85.** -22.0369 **87.** -729.40 **89.** -3 **91.** 1 **93.** $-\frac{3}{4}$ **95.** -17

Problem Set 1.3

1. 6 **3.** 35 **5.** 63 **7.** -20 **9.** -56 **11.** -8 **13.** -6.5 **15.** $-\frac{6}{7}$ **17.** $\frac{51}{8}$ **19.** 0 **21.** -84 **23.** 48 **25.** -3 **27.** 4 **29.** -5 **31.** 9 **33.** -3 **35.** 0 **37.** Undefined **39.** $\frac{7}{2}$ **41.** -3 **43.** $\frac{2}{11}$ **45.** $\frac{1}{12}$ **47.** $\dfrac{1}{-9}$ **49.** $\dfrac{-33}{40}$ **51.** $\dfrac{-9}{14}$ **53.** $\frac{2}{3}$ **55.** -19.4 **57.** 5 **59.** -40 **61.** $-\frac{7}{5}$ **63.** $\frac{7}{18}$ **65.** -6 **67.** Undefined **69.** $-\frac{3}{8}, \dfrac{3}{-8}$ **71.** $-\frac{11}{7}, \dfrac{-11}{7}$ **73.** $\dfrac{-5}{4}, \dfrac{5}{-4}$ **75.** Associative **77.** Invert **79.** 35.755 **81.** -396.610 **83.** -13.840 **85.** 382,097.56 **87.** -1.865 **89.** 0.554 **91.** -1.554 **93.** 2307.946 **95.** -1727.749 **97.** $\frac{4}{3}$ **99.** 0 **101.** $\frac{25}{102}$ **103.** $-\frac{21}{16}$

Problem Set 1.4

1. 1 **3.** -10 **5.** 24 **7.** $2x + 6$ **9.** $5x - 10$ **11.** $-12x - 20$ **13.** $-16y + 56$ **15.** $5ax + 10ay - 5az$ **17.** Terms: $-2y$, 4; coefficients: -2, 4 **19.** Terms: 7, $-\dfrac{y}{4}$; coefficients: 7, $-\dfrac{1}{4}$ **21.** Terms: $4x$, $-6y$, 10; coefficients: 4, -6, 10 **23.** $(12 + 7)x$ **25.** $(15 - 4)y$ **27.** $(-5 + 9)a$ **29.** $(6 - 11)x$ **31.** $(-8 - 9)a$ **33.** $(5 - 3 - 8)x$ **35.** $(\frac{1}{2} - \frac{3}{8})x$ **37.** $6x$ **39.** $12a$ **41.** $9x$ **43.** $-4y$ **45.** $-5y$ **47.** $9y$ **49.** $-4x$ **51.** $10a + 2b$ **53.** $4x - 3$ **55.** $8z - 5$ **57.** $-x - 2$ **59.** $-y + 2$ **61.** $-2a + 3b - c$ **63.** $1 + 5a$ **65.** $3x - 4$ **67.** $10a - 5$ **69.** $-x + 10$ **71.** 12 **73.** $3x + 6$ **75.** $-44 + 48y$

77. $20x + 32$ **79.** $-26 + 8x$ **81.** $5x - 10$ **83.** Numerical coefficient **85.** Negative 1 **87.** 74.074
89. 178.125 **91.** $12.99x - 1.688y$ **93.** $\frac{15}{52}$ **95.** $-\frac{8}{9}$

Problem Set 1.5

1. 7^4 **3.** x^5 **5.** $(2a)^3$ **7.** 8 **9.** -1 **11.** $9y^2$ **13.** $8x^3y^3$ **15.** 2 **17.** 1 **19.** $1, x \neq 0$ **21.** -17
23. 14 **25.** 58 **27.** 3 **29.** 6 **31.** 5 **33.** $\frac{7}{4}$ **35.** $\frac{1}{2}$ **37.** $-\frac{17}{3}$ **39.** $\frac{-5}{2}$ **41.** $\frac{1}{4}$ **43.** 1 **45.** 20
47. 4 **49.** 413 **51.** -38 **53.** 36 **55.** -78 **57.** -158 **59.** Base, exponent **61.** Cubed, power
63. 467.876 **65.** 16.728 **67.** -143.489 **69.** -0.389 **71.** 0.060 **73.** 115.743 **75.** -2.78 **77.** -7.28
79. 0.876 **81.** 6.104 **83.** -0.702 **85.** -1.653 **87.** $(-4)^3$ **89.** $\frac{-1}{64}$ **91.** -16 **93.** 9 **95.** $\frac{53}{20}$

Problem Set 1.6

1. $\frac{1}{4^3}$ **3.** $\frac{1}{y^4}$ **5.** $\frac{1}{-3x}$ **7.** 3^{-4} **9.** $(-3)^{-5}$ **11.** $(6y)^{-7}$ **13.** 3^{12} **15.** $\frac{1}{6^6}$ **17.** $\frac{1}{x^5}$ **19.** $\frac{1}{a^3}$ **21.** $6x^3$
23. $-21x^7y^4$ **25.** 4^3 or 64 **27.** x^{11} **29.** a^2 **31.** $\frac{1}{10^5}$ **33.** $-2x$ **35.** $\frac{-5y^3}{x^5}$ **37.** $\frac{-x^2}{4y^3}$ **39.** $\frac{1}{x^{12n}}$ **41.** a^{13p}
43. x^t **45.** $\frac{2y^{10}}{3x^9}$ **47.** $-\frac{4}{3a^{12}b}$ **49.** 5^{n-1} **51.** 2^{12} **53.** x^{12} **55.** $\frac{1}{7^{12}}$ **57.** $9x^4y^6$ **59.** $\frac{64y^2}{x^4}$ **61.** $\frac{-y^{25}}{x^{15}}$
63. $\frac{1}{a^{18}b^{30}}$ **65.** $\frac{8y^{12}}{x^{24}}$ **67.** $\frac{4x^6}{y^{10}}$ **69.** 7^{4x} **71.** 4^{12pq} **73.** $\frac{49y^{16}}{25x^{16}}$ **75.** $\frac{8^3x^{24}}{4^6y^{30}}$ **77.** $(-60)^8a^{32}b^{16}$ **79.** $4^9x^{54}y^{18}$
81. Subtract **83.** Numerator, denominator **85.** 390,625 **87.** 729 **89.** 384.16 **91.** $(\frac{1}{5})^5$ **93.** $\frac{y^9}{9}$
95. $\frac{a^9c^3}{b^3}$ **97.** $\frac{b^{12}}{27a^{27}}$ **99.** $\left(\frac{a}{b}\right)^{-n} = \frac{1}{\left(\frac{a}{b}\right)^n} = \frac{1}{\frac{a^n}{b^n}} = 1 \cdot \frac{b^n}{a^n} = \frac{b^n}{a^n} = \left(\frac{b}{a}\right)^n$

Problem Set 1.7

1. 3.72×10^3 **3.** 5.04×10^4 **5.** 1.34×10^{-2} **7.** 1.38×10^{-1} **9.** 3.9×10^7 **11.** 2.3×10^{-9}
13. 5×10^{-11} **15.** 452 **17.** 66.9 **19.** 0.03356 **21.** 0.203 **23.** 523,000,000 **25.** 0.0000009408
27. 2.496×10^7 **29.** 1.24×10^6 **31.** 8.066×10^{-10} **33.** 3×10^2 **35.** 5×10^{-3} **37.** 3×10
39. 9.3×10^7 **41.** 0.000066 **43.** 1.8784×10^{14} miles **45.** 16,000 light years **47.** 1.822×10^{-21} grams
49. Scientific notation **51.** Right **53.** 4.4×10^{-2} **55.** 2.1×10^6 **57.** 2.6×10^2 **59.** 4×10^{-8}
61. 4×10^5 **63.** 2×10^2 **65.** 4.5×10^3 **67.** 1.06×10^5 (rounded)

Chapter 1 Additional Exercises

1. 0, 9 **3.** $-4, \frac{11}{8}, -\frac{3}{5}, 0, 9$ **5.** -7 **7.** 4 **9.** -10 **11.** 7 **13.** 8 **15.** $\frac{3}{5}$ **17.** $\frac{1}{15}$ **19.** $-\frac{11}{16}$
21. 7.6 **23.** -4 **25.** 0 **27.** -8 **29.** Associative law of addition **31.** Commutative law of addition
33. 30.72 **35.** -30 **37.** 3 **39.** $-\frac{1}{4}$ **41.** 2 **43.** $\frac{1}{6}$ **45.** -2 **47.** 0 **49.** $x, -2y, 4$ **51.** $x, 3y, -2$
53. $4x + 17$ **55.** $48y + 18$ **57.** $-8x^3y^3$ **59.** $16x^2y^2$ **61.** $25y^2$ **63.** $\frac{8}{7}$ **65.** 17 **67.** $\frac{9}{x^4}$ **69.** 3^4
71. $\frac{9}{8x^{10}y^6}$ **73.** 1.86×10^5 **75.** 886,000,000

Chapter 1 Cooperative Learning

1. $-\frac{11}{4}, \frac{11}{4}$ **3.** $-\$2,360,000$, $-\$472,000$ **5.** -4°F **7.** $7p + 7$ **9.** 9.855×10^7

Chapter 1 Practice Test

1. 5.3 **2.** $\frac{5}{6}$ **3.** $<$ **4.** -7 **5.** -3.1 **6.** $\frac{1}{12}$ **7.** -16 **8.** 0.8 **9.** $-\frac{10}{21}$ **10.** 17.02 **11.** $-\frac{7}{10}$
12. -144 **13.** 3 **14.** -5 **15.** -8 **16.** $\frac{14}{27}$ **17.** -6 **18.** 0 **19.** $\frac{8}{3}$ **20.** $-\frac{1}{4}$ **21.** -26
22. $14x - 28$ **23.** $-6y + 8$ **24.** $-2b - 3c$ **25.** $x - 7y$ **26.** $6a - 4$ **27.** $x - 2$ **28.** $24x + 4$ **29.** $-\frac{5}{4}$
30. $\frac{8x^5}{y^7}$ **31.** $\frac{-4x}{3y^5}$ **32.** $\frac{x^6}{16y^4}$ **33.** $\frac{y^8}{4x^{14}}$ **34.** 8.325×10^6 **35.** 0.0047 **36.** 5.019×10^{-21} gram

CHAPTER 2

Chapter 2 Pretest

1. 56 **2.** $-\frac{7}{3}$ **3.** 7 **4.** $-\frac{11}{2}$ **5.** 19 m, 23 m **6.** 87, 89, 91 **7.** $4000 **8.** $y = \dfrac{3x + 7}{5}$

9. $\{x \mid x \leq 5\}$

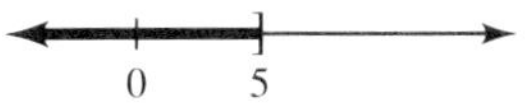

10. $\{x \mid x < -5\}$ **11.** $\{x \mid x \leq -\frac{19}{8}\}$ **12.** $2, -\frac{10}{3}$ **13.** $\varnothing$ **14.** $\{\frac{7}{2}\}$

15. $\{n \mid n < 100\}$

16. $\{x \mid -1 < x < 7\}$

17. $\{x \mid x \leq -\frac{2}{5} \text{ or } x \geq 2\}$

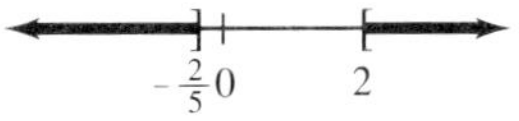

Problem Set 2.1

1. 12 **3.** 42 **5.** −23 **7.** −3.6 **9.** $\frac{11}{8}$ **11.** 6 **13.** $-\frac{21}{2}$ **15.** 2 **17.** $-\frac{11}{3}$ **19.** 20 **21.** 25 **23.** −3 **25.** −13 **27.** 5 **29.** 0 **31.** 4 **33.** $-\frac{9}{5}$ **35.** 3 **37.** $\frac{5}{3}$ **39.** 8 **41.** 6 **43.** 1 **45.** −3 **47.** 2 **49.** $-\frac{10}{3}$ **51.** 5 **53.** −16 **55.** $-\frac{1}{2}$ **57.** −2.5 **59.** −2 **61.** 5 **63.** −6 **65.** Equation **67.** Linear **69.** Add **71.** 3.66 **73.** 3 **75.** 5.27 **77.** 7 **79.** −6 **81.** $\frac{23}{66}$ **83.** 0, 3 **85.** $-\frac{11}{5}, -1, 0, 3, \frac{15}{4}$ **87.** > **89.** <

Problem Set 2.2

1. $x + 5$ **3.** $x + 10$ **5.** $x - 8$ **7.** $-5x$ **9.** $\frac{1}{5}x$ **11.** $\dfrac{x}{8}$ **13.** $4x - 7$ **15.** $4(x - 7)$ **17.** $\dfrac{x}{24} - 10$

19.
$$\begin{aligned} x - 9 &= -4 \\ x &= -4 + 9 \\ x &= 5 \end{aligned}$$
The number is 5.

21.
$$\begin{aligned} 5x + 16 &= 13x \\ 16 &= 13x - 5x \\ 16 &= 8x \\ 2 &= x \end{aligned}$$
The number is 2.

23.
$$\begin{aligned} 8 - \tfrac{3}{4}x &= 14 \\ -\tfrac{3}{4}x &= 14 - 8 \\ -\tfrac{3}{4}x &= 6 \\ x &= -8 \end{aligned}$$
The number is −8.

25.
$$\begin{aligned} 2L + 2W &= P \\ L &= 3W \\ 2(3W) + 2W &= 128 \\ 6W + 2W &= 128 \\ 8W &= 128 \\ W &= 16 \end{aligned}$$
$L = 3W = 3(16) = 48$
The width is 16 m and the length is 48 m.

27.
$$\begin{aligned} 2L + 2W &= P \\ W &= L - 20 \\ 2L + 2(L - 20) &= 248 \\ 2L + 2L - 40 &= 248 \\ 4L &= 248 + 40 \\ 4L &= 288 \\ L &= 72 \end{aligned}$$
$W = L - 20 = 72 - 20 = 52$
The width is 52 cm and the length is 72 cm.

29.
$$\begin{aligned} x + (x + 1) + (x + 2) &= 102 \\ x + x + 1 + x + 2 &= 102 \\ 3x + 3 &= 102 \\ 3x &= 102 - 3 \\ 3x &= 99 \\ x &= 33 \\ x + 1 &= 34 \\ x + 2 &= 35 \end{aligned}$$
The integers are 33, 34, and 35.

31.
$$\begin{aligned} 7x &= 6(x + 2) \\ 7x &= 6x + 12 \\ 7x - 6x &= 12 \\ x &= 12, x + 2 = 14 \end{aligned}$$
The integers are 12 and 14.

33.
$$\begin{aligned} x + 0.08x &= 77{,}760 \\ 1.08x &= 77{,}760 \\ x &= 72{,}000 \end{aligned}$$
The cost of the house was $72,000.

35.
$$\begin{aligned} x + (x + 4660) &= 6740 \\ x + x + 4660 &= 6740 \\ 2x + 4660 &= 6740 \\ 2x &= 2080 \\ x &= 1040 \end{aligned}$$
The cost of electricity was $1040.

37.
$$\begin{aligned} x - 0.16x &= 11{,}760 \\ 0.84x &= 11{,}760 \\ x &= 14{,}000 \end{aligned}$$
The price of the car was $14,000.

39.
$$\begin{aligned} x\% \cdot 73 &= 24.82 \\ x(0.01)73 &= 24.82 \\ 0.73x &= 24.82 \\ x &= 34 \end{aligned}$$
34% of 73 is 24.82.

41.
$$\begin{aligned} 7x - 18 &= 6x + 8 \\ x - 18 &= 8 \\ x &= 26 \end{aligned}$$
The number is 26.

43.
$$\begin{aligned} x + x + 2 + 2x &= 26 \\ 4x + 2 &= 26 \\ 4x &= 24 \\ x &= 4 \\ x + 2 &= 6, 2x = 8 \end{aligned}$$
The lengths are 4 ft, 6 ft, and 8 ft.

45.
$$\begin{aligned} 3x - 12 &= 2(x + 2) - 1 \\ 3x - 12 &= 2x + 4 - 1 \\ 3x - 12 &= 2x + 3 \\ x - 12 &= 3 \\ x &= 15 \\ x + 2 &= 17 \end{aligned}$$
The numbers are 15 and 17.

47.
$$\begin{aligned} x - 33\% \cdot x &= 43.55 \\ x - 33(0.01)x &= 43.55 \\ x - 0.33x &= 43.55 \\ 0.67x &= 43.55 \\ x &= 65 \end{aligned}$$
The original price was $65.

49. $151.04 = x\% \cdot 236$
$151.04 = x(0.01)236$
$151.04 = 2.36x$
$64 = x$
151.04 is 64% of 236.

51. Consecutive **53.** 16.285 m by 19.785 m **55.** \$258,064.52 **57.** 9

59. 42 ft by 48 ft **61.** 10 **63.** Commutative law of addition **65.** Associative law of addition
67. Commutative law of addition **69.** Associative law of addition **71.** Associative law of addition

Problem Set 2.3

1.
$3x = 9(16 - x)$
$3x = 144 - 9x$
$12x = 144$
$x = 12$
$16 - x = 16 - 12 = 4$
The numbers are 4 and 12.

3. $3x - (1 - x) = 55$
$3x - 1 + x = 55$
$4x - 1 = 55$
$4x = 56$
$x = 14$
$1 - x = 1 - 14 = -13$
The numbers are 14 and -13.

5. $0.05x + 0.25(29 - x) = 4.45$
$5x + 25(29 - x) = 445$
$5x + 725 - 25x = 445$
$-20x + 725 = 445$
$-20x = -280$
$x = 14$
$29 - x = 29 - 14 = 15$
Susan has 14 nickels and 15 quarters.

7. $0.05x + 0.10(4x) = 6.75$
$5x + 10(4x) = 675$
$5x + 40x = 675$
$45x = 675$
$x = 15$
$4x = 4(15) = 60$
Ken has 15 nickels and 60 dimes.

9. $0.06x + 0.07(14{,}000 - x) = 900$
$6x + 7(14{,}000 - x) = 90{,}000$
$6x + 98{,}000 - 7x = 90{,}000$
$-x + 98{,}000 = 90{,}000$
$-x = -8000$
$x = 8000$
$14{,}000 - x = 14{,}000 - 8000 = 6000$
Kevin invested \$8000 at 6% and \$6000 at 7%.

11. $0.09x + 0.06(9000 - x) = 630$
$9x + 6(9000 - x) = 63{,}000$
$9x + 54{,}000 - 6x = 63{,}000$
$3x + 54{,}000 = 63{,}000$
$3x = 9000$
$x = 3000$
$9000 - x = 9000 - 3000 = 6000$
\$3000 was invested at 9%. \$6000 was invested at 6%.

13. $0.10x + 0.06(8000 - x) = 600$
$10x + 6(8000 - x) = 60{,}000$
$10x + 48{,}000 - 6x = 60{,}000$
$4x + 48{,}000 = 60{,}000$
$4x = 12{,}000$
$x = 3000$
$8000 - x = 8000 - 3000 = 5000$
\$3000 was invested at 10%. \$5000 was invested at 6%.

15. $x - (71 - x) = 3$
$x - 71 + x = 3$
$2x - 71 = 3$
$2x = 74$
$x = 37$
$71 - x = 71 - 37 = 34$
Maria is 34.

17. $2(45 - x) - 6 = x$
$90 - 2x - 6 = x$
$-2x + 84 = x$
$84 = 3x$
$28 = x$
$45 - x = 45 - 28 = 17$
The larger number is 28.

19. $0.10x + 0.25(3x) = 10.20$
$10x + 25(3x) = 1020$
$10x + 75x = 1020$
$85x = 1020$
$x = 12$
$3x = 36$
Karen has 36 quarters.

21. $0.10x + 0.05(47 - x) = 3.90$
$10x + 5(47 - x) = 390$
$10x + 235 - 5x = 390$
$5x + 235 = 390$
$5x = 155$
$x = 31$
There are 31 dimes in the collection.

23. $0.09x + 0.08(3x) = 495$
$9x + 8(3x) = 49{,}500$
$9x + 24x = 49{,}500$
$33x = 49{,}500$
$x = 1500$
\$1500 was invested at 9%.

25. $3x + 8 = 4(15 - x) - 3$
$3x + 8 = 60 - 4x - 3$
$3x + 8 = 57 - 4x$
$7x = 49$
$x = 7$
$15 - x = 8$
The smaller number is 7.

27. $x + 2x + (2x + 5) = 180$
$x + 2x + 2x + 5 = 180$
$5x + 5 = 180$
$5x = 175$
$x = 35$
$2x = 70,\ 2x + 5 = 2(35) + 5 = 75$
The measures of the angles are 35°, 70°, and 75°.

29. $0.01x + 0.05(79 - x) = 1.99$
$x + 5(79 - x) = 199$
$x + 395 - 5x = 199$
$-4x + 395 = 199$
$-4x = -196$
$x = 49$
He has 49 pennies.

31. $0.05x + 0.10(2x) + 0.25(48 - 3x) = 5.50$
$5x + 10(2x) + 25(48 - 3x) = 550$
$5x + 20x + 1200 - 75x = 550$
$-50x + 1200 = 550$
$-50x = -650$
$x = 13$
$2x = 26$
There are 26 dimes in the bank.

33. $0.06x + 0.08(x + 3000) + 0.09[13{,}000 - (2x + 3000)] = 980$
$6x + 8(x + 3000) + 9[13{,}000 - 2x - 3000] = 98{,}000$
$6x + 8(x + 3000) + 9[10{,}000 - 2x] = 98{,}000$
$6x + 8x + 24{,}000 + 90{,}000 - 18x = 98{,}000$
$-4x + 114{,}000 = 98{,}000$
$-4x = -16{,}000$
$x = 4000$
$x + 3000 = 7000$
\$7000 was invested at 8% interest per year.

35. $I = Prt$ **37.** \$128,000 **39.** 2 pounds of turkey, $2\frac{1}{2}$ pounds of roast beef **41.** 12% **43.** 24 **45.** 5 **47.** 8 **49.** 3 **51.** 6 **53.** 4.5 **55.** 4

Problem Set 2.4

1. $N = L - D$ or $-D + L = N$ **3.** $r = \dfrac{A - P}{Pt}$ **5.** $t = \dfrac{d}{r}$ **7.** $R = \dfrac{P}{B}$ **9.** $L = \dfrac{A}{W}$ **11.** $r = \dfrac{C}{2\pi}$ **13.** $P = \dfrac{I}{rt}$ **15.** $L = \dfrac{V}{WH}$ **17.** $b = \dfrac{2A}{h}$ **19.** $b = \dfrac{2A - hc}{h}$ **21.** $S = C - nR$ **23.** $y = \dfrac{c - 9}{b}$ **25.** $x = \dfrac{a + c}{-b}$ **27.** $x = \dfrac{15 - h}{g}$ **29.** $y = \dfrac{-3x + 7}{4}$ **31.** $y = \dfrac{-7x + 5}{-2}$ **33.** $y = 2x - 8$ **35.** $y = \dfrac{-Ax + c}{B}$ **37.** $a^2 = c^2 - b^2$ **39.** $r^3 = \dfrac{\frac{3}{4}V}{\pi}$ or $r^3 = \dfrac{3V}{4\pi}$ **41.** $\frac{18}{5}$ or $3\frac{3}{5}$ hours **43.** $11\frac{5}{7}$ in. **45.** Formula **47.** 7.4 **49.** 3.64 **51.** $F = \dfrac{S - PS}{P}$ **53.** $M = \dfrac{6V - bh - Bh}{4h}$ **55.** $h = \dfrac{A - 2\pi r^2}{2\pi r}$ **57.** $-5x - 4$ **59.** $-70a + 15$ **61.** $28x - 16$ **63.** 9 **65.** -11

Problem Set 2.5

1. $\{x \mid x > 3\}$;

3. $\{x \mid x \le 4\}$;

5. $\{x \mid x < 4\}$;

7. $\{x \mid x \ge -\frac{1}{2}\}$;

9. $\{x \mid -5 \le x \le -1\}$;

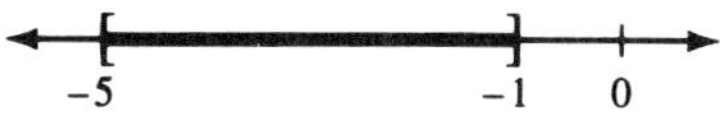

11. $\{x \mid 3 < x \le 5\}$;

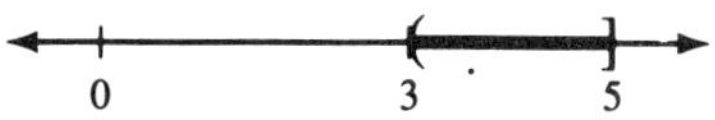

13. $\{x \mid x < -3 \text{ or } x > 2\}$;

15. $\{x \mid x \le 1 \text{ or } x > 4\}$;

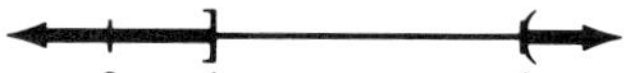

17. $\{x \mid x < -2 \text{ or } x > 5\}$;

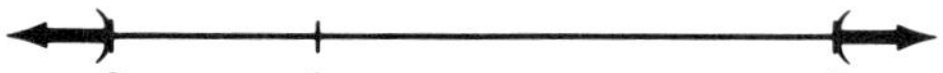

19. $\{x \mid -8 \le x \le -6\}$;

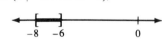

21. $\{x \mid x < -2 \text{ or } x \ge 3\}$;

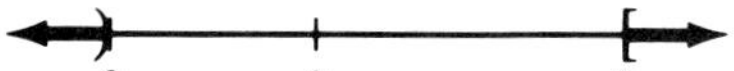

23. $\{x \mid -3 < x < 5\}$;

25. $\varnothing$;

27. $\{x \mid -4.5 < x < 5.5\}$;

29. $\{x \mid -6 \le x < 5\}$;

31. $\{x \mid x < -4 \text{ or } x > -60\}$;

33. $\{x \mid x \le -13.9 \text{ or } x > 6.5\}$;

35. $\varnothing$;

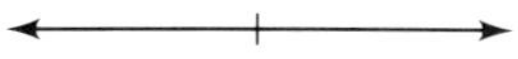

37. Greater

39. Added **41.** $\{x \mid x < 17.5554\}$ **43.** $\{x \mid -13.627 \le x < -3.4546\}$ **45.** $\{x \mid x < 6.98643 \text{ or } x > 9.06565\}$ **47.** $\{x \mid -\frac{11}{24} < x < \frac{11}{48}\}$ **49.** $\{x \mid x < -4 \text{ or } x > -3\}$ **51.** 6 **53.** 8 **55.** 2 **57.** $-\frac{15}{4}$ **59.** $-\frac{7}{4}$

Problem Set 2.6

1. $\{x \mid x < 35\}$ **3.** $\{x \mid x \ge -18\}$ **5.** $\{x \mid x < 7\}$ **7.** $\{y \mid y < -4\}$ **9.** $\{y \mid y \le 2\}$ **11.** $\{x \mid x \ge \frac{6}{5}\}$ **13.** $\{x \mid x < \frac{1}{12}\}$ **15.** $\{x \mid x < 1\}$ **17.** $\{x \mid x \ge \frac{23}{5}\}$ **19.** $\{x \mid 4 \le x < 6\}$ **21.** $\{x \mid -4 < x < 5\}$ **23.** $\{x \mid x \le 2 \text{ or } x > 3\}$ **25.** \$22,000 **27.** 15 **29.** $\{n \mid n < 200\}$ **31.** $-3 < x < 2$ **33.** \$600 **35.** 768 pounds **37.** 18 days **39.** \$20,000 **41.** Positive **43.** $\{y \mid y \le 2.21\}$ **45.** $\{y \mid y \ge 4.26\}$ **47.** $\{x \mid x \le 0.7\}$ **49.** $\{x \mid x < \frac{36}{11}\}$ **51.** $\{x \mid x < 1 \text{ or } x > 0\}$ **53.** False; for example, $-6 < -2$ and $-3 < -1$ but $18 < 2$ is false. **55.** 3 **57.** $-\frac{3}{4}$ **59.** -2.4 **61.** $-\frac{12}{5}$ **63.** -15.6

Problem Set 2.7

1. 3 **3.** 7 **5.** 76 **7.** 8 **9.** $\{8, -8\}$ **11.** $\{8, -10\}$ **13.** $\{-1, -4\}$ **15.** $\{2, -\frac{8}{5}\}$ **17.** $\{\frac{9}{2}, -2\}$ **19.** $\{-14, 2\}$ **21.** $\varnothing$ **23.** $\{\frac{2}{5}\}$ **25.** $\varnothing$ **27.** $\{5, -\frac{3}{5}\}$ **29.** $\{-8, -\frac{2}{3}\}$ **31.** $\{4, \frac{4}{3}\}$

33. $\{x \mid -5 < x < 5\}$;

35. $\{x \mid -13 \leq x \leq 3\}$;

37. $\{x \mid \frac{5}{3} \leq x \leq 3\}$;

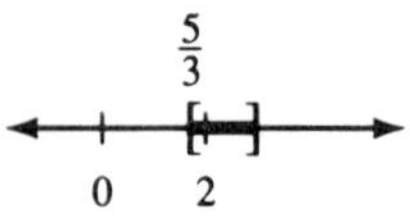

39. $\{x \mid x \geq 5 \text{ or } x \leq -5\}$;

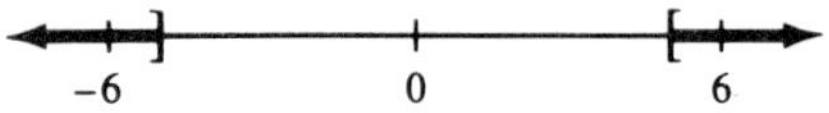

41. $\{x \mid x > 6 \text{ or } x < -2\}$;

43. $\{x \mid x \geq 3 \text{ or } x \leq -1\}$;

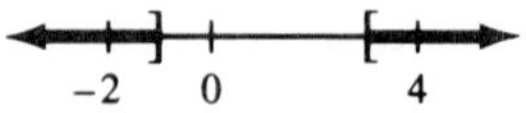

45. $\{x \mid -\frac{1}{2} < x < \frac{7}{2}\}$;

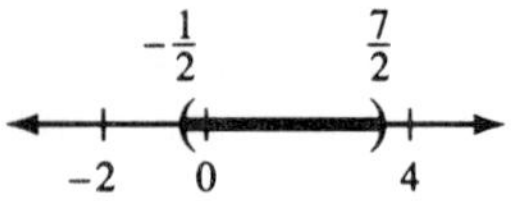

47. $\{x \mid x \geq 1 \text{ or } x \leq -\frac{3}{2}\}$;

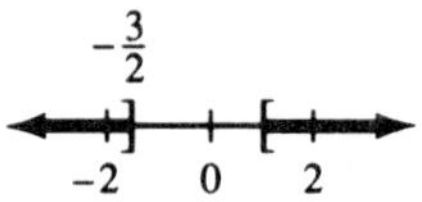

49. $\{y \mid -1 \leq y \leq \frac{7}{4}\}$;

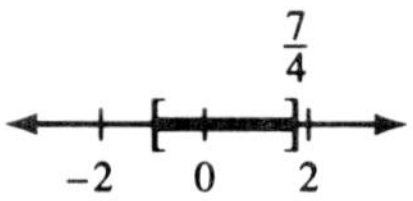

51. $\{x \mid x \text{ is a real number}\}$

53. $\varnothing$

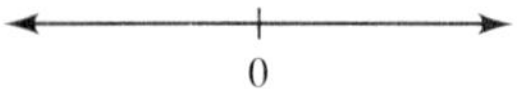

55. $\{y \mid y < -\frac{4}{3} \text{ or } y > 4\}$

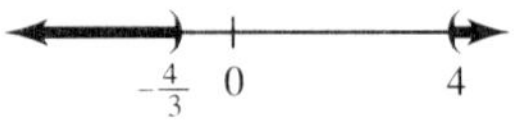

57. $\varnothing$

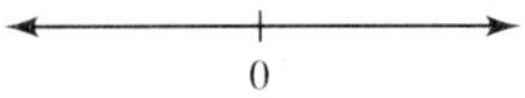

59. $\{x \mid x \text{ is a real number}\}$

61. Absolute value **63.** -1.28 or -1.80 **65.** $0.12 < x < 3.20$

67. $x \geq 8.70$ or $x \leq -22.52$ **69.** $\frac{1}{2}, \frac{1}{4}$ **71.** All real numbers except -6 **73.** $\frac{7}{3}$ **75.** -7 **77.** -4 **79.** -32

Chapter 2 Additional Exercises

1. $-\frac{29}{72}$ **3.** $-\frac{3}{2}$ **5.** $\frac{5}{6}$ **7.** $\frac{9}{10}$ **9.** 2 **11.** 2 **13.** $-\frac{31}{5}$ **15.** 18 **17.** 11 ft by 15 ft **19.** $12,000

21. 38 **23.** 9, 18 **25.** 16 nickels, 13 dimes **27.** $3000 at 6%, $6000 at 7% **29.** $h = \dfrac{S - \pi r^2}{2\pi r}$ **31.** $y = \dfrac{4 - 2x}{3}$

33. $x = \dfrac{8 - 5y}{-3}$ **35.** $y = \dfrac{7 - 4x}{-2}$ **37.** $\{x \mid x \leq -9\}$ **39.** $\{x \mid x > \frac{19}{20}\}$ **41.** $\{x \mid 5 < x \leq 6\}$

43. $\{x \mid x < -2 \text{ or } x > 9\}$ **45.** $\{x \mid x \leq -4\}$;

47. $\{x \mid x > 3\}$;

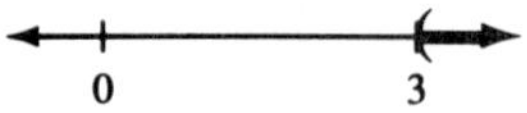

49. $\{x \mid x < 20\}$;

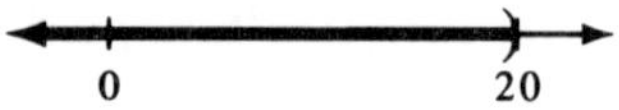

51. $\{x \mid -\frac{7}{2} \leq x < 1\}$;

53. $\{x \mid -2 < x \leq 3\}$;

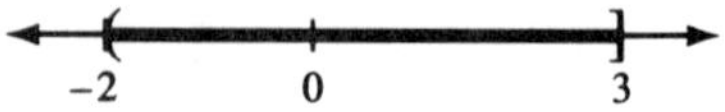

55. $\{x \mid x < 2 \text{ or } x > 6\}$;

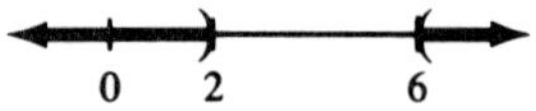

57. 96 **59.** $\{1, -5\}$ **61.** $\varnothing$ **63.** $\{4, \frac{4}{3}\}$ **65.** $\{x \mid x > 4 \text{ or } x < 1\}$

67. $\{x \mid -\frac{32}{3} < x < -\frac{8}{3}\}$ **69.** $\{x \mid x < \frac{3}{2}\}$ **71.** $\varnothing$ **73.** $\{x \mid x \text{ is a real number}\}$ **75.** $\{-\frac{3}{2}\}$

Chapter 2 Cooperative Learning

1. Example: $x + 4 = 7$ (Answers may vary.) **3.** (a) $x + \frac{23}{4} = 0$ (Answers may vary.); (b) $x - 59.6 = 0$ (Answers may vary.)

5. $C = \dfrac{N^2 - N}{2}$ **7.** 133

Chapter 2 Practice Test

1. 25 **2.** −6 **3.** −2.7 **4.** 0 **5.** 16 m, 32 m **6.** 10, 12 **7.** \$7000 **8.** $F = \frac{9C}{5} + 32$

9. $\{x \mid x \le -5\}$;

10. $\{y \mid y < -6\}$ **11.** $\{x \mid x \le -11\}$ **12.** $\{\frac{14}{3}, -4\}$ **13.** $\varnothing$ **14.** $\{-\frac{7}{2}, \frac{1}{12}\}$

15. \$1300 **16.** $\{x \mid -2 < x < 6\}$;

17. $\{x \mid x \ge \frac{7}{2} \text{ or } x \le -\frac{1}{2}\}$;

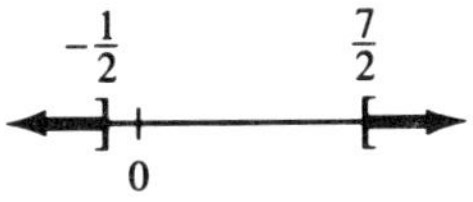

Cumulative Review Chapters 1 and 2

1. $\frac{3}{8}$ **2.** −6.4 **3.** > **4.** −6 **5.** −12.1 **6.** −3 **7.** $\frac{13}{20}$ **8.** −54 **9.** 26.52 **10.** −56 **11.** −5 **12.** 8 **13.** $-\frac{9}{7}$ **14.** 0 **15.** $\frac{9}{-5}$ **16.** $\frac{1}{7}$ **17.** −45 **18.** $-3x - 17y$ **19.** $5x - 8$ **20.** $\frac{19}{14}$ **21.** $\frac{-15x^4}{y^3}$ **22.** $\frac{4y}{x}$ **23.** $\frac{8}{x^6y^3}$ **24.** 3.54×10^{-3} **25.** $\frac{4}{7}$ **26.** 4 **27.** 7 ft by 11 ft **28.** 14 quarters, 11 dimes **29.** $\frac{7 - 3x}{2}$ **30.** $\{x \mid x \ge -2\}$ **31.** $\{6, -10\}$ **32.** $\{x \mid x \text{ is a real number}\}$ **33.** $\{1\}$ **34.** \$74,000

35. $\{x \mid -\frac{1}{5} < x < 1\}$;

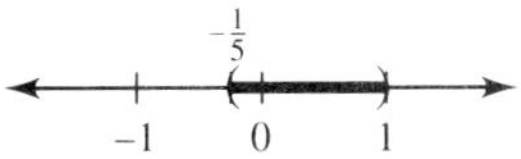

36. $\{x \mid x \ge 1 \text{ or } x \le -5\}$;

CHAPTER 3

Chapter 3 Pretest

1.

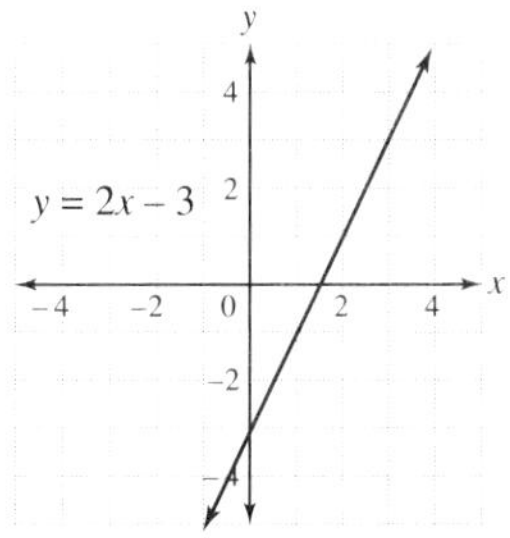

2.

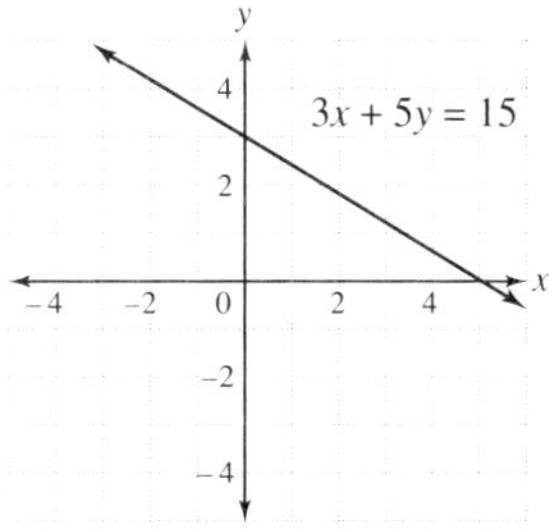

3. 4 **4.** Slope: $\frac{3}{7}$; y-intercept: −2

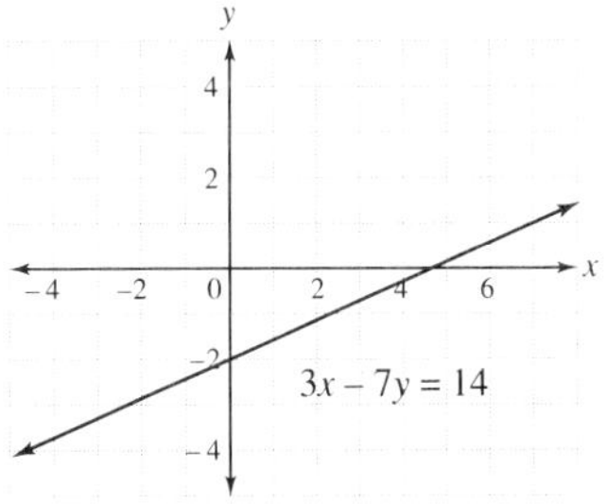

5. $y = -\frac{2}{3}x - 4$ **6.** $y = 11x + 41$ **7.** $y = 4x - 3$ **8.** $y = -\frac{3}{5}x + 1$ **9.** (a) $y = 7x - 1000$; (b) \$4950

10.

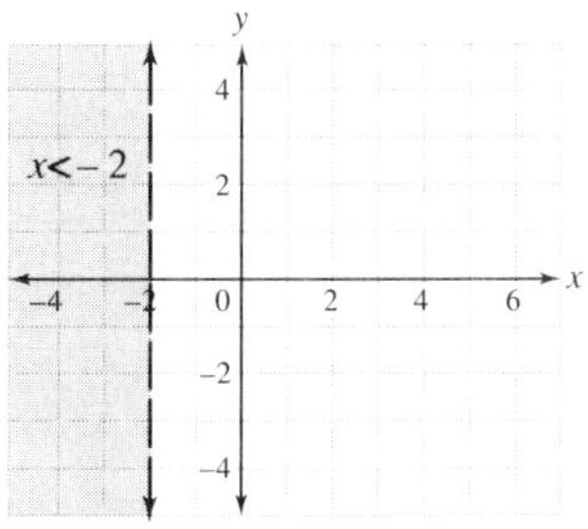

11.

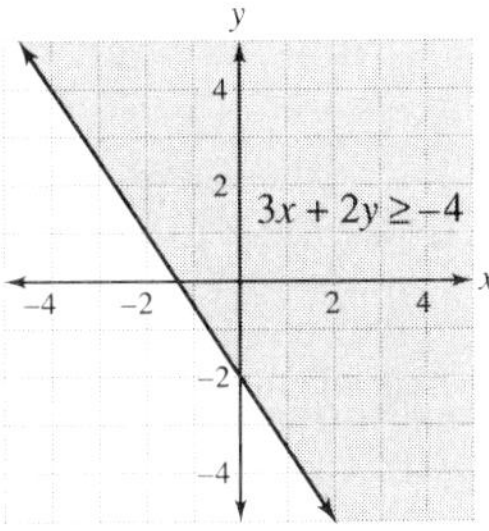

Problem Set 3.1

1, 3, 5, 7.

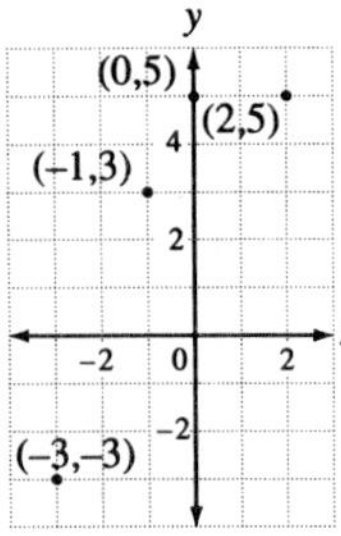

9, 11, 13, 15.

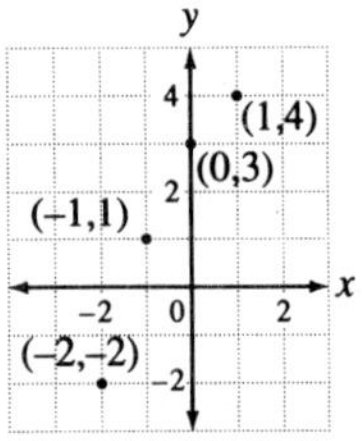

17. II
19. III
21. Yes
23. No
25. Yes

27.

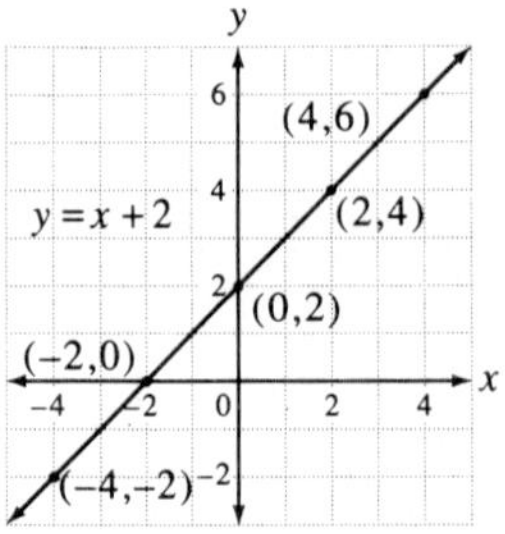

29.

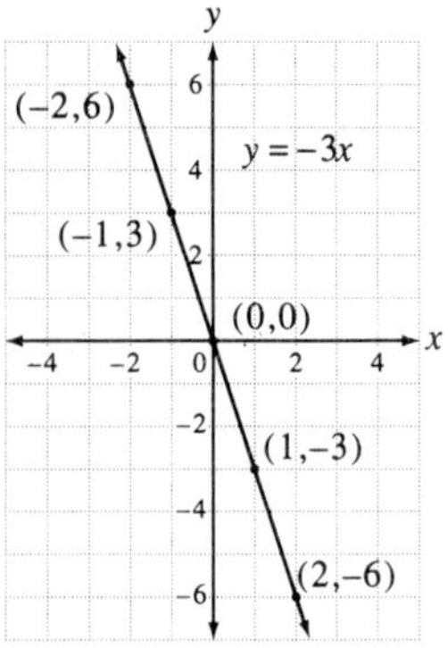

31.

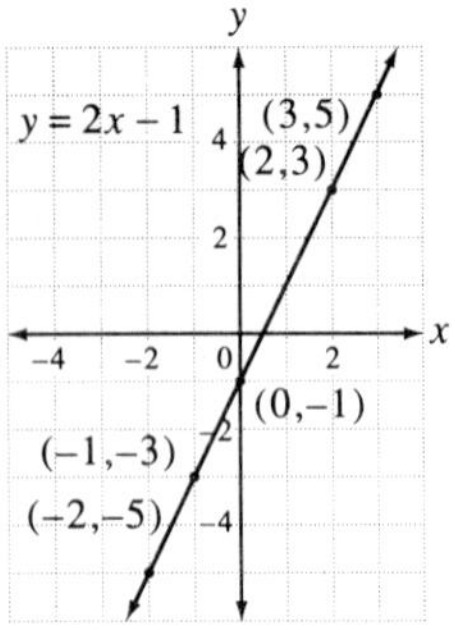

33.

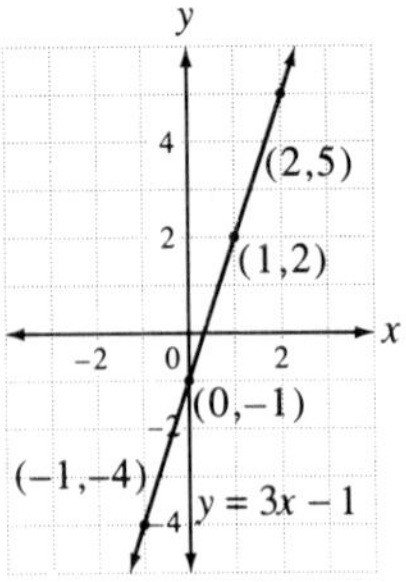

35.

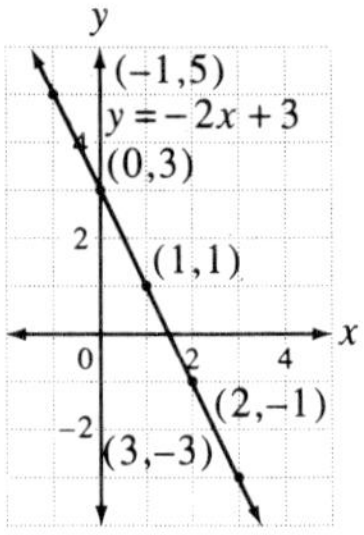

37.

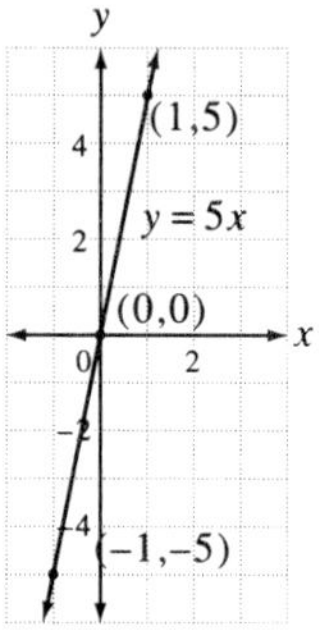

39.

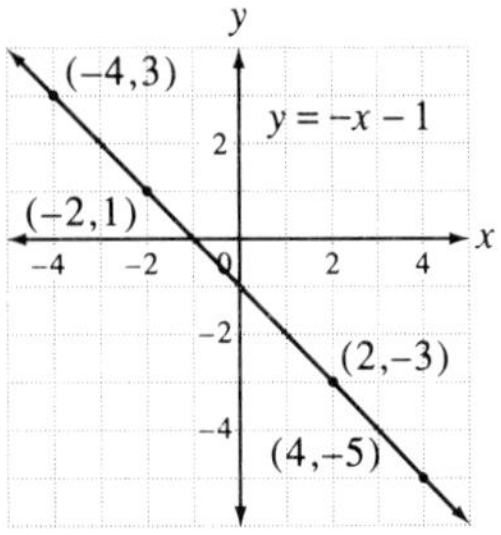

41.

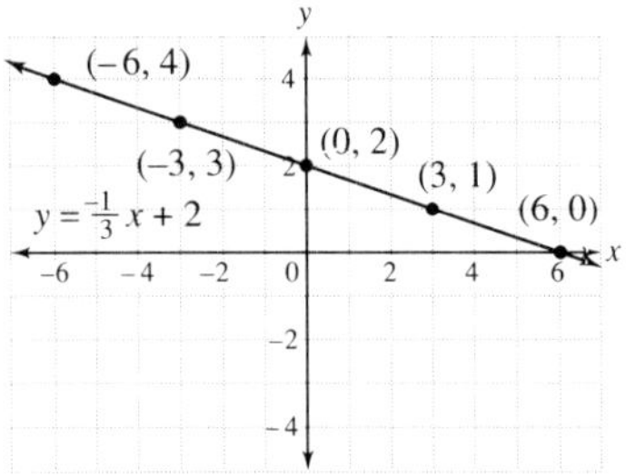

43.

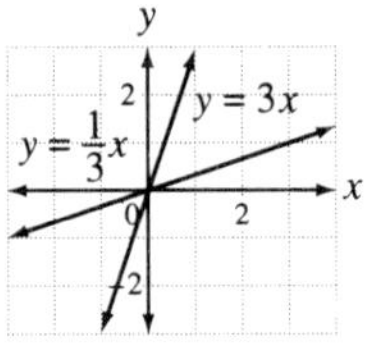

The greater the absolute value of the coefficient, the steeper the line.

45. (a) 20 psi, 30 psi, 40 psi, 50 psi; (b)

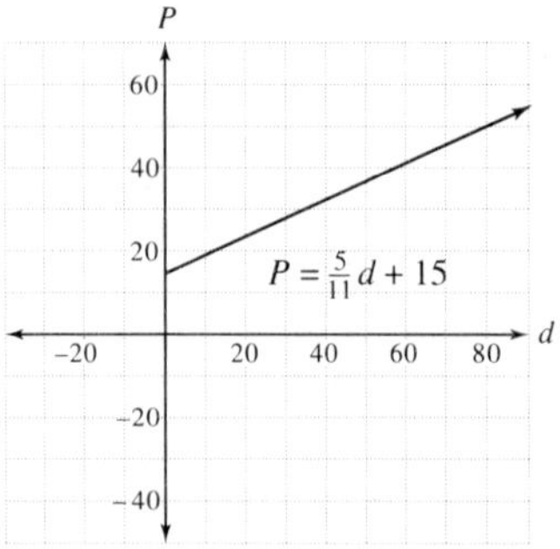

47. Rectangular coordinate
49. Origin
51. Ordered pairs
53. Yes
55. No
57. (a), (d)

59. $2y = x + 4$

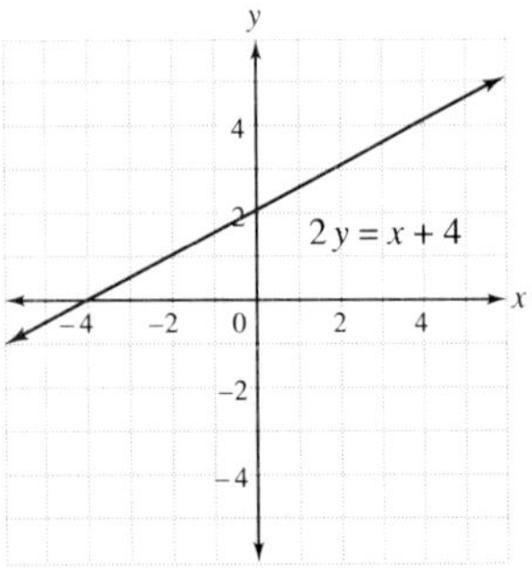

61. 27 cm
63. 6
65. 71, 72, 73
67. 15
69. 8 ft, 12 ft, 12 ft

Problem Set 3.2

1. Linear **3.** Not linear **5.** Not linear **7.** Linear **9.** Linear

11.

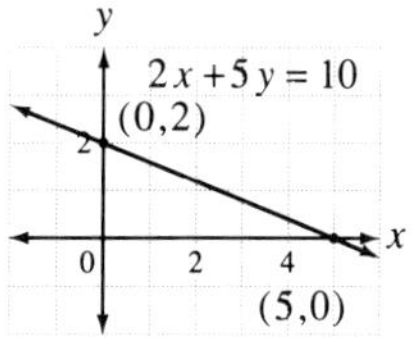

13.

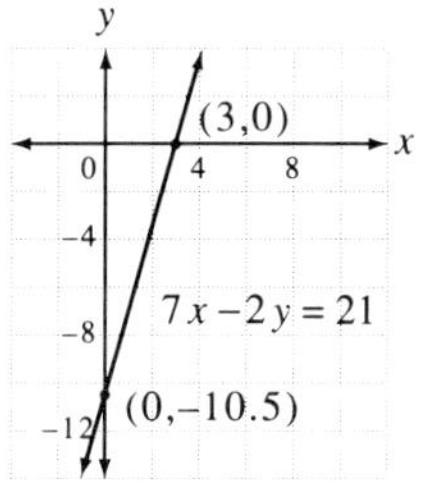

15.

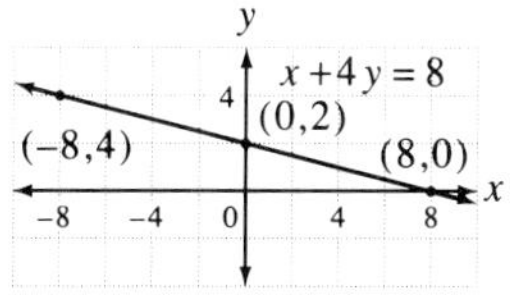

17.

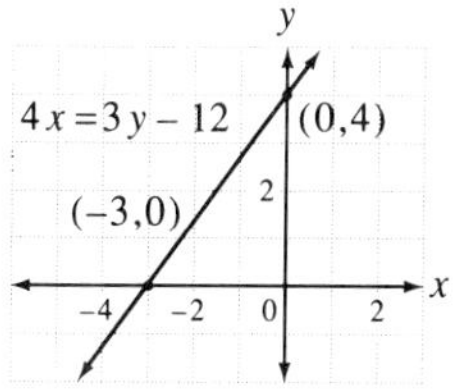

19.

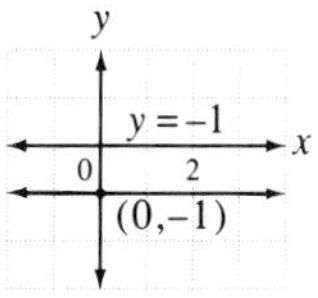

21.

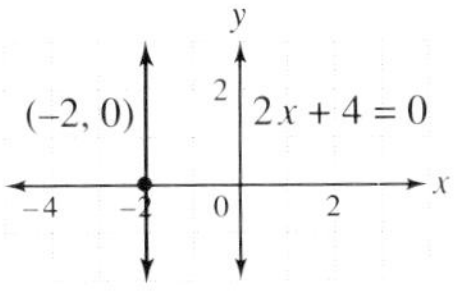

23.

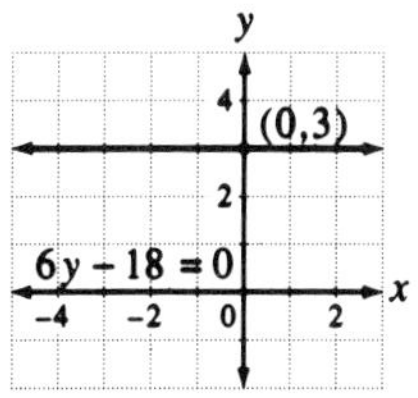

25.

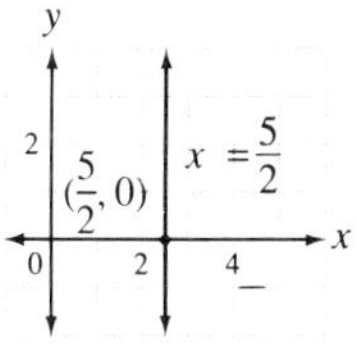

27.

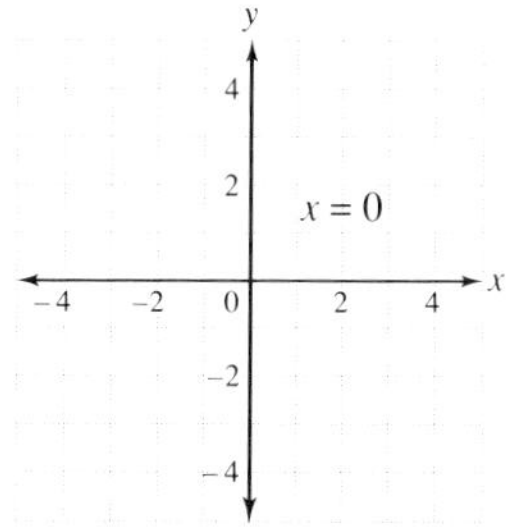

29.

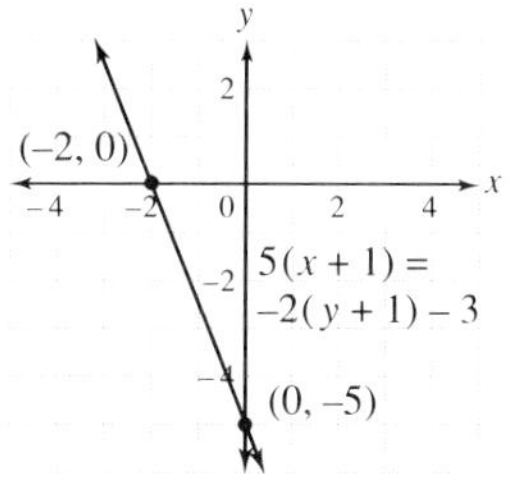

31. Vertical line **33.** Neither **35.** Horizontal line **37.** Vertical line **39.** Neither **41.** Straight line **43.** y-intercept **45.** Linear

47. $y = x - 2, x \geq 0$

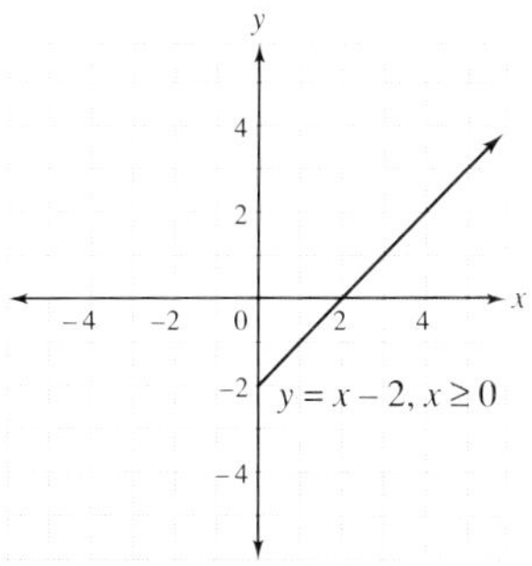

49. $y = |x|$

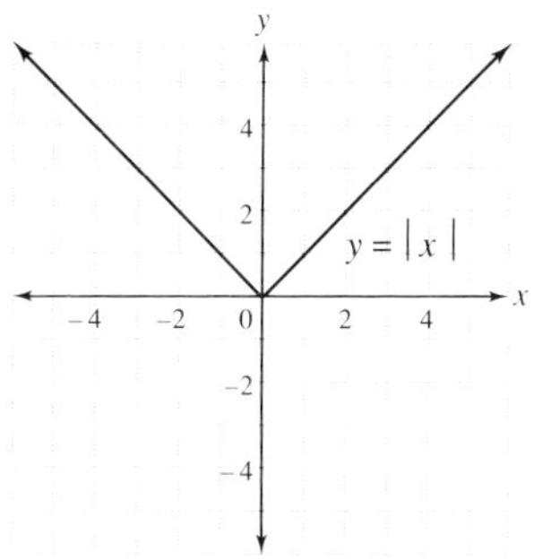

51. $y = |x - 3|$

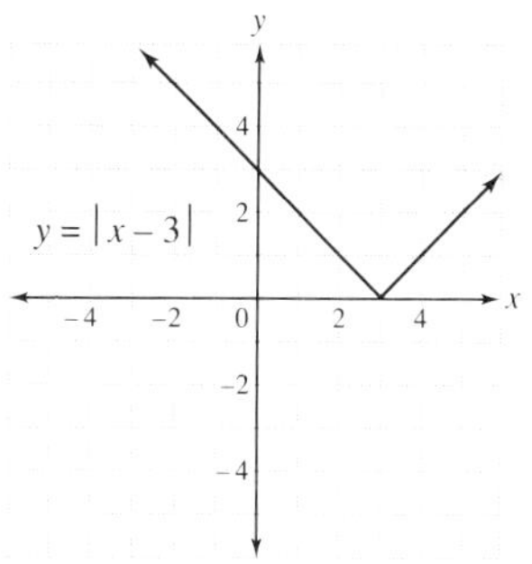

53. $\frac{3}{-5}, -\frac{3}{5}$ **55.** $\frac{-2}{3}, \frac{2}{-3}$ **57.** $\frac{-7}{9}, \frac{7}{-9}$ **59.** $-\frac{11}{4}, \frac{-11}{4}$

Problem Set 3.3

1. -8 **3.** -1 **5.** 5 **7.** Undefined **9.** $-\frac{4}{5}$ **11.** Slope: $\frac{2}{3}$; y-intercept: 4 **13.** Slope: $\frac{3}{2}$; y-intercept: -4 **15.** Slope: $-\frac{3}{2}$; y-intercept: $\frac{3}{4}$ **17.** Slope: $\frac{1}{5}$; y-intercept: $-\frac{8}{5}$ **19.** Slope: $-\frac{5}{4}$; y-intercept: $\frac{1}{2}$ **21.** Slope: $-\frac{4}{3}$; y-intercept: 3 **23.** Slope: $\frac{5}{2}$; y-intercept: 1 **25.** Slope: $-\frac{6}{5}$; y-intercept: 6

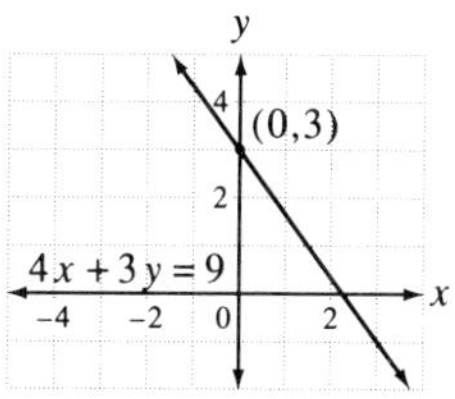

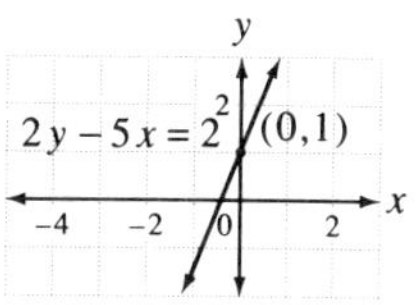

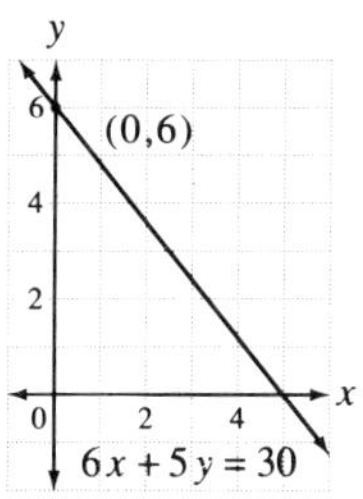

27. Slope: $\frac{1}{3}$; y-intercept: 2

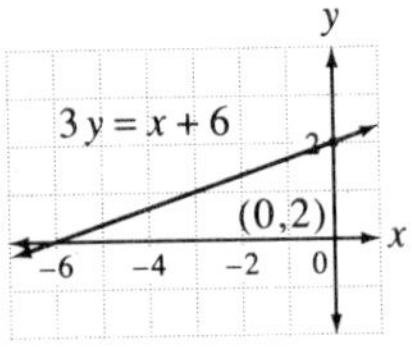

29. Slope: 3; y-intercept: 2

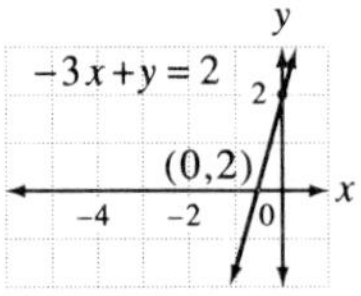

31.

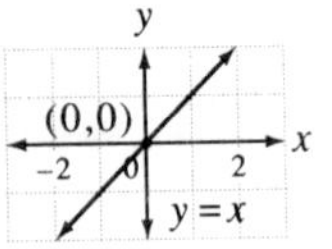

33.

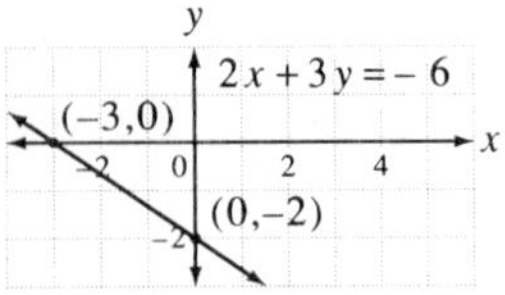

35.

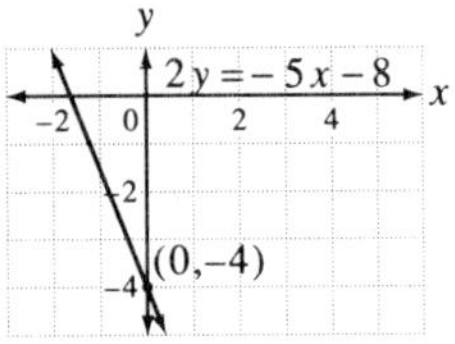

37.

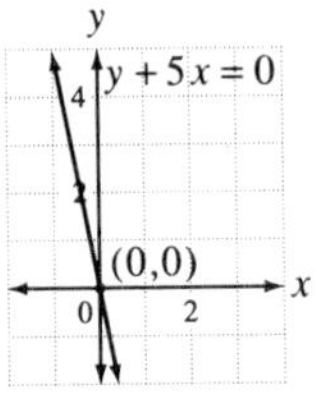

39.

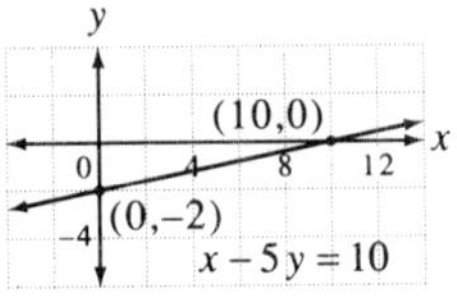

41. 6 **43.** 6 ft **45.** $\frac{2}{25}$ **47.** 94 ft **49.** Negative **51.** $(0, b)$ **53.** 5 **55.** $\frac{1}{3}$ **57.** $-\frac{2}{9}$ **59.** $-\frac{7d}{6c}$

61. Slope: $\frac{-3}{25}$; y-intercept: $-\frac{6}{5}$ **63.** $\frac{2}{5}, -\frac{3}{7}, -\frac{5}{2}$ **65.** 8 **67.** 4 **69.** $\frac{8}{3}$ **71.** $\frac{1}{-2}$

Problem Set 3.4

1. $y = 4x + 1$ **3.** $y = -\frac{5}{8}x + \frac{1}{4}$ **5.** $y = \frac{1}{3}x + 5$ **7.** $y = 3$ **9.** $y = 3.4x - 7.8$ **11.** $y = 3x - 11$
13. $y = -2x + 14$ **15.** $y = -5x - 6$ **17.** $y = -\frac{3}{7}x - \frac{32}{7}$ **19.** $y = -2$ **21.** $y = \frac{3}{2}x - \frac{3}{2}$ **23.** $y = -x + 2$
25. $y = \frac{8}{11}x + \frac{18}{11}$ **27.** $y = \frac{3}{4}x - 1$ **29.** $y = 5$ **31.** $y = 5x - 1000$ **33.** $d = 300t$ **35.** $V = -1200t + 12{,}000$
37. $C = 25r + 2000$; \$12,000 **39.** $d = -\frac{1}{10}c + 5$; 1 **41.** $y = \frac{4}{3}x + \frac{7}{3}$ **43.** $y = -\frac{1}{4}x - 4$ **45.** $y = 5x + 1$
47. $y = 3x + 19$ **49.** $y = \frac{2}{3}x + 3$ **51.** $y = -\frac{1}{6}x - \frac{13}{6}$ **53.** $y = -5x - 22$ **55.** $y = -\frac{1}{4}x - \frac{19}{4}$
57. $y = \frac{3}{2}x - \frac{9}{2}$ **59.** $y = -\frac{3}{2}x + 1$ **61.** Point-slope **63.** Negative reciprocal **65.** $4y + x = -14.823$
67. The slopes of two sides of the triangle are 2 and $-\frac{1}{2}$. Therefore, the sides are perpendicular and the triangle is a right triangle.
69. $3x + 4y = 12$ **71.** $\{x \mid x > 13\}$ **73.** $\{x \mid 1 \le x \le 2\}$ **75.** $\{x \mid x \le -2 \text{ or } x > 2\}$

$$\frac{3x}{12} + \frac{4y}{12} = \frac{12}{12}$$

$$\frac{x}{4} + \frac{y}{3} = 1$$

x-intercept: 4; y-intercept: 3

77.

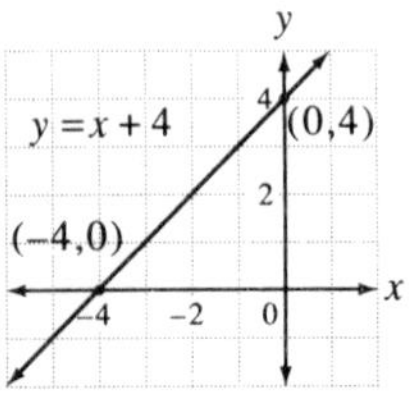

79.

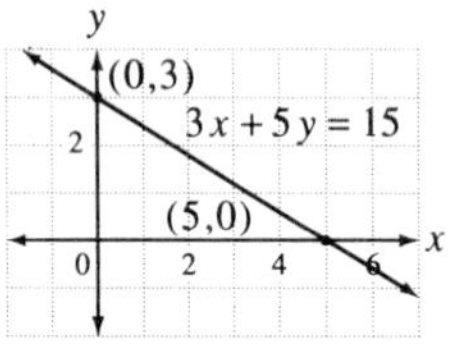

Problem Set 3.5

1. Yes **3.** No **5.**

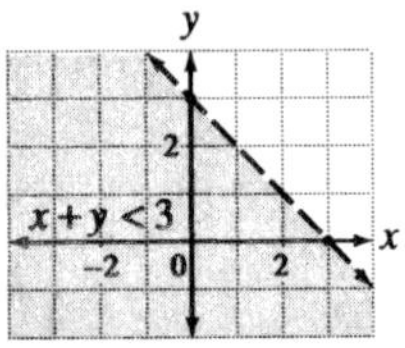

7.

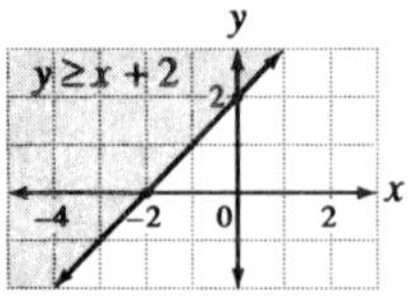

9.

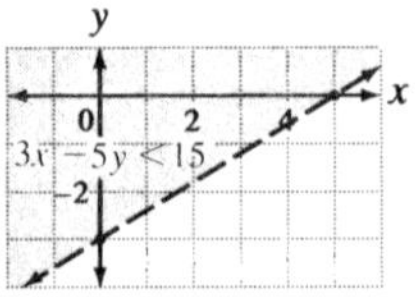

11.

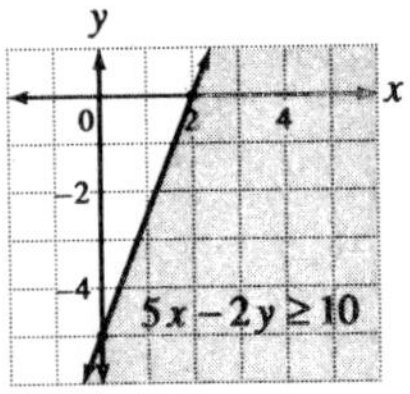

13.

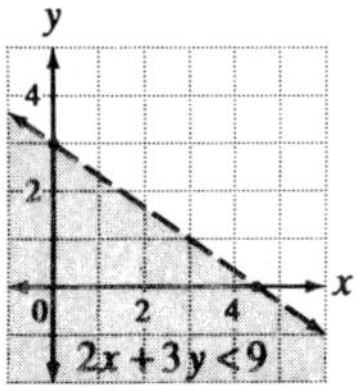

15.

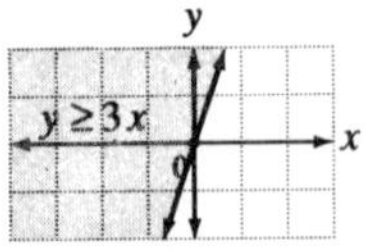

17.

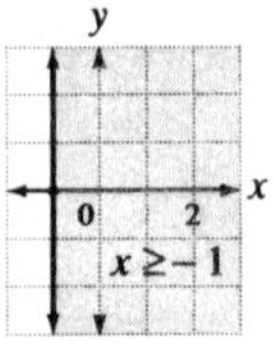

19. 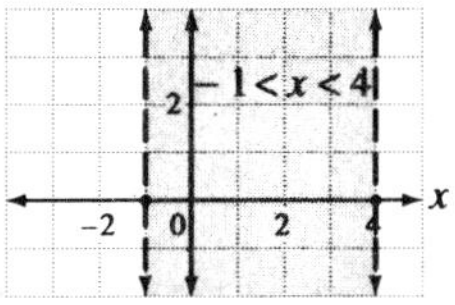

21. 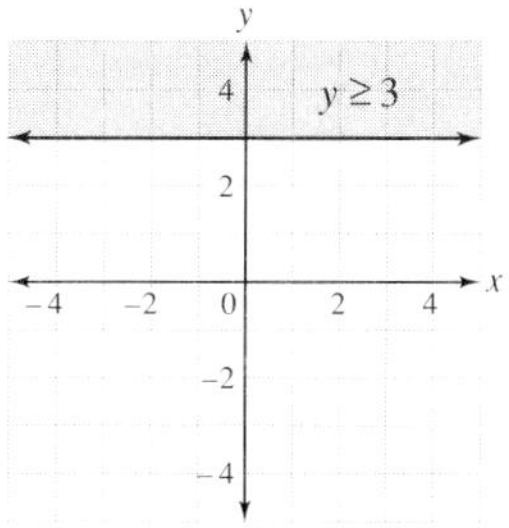

23. 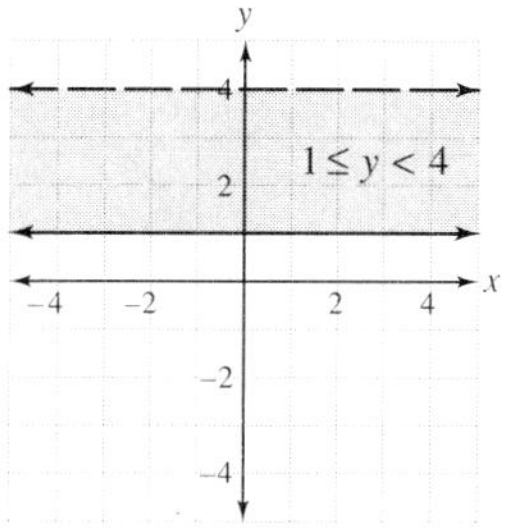

25. 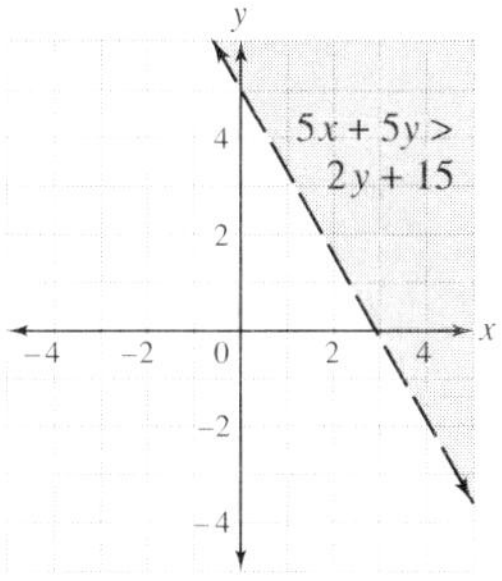

27. 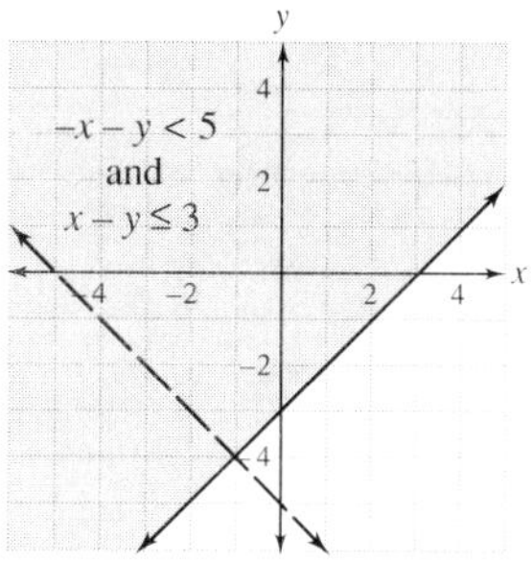

29. 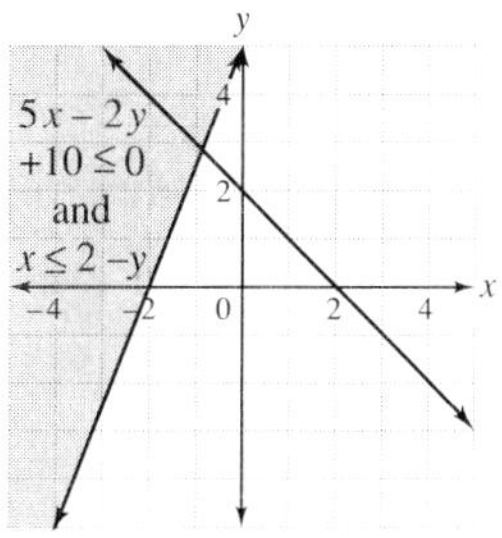

31. 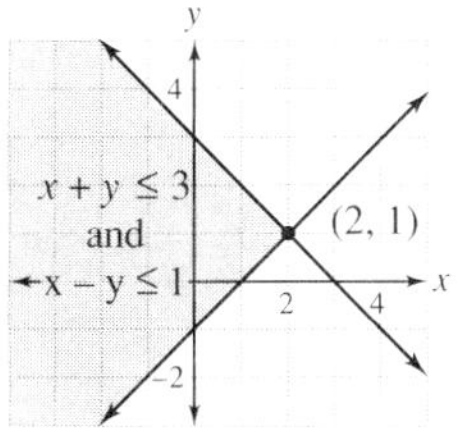

33. 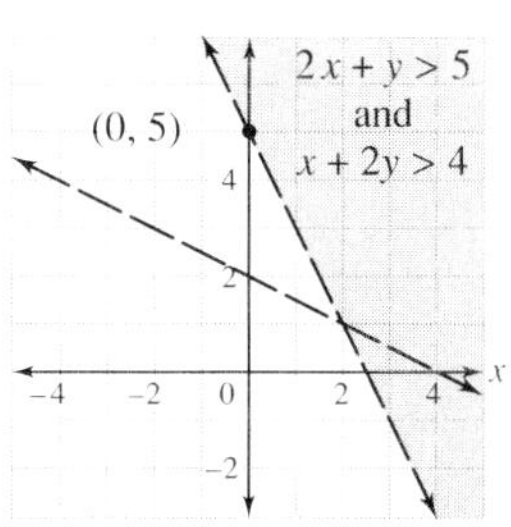

35. Ordered pairs
37. Inequality
39. No
41. Yes

43. Yes

45. $|x| > 3$

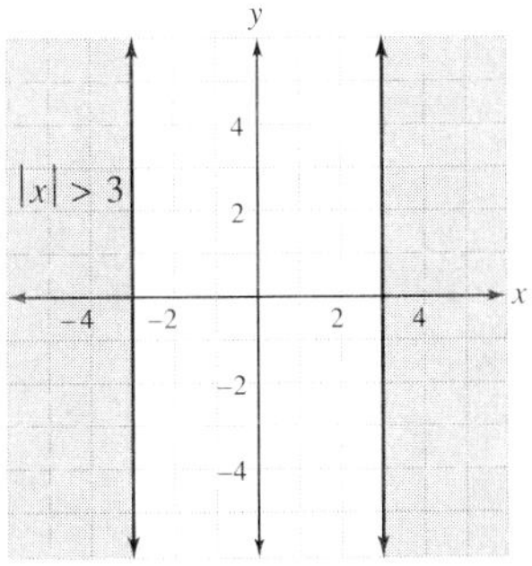

47. $|x + 2| \leq 3$

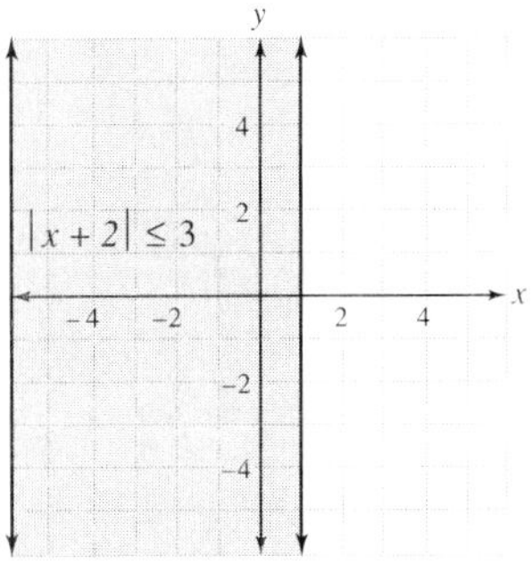

49. $\{6, -12\}$ 51. $\{x \mid -3 \leq x \leq \frac{1}{3}\}$ 53. $\{x \mid x > \frac{14}{5} \text{ or } x < -2\}$ 55. $\{3, -\frac{5}{3}\}$ 57. $\{x \mid -\frac{1}{3} < x < \frac{5}{9}\}$
59. $\{x \mid x \text{ is a real number}\}$

Chapter 3 Additional Exercises

1. II
3. No quadrant
5. III
7. II
9. Yes
11. No

13. 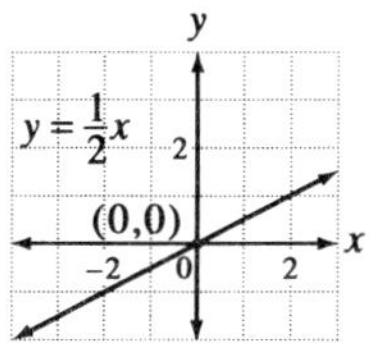

15. Linear
17. Not linear
19. Not linear
21. Not linear

23. 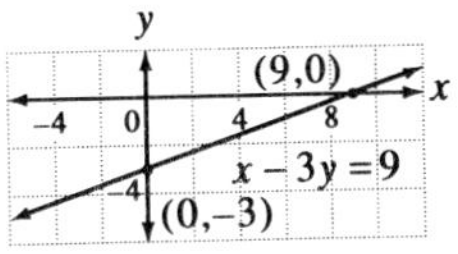

25. 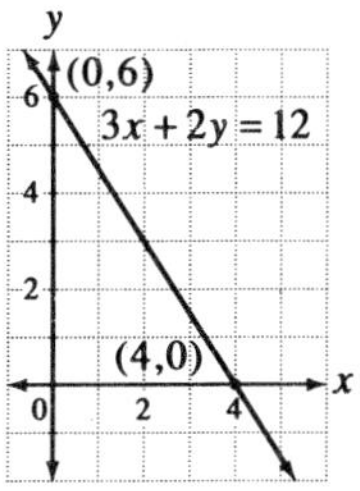

27.

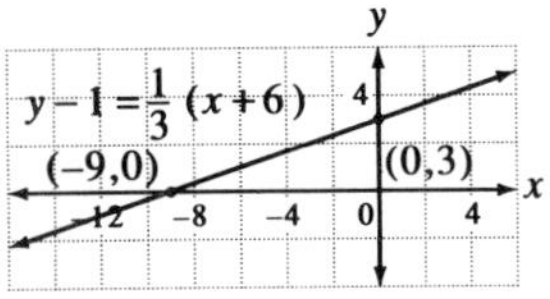

29.

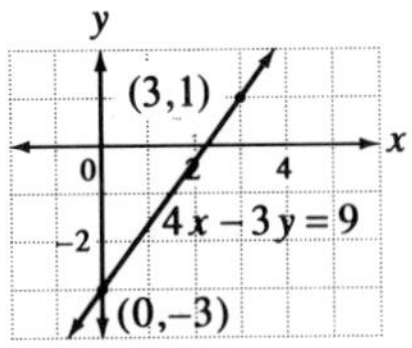

31.

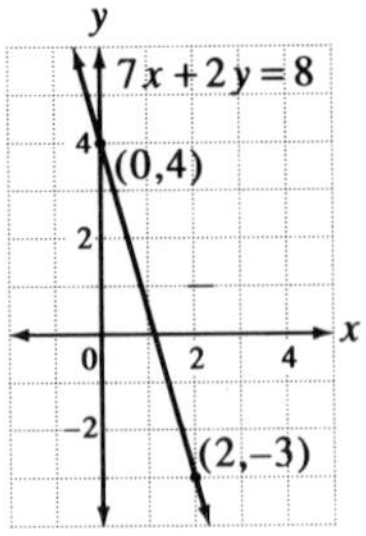

33.

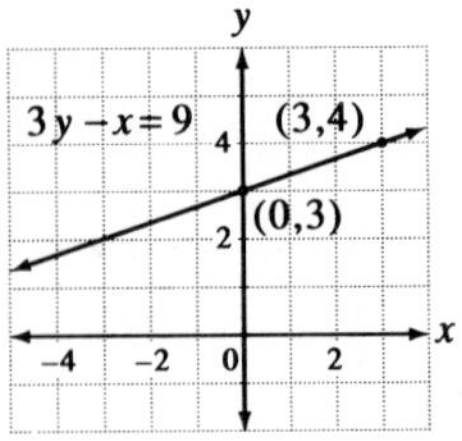

35.

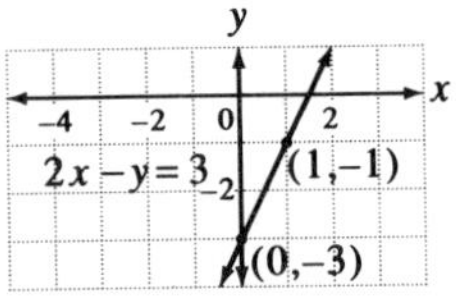

37.

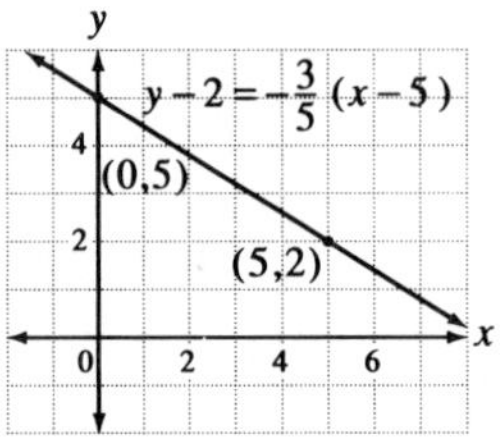

39.

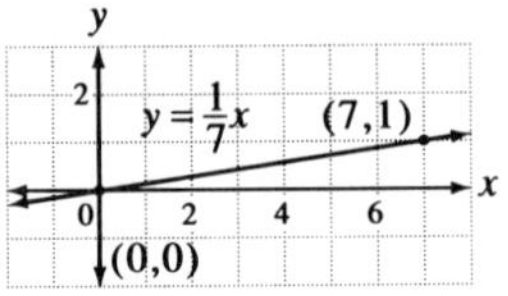

41. $y = 2x - 3$ **43.** $y = \frac{2}{3}x + \frac{1}{2}$ **45.** $y = 0.5x + 1.05$ **47.** $y = \frac{6}{5}$ **49.** $y = \frac{1}{5}x + \frac{26}{5}$ **51.** $y = \frac{3}{4}x - \frac{3}{4}$
53. $y = 7x - 10.2$ **55.** $c = \frac{1}{4}p + 9$ **57.** (a) and (b)

59.

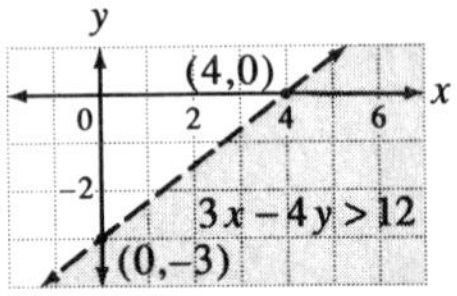

61.

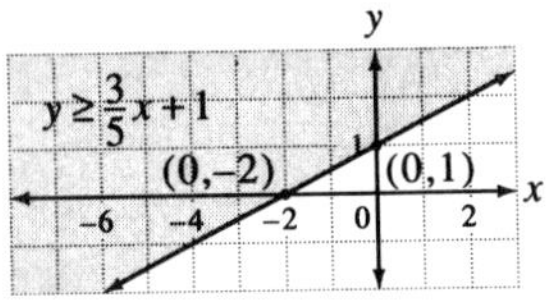

63.

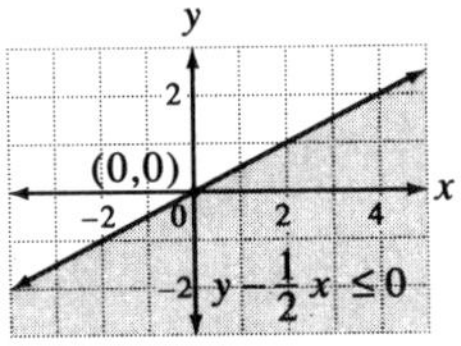

65.

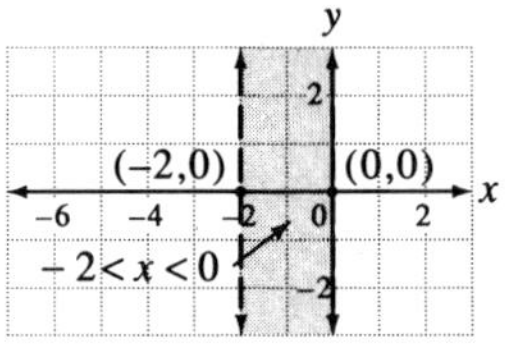

Chapter 3 Cooperative Learning

1. $y = -\frac{3}{2}x - 3$
2. $y = -3$
3. $y = \frac{3}{5}x - 2$
4. $y = 3x - 1$

5.

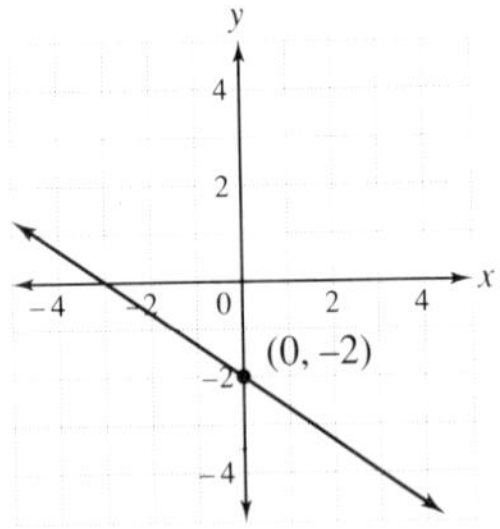

6. (a) $y = \frac{1}{3}x - \frac{10}{3}$; (b) $y = -3x - 10$
7. $y \geq -\frac{2}{3}x + 3$
8. \$120,000, \$4000 per year

Chapter 3 Practice Test

1.

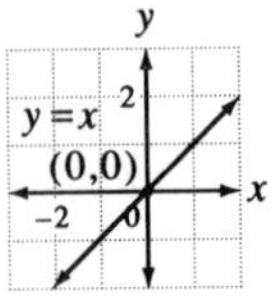

2.

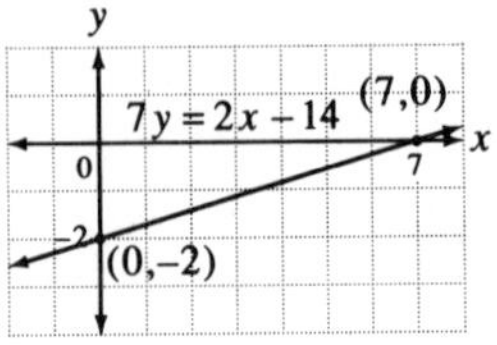

3. 7 **4.** Slope: $-\frac{5}{4}$; y-intercept: 5

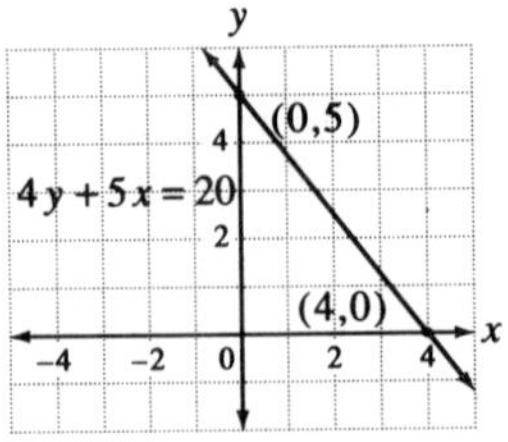

5. $y = \frac{3}{5}x - 2$
6. $y = \frac{1}{5}x - \frac{13}{5}$
7. $y = -\frac{2}{5}x - \frac{4}{5}$
8. $y = -\frac{1}{3}x + \frac{13}{3}$
9. (a) $C = 3d + 450$; (b)\$1350

10.

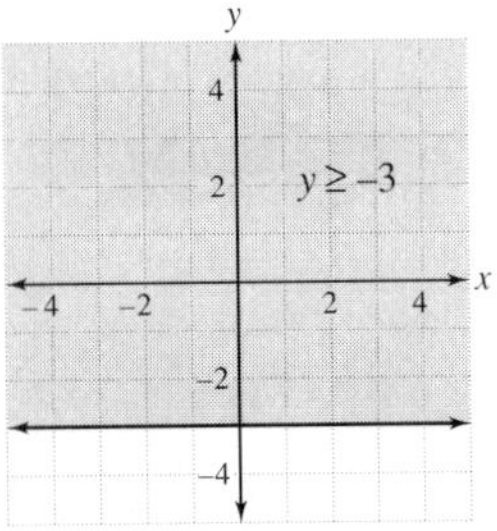

11.

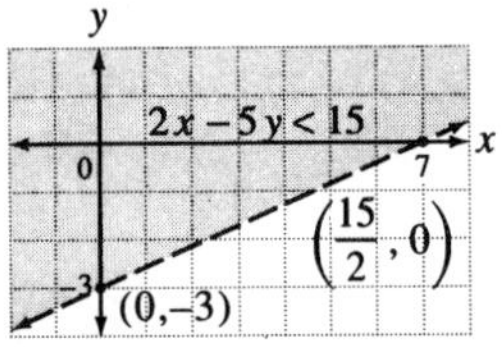

CHAPTER 4

Chapter 4 Pretest

1. (1, −2); consistent, independent **2.** (7, −5) **3.** Infinite number of solutions **4.** 55 ft by 67 ft
5. 12.375 liters of 7%, 5.625 liters of 15% **6.** (−1, −2, −1) **7.** 3 lb of bananas, 6 lb of apples, 6 lb of grapes
8. 92 **9.** 262 **10.** (0, −3, 7)

Problem Set 4.1

1. Yes **3.** No **5.** Yes **7.** No

9.
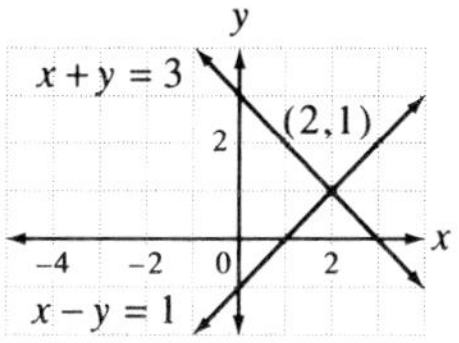

(2, 1); consistent, independent

11.
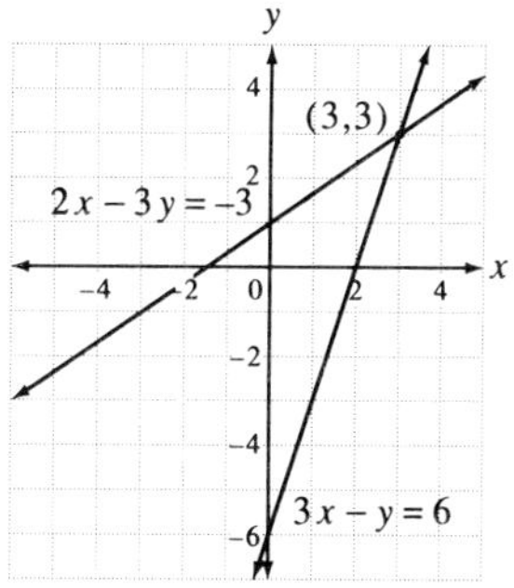

(3, 3); consistent, independent

13.
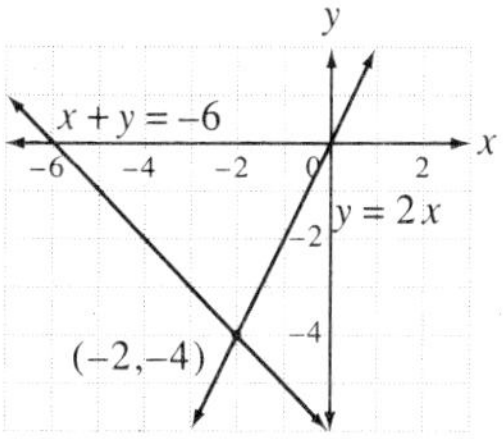

(−2, −4); consistent, independent

15.
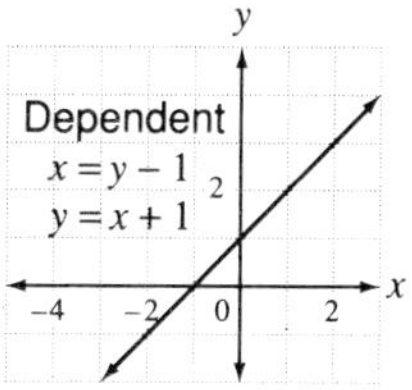

Infinite numbers of solutions; consistent, dependent

17.
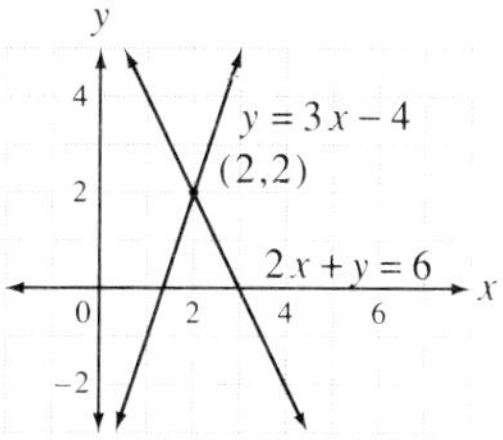

(2, 2); consistent, independent

19.
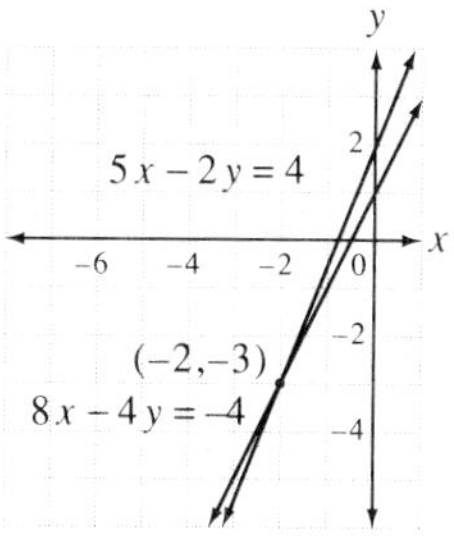

(−2, −3); consistent, independent

21.
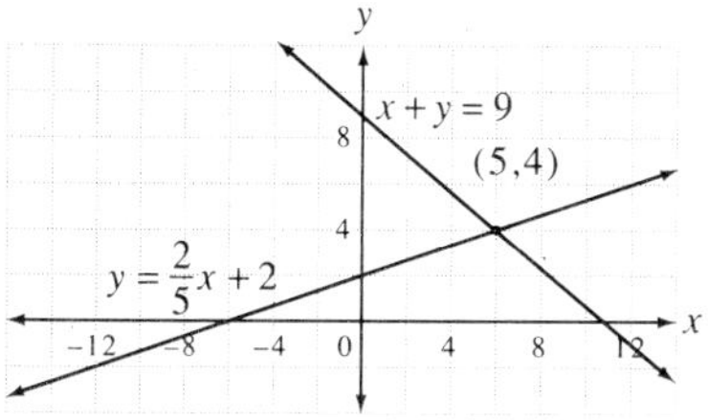

(5, 4); consistent, independent

23.
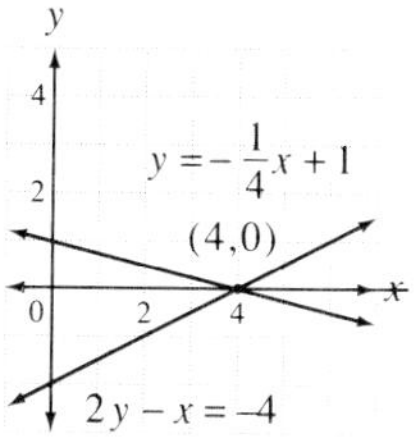

(4, 0); consistent, independent

25.
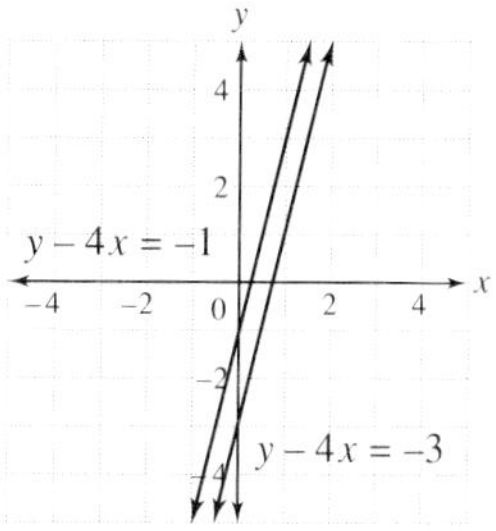

No solution; inconsistent, independent

27.
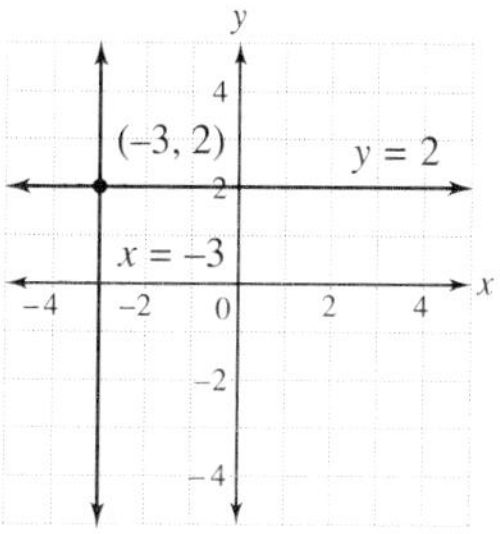

(−3, 2); consistent, independent

29.
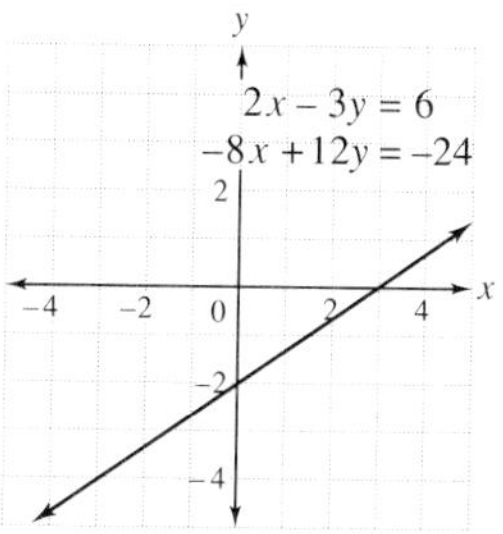

Infinite number of solutions; consistent, dependent

31.
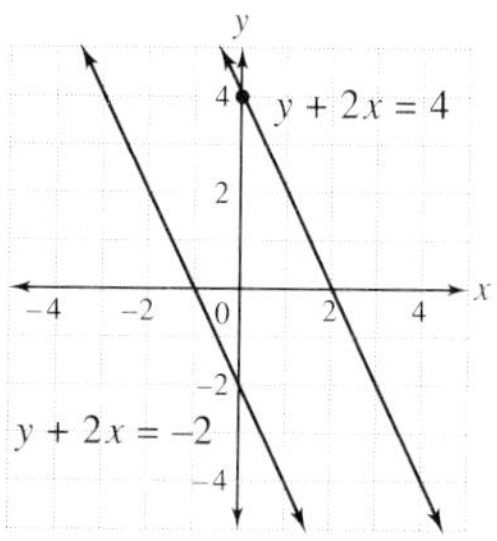

No solution, inconsistent, independent

33.

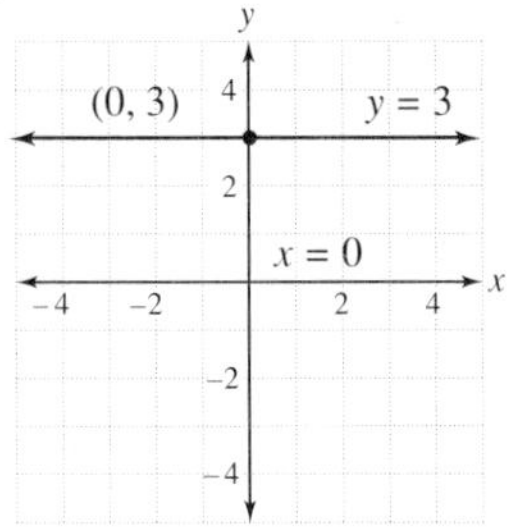

(0, 3); consistent, independent

35. Ordered pairs
37. Inconsistent
39. Yes
41. Yes
43. No

45. $(\frac{5}{2}, -2)$

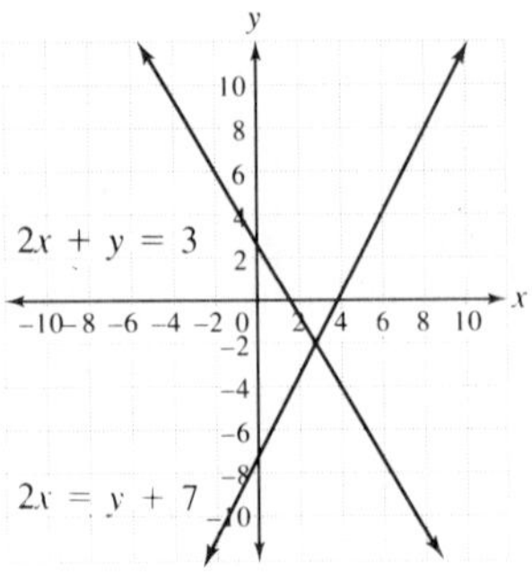

47. $(-3, -2)$

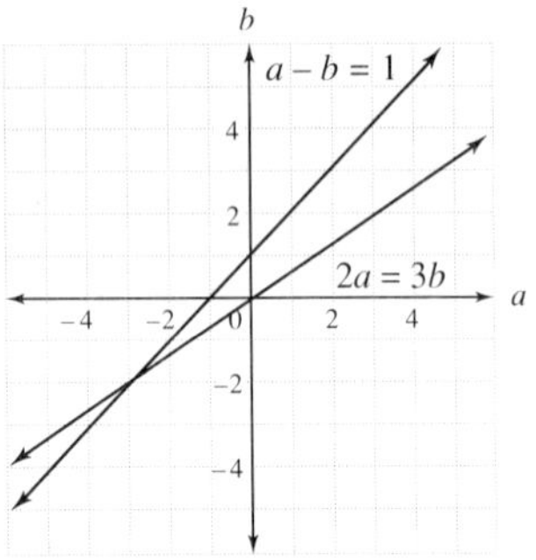

49. $\frac{9}{2}$ **51.** 2 **53.** Undefined **55.** $\frac{1}{2}$ **57.** $\frac{12}{11}$

Problem Set 4.2

1. (3, 1) **3.** $(-2, -1)$ **5.** (1, 2) **7.** (5, 0) **9.** Infinite number of solutions **11.** (3, 2) **13.** $(-1, 3)$ **15.** No solution **17.** $(2, -4)$ **19.** (4, 5) **21.** $(\frac{1}{2}, 1)$ **23.** $(-3, -15)$ **25.** $(-2, 5)$ **27.** $(2, -2)$ **29.** $(1, -1)$ **31.** No solution **33.** $(-2, -2)$ **35.** (9, 6) **37.** (1, 1) **39.** (11, 7) **41.** (6, 2) **43.** Infinite number of solutions **45.** (5, 4) **47.** $(3, \frac{9}{2})$ **49.** Adding **51.** Infinite number **53.** Variable **55.** $(-2, 3)$ **57.** (3, 0.8) **59.** $(-12, -60)$ **61.** $\left(\frac{1}{a}, \frac{1}{b}\right)$ **63.** -4 **65.** 5 **67.** \$4000 **69.** 45 **71.** \$3000

Problem Set 4.3

1. $L = 3W$
$$2L + 2W = 144$$
$$2(3W) + 2W = 144$$
$$6W + 2W = 144$$
$$8W = 144$$
$$W = 18$$
$L = 3W = 3(18) = 54$
The width is 18 cm and the length is 54 cm.

3. $W = L - 16$
$$2L + 2W = 182$$
$$2L + 2(L - 16) = 182$$
$$2L + 2L - 32 = 182$$
$$4L - 32 = 182$$
$$4L = 214$$
$$L = 53.5$$
$W = L - 16 = 53.5 - 16 = 37.5$
The length is 53.5 ft and the width is 37.5 ft.

5. $L = 2W + 3$
$$2L + 2W = 108$$
$$2(2W + 3) + 2W = 108$$
$$4W + 6 + 2W = 108$$
$$6W + 6 = 108$$
$$6W = 102$$
$$W = 17$$
$L = 2W + 3 = 2(17) + 3 = 37$
The width is 17 m and the length is 37 m.

7. $x + y = 8$
$$0.65x + 0.59y = 5.08$$
$$65x + 59y = 508$$
$$\underline{-65x - 65y = -520}$$
$$-6y = -12$$
$$y = 2$$
$x = 8 - y = 8 - 2 = 6$
Kevin bought 6 lb of apples and 2 lb of oranges.

9. $x = 4y$
$$16x + 11y = 1500$$
$$16(4y) + 11y = 1500$$
$$64y + 11y = 1500$$
$$75y = 1500$$
$$y = 20$$
$x = 4y = 80$
Twenty children's tickets and 80 adults' tickets were sold.

11. $x + y = 1800$
$$0.09x + 0.10y = 173$$
$$9x + 10y = 17{,}300$$
$$\underline{-9x - 9y = 16{,}200}$$
$$y = 1100$$
$x = 1800 - y = 700$
\$1100 was invested at 10% and \$700 was invested at 9%.

13. $x = y - 700$
$0.08x + 0.09y = 148$
$8x + 9y = 14{,}800$
$8(y - 700) + 9y = 14{,}800$
$8y - 5600 + 9y = 14{,}800$
$17y = 20{,}400$
$y = 1200$
$x = y - 700 = 500$
\$1200 was invested at 9%.
\$500 was invested at 8%.

15. $x + y = 5$
$0.15x + 0.40y = 0.25(5)$
$15x + 40y = 25(5)$
$-15x - 15y = -75$
$25y = 50$
$y = 2$
$x = 5 - y = 3$
Two gallons of 40% solution and 3 gallons of 15% solution should be used.

17. $x + y = 12$
$0.20x + 40y = 0.25(12)$
$20x + 40y = 25(12)$
$20x + 40y = 300$
$-20x - 20y = -240$
$20y = 60$
$y = 3$
$x = 12 - y = 9$
He should use 3 gallons of 40% solution and 9 gallons of 20% solution.

19. $y = 3x$
$0.48x + 0.45y = 5.49$
$48x + 45y = 549$
$48x + 45(3x) = 549$
$48x + 135x = 549$
$183x = 549$
$x = 3$
$y = 3x = 3(3) = 9$
She bought 3 cans of pop and 9 candy bars.

21. $x + y = 6000$
$0.09x = 0.11y$
$0.09x = 0.11(6000 - x)$
$9x = 11(6000 - x)$
$9x = 66{,}000 - 11x$
$20x = 66{,}000$
$x = 3300$
$y = 6000 - x = 2700$
\$3300 was invested at 9% and \$2700 was invested at 11%.

23. $y = 20 + x$
$0.05x + 0.30(20) = 0.25y$
$0.05x + 0.30(20) = 0.25(20 + x)$
$5x + 30(20) = 25(20 + x)$
$5x + 600 = 500 + 25x$
$600 = 500 + 20x$
$100 = 20x$
$5 = x$
Five liters of 5% acid solution should be added.

25. $x + y = 8$
$3x - 2y = 36$
$2x + 2y = 16$
$5x = 52$
$x = \frac{52}{5} = 10\frac{2}{5}$
$x + y = 8,$
$y = 8 - x = 8 - 10\frac{2}{5} = -2\frac{2}{5}.$
The numbers are $10\frac{2}{5}$ and $-2\frac{2}{5}$.

27. $2L + 2W = 28$
$L = 2W + 8$
$2(2W + 8) + 2W = 28$
$4W + 16 + 2W = 28$
$6W + 16 = 28$
$6W = 12$
$W = 2$
$L = 2W + 8 = 2(2) + 8 = 12$

29. $x + y = 90$
$y = 4x - 4$
$x + (4x - 4) = 90$
$x + 4x - 4 = 90$
$5x - 4 = 90$
$5x = 94$
$x = 18.8$
$y = 4x - 4 = 4(18.8) - 4 = 71.2$
One angle measures 18.8° and the other angle measures 71.2°.

31. $x + y = 275$
$15x + 8.50y = 3800$
$-15x - 15y = -4125$
$15x + 8.50y = 3800$
$-6.50y = -325$
$y = 50$
$x + y = 275,\ x = 225$
There were 50 children's dinners and 225 adult dinners served.

33. $0.085x + 0.071y = 1433$
$y = 2x + 1000$
$85x + 71y = 1{,}433{,}000$
$85x + 71(2x + 1000) = 1{,}433{,}000$
$85x + 142x + 71{,}000 = 1{,}433{,}000$
$227x + 71{,}000 = 1{,}433{,}000$
$227x = 1{,}362{,}000$
$x = 6000$
$y = 2x + 1000 = 2(6000) + 1000 = 13{,}000$
She invested \$13,000 at 7.1%

35. $x + y = 20$
$0.06x + 0.14y = 0.084(20)$
$6x + 14y = 168$
$-6x - 6y = -120$
$8y = 48$
$y = 6$
$x + y = 20,\ x = 14$
She should use 6 ml of 14% solution and 14 ml of 6% solution.

37. $x + y = 180$
$y = 2x + 24$
$x + (2x + 24) = 180$
$x + 2x + 24 = 180$
$3x + 24 = 180$
$3x = 156$
$x = 52$
$y = 2x + 24 = 2(52) + 24 = 128$
The measures of the angles are 52° and 128°.

39. Twice, twice **41.** Final **43.** 5 pizzas, 8 soft drinks
45. 12.6 liters **47.** \$0.31 **49.** \$50,000, \$50,000 **51.** $(\frac{5}{7}, \frac{20}{7})$
53. 25 C. Fashon suits, 35 D. Stile suits **55.** $y = 4x - \frac{1}{2}$
57. $y = 3x - 8.7$ **59.** $y = -5x + 1$ **61.** $y = \frac{8}{5}x + \frac{7}{5}$
63. $y = x + 3$

Problem Set 4.4

1. Yes **3.** No **5.** (1, 2, 3) **7.** (3, −1, −2) **9.** (2, −2, 2) **11.** (3, −1, 4) **13.** (7, −3, −4) **15.** (0, −1, 4) **17.** No solution **19.** (4, 3, −2) **21.** (−1, 2, −2) **23.** Dependent, no unique solution **25.** Dependent, no unique solution **27.** (0, 0, 0) **29.** (−1, 2, 3) **31.** (2, 2, 2) **33.** (2, 1, −1) **35.** No solution **37.** Ordered triple **39.** Point, line, plane **41.** Dependent **43.** Dependent, no unique solution **45.** (1.37, −4.76, 2.66) **47.** $y = -3x - 1,\ z = \dfrac{14 + 8x}{3}$ **49.** Yes **51.** Yes

53.

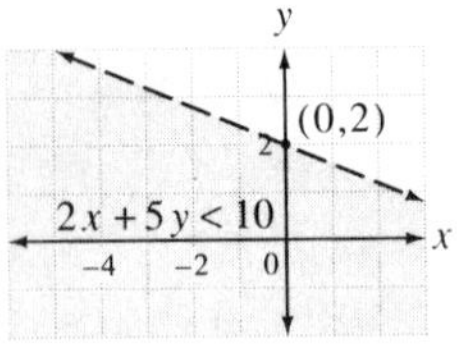

55.

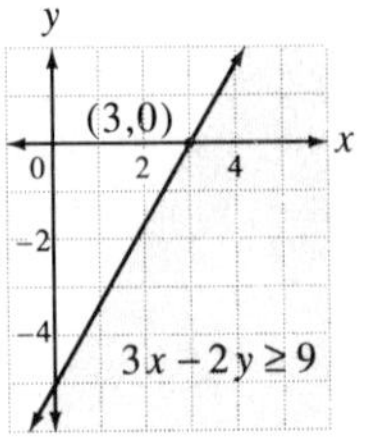

57.

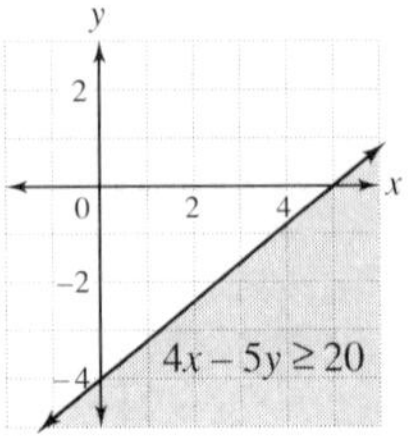

Problem Set 4.5

1. $x + y + z = -1$
$5x + y - z = 1$
$x - 2y - 3z = 5$
The first number is 2, the second number is −6, and the third number is 3.

3. $x + y + z = 53$
$5y = 3x - 3$
$z = 2y - 4$
The first number is 21, the second number is 12, and the third number is 20.

5. $x + y + z = 180$
$y = 3x$
$z = x + y + 20$
The first angle is 20°, the second angle is 60°, and the third angle is 100°.

7. $y = 4x + 5$
$z = x - 5$
$x + y + z = 180$
The first angle is 30°, the second angle is 125°, and the third angle is 25°.

9. $x + y + z = 19$
$y = x - 1$
$z = 2x$
She picked 5 quarts on Tuesday, 4 quarts on Wednesday, and 10 quarts on Thursday.

11. $x + y + z = 100$
$y = z + 10$
$x = 7z$
They made 70 gallons of flat paint, 20 gallons of semigloss, and 10 gallons of enamel.

13. $-2 = a + b + c$
$-5 = 4a - 2b + c$
$10 = 9a + 3b + c$
$a = 1, b = 2, c = -5$

15. $1 = 9a - 3b + c$
$-1 = c$
$3 = 9a + 3b + c$
$a = \frac{1}{3}, b = \frac{1}{3}, c = -1$
$y = \frac{1}{3}x^2 + \frac{1}{3}x - 1$

17. $x + y + z = 9$
$y - 2z = 0$
$0.70x + 0.60y + 0.40x = 5.30$
She bought 4 lb of oranges.

19. $x + y + z = 60{,}000$
$0.12x + 0.05y + 0.04z = 4{,}950$
$y = z$
$30,000 was invested at 12% and $15,000 each at 4% and 5%.

21. $14{,}745 + x + y = 15{,}295$
$14{,}745 + y + z = 16{,}235$
$14{,}745 + x + z = 16{,}015$
The radio costs $120, air conditioning costs $430, and four-wheel drive is $1150.

23. $x + y + z = 140$
$3x + 2y = 159$
$4x + 2z = 286$
Each serving contains the following: whole milk, 33 mg; cheddar cheese, 30 mg; lean beef, 77 mg.

25. $12 = 4a + 2b + c$
$9 = a - b + c$
$5 = a + b + c$
$a = 3, b = -2, c = 4$

27. $-19 = 9a - 3b + c$
$-4 = c$
$-4 = 4a + 2b + c$
$a = -1, b = 2, c = -4$
$y = -x^2 + 2x - 4$

29. 64 **31.** $6.00 per pound **33.** 6 **35.** 12 **37.** $\frac{9}{8}$ **39.** 2 **41.** 106 **43.** −60 **45.** 42

Problem Set 4.6

1. −16 **3.** 13 **5.** 0 **7.** −30 **9.** −41 **11.** −36 **13.** 44 **15.** 12 **17.** 0 **19.** 70 **21.** 120 **23.** (3, 1) **25.** (2, −3) **27.** $(\frac{1}{2}, \frac{1}{3})$ **29.** $(\frac{5}{2}, -\frac{3}{2})$ **31.** $(\frac{1}{2}, \frac{5}{2}, 1)$ **33.** $(-1, -\frac{6}{7}, \frac{11}{7})$ **35.** (1, −2, 3) **37.** (−2, 1, 3) **39.** (1, 1, 1) **41.** (3, 0, 0) **43.** 11 **45.** 8, 6, and −12 **47.** 15 and 8

49. Dan, 90; Megan, 75; Scott, 84 **51.** Second-order **53.** 66.134 **55.** -304.002 **57.** $(2, -3)$ **59.** $(1, 1, 1)$ **61.** $8p$ **63.** $\begin{vmatrix} y & x \\ m & 1 \end{vmatrix} = b$, $y - mx = b$ or $y = mx + b$ **65.** 8 **67.** -9 **69.** 9 **71.** 1 **73.** -7 **75.** 1

Chapter 4 Additional Exercises

1.

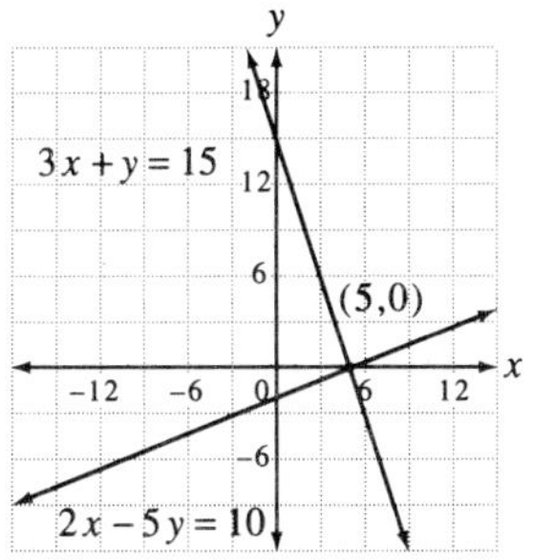

(5, 0); consistent, independent

3.

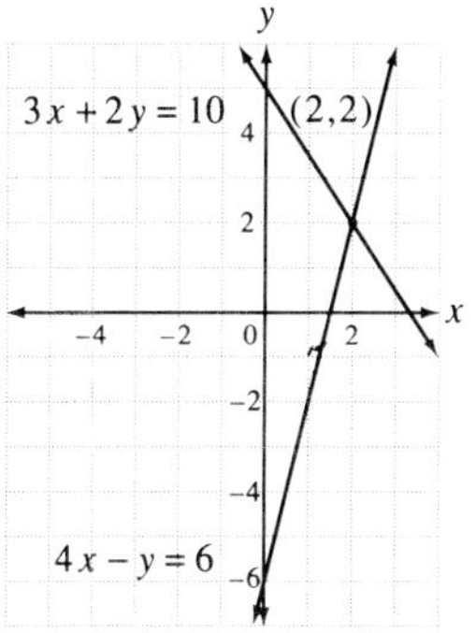

(2, 2); consistent, independent

5.

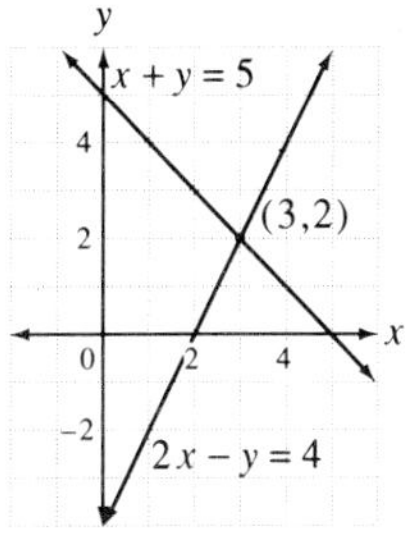

(3, 2); consistent independent

7.

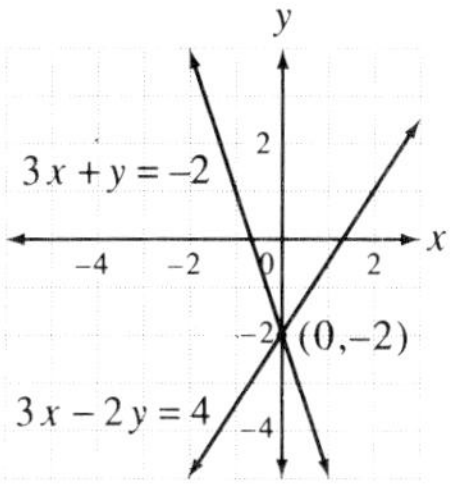

(0, -2); consistent, independent

9.

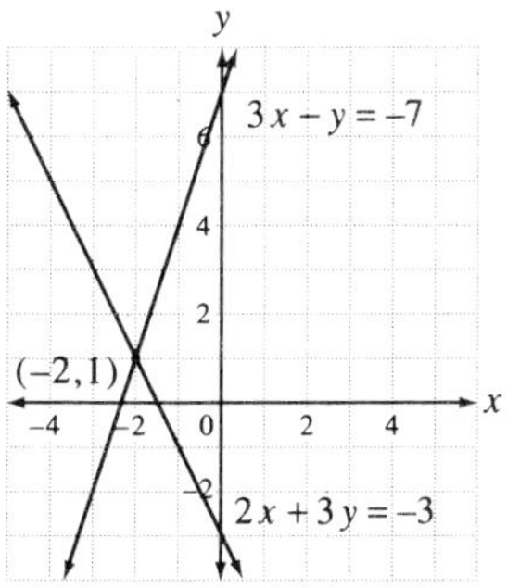

(-2, 1); consistent, independent

11. $(1, 1)$ **13.** $(1, -3)$ **15.** $(5, -2)$ **17.** $(-5, -\frac{10}{3})$ **19.** $(2, -4)$ **21.** $(2, 3)$ **23.** 9 ft by 11 ft **25.** 5 lb **27.** \$7000 at 6%, \$3500 at 8% **29.** 5 gallons **31.** $(1, 2, 3)$ **33.** $(0, 0, 1)$ **35.** $(2, -1, 1)$ **37.** $(3, 0, -2)$ **39.** No solution **41.** \$2500 at 6%, \$5000 at 8%, \$3500 at 9% **43.** 9 dimes, 8 nickels, 5 quarters **45.** (20°, 10°, 150°) **47.** -31 **49.** 14 **51.** -43 **53.** 22 **55.** $(3, 4)$ **57.** $(-3, 2)$ **59.** $(2, -1, 4)$ **61.** $(2, -1, 1)$

Chapter 4 Cooperative Learning

1. $y = 3x + 2$, $y = \frac{3}{2}x - 1$ **3.** $x + y = -81$, $y = 2x$ (Answers may vary.)

5. $x + y + z = 28$
$x + y - z = -26$
$z = 3x$
(Answers may vary.)

7. $\begin{vmatrix} 1 & 9 \\ 4 & -1 \end{vmatrix}$
(Answers may vary.)

Chapter 4 Practice Test

1.

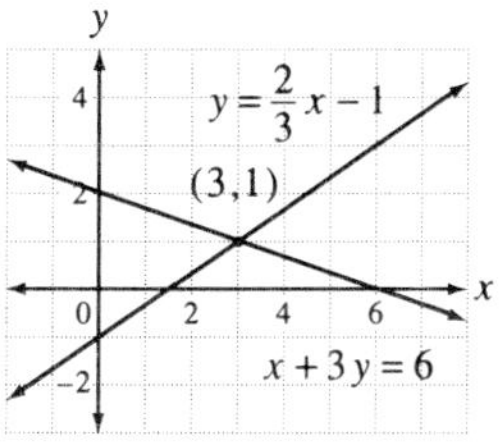

(3, 1); consistent, independent

2. $(1, 2)$ **3.** $(-1, 3)$ **4.** 7 boxes of strawberries, 4 boxes of blueberries **5.** 3 gallons of 30% alcohol, 2 gallons of 80% alcohol **6.** $(1, -1, 2)$ **7.** 7 on Monday, 5 on Tuesday, and 8 on Wednesday **8.** 19 **9.** -7 **10.** $(\frac{3}{2}, -4, 3)$

Cumulative Review Chapters 3 and 4

1.

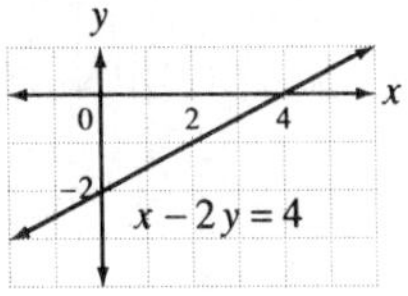

2. -1 **3.** Slope: $-\frac{3}{2}$, y-intercept: 2

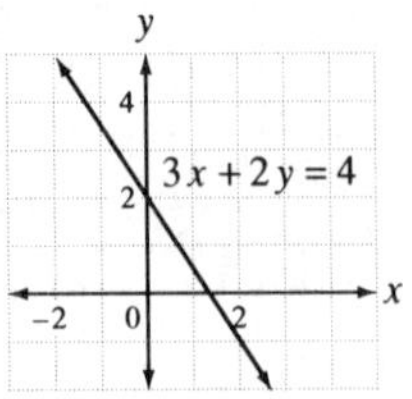

4. $y = -\frac{4}{5}x + 2$
5. $y = \frac{3}{4}x + \frac{13}{4}$
6. $y = -3x + 17$
7. $y = -\frac{1}{4}x + \frac{11}{4}$

8.

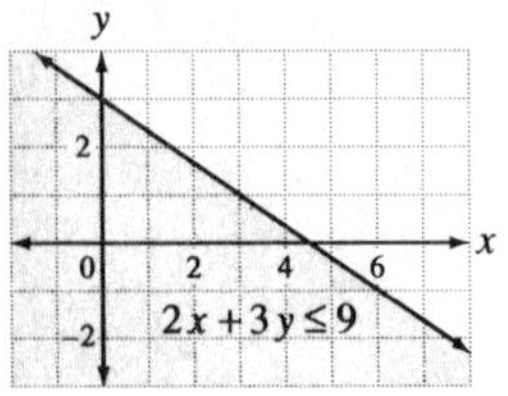

9. $V = -1500t + 18{,}000$ **10.** (4, 2) **11.** $(-1, -4)$ **12.** 17 lb of chocolates, 7 lb of caramels
13. \$1500 at 7%, \$800 at 8% **14.** $(2, -1, -2)$ **15.** -1 **16.** $-\frac{10}{91}, -\frac{9}{13}, \frac{107}{91}$

CHAPTER 5

Chapter 5 Pretest

1. -30 **2.** 7 **3.** $-16a^2b^2 - 6ab^2 + 4$ **4.** $17x^2 - 13xy - 6y^2$ **5.** $-5x^3 + 7x^2 - 5x - 26$
6. $5x^3 - 27x^2 + 25x + 12$ **7.** $27x^2 - 84x + 56$ **8.** $25y^2 - 81$ **9.** $49r^2 - 84r + 36$ **10.** $9p - 7 - \dfrac{1}{2p^2}$
11. $x + 10 + \dfrac{31}{x - 3}$ **12.** $2x^2 + 4x + 8 + \dfrac{15}{x - 2}$ **13.** $(a + b)(x - y)$ **14.** $2x(x - 4)(x + 7)$ **15.** $(9x - 7)^2$
16. $4b^2(5a - 3b)(5a + 3b)$ **17.** $(7x - 3y)(49x^2 + 21xy + 9y^2)$ **18.** $\frac{2}{3}, \frac{1}{5}$ **19.** 0, 19 **20.** 15, -6

Problem Set 5.1

1. 9 **3.** 10 **5.** 7 **7.** -25 **9.** 2 **11.** 1 **13.** 0 **15.** 5 **17.** 5 **19.** $5x^2$ **21.** $-9x^3y$ **23.** $-3x^2$
25. $x - 2y$ **27.** $-6p^2$ **29.** $12x^2 - 2x$ **31.** $8a^2$ **33.** $6a^2b - 9ab^2$ **35.** $-2x^2y^2 + 6xy$ **37.** $x^3 + 3x^2 - 4x - 1$
39. $7y^4 - 9y^3 + y + 3$ **41.** $-8x^4 + 13x^3 + 9x^2 - 6x + 3$ **43.** Binomial **45.** Trinomial **47.** Binomial
49. Monomial **51.** None of these **53.** Trinomial **55.** \$270 **57.** 796 **59.** 24,416.64 m^3 **61.** \$1110
63. 87.92 $in.^3$ **65.** Whole **67.** Degree **69.** Two **71.** -316.18 **73.** 37.63 **75.** $-0.6135x - 5.955y$
77. $9.6389a^2 - 0.8918$ **79.** $\frac{114}{25}$ **81.** $9y - 11$ **83.** $-10x - 2$ **85.** $10y - 9$ **87.** $-12a + 4$ **89.** $3y + 2$

Problem Set 5.2

1. $-2x - 4$ **3.** $-2x^2 - 8x + 2$ **5.** $4x^2y + 5x^2 + 3x - 4$ **7.** $16a - 1$ **9.** $7x^2 - 2x + 1$
11. $-2a^4 + 8a^2 - 14$ **13.** $-9x^2 + x - 3$ **15.** $4x^2 + 2xy + 10y^2$ **17.** $-x^2y - 8xy^2 - 15$ **19.** $2y - 7$
21. $2x + 2y + 4z$ **23.** $-3x^2 - 8x + 7$ **25.** $6x^3 - 10x^2 - 4$ **27.** $-2x^2y^2 - 6xy - 2$ **29.** -12
31. $x^2 + 3x - 7$ **33.** $-6x^2 - 11x + 3$ **35.** $5x^2 - 4x$ **37.** $-2a^2 + a + 1$ **39.** $36x^2y - 20x^2y^2 - 2xy^2$
41. $-3ab^2 + 2ab - 12a^2b$ **43.** $-2x^2 + 7x + 21$ **45.** $13y^2 + 14y - 5$ **47.** $P = 0.6n^2 + 150n - 350$
49. Like terms **51.** Opposite **53.** $16.347x^2 - 12.481$ **55.** $-11.629a^3 + 1.195a^2 - 4.108a$ **57.** $-11x^3 - 9x^2 + 4x$
59. $7x^3 - 3x^2 + 21x + 11$ **61.** x^6 **63.** x^2 **65.** $6x^3$ **67.** $-15x^4$ **69.** 2.166×10^{-8} **71.** 3.33×10^3 (rounded)

Problem Set 5.3

1. $12x^4$ **3.** $10x^8$ **5.** $72x^5y^8$ **7.** $5x^4y^5z^4$ **9.** $-6x^2 + 12x$ **11.** $12x^3 + 32x^2$ **13.** $-10x^5 + 20x^4$
15. $4x^3 - 28x^2 + 8x$ **17.** $-6y^4 + 4y^3 - 16y^2$ **19.** $x^2 - x - 6$ **21.** $x^2 + 2xy + y^2$ **23.** $8a^2 - 2a - 3$
25. $2x^3 - 8x^2 - 3x + 12$ **27.** $12a^4 + a^2 - 6$ **29.** $x^3 - x^2 - 5x + 2$ **31.** $2y^3 - 7y^2 + 10y - 8$
33. $y^5 - 3y^4 - y^2 + 6y - 9$ **35.** $a^3 - b^3$ **37.** $x^4 + 5x^3 - x^2 - 15x + 6$
39. $3p^2q^3 + 4pq^3 - 7p^2q^2 - 7pq^2 + 4p^2q - 15q^3$ **41.** $x^2 - \frac{3}{2}x + \frac{9}{16}$ **43.** $x^4 + 3x^3 - 3x^2 - 7x + 6$
45. $6x^4 - 11x^3 + 12x^2 - 9x + 2$ **47.** $-4x^4 + 11x^3 - 8x^2 + 7x - 6$ **49.** $A = x^2 - 2x$ **51.** Coefficients, adding
53. Distributive **55.** $-10.95x^3 + 68.48x^2 - 4.09x$ **57.** $38.71x^2 - 78.40xy + 26.27y^2$ **59.** $\frac{15}{32}x^2 - \frac{1}{16}x - \frac{1}{6}$
61. $x^4 - \frac{1}{9}x^3 - 6x^2 + \frac{2}{27}$ **63.** $V = 10x^3 + 2x^2 - 8x$ **65.** $(3, -1)$ **67.** $(-2, -8)$ **69.** $(\frac{1}{2}, -1)$ **71.** No solution
73. Infinite number of solutions

Problem Set 5.4

1. $x^2 - 10x + 16$ **3.** $y^2 + 15y + 56$ **5.** $6x^2 - x - 12$ **7.** $16a^2 - 26a + 3$ **9.** $12x^2 + 7xy - 10y^2$
11. $x^2 - 1$ **13.** $x^2 - 100$ **15.** $9x^2 - 16$ **17.** $16a^2 - 9$ **19.** $64x^2 - 81y^2$ **21.** $x^2 + 6x + 9$
23. $y^2 - 10y + 25$ **25.** $16x^2 - 8x + 1$ **27.** $4s^2 - 20s + 25$ **29.** $9y^2 + 24yz + 16z^2$ **31.** $64x^2 - 9$

33. $6x^2 - 13x - 28$ **35.** $x^2 - \frac{5x}{6} + \frac{1}{6}$ **37.** $4x^2 - 28x + 49$ **39.** $3m^2 - 5m - 8$ **41.** $36 - x^2$ **43.** $x^2 - x + \frac{1}{4}$
45. $6x^2 + 5xy - 4y^2$ **47.** $10p^2 + 19pq - 15q^2$ **49.** $x^4 - 16$ **51.** $x^2 - \frac{1}{36}$ **53.** $6y^4 - y^2 - 2$ **55.** $y^4 - 49$
57. $4a^4 - 12a^2 + 9$ **59.** $49 - 9x^2$ **61.** $a^4 - a^2c^2$ **63.** $x^4 - 16$ **65.** $16a^4 - 8a^2b^2 + b^4$
67. $x^3 - 9x^2 + 27x - 27$ **69.** First, inner **71.** Twice **73.** $22.09a^2 - 116.64$ **75.** $96.04x^2 + 72.52xy + 13.69y^2$
77. $\frac{16}{25}y^2 - \frac{6}{5}y + \frac{9}{16}$ **79.** $\frac{49}{400}a^{4n} - \frac{9}{16}b^{4n}$ **81.** $81a^4 - 36a^2b^2$ **83.** $A = 4500 - 280x + 4x^2$ **85.** 18
87. 33.75 liters **89.** 27°, 63° **91.** −34, −51 **93.** 56°, 124°

Problem Set 5.5

1. $x^2 + 2x - 3$ **3.** $3x - 2 + \frac{1}{x}$ **5.** $\frac{8z^2}{5} - z + \frac{2}{z}$ **7.** $a^2b^3 + 4ab^2 - 7b$ **9.** $x + 6$ **11.** $y - 2$
13. $z - 1 + \frac{5}{z - 3}$ **15.** $x + 3$ **17.** $z^2 + 2z - 3 + \frac{3}{2z + 3}$ **19.** $2a^2 - a - 9 + \frac{3a + 12}{a^2 + 2}$ **21.** $2x^2 + 3x - 1$
23. $p^2 + p + 1$ **25.** $8x^3 + 4x^2 + 2x + 1 + \frac{2}{2x - 1}$ **27.** $2b^2 + 2b - 1 + \frac{8}{5b - 2}$ **29.** $x - 8$ **31.** $5x^2 - 11x + 14$
33. $x^2 + 4x + 11 + \frac{26}{x - 2}$ **35.** $3p^2 - 2p + 2 + \frac{-3}{p + 3}$ **37.** $x^5 + 2x^4 + 6x^3 + 12x^2 + 24x + 43 + \frac{97}{x - 2}$
39. $x + 2 + \frac{3}{x - 1}$ **41.** $x^3 - 2x^2 + 4x - 8$ **43.** $x^3 + x^2 + x + 1$ **45.** $6a^4 + 12a^3 + 22a^2 + 40a + 83 + \frac{164}{a - 2}$
47. $a^2 - 2a + 4$ **49.** Term **51.** Long division **53.** Synthetic division **55.** $3x^2 + x + \frac{2}{3} + \frac{4/3x - 10/3}{3x^2 + x - 1}$
57. $y^2 + 2yz - z^2$ **59.** $x^6 - x^5y + x^4y^2 - xy^3 + x^2y^4 - xy^5 + y^6$ **61.** $\frac{3}{4}q - 2 + \frac{1}{4q + 3}$ **63.** $4x + 32$
65. $y^8 - y^5$ **67.** $6x^6 + 3x^8$ **69.** $15x^4 + 10x^3 - 15x^2$ **71.** $48a^3b^4 - 56a^3b^3 - 40a^2b^6$

Problem Set 5.6

1. $12(x + 2)$ **3.** $9(y - 2)$ **5.** $x^2(1 + x)$ **7.** $a(3a + 4)$ **9.** $p^4(9p^4 - 8)$ **11.** $5p^2(4 - 5p^2)$ **13.** $6(4x + 3)$
15. $5x^2(3 - 2x^2 + 5x)$ **17.** $8z^3(4 + 3z^2 - 2z)$ **19.** $15xy^2(xy - 4y^2 + 3x)$ **21.** $-4(x - 5)$ **23.** $-5(y + 6)$
25. $-3(x^2 - 2x + 8)$ **27.** $-5a(a^2 + 3a + 7)$ **29.** $-(x^3 + 6x - 5)$ **31.** $(a^2 + b)(x - y)$ **33.** $(x - 6)(2x + 9)$
35. $(x + y)(a + 3b)$ **37.** $(x - 6)(x + 3)$ **39.** $(x + 4)(x + 3)$ **41.** $(x - 8)(x - 1)$ **43.** $(y - 3)(y - 4)$
45. $a^3(a + 1)(a^3 - 1)$ **47.** $8(2x - 3)(x^2 + 3)$ **49.** $S = \frac{1}{2}x(x + 1)$, 36 **51.** Multiplication **53.** Binomial
55. $(y - 2)(y^2 - 3y + 6)$ **57.** $(4x - y^2)(4x^2 - y)$ **59.** $x^2 + 7x + 12$ **61.** $x^2 - x - 20$ **63.** $6x^2 + 7x + 2$
65. $12x^2 + 16x - 3$

Problem Set 5.7

1. $(x + 3)(x + 3)$ **3.** $(y + 6)(y - 2)$ **5.** $(x - 8)(x - 2)$ **7.** $(x - 5)(x + 4)$ **9.** $2(x - 4)(x + 3)$
11. $x(x + 2)(x + 2)$ **13.** $(2p - 1)(p - 3)$ **15.** $(5x + 6)(x - 1)$ **17.** $(6x - 1)(2x - 3)$ **19.** $(8x + 5)(3x - 2)$
21. $2(x - 5)(x - 4)$ **23.** $(x + 1)(x + 8)$ **25.** $(3x - 5)(2x + 5)$ **27.** $2(4x + 3)(2x - 1)$ **29.** $x(4x - 5)(x + 3)$
31. $(3x - 4)(3x + 5)$ **33.** $2(2x - 5)(2x - 5)$ **35.** $(7p + 4)(5p + 2)$ **37.** $(3x - 1)(2x - 1)$ **39.** $(x^2 - 2)(x^2 + 8)$
41. $(p^2 - 2)(p^2 - 6)$ **43.** $(6x - 5y)(2x + 3y)$ **45.** $3(3x + 2)(2x - 3)$ **47.** $(7y + 3)(5y - 8)$
49. $(2x + 3y)(4x - 13y)$ **51.** $4(2x + 1)(x - 4)$ **53.** $(3a + 1)(3a + 4)$ **55.** $-(2x - 5)(2x - 7)$
57. $4(10a^4 + 4a^2 - 3)$ **59.** $(3x - 4)(x + 1)$ **61.** $(x - 8)(x + 2)$ **63.** $3a^2(a - 4)(a + 6)$ **65.** $(3y^2 + 2)(y^2 + 4)$
67. $(7a + 3b)(5a - 8b)$ **69.** $P = (x - 20)(x - 20)$ **71.** Sum **73.** Greatest common **75.** $2ab^3(a - 12b)^2$
77. $(7p + 30)(9p - 8)$ **79.** $18(x^n - y^n)^2$ **81.** $(3x + 14)(2x + 7)$ **83.** $(6p + 6k + 5)(p + k - 1)$ **85.** $x^2 - 25$
87. $4y^2 - 49$ **89.** $x^2 - 14x + 49$ **91.** $9x^4 + 12x^2y + 4y^2$

Problem Set 5.8

1. $(x - 5)(x + 5)$ **3.** $(7x - 9)(7x + 9)$ **5.** $(p - 8)(p + 8)$ **7.** $(xy - 1)(xy + 1)$ **9.** $y^2(3y - 5)(3y + 5)$
11. $(\frac{1}{6} - p)(\frac{1}{6} + p)$ **13.** $(0.1x - 0.5y)(0.1x + 0.5y)$ **15.** $4(5y - z)(5y + z)$ **17.** $3(x - y^2)(x + y^2)$
19. $9x(x - 2)(x + 2)$ **21.** $8x^2y(2x - y)(2x + y)$ **23.** $(x + 2)^2$ **25.** $(y - 3)^2$ **27.** $a(a - 10)^2$ **29.** $(9x + 1)^2$
31. $3(2x - 3)^2$ **33.** $(5xy - 2)^2$ **35.** $(x + 5)^2$ **37.** $(x + y)^2$ **39.** $(9a - 4b)^2$ **41.** $(6 - 5a)^2$ **43.** $(x^3 + 12)^2$
45. $(0.3a - 0.4)^2$ **47.** $3y^2(y - 5)^2$ **49.** $(\frac{1}{4} - y)^2$ **51.** $(2a^4 + 5)^2$ **53.** $(x + 1 - y)(x + 1 + y)$
55. $(a - 8 - b)(a - 8 + b)$ **57.** $(x + y - 10)(x + y + 10)$ **59.** $(4 - x - y)(4 + x + y)$
61. $(a + 2b - 4x)(a + 2b + 4x)$ **63.** $(10 - p - q)(10 + p + q)$ **65.** $(8 - x + 3)(8 + x - 3)$
67. Two squares **69.** Perfect square **71.** $x^{68}(x - 2)(x + 2)$ **73.** (−2, 3, 1) **75.** (1, −2, 1) **77.** (−1, 1, 1)
79. Dependent, no unique solution **81.** (1, 3, 1) **83.** (3, −4, 0)

Problem Set 5.9

1. $(x+1)(x^2-x+1)$ **3.** $(y-1)(y^2+y+1)$ **5.** $(c+4)(c^2-4c+16)$ **7.** $(3a-1)(9a^2+3a+1)$
9. $(5-3x)(25+15x+9x^2)$ **11.** $(p-q)(p^2+pq+q^2)$ **13.** $(x+\frac{1}{3})(x^2-\frac{1}{3}x+\frac{1}{9})$
15. $(y-0.2)(y^2+0.2y+0.04)$ **17.** $(a-b)(a^2+ab+b^2)(a+b)(a^2-ab+b^2)$ **19.** $(y-\frac{1}{5})(y^2+\frac{1}{5}y+\frac{1}{25})$
21. $(2x+5)(4x^2-10x+25)$ **23.** $(y-2)(y^2+2y+4)$ **25.** $(7p-3q)(49p^2+21pq+9q^2)$
27. $9(y+2)(y^2-2y+4)$ **29.** $(8-x)(64+8x+x^2)$ **31.** $2x(5x+3)(25x^2-15x+9)$
33. $16(x-5)(x^2+5x+25)$ **35.** $(x^2+y^2)(x^4-x^2y^2+y^4)$ **37.** $x^2(5x^2-y^2)(25x^4+5x^2y^2+y^4)$
39. $a(x^2+3x+4)$ **41.** $5(x^2+2x-5)$ **43.** $(x-10)(x+10)$ **45.** $8(x-2y)(x^2+2xy+4y^2)$
47. $(3x-2)(2x+1)$ **49.** $2(3x-1)^2$ **51.** $(x+5)(y+5)$ **53.** $x(7x+2)(x-4)$
55. $(2y+1)(4y^2-2y+1)(2y-1)(4y^2+2y+1)$ **57.** $2x(7x-2y)(7x+2y)$ **59.** $(5x-2)^2$ **61.** $(6x-1)(2x+5)$
63. $2x(4x-15)(x+1)$ **65.** $x(3x+10y)(9x^2-30xy+100y^2)$ **67.** $(3a+4b)^2$ **69.** $(x+8)(x^2-8x+64)$
71. $(4x-3)(2x-5)$ **73.** $12(x-2y)(x+2y)$ **75.** $(x-13-y)(x-13+y)$ **77.** $(3ab+7)(5ab-4)$
79. $(x-3)(2x+11)$ **81.** $(-5+x)(11+x)$ **83.** Greatest common factor **85.** Perfect square trinomial
87. $2x(x^2+3y^2)$ **89.** $(x^q-3)(x^{2q}+3x^q+9)$ **91.** $(x^2-5)(x^2-6)$ **93.** $(x-y+w-z)(x-y-w+z)$
95. $(3x+5y-2)(3x+5y+2)$ **97.** $(3y^n-7)(3y^n+7)$ **99.** $xy(xy+1)(x^2y^2-xy+1)$ **101.** $(xz^2+3)^2$
103. $(x-2)(x+4)$ **105.** $(x+5)(x+2)$ **107.** $(3x-5)(x+2)$ **109.** $(5x-1)(4x-3)$

Problem Set 5.10

1. 4, −9 **3.** $\frac{1}{2}, \frac{4}{3}$ **5.** 0, 5 **7.** −7, 2 **9.** 8, 9 **11.** −6, −5 **13.** 0, −3 **15.** 0, 2 **17.** $\frac{3}{4}, -3$
19. $\frac{5}{3}, -\frac{3}{2}$ **21.** −4, 4 **23.** 4, 3 **25.** −6, 5 **27.** 9, −8 **29.** 0, 8 **31.** $\frac{1}{8}, 5$ **33.** 8, −3 **35.** 0, 4, −8
37. 0, 4, −4

39.
$$\begin{aligned} x(x+2) &= 35 \\ x^2+2x &= 35 \\ x^2+2x-35 &= 0 \\ (x+7)(x-5) &= 0 \\ x+7 = 0 \quad &\text{or} \quad x-5=0 \\ x=-7 \quad &\quad\quad x=5 \\ \text{reject} \quad &\quad\quad x+2=7 \end{aligned}$$
The width is 5 m and the length is 7 m.

41.
$$\begin{aligned} x^2+2x &= 80 \\ x^2+2x-80 &= 0 \\ (x-8)(x+10) &= 0 \\ x-8=0 \quad &\text{or} \quad x+10=0 \\ x=8 \quad &\quad\quad x=-10 \end{aligned}$$
The number is 8 or −10.

43.
$$\begin{aligned} x^2+(x+2)^2 &= 100 \\ x^2+x^2+4x+4 &= 100 \\ 2x^2+4x+4 &= 100 \\ 2x^2+4x-96 &= 0 \\ x^2+2x-48 &= 0 \\ (x+8)(x-6) &= 0 \\ x+8=0 \quad &\text{or} \quad x-6=0 \\ x=-8 \quad &\quad\quad x=6 \\ x+2=-6 \quad &\quad\quad x+2=8 \end{aligned}$$
The integers are −6 and −8 or 6 and 8.

45.
$$\begin{aligned} x(x-5) &= 176 \\ x^2-5x-176 &= 0 \\ (x+11)(x-16) &= 0 \\ x+11=0 \quad &\text{or} \quad x-16=0 \\ x=-11 \quad &\quad\quad x=16 \\ \text{reject} \quad &\quad\quad x-5=11 \end{aligned}$$
The width is 11 m.

47.
$$\begin{aligned} x^2+(8-x)^2 &= 104 \\ x^2+64-16x+x^2 &= 104 \\ 2x^2-16x+64 &= 104 \\ 2x^2-16x-40 &= 0 \\ x^2-8x-20 &= 0 \\ (x+2)(x-10) &= 0 \\ x+2=0 \quad &\text{or} \quad x-10=0 \\ x=-2 \quad &\quad\quad x=10 \\ 8-x=10 \quad &\quad\quad 8-x=-2 \end{aligned}$$
The numbers are −2 and 10.

49.
$$\begin{aligned} x(3x) &= 768 \\ 3x^2 &= 768 \\ 3x^2-768 &= 0 \\ x^2-256 &= 0 \\ (x-16)(x+16) &= 0 \\ x-16=0 \quad &\text{or} \quad x+16=0 \\ x=16 \quad &\quad\quad x=-16 \\ &\quad\quad \text{reject} \end{aligned}$$
$3x = 3(16) = 48$
The dimensions are 16 m by 48 m.

51.
$$\begin{aligned} x(4-x) &= -96 \\ 4x-x^2 &= -96 \\ 0 &= x^2-4x-96 \\ 0 &= (x-12)(x+8) \\ x-12=0 \quad &\text{or} \quad x+8=0 \\ x=12 \quad &\quad\quad x=-8 \\ 4-x=4-12 \quad &\quad\quad 4-x=4-(-8) \\ =-8 \quad &\quad\quad =12 \end{aligned}$$
The numbers are 12 and −8.

53. Zero **55.** Zero

57. −9.2, −3.4 **59.** 2.8, −2.35 **61.** 5, −1 **63.** 3 **65.** 20 feet **67.** 2, −8 **69.** $\frac{8}{3}$ **71.** $\frac{1}{9}$
73. $8(y-1)$ **75.** $(x+4)(x-4)$ **77.** $(2x-3)^2$ **79.** $(3x-1)(2x+5)$

Chapter 5 Additional Exercises

1. 32 **3.** −10 **5.** $\frac{57}{5}$ **7.** (a) 3; (b) $-4a^3+3a^2+8$; (c) trinomial **9.** (a) 4; (b) z^4+7; (c) binomial
11. $-2x-9y+5z$ **13.** $-11x^2+7x-5$ **15.** $4x^3-5x^2-2x+8$ **17.** $-2.02a^2-4.57ab$ **19.** $18b^8$
21. $-y^{9n}$ **23.** $-8x^{4n}$ **25.** $b^{3n}+b^{2n+1}$ **27.** $12x^2-20x+3$ **29.** $a^{2n}-b^{2n}$ **31.** $3y^3-2y^2-3y+2$
33. $4x^4-6x^3+13x^2-5x-3$ **35.** $6x^2-17x+12$ **37.** $25x^2-49$ **39.** $a^{2n}+2a^nb^n+b^{2n}$ **41.** $x^{4n}-y^{6n}$

43. $\frac{7p}{3}+\frac{2}{p}-\frac{1}{p^2}$ **45.** $x-3$ **47.** $3x-4+\frac{-2}{x-2}$ **49.** $5x+7+\frac{-3}{3x-2}$ **51.** $2x^3+6x^2+17x+45+\frac{116x-48}{x^2-3x+1}$
53. $16x^2(1+6x^2-4x)$ **55.** $(a-5b)(x+y)$ **57.** $(x-7)(x-1)$ **59.** $(m+1)(m+n)$ **61.** $(x-2)(3x-2)$
63. $(8x-9)(3x+7)$ **65.** $7(5x-4)(x-1)$ **67.** $x(6x-5)(7x+3)$ **69.** $(6x^n-1)(3x^n-4)$
71. $(8x-9y)(8x+9y)$ **73.** $(5x^n-4y^n)(5x^n+4y^n)$ **75.** $4(x-3)^2$ **77.** $x^n(x^n+5)(x^n+5)$
79. $(4x+5)(16x^2-20x+25)$ **81.** $(x-y)(x^2+xy+y^2)(x+y)(x^2-xy+y^2)$ **83.** $(x-\frac{2}{5})(x^2+\frac{2}{5}x+\frac{4}{25})$
85. $2(7x+10)(49x^2-70x+100)$ **87.** 3, 7 **89.** 0, 3 **91.** $-\frac{5}{6}, \frac{4}{9}$ **93.** 11 ft, 14 ft **95.** 20 in., 14 in.

Chapter 5 Cooperative Learning

1. x^9-3 (Answers may vary.) **3.** $72y^5+128y^4-63y^3+208y^2-84$ **5.** $8x-3$ **7.** x^2-y^2
9. $3y-7$ and $27y-12$ **11.** $56x^2+51x-27=0$

Chapter 5 Practice Test

1. 20 **2.** 7 **3.** $-5x^2-7x$ **4.** $11a^2-8ab-8b^2$ **5.** $2y^2-6y+6$ **6.** $3x^3-13x^2+10x+6$
7. $6x^2-23x+20$ **8.** $9y^2-64$ **9.** $25p^2-90p+81$ **10.** $3a-2+\frac{1}{a^2}$ **11.** $x+8+\frac{12}{x-2}$ **12.** $x+1+\frac{11}{x-4}$
13. $(x+5)(x+2)$ **14.** $3(5x+3)(2x-1)$ **15.** $(2x-5)^2$ **16.** $x^2(7x-9y)(7x+9y)$
17. $(4y-5z)(16y^2+20yz+25z^2)$ **18.** $\frac{3}{2}, -4$ **19.** 0, 15 **20.** 6 m, 10 m

CHAPTER 6

Chapter 6 Pretest

1. $\frac{3x-2}{2(x+2)}$ **2.** $\frac{(x-3)^2}{3}$ **3.** $\frac{1}{x-y}$ **4.** $\frac{17x+42}{(3x-2)(x+6)}$ **5.** $\frac{5b+4a}{b-2a}$ **6.** $\frac{16}{7}$ **7.** 18 hours **8.** 5 mph
9. $r=\frac{Re}{E-e}$ **10.** 272.2 ft **11.** 1920 kg

Problem Set 6.1

1. $2x^2$ **3.** $\frac{1}{5x}$ **5.** $x+2$ **7.** $\frac{3}{4}$ **9.** 3 **11.** $\frac{x-4}{2}$ **13.** $\frac{a+1}{a-2}$ **15.** $\frac{2x-1}{3x+1}$ **17.** $\frac{a-3}{4a-1}$
19. $\frac{x^2-xy+y^2}{x-y}$ **21.** -1 **23.** -1 **25.** $-5-b$ **27.** $-x-3$ **29.** -1 **31.** $\frac{45}{x}$ **33.** $\frac{7x}{6}$ **35.** 1
37. $\frac{(x-5)(3x+1)}{x(x+1)}$ **39.** $\frac{(3x-1)(5x-2)}{(4x-1)(3x+1)}$ **41.** $\frac{(p+3)(p-2)}{p+2}$ **43.** $\frac{4+y}{4-y}$ **45.** $\frac{1}{2}$ **47.** $\frac{12}{x-2}$ **49.** $4a-3$
51. $\frac{x(x+2)(x-3)}{(x+4)(x-4)}$ **53.** $\frac{2x+3}{2x-3}$ **55.** $\frac{x^2+2x+4}{(x+2)^2}$ **57.** $\frac{1}{a+b}$ **59.** $\frac{a+b}{a-b}$ **61.** Polynomials **63.** Factors
65. Reciprocal **67.** -1 **69.** $\frac{x^2-50{,}310.49}{542.52x^2}$ **71.** $\frac{341.42x^2-633.08x+24.96}{811.2x^2+431.6x}$ **73.** $\frac{x-4}{4x^3-1}$
75. $\frac{(x-2)^2}{(x+2)(x-1)(x+1)}$ **77.** $\frac{8}{3(x-2)}$ **79.** $\frac{y^2}{y-x}$ **81.** $\frac{a+5}{3a+1}$ **83.** $2x+3$ **85.** $-5x^2-6x-9$
87. $x^2-10x+5$ **89.** $-5x^2+10x-1$

Problem Set 6.2

1. 24 **3.** $144a^2$ **5.** $75x^2y^4$ **7.** $(x-3)(x-5)$ **9.** $3(x+3)(x-3)$ **11.** $(2x+3)(x-5)x$ **13.** $2(y-2)$
15. $(z-3)(z-4)(z-1)$ **17.** $\frac{14x^2}{10x^4}$ **19.** $\frac{5x+20}{x^2-16}$ **21.** $\frac{14}{6z-8}$ **23.** $\frac{3k^2+15k}{k^2+6k+5}$ **25.** $\frac{-8}{x-7}$ **27.** $\frac{-3x-2}{y^2-x^2}$
29. $\frac{9-x}{x}$ **31.** 1 **33.** $\frac{6y^2-4y+3}{14y^2}$ **35.** $\frac{-z^2+7z+22}{(z-2)(z+2)}$ **37.** $\frac{12+7y}{2(3y-2)}$ **39.** $\frac{4x+15}{(x+5)(x-5)}$ **41.** $\frac{2}{3}$
43. $\frac{z^2+3z-7}{(z+3)(z+3)}$ **45.** $\frac{3x^2+13x-10}{(x-1)(x-5)(x+2)}$ **47.** $\frac{4x}{x-5}$ **49.** $\frac{24y+3xy+3y^2}{4(x+y)(x-y)}$ **51.** $\frac{-3x^2-5x+35}{(x+4)(x-4)(x+5)}$
53. $\frac{x-5}{x^2-y^2}$ **55.** $\frac{5y^2-16y+4}{(2y-3)(y-1)(y-2)}$ **57.** $\frac{5y^2-10y+4}{(y+4)(y-1)(y-2)}$ **59.** $\frac{16}{x}$ **61.** $\frac{-2x-5}{x^2-16}$ **63.** $\frac{2p+1}{(p-1)(p+1)}$
65. $\frac{5x}{(x+2)(x-3)}$ **67.** $\frac{-p^2-4p+35}{(p-4)(p-4)}$ **69.** $\frac{-p-4q}{(p+q)(p-q)}$ **71.** $\frac{3t+7}{(2t-1)(t+2)}$ **73.** $\frac{4a-b}{(a+b)(a-b)}$

75. Same denominator **77.** 1 **79.** $\dfrac{-11y^2 - y + 4}{3(y+2)(y-2)}$ **81.** $\dfrac{-45}{(x+2)(x-2)}$ **83.** $\dfrac{2}{a-2b}$ **85.** $\frac{13}{8}$ **87.** 6
89. $\frac{28}{11}$ **91.** $-\frac{1}{2}$

Problem Set 6.3

1. $\dfrac{5+2x}{3}$ **3.** $\dfrac{4x-12}{x+3}$ **5.** $\dfrac{1+5b}{1-3b}$ **7.** $\dfrac{x^2-3x}{x+4}$ **9.** $\dfrac{a^2-1}{a^2+1}$ **11.** $\dfrac{2}{p^2+2p}$ **13.** $\dfrac{7}{8x+3}$ **15.** $\dfrac{3x^2+x}{5x^2-6x+1}$
17. $\dfrac{2x-1}{-1}$ or $1-2x$ **19.** $\dfrac{2b+3a}{3b-2a}$ **21.** $\dfrac{x-y}{y}$ **23.** $\dfrac{3a^2-a^2b}{b^2-ab^2}$ **25.** $\dfrac{1}{y+4x}$ **27.** $-\dfrac{1}{2x(x+h)}$ **29.** $\dfrac{x^2y^2}{x^2-xy+y^2}$
31. Numerator, denominator **33.** Dividing **35.** $\dfrac{9x^2+y^2}{-6xy}$ **37.** $\dfrac{x^2-2xy+y^2}{x^2-xy+y^2}$ **39.** $\dfrac{17a-5}{4a-1}$ **41.** $\dfrac{a^2b-a^2-b}{ab^2-a^2+b}$
43. The right side is $\dfrac{-2x-h}{x^2(x+h)^2}$. **45.** 0 **47.** 0 **49.** 5 **51.** 2 **53.** 1

Problem Set 6.4

1. 0 **3.** $\frac{5}{3}$ **5.** 7, 1 **7.** None **9.** -4 **11.** $\frac{5}{2}$, 2 **13.** 16 **15.** -8 **17.** -6 **19.** -3 **21.** 4
23. 5 **25.** -1, -3 **27.** 6 **29.** -3 **31.** -6, 4 **33.** No solution **35.** No solution **37.** $\frac{1}{2}$, -4 **39.** 5
41. 6 **43.** No solution **45.** -6, 5 **47.** $-\frac{28}{3}$ **49.** $\frac{22}{3}$ **51.** -2 **53.** $\dfrac{-y+2}{2y}$ **55.** 11 **57.** $\dfrac{2x-22}{(x+4)(x-2)}$
59. No solution **61.** $\frac{5}{2}$ **63.** No solution **65.** Rational **67.** Zero **69.** Linear **71.** Denominator
73. No solution **75.** -1 **77.** $x + 5 + \dfrac{-8}{x+3}$ **79.** $2z - 1 + \dfrac{3}{3z-2}$ **81.** $x - 4 + \dfrac{1}{3x^2-2}$ **83.** $\frac{1}{4}$ **85.** $\dfrac{-5}{11}$

Problem Set 6.5

1.
$$\begin{aligned} 2 - \frac{1}{4x} &= \frac{1}{x} \\ 4x(2) - 4x\left(\frac{1}{4x}\right) &= 4x\left(\frac{1}{x}\right) \\ 8x - 1 &= 4 \\ 8x &= 5 \\ x &= \frac{5}{8} \end{aligned}$$
The number is $\frac{5}{8}$.

3.
$$\begin{aligned} \frac{7+x}{9+x} &= \frac{5}{6} \\ (9+x)(6)\left(\frac{7+x}{9+x}\right) &= (9+x)(6)\left(\frac{5}{6}\right) \\ 6(7+x) &= (9+x)5 \\ 42 + 6x &= 45 + 5x \\ 6x - 5x &= 45 - 42 \\ x &= 3 \end{aligned}$$
The number is 3.

5.
$$\begin{aligned} \frac{1}{x} + 14 &= \frac{1}{2x} \\ 2x\left(\frac{1}{x}\right) + 2x(14) &= 2x\left(\frac{1}{2x}\right) \\ 2 + 28x &= 1 \\ 28x &= -1 \\ x &= \frac{-1}{28} \end{aligned}$$
The number is $\frac{-1}{28}$.

7.
$$\begin{aligned} \tfrac{1}{5}t + \tfrac{1}{4}t &= 1 \\ 20(\tfrac{1}{5}t) + 20(\tfrac{1}{4}t) &= 20(1) \\ 4t + 5t &= 20 \\ 9t &= 20 \\ t &= \tfrac{20}{9} = 2\tfrac{2}{9} \end{aligned}$$
It takes them $2\frac{2}{9}$ hours.

9.
$$\begin{aligned} \tfrac{1}{4}t + \tfrac{1}{3}t &= 1 \\ 12(\tfrac{1}{4}t) + 12(\tfrac{1}{3}t) &= 12(1) \\ 3t + 4t &= 12 \\ 7t &= 12 \\ t &= \tfrac{12}{7} = 1\tfrac{5}{7} \end{aligned}$$
It takes them $1\frac{5}{7}$ hours.

11.
$$\begin{aligned} \frac{1}{\frac{7}{2}}t + \frac{1}{\frac{17}{4}}t &= 1 \\ \tfrac{2}{7}t + \tfrac{4}{17}t &= 1 \\ 119(\tfrac{2}{7}t) + 119(\tfrac{4}{17}t) &= 1(119) \\ 34t + 28t &= 119 \\ 62t &= 119 \\ t &= \tfrac{119}{62} = 1\tfrac{57}{62} \end{aligned}$$
It takes them $1\frac{57}{62}$ hours.

13.
$$\begin{aligned} \tfrac{1}{8}t - \tfrac{1}{12}t &= 1 \\ 24(\tfrac{1}{8}t) - 24(\tfrac{1}{12}t) &= 24(1) \\ 3t - 2t &= 24 \\ t &= 24 \end{aligned}$$
It takes both pipes 24 hours to fill the barrel.

15.
$$\begin{aligned} \frac{8}{x} + \frac{8}{2x} &= 1 \\ 2x\left(\frac{8}{x}\right) + 2x\left(\frac{8}{2x}\right) &= 2x(1) \\ 16 + 8 &= 2x \\ 24 &= 2x \\ 12 &= x \\ 2x &= 24 \end{aligned}$$
It takes the electrician 12 hours and the apprentice 24 hours.

17.
$$\begin{aligned} \frac{225}{r+20} &= \frac{125}{r} \\ (r+20)(r)\left(\frac{225}{r+20}\right) &= (r+20)(r)\left(\frac{125}{r}\right) \\ 225r &= (r+20)125 \\ 100r &= 2500 \\ r &= 25 \end{aligned}$$
Marilyn drives 25 mph.

19. $$\frac{72}{x} + 1 = \frac{810}{5x}$$
$$5x\left(\frac{72}{x}\right) + 5x(1) = 5x\left(\frac{810}{5x}\right)$$
$$360 + 5x = 810$$
$$5x = 450$$
$$x = 90$$
$5x = 450$

The rate of the jet was 450 mph.

21. $$\frac{20}{x} + \frac{40}{3x} = \frac{10}{9}$$
$$9x\left(\frac{20}{x}\right) + 9x\left(\frac{40}{3x}\right) = 9x\left(\frac{10}{9}\right)$$
$$180 + 120 = 10x$$
$$300 = 10x$$
$$30 = x$$
$3x = 90$

The speed of the train was 90 mph.

23. $$\frac{d}{8} = \frac{d}{12} + \frac{1}{4}$$
$$24\left(\frac{d}{8}\right) = 24\left(\frac{d}{12}\right) + 24\left(\frac{1}{4}\right)$$
$$3d = 2d + 6$$
$$d = 6$$
It is 6 miles from her home to school.

25. $$\frac{6}{12 - x} = \frac{10}{12 + x}$$
$$(12 - x)(12 + x)\left(\frac{6}{12 - x}\right) = (12 - x)(12 + x)\left(\frac{10}{12 + x}\right)$$
$$(12 + x)6 = (12 - x)10$$
$$72 + 6x = 120 - 10x$$
$$16x = 48$$
$$x = 3$$
The speed of the current is 3 mph.

27. $$\frac{7}{18 + x} = \frac{5}{18 - x}$$
$$(18 + x)(18 - x)\left(\frac{7}{18 + x}\right) = (18 + x)(18 - x)\left(\frac{5}{18 - x}\right)$$
$$(18 - x)7 = (18 + x)5$$
$$126 - 7x = 90 + 5x$$
$$36 = 12x$$
$$3 = x$$
The speed of the current is 3 mph.

29. $$\frac{22}{x + 3} = \frac{16}{x - 3}$$
$$(x + 3)(x - 3)\left(\frac{22}{x + 3}\right) = (x + 3)(x - 3)\left(\frac{16}{x - 3}\right)$$
$$22(x - 3) = (x + 3)16$$
$$22x - 66 = 16x + 48$$
$$6x = 114$$
$$x = 19$$
The speed of the boat is 19 mph.

31. $$\frac{5}{t} + \frac{5}{2t} = 1$$
$$2t\left(\frac{5}{t}\right) + 2t\left(\frac{5}{2t}\right) = 2t(1)$$
$$10 + 5 = 2t$$
$$15 = 2t$$
$$2t = 15$$
It takes Mrs. Ming 15 hours.

33. $$\frac{2}{5} + \frac{2}{t} = 1$$
$$5t\left(\frac{2}{5}\right) + 5t\left(\frac{2}{t}\right) = 5t(1)$$
$$2t + 10 = 5t$$
$$10 = 3t$$
$$\tfrac{10}{3} = 3t$$
The second person takes $\frac{10}{3}$ or $3\frac{1}{3}$ hours to clean the house.

35. Rate of work **37.** 20 mph **39.** $28\frac{1}{8}$ miles **41.** $37\frac{5}{7}$ miles

43. $I = \dfrac{E}{R}$ **45.** $F = \dfrac{S - PS}{P}$ **47.** $A = PRT + P$ **49.** $y = \dfrac{-5x + 6}{-2}$ **51.** $x = \dfrac{5Q}{y}$

Problem Set 6.6

1. $\frac{17}{4}$ **3.** $\frac{24}{5}$ **5.** $R = \dfrac{E}{I}$ **7.** $c = \dfrac{2A - bh}{h}$ **9.** $d_1 = \dfrac{fd_2}{d_2 - f}$ **11.** $S = \dfrac{PF}{1 - P}$ **13.** $R = \dfrac{R_1R_2}{R_2 + R_1}$
15. $n = \dfrac{p - A}{-Ar}$ or $n = \dfrac{-p + A}{Ar}$ **17.** $p = \dfrac{A}{RT + 1}$ **19.** $T_1 = \dfrac{P_1V_1T_2}{P_2V_2}$ **21.** $d_2 = \dfrac{W_2d_1}{W_1}$ **23.** $g = \dfrac{-Rs}{R - s}$
25. $e = \dfrac{Er}{R + r}$ **27.** $y = \dfrac{-4x + 24}{3}$ **29.** $f = \dfrac{dn - d - L}{-1}$ or $f = -dn + d + L$
31. $V = \dfrac{gt^2 - 2s + 2x}{-2t}$ or $V = \dfrac{-gt^2 + 2s - 2x}{2t}$ **33.** 3 atmospheres **35.** Equations **37.** 5.2 **39.** 5.8
41. $t = \dfrac{\text{mpv}}{k} - m$ **43.** $d = \dfrac{2S - 2an}{n(n - 1)}$ **45.** $2, \frac{1}{4}$ **47.** $0, -5$ **49.** $-5, 3$ **51.** 15 ft **53.** $8, -5$

Problem Set 6.7

1. Direct **3.** Joint **5.** Inverse **7.** $\frac{21}{5}$ **9.** 24 **11.** $\frac{5}{2}$ **13.** 63 **15.** \$113.75 **17.** $\frac{3}{32}$ ampere
19. 1800 lb/ft^2 **21.** 0.75 ohm **23.** 160 lb **25.** 64 kg **27.** 1200 kg **29.** 400 units **31.** \$144,378
33. Quotient **35.** Directly **37.** Jointly **39.** 2.26 **41.** 11.34 lb **43.** Multiplied by $\frac{1}{2}$ **45.** 121 **47.** $\frac{49}{81}$
49. 125 **51.** 81 **53.** y^{12} **55.** $16x^{12}$

Chapter 6 Additional Exercises

1. $\dfrac{6y^2}{5x}$ **3.** $\dfrac{3x-2}{4}$ **5.** $\dfrac{a-2}{a+5}$ **7.** $-\dfrac{6p^3}{q^2}$ **9.** $\dfrac{1}{x+y}$ **11.** $-3x-y$ **13.** 1 **15.** $\dfrac{6y^3+4x^2y^2-12x}{48x^2y^5}$
17. $\dfrac{x^2+3x-3}{(x-5)(x+3)}$ **19.** $\dfrac{3y^2+7y-8}{(y-4)(y-3)(y+2)}$ **21.** $\dfrac{-18x^3-51x^2y-10x-140y+42xy^2}{5(3x-2y)(2x+7y)}$ **23.** $\dfrac{-14x^2+78x-100}{(2x-5)^2}$
25. $\dfrac{5x+20}{x-4}$ **27.** $\dfrac{1+4x}{2-3x}$ **29.** $\dfrac{y^2-2y}{y+2}$ **31.** $\dfrac{10}{12x-5}$ **33.** $\dfrac{-x-3}{5x^2-4x}$ **35.** 3 **37.** $-\frac{4}{3}, \frac{1}{2}$ **39.** $-2, 2$ **41.** 3
43. $\dfrac{7}{y-1}$ **45.** 0 **47.** $\frac{1}{3}$ **49.** $3\frac{3}{13}$ hours **51.** 2 mph **53.** $G=\dfrac{Fd^2}{m_1m_2}$ **55.** $R=\dfrac{-Ar}{A-r}$ **57.** 4 **59.** $\frac{45}{4}$
61. $\frac{27}{4}$ **63.** $-\frac{25}{3}$ **65.** 15 ft^3

Chapter 6 Cooperative Learning

1. A rational expression is the quotient of two polynomials. **3.** $\dfrac{12x^2-37x+28}{21x^2-29x-10}$ **5.** (a) The sum should be $\dfrac{8}{9-x^2}$; (b) the difference should be $\dfrac{8x^2-58x-18}{(3x-2)(x+3)(3x-2)}$ **7.** Neither

Chapter 6 Practice Test

1. $\dfrac{y+3}{3(2y+1)}$ **2.** $\dfrac{x-1}{2}$ **3.** $\dfrac{1}{a+b}$ **4.** $\dfrac{2}{x-3}$ **5.** $\dfrac{x^2+2}{x^2-2x}$ **6.** -8 **7.** $1\frac{5}{7}$ hours **8.** 125 mph
9. $n=\dfrac{IR}{E-IR}$ **10.** 393 **11.** 9.12 kg

Cumulative Review Chapters 5 and 6

1. 17 **2.** 2 **3.** $8a^2+6ab-17b^2$ **4.** $-2x^2-15x-15$ **5.** $12x^2+16x-16$ **6.** $16y^2-81$
7. $2x^3-9x^2+7x-12$ **8.** $64p^2-48pq+9q^2$ **9.** $b-3-\dfrac{1}{2b^2}$ **10.** $3x+5+\dfrac{5}{x-4}$ **11.** $3x(2x+1)(3x-2)$
12. $2(3x-5)^2$ **13.** $x(x-2y)(x+2y)$ **14.** $(3x-7y)(9x^2+21xy+49y^2)$ **15.** $-\frac{8}{3}, 5$ **16.** 0, 4
17. $\dfrac{2(2x-3)(2x-1)}{x-3}$ **18.** $\dfrac{6x^2-16x-6}{(5x+2)^2(x-3)}$ **19.** $\dfrac{2x^2+6x+11}{(x-7)(2x+3)}$ **20.** $\dfrac{3x}{4x^2-14x-8}$ **21.** $-1, \frac{23}{5}$ **22.** $1\frac{1}{5}$ hours
23. $4\frac{2}{3}$ mph **24.** $R=\dfrac{nE}{I+In}$ **25.** \$197.75 **26.** 0.074 calorie per square centimeter per minute **27.** $2\frac{10}{13}$ hours

CHAPTER 7

Chapter 7 Pretest

1. $8|y|$ **2.** $9|x+7|$ **3.** $-4x$ **4.** $x-5y$ **5.** $2y^2\sqrt[3]{y}$ **6.** $3xy^2$ **7.** 13.2 **8.** $-13\sqrt{2}$ **9.** $146-54\sqrt{10}$
10. $\dfrac{5\sqrt{7}}{21}$ **11.** $\dfrac{16\sqrt{6}+72\sqrt{2}-6\sqrt{3}-27}{-69}$ **12.** $(xyz)^{5/4}$ **13.** $\dfrac{1}{7^{1/15}}$ **14.** $x^2y^4\sqrt{x}$ **15.** 4 **16.** $2i\sqrt{7}$
17. $-16-13i$ **18.** $51-83i$ **19.** -4 **20.** $\frac{11}{26}+\frac{3}{26}i$ **21.** 29.698 m

Problem Set 7.1

1. 3 **3.** -5 **5.** $\frac{2}{7}$ **7.** $-\frac{9}{8}$ **9.** 0.4 **11.** $3|y|$ **13.** $8|b|$ **15.** $|p+2|$ **17.** $7|x+4|$ **19.** $|3x-2|$
21. 4 **23.** $-10x$ **25.** 3 **27.** -2 **29.** -5 **31.** p **33.** $|x|$ **35.** 4 **37.** Not defined **39.** $2|a|$
41. -4 **43.** $|x+y|$ **45.** Not defined **47.** $3x-4$ **49.** $2|a|$ **51.** x^4 **53.** $|x^3|$ **55.** x^2 **57.** $\frac{2}{3}$
59. $|x+y|$ **61.** $4ab$ **63.** 10 mph **65.** Nonnegative **67.** Index **69.** $|x^2-y^2|$ **71.** (14, 8, 19) **73.** 12 lb
75. \$10,000 at 12%, \$25,000 at 5%, \$5,000 at 3% **77.** 15 A, 10 B, and 20 C **79.** $a=-7, b=1, c=-1$

Problem Set 7.2

1. $\sqrt{15}$ **3.** $\sqrt{77xy}$ **5.** $\sqrt[3]{20}$ **7.** $\sqrt[4]{63}$ **9.** Cannot be multiplied **11.** $\sqrt[5]{27t^4}$ **13.** $\frac{5}{3}$ **15.** $\dfrac{\sqrt{7}}{2}$ **17.** $\frac{1}{2}$
19. $\frac{3}{2}$ **21.** $\dfrac{x^3}{8}$ **23.** $\dfrac{y^2}{2}$ **25.** $\dfrac{8a\sqrt{a}}{b^2}$ **27.** $\dfrac{3a\sqrt[3]{a^2}}{4b}$ **29.** $\dfrac{a^2b\sqrt[4]{b^3}}{c^2}$ **31.** 5 **33.** $\sqrt{5}$ **35.** 2 **37.** $\sqrt[5]{y^4}$

39. $3xy^2$ **41.** $x + y$ **43.** $2\sqrt{3}$ **45.** $2\sqrt{6}$ **47.** $-3\sqrt{10}$ **49.** $2\sqrt[3]{9}$ **51.** $2x^4\sqrt{19}$ **53.** $3x^2\sqrt[3]{2x}$
55. $2\sqrt[4]{6}$ **57.** $3xy^2\sqrt[4]{2x^2}$ **59.** $(a - b)\sqrt[3]{a - b}$ **61.** $-2pq^2\sqrt[3]{4pq}$ **63.** $2xy^2\sqrt[5]{2x}$ **65.** $4\sqrt{5}$ **67.** $2\sqrt[3]{5}$
69. $10\sqrt{7}$ **71.** $x\sqrt{15y}$ **73.** $2y\sqrt[3]{6}$ **75.** $9\sqrt{xy}$ **77.** $4a^2b\sqrt{6ab}$ **79.** Rational **81.** Rational **83.** Rational
85. 11.6 **87.** 4.7 **89.** 6.0 **91.** -0.2 **93.** Indexes **95.** Radicands **97.** Integers **99.** 2.2 **101.** 9.7
103. -13.0 **105.** 19.4 **107.** 2.1 **109.** 2.3 **111.** 54 **113.** $\frac{3}{x + 1}$ **115.** $24a^4b^3x\sqrt{y}$ **117.** $\frac{12}{17}$
119. $11a - 3$ **121.** $\frac{8 - y}{y}$ **123.** $\frac{x + 12}{(x - 3)(x + 3)}$ **125.** $\frac{5x^2 - 4x + 2}{(x - 3)(x - 1)(x + 4)}$

Problem Set 7.3

1. $11\sqrt{3}$ **3.** $-2\sqrt[3]{6}$ **5.** $17\sqrt[3]{3x}$ **7.** $-\sqrt{5}$ **9.** $5\sqrt[4]{5} - 2\sqrt{6}$ **11.** $10\sqrt{2}$ **13.** $-13\sqrt{3}$ **15.** $80\sqrt{2}$
17. $-5\sqrt[3]{2}$ **19.** $9\sqrt{10}$ **21.** $-9\sqrt{2x}$ **23.** $32x\sqrt{2}$ **25.** $\sqrt[3]{x}$ **27.** $3\sqrt[3]{a^2b}$ **29.** $-9\sqrt{2}$ **31.** $21m\sqrt{2}$
33. $-x\sqrt[3]{xy^2}$ **35.** $-9\sqrt[3]{2}$ **37.** $3\sqrt{3a - 3}$ **39.** $\sqrt{x - 1}\,(x - 4)$ **41.** $-\sqrt[3]{4}$ **43.** Index **45.** $-3x\sqrt[3]{x}$
47. $6\sqrt{x - 1}$ **49.** $\frac{157}{21}ab\sqrt{6b} - \frac{2}{3}a\sqrt{6b}$ **51.** $6x^2 - 14x$ **53.** $8x^2 + 26x - 45$ **55.** $9x^2 - 16$
57. $x^2 - 4xy + 4y^2$ **59.** $\frac{x^2}{2}$ **61.** $\frac{15}{x - 2}$ **63.** $\frac{x}{x + 5}$

Problem Set 7.4

1. $4\sqrt{7} - 7$ **3.** $\sqrt{15} + \sqrt{6}$ **5.** $3\sqrt{14} - 4\sqrt{3}$ **7.** $\sqrt[3]{12} - 12$ **9.** $2 + 3\sqrt[4]{10}$ **11.** $-b\sqrt[3]{3}$ **13.** 22
15. -62 **17.** 7 **19.** 16 **21.** $x - z$ **23.** $-1 + \sqrt{5}$ **25.** $17 + 21\sqrt{3}$ **27.** $-22 - 2\sqrt{15}$
29. $x - \sqrt{2x} - \sqrt{3x} + \sqrt{6}$ **31.** $2\sqrt[3]{2} - 7\sqrt[3]{12} + 10\sqrt[3]{9}$ **33.** $3 - 2x\sqrt{3} + x^2$ **35.** $51 + 14\sqrt{2}$
37. $12 - 4\sqrt{3}a + a^2$ **39.** $a + \sqrt{2a} + \sqrt{5a} + \sqrt{10}$ **41.** $\sqrt[5]{48} - 2\sqrt[5]{4} + \sqrt[5]{12} - 2$ **43.** Binomials
45. $5 - 6\sqrt{x - 4} + x$ **47.** $(\sqrt[3]{x})^2 - (\sqrt[3]{y})^2$ **49.** $6\sqrt{6} + 18\sqrt{14} - 8\sqrt{15} - 24\sqrt{35}$ **51.** $x^2 + 2x\sqrt{y - 3} + y - 3$
53. -19 **55.** $(A - B)(A^2 + AB + B^2) = A^3 - B^3$, Let $A = \sqrt{9}, B = \sqrt{4}$.
$(\sqrt{9} - \sqrt{4})[(\sqrt{9})^2 + \sqrt{9}\sqrt{4} + (\sqrt{16})^2] = (\sqrt{9})^3 - (\sqrt{4})^3$
$= (\sqrt{9} - \sqrt{4})[\sqrt{81} + \sqrt{36} + \sqrt{16}] = 27 - 8 = 19$
57. 7 **59.** $5y$ **61.** 5 **63.** $x\sqrt[3]{9}$
65. $2\sqrt{6}$ **67.** $3a^2\sqrt[3]{2a}$

Problem Set 7.5

1. $\frac{4\sqrt{3}}{3}$ **3.** $\frac{4\sqrt{10}}{5}$ **5.** $\frac{\sqrt{35}}{7}$ **7.** $\frac{2\sqrt{15}}{15}$ **9.** $\frac{\sqrt[3]{2}}{2}$ **11.** $\frac{5\sqrt[3]{2x^2}}{2x}$ **13.** $\frac{-21 + 3\sqrt{6}}{43}$ **15.** $\frac{\sqrt{21} + \sqrt{30}}{-3}$
17. $\frac{3\sqrt{2} - \sqrt{6} + \sqrt{30} - \sqrt{10}}{4}$ **19.** $\frac{8x - 8\sqrt{10} + x\sqrt{5} - 5\sqrt{2}}{x^2 - 10}$ **21.** $\frac{5\sqrt{6} + 3\sqrt{2} - 5 - \sqrt{3}}{22}$
23. $\frac{2\sqrt{15} - 3\sqrt{35} - 4\sqrt{2} + 2\sqrt{42}}{-17}$ **25.** $\frac{a - 2\sqrt{ab} + b}{a - b}$ **27.** $\frac{3p + 10\sqrt{pq} + 3q}{9p - q}$ **29.** $\frac{\sqrt{35}}{5}$ **31.** $\frac{2\sqrt{3ab}}{3b}$
33. $\frac{\sqrt[3]{35}}{5}$ **35.** $\frac{\sqrt[3]{15x^2}}{3x}$ **37.** $\frac{\sqrt[4]{2x^2y}}{2x^2y}$ **39.** $\frac{\sqrt[5]{12x^3y^3}}{2x^2y^2}$ **41.** $\frac{11}{9\sqrt{11}}$ **43.** $\frac{5}{\sqrt{35}}$ **45.** $\frac{3}{\sqrt{3}}$ **47.** $\frac{2a}{\sqrt{10a}}$
49. $\frac{5}{8\sqrt[3]{25}}$ **51.** $\frac{5a}{\sqrt[3]{15ab}}$ **53.** $\frac{xy}{2\sqrt{xy}}$ **55.** $\frac{-79}{7\sqrt{2} + 63}$ **57.** $\frac{78}{81 - 9\sqrt{3} - 9\sqrt{2} + \sqrt{6}}$ **59.** $\frac{-4}{3 - \sqrt{21} + 3\sqrt{2} - \sqrt{42}}$
61. $\frac{-129}{30 + 35\sqrt{6} - 6\sqrt{10} - 14\sqrt{15}}$ **63.** $\frac{3 - x}{3 + 2\sqrt{3x} + x}$ **65.** $\frac{a^2b - c^2}{ab + c\sqrt{b} - ac\sqrt{b} - c^2}$ **67.** 1 **69.** $\frac{2x - 7\sqrt{xy} + 3y}{4x - y}$
71. $\frac{\sqrt{7x} + \sqrt{35} - \sqrt{21}}{7}$ **73.** $\frac{20 + x}{4(5 + \sqrt{5 - x})}$ **75.** $\frac{x\sqrt{x} + 3\sqrt{x}}{x}$ **77.** $\frac{-37\sqrt{10}}{10}$ **79.** 2^8 **81.** x^2 **83.** $\frac{1}{8}$
85. $\frac{y^4}{x^3}$ **87.** $\frac{27x^6}{y^6}$ **89.** $\frac{2}{x(x + 2)}$ **91.** $\frac{(2x + 3)x}{(x - 1)(3x - 1)}$

Problem Set 7.6

1. $\sqrt[3]{y}$ **3.** 5 **5.** 3 **7.** 512 **9.** 625 **11.** 16 **13.** $\sqrt[3]{x^2yz}$ **15.** $\sqrt[9]{(7xy)^4}$ or $(\sqrt[9]{7xy})^4$ **17.** $a^{1/2}$ **19.** $15^{1/4}$
21. $7^{5/4}$ **23.** $5^{7/3}$ **25.** $(xyz)^{9/2}$ **27.** $(pq)^{8/9}$ **29.** $\frac{1}{y^{1/5}}$ **31.** $\frac{1}{3^{2/5}}$ **33.** $x^{2/3}$ **35.** $3a^{4/3}$ **37.** 7^2 **39.** $8^{7/10}$
41. $3^{2/3}$ **43.** $6^{5/12}$ **45.** $5^{1/3}$ **47.** $x^{1/2}$ **49.** $a^{2/5}$ **51.** $\frac{1}{x^3}$ **53.** $\sqrt{x}$ **55.** 2^3 or 8 **57.** $\sqrt{7}$ **59.** $x^2\sqrt[8]{x}$
61. $x^2\sqrt[3]{x^2}$ **63.** $\sqrt{2}$ **65.** x^3y^2 **67.** $\sqrt[4]{x^3y}$ **69.** $2pq^3$ **71.** $\sqrt[3]{a}$ **73.** $x\sqrt[10]{x^3}$ **75.** $\sqrt[6]{2^5}$ **77.** $x\sqrt[6]{x^5}$

79. $\sqrt[4]{5}$ **81.** $\sqrt[10]{x}$ **83.** $\sqrt[8]{27}$ **85.** Nonnegative **87.** 6 **89.** 343 **91.** 8 **93.** 5.08 **95.** $ab^{1/6}$ **97.** $\frac{d^2}{2c}$
99. $x^{1/6}$ **101.** 9^5x^4 **103.** $\frac{10}{3}$ **105.** 5 **107.** -3 **109.** No solution

Problem Set 7.7

1. 16 **3.** 4 **5.** 11 **7.** 4 **9.** 5 **11.** No solution **13.** 18 **15.** $\frac{33}{4}$ **17.** 6 **19.** $\frac{1}{49}$ **21.** 4 **23.** 5
25. No solution **27.** 8 **29.** 7 **31.** $-\frac{5}{3}$ **33.** No solution **35.** 17 **37.** $0, \frac{64}{9}$ **39.** 16 **41.** $\frac{9}{16}$
43. $-4, 2$ **45.** Extraneous **47.** 32.5 **49.** 17.1 **51.** 969.7 cm **53.** 2 **55.** $\frac{3}{4}$ **57.** $5, -2$ **59.** 4
61. 9 **63.** $\frac{4}{5}$ **65.** $y = \frac{xz}{1-x}$ **67.** $z = \frac{xy}{y-x}$ **69.** $\frac{5\sqrt{2}}{2}$ **71.** $\frac{3\sqrt[3]{4}}{2}$ **73.** $\frac{15 + 3\sqrt{3}}{11}$ **75.** $\frac{13 + 2\sqrt{2}}{23}$

Problem Set 7.8

1. 10 ft **3.** 8 m **5.** 6.403 ft **7.** 6.928 m **9.** 12.806 ft **11.** 1 cm **13.** 3.66 miles **15.** 16 in.
17. 22.6 cm **19.** 150 ft **21.** 1000 cm **23.** \$8988.80 **25.** 13.0 ft **27.** 20.5 ft **29.** 92.6 in.
31. 86.023 ft **33.** $\sqrt{x^2 + 9}$ cm **35.** $\sqrt{16 + x^2}$ ft **37.** 4 **39.** 1.389 footcandles **41.** 272.2 ft

Problem Set 7.9

1. $i\sqrt{5}$ **3.** $3i$ **5.** $-3i\sqrt{2}$ **7.** $10 + 7i$ **9.** $12 - 10i$ **11.** $-11 - 2i$ **13.** 0 **15.** $1 + 4i$ **17.** $-4 - 5i$
19. $7 + 9i$ **21.** $-1 - 4i$ **23.** $-6 - 2i$ **25.** $8 + 24i$ **27.** $12 - 21i$ **29.** $7 + 11i$ **31.** $23 - 2i$
33. $6 - 38i$ **35.** $-8 + 26i$ **37.** 13 **39.** 5 **41.** $32 + 24i$ **43.** $-7 - 24i$ **45.** $75 + 15i$ **47.** 29
49. $-\sqrt{15}$ **51.** -7 **53.** $-2\sqrt{3}$ **55.** $\frac{12}{17} - \frac{3}{17}i$ **57.** $\frac{23}{37} + \frac{10}{37}i$ **59.** $\frac{14}{17} - \frac{5}{17}i$ **61.** $-\frac{2}{3} - 3i$ **63.** $-\frac{36}{41} + \frac{45}{41}i$
65. $\frac{2}{3} - \frac{5}{9}i$ **67.** $-i$ **69.** i **71.** -1 **73.** $-i$ **75.** i **77.** $-i$ **79.** Imaginary **81.** Imaginary
83. Imaginary **85.** $-2.5641 + 3i$ **87.** $13.469 - 3.413i$ **89.** $40.35 - 33.14i$ **91.** $32.04 - 65.00i$
93. 159.12 **95.** -1 **97.** $\frac{1 - 2i}{4}$ **99.** $-\frac{6}{13} - \frac{17}{13}i$ **101.** $\frac{1}{i}$ **103.** $\frac{33}{4225} + \frac{56}{4225}i$ **105.** $\frac{7}{3} - \frac{7}{12}i$ **107.** 10
109. $\frac{2}{5}$ **111.** $\frac{11}{8}$ **113.** 3 **115.** y^5

Chapter 7 Additional Exercises

1. $10|y|$ **3.** $5|x - 4|$ **5.** $|4x + 5|$ **7.** $-3x^5$ **9.** Not defined **11.** $2|x^3|$ **13.** x^2 **15.** $\sqrt[3]{15}$ **17.** 10
19. $3y\sqrt{7}$ **21.** $2x^2$ **23.** $5x^2y$ **25.** $6\sqrt[4]{5}$ **27.** $-8\sqrt{5}$ **29.** $5\sqrt[3]{2}$ **31.** $x^4y\sqrt{y} + 5x^3y^4\sqrt{y}$ **33.** $5 - 4\sqrt{5}$
35. $2 + \sqrt[4]{14}$ **37.** 5 **39.** $82 - 10\sqrt{21}$ **41.** $59 + \sqrt{78}$ **43.** $x - 8\sqrt{x + 3} + 19$ **45.** $\sqrt{2}$ **47.** $\frac{\sqrt[3]{14}}{2}$
49. $\frac{\sqrt[5]{4}}{2}$ **51.** $\frac{y - \sqrt{3y} - \sqrt{5y} + \sqrt{15}}{y - 5}$ **53.** $\frac{-2\sqrt{x + 3} - 2\sqrt{x}}{3}$ **55.** $\frac{\sqrt{3y}}{y^2}$ **57.** $\frac{\sqrt[3]{3x^2}}{x}$ **59.** $\frac{-4}{7\sqrt{5} - 21}$
61. $\frac{1}{\sqrt{x + h + 3} + \sqrt{x + 3}}$ **63.** $\frac{27}{8}$ **65.** $\frac{27}{8}$ **67.** $\frac{1}{x^{5/8}}$ **69.** 1 **71.** $3y^4\sqrt[4]{y^3}$ **73.** $\frac{5\sqrt{y}}{y^6}$ **75.** 100 **77.** 6
79. 1 **81.** 3 **83.** $\frac{5}{2}$ **85.** -4 **87.** 12.649 cm **89.** 10.583 in. **91.** 3.317 ft **93.** 186.0 ft **95.** 5 in.
97. $38 - 34i$ **99.** $-3\sqrt{7}$ **101.** Real **103.** Imaginary **105.** Imaginary **107.** Imaginary **109.** 1
111. $-i$ **113.** $\frac{79}{34} + \frac{95}{34}i$ **115.** $-\frac{1}{3} - 7i$

Chapter 7 Cooperative Learning

1. $64y^2$ **3.** $-343x^9$ **5.** $\sqrt{3} - 1$ and $\sqrt{3} + 2$ (Answers may vary.) **7.** $\sqrt{2} + \sqrt{3}$ and $\sqrt{2} - 4\sqrt{3}$ (Answers may vary.) **9.** $5 + i$ and $6 - i$ (Answers may vary.)
11. $i\sqrt{7}$ and $i\sqrt{2}$ or $i\sqrt{14}$ and i

Chapter 7 Practice Test

1. $9|x|$ **2.** $5|x + 3|$ **3.** $-3y$ **4.** $|x - 3y|$ **5.** $3y\sqrt[3]{y^2}$ **6.** x **7.** 13.0 **8.** $21\sqrt{2}$ **9.** $18 - 17\sqrt{6}$
10. $\sqrt{3}$ **11.** $\frac{3\sqrt{6} - 9\sqrt{3} - 5\sqrt{2} + 15}{-14}$ **12.** $(ab^2c)^{4/3}$ **13.** $\frac{1}{6^{5/12}}$ **14.** x^2y^3 **15.** 4 **16.** $2i\sqrt{6}$ **17.** $-6 - 12i$
18. $45 + 11i$ **19.** $-3\sqrt{2}$ **20.** $\frac{27}{61} + \frac{8}{61}i$ **21.** 19.2 ft

CHAPTER 8

Chapter 8 Pretest

1. $\frac{-2+2\sqrt{2}}{5}, \frac{-2-2\sqrt{2}}{5}$ **2.** 5 m **3.** $-1+\frac{\sqrt{6}}{3}i, -1-\frac{\sqrt{6}}{3}i$ **4.** $\frac{1+\sqrt{57}}{2}, \frac{1-\sqrt{57}}{2}$ **5.** 3.4, 0.6

6. 1 (a double root) **7.** $x^2 - 3\sqrt{3}x + 6 = 0$ **8.** $r = \sqrt{\frac{V - \pi R^2 h}{\pi h}}$ **9.** 4.8 mph **10.** $\sqrt{3}, -\sqrt{3}, 2i, -2i$

11.

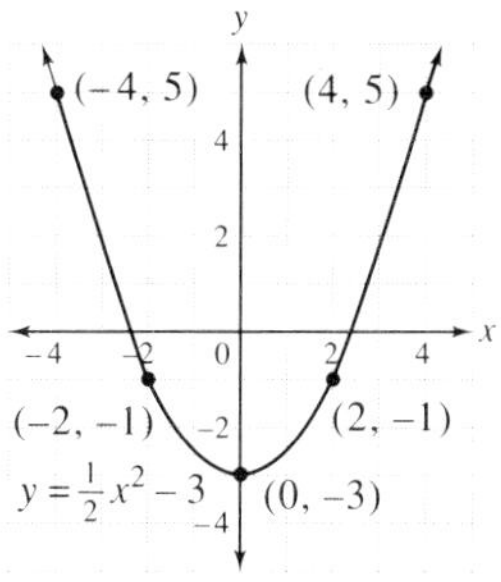

12.

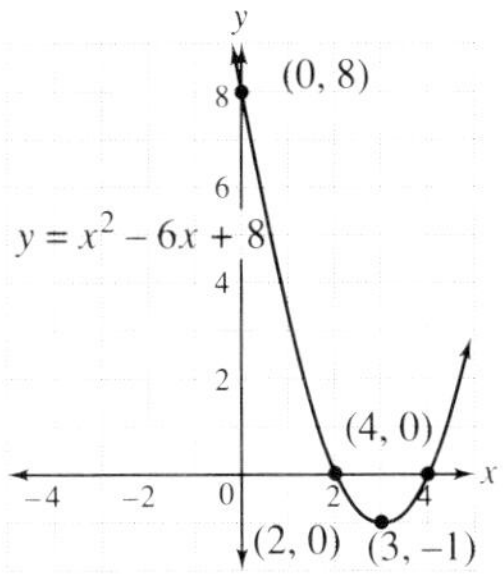

13.

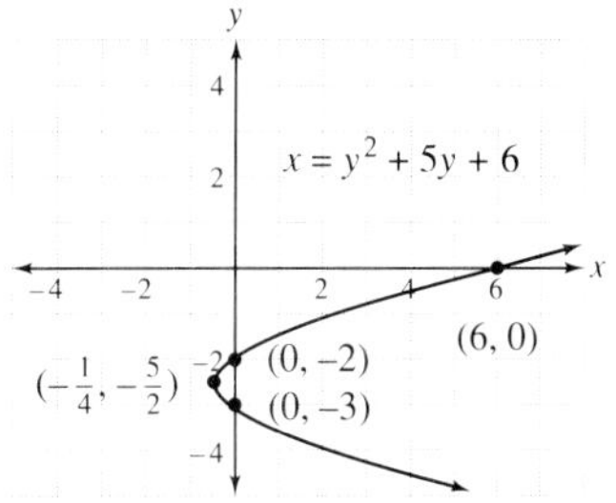

14. 36th week, 1074 **15.** $\{x \mid -5 < x < 1\}$;

16. $\{x \mid x \leq 0 \text{ or } x \geq 2\}$;

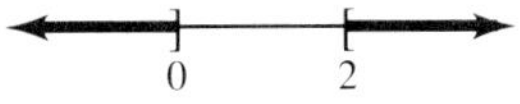

17. $\{x \mid x < 5 \text{ or } x \geq 11\}$

Problem Set 8.1

1. $\sqrt{5}, -\sqrt{5}$ **3.** $2\sqrt{2}, -2\sqrt{2}$ **5.** $\frac{\sqrt{42}}{7}, \frac{-\sqrt{42}}{7}$ **7.** 9, −1 **9.** $\frac{-3+\sqrt{3}}{2}, \frac{-3-\sqrt{3}}{2}$

11. $\frac{-2+2\sqrt{3}}{5}, \frac{-2-2\sqrt{3}}{5}$ **13.** $3 - i, 3 + i$ **15.** $3 + \frac{\sqrt{2}}{2}i, 3 - \frac{\sqrt{2}}{2}i$ **17.** 3, −2 **19.** $\frac{1}{4}, -\frac{1}{2}$ **21.** $\frac{4}{3}$, 1

23. $-\frac{7}{4}, \frac{1}{2}$ **25.** $0, \frac{-7}{10}$ **27.** $0, \frac{7}{3}$ **29.** $0, \frac{5}{6}$ **31.** $\frac{3}{4}i, -\frac{3}{4}i$ **33.** $\frac{-4+2\sqrt{2}}{3}, \frac{-4-2\sqrt{2}}{3}$ **35.** $\frac{3}{7}$

37. $\frac{3}{4} + \frac{\sqrt{5}}{4}i, \frac{3}{4} - \frac{\sqrt{5}}{4}i$ **39.** 1, −5 **41.** $4, -\frac{8}{3}$ **43.** 9 m, 12 m **45.** 80 ft, 60 ft **47.** 4.9 m **49.** 50 ft

51. 6, −9 **53.** 5, 3 **55.** Quadratic **57.** Principal square **59.** Squares **61.** ±6.164 **63.** ±2.121

65. 1.466, −0.277 **67.** $\frac{1}{24}, -\frac{17}{24}$ **69.** $0, \frac{1}{4}, \frac{3}{2}, \frac{15}{2}$ **71.** 30 **73.** $\frac{2x-4}{x}$ **75.** $\frac{4y+7}{4}$ **77.** $\sqrt{15} + 5$ **79.** 5

81. $8\sqrt[3]{49} - 14\sqrt[3]{7} + 5$ **83.** $10 + 2y\sqrt{10} + y^2$

Problem Set 8.2

1. $x^2 + 14x + 49$ **3.** $y^2 - \frac{3}{4}y + \frac{9}{64}$ **5.** $x^2 - \frac{2}{5}x + \frac{1}{25}$ **7.** 1, −3 **9.** $3 + \sqrt{5}, 3 - \sqrt{5}$ **11.** $\frac{5}{2} + \frac{\sqrt{3}}{2}i, \frac{5}{2} - \frac{\sqrt{3}}{2}i$

13. $1 + 2i, 1 - 2i$ **15.** $\frac{-9+\sqrt{129}}{6}, \frac{-9-\sqrt{129}}{6}$ **17.** $-3 + \sqrt{2}, -3 - \sqrt{2}$ **19.** $\frac{-5+\sqrt{33}}{4}, \frac{-5-\sqrt{33}}{4}$

21. $1 + 2\sqrt{2}i, 1 - 2\sqrt{2}i$ **23.** $3 + \sqrt{22}, 3 - \sqrt{22}$ **25.** $3, -\frac{1}{3}$ **27.** 11, −3 **29.** $\frac{-7+\sqrt{13}}{2}, \frac{-7-\sqrt{13}}{2}$

31. $\frac{1}{2} + \frac{\sqrt{19}}{2}i, \frac{1}{2} - \frac{\sqrt{19}}{2}i$ **33.** 8, −2 **35.** $\frac{-3+\sqrt{21}}{3}, \frac{-3-\sqrt{21}}{3}$ **37.** $-\frac{3}{4} + \frac{\sqrt{39}}{4}i, -\frac{3}{4} - \frac{\sqrt{39}}{4}i$ **39.** $2, -\frac{1}{2}$

41. Factoring **43.** Divide **45.** $3, -\frac{3}{2} + \frac{3}{2}\sqrt{3}i, -\frac{3}{2} - \frac{3}{2}\sqrt{3}i$ **47.** $\frac{-a \pm \sqrt{a^2 - 4b}}{2}$ **49.** $2\sqrt{6}$ **51.** $2\sqrt{10}$

53. $-\sqrt[3]{3}$ **55.** $-5\sqrt{5}$ **57.** $\sqrt{2}$

Problem Set 8.3

1. $\frac{-7+\sqrt{41}}{2}, \frac{-7-\sqrt{41}}{2}$ **3.** $\frac{3+\sqrt{41}}{8}, \frac{3-\sqrt{41}}{8}$ **5.** $-\frac{1}{2} + \frac{\sqrt{11}}{2}i, -\frac{1}{2} - \frac{\sqrt{11}}{2}i$ **7.** $-\frac{3}{4} + \frac{\sqrt{7}}{4}i, -\frac{3}{4} - \frac{\sqrt{7}}{4}i$

9. $-\frac{1}{2} + \frac{\sqrt{31}}{2}i, -\frac{1}{2} - \frac{\sqrt{31}}{2}i$ **11.** $\frac{1}{3}, -1$ **13.** $1 + i, 1 - i$ **15.** $\frac{1+\sqrt{17}}{2}, \frac{1-\sqrt{17}}{2}$ **17.** $-\frac{1}{2} + \frac{3}{2}i, -\frac{1}{2} - \frac{3}{2}i$

19. $0, -\frac{1}{5}$ **21.** $\frac{7 + \sqrt{41}}{4}, \frac{7 - \sqrt{41}}{4}$ **23.** $\frac{3}{2} + \frac{\sqrt{19}}{2}i, \frac{3}{2} - \frac{\sqrt{19}}{2}i$ **25.** $\frac{5}{12} + \frac{\sqrt{119}}{12}i, \frac{5}{12} - \frac{\sqrt{119}}{12}i$

27. $4 + \sqrt{26}, 4 - \sqrt{26}$ **29.** $1, -\frac{1}{2} + \frac{\sqrt{3}}{2}i, -\frac{1}{2} - \frac{\sqrt{3}}{2}i$ **31.** 8.4, −0.4 **33.** 1.2, −0.5 **35.** 0.9, −5.9

37. 0.5, −1.2 **39.** 15.1 hours **41.** 11.1 hours **43.** Quadratic **45.** 1.02, −0.81 **47.** −0.29, −2.75 **49.** $-3 \pm i$

51. i **53.** $\frac{1 + \sqrt{1 + 4i}}{2i}, \frac{1 - \sqrt{1 + 4i}}{2i}$ **55.** −3 **57.** 2, −5 **59.** $\frac{3}{2}, -1$ **61.** $\frac{2}{3}, -4$ **63.** $-\frac{1}{4}, -\frac{9}{2}$

Problem Set 8.4

1. Two real **3.** Two imaginary **5.** Two real **7.** One real **9.** Two imaginary **11.** Two imaginary
13. One real **15.** Two real **17.** Two imaginary **19.** Two real **21.** Two real **23.** Two imaginary
25. Two imaginary **27.** $x^2 + 3x - 10 = 0$ **29.** $x^2 + 4 = 0$ **31.** $x^2 - 8x + 16 = 0$ **33.** $x^2 - 10 = 0$
35. $8x^2 + 2x - 3 = 0$ **37.** $x^2 - 2x + 10 = 0$ **39.** $x^2 - \frac{1}{3}ax - \frac{1}{3}bx + \frac{1}{9}ab = 0$ **41.** Discriminant **43.** Double
45. Two imaginary **47.** Two real **49.** $9x^2 - 24x + 11 = 0$ **51.** $\frac{3}{8}$ **53.** Rational solutions **55.** $H = \frac{V}{LW}$

57. $n = \frac{C - S}{R}$ **59.** $n = -\frac{IR}{IR - E}$ **61.** $R = \frac{Er - er}{e}$ **63.** $d_2 = \frac{fd_1}{d_1 - f}$

Problem Set 8.5

1. $h = \frac{D^2}{c}$ **3.** $g = \frac{kL}{p^2}$ **5.** $b = \sqrt{c^2 - a^2}$ **7.** $r = \sqrt{\frac{3V}{\pi h}}$ **9.** $d = \sqrt{\frac{s}{kw}}$ **11.** $h = d^2\sqrt{\frac{k}{L}}$

13. $D = \sqrt{\frac{kLV}{g(P_1 - P_2)}}$ **15.** $t = \frac{-B + \sqrt{B^2 - 4AC}}{2A}$ **17.** $d = \sqrt{\frac{k}{R}}$ **19.** $r = \frac{-2\pi h + \sqrt{4\pi^2h^2 + 4\pi S}}{2\pi}$

21. 5 seconds

23.
$$\begin{aligned}
(30 + 2x)(40 + 2x) &= 1496 \\
1200 + 140x + 4x^2 &= 1496 \\
4x^2 + 140x - 296 &= 0 \\
x^2 + 35x - 74 &= 0 \\
(x - 2)(x + 37) &= 0
\end{aligned}$$
$x - 2 = 0$ or $x + 37 = 0$
$x = 2$ $\quad x = -37$ reject
The width of the strip is 2 ft.

25.
$$\begin{aligned}
\frac{300}{r} + \frac{300}{r - 10} &= 11 \\
r(r - 10)\frac{300}{r} + r(r - 10)\frac{300}{r - 10} &= r(r - 10)(11) \\
(r - 10)300 + r(300) &= r(r - 10)(11) \\
300r - 3000 + 300r &= 11r^2 - 110r \\
600r - 3000 &= 11r^2 - 110r \\
0 &= 11r^2 - 710r + 3000 \\
0 &= (11r - 50)(r - 60)
\end{aligned}$$
$11r - 50 = 0$ or $r - 60 = 0$
$11r = 50$ $\quad r = 60$
$r = \frac{50}{11} = 4\frac{6}{11}$
reject, $r - 10$ is negative
The speed of the car going to the city was 60 mph.

27.
$$\begin{aligned}
\frac{60}{r} - \frac{60}{r + 20} &= \frac{1}{4} \\
4r(r + 20)\frac{60}{r} - 4r(r + 20)\frac{60}{r + 20} &= 4r(r + 20)\frac{1}{4} \\
4(r + 20)60 - 4r(60) &= r(r + 20) \\
240r + 4800 - 240r &= r^2 + 20r \\
0 &= r^2 + 20r - 4800 \\
0 &= (r - 60)(r + 80)
\end{aligned}$$
$r - 60 = 0$ or $r + 80 = 0$
$r = 60$ $\quad r = -80$ (reject)
Kevin's average speed is 60 mph.

29.
$$\begin{aligned}
\frac{8}{6 + x} + \frac{8}{6 - x} &= 8 \\
(6 + x)(6 - x)\frac{8}{6 + x} + (6 + x)(6 - x)\frac{8}{6 - x} &= (6 + x)(6 - x)8 \\
(6 - x)8 + (6 + x)8 &= (6 + x)(6 - x)8 \\
48 - 8x + 48 + 8x &= 288 - 8x^2 \\
96 &= 288 - 8x^2 \\
8x^2 &= 192 \\
x^2 &= 24
\end{aligned}$$
$x = \sqrt{24}$ or $x = -\sqrt{24}$
$x \approx 4.9$ $\quad x \approx -4.9$ (reject)
The speed of the current is approximately 4.9 mph.

31.
$$\begin{aligned}
600 &= 100x^2 - 500x + 600 \\
0 &= 100x^2 - 500x \\
0 &= 100x(x - 5)
\end{aligned}$$
$100x = 0$ or $x - 5 = 0$
$x = 0$ $\quad x = 5$
The company could sell 0 sets or 5 sets.

33.
$$\begin{aligned}
5r^2 + 14r - 3 &= 0 \\
(5r - 1)(r + 3) &= 0
\end{aligned}$$
$5r - 1 = 0$ or $r + 3 = 0$
$5r = 1$ $\quad r = -3$
$r = \frac{1}{5}$ $\quad$ reject
r is $\frac{1}{5}$.

35.

$$(8 - 2x)(14 - 2x) = 40$$
$$112 - 44x + 4x^2 = 40$$
$$72 - 44x + 4x^2 = 0$$
$$18 - 11x + x^2 = 0$$
$$(9 - x)(2 - x) = 0$$
$$9 - x = 0 \quad \text{or} \quad 2 - x = 0$$
$$-x = -9 \qquad -x = -2$$
$$x = 9 \text{ (reject)} \qquad x = 2$$

The width is 2 in.

37.

$$2\left(\frac{18}{r+4}\right) = \frac{18}{r-4}$$
$$(r+4)(r-4)(2)\frac{18}{r+4} = (r+4)(r+4)\frac{18}{r-4}$$
$$(r-4)(2)(18) = (r+4)(18)$$
$$(r-4)2 = r+4$$
$$2r - 8 = r + 4$$
$$r = 12$$

The speed of the houseboat is 12 mph.

39.

$$\frac{400}{360+w} + \frac{400}{360-w} = \frac{9}{4}$$
$$(360+w)(360-w)(4)\frac{400}{360+w} + (360+w)(360-w)(4)\frac{400}{360-w} = (360+w)(360-w)(4)\frac{9}{4}$$
$$(360-w)(4)(400) + (360+w)(4)(400) = (360+w)(360-w)(9)$$
$$576{,}000 - 1600w + 576{,}000 + 1600w = (129{,}600 - w^2)9$$
$$1{,}152{,}000 = 1{,}166{,}400 - 9w^2$$
$$-14{,}400 = -9w^2$$
$$1600 = w^2$$
$$\pm 40 = w \qquad \text{reject } w = -40$$

The speed of the wind is 40 mph.

41. Equations **43.** 3.7 seconds **45.** $-b, 3a$ **47.** $P_1 = \dfrac{kLV + D^2gP_2}{D^2g}$ **49.** 8 cm, 4 cm **51.** $x^{14/15}$ **53.** $\dfrac{1}{6^{1/8}}$

55. $x^{1/3}$ **57.** $\dfrac{1}{2^{3/5}}$ **59.** $\sqrt{5}, -\sqrt{5}$

Problem Set 8.6

1. $1, -1, 2i, -2i$ **3.** 1, 9 **5.** $\sqrt{5}, -\sqrt{5}, \sqrt{2}, -\sqrt{2}$ **7.** 25 **9.** $\sqrt{7}, -\sqrt{7}, i\sqrt{2}, -i\sqrt{2}$ **11.** 7, −1, 5, 1

13. −64, 1 **15.** $-\frac{243}{32}, 32$ **17.** $\frac{1}{2}, \frac{1}{4}$ **19.** 9 **21.** $\frac{3}{4}, 1$ **23.** $\dfrac{\sqrt{5}}{2}, -\dfrac{\sqrt{5}}{2}, \dfrac{\sqrt{6}}{2}, -\dfrac{\sqrt{6}}{2}$ **25.** $\dfrac{-4+\sqrt{41}}{2}, \dfrac{-4-\sqrt{41}}{2}$

27. $\frac{10}{3}, \frac{2}{3}$ **29.** $2, -\frac{4}{3}$ **31.** Substitution **33.** $1, -1, \dfrac{1}{2} \pm \dfrac{\sqrt{3}}{2}i, -\dfrac{1}{2} \pm \dfrac{\sqrt{3}}{2}i$ **35.** $8, -1, -3 \pm \sqrt{17}$

37.

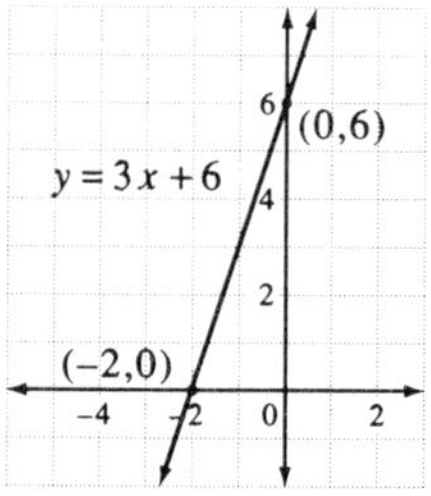

39.

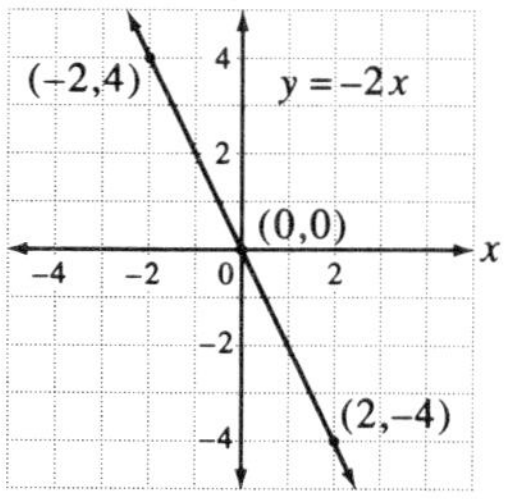

41.

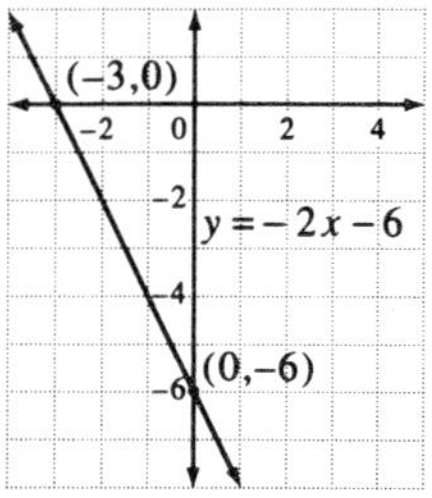

43.

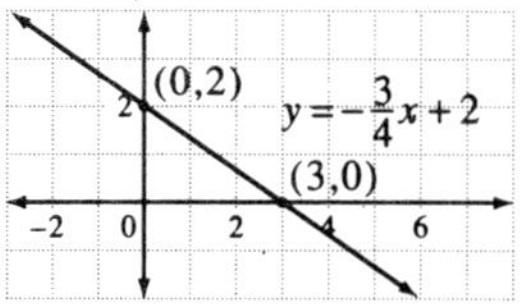

45.

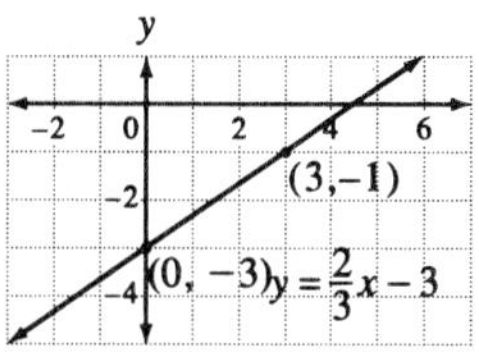

Problem Set 8.7

1.

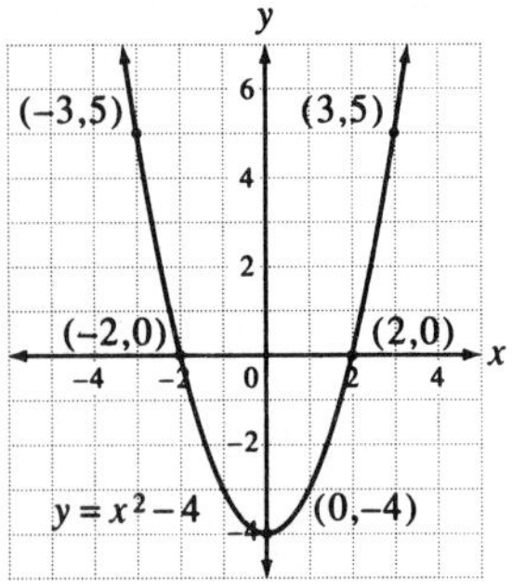

3.

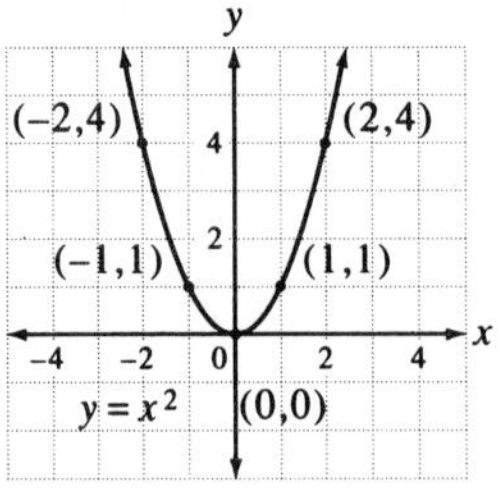

5.

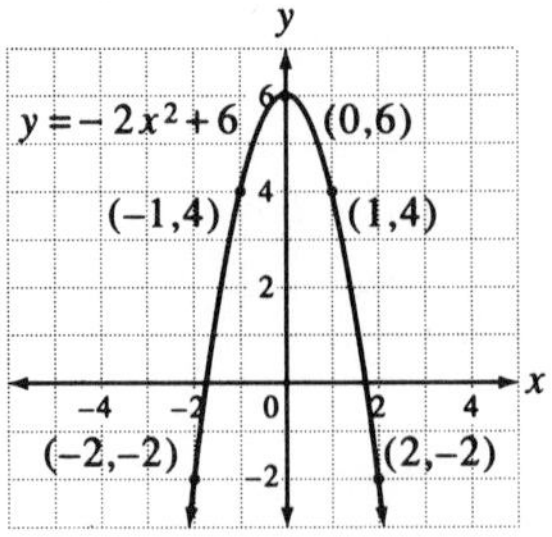

7.

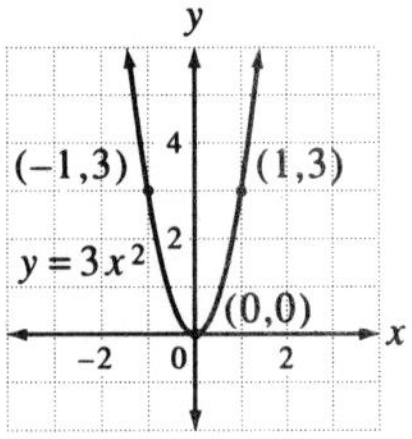

9.

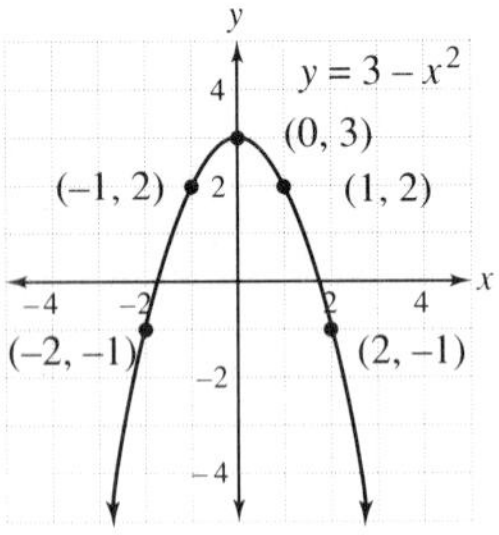

11.

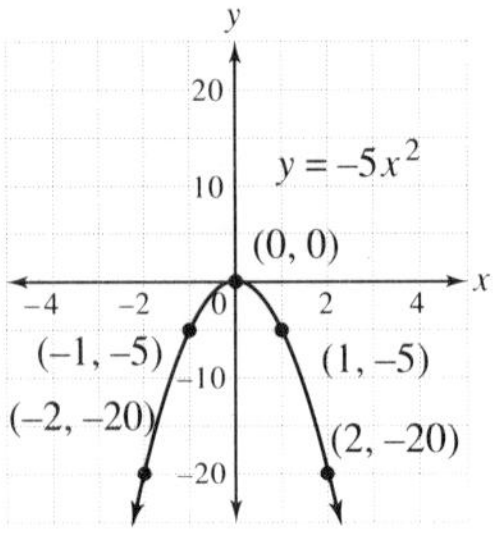

13.

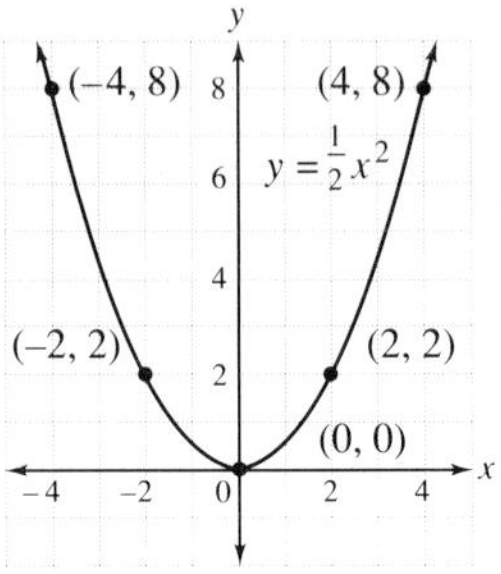

15.

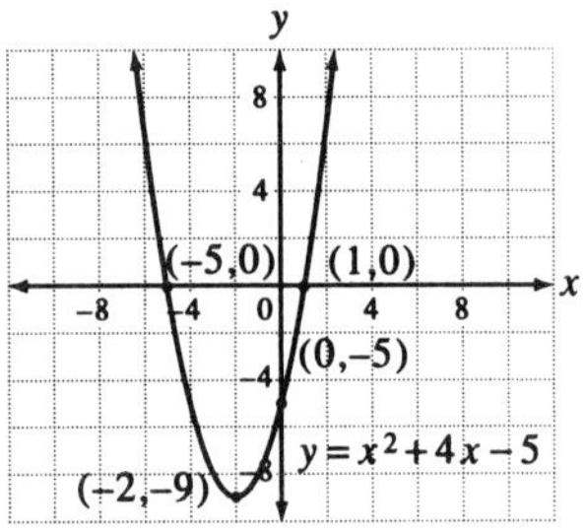

17.

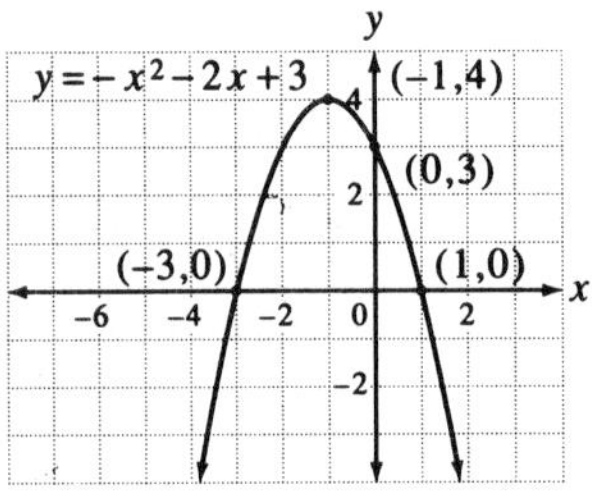

19.

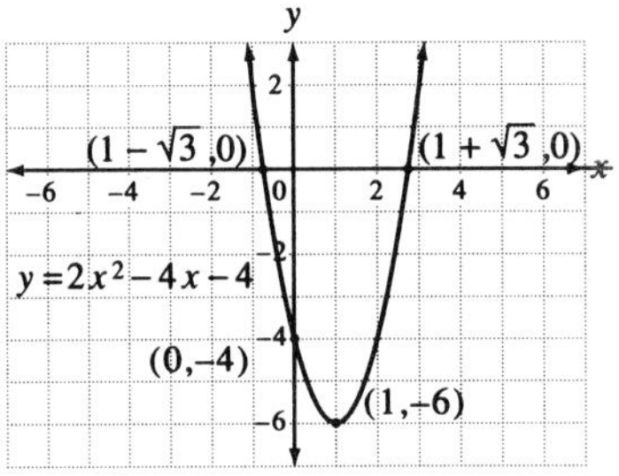

21.

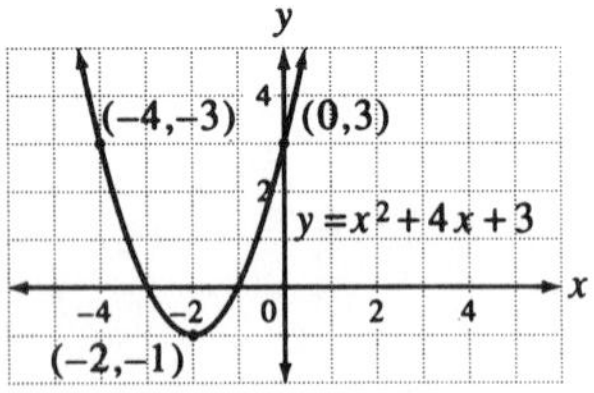

23.

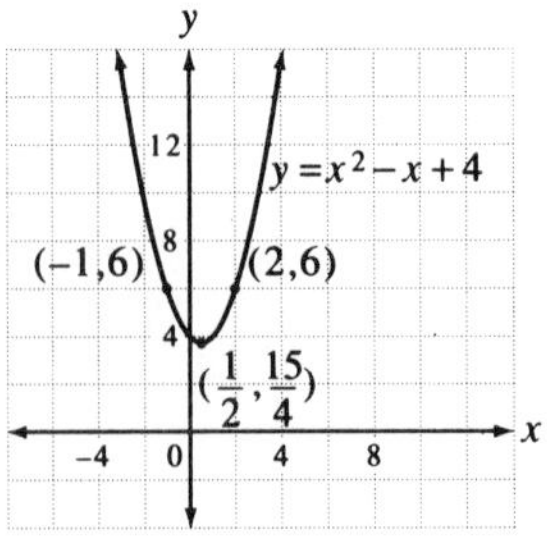

25.

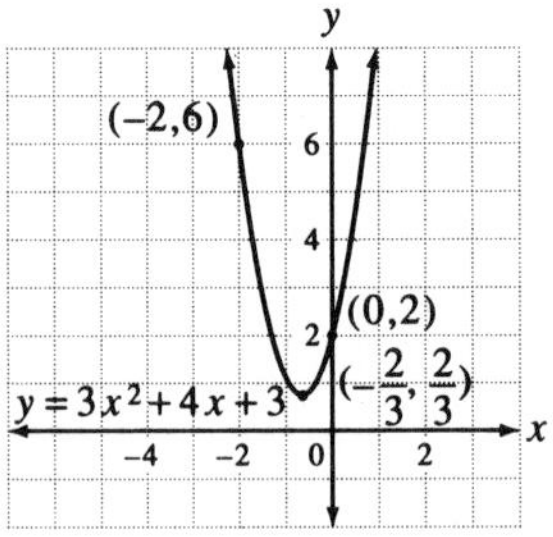

27.

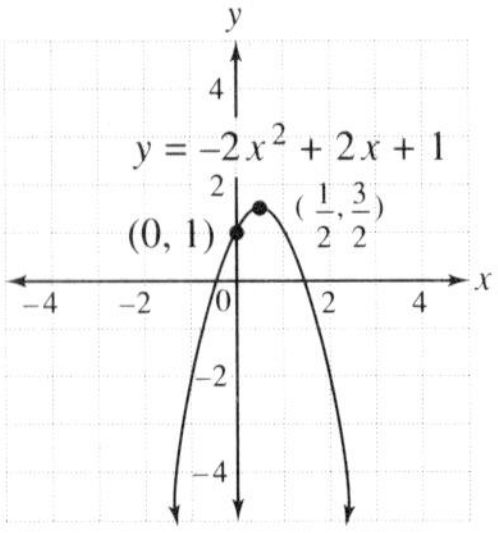

29.

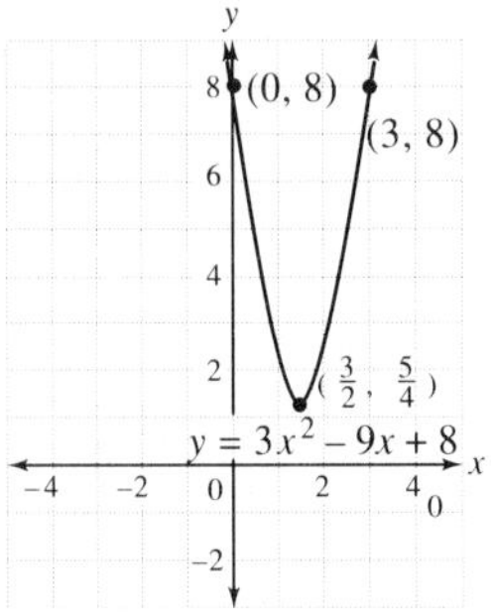

31. Ordered pairs
33. Parabola
35. Axis of symmetry

37. $y = 1.86x^2 + 3.72x - 3$

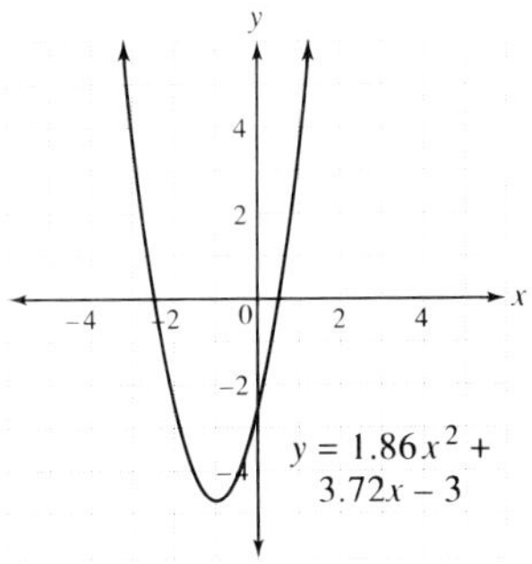

39. $y = -3(x + 1)^2 + 2$

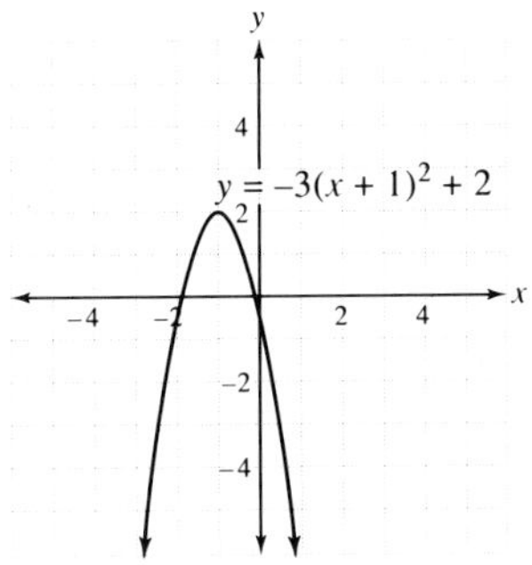

41. 3 **43.** 40 mph **45.** 3 mph **47.** 19.209 m **49.** 25.456 cm **51.** $6998.40

Problem Set 8.8

1. 64 ft **3.** 256 **5.** 441 ft²; 21 ft by 21 ft **7.** $-\frac{9}{4}, \frac{3}{2}, -\frac{3}{2}$ **9.** 400 cars, 40 robots **11.** 36th week, 2148 games

13.

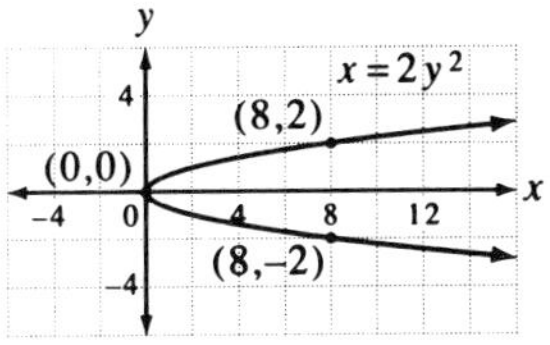

15.

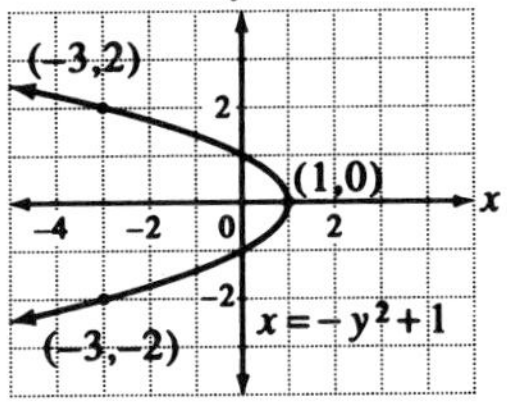

17.

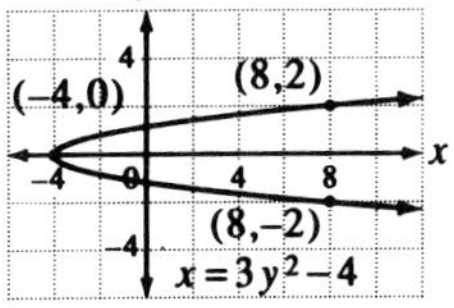

19.

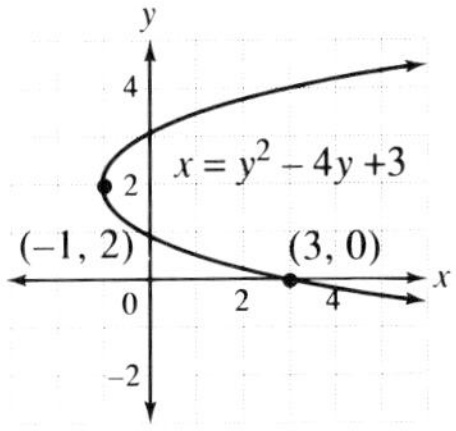

21.

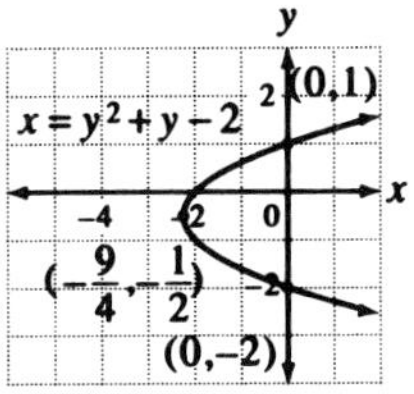

23.

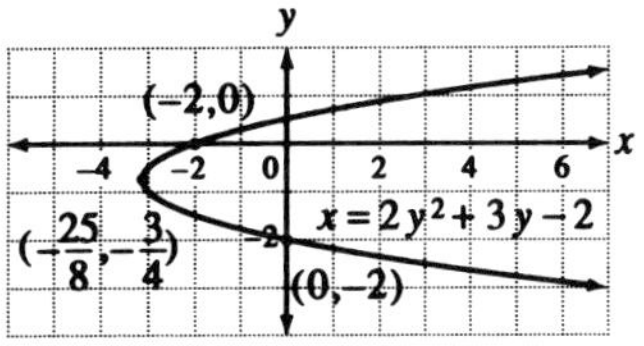

25.

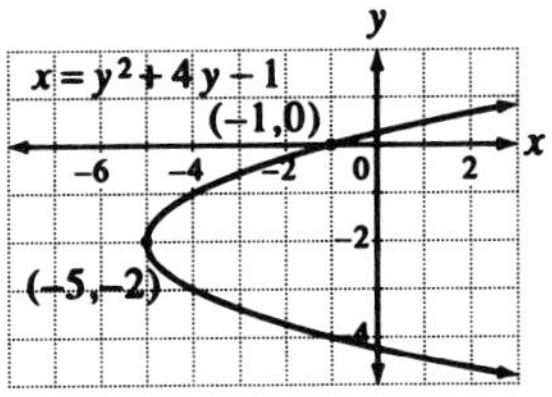

27.

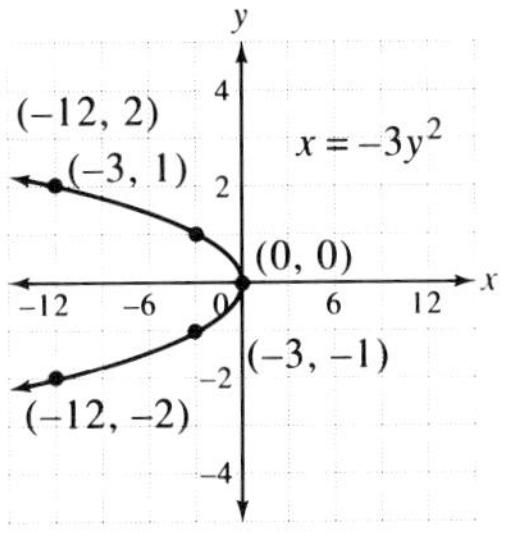

29.

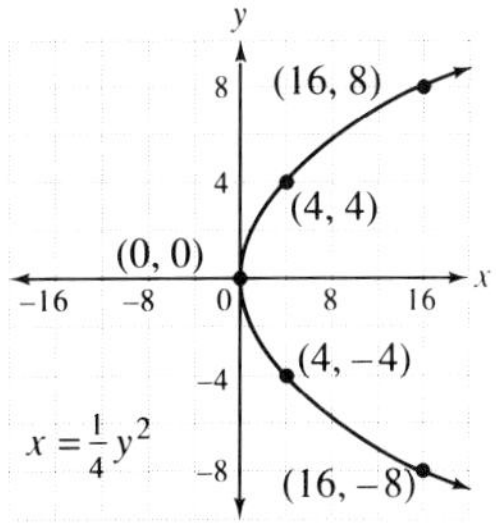

31.

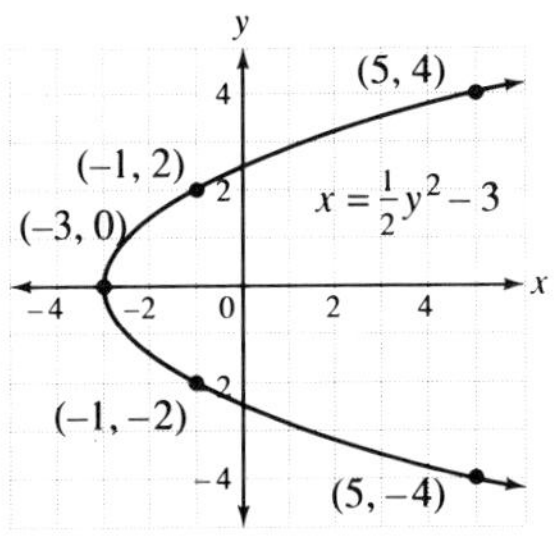

33.

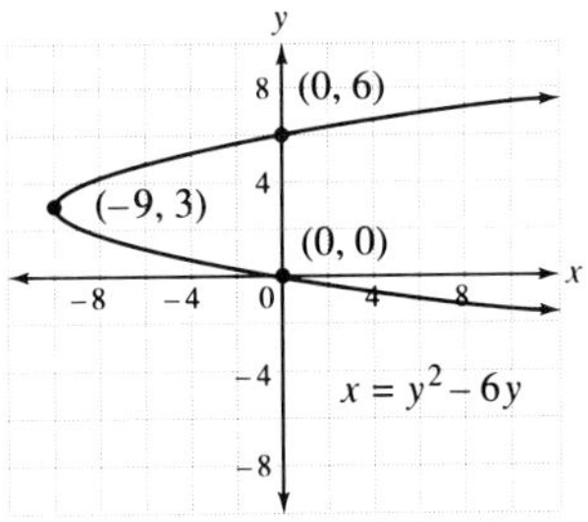

35.

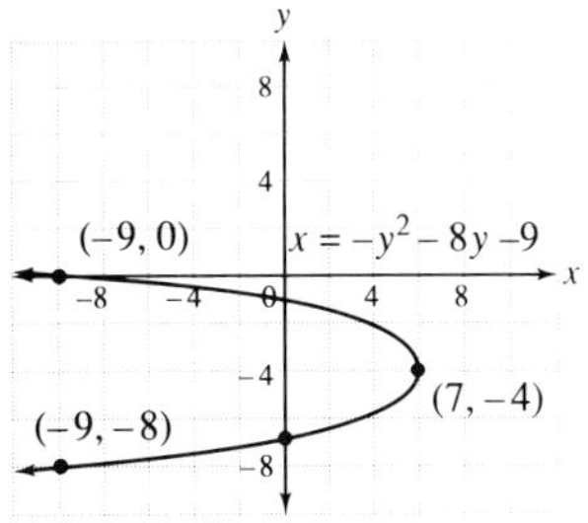

37. Right
39. Parabolas
41. 1968.8 ft^2

43. $3x = -(y + 1)^2$

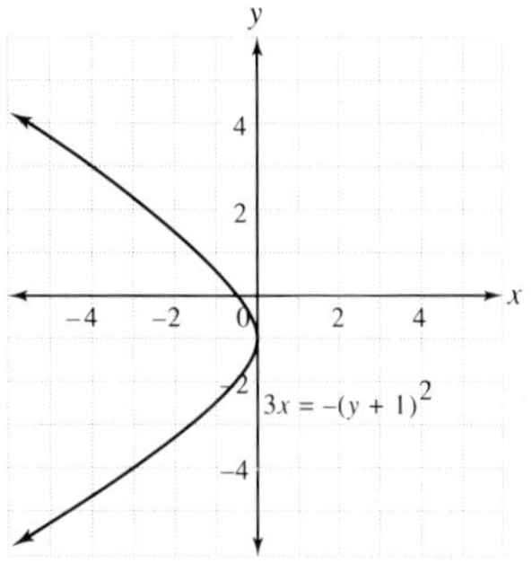

45. $\frac{1}{4}$ **47.** 22 cm, 484 cm^2 **49.** $-1 - 9i$ **51.** $5 - i$ **53.** $2 - 11i$ **55.** 85 **57.** -4 **59.** $-\frac{14}{25} - \frac{2}{25}i$

Problem Set 8.9

1. $\{x \mid x < -4 \text{ or } x > 3\}$

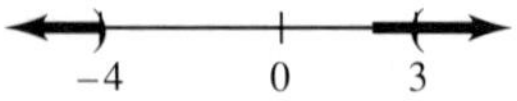

3. $\{x \mid -5 \le x \le -1\}$

5. $\{x \mid -4 < x < 6\}$

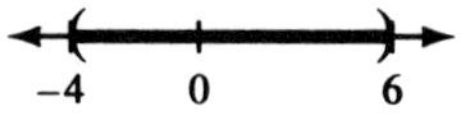

7. $\{x \mid x \le -4 \text{ or } x \ge 0\}$

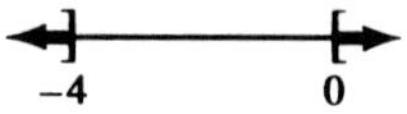

9. $\{x \mid x < 0 \text{ or } x > 1\}$

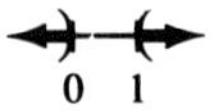

11. $\{x \mid -4 \le x \le \frac{2}{3}\}$

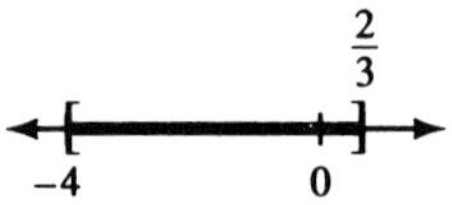

13. $\{x \mid x \text{ is a real number}\}$

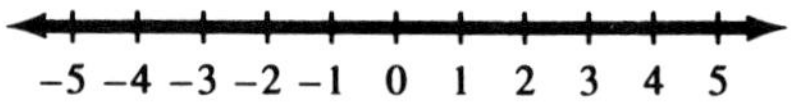

15. $\{x \mid -4 < x < 4\}$

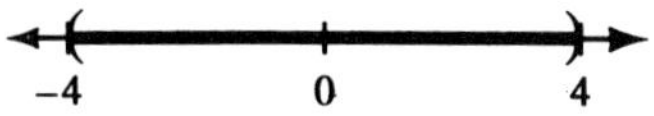

17. $\{x \mid x \le -1 \text{ or } x \ge -\frac{3}{4}\}$

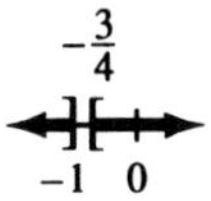

19. $\{x \mid -2 < x < 1 \text{ or } x > 3\}$

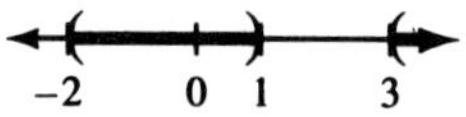

21. $\{x \mid 2 < x \le 6\}$ **23.** $\{x \mid x < 3\}$ **25.** $\{x \mid x < -1 \text{ or } x > 5\}$ **27.** $\{q \mid q \le 1 \text{ or } q > 4\}$
29. $\{x \mid x < -1 \text{ or } x > 2\}$ **31.** $\{x \mid x > \frac{1}{2}\}$ **33.** $\{p \mid p < -6 \text{ or } p > \frac{8}{3}\}$ **35.** $\{x \mid -2 < x \le 0 \text{ or } x > 1\}$
37. Negative **39.** Positive **41.** $\frac{5 - \sqrt{37}}{2} \le x \le \frac{5 + \sqrt{37}}{2}$ **43.** $40 < n < 60$ **45.** 9 **47.** 4 **49.** No solution
51. 7 **53.** 2

Chapter 8 Additional Exercises

1. $i\frac{\sqrt{7}}{3}, -i\frac{\sqrt{7}}{3}$ **3.** $7, -1$ **5.** $\frac{3}{2} + \frac{\sqrt{7}}{2}i, \frac{3}{2} - \frac{\sqrt{7}}{2}i$ **7.** $\frac{5}{8}, -\frac{5}{9}$ **9.** $0, \frac{3}{4}$ **11.** $-2 + \sqrt{6}, -2 - \sqrt{6}$
13. $\frac{-3 + \sqrt{3}}{3}, \frac{-3 - \sqrt{3}}{3}$ **15.** $\frac{1}{8}, -\frac{5}{8}$ **17.** $\frac{15 + \sqrt{165}}{15}, \frac{15 - \sqrt{165}}{15}$ **19.** $\frac{3}{4} + \frac{\sqrt{23}}{4}i, \frac{3}{4} - \frac{\sqrt{23}}{4}i$
21. $-\frac{3}{2} + \frac{\sqrt{19}}{2}i, -\frac{3}{2} - \frac{\sqrt{19}}{2}i$ **23.** $\frac{-3 + \sqrt{41}}{4}, \frac{-3 - \sqrt{41}}{4}$ **25.** $-\frac{1}{L}, -\frac{2}{L}$ **27.** Two real **29.** Two imaginary
31. One real **33.** $x^2 - 2x - 15 = 0$ **35.** $x^2 - \frac{3}{2}x + \frac{7}{16} = 0$ **37.** $y = \frac{2z^2}{x}$ **39.** $V = \sqrt{\frac{rf}{kw}}$
41. $S = \frac{3 + \sqrt{9 + 8N}}{2}$ **43.** 12 **45.** $i\sqrt{2}, -i\sqrt{2}, 2i, -2i$ **47.** 36 **49.** $-\frac{1}{3}, \frac{1}{2}$ **51.** $-\frac{3}{4}, -2$

53.

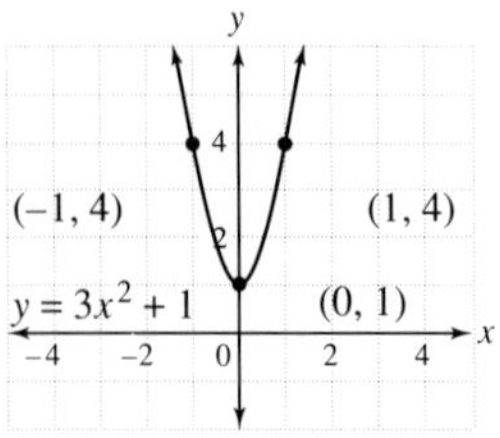

55.

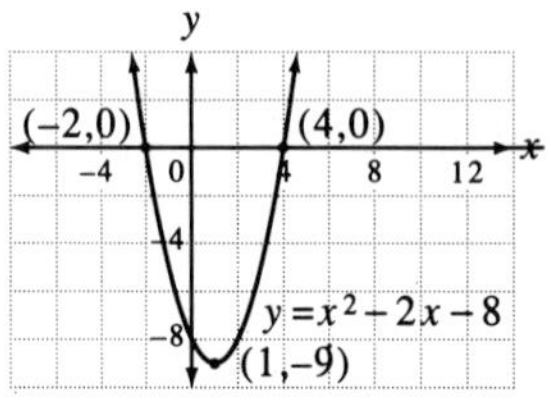

57.

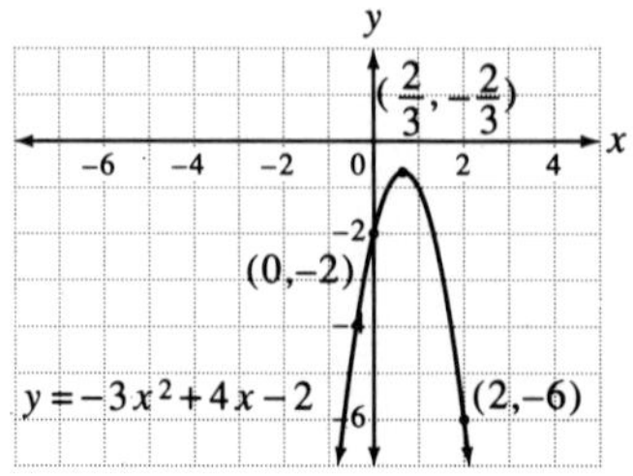

59. 40

61.

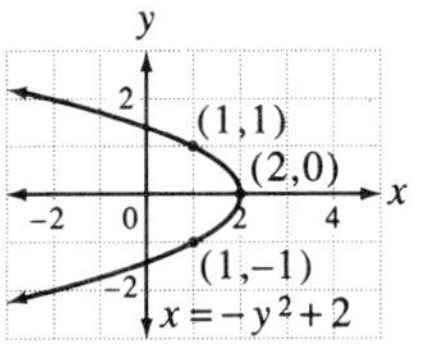

63.

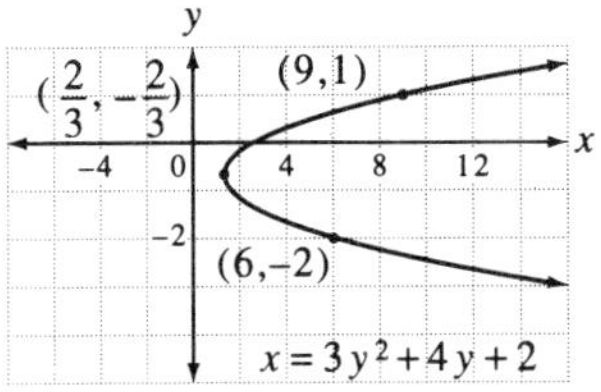

65.

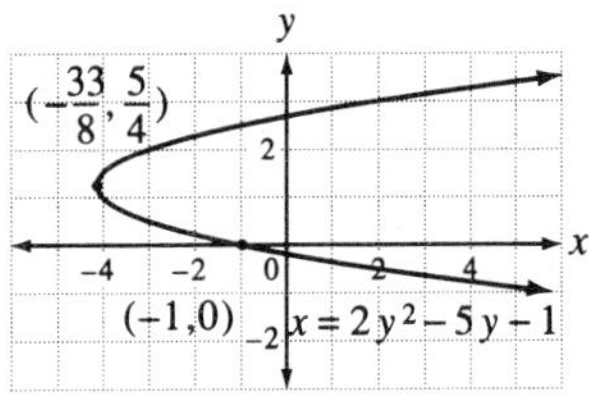

67. $\{x \mid -2 < x < 3\}$;

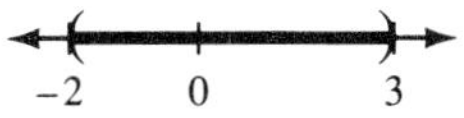

69. $\{x \mid x < 0 \text{ or } x > 3\}$;

71. $\varnothing$

73. $\{y \mid -4 \le y \le -\frac{3}{2} \text{ or } y \ge 1\}$;

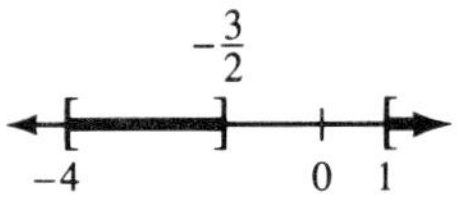

75. $\{x \mid x > -1\}$ **77.** $\{p \mid -3 < p \le \frac{2}{5} \text{ or } p > 2\}$

Chapter 8 Cooperative Learning

1. $\frac{1}{3}, -2$ **3.** $\frac{3+\sqrt{15}}{5}, \frac{3-\sqrt{15}}{5}$ **5.** $\frac{3+\sqrt{15}}{2}, \frac{3-\sqrt{15}}{2}$ **7.** 110 miles at 55 mph, 260 miles at 65 mph

9. $y = -x^2 + 3$ **11.** $y = -x^2 + 4x - 4$ **13.** $x = -y^2 + 1$

Chapter 8 Practice Test

1. $-\frac{1}{3} + \frac{\sqrt{7}}{3}i, -\frac{1}{3} - \frac{\sqrt{7}}{3}i$ **2.** 12 cm **3.** $-3 + \sqrt{14}, -3 - \sqrt{14}$ **4.** $-\frac{3}{2} + \frac{\sqrt{3}}{2}i, -\frac{3}{2} - \frac{\sqrt{3}}{2}i$ **5.** 0.4, −3.4

6. Two real **7.** $x^2 - 3x - 10 = 0$ **8.** $r = \frac{1}{2}\sqrt{\frac{S}{\pi}}$ **9.** 1 inch **10.** $1, -1, i\sqrt{7}, -i\sqrt{7}$

11.

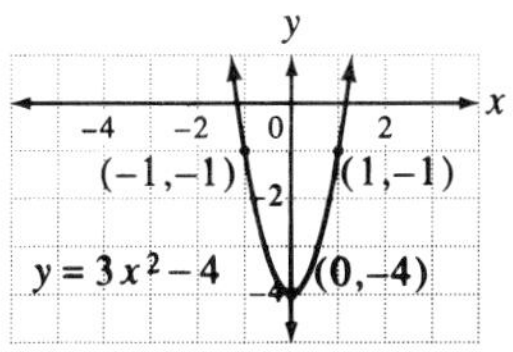

12.

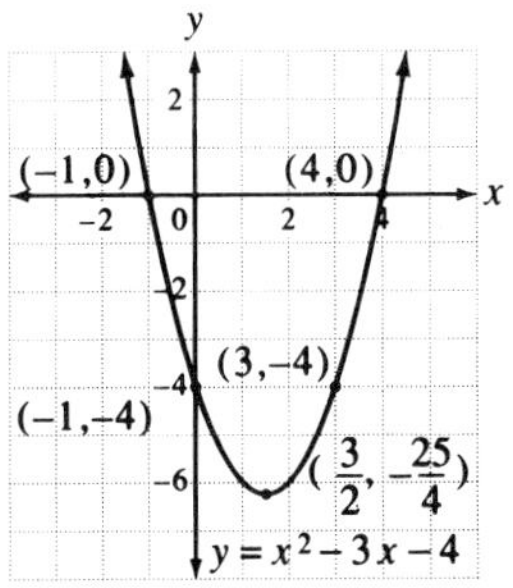

13.

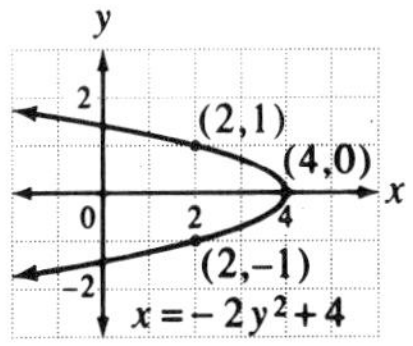

14. 36

15. $\{x \mid x \le -1 \text{ or } x \ge 2\}$ **16.** $\{x \mid -0 < x < 1\}$ **17.** $\{x \mid -6 \le x < -5\}$

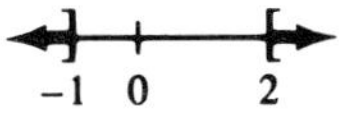

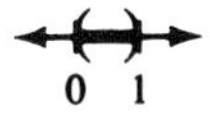

Cumulative Review Chapters 7 and 8

1. $7|x|$ **2.** $-4y^2$ **3.** $8x^2$ **4.** $3y$ **5.** $34\sqrt{3}$ **6.** $19 - 14\sqrt{2}$ **7.** $\frac{x^2 - x\sqrt{2} - x\sqrt{5} + \sqrt{10}}{x^2 - 5}$ **8.** $\sqrt[3]{(a^2b)^2}$

9. $\frac{1}{x^8}$ **10.** 8 **11.** $4i\sqrt{2}$ **12.** $-2 - 9i$ **13.** $7 - 26i$ **14.** $-\sqrt{30}$ **15.** $\frac{1}{3} - \frac{4}{3}i$ **16.** 3.5 miles

17. 4, −4 **18.** $\frac{3+\sqrt{57}}{4}, \frac{3-\sqrt{57}}{4}$ **19.** Two imaginary **20.** $L = \frac{1}{CW^2}$ **21.** 12 in. by 16 in.

22. 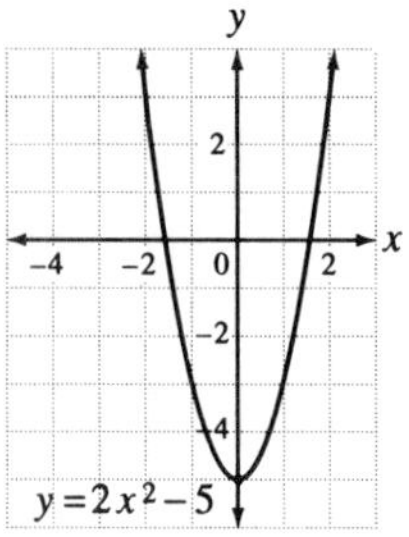

23. 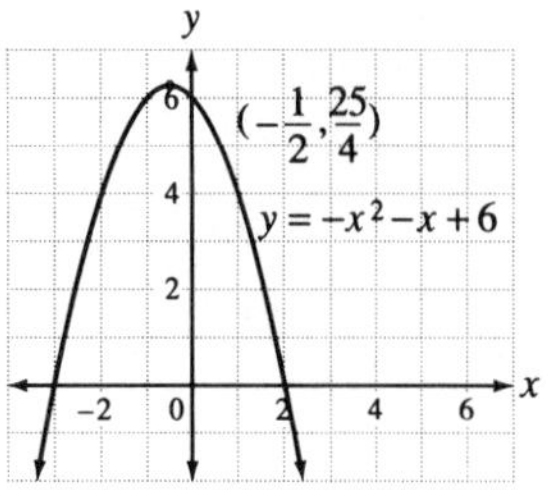

24. 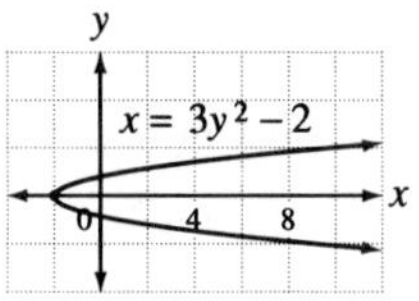

25. $-5 < x < 3$;

26. $\{x \mid x \leq 0 \text{ or } x \geq 3\}$

27. $\{b \mid -2 < b < -1 \text{ or } b \geq 1\}$

CHAPTER 9

Chapter 9 Pretest

1. $\sqrt{58}$ **2.** $(\frac{2}{3}, \frac{71}{80})$ **3.** $(x + 5)^2 + (y + 3)^2 = 54$ **4.** Center: (5, 7), radius: 8

5. 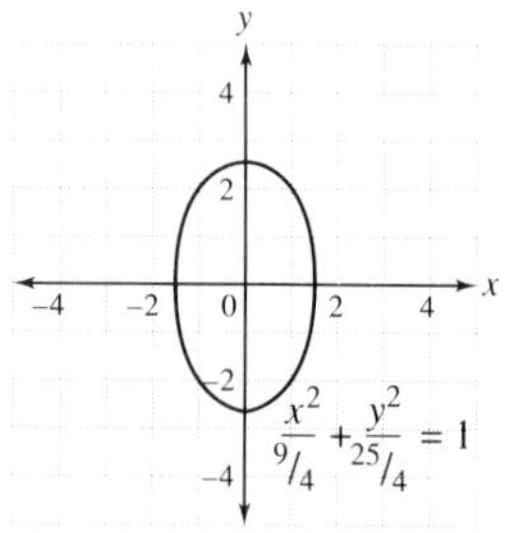

6. 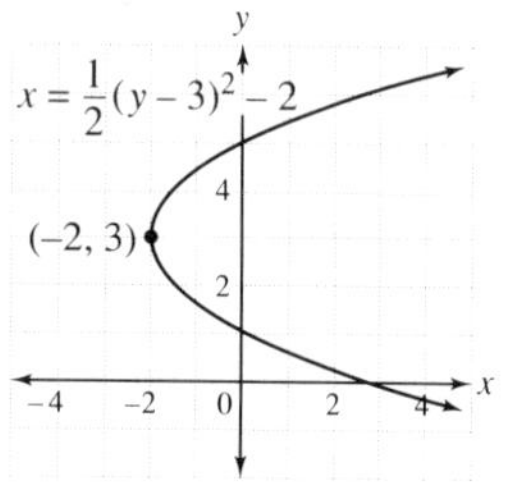

7. 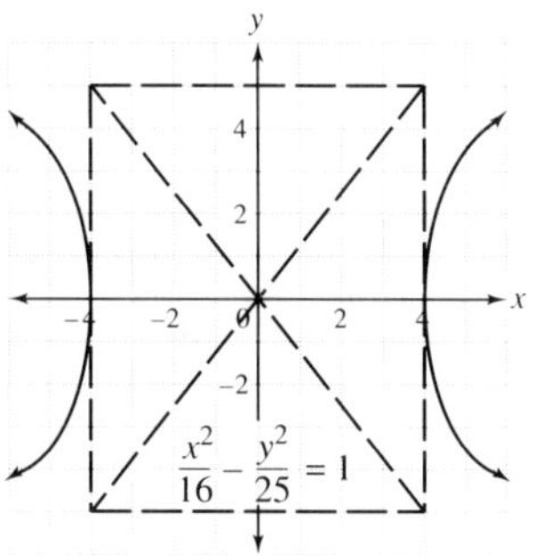

8. Circle **9.** (7, 8), (8, 7) **10.** (4, 2), $(-4, -2)$, $(4, -2)$, $(-4, 2)$ **11.** 11 m by 14 m

Problem Set 9.1

1. 5.099 **3.** 6 **5.** 7.810 **7.** 9.055 **9.** 0.417 **11.** 29.411 **13.** 6.083 **15.** 805.233 **17.** $(-5, \frac{9}{2})$
19. $(\frac{1}{2}, -\frac{5}{2})$ **21.** (3, 1) **23.** $(-4.7, 1.25)$ **25.** $(-\frac{3}{16}, \frac{21}{20})$ **27.** $\left(\frac{c}{2}, \frac{d}{2}\right)$ **29.** Yes **31.** Legs **33.** Negative
35. 1.7 **37.** 32.472 **39.** No **41.** $-1 + \sqrt{5}, -1 - \sqrt{5}$ **43.** $-1 + i\sqrt{7}, -1 - i\sqrt{7}$
45. $\frac{-5 + \sqrt{21}}{2}, \frac{-5 - \sqrt{21}}{2}$ **47.** $\frac{-3 + \sqrt{3}}{2}, \frac{-3 - \sqrt{3}}{2}$

Problem Set 9.2

1. $(x + 3)^2 + (y - 2)^2 = 4$ **3.** $x^2 + y^2 = 3$ **5.** $(x + 4)^2 + (y + 6)^2 = 25$ **7.** $(x - 5)^2 + (y + 1)^2 = 6$
9. Center: (0, 0); radius: 5 **11.** Center: (2, 3); radius: 2 **13.** Center: $(-2, 4)$; radius: $2\sqrt{2} \approx 2.8$

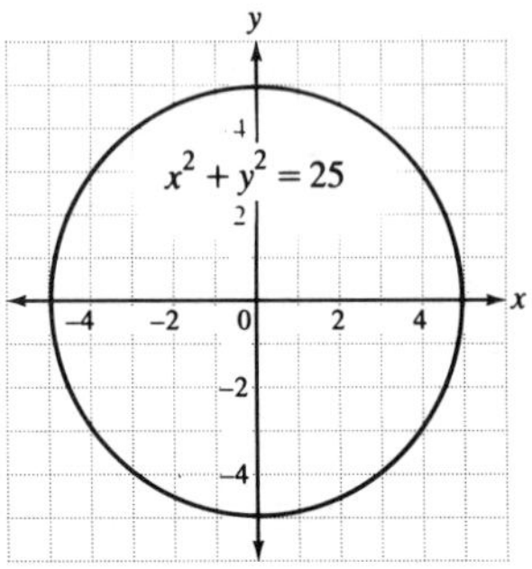

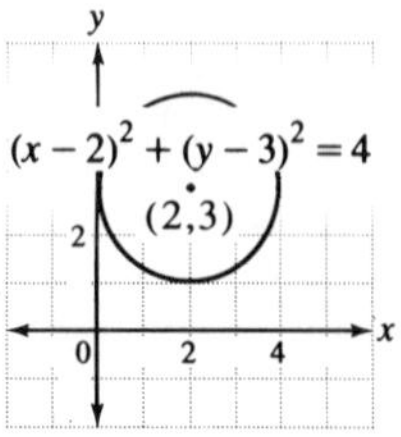

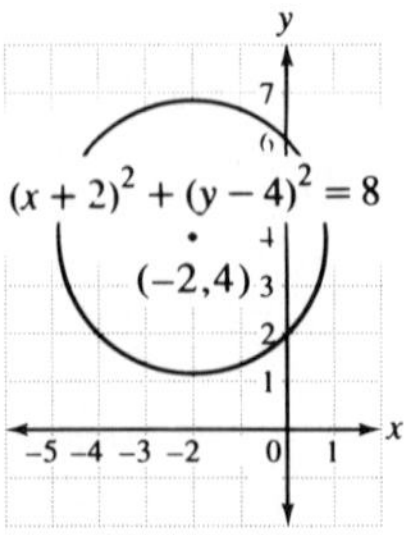

15. Center: $(0, -3)$; radius: $\sqrt{5} \approx 2.2$

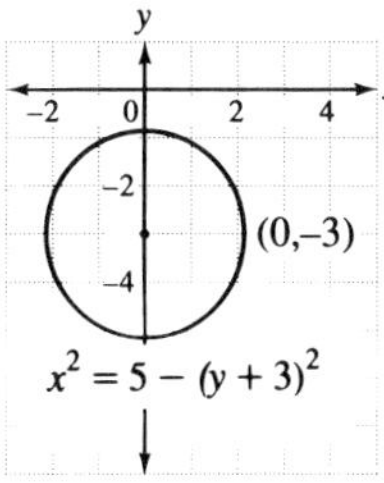

17. Center: $(-2, -6)$; radius: 2
19. Center: $(4, 1)$; radius: $2\sqrt{6} \approx 4.9$
21. Center: $(4, 0)$; radius: $\sqrt{21} \approx 4.6$
23. Center: $(-2, 4)$; radius: 3

25.

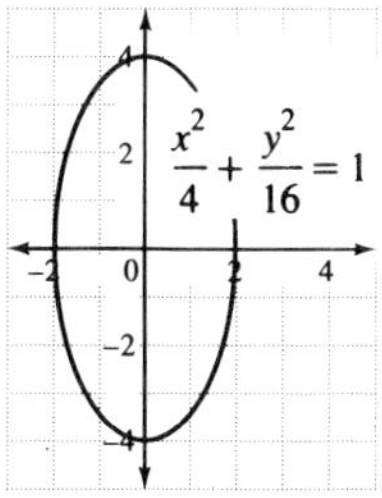

27.

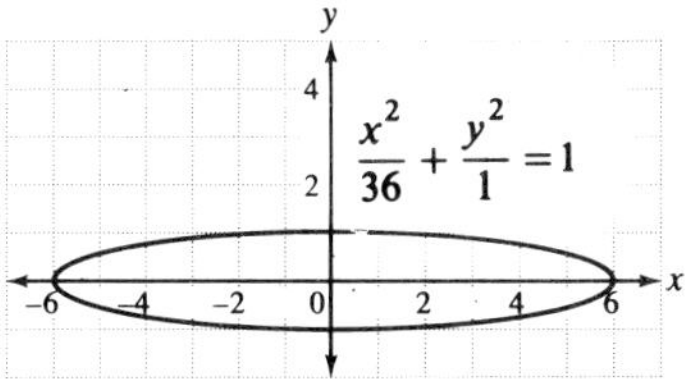

29.

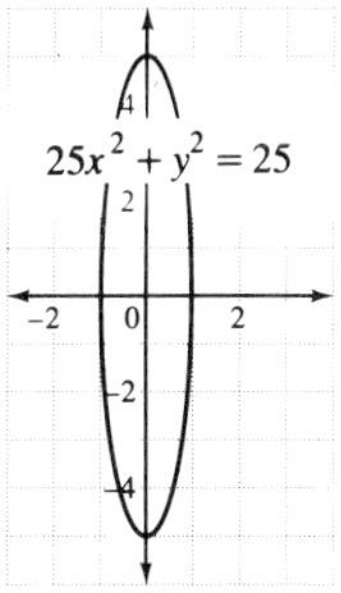

31.

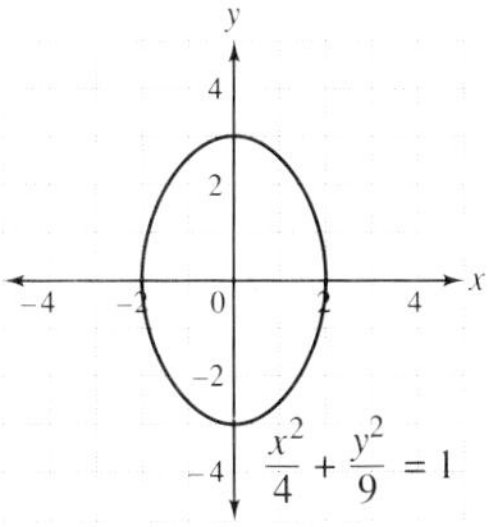

33.

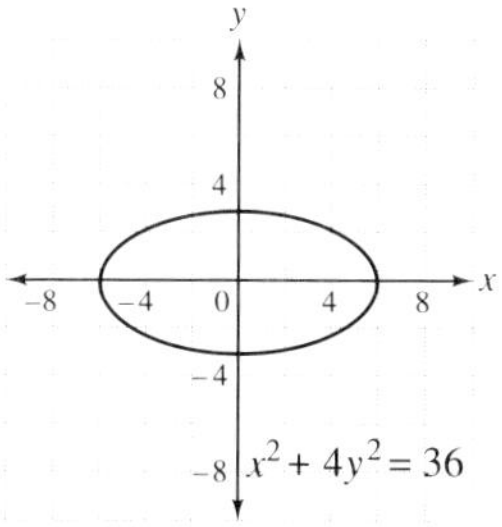

35.

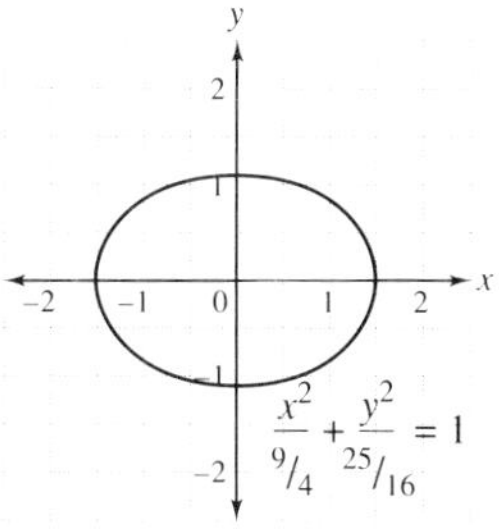

37. Origin **39.** Constant **41.** $x^2 + y^2 = 9$

43. Center: $\left(-\frac{3}{4}, -\frac{5}{4}\right)$; radius $\frac{\sqrt{2}}{4}$

45. Center: $(\frac{1}{2}, \frac{1}{3})$; radius: $\frac{\sqrt{6}}{6}$

47.

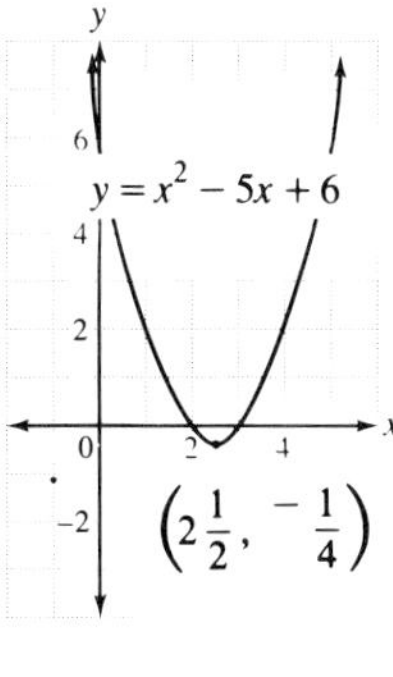

49.

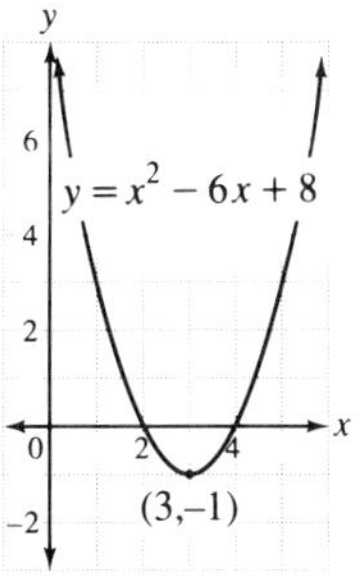

51.

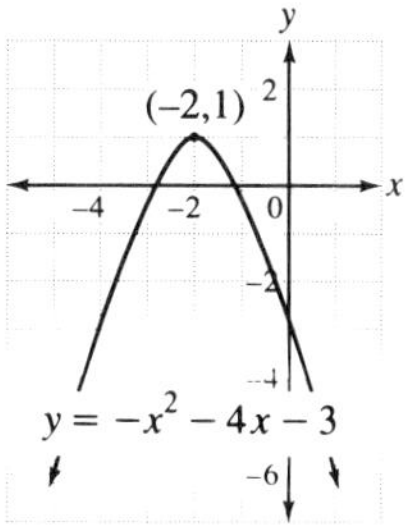

53.

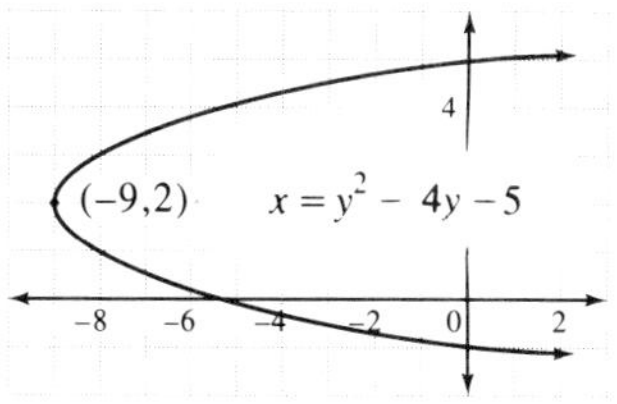

55.

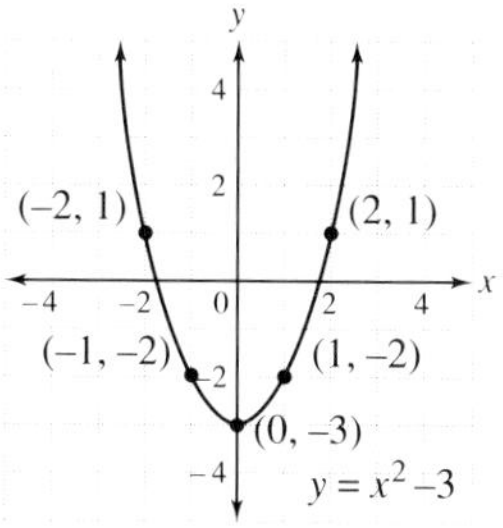

Problem Set 9.3

1.
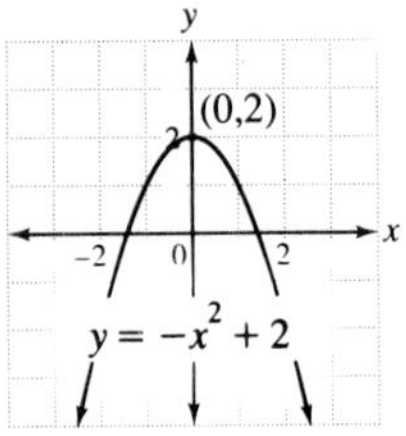

3.
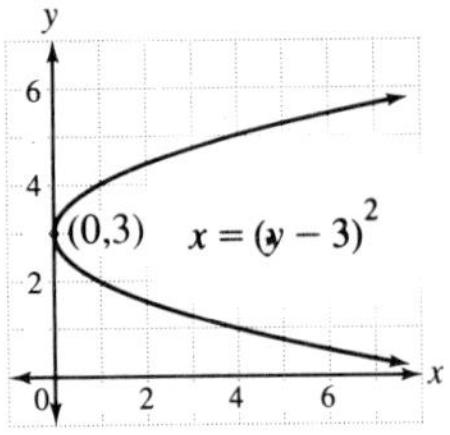

5.
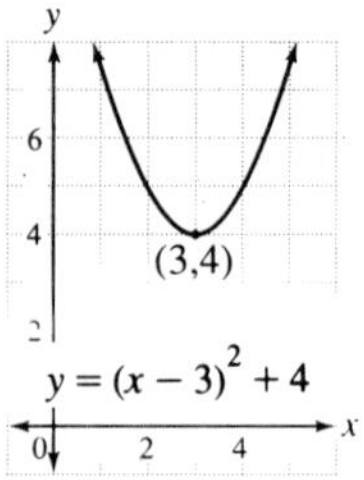

7.
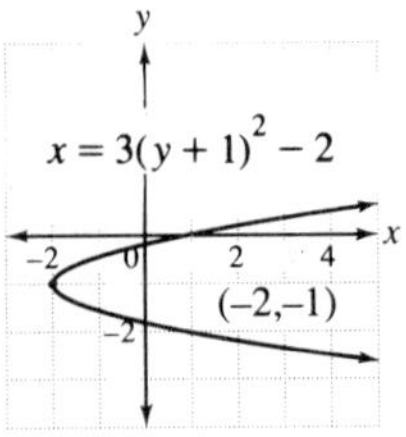

9.
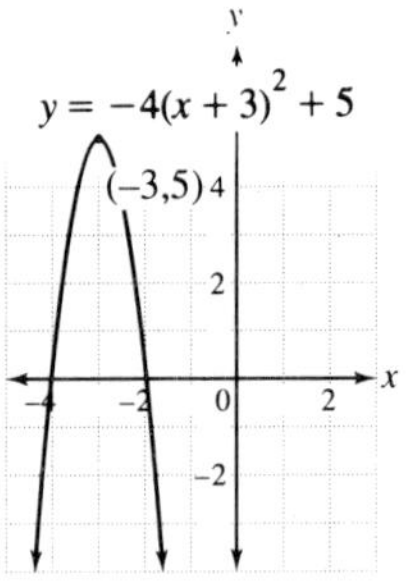

11.
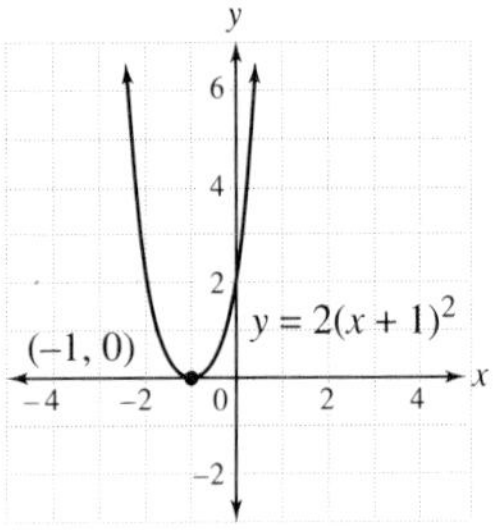

13.
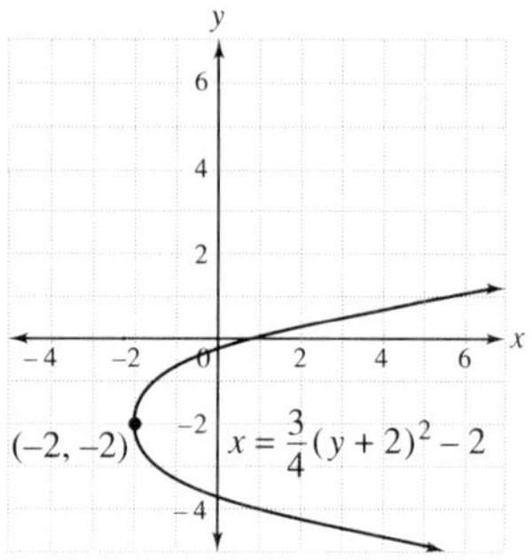

15.
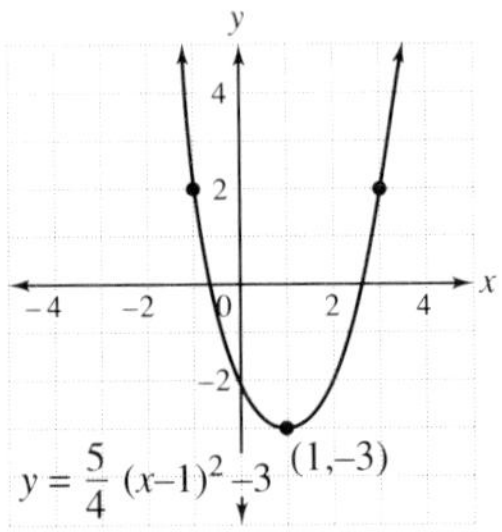

17.
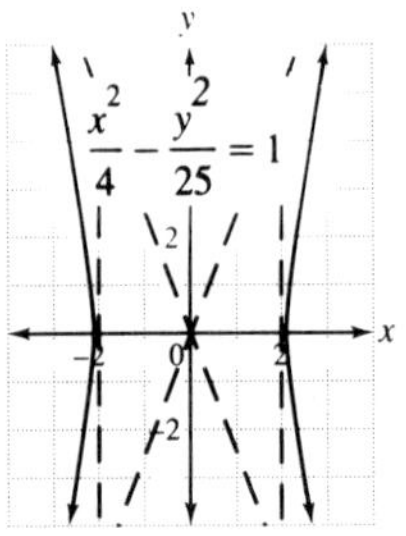

19.
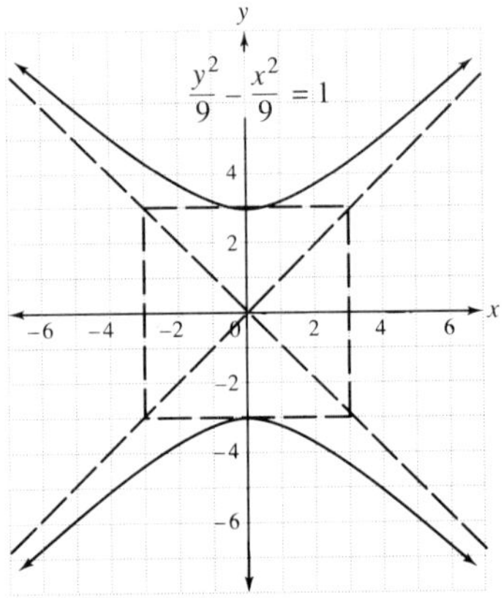

21.
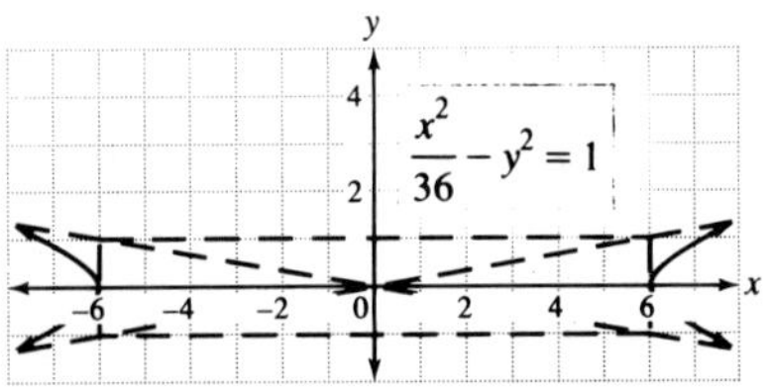

23.
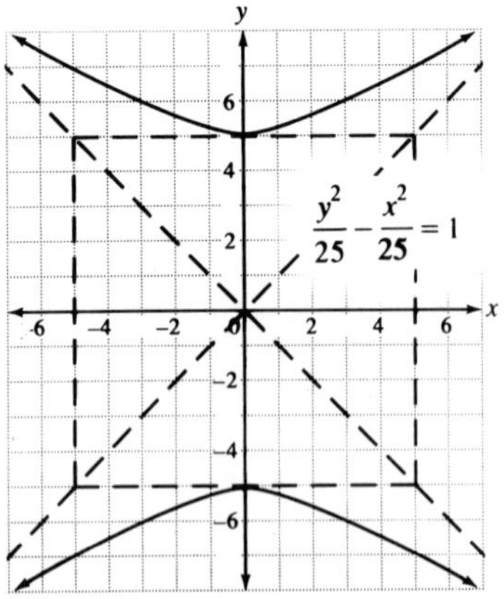

25.
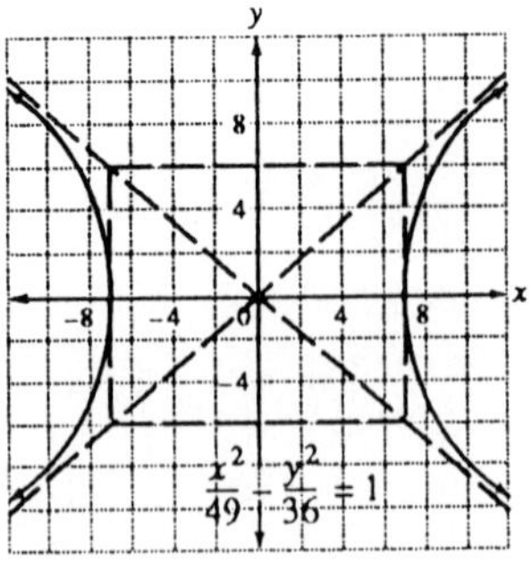

27.

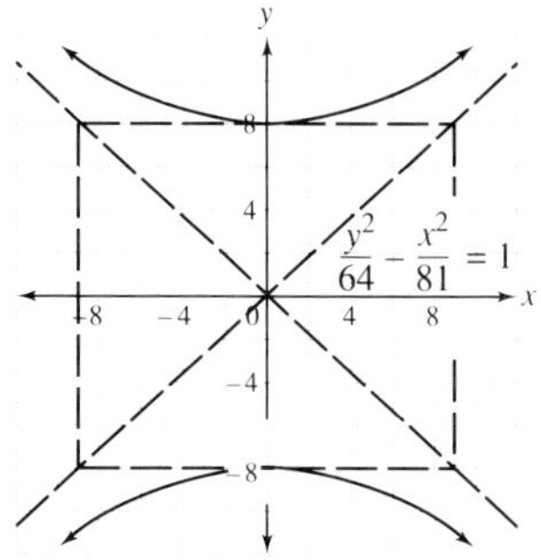

29.

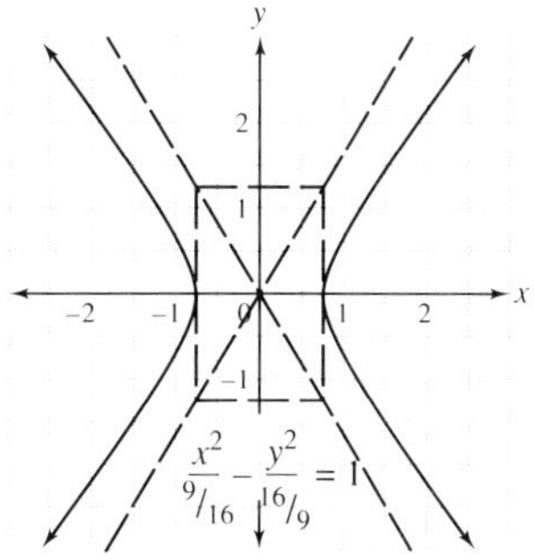

31. Hyperbola **33.** Circle **35.** Parabola **37.** Hyperbola **39.** Parabola **41.** Upward **43.** Negative **45.** Conic

47. $y = \frac{1}{2}x^2 - 3x + 1$

Vertex: $\left(3, -\frac{7}{2}\right)$

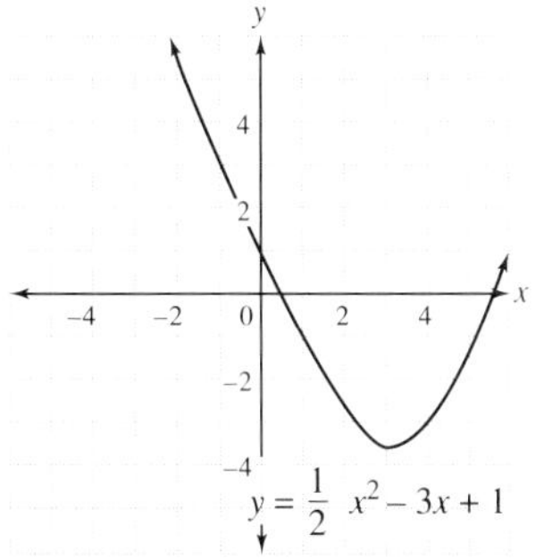

49. Vertex $(-4, -3)$

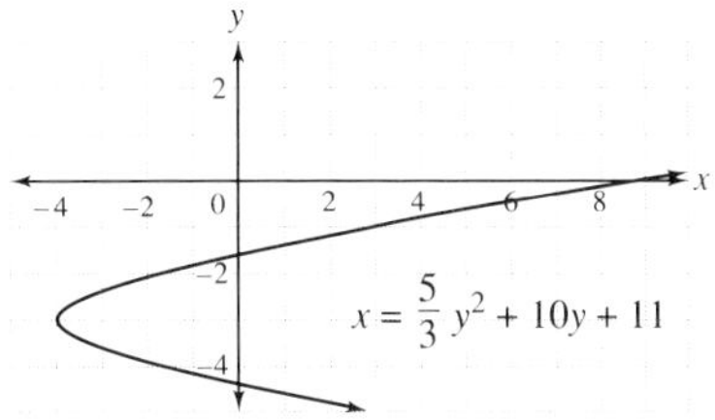

51. $xy = -1$, hyperbola

x	y
-3	$\frac{1}{3}$
-1	1
$\frac{-1}{3}$	3
$\frac{-1}{7}$	7
$\frac{1}{7}$	-7
$\frac{1}{3}$	-3
1	-1
3	$-\frac{1}{3}$

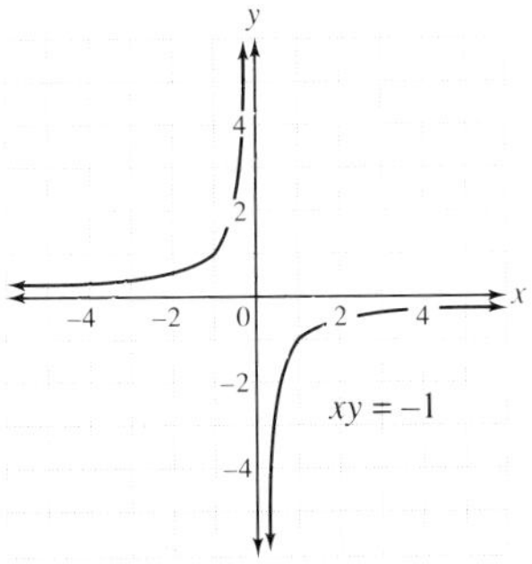

53. $\frac{y^2}{6.25} - \frac{x^2}{0.81} = 1$

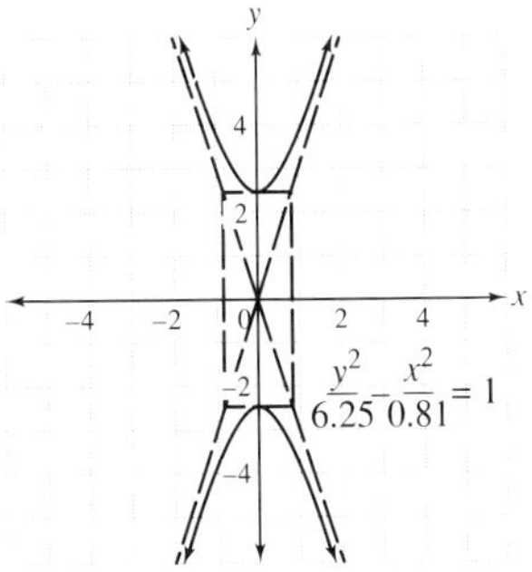

55. $(1, 3)$ **57.** $(-2, -5)$ **59.** $(-3, 4)$ **61.** $(-4, 0)$ **63.** $(5, -7)$

Problem Set 9.4

1. $(\frac{8}{5}, -\frac{6}{5})$, $(0, -2)$ **3.** $(-3, -\frac{2}{3})$, $(-2, -1)$ **5.** $(\frac{3}{2}, \frac{1}{2})$, $(-\frac{3}{2}, -\frac{1}{2})$ **7.** $(-2, 0)$, $(-1, 1)$ **9.** $(0, 2)$, $(3, 0)$ **11.** $(-3, 9)$, $(-1, 1)$ **13.** $(\frac{7}{3}, \frac{1}{3})$, $(1, -1)$ **15.** $(9, -\frac{4}{3})$, $(-4, 3)$ **17.** 7, 9 **19.** 9 ft, 12 ft **21.** 8 ft, 14 ft **23.** 64 mixers, \$512 **25.** $(-2, 1)$, $(1, 4)$ **27.** Substitution **29.** $(2.5, 6.00)$, $(-2.5, -6.00)$ **31.** $(-1.30, 4.55)$, $(8.43, -0.70)$ **33.** 2 and 8 **35.** $(-2, -5)$ **37.** $(-1, 5)$ **39.** $(-2, -3)$ **41.** $W_1 = A^2W_2$ **43.** 2 ft **45.** 9 mph

Problem Set 9.5

1. $(6, \frac{4}{5})$, $(6, -\frac{4}{5})$, $(-6, \frac{4}{5})$, $(-6, -\frac{4}{5})$ **3.** $(\sqrt{3}, 0)$, $(-\sqrt{3}, 0)$ **5.** $(-2, 0)$, $(-5, 30)$ **7.** $(-\frac{2}{3}, \frac{8}{9})$, $(1, 2)$ **9.** $(0, -3)$, $(\sqrt{5}, 2)$, $(-\sqrt{5}, 2)$ **11.** $(1, 3)$, $(-1, -3)$, $(3, 1)$ $(-3, -1)$ **13.** $(1, 4)$, $(-1, -4)$, $(4, 1)$, $(-4, -1)$ **15.** $(1, 0)$ **17.** $(2, 1)$, $(-2, -1)$ **19.** 7 and 3, 7 and -3, -7 and 3, -7 and -3 **21.** 1 m by 3 m **23.** 11 ft by 15 ft **25.** 4 ft, 2 ft **27.** 14 cm, 12 cm **29.** 5 and 8, -5 and -8, -5 and 8, 5 and -8 **31.** Elimination **33.** $(3.77, 3.64)$, $(3.71, -3.64)$, $(-3.77, 3.64)$, $(-3.77, -3.64)$ **35.** $(2.20, 2.77)$, $(2.20, -2.77)$, $(-2.54, 1.29)$, $(-2.54, -1.29)$ **37.** 24 ft by 64 ft **39.** Height: 6 cm, base: 16 cm or height: 8 cm, base: 12 cm **41.** -2 **43.** $\frac{1}{2}$ **45.** 5, -1 **47.** $\frac{5 + \sqrt{17}}{2}$, $\frac{5 - \sqrt{17}}{2}$ **49.** $\frac{3}{4}$, -1

Chapter 9 Additional Exercises

1. 5 **3.** $\sqrt{17}$ **5.** $2\sqrt{13}$ **7.** $2\sqrt{0.17}$ **9.** $2\sqrt{2}$ **11.** $7\sqrt{a^2 + b^2}$ **13.** $(-6, -\frac{11}{2})$ **15.** $(-3.15, 3.85)$
17. $(\frac{29}{48}, -\frac{13}{12})$ **19.** Yes **21.** Yes **23.** $(x + 2)^2 + (y + 3)^2 = 7$ **25.** $x^2 + y^2 = 5$
27. Center: $(4, -3)$; radius: 3 **29.** Center: $(-3, 1)$; radius: $3\sqrt{2}$

31.

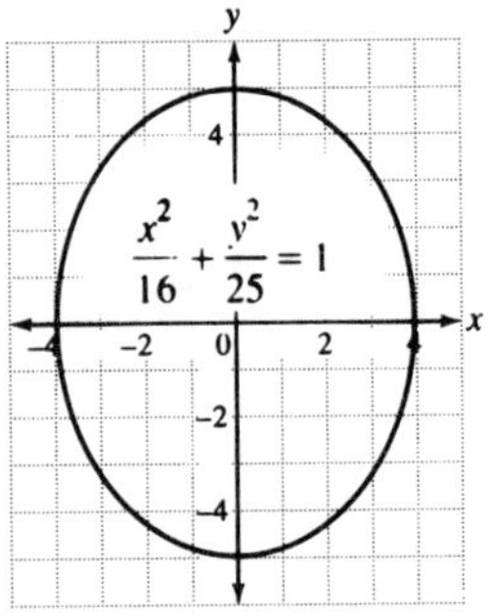

33. Center: (3, 2); vertices: (7, 2), (−1, 2), (3, 4), (3, 0)

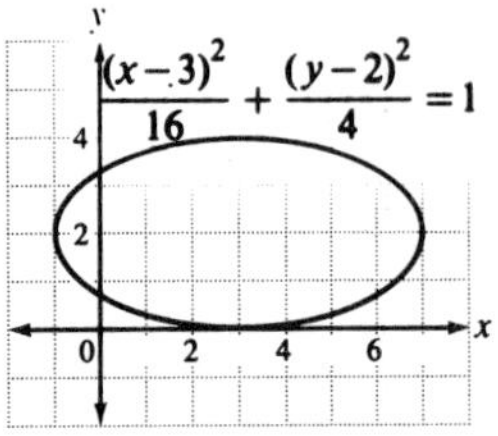

35. Center: (2, 1); vertices: (5, 1), (−1, 1), (2, 3), (2, −1)

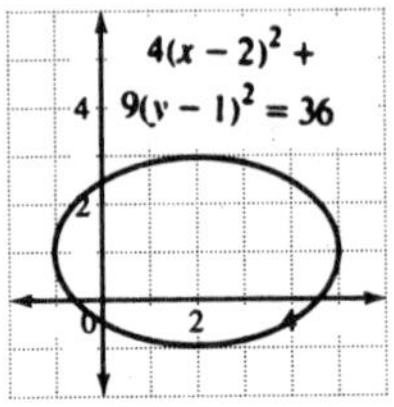

37. Center: (1, 2); vertices: (3, 2), (−1, 2), (1, 3), (1, 1)

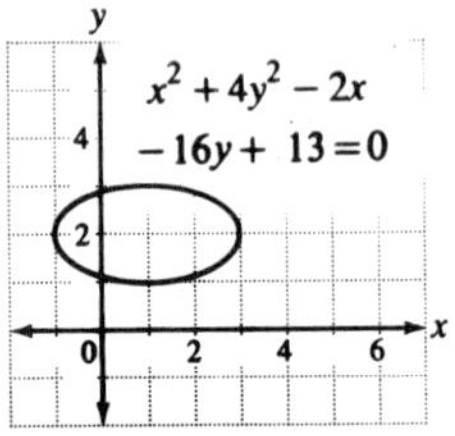

39. Vertex: (2, −5)

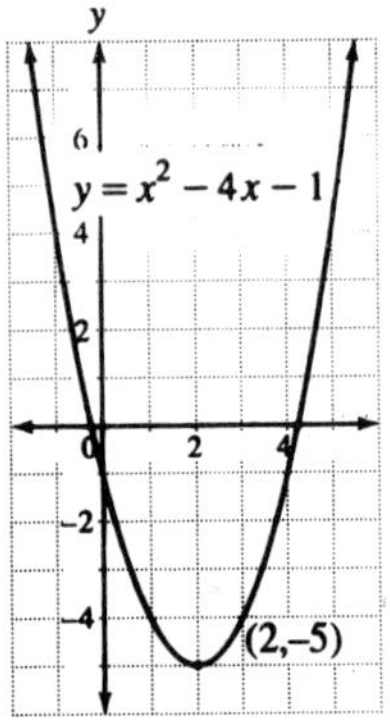

41. Vertex: (−7, 2)

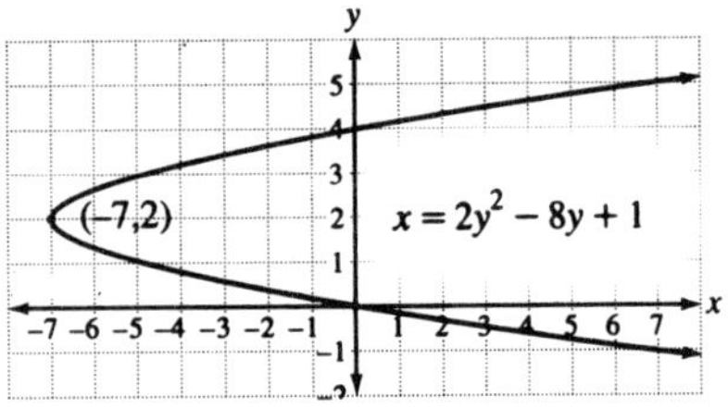

43. Vertex: (1, 3)

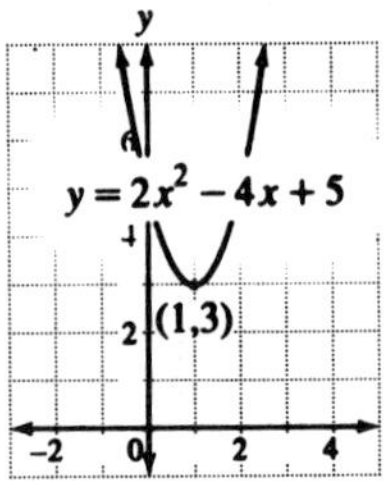

45.

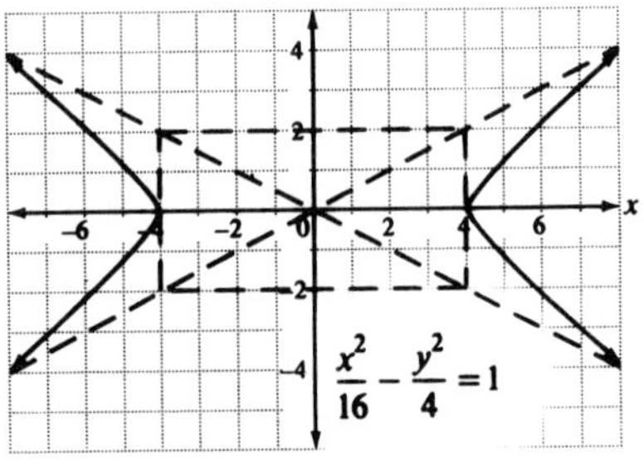

47. Center: (1, 1); vertices: (1, 3), (1, −1); corners of the rectangle: (2, 3), (2, −1), (0, 3), (0, −1)

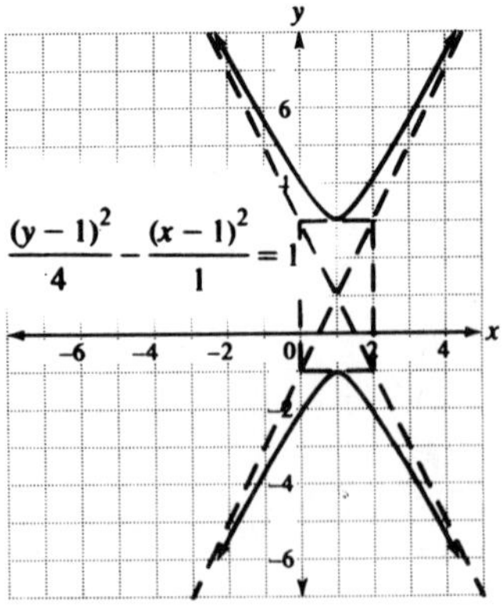

49. Center: (5, 4); vertices: (10, 4), (0, 4); corners of the rectangle: (10, 6), (10, 2), (0, 6), (0, 2)

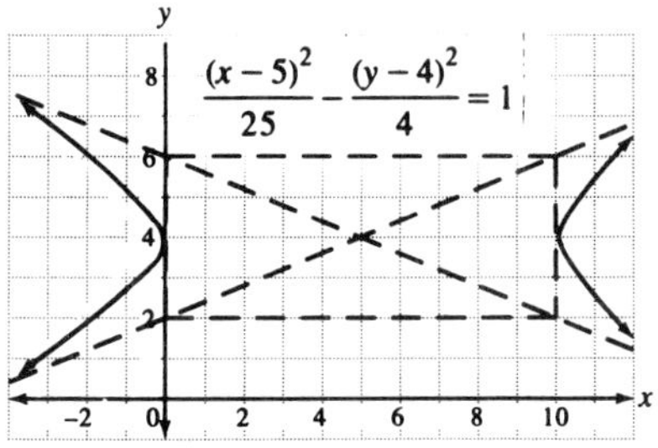

51. $(-7, -5), (5, 1)$ **53.** $(3, 2), (4, \frac{3}{2})$ **55.** $(-1, -2), (1, 2)$ **57.** $(2, 6), (1, 5)$ **59.** 8 cm by 12 cm **61.** $(0, -3), (0, 3)$ **63.** $(3, -1), (\frac{3}{4}, \frac{1}{2})$ **65.** $(2, 3), (-2, 3), (i\sqrt{3}, -4), (-i\sqrt{3}, -4)$ **67.** $(4, 1), (-4, -1), (2, 2), (-2, -2)$

Chapter 9 Cooperative Learning

1. $15|a|$ **3.** $x^2 + y^2 = 9$ **5.** $\frac{x^2}{4} + \frac{y^2}{9} = 1$ **7.** $y = \sqrt{25 - x^2}, y = -\sqrt{25 - x^2}$ **9.** $x^2 + y^2 = 16$

Chapter 9 Practice Test

1. $\sqrt{53}$ **2.** $(-0.3, -2.55)$ **3.** $(x - 4)^2 + (y + 2)^2 = 7$ **4.** Center: $(-2, 4)$; radius: 5

5.

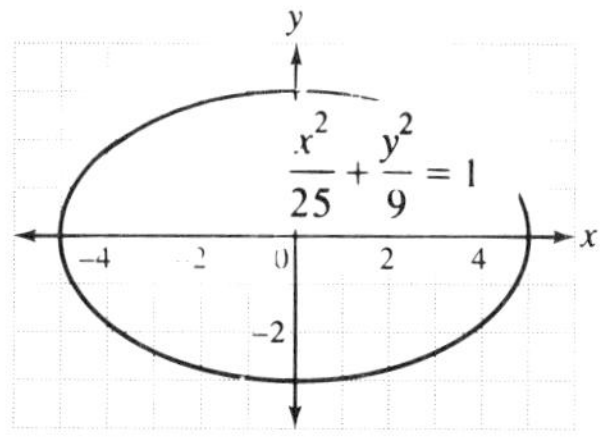

6.

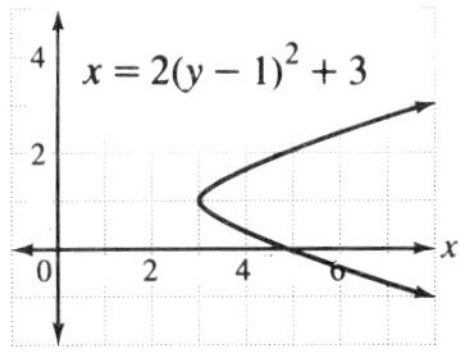

7.

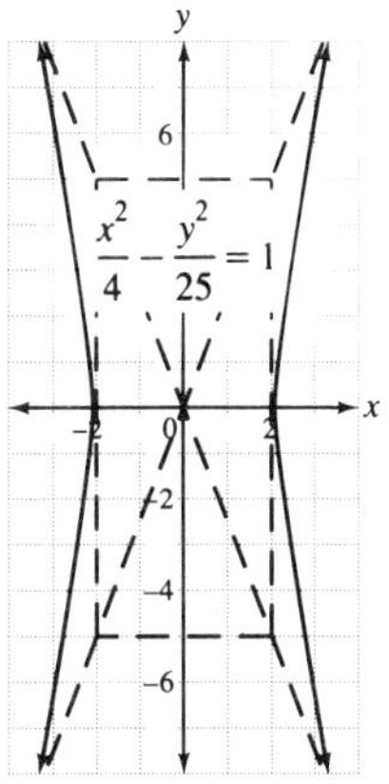

8. Ellipse
9. $(0, 4), (\frac{16}{5}, \frac{12}{5})$
10. $(1, 3), (1, -3), (-1, 3), (-1, -3)$
11. 5 cm by 4 cm

CHAPTER 10

Chapter 10 Pretest

1. Yes
2. $\{-4, -3, 5, 6, -7\}$
3. $\{0, -11, 8, -12, -3\}$
4. All real numbers except 4

5. Not a function

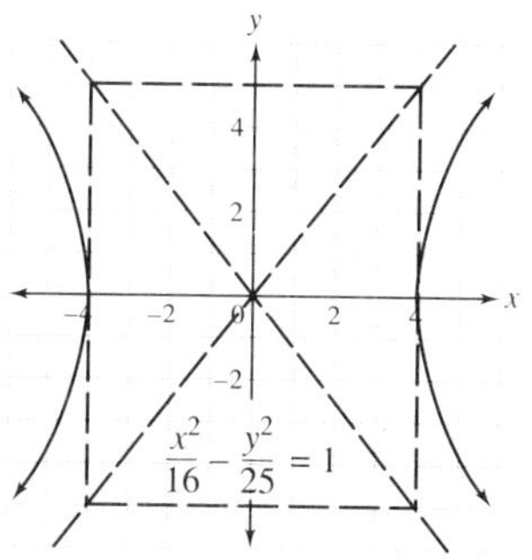

6. 46 **7.** $3x^2 - 16x + 28$ **8.** $-4x^2 - 13x + 17$
9. $x + 5 + \frac{-5}{4x - 3}$ **10.** Linear function **11.** No
12. $f^{-1}(x) = \sqrt{\frac{x + 3}{2}}$

Problem Set 10.1

1. Yes; domain: $\{1, 3, -1\}$; range: $\{5, 7, 8\}$ **3.** No **5.** Yes; domain: $\{-5, 3, -4\}$; range: $\{2, -7\}$ **7.** No **9.** Yes; domain: $\{3, 0, 4, -5, -9\}$; range: $\{-8, 10, 7, 3\}$ **11.** Yes; domain: $\{5, -3, 27, 15, 7\}$; range: $\{12, 8, 4, -32\}$

13. Yes; domain: {300, 400, −100, 325}; range: {−200, 600, −450, −650} **15.** Yes; all real numbers
17. Yes; all real numbers **19.** No **21.** Yes; all real numbers except 3 **23.** No **25.** Yes; all real numbers except $\frac{4}{3}$
27. Yes; $\{x \mid x \geq 0\}$ **29.** Yes; $\{x \mid x \geq \frac{1}{3}\}$ **31.** Function **33.** Not a function **35.** Function **37.** Not a function
39. Function **41.** Yes **43.** Relation **45.** Range **47.** Function **49.** All real numbers except 0 **51.** $\{y \mid y \geq 0\}$
53. $-4 \leq y \leq 4$ **55.** (a) $P = -x^2 + 26x - 160$; (b) yes; (c) 13 **57.** $\{x \mid x \leq -1 \text{ or } x \geq 8\}$ **59.** $\{x \mid \frac{2}{3} < x < 4\}$
61. $\{y \mid -3 < y < 0\}$ **63.** $\{x \mid x > -3\}$
65. $\{x \mid -4 < x \leq -1\}$

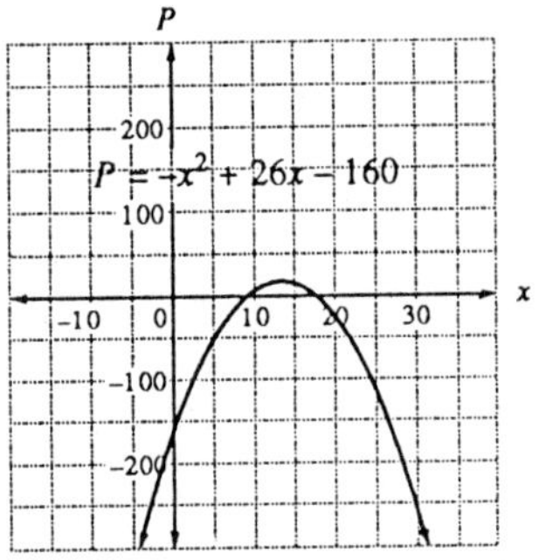

Problem Set 10.2

1. 1 **3.** 10 **5.** −13 **7.** 10 **9.** −5 **11.** $3a^2 - 2$ **13.** $3x^2 + 6hx + 3h^2 - 2$ **15.** 19 **17.** −2
19. 4 **21.** $2x + h$ **23.** $4x + 2h$ **25.** $2x + h + 3$ **27.** $f[g(x)] = 2x^2 + 5$; $g[f(x)] = 4x^2 - 4x + 4$
29. $f[g(x)] = 3x^2 - 24x + 47$; $g[f(x)] = 3x^2 - 5$ **31.** $f[g(x)] = x^2 - x - 10$; $g[f(x)] = x^2 + 3x - 10$
33. $f[g(x)] = \dfrac{3 + x^2}{x^2}$; $g[f(x)] = \dfrac{1}{3x^2 + 1}$ **35.** $f[g(x)] = \dfrac{5}{(x - 4)^2}$; $g[f(x)] = \dfrac{5 - 4x^2}{x^2}$ **37.** $9200
39. The amount paid for 10 gallons of gasoline **41.** of **43.** 42.52 **45.** 10.62 **47.** $\dfrac{-1}{x(x + h)}$
49. $\dfrac{1}{\sqrt{x + h} + \sqrt{x}}$ **51.** $\dfrac{1}{\sqrt{x + h - 3} + \sqrt{x - 3}}$ **53.** Two imaginary **55.** One real **57.** Two real
59. Two imaginary **61.** $x^2 - 5 = 0$ **63.** $24x^2 - 143x + 80 = 0$

Problem Set 10.3

1. $2x^2 - 5x + 2$ **3.** $-2x^2 + 9x - 4$ **5.** $4x^3 - 16x^2 + 13x - 3$ **7.** $x - 3$ **9.** $4x^4 - 28x^3 + 61x^2 - 42x + 9$
11. $-x^2 - 13x + 7$ **13.** $3x^2 + x + 1$ **15.** $x^3 - 10x^2 + 28x - 16$ **17.** $3x^2 - 5x - 28$ **19.** $x - 2 + \dfrac{-4}{x - 4}$
21. $9x^2 + 42x + 49$ **23.** $2x + 11$ **25.** $\dfrac{x - 4}{3x + 7}$ **27.** $x^2 - 7x + 8$ **29.** $3x^3 - 11x^2 - 30x + 28$
31. $2 + \dfrac{19x - 11}{x^2 - 6x + 4}$ **33.** $x^4 - 12x^3 + 44x^2 - 48x + 16$ **35.** $P(x) = -0.06x^2 + 11x - 200$ **37.** Minus
39. $-3.058x^2 - 0.86x - 3.79$ **41.** $24.51x^2 - 0.48x - 66.03$ **43.** $-10.642x^2 - 17.54x + 11.99$
45. $10x^2 - 6x + 24$
47. $x^3 - 2x^2 + 5x - 4$
49. $3 + a^2 - 2ab + b^2$

51.

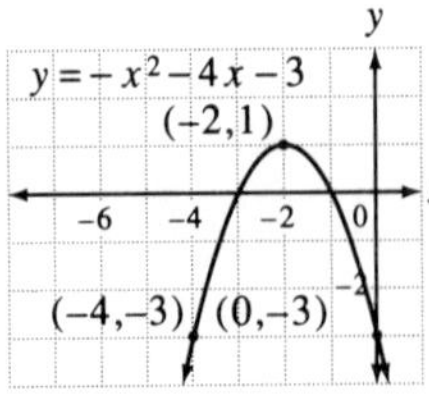

53.

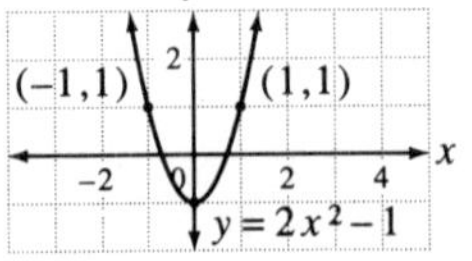

55.

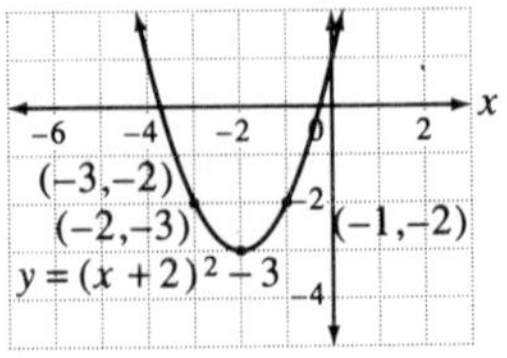

57.

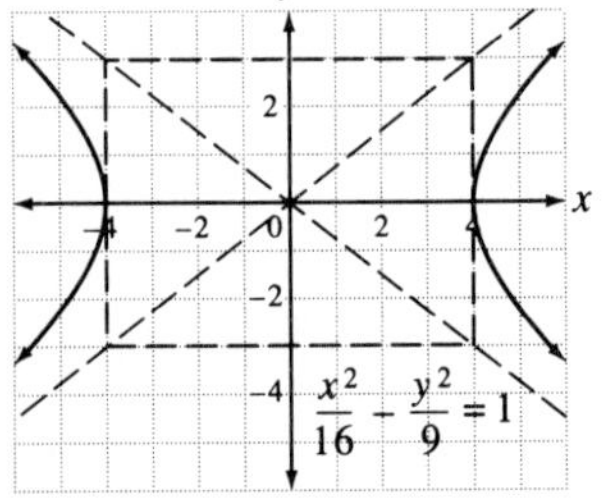

59.

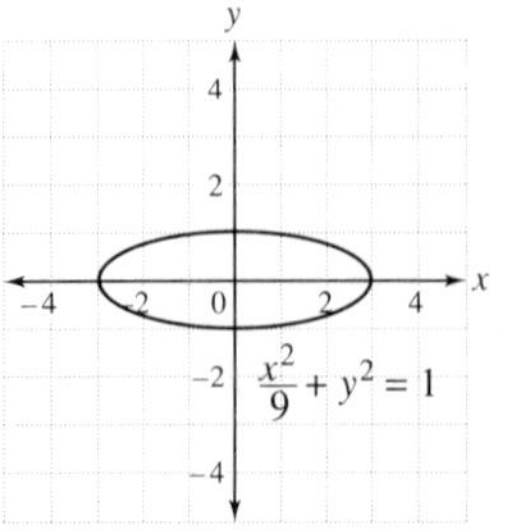

Problem Set 10.4

1. Linear function and constant function

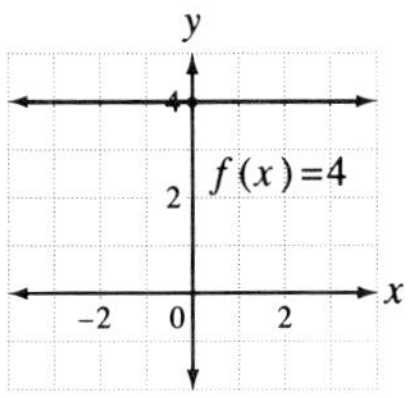

3. Linear function and identity function

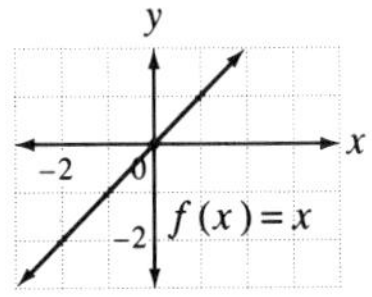

5. Quadratic function

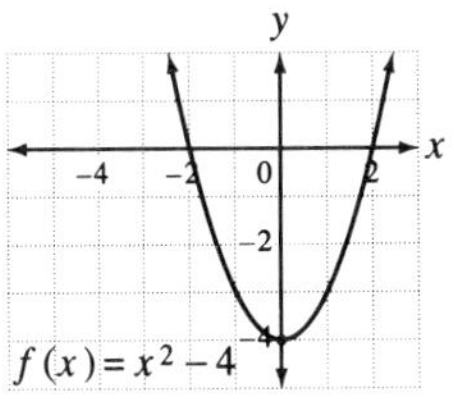

7. Linear function

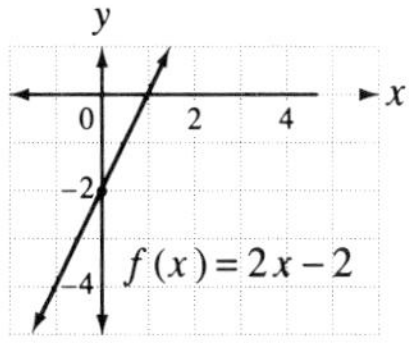

9. Quadratic function

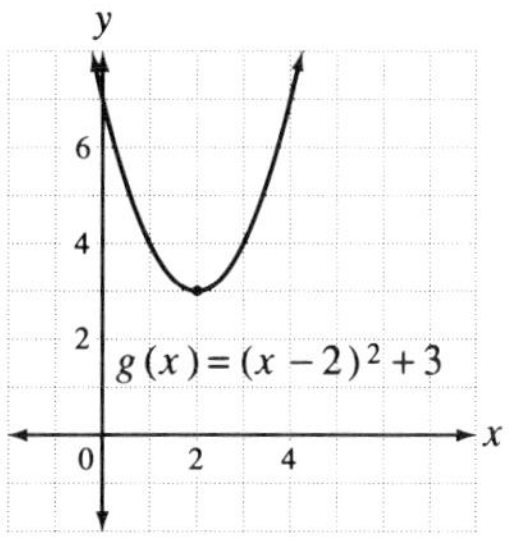

11. Linear function

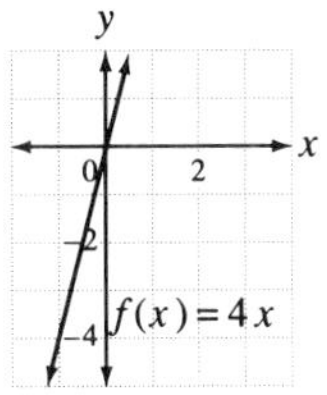

13. Power function

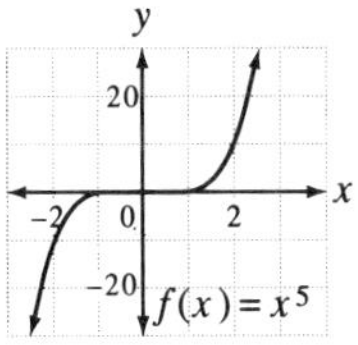

15. Linear function

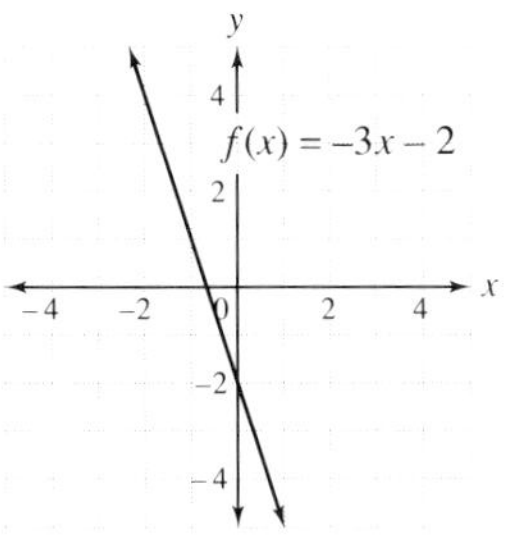

17. Absolute value function

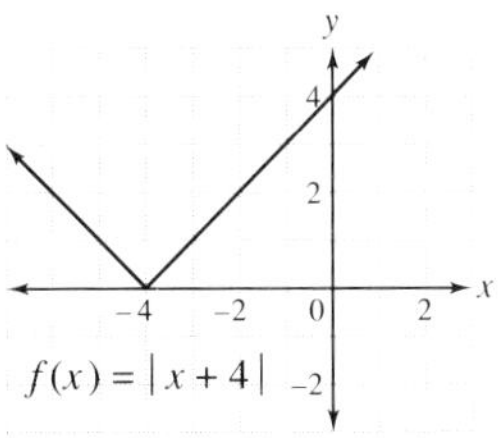

19. Quadratic function

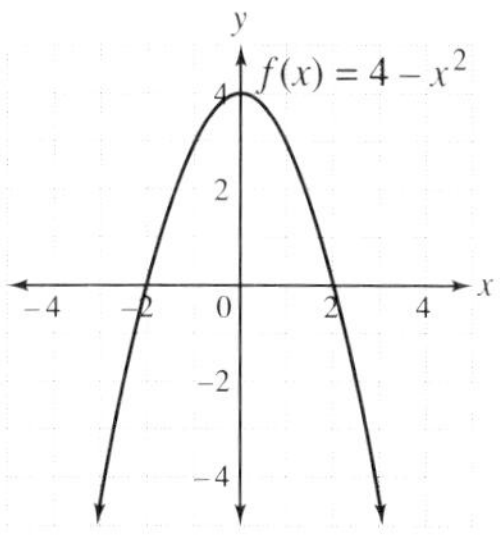

21. Absolute value function

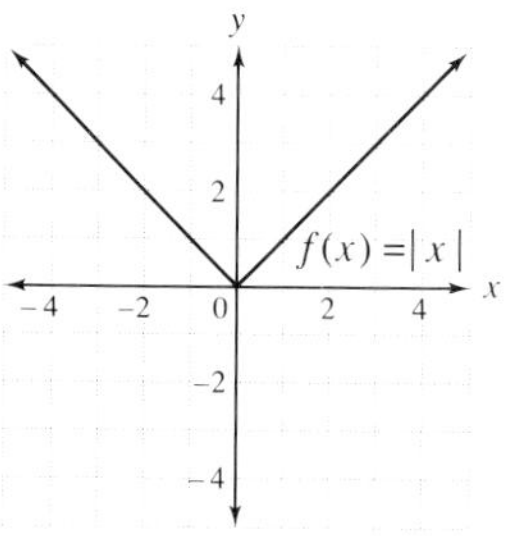

23. Quadratic function

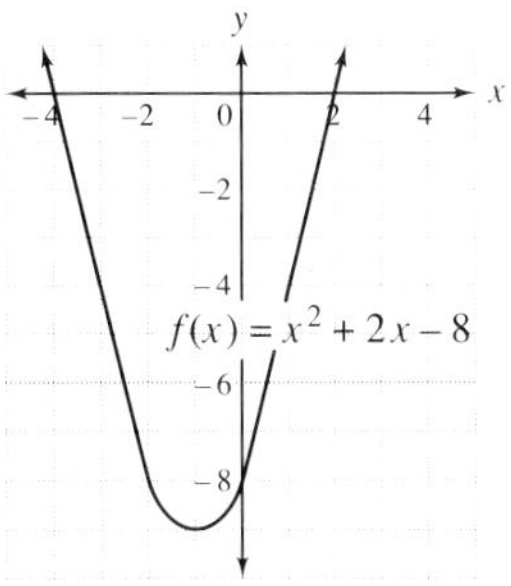

25. Absolute value function

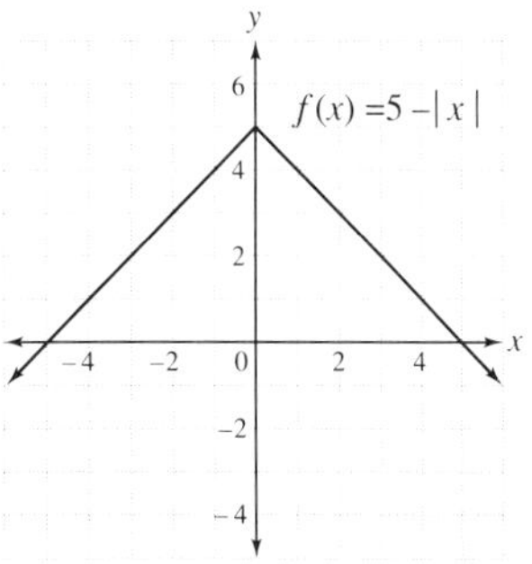

27. 452.16 in.^{2} **29.** Straight line **31.** Quadratic **33.** 30,098.185 cm^3
35. $f(x) = x$ **37.** $f(x) = 5x - 7$ **39.** $f(x) = -3x + 4$ **41.** 3.606 **43.** 7.810 **45.** 7.810

47.

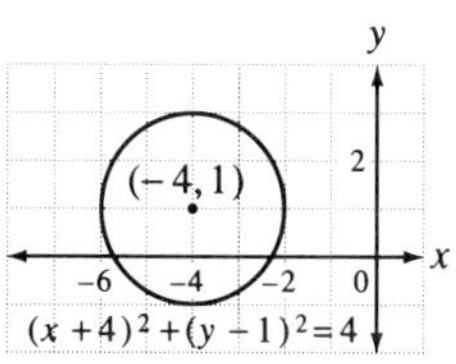

49.

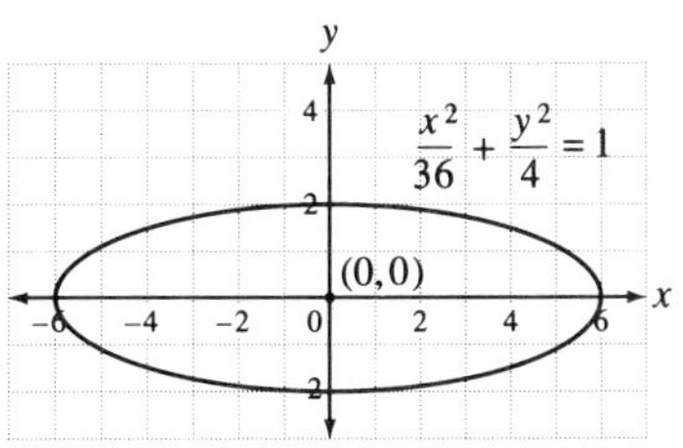

Problem Set 10.5

1. Yes; $f^{-1} = \{(8, 7), (4, -5), (0, 3), (-2, -7)\}$ **3.** No **5.** Yes **7.** No **9.** Yes **11.** No **13.** No

15. $f^{-1}(x) = \dfrac{x}{5}$ **17.** $g^{-1}(x) = \dfrac{x-6}{-5}$ **19.** $f^{-1}(x) = \dfrac{5x+2}{4}$ **21.** $H^{-1}(x) = \sqrt[3]{\dfrac{x+3}{2}}$ **23.** $f^{-1}(x) = x^3 + 3$

25. $F^{-1}(x) = \sqrt{\dfrac{x+1}{3}}$ **27.** $f^{-1}(x) = \dfrac{x-2}{3}$ **29.** $f^{-1}(x) = 4 - x$ **31.** $f^{-1}(x) = \sqrt[3]{x+1}$

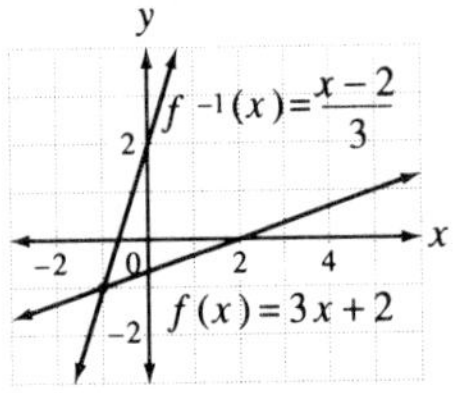

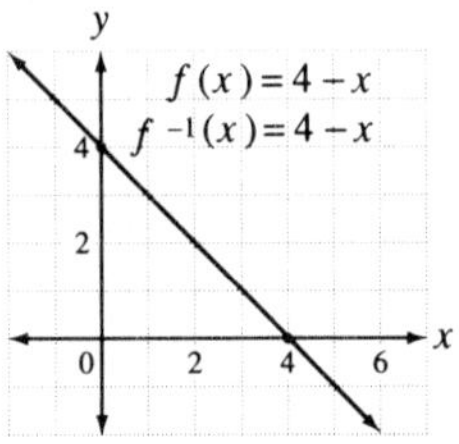

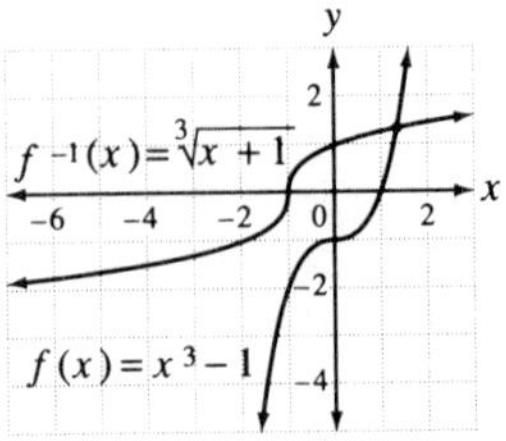

33. $f^{-1}(x) = \dfrac{x-2}{-3}$ **35.** $f^{-1}(x) = \sqrt[3]{x+2}$ **37.** $f^{-1}(x) = \sqrt{x-1}$

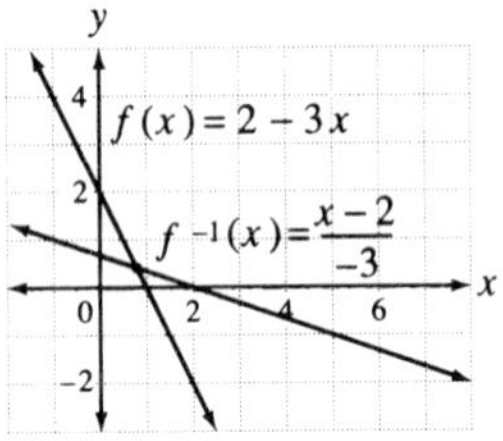

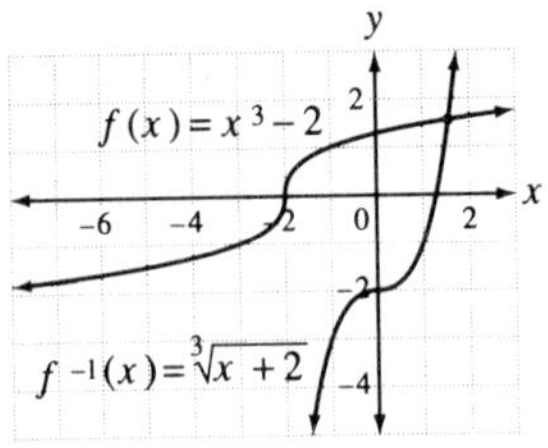

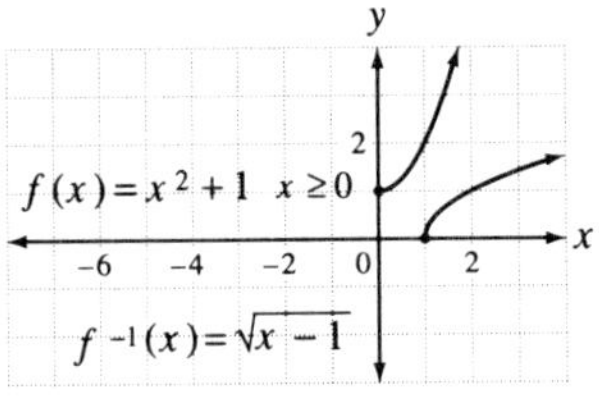

39. $f^{-1}(x) = \sqrt{\dfrac{1-x}{2}}$

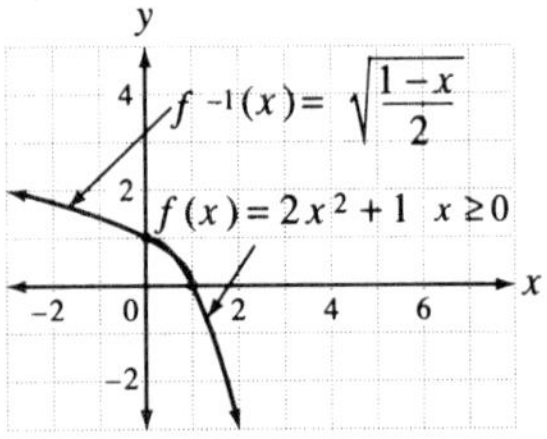

41. Range **43.** Domain **45.** $f[f^{-1}(x)] = x$; $f^{-1}[f(x)] = x$ **47.** $x \geq 1$ (Answers may vary.)

49. $(-3, 0), (\frac{9}{5}, -\frac{12}{5})$ **51.** $(2, -2), (0, 2)$ **53.** $(2, 1), (2, -1), (-2, 1), (-2, -1)$

55. $(1, 4), (-\frac{5}{3}, \frac{100}{9})$

Chapter 10 Additional Exercises

1. Yes; $\{-3, -1, 0, 2\}$ **3.** Yes; all real numbers **5.** No **7.** Yes; all real numbers **9.** No **11.** No
13. Not a function **15.** -3 **17.** -8 **19.** $\frac{1}{64}$ **21.** $2\sqrt{2}$ **23.** 343 **25.** 19 **27.** $8x^3 + 36x^2 + 54x + 27$
29. (a) $2x - 10$; (b) -4; (c) $x^2 - 10x + 21$; (d) $\dfrac{x-7}{x-3}$ **31.** (a) $x^2 - x$; (b) $x^2 + x$; (c) $-x^3$; (d) $-x$
33. (a) $4x^2 + 2x - 30$; (b) $4x^2 - 2x - 20$; (c) $8x^3 - 20x^2 - 50x + 125$; (d) $2x + 5$
35. (a) $\dfrac{13x+25}{(x+4)(2x-1)}$; (b) $\dfrac{-x-31}{(x+4)(2x-1)}$; (c) $\dfrac{21}{(x+4)(2x-1)}$; (d) $\dfrac{6x-3}{7(x+4)}$
37. (a) $6x^2 + 9x$; (b) $6x^2 + 5x - 6$; (c) $12x^3 + 32x^2 + 15x - 9$; (d) $3x - 1$ **39.** Linear function
41. Constant function and linear function **43.** Linear function **45.** Quadratic function **47.** Linear function
49. Absolute value function **51.** Yes; $\{(-5, 0), (-8, -6), (9, 5), (2, -7), (8, 3)\}$ **53.** No **55.** $f^{-1}(x) = \dfrac{x+1}{4}$
57. $f^{-1}(x) = \sqrt[3]{x+4}$
59. $f^{-1}(x) = \sqrt{-x+4}$
61.

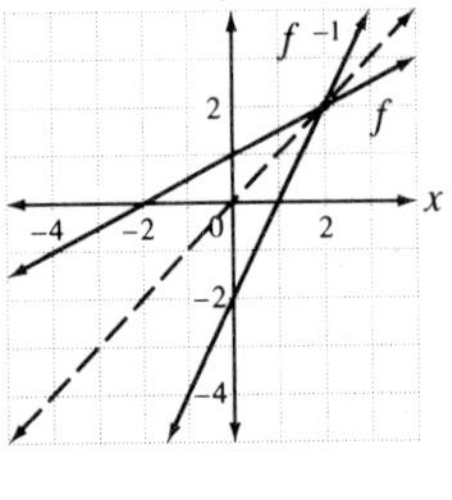

Chapter 10 Cooperative Learning

1. Change one of the −3's in the domain to another number. **3.** 2 **5.** $f(x) = \frac{1}{x}$ **7.** $2x - 5$

Chapter 10 Practice Test

1. Yes
2. $\{1, -2, 4, 0\}$
3. $\{3, 5, 6\}$
4. $\{x \mid x \geq 3\}$
5. Not a function

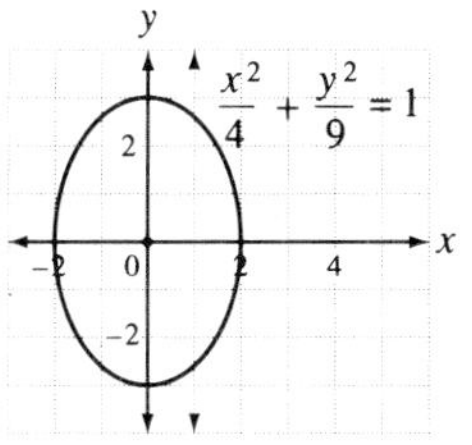

6. 17 **7.** $18x^2 - 12x + 5$ **8.** $-3x^2 + 3x - 12$
9. $6x^3 - 17x^2 + 19x - 35$ **10.** Quadratic function **11.** No
12. $f^{-1}(x) = \sqrt[3]{x + 7}$

Cumulative Review Chapters 9 and 10

1. $2\sqrt{13}$
2. $(\frac{11}{32}, \frac{17}{24})$
3. $(x + 3)^2 + (y - 5)^2 = 16$

4.

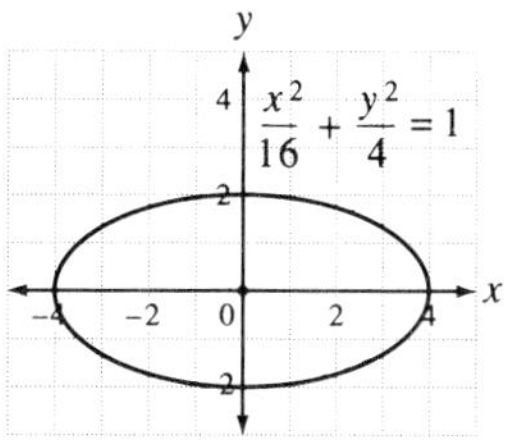

5.

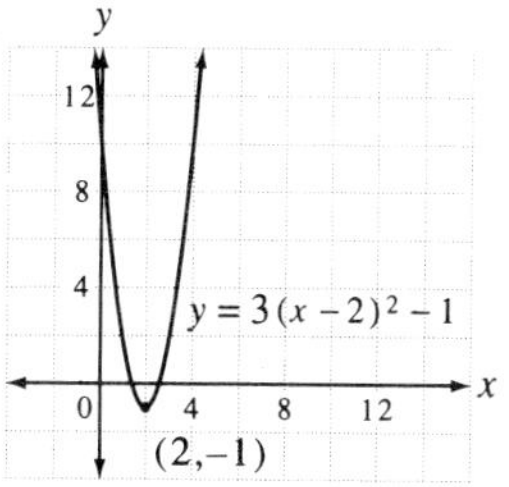

6.

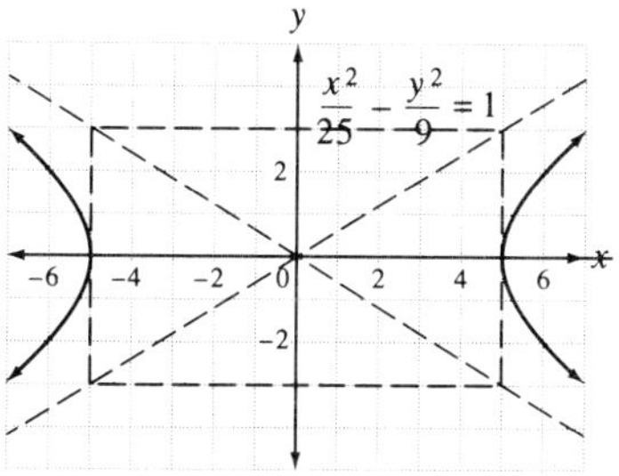

7. Circle **8.** 12 ft by 20 ft **9.** (2, −1), (−2, −1), (2, 1), (−2, 1)
10. Yes **11.** $\{-1, 2, -3\}$ **12.** $\{4, 5\}$ **13.** All real numbers except 8
14. Function **15.** 8 **16.** $x^2 + 8x + 13$ **17.** $2x^2 - 10x + 8$
18. $2x - 3$ **19.** Linear function **20.** $f^{-1}(x) = \frac{x + 2}{7}$ **21.** No

CHAPTER 11

Chapter 11 Pretest

1.

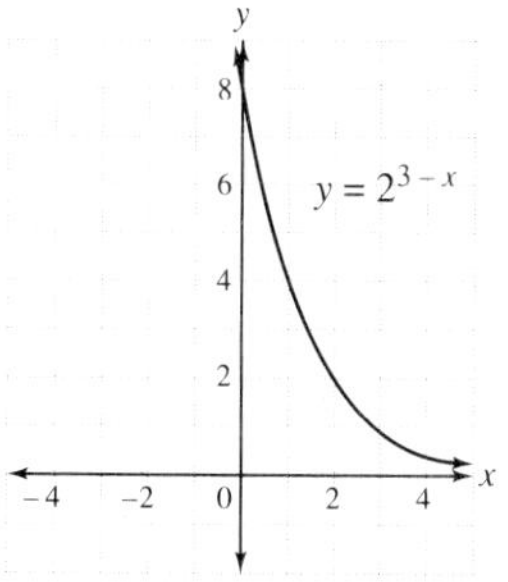

2.

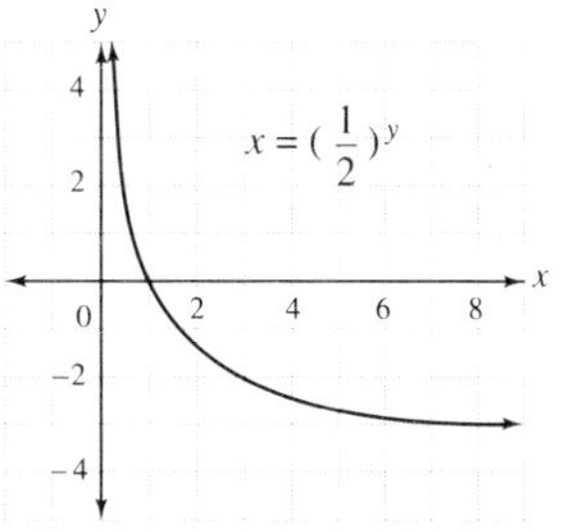

3. $e^{3.9512} = 52$

4. $\log_p y = q$ **5.**

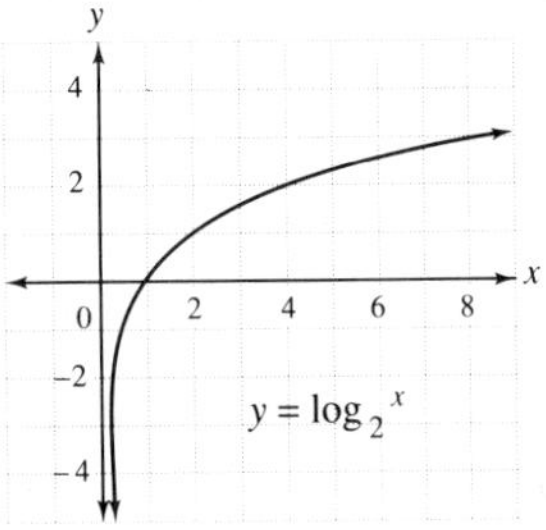

6. $\log_b V^{1/2} W^{2/3}$ **7.** $\log_b x + 3 \log_b y - 2 \log_b p - 4 \log_b q$
8. -1.3098 **9.** 2.1041 **10.** 5223.96 **11.** 0.0138 **12.** 1
13. 0 **14.** 21 **15.** 2.3789 **16.** 2 **17.** 1.9735 **18.** $\frac{31}{3}$
19. 8.5 years **20.** 20.25 lb

Problem Set 11.1

1.

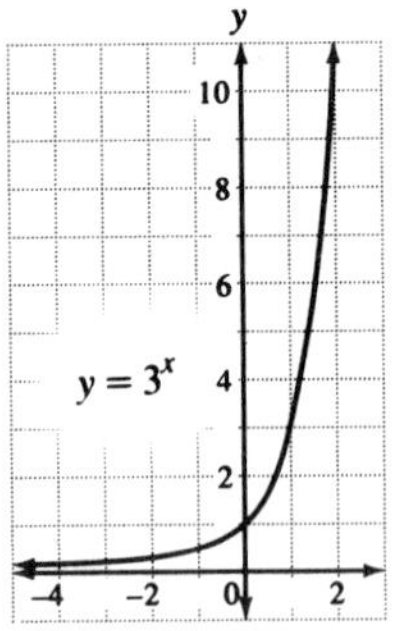

3.

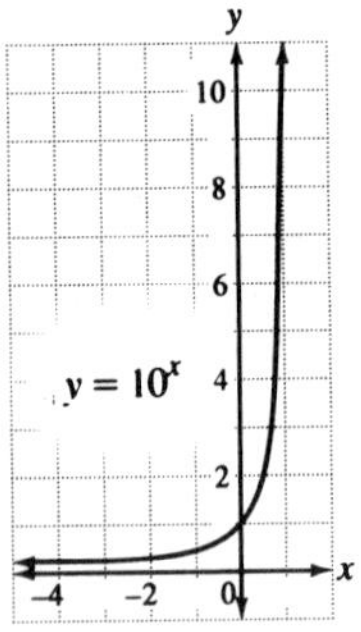

5.

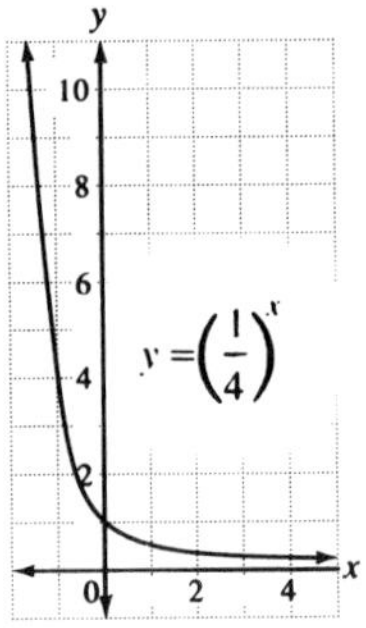

7.

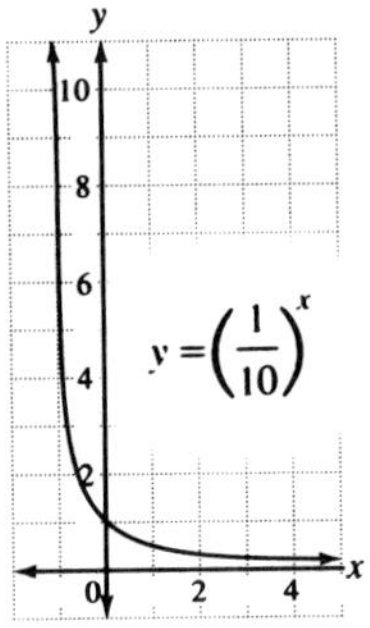

9.

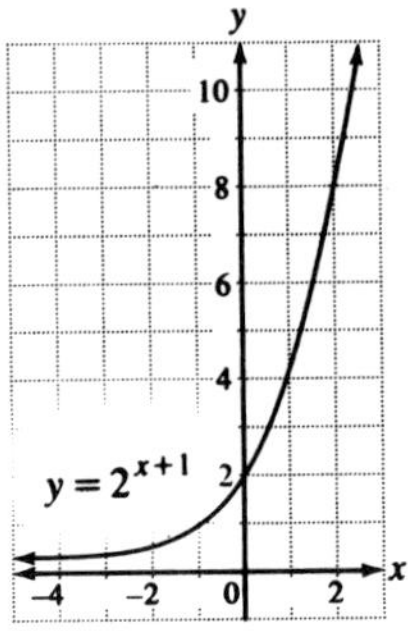

11.

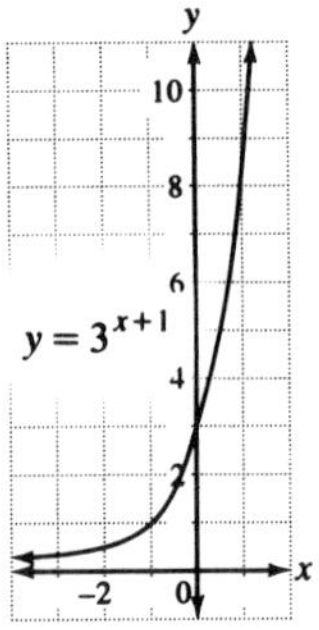

13.

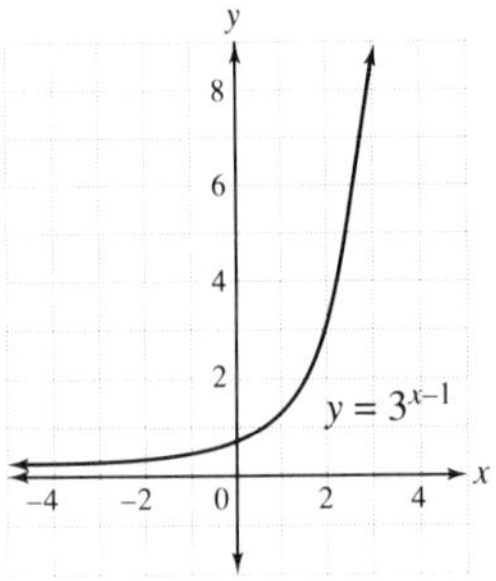

15.

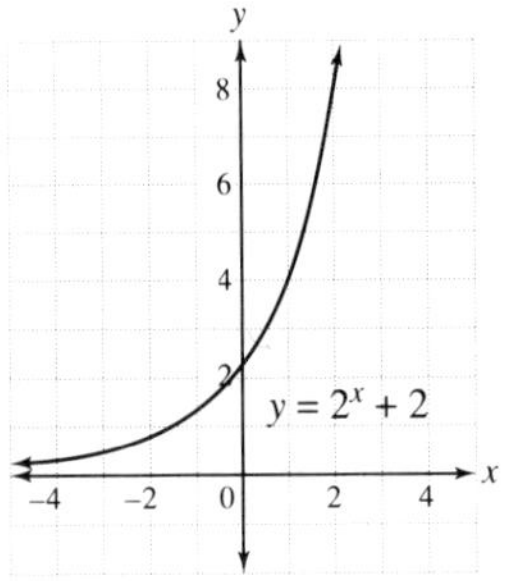

17.

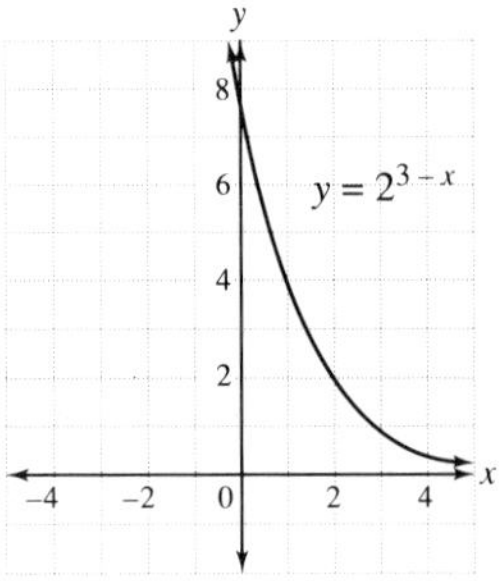

19.

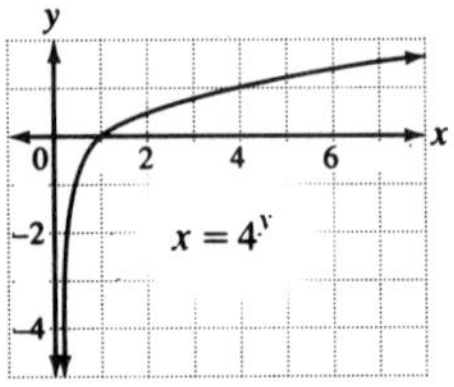

21.

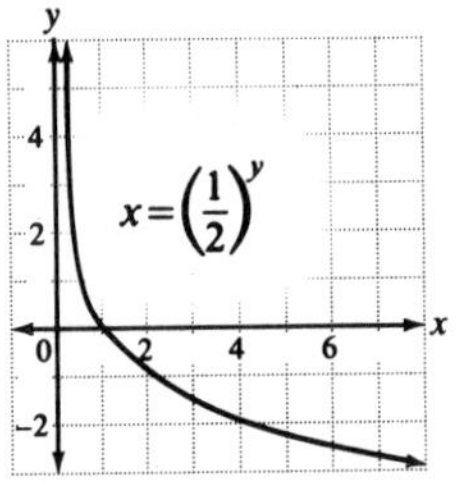

23.

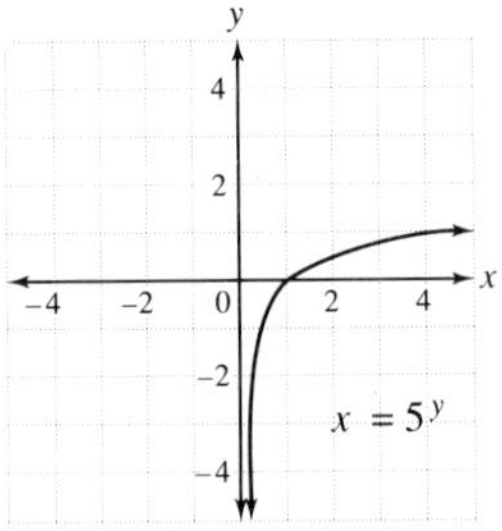

25.

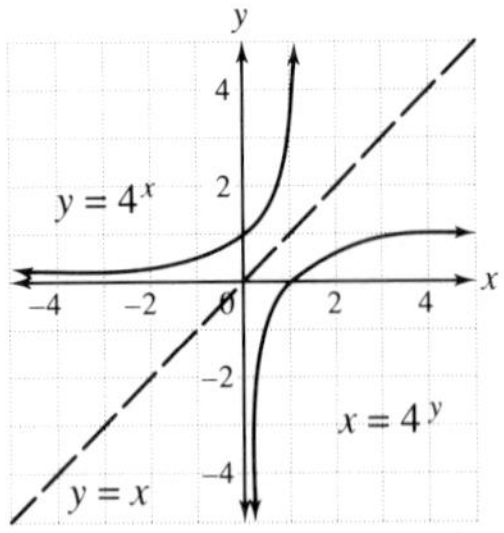

27.

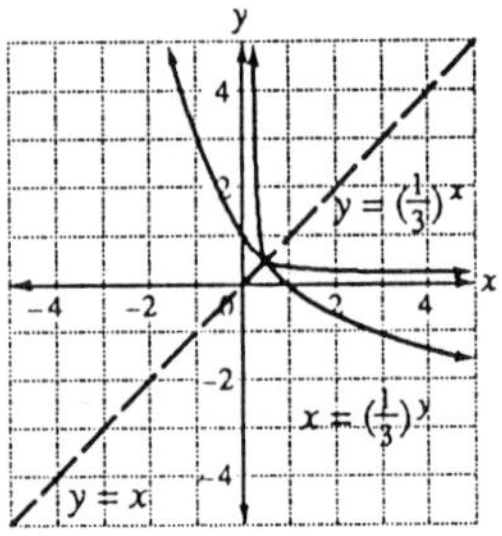

29. Increase **31.** $y = -3^x$

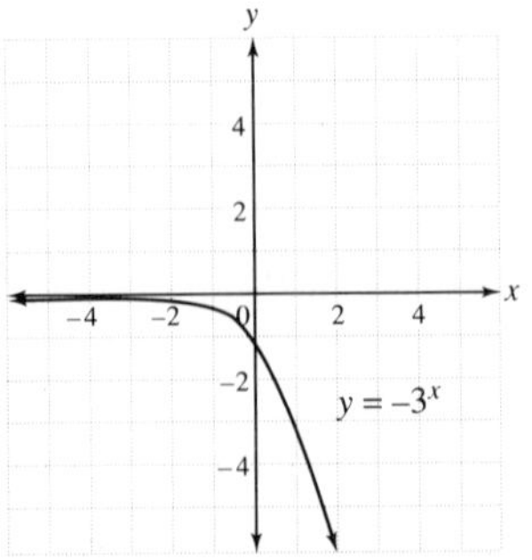

33. 3 **35.** $\frac{1}{3}$
37. Yes, all real numbers
39. Yes, all real numbers except 2
41. No **43.** Yes, $x \geq \frac{5}{2}$
45. No

Problem Set 11.2

1. $4 = 2^2$ **3.** $49 = 7^2$ **5.** $1 = 8^0$ **7.** $\frac{1}{2} = 2^{-1}$ **9.** $6 = 6^1$ **11.** $9 = 10^{0.9542}$ **13.** $15 = e^{2.7081}$ **15.** $R = b^a$
17. $2 = \log_8 64$ **19.** $3 = \log_5 125$ **21.** $0 = \log_2 1$ **23.** $-2 = \log_3 \frac{1}{9}$ **25.** $1 = \log_8 8$ **27.** $0.6990 = \log_{10} 5$
29. $3 = \log_e 20.0855$ **31.** $t = \log_r x$ **33.** 81 **35.** 49 **37.** 16 **39.** 5 **41.** 17 **43.** 1 **45.** 2 **47.** 2

49.

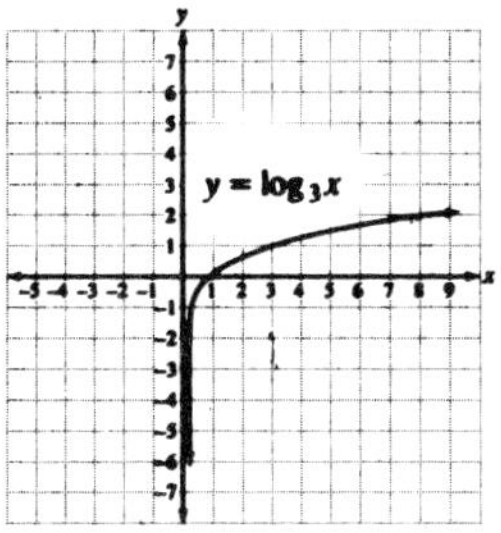

51.

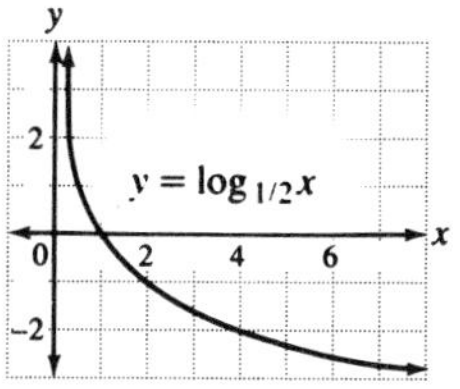

53.

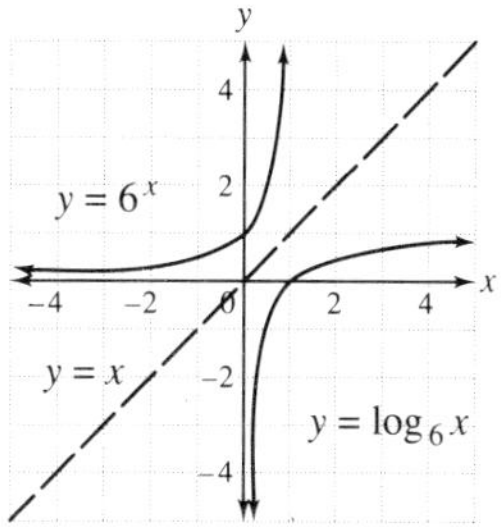

55. Exponents
57. Interchange

59. $y = -\log_{1/4} x$

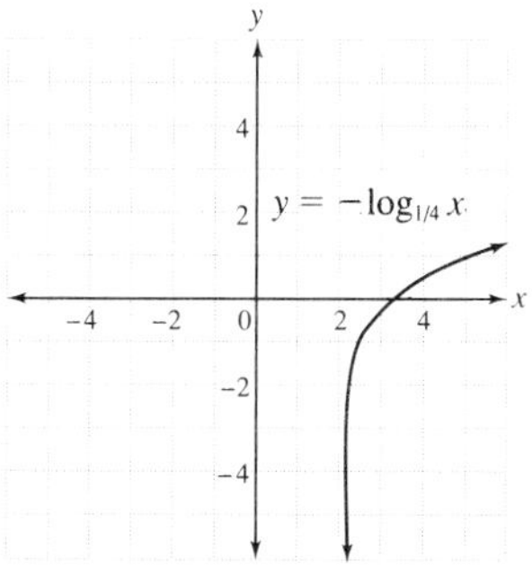

61. $y = \log_3(x - 2)$

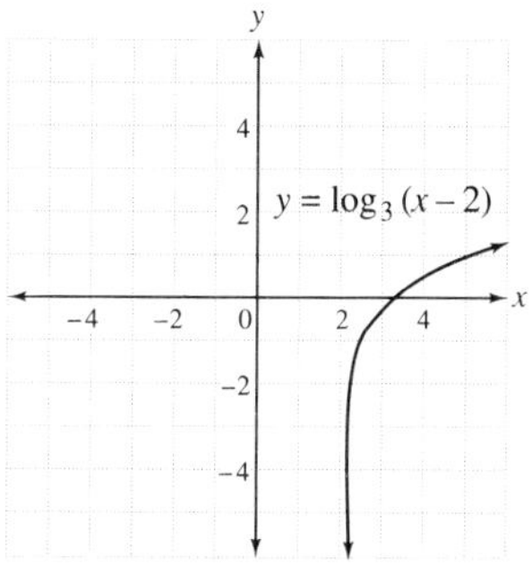

63. 5 **65.** $\frac{1}{2}$ **67.** $\frac{1}{2}$ **69.** 2 **71.** $\frac{4}{7}$
73. −14 **75.** −6 **77.** $-2x^2 - 11$
79. $18x^2 - 12x - 1$

Problem Set 11.3

1. 1 **3.** 0 **5.** 1 **7.** $\log_4 27 + \log_4 5$ **9.** $1 + \log_6 x$ **11.** $\log_a P + \log_a Q$ **13.** $\log_6 7 - \log_6 15$
15. $\log_c 32 - \log_c V$ **17.** $2 \log_8 7$ **19.** $5 \log_a x$ **21.** $-3 \log_a 4$ **23.** $\frac{1}{3} \log_3 4$ **25.** $\log_2 63$ **27.** $\log_a 360$
29. $\log_c My$ **31.** $\log_a 5$ **33.** $\log_b \frac{25}{64}$ **35.** $\log_b M^5 N$ **37.** $\log_a \frac{x^2}{\sqrt[5]{y}}$ **39.** $3 \log_b x + 2 \log_b y$
41. $\frac{1}{4}(\log_a x - 3 \log_a y)$ **43.** $\log_b p + 3 \log_b q - 2 \log_b r$ **45.** $\frac{1}{5}(9 \log_a x + 11 \log_a y - 2 \log_a b - 3 \log_a c)$ **47.** 4
49. −5 **51.** 8 **53.** $-t$ **55.** Zero **57.** Add **59.** False **61.** True **63.** $\log_a \frac{54x^8}{y^2}$ **65.** 4.25×10^3
67. 2.4×10^{-3} **69.** $4x^2 - 3x + 4$ **71.** $8x^3 - 22x^2 + 29x - 21$ **73.** $2x + \frac{1}{2} + \frac{17}{2(2x - 3)}$

Problem Set 11.4

1. 0.8561 **3.** 0.3784 **5.** 2.3979 **7.** 1.8531 **9.** 5.8531 **11.** −1.4584 **13.** −0.4622 **15.** −0.5331
17. −1.5229 **19.** 344 **21.** 40 **23.** 0.0001 **25.** 0.7971 **27.** 0.0206 **29.** 0.6931 **31.** 2.3026
33. 3.5553 **35.** −2.4079 **37.** 5.9140 **39.** −5.3817 **41.** 274.02 **43.** 0.0447 **45.** 1.0618
47. 29, 733 **49.** 0.0169 **51.** 3.8489 **53.** 2.7268 **55.** 1.6131 **57.** 0.8124 **59.** −3.5076 **61.** 2.8836
63. 1.5155 **65.** 1.4 **67.** Common **69.** Antilogarithm **71.** 1.1485 **73.** 1.3×10^{-3} **75.** Power function
77. Constant and linear function **79.** Linear function **81.** Quadratic function **83.** Absolute value function

Problem Set 11.5

1. 3 **3.** −2 **5.** 1 **7.** 2 **9.** 1.4854 **11.** 2.4022 **13.** 0.4679 **15.** 0.3199 **17.** 2 **19.** 2
21. 3.2189 **23.** 3.2189 **25.** −4.1589 **27.** 0.3326 **29.** 7 **31.** 2 **33.** 8 **35.** $\frac{10}{3}$ **37.** 7 **39.** 54.598
41. $\frac{27}{10}$ **43.** Variable **45.** Greater **47.** −3, −1 **49.** $-\frac{1}{3}$ **51.** 0 **53.** 4 **55.** 4 **57.** $y = \frac{x - 7}{-3}$
59. $y = \sqrt[3]{x - 5}$ **61.** $y = \sqrt{4 - x}$ **63.** Yes **65.** No

Problem Set 11.6

1. 147,033 **3.** 4.4 days **5.** \$11,102.59 **7.** \$8141.15 **9.** 20.4 years **11.** 11.9 years **13.** 6.3 years **15.** 29 years **17.** 3.9 grams **19.** \$1201.20 **21.** 25 years **23.** 44 decibels **25.** 7% **27.** $\sqrt{2}, \sqrt{3}, -\sqrt{2}, -\sqrt{3}$ **29.** 1 **31.** $i\sqrt{3}, 0, -i\sqrt{3}$ **33.** $\frac{1}{5}, -\frac{1}{3}$ **35.** 9, 6

Chapter 11 Additional Exercises

1.

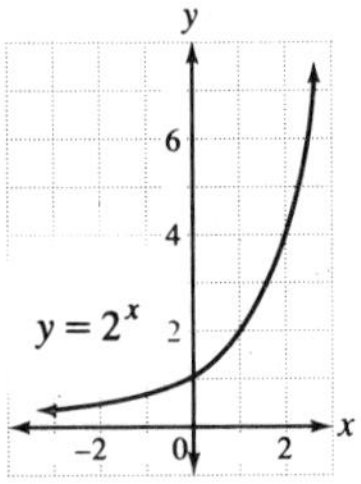

3.

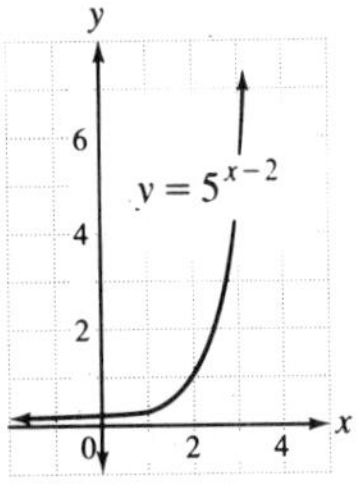

5.

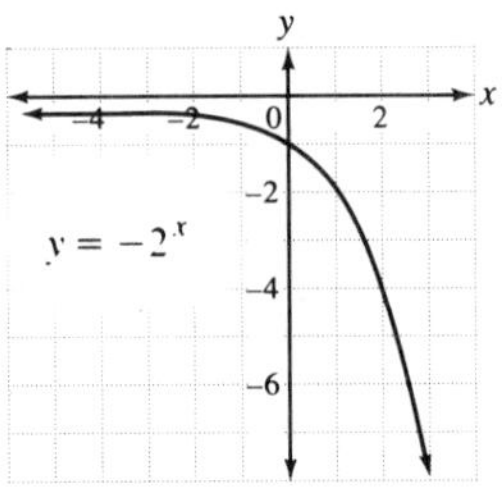

7.

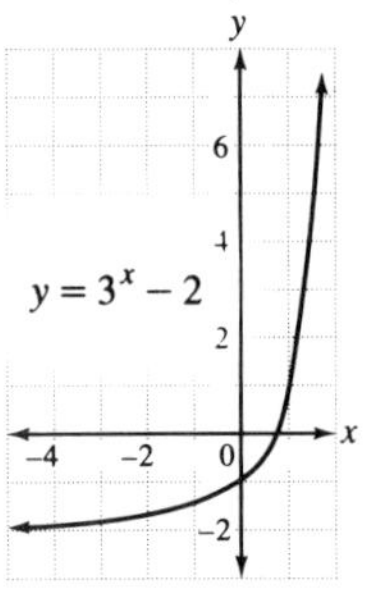

9.

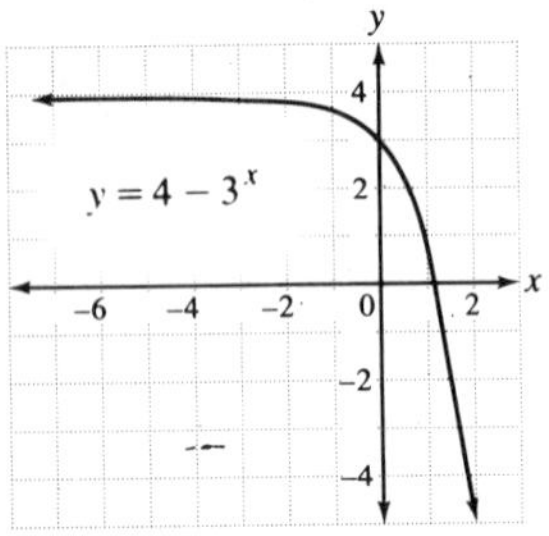

11. $10^0 = 1$ **13.** $3^{-2} = \frac{1}{9}$ **15.** $10^{0.4771} = 3$ **17.** $5 = \log_3 y$ **19.** $5 = \log_b 32$ **21.** $2.9425 = \log_{10} 876$ **23.** 7 **25.** $\frac{1}{10,000}$ **27.** 81 **29.** 0 **31.** 1 **33.** 1 **35.** $\log_5 (x - 1) - \log_5 (x + 1)$ **37.** $4 \log_a (x - 2)$ **39.** $\log_7 (3x + 2) + \log_7 4 + \log_7 y$ **41.** $\log_5 \frac{6}{5}$ **43.** $\log_b x^2y^6$ **45.** $\log_b \sqrt{3x - 1}\,(x + 4)^3$ **47.** 3.7226 **49.** −2.1079 **51.** 5.7683 **53.** −5.8091 **55.** 55,296.8 **57.** 0.117 **59.** 1.0513 **61.** 0.0165 **63.** 3.9069 **65.** 1.2143 **67.** 3 **69.** 1.4270 **71.** 3, −2 **73.** 14.425 **75.** $\pm\sqrt{10}$ **77.** 8, −10 **79.** $\frac{1}{8}$ **81.** 4 **83.** 8.3 **85.** \$7609.81

Chapter 11 Cooperative Learning

1. $y = 2^{x+1}$ **3.** $y = \log_5 x$ **5.** $y = \ln x$ **7.** $t = \dfrac{-\ln b}{c - a}$

Chapter 11 Practice Test

1.

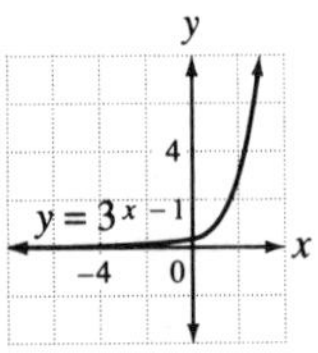

2.

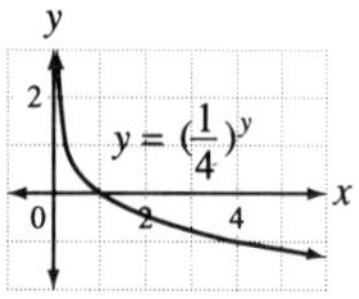

3. $10^{0.8261} = 6.7$ **4.** $\log_4 \frac{1}{16} = -2$

5.

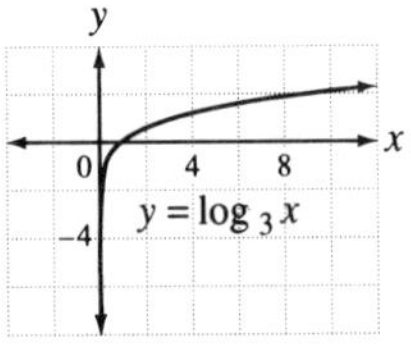

6. $\log_b \dfrac{M^6}{N^4}$ **7.** $\frac{1}{3}(2 \log_a x + \log_a y)$ **8.** 2.1367 **9.** 807.0492 **10.** −3.0262 **11.** 0.0317

12. 1 **13.** 0 **14.** 14 **15.** 1.465 **16.** 3.5 **17.** 0.8371 **18.** 2 **19.** 9 years **20.** 4257 years

FINAL EXAMINATION

1. 48 **2.** $12x + 11$ **3.** $\dfrac{4y^{12}}{x^4}$ **4.** 1 **5.** $22\frac{1}{4}, 30\frac{1}{4}$ **6.** $c = \dfrac{2A - bh}{h}$ **7.** 2, −3 **8.** $\{x \mid -1 < x < 4\}$;

9. \$22,500 **10.**

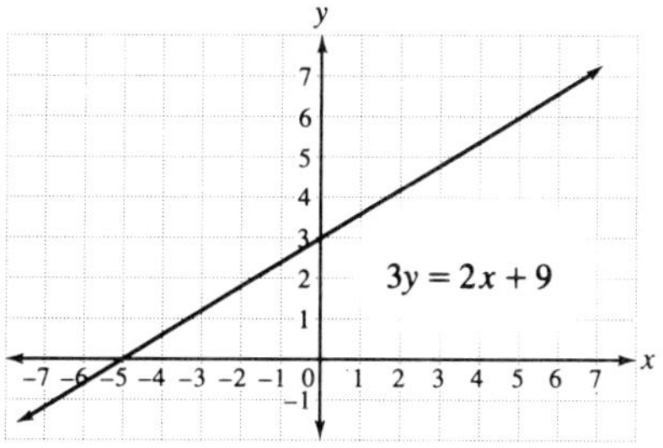

11. $2y + x = 4$
12. $y + 3x = 3$
13. $C = 25n + 1500$; \$10,250

14.

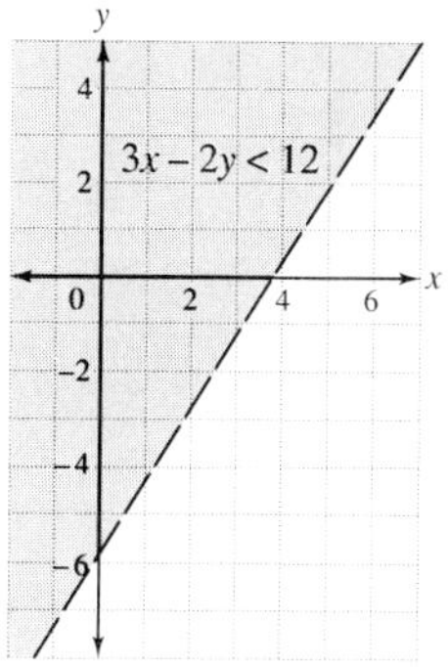

15. (1, 2) **16.** 4.57 gallons **17.** (0, 2, −2) **18.** −33 **19.** 5 **20.** $11x^2 - 3xy - 9y^2$

21. $8x^3 - 8x^2 - 3x + 11$ **22.** $2y^3 - 7y^2 + 10y - 8$ **23.** $6x^2 + 7x - 24$ **24.** $x + 9 + \frac{28}{x - 4}$

25. $2(3x + 1)(2x - 3)$ **26.** $(4x^2 + y^2)(2x + y)(2x - y)$ **27.** $(2y - 5z)(4y^2 + 10yz + 25z^2)$ **28.** $\frac{(x - 4)(2x - 3)}{3}$

29. $\frac{-x - 11}{(x + 4)(x - 2)}$ **30.** $\frac{7}{6x + 5}$ **31.** $3, -\frac{8}{3}$ **32.** $r = \frac{P - A}{-An}$ or $r = \frac{A - P}{An}$ **33.** 108 tons **34.** 45 hours

35. 384 mph **36.** $4xy\sqrt[3]{x}$ **37.** $-14\sqrt{6}$ **38.** $\frac{\sqrt{15} + x\sqrt{5} + x\sqrt{3} + x^2}{5 - x^2}$ **39.** 5 **40.** $-\frac{7}{29} - \frac{26}{29}i$ **41.** 53.7 m

42. $\frac{-5 + \sqrt{33}}{2}, \frac{-5 - \sqrt{33}}{2}$
43. $\sqrt{5}, -\sqrt{5}, i, -i$

44.

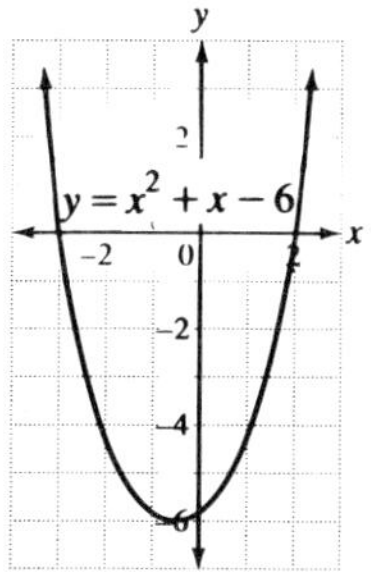

45. −4 **46.** $\{x \mid -6 < x < 3\}$;

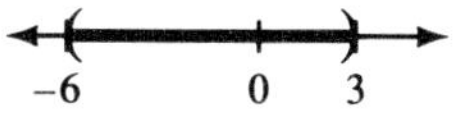

47. $\sqrt{10}$

48. $(x + 1)^2 + (y - 4)^2 = 5$ **49.**

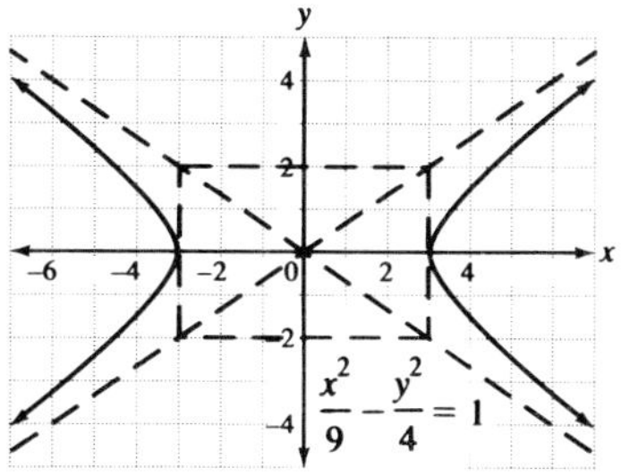

50. (2, 1), (2, −1), (−2, 1), (−2, −1)
51. All real numbers, $x \neq 3$ **52.** 14

53.

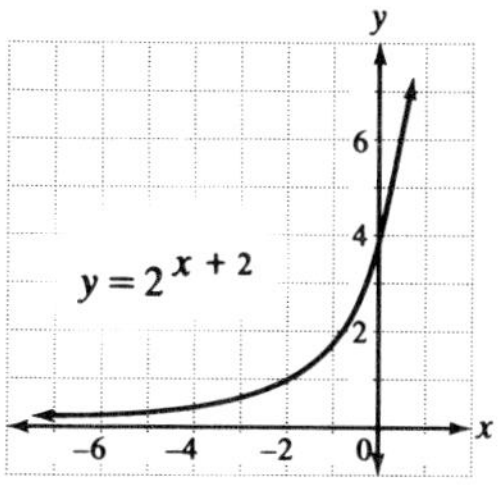

54. $\log_5 \frac{1}{25} = -2$ **55.** 2.7712 **56.** \$7057.40

APPENDIX A

Problem Set

1. True **3.** False **5.** True **7.** False **9.** False
11. True **13.** True **15.** {5, 7}, {5}, {7}, ∅ **17.** {a, b, c}, {a}, {b}, {c}, {a, b}, {a, c}, {b, c}, ∅
19. {3, 8, 9, 5}, {3}, {8}, {9}, {5}, {3, 8}, {3, 9}, {3, 5}, {8, 9}, {8, 5} {9, 5}, {3, 8, 9}, {3, 8, 5}, {8, 9, 5}, {3, 9, 5}, ∅
21. {u, v, w, x}, {u}, {v}, {w}, {x}, {u, v}, {u, w}, {u, x}, {v, w}, {v, x}, {w, x}, {u, v, w}, {u, v, x}, {v, w, x}, {u, w, x}, ∅
23. {1, 2, 3, 4, 6, 8} **25.** {1, 2, 3, 4}

Index